BIOLOGY HOW LIFE WORKS

Volume 2: Evolution, Organisms, and Ecology

Chapters 21–49

SECOND EDITION

James Morris
BRANDEIS UNIVERSITY

Daniel Hartl
HARVARD UNIVERSITY

Andrew Knoll
HARVARD UNIVERSITY

Robert Lue
HARVARD UNIVERSITY

Melissa Michael
UNIVERSITY OF ILLINOIS AT URBANA-CHAMPAIGN

ANDREW BERRY, ANDREW BIEWENER,
BRIAN FARRELL, N. MICHELE HOLBROOK
HARVARD UNIVERSITY

W. H. FREEMAN & COMPANY
A Macmillan Education Imprint

PUBLISHER Kate Ahr Parker
ACQUISITIONS EDITOR Beth Cole
LEAD DEVELOPMENTAL EDITOR Lisa Samols
SENIOR DEVELOPMENTAL EDITOR Susan Moran
DEVELOPMENTAL EDITOR Erica Champion
EDITORIAL ASSISTANTS Jane Taylor, Alexandra Garrett, Abigail Fagan
REVIEW COORDINATOR Donna Brodman
PROJECT MANAGER Karen Misler
DIRECTOR OF MARKET DEVELOPMENT Lindsey Jaroszewicz
EXECUTIVE MARKETING MANAGER Will Moore
ART AND MEDIA DIRECTOR Robert Lue, Harvard University
EXECUTIVE MEDIA EDITOR Amanda Nietzel
SENIOR DEVELOPMENT EDITOR FOR TEACHING & LEARNING STRATEGIES Elaine Palucki
SENIOR MEDIA PRODUCER Keri DiManigold
PROJECT EDITOR Robert Errera
MANUSCRIPT EDITOR Nancy Brooks
DIRECTOR OF DESIGN, CONTENT MANAGEMENT Diana Blume
DESIGN Tom Carling, Carling Design Inc.
ART MANAGER Matt McAdams
ILLUSTRATIONS Imagineering
CREATIVE DIRECTOR Mark Mykytiuk, Imagineering
PHOTO EDITOR Christine Buese
PHOTO RESEARCHERS Lisa Passmore and Richard Fox
PRODUCTION MANAGER Paul Rohloff
COMPOSITION Sheridan Sellers
PRINTING AND BINDING RR Donnelley
COVER IMAGE: Thorsten Henn/Getty Images

Library of Congress Control Number: 2015950976

ISBN-13: 978-1-319-04888-4
ISBN-10: 1-319-04888-9

Printed in the United States of America

First printing

Macmillan Learning
W. H. Freeman and Company
One New York Plaza
Suite 4500
New York, NY 10004-1562
www.macmillanhighered.com

DEDICATION

To all who are curious about life and how it works

ABOUT THE AUTHORS

JAMES R. MORRIS is Professor of Biology at Brandeis University. He teaches a wide variety of courses for majors and non-majors, including introductory biology, evolution, genetics and genomics, epigenetics, comparative vertebrate anatomy, and a first-year seminar on Darwin's *On the Origin of Species*. He is the recipient of numerous teaching awards from Brandeis and Harvard. His research focuses on the rapidly growing field of epigenetics, making use of the fruit fly *Drosophila melanogaster* as a model organism. He currently pursues this research with undergraduates in order to give them the opportunity to do genuine, laboratory-based research early in their scientific careers. Dr. Morris received a PhD in genetics from Harvard University and an MD from Harvard Medical School. He was a Junior Fellow in the Society of Fellows at Harvard University, and a National Academies Education Fellow and Mentor in the Life Sciences. He also writes short essays on science, medicine, and teaching at his Science Whys blog (http://blogs.brandeis.edu/sciencewhys).

DANIEL L. HARTL is Higgins Professor of Biology in the Department of Organismic and Evolutionary Biology at Harvard University and Professor of Immunology and Infectious Diseases at the Harvard Chan School of Public Health. He has taught highly popular courses in genetics and evolution at both the introductory and advanced levels. His lab studies molecular evolutionary genetics and population genetics and genomics. Dr. Hartl is the recipient of the Samuel Weiner Outstanding Scholar Award as well as the Gold Medal of the Stazione Zoologica Anton Dohrn, Naples. He is a member of the National Academy of Sciences and the American Academy of Arts and Sciences. He has served as President of the Genetics Society of America and President of the Society for Molecular Biology and Evolution. Dr. Hartl's PhD is from the University of Wisconsin, and he did postdoctoral studies at the University of California, Berkeley. Before joining the Harvard faculty, he served on the faculties of the University of Minnesota, Purdue University, and Washington University Medical School. In addition to publishing more than 400 scientific articles, Dr. Hartl has authored or coauthored 30 books.

ANDREW H. KNOLL is Fisher Professor of Natural History in the Department of Organismic and Evolutionary Biology at Harvard University. He is also Professor of Earth and Planetary Sciences. Dr. Knoll teaches introductory courses in both departments. His research focuses on the early evolution of life, Precambrian environmental history, and the interconnections between the two. He has also worked extensively on the early evolution of animals, mass extinction, and plant evolution. He currently serves on the science team for NASA's mission to Mars. Dr. Knoll received the Phi Beta Kappa Book Award in Science for *Life on a Young Planet*. Other honors include the Paleontological Society Medal and Wollaston Medal of the Geological Society, London. He is a member of the National Academy of Sciences and a foreign member of the Royal Society of London. He received his PhD from Harvard University and then taught at Oberlin College before returning to Harvard.

ROBERT A. LUE is Professor of Molecular and Cellular Biology at Harvard University and the Richard L. Menschel Faculty Director of the Derek Bok Center for Teaching and Learning. Dr. Lue has a longstanding commitment to interdisciplinary teaching and research, and chaired the faculty committee that developed the first integrated science foundation in the country to serve science majors as well as pre-medical students. The founding director of Life Sciences Education at Harvard, Dr. Lue led a complete redesign of the introductory curriculum, redefining how the university can more effectively foster new generations of scientists as well as science-literate citizens. Dr. Lue has also developed award-winning multimedia, including the animation "The Inner Life of the Cell." He has coauthored undergraduate biology textbooks and chaired education conferences on college biology for the National Academies and the National Science Foundation and on diversity in science for the Howard Hughes Medical Institute and the National Institutes of Health. In 2012, Dr. Lue's extensive work on using technology to enhance learning took a new direction when he became faculty director of university-wide online education initiative HarvardX; he now helps to shape Harvard's engagement in online learning to reinforce its commitment to teaching excellence. Dr. Lue earned his PhD from Harvard University.

MELISSA MICHAEL is Director for Core Curriculum and Assistant Director for Undergraduate Instruction for the School of Molecular and Cellular Biology at the University of Illinois at Urbana-Champaign. A cell biologist, she primarily focuses on the continuing development of the School's undergraduate curricula. She is currently engaged in several projects aimed at improving instruction and assessment at the course and program levels. Her research focuses primarily on how creative assessment strategies affect student learning outcomes, and how outcomes in large-enrollment courses can be improved through the use of formative assessment in active classrooms.

ANDREW BERRY is Lecturer in the Department of Organismic and Evolutionary Biology and an undergraduate advisor in the Life Sciences at Harvard University. With research interests in evolutionary biology and history of science, he teaches courses that either focus on one of the areas or combine the two. He has written two books: *Infinite Tropics*, a collection of the writings of Alfred Russel Wallace, and, with James D. Watson, *DNA: The Secret of Life*, which is part history, part exploration of the controversies surrounding DNA-based technologies.

ANDREW A. BIEWENER is Charles P. Lyman Professor of Biology in the Department of Organismic and Evolutionary Biology at Harvard University and Director of the Concord Field Station. He teaches both introductory and advanced courses in anatomy, physiology, and biomechanics. His research focuses on the comparative biomechanics and neuromuscular control of mammalian and avian locomotion, with relevance to biorobotics. He is currently Deputy Editor-in-Chief for the Journal of Experimental Biology. He also served as President of the American Society of Biomechanics.

BRIAN D. FARRELL is Director of the David Rockefeller Center for Latin American Studies and Professor of Organismic and Evolutionary Biology and Curator in Entomology at the Museum of Comparative Zoology at Harvard University. He is an authority on coevolution between insects and plants and a specialist on the biology of beetles. He is the author of many scientific papers and book chapters on the evolution of ecological interactions between plants, beetles, and other insects in the tropics and temperate zone. Professor Farrell also spearheads initiatives to repatriate digital information from scientific specimens of insects in museums to their tropical countries of origin. In 2011–2012, he was a Fulbright Scholar to the Universidad Autónoma de Santo Domingo in the Dominican Republic. Professor Farrell received a BA degree in Zoology and Botany from the University of Vermont and MS and PhD degrees from the University of Maryland.

N. MICHELE HOLBROOK is Charles Bullard Professor of Forestry in the Department of Organismic and Evolutionary Biology at Harvard University. She teaches an introductory course on biodiversity as well as advanced courses in plant biology. She studies the physics and physiology of vascular transport in plants with the goal of understanding how constraints on the movement of water and solutes between soil and leaves influences ecological and evolutionary processes.

ASSESSMENT AUTHORS

JEAN HEITZ is a Distinguished Faculty Associate at the University of Wisconsin in Madison, WI. She has worked with the two-semester introductory sequence for biological sciences majors for over 30 years. Her primary roles include developing both interactive discussion/recitation activities designed to uncover and modify misconceptions in biology and open-ended process-oriented labs designed to give students a more authentic experience with science. The lab experience includes engaging all second-semester students in independent research, either mentored research or a library-based meta-analysis of an open question in the literature. She is also the advisor to the Peer Learning Association and is actively involved in TA training. She has taught a graduate course in "Teaching College Biology," has presented active-learning workshops at a number of national and international meetings, and has published a variety of lab modules, workbooks, and articles related to biology education.

MARK HENS is Associate Professor of Biology at the University of North Carolina Greensboro, where he has taught introductory biology since 1996. He is a National Academies Education Mentor in the Life Sciences and is the director of his department's Introductory Biology Program. In this role, he guided the development of a comprehensive set of assessable student learning outcomes for the two-semester introductory biology course required of all science majors at UNCG. In various leadership roles in general education, both on his campus and statewide, he was instrumental in crafting a common set of assessable student learning outcomes for all natural science courses for which students receive general education credit on the sixteen campuses of the University of North Carolina system.

JOHN MERRILL is Director of the Biological Sciences Program in the College of Natural Science at Michigan State University. This program administers the core biology course sequence required for all science majors. He is a National Academies Education Mentor in the Life Sciences. In recent years he has focused his research on teaching and learning with emphasis on classroom interventions and enhanced assessment. A particularly active area is the NSF-funded development of computer tools for automatic scoring of students' open-ended responses to conceptual assessment questions, with the goal of making it feasible to use open-response questions in large-enrollment classes.

RANDALL PHILLIS is Associate Professor of Biology at the University of Massachusetts Amherst. He has taught in the majors introductory biology course at this institution for 19 years and is a National Academies Education Mentor in the Life Sciences. With help from the Pew Center for Academic Transformation (1999), he has been instrumental in transforming the introductory biology course to an active learning format that makes use of classroom communication systems. He also participates in an NSF-funded project to design model-based reasoning assessment tools for use in class and on exams. These tools are being designed to develop and evaluate student scientific reasoning skills, with a focus on topics in introductory biology.

DEBRA PIRES is an Academic Administrator at the University of California, Los Angeles. She teaches the introductory courses in the Life Sciences Core Curriculum. She is also the Instructional Consultant for the Center for Education Innovation & Learning in the Sciences (CEILS). Many of her efforts are focused on curricular redesign of introductory biology courses. Through her work with CEILS, she coordinates faculty development workshops across several departments to facilitate pedagogical changes associated with curricular developments. Her current research focuses on what impact the experience of active learning pedagogies in lower division courses may have on student performance and concept retention in upper division courses.

ASSESSMENT CONTRIBUTORS

ELENA R. LOZOVSKY, Principal Staff Scientist, Department of Organismic and Evolutionary Biology, Harvard University

FULTON ROCKWELL, Research Associate, Department of Organismic and Evolutionary Biology, Harvard University

VISION AND STORY OF *BIOLOGY: HOW LIFE WORKS*

Dear students and instructors,

One of the most frequent questions we get about the second edition of *Biology: How Life Works* is, "Has science really changed that much in three years?"

Ongoing discoveries in biology mean that a new edition of an introductory biology textbook will certainly have some new science content. But, more importantly, our second edition is new in the sense that we had the opportunity, for the first time, to listen to students and instructors who used *HLW* in the classroom. The second edition is responsive to this group and their input has proven invaluable.

What we heard from this community is that the philosophy of *HLW* resonates with students and instructors. They appreciate a streamlined text that rigorously focuses on introductory biology, an emphasis on integration, a modern treatment of biology, and equal attention to text, assessment, and media. These elements haven't changed—they are the threads that connect the first and second editions. In fact, all of the changes of the second edition are integrated within the framework of the first edition; they are not simply add-ons.

We are particularly excited about the work we've done in assessment. In the first edition, we worked with a creative and dedicated team of assessment authors to create something wholly new: not a standard test bank, but a thoughtful, curated, well-aligned set of questions that can be used for teaching as well as testing. These questions are written at a variety of cognitive levels. In addition, they can be used in a variety of contexts, including pre-class, in-class, homework, and exam, providing a learning path for students.

Our approach was so well received that we took it a step further in the second edition. The *HLW* team is excited to have Melissa Michael, our lead assessment author in the first edition, join us as a lead author in the second edition. Her new role allows her to work more closely with the text and media, which makes for an even tighter alignment among these various components.

Instructors have told us that they especially like the activities that can be used in class to foster active learning among students. In response, the second edition includes a rich set of activities across the introductory biology curriculum. Some are short, taking just a couple of minutes to explore a specific topic or concept. Others are longer, spanning several class periods and exploring topics and concepts across many chapters.

Although students and instructors appreciated a streamlined text, we also heard that more attention was needed in ecology. In response, we added a chapter that focuses on physical processes and global ecology. This new chapter also had a ripple effect throughout the later chapters of Part 2, giving us more space to explore other ecological concepts more deeply as well.

The media in *HLW* is many layered, so that a static visual synthesis on the page becomes animated online — and even interactive — through visual synthesis maps and simulations, using a consistent visual language and supported by assessment. Media resources for this new edition have been expanded to reflect its increased emphasis on global ecology — for example, there is a new Visual Synthesis figure and online map on the flow of matter and energy in ecosystems. We also developed additional media resources that focus on viruses, cells, and tissues.

We feel that this edition is a wonderful opportunity for us to continue to develop an integrated set of resources to support instructor teaching and student learning in introductory biology. Thank you for taking the time to use it in the classroom.

Sincerely,
The *Biology: How Life Works* Author Team

RETHINKING **BIOLOGY**

The *Biology: How Life Works* team set out to create a resource for today's biology students that would reimagine how content is created and delivered. With this second edition, we've refined that vision using feedback from the many dedicated instructors and students who have become a part of the *How Life Works* community.

We remain committed to the philosophy of *How Life Works*: a streamlined, integrated, and modern approach to introductory biology.

Thematic

We wrote *How Life Works* with six themes in mind. We used these themes as a guide to make decisions about which concepts to include and how to organize them. The themes provide a framework that helps students see biology as a set of connected concepts rather than disparate facts.

- The scientific method is a deliberate way of asking and answering questions about the natural world.
- Life works according to fundamental principles of chemistry and physics.
- The fundamental unit of life is the cell.
- Evolution explains the features that organisms share and those that set them apart.
- Organisms interact with one another and with their physical environment, shaping ecological systems that sustain life.
- In the 21st century, humans have become major agents in ecology and evolution.

Selective

It is unrealistic to expect the majors course to cover everything. We envision *How Life Works* not as a reference for all of biology, but as a resource focused on foundational concepts, terms, and experiments. We explain fundamental topics carefully, with an appropriate amount of supporting detail, so that students leave an introductory biology class with a framework on which to build.

Integrated

How Life Works moves away from minimally related chapters to provide guidance on how concepts connect to one another and the bigger picture. Across the book, key concepts such as chemistry are presented in context and Cases and Visual Synthesis figures throughout the text provide a framework for connecting and assimilating information.

WHAT'S NEW IN **THE SECOND EDITION?**

Expanded ecology coverage on physical processes and global ecology provides additional emphasis on ecological concepts, while ensuring that content is integrated into the larger theme of evolution. *Learn more about the expanded ecology coverage on page xviii.*

Visual Synthesis Figures and Online Maps on the Flow of Matter and Energy through Ecosystems, Cellular Communities, and Viruses allow students to explore connections between concepts through dynamic visualizations. *Learn more about the new media on page xix.*

Lead Author Melissa Michael guides our assessment team in refining and expanding our collection of thoughtful, well-curated assessments. Dr. Michael's role ensures a tight alignment between the assessments and the media and text. *Learn more on page xvi.*

A rich collection of in-class activities provides active learning materials for instructors to use in a variety of settings. *Learn more about the new in-class activities on page xvii.*

Improved LaunchPad functionality makes it easier to search and filter within our expansive collection of assessment questions. *Learn more about new LaunchPad functionality on page xii.*

TABLE OF CONTENTS

The table of contents is arranged in a familiar way to allow its easy use in a range of introductory biology courses. On closer look, there are significant differences that aim to help biology teachers incorporate the outlooks and research of biology today. **Key differences** are identified by ● and **unique chapters** by ✪.

EVOLUTION COVERAGE: Chapter 1 introduces evolution as a major theme of the book before discussing gene expression in Chapters 3 and 4 as a foundation for later discussions of the conservation of metabolic pathways and enzyme structure (Chapters 6–8) and genetic and phenotypic variation (Chapters 14 and 15). After the chapters on the mechanisms and patterns of evolution (Chapters 21–24), we discuss the diversity of all organisms in terms of adaptations and comparative features, culminating in ecology as the ultimate illustration of evolution in action.

CHEMISTRY: Chemistry is taught in the context of biological processes, emphasizing the key principle that structure determines function.

THE CELL: The first set of chapters emphasizes three key aspects of a cell—information flow (Chapters 3 and 4), actively maintaining a constant internal environment (Chapter 5), and harnessing energy (Chapters 6–8). Placing these basic points at the start of the textbook gives them emphasis and helps students build their knowledge of biology naturally.

CASE STUDIES: Biology is best understood when presented using real and engaging examples as a framework for synthesizing information. Eight carefully positioned Cases help provide this framework. For example, the Case about your personal genome is introduced before the set of chapters on genetics and is revisited in each of these chapters where it serves to reinforce important concepts.

GENETICS: The genetics chapters start with genomes and move to inheritance to provide a modern, molecular look at genetic variation and how traits are transmitted.

UNIQUE CHAPTERS: *Biology: How Life Works*, Second Edition includes chapters that don't traditionally appear in introductory biology texts, one in almost every major subject area. These novel chapters represent shifts toward a more modern conception of certain topics in biology and are identified by ✪.

"The approach to teaching is something my colleagues and I had been waiting for in a textbook. However, the text is flexible enough to accommodate a traditional teaching style."

–Steve Uyeda, *Pima Community College*

To hear the authors talk about the table of contents in more depth, visit **biologyhowlifeworks.com**

BIOGEOCHEMICAL CYCLES: We present the carbon cycle as a bridge between the molecular and organismal parts of the book, showing how different kinds of organisms use the biochemical processes discussed in the first half of the book to create a cycle that drives life on Earth and creates ecosystems. The carbon cycle along with other biogeochemical cycles—sulfur and nitrogen—provides the conceptual backbone around which prokaryotic diversity is organized.

PLANT DEFENSE: The chapter on plant defense provides a strong ecological and case-based perspective on the strategies plants use to survive their exploitation by pathogens and herbivores.

DIVERSITY AND PHYSIOLOGY: Diversity follows physiology in order to provide a basis for understanding the groupings of organisms and to avoid presenting diversity as a list of names to memorize. When students understand how organisms function, they can understand the different groups in depth and organize them intuitively. To give instructors maximum flexibility, brief descriptions of unfamiliar organisms and the major groups of organisms have been layered in the physiology chapters, and the diversity chapters include a brief review of organismal form and function.

NEW CHAPTER 48: BIOMES AND GLOBAL ECOLOGY is part of the greatly expanded ecology coverage on physical processes and global ecology. The new coverage broadens connections between ecological concepts, and is carefully integrated into the larger theme of evolution.

RETHINKING THE TEXTBOOK **THROUGH LAUNCHPAD**

Ordinarily, textbooks are developed by first writing chapters, then making decisions about art and images, and finally assembling a test bank and ancillary media. *Biology: How Life Works* develops the text, visual program, and assessment at the same time. These three threads are tied to the same set of core concepts, share a common language, and use the same visual palette, which ensures a seamless learning experience for students throughout the course.

The text, visuals, and assessments come together most effectively through LaunchPad, Macmillan's integrated learning management system. In LaunchPad, students and instructors can access all components of *Biology: How Life Works.*

LaunchPad resources for *How Life Works* are flexible and aligned. Instructors have the ability to select the visuals, assessments, and activities that best suit their classroom and students. All resources are aligned to one another as well as to the text to ensure effectiveness in helping students build skills and develop knowledge necessary for a foundation in biology.

NEW IN LAUNCHPAD FOR BIOLOGY: *HOW LIFE WORKS*, SECOND EDITION

Functionality to search the question database and filter questions for a number of variables including Core Concept, difficulty level, Bloom's level, and class setting allows instructors to make best use of the robust assessment assets of *How Life Works.*

New question types include sequenced questions and multiple true-false. *Learn more on page xviii.*

Metadata tags for each question show at a glance information, including instructional guidance for select questions.

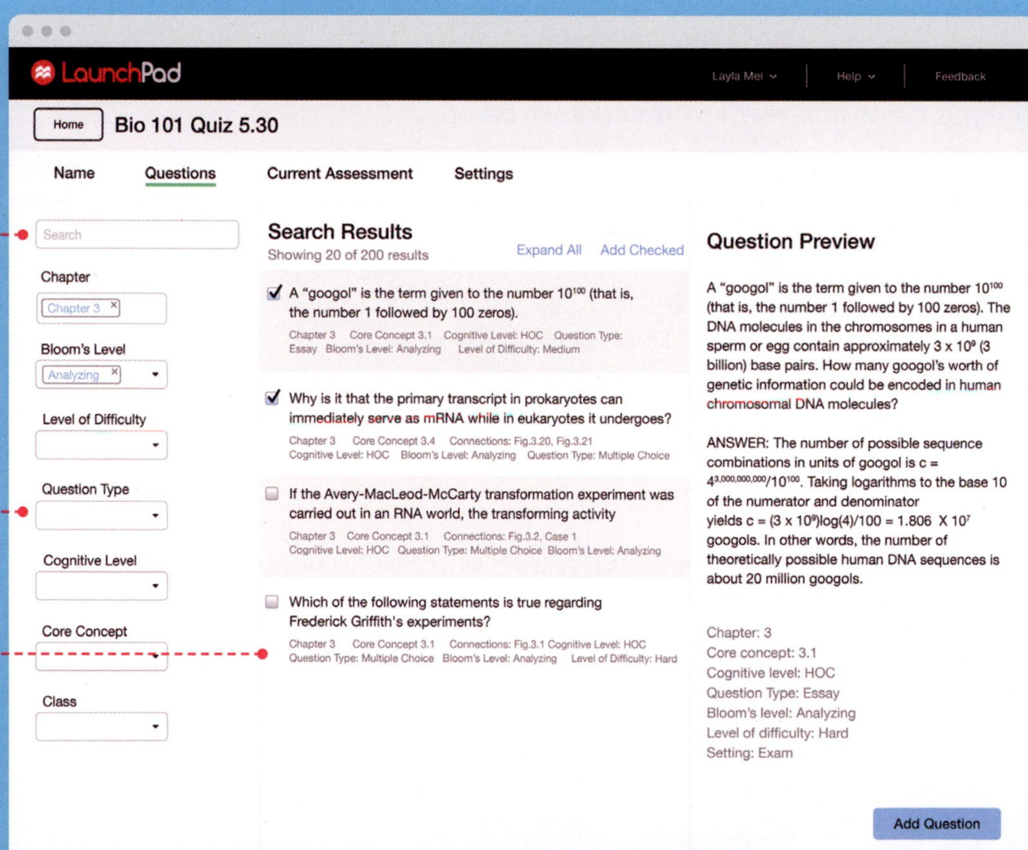

LAUNCHPAD

Where content counts.
Where service matters.
Where students learn.

LAUNCHPADWORKS.COM

Powerful, Simple, and Inviting

LaunchPad for *How Life Works* includes:

The complete *Biology: How Life Works* interactive e-Book

Carefully curated multimedia visuals and assessments, assignable by the instructor and easily accessible by students.

LearningCurve adaptive quizzing that puts "testing to learn" into action, with individualized question sets and feedback for each student based on his or her correct and incorrect responses. All the questions are tied back to the e-Book to encourage students to use the resources at hand.

Pre-built units that are easy to adapt and augment to fit your course.

A Gradebook that provides clear feedback to students and instructors on performance in the course as a whole and individual assignment.

LMS integration allows LaunchPad to be easily integrated into your school's learning management system so your Gradebook and roster are always in sync.

LEARNINGCURVE

LEARNINGCURVEWORKS.COM

Students agree that LearningCurve is extremely helpful in their studies.

Macmillan's LearningCurve adaptive quizzing is part of almost every LaunchPad and has been enthusiastically embraced by students and instructors alike. LearningCurve provides specific feedback for every question and includes links to relevant sections of the e-Book. Questions are written exclusively for the LaunchPad in which they are offered and the question banks are robust and varied.

LearningCurve is **BENEFICIAL**

OVER 90% of students said LearningCurve helped them get better grades

95% of students would recommend LearningCurve to another student

LearningCurve is **APPROVED** by students

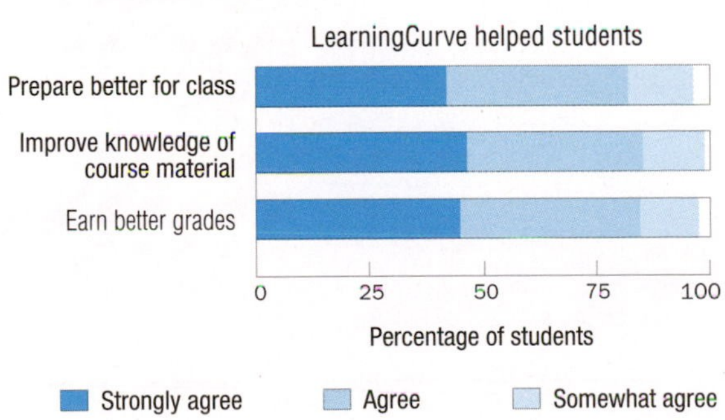

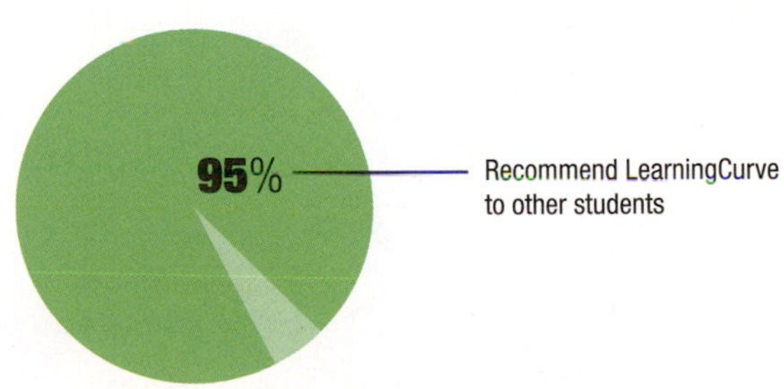

RETHINKING **THE VISUAL PROGRAM**

The art in the text of *How Life Works* and the associated media in LaunchPad were developed in coordination with the text and assessments to present an integrated and engaging visual experience for students.

Two of the biggest challenges introductory biology students face are connecting concepts across chapters and building a contextual picture, or visual framework, of a complex process. To help students think like biologists, we provide **Visual Synthesis** figures at twelve key points in the book. These figures bring together multiple images students have already seen into a visual summary, helping students see how individual concepts connect to tell a single story.

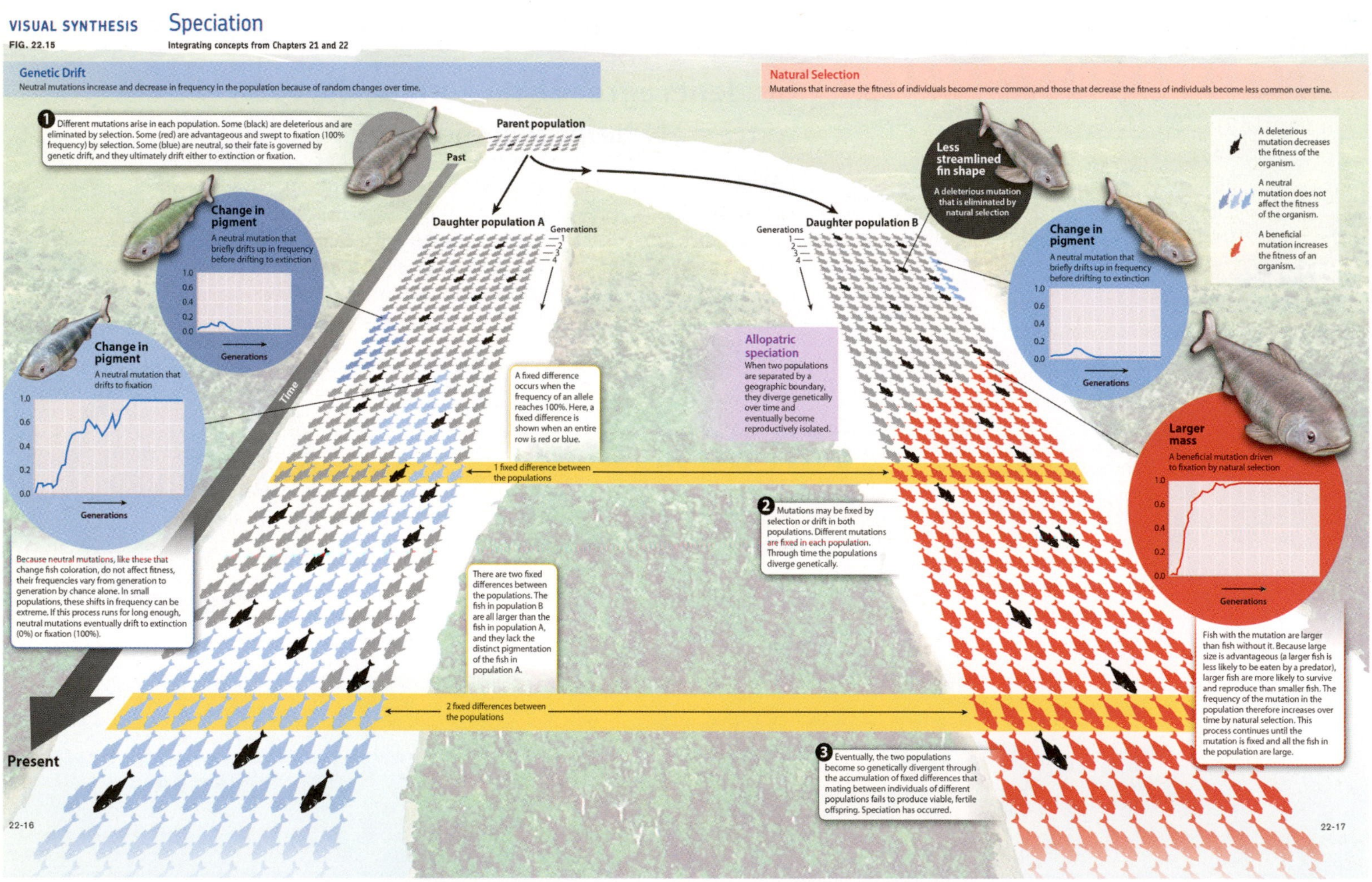

This Visual Synthesis figure on Speciation brings together multiple concepts from the chapters on evolution.

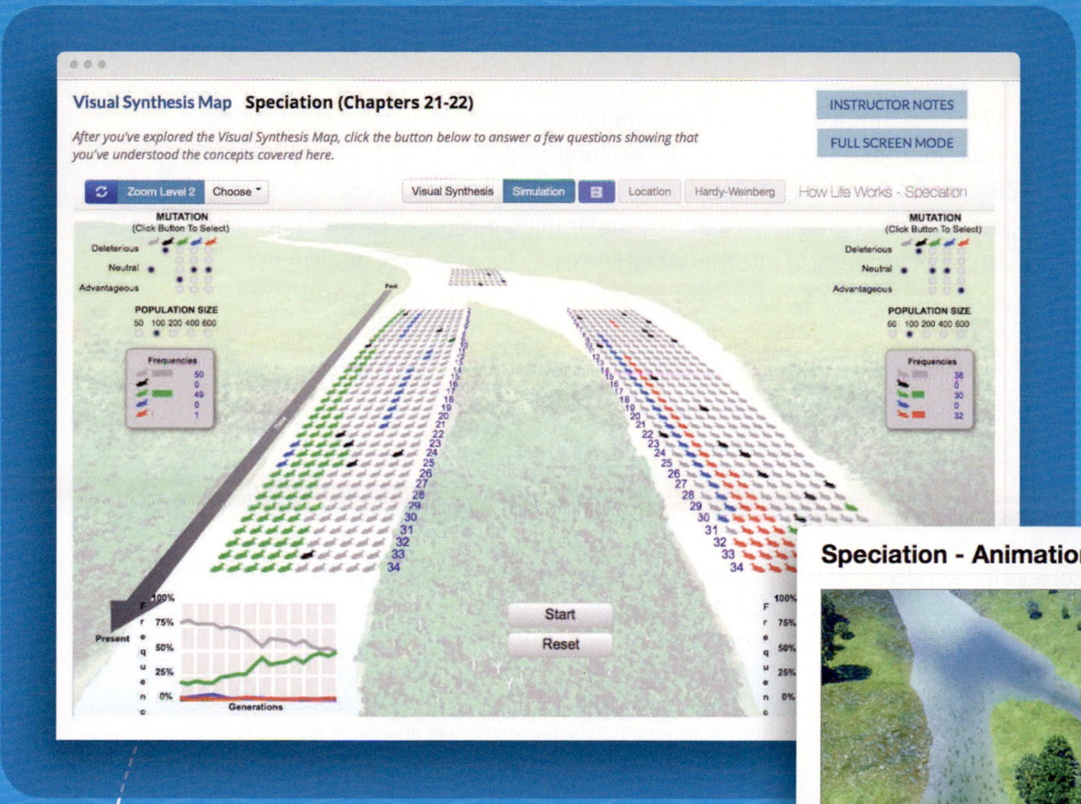

Online, the Visual Synthesis figures turn into zoomable, dynamic **Visual Synthesis Maps** where students can explore both the big picture and the details.

Many of the Visual Synthesis Maps also include **3D animations** (for concepts that are best learned by ***seeing*** them in action) and **simulations** (for concepts best learned by ***doing***).

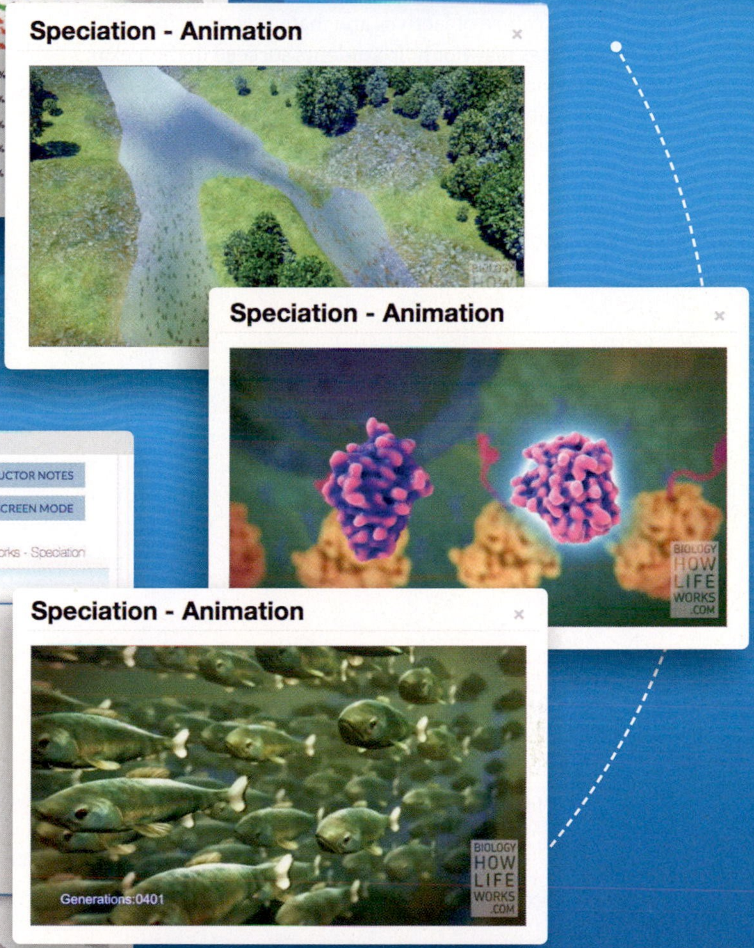

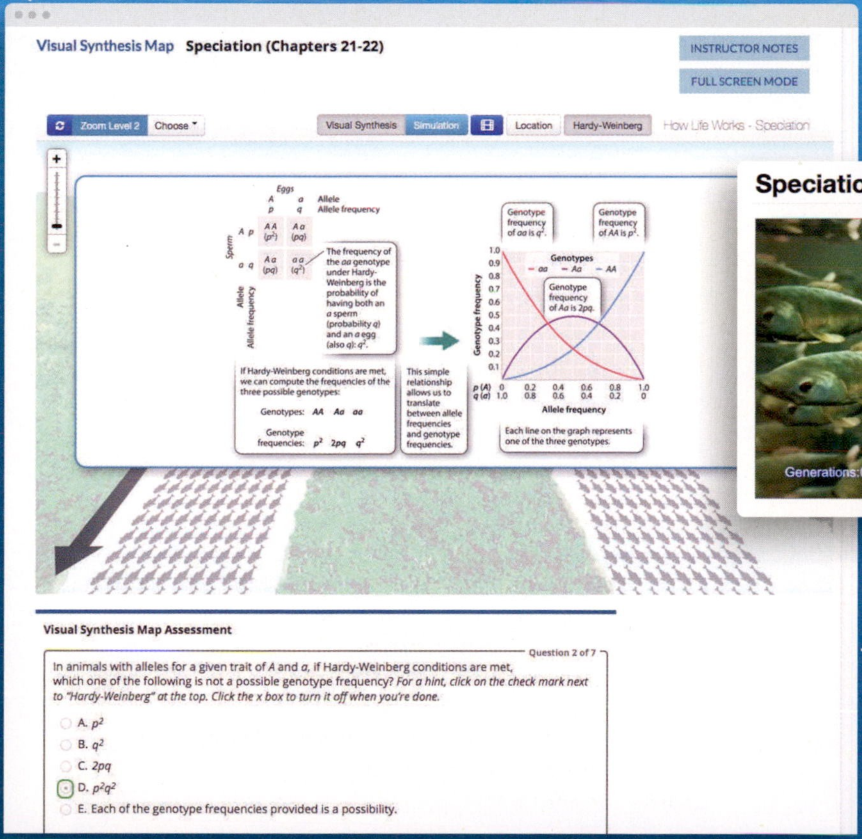

Assessment questions accompany all Visual Synthesis Maps, 3D animation activities, and simulations to help students work through a core concept in a variety of ways.

RETHINKING ASSESSMENT

Well-designed assessment is a tremendous tool for instructors in gauging student understanding, actively teaching students, and preparing students for exams. The *Biology: How Life Works* assessment author team applied decades of experience researching and implementing assessment practices to create a variety of questions and activities for pre-class, in-class, homework, and exam settings. All assessment items are carefully aligned with the text and media and have the flexibility to meet the needs of instructors with any experience level, classroom size, or teaching style.

Alignment

If the questions, exercises, and activities in the course aren't aligned with course objectives and materials, practice with these resources may not help students succeed in their exam or in future biology courses. Each *How Life Works* assessment item is carefully aligned to the goals and content of the text, and to the assessment items used in other parts of the course. Students are guided through a learning path that provides them with repeated and increasingly challenging practice with the important concepts illustrated in the text and media.

Flexibility

The *How Life Works* assessment authors teach in a variety of classroom sizes and styles, and recognize that there is a wide diversity of course goals and circumstances. Each set of materials, from in-class activities to exam questions, includes a spectrum of options for instructors. All the assessment items are housed in the LaunchPad platform, which is designed to allow instructors to assign and organize assessment items to suit the unique needs of their course and their students.

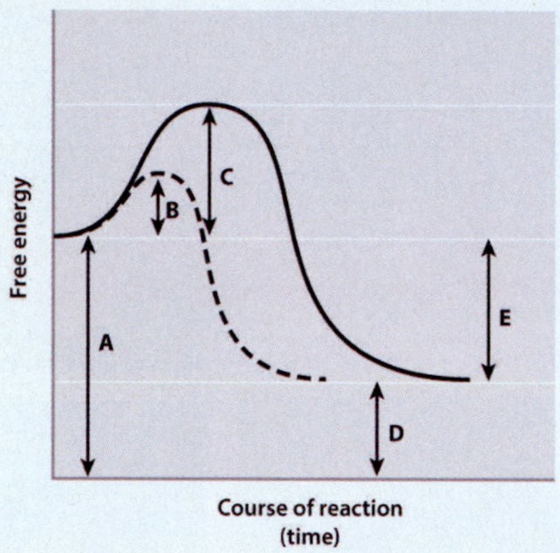

Sample assessment questions, including a sequenced question based on a graph.

Answer the following questions about the reactions shown in the graph.

Which arrow indicates the activation energy of the catalyzed reaction?

❑ A | ■ **B** | ❑ C | ❑ D | ❑ E

Which arrow indicates the activation energy of the uncatalyzed reaction?

❑ A | ❑ B | ■ **C** | ❑ D | ❑ E

Which arrow represents the free energy of the substrate?

■ **A** | ❑ B | ❑ C | ❑ D | ❑ E

Which arrow indicates the free energy of the products?

❑ A | ❑ B | ❑ C | ■ **D** | ❑ E

Which arrow indicates the change in free energy (ΔG) of the reaction?

❑ A | ❑ B | ❑ C | ❑ D | ■ **E**

Is the reaction illustrated by the solid line endergonic or exergonic?

a.) endergonic **b.) exergonic**

Is the reaction illustrated by the dashed line endergonic or exergonic?

a.) endergonic **b.) exergonic**

Which of the following reactions would you predict could be coupled to ATP synthesis from ADP + Pi? Select all that apply.

a.) creatine phosphate + H_2O → creatine + Pi, ΔG -10.3 kcal/mol
b.) phosphoenolpyruvate + H_2O → pyruvate + Pi, ΔG -14.8 kcal/mol
c.) glucose 6-phosphate + H_2O → glucose + Pi, ΔG -3.3 kcal/mol
d.) glucose 1-phosphate + H_2O → glucose + Pi, ΔG -5.0 kcal/mol
e.) glutamic acid + NH_3 → glutamine, ΔG +3.4 kcal/mol

The emperor penguins of Antarctica live on a diet of fish and crustaceans obtained from the cold Antarctic seawaters. During their annual breeding cycle, however, they migrate across the frozen continent to their breeding grounds 50 miles away from the sea (and 50 miles away from their source of food). For over two months the male emperor penguins care for and incubate the eggs while the females return to the sea to feed. During this time a male penguin can lose up to 50% of its biomass (by dry weight). Where did this biomass go?

a.) It was converted to CO_2 and H_2O and then released.
b.) It was converted to heat and then released.
c.) It was converted to ATP molecules.

RETHINKING **ACTIVITIES**

Active learning exercises are an important component of the learning pathway and provide students with hands-on exploration of challenging topics and misconceptions. The second edition of *Biology: How Life Works* includes a new collection of over 40 in-class activities crafted to address the concepts that students find most challenging.

The activities collection was designed to cover a range of classroom sizes and complexity levels, and many can be easily adapted to suit the available time and preferred teaching style. Each activity includes a detailed activity guide for instructors. The activity guide introduces the activity, outlines learning objectives, and provides guidance on how to implement and customize the activity.

Many of the assessment questions and activities throughout *How Life Works* incorporate experimental thinking and data analysis. In addition, two activity types round out the assessment collection by providing applied practice with the data sets and examples from the text. Through ***Working with Data*** activities, students explore and analyze the experiment from a *How Do We Know?* figure from the text. ***Mirror Experiment*** activities introduce students to a new scientific study that relates to, or "mirrors," one of the *How Do We Know?* experiments and ask them to apply what they have learned about data analysis to this new scenario.

List of Materials

1. Activity Guide (file name: 1 Activity Guide Ch11 Double Double Double)
2. In-class Presentation (file name: 2 In-class Presentation Ch11 Double Double Double)
3. Exam Questions (file name: 3 Exam Qs Ch11 Double Double Double)

Description

Students are asked to identify corresponding depictions of chromosomes and DNA pre- and post-replication. The questions take the form shown in this sample, in which students must select one of the DNA representations that most closely depicts the chromosome labeled with the question mark.

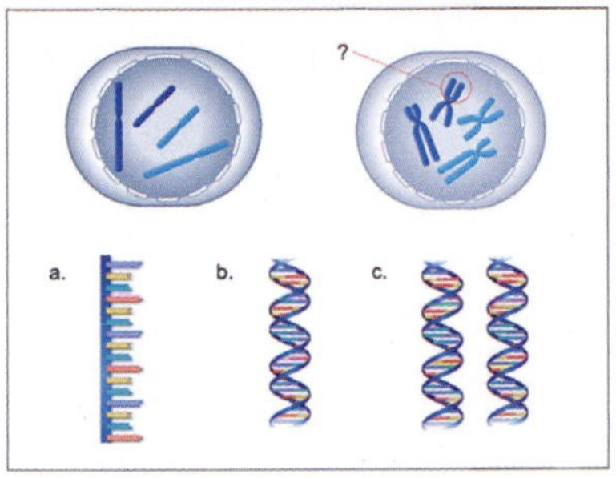

The first clicker question, showing only the mitosis figures, will verify that the students recognize these image types. This might be skipped if the activity comes soon after covering mitosis.

This sequence works very well if students are encouraged to discuss the questions with each other, but with essentially no other instructor guidance or input. Furthermore, clicker results are **not** shown until the final clicker is completed.

Class size and timing

This activity can be used in essentially any class size since it is based on clicker questions. It can be quite brief, 5-10 minutes, even with summary discussion.

Connection to Vision and Change:

This activity is aimed at the Core Concept 3. Information Flow, Exchange, and Storage.

Excerpts of activities from Chapters 11 and 36

Mirror Experiment Activity 36.20

The experiment described below explored the same concepts as the one described in Figure 36.20 in the textbook. Read the description of the experiment and answer the questions below the description to practice interpreting data and understanding experimental design.

Mirror Experiment activities practice skills described in the brief Experiment and Data Analysis Primers, which can be found by clicking on the "Resources" button on the upper right of your LaunchPad homepage. Certain questions in this activity draw on concepts described in the **Experimental Design** and **Data and data Presentation** primers. Click on the "Key Terms" buttons to see definitions of terms used in the question, and click on the "Primer Section" button to pull up a relevant section from the primer.

Experiment

Background

As you have learned, the somatosensory cortex is responsible for processing "touch" stimuli. If someone were tickling the bottom of your foot, mechanoreceptors in the skin of your foot would fire action potentials. These signals would (ultimately) be relayed to the somatosensory cortex portion of your brain, and then your motor cortex. Following this chain of events, you might jerk your foot away from the tickler.

If you were to take a cross section of the somatosensory cortex, you would find that neurons are arranged in six distinct layers; the first layer would be composed of superficial neurons located near the brain surface, and the sixth layer would be composed of the "deepest" neurons (that is, those closest to the white matter). How are neurons that respond to touch stimuli organized in the somatosensory cortex? Do neurons in the six different layers of the somatosensory cortex respond to different types of stimuli?

Hypothesis

Vernon Mountcastle hypothesized that researchers could create a diagram of the somatosensory cortex by tracking which neurons responded to different types of touch stimuli.

Experiment

Mountcastle exposed cats to two types of stimuli: (1) cutaneous or superficial stimuli, which included touching hairs or touching the skin; and (2) deep stimuli, which included bending and extending joints or touching the connective tissue surrounding muscles. He was able to track which neurons in the somatosensory cortex fired action potentials in response to these two types of stimuli, and measured their firing rates (Figure 1).

Results

Mountcastle determined that neurons involved in processing the same type of stimuli are organized in "vertical columns." These columns are composed of cells belonging to different layers of the somatosensory cortex stacked one on top of another. These results demonstrated that just because a neuron responds to deep stimuli does not mean that this neuron will be found deep within the brain; similarly, a neuron that responds to cutaneous stimuli will not necessarily be located near the brain surface. In addition to identifying vertical columns of neurons, Mountcastle

Question 1 of 8

Mountcastle noted that within the somatosensory cortex, neurons that respond to the same type of stimulus are arranged in vertical columns. One way Mountcastle discovered this vertical arrangement was by inserting a measuring device (to record action potentials) into the somatosensory cortex at different angles. This approach is depicted in Figure 2 below. Which of the following statements is true regarding measuring devices that were inserted into the brain at angles of 45° and 90°?

- A. The measuring device inserted into the brain at a 90° angle will likely encounter neurons belonging to the same vertical column.
- B. The measuring device inserted into the brain at a 45° angle will likely encounter neurons that belong to different vertical columns.
- C. All of the neurons encountered by the measuring device inserted into the brain at a 90° angle will likely respond to the same type of stimuli.
- D. Neurons encountered by the measuring device inserted into the brain at a 45° angle will likely respond to different types of stimuli.
- E. All of the answer options are true.

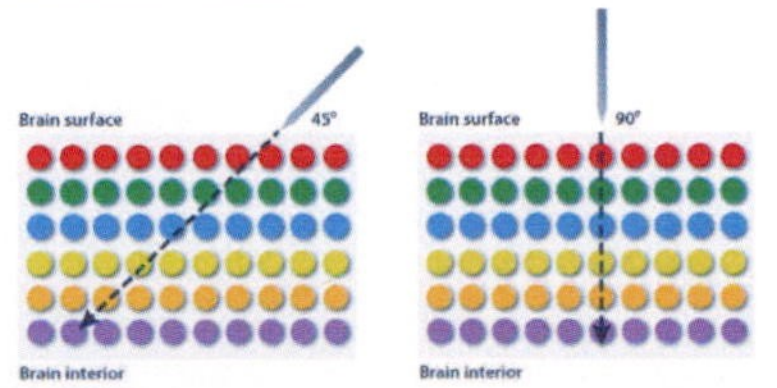

Figure 2

Data from Mountcastle, V. B., 1957. Modality and topographic properties of single neurons of cat's somatic sensory cortex. *J Neurophysiol.* 20, 408-34.

Submit

WHAT'S NEW IN THE SECOND EDITION?

From the start, *Biology: How Life Works* was envisioned not as a reference book for all of biology, but as a resource focused on foundational concepts, terms, and experiments, all placed in a framework that motivates student interest through a coherent and authentic presentation of current science. In preparing this edition, we carefully considered the latest breakthroughs and incremental, but nevertheless significant, changes across the fields of biology. We also reached out to adopters, instructors not using our book, and primary literature to determine what concepts and details are relevant, important, and necessary additions. Our integrated approach to text, media, and assessment means that all changes are carefully reflected in each of these areas.

MAJOR CHANGES AND UPDATES

We've greatly expanded the coverage of ecology in the second edition of *Biology: How Life Works*.

A new ecology chapter, Chapter 48: Biomes and Global Ecology, takes a broad look at ecology on the largest scale. It begins with how and why climates are distributed as they are around the world and introduces Earth's major biomes. Biomes crystallize the relationships among ecology, evolution, and physical environment—landscapes look different in different parts of the world because of the morphological and physiological adaptations that plants and animals have made to different climates. The chapter is distinguished by extensive discussion of biomes in the aquatic realm, especially in the oceans.

Chapter 47: Species Interactions, Communities, and Ecosystems **includes expanded coverage of the ways species interact with one another in communities**. This chapter now has more detail on facilitation, herbivory, and biodiversity.

Chapter 49: The Anthropocene: Humans as a Planetary Force **includes new discussions exploring how human activities affect ecology**. The chapter now examines fracking and its effects on the carbon cycle, habitat loss and its effects on biodiversity, and the overexploitation of resources and its effects on community ecology. The chapter ends with a new section on conservation biology that explores how conservationists are working to preserve natural habitats.

A new Visual Synthesis figure on the flow of matter and energy through ecosystems illustrates, explores, and physically situates the relationships among concepts from Chapters 25, 26, 47, 48, and 49. In LaunchPad, students and instructors can interact with an accompanying dynamic, zoomable, and interactive Visual Synthesis Map based on this figure.

Our new collection of over 40 in-class activities provides tools for instructors to engage their students in active learning. In-class activities are designed to address difficult concepts, and can be used with a variety of classroom sizes and teaching styles. Each activity includes a detailed activity guide for instructors.

We've expanded our collection of high-quality assessment questions by adding over 1000 new questions. New questions are particularly focused on higher-order thinking, including questions based on figures or data, and questions that ask students to consider how perturbing a system would affect outcomes. As in the first edition, questions are carefully aligned with core concepts from the text. New and revised assessment questions also accompany Visual Synthesis Maps, simulations, animations, and other visual media, to more effectively probe student understanding of the media tools they've explored.

The second edition also includes several new question types. **Sequenced questions** ask students several, individually scored questions about a single scenario or system. These questions often build on one another to guide students from lower-order thinking to higher-order thinking about the same concept. **Multiple true–false questions** ask students several, individually scored true–false questions about a single scenario or system.

Improved functionality in LaunchPad allows instructors to search the question database and filter questions by a number of variables, including core concept, difficulty level, Bloom's level, and class setting. Metadata tags for each question show additional information at a glance, including instructional guidance for select questions.

NEW MEDIA

Cell Communities Visual Synthesis Map to accompany the printed Visual Synthesis figure

Virus Visual Synthesis Map to accompany the printed Visual Synthesis figure

New Visual Synthesis figure and map on the Flow of Matter and Energy in Ecosystems

Virus Video featuring author Rob Lue

Cell Membrane simulation

New Animations

Chapter 9: Basic Principles of Cell Signaling

Chapter 9: G protein-coupled Receptor Signaling

Chapter 9: Signal Amplification

Chapter 10: Dynamic Nature of Microtubules

Chapter 10: Motor Proteins

Chapter 10: Dynamic Nature of Actin Filaments

Chapter 19: Lac Operon

Chapter 20: ABC Model of Floral Development

Chapter 40: Glucose Absorption in the Small Intestine

Chapter 42: Gastrulation

NEW TOPICS AND OTHER REVISIONS

The following is a detailed list of content changes in this edition. These range from the very small (nucleotides shown at physiological pH) to quite substantial (an entire new chapter in the ecology section). Especially important changes are indicated with an asterisk (✪).

- ✪ New coverage of functional groups (Chapter 2)
- Nucleotides now shown at physiological pH (Chapter 3)
- Amino acids now shown at physiological pH (Chapter 4)
- The story of the evolution of photosynthesis now brought together in a single major section at the end of Chapter 8 (section 8.5)
- ✪ Chapters 9 and 10 streamlined to better match our mission statement
- A new discussion of cellular response and what determines it (Chapter 9)
- ✪ New inclusion of the trombone model of DNA replication (Chapter 12)
- ✪ Addition of CRISPR technology (Chapter 12)
- Expanded coverage of retrotransposons and reverse transcriptase (Chapter 13)
- A new *How Do We Know?* figure explaining Mendel's experimental results (Chapter 16)
- New coverage of the mechanism of *X*-inactivation (Chapter 19)
- An expanded discussion of non-random mating and inbreeding depression (Chapter 21)
- ✪ Addition of the effect of mass extinctions on species diversity (Chapter 23)
- Updated discussion of the relationship between Neanderthals and *Homo sapiens*, as well as Denisovans (Chapter 24)
- Significantly revised link between the carbon cycle, biodiversity, and ecology (Chapter 25)
- New branching order of the eukaryote tree to reflect new research in the past three years (Chapter 27 and onward)
- A new paragraph on ciliates (Chapter 27)
- A new explanation of protist diversity (Chapter 27)
- ✪ A new discussion of plant nutrients with a table (Chapter 29)
- ✪ An enhanced discussion of seeds, including the development of the embryo and dispersal structures (Chapter 30)
- New coverage of the genetic advantages of alternation of generations, and how it allows inbreeding (Chapter 30)
- Addition of apomixis (Chapter 30)
- The section on the role of plant sensory systems in the timing of plant reproduction moved from Chapter 30 to Chapter 31
- Completely revised explanation of the basis for angiosperm diversity (Chapter 33)
- Brief descriptions of unfamiliar organisms and the major groups of organisms layered in the animal physiology chapters to make it easier to teach physiology before diversity (Chapters 35-42)
- ✪ Brief review of organismal form and function in the plant and animal diversity chapters (Chapters 33 and 44), allowing these chapters to be used on their own or before the physiology chapters and giving instructors maximum flexibility
- ✪ A new section on the composition of blood (Chapter 39)
- New diagrams of hormone feedback loops in the menstrual cycle (Chapter 42)
- A new introduction to the immune system (Chapter 43)
- ✪ A new discussion of nematodes (Chapter 44)
- Introduction of a newly discovered species, *Dendrogramma enigmatica* (Chapter 44)
- A simplified population growth equation (Chapter 46)
- A new discussion of facilitation (Chapter 47)
- An expanded discussion of herbivory (Chapter 47)
- A new example of microbial symbionts (Chapter 47)
- A new discussion of biodiversity and its importance (Chapter 47)
- ✪ An entirely new chapter on physical processes that underlie different biomes (Chapter 48)
    - Differential solar energy around the globe and seasonality
    - Wind and ocean currents
    - Effects of circulation and topography on rainfall
    - Expanded discussion of terrestrial biomes
    - Freshwater and marine biomes
    - Integration of concepts of biogeochemical cycles from Chapters 25 and 26 with ecological concepts
    - Global patterns of primary production
    - Global biodiversity
- ✪ A new exploration of the effect of fracking on the carbon cycle (Chapter 49)
- ✪ New coverage of habitat loss and biodiversity (Chapter 49)
- ✪ New coverage of overexploitation of resources and its effects on community ecology (Chapter 49)
- ✪ A new Core Concept and discussion of conservation biology (Chapter 49)

PRAISE FOR *HOW LIFE WORKS*

I have taught botany and then Biology II for over 20 years and have been very frustrated when I have realized how little knowledge students retained. Since we have gone to this textbook, I find that the questions students are asking in class are much more probing than those in the past, and the students seem much more engaged in the topics. I am hopeful that this approach will help our students be deeper thinkers and better scientists.

– **GLORIA CADDELL, University of Central Oklahoma**

One of the things that really sold me on this text was the LaunchPad system: easy to use; intuitive navigation; really good questions that match the sophistication of the text; love the LearningCurve activities; use most of the animations in my lectures!

– **SARA CARLSON, University of Akron**

This is the best set of questions I've ever seen in a textbook. They are thorough and the right mix of challenging the student with requiring memorization of important details.

– **KURT ELLIOT, Northwest Vista College**

We have all seen an improvement in our students' understanding of the material this year, the first year that we used the Morris text.

– **ANUPAMA SESHAN, Emmanuel College**

I like the figures, especially the 3D ones — we focus on "perceptual ability" training in our classes and figures that encourage students to think about cells in 3D are excellent!

– **KIRKWOOD LAND, University of the Pacific**

These chapters all seem to draw students through the course by referencing what they have learned previously and then adding new information. This makes the course seem like a complete story instead of a series of encyclopedia entries to be learned in isolation.

– **TIM KROFT, Auburn University at Montgomery**

We used this book last year and overall felt that it represented a major improvement from our previous text.

– **PETER ARMBRUSTER, Georgetown University**

Good questions are just as important as a good textbook. The available variety of assessment tools was very important for our adoption of this text.

– **MATTHEW BREWER, Georgia State University**

If the whole book reads like this I would love to use it! This is the way I like to teach! I want students to understand rather than memorize and this chapter seems aimed at this.

– **JENNIFER SCHRAMM, Chemeketa Community College**

The artwork seems very clear-cut and geared to giving the students a very specific piece of information with a very simple example. This should greatly help students with forming a visual image of the various subjects.

– **CHRIS PETRIE, Eastern Florida State College**

The writing style is excellent, it makes a great narrative and incorporates key scientific experiments into the explanation of photosynthesis.

– **DIANNE JENNING, Virginia Commonwealth University**

I think *HLW* does a better job of presenting introductory material than our current text, which tends to overwhelm students.

– **LAURA HILL, University of Vermont**

With the quick checks and the experiments the first chapter already has the learners thinking about experiments and critical analysis.

– **JOHN KOONTZ, University of Tennessee Knoxville**

This book moves teaching away from merely understanding all of the bold terms in a textbook in order to spit them back on a multiple-choice test. I can use this text in order to prepare my students to understand and learn the general principles and concepts in biology and how those concepts translate across different levels of biology. I would not trade this textbook for any other book on the market.

– **PAUL MOORE, Bowling Green State University**

ACKNOWLEDGMENTS

Biology: How Life Works is not only a book. Instead, it is an integrated set of resources to support student learning and instructor teaching in introductory biology. As a result, we work closely with an entire community of authors, publishers, instructors, reviewers, and students. We would like to thank this dedicated group.

First and foremost, we thank the thousands of students we have collectively taught. Their curiosity, intelligence, and enthusiasm have been sources of motivation for all of us.

Our teachers and mentors have provided us with models of patience, creativity, and inquisitiveness that we strive to bring into our own teaching and research. They encourage us to be life-long learners, teachers, and scholars.

We feel very lucky to be a partner with W. H. Freeman and Macmillan Learning. From the start, they have embraced our project, giving us the space and room to achieve something unique, while at the same time providing guidance, support, and input from the broader community of instructors and students.

Beth Cole, our acquisitions editor, deserves thanks for taking on the second edition and becoming our leader. She keeps a watchful eye on important trends in science, education, and technology, carefully listens to what we want to do, and helps us put our aspirations in a larger context.

Lead developmental editor Lisa Samols continues to have just the right touch—the ability to listen as well as offer intelligent suggestions, serious with a touch of humor, quiet but persistent. Senior developmental editor Susan Moran has an eye for detail and the uncanny ability to read the manuscript like a student. Developmental editor Erica Champion brings intelligence and thoughtfulness to her edits.

Karen Misler kept us all on schedule in a clear and firm but always understanding and compassionate way.

Lindsey Jaroszewicz, our market development manager, is remarkable for her energy and enthusiasm, her attention to detail, and her creativity in ways to reach out to instructors and students. Will Moore, our marketing manager, refined the story of *How Life Works* 2e and works tirelessly with our sales teams to bring the second edition to instructors and students everywhere.

We thank Robert Errera for coordinating the move from manuscript to the page, and Nancy Brooks for helping to even out the prose. We also thank Diana Blume, our design director, Tom Carling, our text and cover designer, and Sheridan Sellers, our compositor. Together, they managed the look and feel of the book, coming up with creative solutions for page layout.

In digital media, we thank Amanda Nietzel for her editorial insight in making pedagogically useful media tools, and Keri deManigold and Chris Efstratiou for managing and coordinating the media and websites. They each took on this project with dedication, persistence, enthusiasm, and attention to detail that we deeply appreciate.

We are extremely grateful to Elaine Palucki for her insight into teaching and learning strategies, Donna Brodman for coordinating the many reviewers, and Jane Taylor, Alexandra Garrett, and Abigail Fagan for their consistent and tireless support.

Imagineering under the patient and intelligent guidance of Mark Mykytuik provided creative, insightful art to complement, support, and reinforce the text. We also thank our illustration coordinator, Matt McAdams, for skillfully guiding our collaboration with Imagineering. Christine Buese, our photo editor, and Lisa Passmore and Richard Fox, our photo researchers, provided us with a steady stream of stunning photos, and never gave up on those hard-to-find shots. Paul Rohloff, our production manager, ensured that the journey from manuscript to printing was seamless.

We would also like to acknowledge Kate Parker, Publisher of Sciences, Chuck Linsmeier, Vice President of Editorial, Susan Winslow, Managing Director, and Ken Michaels, Chief Executive Officer, for their support of *How Life Works* and our unique approach.

We also sincerely thank Erin Betters, Jere A. Boudell, Donna Koslowsky, and Jon Stoltzfus for thoughtful and insightful contributions to the assessment materials.

We are extremely grateful for all of the hard work and expertise of the sales representatives, regional managers, and regional sales specialists. We have enjoyed meeting and working with this dedicated sales staff, who are the ones that ultimately put the book in the hands of instructors.

Countless reviewers made invaluable contributions to this book and deserve special thanks. From catching mistakes to suggesting new and innovative ways to organize the content, they provided substantial input to the book. They brought their collective years of teaching to the project, and their suggestions are tangible in every chapter.

Finally, we would like to thank our families. None of this would have been possible without their support, inspiration, and encouragement.

Contributors, First Edition

Thank you to all the instructors who worked in collaboration with the authors and assessment authors to write Biology: How Life Works *assessments, activities, and exercises.*

Allison Alvarado, University of California, Los Angeles
Peter Armbruster, Georgetown University
Zane Barlow-Coleman, formerly of University of Massachusetts, Amherst
James Bottesch, Brevard Community College
Jessamina Blum, Yale University
Jere Boudell, Clayton State University
David Bos, Purdue University
Laura Ciaccia West, Yale University*
Laura DiCaprio, Ohio University
Tod Duncan, University of Colorado, Denver
Cindy Giffen, University of Wisconsin, Madison
Paul Greenwood, Colby College
Stanley Guffey, The University of Tennessee, Knoxville
Alison Hill, Duke University
Meg Horton, University of North Carolina at Greensboro
Kerry Kilburn, Old Dominion University
Jo Kurdziel, University of Michigan
David Lampe, Duquesne University
Brenda Leady, University of Toledo
Sara Marlatt, Yale University*
Kelly McLaughlin, Tufts University
Brad Mehrtens, University of Illinois at Urbana-Champaign
Nancy Morvillo, Florida Southern College
Jennifer Nauen, University of Delaware
Kavita Oommen, Georgia State University
Patricia Phelps, Austin Community College
Melissa Reedy, University of Illinois at Urbana-Champaign
Lindsay Rush, Yale University*
Sukanya Subramanian, Collin College
Michelle Withers, West Virginia University

*Graduate student, Yale University Scientific Teaching Fellow

Reviewers, Class Testers, and Focus Group Participants

Thank you to all the instructors who reviewed and/or class tested chapters, art, assessment questions, and other Biology: How Life Works *materials.*

First Edition

Thomas Abbott, University of Connecticut
Tamarah Adair, Baylor University
Sandra Adams, Montclair State University
Jonathon Akin, University of Connecticut
Eddie Alford, Arizona State University
Chris Allen, College of the Mainland
Sylvester Allred, Northern Arizona University
Shivanthi Anandan, Drexel University
Andrew Andres, University of Nevada, Las Vegas
Michael Angilletta, Arizona State University
Jonathan Armbruster, Auburn University
Jessica Armenta, Lone Star College System
Brian Ashburner, University of Toledo
Andrea Aspbury, Texas State University
Nevin Aspinwall, Saint Louis University
Felicitas Avendano, Grand View University
Yael Avissar, Rhode Island College
Ricardo Azpiroz, Richland College
Jessica Baack, Southwestern Illinois College
Charles Baer, University of Florida
Brian Bagatto, University of Akron
Alan L. Baker, University of New Hampshire
Ellen Baker, Santa Monica College
Mitchell Balish, Miami University
Teri Balser, University of Florida
Paul Bates, University of Minnesota, Duluth
Michel Baudry, University of Southern California
Jerome Baudry, The University of Tennessee, Knoxville
Mike Beach, Southern Polytechnic State University
Andrew Beall, University of North Florida
Gregory Beaulieu, University of Victoria
John Bell, Brigham Young University
Michael Bell, Richland College
Rebecca Bellone, University of Tampa
Anne Bergey, Truman State University
Laura Bermingham, University of Vermont
Aimee Bernard, University of Colorado, Denver
Annalisa Berta, San Diego State University
Joydeep Bhattacharjee, University of Louisiana, Monroe

Arlene Billock, University of Louisiana, Lafayette
Daniel Blackburn, Trinity College
Mark Blackmore, Valdosta State University
Justin Blau, New York University
Andrew Blaustein, Oregon State University
Mary Bober, Santa Monica College
Robert Bohanan, University of Wisconsin, Madison
Jim Bonacum, University of Illinois at Springfield
Laurie Bonneau, Trinity College
David Bos, Purdue University
James Bottesch, Brevard Community College
Jere Boudell, Clayton State University
Nancy Boury, Iowa State University
Matthew Brewer, Georgia State University
Mirjana Brockett, Georgia Institute of Technology
Andrew Brower, Middle Tennessee State University
Heather Bruns, Ball State University
Jill Buettner, Richland College
Stephen Burnett, Clayton State University
Steve Bush, Coastal Carolina University
David Byres, Florida State College at Jacksonville
James Campanella, Montclair State University
Darlene Campbell, Cornell University
Jennifer Campbell, North Carolina State University
John Campbell, Northwest College
David Canning, Murray State University
Richard Cardullo, University of California, Riverside
Sara Carlson, University of Akron
Jeff Carmichael, University of North Dakota
Dale Casamatta, University of North Florida
Anne Casper, Eastern Michigan University
David Champlin, University of Southern Maine
Rebekah Chapman, Georgia State University
Samantha Chapman, Villanova University
Mark Chappell, University of California, Riverside
P. Bryant Chase, Florida State University
Young Cho, Eastern New Mexico University
Tim Christensen, East Carolina University
Steven Clark, University of Michigan
Ethan Clotfelter, Amherst College
Catharina Coenen, Allegheny College
Mary Colavito, Santa Monica College
Craig Coleman, Brigham Young University
Alex Collier, Armstrong Atlantic State University
Sharon Collinge, University of Colorado, Boulder
Jay Comeaux, McNeese State University
Reid Compton, University of Maryland
Ronald Cooper, University of California, Los Angeles
Victoria Corbin, University of Kansas
Asaph Cousins, Washington State University
Will Crampton, University of Central Florida
Kathryn Craven, Armstrong Atlantic State University
Scott Crousillac, Louisiana State University
Kelly Cude, College of the Canyons
Stanley Cunningham, Arizona State University
Karen Curto, University of Pittsburgh
Bruce Cushing, The University of Akron
Rebekka Darner, University of Florida
James Dawson, Pittsburg State University
Elizabeth De Stasio, Lawrence University
Jennifer Dechaine, Central Washington University
James Demastes, University of Northern Iowa
D. Michael Denbow, Virginia Polytechnic Institute and State University
Joseph Dent, McGill University
Terry Derting, Murray State University
Jean DeSaix, University of North Carolina at Chapel Hill
Donald Deters, Bowling Green State University
Hudson DeYoe, The University of Texas, Pan American
Leif Deyrup, University of the Cumberlands
Laura DiCaprio, Ohio University
Jesse Dillon, California State University, Long Beach
Frank Dirrigl, The University of Texas, Pan American
Kevin Dixon, Florida State University
Elaine Dodge Lynch, Memorial University of Newfoundland
Hartmut Doebel, George Washington University
Jennifer Doll, Loyola University, Chicago
Logan Donaldson, York University
Blaise Dondji, Central Washington University
Christine Donmoyer, Allegheny College
James Dooley, Adelphi University
Jennifer Doudna, University of California, Berkeley
John DuBois, Middle Tennessee State University
Richard Duhrkopf, Baylor University
Kamal Dulai, University of California, Merced
Arthur Dunham, University of Pennsylvania
Mary Durant, Lone Star College System
Roland Dute, Auburn University
Andy Dyer, University of South Carolina, Aiken
William Edwards, Niagara University
John Elder, Valdosta State University
William Eldred, Boston University
David Eldridge, Baylor University
Inge Eley, Hudson Valley Community College
Lisa Elfring, University of Arizona
Richard Elinson, Duquesne University
Kurt Elliott, Northwest Vista College
Miles Engell, North Carolina State University
Susan Erster, Stony Brook University
Joseph Esdin, University of California, Los Angeles
Jean Everett, College of Charleston
Brent Ewers, University of Wyoming
Melanie Fierro, Florida State College at Jacksonville
Michael Fine, Virginia Commonwealth University
Jonathan Fingerut, St. Joseph's University
Ryan Fisher, Salem State University
David Fitch, New York University

Paul Fitzgerald, Northern Virginia Community College
Jason Flores, University of North Carolina at Charlotte
Matthias Foellmer, Adelphi University
Barbara Frase, Bradley University
Caitlin Gabor, Texas State University
Michael Gaines, University of Miami
Jane Gallagher, The City College of New York, The City University of New York
Kathryn Gardner, Boston University
J. Yvette Gardner, Clayton State University
Gillian Gass, Dalhousie University
Jason Gee, East Carolina University
Topher Gee, University of North Carolina at Charlotte
Vaughn Gehle, Southwest Minnesota State University
Tom Gehring, Central Michigan University
John Geiser, Western Michigan University
Alex Georgakilas, East Carolina University
Peter Germroth, Hillsborough Community College
Arundhati Ghosh, University of Pittsburgh
Carol Gibbons Kroeker, University of Calgary
Phil Gibson, University of Oklahoma
Cindee Giffen, University of Wisconsin, Madison
Matthew Gilg, University of North Florida
Sharon Gillies, University of the Fraser Valley
Leonard Ginsberg, Western Michigan University
Florence Gleason, University of Minnesota
Russ Goddard, Valdosta State University
Miriam Golbert, College of the Canyons
Jessica Goldstein, Barnard College, Columbia University
Steven Gorsich, Central Michigan University
Sandra Grebe, Lone Star College System
Robert Greene, Niagara University
Ann Grens, Indiana University, South Bend
Theresa Grove, Valdosta State University
Stanley Guffey, The University of Tennessee, Knoxville
Nancy Guild, University of Colorado, Boulder
Lonnie Guralnick, Roger Williams University
Laura Hake, Boston College
Kimberly Hammond, University of California, Riverside
Paul Hapeman, University of Florida
Luke Harmon, University of Idaho
Sally Harmych, University of Toledo
Jacob Harney, University of Hartford
Sherry Harrel, Eastern Kentucky University
Dale Harrington, Caldwell Community College and Technical Institute
J. Scott Harrison, Georgia Southern University
Diane Hartman, Baylor University
Mary Haskins, Rockhurst University
Bernard Hauser, University of Florida
David Haymer, University of Hawaii
David Hearn, Towson University
Marshal Hedin, San Diego State University
Paul Heideman, College of William and Mary
Gary Heisermann, Salem State University
Brian Helmuth, University of South Carolina
Christopher Herlihy, Middle Tennessee State University
Albert Herrera, University of Southern California
Brad Hersh, Allegheny College
David Hicks, The University of Texas at Brownsville
Karen Hicks, Kenyon College
Alison Hill, Duke University
Kendra Hill, South Dakota State University
Jay Hodgson, Armstrong Atlantic State University
John Hoffman, Arcadia University
Jill Holliday, University of Florida
Sara Hoot, University of Wisconsin, Milwaukee
Margaret Horton, University of North Carolina at Greensboro
Lynne Houck, Oregon State University
Kelly Howe, University of New Mexico
William Huddleston, University of Calgary
Jodi Huggenvik, Southern Illinois University
Melissa Hughes, College of Charleston
Randy Hunt, Indiana University Southeast
Tony Huntley, Saddleback College
Brian Hyatt, Bethel College
Jeba Inbarasu, Metropolitan Community College
Colin Jackson, The University of Mississippi
Eric Jellen, Brigham Young University
Dianne Jennings, Virginia Commonwealth University
Scott Johnson, Wake Technical Community College
Mark Johnston, Dalhousie University
Susan Jorstad, University of Arizona
Stephen Juris, Central Michigan University
Julie Kang, University of Northern Iowa
Jonghoon Kang, Valdosta State University
George Karleskint, St. Louis Community College at Meramec
David Karowe, Western Michigan University
Judy Kaufman, Monroe Community College
Nancy Kaufmann, University of Pittsburgh
John Kauwe, Brigham Young University
Elena Keeling, California Polytechnic State University
Jill Keeney, Juniata College
Tamara Kelly, York University
Chris Kennedy, Simon Fraser University
Bretton Kent, University of Maryland
Jake Kerby, University of South Dakota
Jeffrey Kiggins, Monroe Community College
Scott Kight, Montclair State University
Stephen Kilpatrick, University of Pittsburgh, Johnstown
Kelly Kissane, University of Nevada, Reno
David Kittlesen, University of Virginia
Jennifer Kneafsey, Tulsa Community College
Jennifer Knight, University of Colorado, Boulder
Ross Koning, Eastern Connecticut State University
David Kooyman, Brigham Young University
Olga Kopp, Utah Valley University
Anna Koshy, Houston Community College
Todd Kostman, University of Wisconsin, Oshkosh
Peter Kourtev, Central Michigan University
William Kroll, Loyola University, Chicago
Dave Kubien, University of New Brunswick
Allen Kurta, Eastern Michigan University
Ellen Lamb, University of North Carolina at Greensboro

Troy Ladine, East Texas Baptist University
David Lampe, Duquesne University
Evan Lampert, Gainesville State College
James Langeland, Kalamazoo College
John Latto, University of California, Santa Barbara
Brenda Leady, University of Toledo
Jennifer Leavey, Georgia Institute of Technology
Hugh Lefcort, Gonzaga University
Brenda Leicht, University of Iowa
Craig Lending, The College at Brockport, The State University of New York
Nathan Lents, John Jay College of Criminal Justice, The City University of New York
Michael Leonardo, Coe College
Army Lester, Kennesaw State University
Cynthia Littlejohn, University of Southern Mississippi
Zhiming Liu, Eastern New Mexico University
Jonathan Lochamy, Georgia Perimeter College
Suzanne Long, Monroe Community College
Julia Loreth, University of North Carolina at Greensboro
Jennifer Louten, Southern Polytechnic State University
Janet Loxterman, Idaho State University
Ford Lux, Metropolitan State College of Denver
José-Luis Machado, Swarthmore College
C. Smoot Major, University of South Alabama
Charles Mallery, University of Miami
Mark Maloney, Spelman College
Carroll Mann, Florida State College at Jacksonville
Carol Mapes, Kutztown University of Pennsylvania
Nilo Marin, Broward College
Diane Marshall, University of New Mexico
Heather Masonjones, University of Tampa
Scott Mateer, Armstrong Atlantic State University
Luciano Matzkin, The University of Alabama in Huntsville
Robert Maxwell, Georgia State University
Meghan May, Towson University
Michael McGinnis, Spelman College
Kathleen McGuire, San Diego State University
Maureen McHale, Truman State University
Shannon McQuaig, St. Petersburg College
Susan McRae, East Carolina University
Lori McRae, University of Tampa
Mark Meade, Jacksonville State University
Brad Mehrtens, University of Illinois at Urbana-Champaign
Michael Meighan, University of California, Berkeley
Douglas Meikle, Miami University
Richard Merritt, Houston Community College
Jennifer Metzler, Ball State University
James Mickle, North Carolina State University
Brian Miller, Middle Tennessee State University
Allison Miller, Saint Louis University
Yuko Miyamoto, Elon University
Ivona Mladenovic, Simon Fraser University
Marcie Moehnke, Baylor University
Chad Montgomery, Truman State University
Jennifer Mook, Gainesville State College
Daniel Moon, University of North Florida
Jamie Moon, University of North Florida
Jeanelle Morgan, Gainesville State College
David Morgan, University of West Georgia
Julie Morris, Armstrong Atlantic State University
Becky Morrow, Duquesne University
Mark Mort, University of Kansas
Nancy Morvillo, Florida Southern College
Anthony Moss, Auburn University
Mario Mota, University of Central Florida
Alexander Motten, Duke University
Tim Mulkey, Indiana State University
John Mull, Weber State University
Michael Muller, University of Illinois at Chicago
Beth Mullin, The University of Tennessee, Knoxville
Paul Narguizian, California State University, Los Angeles
Jennifer Nauen, University of Delaware
Paul Nealen, Indiana University of Pennsylvania
Diana Nemergut, University of Colorado, Boulder
Kathryn Nette, Cuyamaca College
Jacalyn Newman, University of Pittsburgh
James Nienow, Valdosta State University
Alexey Nikitin, Grand Valley State University
Tanya Noel, York University
Fran Norflus, Clayton State University
Celia Norman, Arapahoe Community College
Eric Norstrom, DePaul University
Jorge Obeso, Miami Dade College
Kavita Oommen, Georgia State University
David Oppenheimer, University of Florida
Joseph Orkwiszewski, Villanova University
Rebecca Orr, Collin College
Don Padgett, Bridgewater State College
Joanna Padolina, Virginia Commonwealth University
One Pagan, West Chester University
Kathleen Page, Bucknell University
Daniel Papaj, University of Arizona
Pamela Pape-Lindstrom, Everett Community College
Bruce Patterson, University of Arizona, Tucson
Shelley Penrod, Lone Star College System
Roger Persell, Hunter College, The City University of New York
John Peters, College of Charleston
Chris Petrie, Brevard Community College
Patricia Phelps, Austin Community College
Steven Phelps, The University of Texas at Austin
Kristin Picardo, St. John Fisher College
Aaron Pierce, Nicholls State University
Debra Pires, University of California, Los Angeles
Thomas Pitzer, Florida International University
Nicola Plowes, Arizona State University
Crima Pogge, City College of San Francisco
Darren Pollock, Eastern New Mexico University

Kenneth Pruitt, The University of Texas at Brownsville
Sonja Pyott, University of North Carolina at Wilmington
Rajinder Ranu, Colorado State University
Philip Rea, University of Pennsylvania
Amy Reber, Georgia State University
Ahnya Redman, West Virginia University
Melissa Reedy, University of Illinois at Urbana-Champaign
Brian Ring, Valdosta State University
David Rintoul, Kansas State University
Michael Rischbieter, Presbyterian College
Laurel Roberts, University of Pittsburgh
George Robinson, The University at Albany, The State University of New York
Peggy Rolfsen, Cincinnati State Technical and Community College
Mike Rosenzweig, Virginia Polytechnic Institute and State University
Doug Rouse, University of Wisconsin, Madison
Yelena Rudayeva, Palm Beach State College
Ann Rushing, Baylor University
Shereen Sabet, La Sierra University
Rebecca Safran, University of Colorado
Peter Sakaris, Southern Polytechnic State University
Thomas Sasek, University of Louisiana, Monroe
Udo Savalli, Arizona State University
H. Jochen Schenk, California State University, Fullerton
Gregory Schmaltz, University of the Fraser Valley
Jean Schmidt, University of Pittsburgh
Andrew Schnabel, Indiana University, South Bend
Roxann Schroeder, Humboldt State University
David Schultz, University of Missouri, Columbia
Andrea Schwarzbach, The University of Texas at Brownsville
Erik Scully, Towson University
Robert Seagull, Hofstra University
Pramila Sen, Houston Community College
Alice Sessions, Austin Community College
Vijay Setaluri, University of Wisconsin
Jyotsna Sharma, The University of Texas at San Antonio
Elizabeth Sharpe-Aparicio, Blinn College
Patty Shields, University of Maryland
Cara Shillington, Eastern Michigan University
James Shinkle, Trinity University
Rebecca Shipe, University of California, Los Angeles
Marcia Shofner, University of Maryland
Laurie Shornick, Saint Louis University
Jill Sible, Virginia Polytechnic Institute and State University
Allison Silveus, Tarrant County College
Kristin Simokat, University of Idaho
Sue Simon-Westendorf, Ohio University
Sedonia Sipes, Southern Illinois University, Carbondale
John Skillman, California State University, San Bernardino
Marek Sliwinski, University of Northern Iowa
Felisa Smith, University of New Mexico
John Sollinger, Southern Oregon University
Scott Solomon, Rice University
Morvarid Soltani-Bejnood, The University of Tennessee
Vladimir Spiegelman, University of Wisconsin, Madison
Chrissy Spencer, Georgia Institute of Technology
Kathryn Spilios, Boston University
Ashley Spring, Brevard Community College
Bruce Stallsmith, The University of Alabama in Huntsville
Jennifer Stanford, Drexel University
Barbara Stegenga, University of North Carolina, Chapel Hill
Patricia Steinke, San Jacinto College, Central Campus
Asha Stephens, College of the Mainland
Robert Steven, University of Toledo
Eric Strauss, University of Wisconsin, La Crosse
Sukanya Subramanian, Collin College
Mark Sugalski, Southern Polytechnic State University
Brad Swanson, Central Michigan University
Ken Sweat, Arizona State University
David Tam, University of North Texas
Ignatius Tan, New York University
William Taylor, University of Toledo
Christine Terry, Lynchburg College
Sharon Thoma, University of Wisconsin, Madison
Pamela Thomas, University of Central Florida
Carol Thornber, University of Rhode Island
Patrick Thorpe, Grand Valley State University
Briana Timmerman, University of South Carolina
Chris Todd, University of Saskatchewan
Gail Tompkins, Wake Technical Community College
Martin Tracey, Florida International University
Randall Tracy, Worcester State University
James Traniello, Boston University
Bibit Traut, City College of San Francisco
Terry Trier, Grand Valley State University
Stephen Trumble, Baylor University
Jan Trybula, The State University of New York at Potsdam
Alexa Tullis, University of Puget Sound
Marsha Turell, Houston Community College
Mary Tyler, University of Maine
Marcel van Tuinen, University of North Carolina at Wilmington
Dirk Vanderklein, Montclair State University
Jorge Vasquez-Kool, Wake Technical Community College
William Velhagen, New York University
Dennis Venema, Trinity Western University
Laura Vogel, North Carolina State University
Jyoti Wagle, Houston Community College
Jeff Walker, University of Southern Maine
Gary Walker, Appalachian State University
Andrea Ward, Adelphi University
Fred Wasserman, Boston University
Elizabeth Waters, San Diego State University
Douglas Watson, The University of Alabama at Birmingham

Matthew Weand, Southern Polytechnic State University
Michael Weber, Carleton University
Cindy Wedig, The University of Texas, Pan American
Brad Wetherbee, University of Rhode Island
Debbie Wheeler, University of the Fraser Valley
Clay White, Lone Star College System
Lisa Whitenack, Allegheny College
Maggie Whitson, Northern Kentucky University
Stacey Wild East, Tennessee State University
Herbert Wildey, Arizona State University and Phoenix College
David Wilkes, Indiana University, South Bend
Lisa Williams, Northern Virginia Community College
Elizabeth Willott, University of Arizona
Mark Wilson, Humboldt State University
Ken Wilson, University of Saskatchewan
Bob Winning, Eastern Michigan University
Candace Winstead, California Polytechnic State University
Robert Wise, University of Wisconsin, Oshkosh
D. Reid Wiseman, College of Charleston
MaryJo Witz, Monroe Community College
David Wolfe, American River College
Kevin Woo, University of Central Florida
Denise Woodward, Penn State
Shawn Wright, Central New Mexico Community College
Grace Wyngaard, James Madison University
Aimee Wyrick, Pacific Union College
Joanna Wysocka-Diller, Auburn University
Ken Yasukawa, Beloit College
John Yoder, The University of Alabama
Kelly Young, California State University, Long Beach
James Yount, Brevard Community College
Min Zhong, Auburn University

Second Edition

Barbara J. Abraham, Hampton University
Jason Adams, College of DuPage
Sandra D. Adams, Montclair State University
Richard Adler, University of Michigan, Dearborn
Nancy Aguilar-Roca, University of California, Irvine
Shivanthi Anandan, Drexel University
Lynn Anderson-Carpenter, University of Michigan
Christine Andrews, The University of Chicago
Peter Armbruster, Georgetown University
Jessica Armenta, Austin Community College
Brian Ashburner, University of Toledo
Ann J. Auman, Pacific Lutheran University
Nicanor Austriaco, Providence College
Felicitas Avendano, Grand View University
J. P. Avery, University of North Florida
Jim Bader, Case Western Reserve University
Ellen Baker, Santa Monica College
Andrew S. Baldwin, Mesa Community College
Stephen Baron, Bridgewater College
Paul W. Bates, University of Minnesota, Duluth
Janet Batzli, University of Wisconsin, Madison
David Baum, University of Wisconsin, Madison
Kevin S. Beach, The University of Tampa
Philip Becraft, Iowa State University
Alexandra Bely, University of Maryland
Lauryn Benedict, University of Northern Colorado
Anne Bergey, Truman State University
Joydeep Bhattacharjee, University of Louisiana, Monroe
Todd Bishop, Dalhousie University
Catherine Black, Idaho State University
Andrew R. Blaustein, Oregon State University
James Bolton, Georgia Gwinnett College
Jim Bonacum, University of Illinois at Springfield
Laurie J. Bonneau, Trinity College
James Bottesch, Eastern Florida State College
Lisa Boucher, The University of Texas at Austin
Nicole Bournias-Vardiabasis, California State University, San Bernardino
Nancy Boury, Iowa State University
Matthew Brewer, Georgia State University
Christoper G. Brown, Georgia Gwinnett College
Jill Buettner, Richland College
Sharon K. Bullock, University of North Carolina at Charlotte
Lisa Burgess, Broward College
Jorge Busciglio, University of California, Irvine
Stephen Bush, Coastal Carolina University
David Byres, Florida State College at Jacksonville
Gloria Caddell, University of Central Oklahoma
Guy A. Caldwell, The University of Alabama
Kim A. Caldwell, The University of Alabama
John S. Campbell, Northwest College
Jennifer Capers, Indian River State College
Joel Carlin, Gustavus Adolphus College
Sara G. Carlson, University of Akron
Dale Casamatta, University of North Florida
Merri Lynn Casem, California State University, Fullerton
Anne Casper, Eastern Michigan University
David Champlin, University of Southern Maine
Rebekah Chapman, Georgia State University
P. Bryant Chase, Florida State University
Thomas T. Chen, Santa Monica College
Young Cho, Eastern New Mexico University
Sunita Chowrira, University of British Columbia
Tim W. Christensen, East Carolina University
Steven Clark, University of Michigan
Beth Cliffel, Triton College
Liane Cochran-Stafira, Saint Xavier University
John G. Cogan, The Ohio State University
Reid Compton, University of Maryland
Ronald H. Cooper, University of California, Los Angeles
Janice Countaway, University of Central Oklahoma

Joseph A. Covi, University of North Carolina at Wilmington
Will Crampton, University of Central Florida
Kathryn Craven, Armstrong State University
Lorelei Crerar, George Mason University
Kerry Cresawn, James Madison University
Richard J. Cristiano, Houston Community College Northwest
Cynthia K. Damer, Central Michigan University
David Dansereau, Saint Mary's University
Mark Davis, Macalester College
Elizabeth A. De Stasio, Lawrence University
Tracy Deem, Bridgewater College
Kimberley Dej, McMaster University
Terrence Delaney, University of Vermont
Tracie Delgado, Northwest University
Mark S. Demarest, University of North Texas
D. Michael Denbow, Virginia Polytechnic Institute and State University
Jonathan Dennis, Florida State University
Brandon S. Diamond, University of Miami
AnnMarie DiLorenzo, Montclair State University
Frank J. Dirrigl, Jr., University of Texas, Pan American
Christine Donmoyer, Allegheny College
Samuel Douglas, Angelina College
John D. DuBois, Middle Tennessee State University
Janet Duerr, Ohio University
Meghan Duffy, University of Michigan
Richard E. Duhrkopf, Baylor University
Jacquelyn Duke, Baylor University
Kamal Dulai, University of California, Merced
Rebecca K. Dunn, Boston College
Jacob Egge, Pacific Lutheran University
Kurt J. Elliott, Northwest Vista College
Miles Dean Engell, North Carolina State University
Susan Erster, Stony Brook University
Barbara I. Evans, Lake Superior State University
Lisa Felzien, Rockhurst University
Ralph Feuer, San Diego State University
Ginger R. Fisher, University of Northern Colorado
John Flaspohler, Concordia College
Sam Flaxman, University of Colorado, Boulder
Nancy Flood, Thompson Rivers University
Arthur Frampton, University of North Carolina at Wilmington
Caitlin Gabor, Texas State University
Tracy Galarowicz, Central Michigan University
Raul Galvan, South Texas College
Deborah Garrity, Colorado State University
Jason Mitchell Gee, East Carolina University
T.M. Gehring, Central Michigan University
John Geiser, Western Michigan University
Carol A. Gibbons Kroeker, Ambrose University
Susan A. Gibson, South Dakota State University
Cynthia J. Giffen, University of Michigan
Matthew Gilg, University of North Florida
Sharon L. Gillies, University of the Fraser Valley
Leslie Goertzen, Auburn University
Marla Gomez, Nicholls State University
Steven Gorsich, Central Michigan University
Daniel Graetzer, Northwest University
James Grant, Concordia University
Linda E. Green, Georgia Institute of Technology
Sara Gremillion, Armstrong State University
Ann Grens, Indiana University, South Bend
John L. Griffis, Joliet Junior College
Nancy A. Guild, University of Colorado, Boulder
Lonnie Guralnick, Roger Williams University
Valerie K. Haftel, Morehouse College
Margaret Hanes, Eastern Michigan University
Sally E. Harmych, University of Toledo
Sherry Harrel, Eastern Kentucky University
J. Scott Harrison, Georgia Southern University
Pat Harrison, University of the Fraser Valley
Diane Hartman, Baylor University
Wayne Hatch, Utah State University Eastern
David Haymer, University of Hawaii at Manoa
Chris Haynes, Shelton State Community College
Christiane Healey, University of Massachusetts, Amherst
David Hearn, Towson University
Marshal Hedin, San Diego State University
Triscia Hendrickson, Morehouse College
Albert A. Herrera, University of Southern California
Bradley Hersh, Allegheny College
Anna Hiatt, East Tennessee State University
Laura Hill, University of Vermont
Jay Hodgson, Armstrong State University
James Horwitz, Palm Beach State College
Sarah Hosch, Oakland University
Kelly Howe, University of New Mexico
Kimberly Hruska, Langara College
William Huddleston, University of Calgary
Carol Hurney, James Madison University
Brian A. Hyatt, Bethel University
Bradley C. Hyman, University of California, Riverside
Anne Jacobs, Allegheny College
Robert C. Jadin, Northeastern Illinois University
Rick Jellen, Brigham Young University
Dianne Jennings, Virginia Commonwealth University
L. Scott Johnson, Towson University
Russell Johnson, Colby College
Susan Jorstad, University of Arizona
Matthew Julius, St. Cloud State University
John S. K. Kauwe, Brigham Young University
Lori J. Kayes, Oregon State University
Todd Kelson, Brigham Young University, Idaho
Christopher Kennedy, Simon Fraser University
Jacob Kerby, University of South Dakota
Stephen T. Kilpatrick, University of Pittsburgh, Johnstown
Mary Kimble, Northeastern Illinois University

Denice D. King, Cleveland State Community College
David Kittlesen, University of Virginia
Ann Kleinschmidt, Allegheny College
Kathryn Kleppinger-Sparace, Tri-County Technical College
Daniel Klionsky, University of Michigan
Ned Knight, Reed College
Benedict Kolber, Duquesne University
Ross Koning, Eastern Connecticut State University
John Koontz, The University of Tennessee, Knoxville
Peter Kourtev, Central Michigan University
Elizabeth Kovar, The University of Chicago
Nadine Kriska, University of Wisconsin, Whitewater
Tim L. Kroft, Auburn University at Montgomery
William Kroll, Loyola University of Chicago
Dave Kubien, University of New Brunswick
Jason Kuehner, Emmanuel College
Josephine Kurdziel, University of Michigan
Troy A. Ladine, East Texas Baptist University
Diane M. Lahaise, Georgia Perimeter College
Janice Lai, Austin Community College, Cypress Creek
Kirk Land, University of the Pacific
James Langeland, Kalamazoo College
Neva Laurie-Berry, Pacific Lutheran University
Brenda Leady, University of Toledo
Adrienne Lee, University of California, Fullerton
Chris Levesque, John Abbott College
Bai-Lian Larry Li, University of California, Riverside
Cynthia Littlejohn, University of Southern Mississippi
Jason Locklin, Temple College
Xu Lu, University of Findlay
Patrice Ludwig, James Madison University
Ford Lux, Metropolitan State University of Denver
Morris F. Maduro, University of California, Riverside
C. Smoot Major, University of South Alabama
Barry Margulies, Towson University
Nilo Marin, Broward College
Michael Martin, John Carroll University
Heather D. Masonjones, University of Tampa
Scott C. Mateer, Armstrong State University
Robert Maxwell, Georgia State University
Joseph McCormick, Duquesne University
Lori L. McGrew, Belmont University
Peter B. McIntyre, University of Wisconsin, Madison
Iain McKinnell, Carleton University
Krystle McLaughlin, Lehigh University
Susan B. McRae, East Carolina University
Mark Meade, Jacksonville State University
Richard Merritt, Houston Community College Northwest
James E. Mickle, North Carolina State University
Chad Montgomery, Truman State University
Scott M. Moody, Ohio University
Daniel Moon, University of North Florida
Jamie Moon, University of North Florida
Jonathan Moore, Virginia Commonwealth University
Paul A. Moore, Bowling Green State University
Tsafrir Mor, Arizona State University
Jeanelle Morgan, University of North Georgia
Mark Mort, University of Kansas
Anthony Moss, Auburn University
Karen Neal, Reynolds Community College
Kimberlyn Nelson, Pennsylvania State University
Hao Nguyen, California State University, Sacramento
John Niedzwiecki, Belmont University
Alexey G. Nikitin, Grand Valley State University
Matthew Nusnbaum, Georgia State University
Robert Okazaki, Weber State University
Tiffany Oliver, Spelman College
Jennifer S. O'Neil, Houston Community College
Kavita Oommen, Georgia State University
Nathan Opolot Okia, Auburn University at Montgomery
Robin O'Quinn, Eastern Washington University
Sarah A. Orlofske, Northeastern Illinois University
Don Padgett, Bridgewater State University
Lisa Parks, North Carolina State University
Nilay Patel, California State University, Fullerton
Markus Pauly, University of California, Berkeley
Daniel M. Pavuk, Bowling Green State University
Marc Perkins, Orange Coast College
Beverly Perry, Houston Community College
John S. Peters, College of Charleston
Chris Petrie, Eastern Florida State College
John M. Pleasants, Iowa State University
Michael Plotkin, Mt. San Jacinto College
Mary Poffenroth, San Jose State University
Dan Porter, Amarillo College
Sonja Pyott, University of North Carolina at Wilmington
Mirwais Qaderi, Mount Saint Vincent University
Nick Reeves, Mt. San Jacinto College
Adam J. Reinhart, Wayland Baptist University

Stephanie Richards, Bates College
David A. Rintoul, Kansas State University
Trevor Rivers, University of Kansas
Laurel Roberts, University of Pittsburgh
Casey Roehrig, Harvard University
Jennifer Rose, University of North Georgia
Michael S. Rosenzweig, Virginia Polytechnic Institute and State University
Caleb M. Rounds, University of Massachusetts, Amherst
Yelena Rudayeva, Palm Beach State College
James E. Russell, Georgia Gwinnett College
Donald Sakaguchi, Iowa State University
Thomas Sasek, University of Louisiana at Monroe
Leslie J. Saucedo, University of Puget Sound
Udo M. Savalli, Arizona State University West Campus
Smita Savant, Houston Community College Southwest
Leena Sawant, Houston Community College Southwest
H. Jochen Schenk, California State University, Fullerton
Aaron E. Schirmer, Northeastern Illinois University
Mark Schlueter, Georgia Gwinnett College
Gregory Schmaltz, University of the Fraser Valley
Jennifer Schramm, Chemeketa Community College
Roxann Schroeder, Humboldt State University
Tim Schuh, St. Cloud State University
Kevin G. E. Scott, University of Manitoba
Erik P. Scully, Towson University
Sarah B. Selke, Three Rivers Community College
Pramila Sen, Houston Community College
Anupama Seshan, Emmanuel College
Alice Sessions, Austin Community College
Vijay Setaluri, University of Wisconsin
Timothy E. Shannon, Francis Marion University
Wallace Sharif, Morehouse College
Mark Sherrard, University of Northern Iowa
Cara Shillington, Eastern Michigan University
Amy Siegesmund, Pacific Lutheran University
Christine Simmons, Southern Illinois University, Edwardsville
S. D. Sipes, Southern Illinois University, Carbondale
John Skillman, California State University, San Bernardino
Daryl Smith, Langara College
Julie Smith, Pacific Lutheran University
Karen Smith, University of British Columbia
Leo Smith, University of Kansas
Ramona Smith-Burrell, Eastern Florida State College
Joel Snodgrass, Towson University
Alan J. Snow, University of Akron–Wayne College
Judith Solti, Houston Community College, Spring Branch
Ann Song, University of California, Fullerton
C. Kay Song, Georgia State University
Chrissy Spencer, Georgia Institute of Technology
Rachel Spicer, Connecticut College
Ashley Spring, Eastern Florida State College
Bruce Stallsmith, The University of Alabama in Huntsville
Maria L. Stanko, New Jersey Institute of Technology
Nancy Staub, Gonzaga University
Barbara Stegenga, University of North Carolina
Robert Steven, University of Toledo
Lori Stevens, University of Vermont
Mark Sturtevant, Oakland University
Elizabeth B. Sudduth, Georgia Gwinnett College
Mark Sugalski, Kennesaw State University, Southern Polytechnic State University
Fengjie Sun, Georgia Gwinnett College
Bradley J. Swanson, Central Michigan University
Brook O. Swanson, Gonzaga University
Ken Gunter Sweat, Arizona State University West Campus
Annette Tavares, University of Ontario Institute of Technology
William R. Taylor, University of Toledo
Samantha Terris Parks, Georgia State University
Jessica Theodor, University of Calgary
Sharon Thoma, University of Wisconsin, Madison
Sue Thomson, Auburn University at Montgomery
Mark Tiemeier, Cincinnati State Technical and Community College
Candace Timpte, Georgia Gwinnett College
Nicholas Tippery, University of Wisconsin, Whitewater
Chris Todd, University of Saskatchewan
Kurt A. Toenjes, Montana State University, Billings
Jeffrey L. Travis, The University at Albany, The State University of New York
Stephen J. Trumble, Baylor University
Jan Trybula, The State University of New York at Potsdam
Cathy Tugmon, Georgia Regents University
Alexa Tullis, University of Puget Sound
Marsha Turell, Houston Community College
Pat Uelmen Huey, Georgia Gwinnett College
Steven M. Uyeda, Pima Community College

Rani Vajravelu, University of Central Florida
Moira van Staaden, Bowling Green State University
Dirk Vanderklein, Montclair State University
William Velhagen, Caldwell University
Sara Via, University of Maryland
Christopher Vitek, University of Texas, Pan American
Neal J. Voelz, St. Cloud State University
Mindy Walker, Rockhurst University
Andrea Ward, Adelphi University
Jennifer Ward, University of North Carolina at Asheville
Pauline Ward, Houston Community College
Alan Wasmoen, Metropolitan Community College
Elizabeth R. Waters, San Diego State University
Matthew Weand, Southern Polytechnic State University
K. Derek Weber, Raritan Valley Community College
Andrea Weeks, George Mason University
Charles Welsh, Duquesne University
Naomi L. B. Wernick, University of Massachusetts, Lowell
Mary E. White, Southeastern Louisiana University
David Wilkes, Indiana University, South Bend
Frank Williams, Langara College
Kathy S. Williams, San Diego State University
Lisa D. Williams, Northern Virginia Community College
Christina Wills, Rockhurst University
Brian Wisenden, Minnesota State University, Moorhead
David Wolfe, American River College
Ramakrishna Wusirika, Michigan Technological University
G. Wyngaard, James Madison University
James R. Yount, Eastern Florida State College
Min Zhong, Auburn University

BIOLOGY

HOW LIFE WORKS

CONTENTS

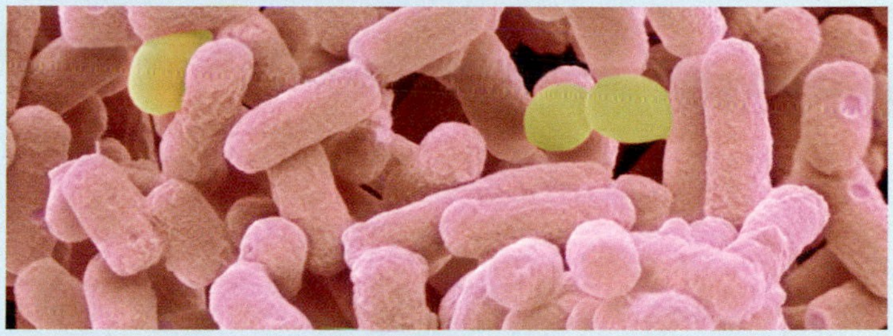

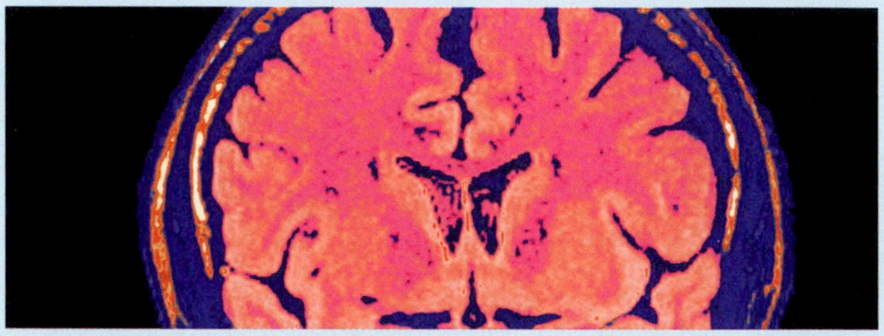

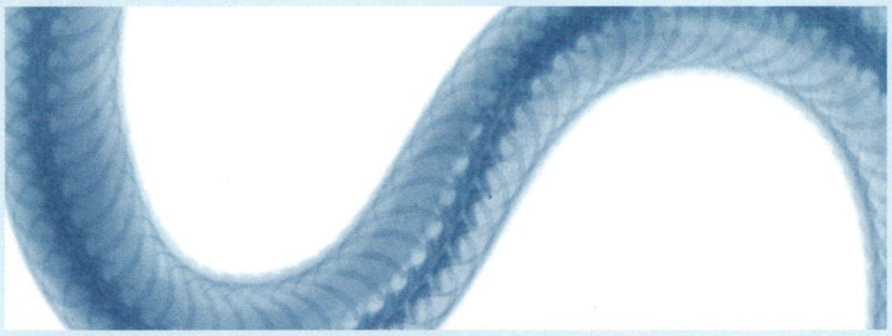

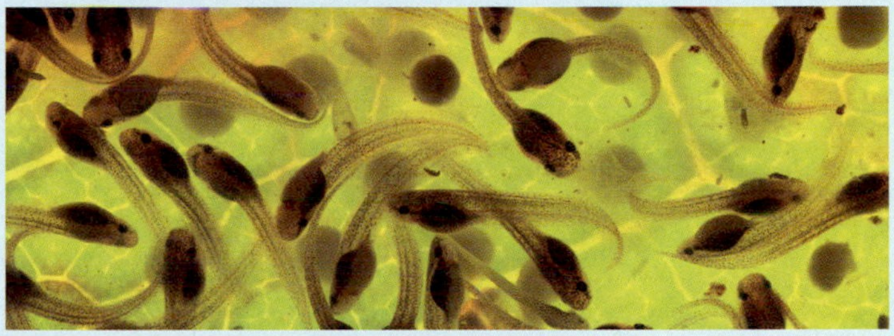

CHAPTER 39 **ANIMAL CARDIOVASCULAR AND RESPIRATORY SYSTEMS** **827**

CHAPTER 40 **ANIMAL METABOLISM, NUTRITION, AND DIGESTION** **851**

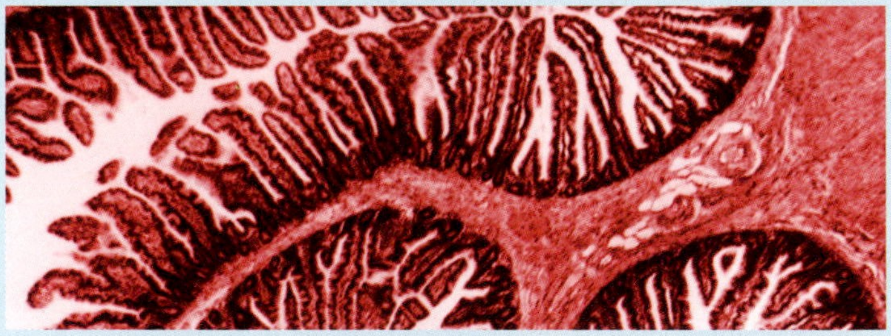

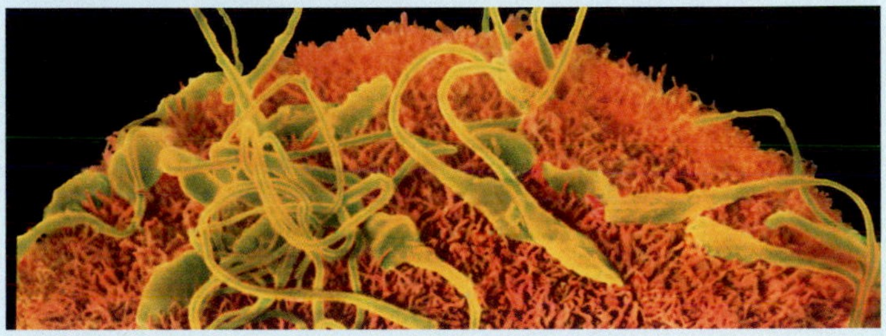

BIOLOGY

HOW LIFE WORKS

CASE 4

Malaria

Coevolution of Humans and a Parasite

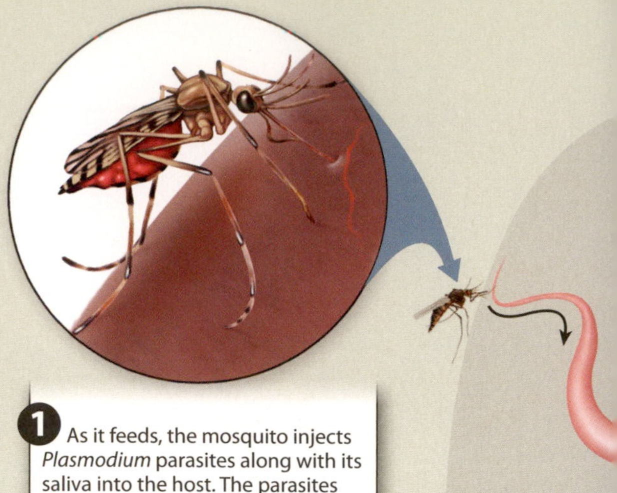

1 As it feeds, the mosquito injects *Plasmodium* parasites along with its saliva into the host. The parasites migrate through the bloodstream to the host's liver.

Most people would say the world has more than enough mosquitoes, but in 2010 scientists at the University of Arizona conjured up a new variety. The high-tech bloodsuckers were genetically engineered to resist *Plasmodium,* the single-celled eukaryote that causes malaria.

Normally, the parasite grows in the mosquito's gut and is spread to humans by the insect's bite. By altering a single gene in the mosquito's genome, the researchers had made the insects immune to the malaria parasite. The accomplishment is a noteworthy advance, the latest in a long line of efforts to stop *Plasmodium* in its tracks.

As humans have tried a succession of weapons to defeat P. falciparum, the parasite has evolved, thwarting their efforts.

Malaria is one of the most devastating diseases on the planet. The World Health Organization estimates that 500 million people contract malaria annually, primarily in tropical regions. The disease is thought to claim about a million lives each year. Of those deaths, 85% to 90% occur in sub-Saharan Africa, mostly among children under 5.

Five species of *Plasmodium* can cause malaria in humans. One of them, *P. falciparum,* is particularly dangerous, accounting for the vast majority of malaria fatalities. Over thousands of years, this parasite and its human host have played a deadly game of tug-of-war. As humans have tried a succession of weapons to defeat *P. falciparum,* the parasite has evolved, thwarting their efforts. And in turn, the tiny organism has helped to shape human evolution.

Plasmodium is a wily and complicated parasite, requiring both humans and mosquitoes to complete its life cycle. The human part of the cycle begins with a single bite from an infected mosquito. As the insect draws blood, it releases *Plasmodium*-laden saliva into the bloodstream. Once inside their human host, the *Plasmodium* parasites invade the liver cells. There they undergo cell division for several days, their numbers increasing. Eventually, the parasites infect red blood cells, where they continue to grow and multiply. The mature parasites burst from the red blood cells at regular intervals, triggering malaria's telltale cycle of fever and chills.

Some of the freed *Plasmodium* parasites go on to infect new red blood cells; others divide to form gametocytes, precursor cells to male and female gametes, which travel through the victim's blood vessels. When another mosquito bites the infected individual, it takes up the gametocytes with its blood meal. Inside the insect, the parasite completes its life cycle. The gametocytes fuse to form zygotes. Those zygotes bore into the mosquito's stomach, where they form cells called oocysts that give rise to a new generation of parasites. When the infected mosquito sets out to feed, the cycle begins again.

The battle between malaria and humankind has raged through the ages. Scientists have recovered *Plasmodium* DNA from the bodies of 3500-year-old Egyptian mummies—evidence that those ancient humans were infected with the malaria parasite. The close connection between humans, mosquitoes, and the malaria parasite almost certainly extends back much further.

In fact, people who hail from regions where malaria is endemic are more likely than others to have certain genetic signatures that offer some degree of protection from the parasite. That indicates that *Plasmodium* has been exerting evolutionary pressure on humankind for quite some time.

Meanwhile, we've done our best to fight back. For centuries, humans fought the infection with quinine, a chemical found in the bark of the South American cinchona tree. In the 1940s, scientists developed a more sophisticated drug based on the cinchona compound. That drug, chloroquine, was effective, inexpensive, easy to administer, and caused few side effects. As a result, it was widely used, and in the late 1950s, *P. falciparum* began showing signs of resistance to chloroquine. Within 20 years, resistance had spread to Africa, and today most strains of *P. falciparum* have evolved resistance to the once-potent medication.

Just as bacteria develop resistance to antibiotics, *Plasmodium* evolves resistance to the antiparasitic drugs designed to fight it. In poor, rural areas where malaria is prevalent, people often can't follow the recommended protocols for antimalarial treatment. Sick individuals may be able to afford only a few pills rather than the full recommended dose. The strength and quality of those pills may be questionable, and the drugs are often taken without oversight from a medical professional.

Unfortunately, inadequate use of the drugs fuels resistance. When the pills are altered or the course of treatment is abbreviated, not all *Plasmodium* parasites are

The life cycle of *Plasmodium falciparum*.

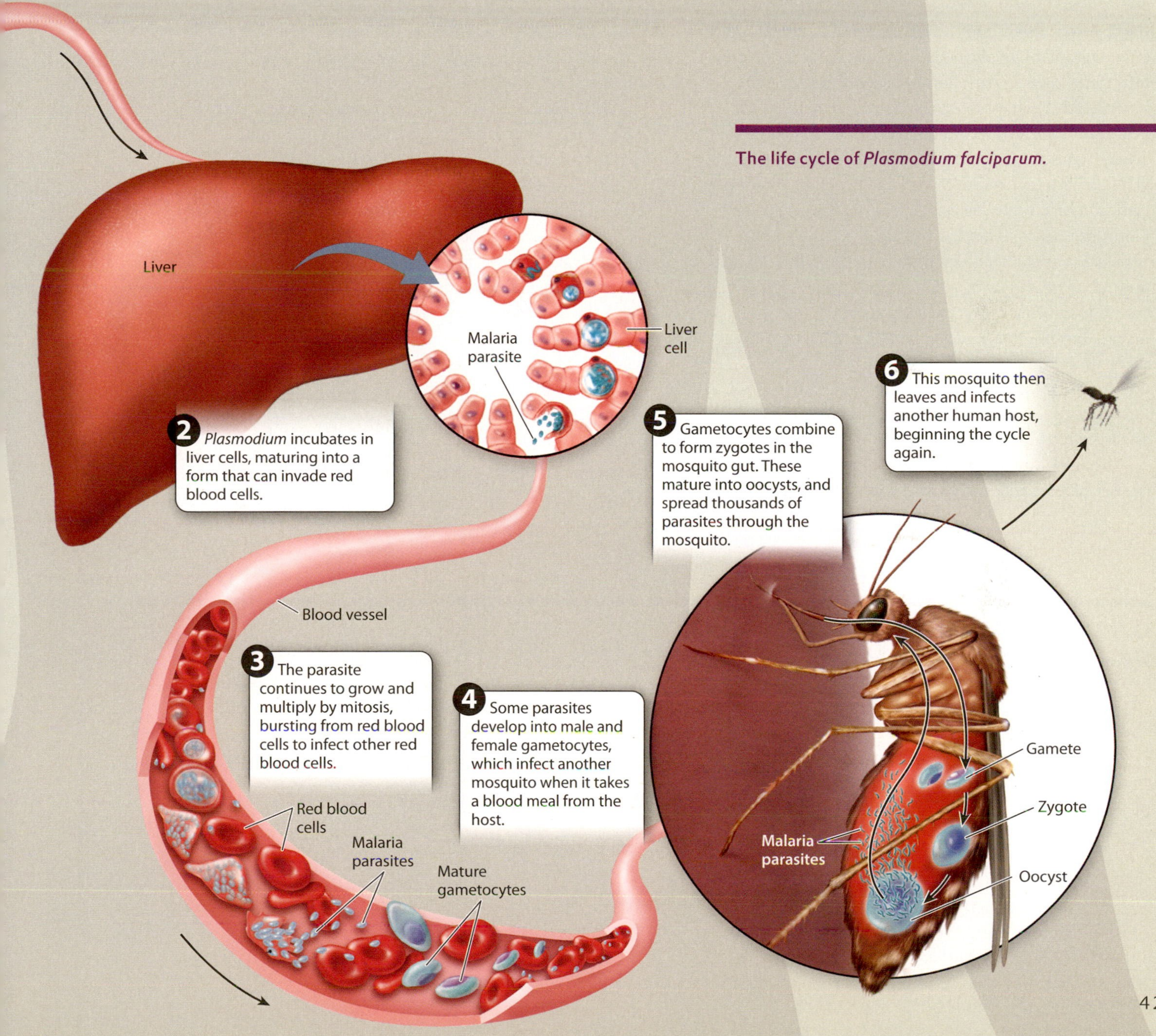

wiped out. Those that survive in the presence of the drug are likely to evolve resistance to the drug. Because the resistant parasites have a survival advantage, the genes for drug resistance spread quickly through the population.

Since chloroquine resistance emerged, pharmaceutical researchers have developed a variety of new medications to prevent or treat malaria. Most, however, are far too expensive for people in the poverty-stricken regions where malaria is rampant. And just as was the case with chloroquine, almost as quickly as new drugs are developed, *Plasmodium* begins evolving resistance.

One of the weapons more recently added to the drug arsenal is artemisinin, a compound derived from the Chinese wormwood tree. It has turned out to be an effective and relatively inexpensive way to treat malaria infections. However, pockets of artemisinin resistance have already been uncovered in Southeast Asia. Public health workers now recommend that artemisinin be given in combination with other drugs. Treating infected patients with multiple drugs is more likely to wipe out *Plasmodium* in their bodies, reducing the chances that more drug-resistant strains will emerge.

Given the challenges of developing practical drugs to prevent or treat malaria, some scientists have turned their attention to other approaches. One goal is to produce a malaria vaccine. The parasite's complex life cycle makes that a complicated endeavor, though. Several vaccines are now in various stages of testing, and some show promise. But researchers expect it will be years before a safe, effective vaccine for malaria could be available.

Other researchers are focusing their efforts on the mosquitoes that carry the parasite, rather than on *Plasmodium* itself. The genetically engineered insects created by the team in Arizona are a promising step in that direction.

The Arizona researchers set out to alter a cellular signaling gene that plays a role in the mosquito's life cycle. Mosquitoes normally live 2 to 3 weeks, and *Plasmodium* takes about 2 weeks to mature in the mosquito's gut. The researchers hoped to create mosquitoes that would die prematurely, before the parasite is mature. The genetic modification worked as planned. The engineered mosquitoes' life-spans were shortened by 18% to 20%. The genetic tweak also had a surprising side effect. The altered gene completely blocked the development of *Plasmodium* in the mosquitoes' guts. The engineered mosquitoes are incapable of spreading malaria to humans, regardless of how long they live.

While the finding was a laboratory success, it will be much harder to translate the results to the real world. To create malaria-free mosquitoes in the wild, scientists would have to release the genetically modified mosquitoes and hope that their altered gene spreads through the wild mosquito population. But that gene would be passed on only if it gave the insects a distinct evolutionary advantage. The engineered mosquitoes may be malaria free, but so far they have no advantage over their wild counterparts.

It's clear that slashing malaria rates will not be an easy task. Despite decades of research, insecticide-treated bed nets are still the best method for preventing the disease. For millennia, the malaria parasite has managed to withstand our efforts to squelch it, yet science continues to push the boundaries. Who will emerge the victor? Stay tuned.

? CASE 4 QUESTIONS

Special sections in Chapters 21–24 discuss the following questions related to Case 4.

1. **What genetic differences make some individuals more and some less susceptible to malaria?** See page 434.
2. **How did malaria come to infect humans?** See page 455.
3. **What human genes are under selection for resistance to malaria?** See page 501.

Roc Canals Photography/Getty Images.

CHAPTER 21

Evolution

How Genotypes and Phenotypes Change over Time

Core Concepts

21.1 Genetic variation refers to differences in DNA sequences.

21.2 Patterns of genetic variation can be described by allele frequencies.

21.3 Evolution is a change in the frequency of alleles or genotypes over time.

21.4 Natural selection leads to adaptations, which enhance the fit between an organism and its environment.

21.5 Migration, mutation, genetic drift, and nonrandom mating are nonadaptive mechanisms of evolution.

21.6 Molecular evolution is a change in DNA or amino acid sequences over time.

Variation is a fact of nature. A walk down any street reveals how variable our species is: Skin color and hair color, for example, vary from person to person. Until the publication in 1859 of Charles Darwin's *On the Origin of Species,* scientists tended to view all the variation we see in humans and other species as biologically unimportant. According to the traditional view, not only were species individually created in their modern forms by a divine Creator, but, because the Creator had a specific design in mind for each species, they were fixed and unchanging. Departures or variations from this divinely ordained type were therefore ignored.

Since Darwin, however, we have appreciated that a species does not conform to a type. Rather, a species consists of a range of variants. In our own species, people may be tall, short, dark-haired, fair-haired, and so on. Furthermore, variation is an essential ingredient of Darwin's theory because the mechanism he proposed for evolution, natural selection, depends on the differential success—in terms of surviving and reproducing—of variants. Darwin changed how we view variation. Before Darwin, variation was irrelevant, something to be ignored; after Darwin, it was recognized as the key to the evolutionary process.

21.1 GENETIC VARIATION

Variation is a major feature of the natural world. We humans are particularly good at noticing phenotypic variation among individuals of our own species. As we discussed in Chapter 16, a phenotype is an observable trait, such as human height or wing color in butterflies. As we saw in Chapter 18, two factors contribute to phenotype: an individual's genotype, which is the set of alleles possessed by the individual, and the environment in which the individual lives. We can take the environment out of the equation by looking directly at genotypic differences through sequencing DNA regions in multiple individuals. We now explore genetic variation directly, in terms of differences at the DNA sequence level.

Population genetics is the study of patterns of genetic variation.

Remarkably, in spite of a high degree of phenotypic variation, humans actually rank low in terms of overall genetic variation compared with other species. Any two randomly selected humans differ from each other on average by one DNA base per thousand (the two genomes are 99.9% identical), while two fruit flies differ by ten bases per thousand (the two genomes are 99% identical). Even one of the most seemingly uniform species on the planet, the Adélie penguins seen in **Fig. 21.1,** is two to three times more genetically variable than we are.

As we discuss in Chapter 22, a **species** consists of individuals that can exchange genetic material through interbreeding. From a genetic perspective, a species is therefore a group of individuals capable, through reproduction, of sharing alleles with one another. Individuals represent different combinations of alleles drawn from the species' **gene pool,** that is, from all the alleles present in all individuals in the species. The human gene pool includes alleles that cause differences in skin color, hair type, eye color, and so on. Each one of us has a different set of those alleles—alleles that cause brown hair and brown eyes, for example, or black hair and blue eyes—drawn from that gene pool.

Population genetics is the study of genetic variation in natural **populations,** which are interbreeding groups of organisms of the same species living in the same geographical area. What factors determine the amount of variation in a population and in a species? Why are humans genetically less variable than penguins? What factors affect the distribution of particular variations? Population genetics addresses detailed questions about patterns of variation. And small differences, given enough time, can lead to the major differences we see among organisms today.

FIG. 21.1 Genetic diversity in Adélie penguins. Adélie penguins are uniform in appearance but are actually more genetically diverse than humans. *Source: Tim Davis/Corbis.*

Mutation and recombination are the two sources of genetic variation.

Genetic variation has two sources: Mutation generates new variation, and recombination followed by segregation of homologous chromosomes during meiotic cell division shuffles mutations to create new combinations. In both cases, new alleles are formed, as shown in **Fig. 21.2.**

As we saw in Chapter 14, mutations can be **somatic,** occurring in the body's tissues, or **germline,** occurring in the reproductive cells and therefore passed on to the next generation. From an evolutionary viewpoint, we are primarily interested in germ-line mutations. A somatic mutation affects only the cells descended from the one cell in which the mutation originally arose, and thus affects only that one individual. However, a germ-line mutation appears in every cell of an individual derived from the fertilization involving the mutation-bearing gamete, and thus appears in its descendants.

Mutations can also be classified by their effects on an organism (Chapter 15). Mutations occur randomly throughout the genome, and, because most of the genome consists of noncoding DNA, most mutations are **neutral,** having little or no effect on the organism. Most mutations that do occur in protein-coding regions of the genome, however, have a **deleterious,** or harmful, effect on an organism. Rarely, a mutation occurs that has a beneficial effect. Mutations like these are **advantageous** if they improve their carriers' chances of survival or reproduction.

FIG. 21.2 Mutation and recombination. The formation of new alleles occurs by mutation and recombination.

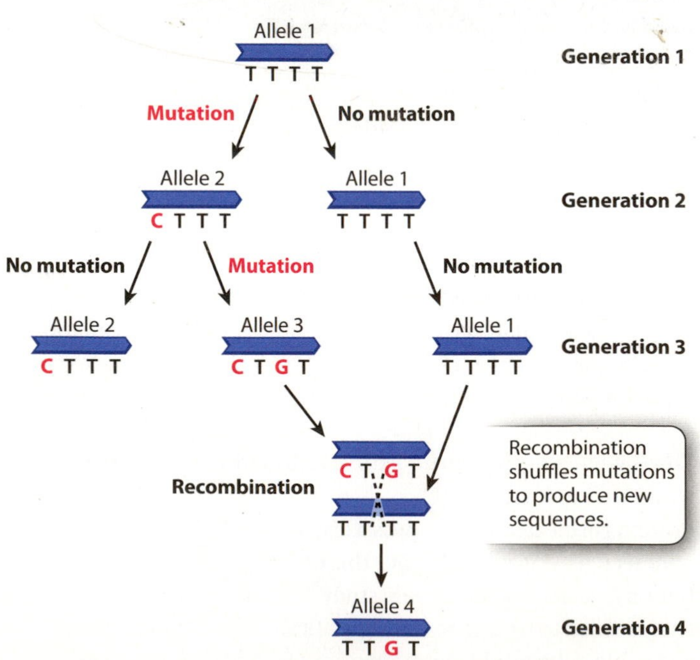

Advantageous mutations, as we will see, can increase in frequency in a population until eventually they are carried by every member of a species. These mutations are the ones that result in a species that is adapted to its environment—better able to survive and reproduce in that environment.

21.2 MEASURING GENETIC VARIATION

Mutations, whether deleterious, neutral, or advantageous, are sources of genetic variation. The goal of population genetics is to make inferences about the evolutionary process from patterns of genetic variation in nature. The raw information for this comes from the rates of occurrence of alleles in populations, or **allele frequencies.**

To understand patterns of genetic variation, we require information about allele frequencies.

The allele frequency of an allele *x* is simply the number of *x*'s present in the population divided by the total number of alleles. Consider, for example, pea color in Mendel's pea plants. In Chapter 16, we discussed how pea color (yellow or green) results from variation at a single gene. Two alleles of this gene are the dominant *A* (yellow) allele and the recessive *a* (green) allele. *AA* homozygotes and *Aa* heterozygotes produce yellow peas, and *aa* homozygotes produce green peas. Imagine that in a population every pea plant produces green peas, meaning that only one allele, *a*, is present: The allele frequency of *a* is 100%, and the allele frequency of *A* is 0%. When a population exhibits only one allele at a particular gene, we say that the population is **fixed** for that allele.

Now consider another population of 100 pea plants with genotype frequencies of 50% *aa*, 25% *Aa*, and 25% *AA*. A genotype frequency is the proportion in a population of each genotype at a particular gene or set of genes. These genotype frequencies give us 50 green-pea pea plants (*aa*), 25 yellow-pea heterozygotes (*Aa*), and 25 yellow-pea homozygotes (*AA*). What is the allele frequency of *a* in this population? Each of the 50 *aa* homozygotes has two *a* alleles and each of the 25 heterozygotes has one *a* allele. Of course, there are no *a* alleles in *AA* homozygotes. The total number of *a* alleles is thus $(2 \times 50) + 25 = 125$. To determine the allele frequency of *a*, we divide the number of *a* alleles by the total number of alleles in the population, 200 (because each pea plant is diploid, meaning that it has two alleles): $125/200 = 62.5\%$. Because we are dealing with only two alleles in this example, the allele frequency of *A* is $100\% - 62.5\% = 37.5\%$

Thus, the allele frequencies of *A* and *a* provide a measure of genetic variation at one gene in a given population. In this example, we were given the genotype frequencies, and from this information we determined the allele frequencies. But how are genotype and allele frequencies measured? We consider three ways to measure genotype and allele frequencies in populations: observable traits, gel electrophoresis, and DNA sequencing.

→ **Quick Check 1** Using the example of pea color in Mendel's pea plants, can you devise equations to determine the allele frequencies of *A* and *a* from the genotype frequencies of *aa*, *Aa*, and *AA*?

Early population geneticists relied on observable traits and gel electrophoresis to measure variation.

It would be a simple matter to measure genetic variation in a population if we could use observable traits. Then we could simply count the individuals displaying variant forms of a trait and have a measure of the variation of that trait's gene. However, as we saw in Chapter 18, this approach can work only rarely, for two important reasons. First, many traits are encoded by a large number of genes. In these cases, it is difficult, if not impossible, to make direct inferences from a phenotype to the underlying genotype. Even traits that seem to have a simple set of phenotypes often prove to have a complicated genetic basis. For instance, human skin color is determined by at least six different genes. Second, the phenotype is a product of both the genotype and the environment.

Until the 1960s, there was only one workable solution: to limit population genetics to the study of phenotypes that are encoded by a single gene. As these are few, the number of genes that population geneticists could study was extremely small. Human blood groups, including the ABO system, provided an early example of a trait encoded by a single gene with multiple alleles. At this gene, there are three alleles in the population—*A*, *B*, and *O*—and therefore six possible genotypes, which result in four different phenotypes (**Table 21.1**).

TABLE 21.1 The ABO blood system.

PHENOTYPE	GENOTYPE
A	*AA* or *AO*
B	*BB* or *BO*
AB	*AB*
O	*OO*

Other instances in which phenotypic variation can be readily correlated with genotype include certain markings in invertebrates. For example, the coloring of the two-spot ladybug *Adalia bipunctata* is controlled by a single gene (**Fig. 21.3**). However, the genetic basis of most traits is not so simple.

Single-gene variation became much easier to detect in the 1960s with the application of gel electrophoresis. In Chapter 12, we saw how gel electrophoresis separates segments of DNA according to their size. Before DNA technologies were developed, the same basic process was applied to proteins to separate them according to their electrical charge and their size. In gel electrophoresis, the proteins being studied migrate through a gel when an electrical charge is applied. The rate at which the proteins move from one end of the gel to the other is determined by their charge and their size.

FIG. 21.3 **Single-gene variation.** A genetic difference in color in the two-spot ladybug, *Adalia bipunctata*, results from variation in a single gene. *Sources: (top) © Biopix: G. Drange; (bottom) Howard Marsh/Shutterstock.*

Early studies of protein electrophoresis focused on enzymes that catalyze reactions that can be induced to produce a dye when the substrate for the enzyme is added. If we add the substrate, we can see the locations of the proteins in the gel. The bands in the gel provide a visual picture of genetic variation in the population, revealing what alleles are present and what their frequencies are. **Fig. 21.4** shows this sort of experiment.

DNA sequencing is the gold standard for measuring genetic variation.

Protein gel electrophoresis was a leap forward in our ability to detect genetic variation, but this technique had significant limitations. Researchers could study only enzymes because they needed to be able to stain specifically for enzyme activity and could detect only mutations that resulted in amino acid

FIG. 21.4

How is genetic variation measured?

BACKGROUND The introduction of protein gel electrophoresis in 1966 gave researchers the opportunity to identify differences in amino acid sequence in proteins both among individuals and, in the case of heterozygotes, within individuals. Proteins with different amino acid sequences run at different rates through a gel in an electric field. Often, a single amino acid difference is enough to affect the mobility of a protein in a gel.

METHOD Starting with crude tissue—in this case, the ground-up whole body of a fruit fly—we load the material on a gel and turn on the current. The rate at which a protein migrates depends on its size and charge, both of which may be affected by its amino acid sequence. To visualize the protein at the end of the gel run, we use a biochemical indicator that produces a stain when the protein of interest is active. Here, we use an indicator that detects the activity of alcohol dehydrogenase (Adh). The result is a series of bands on the gel showing where the Adh proteins in each lane migrated.

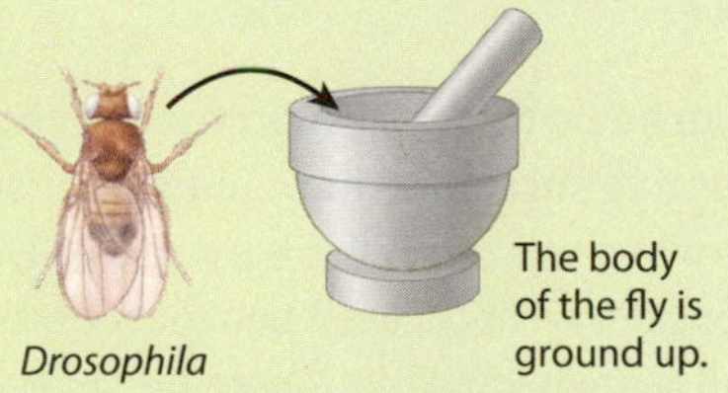

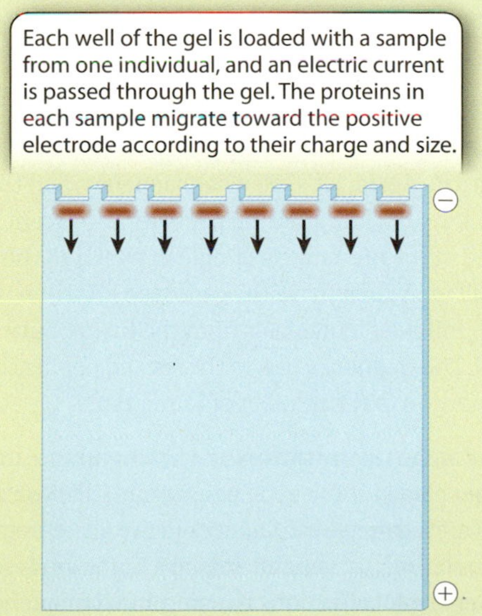

RESULTS The *Adh* gene has two common alleles, distinguished by a single amino acid difference that changes the charge of the protein. One allele, *Fast* (*F*), accordingly runs faster than the other, *Slow* (*S*). Four individuals are *S* homozygotes; two are *F* homozygotes; and two are *FS* heterozygotes. Note that the heterozygotes do not stain as strongly on the gel because each band has half the intensity of the single band in the homozygote. We can measure the allele frequencies simply by counting the alleles. Each homozygote has two of the same allele, and each heterozygote has one of each.

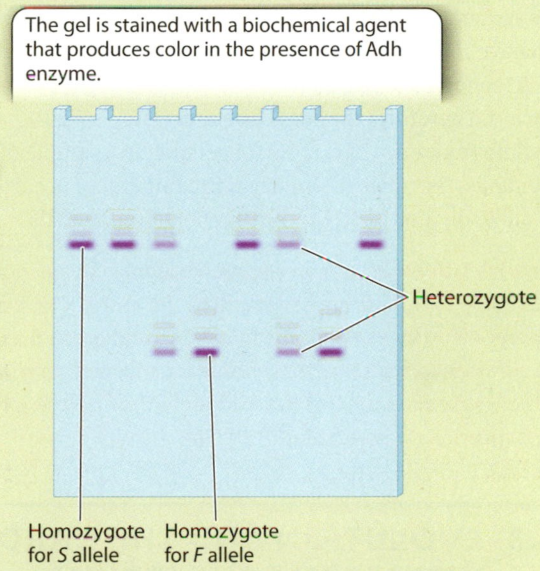

Total number of alleles in the population of 8 individuals = $8 \times 2 = 16$

Number of *S* in the population = $2 \times$ (number of *S* homozygotes) + (number of heterozygotes) = $8 + 2 = 10$

Frequency of $S = \frac{10}{16} = \frac{5}{8}$

Number of *F* in the population = $2 \times$ (number of *F* homozygotes) + (number of heterozygotes) = $4 + 2 = 6$

Frequency of $F = \frac{6}{16} = \frac{3}{8}$

Note that the two allele frequencies add to 1.

CONCLUSION We now have a profile of genetic variation at this gene for these individuals. Population genetics involves comparing data such as these with data collected from other populations to determine the forces shaping patterns of genetic variation.

FOLLOW-UP WORK This technique is seldom used these days because it is easy now to recover much more detailed genetic information about genetic variation from DNA sequencing.

SOURCE Lewontin, R. C., and J. L. Hubby. 1966. "A Molecular Approach to the Study of Genic Heterozygosity in Natural Populations. II. Amount of Variation and Degree of Heterozygosity in Natural Populations of *Drosophila pseudoobscura*." *Genetics* 54:595–609.

substitutions that changed a protein's mobility in the gel. Only with DNA sequencing did researchers finally have an unambiguous means of detecting all genetic variation in a stretch of DNA, whether in a coding region or not. The variations studied by modern population geneticists are differences in DNA sequence, such as a T rather than a G at a specified nucleotide position in a particular gene.

Calculating allele frequencies, then, simply involves collecting a population sample and counting the number of occurrences of a given mutation. We can look even closer at the example of the *Drosophila Adh* gene from Fig. 21.4 to focus not on the amino acid difference between the Fast and Slow phenotypes, but on the A or G nucleotide difference corresponding to the two phenotypes. If we sequence the *Adh* gene from 50 individual flies, we will then have 100 gene sequences from these diploid individuals. We find 70 sequences have an A and 30 have a G at the position in question. Therefore, the allele frequency of A is $70/100 = 0.7$ and the allele frequency of G is 0.3. In general, in a sample of *n* diploid individuals, the allele frequency is the number of occurrences of that allele divided by twice the number of individuals.

→ **Quick Check 2** Data on genetic variation in populations have become ever more precise over time, from phenotypes that are determined by a single gene to gel electrophoresis that looks at variation among genes that encode for enzymes, to analysis of the DNA sequence. Has this increase in precision resulted in the uncovering of more genetic variation or less?

21.3 EVOLUTION AND THE HARDY–WEINBERG EQUILIBRIUM

Determining allele frequencies gives us information about genetic variation. Following and measuring change in that variation over time is key to understanding the genetic basis of evolution.

Evolution is a change in allele or genotype frequency over time.

At the genetic level, evolution is simply a change in the frequency of an allele or a genotype from one generation to the next. For example, if there are 200 copies of an allele that causes blue eye color in a population in generation 1 and there are 300 copies of that allele in a population of the same size in generation 2, evolution has occurred. In principle, evolution may occur without allele frequencies changing. For instance, even if, in our fruit fly example, the A/G allele frequencies stay the same from one generation to the next, the frequencies of the different genotypes (that is, of AA, AG, and GG) may change. This would be evolution *without* allele frequency change.

Evolution is therefore a change in the genetic makeup of a population over time. Note an important and often misunderstood aspect of this definition: *Populations* evolve, not *individuals.* Note, too, that this definition does not specify a mechanism for this change. As we will see, many mechanisms can cause allele or genotype frequencies to change. Regardless of which mechanisms are involved, any change in allele frequencies, genotype frequencies, or both constitutes evolution.

The Hardy–Weinberg equilibrium describes situations in which allele and genotype frequencies do not change.

Allele and genotype frequencies change over time only if specific forces act on the population. This principle was demonstrated independently in 1908 by the English mathematician G. H. Hardy and the German physician Wilhelm Weinberg, and has become known as the **Hardy–Weinberg equilibrium.** In essence, the Hardy–Weinberg equilibrium describes the situation in which evolution does *not* occur. In the absence of evolutionary forces (such as natural selection), allele and genotype frequencies do not change.

To determine whether or not evolutionary forces are at work, we need to determine whether or not a population is in Hardy–Weinberg equilibrium. The Hardy–Weinberg equilibrium specifies the relationship between allele frequencies and genotype frequencies when a number of key conditions are met. In these cases, we can conclude that evolutionary forces are not acting on the gene in the population we are studying. In many ways, then, the Hardy–Weinberg equilibrium is most interesting when we find instances in which allele or genotype frequencies *depart* from expectations. This finding implies that one or more of the conditions are *not* met and that evolutionary mechanisms are at work.

A population that is in Hardy–Weinberg equilibrium meets these conditions:

1. **There can be no differences in the survival and reproductive success of individuals.** Let's examine what happens when this condition is not met. Given two alleles, *A* and *a,* consider what occurs when *a,* a recessive mutation, is lethal. All *aa* individuals die. Therefore, in every generation, there is a selective elimination of *a* alleles, meaning that the frequency of *a* will gradually decline (and the frequency of *A* correspondingly increase) over the generations. As we discuss below, we call this differential success of alleles **selection.**

2. **Populations must not be added to or subtracted from by migration.** Again, let's see what happens when this condition is not met. Consider a second population adjacent to the one we used in the preceding example in which all the alleles are *A* and all individuals have the genotype *AA.* Then there is a sudden influx of individuals from the first population into the second. The frequency of *A* in the second population changes in proportion to the number of immigrants.

3. **There can be no mutation.** If *A* alleles mutate into *a* alleles (or other alleles, if the gene has multiple alleles), and vice versa, then again we see changes in the allele frequencies over the generations. In general, because mutation is so rare, it has a very small effect on changing allele frequencies on the timescales studied by population geneticists.

4. **The population must be sufficiently large to prevent sampling errors.** Small samples are likely to be more misleading than large ones. Campus-wide, a college's sex ratio may be close to 50:50, but in a small class of 8 individuals it is not improbable that we would have 6 women and 2 men (a 75:25 ratio). Sample size, in the form of population size, also affects the Hardy–Weinberg equilibrium such that it technically holds only for infinitely large populations. A change in the frequency of an allele due to the random effects of limited population size is called **genetic drift.**

5. **Individuals must mate at random.** For the Hardy–Weinberg equilibrium to hold, mate choice must be made without regard to genotype. For example, an *AA* homozygote when offered a choice of mate from among *AA*, *Aa*, or *aa* individuals should choose at random. In contrast, **non-random mating** occurs when individuals do not mate randomly. For example, *AA* homozygotes might preferentially mate with other *AA* homozygotes. Non-random mating affects genotype frequencies from generation to generation, but does not affect allele frequencies.

The Hardy–Weinberg equilibrium relates allele frequencies and genotype frequencies.

Now that we have established the conditions required for a population to be in Hardy–Weinberg equilibrium, let us explore the idea in detail. In the *Drosophila* example we looked at earlier, we know the frequency of the two alleles, one with A and the other with G in the *Adh* gene. What are the genotype frequencies? That is, how many AA homozygotes, AG heterozygotes, and GG homozygotes do we see? The Hardy–Weinberg equilibrium predicts the expected genotype frequencies from allele frequencies.

The logic is simple. Random mating is the equivalent of putting all the population's gametes into a single pot and drawing out pairs of them at random to form a zygote, which is the same principle we saw in action in the discussion of independent assortment in Chapter 16. We therefore put in our 70 A alleles and 30 G alleles and pick pairs at random. What is the probability of picking an AA homozygote (that is, what is the probability of picking an A allele followed by another A allele)? The probability of picking an A allele is its frequency in the population, so the probability of picking the first A is 0.7. What is the probability of picking the second A? Also 0.7. What then is the probability of picking an A followed by another A? It is the product of the two probabilities: $0.7 \times 0.7 = 0.49$. Thus, the frequency of an AA genotype is 0.49. We take the same approach to determine the genotype frequency for the GG genotype: Its frequency is 0.3×0.3, or 0.09.

What about the frequency of the heterozygote, AG? This is the probability of drawing G followed by A, or A followed by G. There are thus two separate ways in which we can generate the heterozygote. Its frequency is therefore $(0.7 \times 0.3) + (0.3 \times 0.7) = 0.42$.

As seen in the Punnett square on the left in **Fig. 21.5,** we can generalize these calculations algebraically by substituting letters for the numbers we have computed to derive the relation defined by the Hardy–Weinberg equilibrium. If the allele frequency of one allele, A, is p, and the other, G, is q, then $p + q = 1$ (because there are no other alleles at this site).

Genotypes	AA	AG	GG
Frequencies	p^2	$2pq$	q^2

In the graph on the right hand side of Fig 21.5, we have replaced the p's and q's with numbers. If no *a* alleles are present, then $q = 0$ and $p = 1$ and the frequency of *AA* is 1. In this case, all the genotypes in the population are *AA* and, accordingly, the blue line representing the frequency of *AA* in the population is at 1, and the red line representing the frequency of *aa* and the purple line representing the frequency of *Aa* are both at 0. When no A alleles are present, all individuals have genotype *aa*, and the red line is at 1 and the others at 0.

Not only does the Hardy–Weinberg relation predict genotype

FIG. 21.5 **Hardy–Weinberg relation.** The Hardy–Weinberg relation predicts genotype frequencies from allele frequencies, and vice versa.

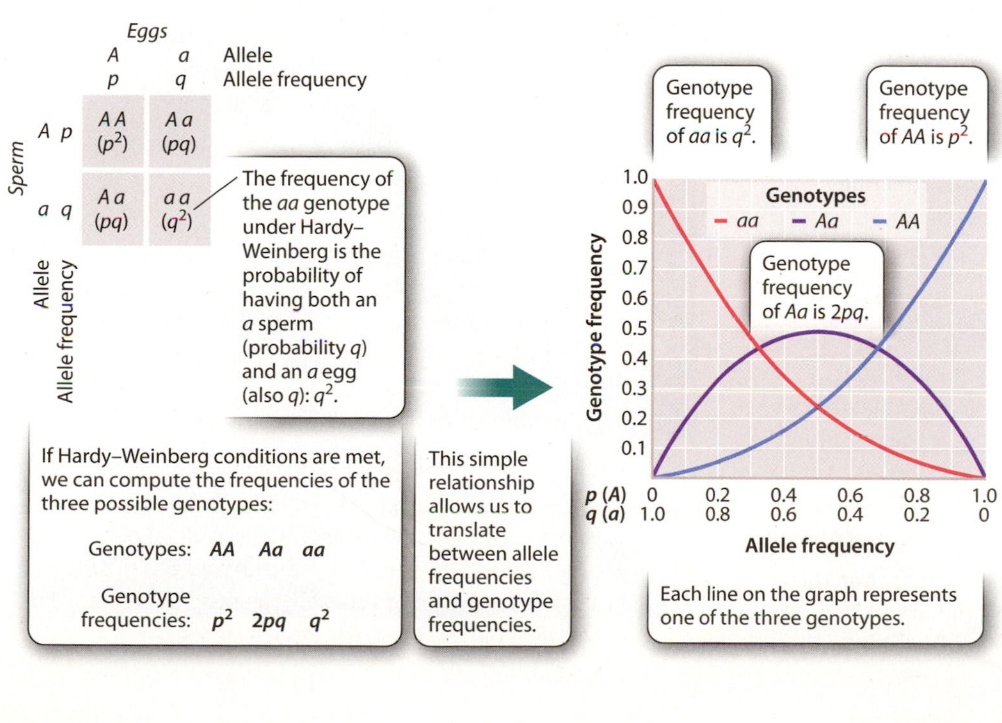

frequency from allele frequencies, but it works in reverse, too: Genotype frequencies predict allele frequencies.

The graph in Fig. 21.5 also shows how allele and genotype frequencies are related. We can use the graph to determine allele frequencies for given genotype frequencies and genotype frequencies for given allele frequencies. For example, if the population we are examining has a 0.5 frequency of heterozygotes, *Aa*, the purple line in the graph indicates that both p and q are 0.5. If we know that $p = 0.5$ (and therefore $q = 0.5$), we can look at the lines to infer that the frequency of heterozygotes is 0.5 and the frequency of both homozygotes, *AA* and *aa*, is the same, 0.25.

We can do this mathematically as well. Knowing the genotype frequency of *AA*, for example, permits us to calculate allele frequencies: if, as in our *Adh* example, p^2 is 0.49 (that is, 49% of the population has genotype *AA*), then p, the allele frequency of *A*, is $\sqrt{0.49} = 0.7$. Because $p + q = 1$, then q, the allele frequency of *a*, is $1 - 0.7 = 0.3$.

Note that these relationships hold only if the Hardy–Weinberg conditions are met. If not, then allele frequencies can be determined only from genotype frequencies, as described earlier in section 21.2.

The Hardy–Weinberg equilibrium is the starting point for population genetic analysis.

Recall our definition of evolution: a change in allele or genotype frequency from one generation to the next. Given this definition, it might seem odd to be discussing factors necessary for allele frequencies to stay the same. The Hardy–Weinberg equilibrium not only provides a means of converting between allele and genotype frequencies, but, critically, it also serves as an indicator that something interesting is happening in a population when it is not upheld.

If we find a population whose allele or genotype frequencies are not in Hardy–Weinberg equilibrium, we can infer that evolution has occurred. With further study of the population, we can then consider, for the gene in question, whether the population is subject to selection, migration, mutation, genetic drift, or non-random mating. These are the primary mechanisms of evolution. The Hardy–Weinberg equilibrium gives us a baseline from which to explore the evolutionary processes affecting populations. We will start by considering one of the most important evolutionary mechanisms: natural selection.

→ **Quick Check 3** When we find a population whose allele frequencies are *not* in Hardy–Weinberg equilibrium, what can and can't we conclude about that population?

21.4 NATURAL SELECTION

Natural selection results in allele frequencies changing from generation to generation according to the allele's impact on the survival and reproduction of individuals. New mutations that are deleterious and eliminated by natural selection have no long-term evolutionary impact; ones that are beneficial, however, can result in adaptation to the environment over time.

Natural selection brings about adaptations.

The adaptations we see in the natural world—the exquisite fit of organisms to their environment—were typically taken by pre-Darwinian biologists as evidence of a divine Creator's existence. Each species, they argued, was so well adapted—the desert plant so physiologically adept at coping with minimal levels of rainfall and the fast-swimming fish so hydrodynamically streamlined—that it must have been designed by a Creator.

With the publication of *On the Origin of Species* in 1859, Darwin, pictured in **Fig. 21.6,** overturned the biological convention of his day on two fronts. First, he showed that species are not unchanging; they have evolved over time. Second, he suggested a mechanism, natural selection, that brings about adaptation. Natural selection was a brilliant solution to the central problem of biology: how organisms come to fit so well in their environments. From where does the woodpecker get its powerful chisel of a bill? And the hummingbird its long delicate bill for probing the nectar stores

FIG. 21.6 **Charles Darwin.** This photograph was taken at about the time Darwin was writing *On the Origin of Species. Source: akg-images.*

FIG. 21.7 **Alfred Russel Wallace.** This photograph was taken in Singapore during Wallace's expedition to Southeast Asia. *Source: A. R. Wallace Memorial Fund & G. W. Beccaloni.*

in flowers? Darwin showed how a simple mechanism, without foresight or intentionality, could result in the extraordinary range of adaptations that all of life is testimony to.

For 20 years after first conceiving the essence of his theory, Darwin collected supporting evidence. In 1858, however, he was spurred to begin writing *On the Origin of Species* by a letter from a little-known naturalist collecting specimens in what is today Indonesia. By a remarkable coincidence, Alfred Russel Wallace, shown in **Fig. 21.7,** had also developed the theory of evolution by natural selection. Aware that Darwin was interested in the problem but having no idea that Darwin was working on the same theory, Wallace wrote to Darwin in 1858 to see what he thought of his idea.

Suddenly, Darwin was confronted with the prospect of losing his claim on the theory that he had been quietly nurturing for 20 years. But all was not lost. Darwin's colleagues arranged for the publication of a joint paper by Wallace and Darwin in 1858. This was done without consulting Wallace, who nonetheless never resented Darwin and afterward was careful to insist that the idea rightly belonged to the older man. It was Darwin's publication of *On the Origin of Species* in 1859 that brought both evolution and natural selection, its underlying mechanism of adaptation, to public attention. Wallace is only fleetingly mentioned in Darwin's great work, and Darwin, not Wallace, is now the name associated with the discovery.

Both Darwin and Wallace recognized their debt to the writings of a British clergyman, Thomas Malthus. In his *Essay on the Principle of Population,* first published in 1798, Malthus pointed out that natural populations have the potential to increase in size geometrically, meaning that populations get larger at an ever-increasing rate. Imagine that human couples can have just four children (two males and two females), so the population doubles every generation. Starting with a single couple, by the twentieth generation, the population will have grown to over a million—1,048,576, to be precise.

However, this geometric expansion of populations does not occur. In fact, population sizes are typically stable from generation to generation. This is because the resources upon which populations are dependent—food, water, places to live—are limited. In each generation many fail to survive or reproduce; there simply is not enough food and other resources to go around. This implies in turn that individuals within a population must compete for resources.

Which individuals will win the competition? Darwin and Wallace suggested that those that are best adapted would most likely survive and leave more offspring. Genetic variation among individuals results in some that are more likely than others to survive and reproduce, passing their genetic material to the next generation. As a result, the next generation will have a higher proportion of these same advantageous alleles. Darwin used the term "natural selection" for the filtering process that acts against deleterious alleles and in favor of advantageous ones.

Competitive advantage is a function of how well an organism is adapted to its environment. A desert plant that is more efficient at minimizing water loss than another plant is better adapted to the desert environment. An organism that is better adapted to its environment is more fit. **Fitness,** in this context, is a measure of the extent to which the individual's genotype is represented in the next generation. We say that the first plant's fitness is higher than the second's if it leaves more surviving offspring either because the plant itself survives for longer, giving it greater opportunity to reproduce, or because it has some other reproductive advantage, such as the ability to produce more seeds. Assuming that the trait maintains its advantage, natural selection then acts over generations to increase the overall fitness of a population. A plant population newly arrived in a desert may be poorly adapted to its environment, but, over time, alleles that minimize water loss increase under natural selection, resulting in a population that is better adapted to the desert.

Such changes in populations take time. Borrowing from the geologists of his day, Darwin recognized that time was a critical ingredient of his theory. Geologists had put forward a view of Earth's history that argued that large geological changes—like the carving of the Grand Canyon—can be explained by simple day-to-day processes operating over vast timescales. Darwin applied this worldview to biology. He recognized that small changes, like subtle shifts in the frequencies of alleles, could add up to major changes given long enough time periods. What might seem to us to be a trivial change over the short term can, over the long term, result in substantial differences among populations.

The Modern Synthesis combines Mendelian genetics and Darwinian evolution.

Darwinian evolution involves the change over time of the genetic composition of populations and is thus a genetic theory. Although Mendel published his genetic studies of pea plants in 1866, not long after *The Origin,* Darwin never saw them, so a key component of the theory was missing. The rediscovery of Mendel's work in 1900 unexpectedly provoked a major controversy among evolutionary biologists. Some argued that Mendel's discoveries did not apply to most genetic variation because the traits studied by Mendel were discrete, meaning that they had clear alternative states, such as either yellow or green color in peas. Most of the variation we see in natural populations, in contrast, is continuous, meaning that variation occurs across a spectrum (Chapter 18). Human height, for example, does not come in discrete classes. People are not either 5 feet tall or 6 feet tall and of no height in between. Instead, they may be any height within a certain range.

How could the factors that controlled Mendel's discrete traits account for the continuous variation seen in natural populations? This question was answered by the English theoretician Ronald Fisher, who realized that, instead of a single gene contributing to a trait like human height, there could be several genes that contribute to the trait. He argued that extending Mendel's theory to include multiple genes per trait could account for patterns of continuous variation that we see all around us.

Fisher's insight formed the basis of a synthesis between Darwin's theory of natural selection and Mendelian genetics that was forged during the middle part of the twentieth century. The product of this **Modern Synthesis** is our current theory of evolution.

Natural selection increases the frequency of advantageous mutations and decreases the frequency of deleterious mutations.

Natural selection increases the frequency of advantageous alleles, resulting in adaptation. In some cases, it can promote the fixation of advantageous alleles, meaning the allele has a frequency of 1. To start with, a new advantageous allele will exist as a single copy in a single individual (that is, as a heterozygote), but, under the influence of natural selection, the advantageous allele can eventually replace all the other alleles in the population. Natural selection that increases the frequency of a favorable allele is called **positive selection.**

As we have seen, most mutations to functional genes are deleterious. In extreme cases, they are lethal to the individuals carrying them and are thus eliminated from the population. Sometimes, however, natural selection is inefficient in getting rid of a deleterious allele. Consider a recessive lethal mutation, *b* (that is, one that is lethal only as a homozygote, *bb,* and has no effect as a heterozygote, *Bb*). When it first arises, all the other alleles in the population are *B*, which means that the first *b* allele that appears in the population must be paired with a *B* allele, resulting in a *Bb* heterozygote. Because natural selection does not act against heterozygotes in this case, the *b* allele may increase in frequency by chance alone (we discuss below how this happens). Only when two *b* alleles come together to form a *bb* homozygote does natural selection act to rid the population of the allele. Natural selection that decreases the frequency of a deleterious allele is called **negative selection.**

Many human genetic diseases show this pattern: The deleterious allele is rare and recessive. Because it is rare, homozygotes for it are formed only infrequently. Remember that the expected frequency of homozygotes in a population under the Hardy–Weinberg equilibrium is the square of the frequency of the allele in the population. Therefore, if the allele frequency is 0.01, we expect 0.01×0.01, or 1 in every 10,000 individuals, to be homozygous for it. Thus, the genetic disease occurs rarely, and the allele remains in the population because it is recessive and not expressed as a heterozygote.

? CASE 4 MALARIA: CO-EVOLUTION OF HUMANS AND A PARASITE

What genetic differences have made some individuals more and some less susceptible to malaria?

In addition to allowing alleles to be either eliminated or fixed, natural selection can also maintain an allele at some intermediate frequency between 0% and 100%. This form of natural selection is called **balancing selection,** and it acts to maintain two or more alleles in a population. A simple case is members of a species that face different conditions depending upon where they live. One allele might be favored by natural selection in a dry area, but a different one favored in a wet area. Taking the species as a whole, these alleles are maintained by natural selection at intermediate frequencies.

Another example of balancing selection occurs when the heterozygote's fitness is higher than that of either of the homozygotes, resulting in selection that ensures that both alleles remain in the population at intermediate frequencies. This form of balancing selection is called **heterozygote advantage,** and it is exemplified by human populations in Africa, where malaria has been a long-standing disease. Because the malaria parasite spends part of its life cycle in human red blood cells, mutations in the hemoglobin molecule that affect the structure of the red blood cells have a negative impact on the parasite and can reduce the severity of malarial attacks.

Two alleles of the gene for one of the subunits of hemoglobin are *A* and *S* (Chapter 15). The *A* allele codes for normal hemoglobin, resulting in fully functional, round red blood cells. The *S* allele encodes a polypeptide that differs from the *A* allele's product in just a single amino acid, which is enough to cause the molecules to aggregate end to end, so the red blood cell is distorted into a sickle.

In regions of the world with malaria, heterozygous individuals (*SA*) have an advantage over homozygous individuals (*SS* and *AA*). *SS* homozygotes are protected against malaria, but they are burdened with severe sickling disease. Sickle-shaped red blood cells can block capillaries, and therefore people with the *SS* genotype are prone to debilitating, painful, and sometimes fatal episodes resulting from capillary blockage. *AA* homozygotes lack sickling disease but are vulnerable to malaria. *SA* heterozygotes, however, do not have severe sickling disease and have some protection from malaria. As a result, natural selection maintains both the *S* and *A* alleles in the population at intermediate frequencies.

In areas where there is no malaria, this balance is shifted. Many African Americans, descended from Africans upon whom natural selection operated in favor of the heterozygote, still carry the *S* allele, even though the allele is no longer useful to them in their malaria-free environment. If natural selection were to run its course among African-Americans, the *S* allele would gradually be eliminated. However, this is a slow process, and many more people will continue to suffer from sickle-cell anemia before it is complete.

Natural selection can be stabilizing, directional, or disruptive.

Up to this point, we have followed the fate of individual mutations, which can increase, decrease, or be maintained at an intermediate frequency under the influence of natural selection. We can also look at the consequence of natural selection from a different perspective. Instead of following individual mutations, we can look at changes over time of a particular trait of an organism. For example, we might track the evolution of height in a population, despite not knowing the specifics of the genetic basis of height differences. When we look at natural selection from this perspective, we see three types of patterns: stabilizing, directional, and disruptive.

Stabilizing selection maintains the status quo and acts against extremes. A good example is provided by human birth weight, a trait affected by a number of factors, including many fetal genes (**Fig. 21.8**). If a baby is too small, then its chances of survival after birth are low. However, if it is too big, there may be complications during delivery that endanger both mother and baby. Thus, the optimum birth weight is between these two extremes. In this case, natural selection acts against the extremes. The vast majority of natural selection is of this kind as deleterious mutations that cause a departure from the optimal phenotype are selected against.

Whereas stabilizing selection keeps a trait the same over time, **directional selection** leads to a change in a trait over time. A well-documented case of directional selection is found in Darwin's finches on the Galápagos Islands (**Fig. 21.9**). A severe drought in 1977 killed a significant amount of the vegetation that provided food for one island's population of seed-eating ground finches, *Geospiza fortis*, reducing that population's numbers from 750 to 90. Because plant species that produce big seeds fared better in drought conditions than other plant species did, the average size of seeds available to the remaining birds increased. Birds with bigger bills were better at handling the big seeds and therefore

FIG. 21.8 Stabilizing selection. Stabilizing selection on human birth weight results in selection against babies that are either too small or too large. *Source: Data from L. L. Cavalli-Sforza and W. F. Bodmer, 1971, The Genetics of Human Populations, San Francisco: W. H. Freeman, p. 613.*

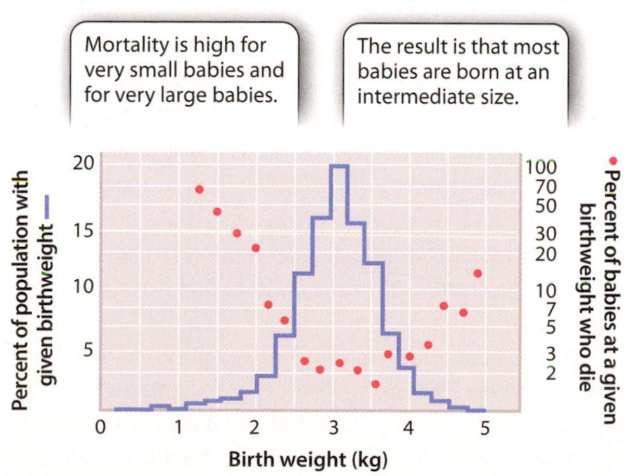

FIG. 21.9 Directional selection. Directional selection in a population of Galápagos finches caused a shift toward larger bills in a single drought year, 1977. *Data from Freeman & Herron Evolutionary Analysis 2004.*

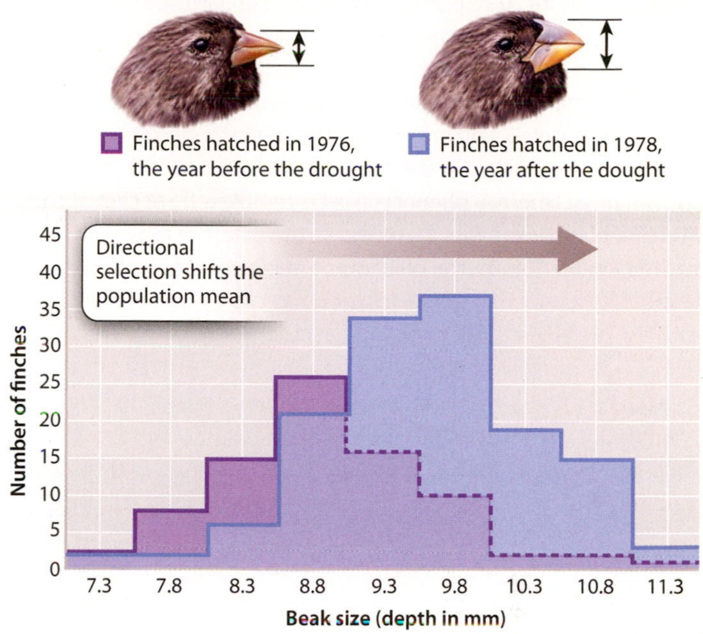

had higher survival rates than birds with smaller bills. Because bill size is genetically determined, the drought resulted in directional selection for increased bill size.

Artificial selection, which has been practiced by humans since at least the dawn of agriculture, is a form of directional selection. Artificial selection is analogous to natural selection, but the competitive element is removed. Successful genotypes are selected by the breeder, not through competition. Because it can be carefully controlled by the breeder, artificial selection is astonishingly efficient at generating genetic change. Practiced over many generations, artificial selection can create a population in which the selected phenotype is far removed from that of the starting population. **Fig. 21.10** shows the result of long-continued artificial selection for the oil content in kernels of corn.

A third mode of selection, known as **disruptive selection,** operates in favor of extremes and against intermediate forms. Apple maggot flies of North America, *Rhagotletis pomonella,* provide an example (**Fig. 21.11**). The larvae of these flies feed on the fruit of hawthorn trees. However, with the introduction of apples from Europe about 150 years ago, these flies have become pests of apples. Apple trees flower and produce fruit earlier every summer than hawthorns, so disruptive selection has resulted in the production of two genetically distinct groups of flies, one specializing on apple trees and the other on hawthorn trees. Disruptive selection acts against intermediates between the two groups, which miss the peaks of both the apple and hawthorn seasons. We explore this mechanism, which can lead to the evolution of new species, in more detail in the next chapter.

HOW DO WE KNOW?

FIG. 21.10

How far can artificial selection be taken?

BACKGROUND Begun in the 1890s and continuing to this day at the University of Illinois as an attempt to manipulate the properties of corn, this experiment has become one of the longest-running biological experiments in history.

HYPOTHESIS Researchers hypothesized that there is a limit to the extent to which a population can respond to continued directional selection.

EXPERIMENT Corn was artificially selected for either high oil content or low oil content: Every generation, researchers bred together just the plants that produced corn with the highest oil content, and did the same for the plants that produced corn with the lowest oil content. Every generation, kernels showed a range of oil levels, but only the 12 kernels with the highest or the lowest oil content were used for the next generation.

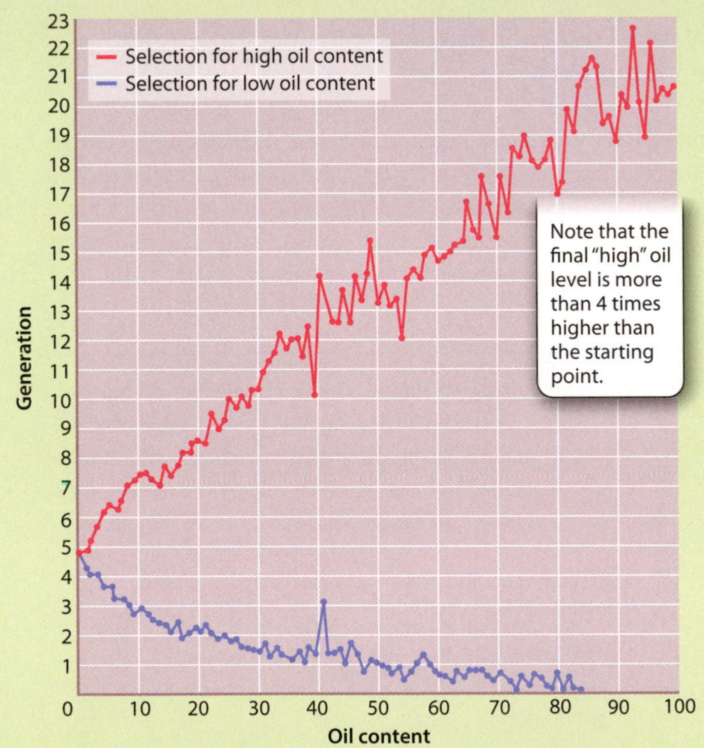

RESULTS In the line selected for high oil content, the percentage of oil more than quadrupled, from about 5% to more than 20%. In the line selected for low oil content, the oil content fell so close to zero that it could no longer be measured accurately, and the selection was terminated. Both selected lines are completely outside the range of any phenotype observed at the beginning of the experiment.

FOLLOW-UP WORK Genetic analysis of the selected lines indicates that the differences in oil content are due to the effects of at least 50 genes.

SOURCE Moose, S. P., J. W. Dudley, and T. R. Rocheford. 2004. "Maize Selection Passes the Century Mark: A Unique Resource for 21st Century Genomics." *Trends in Plant Science* 9:358–364.

FIG. 21.11 **Disruptive selection.** Disruptive selection has produced two genetically distinct populations of apple maggot fly, each one coordinated with fruiting times of two different species of tree. *Sources: (photo) Rob Oakleaf (National Science Foundation); Data from Filchak et al. 2000 Nature 407:739–42.*

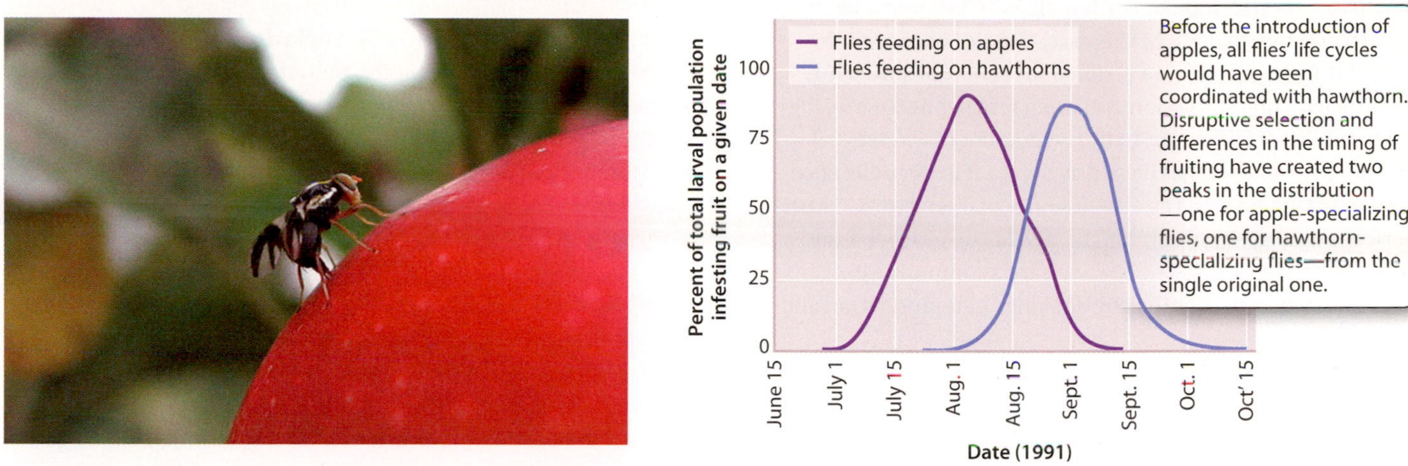

Sexual selection increases an individual's reproductive success.

Initially, Darwin was puzzled by features of organisms that seemed to reduce an individual's chances of survival. In a letter a few months after the publication of *The Origin,* he wrote, "The sight of a feather in a peacock's tail, whenever I gaze at it, makes me sick!" The tail is metabolically expensive to produce; it is an advertisement to potential predators; and it is an encumbrance in any attempt to escape a predator. How could such a feature evolve under natural selection?

In his 1871 book, *The Descent of Man, and Selection in Relation to Sex,* Darwin introduced a solution to this problem. Natural selection is indeed acting to reduce the showiness and size of the peacock's tail, but another form of selection, **sexual selection,** is acting in the opposite direction. Sexual selection promotes traits that increase an individual's access to reproductive opportunities.

Darwin recognized that this could occur in two different ways (**Fig. 21.12**). In one form of sexual selection, members of one sex (usually the males) compete with one another for access to the other sex (usually the females). This form is called **intrasexual selection** since it focuses on interactions between individuals of one sex. Because competition typically occurs among males, it is in males that we see physical traits such as large size and horns and other elaborate weaponry, as well as fighting ability. Larger, more

FIG. 21.12 **Sexual selection.** (a) Intrasexual selection often involves competition between males, as in this battle between two male elk. (b) Intersexual selection often involves bright colors and displays by males to attract females, as shown by this male and female Japanese Red-crowned Crane. *Source: (a) Kelly Funk/All Canada Photos/Corbis; (b) Steven Kaufman/Getty Images.*

powerful males tend to win more fights, hold larger territories, and have access to more females.

Darwin also recognized a second form of sexual selection. Here, males (typically) do not fight with one another, but instead compete for the attention of the female with bright colors or advertisement displays. In this case, females choose their mates. This form of selection is called **intersexual selection** since it focuses on interactions between females and males. The peacock's tail is thought to be the product of intersexual selection: Its evolution has been driven by a female preference for ever-showier tails. In the absence of sexual selection, natural selection would act to minimize the size of the peacock's tail. Presumably, the peacocks' tails we see are a compromise, a trade-off between the conflicting demands of reproduction and survival.

→ **Quick Check 4** Sexual selection tends to cause bigger size, more elaborate weaponry, or brighter colors in males. Is this an example of stabilizing, directional, or disruptive selection?

21.5 MIGRATION, MUTATION, GENETIC DRIFT, AND NON-RANDOM MATING

Selection is evolution's major driving force, enriching each new generation for the mutations that best fit organisms to their environments. However, as we have seen, it is not the only evolutionary mechanism. There are other mechanisms that can cause allele and genotype frequencies to change. These are migration, mutation, genetic drift, and non-random mating. Like natural selection, these mechanisms can cause allele frequencies to change. Unlike natural selection, they do not lead to adaptations.

Migration reduces genetic variation between populations.

Migration is the movement of individuals from one population to another, resulting in **gene flow,** the movement of alleles from one population to another. It is relatively simple to see how movements of individuals and alleles can lead to changes in allele frequencies. Consider two isolated island populations of rabbits, one white and the other black. Now imagine that the isolation breaks down—a bridge is built between the islands—and migration occurs. Over time, black alleles enter the white population and vice versa, and the allele frequencies of the two populations gradually become the same.

The consequence of migration is therefore the homogenizing of populations, making them more similar to each other and reducing genetic differences between them. Because populations are often adapted to their particular local conditions (think of dark-skinned humans in regions of high sunlight versus fair-skinned humans in regions of low sunlight), migration may be worse than merely non-adaptive—it may be maladaptive, in that it causes a decrease in a population's average fitness. Fair-skinned people arriving in an equatorial region, for example, are at risk of sunburn and skin cancer.

Mutation increases genetic variation.

As we saw earlier in this chapter, mutation is a rare event. This means that it is generally not important as an evolutionary mechanism that leads allele frequencies to change. However, as we have also seen, it is the source of new alleles and the raw material on which the other forces act. Without mutation, there would be no genetic variation and no evolution.

Genetic drift has a large effect in small populations.

Genetic drift is the random change in allele frequencies from generation to generation. By "random," we mean that frequencies can either go up or down simply by chance. An extreme case is a population **bottleneck,** which occurs when an originally large population falls to just a few individuals.

Consider a rare allele, *A*, with a frequency of 1/1000. Habitat destruction then reduces the population to just one pair of individuals, one of which is carrying *A*. The frequency of *A* in this new population is 1/4 because each individual has two alleles, giving a total of four alleles. In other words, the bottleneck resulted in a dramatic change in allele frequencies. It also caused a loss of genetic variation as much of the variation present in the original population was not present in the surviving pair. That is, the surviving pair carries only a few of the alleles that were present in their original population. A population of Galápagos tortoises that has very low levels of genetic diversity probably went through just such a bottleneck about 100,000 years ago when a volcanic eruption eliminated most of the tortoises' habitat.

Genetic drift also occurs when a few individuals start a new population, in what is called a **founder event.** Such events occur, for example, when a small number of individuals arrive on an island and colonize it. Once again, relative to the parent population, allele frequencies are changed and genetic variation is lost.

Earlier, we considered the fate of beneficial and harmful mutations under the influence of natural selection. What about neutral mutations? Natural selection, by definition, does not govern the fate of neutral mutations. Consider a neutral mutation, *m*, which occurs in a noncoding region of DNA and therefore has no effect on fitness. At first, it is in just a single heterozygous individual. What happens if that individual fails to reproduce (for reasons unrelated to *m*)? In this case, *m* is lost from the population, but not by natural selection (which does not select against *m*). Alternatively, the *m*-bearing individual might by chance leave many offspring (again for reasons unrelated to *m*), in which case the frequency of *m* increases. In principle, it is possible over a long period of time for *m* to become fixed in the population. At the end of the process, every member of the population is homozygous *mm*.

Like natural selection, genetic drift leads to allele frequency changes and therefore to evolution. Unlike natural selection, however, it does not lead to adaptations since the alleles whose frequencies are changing as a result of drift do not affect an individual's ability to survive or reproduce.

The impact of genetic drift depends on population size (**Fig. 21.13**). If *m* arises in a very small population, its frequency will change rapidly, as shown in Figs. 21.13a and 21.13b. Imagine *m* arising in a population of just six individuals (or three pairs). Its initial frequency is 1 in 12, or about 8% (there are a total of 12 alleles because each individual is diploid). If, by chance, one pair fails to breed and the other two (including the one who is an *Mm* heterozygote) each produce three offspring, and all three of the *Mm* individual's offspring happen to inherit the *m* allele, then the frequency of *m* will increase to 3 in 12 (25%) in a single generation. In effect, genetic drift is equivalent to a sampling error. In a small sample, extreme departures from the expected outcome are common.

On the other hand, if the population is large, as in Figs. 21.13c and 21.13d, then changes in allele frequency from generation to generation are much smaller, typically less than 1%. A large population is analogous to a large sample size, in which we tend not to see large departures from expectation. Toss a coin 1000 times, and you will end up with approximately 500 heads. Toss a coin 5 times, and you might well end up with zero heads. In other words, in a small sample of coin tosses, we are much more likely to see departures from our 50:50 expectation than in a large sample. The same is true of genetic drift. It is likely to be much more significant in small populations than in large ones.

→ **Quick Check 5** Why, of all the evolutionary mechanisms, is selection the only one that can result in adaptation?

FIG. 21.13 **Genetic drift.** The fate of neutral mutations is governed by genetic drift, the effect of which is more extreme in small populations than in large populations.

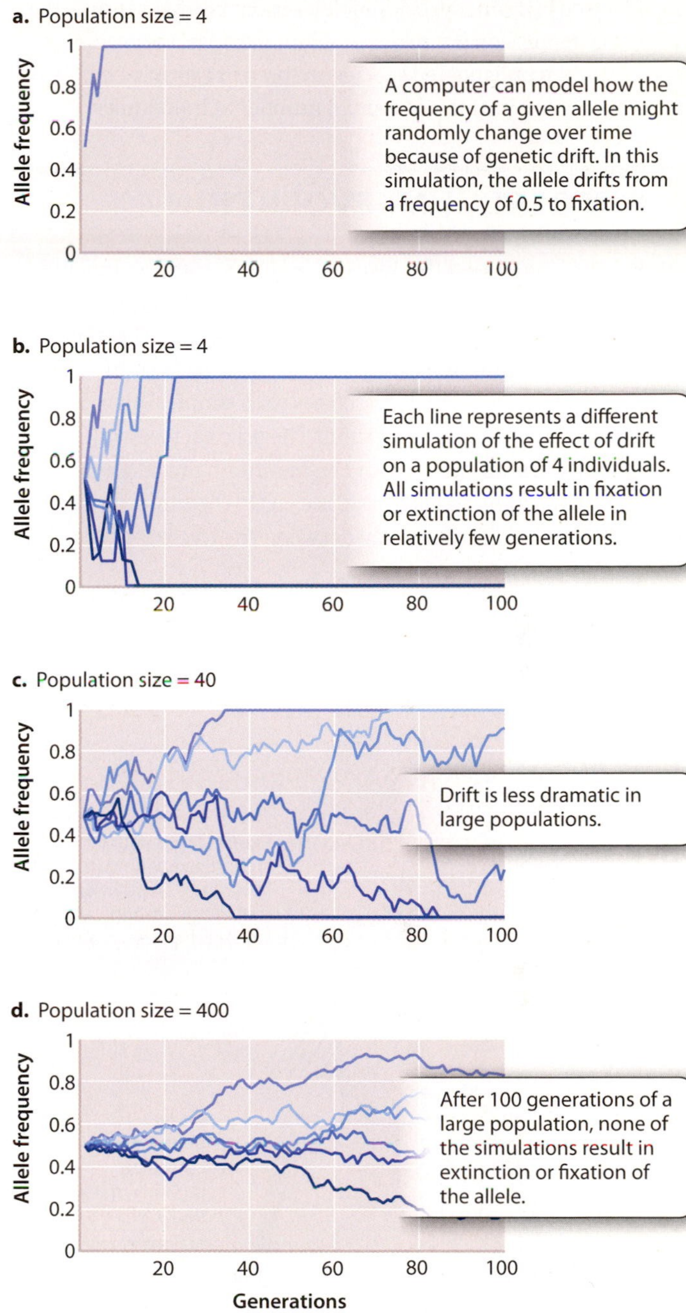

Non-random mating alters genotype frequencies without affecting allele frequencies.

As we saw earlier, another way that a population can evolve is through non-random mating. In random mating, individuals select mates without regard for genotype. In non-random mating, by contrast, individuals preferentially choose mates according to their genotypes. The result is that certain phenotypes increase and others decrease. Because non-random mating just rearranges alleles already in the gene pool and, unlike migration or mutation, does not add new alleles to the population, the genotype frequencies change, whereas the allele frequencies do not.

Probably the most evolutionarily significant form of non-random mating is inbreeding, in which mating occurs between close relatives. Inbreeding increases the frequency of homozygotes and decreases the number of heterozygotes in a population without affecting allele frequencies.

Consider a rare allele *b* that is at frequency 0.001 in a population. According to the Hardy–Weinberg equilibrium, the expected frequency of *bb* homozygotes is $0.001^2 = 0.000001$. Now let's see what happens to the frequency of *bb* homozygotes when there is inbreeding. A father with one *b* allele has a son and daughter who mate and have offspring. We know that the probability that the brother inherited the allele from his father is 0.5, and the same is true for his sister. The probability that the siblings' child is homozygous *bb* is the probability each sibling inherited a copy from their father (0.5×0.5) multiplied by the probability that their child inherited *b* from both of them (0.5 × 0.5), or $(0.5)^4 = 0.0625$. Needless to say, 0.0625 is a considerably higher probability than 0.000001.

If *b* is a deleterious recessive mutation, it may contribute to **inbreeding depression** in the child, a reduction in the child's fitness caused by homozygosity of deleterious recessive mutations. Inbreeding depression is a major problem in conservation biology, especially when endangered species are bred in captivity in programs starting with a just a small number of individuals.

21.6 MOLECULAR EVOLUTION

How do DNA sequence differences arise among species? Imagine starting with two pairs of identical twins, one pair male and the other female. Now we place one member of each pair together on either side of a mountain range (**Fig. 21.14**). Let's assume the mountain range completely isolates each couple. What, in genetic terms, will happen over time? The original pairs will found populations on each side of the mountain range. The genetic starting point, in each case, is exactly the same, but, over time, differences will accumulate between the two populations.

FIG. 21.14 Genetic divergence in isolated populations.

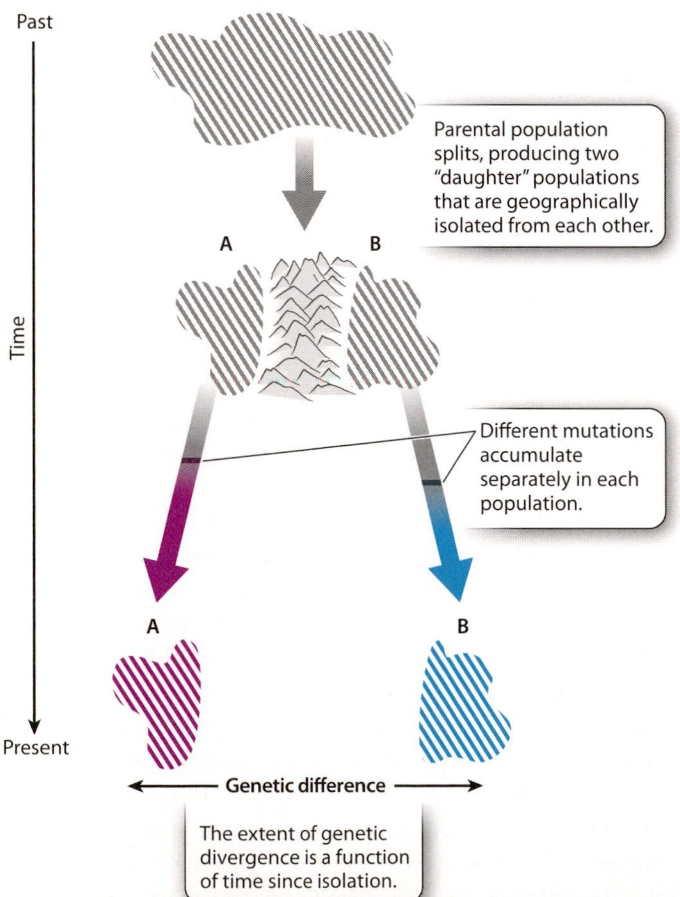

Mutations will occur in one population that will not have arisen in the other population, and vice versa.

A mutation in either population has one of three fates: It goes to fixation (either through genetic drift or through positive selection); it is maintained at intermediate frequencies (by balancing selection); or it is eliminated (either through natural selection or genetic drift). Different mutations will be fixed in each population. When we come back thousands of generations later and sequence the DNA of our original identical individuals' descendants, we will find that many differences have accumulated. The populations have diverged genetically. What we are seeing is evidence of **molecular evolution.**

Species are the biological equivalents of islands because they, too, are isolated. They are *genetically* isolated because, by definition, members of one species cannot exchange genetic material with members of another (Chapter 22). The amount of time that two species have been isolated from each other is the time since their most recent common ancestor. Thus, humans and chimpanzees, whose most recent common ancestor lived about 6–7 million years ago, have been isolated from each other for about 6–7 million years. Mutations arose and were fixed in the human lineage over that period; mutations, usually different ones, also arose and were fixed in the chimpanzee lineage over the same period. The result is the genetic difference between humans and chimpanzees.

The molecular clock relates the amount of sequence difference between species and the time since the species diverged.

The extent of genetic difference, or genetic divergence, between two species is a function of the time they have been genetically isolated from each other. The longer they have been apart, the greater the opportunity for mutation and fixation to occur in each population. This correlation between the time two species have been evolutionarily separated and the amount of genetic divergence between them is known as the **molecular clock.**

For a clock to function properly, it not only needs to keep time, but it also needs to be set. We set the clock using dates from the fossil record. For example, in a 1967 study, Vince Sarich and Allan Wilson determined from fossils that the lineages that gave rise to the Old and New World monkeys separated about 30 million years ago. Finding that the amount of genetic divergence between humans and chimpanzees was about one-fifth of that between Old and New World monkeys, they concluded that humans and chimpanzees had been separated one-fifth as long, or about 6 million years. Although this is the generally accepted number today, Sarich and Wilson's result was revolutionary at the time, when it was thought that the two species had been separated for as long as 25 million years.

The rate of the molecular clock varies.

Molecular clocks can be useful for dating evolutionary events like the separation of humans and chimpanzees. However, because

FIG. 21.15 The molecular clock. Different genes evolve at different rates because of differences in the intensity of negative selection.

After Fig. 20-3, p. 733, in A. J. F. Griffiths, S. R. Wessler, S. B. Carroll, and J. Doebley, 2012, Introduction to Genetic Analysis, 10th ed., New York: W. H. Freeman.

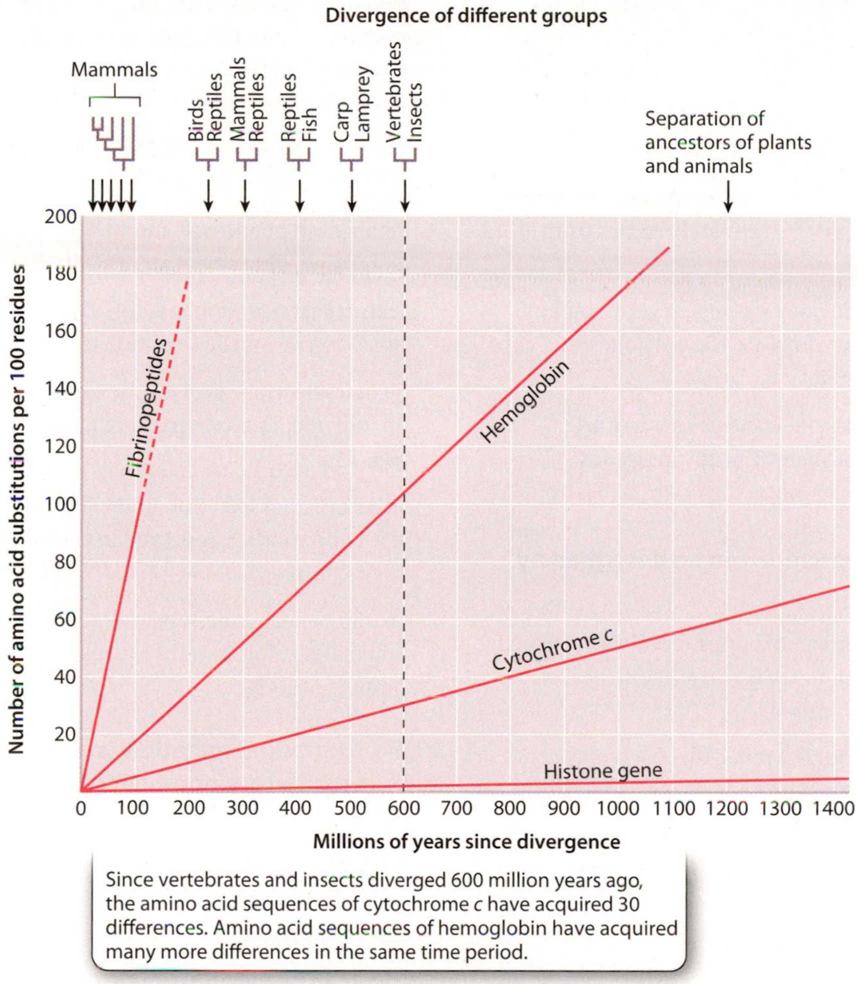

the rates of molecular clocks vary from gene to gene, clock data should be interpreted cautiously. These rate differences can be attributed largely to differences in intensity of negative selection (which results in the elimination of harmful mutations) among different genes. The slowest molecular clock on record belongs to the histone genes, which encode the proteins around which DNA is wrapped to form chromatin (Chapters 3 and 13). These proteins are exceptionally similar in all organisms; only 2 amino acids (in a chain of about 100) distinguish plant and animal histones. Plants and animals last shared a common ancestor more than 1 billion years ago, which means, because each evolutionary lineage is separate, that there have been at least 2 billion years of evolution since they were in genetic contact. And yet the histones have hardly changed at all. Almost any amino acid change fatally disrupts the histone protein, preventing it from carrying out its proper function. Negative selection has thus been extremely effective in eliminating just about every amino acid–changing histone mutation over 2 billion years of evolution. The histone molecular clock is breathtakingly slow.

Other proteins are less subject to such rigorous negative selection. Occasional mutations may therefore become fixed, either through drift (if they are neutral) or selection (if beneficial). The extreme case of a fast molecular clock is that derived from a **pseudogene,** a gene that is no longer functional. Because *all* mutations in a pseudogene are by definition neutral—there is no function for a mutation to disrupt, so a mutation is neither deleterious nor beneficial—we expect to see a pseudogene's molecular clock tick at a very fast rate. In the histone genes, virtually all mutations are selected against, constraining the rate of evolution; in pseudogenes, none is. **Fig. 21.15** shows the varying rates of the molecular clock for different genes.

In the next chapter, we examine how genetic divergence between populations can lead to the evolution of new species. ■

Core Concepts Summary

21.1 Genetic variation refers to differences in DNA sequences.

Visible differences among members of a species (phenotypic variation) are the result of differences at the DNA level (genetic variation) as well as the influence of the environment. page 426

Mutation and recombination are the two sources of genetic variation, but all genetic variation ultimately comes from mutation. page 427

Mutations can be somatic (in body tissues) or germ line (in gametes), but germ-line mutations are the only ones that can be passed on to the next generation. page 427

When a mutation occurs in a gene, it creates a new allele. Mutations can be deleterious, neutral, or advantageous. page 427

21.2 Patterns of genetic variation can be described by allele frequencies.

An allele frequency is the number of occurrences of a particular allele divided by the total number of occurrences of all alleles of that gene in a population. page 427

In the past, population geneticists relied on observable traits determined by a single gene and protein gel electrophoresis to measure genetic variation. page 428

DNA sequencing is now the standard technique for measuring genetic variation. page 428

21.3 Evolution is a change in the frequency of alleles or genotypes over time.

The Hardy–Weinberg equilibrium describes situations in which allele frequencies do not change. By seeing if a population is in Hardy–Weinberg equilibrium, we can determine whether or not evolution is occurring in a population. page 430

The Hardy–Weinberg equilibrium makes five assumptions. These assumptions are that the population experiences no selection, no migration, no mutation, no sampling error due to small population size, and random mating. page 430

The Hardy–Weinberg equilibrium allows allele frequencies and genotype frequencies to be calculated from each other. page 431

21.4 Natural selection leads to adaptation, which enhances the fit between an organism and its environment.

Independently conceived by Charles Darwin and Alfred Russel Wallace, natural selection is the differential reproductive success of genetic variants. page 432

Under natural selection, a harmful allele decreases in frequency, and a beneficial one increases in frequency. Natural selection does not affect the frequency of neutral mutations. page 434

Natural selection can maintain alleles at intermediate frequencies by balancing selection. page 434

Changes in phenotype show that natural selection can be stabilizing, directional, or disruptive. page 435

In artificial selection, a form of directional selection, a breeder governs the selection process. page 436

Sexual selection involves the evolution of traits that increase an individual's access to members of the opposite sex. page 436

In intrasexual selection, individuals of the same sex compete with one another, resulting in traits like large size and horns. page 437

In intersexual selection, interactions between females and males result in traits like elaborate plumage in male birds. page 438

21.5 Migration, mutation, genetic drift, and non-random mating are non-adaptive mechanisms of evolution.

Migration involves the movement of alleles between populations (gene flow) and tends to have a homogenizing effect. page 438

Mutation is the ultimate source of variation, but it also can change allele frequencies on its own. page 438

Genetic drift is a kind of sampling error, which acts more strongly in small populations than in large ones. page 438

Non-random mating, such as inbreeding, results in an increase in homozygotes and a decrease in heterozygotes, but does not change allele frequencies. page 439

21.6 Molecular evolution is a change in DNA or amino acid sequences over time.

The extent of sequence difference between two species is a function of the time they have been genetically isolated from each other. page 440

Correlation between sequence differences among species and time since common ancestry of those species is known as the molecular clock. page 440

The rate of the molecular clock varies among genes because some genes are more selectively constrained than others. page 441

Self-Assessment

1. Why does the relation between phenotype and genotype matter in evolution?
2. What is a neutral mutation and what evolutionary mechanism causes it to change in frequency over time?
3. Define genetic variation and explain how it can be measured.
4. How would you calculate the allele frequencies for a two-allele trait in a population if given the genotype frequencies?
5. Can evolution occur without allele frequency changes? If not, why not? If so, how?
6. Describe what happens to allele and genotype frequencies under the Hardy–Weinberg equilibrium.
7. Name the five assumptions of the Hardy–Weinberg equilibrium and, for each one, explain what happens in a population in which that condition is not met.
8. How would you calculate genotype frequencies of a population in Hardy–Weinberg equilibrium, given the allele frequencies of that trait?
9. Define natural selection and explain how it is different from other mechanisms of evolution.
10. Explain how a molecular clock can be used to determine the time of divergence of two species.

Log in to LaunchPad to check your answers to the Self-Assessment questions, and to access additional learning tools.

CHAPTER 22

Species and Speciation

Core Concepts

22.1 Reproductive isolation is the key to the biological species concept.

22.2 Reproductive isolation is caused by barriers to reproduction before or after egg fertilization.

22.3 Speciation underlies the diversity of life on Earth.

22.4 Speciation can occur with or without natural selection.

Rangzen/Shutterstock.

Imagine for a moment a world without **speciation,** the process that produces new and distinct forms of life. Life would have originated and natural selection would have done its job of winnowing advantageous mutations from disadvantageous ones, but the planet would be inhabited by a single kind of generally adapted organism. Instead of the staggering biological diversity we see around us—by current estimates, between 10 and 100 million species call Earth home—there would be just a single life-form. From a biological perspective, the planet would be a decidedly dull place. Evolution is as much about speciation, the engine that generates this breathtaking biodiversity, as it is about adaptation, the result of natural selection.

22.1 THE BIOLOGICAL SPECIES CONCEPT

The definition of **species** has been a long-standing problem in biology. Many biologists respond to the problem in the same way that Darwin himself did. In *On the Origin of Species,* Darwin wrote, "No one definition has as yet satisfied all naturalists; yet every naturalist knows vaguely what he means when he speaks of a species." The difficulty of defining species has come to be called the species problem.

Here is the problem in a nutshell: The species, as an evolutionary unit, must by definition be fluid and capable of changing, giving rise through evolution to new species. The whole point of the Darwinian revolution is that species are not fixed. How, then, can we define something that changes over time, and, by the process of speciation, even gives rise to two species from one?

Species are reproductively isolated from other species.

We can plainly see biodiversity, but are what we call "species" real biological entities? Or was the term coined by biologists to simplify their description of the natural world? To test whether or not species are real, we can examine the natural world, measure some characteristic of the different living organisms we see, and then plot these measurements on a graph. **Fig. 22.1** shows such a plot, graphing antenna length and wing length of three different types of butterfly. Note that the dots, representing individual organisms, fall into non-overlapping clusters. Today, we can add a molecular dimension to this kind of analysis. When we compare genomes of multiple organisms, they, too, cluster on the basis of similarity. Each cluster is a species, and the fact that the clusters are distinct implies that species are biologically real.

The distances we see between the dots within a cluster reflect variation from one individual to the next within a species (Chapter 21). Humans are highly variable, but overall we are more similar to one another than to our most humanlike relative, the chimpanzee. We form a messy cluster, but that cluster does not overlap with the chimpanzee cluster.

Species, then, are real biological entities, not just a convenient way to group organisms. Whether or not two individuals are

FIG. 22.1 Species clusters on the basis of two characteristics.
Source: Museum of Comparative Zoology, Harvard University.

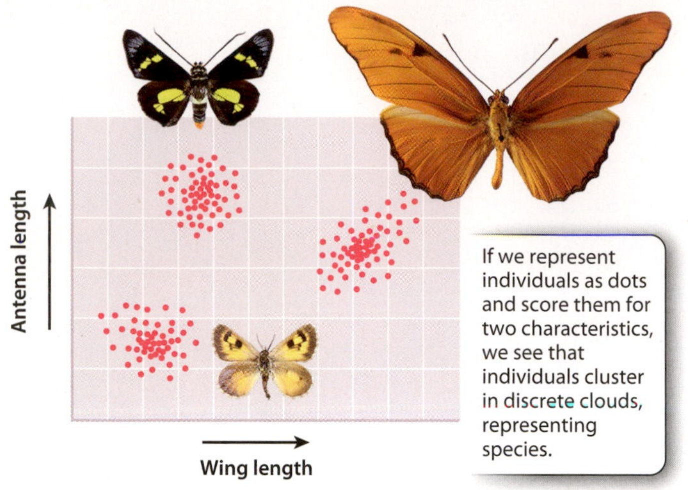

members of the same species is not a matter of judgment, but rather a reflection of their ability (or inability) to exchange genetic material by producing fertile offspring. Consider a new advantageous mutation that appears initially in a single individual. That individual and its offspring inheriting the mutation will have a competitive advantage over other members of the population, and the mutation will increase in frequency until it reaches 100%. Migration among populations causes the mutation to spread further until all individuals within the species have it. The mutation spreads *within* the species but, with some exceptions, cannot spread beyond it.

Therefore, a species represents a closed gene pool, with alleles being shared among members of that species but usually not with members of others. As a result, it is species that become extinct and it is species that, through genetic divergence, give rise to new species.

The definition of "species" continues to be debated to this day. The most widely used and generally accepted definition of a species is known as the **biological species concept (BSC).** The BSC was described by the great evolutionary biologist Ernst Mayr (1904–2005) as follows:

> Species are groups of actually or potentially interbreeding populations that are reproductively isolated from other such groups.

Let us look at this definition closely. At its heart is the idea of reproductive compatibility. Members of the same species are capable of producing offspring together, whereas members of different species are incapable of producing offspring together. In other words, members of different species are **reproductively isolated** from one another.

FIG. 22.2 ***Elephas maximus* in Sri Lanka and in India.** *Sources: (top) Louise Morgan/Getty Images; (bottom) Mike Smith/age footstock.*

As Mayr and many others realized, however, reproductive compatibility entails more than just the ability to produce offspring. The offspring must be fertile and therefore capable of passing their genes on to their own offspring. For example, although a horse and a donkey, two different species, can mate to produce a mule, the mule is infertile. If a hybrid offspring is infertile, it is a genetic dead end.

Note, too, the "actually or potentially" part of Mayr's definition. An Asian elephant living on the island of Sri Lanka and one living in nearby India are considered members of the same species, *Elephas maximus,* even though Sri Lankan and Indian Asian elephants never have a chance to mate with each other in nature because they are geographically separated (**Fig. 22.2**).

The BSC is more useful in theory than in practice.

Species, then, are real biological entities and the BSC is the most useful way to define them. However, the BSC has shortcomings, the most important of which is that it can be difficult to apply. Imagine you are on a field expedition in a rain forest and you find two insects that look reasonably alike. Are they members of the same species or not? To use the BSC, you would need to test whether or not they are capable of producing fertile offspring. However, in practice, you probably will not have the time and resources to perform such a test. And even if you do, it is possible that members of the same species will not reproduce when you place them together in a laboratory setting or in your field camp. The conditions may be too unnatural for the insects to behave in a normal way. And even if they do mate, you may not have time to determine whether or not the offspring are fertile.

Thus, we should consider the BSC as a valid framework for thinking about species, but one that is difficult to test. On a day-to-day basis, therefore, biologists often use a rule of thumb called the **morphospecies concept.** Stated simply, the morphospecies concept holds that members of the same species usually look alike. For example, the shape, size, and coloration of the bald eagle make it easy to determine whether or not a bird observed in the wild is a member of the bald eagle species.

Today, the morphospecies concept has been extended to the molecular level: Members of the same species usually have similar DNA sequences that are distinct from those of other species. A remarkable project called the Barcode of Life has established a database linking DNA sequences to species. Scientists can therefore identify species from DNA alone by sequencing the relevant segment of DNA and finding the matching sequence (and therefore the species) in the Barcode of Life database. Here, at both morphological and molecular levels, we are returning to those biological clusters of Fig. 22.1 in which members of a species fall close to one another within a single, discrete group.

Although the morphospecies concept is useful and generally applicable, it is not infallible. Members of a species may not always look alike, but instead show different phenotypes called polymorphisms (Chapter 15)—for example, color difference in some species of birds. Sometimes, males of a species may look different from females—think of the showy peacock, which differs dramatically from the relatively drab peahen. Young can be very different from old: Caterpillars that mature into butterflies are a striking example.

So members of the *same* species can look quite different from one another. It is also true that members of *different* species can look quite similar. Some species of butterfly, for example, appear similar but have chromosomal differences that can be observed only with the aid of a microscope (**Fig. 22.3**). These chromosomal differences prevent the formation of interspecies hybrids, so those butterflies meet the BSC criterion—they are different species—but they do not meet the morphospecies criterion because they look so much alike.

With the application of ever more powerful DNA-based approaches to studies of natural populations, scientists frequently uncover what are called cryptic species. These are composed of organisms that had been traditionally considered as belonging to one species because they look similar, but turn out to belong to two species because of a distinction at the DNA sequence level.

The BSC does not apply to asexual or extinct organisms.

In addition to the difficulties of putting the BSC into practice, a second problem is that it overlooks some organisms. For example, because it is based on the sexual exchange of genetic information, the BSC cannot apply to species that reproduce asexually, such as bacteria. Although some asexual species do occasionally exchange genetic information in a process known as conjugation (Chapter 26), true asexual organisms do not fit the BSC.

Furthermore, because it depends on reproduction, the BSC obviously cannot be applied to species that have become extinct

FIG. 22.3 Three species of *Agrodiaetus* butterflies. These similar-appearing butterflies are identifiable only by differences in their chromosome numbers. *Source: Vladimir Lukhtanov, Zoological Institute of Russian Academy of Sciences, St. Petersburg, Russia.*

and are known only through the fossil record, like different "species" of trilobites and dinosaurs.

Ring species and hybridization complicate the BSC.

An unusual but interesting geographic pattern shown by **ring species,** species with populations that are reproductively but not genetically isolated, highlights another shortcoming of the BSC. Here, we find that some populations within a species are reproductively isolated from each other, but others are not.

For example, populations of the greenish warbler, *Phylloscopus trochiloides,* are distributed in a large geographic loop around the Himalayas (**Fig. 22.4**). In Russia to the north, members of two neighboring populations do not interbreed. Therefore, according to the BSC, they are different species. However, the more western of the two Russian populations is capable of reproducing with the population to the south of it, and that population is capable of reproducing with the population to the east of it, and so on. Eventually, the loop of genetically exchanging populations comes all the way round to the more eastern of the two Russian populations. Thus, though members of the two Russian populations cannot exchange genes directly, they can do so indirectly, with the genetic material passing through many intermediate populations. This situation is more complicated than anything predicted by the BSC. The two Russian populations are reproductively isolated from each other but they are not genetically isolated from each other because of gene flow around the ring.

We also see a complicated situation in some groups of closely related species of plants. Despite apparently being good morphospecies that can be distinguished by appearance alone, many different species of willow (*Salix*), oak (*Quercus*), and dandelion (*Taraxacum*) are still capable of exchanging genes with other species in their genera through **hybridization,** or interbreeding, between species. By the BSC, these different forms should be considered one large species because they are able to reproduce and produce fertile offspring. However, because they maintain their distinct appearances, natural selection must work against the hybrid offspring.

This unusual phenomenon seems to occur mainly in plants, but, with the application of powerful comparative genomic approaches, we are discovering that the boundaries between closely related animal species are also not as strictly drawn as traditionally supposed. As we will see in Chapter 24, for example, we now know that our own species, *Homo sapiens,* interbred in the past with another species, *Homo neanderthalensis.*

Ecology and evolution can extend the BSC.

Because of these problems, much work has been put into modifying and improving the BSC and many alternative definitions of species have been suggested. In general, these efforts highlight how difficult it is to make all species fit easily into one definition. The natural world truly defies neat categorization!

Several useful ideas, however, have come out of this literature. One of these is the notion that a species can sometimes be characterized by its **ecological niche,** which, as we will discuss further in Chapter 47, is a complete description of the role the species plays in its environment—its habitat requirements, its nutritional and water needs, and the like. It turns out that it is impossible for two species to coexist in the same location if their niches are too similar because competition between them for resources will inevitably lead to the extinction of one of them. This observation has given rise to the **ecological species concept (ESC),** the idea that there is a one-to-one correspondence between a species and its niche. Thus, we can determine whether or not asexual bacterial lineages are distinct species on the basis of differences or similarities in their ecological requirements. If two lineages have very different nutritional needs, for example, we can infer on ecological grounds that they are separate species.

Another species concept is the **phylogenetic species concept (PSC),** which is the idea that members of a species all

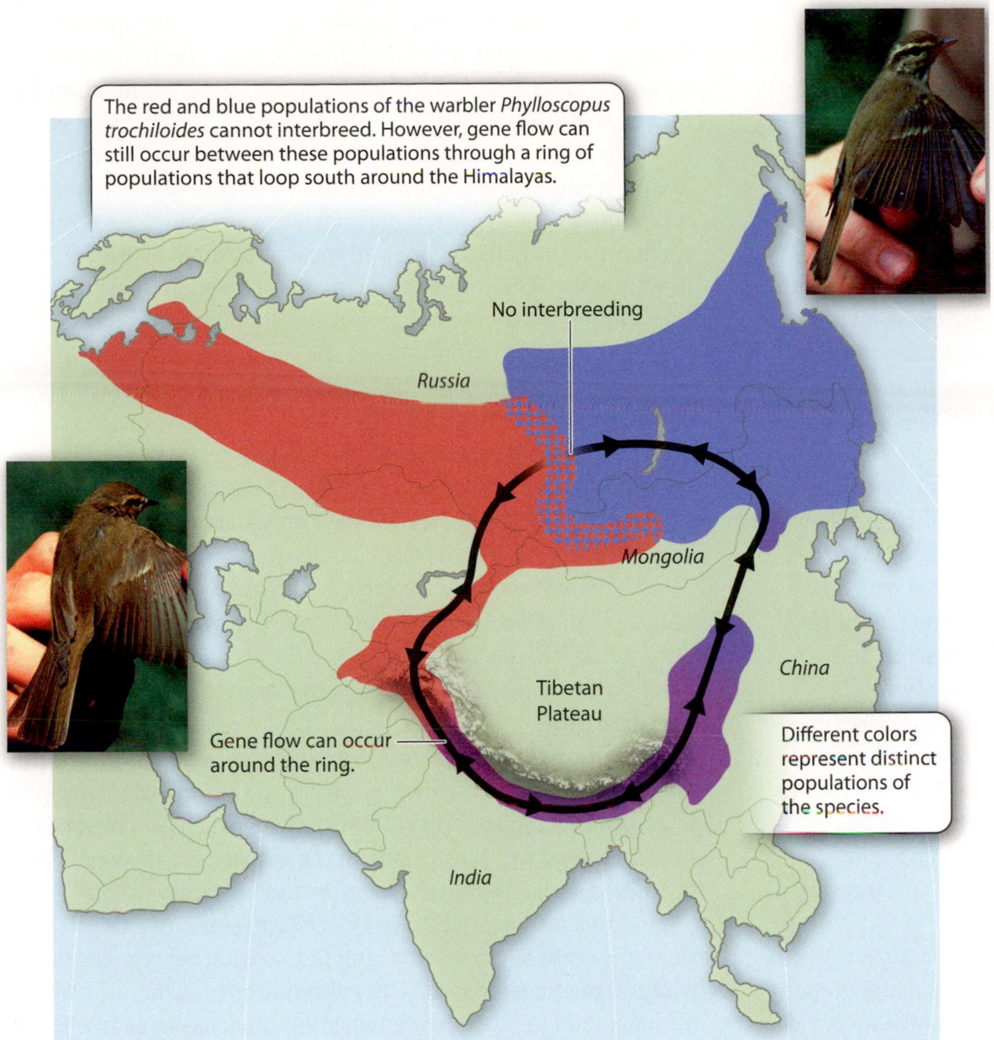

FIG. 22.4 **Ring species.** Ring species contain populations that are reproductively isolated from each other but can still exchange genetic material through other linking populations. *Source: Courtesy of Darren E. Irwin.*

share a common ancestry and a common fate. It is, after all, species rather than individuals that become extinct. The PSC requires that all members of a species are descended from a single common ancestor. It does not specify, however, on what scale this idea should be applied. All mammals derive from a single common ancestor that lived about 200 million years ago, but there are thousands of what we recognize as species of mammals that have evolved since that long-ago common ancestor. But, under a strict application of the PSC, would we consider all mammals to be a single species? Similarly, siblings and cousins are all descended from a common ancestor, a grandmother, but that is surely not sufficient grounds to classify more-distant relatives as a distinct species. The PSC can be useful when thinking about asexual species, but, given the arbitrariness of the decisions involved in assessing whether or not the descendants of a single ancestor warrant the term "species," its utility is limited.

It is worth bearing these ecological and phylogenetic considerations in mind when thinking about species. Although the BSC remains our most useful definition of species, the ESC and PSC broaden and generalize the concept. For example, in the case of a normally asexual species that is difficult to study because conjugation is so infrequent, we can jointly apply the ESC and PSC. We can use the ESC to loosely define the species in terms of its ecological characteristics (for example, its nutritional requirements), and we can refine that definition by using genetic analyses to determine whether the group is indeed a species by the PSC's standard (that is, that all its members derive from a single common ancestor).

Despite the shortcomings of the BSC and the usefulness of alternative ideas, we stress that the BSC is the most constructive way to think about species. In particular, by focusing on reproductive isolation—the inability of two different species to produce viable, fertile offspring—the BSC gives us a means of studying and understanding speciation, the process by which two populations, originally members of the same species, become distinct.

→ **Quick Check 1** Why haven't we been able to come up with a single, comprehensive, and agreed-upon species concept?

22.2 REPRODUCTIVE ISOLATION

Factors that cause reproductive isolation are generally divided into two categories, depending on when they act. **Pre-zygotic** isolating factors act before the fertilization of an egg, and **post-zygotic** factors come into play after fertilization. In other words, pre-zygotic factors prevent fertilization from taking place, whereas post-zygotic factors result in the failure of the fertilized egg to develop into a fertile individual.

Pre-zygotic isolating factors occur before egg fertilization.

Most species are reproductively isolated by pre-zygotic isolating factors, which can take many forms. Among animals, species are often **behaviorally isolated,** meaning that individuals mate only with other individuals based on specific courtship rituals, songs, or other behaviors. Chimpanzees may be our closest relative—and therefore the species we are most likely to confuse with our own—but a chimpanzee of the appropriate sex, however attractive to a chimpanzee of the opposite sex, fails to provoke even the faintest reproductive impulse in a human. In this case, the pre-zygotic reproductive isolation of humans and chimpanzees is behavioral.

Behavior does not play a role in plants, but pre-zygotic factors can still be important in their reproductive isolation. Pre-zygotic isolation in plants can take the form of incompatibility between the incoming pollen and the receiving flower, so fertilization fails to take place. We see similar forms of isolation between members of marine species, such as abalone, which simply discharge their gametes into the water. In these cases, membrane-associated proteins on the surface of sperm interact specifically with membrane-associated proteins on the surface of eggs of the same species but not with those of different species. These specific interactions ensure that a sperm from one abalone species, *Haliotis rufescens,* fertilizes only an egg of its own species and not an egg from *H. corrugata,* a closely related species. Incompatibilities between the gametes of two different species is called **gametic isolation.**

In some animals, especially insects, incompatibility arises earlier in the reproductive process. The genitalia of males of the fruit fly *Drosophila melanogaster* are configured in such a way that they fit only with the genitalia of females of the same species. Attempts by males of *D. melanogaster* to copulate with females of another species of fruit fly, *D. virilis,* are prevented by **mechanical incompatibility.**

Both plants and animals may also be pre-zygotically isolated in time (**temporal isolation**). For example, closely related plant species may flower at different times of the year, so there is no chance that the pollen of one will come into contact with the flowers of the other. Similarly, members of a nocturnal animal species simply will not encounter members of a closely related species that are active only during the day.

Plants and animals can also be isolated in space (**geographic** or **ecological isolation**). This type of isolation can be subtle. For example, the two Japanese species of ladybug beetle shown in **Fig. 22.5** can be found living side by side in the same field, but they feed on different plants. Because their life cycles are so intimately associated with their host plants (adults even mate on their host plants), these two species never breed with each other. This ecological separation is what leads to their pre-zygotic isolation.

Post-zygotic isolating factors occur after egg fertilization.

Post-zygotic isolating factors involve mechanisms that come into play after fertilization of the egg. Typically, they involve some kind of **genetic incompatibility.** One example, which we saw earlier in Fig. 22.3 and will explore later in the chapter, is the case of two organisms with different numbers of chromosomes.

In some instances, the effect can be extreme. For example, the zygote may fail to develop after fertilization because the two parental genomes are sufficiently different to prevent normal development. In others, the effect is less obvious. Some matings between different species produce perfectly viable adults, as in the case of the horse–donkey hybrid, the mule. As we have seen, though, all is not well with the mule from an evolutionary perspective. The horse and donkey genomes are different enough to cause the mule to be infertile. As a general rule, the more closely related—and therefore genetically similar (Chapter 21)—a pair of species, the less extreme the genetic incompatibility between their genomes.

FIG. 22.5 Ecological isolation. The ladybugs (a) *Henosepilachna yasutomii* and (b) *H. niponica* are reproductively isolated from each other because they feed and mate on different host plants. *Source: Courtesy Dr. Haruo Katakura.*

a. b.

22.3 SPECIATION

Recognizing that species are groups of individuals that are reproductively isolated from other such groups, we are now in a position to recast the key question: How does speciation occur?

FIG. 22.6 Speciation. Speciation occurs when two populations that are genetically diverging become reproductively isolated from each other.

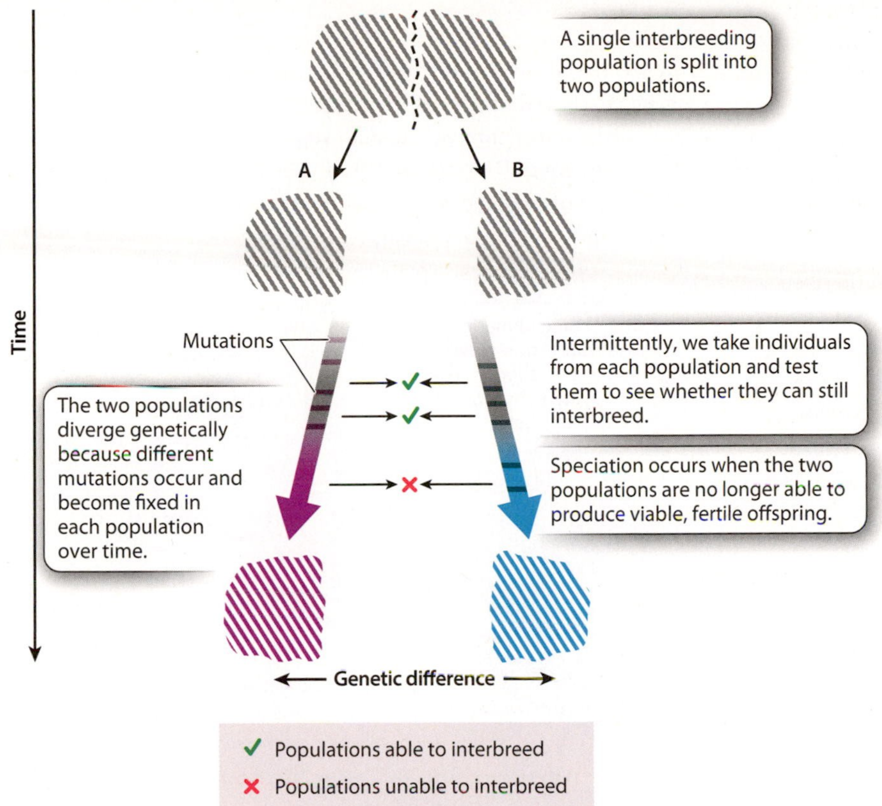

Instead of asking, "How do new species arise?" we can ask, "How does reproductive isolation arise between populations?"

Speciation is a by-product of the genetic divergence of separated populations.

The key to speciation is the fundamental evolutionary process of genetic divergence between genetically separated populations. As we saw in Chapter 21, if a single population is split into two populations that are unable to interbreed, different mutations will appear by chance in the two populations. Like all mutations, these will be subject to genetic drift or natural selection (or both), resulting over time in the genetic divergence of the two populations. Two separate populations that are initially identical will, over long periods of time, gradually become distinct as different mutations are introduced and propagated in each population. At some stage in the course of divergence, changes occur in one population that lead to its members' being reproductively isolated from members of the other population (**Fig. 22.6**). It is this process that results in speciation.

Speciation—the development of reproductive isolation between populations—is, therefore, typically just a by-product of the genetic divergence of separated populations.

As Fig. 22.6 shows, speciation is typically a gradual process. If we try to cross members of two populations that have genetically diverged but not diverged far enough for full reproductive isolation (that is, speciation) to have arisen, we may find that the populations are **partially reproductively isolated.** They are not yet separate species, but the genetic differences between them are extensive enough that the hybrid offspring they produce have reduced fertility or viability compared to offspring produced by crosses between individuals within each population.

Allopatric speciation is speciation that results from the geographical separation of populations.

Speciation is the process by which two groups of organisms become reproductively isolated from each other. Because this process requires genetic isolation between the diverging populations and because geography is the easiest way to ensure physical and therefore genetic isolation, many models of speciation focus on geography.

The process usually begins with the creation of **allopatric** (literally, "different place") populations, populations that are geographically separated from each other. Clearly, physical separation will not immediately cause reproductive isolation. If a single population is split in two by a geographic barrier, and, after a couple of generations, individuals from each population are allowed to interbreed again, they will still be capable of producing fertile offspring and therefore still be members of the same species. What's also required for speciation to occur is time. Time allows for different mutations to become fixed in the two separated populations so that eventually they become reproductively isolated from each other.

Because genetic divergence is typically gradual, we often find allopatric populations that have yet to evolve even partial reproductive isolation but which have accumulated a few population-specific traits. This genetic distinctness is sometimes recognized by taxonomists, who call each geographic form a **subspecies** by adding a further designation after its species name. For example, Sri Lankan Asian elephants, subspecies *Elephas maximus maximus,* are generally larger and darker than Indian ones, subspecies *Elephas maximus indicus* (see Fig. 22.2).

Dispersal and vicariance can isolate populations from each other.

How do populations become allopatric? There are two ways. The first is by **dispersal,** in which some individuals colonize a distant

FIG. 22.7

Can vicariance cause speciation?

BACKGROUND Three and a half million years ago, the Isthmus of Panama was not completely formed. Several marine corridors remained open, allowing interbreeding between marine populations in the Caribbean and the eastern Pacific. Subsequently, the gaps in the isthmus were plugged, separating the Caribbean and eastern Pacific populations and preventing gene flow between them.

HYPOTHESIS Nancy Knowlton and her colleagues hypothesized that patterns of speciation would reflect the impact of the vicariance resulting from the closing of the direct marine connections between the Pacific and the Caribbean. Specifically, they predicted that each ancestor species (from the time before the formation of the isthmus) split into two "daughter" species, one in the Caribbean and the other in the Pacific. The closest relative of each current Pacific species, then, is predicted to be a Caribbean species (and vice versa).

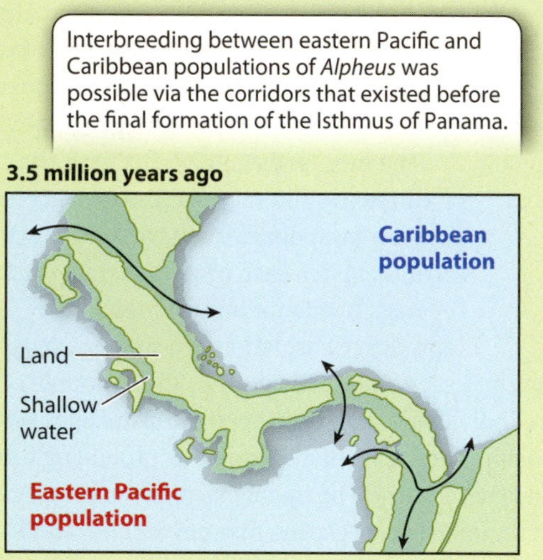

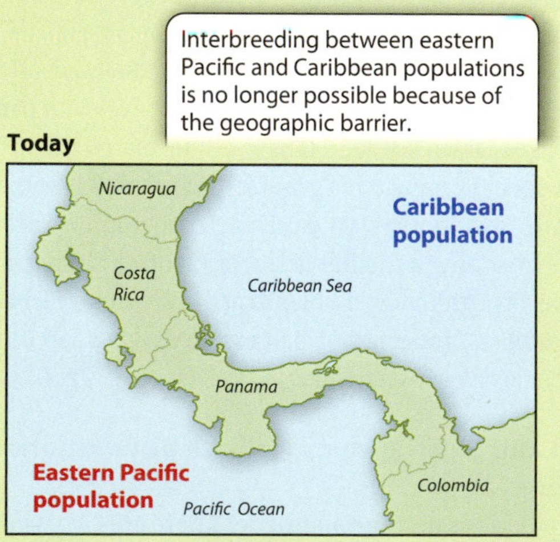

EXPERIMENT This study focused on 17 species of snapping shrimp in the genus *Alpheus,* a group that is distributed on both sides of the isthmus. The first step was to sequence the same segment of DNA from each species. The next step was to compare those sequences in order to reconstruct the phylogenetic relationships among the species.

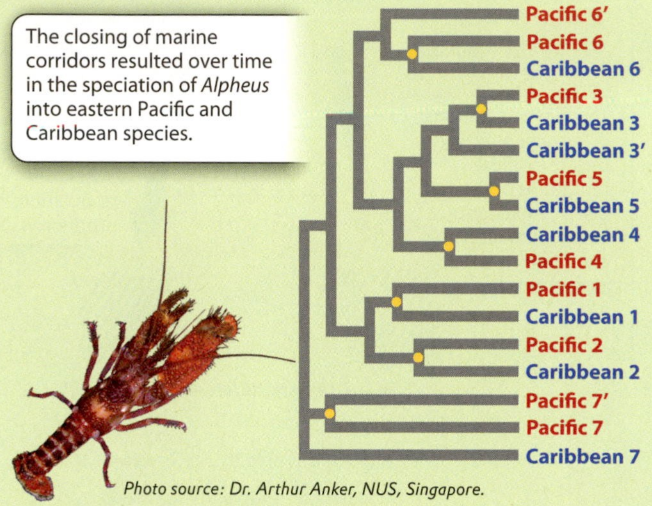

Photo source: Dr. Arthur Anker, NUS, Singapore.

RESULTS The phylogeny reveals that species show a distinctly paired pattern of relatedness: The closest relative of each species is one from the other side of the isthmus.

CONCLUSION That we see these consistent Pacific/Caribbean sister species pairings strongly supports the hypothesis that the vicariance caused by the formation of the isthmus has driven speciation in *Alpheus.* Each Pacific/Caribbean pairing is derived from a single ancestral species (indicated by a yellow dot) whose continuous distribution between the Caribbean and eastern Pacific was disrupted by the formation of the isthmus. Here, we see striking evidence of the role of vicariance in multiple speciation events.

FOLLOW-UP WORK Speciation is about more than just genetic differences between isolated populations, as shown here. Knowlton and colleagues also tested the different species for reproductive isolation and found that there were high levels of isolation between Caribbean/Pacific pairs. Given that we know that each species pair has been separated for at least 3 million years, we can then calibrate the rate at which new species are produced.

SOURCE Knowlton, N., et al. 1993. "Divergence in Proteins, Mitochondrial DNA, and Reproductive Compatibility Across the Isthmus of Panama." *Science* 260 (5114): 1629–1632.

place, such as an island, far from the main source population. The second is by **vicariance,** in which a geographic barrier arises within a single population, separating it into two or more isolated populations. For example, when sea levels rose at the end of the most recent ice age, new islands formed along the coastline as the low-lying land around them was flooded. The populations on those new islands suddenly found themselves isolated from other populations of their species. This kind of island formation is a vicariance event.

Regardless of how the allopatric populations came about—whether through dispersal or vicariance—the outcome is the same. The two separated populations will diverge genetically until speciation occurs.

Often, vicariance-derived speciation is the easier to study because we can date the time at which the populations were separated if we know when the vicariance occurred. One such event whose history is well known is the formation of the Isthmus of Panama between Central and South America, shown in **Fig. 22.7.** This event took place about 3.5 million years ago. As a result, populations of marine organisms in the western Caribbean and eastern Pacific that had formerly been able to interbreed freely were separated from each other. After a period of time, the result was the formation of many distinct species, with each one's closest relative being on the other side of the isthmus.

Dispersal is important in a specific kind of allopatric speciation known as **peripatric speciation** (that is, in a peripheral place). In this model, a few individuals from a **mainland population** (the central population of a species) disperse to a new location remote from the original population and evolve separately. This may be an intentional act of dispersal, such as young mammals migrating away from where they were raised, or it could be an accident brought about by, for example, an unusual storm that blows migrating birds off their normal route. The result is a distant, isolated **island population.** "Island" in this case may refer to a true island—like Hawaii—or may simply refer to a patch of habitat on the mainland that is appropriate for the species but is geographically remote from the mainland population's habitat area. For a species adapted to life on mountaintops, a new island might be another previously uninhabited mountaintop. For a rain forest tree species, that new island might be a patch of lowland forest on the far side of a range of mountains that separates it from its mainland forest population.

The island population is classically small and often in an environment that is slightly different from that of the mainland population. The peripatric speciation model suggests that change accumulates faster in these peripheral isolated populations than in the large mainland populations, both because genetic drift is more pronounced in smaller populations than in larger ones and because the environment may differ between the mainland and island in a way that results in natural selection driving differences between the two populations. These mechanisms cause genetic divergence of the island population from the mainland one, ultimately leading to speciation.

It is possible to glimpse peripatric differentiation in action (**Fig. 22.8**). Studies of a kingfisher, *Tanysiptera galatea,* in New

FIG. 22.8 Peripatric speciation in action among populations of New Guinea kingfishers. *Sources: Data from D. J. Futuyma, 2009, Evolution, 2nd ed., Sunderland, MA: Sinauer Associates, Fig. 18.7, p. 484; Photo from C. H. Greenewalt/VIREO.*

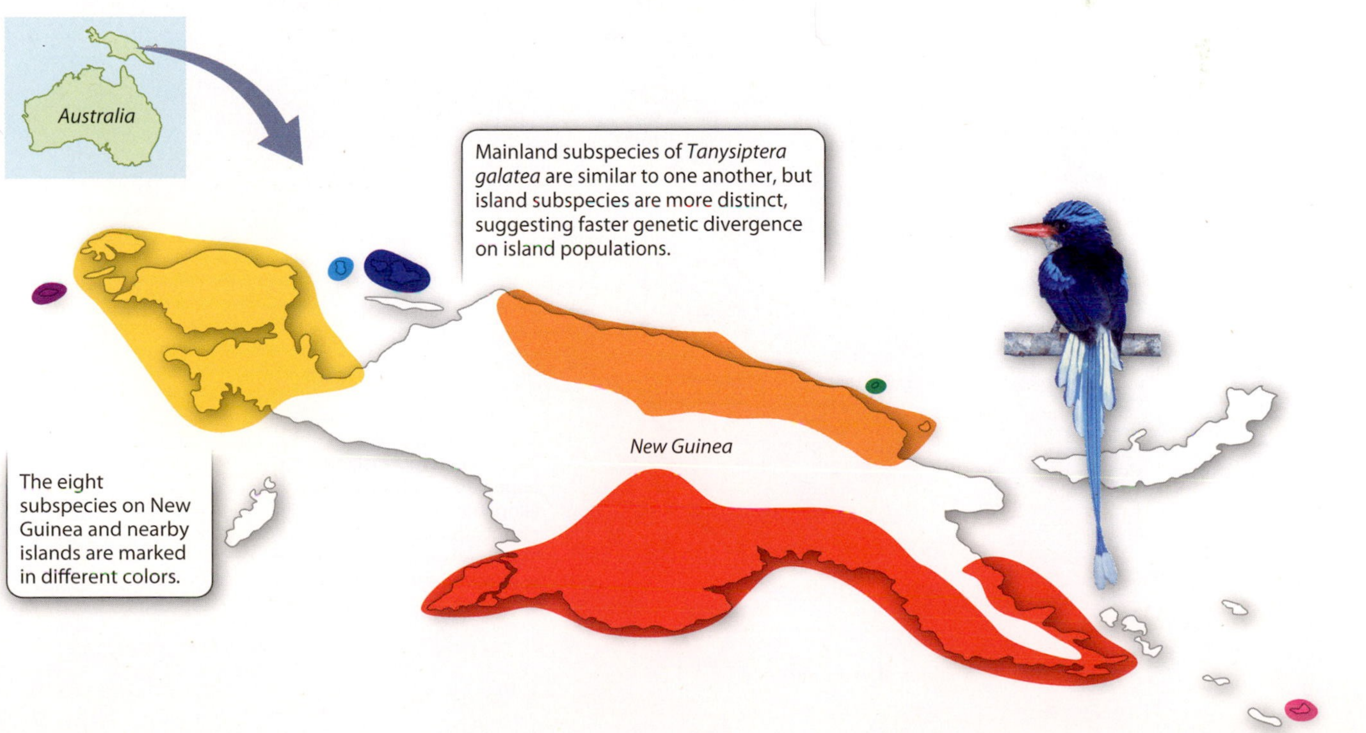

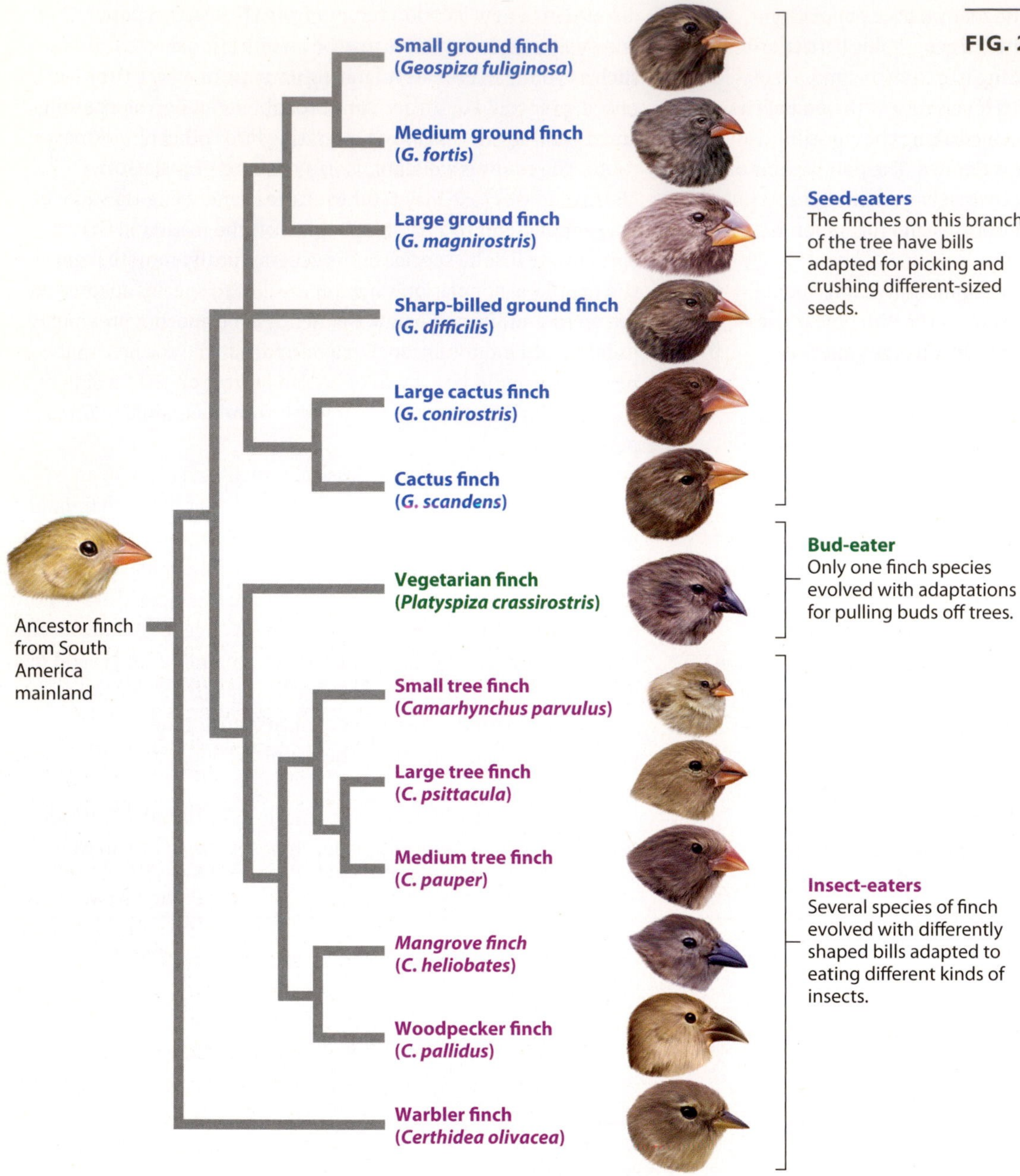

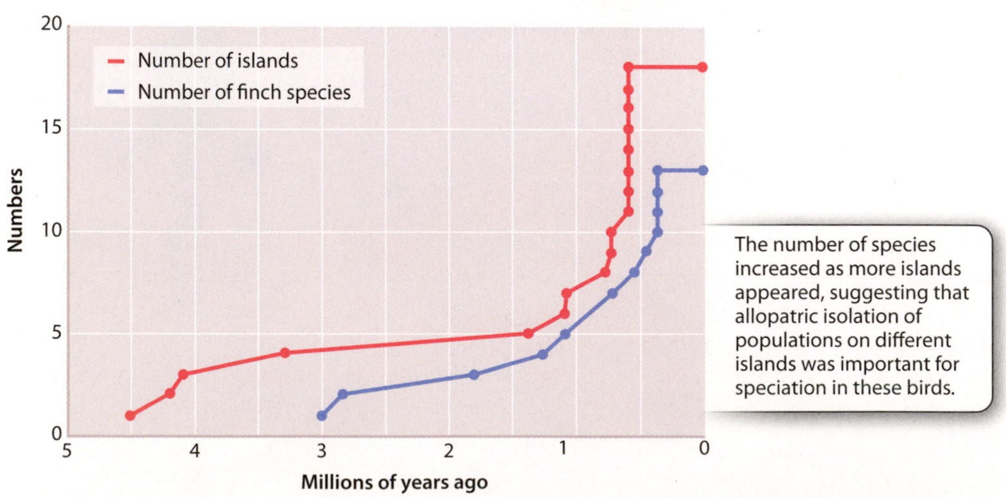

FIG. 22.9 **Adaptive radiation in Darwin's finches.** Today, there are 13 different species of Darwin's finches. *Source: Phylogenetic tree data from Fig. 2.1, p. 15, finch species plot data from Fig. 2.4, p. 23, in P. R. Grant and B. R. Grant, 2008, How and Why Species Multiply, Princeton, NJ: Princeton University Press.*

Guinea and nearby islands show the process under way. There are eight recognized subspecies of *T. galatea,* three on mainland New Guinea (where they exist in large populations separated by mountain ranges) and five on nearby islands (where, because the islands are small, the populations are correspondingly small). The mainland subspecies are still quite similar to one another, but the island subspecies are much more distinct, suggesting that genetic divergence is occurring faster in the small island populations. If we wait long enough, these subspecies will probably diverge into new species.

Because dating such dispersal events is tricky—newcomers on an island tend not to leave a record of when they arrived—we can sometimes use vicariance information to study the timing of peripatric speciation. For instance, we know that the oldest of the Galápagos Islands were formed 4–5 million years ago by volcanic action. Sometime early in the history of the Galápagos, individuals of a small South American finch species arrived there. Conditions on the Galápagos are very different from those on the South American continent, where the mainland population of this ancestral finch lived, and so the isolated island population evolved to become distinct from its mainland ancestor and eventually became a new species.

The finches' subsequent dispersal among the other islands of the Galápagos has promoted further peripatric speciation (**Fig. 22.9**). In the graph at the bottom of the figure, we see that the number of finch species is correlated with the number of islands in the archipelago. This is clear indication of the importance of geographic separation (allopatry) in speciation: The availability of islands provided opportunities for populations to become isolated from one another, allowing speciation. The result was the evolution of 13 different species of finches, collectively known today as Darwin's finches.

The Galápagos finches and their frenzy of speciation illustrate the important evolutionary idea of **adaptive radiation,** a bout of unusually rapid evolutionary diversification in which natural selection accelerates the rates of both speciation and adaptation. Adaptive radiation occurs when there are many ecological opportunities available for exploitation. Consider the ancestral finch immigrants arriving on the Galápagos. A wealth of ecological opportunities was open and available. Until the arrival of the colonizing finches, there were no birds on the islands to eat the plant seeds, or to eat the insects on the plants, and so on. Suppose that the ancestral finches fed specifically on medium-sized seeds on the South American mainland (that is, they were medium-seed specialists, with bills that are the right size for handling medium-sized seeds). On the mainland, they were constrained to that size of seed because any attempt to eat larger or smaller ones brought them into conflict with other species—a large-seed specialist and a small-seed specialist—that already used these resources. In effect, stabilizing selection (Chapter 21) was operating on the mainland to eliminate the extremes of the bill-size spectrum in the medium-seed specialists.

When the medium-billed immigrant finches first arrived on the Galápagos, however, no such competition existed. Therefore, natural selection promoted the formation of new species of small- and large-seed specialists from the original medium-seed-eating ancestral stock. With the elimination of the stabilizing selection that kept the medium-billed finches medium-billed on the mainland, selection actually operated in the opposite direction, favoring the large and small extremes of the bill-size spectrum because these individuals could take advantage of the abundance of unused resources, small and large seeds. It is this combination of emptiness—the availability of ecological opportunity—and the potential for allopatric speciation that results in adaptive radiation.

→ **Quick Check 2** Why do we see so many wonderful examples of adaptive radiation on mid-ocean volcanic archipelagos like the Galápagos?

Co-speciation is speciation that occurs in response to speciation in another species.

As we have seen, physical separation is often a critical ingredient in speciation. Two populations that are not fully separated from each other—that is, there is gene flow between them—will typically not diverge from each other genetically, because genetic exchange homogenizes them. This is why most speciation is allopatric. Allopatric speciation brings to mind populations separated from each other by stretches of ocean or deserts or mountain ranges. However, separation can be just as complete even in the absence of geographic barriers.

Consider an organism that parasitizes a single host species. Suppose that the host undergoes speciation, producing two daughter species. The original parasite population will also be split into two populations, one for each host species. Thus, the two new parasite populations are physically separated from each other and will diverge genetically, ultimately undergoing speciation. This divergence results in a pattern of coordinated host–parasite speciation called **co-speciation,** a process in which two groups of organisms speciate in response to each other and at the same time.

Phylogenetic analysis of lineages of parasites and their hosts that undergo co-speciation reveals trees that are similar for each group. Each time a branching event—that is, speciation—has occurred in one lineage, a corresponding branching event occurred in the other (**Fig. 22.10**).

? CASE 4 MALARIA: CO-EVOLUTION OF HUMANS AND A PARASITE

How did malaria come to infect humans?

Now let's look at a human parasite, *Plasmodium falciparum,* the single-celled eukaryote that causes malaria. It had been suggested that *P. falciparum*'s closest relative is another *Plasmodium* species, *P. reichenowi,* found in chimpanzees, our closest living relative. Were *P. falciparum* and *P. reichenowi* the products of co-speciation? When the ancestral population split millions of years ago to give

FIG. 22.10 Co-speciation. Parasites and their hosts often evolve together, and the result is similar phylogenies. *Sources: Data from J. P. Huelsenbeck and B. Rannala, 1997, "Phylogenetic Methods Come of Age: Testing Hypotheses in an Evolutionary Context," Science 276:227, doi: 10.1126/science.276.5310.227; Louse photo by Alex Popinga, courtesy of James Demastes; pocket gopher photo by Richard Ditch @ richditch.com.*

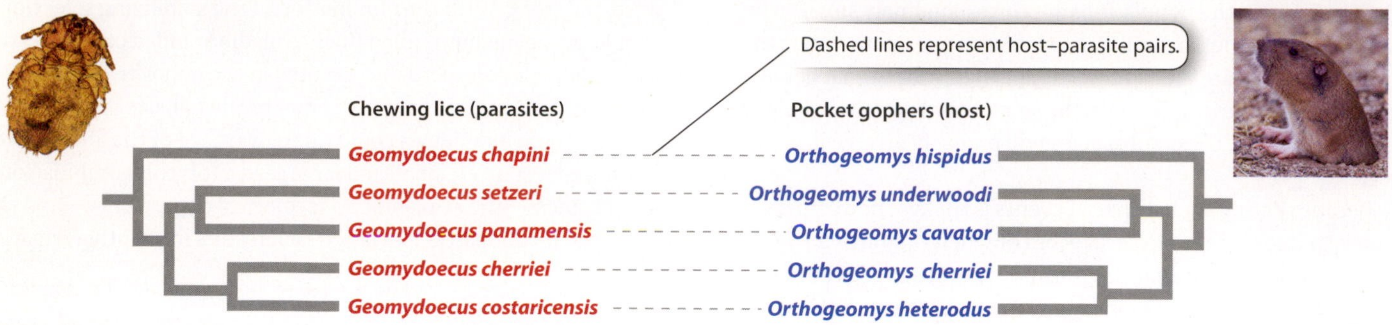

rise to human and chimpanzee lineages, that population's parasitic *Plasmodium* population could also have been split, ultimately yielding *P. falciparum* and *P. reichenowi.*

Recent studies, however, have disproved this hypothesis. We now know that *P. falciparum* was introduced to humans relatively recently from gorillas. Why doesn't the evolutionary history of *Plasmodium* follow the classical host–parasite co-speciation pattern? We know that this history is complex and is still being unraveled. However, we also know that the mosquito-borne phase of its life cycle facilitates transfer to new hosts. Malaria parasites are thus not as inextricably tied to their hosts as the pocket gopher lice in Fig 22.10 and so their evolutionary history is not as parallel to their hosts.

Sympatric populations—those not geographically separated—may undergo speciation.

Can speciation occur *without* complete physical separation of populations? Yes, though evolutionary biologists are still exploring how common this phenomenon is. Recall how separated populations inevitably diverge genetically over time (see Fig. 22.6). If a mutation arises in population A after it has separated from population B, that mutation is present only in population A and may eventually become fixed (100% frequency) in that population, either through natural selection, if it is advantageous, or through drift, if it is neutral. Once the mutation is fixed in population A, it represents a genetic difference between populations A and B. Repeated independent fixations of different mutations in the two populations result over time in the genetic divergence of separated populations.

Now imagine that populations A and B are not completely separated and there is some gene flow between them. The mutation that arose in population A can, in principle, appear in population B as members of the two populations interbreed. The new mutation may indeed have become fixed in population A, but a migrant from population A to population B may introduce the mutation to population B as well. Gene flow effectively negates the genetic divergence of populations. If there is gene flow, a pair of populations may change over time, but they do so together. How, then, can speciation occur if gene flow exists? The term we use to describe populations that are in the same geographic location is **sympatric** (literally, "same place"). So we can rephrase the question in technical terms: How can speciation occur sympatrically?

FIG. 22.11 Sympatric speciation by disruptive selection. Natural selection eliminates individuals in the middle of the spectrum.

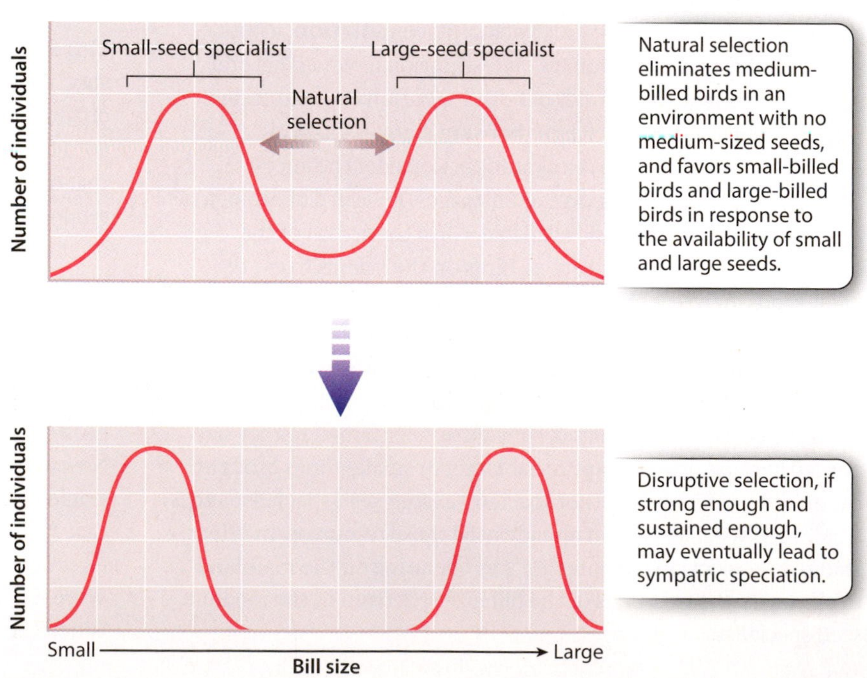

For speciation to occur sympatrically, natural selection must act strongly to counteract the homogenizing effect of gene flow. Consider two sympatric populations of finch-like birds, represented in the graph in **Fig. 22.11.** One population begins to specialize on small seeds and the other on large seeds. If the two populations freely interbreed, no genetic differences between the two will occur and speciation will not take place. Now suppose that the offspring produced by the pairing of a big-seed specialist with a small-seed specialist is an individual best adapted to eat medium seeds, and there are no medium-sized seeds available in the environment. Natural selection will act against the hybrids, which will starve to death because there are no medium-sized seeds for them to eat and they are not well adapted to compete with the big- or small-seed specialists. Natural selection would then, in effect, eliminate the products of gene flow. So, although gene flow is occurring, it does not affect the divergence of the two populations because the hybrid individuals do not survive to reproduce. As discussed in Chapter 21, this form of natural selection, which operates against the middle of a spectrum of variation, is called disruptive selection.

However, it turns out to be difficult to find evidence of sympatric speciation in nature, though it may not be especially rare in plants, as we will see. We might find two very closely related species in the same location and argue that they must have arisen through sympatric speciation. However, there is an alternative explanation. One species could have arisen elsewhere by, for example, peripatric speciation and subsequently moved into the environment of the other one. In other words, the speciation occurred in the past by allopatry, and the two species are only currently sympatric because migration after speciation occurred.

However, recent studies of plants on an isolated island have provided strong, if not definite, evidence of sympatric speciation. Lord Howe Island is a tiny island (about 10 km long and 2 km across at its widest) about 600 km east of Australia. Two species of palm tree that are only found on the island are each other's closest relatives. Because the island is so small, there is little chance that the two species could be geographically separated from each other, meaning that the species are and were sympatric, and, because of the distance of Lord Howe from Australia (or other islands), it is extremely likely that they evolved and speciated from a common ancestor on the island.

This and other demonstrations show that sympatric speciation can occur. We must recognize, however, that we do not yet know just how much of all speciation is sympatric and how much is allopatric. **Fig. 22.12** summarizes modes of speciation based on geography.

FIG. 22.12 Modes of speciation.

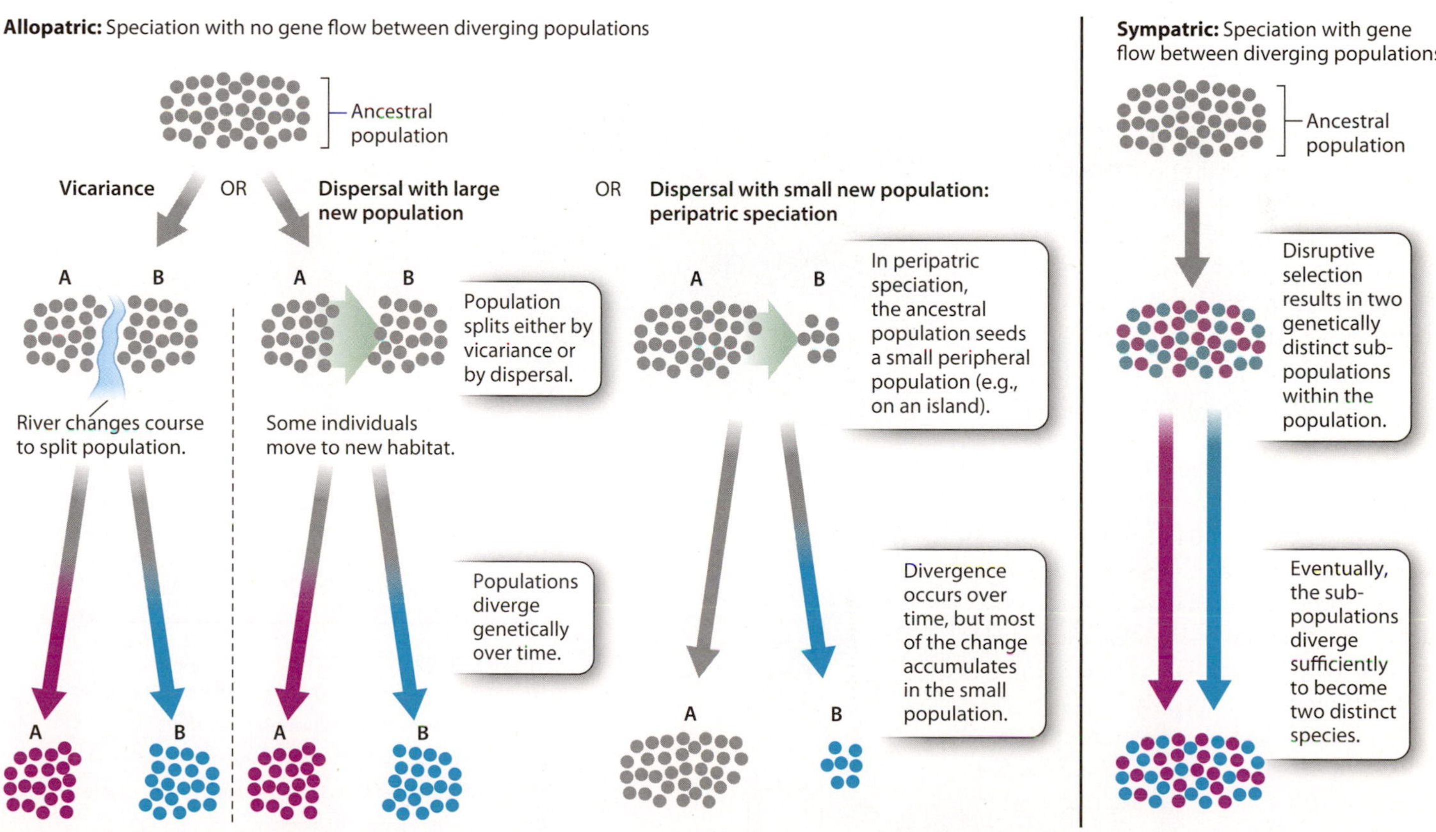

→ **Quick Check 3** There are hundreds of species of cichlid fish in Lake Victoria in Africa. Some scientists argue that they evolved sympatrically, but recent studies of the lake suggest that it periodically dried out, leaving a series of small ponds. Why is this observation relevant to evaluating the hypothesis that these species arose by sympatric speciation?

Speciation can occur instantaneously.

Although speciation is typically a lengthy process, it can occasionally occur in a single generation, making it sympatric by definition. Typically, cases of such **instantaneous speciation** are caused by hybridization between two species in which the offspring are reproductively isolated from both parents.

For example, hybridization in the past between two sunflower species, *Helianthus annuus* and *H. petiolaris* (the ancestor of the cultivated sunflower) has apparently given rise to three new sunflower species, *H. anomalus, H. paradoxus,* and *H. deserticola* (**Fig. 22.13**). *H. petiolaris* and *H. annuus* have probably formed innumerable hybrids in nature, virtually all of them inviable. However, a few—the ones with a workable genetic complement—are the ones that survived to yield these three daughter species. In this case, each one of these new species has acquired a different mix of parental chromosomes. It is this species-specific chromosome complement that makes all three distinct and reproductively isolated from the parent species and from one another.

In many cases of hybridization, chromosome numbers may change. Two diploid parent species with 5 pairs of chromosomes, for a total of 10 chromosomes each, may produce a hybrid with double the number of chromosomes (that is, the hybrid inherits a full paired set of chromosomes from each parental species)—a total of 20. In this case, the hybrid has four genomes rather than the diploid number of two. We call such a double diploid a tetraploid. In general, animals cannot sustain this kind of expansion in chromosome complement, but plants are more likely to do so. As a result, the formation of new species through **polyploidy**—multiple chromosome sets (Chapter 13)—has been relatively common in plants.

Polyploids may be allopolyploids, meaning that they are produced from hybridization of two different species. For example, related species of *Chrysanthemum* appear, on the basis of their chromosome numbers, to be allopolyploids. Alternatively, polyploids may be autopolyploids, meaning that they are derived from an unusual reproductive event between members of a single species. In this case, through an error of meiosis in one or both parents in which homologous chromosomes fail to separate, a gamete may be produced that is not haploid. For example, *Anemone rivularis,* a plant in the buttercup family, has 16 chromosomes, and its close relative *A. quinquefolia* has 32. *Anemone quinquefolia* appears to be an autopolyploid derived from the joining of *A. rivularis* gametes with two full sets of chromosomes each.

So rampant is speciation by polyploidy in plants that it affects the pattern of chromosome numbers across all plants. In **Fig. 22.14,** we see the haploid chromosome numbers of thousands of plant species plotted on a graph. Note that as the numbers get higher, even numbers tend to predominate, suggesting that the doubling of the total number of chromosomes (a form

FIG. 22.13 **Speciation by hybridization.** In sunflowers, *Helianthus anomalus* (bottom) is the product of natural hybridization between *H. annuus* (top left) and *H. petiolaris* (top right). *Source: (top left) Gary A. Monroe @ USDA-NRCS PLANTS Database, (top right) Jason Rick, (bottom) Gerald J. Seiler, USDA-ARS.*

FIG. 22.14 Plant chromosome numbers. Plant chromosome numbers suggest that polyploidy has played an important role in plant evolution. *Data from Fig. 20-18, p. 752, in A. J. F. Griffiths, S. R. Wessler, S. B. Carroll, and J. Doebley, 2012, Introduction to Genetic Analysis, 10th ed., New York: W. H. Freeman.*

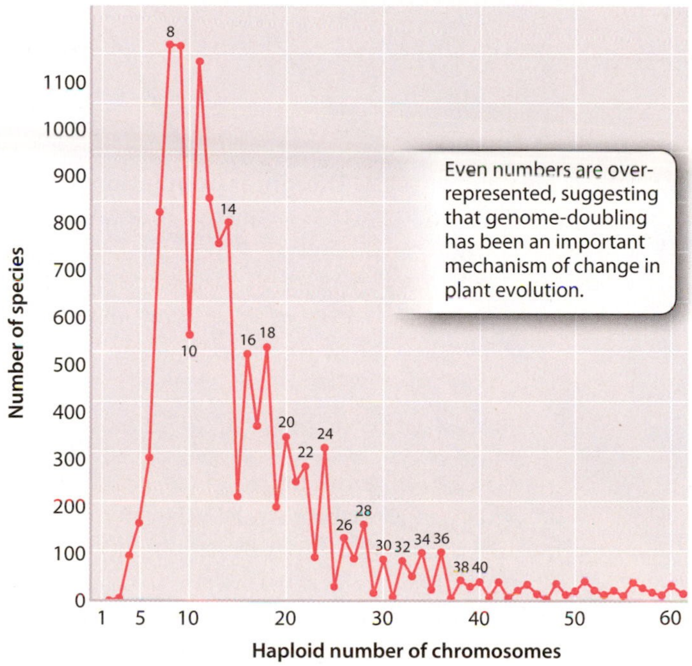

of polyploidy that will always result in an even number of chromosomes) is an important factor in plant evolution.

22.4 SPECIATION AND SELECTION

The association of both the origin of species and natural selection with Charles Darwin may make it seem that one cannot occur without the other. However, speciation can occur in the presence or absence of natural selection, and natural selection does not always lead to speciation.

Speciation can occur with or without natural selection.

Natural selection may or may not play a role in speciation. The genetic divergence of two populations can be entirely due to genetic drift, for example, with no role for natural selection.

We have also seen two ways in which natural selection can be involved in speciation. First, sympatric speciation requires some form of disruptive natural selection, as when hybrid offspring are competitively inferior. Second, allopatric speciation (and adaptive radiation) may be facilitated by natural selection. For example, when a peripheral population is in a new environment, natural selection will act to promote its adaptation to the new conditions, accelerating in the process the rate of genetic divergence between it and its parent population.

Natural selection can enhance reproductive isolation.

There is a third way in which natural selection may play an important role in speciation. In this case, natural selection contributes directly to the process of speciation (rather than contributing indirectly by causing differences among the diverging populations), when individuals better at choosing mates from their own group are selectively favored over those that frequently mate with members of the "wrong" group.

Recall the two hypothetical bird populations, one with large bills and the other with small bills, whose medium-billed hybrid offspring are at a disadvantage because of a lack of medium-sized seeds in their environment. If a large-billed individual cannot distinguish between large- and small-billed individuals as potential mates (that is, there is a lack of pre-zygotic isolation between them), it will frequently make the "wrong" mate choice, picking a small-billed individual and paying a considerable evolutionary cost of producing poorly adapted hybrid offspring.

Now imagine a new mutation in the large-billed group that permits individuals carrying it to distinguish between the two groups of birds and to mate only with other large-billed individuals. Such a mutation would spread under natural selection because it would prevent the wasted reproductive effort of producing disadvantaged hybrid offspring. This is an example of **reinforcement of reproductive isolation,** or **reinforcement** for short. Reinforcement is the process by which diverging populations undergo natural selection in favor of traits that enhance pre-zygotic isolation, thereby preventing the production of less fit hybrid offspring. In this case, enhanced pre-zygotic isolation takes the form of mating discrimination, an increased ability to recognize and mate with members of one's own population.

The best evidence in support of reinforcement comes from a study of related fruit-fly species living either in allopatry (geographically separated) or sympatry (not geographically separated). Sympatric species evolve pre-zygotic isolating mechanisms more rapidly than allopatric species. Why would this be? When the two populations are geographically separated, the formation of less fit hybrids is impossible since the two populations do not interbreed, so a mutation that increases mating discrimination between the two groups would not be favored by natural selection. In sympatry, however, where the production of less fit hybrids is a problem, such a mutation provides a fitness benefit, so natural selection favors its spread in the population and reinforces reproductive isolation between the two groups.

Fig. 22.15 summarizes the evolutionary mechanisms that lead to speciation. Speciation is caused by the accumulation of genetic differences between populations. Mutation is therefore a key component of the process, but so, too, is the fixation process whereby mutations go to 100% in a population (Chapter 21). A mutation can go to fixation by selection if it is advantageous and by genetic drift. The differences that accumulate between diverging populations are therefore produced by multiple forces acting over long periods on each population. ■

VISUAL SYNTHESIS # Speciation

FIG. 22.15 **Integrating concepts from Chapters 21 and 22**

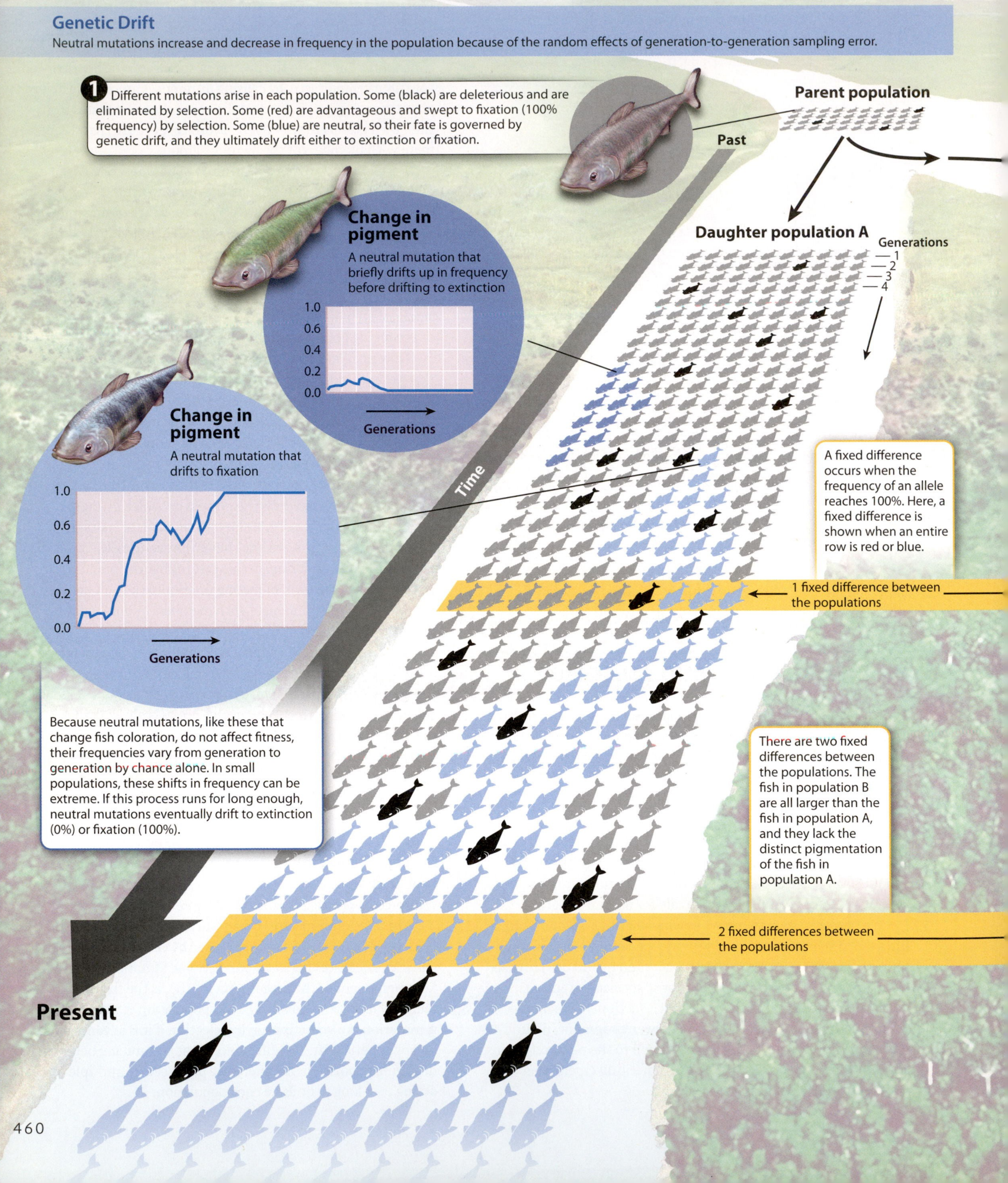

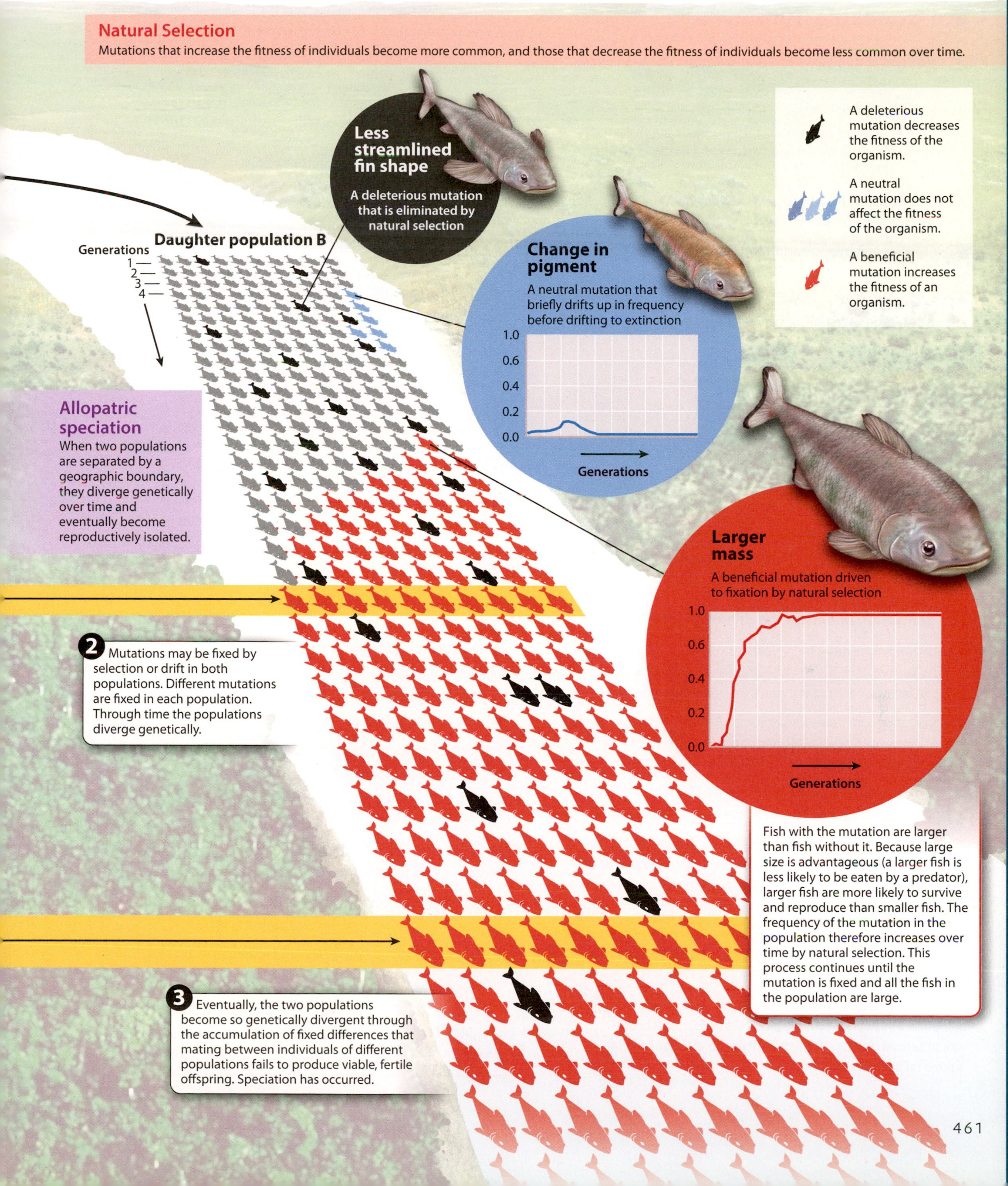
Natural Selection
Mutations that increase the fitness of individuals become more common, and those that decrease the fitness of individuals become less common over time.
Less streamlined fin shape
A deleterious mutation that is eliminated by natural selection
A deleterious mutation decreases the fitness of the organism.
A neutral mutation does not affect the fitness of the organism.
A beneficial mutation increases the fitness of an organism.
Daughter population B
Generations
1
2
3
4
Change in pigment
A neutral mutation that briefly drifts up in frequency before drifting to extinction
1.0
0.6
0.4
0.2
0.0
Generations
Allopatric speciation
When two populations are separated by a geographic boundary, they diverge genetically over time and eventually become reproductively isolated.
Larger mass
A beneficial mutation driven to fixation by natural selection
1.0
0.6
0.4
0.2
0.0
Generations
2 Mutations may be fixed by selection or drift in both populations. Different mutations are fixed in each population. Through time the populations diverge genetically.
Fish with the mutation are larger than fish without it. Because large size is advantageous (a larger fish is less likely to be eaten by a predator), larger fish are more likely to survive and reproduce than smaller fish. The frequency of the mutation in the population therefore increases over time by natural selection. This process continues until the mutation is fixed and all the fish in the population are large.
3 Eventually, the two populations become so genetically divergent through the accumulation of fixed differences that mating between individuals of different populations fails to produce viable, fertile offspring. Speciation has occurred.

Core Concepts Summary

22.1 Reproductive isolation is the key to the biological species concept.

The biological species concept (BSC) states that species are groups of actually or potentially interbreeding populations that are reproductively isolated from other such groups. page 446

We cannot apply the BSC to asexual or extinct organisms. page 447

Ring species and hybridization further demonstrate that the BSC is not a comprehensive definition of species. page 448

The BSC is nevertheless especially useful because it emphasizes reproductive isolation. page 449

22.2 Reproductive isolation is caused by barriers to reproduction before or after egg fertilization.

Reproductive barriers can be pre-zygotic, occurring before egg fertilization, or post-zygotic, occurring after egg fertilization. page 449

Pre-zygotic isolation may be behavioral, gametic, temporal, or ecological. page 450

In post-zygotic isolation, mating occurs but genetic incompatibilities prevent the development of a viable, fertile offspring. page 450

22.3 Speciation underlies the diversity of life on Earth.

Speciation is typically a by-product of genetic divergence that occurs as a result of the fixation of different mutations in two populations that are not regularly exchanging genes. page 451

If divergence continues long enough, chance differences will arise that result in reproductive barriers between the two populations. page 451

Most speciation is thought to be allopatric, involving two geographically separated populations. page 451

Geographic separation may be caused by dispersal, resulting in the establishment of a new and distant population, or by vicariance, in which the range of a species is split by a change in the environment. page 451

A special case of allopatric speciation by dispersal is peripatric speciation, in which the new population is small and outside the species' original range. page 453

Adaptive radiation, in which speciation occurs rapidly to generate a variety of ecologically diverse forms, is best documented on oceanic islands after the arrival of a single ancestral species. page 455

Co-speciation occurs when one species undergoes speciation in response to speciation in another. In parasites and their hosts, co-speciation can result in host and parasite phylogenies that have the same branching patterns. page 455

Speciation may be sympatric, meaning that there is no geographic separation between the diverging populations. For this type of speciation to occur, natural selection for two or more different types within the population must act so strongly that it overcomes the homogenizing effect of gene flow. page 456

22.4 Speciation can occur with or without natural selection.

Separated populations can diverge as a result of genetic drift, natural selection, or both. page 459

Natural selection may act on mutations that allow individuals to identify and mate with individuals that are more like themselves. This process is called reinforcement of reproductive isolation. page 459

Self-Assessment

1. Define the term "species."
2. Given a group of organisms, describe how you would test whether they all belong to one species or whether they belong to two separate species.
3. Name two types of organism that do not fit easily into the biological species concept. What species concept would work best for these organisms?
4. Explain how ecological and phylogenetic considerations can help inform whether or not a group of organisms represents a single species.
5. Name four reproductive barriers and indicate whether each is pre- or post-zygotic.
6. Describe how genetic divergence and reproductive isolation are related to each other.
7. Differentiate between allopatric and sympatric speciation, and state which is thought to be more common and why.
8. Differentiate between allopatric speciation by dispersal and by vicariance, and give an example of each.
9. Describe how genetic drift can result in speciation.
10. Describe how natural selection can result in speciation.

Log in to LaunchPad to check your answers to the Self-Assessment questions, and to access additional learning tools.

Dudley Museums Service.

CHAPTER 23

Evolutionary Patterns

Phylogeny and Fossils

Core Concepts

23.1 A phylogenetic tree is a reasoned hypothesis of the evolutionary relationships among organisms.

23.2 A phylogenetic tree is built on the basis of shared derived characters.

23.3 The fossil record provides direct evidence of evolutionary history.

23.4 Phylogeny and fossils provide independent and corroborating evidence of evolution.

All around us, nature displays nested patterns of similarity among species. For example, as noted in Chapter 1, humans are more similar to chimpanzees than either humans or chimpanzees are to monkeys. Humans, chimpanzees, and monkeys, in turn, are more similar to one another than any one of them is to a mouse. And humans, chimpanzees, monkeys, and mice are more similar to one another than any of them is to a catfish. This pattern of nested similarity was recognized more than 200 years ago and used by the Swedish naturalist Carolus Linnaeus to classify biological diversity. A century later, Charles Darwin recognized this pattern as the expected outcome of a process of "descent with modification," or evolution.

Evolution produces two distinct but related patterns, both evident in nature. First is the nested pattern of similarities found among species on present-day Earth. The second is the historical pattern of evolution recorded by fossils. Life, in its simplest form, originated more than 3.5 billion years ago. Today, an estimated 10 million species inhabit the planet. Short of inventing a time machine, how can we reconstruct those 3.5 billion years of evolutionary history in order to understand the extraordinary events that have ultimately resulted in the biological diversity we see around us today? These two great patterns provide the answer.

Darwin recognized that the species he observed must be the modified descendants of earlier ones. Distinct populations of an ancestral species separate and diverge through time, again and again, giving rise to multiple descendant species. The result is the pattern of nested similarities observed in nature (see Fig. 1.17). This history of descent with branching is called **phylogeny,** and is much like the genealogy that records our own family histories.

The evolutionary changes inferred from patterns of relatedness among present-day species make predictions about the historical pattern of evolution we should see in the fossil record. For example, groups with features that we infer to have evolved earlier than others should appear earlier in time as fossils. Paleontological research reveals that the history of life is indeed laid out in the chronological order predicted on the basis of comparative biology. How do we reconstruct the history of life from evolution's two great patterns, and how do they compare?

23.1 READING A PHYLOGENETIC TREE

Chapter 22 introduced the concept of speciation, the set of processes by which physically, physiologically, or ecologically isolated populations diverge from one another to the point where they can no longer produce fertile offspring. As illustrated in **Fig. 23.1,** speciation can be thought of as a process of branching. Now consider what happens as this process occurs over and over in a group through time. As species proliferate, their evolutionary relationships to one another unfold in a treelike pattern, with individual species at the twig tips and their closest relatives connected to them at the nearest fork in the branch, called a **node.** A node thus represents the most recent common ancestor of two descendant species.

FIG. 23.1 The relationship between speciation and a phylogenetic tree. The phylogenetic tree, on the right, depicts the evolutionary relationships that result from the two successive speciation events diagrammed on the left.

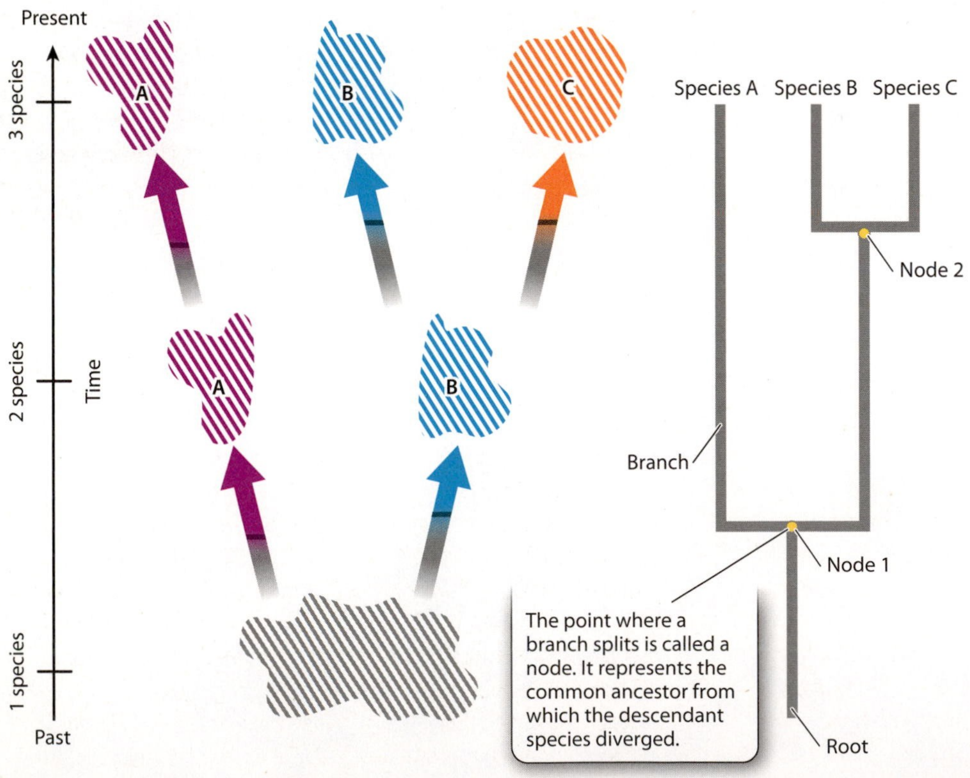

Phylogenetic trees provide hypotheses of evolutionary relationships.

Phylogenetics is one of two related disciplines within systematics, the study of evolutionary relationships among organisms. The other is taxonomy, the classification of organisms.

The aim of taxonomy is to recognize and name groups of individuals as species, and, subsequently, to group closely related species into the more inclusive taxonomic group of the genus, and so on up through the taxonomic ranks—species, genus, family, order, class, phylum, kingdom, domain. Taxonomy, then, provides us with a hierarchical classification of species in groups that are more and more inclusive, giving us a convenient way to communicate information about the features each group possesses. So, if we want to tell someone about a small animal we have seen with fur, mammary glands, and extended finger bones that permit it to fly, we can give them this long description, or we can just say we saw a bat, or a member of Order Chiroptera. All the rest is understood (or can be looked up in a reference).

Phylogenetics, on the other hand, aims to discover the pattern of evolutionary relatedness among groups of species or other groups by comparing their anatomical or molecular features, and to depict these relationships as a **phylogenetic tree.** A phylogenetic tree is a hypothesis about the evolutionary history, or phylogeny, of the species. Phylogenetic trees are hypotheses because they represent the best model, or explanation, of the relatedness of organisms on the basis of all the existing data. As with any model or hypothesis, new data may provide evidence for alternative relationships, leading to changes in the hypothesized pattern of branching on the tree.

Many phylogenetic trees explore the relatedness of particular groups of individuals, populations, or species. We may, for example, want to understand how wheat is related to other, non-commercial grasses, or how disease-causing populations of *Escherichia coli* relate to more benign strains of the bacterium. At a much larger scale, universal similarities of molecular biology indicate that all living organisms are descended from a single common ancestor. This insight inspires the goal of reconstructing phylogenetic relationships for all species in order to understand how biological diversity has evolved since life originated. This universal tree is commonly referred to as the tree of life (Chapter 1). In Part 2 of this book, we will make use of the tree of life and many smaller-scale phylogenetic trees to understand our planet's biological diversity.

Fig. 23.2 shows a phylogenetic tree for vertebrate animals. The informal name at the end of each branch represents a group of organisms, many of them familiar. We sometimes find it useful to refer to groups of species this way (for example, "frogs," or "Class Anura") rather than name all the individual species or list the characteristics they have in common. It is important, however, to remember that such named groups represent a number of member species. If, for example, we were able to zoom in on the branch labeled "Frogs," we would see that it consists of many smaller branches, each representing a distinct species of frog, either living or extinct.

This tree provides information about evolutionary relationships among vertebrates. For example, it proposes that the closest living relatives of birds are crocodiles and alligators. The tree also proposes that the closest relatives of all tetrapod (four-legged) vertebrates are lungfish, which are fish with lobed limbs and the ability to breathe air. Phylogenetic trees are built from careful analyses of the morphological and molecular attributes of the species or other groups under study. A tree is therefore a hypothesis about the order of branching events in evolution, and it can be tested by gathering more information about anatomical and molecular traits.

A phylogenetic tree does not in any way imply that more recently evolved groups are more advanced than groups that arose earlier. A modern lungfish, for example, is not more primitive or "less evolved" than an alligator, even though its group branches off the trunk of the vertebrate tree earlier than the alligator group does. After all, both species are the end products of the same interval of evolution since their divergence from a common ancestor more than 370 million years ago.

FIG. 23.2 **A phylogeny of vertebrate animals.** The branching order constitutes a hypothesis of evolutionary relationships within the group.

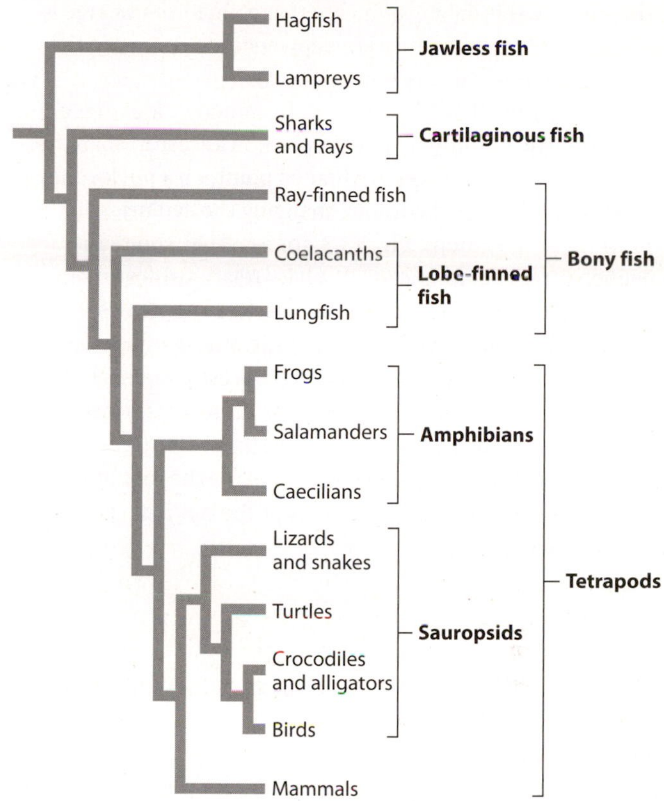

The search for sister groups lies at the heart of phylogenetics.

Two species, or groups of species, are considered to be closest relatives if they share a common ancestor not shared by any other species or group. In Fig. 23.2, for example, we see that frogs are more closely related to salamanders than to any other group of organisms because frogs and salamanders share a common amphibian ancestor not shared by any other group. Similarly, lungfish are more closely related to tetrapods than to any other group. A lungfish may look more like a fish than it does an amphibian, but lungfish are more closely related to amphibians than they are to other fish because lungfish share a common ancestor with amphibians (and other tetrapods) that was more recent than their common ancestor with other fish (Fig. 23.2).

Groups that are more closely related to each other than either of them is to any other group, like lungfish and tetrapods, are called **sister groups.** Simply put, phylogenetic hypotheses amount to determining sister-group relationships because the simplest phylogenetic question we can ask is which two of any three species (or other groups) are more closely related to each other than either is to the third. In this light, we can see that a phylogenetic tree is simply a set of sister-group relationships; adding a species to the tree entails finding its sister group in the tree.

Closeness of relationship is then determined by looking to see how recently two groups share a common ancestor. Shared ancestry is indicated by a node, or branch point, on a phylogenetic tree. Nodes can be rotated without changing the evolutionary relationships of the groups. **Fig. 23.3,** for example, shows four phylogenetic trees depicting evolutionary relationships among birds, crocodiles and alligators, and turtles. In all four trees, birds are a sister group to crocodiles and alligators because birds, crocodiles, and alligators share a common ancestor not shared by turtles. The more recent a common ancestor, the more closely related two groups are. Evolutionary relatedness therefore is determined by following nodes from the tips to the root of the tree, and is not determined by the order of the tips from the top to bottom of a page.

→ **Quick Check 1** Does either of these two phylogenetic trees indicate that humans are more closely related to lizards than to mice?

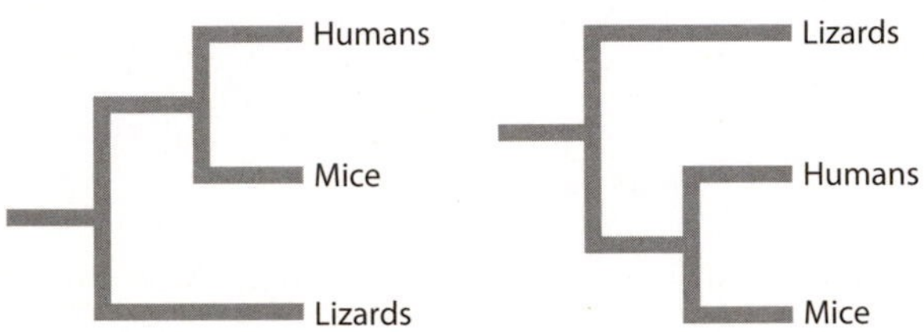

A monophyletic group consists of a common ancestor and all its descendants.

Up to this point, we have used the word "group" to mean all the species in some taxonomic entity under discussion. A more technical word is **taxon** (plural, **taxa**), with taxonomy providing a formal means of naming groups. Recently, biologists have worked to integrate evolutionary history with taxonomic classification. The resulting classification emphasizes groups that are **monophyletic,** meaning that all members share a single common ancestor not shared with any other species or group of species. In Fig. 23.2, the tetrapods are monophyletic because they all share a common ancestor not shared by any other taxa. Similarly, amphibians are monophyletic.

In contrast, consider the group of animals traditionally recognized as reptiles, which includes turtles, snakes, lizards, crocodiles, and alligators (Fig. 23.2). The group "reptiles" excludes birds, although they share a common ancestor with the included animals. Such a group is **paraphyletic.** A paraphyletic group includes some, but not all, of the descendants of a common ancestor. Early zoologists separated birds from reptiles because they are so distinctive. However, many features of skeletal anatomy and DNA sequence strongly support the placement of birds as a sister group to the crocodiles and alligators.

There is a simple way to distinguish between monophyletic and paraphyletic groups, illustrated in **Fig. 23.4.** If in order to separate a group from the rest of the phylogenetic tree you need only to make one cut, the group is monophyletic. If you need a second cut to trim away part of the separated branch, the group is paraphyletic.

Groupings that do not include the last common ancestor of all members are called **polyphyletic.** For example, clustering bats and birds together as flying tetrapods results in a polyphyletic group (Fig. 23.4).

Identifying monophyletic groups is a main goal of phylogenetics because monophyletic groups include *all* descendants of a common ancestor and *only* the descendants of that common ancestor. This means that monophyletic groups alone show the evolutionary path a given group has taken since its origin. Omitting some members of a group, as in the case of reptiles and other paraphyletic groups, can provide a misleading sense of evolutionary history. By using monophyletic groups in taxonomic classification, we effectively convey our knowledge of their evolutionary history.

→ **Quick Check 2** Look at Fig. 23.4. Are fish a monophyletic group?

FIG. 23.3 Sister groups. The four trees illustrate the same set of sister-group relationships.

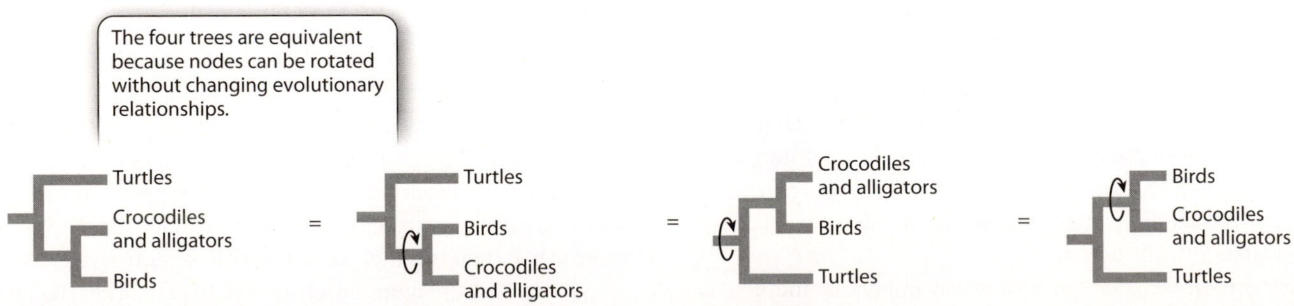

FIG. 23.4 Monophyletic, paraphyletic, and polyphyletic groups. Only monophyletic groups reflect evolutionary relationships because only they include all the descendants of a common ancestor.

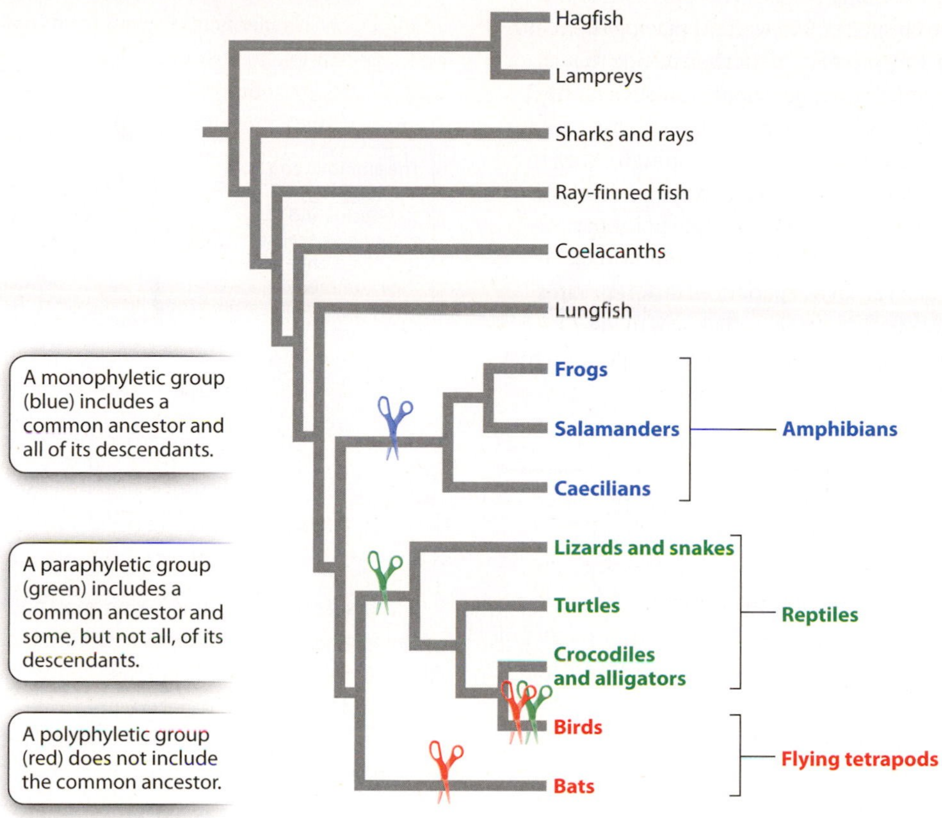

Taxonomic classifications are information storage and retrieval systems.

The nested pattern of similarities among species has been recognized by naturalists for centuries. In the vocabulary of formal classification, closely related species are grouped into a **genus** (plural, **genera**). Closely related genera, in turn, belong to a larger, more inclusive branch of the tree, as a **family.** Closely related families, in turn, form an **order,** orders form a **class,** classes form a **phylum** (plural, **phyla**), and phyla form a **kingdom,** each more inclusive taxonomic level occupying a successively larger limb on the tree (**Fig. 23.5**). Biologists today commonly refer to the three largest limbs of the entire tree of life as **domains** (Eukarya, or eukaryotes; Bacteria; and Archaea).

The ranks of classification form a nested hierarchy, but the boundaries of ranks above the species level are arbitrary

FIG. 23.5 Classification. Classification reflects our understanding of phylogenetic relationships, and the taxonomic hierarchy reflects the order of branching.

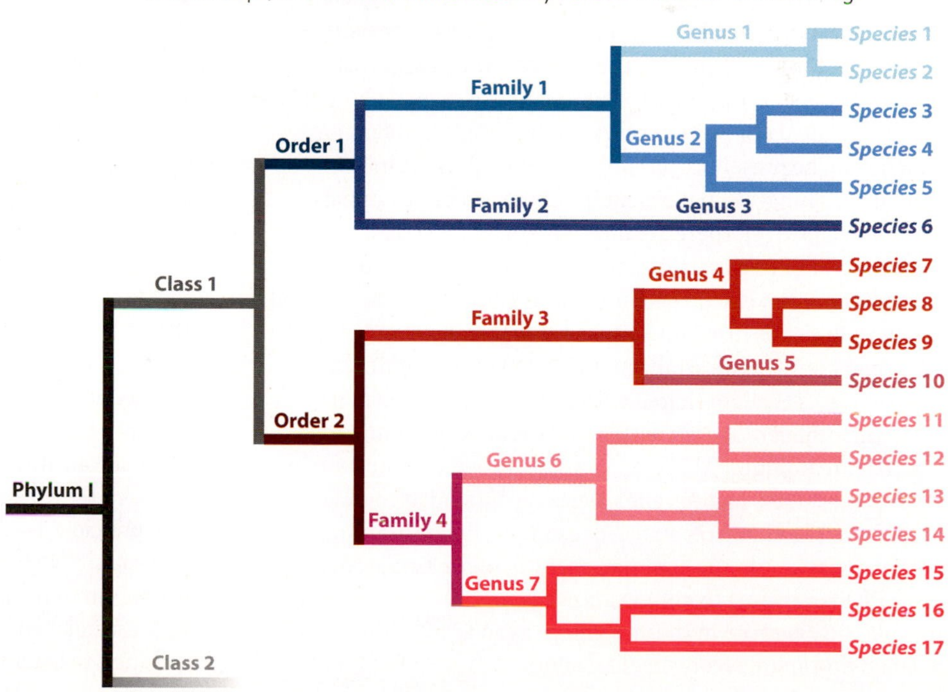

in that there is nothing particular about a group that makes it, for example, a class rather than an order. A taxonomist examining the 17 species included in Fig. 23.5 might decide that species 3 and 4 are sufficiently distinct from species 5 to warrant placing them into a distinct genus. The same taxonomist might then decide that the grouping of the two new genera together should rank as a family. For this reason, it is not necessarily true that orders or classes of, for example, birds and ferns are equivalent in any meaningful way. In contrast, sister groups are equivalent in several ways—notably, in that they diverged from a single ancestor at a single point in time. Therefore, if one branch is a sister group that contains 500 species and another has 6, the branches have experienced different rates of speciation, extinction, or both since they diverged. In Fig 23.5, Families 1 and 2 are sister groups, but Family 1 has five species while Family 2 has just one.

23.2 BUILDING A PHYLOGENETIC TREE

Up to this point, we have focused on how to interpret a phylogenetic tree, a diagram that depicts the evolutionary history of organisms. But how do we infer evolutionary history from a group of organisms? That is, how do we actually construct a phylogenetic tree? Biologists use characteristics of organisms to figure out their relationships. Similarities among organisms are particularly important in that similarities sometimes suggest shared ancestry. However, a key principle of constructing trees is that only *some* similarities are actually useful. Others can in fact be misleading.

Homology is similarity by common descent.

Phylogenetic trees are constructed by comparison of character states shared among different groups of organisms. **Characters** are the anatomical, physiological, or molecular features that make up organisms. In general, characters have several observed conditions, called **character states.** In the simplest case, a character can be present or absent—lungs are present in tetrapods and lungfish, but absent in other vertebrate animals. Commonly, however, there are multiple character states. Petals are a character of flowers, for example, and each observed arrangement—petals arranged in a helical pattern, petals arranged in a whorl, or petals fused into a tube—can be considered a state of the character of petal arrangement. All species contain some character states that are shared with other members of their group, some that are shared with members of other groups, and some that are unique.

Character states in different species can be similar for one of two reasons: The character state (for example, helically arranged petals) was present in the common ancestor of the two groups and retained over time (common ancestry), or the character state independently evolved in the two groups as an adaptation to similar environments (convergent evolution).

FIG. 23.6 Homology and analogy. A homology is a similarity that results from shared ancestry, whereas an analogy is a similarity that results from convergent evolution.

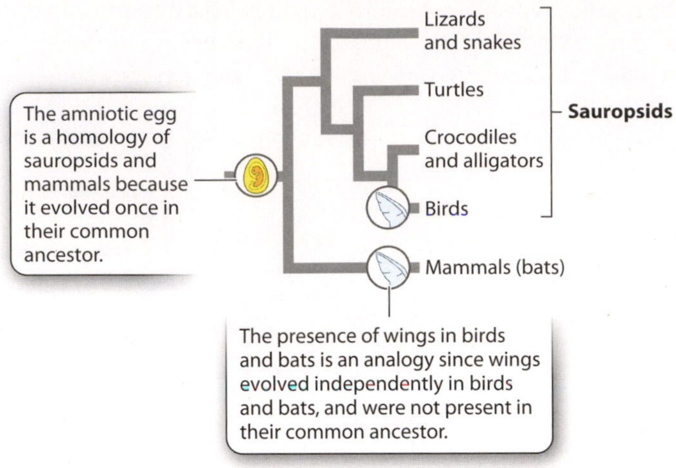

Consider two examples. Mammals and birds both produce amniotic eggs. Amniotic eggs occur only in groups descended from the common ancestor at the node connecting the mammal and sauropsid branches of the tree, and so we reason that birds and mammals each inherited this character from a common ancestor in which the trait first evolved (**Fig. 23.6**). Characters that are similar because of descent from a common ancestor are said to be **homologous.**

Not all similarities arise in this way, however. Think of wings, a character exhibited by both birds and bats. Much evidence supports the view that wings in these two groups do not reflect descent from a common, winged ancestor but rather evolved independently in the two groups. Similarities due to independent adaptation by different species are said to be **analogous.** They are the result of convergent evolution.

Innumerable examples of convergent evolution less dramatic than wings are known. In some, we even understand the genetic basis of the convergence. For example, echolocation has evolved in bats and in dolphins, but not in other mammals. Prestin is a protein in the hair cells of mammalian ears that is involved in hearing ultrasonic frequencies. Both bats and dolphins independently evolved similar changes in their *Prestin* genes, apparently convergent adaptations for echolocation. Similarly, unrelated fish that live in freezing water at the poles, Arctic and Antarctic, have evolved similar glycoproteins that act as molecular "antifreeze," preventing the formation of ice in their tissues.

In principle, two characters or character states are homologous if they are similar because of descent from a common ancestor with the same character or character state; they are analogous if they arose independently because of similar selective pressures. In practice, to determine if characters observed in two organisms are homologous or analogous, we can weigh evidence from where other traits place the two organisms on a phylogenetic tree, we can look at where on the organisms the trait occurs, and we can look at the anatomical or genetic details of how the trait is constructed. Wings in birds and bats are similar in morphological position (both are modified forelimbs) but differ in details of construction (the bat wing is supported by long fingers), and all other traits of birds and bats place them at the tips of different lineages, with many nonwinged species between them and their most recent common ancestor.

→ **Quick Check 3** Fish and dolphins have many traits in common, including a streamlined body and fins. Are these traits homologous or analogous?

Shared derived characters enable biologists to reconstruct evolutionary history.

Because homologies result from shared ancestry, only homologies, and not analogies, are useful in constructing phylogenetic trees. However, it turns out that only some homologies are useful. For example, character states that are unique to a given species or other monophyletic group can't tell us anything about its sister group. They evolved after the divergence of the group from its sister group and so can be used to characterize a group but not to relate it to other groups. Similarly, homologies formed in the common ancestor of the entire group and therefore present in all its descendants do not help to identify sister-group relationships among the descendants of that common ancestor.

What we need to build phylogenetic trees are homologies that are shared by some, but not all, of the members of the group under consideration. These shared derived characters are called **synapomorphies.** A derived character state is an evolutionary innovation (for example, the change from five toes to a single toe—the hoof—in the ancestor of horses and donkeys). When such a novelty arises in the common ancestor of two taxa, it is shared by both (thus, the hoof is a synapomorphy defining horses and donkeys as sister groups).

In **Fig. 23.7,** we indicate the major synapomorphies that have helped us construct the phylogeny of vertebrates. For example, the lung is a character present in lungfish and tetrapods, but absent in other vertebrates. Thus, the presence of lungs provides one piece of evidence that lungfish are the sister group of tetrapods. Phylogenetic reconstruction on the basis of synapomorphies is called **cladistics.**

The simplest tree is often favored among multiple possible trees.

To show how synapomorphies help us chart out evolutionary relationships, let's consider the simple example in **Fig. 23.8.** We begin with four species of animals (labeled "A" through "D") in a group we wish to study that we will call our ingroup; we believe the species to be closely related to each other. For comparison, we have a species that we believe is outside this ingroup—that is, it falls on a branch that splits off nearer the root of the tree—and so is called an outgroup (labeled "OG" in Fig. 23.8). Each species in the ingroup and the outgroup has a different combination of characters, such as leg number, presence or absence of wings, and whether development of young to adult is direct or goes through a pupal stage (Fig. 23.8a).

We are interested in the relationships among species A–D and so focus on potential synapomorphies, character states shared by some but not all species within the group. For example, only C and D have

FIG. 23.7 Synapomorphies, or shared derived characters. Homologies that are present in some, but not all, members of a group help us to construct phylogenetic trees.

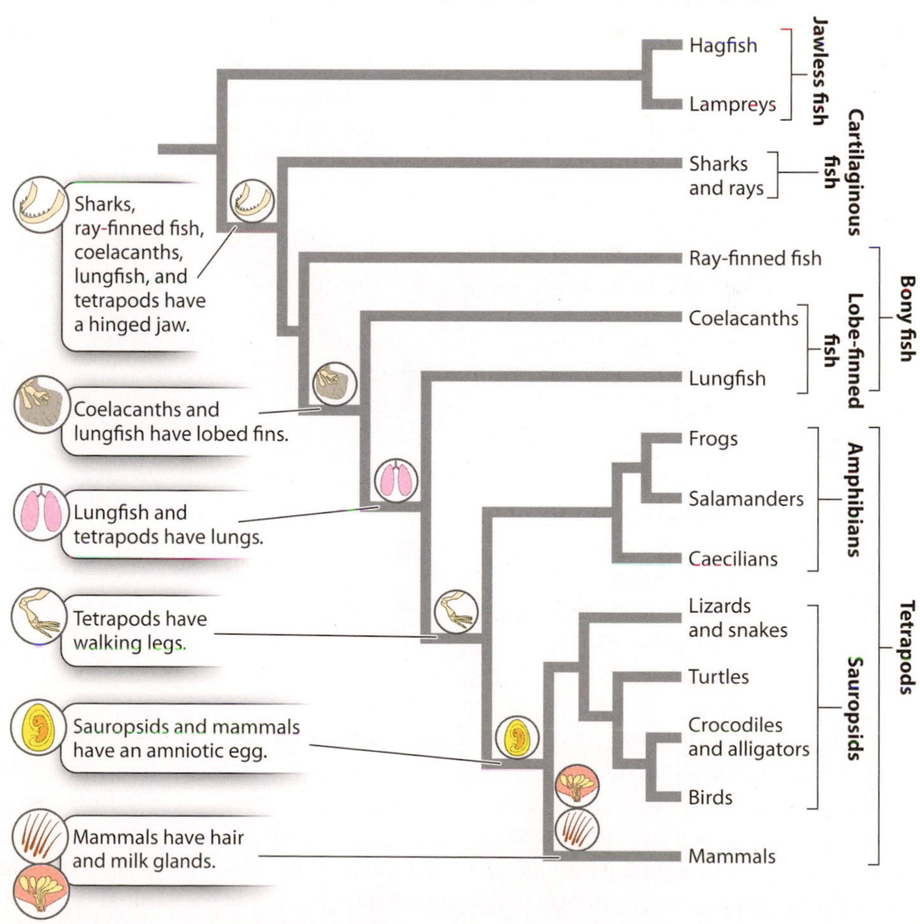

FIG. 23.8 Constructing a phylogenetic tree from shared derived traits. The strongest hypothesis of evolutionary relationships overall is the tree with the fewest number of changes because it minimizes the total number of independent origins of character states.

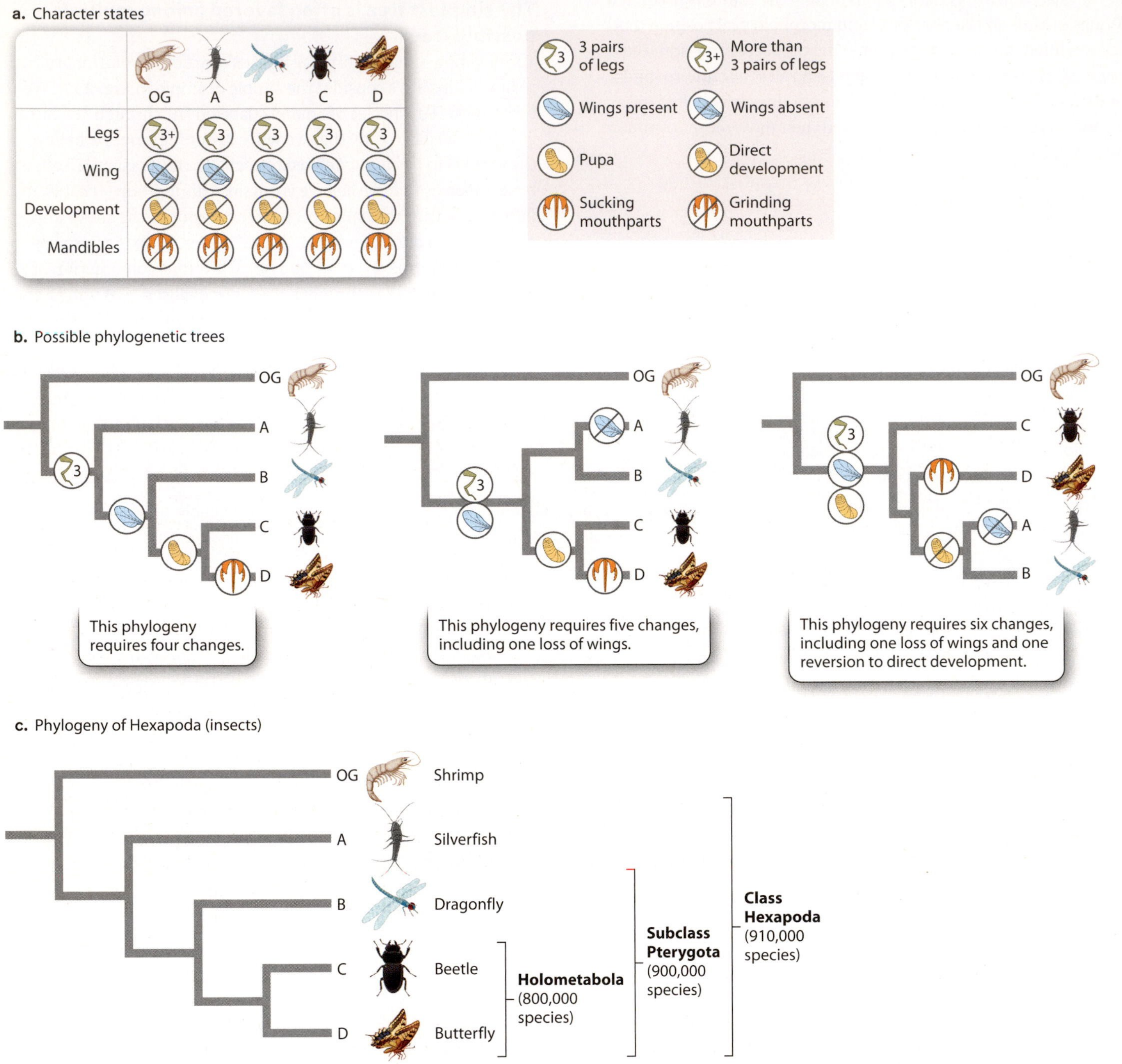

pupae. This character suggests that C and D are more closely related to each other than either is to the other species. The alternatives are that C and D each evolved pupae independently or that pupae were present in the common ancestor of A–D but were lost in A and B.

How do we choose among the alternatives? Studies of the outgroup show that it does not form pupae, supporting the hypothesis that pupal development evolved within the ingroup. In practice, biologists examine multiple characters and choose the phylogenetic hypothesis that best fits all of the data.

How do we determine "best fit"? Fig. 23.8b illustrates three different hypotheses for the relationships among the species based on four characters and their various character states. Each reflects the sister-group relationship between C and D proposed earlier. The leftmost tree requires exactly four character-state changes during the evolution of these species: reduction of leg pairs to three in the group ABCD, wings in the group BCD, and pupae in group CD, plus a change of form of the mandible, or jaw, in species D.

Now consider the middle tree in Fig. 23.8b. It groups A and B together, and so differs from the tree on the left in requiring either a loss of wings in A, or an additional origin of wings in B, independent of that in the common ancestor of C and D—five changes in all. The tree on the right groups A and B together, and requires two extra steps for a total of six steps. No tree that we can construct from species A–D requires fewer than four evolutionary changes, so the left-hand version in Fig. 23.8b is the best available hypothesis of evolutionary relatedness. In fact, this is the phylogeny for a sample of species from the largest group of animals on Earth, the Hexapoda—insects and their closest relatives (Fig. 23.8c).

In general, trees with fewer character changes are preferred to ones that require more because they provide the simplest explanation of the data. This approach is an example of **parsimony,** that is, choosing the simpler of two or more hypotheses to account for a given set of observations. When we use parsimony in phylogenetic reconstruction, we make the implicit assumption that evolutionary change is typically rare. Over time, most features of organisms stay the same—we have the same number of ears as our ancestors, the same number of fingers, and so on. Thus, biologists commonly prefer the phylogenetic tree requiring the fewest evolutionary steps.

In systematics, parsimony suggests counting character changes on a phylogenetic tree to find the simplest tree for the data (the one with the fewest number of changes). Each change corresponds to a mutation (or mutations) in an ancestral species, and the more changes or steps we propose, the more independent mutations we must also hypothesize.

Note also that it isn't necessary to make decisions in advance about which characters are homologies and which are analogies. We can construct all possible trees and then choose the one requiring the fewest evolutionary changes. This is a simple matter for the example in Fig. 23.8 because four species can be arranged into only 15 possible different trees. As the number of groups increases, however, the number of possible trees connecting them increases as well, and dramatically so. There are 105 trees for 5 groups and 945 trees for 6 groups, and there are nearly 2 million possible trees for 10. For 50 groups the possibilities balloon to 3×10^{76}! Clearly, computers are required to sort through all the possibilities.

As is true for all hypotheses, phylogenetic hypotheses can be supported strongly or weakly. Biologists use statistical methods to evaluate a given phylogenetic hypothesis. Available character data may not strongly favor any hypothesis. When support for a specific branching pattern is weak, biologists commonly depict the relationships as unresolved and show multiple groups diverging from one node, rather than just two. Such branching patterns are not meant to suggest that multiple species diverged simultaneously, but rather to indicate that we lack the data to choose unequivocally among several different hypotheses of relationship. In Part 2, we show unresolved branches in a number of groups. These shouldn't be read as admissions of defeat, but instead as problems awaiting resolution and opportunities for future research.

Molecular data complement comparative morphology in reconstructing phylogenetic history.

Trees can be built using anatomical features, but increasingly tree construction relies on molecular data. The amino acids at particular positions in the primary structure of a protein can be used, as can the nucleotides at specific positions along a strand of DNA.

From genealogy to phylogeny, tracing mutations in DNA or RNA sequences has revolutionized the reconstruction of historical genetic connections. Whether we are tracing the paternity of the children of Sally Hemings, mistress of Thomas Jefferson, identifying the origin of a recent cholera epidemic in Haiti, or placing baleen whales near the hippopotamus family in a phylogenetic tree of mammals, molecular data are a rich source of phylogenetic insight.

There is nothing about molecular data that provides a better record of history than does anatomical data; molecular data simply provide more details because there are more characters that can vary among the species. A sequence of DNA with hundreds or thousands of nucleotides can represent that many characters, as opposed to the tens of characters usually visible in morphological studies. Indeed, for microbes and viruses there is very little morphology available, so molecular information is critical for phylogenetic reconstruction. Once a gene or other stretch of DNA or RNA is identified that seems likely (based on previous studies of other species) to vary among the species to be studied, sequences are obtained and aligned to identify homologous nucleotide sites. Analyses of this kind commonly involve comparisons of sequences of about 1000 nucleotides from one or more genes. Increasingly, though, the availability of whole-genome sequences is changing the way we do molecular phylogenetics. Rather than comparing the sequences of a few genes, we compare the sequences of entire genomes.

The process of using molecular data is conceptually similar to the process described earlier for morphological data. Through comparison to an outgroup, we can identify derived and ancestral molecular characters (whether DNA nucleotides or amino acids in proteins) and generate the phylogeny on the basis of synapomorphies as before.

An alternative method of reconstruction is based on overall similarity rather than synapomorphies. Here, the premise is simple: The descendants of a recent common ancestor will have had relatively little time to evolve differences, whereas the descendants of an ancient common ancestor have had a lot of time to evolve differences. Thus, the extent of similarity (or distance) indicates how recently two groups shared a common ancestor.

Underpinning this approach is the assumption that the rate of evolution is constant. (Otherwise, a pair of taxa with a recent common ancestor could be more different than expected because of an unusually fast rate of evolution.) This rate-constancy assumption is less likely to be violated when we are using molecular data than when we are using morphological data. Recall from Chapter 21

FIG. 23.9 Phylogenetic trees of DNA sequences based on (a) synapomorphies and (b) overall similarity.

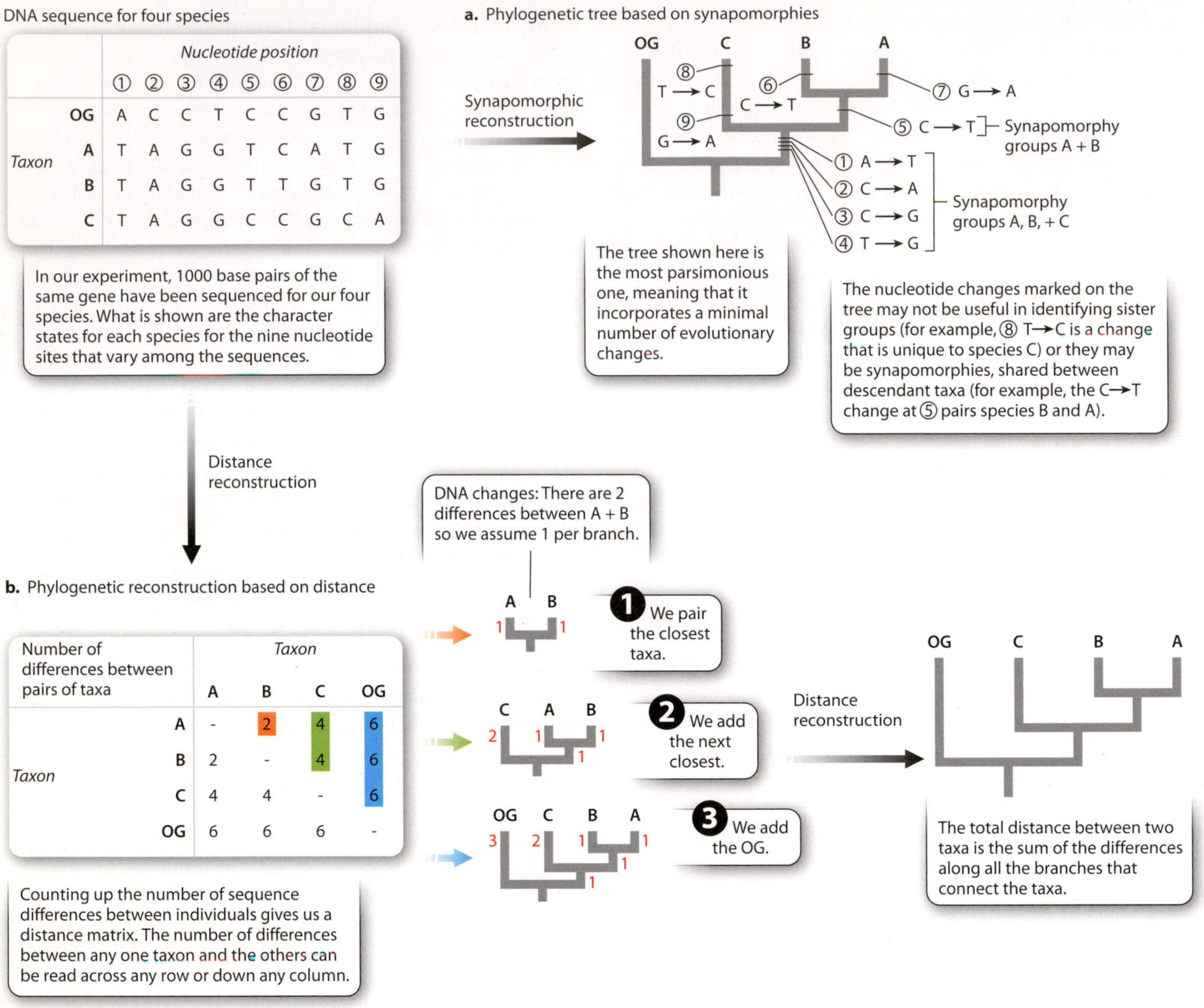

		Nucleotide position								
		①	②	③	④	⑤	⑥	⑦	⑧	⑨
	OG	A	C	C	T	C	C	G	T	G
Taxon	A	T	A	G	G	T	C	A	T	G
	B	T	A	G	G	T	T	G	T	G
	C	T	A	G	G	C	C	G	C	A

Number of differences between pairs of taxa		Taxon			
		A	B	C	OG
Taxon	A	-	2	4	6
	B	2	-	4	6
	C	4	4	-	6
	OG	6	6	6	-

that the molecular clock is based on the observation of constant accumulation of genetic divergence through time. **Fig. 23.9** shows a simple DNA sequence dataset that we can analyze either on the basis of synapomorphies (Fig. 23.9a) or on the basis of distance (Fig. 23.9b). Note that both give the same result in this case.

Molecular data are often combined with morphological data, and each can also serve as an independent assessment of the other. Not surprisingly, results from analyses of each kind of data are commonly compatible, at least for plants and animals rich in morphological characters.

The single largest library of taxonomic information is GenBank, the National Institutes of Health's genetic data storage facility. As of this writing, GenBank gives users access to more than 100 billion observations (mostly nucleotides) collected under more than 430,000 taxonomic names. A growing internet resource is the Encyclopedia of Life, which is gathering additional biological information about species, including ecology, geographic distributions, photographs, and sounds in pages for individual species that are easy to navigate. Another web resource, the Tree of Life, provides information on phylogenetic trees for many groups of organisms.

Phylogenetic trees can help solve practical problems.

The sequence of changes on a tree from its root to its tips documents evolutionary changes that have accumulated through time. Trees suggest which groups are older than others, and which

FIG. 23.10

Did an HIV-positive dentist spread the AIDS virus to his patients?

BACKGROUND In the late 1980s, several patients of a Florida dentist contracted AIDS. Molecular analysis showed that the dentist was HIV-positive.

HYPOTHESIS It was hypothesized that the patients acquired HIV during dental procedures carried out by the infected dentist.

METHOD Researchers obtained two HIV samples each (denoted 1 and 2 in the figure) from several people, including the dentist (Dentist 1 and Dentist 2), several of his patients (Patients A through G), and other HIV-positive individuals chosen at random from the local population (LP). In addition, a strain of HIV from Africa (HIVELI) was included in the analysis.

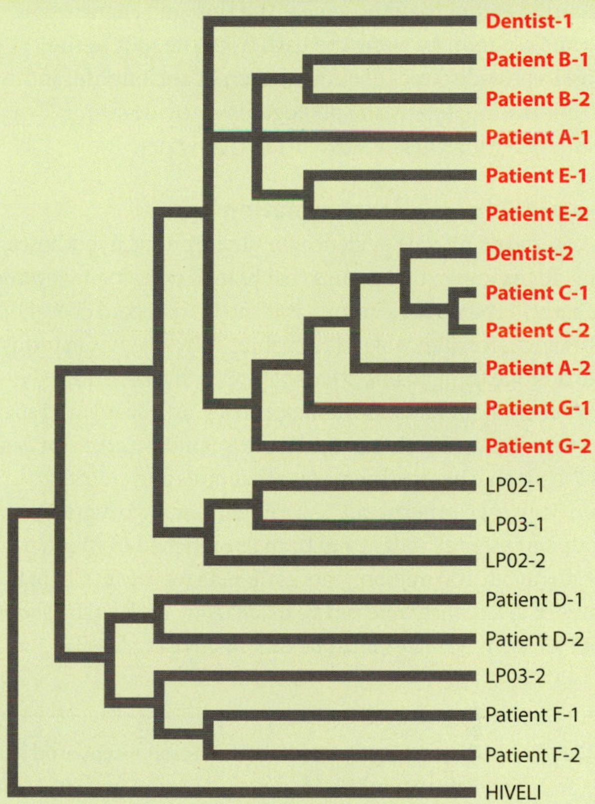

RESULTS Biologists constructed a phylogeny based on the nucleotide sequence of a rapidly evolving gene in the genome of HIV. Because the gene evolves so quickly, its mutations preserve a record of evolutionary relatedness on a very fine scale. HIV in some of the infected patients—patients A, B, C, E, and G—were more similar to the dentist's HIV than they were to samples from other infected individuals. Some patients' sequences, however, did not cluster with the dentist's, suggesting that these patients, D and F, had acquired their HIV infections from other sources.

CONCLUSION HIV phylogeny makes it highly likely that the dentist infected several of his patients. The details of how the patients were infected remain unknown, but rigidly observed safety practices make it unlikely that such a tragedy could occur again.

FOLLOW-UP WORK Phylogenies based on molecular sequence characters are now routinely used to study the origin and spread of infectious diseases, such as swine flu and Ebola.

SOURCE Hillis, D. M., J. P. Huelsenbeck, and C. W. Cunningham. 1994. "Application and Accuracy of Molecular Phylogenies." *Science* 264: 671–677.

traits came first and which followed later. Proper phylogenetic placement thus reveals a great deal about evolutionary history, and it can have practical consequences as well. For example, oomycetes, microorganisms responsible for potato blight and other important diseases of food crops, were long thought to be fungi because they look like some fungal species. The discovery, using molecular characters, that oomycetes belong to a very different group of eukaryotic organisms, has opened up new possibilities for understanding and controlling these plant pathogens. Similarly, in 2006, researchers used DNA sequences to identify the Malaysian parent population of a species of butterfly called lime swallowtails that had become an invasive species in the Dominican Republic, pinpointing the source populations from which natural enemies of this pest can be sought.

Phylogenetics solved a famous case in which an HIV-positive dentist in Florida was accused of infecting his patients (**Fig. 23.10**). HIV nucleotide sequences evolve so rapidly that biologists can build phylogenetic trees that trace the spread of specific strains from one individual to the next. Phylogenetic study of HIV present in samples from several infected patients, the dentist, and other individuals provided evidence that the dentist had, indeed, infected his patients.

Similarly, phylogenetic studies of influenza virus strains show their origins and subsequent movements among geographic regions and individual patients. Today, there is a growing effort to use specific DNA sequences as a kind of fingerprint or barcode for tracking biological material. Such information could quickly identify samples of shipments of meat as being from endangered species, or track newly emerging pests. The Consortium for the Barcode of Life has already accumulated species-specific DNA barcodes for more than 100,000 species. Phylogenetic evidence provides a powerful tool for evolutionary analysis and is useful across timescales ranging from months to the entire history of life, from the rise of epidemics to the origins of metabolic diversity.

23.3 THE FOSSIL RECORD

Phylogenies based on living organisms provide hypotheses about evolutionary history. Branches toward the root of the tree occurred earlier than those near the tips, and characters change and accumulate along the path from the root to the tips. Fossils provide direct documentation of ancient life, and so, in combination, fossils and phylogenies provide strong complementary insights into evolutionary history.

Fossils provide unique information.

Fossils can and do provide evidence for phylogenetic hypotheses, showing, for example, that groups that branch early in phylogenies appear early in the geologic record. But the fossil record does more than this. First, fossils enable us to calibrate phylogenies in terms of time. It is one thing to infer that mammals diverged from the common ancestor of birds, crocodiles, turtles, and lizards and snakes before crocodiles and birds diverged from a common ancestor (see Fig. 23.7), but another matter to state that birds and crocodiles diverged from each other about 220 million years ago, whereas the group represented today by mammals branched from other vertebrates about 100 million years earlier. As we saw in Chapter 21, estimates of divergence time can be made using molecular sequence data, but all such estimates must be calibrated using fossils.

The evolutionary relationship between birds and crocodiles highlights a second kind of information provided by fossils. Not only do fossils record past life, they also provide our only record of extinct species. The phylogeny in Fig. 23.7 contains a great deal of information, but it is silent about dinosaurs. Fossils demonstrate that dinosaurs once roamed Earth, and details of skeletal structure place birds among the dinosaurs in the vertebrate tree. Indeed, some remarkable fossils from China show that the dinosaurs most closely related to birds had feathers (**Fig. 23.11**).

A third, and also unique, contribution of fossils is that they place evolutionary events in the context of Earth's dynamic environmental history. Again, dinosaurs illustrate the point. As discussed in Chapter 1, geologic evidence from several continents suggests that a large meteorite triggered drastic changes in the global environment 66 million years ago, leading to the extinction of dinosaurs (other than birds). In fact, at five times in the past, large environmental disturbances sharply decreased Earth's biological diversity. These events, called mass extinctions, have played a major role in shaping the course of evolution.

Fossils provide a selective record of past life.

Fossils are the remains of once-living organisms, preserved through time in sedimentary rocks. If we wish to use fossils to complement phylogenies based on modern organisms, we must understand how fossils form and how the processes of formation govern what is and is not preserved.

For all its merits, the fossil record should not be thought of as a complete dictionary of everything that ever walked, crawled, or swam across our planet's surface. Fossilization requires burial, as when a clam dies on the seafloor and is quickly covered by sand, or a leaf falls to the forest floor and ensuing floods cover it in mud. Through time, accumulating sediments harden into sedimentary

FIG. 23.11 ***Microraptor gui*, a remarkable fossil discovered in approximately 125-million-year-old rocks from China.** The structure of its skeleton identifies *M. gui* as a dinosaur, yet it had feathers on its arms, tail, and legs. *Source: Nature/Xing Xu/Getty Images.*

FIG. 23.12 **The Grand Canyon.** Erosion has exposed layers of sedimentary rock that record Earth history. *Source: National Park Service.*

rocks such as those exposed so dramatically in the walls of the Grand Canyon (**Fig. 23.12**). If they are not buried, the remains of organisms are eventually recycled by biological and physical processes, and no fossil forms. In general, the fossil record of marine life is more completely sampled than that for land-dwelling creatures because marine habitats are more likely than those on land to be places where sediments accumulate and become rock. Thus, trees and elks living high in the Rocky Mountains have a low probability of fossilization, whereas clams and corals on the shallow seafloor are commonly buried and become fossils.

Biological factors contribute to the incompleteness of the fossil record. Most fossils preserve the hard parts of organisms, those features that resist decay after death. For animals, this usually means mineralized skeletons. Clams and snails that secrete shells of calcium carbonate have excellent fossil records. More than 80% of the clam species found today along California's coast also occur as fossils in sediments deposited during the past million years. In contrast, nematodes, tiny worms that may be the most abundant animals on Earth, have no mineralized skeletons and almost no fossil record. The wood and pollen of plants, which are made in part of decay-resistant organic compounds, enter the fossil record far more commonly than do flowers. And, among unicellular organisms, the skeleton-forming diatoms, radiolarians, and foraminiferans have exceptionally good fossil records, whereas most amoebas are unrepresented.

Together, then, the properties of organisms (do they make skeletons or other features that resist decay after death?) and environment (did the organisms live in a place where burial was likely?) determine the probability that an ancient species will be represented in the fossil record. Fortunately, for species that make mineralized skeletons and live on the shallow seafloor, the fossil record is very good, preserving a detailed history of evolution through time.

Organisms that lack hard parts can leave a fossil record in two other distinctive ways. Many animals leave tracks and trails as they move about or burrow into sediments. These **trace fossils,** from dinosaur tracks to the feeding trails of snails and trilobites, preserve a record of both anatomy and behavior (**Fig. 23.13**).

Organisms can also contribute **molecular fossils** to the rocks. Most biomolecules decay quickly after death. Proteins and DNA, for example, generally break down before they can be preserved, although, remarkably, a sizable fraction of the Neanderthal genome has been pieced together from DNA in 40,000-year-old

FIG. 23.13 **Trace fossils.** These footprints in 150-million-year-old rocks record both the structure and behavior of the dinosaurs that made them. *Source: José Antonio Hernaiz/ age fotostock.*

bones. Other molecules, especially lipids like cholesterol, are more resistant to decomposition. Sterols, bacterial lipids, and some pigment molecules can accumulate in sedimentary rocks, documenting organisms that rarely form conventional fossils, especially bacteria and single-celled eukaryotes.

Rarely, unusual conditions preserve fossils of unexpected quality, including animals without shells, delicate flowers or mushrooms, fragile seaweeds or bacteria, even the embryos of plants and animals. For example, 505 million years ago, during the Cambrian Period, a sedimentary rock formation called the **Burgess Shale** accumulated on a relatively deep seafloor covering what is now British Columbia. Waters just above the basin floor contained little or no oxygen, so that when mud swept into the basin, entombed animals were sealed off from scavengers, disruptive burrowing activity, and even bacterial decay. For this reason, Burgess rocks preserve a remarkable sampling of marine life during the initial diversification of animals (**Fig. 23.14**).

FIG. 23.14 **A Burgess Shale fossil.** This fossil is *Opabinia regalis,* an extinct early relative of the arthropods, from the 505-million-year-old Burgess Shale. *Source: Courtesy of Smithsonian Institution. Photo by Chip Clark.*

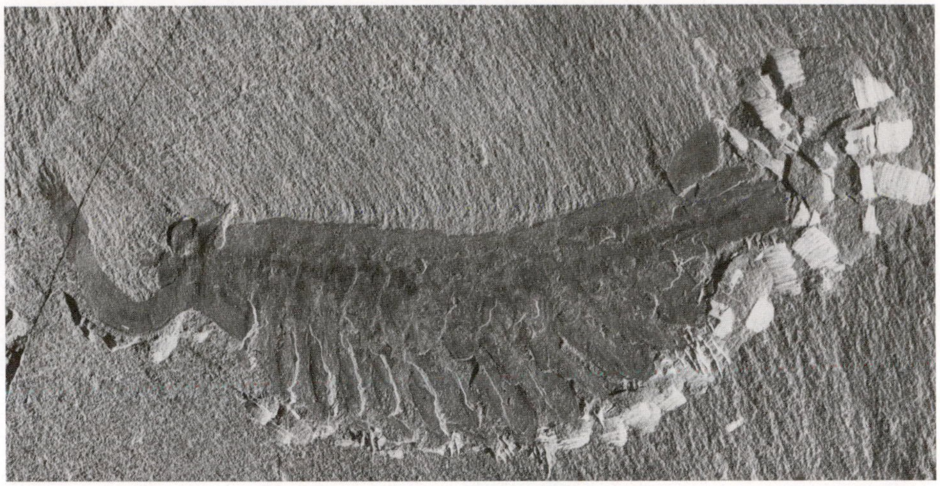

The **Messel Shale** formed more recently—about 50 million years ago—in a lake in what is now Germany. Release of toxic gases from deep within the Messel lake suffocated local animals, and their carcasses settled into oxygen-poor muds on the lake floor. Fish, birds, mammals, and reptiles are preserved as complete and articulated skeletons, and mammals retain impressions of fur and color patterning (**Fig. 23.15**). Plants were also beautifully preserved, as were insects, some with a striking iridescence still intact. Messel rocks provide a truly outstanding snapshot of life on land as the age of mammals began.

FIG. 23.15 **A Messel Shale fossil.** Even the furry tail of this extinct squirrel-like mammal is preserved in this remarkable 50-million-year-old fossil.

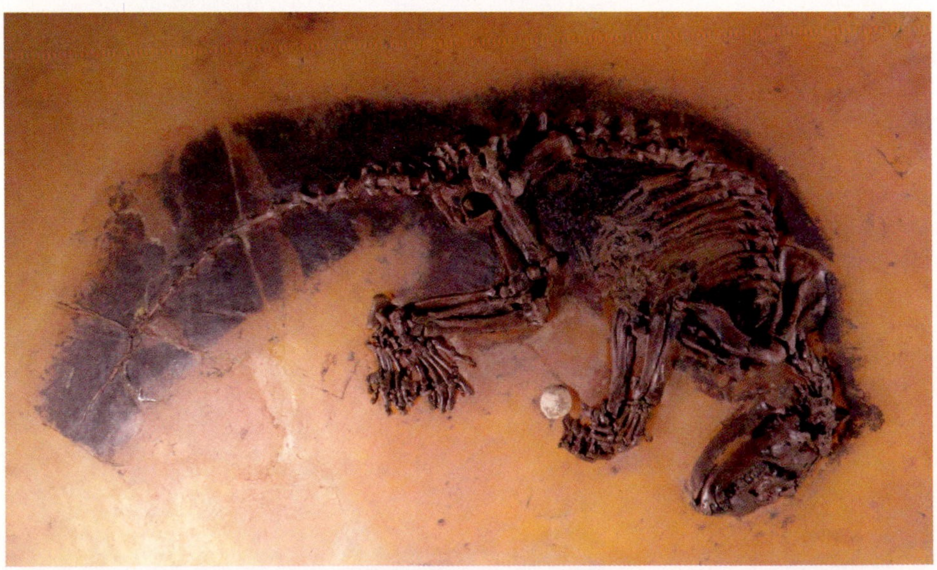

In general, the fossil record preserves some aspects of biological history well and others poorly. Fossils provide a good sense of how the forms, functions, and diversity of skeletonized animals have changed over the past 500 million years. The same is true for land plants and unicellular organisms that form mineralized skeletons. These fossils shed light on major patterns of morphological evolution and diversity change through time; their geographic distributions record the movements of continents over millions of years; and the radiations and extinctions they document show how life responds to environmental change, both gradual and catastrophic.

Geological data indicate the age and environmental setting of fossils.

How do we know the age of a fossil? Beginning in the nineteenth century, geologists recognized that groups of fossils change systematically from the bottom of a sedimentary rock formation to its top. As more of Earth's surface was mapped and studied, it became clear that certain fossils always occur in layers that lie beneath (and so are older than) layers that contain other species. From these patterns, geologists concluded that fossils mark time in Earth history. At first, geologists didn't know why fossils changed from one bed to the next, but after Darwin the reason became apparent:

Fossils record the evolution of life on Earth. They eventually mapped out the **geologic timescale,** the series of time divisions that mark Earth's long history (**Fig. 23.16**).

The layers of fossils in sedimentary rocks can tell us that some rocks are older than others, but they cannot by themselves provide an absolute age. Calibration of the timescale became possible with the discovery of radioactive decay. In Chapter 2, we discussed isotopes, variants of an element that differ from one another in the number of neutrons they contain. Many isotopes are unstable and spontaneously break down to form other, more stable isotopes. In the laboratory, scientists can measure how fast unstable isotopes decay. Then, by measuring the amounts of the

FIG. 23.16 The geologic timescale, showing major events in the history of life on Earth. *Sources: (top, left to right) Scott Orr/iStockphoto; Hans Steur, The Netherlands; T. Daeschler/VIREO; Reconstruction artwork: Mark A. Klingler/Carnegie Museum of Natural History, from the cover of Science, 25 MAY 2001 VOL 292, ISSUE 5521, Reprinted with permission from AAAS; dimair/Shutterstock; DEA/G. Cigolini/Getty Images; (bottom, left to right) Eye of Science/Science Source; Andrew Knoll, Harvard University; Antonio Guillén, Proyecto Agua, Spain; Andrew Knoll, Harvard University.*

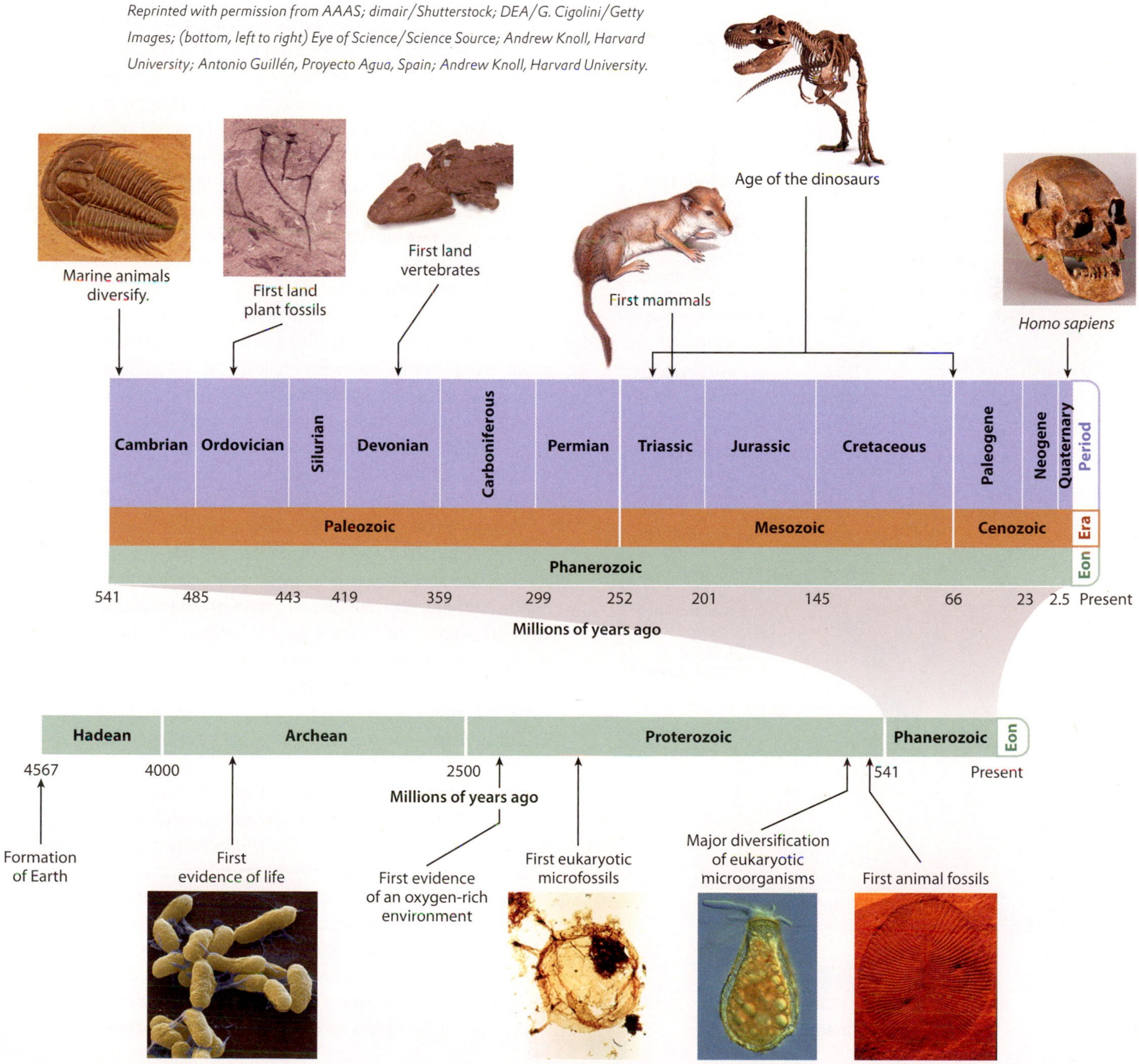

FIG. 23.17 **^{14}C decay.** Scientists can determine the age of relatively young materials such as wood and bone from the amount of ^{14}C they contain.

unstable isotope and its stable daughter inside a mineral, they can determine when the mineral formed.

Archaeologists commonly use the radioactive decay of the isotope carbon-14, or ^{14}C, to date wood and bone, a process called **radiometric dating.** As shown in **Fig. 23.17,** cosmic rays continually generate ^{14}C in the atmosphere, much of which is incorporated into atmospheric carbon dioxide. Through photosynthesis, carbon dioxide that contains ^{14}C is incorporated into wood, and animals incorporate small amounts of ^{14}C into their tissues when they eat plant material. After the organism's death, the unstable ^{14}C in these tissues begins to break down, losing an electron to form ^{14}N, a stable isotope of nitrogen. Laboratory measurements indicate that half of the ^{14}C in a given sample will decay to nitrogen in 5730 years, a period called its **half-life** (Fig. 23.17). Armed with this information, scientists can measure the amount of ^{14}C in an archaeological sample and, by comparing it to the amount of ^{14}C in a sample of known age—annual rings in trees, for example, or yearly growth of coral skeletons—determine the age of the sample.

Because its half-life is so short (by geological standards), ^{14}C is useful only in dating materials younger than 50,000 to 60,000 years. Beyond that, there is too little ^{14}C left to measure accurately. Older geological materials are commonly dated using the radioactive decay of uranium (U) to lead (Pb): ^{238}U, incorporated in trace amounts into the minerals of volcanic rocks, breaks down to ^{206}Pb with a half-life of 4.47 billion years; ^{235}U decays to ^{207}Pb with a half-life of 704 million years. Calibration of the geologic timescale is based mostly on the ages of volcanic ash interbedded with sedimentary rocks that contain key fossils, as well as volcanic rocks that intrude into (and so are younger than) layers of rock containing fossils. In turn, the ages of fossils provide calibration points for phylogenies.

The sedimentary rocks that contain fossils also preserve, encrypted in their physical features and chemical composition, information about the environment in which they formed. Sandstone beds, for example, may have rippled surfaces, like the ripples produced by currents that we see today in the sand of a seashore or lake margin. Pyrite (FeS_2), or fool's gold, forms when H_2S generated by anaerobic bacteria reacts with iron. As these conditions generally occur where oxygen is absent, pyrite enrichment in ancient sedimentary rocks can signal oxygen depletion.

We might think our moment in geologic time is representative of Earth as it has always existed, but nothing could be further from the truth. In the location and sizes of its continents, ocean chemistry, and atmospheric composition, the Earth we experience is unlike any previous state of the planet. Today, for example, the continents are distributed widely over the planet's surface, but 290 million years ago they were clustered in a supercontinent called Pangaea (**Fig. 23.18**). Oxygen gas permeates most surface environments of Earth today, but 3 billion years ago, there was no O_2 anywhere. And,

FIG. 23.18 **Pangaea, 290 million years ago.** Plate tectonics has shaped and reshaped Earth's geography through time. The white area near the South Pole is glacial ice. *Source: © Ron Blakey and Colorado Plateau Geosystems, Inc.*

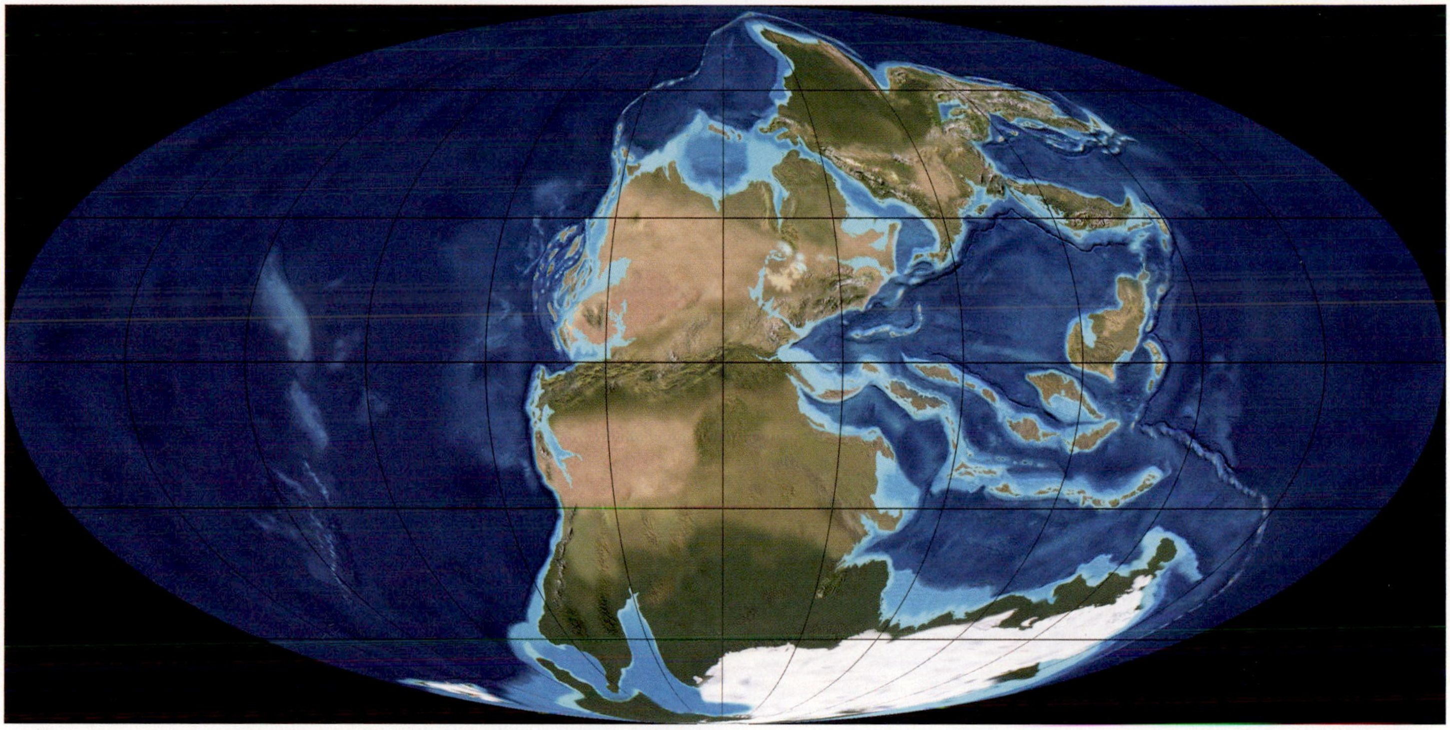

just 20,000 years ago, 2 km of glacial ice stood where Boston lies today. Sedimentary rocks record the changing state of Earth's surface over billions of years and show that life and environment have changed together through time, each influencing the other.

Fossils can contain unique combinations of characters.

Phylogenies hypothesize impressive morphological and physiological shifts through time—amphibians from fish, for example, or land plants from green algae. Do fossils capture a record of these transitions as they took place?

Let's begin with an example introduced earlier in this chapter. Phylogenies based on living organisms generally place birds as the sister group to crocodiles and alligators, but birds and crocodiles are decidedly different from each other in structure—birds have wings, feathers, toothless bills, and a number of other skeletal features distinct from those of crocodiles. In 1861, just two years after publication of *On the Origin of Species,* German quarry workers discovered a remarkable fossil that remains paleontology's most famous example of a transitional form. *Archaeopteryx lithographica,* now known from 11 specimens splayed for all time in fine-grained limestone, lived 150 million years ago. Its skeleton shares many characters with dromaeosaurs, a group of small, agile dinosaurs, but several features—its pelvis, its braincase, and, especially, its winglike forearms—are distinctly birdlike. Spectacularly, the fossils preserve evidence of feathers. *Archaeopteryx* clearly suggests a close relationship between birds and dinosaurs, and phylogenetic reconstructions that include information from fossils show that many of the characters found today in birds accumulated through time in their dinosaur ancestors. As noted earlier, even feathers first evolved in dinosaurs (**Fig. 23.19**).

Tiktaalik roseae and other skeletons in rocks deposited 375 to 362 million years ago record an earlier but equally fundamental transition: the colonization of land by vertebrates. Phylogenies show that all land vertebrates, from amphibians to mammals, are descended from fish. As seen in **Fig. 23.20,** *Tiktaalik* had fins, gills, and scales like other fish of its day, but its skull was flattened, more like that of a crocodile than a fish, and it had a functional neck and ribs that could support its body—features today found only in tetrapods. Along with other fossils, *Tiktaalik* captures key moments in the evolutionary transition from water to land, confirming the predictions of phylogeny.

→ **Quick Check 4** You have just found a novel vertebrate skeleton in 200-million-year-old rocks. How would you integrate this new fossil species into the phylogenetic tree depicted in Fig. 23.7?

FIG. 23.19 Dinosaurs and birds. A number of dinosaur fossils link birds phylogenetically to their closest living relatives, the crocodiles. The fossil at the bottom is *Archaeopteryx*. *Data from C. Zimmer, 2009, The Tangled Bank, Greenwood Village, CO: Roberts and Company; (photo) Jason Edwards/ Getty Images.*

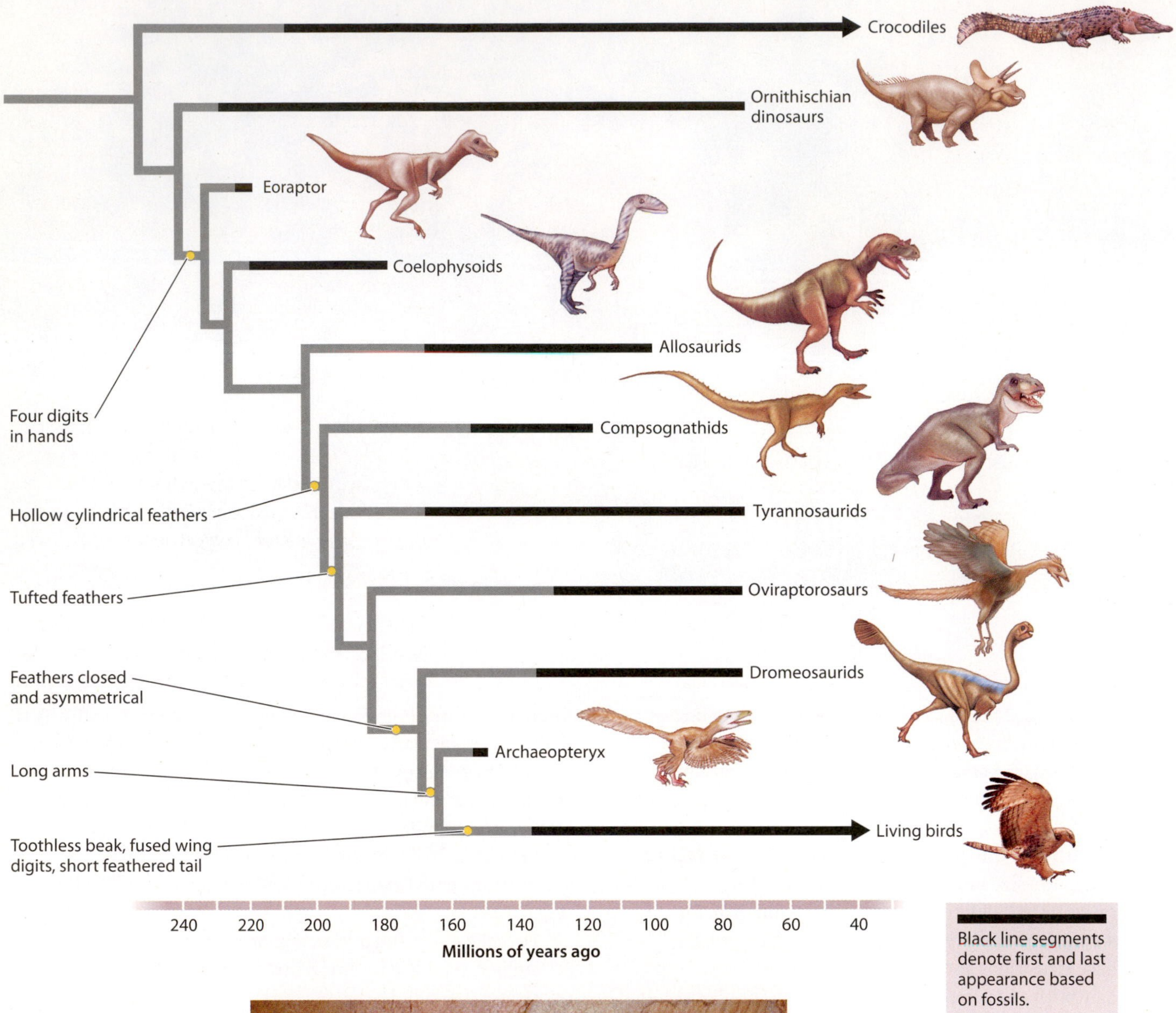

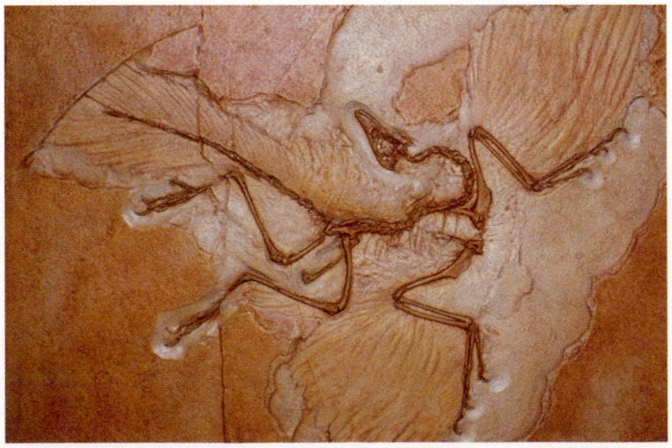

FIG. 23.20

Can fossils bridge the evolutionary gap between fish and tetrapod vertebrates?

BACKGROUND Phylogenies based on both morphological and molecular characters indicate that fish are the closest relatives of four-legged land vertebrates.

HYPOTHESIS Land vertebrates evolved from fish by modifications of the skeleton and internal organs that made it possible for them to live on land.

OBSERVATION Fossil skeletons 390 to 360 million years old show a mix of features seen in living fish and amphibians. Older fossils have fins, fishlike heads, and gills, and younger fossils have weight-bearing legs, skulls with jaws able to grab prey, and ribs that help ventilate lungs. Paleontologists predicted that key intermediate fossils would be preserved in 380–370-million-year-old rocks. In 2004, Edward Daeschler, Neil Shubin, and Farish Jenkins discovered *Tiktaalik,* a remarkable fossil that has fins, scales, and gills like fish, but wrist bones and fingers, an amphibian-like skull, and a true neck (which fish lack).

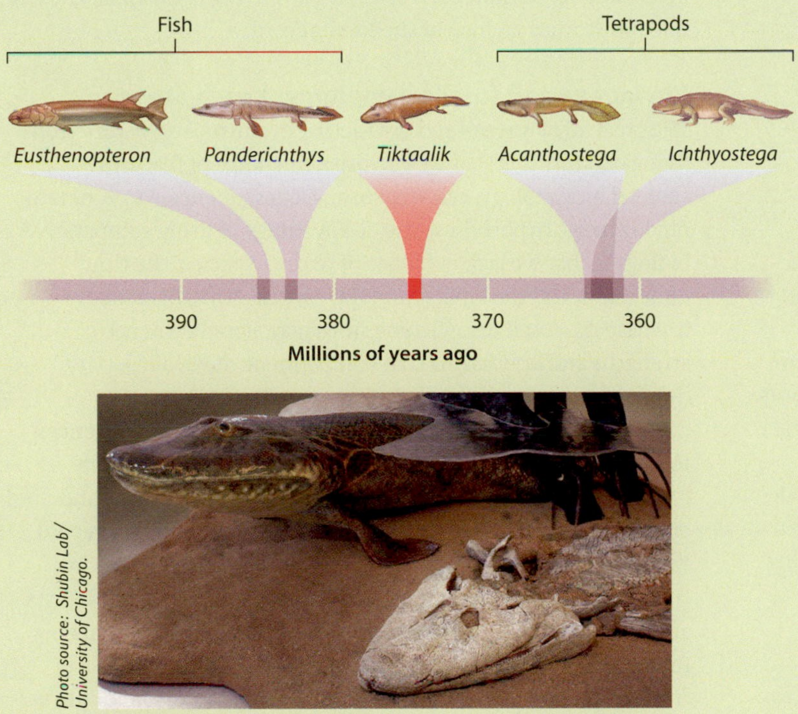

Photo source: Shubin Lab/ University of Chicago.

CONCLUSION Fossils confirm the phylogenetic prediction that tetrapod vertebrates evolved from fish by the developmental modification of limbs, skulls, and other features.

FOLLOW-UP WORK Research into the genetics of vertebrate development shows that the limbs of fish and amphibians are shaped by similar patterns of gene expression, providing further support for the phylogenetic connection between the two groups.

SOURCE Daeschler, E. B., N. H. Shubin, and F. A. Jenkins, Jr. 2006. "A Devonian Tetrapod-like Fish and the Evolution of the Tetrapod Body Plan." *Nature* 440:757–763.

Rare mass extinctions have altered the course of evolution.

Beginning in the 1970s, American paleontologist Jack Sepkoski scoured the paleontological literature, recording the first and last appearances in the geologic record for every genus of marine animals he could find. The results of this monumental effort are shown in **Fig. 23.21.** Sepkoski's diagram shows that animal diversity has increased over the past 540 million years: The biological diversity of animals in today's oceans may well be higher than it has ever been in the past. But it is also clear that animal evolution is not simply a history of gradual accumulation.

Repeatedly during the past 500 million years, animal diversity in the oceans dropped both rapidly and substantially, and extinctions also occurred on land. Known as **mass extinctions,** these events eliminated ecologically important taxa and thereby provided evolutionary opportunities for the survivors. In Chapter 22, we explored how species often radiate on islands where there is little or no competition or predation. For survivors, mass extinctions provided ecological opportunities on a grand scale.

The best-known mass extinction occurred 66 million years ago, at the end of the Cretaceous Period (Fig. 23.21). On land, dinosaurs disappeared abruptly, following more than 150 million years of dominance in terrestrial ecosystems. In the oceans, ammonites, cephalopod mollusks that had long been abundant predators, also became extinct, and most skeleton-forming microorganisms in the oceans disappeared as well. As discussed in Chapter 1, a large body of geologic evidence supports the hypothesis that this biological catastrophe was caused by the impact of a giant meteorite. By eliminating dinosaurs and much more of the biological diversity that had built up over millions of years on land, the mass extinction at the end of the Cretaceous Period also had the closely related effect of generating new evolutionary possibilities for terrestrial animals that survived the event, including mammals.

Before dinosaurs ever walked on Earth, the greatest of all mass extinctions occurred 252 million years ago, at the end of the Permian Period, when environmental catastrophe eliminated half of all families in the oceans and about 80% of all genera. Estimates of marine species loss run as high as 90%. Geologists have hypothesized that this mass extinction resulted from the catastrophic effects

FIG. 23.21 Mass extinctions. Though they eliminate much of the life found on Earth at the time, mass extinctions allow new groups to proliferate and diversify. *Source: Sepkoski's Online Genus Database. http://strata.geology.wisc.edu/jack/. Data from J. J. Sepkoski, Jr., 2002, "A Compendium of Fossil Marine Animal Genera," David Jablonski and Michael Foote (eds.), Bulletin of American Paleontology 363:1–560.*

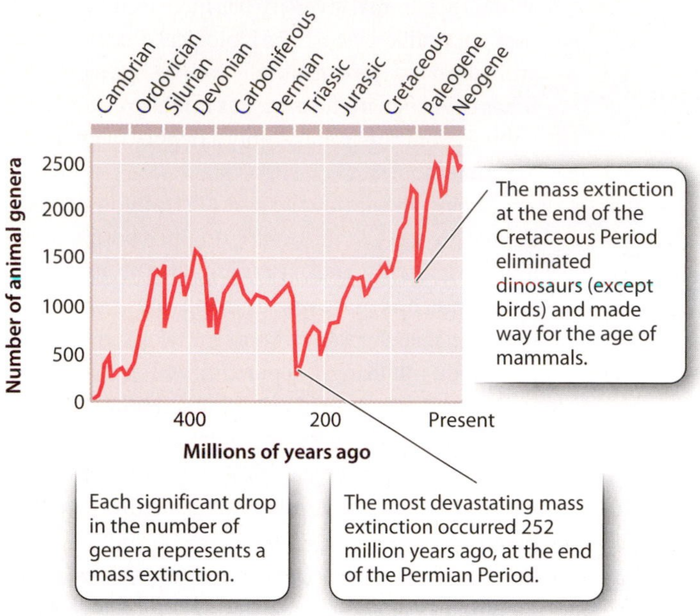

of massive volcanic eruptions. At the end of the Permian Period, most continents were gathered into the supercontinent Pangaea (see Fig. 23.18), and a huge ocean covered more than half of the Earth. Levels of oxygen in the deep waters of this ocean were low, the result of sluggish circulation and warm seawater temperatures. Then, a massive outpouring of ash and lava—a million times larger than any volcanic eruption experienced by humans—erupted across what is now Siberia. Enormous emissions of carbon dioxide and methane from the volcanoes caused global warming (so even less oxygen reached deep oceans) and ocean acidification (making it difficult for animals and algae to secrete calcium carbonate skeletons).

The three-way insult of lack of oxygen, ocean acidification, and global warming doomed many species on land and in the seas. Seascapes dominated for 200 million years by corals and shelled invertebrates called brachiopods disappeared. As ecosystems recovered from this mass extinction, they came to be dominated by new groups descended from survivors of the extinction. Bivalves and gastropods diversified, new groups of arthropods radiated, including the ancestors of the crabs and shrimps we see today, and surviving sea anemones evolved a new capacity to make skeletons of calcium carbonate, resulting in the corals that build modern reefs. In short, mass extinction reset the course of evolution, much as it did and the end of the Cretaceous Period.

When you stroll through a zoo or snorkel above a coral reef, you are not simply seeing the products of natural selection played out over Earth history. You are encountering the descendants of Earth's biological survivors. Current biological diversity reflects the interplay through time of natural selection and rare massive perturbations to ecosystems on land and in the sea. The fossil record provides a silent witness to our planet's long and complex evolutionary history.

23.4 COMPARING EVOLUTION'S TWO GREAT PATTERNS

The diversity of life we see today is the result of evolutionary processes playing out over geologic time. Evolutionary process can be studied by experiment, both in the field and in the laboratory, but evolutionary history is another matter. There is no experiment we can do to determine why the dinosaurs became extinct—we cannot rerun the events of 66 million years ago, this time without the meteorite impact. The history of life must be reconstructed from evolution's two great patterns: the nested similarity observed in the forms and molecular sequences of living organisms, and the direct historical archive of the fossil record.

Phylogeny and fossils complement each other.

The great advantage of reconstructing evolutionary history from living organisms is that we can use a full range of features—skeletal morphology, cell structure, DNA sequence—to generate phylogenetic hypotheses. The disadvantage of using comparative biology is that we lack evidence of extinct species, the time dimension, and the environmental context. This, of course, is where the fossil record comes into play. Fossil evidence has strengths and limitations that complement the evolutionary information in the living organisms.

We can use phylogenetic methods based on DNA sequences to infer that birds and crocodiles are closely related, but only fossils can show that the evolutionary link between birds and crocodiles runs through dinosaurs. And only the geologic record can show that mass extinction removed the dinosaurs, paving the way for the emergence of modern mammals. Paleontologists and biologists work together to understand evolutionary history. Biology provides a functional and phylogenetic framework for the interpretation of fossils, and fossils provide a record of life's history in the context of continual planetary change.

Agreement between phylogenies and the fossil record provides strong evidence of evolution.

Phylogenies based on morphological or molecular comparisons of living organisms make hypotheses about the timing of evolutionary changes through Earth history. We humans are a case in point. As we discuss in Chapter 24, comparisons of DNA sequences suggest that chimpanzees are our closest living relatives. This is hardly surprising, as simple observation shows that chimpanzees and humans share many features. However, among other differences, chimpanzees have smaller stature, smaller brains, long arms that facilitate knuckle-walking (the arms help support the body as the chimpanzee moves forward), a more prominent snout, and larger teeth.

How did these differences accumulate over the 7 million years or so since the human and chimpanzee lineages split? Fossils painstakingly unearthed over the past century show, for example, that the ability to walk upright evolved before the large human brain. Lucy, a famous specimen of *Australopithecus afarensis* dating to about 3.2 million years ago, was fully bipedal but had a brain the size of a chimpanzee's. Fossils like Lucy that are a mix of ancestral (brain size) and derived (bipedalism) characters provide strong support for our phylogenetic tree that shows that humans and chimpanzees are sister taxa.

The agreement of comparative biology and the fossil record can be seen at all scales of observation. Humans form one tip of a larger branch that contains all members of the primate family. Primates, in turn, are nested within a larger branch occupied by mammals, and mammals nest within a still larger branch containing all vertebrate animals, which include fish. This arrangement predicts that the earliest fossil fish should be older than the earliest fossil mammals, the earliest fossil mammals older than the earliest primates, and the earliest primates older than the earliest humans. This is precisely what the fossil record shows (**Fig. 23.22**).

FIG. 23.22 **Phylogeny and fossils of vertebrates.** The branching order of the phylogeny corresponds to the order of appearance of each group in the fossil record.

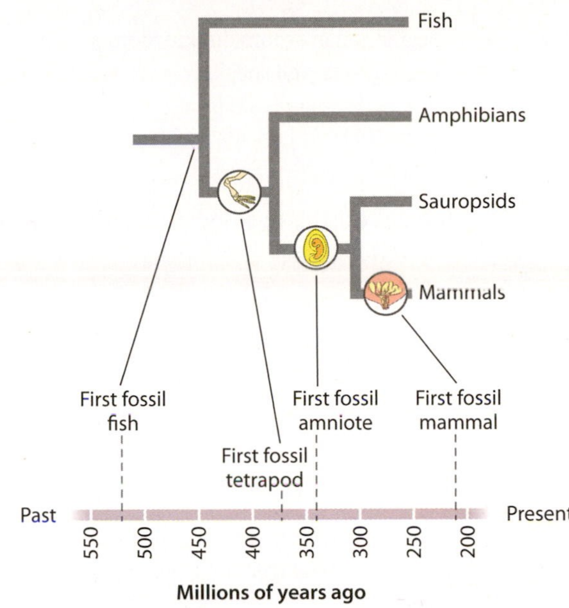

The agreement between fossils and phylogenies can be seen again and again when we examine different branches of the tree of life or, for that matter, the tree as a whole. All phylogenies indicate that microorganisms diverged early in evolutionary history, and mammals, flowering plants, and other large complex organisms diverged more recently. The tree's shape implies that diversity has accumulated through time, beginning with simple organisms and later adding complex macroscopic forms.

The geologic record shows the same pattern. For nearly 3 billion years of Earth history, microorganisms dominate the fossil record, with the earliest animals appearing about 600 million years ago, the earliest vertebrate animals 520 million years ago, the earliest tetrapod vertebrates about 360 million years ago, the earliest mammals 210 million years ago, the earliest primates perhaps 55 million years ago, the earliest fossils of our own species a mere 200,000 years ago. Similarly, if we focus on photosynthetic organisms, we find a record of photosynthetic bacteria beginning at least 3500 million years ago, algae 1200 million years ago, simple land plants 470 million years ago, seed plants 370 million years ago, flowering plants about 140 million years ago, and the earliest grasses 70 million years ago.

In Part 2, we explore the evolutionary history of life in some detail. Here, it is sufficient to draw the key general conclusion: The fact that comparative biology and fossils, two complementary but independent approaches to reconstructing the evolutionary past, yield the same history is powerful evidence of evolution. ■

Core Concepts Summary

23.1 A phylogenetic tree is a reasoned hypothesis of the evolutionary relationships among organisms.

The nested pattern of similarities seen among organisms is a result of descent with modification and can be represented as a phylogenetic tree. page 465

The order of branches on a phylogenetic tree indicates the sequence of events in time. page 465

Sister groups are more closely related to one another than they are to any other group. page 465

A node is a branching point on a tree, and it can be rotated without changing evolutionary relationships. page 466

A monophyletic group includes all the descendants of a common ancestor, and it is considered a natural grouping of organisms based on shared ancestry. page 466

A paraphyletic group includes some, but not all, of the descendants of a common ancestor. page 466

A polyphyletic group includes organisms from distinct groups based on shared characters, but it does not include a common ancestor. page 466

Organisms are classified into domain, kingdom, phylum, class, order, family, genus, and species. page 467

23.2 A phylogenetic tree is built on the basis of shared derived characters.

Characters, or traits, existing in different states are used to build phylogenetic trees. page 468

Homologies are similarities based on shared ancestry, while analogies are similarities based on independent adaptations. page 468

Homologies can be ancestral, unique to a particular group, or present in some, but not all, of the descendants of a common ancestor (shared derived characters). page 468

Only shared derived characters, or synapomorphies, are useful in constructing a phylogenetic tree. page 469

Molecular data provide a wealth of characters that complement other types of information in building phylogenetic trees. page 471

Phylogenetic trees can be used to understand evolutionary relationships of organisms and solve practical problems, such as how viruses evolve over time. page 473

23.3 The fossil record provides direct evidence of evolutionary history.

Fossils are the remains of organisms preserved in sedimentary rocks. page 474

The fossil record is imperfect because fossilization requires burial in sediment, sediments accumulate episodically and discontinuously, and fossils typically preserve only the hard parts of organisms. page 474

Radioactive decay of certain isotopes of elements provides a means of dating rocks. page 477

Archaeopteryx and *Tiktaalik* are two fossil organisms that document, respectively, the bird–dinosaur transition and the fish–tetrapod transition. page 479

The history of life is characterized by five mass extinctions that changed the course of evolution. page 481

The extinction at the end of the Cretaceous Period 66 million years ago led to the extinction of the dinosaurs (other than birds). page 481

The extinction at the end of the Permian Period 252 million years ago is the largest documented mass extinction in the history of Earth. page 481

23.4 Phylogeny and fossils provide independent and corroborating evidence of evolution.

Phylogeny makes use of living organisms, and the fossil record supplies a record of species that no longer exist, absolute dates, and environmental context. page 482

Data from phylogeny and fossils are often in agreement, providing strong evidence for evolution. page 482

Self-Assessment

1. Draw a phylogenetic tree of three groups of organisms and explain how a nested pattern of similarity can be seen in the tree and how it might arise.
2. Distinguish among monophyletic, paraphyletic, and polyphyletic groups, and give an example of each.
3. List the levels of classification, from the least inclusive (species) to the most inclusive (domain).
4. Define "homology" and "analogy" and describe two traits that are homologous and two that are analogous.
5. Name a type of homology that is useful in building phylogenetic trees and explain why this kind of homology, and not others, is useful.
6. Describe three ways that an organism can leave a record in sedimentary rocks and explain why this means that there are gaps in the fossil record.
7. Explain how the fossil record can be used to determine both the relative and the absolute timescales of past events.
8. Describe the significance of *Archeopteryx* and *Tiktaalik*.
9. Describe how mass extinctions have shaped the ecological landscape.

Log in to LaunchPad to check your answers to the Self-Assessment questions, and to access additional learning tools.

Brad Wilson / Getty Images.

CHAPTER 24

Human Origins and Evolution

Core Concepts

24.1 Anatomical, molecular, and fossil evidence shows that the human lineage branches off the great apes tree.

24.2 Phylogenetic analysis of mitochondrial DNA and the Y chromosome shows that our species arose in Africa.

24.3 During the 5–7 million years since the most recent common ancestor of humans and chimpanzees, our lineage acquired a number of distinctive features.

24.4 Human history has had an important impact on patterns of genetic variation in our species.

24.5 Culture is a potent force for change in modern humans.

Charles Darwin carefully avoided discussing the evolution of our own species in *On the Origin of Species*. Instead, he wrote only that he saw "open fields for far more important researches," and that "[l]ight will be thrown on the origin of man and his history." Darwin, an instinctively cautious man, realized that the ideas presented in *On the Origin of Species* were controversial enough without his adding humans to the mix. He presented his ideas on human evolution to the public only when he published *The Descent of Man* 12 years later, in 1871.

As it turned out, Darwin's delicate sidestepping of human origins had little effect. The initial print run of *The Origin* sold out on the day of publication, and the public was perfectly capable of reading between the lines. The Victorians found themselves wrestling with the book's revolutionary message: that humans are a species of ape.

Darwin's conclusions remain controversial to this day among the general public, but they are not controversial among scientists. The evidence that humans are descended from a line of apes whose modern-day representatives include gorillas and chimpanzees is compelling. We know now that about 5–7 million years ago the family tree of the great apes split, one branch ultimately giving rise to chimpanzees and the other to our species. It is those 5–7 million years that hold the key to our humanity. It was over this period—brief by evolutionary standards—that the attributes that make our species so remarkable arose. This chapter discusses what happened over those 5–7 million years and how we came to be the way we are.

24.1 THE GREAT APES

We can approach the question of our place in the tree of life in three different ways: through comparative anatomy, through molecular analysis, and through the fossil record. In this section, we use data from all three sources as we apply the standard methods of phylogenetic reconstruction (Chapter 23) to figure out the evolutionary relationships between humans and other mammals.

Comparative anatomy shows that the human lineage branches off the great apes tree.

There are about 400 species of **primates,** which include prosimians (lemurs, bushbabies), monkeys, and apes (**Fig. 24.1**). All primates share a number of general features that distinguish them from other mammals: nails rather than claws together with a versatile thumb allow objects to be manipulated more dexterously, and eyes on the front of the face instead of the side allow stereoscopic (three-dimensional) vision.

Prosimians are thought to represent a separate primate group from the one that gave rise to humans. Lemurs, which today are confined to the island of Madagascar, are thus only distantly related to humans. Monkeys underwent independent bouts of evolutionary change in the Americas and in Africa and Eurasia, so the family tree is split along geographic lines into New World and Old World monkeys. Though both groups have evolved similar habits, there are basic distinctions. There are differences between the teeth of the two groups, and in New World monkeys the nostrils tend to be widely spaced, whereas in Old World species they are closer together.

One line of Old World monkeys gave rise to the apes, which lack a tail and show more sophisticated behaviors than other monkeys. The apes are split into two groups, the lesser and the great apes (**Fig. 24.2**). Lesser apes include the fourteen species of gibbon, all of which are found in Southeast Asia. The great apes include orangutans, gorillas, chimpanzees, and humans. Taxonomists classify all the descendants of a specified common

FIG. 24.1 The primate family tree. This tree shows the evolutionary relationships of prosimians, monkeys, and apes. *Sources: (left to right) George Holton/Science Source; Penelope Dearman/Getty Images; Daily Mail/Rex/Alamy; Patrick Shyu/Getty Images; Yellow Dog Productions/Getty Images.*

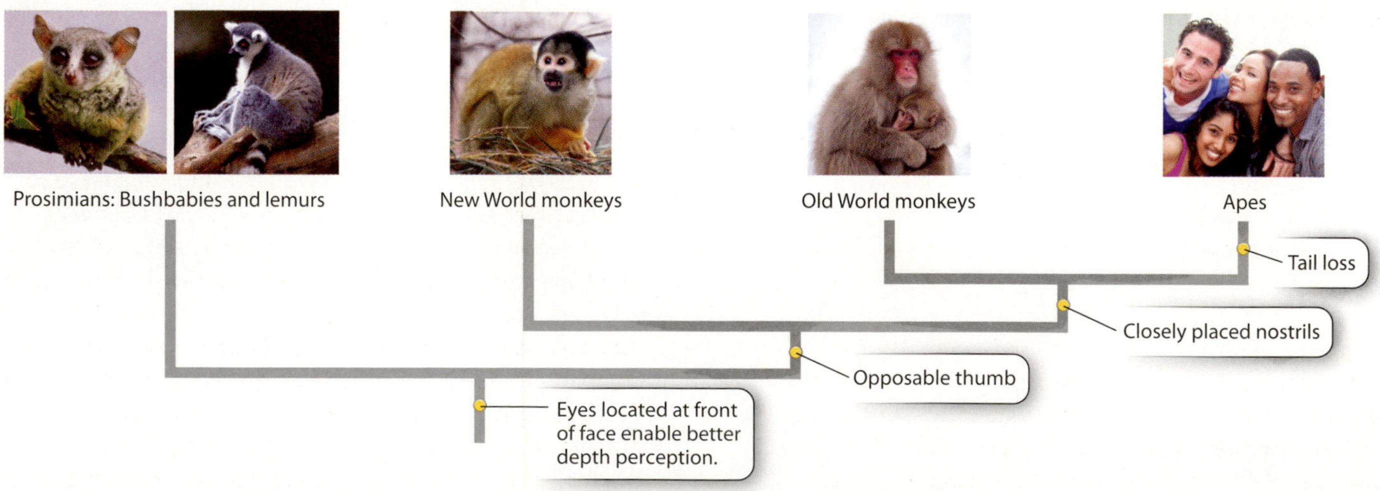

FIG. 24.2 The family tree of the lesser and great apes. The apes consist of two major groups, the lesser apes and great apes. The great apes group includes humans. *Sources: (left to right) Zoonar/K. Jorgensen/age fotostock; S Sailer/A Sailer/age fotostock; J & C Sohns/age fotostock; Michael Dick/Animals Animals–Earth Scenes; FLPA/Jurgen & Christi/age fotostock; Yellow Dog Productions/Getty Images.*

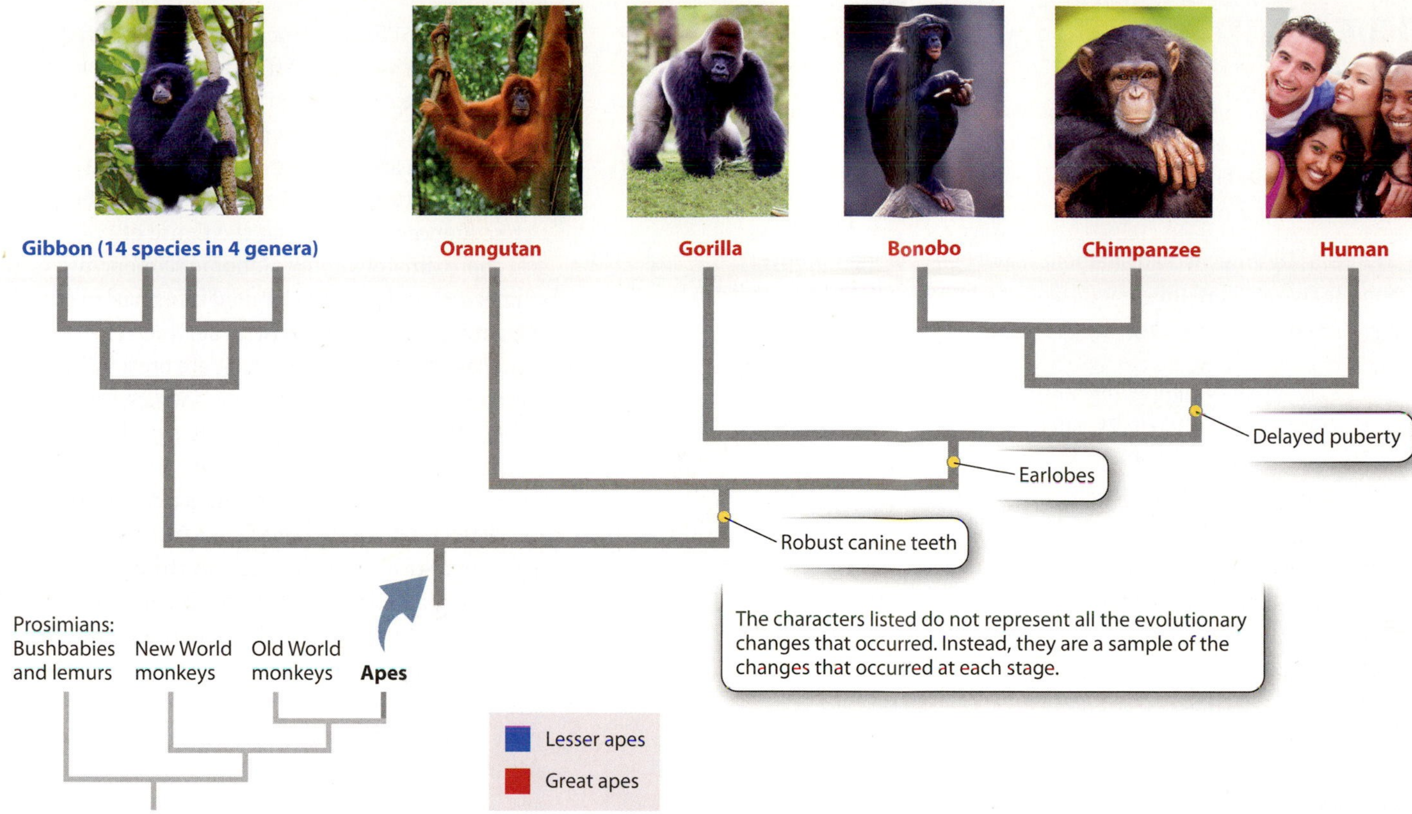

ancestor as belonging to a monophyletic group (Chapter 23). Thus, humans, blue whales, and hedgehogs are all mammals because all three are descended from the first mammal, the original common ancestor of all mammals. Because human ancestry can be traced to the common ancestor of orangutans, gorillas, and chimpanzees, humans, too, are a member of the monophyletic great ape group.

Molecular analysis reveals that the human lineage split from the chimpanzee lineage about 5–7 million years ago.

Which great ape is most closely related to humans? That is, which is our sister group? Traditional approaches of reconstructing evolutionary history by comparing anatomical features failed to determine which of two candidates, gorillas or chimpanzees, is the sister group to humans. It was only with the introduction of molecular methods of assessing evolutionary relationships—through the comparison of DNA and amino acid sequences from the different species—that we had the answer. Our closest relative is the chimpanzee (Fig. 24.2), or, more accurately, the chimpanzees, plural, because there are two closely related chimpanzee species, the smaller of which is often called the bonobo.

Just how closely are humans and chimpanzees related? To answer this question, we need to know the timing of the evolutionary split that led to chimpanzees along one fork and to humans along the other. As we saw in Chapter 21, DNA sequence differences accumulate between isolated populations or species, and they do so at a more or less constant rate. As a result, the extent of sequence difference between two species is a good indication of the amount of time they have been separate, that is, the amount of time since their last common ancestor.

The first thorough comparison of DNA molecules between humans and chimpanzees was carried out before the advent of DNA sequencing methods by Mary-Claire King and Allan Wilson at the University of California at Berkeley (**Fig. 24.3**). One of their methods of measuring molecular differences between species relied on DNA–DNA hybridization (Chapter 12). Two complementary strands of DNA in a double helix can be separated by heating the sample. If the two strands are not perfectly complementary, as is the case if there is a base-pair mismatch (for example, a G paired with a T rather than with a C), less heat is required to separate the strands.

King and Wilson used this observation to examine the differences between a human strand and the corresponding

FIG. 24.3

How closely related are humans and chimpanzees?

BACKGROUND Phylogenetic analysis based on anatomical characteristics had established that chimpanzees are closely related to humans. Mary-Claire King and Allan Wilson used molecular techniques to determine *how* closely related the two species are.

HYPOTHESIS Despite marked anatomical and behavioral differences between the two species, the genetic distance between the two is small, implying a relatively recent common ancestor.

METHOD When two complementary strands of DNA are heated, the hydrogen bonds pairing the two helices are broken at around 95°C and the double helix denatures, or separates (Chapter 12). Two complementary strands with a few mismatches separate at a temperature slightly lower than 95°C because fewer hydrogen bonds are holding the helix together. More mismatches between the two sequences results in an even lower denaturation temperature. Using hybrid DNA double helices with one strand contributed by each species—humans and chimpanzees, in this case—and determining their denaturation temperature, King and Wilson could infer the genetic distance (the extent of genetic divergence) between the two species.

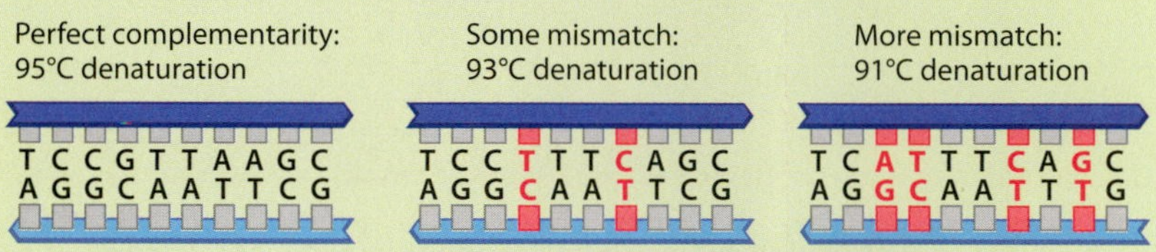

RESULTS King and Wilson found that human–chimpanzee DNA molecules separated at a temperature approximately 1°C lower than the temperature at which human–human DNA molecules separate. This difference could be calibrated on the basis of studies of other species whose genetic distances were known from other methods. The DNA of humans differs from that of chimpanzees by about 1%.

CONCLUSION AND INTERPRETATION King and Wilson noted the discrepancy between the extent of genetic divergence (small) and the extent of anatomical and behavioral divergence (large) between humans and chimpanzees. They suggested that one way in which relatively little genetic change could produce extensive phenotypic change is through differences in gene regulation (Chapter 19). A small genetic change in a control region responsible for switching a gene on and off might have major consequences for the organism.

FOLLOW-UP WORK The sequences of the chimpanzee and human genomes allow us to compare the two sequences directly, and these data confirm King and Wilson's observations.

SOURCE King, M.-C., and A. C. Wilson. 1975. "Evolution at Two Levels in Humans and Chimpanzees." *Science* 188:107–116.

chimpanzee strand. They inferred the extent of DNA sequence divergence from the melting temperature and made a striking discovery: Human and chimpanzee DNA differ in sequence by just 1%.

King and Wilson's conclusions have proved robust as ever more powerful techniques have been applied to the comparison of human and chimpanzee DNA. Today, after sequencing entire human and chimpanzee genomes and literally counting the differences between the two sequences, we find that the figure of 1% still approximately holds. Genome sequencing studies give us superbly detailed information on the differences and similarities between the two genomes, revealing regions that are present in one species but not the other, and which parts of the genomes have evolved more rapidly than others.

Because the amount of sequence difference is correlated with the length of time the two species have been isolated, we can convert the divergence results into an estimate of the timing of the split between the human and chimpanzee lineages. That split occurred about 5–7 million years ago. All the extraordinary characteristics that set our species apart from the rest of the natural world—those attributes that are ours and ours alone—arose in just 5–7 million years.

→ **Quick Check 1** Did humans evolve from chimpanzees? Explain.

The fossil record gives us direct information about our evolutionary history.

Molecular analysis is a powerful tool for comparing species and populations within species. It allows us to compare humans and chimpanzees and look at differences among groups of humans or groups of chimpanzees. However, for a full picture of human evolution, we must turn to fossils (Chapter 23).

For the first several million years of human evolution, all the fossils from the human lineage are found in Africa. This is not surprising. Charles Darwin himself noted that it was likely that the human lineage originated there, as humans' two closest relatives, chimpanzees and gorillas, live only in Africa. The fossil material varies in quality, and a great deal of ingenuity is often required to reconstruct the appearance and attributes of an individual from fragmentary fossil material. It's hard to determine whether two fossil specimens with slight differences belong to the same or different species. And, of course, interbreeding, the criterion that researchers most often use to define a species (Chapter 22), cannot readily be applied to fossils. As a result, experts often disagree over the details of the human fossil record—whether or not, for example, a specimen belongs to a particular species. Regardless, several robust general conclusions may be drawn.

The many different species, including modern humans, that have arisen on the human side of the split since the human/chimpanzee common ancestor are called **hominins.** Let's start with the earliest known hominin, *Sahelanthropus tchadensis* (**Fig. 24.4**). Discovered in Chad in 2002, the skull of *S. tchadensis* combines both modern (human) and ancestral features. *S. tchadensis,* which has been dated to about 7 million years ago and has a chimpanzee-sized brain but hominin-type brow ridges, probably lived shortly after the split between the hominin and chimpanzee lineages, although this conclusion is controversial because of difficulties in interpreting whether a fossil is part of a newly diverged lineage or just another specimen of the ancestral lineage.

An important early hominin, dating from about 4.4 million years ago, is a specimen of *Ardipithecus ramidus* from Ethiopia. This individual, known as **Ardi,** was capable of walking upright, using two legs on the ground but all four limbs in the trees.

An unusually complete early hominin fossil, **Lucy,** found in 1974 at Hadar, Ethiopia, represents the next step in the evolution of hominin gait: She was fully **bipedal,** habitually walking upright (**Fig. 24.5**). Her name comes from the Beatles' song "Lucy in the Sky with Diamonds," which was playing in the paleontologists' field camp when she was unearthed. This fossil dates from around 3.2 million years ago. Lucy was a member of the species *Australopithecus afarensis* and was much smaller than modern humans, less than 4 feet tall, and had a considerably smaller brain (even when the difference in body size is taken into consideration). In many ways, then, Lucy was similar to the common ancestor of humans and chimpanzees, except that she was bipedal.

FIG. 24.4 The skull of *Sahelanthropus tchadensis.* This skull, which is about 7 million years old and was found in Chad, has both human and chimpanzee features. *Source: Richard T. Nowitz/Science Source.*

FIG. 24.5 Lucy, a specimen of *Australopithecus afarensis.* Lucy was fully bipedal, establishing that the ability to walk upright evolved at least 3.2 million years ago. *Source: The Natural History Museum, London/The Image Works.*

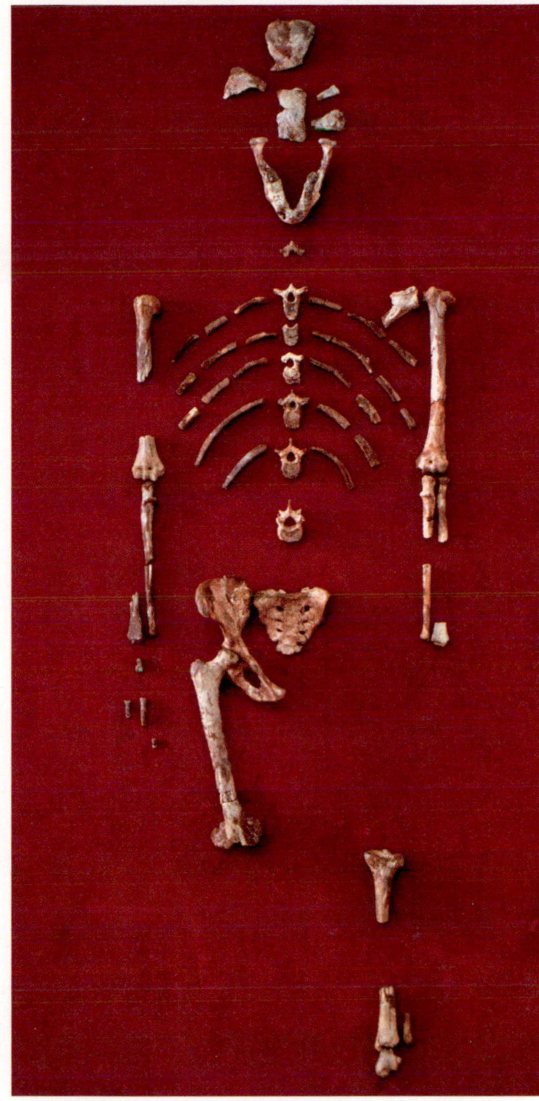

FIG. 24.6 **Hominin lineages.** Three main lineages, *Ardipithecus, Australopithecus,* and *Homo* existed at various and sometimes overlapping times in history. The *Homo* lineage led to modern humans, *Homo sapiens*. *Sources: Data from R. G. Klein, 2009, The Human Career, Chicago: University of Chicago Press, p. 244; Homo floresiensis skull from R. D. Martin, A. M. MacLarnon, J. L. Phillips, L. Dussubieux, P. R. Williams and W. B. Dobyns, 2006, Comment on "The Brain of LB1, Homo floresiensis," Science 312:999.*

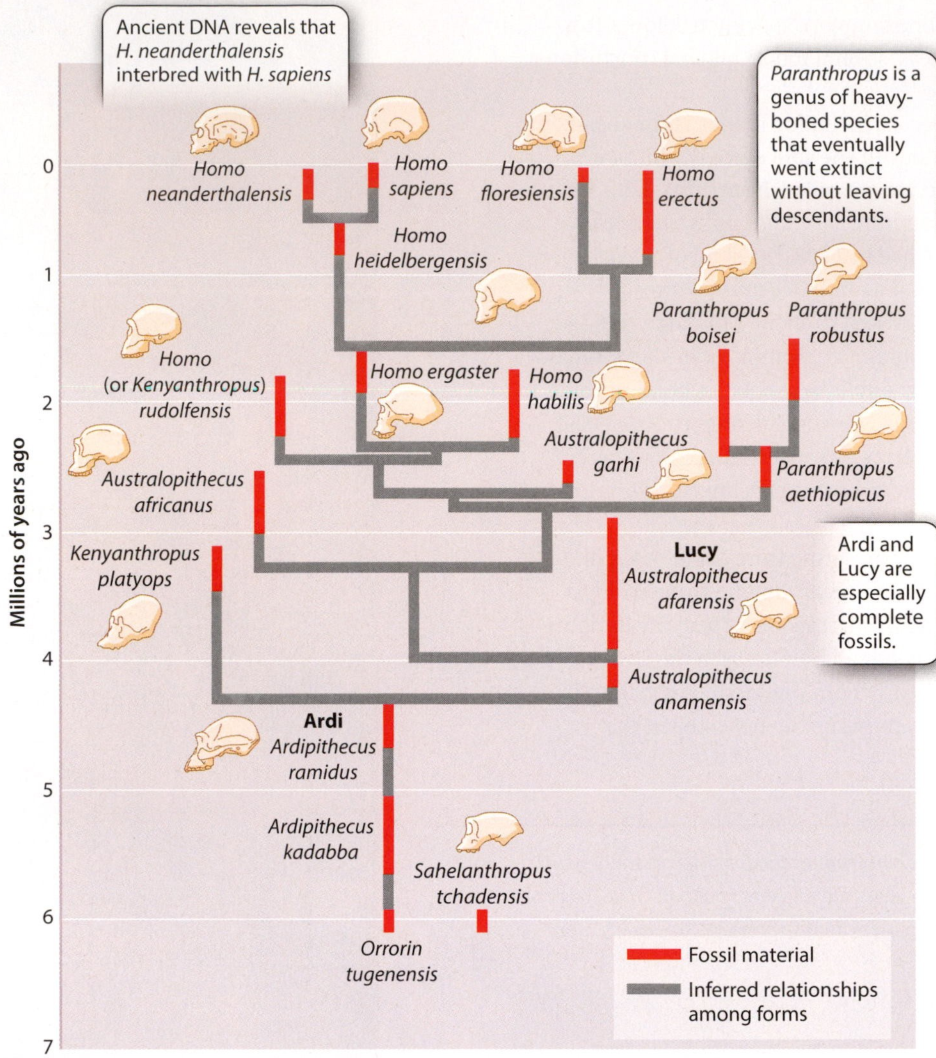

Remarkably, there were many different hominin species living in Africa at the same time. **Fig. 24.6** shows the main hominin species and the longevity of each in the fossil record, along with suggested evolutionary relationships among the species. Note that all hominins have a common ancestor, but not all of these groups lead to modern humans, producing instead other branches of the hominin tree that ultimately went extinct.

A number of trends can be seen when we look over the entire record. Body size increased. Most striking is the increase in size of the cranium and therefore, by inference, of the brain, as shown in **Fig. 24.7.**

The earliest hominin fossils found in Asia are about 2 million years old, indicating that at least one group of hominins ventured out of Africa then. The individuals that first left Africa were members of a species that is sometimes called *Homo ergaster* and sometimes *Homo erectus.* The confusion stems from the controversies that surround the naming of fossil species. Some researchers contend that *H. ergaster* is merely an early form of *H. erectus.* Here, we designate this first hominin *Homo ergaster* and a later descendant species *H. erectus* (see Fig. 24.6). The naming details, however, are relatively unimportant. What matters is that some hominins first colonized areas beyond Africa about 2 million years ago.

FIG. 24.7 Increase in brain size (as inferred from fossil cranium volume) over 3 million years. *Data from T. Deacon, "The Human Brain," pp. 115–123, in J. Jones, R. Martin, and D. Pilbeam (eds.), 1992, The Cambridge Encyclopedia of Human Evolution, Cambridge, England: Cambridge University Press.*

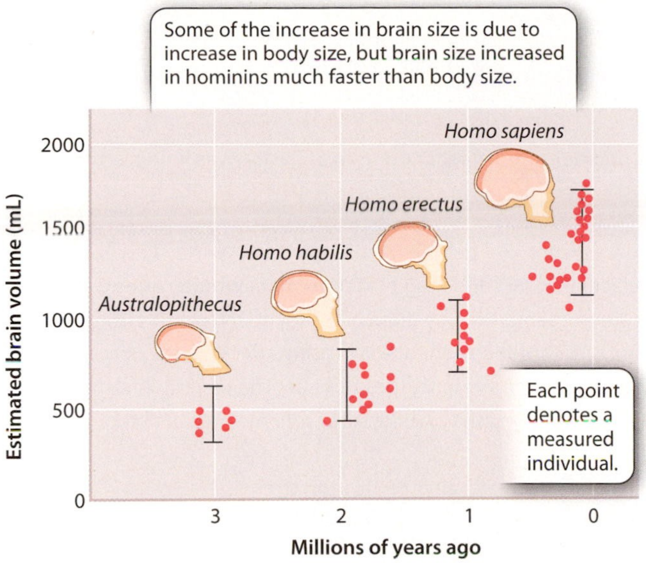

Another hominin species was *Homo neanderthalensis,* whose fossils appear in Europe and the Middle East. **Neanderthals** represent a second hominin exodus from Africa dating from around 600,000 years ago. Thicker boned than us, and with flatter heads that contained brains about the same size as, or slightly larger than, ours, Neanderthals disappeared around 30,000 years ago. As we will see, genetic analysis suggests that this group likely interbred with our own group, *Homo sapiens,* so this disappearance was perhaps not as complete as the fossil record suggests.

Another hominin species became extinct only about 12,000 years ago. This was *H. floresiensis,* known popularly as the Hobbit. *H. floresiensis* is peculiar. Limited to the Indonesian island of Flores, adults were only just over 3 feet tall. Some have suggested that *H. floresiensis* is not a genuinely distinct species, but, rather, is an aberrant *H. sapiens.* Plenty of morphological evidence, however, suggests that *H. floresiensis* is a distinct species derived from an archaic *Homo* species, probably *H. erectus.* Mammals often evolve small body size on islands because of the limited availability of food.

24.2 AFRICAN ORIGINS

From the wealth of fossil evidence, it's clear that modern humans first evolved in Africa. But for a while the timing of this event was unclear. Can all modern humans, *Homo sapiens,* date their ancestry to early hominins in Africa that migrated out and spread around the world about 2 million years ago? Or did the groups that left Africa ultimately go extinct, so that modern humans evolved from a group of hominins in Africa that migrated out much more recently? Molecular studies helped to answer this question.

Studies of mitochondrial DNA reveal that modern humans evolved in Africa relatively recently.

For a long time, it was argued that modern humans derive from the *Homo ergaster* (or *H. erectus*) populations that spread around the world starting from around the time of the early emigration from Africa 2 million years ago. This idea is called the **multiregional hypothesis** of human origins because it implies that different *Homo ergaster* populations throughout Africa and Eurasia evolved in parallel, with some limited gene flow among them, each producing modern *H. sapiens* populations. In short, racial differences among humans would have evolved over 2 million years in different geographic locations.

This idea was overturned in 1987 by another study from Allan Wilson's laboratory, which instead suggested that modern humans arose much more recently from *Homo ergaster* descendants that remained in Africa (sometimes called *Homo heidelbergensis;* see Fig 24.6), and are all descended from an African common ancestor dating from around 200,000 years ago. This newer idea is the **out-of-Africa hypothesis** of human origins.

To test these two hypotheses about human origins, Rebecca Cann, a student in Allan Wilson's laboratory, analyzed DNA sequences to reconstruct the human family tree. Specifically, she studied sequences of a segment of mitochondrial DNA from people living around the world (**Fig. 24.8**).

Mitochondrial DNA (mtDNA) is a small circle of DNA, about 17,000 base pairs long, found in every mitochondrion (Chapter 17). Cann chose to study mtDNA for several reasons. Although a typical cell contains a single nucleus with just two copies of nuclear DNA, each cell has many mitochondria, each carrying multiple copies of mtDNA, making mtDNA much more abundant than nuclear DNA and therefore easier to extract. More important, however, is its mode of inheritance. All your mtDNA is inherited from your mother in the egg she produces because sperm do not contribute mitochondria to the zygote. This means that there is no opportunity for genetic recombination between different mtDNA molecules, so the only way in which sequence variation can arise is through mutation. In nuclear DNA, by contrast, differences between two sequences can be introduced through both mutation and recombination. Recombination obscures genealogical relationships because it mixes segments of DNA with different evolutionary histories.

FIG. 24.8

When and where did the most recent common ancestor of all living humans live?

BACKGROUND With the availability of DNA sequence analysis tools, it became possible to investigate human prehistory by comparing the DNA of living people from different populations.

HYPOTHESIS The multiregional hypothesis of human origins suggests that our most recent common ancestor was living at the time *Homo ergaster* populations first left Africa, about 2 million years ago.

METHOD Rebecca Cann compared mitochondrial DNA (mtDNA) sequences from 147 people from around the world. This approach required a substantial amount of mtDNA from each individual, which she acquired by collecting placentas from women after childbirth. Instead of sequencing the mtDNA, Cann inferred differences among sequences by digesting the mtDNA with different restriction enzymes, each of which cuts DNA at a specific sequence (Chapter 12). If the sequence is present, the enzyme cuts. If any base in the recognition site of the restriction enzyme has changed in an individual, the enzyme does not cut. The resulting fragments were then separated by gel electrophoresis. By using 12 different enzymes, Cann was able to assay a reasonable proportion of all the mtDNA sequence variation present in the sample.

Here, we see a sample dataset for four people and a chimpanzee. There are seven varying restriction sites:

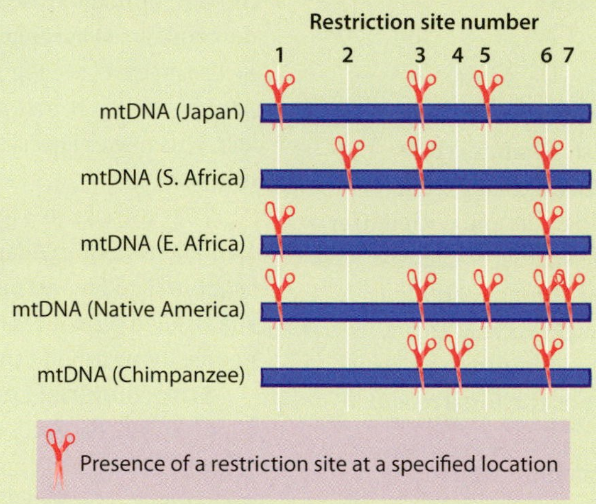

ANALYSIS The data were converted into a family tree by using shared derived characters, described in Chapter 23. By mapping the pattern of changes in this way, the phylogenetic relationships of the mtDNA sequences can be reconstructed. For example, the derived "cut" state at restriction site 1 shared by the East African, Japanese, and Native American sequences implies that the three groups had a common ancestor in which the mutation that created the new restriction site occurred.

Mitochondrial DNA sequences offered Cann a clear advantage. With sequence information from the mtDNA of 147 people from around the world, Cann reconstructed the human family tree (Fig. 24.8). The tree contained two major surprises. First, the two deepest branches—the ones that come off the tree earliest in time—are African. All non-Africans are branches off the African tree. The implication is that *Homo sapiens* evolved in Africa and only afterward did populations migrate out of Africa and become established elsewhere. This finding contradicted the expectations of the multiregional theory, which predicted that *Homo sapiens*

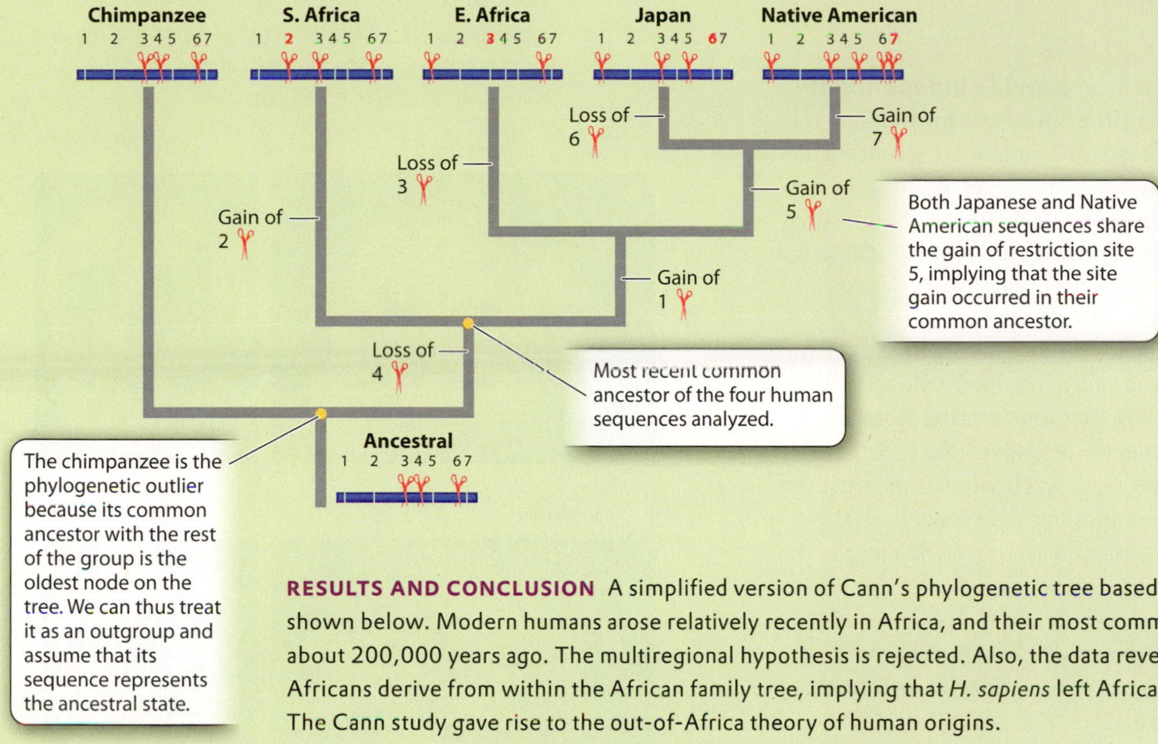

RESULTS AND CONCLUSION A simplified version of Cann's phylogenetic tree based on mtDNA is shown below. Modern humans arose relatively recently in Africa, and their most common ancestor lived about 200,000 years ago. The multiregional hypothesis is rejected. Also, the data revealed that all non-Africans derive from within the African family tree, implying that *H. sapiens* left Africa relatively recently. The Cann study gave rise to the out-of-Africa theory of human origins.

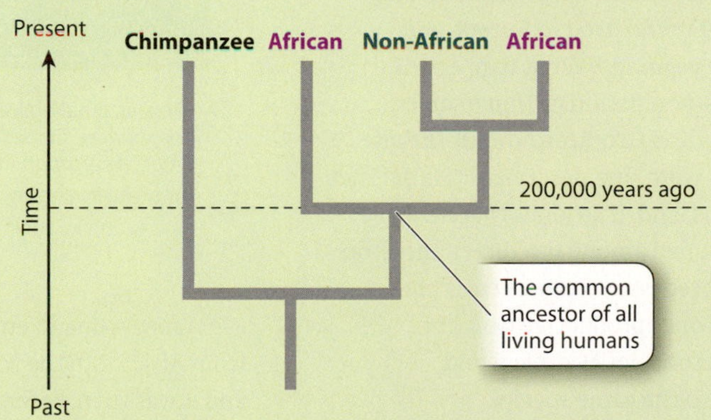

FOLLOW-UP WORK This study set the stage for an explosion in genetic studies of human prehistory that used the same approach of comparing sequences in order to identify migration patterns. Today, with molecular tools vastly more powerful than those available to Cann, studies of the distribution of genetic variation are giving us a detailed pattern, even on relatively local scales, of demographic events.

SOURCE Cann, R. L., M. Stoneking, and A. C. Wilson. 1987. "Mitochondrial DNA and Human Evolution." *Nature* 325:31–36.

evolved independently in different locations throughout the Old World.

Second, the tree is remarkably shallow. That is, even the most distantly related modern humans have a relatively recent common ancestor. By calibrating the rate at which mutations occur in mtDNA, Cann was able to estimate the time back to the common ancestor of all modern humans as about 200,000 years. Subsequent analyses have somewhat refined this estimate, but the key message has not changed: The data contradict the multiregional hypothesis, which predicted a number closer to 2 million years.

→ **Quick Check 2** What do the multiregional and out-of-Africa hypotheses predict about the age of the common ancestor of all humans living today?

Studies of the *Y* chromosome provide independent evidence for a recent origin of modern humans.

Two hundred thousand years ago is 10 times more recent than 2 million years ago, so this change in dating represents a revolution in our understanding of human prehistory. But surely it is risky to make such major claims on the basis of a single study. Maybe there's something peculiar about mtDNA, and it doesn't offer an accurate picture of the human past.

To find independent evidence for a recent origin of modern humans in Africa, another dataset was required. One such dataset comes from studies of the Y chromosome, another segment of human DNA that does not undergo recombination (Chapter 17). When an approach similar to Cann's was used to reconstruct the human family tree using Y chromosome DNA sequences, the result was completely in agreement with the mtDNA result: The human family is young, and it arose in Africa.

Neanderthals disappear from the fossil record as modern humans appear, but have contributed to the modern human gene pool.

If we exclude *H. floresiensis,* the last of the nonmodern humans were the Neanderthals (**Fig. 24.9**), who lived in Europe and Western Asia until about 30,000 years ago. What happened? Were they eliminated by the first population of *Homo sapiens* to arrive in Europe (a group known as **Cro-Magnon** for the site in France from which specimens were first described)? Or did they interbreed with the Cro-Magnons, and, if so, does that mean that Neanderthals should be included among our direct ancestors? Indeed, if modern humans interbred with Neanderthals and our definition of species is based on reproductive isolation (Chapter 22), perhaps we should consider *H. sapiens* and *H. neanderthalensis* as belonging to the same species.

In 1997, we thought we had a clear answer. Matthias Krings, working in Svante Pääbo's lab in Germany, extracted intact stretches of mtDNA from 30,000-year-old Neanderthal bone. If Neanderthals had interbred with our ancestors, we would expect to see evidence of their genetic input in the form of Neanderthal mtDNA in the modern human gene pool. However, the Neanderthal sequence was strikingly different from that of modern humans. That we see nothing even close to the Neanderthal mtDNA sequence in modern humans strongly suggested that Neanderthals did not interbreed with our ancestors (**Fig. 24.10a**).

This conclusion, however, has been reversed following remarkable technological advances that permitted Pääbo and others to sequence the entire Neanderthal genome (rather than just its mtDNA). Careful population genetic analysis revealed that our ancestors did interbreed with Neanderthals, and that 1% to 4% of the genome of every non-African is Neanderthal-derived.

FIG. 24.9 (a) ***Homo sapiens*** **and (b)** ***Homo neanderthalensis.*** Neanderthals disappeared from Europe about 30,000 years ago, about 10,000 years after a group of modern humans called Cro-Magnon first appeared in Europe. *Sources: Javier Trueba/MSF/Science Source.*

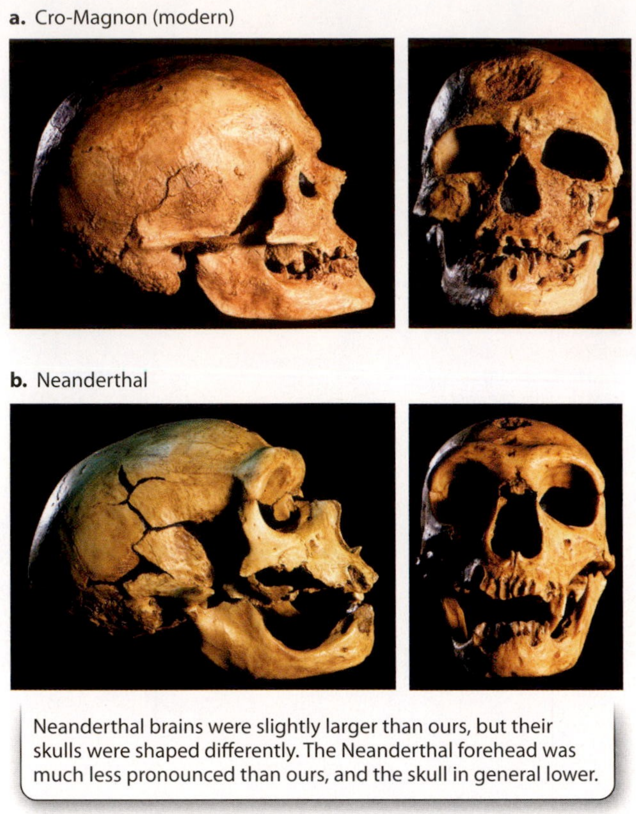

A new scenario emerged: As populations of our ancestors headed from Africa into the Middle East, they encountered Neanderthals, and it was then, before modern humans had spread out across the world, that interbreeding took place (**Fig. 24.10b**). Remarkably, ancient DNA analysis has further revealed that early non-African *Homo sapiens* did not interbreed only with Neanderthals. Another group of Eurasian hominins, relatives of Neanderthals known as Denisovans after the Siberian cave in which their remains were found, also contributed genetically to some modern human populations, especially indigenous people from Australasia.

How can we reconcile the two apparently contradictory results about the contributions of Neanderthals to the modern human gene pool? The mtDNA study indicates there was no genetic input from Neanderthals in modern humans, but the whole-genome study suggests that in fact there was. One possible solution lies in the difference in patterns of transmission between mtDNA and genomic material (Chapter 17). Recall that mtDNA is maternally inherited because sperm do not contribute mitochondrial material. Your mtDNA, whether you are male or female, is derived solely

FIG. 24.10 Evidence for and against early interbreeding between Neanderthals and early humans. (a) Studies using only mtDNA did not support the idea of interbreeding, but (b) later studies using whole genomes did.

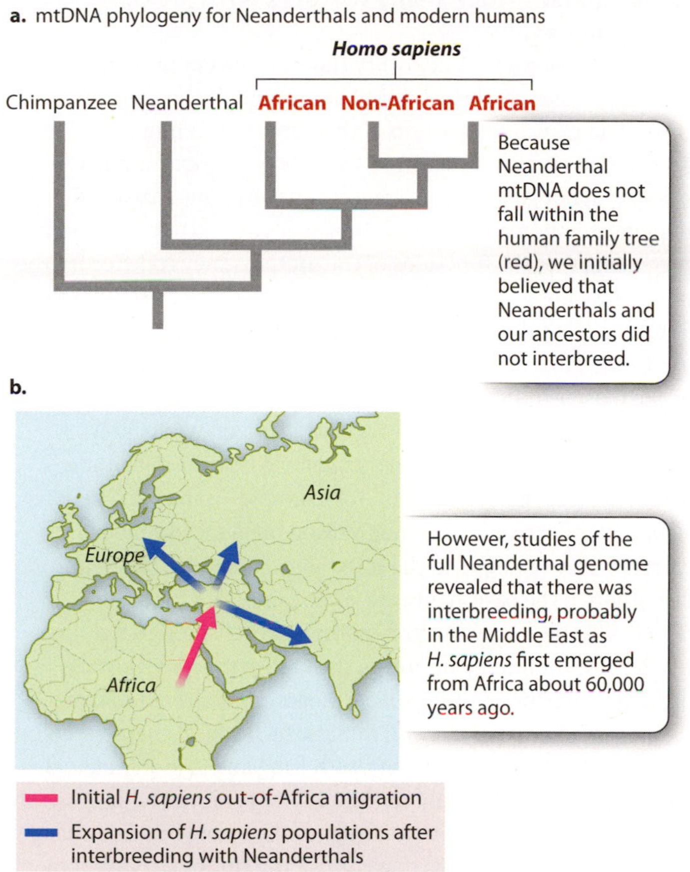

from your mother. Therefore, it is possible that the Neanderthal mtDNA lineage has been lost through genetic drift (Chapter 21).

An alternative is that the discrepancy in the ancient DNA stems from a sex-based difference in interbreeding. If Neanderthal females did not interbreed with our ancestors, Neanderthal mtDNA would not have entered the modern human gene pool. If only male Neanderthals interbred with our ancestors, we would expect exactly the pattern we observe: no Neanderthal mtDNA in modern humans, but Neanderthal genomic DNA in the modern human gene pool.

Regardless of the details of what happened when modern humans, Neanderthals, and Denisovans encountered each other, one thing is clear: We are forced to reimagine the ancestry of a large portion of the human population. Every non-African on the planet is part Neanderthal, albeit a very small part, and many non-Africans can count Denisovans among their ancestors as well.

→ **Quick Check 3** Explain how genetic data are consistent with the possibility that only male Neanderthals interbred with our ancestors.

24.3 DISTINCT FEATURES OF OUR SPECIES

Many extraordinary changes in anatomy and behavior occurred in the 5–7 million years since our lineage split from the lineage that gave rise to the chimpanzees. Fossils tell us a great deal about those changes, especially when high-quality material such as Lucy or Ardi is available, but in general this is an area in which fossils are hard to come by and there is a lot of speculation. Speculation is especially common when we try to explain the reasons behind the evolution of a particular trait. Why, for example, did language evolve? It is easy enough to think of a scenario in which natural selection favors some ability to communicate—maybe language arose to facilitate group hunting. There are plenty of plausible ideas on the subject, but, in most cases, no evidence, so it is impossible to distinguish among competing hypotheses. We can, however, be confident that the events that produced language occurred in Africa, and, through paleontological studies of past environments, we can conclude that humans evolved in an environment similar in many ways to today's East African savanna.

Bipedalism was a key innovation.

The shift from walking on four legs to walking on two was probably one of the first changes in our lineage. Many primates are partially bipedal—chimpanzees, for example, may be observed knuckle-walking—but humans are the only living species that is wholly bipedal. As we've seen, Lucy, dating from about 3.2 million years ago, was already bipedal, and 4.4 million years ago, Ardi was partially bipedal. We can further refine our estimate of when full bipedalism arose from the evidence of a trace fossil. A set of 3.5-million-year-old fossil footprints discovered in Laetoli, Tanzania, reveal a truly upright posture.

Becoming bipedal is not simply a matter of standing up on hind legs. The change required substantial shifts in a number of basic anatomical characteristics, described in **Fig. 24.11.**

Why did hominins become bipedal? Maybe it gave them access to berries and nuts located high on a bush. Maybe it allowed them to scan the vicinity for predators. Maybe it made long-distance travel easier. Whatever the reason, bipedalism freed our ancestors' hands. No longer did they need their hands for locomotion, so for the first time there arose the possibility that specialized hand function could evolve. Most primates have some kind of opposable thumb, but the human version is much more dexterous. The human thumb has three muscles that are not present in the thumb of chimpanzees, and these allow much finer motor control of the thumb. Tool use, present but crude in chimpanzees, can be much more subtle and sophisticated with a human hand.

Bipedalism also made it possible to carry material over long distances. Hominins could then set up complex foraging strategies whereby some individuals supplied others with resources. Exactly when sophisticated tool use arose in our ancestors is controversial, but it is indisputable that bipedalism contributed. Similarly, the

FIG. 24.11 **The shift from four to two legs.** Anatomical changes were required for the shift, especially in the structure of the skull, spine, legs, and feet.

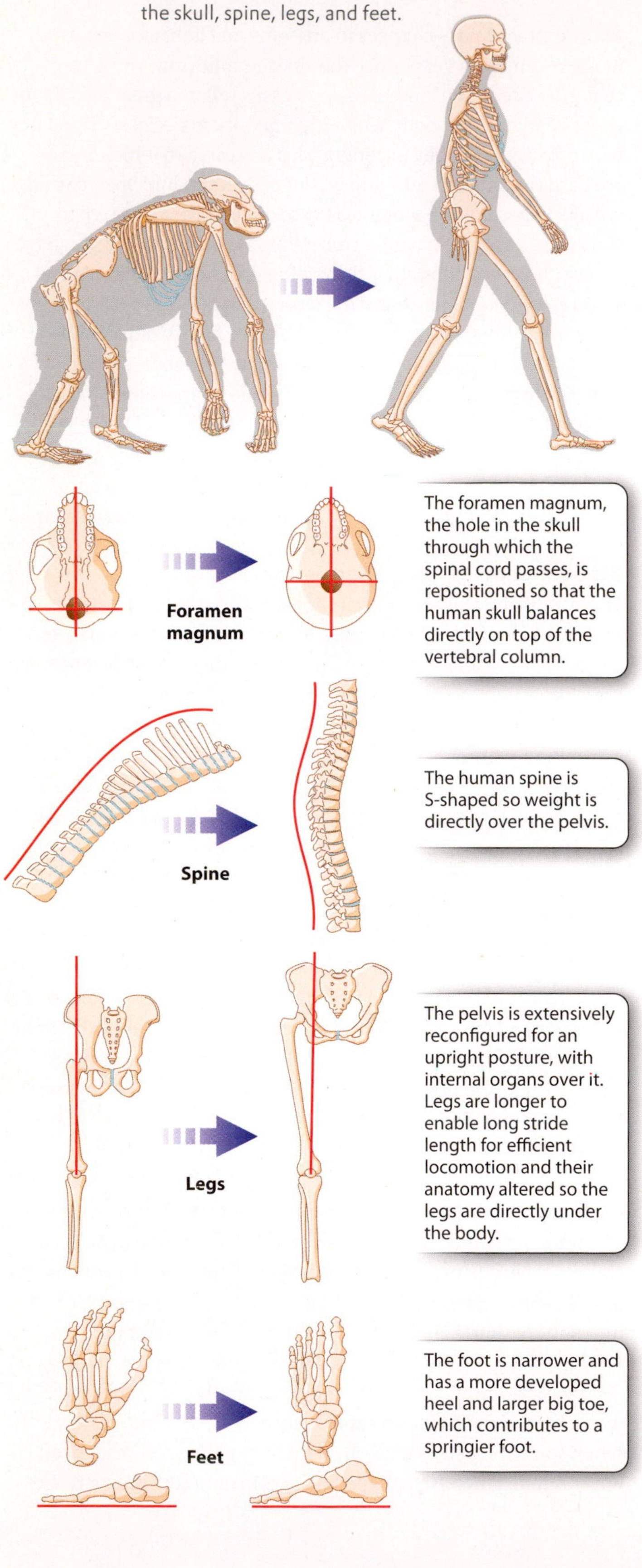

ability to manipulate food with the hands, and to carry material using the hands rather than the mouth, likely permitted the evolution of the human jaw—indeed, the entire facial structure—in such a way that language became a possibility.

Adult humans share many features with juvenile chimpanzees.

King and Wilson's 1975 discovery that the DNA of humans and of chimpanzees are 99% identical has recently been confirmed by DNA sequence comparison of the human and chimpanzee genomes. It turns out that the two genomes are extraordinarily similar: All our genes are also present in the chimpanzee and their sequences in humans and in chimpanzees are extremely similar, suggesting that the functions of the proteins they code for are the same. If we are so similar, how can we account for the extensive differences between the two species?

In their original paper, King and Wilson suggested that much of the most significant evolution along the hominin lineage came about through changes in the regulation of genes (Chapter 19). A small change—one that causes a gene to be transcribed at a different stage of development, for example—can have a major effect. In other words, King and Wilson introduced a model whereby small changes in the software could have a major impact even though the basic hardware is the same.

One of the gene-regulatory pathways that changed may be responsible for human **neoteny,** the long-term evolutionary process in which the timing of development is altered so that a sexually mature organism still retains the physical characteristics of the juvenile form. In keeping with King and Wilson's idea, this shift in development could conceivably be achieved with relatively little genetic change. All that it might take would be a few changes to the regulatory switches that control the timing of development.

As early as 1836, French naturalist Étienne Geoffroy Saint-Hilaire noted that a young orangutan on exhibit in Paris looked considerably more like a human than the adult of its own species (**Fig. 24.12**). Several human attributes support this model, including heads that are large relative to our bodies (and our correspondingly large brains), a feature of juvenile great apes. A second human attribute is our lack of hair. The juveniles of other great apes are not as hairless as humans, but they're considerably less hairy than adult apes. A third attribute is the position of the foramen magnum at the base of the skull. In primate development, the foramen magnum starts off in the position it occupies in the adult human and then, in nonhuman great apes, migrates toward the back of the skull. Adult humans have retained the juvenile great ape foramen magnum position. Finally, it has been suggested that our mentality, with its questioning and playfulness, is equivalent in many ways to that of a juvenile ape rather than to that of the comparatively inflexible adult ape.

Humans have large brains relative to body size.

In mammals, brain size is typically correlated with body size. Humans are relatively large-bodied mammals, but our brains

FIG. 24.12 **A juvenile and an adult orangutan.** Humans look more like juvenile orangutans than like adult orangutans, suggesting that humans may be neotenous great apes. *Sources: (top to bottom) McPHOTO/W. Layer/age fotostock; Russell Watkins/Shutterstock.*

are large even for our body size (**Fig. 24.13**). It is our large brains that have allowed our species' success, extraordinary technological achievements, and at times destructive dominion over the planet.

What factors acted as selective pressure for the evolution of the large human brain? Again, speculation is common, but because a large brain is metabolically expensive to produce and to maintain, we can conclude that the brain has adaptive features that natural selection acted in favor of. What are the selective factors? Here are some possibilities:

- *Tool use.* Bipedalism permitted the evolution of manual dexterity, which in turn requires a complex nervous organization if hands are to be useful.
- *Social living.* Groups require coordination, and coordination requires some form of communication and the means of integrating and acting upon the information conveyed. One scenario, for example, sees group hunting as critical in the evolution of the brain: Natural selection favored those individuals who cooperated best as they pursued large prey.
- *Language.* Did the evolution of language drive the evolution of large brains? Or did language arise as a result of having large brains? Again, we will probably never know, but it is tempting to speculate that the brain and our extraordinary powers of communication evolved in concert.

Probably, as with bipedalism, there was no single factor but a mix of elements that worked together to result in the evolution of large brain size. We tend to focus on brain size because it is a convenient stand-in for mental power and because we can measure it in the fossil record by making the reasonable assumption that the volume of a fossil's cranium reflects the size of its brain. However, it should be emphasized that our minds are not the products solely

FIG. 24.13 **Brain size plotted against body size for different species.** Humans have large brains for their body size. *Data from Fig. 2.4, p. 44, in H. J. Jerison, 1973, Evolution of the Brain and Intelligence, New York: Academic Press.*

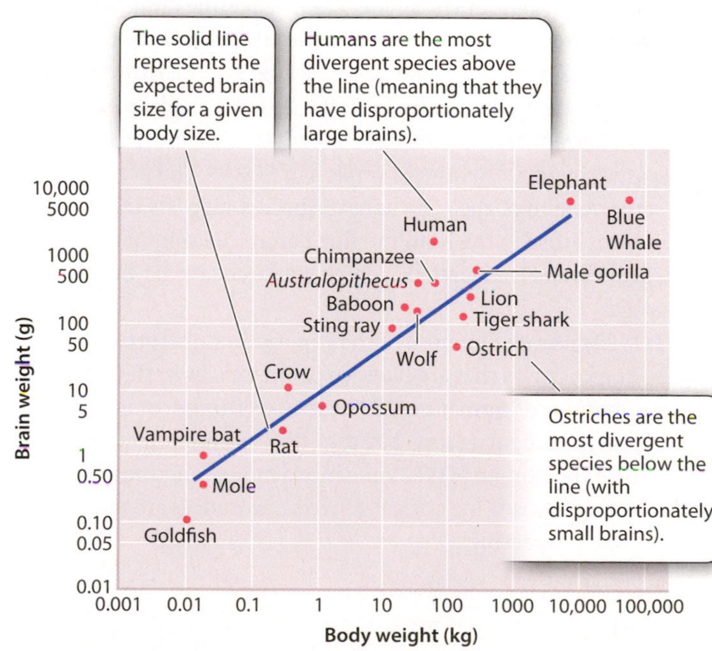

of larger brains. Rather, the reorganization of existing structures and pathways is the more important product of the evolution of the brain. Not only was the brain expanding over those 7 million years of human evolution, it was also being rewired.

The human brain evolved through natural selection. What we have today is a learning machine capable of generating many more skills and abilities than just those that directly enhance evolutionary fitness. For example, let's assume that the brain evolved to make group hunting more efficient. The result is that the brain does indeed enhance our ability to hunt together, but there are numerous by-products of this brain that take its abilities beyond just hunting. Making art, for instance, has nothing to do with group hunting, and yet the brain allows us to do it. A brain evolved for an essential but mundane task like group hunting lies at the heart of much that is wonderful about humanity.

The human and chimpanzee genomes help us identify genes that make us human.

The key genetic differences between humans and chimpanzees must lie in the approximately 1% of protein-coding DNA sequence that differs between the two genomes. What do we see when we compare human and chimpanzee genomes?

A gene that has attracted a lot of interest is *FOXP2,* a member of a large family of evolutionarily conserved genes that encode transcription factors that play important roles in development. Individuals with mutations in *FOXP2* often have difficulty with speech and language. Interestingly, laboratory animals whose *FOXP2* has been knocked out also have communication impairments. Songbirds are less capable of learning new songs, and the high-frequency "songs" of mice are disrupted. In addition to its effects on the brain, *FOXP2* plays a role in the development of many tissues.

Studies of the gene's amino acid sequence reveal a pattern of extreme conservation. The sequences in mice and chimpanzees, whose most recent common ancestor lived about 75 million years ago, differ by a single amino acid (**Fig. 24.14**). However, two amino acids are present in humans but absent in chimpanzees. Studies of Neanderthal DNA have shown that Neanderthals possessed the modern human version of *FOXP2.* The presence of the same version of the gene in both species suggests that the differences arose in the hominin line before the split between our species and Neanderthals, which occurred at least 600,000 years ago.

FOXP2 gives us a glimpse of the genetic architecture of the biological traits that are likely to be important in the determination of "humanness." Those two differences in amino acid sequence are intriguing, but they do not a human make. Substitute the human *FOXP2* gene into a mouse and you will not get a talking mouse. The critical genetic differences are many, subtle, and interacting.

➔ **Quick Check 4** The *FOXP2* gene is sometimes called the "language gene." Why is this name inaccurate?

FIG. 24.14 A family tree of mammals showing amino acid changes in *FOXP2.* The gene is highly conserved, but two changes are seen in the Neanderthals and modern humans.

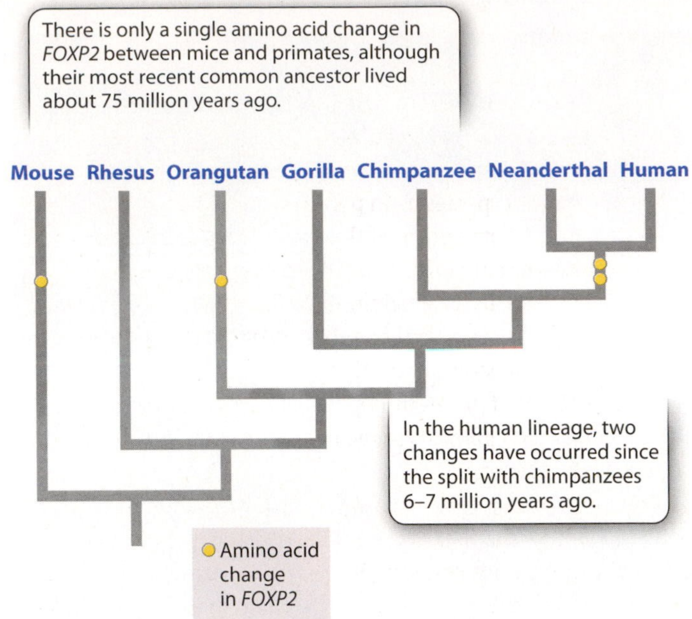

24.4 HUMAN GENETIC VARIATION

So far, we have treated humans as all alike, and in many ways, of course, we are. But there are also many differences from one person to the next. Those differences ultimately have two sources: genetic variation and differences in environment (Chapter 18). A person may be born with dark skin, or a person born with pale skin may acquire darker skin—a tan—in response to exposure to sun.

The differences we see from one person to the next are deceptive. Despite appearances, ours is not the most genetically variable species on Earth. While it is certainly true that everyone alive today (except for identical twins) is genetically unique, our species is actually rather low in overall amounts of genetic variation. Modern estimates based on comparisons of many human DNA sequences indicate that, on average, about 1 in every 1000 base pairs differs among individuals (that is, our level of DNA variation is 0.1%). That's about 10 times less genetic variation than in fruit flies (which nevertheless all look the same to us) and about two to three times less than in Adélie penguins, which look strikingly similar to one another (see Fig. 21.1).

Why, then, are we all so phenotypically different if there is so little genetic variation in our species? Given the large size of our genome, a level of variation of 0.1% translates into a great many genetic differences. Our genome consists of approximately

3 billion base pairs, so 0.1% variation means that 3 million bp differ between any two people chosen at random. Many of those differences are in noncoding DNA, but some fall in regions of DNA that encode proteins and therefore influence the phenotype. When those mutations are reshuffled by recombination, we get the vast array of genetic combinations present in the human population.

The prehistory of our species has had an impact on the distribution of genetic variation.

The reasons for our species' relative lack of genetic variation compared to other species lie in prehistory, and factors affecting the geographical distribution of that variation also lie in the past. Studies of the human family tree initiated by Rebecca Cann's original mtDNA analysis are giving a detailed picture of how our ancestors colonized the planet and how that process affected the distribution of genetic variation across populations today. Detailed analyses of different populations, often using mtDNA or Y chromosomes, allow us to reconstruct the history of human population movements.

As we have seen, *Homo sapiens* arose in Africa. Genetic analyses indicate that perhaps 60,000 years ago, populations started to venture out through the Horn of Africa and into the Middle East (**Fig. 24.15**). The first phase of colonization took our ancestors through Asia and into Australia by about 50,000 years ago. Archaeological evidence indicates that it was not until about 15,000 years ago that the first modern humans crossed from Siberia to North America to populate the New World.

Genetic analyses also indicate that other colonizations were even later. Despite its closeness to the African mainland, the first humans arrived in Madagascar only about 2000 years ago, and the colonists came from Southeast Asia, not Africa. Madagascar populations to this day bear the genetic imprint of this surprising Asian input. The Pacific Islands were among the last habitable places on Earth to be colonized during the Polynesians' extraordinary seaborne odyssey from Samoa, which began about 2000 years ago. Hawaii was colonized about 1500 years ago, and New Zealand only 1000 years ago.

By evolutionary standards, the beginning of the spread of modern humans out of Africa about 60,000 years ago is very recent. There has therefore been relatively little time for differences to accumulate among regional populations, and most of the variation in human populations today arose in ancestral populations before any humans left Africa. When we compare levels of variation in a contemporary African population with that in a non-African population, like Europeans or Asians, we

FIG. 24.15 Human migratory routes. Tracking the spread of mitochondrial DNA mutations around the globe allows us to reconstruct the colonization history of our species. *Data from Fig. 6.18, p. 149, in D. J. Futuyma, 2009, Evolution, 2nd ed., Sunderland, MA: Sinauer Associates.*

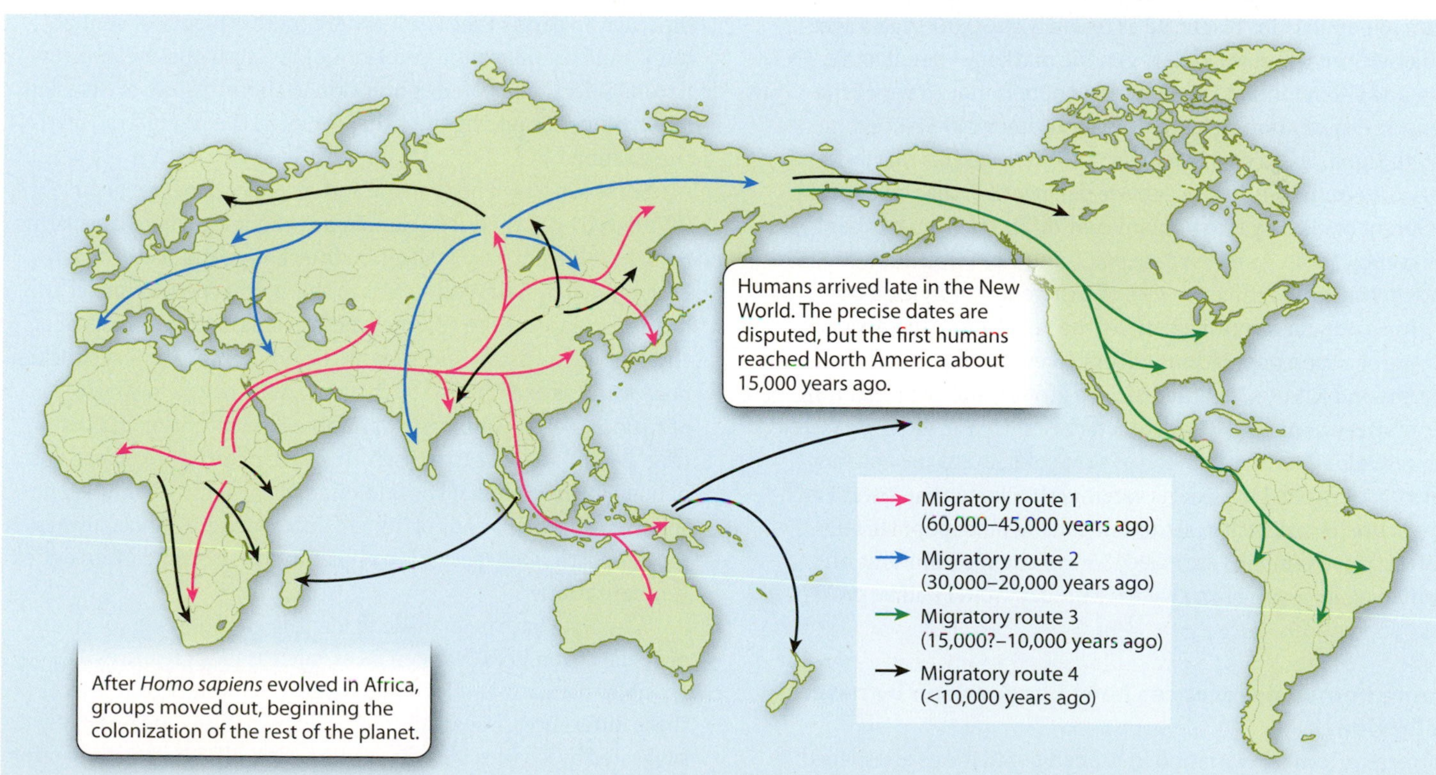

find there is more variation in the African population. This is because the individuals that left Africa 60,000 years ago to found populations in the Middle East and beyond were a relatively small sample of the total amount of genetic variation then present in the human population. Non-African populations therefore began with less genetic variation, and that initial lower variation is reflected in their genetic profiles today.

The recent spread of modern humans means that there are few genetic differences between groups.

Because the out-of-Africa migration was so recent, the genetic differences we see among geographical groups—sometimes called races—are minor. This fact is highly counterintuitive. We see many superficial differences between an African and a Caucasian, such as skin color, facial form, and hair type, and assume that these superficial differences must reflect extensive genetic differences. This assumption made sense when the standard theory about the origin of modern humans was the multiregional one. If European and African populations really had been geographically isolated from each other for as long as 2 million years, then we would expect significant genetic differences among populations.

We expect isolated populations to diverge genetically over time as different mutations occur and are fixed in each population. The longer two populations have been isolated from each other, the more genetic differences between them we expect to see (Chapter 21). Isolation lasting 2 million years implies that the differences are extensive, but isolation of just 60,000 years suggests they are relatively few. Patterns of genetic variation among different human populations support the hypothesis that human populations dispersed as recently as 60,000 years ago. What we see when we look at genetic markers—variable A's, T's, G's, and C's in human DNA—is that there is indeed very little genetic differentiation by what is sometimes called race.

In short, there's a disconnect: Different groups may look very different, but, from a genetic perspective, they're not very different at all. Any two humans may differ from each other by, on average, only 3 million base pairs, and statistical analyses have shown that approximately 85% of that genetic variation occurs within a population (for example, the Yoruba in West Africa); 8% occurs between populations within races (for example, between Yoruba and Kikuyu, another African group); and the remaining 7% occurs between races. The characteristics we use when we assess an individual's ethnicity, such as skin color, eye type, and hair form, are encoded by genetic variants that lie in that 7%. If Earth were threatened with destruction and only one population—Yoruba, for example—survived, 85% of the total amount of human genetic variation that exists today would still be present in that population.

Some human differences have likely arisen by natural selection.

Patterns of genetic variation in human populations are shaped by the set of evolutionary forces discussed in Chapter 21. Neutral variants, for example, are subject to genetic drift, and, given the stepwise global colonization history of our species, it is likely that founder events have played a role as well.

Selection, too, has been important. It is apparent, for example, that the genetic variants affecting traits we can easily see are an especially prominent feature of the 7% of human genetic variation that occurs between so-called races. If we look at other genetic variants, ones that don't affect traits that we can see, there is little or no racial pattern. An African is as likely to have a particular base-pair mutation in a randomly chosen gene as a European. So why are visible traits so markedly different among races and other traits are not? Given the short amount of time (by evolutionary standards) since all *Homo sapiens* were in Africa, it is likely that the differences we see between groups are the product of selection.

People with dark skin tend to originate from areas in lower latitudes with high levels of solar radiation, and people with light skin tend to originate from areas in higher latitudes with low levels of solar radiation. It is likely that natural selection is responsible for the physical differences between these populations. Assuming that the ancestors of non-African populations were relatively dark-skinned, what selective factors can account for the loss of pigmentation?

A likely factor is an essential vitamin, vitamin D, which is particularly important in childhood because it is needed for the production of bone. A deficiency of vitamin D can result in the skeletal malformation known as rickets. The body can synthesize vitamin D, but the process requires ultraviolet radiation. Heavily pigmented skin limits the entry of UV radiation into cells and so limits the production of vitamin D. This does not present a difficulty in parts of the world where there is plenty of sunlight, but it can be problematic in regions of low sunlight. Presumably, natural selection favored lighter skin in the ancestors of Eurasian populations because lighter skin promoted the production of the vitamin.

Some aspects of body shape and size may also have been influenced by natural selection. In hot climates, where dissipating body heat is a priority, a tall and skinny body form has evolved. Exemplified by East African Masai, this body type maximizes the ratio of surface area to volume and thus aids heat loss. In colder climates, by contrast, selection has favored a more robust, stockier body form, as exemplified by the Inuit, who have a low ratio of surface area to volume that promotes the retention of heat (**Fig. 24.16**). In these two cases, these are plausible explanations of body form. We should bear in mind, however, that simple one-size-fits-all explanations of human difference are almost always too simplistic. Our species is complex and diverse and often defies generalizations.

Attempts have been made to identify the adaptive value of visual differences between races, such as facial features. It's possible that natural selection played a role in the evolution of these differences, but an alternative explanation, one originally suggested by Charles Darwin, is more compelling: sexual selection (Chapters 21 and 45).

FIG. 24.16 Evolutionary responses of body shape to climate. (a) A heat-adapted Masai in Kenya and (b) a cold-adapted Inuit in Greenland.
Sources: a. Ryan Heffernan/Aurora/Getty Images; b. B&C Alexander/ArcticPhoto.

As we have seen, there is an apparent mismatch between the extent of difference among groups in visible characters, such as facial features, and the overall level of genetic difference between human groups. Sexual selection can account for this mismatch because it operates solely on characteristics that can readily be seen—think of the peacock's tail. As we learn more about the genetic underpinnings of the traits in question, we will be able to investigate directly the factors responsible for the differences we see among groups.

? CASE 4 MALARIA: CO-EVOLUTION OF HUMANS AND A PARASITE

What human genes are under selection for resistance to malaria?

We see evidence of regional genetic variation in response to local challenges, especially those posed by disease. Malaria, for example, is largely limited to warm climates because it is transmitted by a species of mosquito that can survive only in these regions. Historically, the disease has been devastating in Africa and the Mediterranean. As we saw in Chapter 21, the sickle allele of hemoglobin, *S*, has evolved to be present at high frequencies in these regions because in heterozygotes it confers some protection against the disease. But in homozygotes, the *S* allele is highly detrimental because it causes sickle-cell anemia. Homozygotes for the allele encoding normal hemoglobin are also at a disadvantage because they are entirely unprotected from the parasite.

The *S* allele is beneficial only in the presence of malaria. If there is no malaria in an area, the *S* allele is disadvantageous, so natural selection presumably acted rapidly to eliminate it in the ancestors of Europeans when they arrived in malaria-free regions. The continued high frequency of the *S* allele in Africans, some Mediterranean populations, and in populations descended recently from Africans (such as African-Americans) is, however, a reflection of the response of natural selection to a regional disease.

The hemoglobin genes are not the only genes that are under selection for resistance to malaria. *Glucose-6-phosphate dehydrogenase* (*G6PD*), a gene involved in glucose metabolism, is one of several other genes implicated. People who are heterozygotes for a mutation in the *G6PD* gene—and therefore have a G6PD enzyme deficiency—can develop severe anemia when they eat certain foods (most notably fava beans; hence, the condition is called favism). People who are heterozygotes for a mutation in the *G6PD* gene, however, also have increased resistance to malaria, apparently because they are better at clearing infected red blood cells from their bloodstream. In areas where malaria is common, the advantage of malaria resistance offsets the disadvantage of favism.

Detailed evolutionary analysis of mutations in *G6PD* shows that favism has arisen multiple times, each time selectively favored because of its role in the body's response to the malaria parasite. As expected, favism, like sickle-cell anemia, is mainly a feature of populations in malarial areas or of populations whose evolutionary roots lie in these areas.

24.5 CULTURE, LANGUAGE, AND CONSCIOUSNESS

The most remarkable outcome of the evolutionary process described in this chapter is the human brain. This allows us to do extraordinary things, like appreciate Bach's music, read books, and build skyscrapers. But does this wonderful brain make us qualitatively different from other organisms? Does it in some way take us out of nature? Traditionally, the answer to these questions would have been a resounding yes. However, research into the capabilities of other species is questioning this conclusion: The human brain is certainly remarkable, but, in essence, what we can do is merely an extension of what other animals can do.

Culture changes rapidly.

Culture is generally defined as a body of learned behavior that is socially transmitted among individuals and passed down from one generation to the next. Culture has permitted us in part to transcend our biological limits. To take a simple example, clothing and ingeniously constructed shelters have enabled us to live in extraordinarily inhospitable parts of the planet, like the Arctic (**Fig. 24.17**). The capacity to innovate coupled with the ability to transmit culture is the key to the success of humans.

Culture, of course, changes over time. In many ways, cultural change is responsible for our species' extraordinary achievements. Genetic evolution is slow because it involves mutation followed by changes in allele frequencies that take place over many generations. Cultural change, on the other hand, can occur much more rapidly. Ten years ago, nobody had heard of smartphones, but today, millions of people own them. Or think of the speed at which a change in clothing style—a shift from flared to straight-leg jeans, for example—spreads through a population. Today's human population as a whole is genetically almost identical to the population when your grandparents were young. But think of the cultural changes that have occurred in the 50 years or so between your grandparents' youth and your own.

Despite this clear contrast between biological evolution and cultural change, we should not necessarily think of the two processes as independent of each other. Sometimes cultural change drives biological evolution.

A good example of the interaction between cultural change and biological evolution is the evolution of lactose tolerance in populations for which domesticated animals became an important source of dairy product. Most humans are lactose intolerant. Lactose, a sugar, is a major component of mammalian milk, including human breast milk. We have an enzyme, lactase, that breaks down lactose in the gut, but, typically, the enzyme is produced only in the first years of life, when we are breast-feeding. Once a child is weaned, lactase production is turned off. Lactase, however, is clearly a useful enzyme to have if there is a major dairy component to your diet.

FIG. 24.17 **The power of culture.** Inventions (such as clothing) have allowed our species to expand its geographic range. *Source: B&C Alexander/ArcticPhoto.*

Archaeological and genetic analyses indicate that cattle were domesticated probably three separate times in the past 10,000 years, in three separate places: in the Middle East, in East Africa, and in the Indus Valley. In at least two of these cases, there has been subsequent human biological evolution in favor of lactose tolerance, that is, continued lactase production throughout life. Analysis of the gene region involved in switching lactase production on and off has revealed mutations in European lactose-tolerant people that are different from those in African lactose-tolerant people, implying independent, convergent evolution of this trait in the two populations. Furthermore, we see evidence that these changes have evolved very recently, implying that they arose as a response to the domestication of cattle. Here, we see the interaction between cultural change and biological evolution. Biological

FIG. 24.18 **Nonhuman culture.** (a) English blue tits steal cream from a milk bottle on the doorstep. (b) Adult meerkats teach their young how to handle their prey. (c) Chimpanzees in different populations have devised different ways of using tools to hunt insects. *Sources: a. Colin F. Sargent; b. Dr. Alex Thornton, University of Exeter; c. FLPA/Peter Davey/age fotostock.*

evolution of continued lactase production has resulted from the change in a cultural practice, namely cattle domestication.

Is culture uniquely human?

Even nonprimates are known to be capable of learning from another member of their own species—culture is not uniquely human. Bird songs, for instance, may vary regionally because juveniles learn the song from local adults. In a famous case, shown in **Fig. 24.18a,** small birds called blue tits (*Parus caeruleus*) in Britain learned from each other how to peck through the aluminum caps of milk bottles left on doorsteps by milkmen to reach the rich cream at the top of the milk. Presumably, one individual discovered accidentally how to peck through a milk bottle cap and the others then imitated the action.

Nor is teaching, one way in which culture is transmitted, limited to humans. In teaching, one individual tailors the information available to another in order to facilitate learning. Teaching and imitation together make learning highly efficient. Adult meerkats teach young meerkats how to handle prey by giving them the opportunity to interact with live prey (**Fig. 24.18b**).

The most sophisticated example of nonhuman culture is shown by chimpanzees. Detailed studies of several geographically isolated populations have revealed 39 culturally transmitted behaviors, such as ways of using tools to catch insects, which are specific to a particular population (**Fig. 24.18c**). West African Chimpanzees in Guinea and the Ivory Coast both use tools to help them harvest the same species of army ants for food (and avoid the painful defensive bites of the ants), but each population has its own distinct way of doing this. Chimpanzees, then, are like us in having regional variation in culture.

Despite the cultural achievements of other species, it is clear that culture in *Homo sapiens* is very far removed from anything seen in the natural world. We are amazingly adept at imitation, learning, and acquiring culture. Teaching, learning, and cultural transmission all benefit from another extraordinary human attribute, language.

Is language uniquely human?

Language is a complex form of communication involving the transfer of information encoded in specific ways, such as speech, facial expressions, posture, and the like. Despite previous claims that language is unique to humans, it is not. The waggle dance of a worker bee on its return to the hive communicates information on the direction, distance, and nature of a food resource (Chapter 45). Vervet monkeys use warning vocalizations to specify the identity of a potential predator: They have one call for "leopard," for example, and another for "snake." But nonhuman animal languages are limited. Attempts to teach even our closest relatives, chimpanzees, to communicate using sign language have met with only limited success (**Fig. 24.19**).

Grammar provides a set of rules that allow the combination of words into a virtually infinite array of meanings. Noam Chomsky, father of modern linguistics, has pointed out that all human languages are basically similar from a grammatical viewpoint. Humans have what he has called a "universal grammar" that would lead a visiting linguist from another planet to conclude that all Earth's languages are dialects of the same basic language. That universal grammar is in some way hard wired into the human brain in such a way that every human infant spontaneously strives to acquire language. The specific attributes of the language depend on the baby's environment—a baby in France will learn French, and one in Japan will learn Japanese—but the basic process is similar in every case. Chimpanzees, and the entire natural world, lack the drive toward the acquisition of a grammatical language.

Is consciousness uniquely human?

Descartes famously wrote, "Cogito, ergo sum"—"I think, therefore I am." Can we legitimately rewrite his statement to declare, "Animals think, therefore they are"? With the growth

FIG. 24.19 Dr. Susan Savage Rumbaugh with Panbanisha, a female bonobo who learned to communicate using sign language. Chimpanzees and bonobos are able to learn and use sign language to express words and simple sentences. *Source: Anna Clopet/CORBIS.*

of the animal rights movement, particularly in reference to the treatment of animals in factory farms, this question is of more than academic interest. We now have many examples of animal thinking from a range of species, including, not surprisingly, chimpanzees and gorillas.

Perhaps more remarkable are the examples that come from animal species that are not closely related to us. In experiments carried out by Alex Kacelnik in Oxford, England, a pair of New Caledonian Crows was presented with two pieces of wire, one straight and the other hooked, and offered a food reward that could be obtained only by using the hooked wire. One member of the pair, the male, disregarded the experiment and flew off with the hooked wire. The female, however, having discovered that she could not get the food reward with the straight wire, went to some considerable trouble to bend a hook into the straight wire. She succeeded in getting the food. It is difficult to deny that the crow thought about the problem and was able to solve it, perhaps in the same way as we would. Definitions of consciousness are contested, but, as with language and culture, it seems clear that other species are capable of some form of conscious thought.

The evolutionary biologist Theodosius Dobzhansky once said, "All species are unique, but humans are uniquest." Our "uniquest" status is not derived from having attributes absent in other species, but from the extent to which those attributes are developed in us. Human language, culture, and consciousness are extraordinary products of our extraordinary brains. Nevertheless, as Darwin taught us and as we should never forget, we are fully a part of the natural world. ■

Core Concepts Summary

24.1 Anatomical, molecular, and fossil evidence shows that the human lineage branches off the great apes tree.

Anatomical features indicate that primates are a monophyletic group that includes prosimians, monkeys, and apes. The apes in turn include the lesser and great apes. page 486

The great apes include orangutans, gorillas, chimpanzees, and humans. page 487

Analysis of sequence differences between humans and our closest relatives, chimpanzees, indicate that our lineage split from chimpanzees 5–7 million years ago. page 487

Lucy, an unusually complete specimen of *Australopithecus afarensis,* demonstrates that our ancestors were bipedal by about 3.2 million years ago. page 489

Hominin fossils occur only in Africa until about 2 million years ago, when *Homo ergaster* migrated out of Africa to colonize the Old World. page 490

24.2 Phylogenetic analysis of mitochondrial DNA and the *Y* chromosome shows that our species arose in Africa.

Studies of mitochondrial DNA (mtDNA) suggest that the time back to the common ancestor of modern humans is about 200,000 years, implying that modern humans (*Homo sapiens*) arose in Africa (the out-of-Africa theory). page 491

The mtDNA out-of-Africa pattern is supported by Y chromosome analysis, which also shows a recent African origin of modern humans. page 494

Analysis of Neanderthal DNA from 30,000-year-old material indicates that, as the ancestors of non-African humans emigrated from Africa, they interbred with the Neanderthals. page 494

Our species originated in Africa and subsequently colonized the rest of the planet, starting about 60,000 years ago. page 495

24.3 During the 5–7 million years since the most recent common ancestor of humans and chimpanzees, our lineage acquired a number of distinctive features.

The development of bipedalism involved a wholesale restructuring of anatomy. page 495

Neoteny is the process in which the timing of development is altered over evolution so that a sexually mature organism retains the physical characteristics of the juvenile form; humans are neotenous, exhibiting many traits as adults that chimpanzees exhibit as juveniles. page 496

There are many possible selective factors that explain the evolution of our large brain, including tool use, social living, and language. page 497

FOXP2, a transcription factor involved in brain development, may be important in language, as mutations in the gene that encodes FOXP2 are implicated in speech pathologies. page 498

24.4 Human history has had an important impact on patterns of genetic variation in our species.

Because our ancestors left Africa very recently in evolutionary terms, there has been little chance for genetic differences to accumulate among geographically separated populations. page 499

Humans have very little genetic variation, with only about 1 in every 1000 base pairs varying among individuals. page 499

Most of the variation among humans occurs within populations. As much as 85% of the total amount of genetic variation in humans can be found within a single population. Only about 7% of human genetic variation segregates between groups that are commonly called races. page 500

Some racial differences, such as skin color and resistance to malaria, have probably arisen by natural selection. page 500

Other racial differences have probably arisen by sexual selection. page 501

24.5 Culture is a potent force for change in modern humans.

Cultural evolution and biological evolution may interact, as in the case of the evolution of lactose tolerance in regions where cattle were domesticated. page 502

Other animals possess simple versions of culture, language, and even consciousness, but the capabilities of our species in all three are truly exceptional. page 503

Self-Assessment

1. Describe the evidence suggesting that chimpanzees are the closest living relatives of humans.
2. Explain the out-of-Africa theory of human origins and how studies of mitochondrial DNA and the Y chromosome support it.
3. List four anatomical differences between chimpanzees and humans, and explain how these changes facilitated walking upright.
4. Given the high genetic similarity of humans and chimpanzees, how can we account for the differences we see between the two species?
5. Describe three possible selective factors underlying the evolution of large brains in our ancestors.
6. Explain how differences among different human populations arose by natural and sexual selection.
7. Provide arguments for and against the idea that culture, language, and consciousness are uniquely human.

Log in to LaunchPad to check your answers to the Self-Assessment questions, and to access additional learning tools.

Gallery Stock.

PART 2

FROM ORGANISMS TO THE ENVIRONMENT

"From so simple a beginning endless forms most beautiful and most wonderful have been, and are being, evolved."

— CHARLES DARWIN

Andre Gallant/Getty Images.

CHAPTER 25

Cycling Carbon

Core Concepts

25.1 Photosynthesis and respiration are the key biochemical pathways for the biological, or short-term, carbon cycle.

25.2 Physical processes govern the long-term carbon cycle.

25.3 The carbon cycle can help us understand ecological interactions and the evolution of biological diversity.

Let's imagine that we could tag a carbon atom at its moment of origin and then follow its odyssey through time and space. Formed in a nuclear blast furnace deep within an ancient star and then ejected into space as the star died, our atom was eventually swept up with other materials to form the Earth, a small planet orbiting the newer star we now call the sun. Volcanoes introduced our carbon atom into the early atmosphere as carbon dioxide (CO_2), and slowly, over millions of years, this CO_2 reacted chemically with rocks, transferring the carbon to limestone that accumulated on the sea floor. Here, our atom sat for many millions of years, until earthquakes, erosion, or other geologic activities returned it to the atmosphere as, once again, CO_2. Slowly but surely, geologic processes on the early Earth cycled carbon from atmosphere to rocks, and back again.

Sometime between 4 and 3.5 billion years ago, our carbon atom began to cycle more rapidly—much more rapidly—as the carbon cycle gained new players. Photosynthetic microorganisms converted CO_2 into organic molecules, and respiring microorganisms returned CO_2 to the environment, completing a rapid biological carbon cycle. To this day, the biological carbon cycle endlessly propels our atomic wayfarer from atmosphere and oceans to cells and back again, while the slow burial of organic matter and limestone in sedimentary rocks continues to bring carbon atoms to the solid Earth, returning it to the atmosphere only slowly, on geologic timescales.

In Part 1, we explored the molecular and cellular basis of life, culminating in a discussion of how genetic variation and natural selection shape evolution. Part 2 examines the products of evolution as it has played out through Earth history: our emerging sense of the immense biological diversity of Bacteria and Archaea; the myriad microscopic eukaryotes found in oceans, lakes and soils; and the millions of plant, fungal, and animal species that define our biological landscape. As we will see, function and diversity underpin the ecological interactions that shape communities, ecosystems, and biomes across our planet.

We begin with an introduction to the **carbon cycle,** the intricately linked network of biological and physical processes that shuttles carbon among rocks, soil, oceans, air, and organisms. Why start here? Because the carbon cycle provides a fundamental organizing principle for understanding life on Earth. Indeed, it lies at the heart of everything else we discuss in Part 2. The chemistry of life is, to a first approximation, the chemistry of carbon. How organisms move carbon from one species to another, and between organisms and their surrounding environment, underpins both the efficient functioning of ecosystems and their persistence over an immense span of time. In no small part, biological diversity can be understood in terms of the varied ways that organisms obtain the carbon needed for growth and reproduction. Much of ecology, in turn, concerns the ways that organisms interact to cycle carbon and transfer the energy stored in organic molecules. The carbon cycle focuses our attention on the ways that physical and biological processes together determine the properties of environments. And it provides a basis for assessing the role that humans play in our environmental present and future.

25.1 THE SHORT-TERM CARBON CYCLE

In 1958, American chemist Charles David Keeling began a novel program to monitor the atmosphere. At the time he began his study, scientists had only a vague notion of how carbon dioxide (CO_2) behaves in air. Many thought that CO_2 might vary unpredictably from time to time, and from place to place. Keeling decided to find out if in fact it did. From five towers set high on Mauna Loa in Hawaii, 3400 m above sea level, he sampled the atmosphere every hour and measured its composition with an infrared gas analyzer. Within the first few years of his project, a pattern of seasonal oscillation became apparent: CO_2 concentration in the air reached its annual high point in spring and then declined by about 6 parts per million (ppm, by volume) to a minimum in early fall (**Fig. 25.1**). Such regular variation was unexpected: What could be causing it?

Perhaps the pattern was local. Airplane traffic to Hawaii peaked in winter, and maybe that was affecting the air around Mauna Loa. As measurements continued through a number of years, however, the pattern persisted, even as trends in air traffic changed. Moreover, monitoring stations in many other parts of the globe gave similar results. Knowing that the Mauna Loa measurements are representative of the atmosphere as a whole allows us to appreciate what an annual variation of 6 ppm really means. With calendar-like regularity, approximately 47 billion metric tons of CO_2 were entering and leaving the atmosphere annually—that's 13 billion metric tons of the element carbon (the rest of the mass is oxygen). To explain such a pattern, we must consider processes that affect the planet as a whole.

A second pattern is evident in Fig. 25.1. Summertime removal of CO_2 from the atmosphere does not quite balance winter increase, with the result that the amount of CO_2 in the atmosphere on any given date is greater than it was a year earlier. Atmospheric CO_2 levels have therefore increased steadily through the period of monitoring. For example, mean annual CO_2 levels, about 315 ppm in 1958, reached 404 ppm by late April 2015. This is an increase of more than 25%, and there is no indication that the rise will slow or stop any time soon.

What causes atmospheric CO_2 to vary through the year and over a timescale of decades? To answer this question, we need to ask what processes introduce CO_2 into the atmosphere and what processes remove it. Carbon dioxide is added to the atmosphere by (1) geologic inputs, mainly from volcanoes and mid-ocean ridges; (2) biological inputs, especially respiration; and (3) human activities, including deforestation and the burning of fossil fuels. Processes that remove CO_2 from the atmosphere include (1) geologic removal, especially by chemical weathering in which

FIG. 25.1 The Keeling curve. The Keeling curve provides a record of atmospheric CO_2 concentrations over half a century, measured from an observatory at Mauna Loa, Hawaii. *Data from Dr. Pieter Tans, NOAA/ESRL (www.esrl.noaa.gov/gmd/ccgg/trends/) and Dr. Ralph Keeling, Scripps Institution of Oceanography (scrippsco2.ucsd.edu/).*

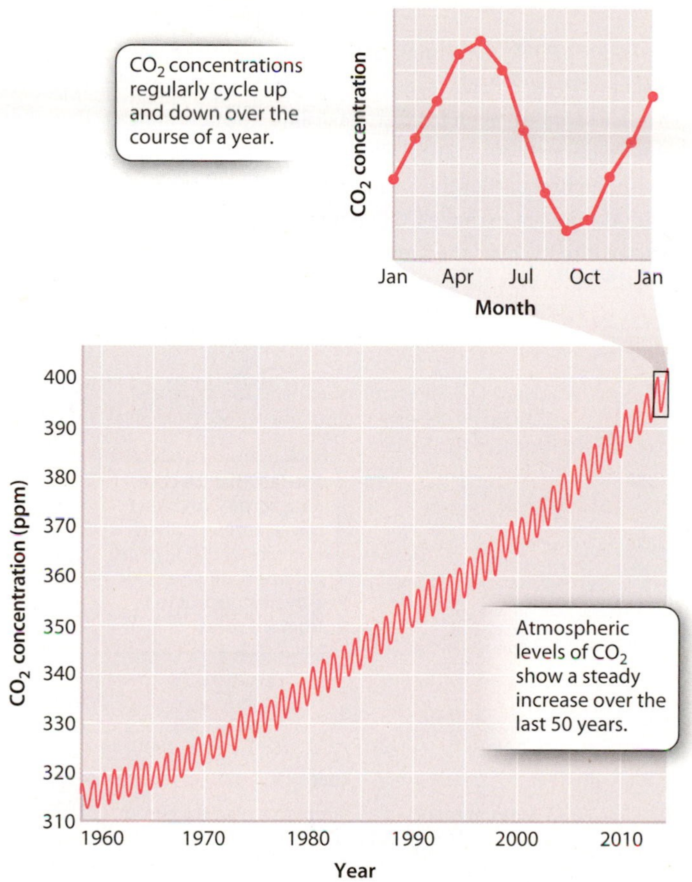

CO_2 in rainwater reacts with exposed rocks, and (2) biological removal, mainly through photosynthesis. As we will see, geologic processes play key roles in the carbon cycle on timescales of centuries and longer. Only biology, however, can account for the annual rhythm of the Keeling curve.

Photosynthesis and respiration are key processes in short-term carbon cycling.

To grow, all organisms require both carbon and energy. As we saw in Chapter 8, photosynthetic organisms use energy from the sun to form ATP and NADPH, which reduce CO_2 to carbohydrates. As photosynthesis pulls CO_2 out of the atmosphere (or water, for aquatic organisms), the carbon is transferred to carbohydrates, and oxygen is given off as a by-product. The overall chemical reaction is shown here:

$$6CO_2 + 6H_2O \rightarrow C_6H_{12}O_6 + 6O_2$$

It is estimated that photosynthetic organisms remove about 210 billion metric tons of carbon (in nearly 770 billion metric tons of CO_2) from the atmosphere each year, about 60% of it by land plants and the rest by phytoplankton and seaweeds in the oceans. This annual photosynthetic removal of CO_2 equals about 25% of the total CO_2 in the atmosphere. In the absence of processes that restore CO_2 to the atmosphere, high rates of photosynthesis would use up the atmospheric reservoir of CO_2 in just a few years. But the Keeling curve tells us that atmospheric CO_2 is *not* decreasing rapidly—it is rising. Moreover, the geologic record indicates that photosynthesis has fueled ecosystems for billions of years without causing CO_2 depletion.

What processes return CO_2 to the atmosphere, balancing photosynthetic removal? In Chapter 7, we discussed aerobic respiration. Humans and many other organisms gain both the energy and carbon needed for growth from organic molecules (that is, by consuming food). As explained in Chapter 7, aerobic respiration uses oxygen to oxidize organic molecules to CO_2, converting chemical energy in the organic compounds to ATP for use in cellular processes. Aerobic respiration can be summarized by this chemical equation:

$$C_6H_{12}O_6 + 6O_2 \rightarrow 6CO_2 + 6H_2O$$

Plants respire, too, converting the chemical energy in carbohydrates into useful work. On the present-day Earth, the amount of CO_2 returned annually to the atmosphere by aerobic respiration and related processes is about equal to the amount that plants and algae remove by photosynthesis. That is why photosynthesis does not deplete the CO_2 in air.

As we discussed in Chapter 8, the equations for oxygenic photosynthesis and aerobic respiration mirror each other. Photosynthesis continually converts CO_2 to organic molecules, using energy from the sun, and respiration harvests the chemical energy stored in organic molecules, releasing CO_2 and water in the process. Photosynthesis and aerobic respiration are complementary metabolic pathways: Each uses the products of the other and generates a new supply of substrates for its complement. The result is a cycle (**Fig. 25.2**).

FIG. 25.2 The short-term carbon cycle. The complementary metabolic processes of photosynthesis and respiration drive the short-term cycling of carbon through the biosphere.

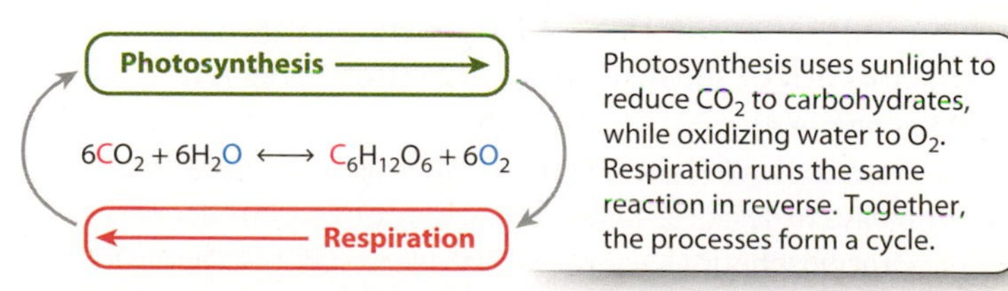

HOW DO WE KNOW?

FIG. 25.3

How much CO_2 was in the atmosphere 1000 years ago?

BACKGROUND Direct measurements of the atmosphere show that CO_2 has increased by 25% over the past 50 years. To know whether or not such a change is unusual, we need to know CO_2 levels over a longer time interval.

OBSERVATION Scientists recognized that the snow that accumulates on glaciers to form new layers of ice is initially full of small spaces that are in contact with the air. Through time, as snow changes to ice, these spaces become sealed off from the surrounding environment and form bubbles, preserving a minute sample of air at the time of ice formation.

Ice core drilled from a glacier in Antarctica

Viewed under the microscope, glacial ice contains small bubbles that trap samples of air.

Photo sources: (left) Courtesy of Ted Scambos & Rob Bauer, National Snow and Ice Data Center, University of Colorado, Boulder, (right) Courtesy Vin Morgan.

MEASUREMENTS Identifying chemical indicators that vary on an annual basis in successive layers of ice enabled scientists to assign ages to ice samples taken from cores drilled through the Law Dome, a large ice dome in Antarctica. The youngest ice samples overlap in age with the first part of the Keeling data, providing a check that air samples trapped in ice have CO_2 concentrations similar to those in air measured directly over the same interval.

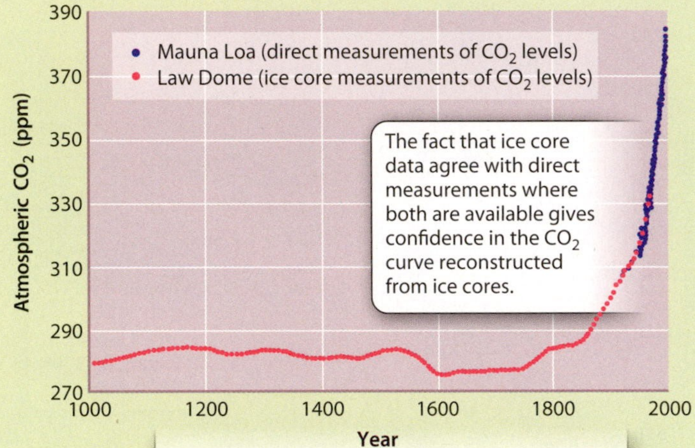

Measurements of air samples in glacial ice from the Law Dome, Antarctica, show that atmospheric CO_2 varied little between 1000 and 1800.

CONCLUSIONS Before the Industrial Revolution, atmospheric CO_2 levels had varied little over 1000 years, generally falling between 270 and 280 ppm. From this, scientists have concluded that current changes in atmospheric CO_2 are unusual on the timescale of the past millennium.

SOURCE Etheridge, D. M., et al. 1996. "Natural and Anthropogenic Changes in Atmospheric CO_2 over the Last 1000 Years from Air in Antarctic Ice and Firn." *Journal of Geophysical Research* 101:4115–4128.

The regular oscillation of CO_2 reflects the seasonality of photosynthesis in the Northern Hemisphere.

Now that we have identified the major processes in the short-term (days to decades) carbon cycle, we can examine what makes atmospheric CO_2 level oscillate seasonally. Except in a narrow band around the equator, photosynthesis is seasonal, with higher rates in the summer and lower rates in the winter. Respiration, in contrast, remains more or less constant through the year.

Even cursory examination of the Earth shows that land is distributed asymmetrically, with more land—and hence more plants—in the Northern Hemisphere than in the Southern Hemisphere. For this reason, global atmospheric CO_2 declines through the northern summer, when the ratio of photosynthesis to respiration is highest, and then increases through fall and winter, when the ratio is reversed. The result is the seasonal oscillation of atmospheric CO_2 levels documented by Keeling.

Human activities play an important role in the modern carbon cycle.

Now let us ask about the second major pattern evident in the Keeling curve. Although CO_2 levels oscillate on an annual basis, the overall pattern is one of sustained increase. During the 1960s, the observed increase was less than 1 ppm each year. In the first decade of the new millennium, it has been closer to 2 ppm. The current level of atmospheric CO_2 is more than 25% higher than it was 50 years ago. Why has CO_2 input into the atmosphere outstripped removal for the past half century? Should we consider this pattern unusual?

Answering the second question helps us to address the first. Conclusions about whether it is unusual for atmospheric CO_2 to increase 25% in 50 years can be answered only if we can compare the Keeling curve with longer records of atmospheric CO_2 levels. Before Keeling, no one systematically collected air samples. Fortunately, however, nature did it for us.

When glacial ice forms, it traps tiny bubbles of air. Each year, snowfall gives rise to a new layer of ice, and as layers accumulate through time their bubbles preserve a history of the atmosphere. The graph in **Fig. 25.3** shows the amount of CO_2 in air bubbles trapped in Antarctic ice that has accumulated over the past 1000 years. Ice samples show that CO_2 levels actually began to increase slowly in the 1800s, the time of major transformations in mining, manufacturing, and transportation we call the Industrial Revolution. Before that, however, atmospheric CO_2 levels had varied little since the Middle Ages, staying at 270–280 ppm for centuries on end. On a 1000-year timescale, then, big changes in CO_2 abundance happened only once, in the last 200 years. Atmospheric CO_2 inputs and outputs were approximately in balance until the Industrial Revolution.

Fig. 25.3 shows that there is a **correlation** between the increasing CO_2 content of the atmosphere and human activities. A correlation simply indicates that two events or processes occur together. Can we actually demonstrate that humans have played a role in recent atmospheric change? By itself, correlation does not establish **causation,** a relationship in which one event leads to another. We might propose instead that recent CO_2 increase reflects natural processes and only matches the period of industrialization by coincidence.

→ **Quick Check 1** How does the graph in Fig. 25.3 suggest that human activities have influenced CO_2 levels in the atmosphere? What might be a plausible alternative hypothesis?

Carbon isotopes show that much of the CO_2 added to air over the past half century comes from burning fossil fuels.

We can test the hypothesis that human activities have contributed to the increases in atmospheric CO_2 measured at Mauna Loa and in ice cores by making careful measurements of a chemical detail: the isotopic composition of atmospheric CO_2. In Chapter 2, we saw that many elements have several isotopes, atoms of the element that vary in atomic mass because they have different numbers of neutrons. Carbon has three isotopes: ^{12}C (with six neutrons, about 99% of all carbon atoms); ^{13}C (with seven neutrons, most of the remaining 1%); and the rare ^{14}C (with eight neutrons, about one part per trillion of atmospheric carbon).

More than 40 years ago, the Austrian-born chemist Hans Suess began to measure the relative abundances of the three isotopes of carbon in atmospheric CO_2, a program that continues today. He observed that the proportion of ^{13}C in atmospheric CO_2 has declined as the total amount of CO_2 has increased (**Fig. 25.4**). This subtle change in isotopic composition has occurred because the CO_2 being added to the atmosphere has less ^{13}C than the CO_2 already in the air.

How does this result influence the hypotheses we developed to account for increasing atmospheric CO_2 levels? In photosynthesis, CO_2 containing the lighter isotope ^{12}C is incorporated into biomolecules preferentially over CO_2 containing ^{13}C, and for this reason organic matter generated by photosynthesis (and the organic matter of organisms that eat photosynthetic organisms) differs in its proportions of ^{13}C and ^{12}C from CO_2 from volcanic gases and inorganic carbon dissolved in the oceans. Volcanic and dissolved marine carbon do not have the right ratio of ^{13}C to ^{12}C to explain the isotopic change observed in the atmosphere over time. In contrast, organic matter in living organisms has just the right ratio of ^{13}C to ^{12}C to account for Suess's measurements. Therefore, processes that cause a net conversion of organic matter to CO_2, for example by burning, must be adding CO_2 to the atmosphere.

Here's where the other isotope of carbon, ^{14}C, comes in. Measurements also show that the sources of added CO_2 are depleted in ^{14}C relative to the CO_2 already in the air. Modern organic matter contains too much ^{14}C to account for the observed pattern, but ancient organic matter—the coal, petroleum, and natural gas burned as fossil fuels—is isotopically just right. Chemical analyses, therefore, support the hypothesis that human activities cycle carbon in amounts high enough to affect the chemical composition of the atmosphere.

That atmospheric CO_2 is increasing is not a hypothesis; it is a measurement. That the burning of fossil fuels plays an important role in this increase is also unambiguously documented by chemical analyses. Current debate focuses not on these observations, but instead on the more difficult problem of understanding how increasing CO_2 will affect climate. We discuss this debate in Chapter 49. For now, however, we can use

HOW DO WE KNOW?

FIG. 25.4

What is the major source of the CO_2 that has accumulated in Earth's atmosphere over the past two centuries?

BACKGROUND Chemical analyses show that CO_2 levels in the current atmosphere are about 40% higher than they were at the time of the American Revolution. This rise coincides with major advances in manufacturing and transportation, which are driven by the burning of fossil fuels. These coincidences in timing suggest that human activities are responsible for increasing CO_2 levels.

HYPOTHESIS The burning of fossil fuels, central to the emergence of modern industrial societies, has been the principal cause of the measured increase of CO_2 in the atmosphere.

EXPERIMENT AND RESULTS Hans Suess measured the abundance of the two stable isotopes of carbon, ^{13}C and ^{12}C, in air samples and demonstrated that the ratio of ^{13}C to ^{12}C in atmospheric CO_2 has decreased over the second half of the twentieth century. Additional measurements of coral skeletons and wood show that the ratio of ^{13}C to ^{12}C in air has been decreasing over the entire period during which atmospheric CO_2 levels have been increasing.

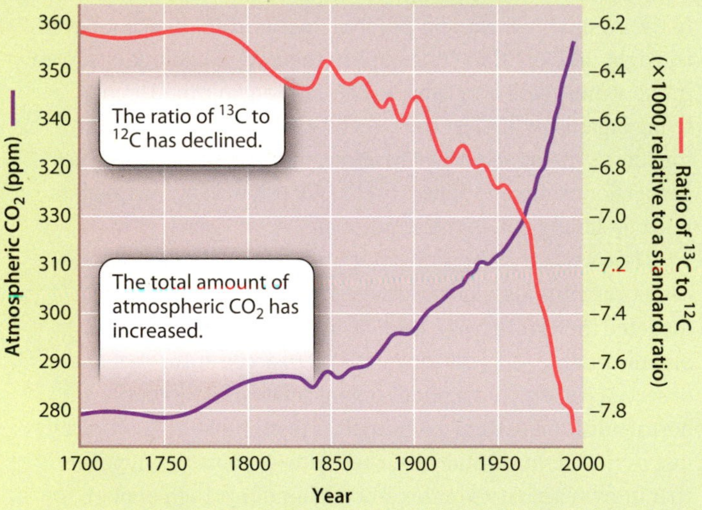

DISCUSSION The ratio of ^{13}C to ^{12}C in CO_2 added to the atmosphere over the past 200 years is lower than that in CO_2 that was already in the air more than 200 years ago. The ratio of ^{13}C to ^{12}C in CO_2 emitted by volcanoes is too high to account for the data, as is the ratio in CO_2 released from the oceans. In contrast, organic matter formed by photosynthesis has just the right ratio of ^{13}C to ^{12}C to account for Suess's measurements. By itself, the changing abundance of ^{13}C in the air could reflect the conversion of plant carbon to CO_2 by burning, or it could reflect the burning of fossil fuels formed by the burial of plant and algal materials in the geologic past.

FOLLOW-UP WORK Further studies, using ^{14}C, the third, unstable isotope of C, also show a strong decrease in the proportional abundance of ^{14}C in atmospheric CO_2 over time. While the burning of vegetation to clear land for agriculture results in CO_2 with the right ratio of ^{13}C to ^{12}C, modern vegetation has far too much ^{14}C to account for observed changes in the ^{14}C of air. Only fossil fuels have the right ratios of all three carbon isotopes—^{12}C, ^{13}C, *and* ^{14}C—to account for the pattern of isotopic change in atmospheric CO_2 measured by Suess and others. In fact, ^{14}C data from the high Rocky Mountains indicate that if fossil fuels were the principal source of added CO_2, CO_2 levels should have increased by a little less than 2 ppm per year between 2004 and 2008—which is very close to the increase actually measured.

Air measurements at Niwot Ridge, Colorado

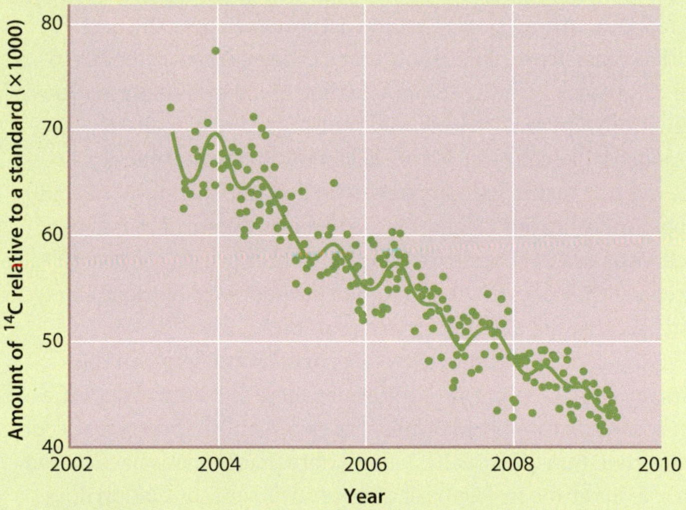

CONCLUSION Fossil fuel burning by industrialized societies has been, and continues to be, a principal source of CO_2 buildup in Earth's atmosphere.

SOURCES Revelle, R., and H. E. Suess. 1957. "Carbon dioxide exchange between atmosphere and ocean and the question of an increase of atmospheric CO_2 during the past decades." *Tellus* 9:18–27; Turnbull, J. C., et al. 2007. "A new high precision $^{14}CO_2$ time series for North American continental air." *Journal of Geophysical Research: Atmospheres* 112, no. D11. Article Number: D11310. doi: 10.1029/2006JD008184.

FIG. 25.5 **Carbon dioxide sources and sinks.** Much of the CO_2 introduced by humans into the atmosphere has not stayed there. About half of it has accumulated as dissolved inorganic carbon in the oceans or as new biomass on the continents. *Data from J. G. Canadell, C. Le Quéré, M. R. Raupach, C. B. Field, E. T. Buitenhuis, P. Ciais, T. J. Conway, N. P. Gillett, R. A. Houghton, and G. Marland, 2007, "Contributions to Accelerating Atmospheric CO_2 Growth from Economic Activity, Carbon Intensity, and Efficiency of Natural Sinks," Proceedings of the National Academy of Sciences USA 104:18866–18870.*

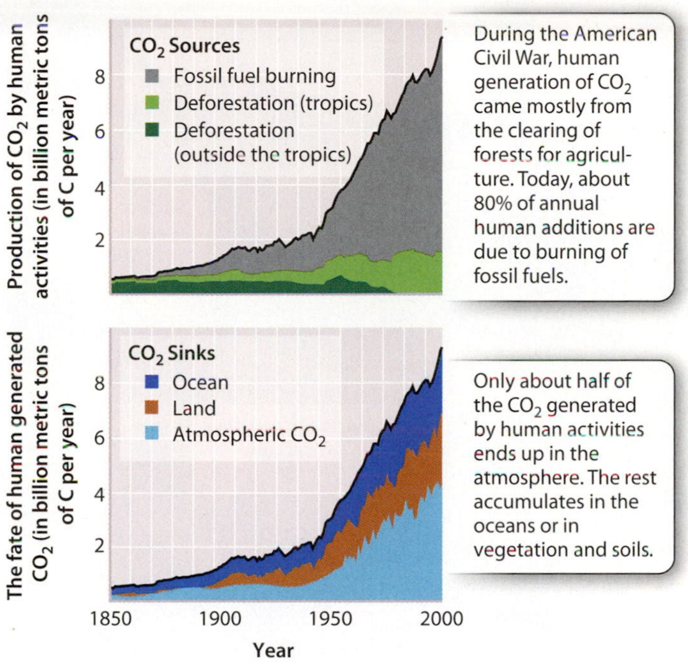

some relevant numbers to reveal another important aspect of the carbon cycle.

As shown in the top graph in **Fig. 25.5,** humans now add almost 9 billion metric tons of carbon to the atmosphere each year, nearly all of it as CO_2. Much of this material comes from power plants, airplanes, ships, and automobiles, which oxidize ancient organic matter to CO_2 on a grand scale. Further important inputs come from the conversion of forests to agriculture or pastureland. Clearing and burning forest trees also oxidizes organic carbon to CO_2.

Interestingly, if we compare our best estimates of human CO_2 generation since 1958 (not including CO_2 added through respiration) with the measured increase in atmospheric CO_2, we find that about half of the CO_2 released by human activities *did not* end up in the air. Where did it go? This has been a major question for years, and it now appears that much of the "missing carbon" has been stored as inorganic carbon in the form of CO_2, bicarbonate (HCO_3^-), and carbonate (CO_3^{2-}) ions dissolved in the oceans. The bottom graph in Fig. 25.5 shows an estimate of the various places where human-produced CO_2 ends up. These observations indicate that Earth's carbon cycle is more complicated than a simple exchange of carbon between photosynthesis and respiration. There are additional storehouses of carbon and additional processes that move carbon from one store to another, topics we discuss next.

25.2 THE LONG-TERM CARBON CYCLE

Up to this point, we have considered only the short-term carbon cycle: exchanges over days, years, and decades driven by the biological processes of photosynthesis and respiration, and altered in recent times by human activities. Geologic evidence, however, tells us that on timescales of thousands to millions of years, CO_2 levels in air have changed even more dramatically. These longer-term changes mean that we must consider additional contributions to the carbon cycle: physical processes, including volcanism and climate change. Indeed, the complete carbon cycle links Earth's physical and biological processes, providing a foundation for understanding the interconnected histories of life and environment through our planet's long history.

Reservoirs and fluxes are key in long-term carbon cycling.

To understand how biological and physical processes interact to govern CO_2 levels in the air, we must first look at how carbon is distributed among its various **reservoirs,** the places where carbon is found on the Earth. As shown in the purple boxes in **Fig. 25.6,** reservoirs of carbon include organisms, the atmosphere, soil, the oceans, and sedimentary rocks, and both biological and geologic processes cycle carbon back and forth among these pools.

How much carbon is stored in each of Earth's major reservoirs? If we add up all the carbon contained in the total mass of organisms living on land, that amount of carbon is just a bit smaller than the amount of carbon stored as CO_2 in the atmosphere. Soil, by comparison, stores as much carbon as do land organisms and the atmosphere combined, mostly as slowly decaying organic compounds. In the oceans, the amount of carbon contained in living organisms is actually very small. Much more resides as inorganic carbon dissolved in the water—most of it in the deep oceans as CO_2 and bicarbonate and carbonate ions. Perhaps fortunately for us, the oceans have sopped up some of the CO_2 generated by human activities.

The biggest carbon reservoir of all, however, lies beneath our feet, within sediments and sedimentary rocks. Calcium carbonate minerals ($CaCO_3$), which form limestone, and organic matter preserved in sedimentary rocks dwarf all other carbon reservoirs combined by three orders of magnitude. Coal, petroleum, and natural gas make up only a small part of the organic matter stored in sedimentary rocks, but as we have seen, they play an important role in the modern carbon cycle.

Fluxes are the rates at which carbon flows from one reservoir to another. The sensitivity of different reservoirs to change depends on the relative sizes of the reservoir and of the amount of movement of material into and out of it. When fluxes are large

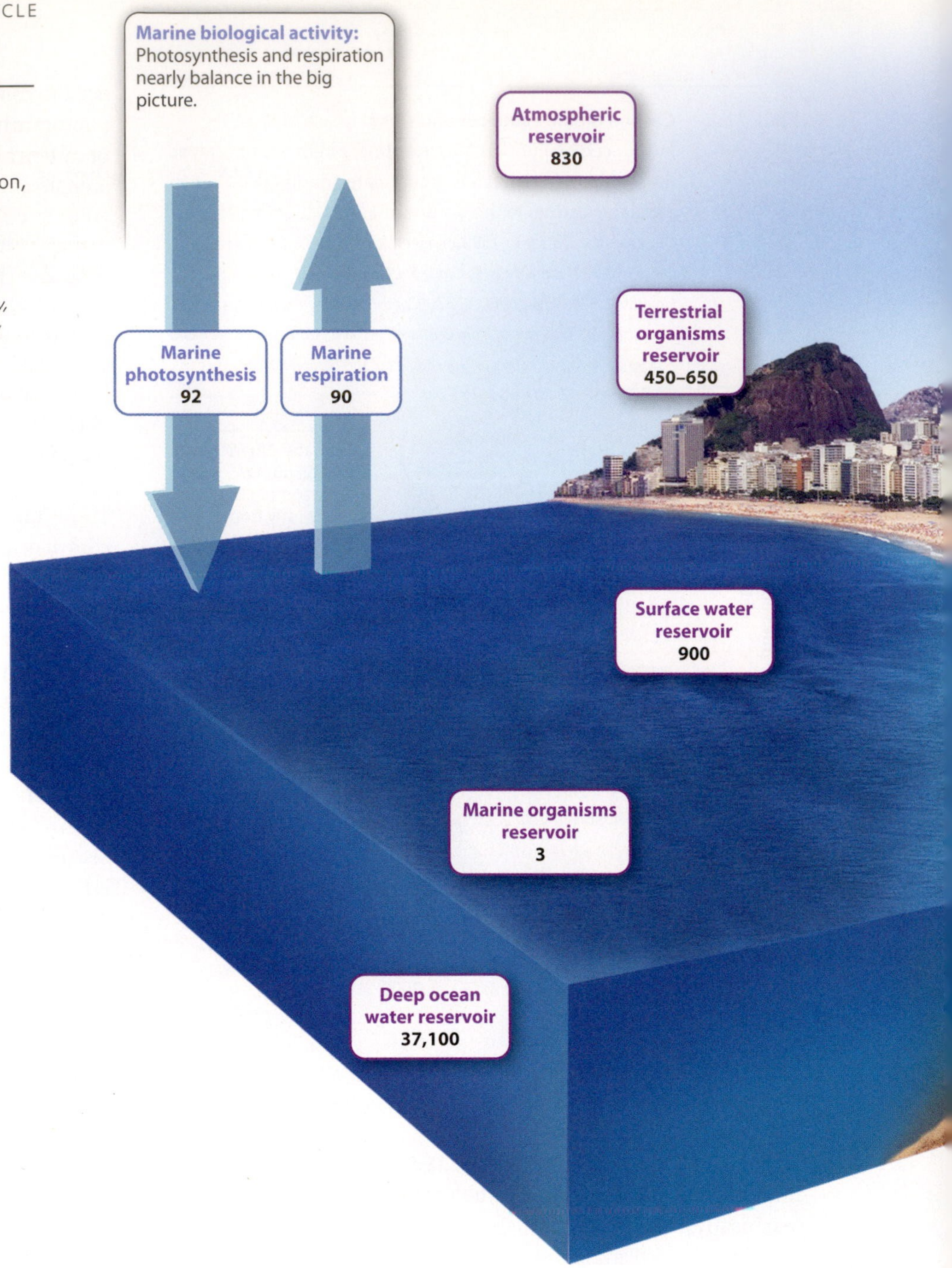

FIG. 25.6 Movement of carbon between reservoirs. The numbers for each reservoir are gigatons (a gigaton equals 1 billion metric tons) of carbon, and those for fluxes are gigatons of carbon per year. *Photo source: Fotosearch Stock Images; Data from P. Rekacewicz, "The Carbon Cycle," from Vital Climate Graphics, UNEP/GRID–Arendal Maps & Graphics Library, JPG file, 2005, http://www.grida.no/graphicslib/detail/the-carbon-cycle_a224.*

relative to the size of the reservoir, reservoir size can change rapidly. As shown by the arrows in Fig. 25.6, the amount of carbon stored as CO_2 in air is not that much larger than the annual fluxes into and out of the atmosphere. For this reason, atmospheric CO_2 abundance can be influenced by a number of processes at work in the carbon cycle.

Physical processes add and remove CO_2 from the atmosphere.

Physical processes, like biological ones, are capable of adding and removing CO_2 from the atmosphere. A key process is volcanism: Volcanoes and mid-ocean ridges release an estimated 0.1 billion metric tons of carbon (as CO_2) into the atmosphere each year. More CO_2, about 0.05 billion metric tons of carbon per year, is released by the slow oxidation of coal, oil, and other ancient organic material in sedimentary rocks exposed at the Earth's surface. In nature, bacteria and fungi accomplish most of this oxidation by respiring old organic molecules. By burning fossil fuels, humans accelerate this process dramatically, increasing a hundredfold the rate at which sedimentary organic carbon is oxidized.

What geologic processes remove carbon from the atmosphere? About 0.43 billion metric tons of CO_2 are removed from the atmosphere by chemical reactions between air and exposed rocks, a process called chemical weathering. To understand how weathering fits into the carbon cycle, we must first understand that CO_2 in the air reacts with rainwater to form carbonic acid (H_2CO_3). This acid slowly reacts with rock-forming minerals, generating (among other substances) calcium and bicarbonate (HCO_3^-) ions that are transported by rivers to the oceans. Within the oceans, the calcium and bicarbonate ions react with each other to form calcium carbonate ($CaCO_3$) minerals that precipitate out of seawater and accumulate on the ocean floor, forming limestone. In essence, carbon moves from the atmosphere to the sediments, where it can be stored for millions of years.

The two linked processes of chemical weathering and mineral precipitation in the oceans can be summarized by a simple chemical reaction (here, $CaSiO_3$ stands in for rock-forming minerals as a whole):

$$CaSiO_3 + CO_2 \rightarrow CaCO_3 + SiO_2$$

In present-day oceans, the calcium carbonate minerals that accumulate on the seafloor do not precipitate directly from seawater, like salt. Instead, they are generated predominantly by organisms, as mineralized skeletons and shells. Clams, corals, and many other organisms form beautiful and functional shells of $CaCO_3$ (**Fig. 25.7**). Skeleton formation is an example of

biomineralization, the precipitation of minerals by organisms. Biomineralization provides yet another link between Earth and life, and one that has varied through evolutionary history.

The annual carbon fluxes associated with geologic processes are small relative to the yearly inputs and outputs from photosynthesis and respiration. As we see in Fig. 25.6, however, photosynthesis and respiration almost cancel each other out. In the long-term carbon cycle, it is the net result of these rapid biological processes—photosynthesis input minus respiratory output—that matters. Is CO_2 produced or consumed when all biological processes are considered together?

There is some uncertainty in estimates of reservoir and flux sizes in the present- day carbon cycle, but by most estimates the amount of carbon incorporated into organic matter by photosynthesis is a bit larger than the amount returned to the atmosphere by respiration. The difference isn't much—about the same magnitude as volcanic and weathering fluxes. The "excess" organic carbon produced by photosynthesis is deposited in sediments as they accumulate on land or the seafloor, sustaining a small but constant leak of carbon from the short-term biological components of the carbon cycle to long-term geologic reservoirs.

One final set of geological processes completes the long-term carbon cycle. Earth's crust is constantly in motion, propelled by heat within the Earth's interior. **Plate tectonics** is the name given to this dynamic movement of our planet's outer layer. New crust forms at spreading centers, the places where molten rock rises upward from the underlying mantle. Old crust is destroyed in subduction zones, where one slab of crust slides beneath another, returning material to the mantle. Plate tectonics provides geology's best explanation for the formation of mountains and ocean basins, and it plays an important role in the long-term carbon cycle. The crust that descends into subduction zones carries with it sediments, including carbonate minerals and organic matter. Subduction removes carbon from Earth's surface, but this carbon will be recycled to the surface as CO_2 emitted from volcanoes and mid-ocean ridges. **Fig. 25.8** shows the physical processes at work in Earth's long-term carbon cycle.

→ **Quick Check 2** If plate tectonic processes form a chain of high mountains, would you expect atmospheric CO_2 to increase or decrease?

Records of atmospheric composition over 400,000 years show periodic shifts in CO_2 content.

Earlier in this chapter, we discussed evidence that before the Industrial Revolution, atmospheric CO_2 levels had not changed

FIG. 25.7 Clams and corals with skeletons made of calcium carbonate. *Source: Sami Sarkis/Getty Images.*

FIG. 25.8 Geological processes that drive the long-term carbon cycle. Numbers are estimates for annual fluxes of carbon, in gigatons.
Source: Data from J. Gaillardet and A. Galy, 2008, "Himalaya—Carbon Sink or Source?" Science 320:1728, doi:10.1126/science.1159279.

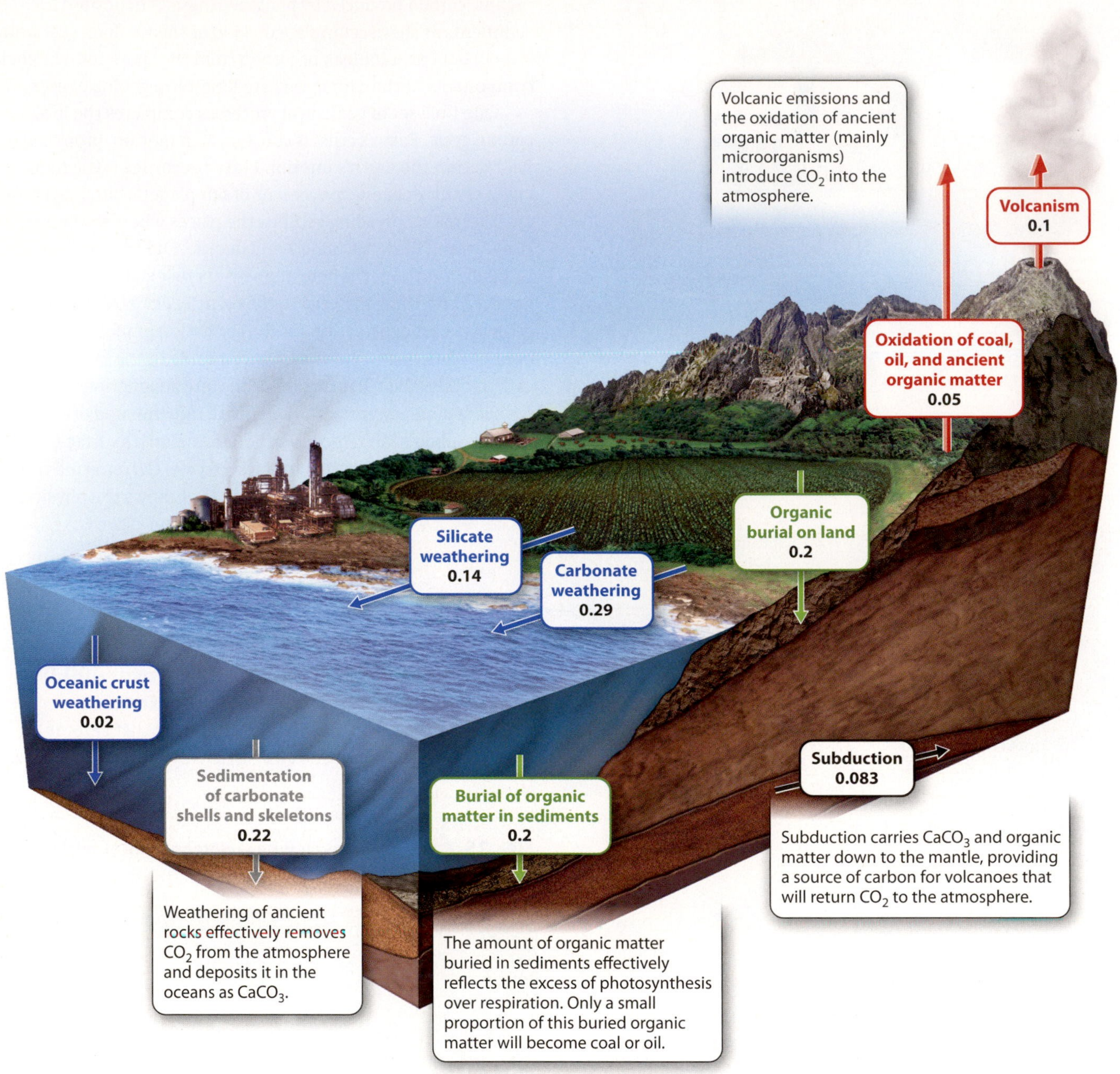

appreciably for 1000 years or more. Longer-term records, however, show that the CO_2 levels in air can change substantially through time.

At Vostok, high on the Antarctic ice sheet, glacial ice records more than 400,000 years of environmental history (**Fig. 25.9**). As shown in the top graph of Fig. 25.9, the youngest samples show about 285 ppm CO_2 in the atmosphere, consistent with direct measurements of air, including the first years of the Keeling curve. Notice, however, that 20,000 years ago, CO_2 levels were much lower—about 180 ppm. In fact, the Vostok ice core in its entirety shows that atmospheric CO_2 has oscillated between 285 ppm and 180 ppm for at least 400,000 years. On long timescales, therefore, the natural variations in the carbon cycle can be large.

The bottom graph in Fig. 25.9 shows an estimate of surface temperature obtained by chemical analysis of oxygen isotopes in ice from the same glacier and how it differed from the average temperature in 1950. Over the last 400,000 years, Antarctic temperature has oscillated between peaks of a few degrees warmer than today and temperatures as much as 6°C to 8°C colder than the present. Interestingly, the temperature and CO_2 curves closely

FIG. 25.9 Atmospheric CO_2 content for the past 400,000 years. These measurements were recorded from air bubbles trapped in glacial ice at Vostok, Antarctica. *Source: Data from P. Rekacewicz, "Temperature and CO_2 Concentration in the Atmosphere over the Past 400,000 Years," from Vital Climate Graphics, UNEP/GRID–Arendal Maps & Graphics Library, JPG file, 2005, http://www.grida.no/publications/vg/climate/page/3057.aspx. Based on J. R. Petit et al., 1999, "Climate and Atmospheric History of the Past 420,000 Years from the Vostok Ice Core, Antarctica," Nature 399:429–436.*

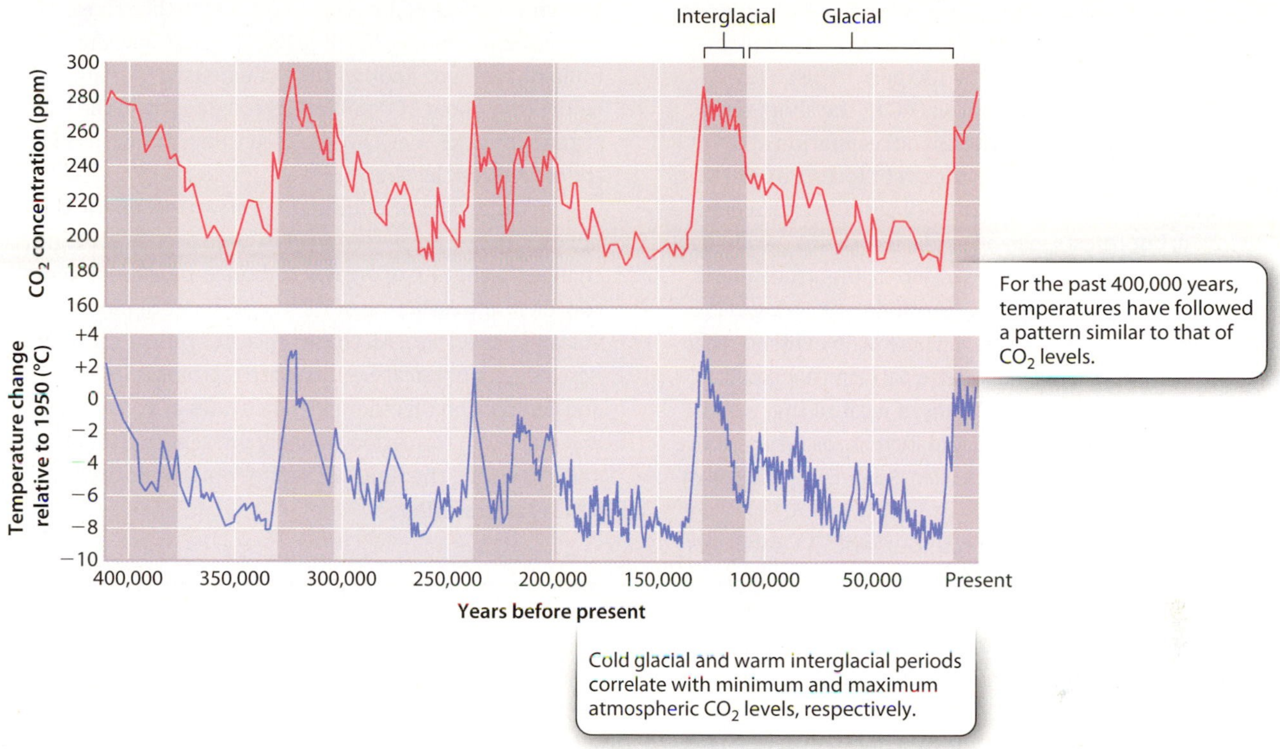

parallel each other. As discussed more fully in Chapter 49, CO_2 is known to be an effective **greenhouse gas,** meaning that it allows incoming solar radiation to reach Earth's surface but traps heat that is re-emitted from land and sea. Higher concentrations of CO_2 result in warmer temperatures. Therefore, it is not surprising that climate and atmospheric CO_2 levels show the parallel history documented in the figure.

The two curves correlate closely with one further phenomenon, the periodic growth and decay of continental ice sheets. Large glaciers expanded in the Northern and Southern hemispheres a few million years ago, ushering in an ice age. Today, we live in an interglacial interval, when climate is relatively mild, but 20,000 years ago thick sheets of ice extended far enough away from the poles to cover the present site of Boston (**Fig. 25.10**).

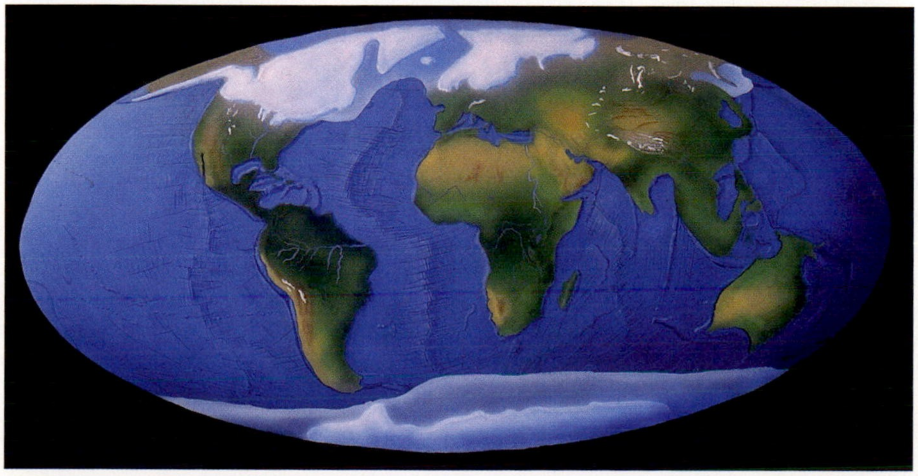

FIG. 25.10 Earth's most recent ice age. Twenty thousand years ago, glacial ice (shown in white) covered much of North America and Europe, as well as mountain ranges such as the Andes and Himalayas. Sea ice (light blue) expanded markedly in both northern and southern oceans. *Source: Plate tectonic maps and continental drift animations by C. R. Scotese, PALEOMAP Project (www.scotese.com).*

The repeated climatic shifts recorded in ice cores reflect periodic variations in the amount and distribution of solar radiation on Earth's surface, which are caused by oscillating changes in Earth's orbit around the sun.

The temperature and CO_2 increases recorded by Vostok ice between 20,000 and 10,000 years ago coincide with the last great retreat of continental ice sheets. What processes might explain how atmospheric CO_2 could increase by 100 ppm in just a few thousand years, as glaciers began to retreat? Can the short-term carbon cycle processes of photosynthesis and respiration account for this much carbon? Certainly, the amount of forests on Earth's surface has varied through the past 500,000 years as ice sheets grew and decayed, but forests expand as glaciers shrink, so changes in forests cannot account for a pattern of *increasing* atmospheric CO_2 with the retreat of glaciers. Volcanism and weathering also fail to account for the observed pattern. There is no evidence that volcanic activity has waxed and waned in a pattern that could explain observed CO_2 variations. And rates of weathering, which remove CO_2 from the atmosphere, should increase as temperature rises, but CO_2 levels have actually increased. Something else must be going on.

Scientists continue to debate why atmospheric CO_2 oscillates in parallel with glacial expansion and retreat, but, increasingly, proposed mechanisms suggest interactions involving the ocean and its large reservoir of inorganic carbon. For example, it has been hypothesized that during glacial advances, the circulation of carbon-rich deep-ocean waters back to the sea surface slows, causing more inorganic carbon to accumulate in the deep sea. With glacial retreat, the oceans circulate more vigorously, returning CO_2 to the surface and then to the atmosphere. Whatever the explanation, the historical record of the past 400,000 years shows that climate can and does change without any input from humans, something we must take into account when considering our climatic future (Chapter 49).

Variations in atmospheric CO_2 over hundreds of millions of years reflect plate tectonics and evolution.

The atmospheric record trapped in glacial ice extends back in time less than 1,000,000 years. To estimate atmospheric CO_2 levels for earlier intervals of Earth history, we must rely on computer models and measurements of chemical or paleontological features of ancient rocks that are thought to reflect the levels of atmospheric CO_2 at their time of formation. One substitute for direct measurements of ancient CO_2 levels is the anatomy of fossil leaves: Experiments show that stomata, the small pores on leaf surfaces (Chapter 29), decrease in density as atmospheric CO_2 levels increase. The chemistry of ancient soils is also thought to reflect atmospheric history. Such indirect observations and analyses come with a large degree of uncertainty, but most Earth scientists accept at least the broad pattern of atmospheric history shown in **Fig. 25.11.**

For the past 30 million years or so, atmospheric CO_2 levels have probably not exceeded those recorded today at Mauna Loa by much (Fig. 25.11). Before that, however, higher levels of CO_2—perhaps four to six times the 1958 level—characterized the Mesozoic Era (252 to 66 million years ago), a generally warm interval commonly known as the age of dinosaurs. Levels more like those of today existed earlier, during much of the late Paleozoic Era (350–252 million years ago), another time of extensive glaciation. And before that, about 500 million years ago, early in the Paleozoic Era, atmospheric CO_2 may have reached values as high as 15 to 20 times the present-day level.

On timescales of millions of years, atmospheric CO_2 and, hence, climate, are determined in large part by geologic processes: changes in the rate of organic carbon burial in sediments, continental weathering of rocks uplifted into mountains, and volcanic gas release. All these processes reflect the action of Earth's great physical engine: plate tectonics. Seafloor formation and destruction together influence rates of volcanic gas release, and mountain formation strongly influences rates of weathering and, therefore, the fluxes of sediments that bury organic carbon beneath the seafloor.

FIG. 25.11 Atmospheric CO_2 content for the past 541 million years. Models and indirect estimates from geological measurements suggest that CO_2 levels have varied substantially through Earth's history. *Data from Fig 3.4 in K. Wallmann and G. Aloisi, 2012, "The Global Carbon Cycle: Geological Processes," in Fundamentals of Geobiology, A. H. Knoll, D. E. Canfield, and K. O. Konhauser (eds.), Chichester, England: Wiley-Blackwell, pp. 20–35.*

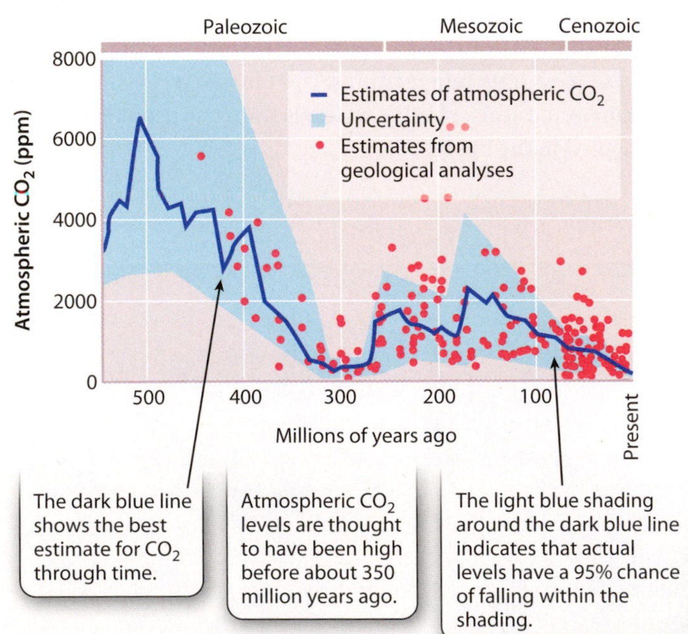

25.3 THE CARBON CYCLE: ECOLOGY, BIODIVERSITY, AND EVOLUTION

So far, we have outlined the biological and physical pathways that cycle carbon through Earth's surface environments. How can the carbon cycle provide a framework for understanding basic features of ecology, biodiversity, and evolution?

Food webs trace the cycling of carbon through communities and ecosystems.

As you sit quietly by a woodland pond, the carbon cycle is at work all around you (**Fig. 25.12**). In the forest, plants generate carbohydrates as they photosynthesize, and algae and photosynthetic bacteria are doing the same in the pond. All of the organisms in a habitat such as a woodland pond constitute an ecological **community.** In Chapter 6, we called photosynthetic organisms autotrophs ("self feeders") because they synthesize the organic molecules needed for growth and reproduction from CO_2. Ecologists commonly refer to these organisms as **primary producers,** reflecting their position within the carbon cycle. Primary producers are the first biological reservoir for carbon once it has been taken out of the physical environment.

An insect grazes on leaves, and ducks feed on both insects and algae. Ducks and insects obtain the carbon they need for growth and reproduction from the foods they eat, and they also gain energy by respiring food molecules. In Chapter 6, we called such organisms heterotrophs ("other feeders"), and ecologists label them **consumers,** again with reference to the carbon cycle. Consumers include all the animals in both pond and forest. Beneath our feet, similar interactions play out on a microscopic scale, especially among the fungi, protists, and bacteria that recycle the organic carbon in soil back to CO_2.

Primary consumers consume primary producers. Biologists call these organisms herbivores or grazers. Primary consumers transfer carbon drawn by the primary producers from the environment to a second biological reservoir and, by respiration, return some of the carbon to surrounding air or water.

FIG. 25.12 **A food web.** Food webs trace the cycling of carbon through communities.

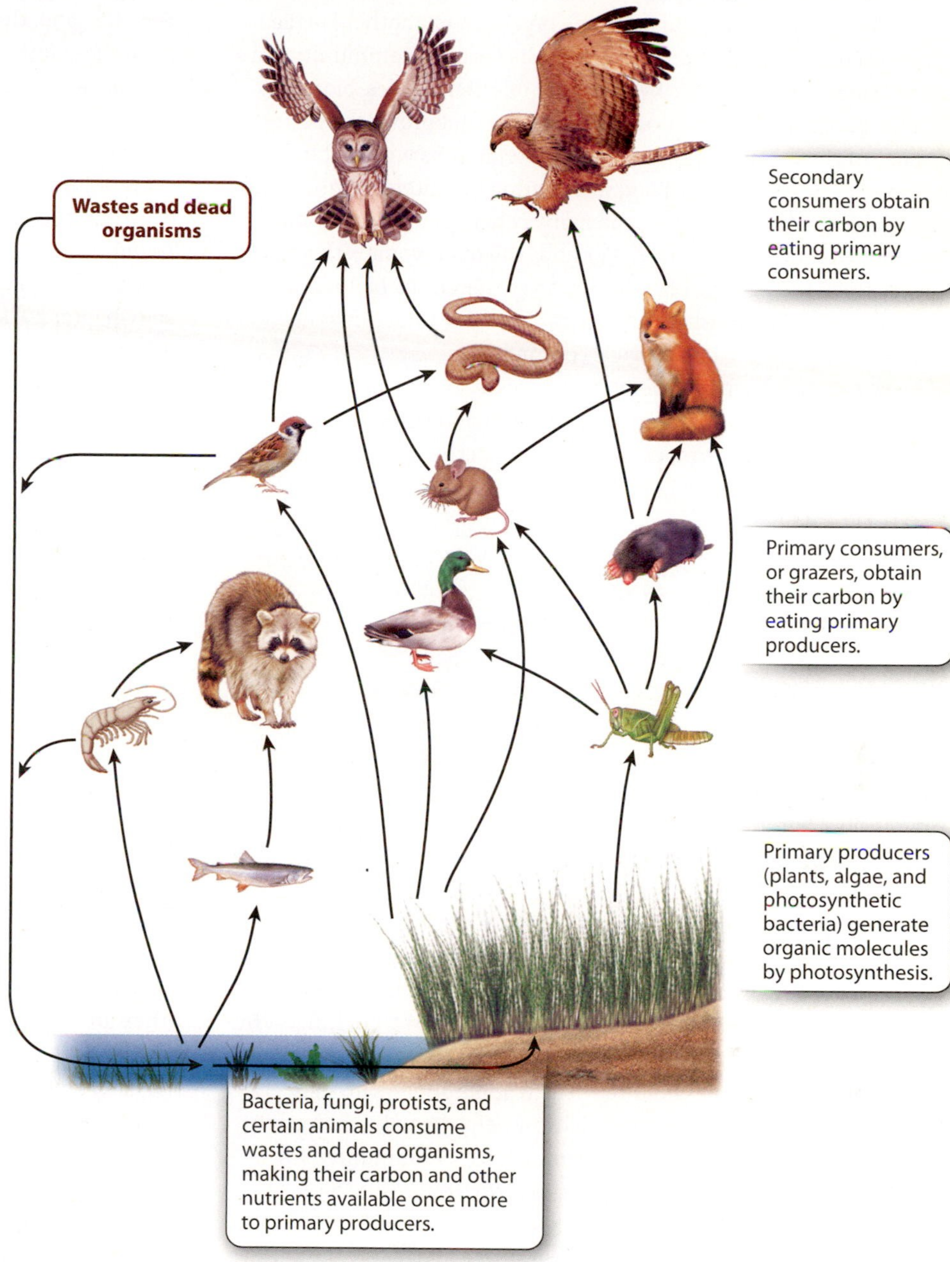

Secondary consumers, in turn, are predators or scavengers that feed on primary consumers. In particular, animals that eat other animals are called **carnivores,** and these organisms may provide food for still other carnivores, generating successive tiers of consumers. Like primary consumers, secondary consumers continue the transfer of carbon from reservoir to reservoir by eating the primary consumers and returning some of the carbon stored in that reservoir to the environment through respiration. Eventually, having passed from one consumer to another, the carbon originally fixed by photosynthesis is returned to the atmosphere by the respiration of fungi, bacteria, and other **decomposers** that break down dead tissues. The cycle has come full circle.

The transfer of carbon from one organism to another is called a food chain. Because most heterotrophs within a community can consume or be consumed by a number of other species, biologists often prefer to speak of **food webs,** a term that provides a better sense of the complexity of biological interactions within the carbon cycle (Fig. 25.12). Food webs define the interactions among organisms in ponds, forests, and many other habitats; we discuss them in detail in Chapter 47. For now, the main point is that food webs track the passage of carbon atoms through the biological carbon cycle. Put another way, the carbon cycle underpins the ecological structure of biological communities.

The complementary metabolic processes of photosynthesis and respiration cycle carbon through forest and pond communities. Furthermore, as we discuss in the next chapter, complementary metabolic processes also cycle nitrogen, sulfur, and other elements required for life. By continually recycling materials, biogeochemical cycles, which involve both biological and physical processes, sustain life over long intervals. In their absence, life could hardly have persisted for 4 billion years.

Biological diversity reflects the many ways that organisms participate in the carbon cycle.

Among the most remarkable features of life is its astonishing diversity. Biological diversity, or **biodiversity,** is the product of evolution, but it is shaped and sustained by ecological interactions among organisms and between organisms and the physical environment.

To begin, an estimated 500,000 species of photosynthetic organisms fuel the carbon cycle in all but a few deep-sea and subterranean environments. In terms of carbon and energy metabolism, all these species do pretty much the same thing. Why, then, is there such a diversity of photosynthetic organisms? Why don't just a few species dominate the world's photosynthesis?

The example of the pond and forest helps to make the basis of photosynthetic diversity clear. In the forest community, the leaves of several different tree species form a photosynthetic canopy above the forest floor. Below, shrubs grow, making use of light not absorbed by the leaves above them. And below the shrubs are grasses, herbs, ferns, and mosses that can grow in the reduced light levels of the forest floor. In the nearby pond, a few species of aquatic plants line the water's edge, but beyond that algae and photosynthetic bacteria dominate photosynthesis, with some species anchored to the pond bottom and others floating in the water column.

Locally, then, the different photosynthetic species that transfer carbon atoms from CO_2 to organic molecules subdivide the forest on the basis of light, water, and nutrient availability—a pattern complicated by grazing and environmental disturbances such as fire and landslides (Chapter 47). On a larger scale, climate and topography vary tremendously from one region to another, and these features, as well, help to explain how plants can build and maintain diversity by performing comparable metabolic functions in different habitats (Chapter 48). In the forests of New England, the tree species that thrive in wet regions differ from those found on well-drained hillsides. And seasonally dry woodlands in southern California support yet another set of plant species. In general, plants of varying size, shape, and physiology inhabit physically and biologically distinct environments, and the same is true of photosynthetic organisms in lakes and oceans.

Thus, the immense diversity of photosynthetic organisms found today does not reflect evolutionary variations in the biochemistry of photosynthesis (although some of that occurs; see Chapter 29) so much as it does structural and physiological adaptations. These adaptations allow the effective gathering of light, nutrients, and—critical to life on land—water, in widely varying local environments. Natural selection, acting on local populations, links the diversity of photosynthetic organisms to the carbon cycle.

If half a million species function as primary producers, an estimated 10 million species help to cycle carbon through respiration. These include the plants, algae, and bacteria that generate carbohydrates by photosynthesis, as well as animals, fungi, and microorganisms that obtain both carbon and energy from organic compounds in photosynthetic organisms or the consumers that eat them. These organisms are essential to the completion of the short-term carbon cycle, returning carbon atoms to the environment as CO_2.

Heterotrophic bacteria, amoebas, and humans may use essentially the same biochemical pathway to respire organic molecules, but they differ markedly in how they feed and, therefore, in what they can eat. Bacteria (and also fungi) absorb molecules from their environment, but amoebas and many other eukaryotic microorganisms can capture and ingest cells—they are capable of predation. Animals capture prey, as well, but commonly feed on organisms far too large for an amoeba to eat. As photosynthetic organisms have adapted structurally and physiologically to local environments across the globe, consumers have adapted by means of locomotion, mouth and limb specialization, perception, and behavior to obtain their food. So, as you read about bacteria, amoebas, and insects in the chapters that follow, think about how each operates within the carbon cycle and so contributes to the operation of Earth's diverse ecosystems.

The carbon cycle weaves together biological evolution and environmental change through Earth history.

The world did not always support the biological diversity we see today. Indeed, 2 billion years ago, the carbon cycle included mostly photosynthetic bacteria and microbial heterotrophs because there were no plants or animals (**Fig. 25.13**). Subsequently, algae gained an ecological foothold as nutrients such as nitrate became more widely available, and single-celled eukaryotic heterotrophs expanded their reach by gathering food in ways not possible for bacteria, ingesting cells and other types of particulate food. With

FIG. 25.13 Microbial communities in hot springs in New Zealand. This environment and others that few eukaryotes tolerate provide a glimpse of how the carbon cycle worked before algae, amoebas, plants, and animals evolved. *Source: Albert Aanensen/Nature Picture Library.*

the evolution of multicellularity, animals and seaweeds added further complexity to carbon cycling in the sea. Eventually, some aquatic green algae evolved the capacity to live on land, and animals soon followed, building the carbon cycle on land and thereby generating unprecedented levels of diversity on our planet.

The history of life is one of accumulating variety and complexity, based mostly on the great biological components of the carbon cycle: photosynthesis and respiration. Indeed, the carbon cycle did more than provide a framework for accumulating diversity; it changed the very nature of Earth's surface environments in ways that enabled new types of organism to evolve.

When life began, the atmosphere and oceans contained little or no oxygen gas. How did our present environment come to be? The key lies in the formulas for photosynthesis and respiration, shown in Fig. 25.2. Photosynthesis and respiration not only cycle carbon, they cycle oxygen (and water) as well. For this reason, the history of atmospheric oxygen is closely tied to the workings of the carbon cycle through time. When oxygen production by photosynthesis and oxygen consumption by respiration are in balance, oxygen levels do not change. However, when some of the organic carbon generated by photosynthesis is buried in sediments, thereby avoiding being returned to the environment through respiration, some of the O_2 also generated by primary producers can accumulate in the atmosphere and oceans. In other words, because sedimentary organic matter burial can break the tight coupling depicted in Fig. 25.2, it can facilitate an increase in O_2 through time. The geologic history of oxygen shown in **Fig. 25.14** results from the interactions through time between the carbon (and sulfur) cycle with plate tectonics. Our oxygen-rich world is the result.

But life does not just respond passively to changing environments—it can drive environmental change through time. For example, the large drop in atmospheric CO_2 suggested for the mid-Paleozoic Era, 400–350 million years ago (see Fig. 25.11), is thought to reflect the evolution of a new player in the carbon cycle: woody plants. The evolution of trees increased the size of the carbon reservoir on land and ushered in an important new mechanism for removing carbon from the air and ultimately to sedimentary rocks. That mechanism was the burial of plant material on land, forming peat and, eventually, coal.

And, in the current century, a single animal species is influencing the carbon cycle in new ways. That species is us. As noted in our opening discussion of the Keeling curve, humans are adding CO_2 to the atmosphere at a high rate, warming Earth's surface by the greenhouse effect. The consequences of human activities for Earth and its living organisms are discussed more fully in Chapter 49. For now, just bear in mind that modern humans are not simply observers of the carbon cycle; we are major participants in it.

In the chapters that follow, we examine life's diversity in terms of function, phylogeny, and ecology, but don't be fooled into thinking that these are discrete topics: They are inextricably intertwined. The remarkable diversity of life on Earth reflects the ecology and evolution of populations that interact to cycle carbon through the biosphere. ■

FIG. 25.14 Oxygen levels in Earth's atmosphere during the history of life.

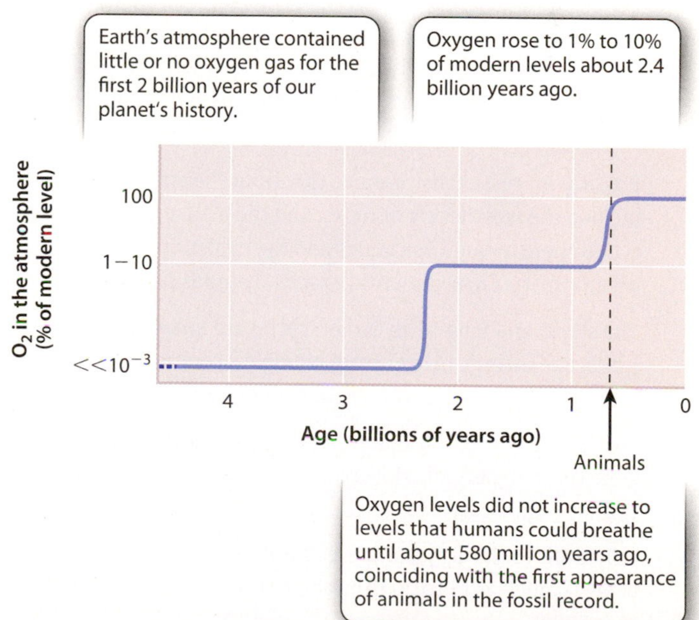

Core Concepts Summary

25.1 Photosynthesis and respiration are the key biochemical pathways of the biological, or short-term, carbon cycle.

Keeling's program to monitor CO_2 levels in the atmosphere revealed two patterns: (1) CO_2 levels oscillate on a yearly basis, and (2) CO_2 levels have increased steadily to the present day since measurements began. page 510

The first pattern (seasonal oscillation) can be explained by the global imbalance between photosynthesis and respiration in the summer and in the winter months. page 512

The second pattern (steady increase) can be extended further back in time by measuring the composition of air trapped in gas bubbles within polar ice. These measurements indicate that CO_2 levels in the atmosphere remained steady for more than a thousand years, but then started to increase in the mid-1800s, the time of the Industrial Revolution. page 513

The ratio of different isotopes of carbon in the atmosphere indicates that most of the carbon added to the atmosphere in recent decades comes from human activities, particularly the burning of fossil fuels. page 513

25.2 Physical processes govern the long-term carbon cycle.

Physical processes, which act at slower rates and over longer timescales, are the major drivers of the long-term carbon cycle. page 515

In following carbon around its cycle, it is important to consider where carbon is stored (reservoirs) and rates at which it is added to or removed from these stores (fluxes). page 515

Important reservoirs of carbon include organisms, the atmosphere, soils, the oceans, and sedimentary rocks. Sedimentary rocks are by far the largest carbon reservoir. page 515

Physical processes that remove CO_2 from the atmosphere include the weathering of rocks, and those that add CO_2 to the atmosphere include volcanoes and the oxidation of ancient organic matter in sedimentary rocks. page 516

The short- and long-term carbon cycles are linked by the slow leakage of organic matter from biological communities to sediments accumulating on the seafloor. page 516

Over the last 400,000 years, CO_2 levels in the atmosphere have gone through periodic shifts that coincide with repeated cycles of glaciation. page 518

CO_2 levels in the atmosphere 500 million years ago may have been as much as 15 to 20 times higher than present-day levels. page 520

Over time spans of hundreds of millions of years, movements of Earth's plates become major influences on the carbon cycle. page 520

25.3 The carbon cycle can help us understand ecological interactions and the evolution of biological diversity.

Photosynthetic and other autotrophic organisms are primary producers, converting CO_2 into organic molecules. page 521

Heterotrophic organisms that eat primary producers are called primary consumers. page 521

Heterotrophs that prey on grazers are called secondary consumers. page 521

Interacting networks of species, linked by predator–prey interactions, are called food webs, and they are an important feature of the carbon cycle. page 522

Primary producers are well adapted to obtaining sunlight, water, and nutrients in different environments. page 522

Primary and secondary consumers are adapted to obtaining different sources and sizes of food. page 522

The evolution of life on Earth can be considered a long history during which organisms have evolved different ways of obtaining energy and carbon from their environment. page 523

The interactions between biological and geologic processes at work in the carbon cycle are also responsible for Earth's oxygen-rich atmosphere and oceans. page 523

Self-Assessment

1. Why does the coupling of photosynthesis and respiration result in the cycling of carbon?
2. Draw and explain the curve representing changing atmospheric levels of CO_2 over the course of a year.
3. Draw and explain the curve representing changing atmospheric levels of CO_2 over the past 150 years.
4. Describe one way by which the source of the modern increase in CO_2 in the atmosphere can be determined.
5. How do geologic processes participate in the long-term carbon cycle?
6. Draw and explain the curve representing changing atmospheric levels of CO_2 over the last 400,000 years.

CASE 5

The Human Microbiome

Diversity Within

Clostridium difficile doesn't grab many headlines. Perhaps it should. This bacterium is thought to be the most common cause of hospital-acquired infections in the United States. It sickens as many as 300,000 U.S. hospital patients each year, causing fevers, chills, diarrhea, and severe intestinal pain—and adds more than $1 billion annually to the nation's health care costs. Altogether, the infection kills about 29,000 people every year, according to the Centers for Disease Control and Prevention.

Now a promising (if stomach-turning) new treatment is gaining traction: fecal transplants. The vast majority of *C. difficile* cases occur after patients have taken antibiotics for other infections. The drugs kill the beneficial bacteria normally found in a healthy gut, allowing *C. difficile* to proliferate. Recently, pioneering researchers have begun to replace the missing microbes by transplanting fecal material from healthy donors into patients sick with *C. difficile*. Some doctors have reported that the procedure alleviates symptoms in more than 80% of patients.

While transplanting fecal bacteria may not yet be mainstream medicine, there is no question that our gut, as well as our skin, mouth, nose, and urogenital tract, is home to a diversity of bacterial species. No one has yet identified all of the many microbes living on and within our bodies, but estimates suggest that each of us contains 10 times as many bacterial cells as human cells. For the most part, the relationship between "us and them" is a type of symbiosis, in which both humans and bacteria benefit from the close association. Indeed, the microbiome, as our assemblage of bacterial houseguests is known, affects human health in many important ways.

Some of the most mysterious, and most intriguing, members of our microbiome live in our gut. Bacteria living in our intestines help us break down food and synthesize vitamins. On the other hand, abnormal communities of microbes in the gut have been linked to Crohn's disease, colon cancer, and nonalcoholic fatty liver disease.

Your own personal gut flora is a unique collection of species—even identical twins can have surprisingly different microbiomes. This variability has made it difficult for scientists to decipher exactly what role individual species play in health and in disease. The sheer number of bacteria in your gut only compounds the problem. One international team of scientists recently reported that each human gut contains at least 160 different bacterial species.

Despite these challenges, researchers have made strides toward understanding the unique bacterial communities within our intestinal tract. In 2011, scientists compared all the DNA sequences from gut microbes of people living in Europe, America, and Japan. They found that each individual fell into one of three enterotypes, or "gut types." Most people had a bacterial mix dominated by organisms from the genus *Ruminococcus*. But others had gut types dominated by the genus *Bacteroides* or the genus *Prevotella*. The three different gut compositions appear to produce different vitamins and may even make a person more, or less, susceptible to certain diseases.

> ***No one has yet identified the many microbes living on and within our bodies, but estimates suggest that each of us contains 10 times as many bacterial cells as human cells.***

Surprisingly, the researchers reported that enterotype was unrelated to gender, body mass index, or nationality. So what determines your gut type? A follow-up study suggested that enterotypes were associated with long-term dietary patterns. People who eat a diet high in animal protein and saturated fats are more likely to contain a gut community dominated by *Bacteroides*. People who stick to a high-carbohydrate diet rich in plant-based nutrition are more likely to host an abundance of *Prevotella*.

The researchers don't know yet what a person's enterotype means for his or her health. But scientists are eager to explore the link between microbiome and physical well-being. Meanwhile, various researchers are exploring the microbiome of other organisms—specifically, cows.

As it happens, there are good reasons to peer into a cow's gut. The cow's rumen (one of its four digestive

The human microbiome. Different groups of bacteria reside in or on each part of the body. When the *Helicocbacter pylori* is present in the stomach (*H. pylori* +), it can overwhelm other bacterial populations (*H. pylori* –). *Source: Mimi Haddon/Getty Images.*

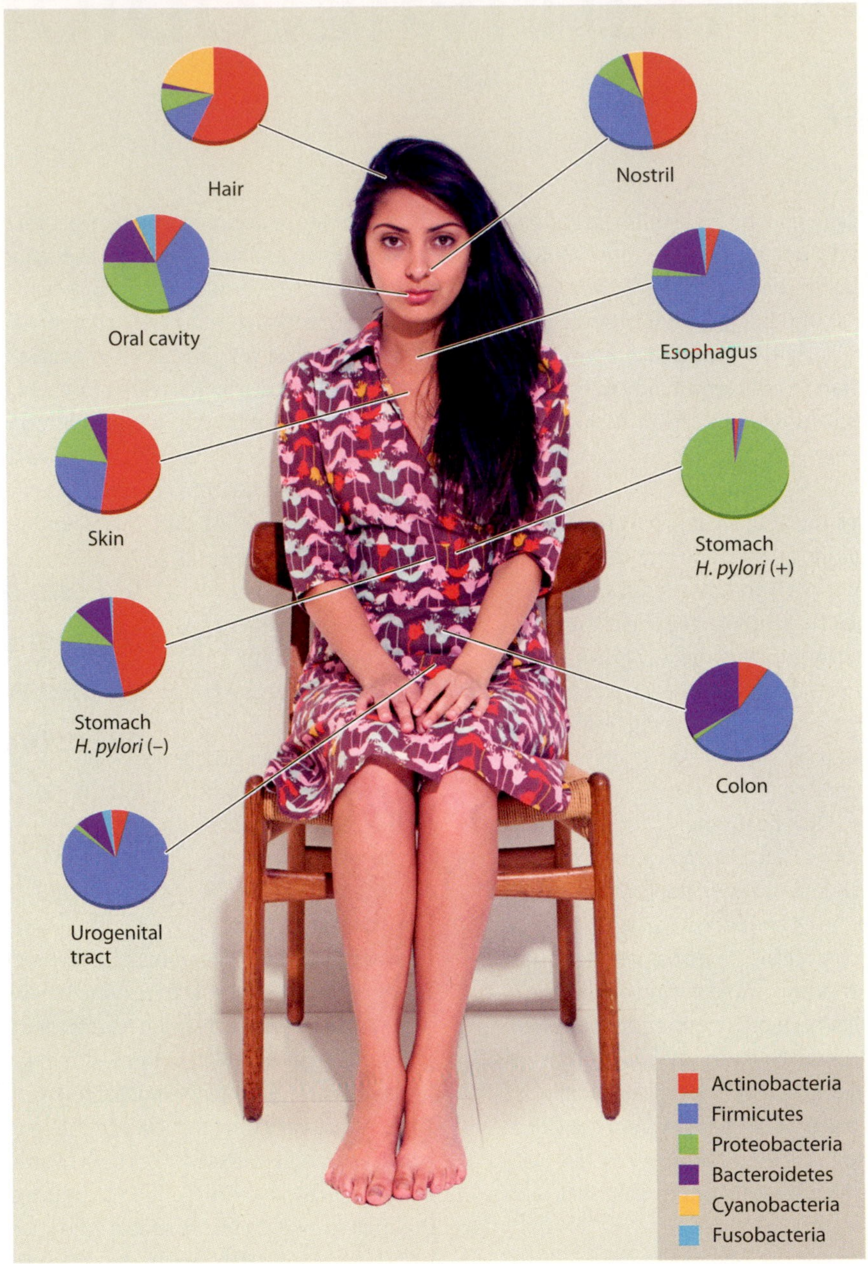

chambers) contains microbes that break down the cellulose in plant material. Without them, cattle wouldn't be able to extract adequate nutrition from the grass they graze. Those microbes are the particular focus of scientists at the Department of Energy (DOE), who are interested in converting material containing cellulose, such as switchgrass, into biofuels.

As researchers look for more sustainable replacements for fossil fuels, many have pinned their hopes on biofuels. But the enzymes currently used commercially to break

The bobtail squid and its luminescent bacteria. This squid contains specialized organs that harbor light-emitting bacteria. *Source: Gary Bell/OceanwideImages.com.*

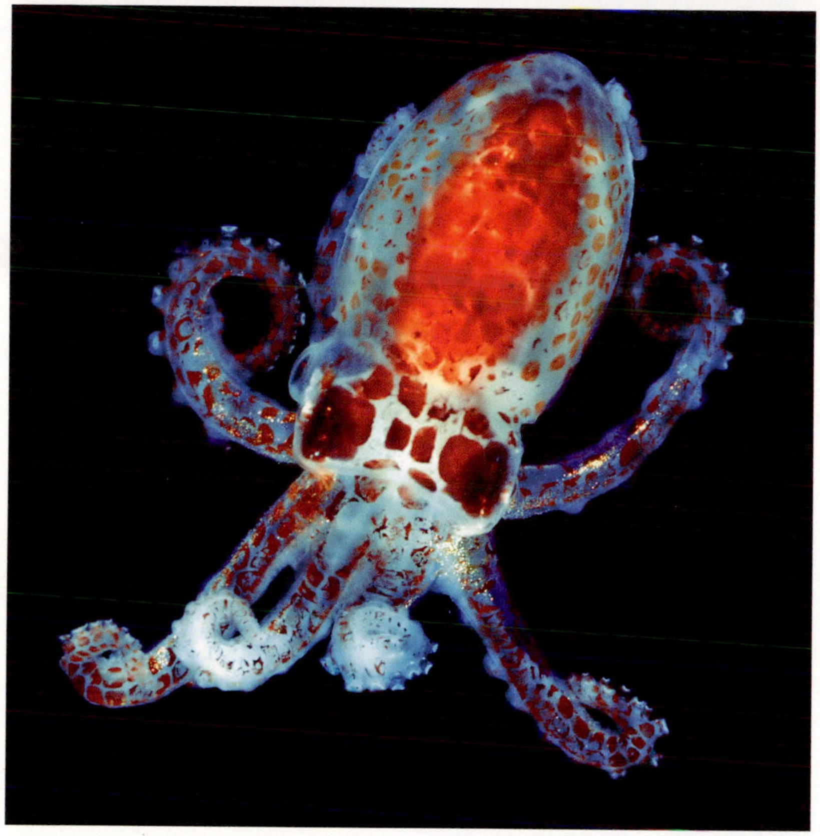

down cellulose aren't efficient or cost effective enough to produce biofuels on an industrial scale. That's where the cows come in. DOE scientists recently analyzed the microbial genes present in the cow rumen. They discovered nearly 28,000 genes for proteins that metabolize carbohydrates, and at least 50 new proteins. The hope is that some of these newly discovered proteins will help break down cellulose in ways that will produce biofuels more efficiently and cheaply.

Mammals such as cows and humans are hardly the only organisms that have symbiotic relationships with bacteria. All living animals have their own microbiomes. Even tiny single-celled eukaryotes can harbor still-tinier bacterial guests.

Symbiotic bacteria provide a variety of benefits to their hosts. Consider the bobtail squid, *Euprymna scolopes,* which possesses a specialized organ to house bacteria. The bacteria, *Vibrio fischeri,* are luminescent—that is, capable of generating light. The squid use their *Vibrio*-filled light organs like spotlights, projecting light downward to match light from the surface as a way to camouflage themselves from would-be predators below.

In many cases, animals and their bacterial symbionts are so intimately connected that it's hard to separate one from the other. Many human gut bacteria are poorly studied because scientists haven't been able to grow them outside the body. But some species have taken their intimate bacterial partnership a step further.

Aphids are small insects that feed on plant sap. They rely on the bacterium *Buchnera aphidicola* to synthesize certain amino acids for them that aren't available from their plant-based diet. The partnership between aphids and *Buchnera* dates back as far as 250 million to 150 million years ago. Over time, the organisms have become completely dependent on one another for survival. The *Buchnera* bacteria actually live inside the aphid's cells. In some ways, the bacteria resemble organelles rather than independent organisms. Intracellular symbionts may not be so unusual. In fact, mitochondria and chloroplasts—key organelles of eukaryotic cells—are thought to have originated from symbiotic bacteria. Over time, the progenitor bacteria became so intertwined with their host cells that they developed into an integral part of the cells themselves. Studying simple sap-sucking aphids, it turns out, may offer insights into the evolutionary history of all eukaryotic cells.

Clearly, there are many good reasons to learn more about our bacterial comrades. From evolution to energy to human health, our microbiomes hold great promise—and great mystery. We've evolved hand in hand with these bacteria for millions of years. Let's hope it takes less time to uncover their secrets.

? CASE 5 QUESTIONS

Special sections in Chapters 26 and 27 discuss the following questions related to Case 5.

1. **How do intestinal bacteria influence human health?** See page 550.
2. **What role did symbiosis play in the origin of chloroplasts?** See page 557.
3. **What role did symbiosis play in the origin of mitochondria?** See page 559.
4. **How did the eukaryotic cell originate?** See page 560.

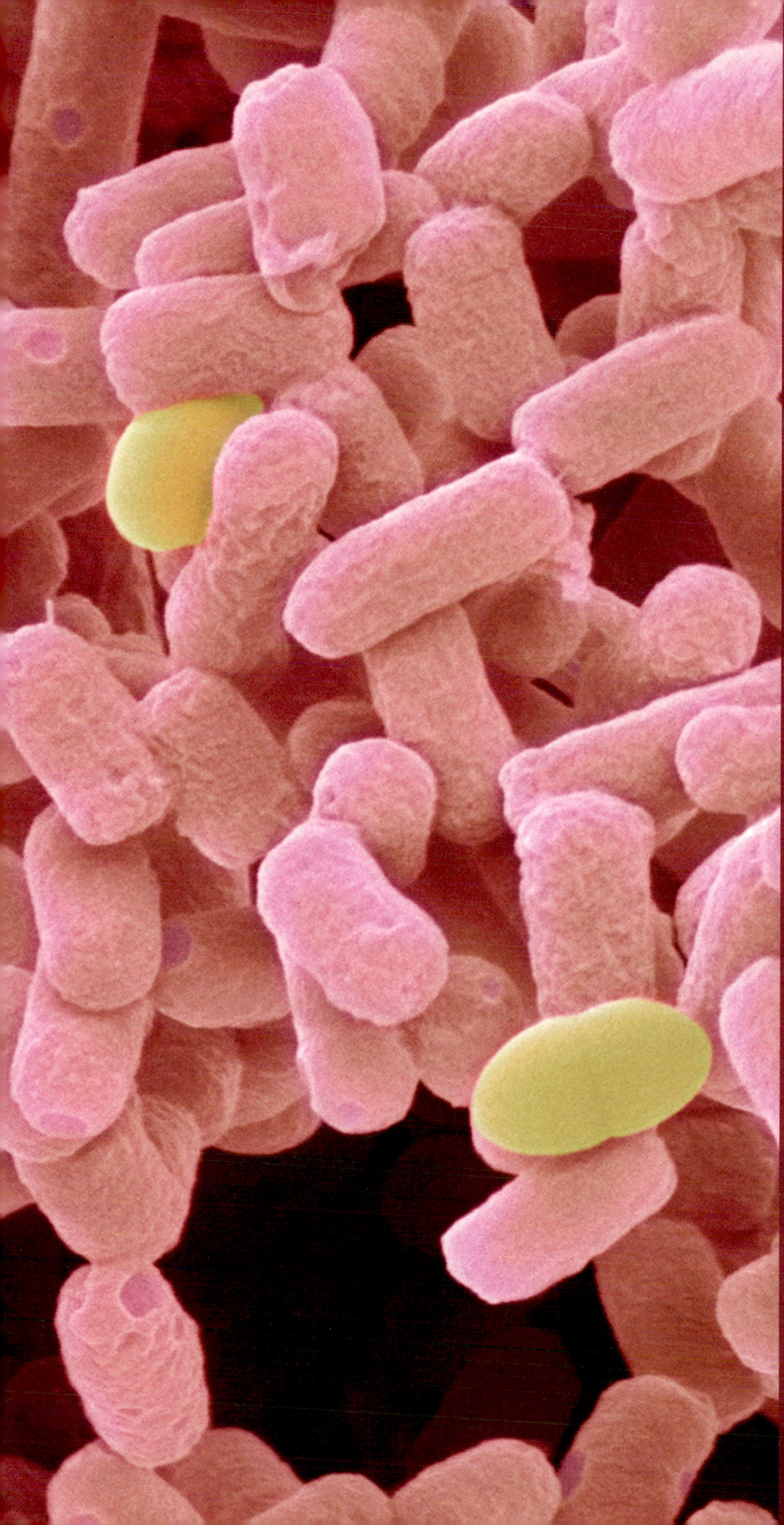

Steve Gschmeissner/SPL/Getty Images.

CHAPTER 26

Bacteria and Archaea

Core Concepts

26.1 The tree of life has three main branches, called domains: Eukarya, Bacteria, and Archaea.

26.2 Bacteria and Archaea are notable for their metabolic diversity.

26.3 In addition to their key roles in the carbon cycle, Bacteria and Archaea are critical to the biological cycling of sulfur and nitrogen.

26.4 The extent of bacterial diversity was recognized when sequencing technologies could be applied to non-culturable bacteria.

26.5 The diversity of Archaea has only recently been recognized.

26.6 The earliest forms of life on Earth were Bacteria and Archaea.

In Chapter 25, we discussed how the cycling of carbon by plants and animals is linked to their production and consumption of oxygen. But can carbon cycle through the deep waters of the Black Sea, within black muds beneath swamps and marshlands, or in other habitats where oxygen is limited or absent? Can biological communities even survive without oxygen? The answer, emphatically, is yes. Indeed, life existed on Earth for more than a billion years before oxygen-rich habitats first appeared on our planet.

The seemingly alien oxygen-poor environments on today's Earth are populated by organisms that neither produce nor consume oxygen, yet are able to cycle carbon. The organisms responsible for this expanded cycle share one fundamental feature: They have prokaryotic cell organization. Prokaryotes, which include bacteria and archaeons, lack a nucleus (Chapter 5). They inhabit the full range of environments present on Earth, including those rich in oxygen, but their presence in habitats where plants and animals cannot live suggests that some of these organisms have biological features quite unlike those of our familiar world. We'll look at the carbon cycle again later in this chapter, but first we discuss **Bacteria** and **Archaea,** the prokaryotic domains of microscopic organisms so critical to the cycling of carbon and other elements essential to life. We examine their diversity and how they have evolved through time.

26.1 TWO PROKARYOTIC DOMAINS

Chapter 5 outlined the two distinct ways that cells are organized internally. Eukaryotic cells, which include the cells that make up our bodies, have a membrane-bounded nucleus and organelles that form separate compartments for distinct cell functions. Prokaryotic cells have a simpler organization. No membrane surrounds the cell's DNA, and there is little in the way of cell compartments. Prokaryotic cell organization is an ancestral character for life as a whole—that is, it is a feature that was present in the last common ancestor of all organisms alive today. The group defined traditionally as the prokaryotes, however, is paraphyletic in that it excludes some descendants of the last common ancestor of all living organisms, namely eukaryotes (Chapter 23).

Bacteria and Archaea are the two domains characterized by prokaryotic cell structure. These organisms are present almost everywhere on Earth. What they lack in complexity of cell structure, these tiny cells more than make up for in their dazzling metabolic diversity. As will become clear over the course of this chapter, Bacteria and Archaea underpin the efficient operation of ecosystems on our planet.

The bacterial cell is small but powerful.

Because of their small size and deceptively simple cell organization, bacteria were long dismissed as primitive organisms, distinguished mostly by the eukaryotic features they lack: They have no membrane-bounded nuclei, no energy-producing organelles, no sex. This point of view turns out to be more than a little misleading. Bacteria are the diverse and remarkably successful products of nearly 4 billion years of evolution. Today, bacterial cells outnumber eukaryotic cells by several orders of magnitude. Even in your own body, bacteria outnumber human cells 10 to 1.

Fig. 26.1 illustrates the bacterial cell, which was briefly introduced in Chapter 5. The cell's DNA is present in a single circular chromosome, in contrast to the multiple linear chromosomes characteristic of eukaryotic cells. Many bacteria carry additional DNA in the form of **plasmids,** small circles of DNA that replicate independently of the cell's circular chromosome. In general, plasmid DNA is not essential for the cell's survival, but it may contain genes that have adaptive value under specific environmental conditions. No nuclear membrane separates DNA from the surrounding cytoplasm, and so transcribed mRNA is immediately translated into proteins by ribosomes.

Bacteria lack the membrane-bounded organelles found in eukaryotic cells. Instead, cell processes such as metabolism are carried out by proteins that float freely in the cytoplasm or are embedded in the plasma membrane. A few bacteria, notably the photosynthetic bacteria, contain internal membranes similar to those found in chloroplasts and mitochondria. The light reactions of photosynthetic bacteria take place in association with membranes distributed within the cytoplasm.

Structural support is provided by a cell wall made of **peptidoglycan,** a complex polymer of sugars and amino acids. Some bacteria have thick walls made up of multiple peptidoglycan layers, while others have thin walls surrounded by an outer layer of lipids. For many years, it was believed that bacteria lacked the cytoskeletal framework that organizes cytoplasm in eukaryotic cells. However, careful studies now show that bacteria do have an internal scaffolding of proteins that plays an important role in determining the shape, polarity, and other spatial properties of bacterial cells (Chapter 10).

Diffusion limits cell size in bacteria.

Most bacterial cells are tiny: The smallest are only 200–300 nanometers (nm) in diameter, and relatively few are more than 1–2 micrometers (μm) long. Why are bacteria so small? The answer has to do with diffusion, a critical process introduced in Chapter 5. If you could watch the movement of any particular molecule in air or water, you would see that its motion is random, sometimes in one direction and sometimes in another. On average, however, more molecules move from a region with a higher concentration of the molecule to a region with a lower concentration of the molecule than move in the other direction. Net movement stops only when the two regions achieve equal concentration, but diffusion continues. Photosynthetic bacteria gain the carbon dioxide they need by the diffusion of CO_2 from the environment into the cell, and that is also how respiring bacteria take in small organic molecules and oxygen.

FIG. 26.1 A bacterial cell. The cells of Bacteria (and Archaea) do not have a membrane-bounded nucleus or other organelles.

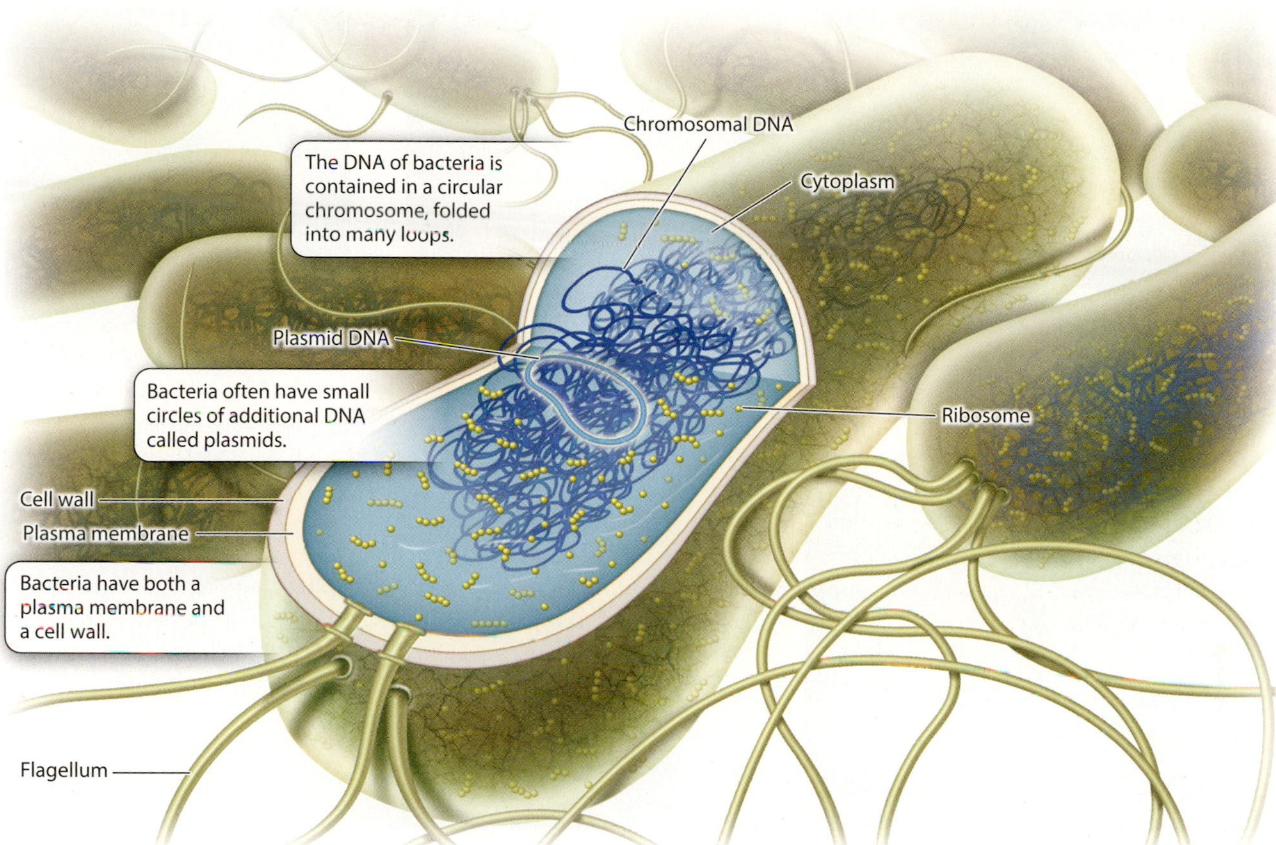

Diffusion explains why bacterial cells tend to be small. A small cell has more surface area in proportion to its volume, and the interior parts of a small cell are closer to the surrounding environment than those of a larger cell. As a consequence, slowly diffusing molecules do not have to travel far to reach every part of a small cell's interior. The surface area of a spherical cell—the area available for taking up molecules from the environment—increases as the square of the radius. However, the cell's volume—the amount of cytoplasm that is supported by diffusion—increases as the cube of the radius. Therefore, as cell size increases, it becomes harder to supply the cell with the materials needed for growth. For this reason, most bacterial cells are tiny spheres, rods, spirals, or filaments—small enough for molecules to diffuse into cell interiors (**Fig. 26.2**).

A few exceptional bacteria exceed 100 μm in maximum dimension. For example, *Thiomargarita namibiensis,* a bacterium that lives in oxygen-poor sediments off the coast of southwestern Africa, has a total volume about 100,000,000 times larger than that of *Escherichia coli* (**Fig. 26.3**). But, in one sense, *T. namibiensis* cheats: 98% of its volume is taken up by a large vacuole, so the metabolically active cytoplasm is restricted to a thin film around the cell's periphery. Thus, the distance through which nutrients move by diffusion is only a few micrometers, as in many other bacteria.

Some bacteria are multicellular, forming simple filaments or sheets of cells. More unusual are myxobacteria, which aggregate to form multicellular reproductive structures that are composed of several distinct cell types (see Fig. 26.2e).

Horizontal gene transfer promotes genetic diversity in bacteria.

Bacterial genomes are generally smaller than the genomes of eukaryotes, in part because bacteria lack the large stretches of noncoding DNA characteristic of eukaryotic chromosomes (Chapter 13). The streamlining of the bacterial genome confers certain benefits. For example, bacteria can reproduce rapidly when nutrients required for growth are available in the local environment.

FIG. 26.2 Cell shape and size in Bacteria and Archaea. *Sources: a. Eye of Science/Science Source; b. Scimat/Getty Images; c. Courtesy Mike Dyall-Smith; d. David Scharf/Science Source; e. © ANIMA RES.*

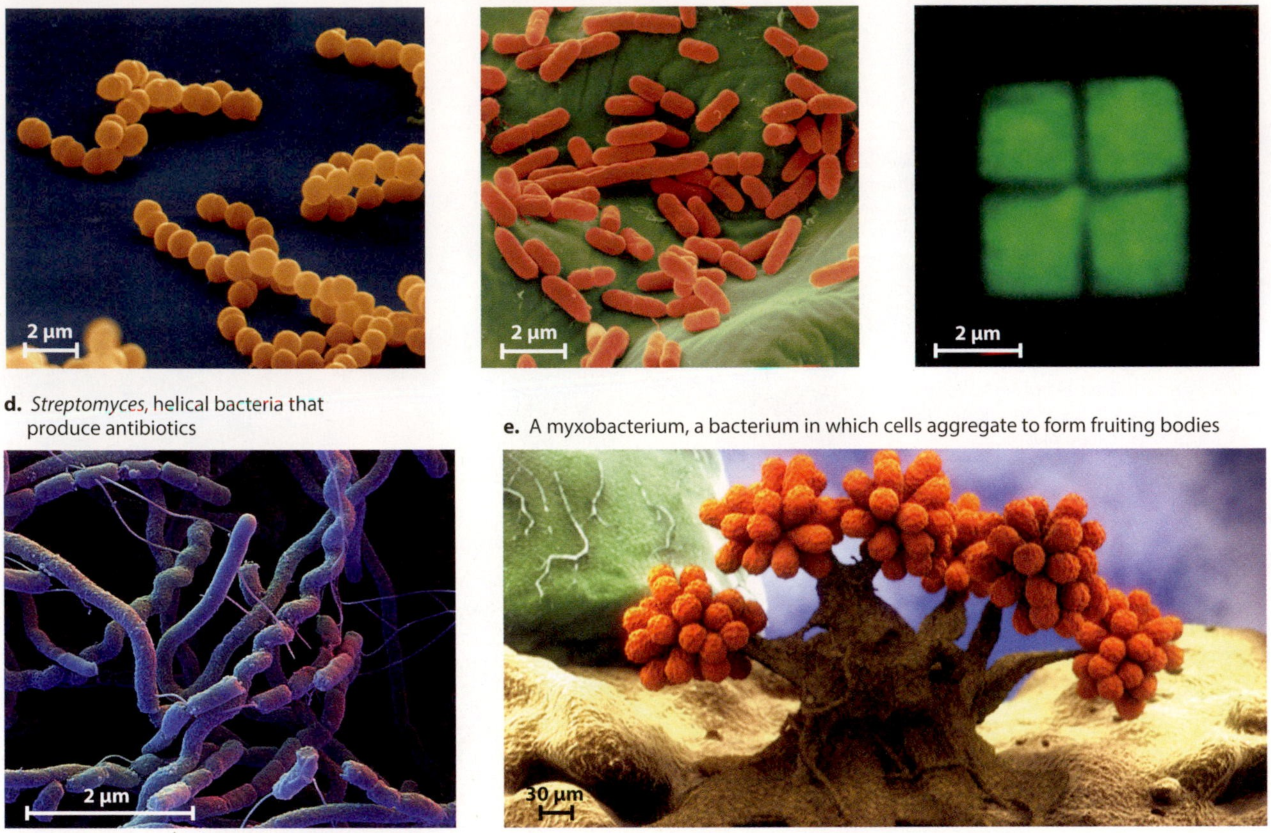

a. *Streptococcus*, strings of spheroidal or coccoidal bacteria

b. *E. coli*, bacterial rods

c. *Haloquadratum walsbyi*, a square archaeon that lives in salt ponds

d. *Streptomyces*, helical bacteria that produce antibiotics

e. A myxobacterium, a bacterium in which cells aggregate to form fruiting bodies

FIG. 26.3 ***Thiomargarita namibiensis*, the largest known bacterial cell.** The volume of the cell is mostly taken up by a vacuole, so the active cytoplasm is only a few micrometers thick. *Source: Courtesy Heide Schulz-Vogt, Leibniz Institute for Baltic Sea Research, Department of Biological Oceanography.*

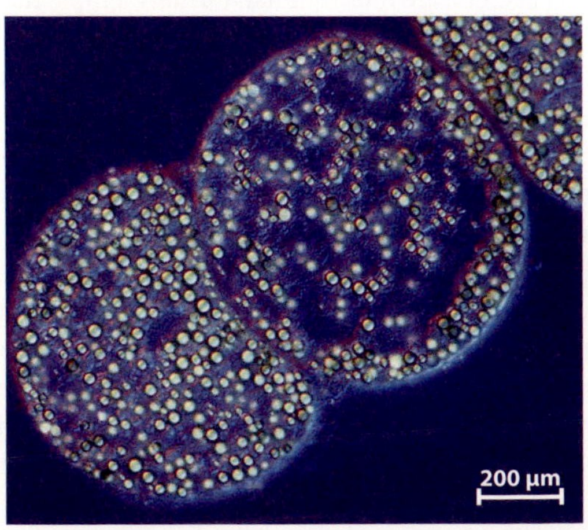

Bacteria replicate their DNA from only one or a small number of initiation sites, so genome size can influence rate of reproduction.

Bacteria do not undergo meiotic cell division and cell fusion. That is, they lack the sexual processes characteristic of eukaryotic organisms (Chapters 11 and 42). Despite this, bacterial populations display remarkable genetic diversity. For example, different strains of *Pseudomonas aeruginosa*, a common disease-causing organism, may differ in genome size by nearly a factor of 2 (3.7 million base pairs compared to 7.1 million base pairs), yet these strains are quite similar in function.

How does this variation arise? In eukaryotic organisms, genes generally pass from parent to offspring. Bacteria also inherit most of their genes from parental cells. However, bacterial cells can also obtain new genes from distant relatives in a process called **horizontal gene transfer.** This process is a major source of genetic diversity in bacteria.

How does DNA move from one bacterial cell into another independently of cell division? Some bacteria synthesize thin strands of membrane-bound cytoplasm called pili (Chapter 5) that connect them to other cells (**Fig. 26.4a**). After joining to a second cell, the pilus contracts, drawing the two cells close together. A pore-like opening develops where the two cells are in close contact, providing a migration route for the direct cell-to-cell transfer of DNA. This process, called **conjugation,** commonly transfers plasmids from one cell to another, spreading novel genes throughout a population. Genes that confer resistance to antibiotics are a well-studied example of horizontal gene transfer by conjugation.

Genes can also be transferred from one cell to another without any direct bridge between cells. DNA released to the environment by cell breakdown can be taken up by other cells, a process called **transformation** (**Fig. 26.4b**). Transformation was revealed when experiments showed that harmless strains of the bacteria causing pneumonia could be transformed into virulent strains by exposure to media containing dead cells of disease-causing strains (Chapter 3). Scientists reasoned that the transformation occurred because some substance was being taken up by the living bacteria. The "transforming substance" was later shown to be DNA. Today, biologists commonly use transformation in the laboratory to introduce genes into cells.

Viruses provide a third mechanism of horizontal gene transfer. Recall from Chapter 19 that viruses in bacterial cells sometimes integrate their DNA into the host bacterial DNA. This viral DNA persists within the cells as they grow and divide. Before the virus leaves the cell to infect others, the viral DNA removes itself from the bacterial genome and is packaged in a protein capsule to produce the complete virus particle. This excision is not always precise, and sometimes genetic material from the bacterial host is incorporated into the virus. Viruses released from their host cell go on to infect others, bringing host-derived genes with them. Horizontal gene transfer by means of viruses is called **transduction** (**Fig. 26.4c**). It is common in nature and is also widely used in the laboratory to introduce novel genes into bacteria for medical research.

Horizontal gene transfer allows bacterial cells to gain beneficial genes from organisms distributed throughout the bacterial domain and beyond. Indeed, bacteria everywhere are constantly reshaping their genomes. Bacteria, therefore, are not the poor

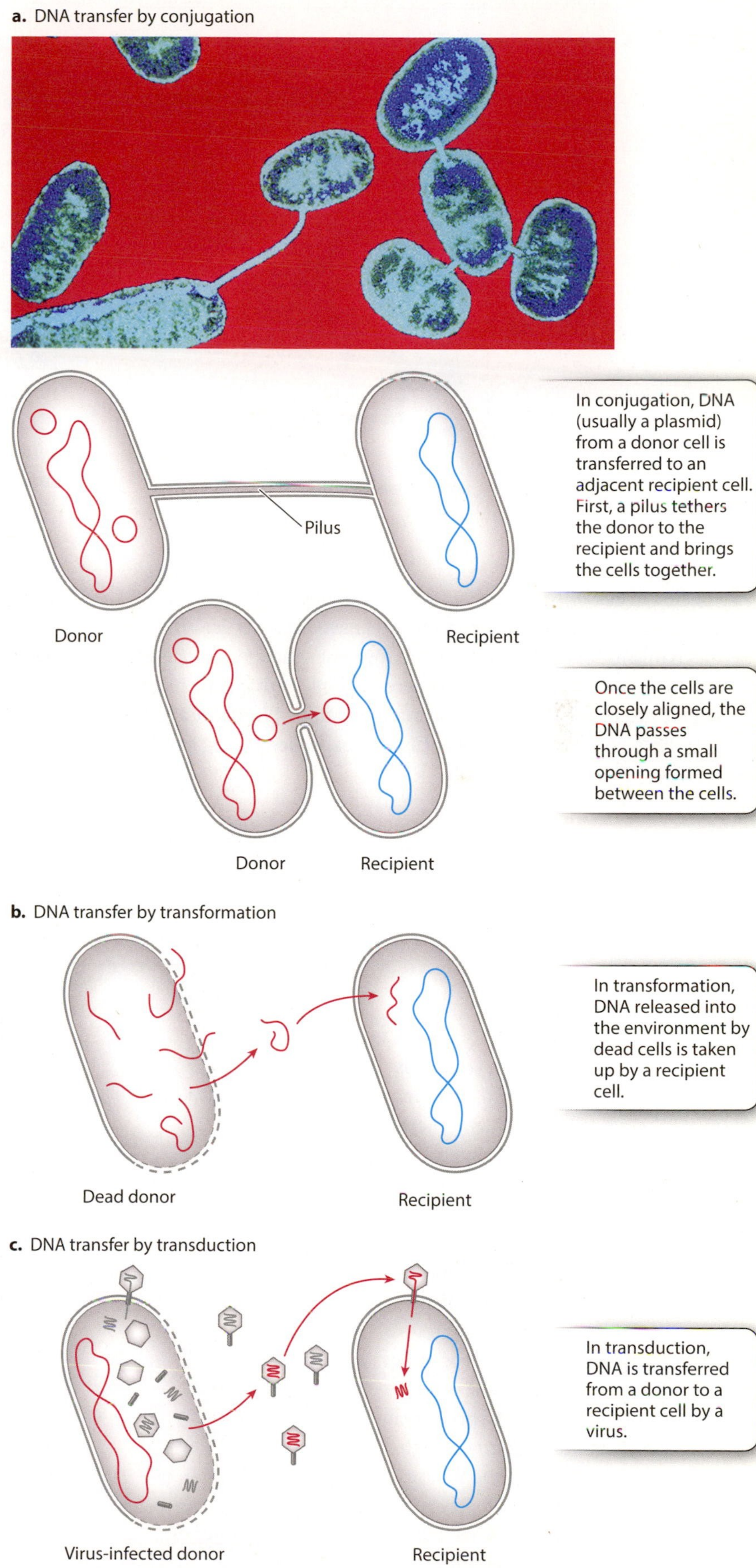

FIG. 26.4 Horizontal gene transfer in bacteria. DNA, shown in red in the diagrams, originates from the donor cell. DNA shown in blue is that of the recipient cell. *Source: Dr. Linda M. Stannard, University of Cape Town/Science Source.*

FIG. 26.5 A three-branched tree of life, based on sequence comparisons of the genes for the small subunit of ribosomal RNA. Archaea and Bacteria form distinct branches, although they inherited prokaryotic cell organization from a common ancestor. Some data suggest that eukaryotes arose within the Archaea, making this domain paraphyletic.

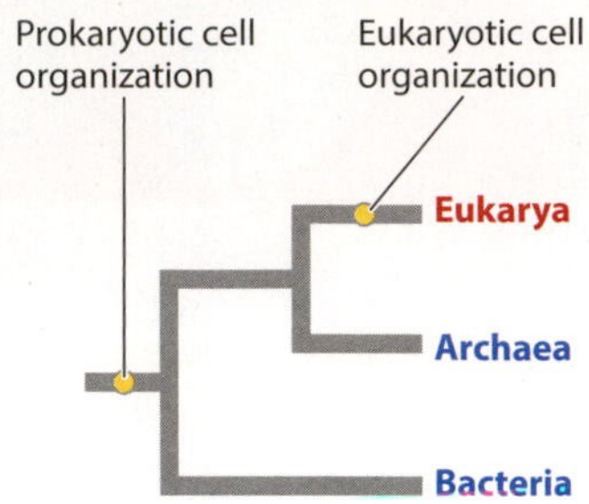

cousins of eukaryotes, unable to generate genetic diversity by sexual recombination. Instead, bacteria have highly efficient mechanisms for adding and subtracting genes that permit them to evolve and adapt rapidly to local conditions. Perhaps the most widely discussed and worrisome manifestation of horizontal gene transfer is the rapid spread of antibiotic resistance among bacteria (Chapter 49).

➔ **Quick Check 1** In eukaryotes, sexual reproduction is the main process by which new gene combinations are generated. How do bacteria generate new gene combinations in the absence of sexual reproduction?

Archaea form a second prokaryotic domain.

For many years, all prokaryotic cells were classified as bacteria. However, in 1977, George Fox and Carl Woese published a revolutionary hypothesis. From comparisons of RNA molecules from the small subunits of ribosomes (and later, comparisons of the genes for these RNAs), Fox and Woese argued that prokaryotic organisms actually fall into two distinct groups, as different from each other as either is from eukaryotes. If this is true, the tree of life must have three great branches: Bacteria, Eukarya, and the limb recognized by Fox and Woese, called Archaea (**Fig. 26.5**). Research since the late 1970s confirms this hypothesis. Like bacterial cells, archaeal cells are prokaryotic—they have no membrane-bounded nucleus and their genes are arrayed along a single circular chromosome. Moreover, cell size in Archaea is also limited by diffusion, and genetic diversity is promoted by horizontal gene transfer, much as in Bacteria.

In detail, however, Archaea are quite distinctive (**Table 26.1**). Their membranes are made from lipids different from the fatty acids found in bacterial and eukaryotic membranes. Archaea also show a diversity of molecules in their cell walls, but none has the peptidoglycan characteristic of Bacteria or the cellulose or chitin found in most eukaryotic cell walls.

Archaea differ from Bacteria in another intriguing way. DNA transcription in archaeons employs RNA polymerase and ribosomes more similar to those of eukaryotes than to bacteria (Chapter 3). Moreover, many of the antibiotics that target protein synthesis in bacteria are ineffective against archaeons, suggesting fundamental differences in translation as well.

Many of the microorganisms first identified as Archaea have unusual physiological properties or inhabit extreme environments. For example, some archaeons live in acid mine water at pH 1 (or less!). Others live in water salty enough to precipitate NaCl (**Fig. 26.6**), or in deep-sea hydrothermal vents where temperatures can exceed 100°C. We now know, however, that many archaeons live under less extreme conditions in soils, lakes, and the sea. Indeed, they may be the most abundant organisms throughout much of the ocean.

➔ **Quick Check 2** Why were Archaea originally thought to be simply unusual forms of Bacteria? What lines of evidence showed this domain to form a distinct branch on the tree of life?

TABLE 26.1 Principal Differences among Archaea, Bacteria, and Eukarya

CHARACTERISTIC	ARCHAEA	BACTERIA	EUKARYA
Cell contains a nucleus and other membrane-bound organelles	No	No	Yes
DNA occurs in a circular form*	Yes	Yes	No
Ribosome size	70s	70s	80s
Membrane lipids ester-linked†	No	Yes	Yes
Photosynthesis with chlorophyll	No	Yes	Yes
Capable of growth at temperatures greater than 80°C	Yes	Yes	No
Histone proteins present in cell	Yes	No	Yes
Operons present in DNA	Yes	Yes	No
Introns present in most genes	No	No	Yes
Capable of methanogenesis	Yes	No	No
Sensitive to the antibiotics chloramphenicol, kanamycin, and streptomycin	No	Yes	No
Capable of nitrogen fixation	Yes	Yes	No
Capable of chemoautotrophy	Yes	Yes	No

*Eukaryote DNA is linear. †Archaea membrane lipids are ether-linked.

FIG. 26.6 **Archaeons in San Francisco Bay.** The ponds are flooded and then evaporated to precipitate table salt. Salt-loving archaeons, made conspicuous by their reddish purple pigment, thrive here. *Source: David Wall/Alamy.*

26.2 AN EXPANDED CARBON CYCLE

In our discussion of energy metabolism in Chapters 7 and 8, we emphasized oxidation–reduction (redox) chemistry. In redox reactions, a pair of molecules reacts, one molecule becoming more oxidized and the other becoming more reduced. If we follow the transfer of electrons in a redox reaction, we see that the molecule that is reduced gains electrons, whereas the molecule that is oxidized loses electrons. Reduction therefore requires a source of electrons, or an electron donor, and oxidation requires a sink for electrons, or an electron acceptor.

In photosynthesis carried out by plants, algae, and cyanobacteria (Chapter 8), carbon dioxide (CO_2) is reduced to form carbohydrates, and water is oxidized to oxygen gas (O_2). Water is the electron donor needed to reduce CO_2, generating O_2 as a by-product. In respiration, as described in Chapter 7, organic molecules are oxidized to CO_2, and O_2 is reduced to water. In this case, O_2 serves as the electron acceptor needed to oxidize organic molecules.

In all known photosynthetic eukaryotes, the photosynthetic reaction is **oxygenic,** or oxygen producing. There is a good reason for this. Water occurs nearly everywhere and, in the oxygen-rich environments where most eukaryotic organisms thrive, no other molecule is available to donate electrons. Nonetheless, where oxygen is limited or absent, other electron donors such as hydrogen sulfide (H_2S) are available for photosynthesis and are utilized by photosynthetic microorganisms. Similarly, essentially all respiration in eukaryotic cells is **aerobic,** or oxygen utilizing. Again, this makes good chemical sense because oxidation of carbohydrates using oxygen generates far more energy than can be obtained using other oxidants. Once again, however, where oxygen is absent, other electron acceptors can be used for respiration. The organisms that can use these alternative electron donors and electron acceptors are Bacteria and Archaea.

We've already noted that Bacteria and Archaea differ from Eukarya in cell structure and genome organization, but it is the distinctive ways in which Bacteria and Archaea gather carbon and harvest energy that make them indispensable parts of nature. Prokaryotic organisms are thus the real foundation of the carbon cycle, capable of building and dismantling organic molecules throughout the full breadth of habitats on Earth.

Many photosynthetic bacteria do not produce oxygen.

In saline bays along the west coast of Australia and many other low-lying tropical coastlines, carpets of deep blue-green cover tidal flats along restricted lagoons where high salinity and periodic exposure limit colonization by algae and animals (**Fig. 26.7**). Called microbial mats, the carpets shown in Fig. 26.7a are densely packed communities of (mostly) bacteria and archaeons that have a predictable structure based on the depths to which light and oxygen penetrate into the mat (Figs. 26.7b and 26.7c). In the upper layer, where both light and oxygen are present, the mat is dominated by cyanobacteria, bacteria that photosynthesize like plants and algae, using water as an electron donor and releasing oxygen gas (the blue-green layer in Fig. 26.7b). It is the photosynthetic pigments in the cyanobacteria that impart the blue-green color to microbial mats. The cyanobacteria share this layer with heterotrophic bacteria, which get their carbon from preformed organic molecules. Both cyanobacteria and the heterotrophic bacteria respire aerobically, using O_2 to oxidize organic molecules to CO_2. Thus, at the mat surface, prokaryotic microorganisms carry out a biological carbon cycle much like the one outlined for plants and animals in Chapter 25.

What happens beneath the mat surface is a different story. Note in Fig. 26.7b that a distinct purple layer lies beneath the blue-green mat surface. Light penetrates into this layer, but oxygen does not because aerobic respiration rapidly consumes O_2 within the mat (Fig. 27.6c). What organisms live where light is present and oxygen absent?

Like the cyanobacteria in the surface zone, these bacteria are photosynthetic, but they do not use water as an electron donor and so do not generate oxygen gas. Called **anoxygenic** because they do not produce oxygen, these photosynthetic bacteria use electron donors such as hydrogen sulfide (H_2S), hydrogen gas (H_2), ferrous iron (Fe^{2+}), and even the arsenic compound arsenite (AsO_3^{3-}). Where O_2 is present, these compounds oxidize quickly and so are not available for photosynthesis. Therefore,

FIG. 26.7 Microbial mats. Mats, like the example in (a) from Western Australia, cover tidal flats and lagoons where salinity or other environmental factors inhibit animals and seaweeds. As shown by the example from (b) Cape Cod, microbial mats have sharp vertical gradients of light, oxygen, and other chemical compounds, and so support nearly all the energy metabolisms found in (c) the expanded carbon cycle. *Sources: a. Andrew Knoll, Harvard University; b. Republished with permission of Springer Science+Business Media, from The Prokaryotes: Vol. 2: Ecophysiology and Biochemistry 2006, The Phototrophic Way of Life Jörg Overmann and Ferrau Garcia-Pichel, figure 5.*

microorganisms that conduct anoxygenic photosynthesis are restricted to sunlit habitats where O_2 is absent.

The purple color in the mat layer comes from **bacteriochlorophyll,** a light-harvesting pigment closely related to the chlorophyll found in plants, algae, and cyanobacteria (**Fig. 26.8a**). Because of its chemical structure, bacteriochlorophyll absorbs most strongly at wavelengths different from those absorbed by the cyanobacteria above them in the mat (**Fig. 26.8b**).

Anoxygenic photosynthesis differs from oxygenic photosynthesis in another important way: It employs only a single photosystem. As discussed in Chapter 8, a single photosystem cannot capture enough energy to oxidize water and then use those electrons to reduce CO_2. Therefore, oxygenic photosynthesis has two photosystems in series. With only a single photosystem, bacteria that carry out anoxygenic photosynthesis use electron donors that donate their electrons more easily than water.

Many bacteria respire without oxygen.

Light but no oxygen penetrates into the purple layer in Fig. 26.7b, and neither light nor oxygen penetrates into the black layer deep within the microbial mat. Yet both layers are inhabited by a diversity of heterotrophic Bacteria and Archaea that metabolize organic molecules originally synthesized by photosynthetic bacteria higher in the mat community.

Because there is no oxygen in these layers, these organisms must use other molecules as electron acceptors in cellular respiration, including oxidized forms of nitrogen (NO_3^-), sulfur (SO_4^{2-}), manganese (Mn^{4+}), iron (Fe^{3+}), and even arsenic (AsO_4^{3-}). Such microorganisms thrive in many environments where oxygen is absent.

Still other types of heterotrophs live in deeper layers of microbial mats. **Fermentation** provides an alternative to cellular respiration as a way of extracting energy from organic molecules. Whereas cellular respiration is the full oxidation of carbon compounds to CO_2, fermentation is the partial oxidation of carbon compounds to molecules that are less oxidized than CO_2 (Chapter 7). Fermentation has an advantage over cellular respiration in that it does not require an external electron acceptor, such as O_2. On the other hand, fermentation yields only a modest amount of energy.

Fermentation plays an important role in oxygen-poor environments that are rich in organic matter, such as landfills and the digestive tracts of animals. Here the breakdown of organic molecules typically requires more than one organism, each able to metabolize a different intermediate. These cooperating groups of fermenters enable the breakdown of substances that could not be metabolized by any one organism alone. Groups of fermenters thus form what is essentially an ecological solution to the metabolically difficult problem of gaining energy from complex organic molecules. Fermentation is widespread in Bacteria and Archaea, but is of only minor importance in most Eukarya, the exception being yeasts, which thrive in the sugar-rich

FIG. 26.8 Structures and absorption spectra of bacteriochlorophyll *a* and chlorophyll *a*. Note that bacteria absorb wavelengths of light that are outside the visible spectrum.

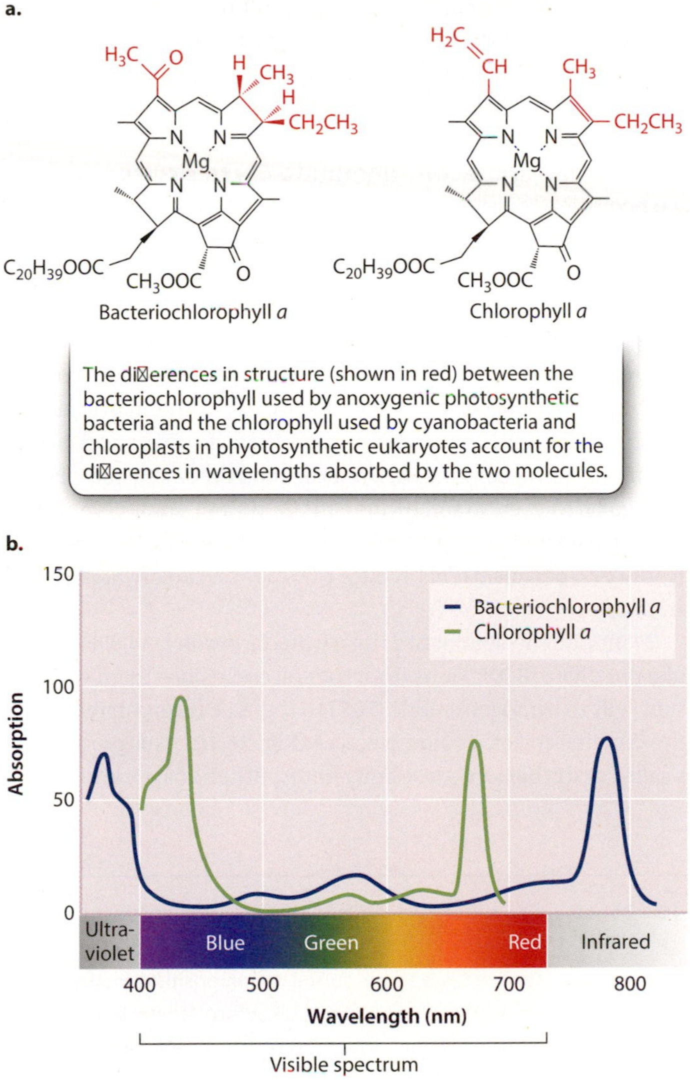

environment of rotting fruit, as well as in breweries throughout the world.

In summary, where O_2 is present, many different kinds of organisms participate in the carbon cycle. Photosynthetic plants, algae, and cyanobacteria all transform CO_2 into organic molecules, and animals, fungi, single-celled eukaryotes, bacteria, and archaeons return CO_2 to the environment by aerobic respiration. In Chapter 25, we noted that some of the organic carbon produced by photosynthesis escapes aerobic respiration and accumulates in oxygen-depleted waters or sediments. Only prokaryotic heterotrophs can complete the recycling of organic carbon in oxygen-poor environments. Thus, in carrying out anaerobic respiration or a diversity of fermentation reactions, prokaryotes play a major role in the carbon cycle.

Photoheterotrophs obtain energy from light but obtain carbon from preformed organic molecules.

In Chapter 6, we saw that organisms have two sources of energy: the sun (phototrophs) and chemical compounds (chemotrophs). Organisms also have two sources of the carbon needed for growth and other functions: inorganic molecules like CO_2 (autotrophs) and organic molecules like glucose (heterotrophs). This means that there are four ways that organisms acquire the energy and carbon they need (see Fig. 6.1).

Up to this point, we have considered only two of these ways. Plants, algae, and cyanobacteria are photoautotrophs, gaining energy from the sunlight and carbon from CO_2, whereas animals, fungi, and many prokaryotes are chemoheterotrophs, gaining both energy and carbon from organic molecules taken up from the environment. Among the Bacteria and Archaea, additional metabolisms are possible. Some use the energy from sunlight to make ATP, just as plants do, but rather than reducing CO_2 to make their own organic molecules, they rely on organic molecules obtained from the environment as the source of carbon for growth and other vital functions. These organisms are **photoheterotrophs.**

Photoheterotrophy can be advantageous in environments rich in dissolved organic compounds. It allows organisms to use all their absorbed light energy to make ATP while directing all absorbed organic molecules toward growth and reproduction. Photoheterotrophic bacteria have a single photosystem and so live where oxygen is limited or absent. The purple layer of a microbial mat very likely contains heliobacteria or green nonsulfur bacteria that take advantage of the large amounts of organic matter generated by photoautotrophs in the mat.

→ **Quick Check 3** How do photoautotrophs such as cyanobacteria differ from photoheterotrophs such as heliobacteria?

Chemoautotrophy is a uniquely prokaryotic metabolism.

On the deep seafloor, animals are relatively uncommon, their abundance limited by the slow descent of organic matter from surface oceans. However, where hydrothermal springs punctuate the deep seafloor, animal populations can be remarkably dense (**Fig. 26.9**). Why are animals so abundant around hydrothermal vents?

The many animals around deep-sea hydrothermal vents subsist on a thriving community of microorganisms. These microorganisms gain carbon by reducing CO_2 to form carbohydrates. However, they obtain the energy to fuel this process not from sunlight but from chemical reactions. Hence, these microorganisms are called **chemoautotrophs.** They oxidize molecules and ions such as H_2, H_2S, and Fe^{2+} to generate the ATP and reducing power required to incorporate CO_2 into organic molecules. In chemoautotrophic metabolism, the electron acceptor is most commonly O_2 or nitrate (NO_3^-). Thus, chemoautotrophy requires access to both oxidized (O_2 and nitrate) and reduced (H_2, H_2S, and Fe^{2+}) substances.

FIG. 26.9 **Deep-sea hydrothermal vents.** Chemoautotrophs fuel the carbon cycle in these environments. *Source: Science Source/Getty Images.*

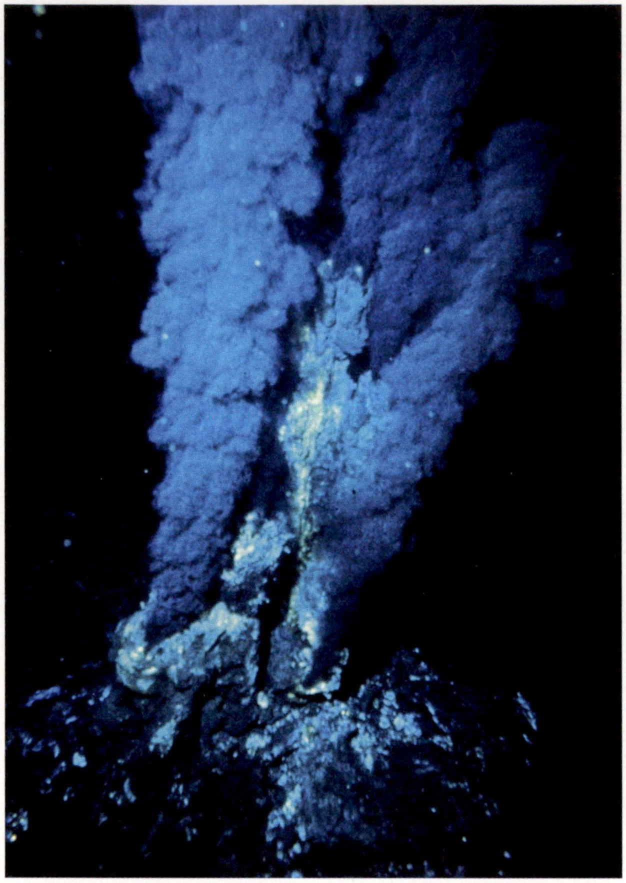

Oxidized and reduced molecules react readily with each other and so are rarely present in the same environment. For this reason, chemoautotrophs tend to live along the interface between an environment where oxygen is present and one in which oxygen is absent. Chemoautotrophic prokaryotes therefore thrive where reduced gases from Earth's interior meet oxygen-rich seawater.

→ **Quick Check 4** How could the biological carbon cycle have worked on the primitive Earth, where oxygen gas was essentially absent from the atmosphere and oceans?

26.3 OTHER BIOGEOCHEMICAL CYCLES

Let's think further about prokaryotic metabolism. We saw that some photosynthetic and chemosynthetic bacteria oxidize reduced sulfur compounds like H_2S to sulfate (SO_4^{2-}), whereas other heterotrophic bacteria reduce sulfate to H_2S. This is reminiscent of the carbon cycle outlined in Chapter 25, where plants generate oxygen and animals consume it. Metabolisms at work where oxygen is absent can also complete the carbon cycle, and they do so by coupling the cycling of carbon to the cycling of other elements—in this case, sulfur. Once again, we see complementary metabolic pathways. The expanded carbon cycle of Bacteria and Archaea promotes the cycling of sulfur, nitrogen, and other biologically important elements.

Bacteria and archaeons dominate Earth's sulfur cycle.

Our bodies are mostly carbon, oxygen, and hydrogen, but we also contain about 0.2% sulfur by weight. Sulfur is a component of the amino acids cysteine and methionine and therefore is present in many proteins (Chapter 2). Sulfur is also present in iron–sulfur clusters that are key components of enzymes fundamental to electron transfer (Chapter 6) and in some vitamins. Where do we get the sulfur we need? From the food we eat—protein in steak or fish provides a rich source of sulfur. Where do cows and fish get the sulfur they need? Like us, from the food they eat. Food chains don't go on forever, so somewhere in the system, some organisms must take up inorganic sulfur from the environment. **Primary producers,** organisms that reduce CO_2 to form carbohydrates, fill this role.

Plants are the dominant primary producers on land. Plants take up sulfate (SO_4^{2-}) ions from the soil and reduce them within their cells to hydrogen sulfide (H_2S) that can be incorporated into cysteine and other biomolecules (**Fig. 26.10**). This process is called **assimilation.** Algae and photosynthetic bacteria do

FIG. 26.10 **The sulfur cycle.** Plants take up sulfur as sulfate ions for incorporation into proteins and other compounds, and fungi release sulfur during decomposition. However, the major role in cycling sulfur through the biosphere is played by microbes that oxidize or reduce sulfur compounds to gain energy or carbon.

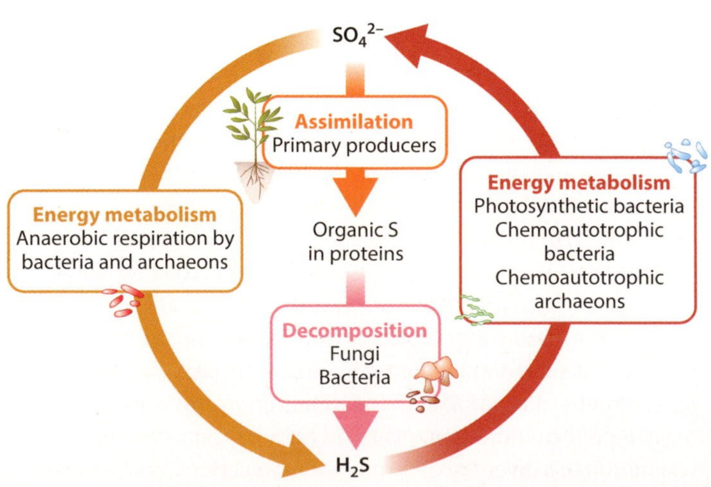

much the same thing in lakes, rivers, and oceans. Why don't primary producers take up H_2S directly? There are two answers to this question. First, H_2S is rapidly oxidized in the presence of oxygen and so does not occur in environments where oxygenic photosynthesis is common. Second, H_2S is generally toxic to eukaryotic organisms, so plants and algae do not thrive where it is abundant. (The H_2S produced within eukaryotic cells has a short lifetime and is restricted to intracellular sites distant from those that are vulnerable to its toxic effects.)

We've now seen half the sulfur cycle, the conversion of sulfate to H_2S within cells. How did sulfate molecules get into the soil in the first place? After death, fungi and bacteria decompose cells, returning carbon, sulfur, and other compounds to the environment. Reduced sulfur compounds released from decomposing cells are oxidized by bacteria and archaeons, completing the cycle. These microbes are chemoautotrophs that obtain energy by oxidizing H_2S or photosynthesizers that use H_2S as the electron donor. In addition to being taken up by plants, the sulfate produced by these processes is consumed by heterotrophic bacteria living in oxygen-free environments found in soil, sediments, and occasionally in lakes and seawater. Sulfate rather than oxygen is used as the final electron acceptor in respiration and is reduced to H_2S. In fact, most of the sulfur cycled biologically is used to drive energy metabolism in oxygen-poor environments, not to build proteins.

Note that eukaryotes use neither H_2S for photo- (or chemo-) synthesis nor sulfate in respiration. Thus, the biological sulfur cycle is completed by bacteria and archaeons alone.

The nitrogen cycle is also driven by bacteria and archaeons.

The model provided by the sulfur cycle can be extended to other biologically important elements. Nitrogen is the fourth most abundant element in the human body, contributing 4% by weight as a component of proteins, nucleic acids, and other compounds. As is true of sulfur, primary producers incorporate inorganic forms of nitrogen into biomolecules, and we get the nitrogen we need from our food. As we'll see, however, there is a twist.

Let's examine the biological nitrogen cycle (**Fig. 26.11**). Carbon is a minor component of the atmosphere and sulfur only a trace constituent, but the air we breathe consists mostly (78%) of nitrogen gas (N_2). Indeed, most of the nitrogen at Earth's surface resides in the atmosphere. Although nitrogen is plentiful, the nitrogen available to primary producers in many environments is limited. The reason is that plants cannot make use of nitrogen gas and neither can algae. Only certain bacteria and archaeons can reduce N_2 to ammonia (NH_3), a form of nitrogen that can be incorporated into biomolecules.

The process of converting N_2 into a biologically useful form such as ammonia is called **nitrogen fixation,** and it is one of the most important reactions ever evolved by organisms. Farmers plant soybeans to regenerate nitrogen nutrients in soils, but it is not the soybean that fixes nitrogen. Rather, soybean roots have nodules that harbor nitrogen-fixing bacteria (**Fig. 26.12**). That is, plants like soybean have entered into an intimate partnership with nitrogen-fixing bacteria to obtain the biologically useful forms of nitrogen needed for growth (Chapter 29).

FIG. 26.11 **The nitrogen cycle.** Unique bacterial and archaeal metabolisms include nitrogen fixation, nitrification and denitrification, as well as anammox.

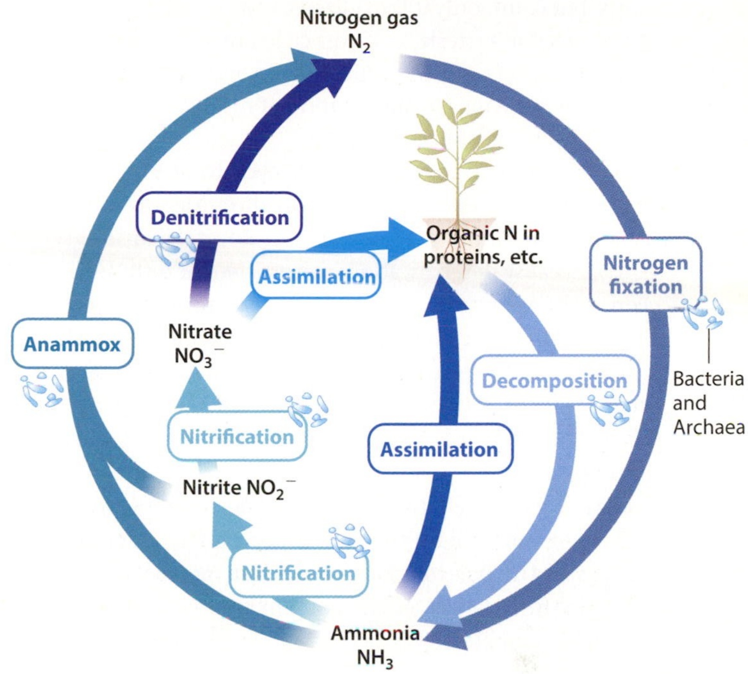

FIG. 26.12 **Roots of a soybean plant.** The nodules contain bacteria capable of reducing nitrogen gas to ammonia. *Source: Scimat/Getty Images.*

The nitrogen cycle, then, begins with nitrogen fixation, and it continues through many additional metabolisms, most of them confined to bacteria and archaeons. Ammonia released by the decomposition of dead cells can be taken up by primary producers, but commonly it is oxidized to nitrate (NO_3^-) by chemoautotrophic bacteria, a process called **nitrification** (see Fig. 26.11). This process works efficiently in both soil and water, and so nitrate is the most common form of nitrogen available to life in both environments.

Anaerobic respiration completes the cycle by returning N_2 to the atmosphere. Some bacteria can use nitrate as an electron acceptor in respiration, a process known as **denitrification** (see Fig. 26.11). Because denitrification is so efficient at returning nitrogen to the atmosphere as nitrogen gas, nitrogen fixation must continually make nitrogen available for organisms. Without nitrogen fixation, the nitrogen cycle would slowly wind down, and the productivity of Earth's biota—that is, of all its organisms—would wind down with it. This is why nitrogen fixation is such an important process.

Until the 1990s, the nitrogen cycle was described—completely, it was thought—in terms of the metabolic reactions just discussed. Then, a new and unexpected form of nitrogen metabolism was discovered in the oceans. A number of prokaryotic microorganisms function as chemoautotrophs by reacting ammonium ion (NH_4^+) with nitrite (NO_2^-) to gain energy. The reaction is shown here:

$$NH_4^+ + NO_2^- \rightarrow N_2 + 2H_2O$$

Note that this is a redox reaction, like other reactions involved in energy metabolism. In this case, the ammonium ion is oxidized to nitrogen gas, and nitrite is reduced to nitrogen gas, with water produced as a by-product. The reaction, called **anammox** (**an**aerobic **amm**onia **ox**idation), provides only a small amount of energy, but it supports many bacteria and archaeons in oxygen-depleted parts of the ocean, providing a major route by which fixed nitrogen returns to the atmosphere as N_2 (see Fig. 26.11).

The biological nitrogen cycle, then, begins and ends with prokaryotic metabolisms. Bacteria and archaeons fix N_2 into a biologically available form, nitrifying bacteria oxidize ammonia to nitrate, and still other prokaryotes gain energy from denitrification and anammox, returning N_2 to the air. Unlike the carbon and sulfur cycles, the nitrogen cycle does not rely at any point on elements stored as minerals in rocks and sediments. Most of Earth's nitrogen was released to the atmosphere as gas early in our planet's history, and so movement into sediments and out of volcanoes contributes relatively little to the cycle. Humans, however, contribute in important ways, most notably through the industrial synthesis of nitrate fertilizer (Chapter 49).

➔ **Quick Check 5** How would the nitrogen cycle operate in the absence of bacteria and archaeons?

26.4 THE DIVERSITY OF BACTERIA

Traditionally, bacterial taxonomy was a laborious process. Microbiologists grew bacteria in beakers or on petri dishes, carefully separating and enriching individual populations until only one type of bacterium—that is, a pure culture—was present. Species were identified on the basis of details of cell structure and division observed under the microscope, and metabolic capabilities established through experiments and reactions to chemical stains or antibiotics. By the 1980s, a few thousand species had been described in this way.

With molecular sequencing technologies, however, bacterial species are now characterized by their DNA sequences, especially the genes for the small subunit of ribosomal RNA (rRNA). A novel application of sequencing technology paved the way for modern studies of bacterial diversity and ecology. Microbiologists reasoned that the same procedures used to extract, amplify, and sequence genes from pure cultures in the lab could be used to study bacteria in nature. That is, extracting and reading gene sequences directly from soil or seawater could reveal the diversity of the microbial communities in these environments.

The results were stunning. Most bacteria in nature—perhaps 99% or more—are species that have not been cultured in the lab. Indeed, species well characterized by studies of pure cultures commonly represent only minor components of natural communities. To give just one illustration, a bacterium called SAR11 was originally identified on the basis of gene sequences amplified from seawater. SAR11 caught the attention of microbiologists because it makes up about a third of all sequences identified in the samples from the surface ocean. How could we be ignorant of a bacterium that must be among the most abundant organisms on Earth? The answer is simple: No one had cultured it. Recently, SAR11 has been cultured. Formally described as *Pelagibacter ubique* (*ubique* is Latin for "everywhere"), this bacterium is now known to be a heterotroph that plays an important role in recycling organic molecules dissolved in the sea.

More recently, the application of shotgun sequencing techniques (Chapter 13) to the environmental sampling of genomes has allowed microbiologists to begin matching uncultured populations with specific metabolisms. That is, we need no longer be content to know from molecular sequences what is living in a given environment. Now we can begin to understand what the different microorganisms are doing there. For this reason, another revolution in our understanding of the microbial world lies just around the corner (**Fig. 26.13**).

Bacterial phylogeny is a work in progress.

In Chapter 23, we introduced molecular phylogenies, in which gene sequences are compared to enable us to draw conclusions about evolutionary relatedness among populations in nature. Gene sequencing made it possible to test traditional bacterial groupings based on form and physiology. Some traditionally defined

HOW DO WE KNOW?

FIG. 26.13

How many kinds of bacterium live in the oceans?

BACKGROUND The oceans are full of bacteria—by some estimates, the sea harbors more than 10^{29} bacterial cells. How diverse are these bacterial communities?

METHOD 1 Different types of bacterium can be recognized by the unique nucleotide sequences of their genes. In fact, individual types of bacterium can be identified on the basis of nucleotide sequence in a relatively small region of the gene for the small subunit of ribosomal RNA (rRNA). This molecular "barcode" can be sequenced quickly and accurately for samples of DNA drawn from the environment.

Using this strategy, scientists associated with the International Census of Marine Life obtained samples of seawater and seafloor sediment from deep-ocean environments. In the laboratory, the rRNA sequences were amplified by PCR, separated by gel electrophoresis, and sequenced, thus providing a library of tags that document bacterial diversity

METHOD 2 In another set of experiments, scientists collected samples of seawater from the Sargasso Sea, and from these they amplified whole genomes—more than a billion nucleotides' worth. They used small-subunit rRNA and other genes to characterize species richness, and also characterized a large assortment of additional sequence data to understand genetic and physiological diversity.

RESULTS Both surveys found that the bacterial diversity of marine environments is much higher than had been thought. The barcode survey (method 1) found as many as 20,000 distinct types of bacteria in a single liter of seawater and estimated that as many as 5–10 million different microbes may live in the world's oceans. Some of these bacteria are abundant and widespread, but most are rare. The genomic survey (method 2) of the Sargasso Sea found more than 1800 distinct bacteria in sea-surface samples and, remarkably, identified more than 1.2 million previously unknown genes within these genomes.

CONCLUSION The bacterial diversity of the oceans is huge and still largely unexplored. Both barcode and genomic surveys are continuing, and promise a more nearly complete accounting of marine diversity. Together, these surveys will help us understand how marine bacteria function and how they sort into communities.

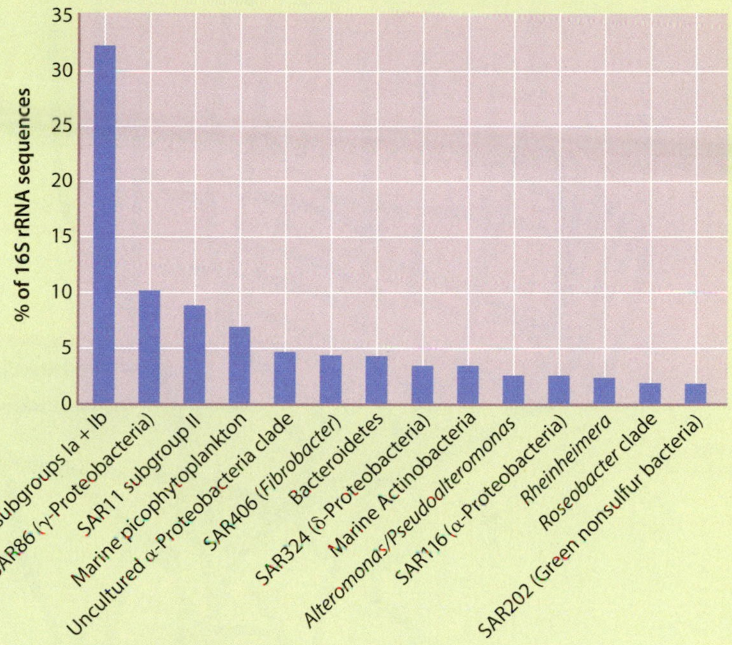

Relative abundance of different bacterial groups in samples from the Sargasso Sea, based on percent representation of small subunit rRNA sequences. Note the abundance of SAR11, only recently characterized.

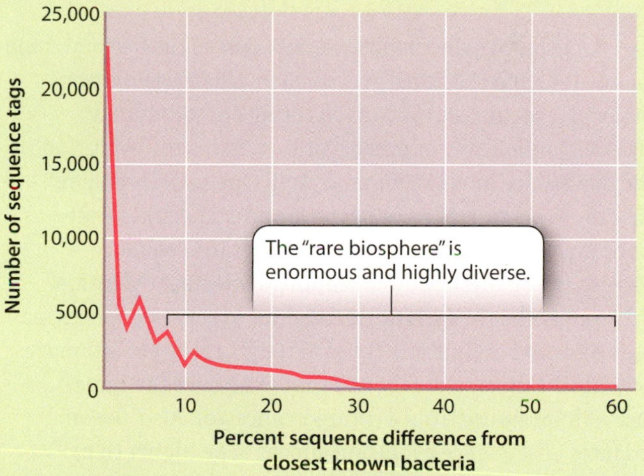

Computer-aided comparison of sequences from the world's oceans shows that about 75% of all sequences collected are similar to other sequences in the library of sequence tags. However, 25% are distinctive, forming a rare but enormously diverse pool of bacterial types.

SOURCES Venter, J. C., et al. 2004. "Environmental Genome Shotgun Sequencing of the Sargasso Sea." *Science* 304:66–74; Sogin, M. L., et al. 2006. "Microbial Diversity in the Deep Sea and the Underexplored 'Rare Biosphere'" *Proceedings of the National Academy of Sciences USA* 103:12115–12120.

FIG. 26.14 Bacterial and archaeal phylogeny. Some scientists argue that the evolution of prokaryotes should not be viewed as a tree but as a series of branches that diverge and then come back together, representing genes that diverge from one another but then come to reside together in new organisms because of horizontal gene transfer. *Source: Adapted from W. F. Doolittle, 2000, "Uprooting the Tree of Life," Scientific American 282:90–95.*

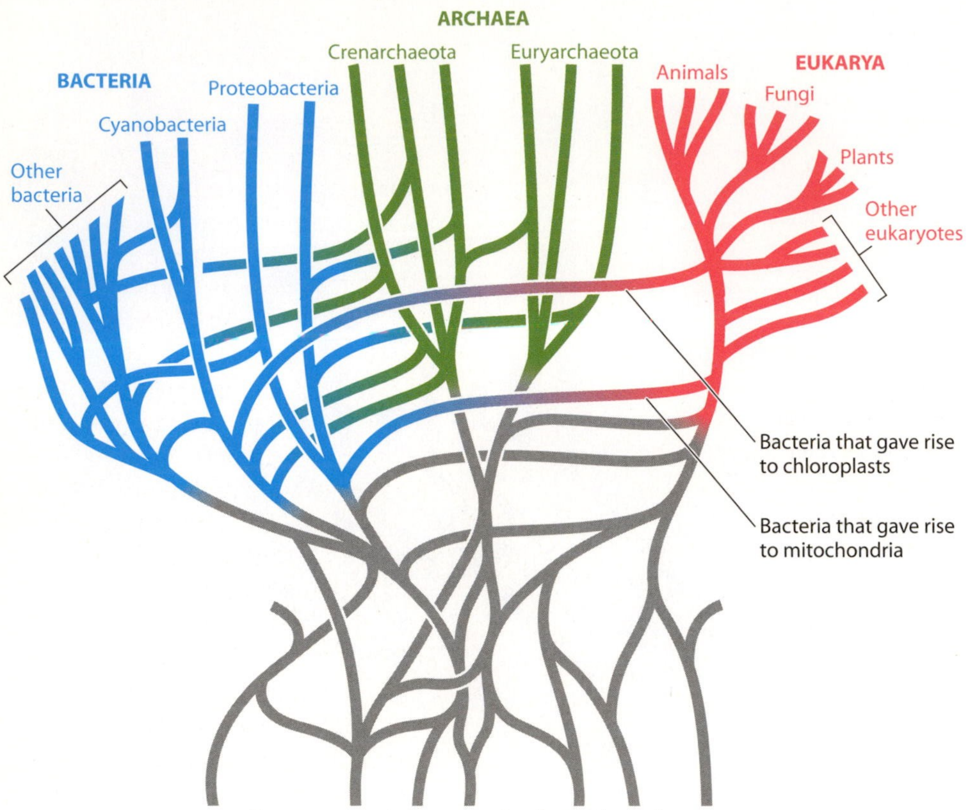

groups—for example, the cyanobacteria— passed the test and stand today as well-defined bacterial groups. Many, however, did not.

Molecular sequence comparisons bring enormous new possibilities to the study of prokaryotic diversity and evolution, but they introduce new problems as well. One set of problems arises because major groups of Bacteria and Archaea separated from one another billions of years ago. Inherited sequence similarities can be masked by continuing molecular evolution within groups. That is, specific nucleotides in gene sequences may change not once but multiple times through a long evolutionary history, erasing molecular features that were once the same because of inheritance from a common ancestor. Also, in some groups rates of sequence evolution appear to be higher than in others. As a result, these groups have especially divergent sequences and so may misleadingly fall toward the bottoms of phylogenetic trees.

Another problem arises because of horizontal gene transfer. Phylogenies may falsely group distantly related bacteria because they contain genes passed on by conjugation, transformation, or transduction. Statistical analyses of complete bacterial genomes suggest that at least 85% of all genes in bacterial genomes have been transferred horizontally at least once. Such apparently rampant gene exchange might well doom attempts to build a bacterial tree of life, and some biologists argue that bacterial evolution is more properly depicted as a complex bush with many connecting branches (**Fig. 26.14**).

Other microbiologists are more optimistic. Small-subunit rRNA genes, for example, show little evidence of horizontal transfer and so may faithfully record the evolutionary history of the organisms in which they occur. Nonetheless, we need to be cautious when using bacterial trees to study the evolution of specific metabolic or physiological characteristics. Nitrogen fixation, photosystems, drug tolerance—such features depend on genes known to jump from branch to branch on the tree.

At present, molecular signatures have been extremely useful in identifying major groups of bacteria, but less clear in establishing branching order among these groups. About 50 major groups of bacteria (each one very roughly equivalent to a eukaryotic phylum) have been recognized. Nearly half of these groups cannot be cultured, but have been identified only by the cloning of genes from environmental samples. To support our discussion of bacterial diversity, we present a phylogeny based on comparisons of whole genomes in **Fig. 26.15.** As you examine this figure, remember that microbiologists continue to debate branching order on the bacterial tree.

What, if anything, is a bacterial species?

What do we mean when we talk about bacterial species? As we saw in Chapter 22, Ernst Mayr defined species as "groups of interbreeding natural populations that are reproductively isolated from other such groups." This widely accepted definition is called the biological species concept, but a moment's reflection will show that it can't be applied to the most abundant organisms on Earth—Bacteria and Archaea. In the absence of sexual reproduction, how do we think about species in the prokaryotic domains?

Some microbiologists argue that we should abandon the concept of species when talking about bacteria, especially since

FIG. 26.15 Evolutionary relationships among bacteria, inferred from data from complete genomes. Photosynthetic groups are shown in green (the Acidobacteria have one known photoheterotrophic member). Other groups discussed in the text are shown in red.

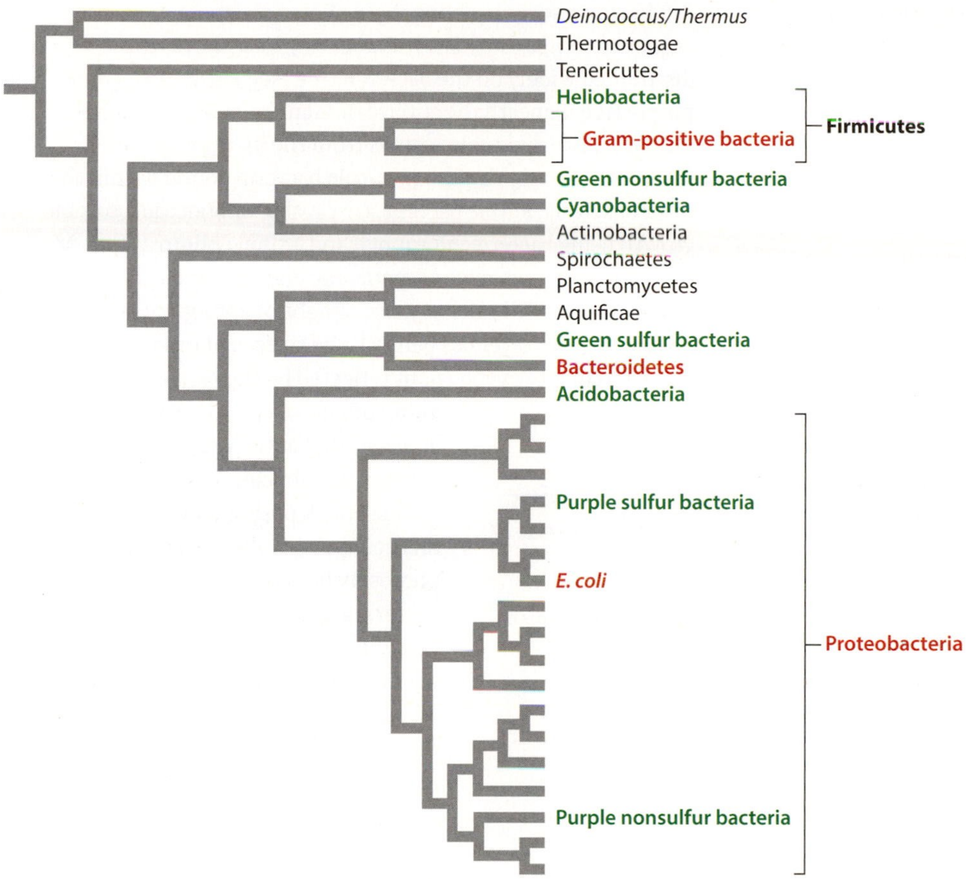

distantly related bacteria can exchange genes. Nonetheless, nature is full of microbial populations with well-defined features of form and function.

In the age of molecular sequence comparisons, bacterial species have sometimes been defined as populations whose small-subunit rRNA sequences are more than 97% identical. It would be useful, however, to conceive of bacterial species in terms of functional processes, much as Ernst Mayr did for eukaryotic organisms. In plants, animals, and single-celled eukaryotes, interbreeding ensures that populations share the same pool of genes, providing a counterforce to mutations and selection pressures that promote divergence.

A similar process exists in bacteria. In early research on the population genetics of bacteria, biologists observed that genetic diversity in laboratory cultures gradually increases through time and then rapidly decreases with the emergence of a successful variant that outcompetes the rest. The episodic loss of diversity is called **periodic selection.** Some biologists argue that this process provides a means of recognizing bacterial species: Populations subject to the same episodes of periodic selection belong to a single species. This is a useful way to define bacterial species, but other definitions are possible, and as yet there is no consensus among microbiologists.

Proteobacteria are the most diverse bacteria.

Proteobacteria are far and away the most diverse of all bacterial groups (Fig. 26.15). Defined largely by similarities in rRNA sequences, Proteobacteria include many of the organisms that populate our expanded carbon cycle and other biogeochemical cycles. For example, they include anoxygenic photosynthetic bacteria and chemoautotrophs that oxidize NH_3, H_2S, and Fe^{2+}, as well as bacteria able to respire using SO_4^{2-}, NO_3^-, or Fe^{3+}.

Many Proteobacteria have evolved intimate ecological relationships with eukaryotic organisms. Some of these are mutually beneficial, such as the nitrogen-fixing bacteria in soybean root nodules. Other Proteobacteria, however, are our worst pathogens: the rickettsias that cause typhus and spotted fever, vibrios that trigger cholera, salmonellas responsible for typhoid fever and food poisoning, and *Pseudomonas aeruginosa* strains that infect burn victims and others whose resistance is reduced. *Escherichia coli,* famous as a laboratory model organism (see Fig. 26.4b), is usually a harmless presence in our intestinal tract, but pathogenic strains can cause serious diarrhea and even death.

The gram-positive bacteria include organisms that cause and cure disease.

In 1884, the Danish scientist Hans Christian Gram developed a diagnostic dye that, after washing, is retained by bacteria with thick peptidoglycan walls, but not by those with thin walls. Molecular sequence comparisons confirm that most **gram-positive bacteria** (those that retain the dye) form a well-defined branch of the bacterial tree (Fig. 26.15). One species in this group, *Bacillus subtilis,* is the focus of many studies about basic molecular function in bacteria. All of us harbor *Streptococcus* species in our mouths, and many of us have suffered from strep throat, produced by streptococcal pathogens. More serious infections, including some forms of meningitis, are also associated with streptococci. *Staphylococcus* bacteria are commonly found on human skin, but only occasionally do we suffer from maladies such as staph infection, toxic shock syndrome, or food poisoning induced by staphylococcal bacteria.

In contrast to these pathogens, gram-positive bacteria called streptomycetes have proved invaluable to medicine because of a remarkable property. More than half of the species in this group secrete compounds that kill other bacteria and fungi. Streptomycetes have provided us with tetracycline, streptomycin, erythromycin, and numerous other antibiotics that combat infectious disease.

Photosynthesis is widely distributed on the bacterial tree.

Molecular studies confirm that all bacteria capable of oxygenic photosynthesis form a single branch of the bacterial tree: the **cyanobacteria.** Cyanobacteria can be found in environments that range from deserts to the open ocean. Their diverse species include unicellular rods and spheroidal cells, as well as multicellular balls and filaments (**Fig. 26.16**). Some filamentous cyanobacteria can even form several different cell types, including specialized cells for nitrogen fixation and resting cells that provide protection when the local environment does not favor growth.

In contrast to the monophyly of cyanobacteria, molecular sequence comparisons show that *anoxygenic* photosynthesis is distributed widely on the bacterial tree (see Fig. 26.15). Think of the purple layer beneath the surface of many microbial mats (see Fig. 26.7). The vivid color comes from the light-sensitive pigments of the aptly named purple bacteria, found within the Proteobacteria. Purple bacteria are capable of photoautotrophic growth using bacteriochlorophyll and a single photosystem. Most, however, show evidence of metabolic diversity and can grow heterotrophically in the absence of light or appropriate electron donors. A green layer lies beneath the surface of other microbial mats, its color caused by light reflected by the pigments of another group of photosynthetic bacteria, called green sulfur bacteria because they commonly gain electrons from hydrogen sulfide and deposit elemental sulfur on their walls. Unlike most groups of photosynthetic bacteria, the green sulfur bacteria are intolerant of oxygen gas. Oddly, a species of green sulfur bacteria has been found in hydrothermal rift vents 2 km beneath the surface of the ocean. Light doesn't penetrate to these depths, so if these organisms harvest light, it must come from incandescent lava as it erupts on the bottom of the ocean.

Light-harvesting bacteria occur on several other branches of the bacterial tree—the photoheterotrophic heliobacteria, for example, and the green nonsulfur bacteria often found in freshwater hot springs. Despite a century of research, our knowledge of photosynthesis within the bacterial tree remains incomplete, as illustrated by the discovery in 2007 of unique light-harvesting bacteria in Yellowstone National Park. Genomic surveys of hot springs turned up evidence of a previously unknown bacteriochlorophyll-containing microorganism, and gene sequencing established that this organism belongs to the phylum Acidobacteria. There is no evidence for carbon fixation, however, suggesting that the Yellowstone bacterium is a photoheterotroph. To date, these unusual microorganisms have not been grown in pure culture, so they remain difficult to study.

FIG. 26.16 Diversity of form among cyanobacteria. *Sources: a. Dr. Ralf Wagner; b. blickwinkel/Alamy; c. Hecker/Sauer/age fotostock; d. Jason Oyadomari.*

a. *Aphanothece*, small single cells

b. *Chroococcus*, larger cells bound into a colony by a common extracellular envelope made of polysaccharides

c. *Oscillatoria*, filaments with no cell differentiation

d. *Anabaena*, strings of undifferentiated cells at either end of the filament, flanking two elongated resting cells, and in the middle, a heterocyst specialized for nitrogen fixation

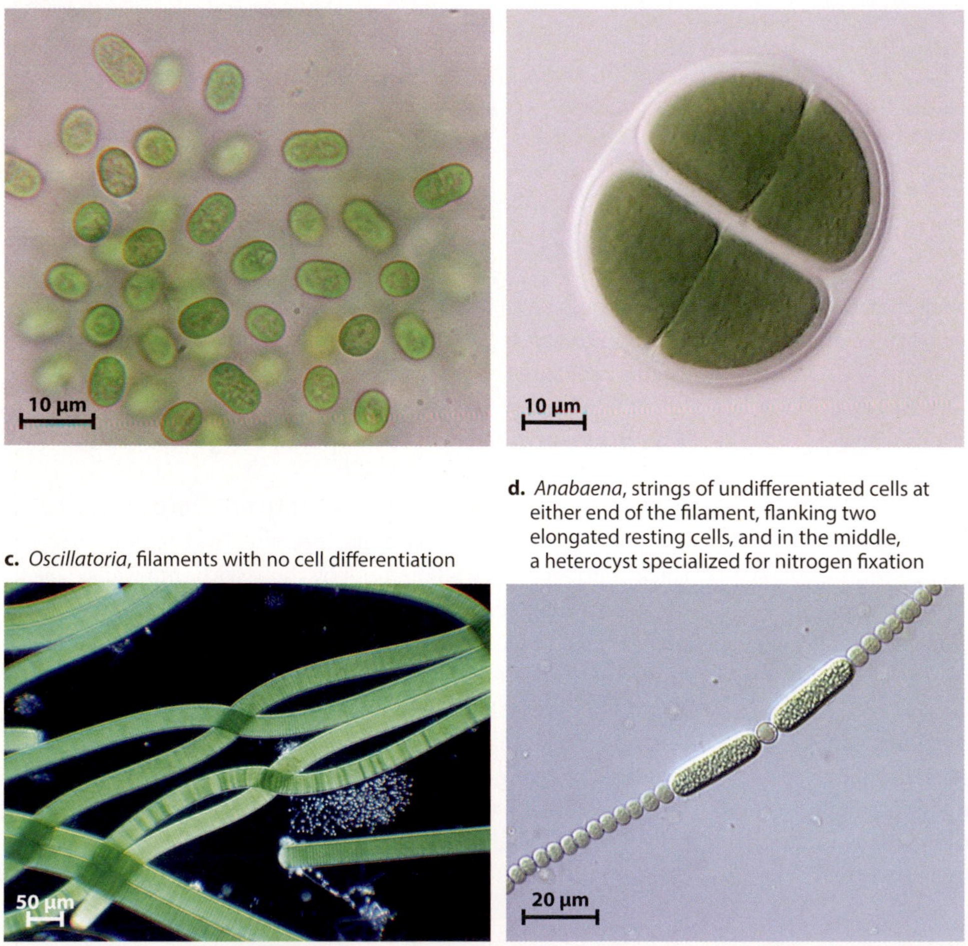

26.5 THE DIVERSITY OF ARCHAEA

We have already noted that Archaea and Bacteria are prokaryotic, but differ in their membrane and wall composition, molecular mechanisms for transcription and translation, and many other biological features (see Table 26.1). As a result, the two groups tend to thrive in very different environments. Many archaeons tolerate environmental extremes such as heat and acidity. In fact, for many environmental conditions, archaeons define the known limits of life. What do these disparate environments have in common? One proposal is that archaeons thrive in environments where the energy available to fuel cell activities is limited—where there is no light, or a limited abundance of fuels and oxidants for chemoautotrophy or respiration, or both. That is, archaeons commonly live where energy resources are too low to support bacteria and eukaryotes.

Bacteria and Archaea diverged from their last common ancestor more than 3 billion years ago. Just as Bacteria have evolved a phylogenetic diversity that defies easy categorization, Archaea also include many distinct types of organism. Many microbiologists recognize three major divisions of Archaea, called the **Crenarchaeota, Euryarchaeota,** and **Thaumarchaeota,** but a few poorly known groups may indicate further major branches within the domain (for example, the Korarchaeota and Nanoarchaeota noted in **Fig. 26.17**). Indeed, some microbiologists now propose that several of these groups together form a distinct clade nicknamed TACK (Thaumarchaeota + Aigarchaeota + Crenarchaeota + Korarchaeota; the upper main branch in Fig. 26.17). Like the Bacteria, Archaea include forms able to respire aerobically or anaerobically, as well as chemoautotrophs and fermenters. There are, however, metabolic differences between the two domains. No Archaea carry out photosynthesis using chlorophyll and related pigments, but some Archaea exhibit forms of energy metabolism unknown in bacteria or eukaryotes.

The archaeal tree has anaerobic, hyperthermophilic organisms near its base.

Archaeons found on lower branches of the archaeal tree exhibit a remarkable ability to grow and reproduce at temperatures of 80°C or more (Fig. 26.17). Called **hyperthermophiles,** these organisms do not simply tolerate high temperature; they require it. At present, the world record for high temperature growth is held by *Methanopyrus kandleri,* which has been shown to grow at 122°C among hydrogen- and CO_2-rich hydrothermal vents on the seafloor beneath the Gulf of California. Most of these heat-loving archaeons are anaerobic, living as anaerobic respirers or as chemoautotrophs that obtain energy from the oxidation of hydrogen gas.

Why do anaerobic hyperthermophiles sit on the lowest branches of the archaeal tree? Many biologists believe that these features characterized the first Archaea and therefore imply that

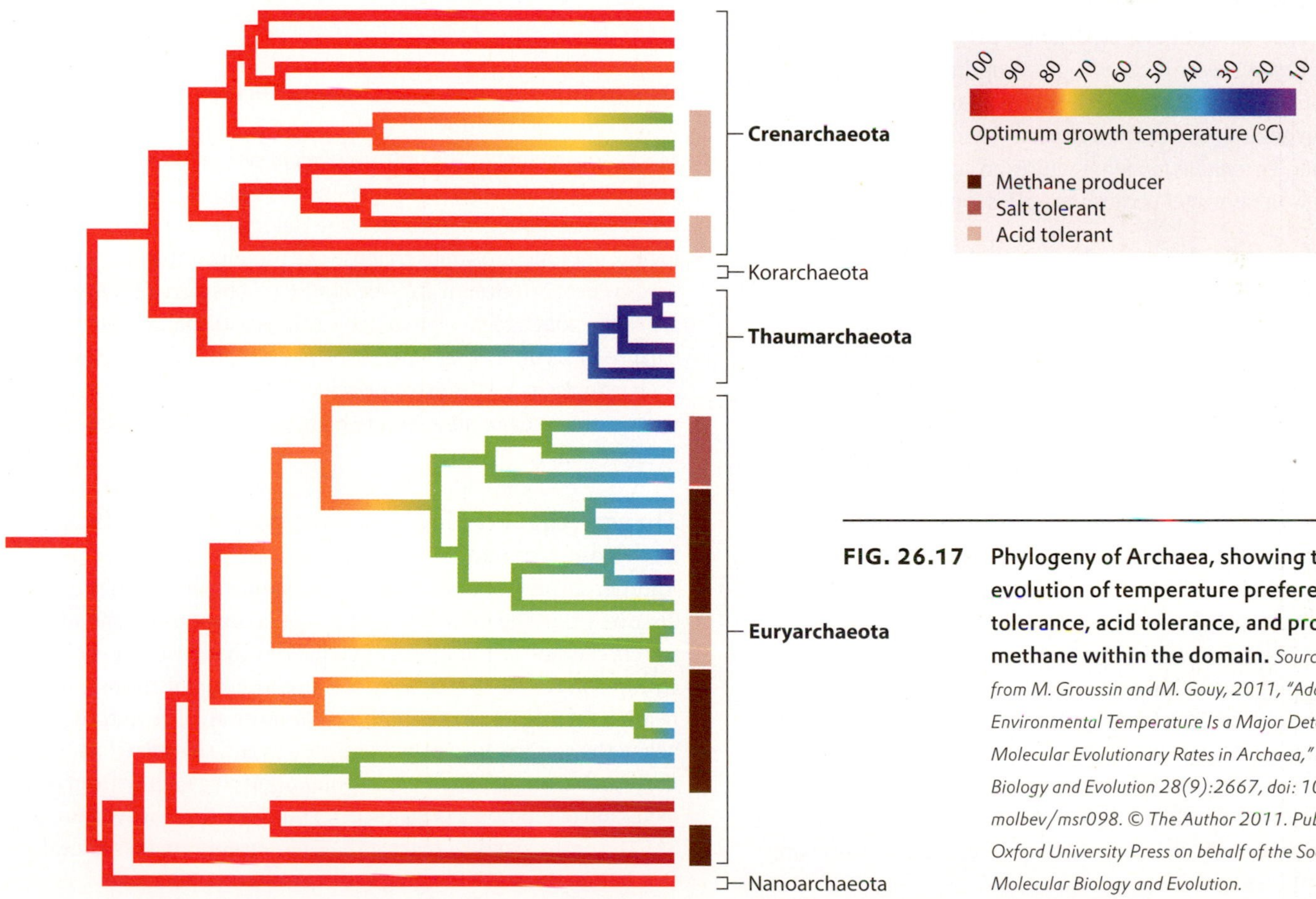

FIG. 26.17 Phylogeny of Archaea, showing the evolution of temperature preference, salt tolerance, acid tolerance, and production of methane within the domain. *Source: Adapted from M. Groussin and M. Gouy, 2011, "Adaptation to Environmental Temperature Is a Major Determinant of Molecular Evolutionary Rates in Archaea," Molecular Biology and Evolution 28(9):2667, doi: 10.1093/molbev/msr098. © The Author 2011. Published by Oxford University Press on behalf of the Society for Molecular Biology and Evolution.*

the Archaea evolved in hot environments with no oxygen. As we discussed in Chapter 23, geologists agree that oxygen gas was not present on the early Earth, consistent with the inference we can make from the archaeal tree that early Archaea were anaerobic. A hot environment for early Archaea can be understood in two ways. Either the entire ocean was hot—for which there is little evidence—or Archaea first evolved in hot springs where local supplies of hydrogen and metals enabled them to live as chemoautotrophs. Because some versions of the bacterial tree also show hyperthermophilic species on their lowest branches, it has been suggested that the last common ancestor of *all* living organisms lived in a hot environment.

➔ **Quick Check 6** If early branches on the bacterial and archaeal trees are dominated by hyperthermophilic microorganisms, does this mean that the early oceans were very hot?

The Archaea include several groups of acid-loving microorganisms.

In many parts of the globe where humans mine coal or metal ores, abandoned mines fill up with very acidic water. Called acid mine drainage, this water has been acidified through the oxidation of pyrite (FeS_2) and other sulfide minerals to produce sulfuric acid (H_2SO_4). The pH of these waters is commonly about 1 to 2, and can actually fall below 0. To us, acid mine drainage is an environmental catastrophe, but a number of archaeons thrive in it (**Fig. 26.18**). Modified protein and membrane structures enable them to tolerate their harsh surroundings, making it possible to use sulfur compounds in the environment for both chemoautotrophic and heterotrophic growth. What is the limit to growth at low pH? *Picrophilus torridus*, a euryarchaeote archaeon, grows optimally at 60°C and in environments with a pH of 0.7—the pH of battery acid. It is capable of growth in environments down to pH 0— a record, as far as we know. In fact, *Picrophilus torridus* requires an acidic environment. It will not grow in rainwater because the pH is too high.

FIG. 26.18 Headwaters of the Rio Tinto, southwestern Spain. Archaeons thrive in these highly acidic waters, pH 1–2. *Source: Andrew Knoll, Harvard University.*

Only Archaea produce methane as a by-product of energy metabolism.

Euryarchaeotes include Earth's only biological source of natural gas, or methane (CH_4). Known as **methanogens,** they generate methane by fermenting acetate (CH_3COO^-) or by chemoautotrophic pathways in which hydrogen gas (H_2) reduces CO_2 to CH_4. Methanogenic Archaea play an important role in the carbon cycle, recycling organic molecules in environments where the supplies of electron acceptors for respiration are limited.

Most methanogens live in soil or sediments, and they are particularly prominent in peats and lake bottoms where sulfate levels are low and bacterial sulfate reduction is therefore limited. Some, however, live in the rumens (a specialized chamber in the gut) of cows (Case 5), helping to maximize the energy yield from ingested grass. Methane doesn't last long in the atmosphere—it is quickly oxidized to carbon dioxide. However, the steady supply of methane generated by archaeons contributes significantly to the greenhouse properties of the atmosphere. For this reason, biologists are interested in how methanogenic Archaea will respond to 21st-century climate change (Chapter 49).

One group of the Euryarchaeota thrives in extremely salty environments.

Methanogens are not the only unique euryarchaeotes. A related group illustrates the limits of tolerance to another environmental condition—salinity, or saltiness. The Haloarchaea are halophilic (salt-loving) photoheterotrophs that use the protein bacteriorhodopsin to absorb energy from sunlight (see Fig. 26.6). They live in waters that are salty enough to precipitate table salt (NaCl), maintaining osmotic balance by accumulating ions (especially K^+) and organic solutes in their cytoplasm. Just as extreme acidophiles require acid conditions, these extreme halophiles require high salt conditions and cannot live in dilute water.

Thaumarchaeota may be the most abundant cells in the deep ocean.

Another surprise of environmental sequencing has been the discovery that the immense volumes of ocean beneath the surface waters contain vast numbers of archaeons that have not yet been cultured in the lab. In the deep ocean, these organisms may be nearly as abundant as the combined number of all bacteria on and below the surface. Recently segregated as the Thaumarchaeota, these archaeons are chemoautotrophs, deriving energy from the oxidation of ammonia (**Fig. 26.19**). Thus, like the discovery of SAR11, the discovery of marine thaumarchaeotes makes it clear that before the still-emerging age of environmental genetics

FIG. 26.19

How abundant are archaeons in the oceans?

BACKGROUND Early exploration of microbial diversity in the oceans revealed that archaeons are tremendously abundant. Just how abundant are they, and what metabolisms do they employ to obtain energy and carbon?

METHOD Scientists sampled seawater throughout the depth of the ocean at a test site in the northern Pacific Ocean. To the samples, they added molecular tags bound to fluorescent molecules that are visible under the microscope. The tags were RNA sequences that bind to small-subunit ribosomal RNA genes known to be useful in identifying different types of bacteria and archaeons. In this way, separate fluorescent markers were attached to thaumarchaeote archaeons, euryarchaeote archaeons, and bacteria.

RESULTS By counting the cells marked by different fluorescent tags in seawater samples, the biologists showed that bacteria dominate microbial communities in near-surface seawater, but that thaumarchaeotes are as numerous as bacteria in deeper waters (Fig. 26.19a).

a.

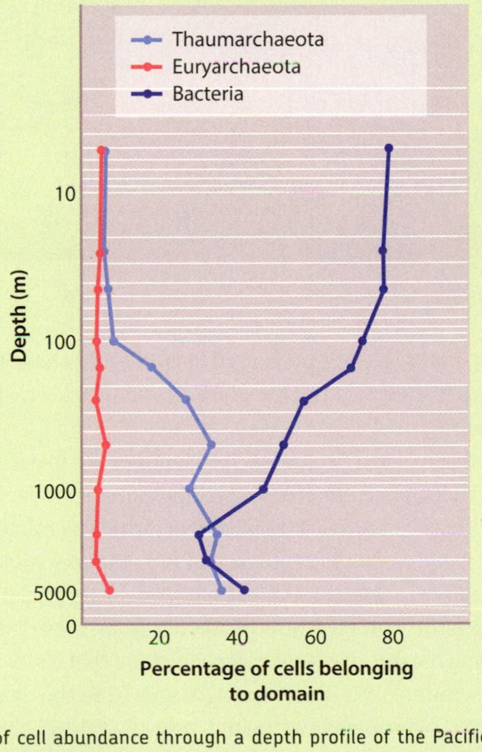

Surveys of cell abundance through a depth profile of the Pacific Ocean show that bacteria dominate cell numbers near the surface, but thaumarchaeotes make up about 40% of all cells in deeper waters.

CONCLUSION Marine thaumarchaeota, unknown in 1990, are now known to be among the most abundant organisms in the oceans.

FOLLOW-UP WORK Archaea are among the most abundant of all cells in the world's oceans. How do these cells live? Microbial samples from water known to be sites of nitrification (that is, the conversion of ammonia to nitrite or nitrate) contained an abundance of thaumarchaeote cells. The high numbers of thaumarchaeotes from these communities supported the hypothesis that they are nitrifiers—the higher the abundance of thaumarchaeote cells (Fig. 26.19b, dark blue line), the higher the amount of nitrite produced (Fig. 26.19b, light blue line). Thaumarchaeotes were grown in pure culture and shown to grow by consuming ammonia (Fig. 26.19b, red line). In other words, these organisms oxidize ammonia to provide the ATP and reducing power needed to incorporate CO_2 into organic molecules. Therefore, they play a major role in the marine nitrogen cycle, thriving where sources of carbon and energy for other types of metabolism are scarce.

b.

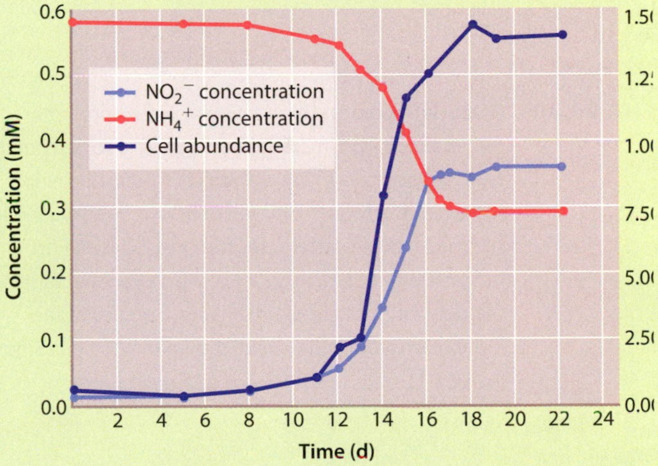

Growth of a marine thaumarchaeote in a medium containing ammonium chloride and bicarbonate as the only sources of energy and carbon, respectively. As cell number increased, ammonia (the ammonium ion NH_4^+) was increasingly converted to nitrite (NO_2^-), supporting the hypothesis that these cells are ammonia-oxidizing chemoautotrophs.

SOURCES Karner, M. B., E. F. DeLong, and D. M. Karl. 2001. "Archaeal Dominance in the Mesopelagic Zone of the Pacific Ocean." *Nature* 409:507–510 (The figure has been modified to reflect more recent nomenclature.); Könneke, K., et al. 2005. "Isolation of an Autotrophic Ammonia-Oxidizing Marine Archaeon." *Nature* 437:543–546.

and genomics, we simply didn't have a very good idea of which microorganisms lived where and did what in the oceans. Fortunately, that situation is changing rapidly.

26.6 THE EVOLUTIONARY HISTORY OF PROKARYOTES

On the tree of life, the plants and animals of our familiar existence populate only the most recent branches. For most of Earth's history, life was entirely microbial. Perhaps surprisingly given their small size, Bacteria and Archaea have left a fossil record that allows paleontologists to reconstruct aspects of life's early evolution. Sedimentary rocks preserve microscopic fossils of ancient bacteria, including the photosynthetic cyanobacteria (**Fig. 26.20**). Limestones and related rocks also preserve **stromatolites,** layered structures that record sediment accumulation by microbial communities (**Fig. 26.21**). Some especially well preserved ancient rocks also contain molecular fossils—biomolecules, usually lipids, that resist decay and so preserve a chemical signature of ancient life.

Life originated early in our planet's history.

Earth is about 4.6 billion years old, but few sedimentary rocks have survived destruction by geologic processes and remain to document our planet's earliest history. Sedimentary rocks deposited nearly 3.5 billion years ago in Western Australia and South Africa provide some of our earliest records of Earth's history. Remarkably, they preserve a scarce but discernible signature of life. Tiny organic structures preserved in these rocks have been interpreted as microfossils, although this conclusion remains controversial.

More pervasive (and persuasive) evidence of life in early oceans is provided by stromatolites. Stromatolites in ancient rocks are strikingly similar to modern structures formed by microbial communities, and so provide evidence of life that goes back about 3.5 billion years (Fig. 26.21). Also, because purely physical processes cannot explain the isotopic composition of carbon in very old limestones and organic matter, we can conclude that a biological carbon cycle existed 3.5 billion years ago (and, from the evidence of 3.8-billion-year-old rocks in Greenland, possibly earlier).

The chemistry of Earth's oldest sedimentary rocks also shows that the early atmosphere and ocean contained little or no free

FIG. 26.20 **Bacterial microfossils.** (a) Bacteria that probably metabolized iron in 1.9-billion-year-old rock from the Gunflint Formation in northwest Ontario, Canada. (b) Cyanobacteria preserved in 800-million-year-old silicified marine carbonates from the Draken Formation in Spitsbergen, Norway. (c) A short cyanobacterial filament from silicified tidal flat carbonates of the 1.5-billion-year-old Bilyakh Group in northern Siberia. *Source: a–c. Andrew Knoll, Harvard University.*

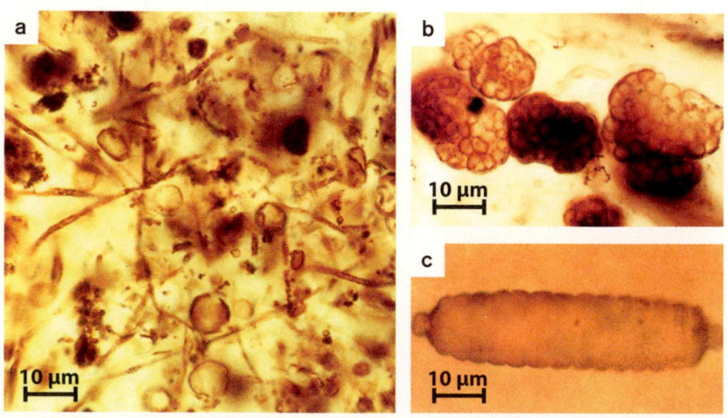

FIG. 26.21 **Stromatolites.** Modern-day stromatolites at Shark Bay, Western Australia (top) show microbial communities building the domed structures exhibited by stromatolites in ancient rocks (bottom). Stromatolites are among the earliest records of life on Earth. *Sources: (top) Andrew Knoll, Harvard University; (bottom) François Gohier/Science Source.*

oxygen. What would the carbon cycle look like on an oxygen-free planet? We know the answer from our discussion of prokaryotic metabolism. Photosynthesis can proceed in oxygen-poor waters, using H_2S, H_2, or Fe^{2+} as electron donors, and fermenters and anaerobic respirers can recycle carbon using electron acceptors such as SO_4^{2-} and Fe^{3+}. Geologic evidence suggests that iron played a particularly important role in cycling carbon for the first billion years of evolutionary history.

The evolution of cyanobacteria, with their capacity for oxygenic photosynthesis, made possible the accumulation of oxygen in the atmosphere and oceans. As discussed in Chapter 25, oxygen began to accumulate in the atmosphere about 2.4 billion years ago, making possible aerobic respiration and, eventually, our present-day carbon cycle. Both microfossils and molecular fossils indicate that cyanobacteria and other photosynthetic bacteria continued to dominate primary production until about 800 million years ago. Thus, our modern carbon cycle, powered overwhelmingly by oxygenic photosynthesis and with major participation by eukaryotic organisms, may have existed for only the last 20% of Earth's history.

Prokaryotes have coevolved with eukaryotes.

Even after the rise of eukaryotic cells and, later, plants and animals, Bacteria and Archaea have remained essential to the functioning of biogeochemical cycles and hence to the maintenance of habitable environments on Earth. Prokaryotic organisms have also radiated into the many novel environments made possible by eukaryotes. We noted earlier that bacteria fix nitrogen in nodules formed on soybean roots. Soybeans and nitrogen-fixing bacteria evolved together, a process called **coevolution.**

Coevolutionary relationships between prokaryotic microorganisms and eukaryotes are common in nature. As noted earlier, cows metabolize grass with the help of methanogenic archaeons in the rumen, a specialized digestive organ. The microbes enable the cow to gain nutrition from plant materials that could otherwise not be digested, and the benefit to the microorganisms is a steady supply of food. Similarly, clams that live near methane-rich vents in the Gulf of Mexico have evolved specialized organs that house methane-oxidizing bacteria, providing them with a reliable source of nutrition.

Bacteria can affect animal behavior as well as nutrition. For example, some squid species attract mates by emitting light from specialized organs full of bioluminescent bacteria (Case 5). The bacteria provide the squid with a mechanism for communication in the dark deep-sea environment, and the squid feeds the bacteria. Interestingly, many different kinds of bacteria live in the waters that surround the squid, but only the bioluminescent species accumulates in the squid's light organ. The squid apparently sends molecular signals that attract just those bacteria.

Bacteria can also influence animal reproduction. Earth supports more than a million species of insects, and perhaps two-thirds of these harbor the proteobacterial parasite *Wolbachia*

FIG. 26.22 A large population of *Wolbachia* bacteria (fluorescing green) in the embryo of a fruit fly. *Wolbachia* can influence the development and reproductive biology of their animal hosts. *Source: Image by Zoe Veneti and Kostas Bourtzis.*

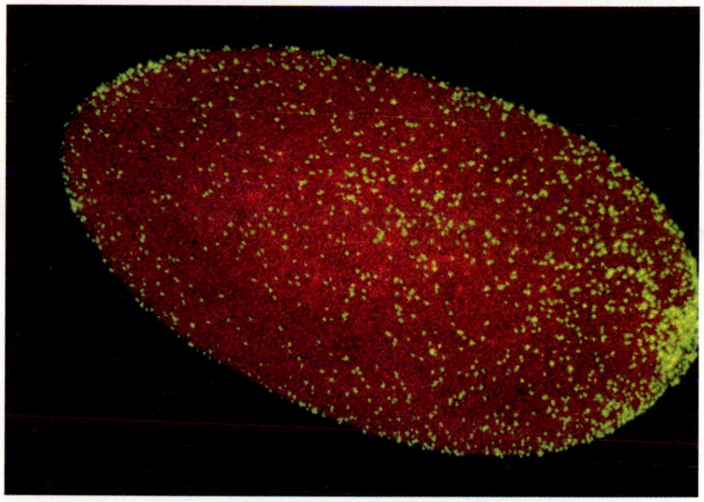

within their tissues. A heterotroph that gains nutrition by metabolizing organic compounds supplied by its host, *Wolbachia* infects its host's reproductive tissues and passes from one generation to the next in the host's eggs (**Fig. 26.22**). Remarkably, many *Wolbachia* strains alter host reproductive cells so that the insect host can successfully reproduce only with partners infected by the same *Wolbachia* strain. Because *Wolbachia* cells are transferred by eggs and not sperm, they have little use for males. Consequently, some cause male host embryos to die, thus increasing the proportion of females. Others cause genetically male embryos to develop as females, or enable eggs to develop into female embryos without fertilization.

Wolbachia have another effect on their insect hosts that is highly relevant to human health: They commonly inhibit infection by viruses and the eukaryotic parasites that cause malaria. Recent research shows that *Wolbachia* populations in some fly species protect their hosts against infection by RNA viruses. Scientists introduced these bacteria into mosquitos that transmit dengue fever, a tropical disease that infects 50–100 million people each year. The *Wolbachia*-infected mosquitos were strongly resistant to the Dengue virus, which causes dengue fever. Moreover, when introduced into the wild, these virus-resistant forms rapidly expanded to dominate regional mosquito populations. *Wolbachia* parasites present a possible path to controlling dengue fever, malaria, and other insect-borne diseases.

All around us, plants, animals, and microorganisms live in intimate association—sometimes to our detriment, but commonly in mutually beneficial relationships that influence how we grow, develop, eat, and behave. Humans are no exception.

? CASE 5 THE HUMAN MICROBIOME: DIVERSITY WITHIN

How do intestinal bacteria influence human health?

It has been estimated that the bacteria (and, to a much lesser extent, archaeons) in and on your body outnumber your own cells by as much as 10 to 1. Estimates of the number of microbial species that inhabit our bodies vary, but about 750 types have been identified in the mouth alone, and the list remains incomplete (**Fig. 26.23**). An equal number resides in the colon, and still more live on the skin and elsewhere. Some are transients, entering and leaving the body within a few bacterial generations. Others are specifically adapted for life within humans, forming complex communities of interacting species.

At present, much research is focused on the bacteria within our intestinal tracts. We commonly think of gut bacteria as harmful, but this perception arises from only a small—albeit devastating—subset of our microbial guests. Illnesses known to be caused by bacteria within our bodies include cholera, dysentery, and tuberculosis. Once the leading causes of death in New York and London, they remain major killers in Africa. Other diseases, not previously thought to be of microbial origin, are now known to be caused by bacteria—ulcers, for example, and stomach cancer, both mediated by the acid-tolerant bacterium *Helicobacter pylori.* More often than not, however, intestinal bacteria have a beneficial effect. They help to break down food in our digestive system and secrete vitamins and other biomolecules into the colon for absorption into our tissues. Molecular signals from bacteria guide the proper development of cells that line the interior of our intestines.

FIG. 26.23 The human microbiome. This chart provides a rough guide to the distribution of the 3000 or more species of bacteria found on and within a healthy human adult, showing the locations of bacteria that had been sequenced as of 2010. *Source: Data from J. Peterson et al., 2009, "The NIH Human Microbiome Project," Genome Research 19:2317–2323; originally published online October 9, 2009.*

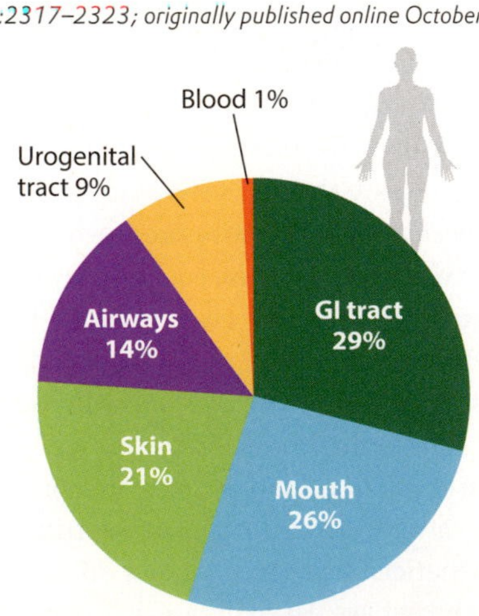

A fertilized human egg has no bacteria attached to it, so our microscopic passengers arrive by colonization—by infection, if you will. Research on children from Europe and rural Africa clearly shows the importance of diet in establishing our gut microbiota (**Fig. 26.24**). The African children, raised on a diet low in fat and animal protein but rich in plant matter, have gut microbiotas

FIG. 26.24 Diet and the intestinal microbiota. Studies of gut bacteria in children from (a) rural Africa and (b) Western Europe show markedly different communities. *Source: Data from C. De Filippo, D. Cavalieri, M. Di Paola, M. Ramazzotti, J. B. Poullet, S. Massart, S. Collini, G. Pieraccini, and P. Lionetti, 2010, "Impact of Diet in Shaping Gut Microbiota Revealed by a Comparative Study in Children from Europe and Rural Africa," Proceedings of the National Academy of Sciences USA 107(33):14693, doi:10.1073/pnas.1005963107.*

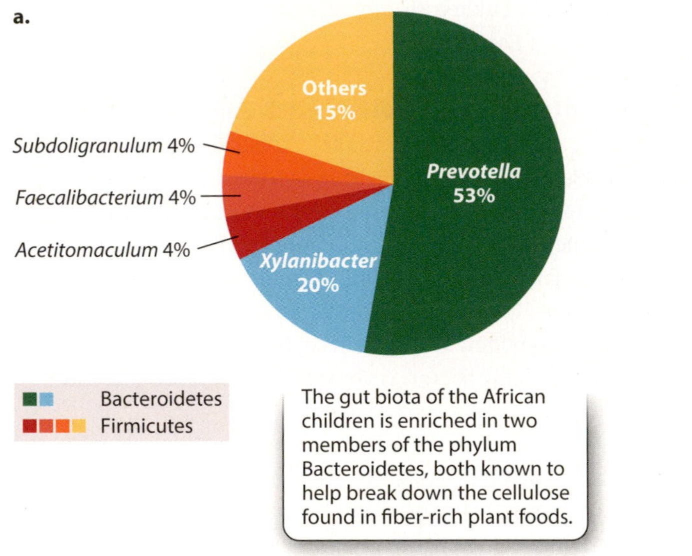

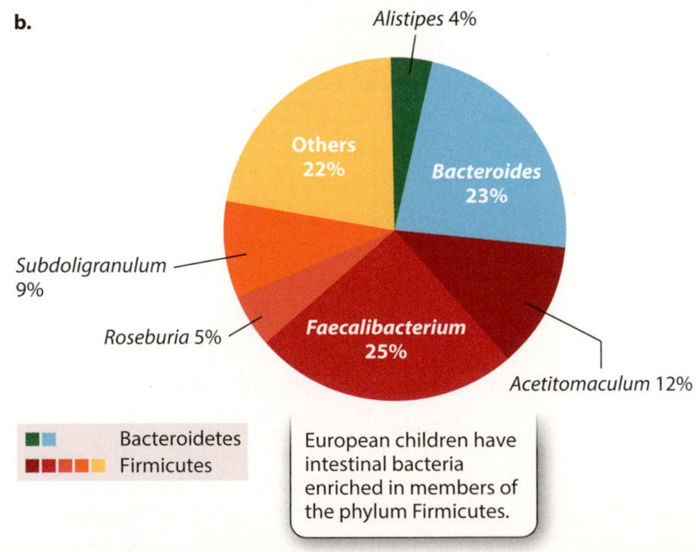

enriched in species of the phylum Bacteroidetes that are known to help digest cellulose. European children raised on a typical Western diet rich in sugar and animal fat lack these bacteria, harboring instead diverse members of the phylum Firmicutes. We do not understand the full ramifications of these differences, but active research suggests that they may help to explain the relative prominence in Western societies of allergies and other disorders involving the immune system.

We also influence our gut biota by ingesting antibiotics. Although prescribed for the control of pathogens, most antibiotics kill a wide range of bacteria and so can change the balance of our intestinal microbiota. Clinical studies show that the incidence of inflammatory bowel disease, a disabling inflammation of the colon, increases in humans who have been treated with antibiotics.

To understand the relationships between the human microbiome and human health, we need to know what constitutes a healthy gut biota, and this means studying the metabolism, ecology, and population genetics of the bacteria in our bodies. We need to know which bacteria have evolved to take advantage of the environments provided by the human digestive system, and which are "just passing through." Future visits to the doctor may include routine genetic fingerprinting of our bacterial biota, as well as treatments designed to keep our microbiome—and, so, ourselves—healthy. ■

Core Concepts Summary

26.1 The tree of life has three main branches, called domains: Eukarya, Bacteria, and Archaea.

Prokaryotic cells are cells that lack a nucleus; they include Bacteria and Archaea. page 530

Bacteria are small and lack membrane-bounded organelles. page 530

The bacterial genome is circular. Some bacteria also carry smaller circles of DNA called plasmids. page 530

Diffusion limits size in bacterial cells. page 530

Bacteria can obtain DNA by horizontal gene transfer from organisms that may be distantly related. page 532

Like Bacteria, Archaea lack a nucleus, but form a second prokaryotic domain distinct from Bacteria. page 534

Some archaeons are extremophiles, living in extreme environments characterized by low pH, high salt, or high temperatures, but others live in less extreme environments like the upper ocean or soil. page 534

26.2 Bacteria and Archaea are notable for their metabolic diversity.

Bacteria are capable of oxygenic photosynthesis, using water as a source of electrons and producing oxygen as a by-product, and of anoxygenic photosynthesis, using electron donors other than water, such as H_2S, H_2, and Fe^{2+}. page 535

Some bacteria and archaeons are capable of anaerobic respiration, in which NO_3^-, SO_4^{2-}, Mn^{4+}, and Fe^{3+} serve as the electron acceptor instead of oxygen gas. page 535

Many bacteria and archaeons obtain energy from fermentation, which involves the partial oxidation of organic molecules and the production of ATP in limited quantities. page 536

Some photosynthetic bacteria are photoheterotrophs, obtaining energy from sunlight but using preformed organic compounds rather than CO_2 as a source of carbon. page 537

Chemoautotrophy, in which chemical energy is used to convert CO_2 to organic molecules, is unique to Bacteria and Archaea. page 537

26.3 In addition to their key roles in the carbon cycle, Bacteria and Archaea are critical to the biological cycling of sulfur and nitrogen.

Plants and algae can take up sulfur and incorporate it into proteins, but bacteria and archaeons dominate the sulfur cycle by means of oxidation and reduction reactions that are linked to the carbon cycle. page 538

Some bacteria and archaeons can reduce nitrogen gas to ammonia in a process called nitrogen fixation. page 539

The nitrogen cycle also involves oxidation and reduction reactions by Bacteria and Archaea that are linked to the carbon cycle. page 540

26.4 The extent of bacterial diversity was recognized when sequencing technologies could be applied to non-culturable bacteria.

Traditionally, bacterial groups were recognized by morphology, physiology, and the ability to take up specific stains in culture. page 540

Direct sequencing of ribosomal RNA genes from organisms in soil and seawater samples revealed new groups of bacteria. page 540

Proteobacteria are the most diverse group of bacteria and are involved in many of the biogeochemical processes that are linked to the carbon cycle. page 543

Gram-positive bacteria include important disease-causing strains as well as species that are principal sources of antibiotics. page 543

Photosynthetic bacteria are not limited to a single branch of the bacterial tree. page 544

26.5 The diversity of Archaea has only recently been recognized.

Archaeons tend to thrive where energy available for growth is limited. page 545

Archaea are commonly divided into three major groups, the Crenarchaeota, Thaumarchaeota, and Euryarchaeota. page 545

Archaeons at the base of the Crenarchaeota and Euryarchaeota are hyperthermophiles, meaning that they grow at high temperatures. page 545

A number of archaeons grow in highly acidic waters, such as those associated with acid mine drainage. page 546

Some archaeons (but no bacteria or eukaryotes) generate methane as a by-product of their energy metabolism, contributing in important ways to the carbon cycle. page 546

Haloarchaea are archaeons that can live only in extremely salty environments. page 546

Thaumarchaeotes are among the most abundant cells in the oceans. page 546

26.6 The earliest forms of life on Earth were Bacteria and Archaea

Evidence for the early history of life comes from microfossils, fossilized structures called stromatolites, and the isotopic composition of rocks and organic matter. page 548

Fossils indicate that life on Earth originated more than 3.5 billion years ago. page 548

The early atmosphere and ocean contained little or no free oxygen. page 548

Oxygen began to accumulate in the atmosphere and oceans about 2.4 billion years ago as a result of the success of cyanobacteria utilizing oxygenic photosynthesis. page 549

Prokaryotic metabolisms were not only essential in the early history of Earth, but are also vital today, as many forms of life depend on biogeochemical cycles and metabolisms unique to Bacteria and Archaea. page 549

Most animals, including humans, live in intimate association with bacteria, which in turn affect health. page 550

Self-Assessment

1. Name and describe the three domains of life.
2. Describe shared and contrasting features of bacterial and archaeal cells.
3. Explain how prokaryotic cells obtain nutrients and how this process puts constraints on their size.
4. Describe how surface area and volume change with size.
5. Explain how photosynthesis can occur without the production of oxygen, and how respiration can occur without requiring oxygen.
6. Describe the roles of bacteria and archaeons in the sulfur and nitrogen cycles.
7. Explain how horizontal gene transfer complicates our understanding of evolutionary relationships among bacteria and archaeons.
8. Name and describe three major groups of Bacteria.
9. Name and describe three major groups of Archaea.
10. State the age of Earth and the time when life is thought to have first originated.

Log in to LaunchPad to check your answers to the Self-Assessment questions, and to access additional learning tools.

Lebendkulturen.de/Shutterstock.

CHAPTER 27

Eukaryotic Cells

Origins and Diversity

Core Concepts

27.1 Eukaryotic cells are defined by the presence of a nucleus, but features like a dynamic cytoskeleton and membrane system explain their success in diversifying.

27.2 The endosymbiotic hypothesis proposes that the chloroplasts and mitochondria of eukaryotic cells were originally free-living bacteria.

27.3 Eukaryotes were formerly divided into four kingdoms, but are now divided into at least seven superkingdoms.

27.4 The fossil record extends our understanding of eukaryotic diversity by providing perspectives on the timing and environmental context of eukaryotic evolution.

Eukaryotic cells have a nucleus—that is their defining characteristic. As we will see, however, it is quite another set of features that distinguishes eukaryotes in terms of function and the capacity for evolutionary innovation. Eukaryotic diversity—the animals, plants, fungi, and protists we are familiar with—does not reflect exceptional versatility in energy metabolism. In fact, compared with Bacteria and Archaea, the Eukarya obtain carbon and energy in only a limited number of ways. For example, oxygenic photosynthesis is widespread among eukaryotes, but anoxygenic photosynthesis is unknown. Furthermore, aerobic respiration is present almost everywhere, but anaerobic respiration has been reported in only a handful of species. Chemoautotrophy has not been documented in eukaryotes at all. Many eukaryotes can ferment organic substrates—yeasts are particularly important in this regard—but most use fermentation only as a supplementary metabolism and lack the diversity of reactions known among prokaryotic organisms.

Why then, have eukaryotes been so successful? Eukaryotic success lies in their remarkable capacity to form cells of diverse shape and size and to produce multicellular structures in which multiple cell types act in concert. Together, these innovations opened up a range of functional possibilities not found among Bacteria and Archaea.

FIG. 27.1 A eukaryotic cell. Eukaryotic cells have a nucleus and extensive internal compartmentalization.

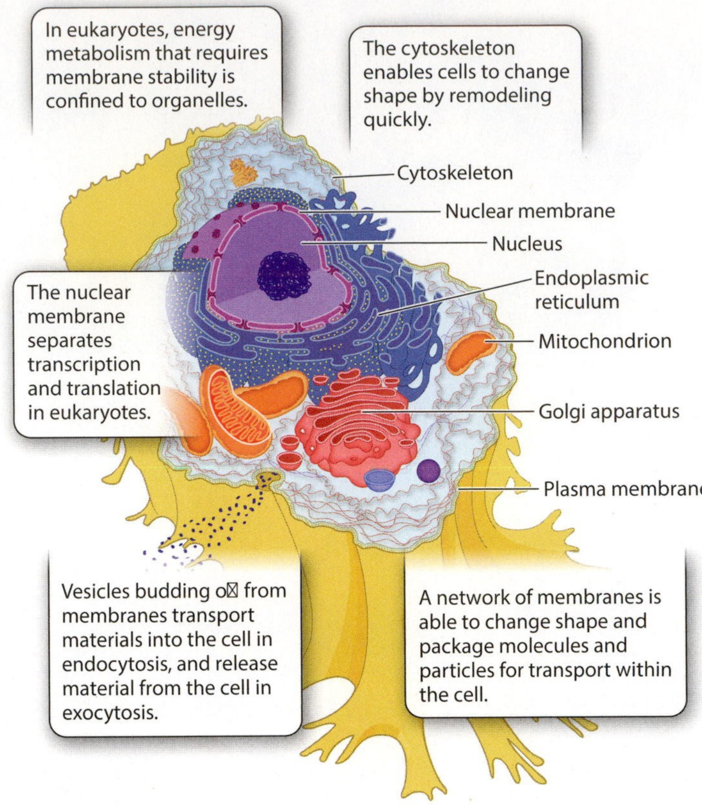

27.1 THE EUKARYOTIC CELL: A REVIEW

Chapter 5 introduced the fundamental features of eukaryotic cells. Here, we revisit those attributes (**Fig. 27.1**) and consider their consequences for function and diversity. How do eukaryotic cells differ from bacteria and archaeons, and how do these differences explain the roles that eukaryotes play in modern ecosystems?

Internal protein scaffolding and dynamic membranes organize the eukaryotic cell.

All cells require a mechanism to maintain spatial order in the cytoplasm. As we saw in Chapter 26, bacteria and archaeons rely primarily on walls that support the cell from the outside, along with a framework of proteins within the cytoplasm. Eukaryotes also use an internal scaffolding of proteins to organize the cell, mostly microtubules composed of the protein tubulin and microfilaments of actin. This cytoskeleton, found in all eukaryotic cells, differs in one key property from the rigid protein framework of bacteria: It can be remodeled quickly, enabling cells to change shape (Chapter 10).

Dynamic cytoskeletons require dynamic membranes that enable the cell to continue functioning even as it changes shape; eukaryotes maintain within their cells a remarkably dynamic network of membranes called the endomembrane system (Chapter 5). This network includes the nuclear envelope, an assembly of membranes that runs through the cytoplasm called the endoplasmic reticulum (ER) and Golgi apparatus, and a plasma membrane that surrounds the cytoplasm. All membranes of the endomembrane system are interconnected, either directly or by the movement of vesicles. Many are also capable of changing shape rapidly. In fact, the different membranes are interchangeable in the sense that material originally added to the endoplasmic reticulum may in time be transferred to the cell membrane or nuclear membrane. Biologists like to say that the membranes of eukaryotic cells are in dynamic continuity. Membranes are stable, as required for energy metabolism, only in the mitochondria and chloroplasts.

In combination, the dynamic cytoskeleton and membrane system provide eukaryotes with new possibilities for movement. For example, amoebas extend finger-like projections that pull the rest of the cell forward. They also enable eukaryotic cells to engulf molecules or particles, including other cells, in a process called endocytosis, which prokaryotic cells simply cannot do. Phagocytosis is a specific form of endocytosis in which eukaryotic cells surround food particles and package them in vesicles that bud off from the cell membrane (**Fig. 27.2**). Thus packaged, particles can be transported into the cell interior for digestion. The same process also works in reverse: In exocytosis, newly formed molecules or cytoplasmic waste are packaged in vesicles and moved to the cell surface for removal (Chapter 5).

FIG. 27.2 Phagocytosis. The flexibility of the eukaryotic cytoskeleton and membrane system enables eukaryotic cells to engulf food particles, including other cells. This sequence shows an amoeba engulfing a yeast cell. Note how the cell membrane (green) folds inward locally to form a vesicle around the red food particle. *Source: V. Mercanti, S. J. Charette, N. Bennett, J. J. Ryckewaert, F. Letourneur, and P. Cosson, 2006, "Selective Membrane Exclusion in Phagocytic and Macropinocytic Cups," Journal of Cell Science 119:4079–4067.*

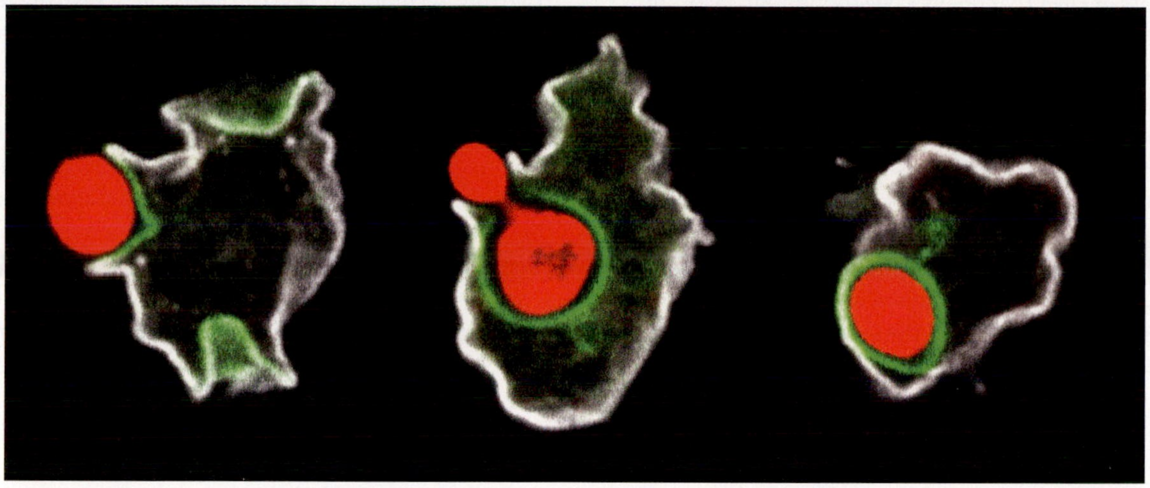

Intracellular vesicles and the molecules they carry are transported through the cytoplasm by means of molecular motors associated with the cytoskeleton (Chapter 10). In this way, both nutrients and signaling molecules move through the cell at speeds much greater than diffusion allows. A major consequence is that eukaryotic cells can be much larger than most bacteria.

The cytoskeleton and membrane system are flexible in another way: A change in the expression of a few genes can change their shape and organization. This makes possible yet another hallmark of eukaryotic evolution—complex multicellularity (Chapter 28). In complex multicellular organisms, patterns of gene expression can modify the cytoskeleton and membranes of individual cells, enabling them to function in different ways.

In eukaryotic cells, energy metabolism is localized in mitochondria and chloroplasts.

Relative to prokaryotic organisms, eukaryotes are fairly limited in the ways they obtain carbon and energy. Moreover, the metabolic processes that power eukaryotic cells take place only in specific organelles—aerobic respiration in the mitochondrion (Chapter 7) and photosynthesis in the chloroplast (Chapter 8). Only limited anaerobic processing of food molecules takes place within the cytoplasm.

As noted above, many eukaryotic cells engulf food particles and package them inside a vesicle, which is then transported into the cytoplasm. Within the cytoplasm, enzymes break down the particles into molecules that can be processed by the mitochondria. Many single-celled eukaryotes feed on bacteria or other eukaryotic cells, and animals, in turn, ingest larger foodstuffs, including other animals and plants. In consequence, eukaryotes can exploit sources of food not readily available to bacterial heterotrophs, which feed on individual molecules. This ability opens up a great new ecological possibility—predation—increasing the complexity of interactions among organisms.

The structural flexibility of eukaryotic cells also allows photosynthetic eukaryotes to interact with their environment in ways that photosynthetic bacteria cannot. Unicellular algae (which are eukaryotes) can move effectively through surface waters vertically as well as horizontally and therefore can seek and exploit local patches of nutrients. Diatoms, discussed shortly, go one step further. Large internal vacuoles allow them to store nutrients for later use when nutrient levels in the environment may become low. Plants have evolved multicellular bodies with many different cell types, allowing them to capture sunlight many meters above the ground. That ability gives plants a tremendous advantage on land, as long as their leaves can obtain water and nutrients from the soil in which the plants are rooted.

The organization of the eukaryotic genome also helps explain eukaryotic diversity.

Bacteria and Archaea absorb available nutrients quickly, and both rapid deployment of metabolic proteins and rapid reproduction are key to the exploitation of patchily distributed nutrients. As the majority of a prokaryote's DNA is arrayed in a single circular chromosome, speed of replication allows for speed of reproduction. As a result, selection favors those strains of Bacteria and Archaea that retain only the genetic material vital to the organism.

Eukaryotes have multiple linear chromosomes and can begin replication from many sites on each one. Eukaryotes are thus able to replicate multiple strands of DNA simultaneously and rapidly (Chapter 12). This ability relieves the evolutionary pressure for streamlining, allowing eukaryotic genomes to build up large amounts of DNA that do not code for proteins. Most of this additional DNA was originally considered to have no function and, indeed, was called "junk DNA." That view, however, has been modified in recent years, as the complete genome has come to be better understood (Chapter 13). Eukaryotic genomes appear to contain truly junky DNA, but at least some of the DNA that does not code for proteins functions in gene regulation (Chapter 19). This regulatory DNA gives eukaryotes the fine control of gene expression required for both multicellular development and complex life cycles, two major features of eukaryotic diversity.

At this point, it can be seen that the evolutionary success of eukaryotes rests on a combination of features. The innovations of dynamic cytoskeletal and membrane systems gave eukaryotes the structure required for larger cells with complex shapes and the ability to ingest other cells. Thus, early unicellular eukaryotes did not gain a foothold in microbial ecosystems by outcompeting bacteria and archaeons. Instead, they succeeded by evolving novel functions. Along with the capacity to remodel cell shape, eukaryotes evolved complex patterns of gene regulation, which in turn enabled unicellular eukaryotes to evolve complex life cycles and multicellular eukaryotes to generate multiple, interacting cell types during growth and development. These abilities opened up still more possibilities for novel functions, which we explore in this and later chapters.

→ **Quick Check 1** How did the evolutionary expansion of eukaryotic organisms change the way carbon is cycled through biological communities?

Sex promotes genetic diversity in eukaryotes and gives rise to distinctive life cycles.

In Chapter 26, we saw how Bacteria and Archaea generate genetic diversity by horizontal gene transfer. Horizontal gene transfer has been documented in eukaryotic species, but it is relatively uncommon. How then do eukaryotes generate and maintain genetic diversity within populations? The answer is sex.

Sexual reproduction involves meiosis and the formation of gametes, and the subsequent fusion of gametes during fertilization (Chapters 11 and 42). Sex promotes genetic variation in two simple ways. First, meiotic cell division results in gametes or spores that are genetically unique. Each gamete has a combination of alleles different from the other gametes and from the parental cell as a result of recombination and independent assortment. Second, in fertilization, new combinations of genes are brought together by the fusion of gametes. Interestingly, a few eukaryotic groups have lost the capacity for sexual reproduction. The best-studied of these eukaryotes are tiny animals called bdelloid rotifers. The genetic diversity of these organisms is actually high, maintained by high rates of horizontal gene transfer.

Meiotic cell division results in cells with one set of chromosomes. Such cells are **haploid.** Sexual fusion brings two haploid ($1n$) cells together to produce a **diploid** ($2n$) cell that has two sets of chromosomes. The life cycle of sexually reproducing eukaryotes, then, necessarily alternates between haploid and diploid states.

Many single-celled eukaryotes normally exist in the haploid stage and reproduce asexually by mitotic cell division (**Fig. 27.3a**). The green alga *Chlamydomonas* is one such organism. Under the right conditions, however—typically, starvation or other environmental stress—two cells fuse, forming a diploid cell, or **zygote.** The zygote formed by these single-celled eukaryotes commonly functions as a resting cell. It covers itself with a protective wall and then lies dormant until environmental conditions improve. In time, further signals from the environment induce meiotic cell division, resulting in four genetically distinct haploid cells that emerge from their protective coating to complete the life cycle.

As shown in **Fig. 27.3b,** some single-celled eukaryotes normally exist as diploid cells. An example is provided by the diatoms, single-celled eukaryotes commonly found in lakes, soils, and the oceans. Diatom cells are mostly diploid and reproduce asexually by mitotic cell division to make more diploid cells. Because their mineralized skeletons constrain growth, diatoms become smaller with each asexual division. Once a critical size is reached, meiotic cell division is triggered, producing haploid gametes that fuse to regenerate the diploid state as a round, thick-walled cell. This cell eventually germinates to form an actively growing, skeletonized cell. In diatoms, then, short-lived gametes constitute the only haploid phase of the life cycle.

The two life cycles just introduced, of *Chlamydomonas* and diatoms, are similar in many ways. In both, haploid cells fuse to form diploid cells, and diploid cells undergo meiotic cell division to generate haploid cells. Both life cycles also commonly include cells capable of persisting in a protected form when the environment becomes stressful. Why some single-celled eukaryotes usually occur as haploid cells and others usually occur as diploid cells remains unknown—a good question for continuing research.

Sexual reproduction has never been observed in some eukaryotes, but most species appear to be capable of sex, even if they reproduce asexually most of the time. Variations on the eukaryotic life cycle can be complex, especially in parasitic microorganisms that have multiple animal hosts. All, however, have the same fundamental components as the two life cycles described here.

In the following chapters, we discuss multicellular organisms in detail. Here we note only that animals, plants, and other complex multicellular organisms have life cycles with the same features as those just discussed. The big difference is that in animals, the zygote divides many times to form a multicellular diploid body before a

FIG. 27.3 Eukaryotic life cycles. (a and b) The life cycles of single-celled eukaryotes differ in the proportion of time spent as haploid (1*n*) versus diploid (2*n*) cells. (c) The life cycles of animals have many mitotic divisions between formation of the zygote and meiosis. (d) Vascular plants have two multicellular phases.

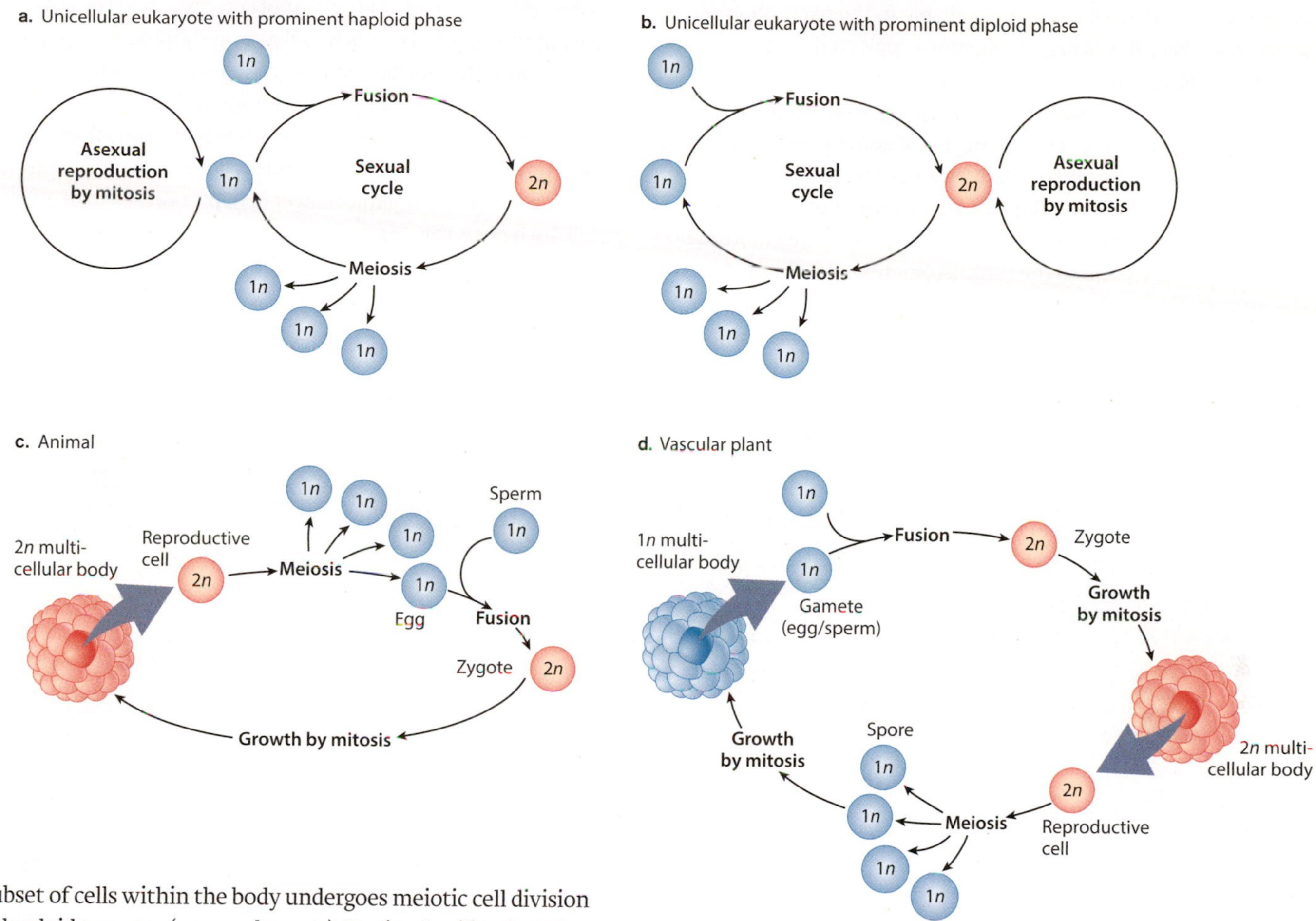

small subset of cells within the body undergoes meiotic cell division to form haploid gametes (eggs and sperm). During fertilization, the egg and sperm combine sexually to form a zygote (**Fig. 27.3c**). In animals, as in diatoms, the only haploid phase of the life cycle is the gamete. As we will see in Chapter 32, plants have two multicellular phases in their life cycle, one haploid and one diploid (**Fig. 27.3d**). Like single-celled eukaryotes, many plants and animals can reproduce asexually. Humans and other mammals are unusual in that we cannot reproduce asexually.

27.2 EUKARYOTIC ORIGINS

In the preceding section, we reviewed the basic features of eukaryotic cells. The distinctive cellular organization of Eukarya differs markedly from the simpler cells of Bacteria and Archaea. How did the complex cells of eukaryotes come to exist? In Case 5: The Human Microbiome, we saw that symbioses are common, even in humans. Is the relationship between humans and bacteria unusual, or are symbioses so fundamental that they played key roles in the evolution of the eukaryotic cell?

? CASE 5 THE HUMAN MICROBIOME: DIVERSITY WITHIN

What role did symbiosis play in the origin of chloroplasts?

The chloroplasts found in plant cells closely resemble certain photosynthetic bacteria, specifically cyanobacteria. The molecular workings of photosynthesis are nearly identical in the two, and internal membranes organize the photosynthetic machinery in similar ways (Chapter 8). The Russian botanist Konstantin Sergeevich Merezhkovsky recognized this similarity more than a century ago. He also understood that corals and some other organisms harbor algae within their tissues as symbionts that aid the growth of their host. A **symbiont** is an organism that lives in closely evolved association with another species. This association, called **symbiosis**, is discussed in Chapter 47.

Putting these two observations together, Merezhkovsky came up with a radical hypothesis. Chloroplasts, he argued, originated as symbiotic cyanobacteria that through time became permanently incorporated into their hosts. Such a symbiosis, in which one partner lives within the other, is called an **endosymbiosis.** Merezkhovsky's hypothesis of chloroplast origin by endosymbiosis was difficult to test with the tools available in the early twentieth century, and his idea was dismissed, more neglected than disproved, by most biologists.

In 1967, American biologist Lynn Margulis resurrected the endosymbiotic hypothesis, supporting her arguments with new types of data made possible by the then-emerging techniques of cell and molecular biology (**Fig. 27.4**). Transmission electron microscopy showed that structural similarities between chloroplasts and cyanobacteria extend to the submicrometer level, for example in the organization of the photosynthetic membranes. It also became clear that the chloroplasts in red and green algae (and in land plants) are separated from the cytoplasm that surrounds them by two membranes. This is expected if a cyanobacterial cell had been engulfed by a eukaryotic cell. The inner membrane corresponds to the cell membrane of the cyanobacterium, and the outer one is part of the engulfing cell's membrane system, as shown in Fig. 27.2. In addition, the biochemistry of photosynthesis was found to be essentially the same in cyanobacteria and chloroplasts. Both use two linked photosystems, a common mechanism for extracting electrons from water, and the same reactions to reduce carbon dioxide (CO_2) into organic matter (Chapter 8).

Such observations kindled renewed interest in the endosymbiotic hypothesis, but the decisive tests were made possible by another, and unexpected, discovery. It turns out

HOW DO WE KNOW?

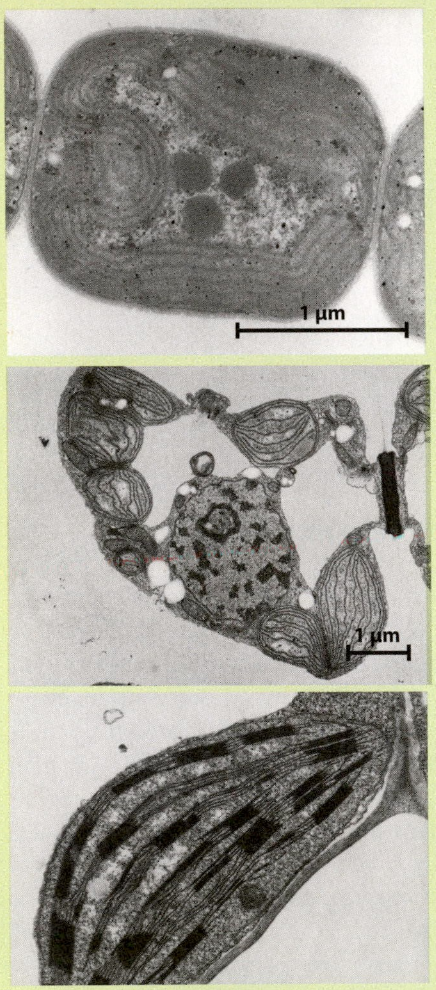

FIG. 27.4

What is the evolutionary origin of chloroplasts?

HYPOTHESIS Chloroplasts evolved from cyanobacteria living as endosymbionts within a eukaryotic cell.

OBSERVATION Chloroplasts and cyanobacteria have closely similar internal membranes that organize the light reactions of photosynthesis. Transmission electron microscopy shows that cyanobacteria (top), red algal chloroplasts (middle), and plant chloroplasts (bottom) have very similar internal membranes that organize the light reactions of photosynthesis.

EXPERIMENT Like cyanobacteria, chloroplasts have DNA organized in a single circular chromosome. Phylogenies based on molecular sequence comparisons place chloroplasts among the cyanobacteria.

CONCLUSION Molecular and electron microscope data support the hypothesis that chloroplasts originated as endosymbiotic cyanobacteria.

SOURCE Giovannoni, S. J., et al. 1988. "Evolutionary Relationships Among Cyanobacteria and Green Chloroplasts." *Journal of Bacteriology* 170:3584–3592. Photos (top to bottom): Dr. Kari Lounatmaa/Science Source; Biology Pics/Getty Images; Dr. Jeremy Burgess/Science Source.

that chloroplasts have their own DNA, organized into a single circular chromosome, like that of bacteria. Armed with this knowledge, investigators could use the tools of molecular sequence comparison (Chapter 23). The sequences of nucleotides in chloroplast genes closely match those of cyanobacterial genes but are strikingly different from sequences of nucleotides in genes within the nuclei of photosynthetic eukaryotes. This finding provides strong support for the hypothesis of Merezhkhovsky and Margulis. Chloroplasts are indeed the descendants of symbiotic cyanobacteria that lived within eukaryotic cells.

Eukaryotes were able to acquire photosynthesis because they could engulf and retain cyanobacterial cells. But engulfing another microorganism was just the beginning. The cyanobacteria most closely related to chloroplasts have 2000 to 3000 genes. In contrast, among photosynthetic eukaryotes, chloroplast gene numbers vary from just 60 to 200. Where did the rest of the cyanobacterial genome go? Some genes may simply have been lost if similar nuclear genes could supply chloroplast's requirements. Many, however, were transported to the nucleus when chloroplasts broke or by hitching a ride with viruses. The nuclear genome of the flowering plant *Arabidopsis* contains several thousand genes of cyanobacterial origin. Some code for proteins destined for use within the chloroplast.

Although foreign to mammals and other vertebrate animals, symbiosis between a heterotrophic host and a photosynthetic partner is common throughout the eukaryotic domain. For example, reef corals harbor photosynthetic cells that live symbiotically within their tissues. Some corals have even lost the capacity to capture food from surrounding waters. Giant clams of the genus *Tridacna,* found in tropical Pacific waters, obtain some or even most of their nutrition from symbiotic algae that live within their tissues.

Until recently, most biologists agreed that chloroplasts originated from endosymbiotic cyanobacteria only once, in a common ancestor of the green algae and red algae. Now, remarkably, a second case of cyanobacterial endosymbiosis has come to light. *Paulinella chromatophora* is a photosynthetic amoeba (**Fig. 27.5**). Gene sequence comparisons show that its chloroplast originated in a branch of the cyanobacteria different from the one that gave rise to chloroplasts in other photosynthetic eukaryotes. About a third to a half of the ancestral genome remains in *P. chromatophora* chloroplasts, suggesting that in this organism chloroplast evolution is still in progress.

FIG. 27.5 A second example of chloroplast endosymbiosis. The photosynthetic amoeba *Paulinella chromatophora* acquired photosynthesis by endosymbiosis independently of and more recently than the endosymbiosis that gave rise to chloroplasts in green plants and other eukaryotes. *Source: Courtesy of Hwan Su Yoon.*

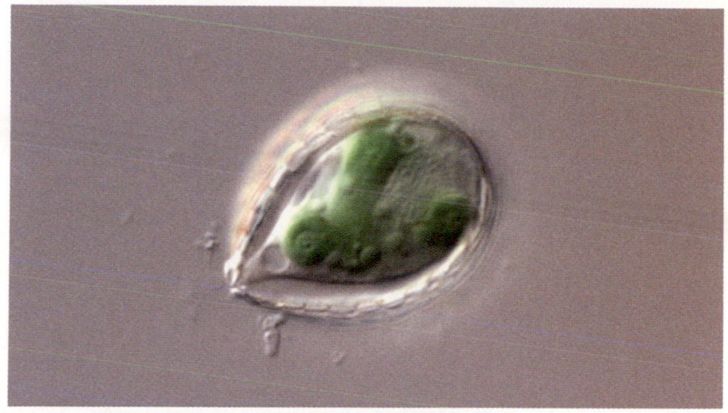

? CASE 5 THE HUMAN MICROBIOME: DIVERSITY WITHIN

What role did symbiosis play in the origin of mitochondria?

Like chloroplasts, mitochondria closely resemble free-living bacteria in organization and biochemistry. And like chloroplasts, mitochondria contain DNA that confirms their close phylogenetic relationship to a form of bacteria—in this case proteobacteria (Chapter 26). Like chloroplasts, then, mitochondria originated as endosymbiotic bacteria.

Mitochondria are also like chloroplasts in having a small genome. Indeed, the mitochondrial genome is dramatically reduced compared to the ancestral proteobacterial genome, in most eukaryotes containing only a handful of functioning genes. Human mitochondria, for example, code for just 13 proteins and 24 RNAs. Once again, many genes from the original bacterial endosymbiont migrated to the nucleus, where they still reside.

Most eukaryotic cells contain mitochondria, but a few single-celled eukaryotic organisms found in oxygen-free environments do not. Biologists earlier hypothesized that these eukaryotes evolved before the endosymbiotic event that established mitochondria in cells having a nucleus, but that proposal turns out to be wrong. We know this because of the propensity for genes to migrate from the endosymbiont to the host's nucleus. Every mitochondria-free eukaryote examined to date has relic mitochondrial genes in its nuclear genome, providing support for a second hypothesis, that eukaryotic cells without mitochondria had them once but have lost them.

In fact, many eukaryotes that lack mitochondria contain small organelles called hydrogenosomes that generate ATP by anaerobic processes (**Fig. 27.6**). These organelles have little or no DNA, but genes of mitochondrial origin in the cells' nuclei code for proteins that function in the hydrogenosome. Thus, hydrogenosomes appear to be highly altered mitochondria adapted to life in oxygen-poor environments. Other anaerobic eukaryotes contain still smaller organelles called mitosomes that also appear to be remnant mitochondria. Interestingly, mitosomes have lost all capacity for metabolism, retaining only genes needed for assembling the clusters of iron and sulfur found in many proteins. Why these genes are retained in a modified mitochondrion remains unknown.

FIG. 27.6 Transmission electron microscope images comparing mitochondria within the aerobic protist *Euplotes* (left) and *Nyctotherus*, a close relative that lives in anoxic environments (right). The organelle in *Nyctotherus* is halfway between a normal mitochondrion and a hydrogenosome, supporting the view that these two organelles are descended from a common ancestor. *Source: Reprinted by permission from Macmillan Publishers Ltd: B. Boxma, et al., 2005, "An Anaerobic Mitochondrion That Produces Hydrogen," Nature 434:74–79. Copyright 2005.*

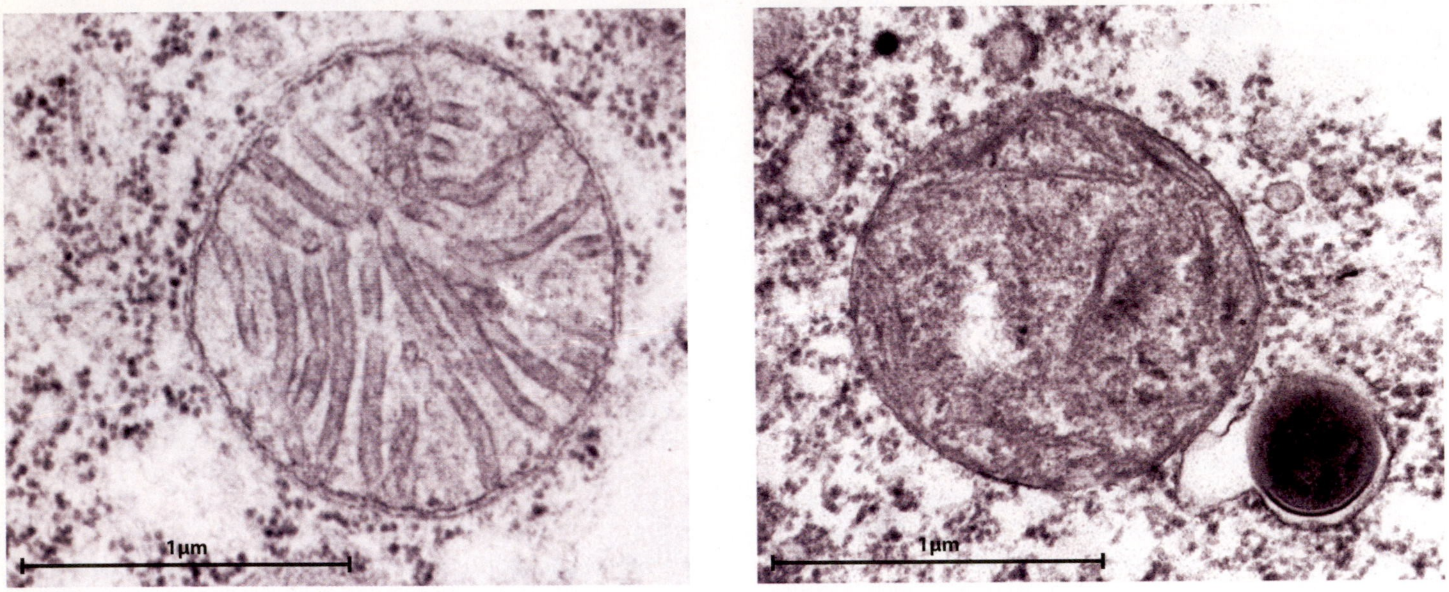

Normal mitochondrion in *Euplotes*

Altered mitochondrion in *Nyctotherus* produces ATP and generates hydrogen. The small dark body in lower right is an endosymbiotic archaeon.

? CASE 5 THE HUMAN MICROBIOME: DIVERSITY WITHIN

How did the eukaryotic cell originate?

If mitochondria originated as proteobacteria and chloroplasts are descended from cyanobacteria, where does the rest of the eukaryotic cell come from? Analysis of the nuclear genome alone provides no clear picture because genes of bacterial, archaeal, and purely eukaryotic origin are all present. As discussed, many nuclear genes originated with the mitochondria and chloroplasts acquired from specific bacteria. However, genes from other groups of bacteria also reside in the eukaryotic nucleus, recording multiple episodes of horizontal gene transfer through evolutionary history. In contrast, some genes are present only in eukaryotes and apparently evolved after the domain originated. Still others, including the genes that govern DNA transcription and translation, are clearly related to the genes of Archaea.

Two starkly different hypotheses have been proposed to explain this mix of genes. Some biologists believe that the host for mitochondrion-producing endosymbiosis was itself a true eukaryotic cell. This cell had a nucleus, cytoskeleton, and endomembrane system, but only limited ability to derive energy from organic molecules (**Fig. 27.7**). In this view, nuclear genes in Eukarya resemble those of Archaea because the primordial eukaryotic host cell was closely related to Archaea.

In striking support of this hypothesis, microbiological exploration of hydrothermal vents beneath the Arctic ocean recently revealed a previously known archaeon whose genome encodes actin-like proteins and other molecules fundamental to cell organization, vesicle trafficking and phagocytosis in eukaryotes. Gene sequence comparisons, in fact, place these novel archaeons as the closest relatives of eukaryotes. An archaeal component of the eukaryotic cell seems assured, but some biologists argue that no eukaryotic cell existed before there were mitochondria. Instead, they propose that the eukaryotic cell as a whole began as a symbiotic association between a proteobacterium and an archaeon (Fig. 27.7). The proteobacterium became the mitochondrion and provided many genes to the nuclear genome. The archaeon provided other genes, including those used to transcribe DNA and translate it into proteins.

Biologists continue to debate these alternatives. Both hypotheses explain the hybrid nature of the eukaryotic genome, but neither fully explains the origins and evolution of the nucleus, linear chromosomes, the eukaryotic cytoskeleton, or a cytoplasm subdivided by ever-changing membranes. There is no consensus on the question of eukaryotic origins—it is one of biology's deepest unanswered questions, awaiting novel observations by a new generation of biologists. Once a dynamic cytoskeleton became coupled to a flexible membrane system, however, the evolutionary possibilities of eukaryotic form were established.

FIG. 27.7 Two hypotheses for the origin of the eukaryotic cell. Both hypotheses lead to the evolution of a mitochondrion-bearing protist.

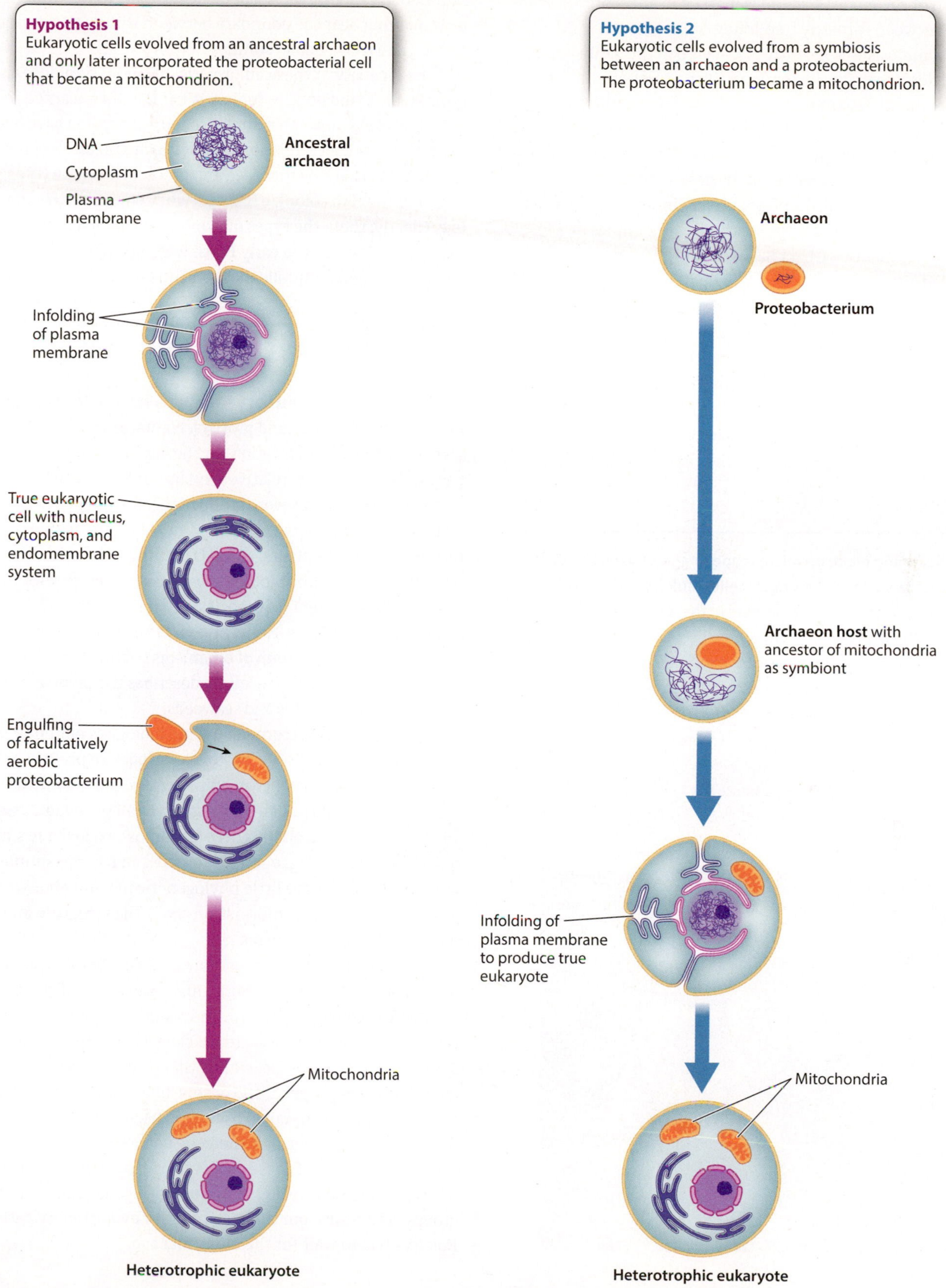

In the oceans, many single-celled eukaryotes harbor symbiotic bacteria.

The evolution of the eukaryotic cell is marked by intimate associations between formerly free-living organisms. An unexpected symbiosis was recently discovered in the Santa Barbara Basin, a local depression in the seafloor off the coast of southern California. Sediments accumulating in this basin contain little oxygen but large amounts of hydrogen sulfide (H_2S) generated by anaerobic bacteria within the sediments. We might predict that eukaryotic cells would be uncommon in these sediments. Not only is oxygen scarce, but also sulfide actively inhibits respiration in mitochondria. It came as a surprise, then, that samples of sediment from the Santa Barbara Basin contain large populations of single-celled eukaryotes.

Some of these cells have hydrogenosomes, mitochondria altered by evolution to generate energy where oxygen is absent. Others, however, appear to thrive by supporting populations of symbiotic bacteria on or within their cells. In one case, rod-shaped bacteria cover the surface of their eukaryotic host (**Fig. 27.8**). Research has shown that the bacteria metabolize sulfide in the local environment, thereby protecting their host from this toxic substance. Scientists hypothesize that the bacteria benefit from this association as well because they get a free ride through the sediments, enabling these chemoautotrophs to maximize growth by remaining near the boundary between waters that contain oxygen and those rich in sulfide.

The diversity of these eukaryotic–bacterial symbioses is remarkable—and poorly studied. What has been learned to date, however, shows that single-celled eukaryotes have evolved numerous symbiotic relationships with chemoautotrophic bacteria, associations that feed and protect the eukaryotes, enabling them to colonize habitats where most eukaryotes cannot live. Clearly, then, the types of symbioses that led to mitochondria and chloroplasts on the early Earth were not rare events, but basic associations between cells that continue to evolve today.

FIG. 27.8 Scanning electron microscope image of bacterial cells on the surface of a single-celled eukaryote found in oxygen-depleted sediments in the Santa Barbara Basin, off the coast of California. The bacteria are hypothesized to metabolize the H_2S in this environment, thereby protecting the eukaryote that they enclose. *Source: Reprinted by permission from Macmillan Publishers Ltd: V. P. Edgcomb, S. A. Breglia, N. Yubuki, D. Beaudoin, D. J. Patterson, B. S. Leander, and J. M. Bernhard, 2011, "Identity of Epibiotic Bacteria on Symbiontid Euglenozoans in O2-Depleted Marine Sediments: Evidence for Symbiont and Host Co-evolution," The ISME Journal 5:231–243. Copyright 2010. Courtesy of Naoji Yubuki.*

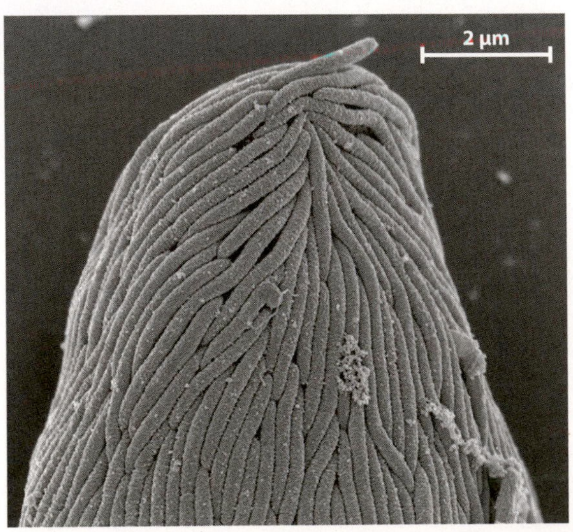

27.3 EUKARYOTIC DIVERSITY

Historically, Domain **Eukarya** was divided into four kingdoms: plants, animals, fungi, and protists. Plants, animals, and fungi received special consideration for the obvious reason that they include large and relatively easily studied species. All remaining eukaryotes were grouped together as **protists,** which were defined as organisms having a nucleus but lacking other features specific to plants, animals, or fungi. The term "protist" is frowned on by some biologists because it doesn't refer to a monophyletic grouping of species—that is, an ancestral form and all its descendants. Nonetheless, as an informal term that draws attention to a group of organisms sharing particular characteristics, "protist" usefully describes the diverse world of microscopic eukaryotes and seaweeds.

Two other terms scorned by some biologists but embraced by ecologists are "algae" and "protozoa." **Algae** are photosynthetic protists. They may be microscopic single-celled organisms or the highly visible, multicelled organisms we call seaweed. **Protozoa** are heterotrophic protists. These are almost exclusively single-celled organisms. "Algae" and "protozoa" in fact are simple and useful terms that have little phylogenetic meaning but which convey a great deal of information about the structure and function of these organisms.

Protist cells exhibit remarkable diversity. Some have cell walls, while others do not. Some make skeletons of silica (SiO_2) or calcium carbonate ($CaCO_3$), and others are naked or live within **tests,** or "houses," constructed exclusively of organic molecules. Some are photosynthetic, while others are heterotrophic. Some move by beating flagella, and others, like *Amoeba,* can extend fingers of cytoplasm, called pseudopodia, to move and capture food. Most are aerobic, but there are anaerobic protists as well. Surprisingly, most of these distinctive features have evolved multiple times and so do not define phylogenetically coherent groups. As a result, our understanding of evolutionary pattern in Eukarya had to wait for the molecular age.

Fig. 27.9 shows a phylogenetic tree of eukaryotes as biologists currently understand it. As in the case of Bacteria, it is easier to identify major groups of Eukarya than to establish their evolutionary relationships to one another. Molecular sequence comparisons consistently recognize seven major groups, called **superkingdoms,** noted in the figure. Animals fall within one of these superkingdoms, the Opisthokonta, whereas plants fall within another, the Archaeplastida. Most, but not all, of the divisions into these major eukaryotic branches are strongly supported by data from genes and genomes. A few groups, notably the cryptophytes, haptophytes, and centrohelid heliozoans, seem to wander from position to position from one phylogenetic analysis to the next, sometimes moving together, sometimes separately. Whether the well-resolved superkingdoms constitute the only major limbs on the eukaryotic tree is less certain. Many microscopic protists remain unstudied, and we may yet be surprised by what they show.

In the following sections, we introduce the major features of eukaryotic diversity. The species numbers we present indicate only the taxonomic diversity that is known to us; most protistan species remain unknown. Many biologists estimate that only about 1 in 10 protist species has been described so far. As this is also the case for animals and fungi (but arguably not for land plants because they are all visible to the naked eye), the relative numbers of species within different superkingdoms may reflect something like the actual distribution of biodiversity among eukaryotes.

FIG. 27.9 The eukaryotic tree of life showing major groups. Branches with photosynthetic species (shown in green) are distributed widely. Additional branches, including animals, foraminiferans, radiolarians, and ciliates, contain species that harbor photosynthetic symbionts. Dashed lines indicate a high degree of uncertainty.

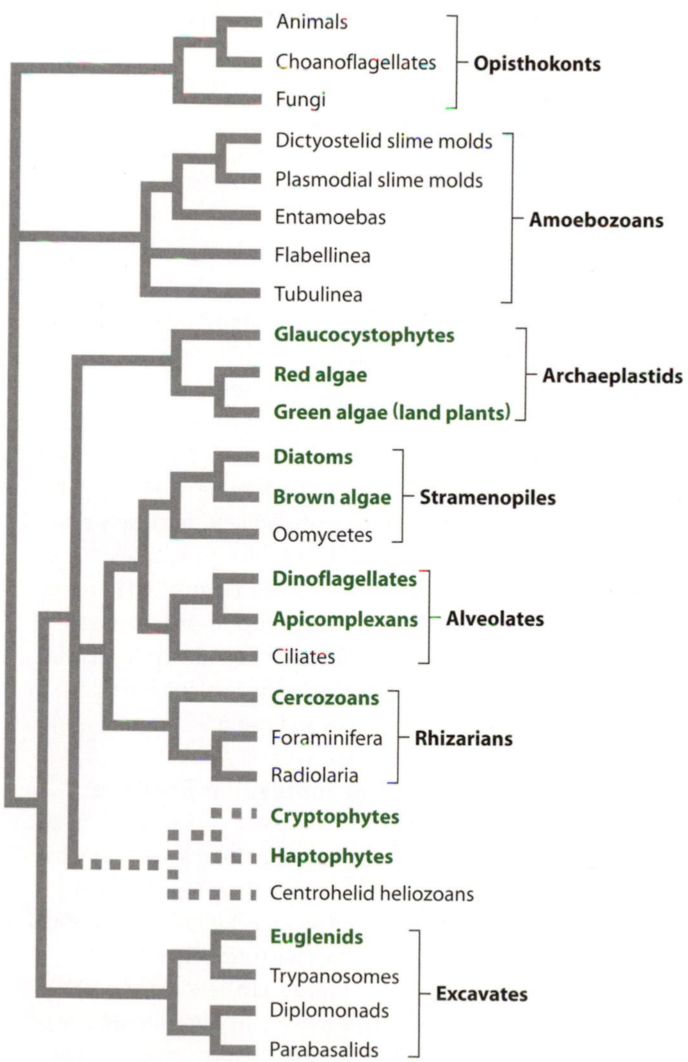

Our own group, the opisthokonts, is the most diverse eukaryotic superkingdom.

Over the past 250 years, biologists have described about 1.8 million species. Of these, 75% or so fall within the **Opisthokonta** (**Fig. 27.10**), a group that encompasses animals, fungi, and some protists. The name for this superkingdom is derived from Greek words meaning "posterior pole," calling attention to the fact that cell movement within this group is propelled by a single flagellum attached to the posterior end of the cell. Not all cells in this superkingdom move around. In humans, for example, the flagellum is limited to sperm. Opisthokonts are heterotrophic, although some species harbor photosynthetic symbionts. Animals are the most diverse and conspicuous opisthokonts. More than 1.3 million animal species, mostly insects and their relatives, have been described. Fungi are also diverse, with more than 75,000 described species, including the visually arresting mushrooms (Fig. 27.10a). Here, however, we focus on opisthokont protists, somewhat poorly known microorganisms that hold clues to the origins of complex multicellularity in this superkingdom (Chapter 28).

It isn't easy to find morphological characters that unite all opisthokonts. One reason is that animals and fungi have diverged so strikingly from what must have been the ancestral condition of the group. However, as molecular sequence comparisons began to reshape our understanding of eukaryotic phylogeny, it soon became clear that fungi and animals are closely related.

Even more closely related to the animals are the **choanoflagellates,** a group of mostly unicellular protists characterized by a ring of microvilli, fingerlike projections that form a collar around the cell's single flagellum (Fig. 27.10b). About 150 choanoflagellate species have been described from marine and freshwater environments, where they prey on bacteria. As early as 1841, the close similarity between choanoflagellates and the collared feeding cells of sponges suggested that these minute organisms might be our closest protistan relatives. This

FIG. 27.10 Opisthokonts. This superkingdom includes animals and fungi, as well as several protistan groups, including choanoflagellates, the closest protistan relatives of animals. *Photo source: Nicole King, Steve Paddock, and Sean Carroll, HHMI, University of Wisconsin.*

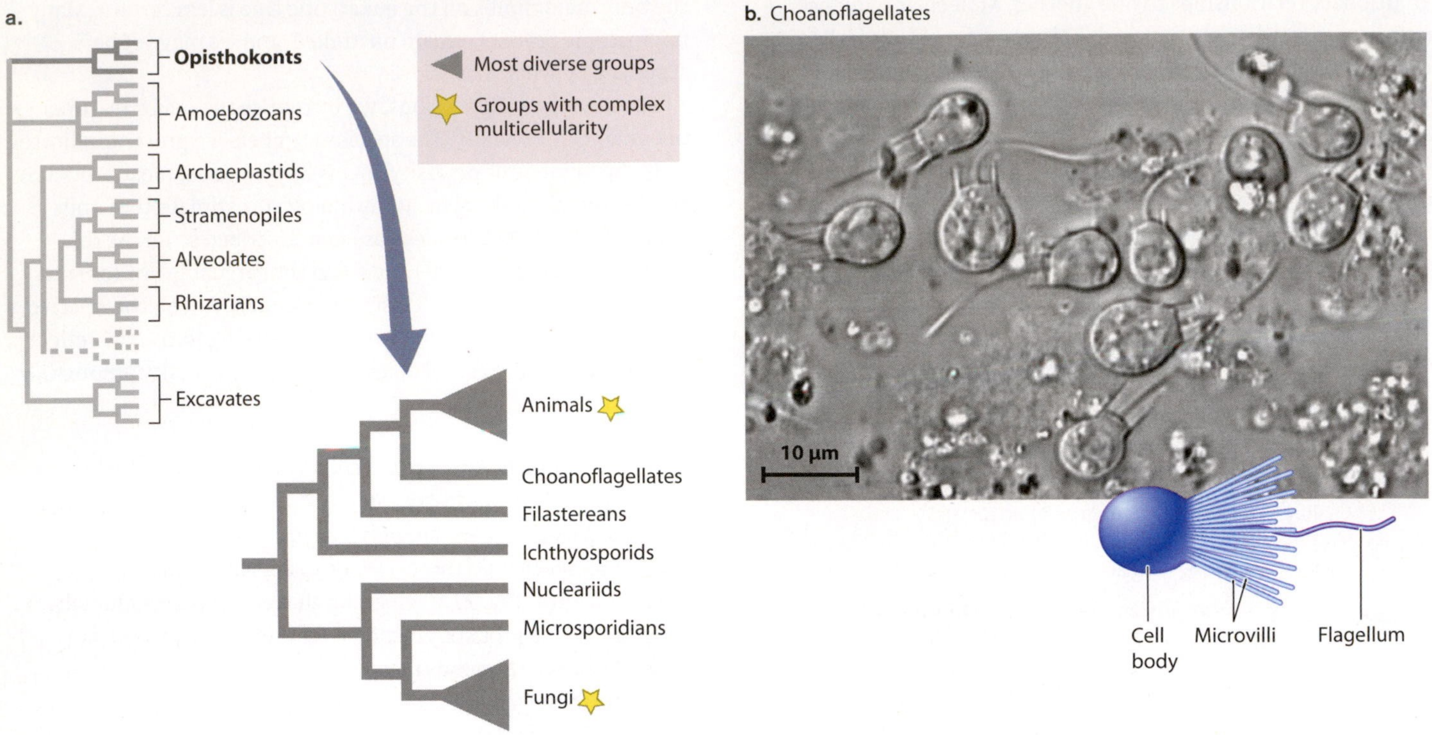

view gained widespread popularity with the discovery in 1880 of *Proterospongia haeckeli*, a choanoflagellate that lives in colonies of cells joined by adhesive proteins.

Molecular sequence comparisons now confirm the close relationship between animals and choanoflagellates. For example, the complete genome of the choanoflagellate *Monosiga brevicollis* shows that a number of signaling molecules known to play a role in animal development are present in our choanoflagellate relatives, although their function in these single-celled organisms remains largely unknown. More generally, many genes once thought to be unique to animals have now been identified in choanoflagellates and other unicellular opisthokonts, underscoring the deep evolutionary roots of animal biology.

Microsporidia form another group of single-celled opisthokonts. Microsporidia are parasites that live inside animal cells. Only their spores survive in the external environment, where they await the opportunity to infect a host and complete their life cycle. Microsporidia infect all animal phyla, so the approximately 1000 known species probably represent only the tip of the iceberg. More than a dozen species have been isolated from human intestinal tissues. Microsporidia infections can be particularly devastating in AIDS patients.

Beyond their importance in human health, microsporidians have attracted the attention of biologists because of features they *lack*: Microsporidian cells have no aerobically respiring mitochondria, no Golgi apparatus, and no flagella. Furthermore, they have a highly reduced metabolism and among the smallest genomes of any known eukaryote. The cellular simplicity of microsporidians does not mean that they are early-evolved organisms, however. Molecular sequencing studies show that microsporidians are the descendants of more complex organisms. Their simplicity is an adaptation for life as an intracellular parasite. It is now widely accepted that microsporidians are closely related to the fungi.

Other protistan opisthokonts have been identified in recent years. These are mostly bacteria-eating unicells or parasites of aquatic animals, but like the choanoflagellates described earlier, their biology is beginning to illuminate the evolutionary path to complex multicellularity in animals and fungi.

Amoebozoans include slime molds that produce multicellular structures.

Previously in this chapter, we learned one thing about amoebas: they move and feed by extending cytoplasmic fingers called pseudopodia. As its name implies, the superkingdom **Amoebozoa** (**Fig. 27.11a**) is a group of eukaryotes with amoeba-like cells (illustrated by *Amoeba proteus* in **Fig. 27.11b**). More than 1000 amoebozoan species have been described. In some amoebozoan species, large numbers of cells aggregate to form multicellular structures as part of their life cycle.

FIG. 27.11 **Amoebozoans.** This group includes protists with an amoeboid stage in their life cycle. *Photo sources: b. M. I. Walker/Science Source; c. AFIP/Science Source.*

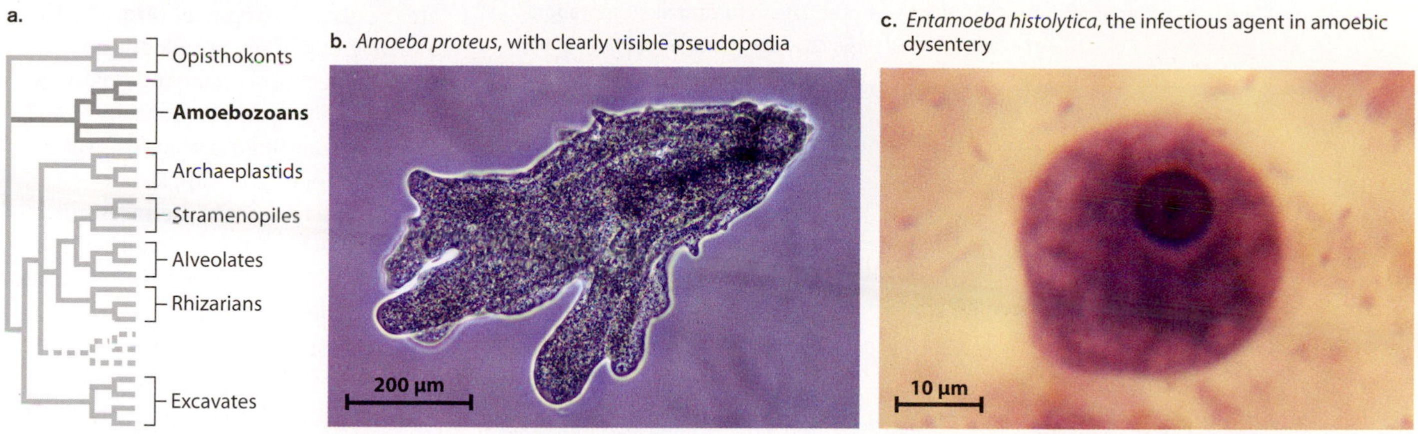

Although there are amoeba-like cells that are not members of the Amoebozoa superkingdom, all members of this group have cells similar to that of *Amoeba*, at least at some stage of their life cycle. Amoebozoans play an important role in soils as predators on other microorganisms. Some amoebozoans cause disease in humans. For example, *Entamoeba histolytica* (**Fig. 27.11c**), an anaerobic protist that causes amoebic dysentery, is responsible for 50,000–100,000 deaths every year.

Beyond considerations of human health, the amoebozoans of greatest biological interest are the slime molds. In plasmodial slime molds (**Fig. 27.12**), haploid cells fuse to form zygotes that subsequently undergo repeated rounds of mitosis but not cell division to form colorful, often lacy structures visible to the naked eye (Fig. 27.12a). These structures, called plasmodia, are **coenocytic,** which means they contain many nuclei within one giant cell. The plasmodia can move and so seek out and feed on the bacteria and small fungi commonly found on bark or plant litter on the forest floor. Triggered by poorly understood environmental signals, plasmodia eventually differentiate to form stalked structures called sporangia that can be 1 to 2 mm high (Fig. 27.12b). Within the mature sporangium, cell walls form around the many nuclei, producing discrete cells. These cells undergo meiosis, generating haploid spores that disperse into the environment. Germination of these spores begins the life cycle anew.

Cellular slime molds (**Fig. 27.13**) are a second type of slime mold. These slime molds spend most of their life cycle as solitary amoeboid cells feeding on bacteria in the soil. Starvation, however, causes these cells to produce the chemical signal cyclic AMP (Chapter 9), which induces as many as 100,000 cells to aggregate into a large multicellular slug-like form (Fig. 27.13a).

FIG. 27.12 **Plasmodial slime molds.** (a) Plasmodia, such as this pretzel slime mold, are coenocytic structures containing many nuclei. (b) Plasmodia generate sporangia, stalked structures that produce spores for dispersal. *Sources: a. Matt Meadows/Science Source; b. Ray Simons/Getty Images.*

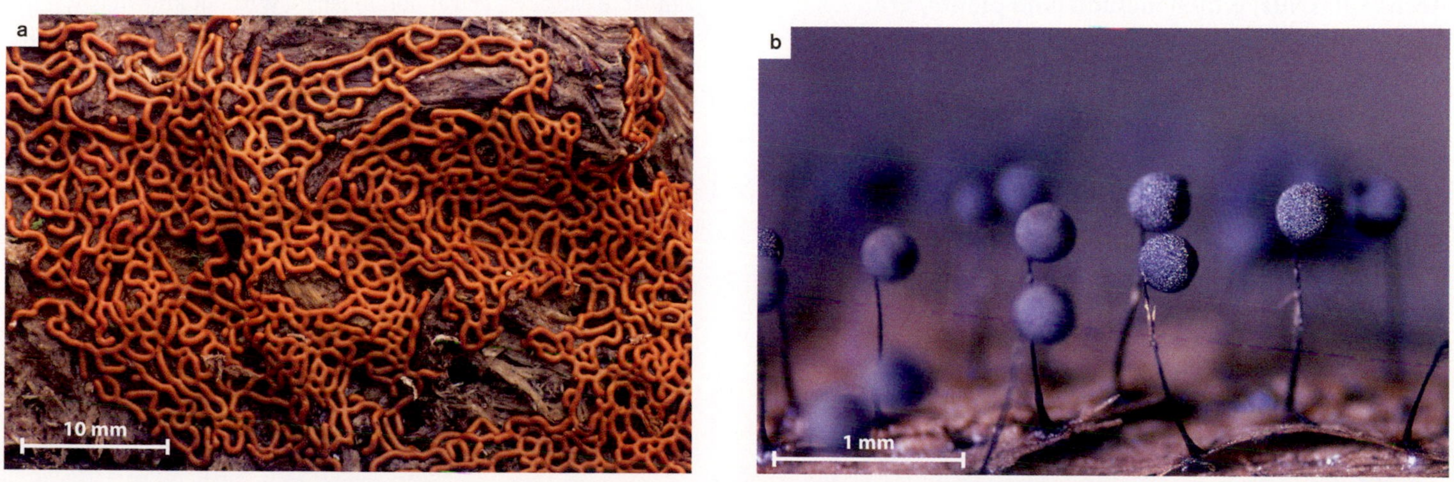

FIG. 27.13 Cellular slime molds. (a) Starvation causes amoeboid feeding cells of cellular slime molds to aggregate into a multicellular "slug" that can migrate along surfaces, leaving behind a trail of slime. (b) "Slugs" differentiate to form stalked sporangia that produce spores. *Source: Carolina Biological Supply Company/Phototake.*

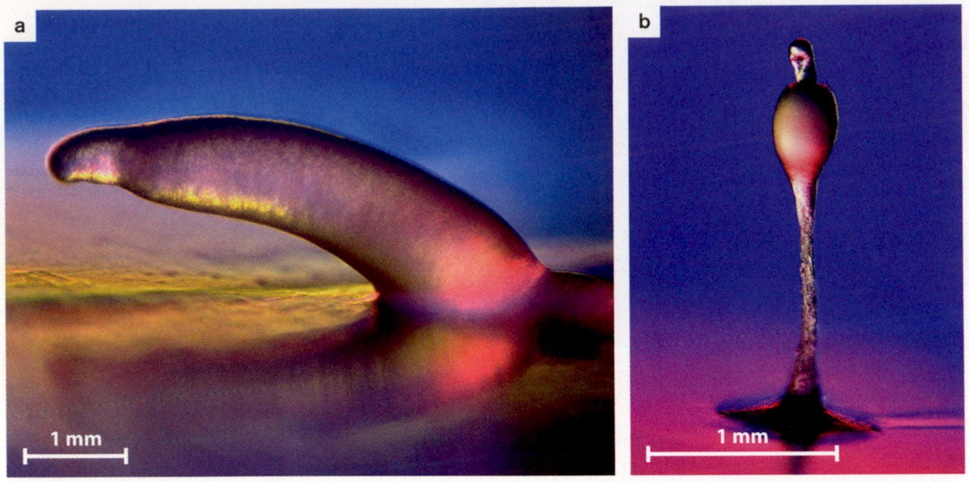

This "slug" can migrate by a coordinated movement of its cells governed by actin and myosin, the same proteins that control muscle movement in animals. Migrating "slugs" can forage for food, leaving behind a trail of mucilage as they move—that's why they're called slime molds. Like plasmodial slime molds, the "slugs" of cellular slime molds eventually differentiate to form stalk-borne sporangia that produce the spores responsible for dispersal of the organism and renewal of the life cycle (Fig. 27.13b).

Slime molds are not true intermediates in the evolution of multicellularity—that is, they are not closely related to animals, fungi, or other groups in which complex multicellularity evolved. Nonetheless, the cellular slime mold *Dictyostelium* has been extensively studied by biologists as a model organism because it has much to teach us about fundamental eukaryotic processes of cell signaling and differentiation, cellular motility, and mitosis.

Archaeplastids, which include land plants, are photosynthetic organisms.

After the opisthokonts, the most conspicuous and diverse eukaryotes fall within the superkingdom **Archaeplastida,** another major branch of the eukaryotic tree (**Fig. 27.14a**). This is where we find the land plants, whose 400,000 described species dominate eukaryotic biomass on this planet (Chapter 33). The archaeplastids contain three major phylogenetic groups, and with the exception of a few parasitic and saprophytic (living on dead matter) species, all are photosynthetic. This indicates that the last common ancestor of living archaeplastids was photosynthetic. In fact, archaeplastids are the direct descendants of the protist that first evolved chloroplasts from endosymbiotic cyanobacteria. Some molecular analyses also place cryptophyte and haptophyte algae within this branch, but this remains a topic of continuing research.

The least conspicuous archaeplastids are the glaucocystophytes (**Fig. 27.14b**), a small group of single-celled algae found in freshwater ponds and lakes. We might skip over them entirely, except for one illuminating observation: Glaucocystophyte chloroplasts retain more features of the ancestral cyanobacterial endosymbiont than any other algae. Their chloroplasts have walls of peptidoglycan, the same molecule found in cyanobacterial walls, and their photosynthetic pigments include biliproteins, also found in cyanobacteria.

More abundant and diverse is the second major archaeplastid group, the red algae (**Fig. 27.14c**). About 5000 species are known, mostly from marine environments. Most are multicellular and some have differentiated tissues and complex morphology. Like other archaeplastids, most have walls made of cellulose. The principal photosynthetic pigments in red algae are a form of chlorophyll called chlorophyll *a* and the biliproteins also found in glaucocystophytes and most cyanobacteria.

Red algae can be conspicuous as seaweeds on the shallow seafloor. One subgroup of red algae, known as the coralline algae, secretes skeletons of calcium carbonate ($CaCO_3$) that deter predators and help the algae withstand the energy of waves in shallow marine environments. If you examine the margin of coral reefs, where waves crash into the reef, you will often see that it is coralline algae and not the corals themselves that resist breaking waves and so enable reefs to expand upward into wave-swept environments.

Red algae are important as food in certain cultures, for example, as laver in the British Isles (particularly in Wales, in cakes similar to oatcakes) and as nori in Japan. Without knowing it, you often ingest carageenan, a polysaccharide synthesized by red algae. Carageenan molecules stabilize mixtures of biomolecules and promote gel formation, and so they are used in products ranging from ice cream to toothpaste. Agar, another red algal polysaccharide, is found universally in microbiology laboratories because the gel it forms at room temperature provides a nearly ideal substrate for bacterial growth.

The land plants and their algal relatives form Viridoplantae, the third branch of Archaeplastida. In this chapter, we are concerned with the protistan members of this group, the green algae (**Fig. 27.14d**). Green algae include a diverse assortment of species (about 10,000 have been described) that live in marine and, especially, freshwater environments. Green algae range from

FIG. 27.14 Archaeplastids. (a) Archaeplastids are a photosynthetic group descended from the protist that acquired photosynthesis from an endosymbiotic cyanobacterium. They include the (b) glaucocystophytes, (c) red algae, and (d) green algae and their descendants, the land plants. *Photo sources: b. Jerome Pickett-Heaps/Science Source; c. Dr. D. P. Wilson/Science Source; d. D. J. Patterson.*

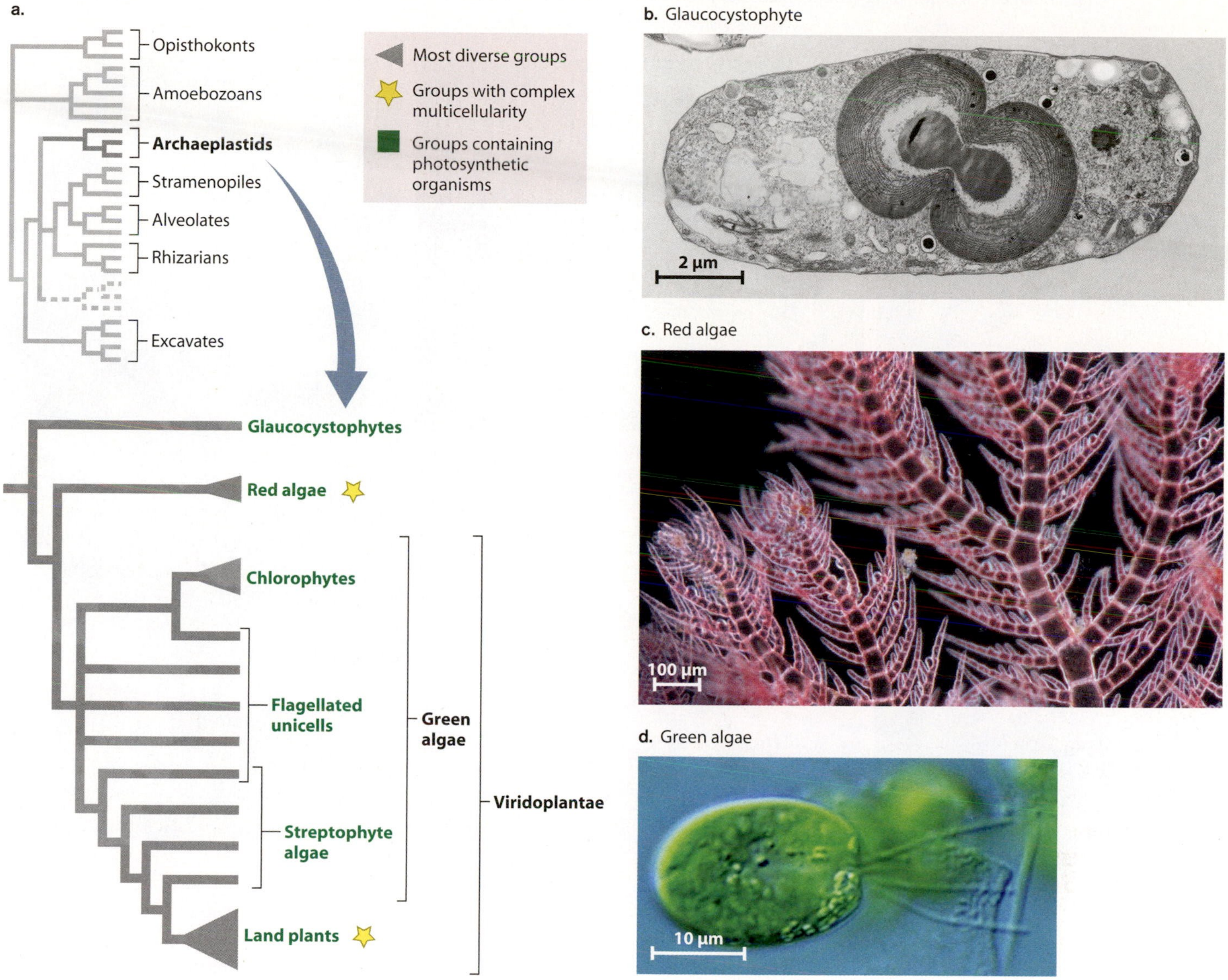

tiny single-celled flagellates to meter-scale seaweeds (**Fig. 27.15**). They all are united by two features: the presence of chlorophyll *a* and chlorophyll *b* in chloroplasts that have two membranes, and a unique attachment for flagella.

It is thought that green algae originated as small flagellated cells, and today most species that branch early on the green algal tree still live as photosynthetic cells in marine or fresh waters (Fig. 27.15a). The larger diversity of green algae, however, is found on two branches that diverged from these flagellated ancestors. One branch, the chlorophytes, radiated mostly in the sea and includes common seaweeds seen globally along seashores (Fig. 27.15e). Chlorophyte green algae have also played a major role in laboratory studies of photosynthesis (*Chlamydomonas*, Fig. 27.15a) and multicellularity (*Volvox*, Fig. 27.15d).

For humans, however, the more important branch of the green algal tree is the one that diversified in freshwater and, eventually, on land. Called streptophytes, the species on this branch show a progression of form from unicells at the base through simple

FIG. 27.15 **Diversity of green algae.** (a) *Chlamydomonas reinhardtii,* a tiny flagellated unicell widely used in laboratory research on photosynthesis; (b) *Spirogyra,* a freshwater green alga with distinctive helical chloroplasts; (c) a desmid, among the most diverse and widespread of all green algae in fresh water; (d) *Volvox,* a simple multicellular organism; and (e) *Acetabularia,* a macroscopic (but single-celled!) green alga found in coastal marine waters in tropical climates. *Sources: a. DeAgostini/Getty Images; b. Eric Grave/Phototake; c. Gerd Guenther/Science Source; d. Frank Fox/Science Source; e. Wolfgang Poelzer/age fotostock.*

cell clusters and filaments on intermediate branches (Figs. 27.15b and 21.15c), to complex multicellular algae that form the closest relatives of land plants (Chapter 33). As was true of choanoflagellates and animals, green algae display features of cell biology and genomics that tie them unequivocally to land plants.

→ **Quick Check 2** Do plants and animals have a common multicellular ancestor?

Stramenopiles, alveolates, and rhizarians dominate eukaryotic diversity in the oceans.

The phylogenetically related superkingdoms called the stramenopiles, alveolates, and rhizarians may not be household names, but perhaps they should be. The surface ocean hosts three species belonging to these groups for every marine animal species, and these protists also play important roles in the ecology of lakes, rivers, and soils.

The superkingdom **Stramenopila** (see Fig. 27.9) includes unicellular organisms and giant kelps, algae and protozoa, free-living cells and parasites. All have an unusual flagellum that bears two rows of stiff hairs (**Fig. 27.16**). Most also have a second, smooth flagellum.

Despite the presence of heterotrophic species, most stramenopiles are photosynthetic. There are about a dozen groups of stramenopile algae, of which the brown algae and diatoms deserve special mention. Brown seaweeds are common along rocky shorelines across the world, and in the Sargasso Sea, huge masses of ropy brown algae (aptly named *Sargassum*) float at the surface. Easily the most impressive brown algae are the kelps, giant seaweeds that form forests above the seafloor

FIG. 27.16 **Stramenopiles.** This electron micrograph of a unicellular stramenopile shows a flagellum with two rows of stiff hairs. Like this cell, most stramenopiles also have a second, smooth flagellum. *Source: D. J. Patterson, image used under license to MBL (micro*scope).*

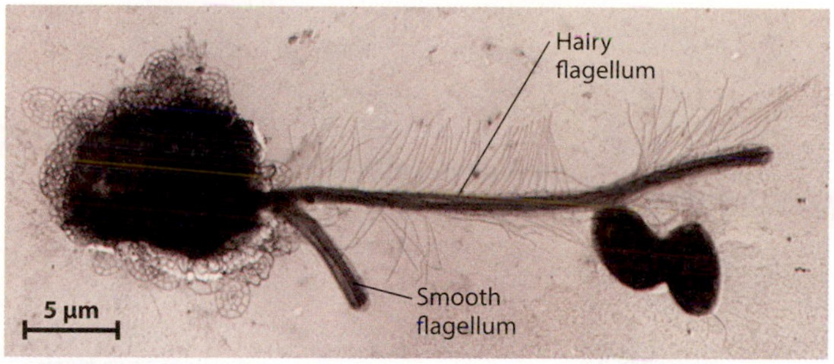

FIG. 27.17 **Kelps.** These complex multicellular brown algae can form "forests" tens of meters high in the ocean. *Source: Mark Conlin/Alamy.*

(**Fig. 27.17**). Kelps not only develop functionally distinct tissues, but also have evolved distinct organs specialized for attachment, photosynthesis, flotation, and reproduction.

The most diverse stramenopiles, however, are the diatoms (**Fig. 27.18**). Recognized by their distinctive skeletons of silica, diatoms account for half of all photosynthesis in the sea (and, so, nearly a quarter of all photosynthesis on Earth). The 10,000 known diatom species thrive in environments that range from wet soil to the open ocean.

Among heterotrophic stramenopiles, the oomycetes, or water molds, are particularly interesting. Originally classified as fungi, these multicellular pathogens cause a number of plant diseases,

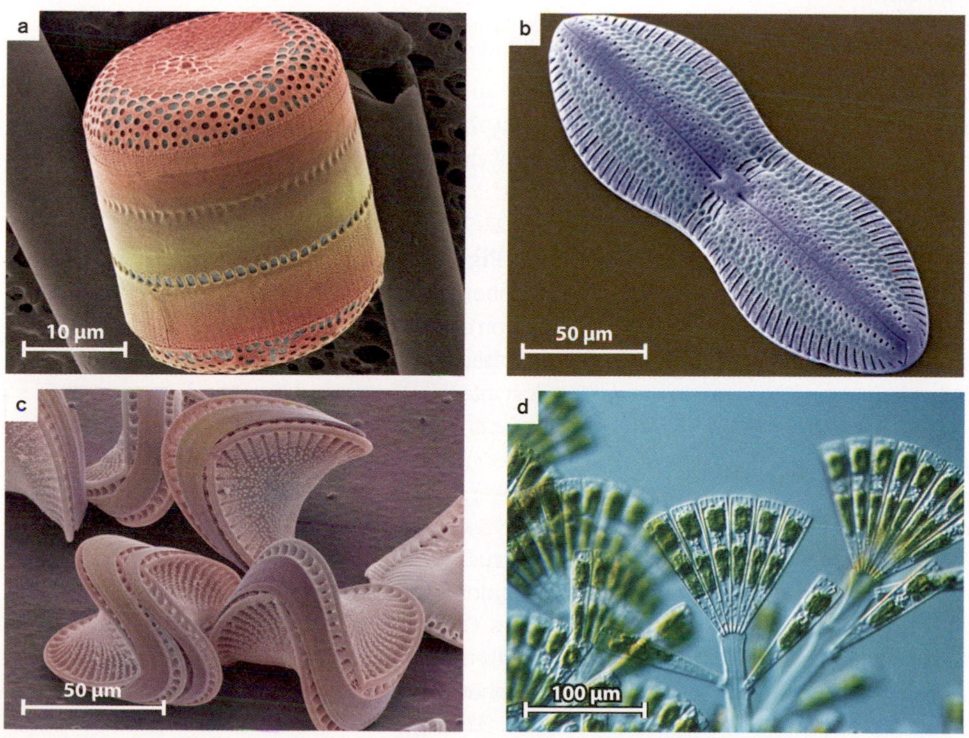

FIG. 27.18 **Diatoms.** Diatoms are the most diverse stramenopiles, and among the most diverse of all protists. Their shapes, outlined by their tiny skeletons made of silica (SiO_2), can be (a) like a pill-box, (b) elongated, or (c) twisted around the long axis of the cell. Some diatom species form colonies, like those shown in (d). *Sources: a. and b. Steve Gschmeissner/Science Source; c. Andrew Syred/Science Source; d. Biophoto Associates/Science Source.*

FIG. 27.19 **Alveolates.** (a) The arrows point to alveoli, flattened vesicles just beneath the surface of the cell that give alveolates their name. (b) Dinoflagellates have cell walls made of interlocking plates of cellulose. (c) A ciliate has many short flagella, or cilia, that line the cell. *Sources: a. Gert Hansen–SCCAP, University of Copenhagen; b. Electron Microscopy Unit, Australian Antarctic Division © Commonwealth of Australia; c. Andrew Syred/Science Source.*

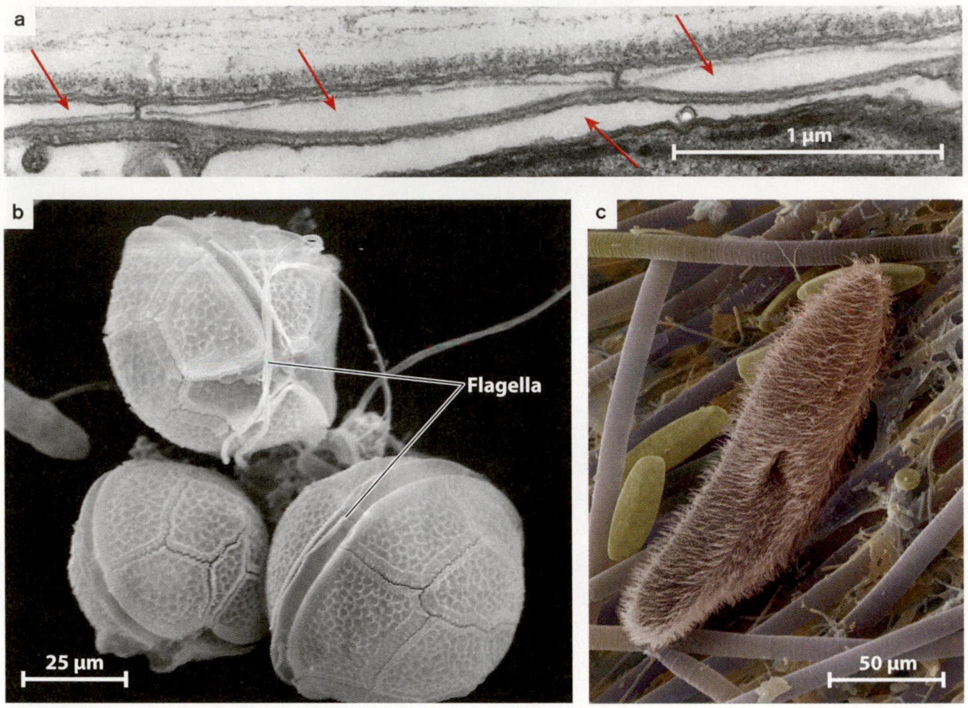

including the potato blight that devastated Ireland during the 1840s (Chapter 33). Other species infect animals, including humans.

Like stramenopiles, the superkingdom **Alveolata** is a diverse group that includes both photosynthetic and heterotrophic species. And, again, like stramenopiles and many other eukaryotic groups, alveolates are united by a feature observed only with careful electron microscopy—small vesicles called cortical alveoli that are packed beneath the cell surface and, in some species at least, store calcium ions for use by the cell (**Fig. 27.19a**).

Most photosynthetic alveolates belong to a large subgroup called dinoflagellates. Dinoflagellates are distinguished by two flagella that produce a rotary motion of the cell as it moves through its environment; many also have a distinctive covering of organic plates (**Fig. 27.19b**). About 4000 dinoflagellate species have been described, but recent genomic surveys of the oceans make it clear that they are much more diverse—by nearly an order of magnitude. About half of the known dinoflagellate species are photosynthetic, but most of these are also capable of heterotrophic growth. Dinoflagellates are important components of the photosynthetic biota in many parts of the oceans, and they affect humans when they form red tides, large and toxic blooms in coastal waters. Red tides occur when nutrients are supplied to coastal waters in large amounts, sometimes for natural reasons but commonly because of human activities such as sewage seeps or runoff from highly fertilized croplands.

The alveolates also include a second diverse (8000 described species) group of protists called ciliates (**Fig. 27.19c**). Ciliates are also distinct at the cellular level, with two nuclei in each cell and numerous short flagella called cilia. They are heterotrophs that feed on bacteria or other protists. In ecological terms, ciliates are among Earth's most important heterotrophic protists.

Dinoflagellates and ciliates are both diverse and ecologically important, but it may be the third principal group of alveolates, the Apicomplexans, that attracts most scientific attention. The reason is simple: The apicomplexan *Plasmodium falciparum* and related species cause malaria, a lethal disease that sickened millions and killed nearly 600,000 people in 2013 (Case 4: Malaria). Although *Plasmodium* cells are heterotrophic, they contain vestigial chloroplasts that suggest evolutionary loss of photosynthesis. Recently, this hypothesis has been confirmed by the discovery in Sydney Harbour, Australia, of photosynthetic cells closely related to parasitic *Plasmodium.*

Photosynthesis spread through eukaryotes by repeated endosymbioses involving eukaryotic algae.

A curious feature of the eukaryotic tree is that photosynthetic eukaryotes are distributed widely but discontinuously among its branches (see Fig. 27.9). How can we explain this pattern?

As explored in **Fig. 27.20,** two hypotheses have been suggested: Either photosynthesis was established early in eukaryotic evolution and then subsequently lost in some lineages, or eukaryotes acquired photosynthesis multiple times by repeated episodes of endosymbiosis. The second hypothesis, it turns out, is correct. The twist is that most of the symbionts involved in the spread of photosynthesis across eukaryotic phylogeny were photosynthetic eukaryotes, not cyanobacteria.

Once more, DNA sequence comparisons provide the decisive data. We can construct phylogenetic trees for photosynthetic eukaryotes using chloroplast genes and then repeat the exercise using nuclear genes. Comparing the two trees, as shown in Figure 27.20, reveals that two groups, chlorarachniophytes and euglenids, gained photosynthesis by establishing green algal cells as endosymbionts. (The algal cells contained chloroplasts derived from an earlier symbiotic assimilation of cyanobacteria.) Other

FIG. 27.20

How did photosynthesis spread through the Eukarya?

BACKGROUND Many different branches on the eukaryotic tree include photosynthetic species, commonly interspersed with non-photosynthetic lineages. Did eukaryotes gain photosynthesis once, by the endosymbiotic incorporation of a cyanobacterium, and then lose this capacity multiple times? Or, is the history of photosynthetic endosymbioses more complicated, involving multiple events of symbiont capture and transformation?

HYPOTHESIS 1 Photosynthesis was established early in eukaryotic evolution and was subsequently lost in some lineages.

HYPOTHESIS 2 Eukaryotes acquired photosynthesis multiple times by repeated episodes of endosymbiosis.

PREDICTIONS The two hypotheses make different predictions. If photosynthesis were acquired in a single common ancestor of all living eukaryotes and then lost in some lineages, we would expect chloroplast and nuclear gene phylogenies to show the same pattern of branching. If photosynthesis were acquired multiple independent times, we would expect chloroplast and nuclear phylogenies to show different patterns.

EXPERIMENT Chloroplasts have DNA, so scientists developed a phylogenetic tree based on molecular sequence comparison of chloroplast genes (Fig. 27.20a). This phylogeny was then compared with a second phylogeny, for all eukaryotes, based on molecular sequence comparison of nuclear genes (Fig. 27.20b).

a. Phylogenetic relationships based on chloroplast genes

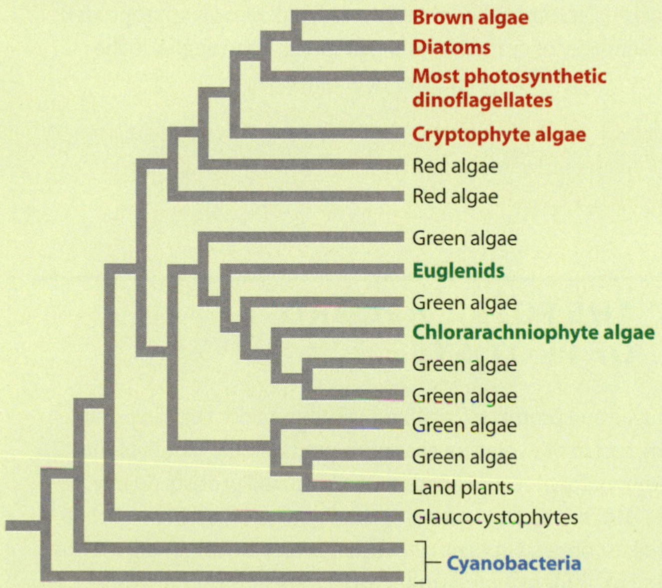

For branches in black, chloroplast relationships match the phylogeny indicated by nuclear genes—these groups diverged from the ancestor in which endosymbiosis first gave rise to chloroplasts.

Nuclear genes place brown algae and diatoms among the stramenopiles, dinoflagellates within the alveolates, and cryptophytes on their own poorly resolved branch. Chloroplast genes, however, relate them to red algae. These groups gained photosynthesis by incorporating red algal cells as endosymbionts.

Nuclear genes place chlorarachniophytes and euglenids among the rhizarians and excavates, respectively, but chloroplast genes relate them to green algae. Both groups gained photosynthesis by incorporating green algal cells as endosymbionts.

Chloroplasts form a monophyletic grouping nested within the cyanobacteria—a principal line of evidence supporting the endosymbiotic origin of chloroplasts.

b. Phylogenetic relationships based on nuclear genes

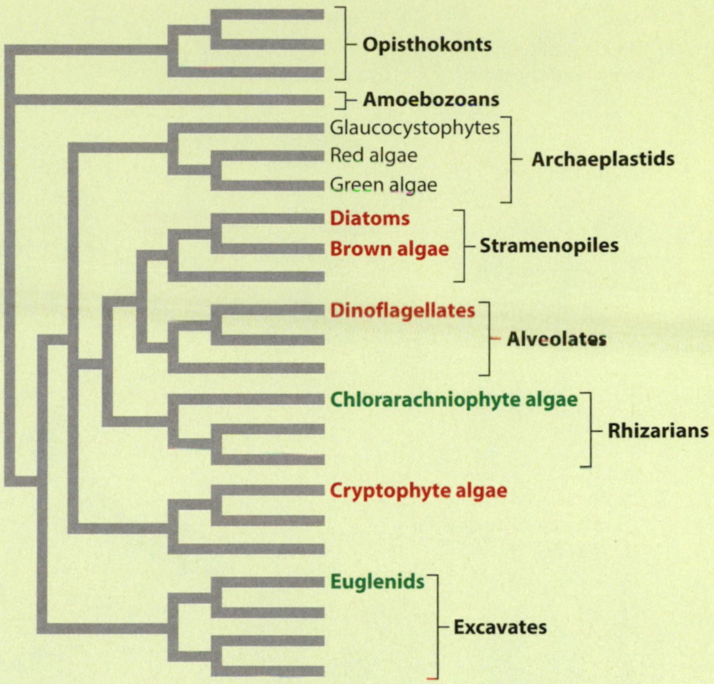

RESULTS Molecular sequence comparisons show that evolutionary relationships among chloroplasts do not mirror those based on nuclear genes. The chlorarachniophyte algae and photosynthetic euglenids (shown in green) fall in a monophyletic group with the green algae when chloroplast DNA is analyzed (Fig. 27.20a). However, these groups lie far from green algae in the phylogeny based on nuclear genes (Fig. 27.20b). Similarly, brown algae, diatoms, most photosynthetic dinoflagellates, and cryptophyte algae (shown in red) form a monophyletic group with the red algae in the phylogeny based on chloroplast DNA (Fig. 27.20a), but analysis of nuclear genes places them in different groups (Fig. 27.20b).

CONCLUSION The chloroplast and nuclear phylogenies show different patterns of branching, supporting the hypothesis that photosynthesis spread through the Eukarya by means of multiple eukaryotic endosymbionts.

SOURCE Hackett, J. D., et al. 2007. "Plastid Endosymbiosis: Sources and Timing of the Major Events." In *Evolution of Primary Producers in the Sea,* edited by P. G. Falkowski and A. H. Knoll, 109–132. Boston: Elsevier Academic Press.

FIG. 27.21 **Successive endosymbiotic events that established photosynthesis in chlorarachniophyte algae.**

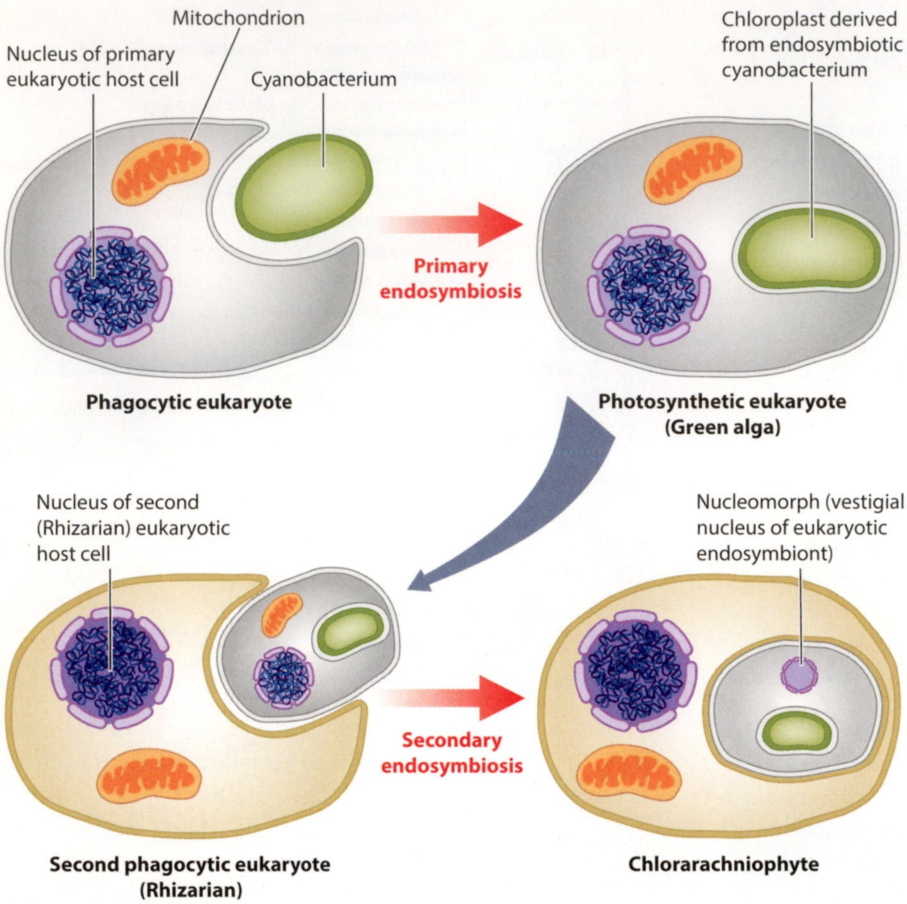

groups (including brown algae, diatoms, most photosynthetic dinoflagellates, and cryptophyte algae) gained photosynthesis as well, but most likely by ingesting a photosynthetic endosymbiont that was a red algal cell.

Several features of cell biology support the view that eukaryotes acquired photosynthesis multiple times by repeated episodes of endosymbiosis. The chloroplasts of euglenids and chloarachniophyte algae contain chlorophyll *a* and *b*, like green algae, whereas cryptophytes contain chlorophyll *c*, a form of chlorophyll not known from either red or green algae, along with chlorophyll *a* and biliproteins, both found in red algae. (Diatoms, brown algae and their relatives contain chlorophyll *a* and *c*, but not biliproteins.)

We noted earlier that chloroplasts in green and red algae are bounded by two membranes, but most of the algae shown in red and green in Fig. 27.20 have chloroplasts bounded by four membranes. The presence of four membranes is expected under the hypothesis of eukaryotic endosymbionts: The inner two membranes are those that bounded the chloroplast in the green or red algal endosymbiont, the third membrane is derived from the outer cell membrane of the symbiont, and the fourth, outer membrane is part of the cell membrane of the host (**Fig. 27.21**).

Chlorarachniophyte algae also retain another feature that supports the eukaryotic origin of their chloroplasts: a small organelle called a nucleomorph (Fig. 27.21). The nucleomorph in chlorarachniophyte algae is, in fact, the remnant nucleus of a green algal symbiont. It retains only a few genes, but those show close phylogenetic affinity with green algae. Nucleomorph genes in cryptophyte algae similarly provide evidence for a red algal symbiont.

How many endosymbiotic events are required to account for the diversity of photosynthetic eukaryotes? There was the primary endosymbiosis that gave rise to green and red algae, two further events that established chloroplasts derived from green algae in euglenids and chlorarachniophytes, and at least one to three events that established chloroplasts derived from red algae in cryptophytes, haptophytes, and stramenopiles.

The dinoflagellates add a remarkable coda to this accounting. Like stramenopiles, most dinoflagellates have chloroplasts derived from red algae. Three dinoflagellate groups, however, have chloroplasts that are clearly derived from green, cryptophyte, and haptophyte algae. Include *Paulinella* and at least 8 to 10 endosymbiotic events are required to explain eukaryotic photosynthesis. Eukaryotes never evolved oxygenic photosynthesis—only the cyanobacteria accomplished that. Instead, eukaryotes appropriated the biochemistry of cyanobacteria and spread it throughout the domain by the horizontal transfer of entire cells.

→ **Quick Check 3** How do the molecular sequences of genes in chloroplasts show how photosynthesis spread throughout the eukaryotic domain?

27.4 THE FOSSIL RECORD OF PROTISTS

Not all algae and protozoa fossilize easily, but those that do provide a striking record of evolution that both predates and parallels the better-known fossil record of animals. Fossilized protists do not contain DNA, nor do they preserve the features of cell biology that define eukaryotes, such as nuclear membranes and organelles. Thus, for a fossil to be identified as eukaryotic, it must preserve other features of morphology found today in Eukarya, but not in Bacteria or Archaea. For this reason, fossils that can be identified with certainty as eukaryotic are limited to those left by eukaryotes that synthesized distinctive and preservable cell walls at some stage of their life cycle.

FIG. 27.22 Eukaryotic fossils in Precambrian sedimentary rocks. (a) Simple protist in 1500-million-year-old rocks from Australia. (b) Simple multicellular red algae in 1100–1200-million-year-old rocks from Canada. (c) Amoebozoan test in 750-million-year-old rocks from the Grand Canyon, Arizona. *Sources: a. Courtesy of Andrew H. Knoll, b. courtesy of N. Butterfield, University of Cambridge (from Butterfield NJ. 2000. Bangiomorpha pubescens n. gen., n. sp.: Implications for the evolution of sex, multicellularity, and the Mesoproterozoic/Neoproterozoic radiation of eukaryotes. Paleobiology 26: 386–404, c. Porter et al. 2003. Vase-shaped microfossils from the Neoproterozoic Chuar Group, Grand Canyon: a classification guided by modern testate amoebae. Journal of Paleontology 77(3): 409–429.) Photo courtesy of Susannah Porter, University of California, Santa Barbara.*

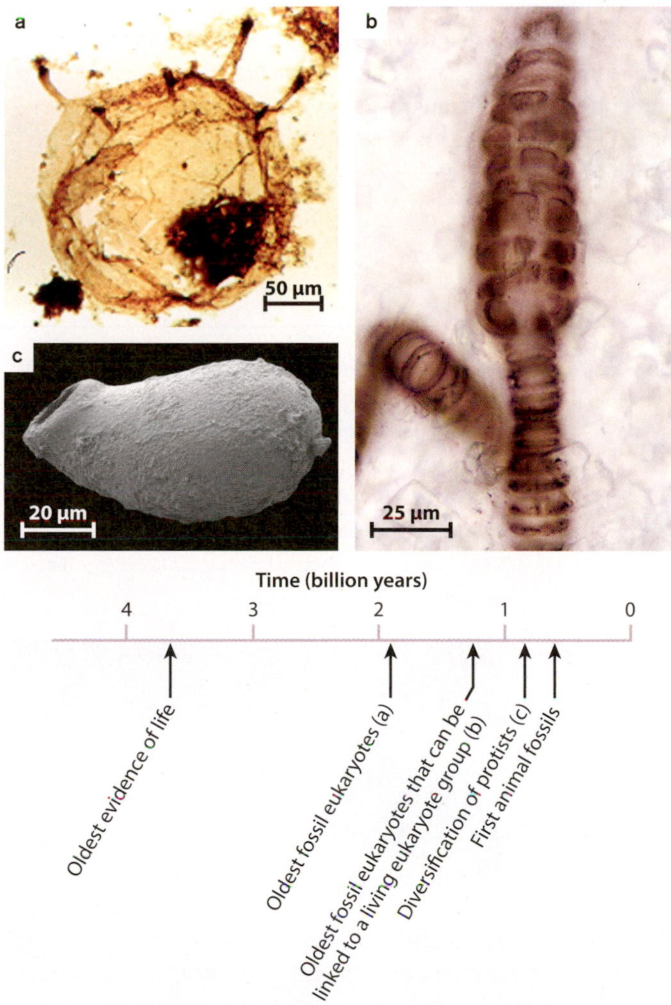

Fossils show that eukaryotes existed at least 1800 million years ago.

Sedimentary rocks deposited in coastal marine environments 1800–1400 million years ago contain microscopic fossils (called **microfossils** for short) up to about 300 μm in diameter. Many of these have complicated wall structures that identify them as eukaryotic. Interlocking plates, long and branching arms, and complex internal layering have been imaged by both scanning and transmission electron microscopes (**Fig. 27.22a**). Comparison with living organisms suggests that such fossils could only be formed by organisms with a cytoskeleton and endomembrane system, the hallmarks of eukaryotic biology.

At present, it appears that the eukaryotic domain first appeared no later than 1800 million years ago and possibly earlier, although the record is difficult to trace in older rocks. None of these earliest eukaryotic microfossils can be placed with confidence into one of the living superkingdoms. Conceivably, at least some of the oldest protistan microfossils could predate the last common ancestor of living eukaryotes.

In sedimentary rocks thought to be 1100–1200 million years old, we find the oldest fossils that can be linked clearly to a living group of eukaryotes (**Fig. 27.22b**). These fossils preserve features found today only in the red algae. We can therefore conclude that by 1200 million years ago Archaeplastida had already diverged from other eukaryotic superkingdoms, photosynthesis had become established in eukaryotes, and simple multicellularity had already evolved within the domain.

Much more eukaryotic diversity is recorded in rocks 800–700 million years old, including green algae and the remains of test-forming amoebozoans (**Fig. 27.22c**). Molecular fossils further indicate the presence of ciliates and dinoflagellates in the emerging eukaryotic world and show that algae were expanding to become major photosynthesizers in the oceans. How can we explain this diversification? One hypothesis is that 800–700 million years ago, protists first evolved (or expanded) the ability to eat other protists rather than preying only on Bacteria and Archaea. This ability initiated an evolutionary arms race that favored divergent means of preying on other organisms and avoiding predation. Both fossils and molecular clocks support this ecological hypothesis for eukaryotic diversification, perhaps facilitated by a small increase in the oxygen content of seawater.

Protists have continued to diversify during the age of animals.

Modern protistan diversity took shape alongside evolving animals during the Phanerozoic Eon, 541 million years ago to the present. For example, foraminiferans and radiolarians, marine protozoans with shells of calcium carbonate and silica, respectively, appear for the first time in rocks deposited near the beginning of the Phanerozoic Eon and have continued to diversify since that time (**Fig. 27.23**). Some algal seaweeds also evolved skeletons of calcium carbonate at this time, and diversified along with animals in Phanerozoic oceans. Why did algae and protozoans evolve mineralized skeletons? Presumably to keep from being eaten.

Marine protists underwent a new round of radiation during the Mesozoic Era, 252–66 million years ago. The most significant changes occurred among photosynthesizers. During the preceding Paleozoic Era, green algae and cyanobacteria were the primary photosynthesizers in the oceans, but as the Mesozoic Era began, new groups of photosynthetic dinoflagellates and coccolithophorid algae appeared, followed by diatoms. Today, these three groups dominate photosynthesis in many parts of the sea.

Within the oceans, diatom and coccolithophorid evolution had another important consequence. The silica skeletons of diatoms are now the principal means by which SiO_2 is transferred from seawater to sediments. Similarly, the $CaCO_3$ scales made by coccolithophorids and the skeletons of planktonic foraminifera are a primary source of carbonate sediments on the seafloor (**Fig. 27.24**). In this way, the ongoing evolution of protists in the age of animals has continued to transform marine ecosystems.

Today, microscopic eukaryotes dominate photosynthetic production in many parts of the oceans and play a major role in biogeochemical cycles. And for thousands of protists, plants and animals are simply habitats, ripe for colonization—we need look no further than the many eukaryotic vectors of human disease. The key conclusion is that a remarkable, and still largely undocumented, diversity of eukaryotic organisms has evolved through time to form an integral part of ecosystems across our planet, shaping in ways both positive and negative the world in which we live. ■

FIG. 27.23 (a) Foraminifera and (b) Radiolaria, showing skeletons of calcium carbonate and silica, respectively. *Sources: a. Astrid & Hanns-Frieder Michler/Science Source; b. M. I. Walker/Science Source.*

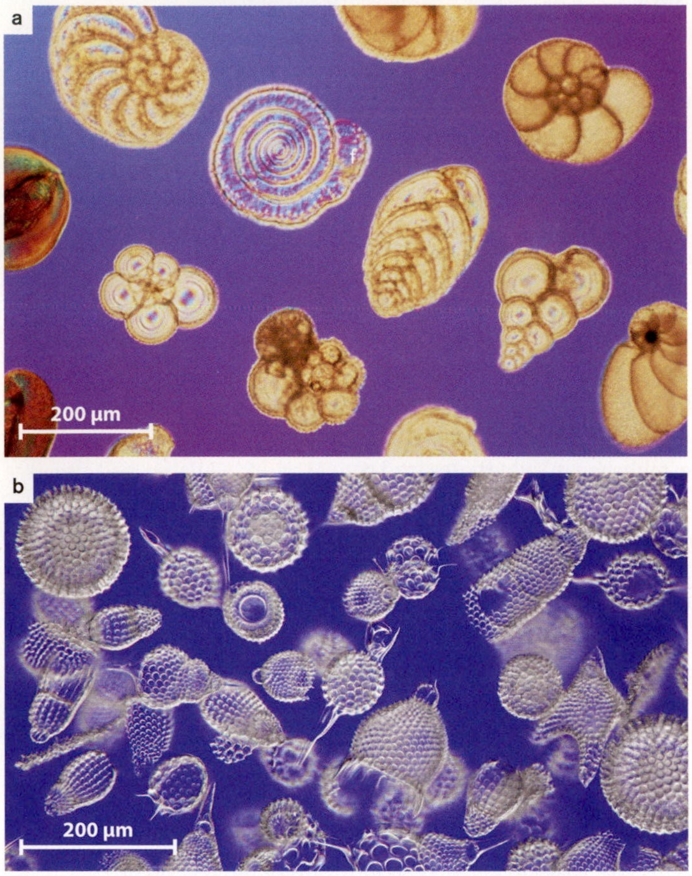

FIG. 27.24 **Coccolithophorids.** (a) Coccolithophorid algae, like *Emiliania huxleyi*, form tiny scales of calcium carbonate. (b) A satellite image captures a coccolithophorid bloom off the coast of Norway, visible as light blue streamers of calcium carbonate in the otherwise dark blue ocean. *Sources: a. ©The Trustees of the Natural History Museum, London; b. NASA image courtesy Jacques Descloitres, MODIS Rapid Response Team at NASA GSFC.*

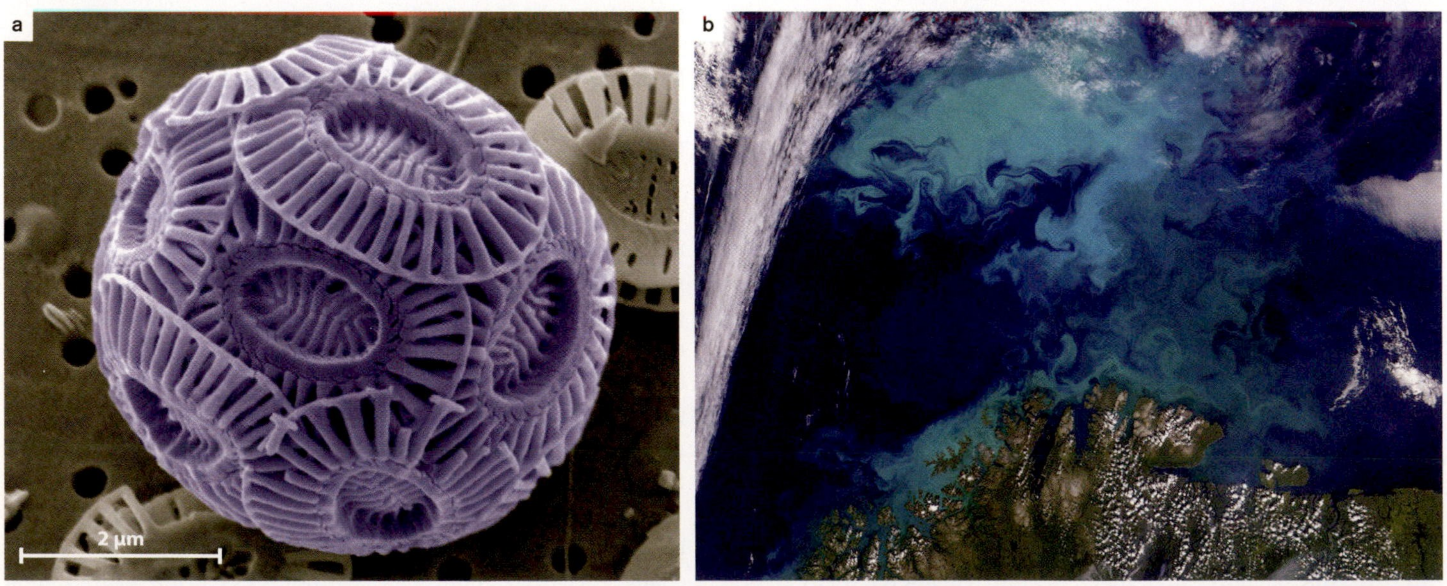

Core Concepts Summary

27.1 Eukaryotic cells are defined by the presence of a nucleus, but features like a dynamic cytoskeleton and membrane system explain their success in diversifying.

Eukaryotic cells have a network of proteins inside the cell that allow them to change shape, move, and transfer substances in and out of the cell. page 554

Eukaryotic cells have dynamic membranes that facilitate movement and feeding. page 554

Eukaryotic cells compartmentalize their machinery for energy metabolism into mitochondria and chloroplasts, freeing the cell to interact with the environment in novel ways not available to prokaryotes. page 555

The eukaryotic genome is larger than that of prokaryotes, allowing new mechanisms of gene regulation. page 556

Sexual reproduction promotes genetic diversity in eukaryotic populations. page 556

27.2 The endosymbiotic hypothesis proposes that the chloroplasts and mitochondria of eukaryotic cells were originally free-living bacteria.

The endosymbiotic hypothesis is based on physical, biochemical, and genetic similarities between chloroplasts and cyanobacteria, and between mitochondria and proteobacteria. page 557

Chloroplasts and mitochondria have their own genomes, but their genomes are small relative to free-living bacteria, to which they are closely related, mostly because of gene migration to the nuclear genome. page 559

Symbiosis between a heterotrophic host and photosynthetic partner is common throughout the eukaryotic domain; reef-forming corals are an example. page 559

Photosynthesis spread through Eukarya by means of multiple independent events involving a protozoan host and a eukaryotic endosymbiont. page 559

Most eukaryotic cells have mitochondria, but a few do not. Evidence suggests that cells lacking mitochondria once had them but lost them. page 559

The eukaryotic nuclear genome contains genes unique to Eukarya, but also genes related to Bacteria and Archaea, suggesting that the ancestor of the modern eukaryotic cell was either a primitive eukaryote descended from archaeal ancestors or an archaeon that engulfed a bacterium. page 560

27.3 Eukaryotes were formerly divided into four kingdoms, but are now divided into at least seven superkingdoms.

The terms "protist," "algae," and "protozoa" are useful, but do not describe monophyletic groups. page 562

Protists are eukaryotes that do not have features of animals, plants, and fungi; algae are photosynthetic protists; and protozoa are heterotrophic protists. page 562

The seven superkingdoms of eukaryotes are Opisthokonta, Amoebozoa, Archaeplastida, Stramenopila, Alveolata, Rhizaria, and Excavata. page 563

Opisthokonts are the most diverse eukaryotic superkingdom. They include animals and fungi, as well as choanoflagellates (our closest protistan relatives). page 563

Amoebozoans produce multicellular structures by the aggregation of amoeba-like cells, and include organisms that cause human disease and those important in biological research. page 564

The archaeplastids, which include land plants, are photosynthetic, and are divided into three major groups. page 566

The related superkingdoms called the stramenopiles, alveolates, and rhizarians dominate eukaryotic diversity in the oceans. page 568

27.4 The fossil record extends our understanding of eukaryotic diversity by providing perspectives on the timing and environmental context of eukaryotic evolution.

Fossils in sedimentary rocks as old as 1800 million years have unmistakable signs of eukaryotic cells, including complicated wall structures. page 573

The earliest fossil eukaryotes that can be placed into one of the present-day superkingdoms are 1100–1200 million years old and belong to the red algae. page 573

Eukaryotic fossils diversified greatly about 800 million years ago, reflecting a new capacity for predation on other protists and, perhaps, a modest increase in oxygen. page 573

Protists continue to diversify to the present. Green algae and cyanobacteria dominated primary production in earlier oceans, but, since about 200 million years ago, dinoflagellates, coccolithophorids, and diatoms have become the primary photosynthetic organisms in the oceans. page 574

Protists have evolved to take advantage of the environments provided by animals and plants; these protists include many that cause disease in humans. page 574

Self-Assessment

1. List key features that distinguish a eukaryotic cell from a prokaryotic cell.
2. Describe the forms of energy metabolism found in eukaryotes.
3. Describe the origin and evolution of the chloroplast and mitochondrion.
4. Present two hypotheses for the origin of the eukaryotic cell.
5. Name the seven superkingdoms of eukaryotes and an organism in each one.
6. State when the eukaryotic cell is first thought to have evolved and the evidence that supports this date.

Log in to LaunchPad to check your answers to the Self-Assessment questions, and to access additional learning tools.

irawansubingarphotography/Getty Images.

CHAPTER 28

Being Multicellular

Core Concepts

28.1 Complex multicellularity arose several times in evolution.

28.2 In complex multicellular organisms, bulk flow circumvents the limitations of diffusion.

28.3 Complex multicellularity depends on cell adhesion, communication, and a genetic program for development.

28.4 Plants and animals evolved complex multicellularity independently of each other, and solved similar problems with different sets of genes.

28.5 The evolution of large and complex multicellular organisms, which required abundant oxygen, is recorded by fossils.

Chapters 26 and 27 introduced the extraordinarily diverse, but largely unseen, world of microorganisms. Among the Bacteria, Archaea, and protists, each cell usually functions as an individual, growing and reproducing, moving from one place to another, taking in nutrients, and both sensing and transmitting molecular signals. In contrast, complex multicellular organisms contain as many as a trillion or more cells that work in close coordination. In your own body, for example, different cells are specialized for specific functions, so that while your body as a whole can perform the broad range of tasks accomplished by microorganisms, individual cells for the most part cannot. Cells lining the intestine absorb food molecules, but nutrients must then be transported to other parts of the body. Lungs take up oxygen from inhaled air, but this, too, must be distributed to other tissues and organs. And while cells at the body's surface sense signals from the environment, the signals affecting interior cells come mostly from surrounding cells.

The biological gulf between microbes and complex multicellular organisms is enormous, but complex multicellularity has evolved a half dozen times. In the chapters that follow, we explore the structure and diversity of plants, fungi, and animals. Here, however, we address the broader question of how complex multicellular organisms evolved in the first place.

28.1 THE PHYLOGENETIC DISTRIBUTION OF MULTICELLULAR ORGANISMS

Most prokaryotic organisms are composed of a single cell, although some form simple filaments or live in colonies. A few types of bacterium, notably some cyanobacteria, differentiate to form several distinct cell types. No bacteria, however, develop macroscopic bodies with functionally differentiated tissues. Only eukaryotes have evolved those. Simple multicellular organisms, composed of multiple similar cells, occur widely within the eukaryotic tree of life, and a few of these evolved into complex multi-cellular organisms characterized by differentiated tissues and organs. How are these different types of organization distributed on the eukaryotic tree of life?

Simple multicellularity is widespread among eukaryotes.

A recent survey of eukaryotic organisms recognized 119 major groups within the superkingdoms discussed in Chapter 27. Of these, 83 contain only single-celled organisms, predominantly cells that engulf other microorganisms or ingest small organic particles, photosynthetic cells that live suspended in the water column, or parasitic cells that live within other organisms. Each of the 36 remaining branches exhibits some cases of simple multicellularity, mostly in the form of filaments, hollow balls, or sheets of little-differentiated cells (**Fig. 28.1**).

Simple multicellular eukaryotes share several properties. In many, cell adhesion molecules cause adjacent cells to stick together, but there are few specialized cell types and relatively little communication or transfer of resources between cells. Most or all of the cells in simple multicellular organisms retain a full range of functions, including reproduction, and so the organism usually pays only a small penalty for individual cell death. Importantly, in simple multicellular organisms, nearly every cell is in direct contact with the external environment, at least during phases of the life cycle when the cells must acquire nutrients.

Simple multicellularity occurs most prominently among algae, although stalklike colonies of particle feeders have evolved in at least three groups of heterotrophic protists, and simple filamentous fungi absorb organic molecules as sources of carbon and energy. In addition, four eukaryotic groups have achieved simple multicellularity by a different route, aggregating during

FIG. 28.1 **Simple multicellularity.** Cells in simple multicellular organisms show little differentiation and remain in close contact with the external environment. Many groups of eukaryotic organisms display simple multicellularity: (a) *Uroglena,* a stramenopile alga found commonly in lakes; (b) *Epistylis,* a stalked ciliate protozoan that lives attached to the surfaces of fish and crabs; and (c) *Prasiola,* a bladelike green alga only a single cell thick, found on tree trunks and rock surfaces. *Sources: a. Jason Oyadomari; b. D. J. Patterson; c. Fabio Rindi.*

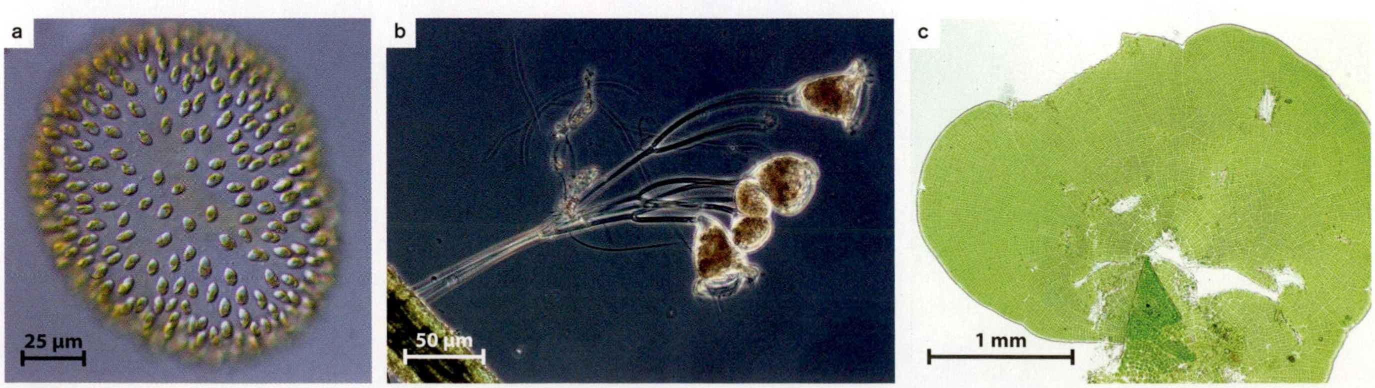

FIG. 28.2 **Coenocytic organization.** Nuclei divide repeatedly but are not partitioned into individual cells. Two different groups of green algae have evolved large size as coenocytic organisms: (a) *Codium,* found abundantly along temperate coastlines, and (b) *Caulerpa,* common in shallow tropical seas and aquaria. *Sources: a. Marevision/age fotostock; (b) Wolfgang Poelzer/WaterFrame/age fotostock.*

just one stage of the life cycle. Slime molds are the best-known example (Chapter 27).

Six groups (two algal, three protozoan, and one fungal) include species characterized by coenocytic organization. In coenocytic organisms, the nucleus divides multiple times, but the cell does not, so the nuclei are not partitioned into individual cells. The result is a large cell—sometimes even visible to the naked eye—with many nuclei. The green algae *Codium* and *Caulerpa,* found along the shorelines of temperate and tropical seas, respectively, are common examples of coenocytic organisms (**Fig. 28.2**). Coenocytic rhizarians on the deep seafloor can be 10 to 20 cm long. There is no evidence that any coenocytic organisms evolved from truly multicellular ancestors, nor have any given rise to complex multicellular descendants.

What selection pressures favored the evolution of simple multicellular organisms from single-celled ancestors? One selective advantage is that multicellularity helps organisms avoid getting eaten. In an illuminating experiment, single-celled green algae were grown in the presence of a protistan predator. Within 10 to 20 generations, most of the algae were living in eight-cell colonies that were essentially invulnerable to predator attack. Another advantage is that multicellular organisms may be able to maintain their position on a surface or in the water column better than their single-celled relatives. Seaweeds, for example, live anchored to the seafloor in places where light and nutrients support growth. Feeding provides a third potential advantage: In colonial heterotrophs such as the stalked ciliate *Epistylis* (see Fig. 28.1b), the coordinated beating of flagella assists feeding by directing currents of food-laden water toward the cells.

There are also costs associated with multicellularity, particularly for complex multicellular organisms with differentiated reproductive tissues. In these organisms, most cells do not reproduce, instead supporting the few that do. This requires cooperation among cells, but it creates opportunities for cells to "cheat"—to use nutrients for their own proliferation rather than the growth and reproduction of the organism as a whole. Cancer is the most conspicuous and lethal example of non-cooperation, perhaps the defining disease of complex multicellular organisms.

As we will see, complex multicellular organisms evolved from simple multicellular ancestors, but most taxonomic groups containing simple multicellular organisms never gave rise to complex descendants.

Complex multicellularity evolved several times.

Complex multicellular organisms are conspicuous parts of our daily existence. Plants, animals, and, if you look a little more closely, mushrooms and complex seaweeds are what we notice when we view a landscape or the coastal ocean (**Fig. 28.3**). Complex multicellular organisms differ from one another in many ways, but they share three general features. They have highly developed molecular mechanisms for adhesion between cells. They display specialized structures that allow cells to communicate with one another. And they display complex patterns of cellular and tissue differentiation, guided by networks of regulatory genes. Without these features, complex multicellularity would be impossible.

For example, plants and animals both have differentiated cells and tissues with specialized functions. Only some tissues photosynthesize or absorb organic molecules; other tissues transport food and oxygen through the body; and still others generate the molecular signals that govern development. Only a small subset of all cells contributes to reproduction. Because of this functional differentiation, cell or tissue loss can be lethal for the entire organism.

FIG. 28.3 Complex multicellular organisms. Complex multicellularity evolved independently in (a) red algae, (b) brown algae, (c) land plants, (d) animals, and (e) fungi (at least twice). *Sources: a. Dr. Keith Wheeler/Science Source; b. Dave Fleetham/Getty Images; c. Filmfoto/Dreamstime.com; d. Jack Milchanowski/age fotostock; e. AntiMartina/Getty Images.*

There is one more feature of complex multicellular organisms that is key to understanding their biology: They have a three-dimensional organization, so only some cells are in direct contact with the environment. Cells that are buried within tissues, relatively far from the exterior of the organism, do not have direct access to nutrients or oxygen. Therefore, interior cells cannot grow as fast as surface cells unless there is a way to transfer resources from one part of the body to another. Similarly, interior cells do not receive signals directly from the environment, even though all cells must be able to respond to environmental signals if the organism is to grow, reproduce, and survive. Complex multicellular organisms, therefore, require mechanisms for transferring environmental signals received by cells at the body's surface to interior cells, where genes will be activated or repressed in response. Development in complex multicellular organisms can, in fact, be defined as increasing or decreasing gene expression in response to molecular signals from surrounding cells (Chapter 20).

Complex multicellularity evolved at least six separate times in different eukaryotic groups (**Fig. 28.4**). Complex multicellularity characterizes animals, but it also evolved at least twice in the fungi, once in the green algal group that gave rise to land plants, once in the red algae, and once in the brown algae, producing the giant kelps that form forests in the sea.

→ **Quick Check 1** How do simple multicellular organisms differ from complex multicellular organisms?

28.2 DIFFUSION AND BULK FLOW

We already mentioned a key functional challenge of complex multicellularity: transporting food, oxygen, and molecular signals rapidly across large distances within the body. How does oxygen get from the air in your lungs into your bloodstream? How does atmospheric carbon dioxide get into leaves? How does ammonia get from seawater into the cells of seaweeds? The answer to all three questions is the same: by diffusion. But oxygen absorbed by your lungs doesn't reach your toes by diffusion alone—it is transported actively, and in bulk, by blood pumped through your

FIG. 28.4 Complex multicellularity within the eukaryotic tree of life. Complex multicellular organisms (found on branches marked by blue) evolved at least six times within three superkingdoms of eukaryotes.

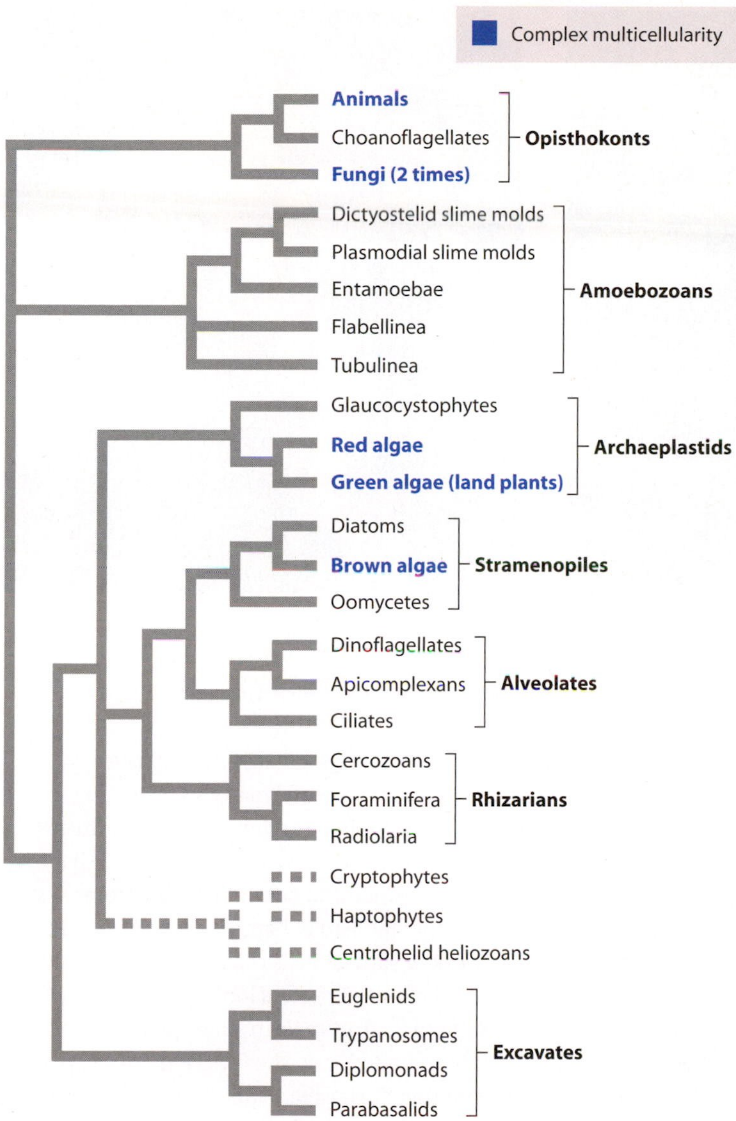

circulatory system. Bulk flow of oxygen, nutrients, and signaling molecules, at rates and across distances far larger than can be achieved by diffusion alone, lies at the functional heart of complex multicellularity.

Diffusion is effective only over short distances.

Diffusion is the random motion of molecules, with net movement from areas of higher to lower concentration (Chapter 5). The distance through which molecules can diffuse quickly depends in part on how great the concentration difference is. Because diffusion supplies key molecules for metabolism, it exerts a strong constraint on the size, shape, and function of cells. Diffusion is effective only over small distances and so it strictly limits the size and shape of bacterial cells, as discussed in Chapter 26. Diffusion also constrains how eukaryotic organisms function.

Oxygen provides a good example. Most eukaryotes require oxygen for respiration. If a cell or tissue must rely on diffusion for its oxygen supply, its thickness is limited by the concentration difference in oxygen between the cell or tissue and its environment. The concentration difference itself depends on the amount of oxygen in the environment and the rate at which oxygen is used for respiration inside the organism. In shallow water that is in direct contact with the atmosphere, diffusion permits animals to reach thicknesses of about 1 mm to 1 cm, depending on their metabolic rate. Of course, a quick swim along the seacoast will reveal many animals much larger than this. And, obviously, you are larger than this. How do you and other large animals get enough oxygen to all of your cells to enable you to survive?

Animals achieve large size by circumventing limits imposed by diffusion.

Sponges can reach overall dimensions of a meter or more, but they actually consist of only a few types of cell that line a dense network of pores and canals and so remain in close contact with circulating seawater (**Fig. 28.5a**). The large size of a sponge is therefore achieved without placing metabolically active cells at any great distance from their environment. Similarly, in jellyfish, active metabolism is confined to thin tissues that line the inner and outer surfaces of the body. Essentially, a large flat surface is folded up to produce a three-dimensional structure (**Fig. 28.5b**). The jellyfish's bell-shaped body is often thicker than the metabolically active tissue, but its massive interior is filled by materials that are not metabolically active. This material constitutes the mesoglea, the jellyfish's "jelly." The mesoglea provides structural support but does not require much oxygen.

But what about us? How does the human body circumvent the constraints of diffusion? Our lungs gather the oxygen we need for respiration, but the lung is a prime example of diffusion in action, not a means of avoiding it. Because lung tissues have a very high ratio of surface area to volume, oxygen can diffuse efficiently from the air you breathe into lung tissue (Chapter 39). A great deal of oxygen can be taken in this way, but how does it get from the lungs to our brains or toes? The distances are far too large for diffusion to be effective. The answer is that oxygen binds to molecules of hemoglobin in red blood cells and then is carried through the bloodstream to distant sites of respiration. We circumvent diffusion by actively pumping oxygen-rich blood through our bodies.

Complex multicellular organisms have structures specialized for bulk flow.

We have seen how humans circumvent the limitations of oxygen diffusion. Many other animals similarly rely on the active

FIG. 28.5 Circumventing limits imposed by diffusion. (a) Sponges can attain a large size because the many pores and canals in their bodies ensure that all cells are in close proximity to the environment. (b) Jellyfish also have thin layers of metabolically active tissue, but their familiar bell can be relatively thick because it is packed with metabolically inert molecules (the mesoglea, or "jelly"). *Sources: a. Andrew J. Martinez/SeaPics.com; b. D. R. Schrichte/SeaPics.com.*

circulation of fluids to transport oxygen and other essential molecules, including food and molecular signals, across distances far larger than those that could be traversed by diffusion alone. Indeed, without a mechanism like bulk flow, animals could not have achieved the range of size, shape, and function familiar to us.

Bulk flow is any means by which molecules move through organisms at rates beyond those possible by diffusion across a concentration gradient (**Fig. 28.6**). In humans and other vertebrate animals, the active pumping of blood through blood vessels supplies oxygen to tissues that may be more than a meter distant from the lungs (Fig. 28.6a). Most invertebrate animals lack well-defined blood vessels but circulate fluids freely throughout the body cavity.

Bulk flow also distributes nutrients through the body, transporting the organic molecules required for respiration large distances from the intestinal cells that absorb these molecules from the digestive tract. (Again, the molecules are transported actively in the bloodstream.) Endocrine signaling molecules such as hormones (Chapters 9 and 38) also move rapidly through the body by means of the blood and other fluids.

Complex organisms other than animals also rely on bulk flow. A redwood tree must transport water upward from its roots to leaves that may be 100 m above the soil. If plants relied on diffusion to transport water, they would be only a few millimeters tall. How, then, do they move water? Plants move water by bulk flow through a system of specialized tissues powered by the evaporation of water from leaf surfaces (Fig. 28.6b; Chapter 29). Vascular plants also have specialized tissues for the transport of nutrients and signaling molecules upward and downward through roots, stems, and leaves.

Still other complex multicellular organisms also have specialized tissues to transport essential materials over long distances at rates faster than could be accomplished by diffusion. Fungi transport nutrients through networks of filaments that may be meters long, relying on osmosis to pump materials from sites of absorption to sites of metabolism (Chapter 34). The giant kelps have an internal network of tubular cells that transports molecules through a body that can be tens of meters long. We will discuss the biology of plants, fungi, and animals in subsequent chapters. For now, it is important to understand the functional similarities among these diverse multicellular organisms. In general, when some cells within an organism are buried within tissues, far from the external environment, bulk flow is required to supply those cells with molecules needed for metabolism.

→ **Quick Check 2** How do mechanisms for bulk flow enable organisms to achieve large size?

28.3 HOW TO BUILD A MULTICELLULAR ORGANISM

We have emphasized three general requirements for complex multicellular life: Cells must stick together; they must communicate with one another; and they must participate in a network of genetic interactions that regulates cell division and

FIG. 28.6 **Bulk flow.** (a) The circulatory system in animals and (b) the vascular system in plants allow these organisms to get around the size limits of diffusion.

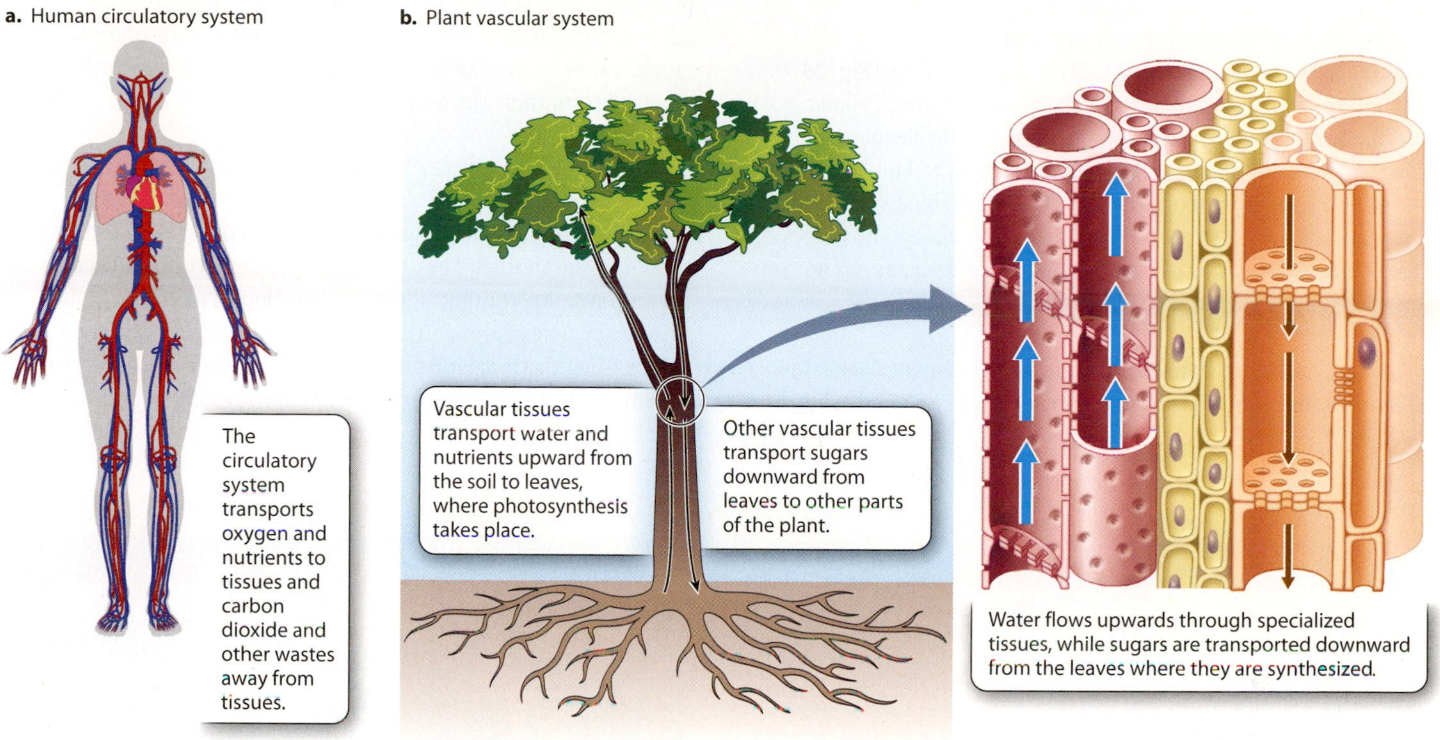

differentiation. Once these are in place, the stage is set for the evolution of specialized tissues and organs observed in plants, animals, and other complex multicellular organisms.

Complex multicellularity requires adhesion between cells.

If a fertilized egg is to develop into a complex multicellular organism, it must divide many times, and the cells produced from those divisions must stick together. In addition, they must retain a specific spatial relationship with one another in order for the developing organism to function. Chapter 10 introduced the molecular mechanisms that hold multicellular organisms together. We begin by reviewing those mechanisms briefly before placing them in evolutionary context.

In animal development, proliferating cells commonly become organized into sheets of cells called epithelia that line the inside and cover the outside of the body. Within epithelia, adjacent cells adhere to each other by means of transmembrane proteins, especially cadherins, which form molecular attachments between cells. Epithelial cells also secrete an extracellular matrix made of proteins and glycoproteins, and these cells attach themselves to this matrix by means of other transmembrane proteins called integrins. Cadherins, integrins, and additional transmembrane proteins thus provide the molecular mechanisms for adhesion in animals.

Cadherins and integrins do not occur in complex multicellular organisms other than animals, but all such organisms require some means of keeping their cells stuck together. How do they do it? Plants also synthesize cell adhesion molecules that bind cells into tissues, but, in this case, the molecules are polysaccharides called pectins (pectins are what make fruit jelly jell). Molecules that control adhesion between adjacent cells are less well known in other complex multicellular organisms, but the general picture is clear. Without adhesion there can be no complex multicellularity, and different groups of eukaryotes have evolved distinct types of molecular "glue" for this purpose.

How did animal cell adhesion originate?

The genome of the choanoflagellate *Monosiga brevicollis* provides fascinating insight into the evolution of cell adhesion molecules. Choanoflagellates, the closest protistan relatives of animals, are unicellular microorganisms. Therefore, it came as a surprise that the genes of *M. brevicollis* code for many of the same protein families that promote cell adhesion in animals. The choanoflagellate genome contains genes for both cadherin and integrin proteins. Clearly, these proteins are not supporting epithelia in *M. brevicollis,* so what are they doing?

One approach to an answer came from the observation that cells stick not only to one another but also to rock or sediment surfaces. Proteins that originally evolved to promote adhesion of individual cells to sand grains or a rock surface may have been modified during evolution for cell–cell adhesion. It is also possible that adhesion molecules were originally used in the capture of bacterial cells, to make prey adhere to the predator.

An intriguing clue to the role of these proteins in choanoflagellates comes from careful laboratory studies (**Fig. 28.7**). Although choanoflagellates are unicellular, simple multicellular structures can be induced in a number of species. Molecular signals induce some choanoflagellates to form a novel multicellular structure, and, unexpectedly, the source of these signals is a bacterium, the preferred prey of the choanoflagellates. The function of the multicellular structures in these choanoflagellates has not yet been determined, but these observations support the hypothesis that they aid predation.

To date, cadherins have been found only in choanoflagellates and animals, but proteins of the integrin complex extend even deeper into eukaryotic phylogeny—they are found in single-celled protists that branch near the base of the opisthokont superkingdom (Chapter 27). Such a phylogenetic distribution provides strong support for the general hypothesis that cell adhesion in animals resulted from the redeployment of protein families that evolved to perform other functions before animals diverged from their closest protistan relatives.

Complex multicellularity requires communication between cells.

In complex multicellular organisms, it is not sufficient for cells to adhere to one another. They must also be able to communicate. Communication is important during development, guiding the patterns of gene expression that differentiate cells, tissues, and organs. The functional integration of cells within tissues and tissues within organs also depends on the flow of information among cells.

HOW DO WE KNOW?

FIG. 28.7

How do bacteria influence the life cycles of choanoflagellates?

BACKGROUND Choanoflagellates are the closest protistan relatives of animals. Cell adhesion, cell differentiation, and multicellularity in these organisms are being studied to increase our understanding of early events in the evolution of animal multicellularity. Nicole King and her colleagues cultured the choanoflagellate species *Salpingoeca rosetta* in the laboratory. High-speed videos revealed that these choanoflagellates differentiate into several distinct cell types in the course of their life cycle. Actively swimming cells can be fast swimmers or slow swimmers that differ in shape and ornamentation (Fig. 28.7a). When swimming cells make contact with a solid surface, they can differentiate to form a stalk that tethers them to the substrate (Fig. 28.7b). Simple multicellular colonies in the shape of rosettes (Fig. 28.7c) were also seen when the choanoflagellates were cultured in the presence of many bacteria, including *Algoriphagus machipongonesis.*

HYPOTHESIS The bacterium *Algoriphagus machipongonesis* induces the choanoflagellate to form rosette-like colonies.

EXPERIMENT The scientists grew the choanoflagellates in a medium containing only *A. machipongonesis,* as well as in media without this prey bacterium.

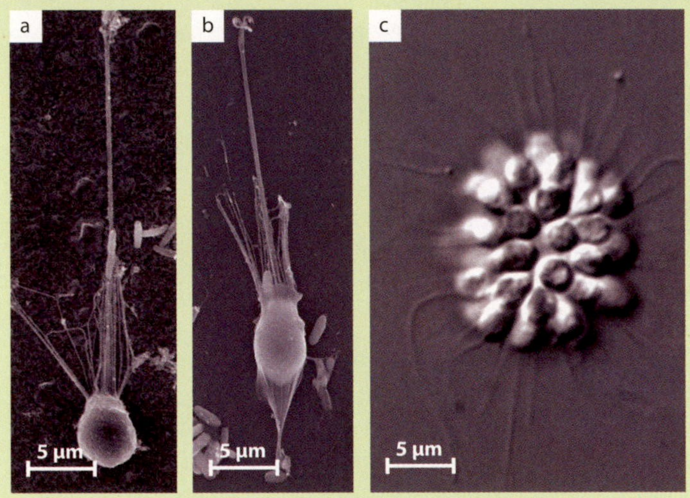

Photo source: Reprinted from Developmental Biology, 357, Mark Dayel and others, Cell differentiation and morphgenesis in the colony-forming choanoflagellate Salpingoeca rosetta, 73–82, Copyright 2011, with permission from Elsevier.

RESULTS The multicellular state formed only when *A. machipongonesis* was present (Fig. 28.7c).

CONCLUSION Choanoflagellates differentiate several cell types upon cues from the environment. In particular, molecular signals from prey bacteria can induce a simple form of multicellularity. The direct ancestors of animals may have gained simple multicellularity by a similar route.

FOLLOW-UP WORK Research is continuing to investigate the molecular mechanisms of cell differentiation, cell adhesion, and signal reception in these distant relatives of ours.

SOURCE Dayel, M., et al. 2011. "Cell Differentiation and Morphogenesis in the Colony-Forming Choanoflagellate *Salpingoeca rosetta*." *Developmental Biology* 357:73–82.

FIG. 28.8 Gap junctions. Gap junctions facilitate the transport of ions and molecules between adjacent cells in animals.

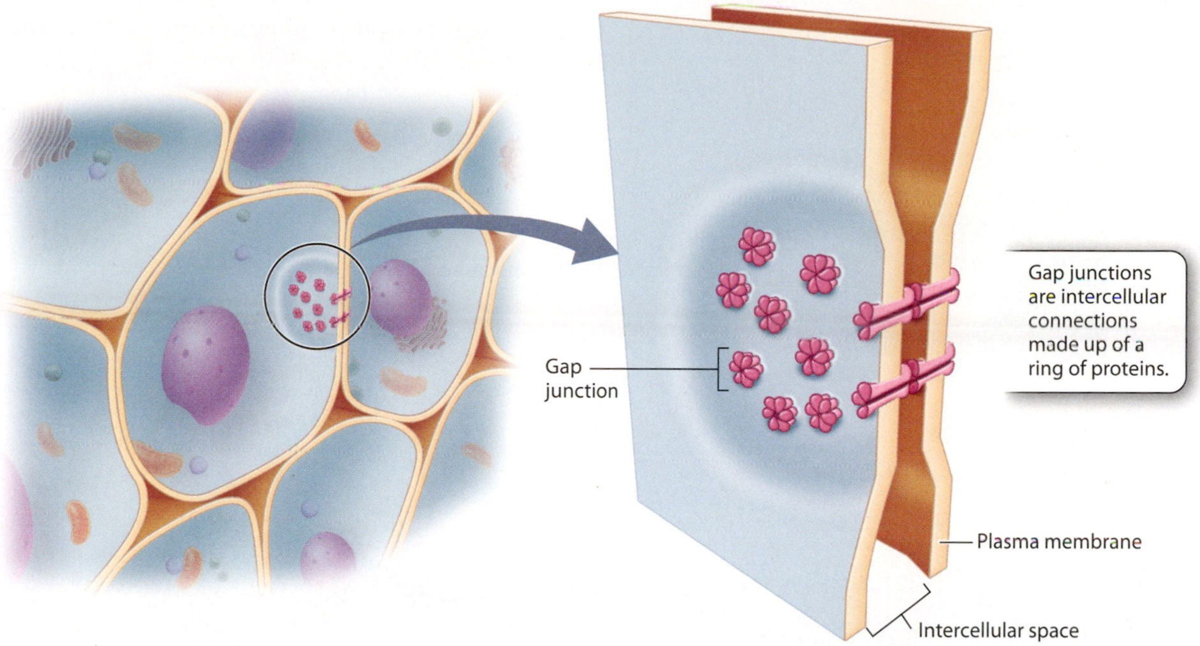

As discussed in Chapter 9, cells communicate by molecular signals. A signaling molecule (generally a protein) synthesized by one cell binds with a receptor protein on the surface of a second cell, essentially flipping a molecular switch that activates or represses gene expression in the receptor cell's nucleus. Plants and animals both have receptors of the receptor kinase type, and these initiate similar signaling pathways within the cell. As in the case of adhesion molecules, these signaling pathways, or components of the pathways, have also been identified in their close protistan relatives. Evidently, many of the signaling pathways used for communication between cells in complex multicellular organisms first evolved in single-celled eukaryotes. Again we can ask, what function did molecular signals and receptors have in the protistan ancestors of complex organisms?

All cells have transmembrane receptors that respond to signals from the environment. In some cases, the signal is a molecule released by a food organism (such as the bacteria that induce simple multicellularity in choanoflagellates), and in some cases the cells sense nutrients, temperature, or oxygen level. Single-celled eukaryotes also communicate with other cells within the same species, for example to ensure that two cells can find each other to fuse in sexual reproduction. Signaling between two cells within an animal's body can be seen as a variation on this more general theme of a cell responding to other cells and the physical environment.

While all eukaryotic cells have molecular mechanisms for communication between cells, complex multicellular organisms have distinct cellular pathways for the movement of molecules from one cell to another. For example, animals more complex than sponges have **gap junctions,** protein channels that allow ions and signaling molecules to move from one cell into another (**Fig. 28.8**). Gap junctions not only help cells to communicate with their neighbors, they allow *targeted* communication between a cell and specific cells adjacent to it (Chapter 10).

Plants, in contrast, have an intrinsic barrier to intercellular communication: the cell wall. Careful observation of plant cell walls with the electron microscope reveals that these walls have tiny holes through which thin strands of cytoplasm extend from one cell to the next (**Fig. 28.9**). Called **plasmodesmata,** these intercellular channels are lined by extensions of the cell membrane (Chapter 10). Tubules running through these channels connect the endomembrane systems of the two cells. Like gap junctions, plasmodesmata permit signaling molecules to pass between cells in such a way that they can be targeted to only one or a few adjacent cells.

Complex red and brown algae also have plasmodesmata, and complex fungi have pores between cells that enable communication by means of cytoplasmic flow. As similar channels do not occur in most other eukaryotic organisms, they appear to represent an important step in the evolution of complex multicellularity.

Complex multicellularity requires a genetic program for coordinated growth and cell differentiation.

All of the cells in your body derive from a single fertilized egg, and most of those cells contain the same genes. Yet your body contains about 200 distinct cell types precisely arranged in a variety of tissues and organs. How can two cells with the same genes become

FIG. 28.9 Plasmodesmata. Plasmodesmata facilitate the movement of ions and molecules between cells in land plants and in complex red and brown algae. *Photo source: Biophoto Associates/Science Source.*

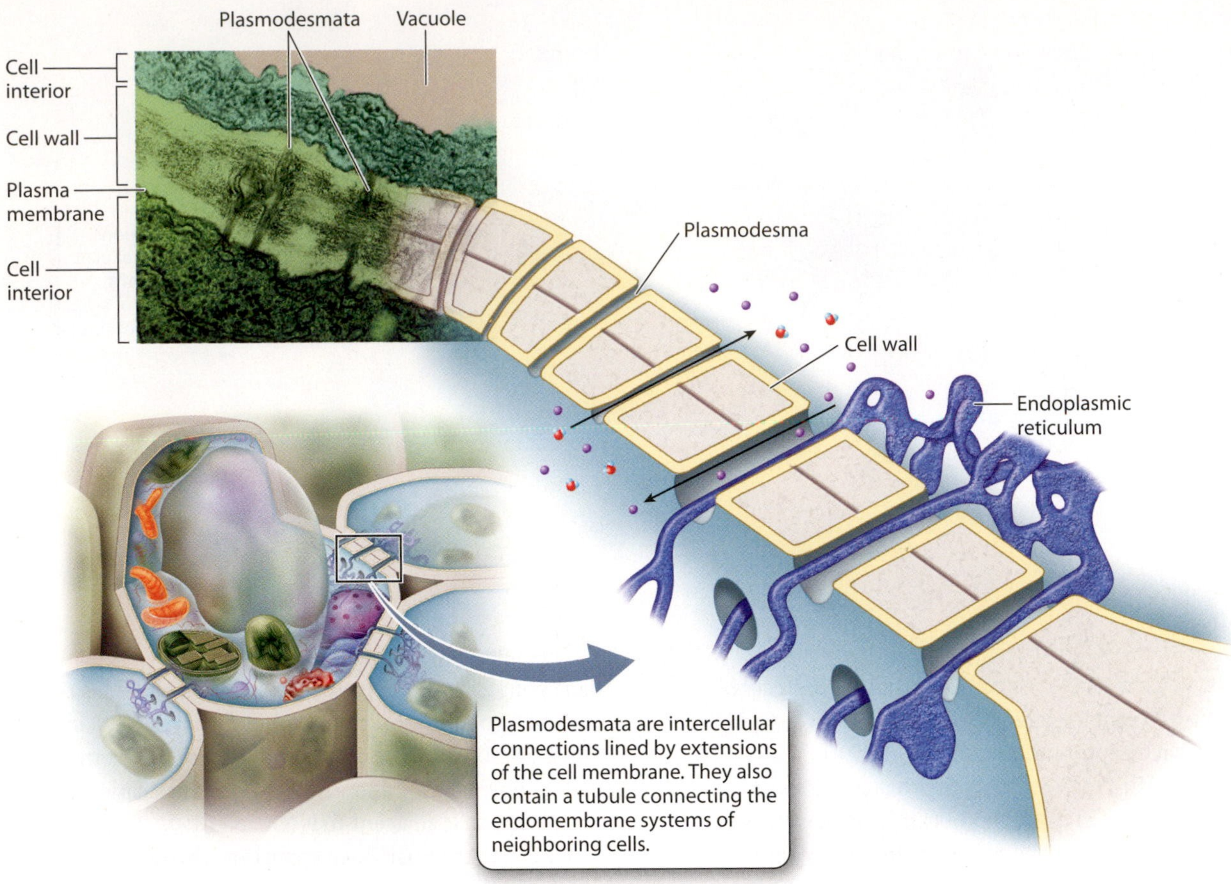

different cell types? The answer lies in development, the system of gene regulation that guides growth from zygote to adult (Chapter 20).

Development is the result of molecular communication between cells. Cells have different fates depending on which genes are switched off or on, and genes are switched off and on by the molecular signals that cells receive. A signal commonly alters the production of proteins—inducing the reorganization of the cytoskeleton, for example. As a result, a stem cell may become an epithelial cell, or a muscle cell, or a neuron. This observation leads to another question: What causes the same gene to be turned on in one cell and off in another? The ultimate answer is that two cells in the same developing organism can be exposed to very different environments.

When we think about development as a process of programmed cell division and differentiation, the link to unicellular ancestors becomes clearer. Many biological innovations accompanied the evolution of complex multicellularity, but, as noted in Chapter 27, the differentiation of distinct cell types is not one of them. Many unicellular organisms have life cycles in which different cell types alternate in time, depending on environmental conditions. For example, if we experimentally starve dinoflagellate cells, two cells undergo sexual fusion to form morphologically and physiologically distinct resting cells protected by thick walls. That is, a nutrient shortage induces a change in gene expression that leads to the formation of resting cells. When food becomes available again, the cells undergo meiotic cell division to form new feeding cells. Many other single-celled eukaryotes form resting cells in response to environmental cues, especially deprivation of nutrients or oxygen.

The innovation of complex multicellularity was to differentiate cells in *space* instead of time. In a three-dimensional multicellular organism, only surface cells are in direct contact with the outside environment. Interior cells are exposed to a different physical and chemical environment because nutrients, oxygen, and light become less abundant with increasing depth within tissues. In effect, there is a gradient of environmental signals within multicellular organisms. We might, therefore, hypothesize that in the earliest organisms with three-dimensional multicellularity, a nutrient or oxygen gradient triggered oxygen- or nutrient-starved interior cells to differentiate, much as happens to trigger the formation of resting cells in their single-celled relatives.

With increasing genetic control of cellular responses to signaling gradients, the seeds of complex development were sown.

Green algae provide a fascinating example that links cell differentiation in unicellular and multicellular organisms. The simple multicellular organism *Volvox* (see Fig. 27.15) has two types of cells, vegetative cells that photosynthesize and control movement of the organism, and reproductive cells. Cell differentiation in *Volvox* is regulated by a gene also involved in the formation of distinct cell types in the life cycle of *Volvox*'s single-celled relative *Chlamydomonas,* supporting the hypothesis that the spatial differentiation of cells in multicellular organisms began with the redeployment of genes that regulate cell differentiation in single-celled ancestors.

Bulk flow, which transports nutrients, oxygen, and water within complex multicellular organisms, also carries developmental signals. Signals carried by bulk flow can travel far greater distances through the body than signals transmitted by diffusion alone. For example, in animals the endocrine system releases hormones directly into the bloodstream, enabling them to affect cells far from those within which they formed (Chapter 38). Thus, the sex hormones estrogen and testosterone are synthesized in reproductive organs but regulate development throughout the body, contributing to the differences between males and females. In this way, signals carried by bulk flow can induce the formation of distinct cell types and tissues along the path of signal transport.

The genome of the choanoflagellate *Monosiga brevicollis,* discussed earlier, has been a treasure trove of information on the antiquity of signaling molecules deployed in animal development. In addition to expressing proteins that govern cell adhesion and epithelial cohesion in animals, *M. brevicollis* expresses a number of proteins active in animal cell differentiation. For example, signaling based on specific receptor kinases was long thought to be restricted to animals, but it also occurs in *M. brevicollis.* In other cases, individual components of signaling proteins are present in the *M. brevicollis* genome, but not the complex multidomain proteins formed by animals. This is true, for example, of several protein complexes that are important in development. Similarly, molecules that play an important role in plant development are also being identified in the genomes of morphologically simple green algae. The key point is that genome sequences interpreted in light of eukaryotic phylogeny are now enabling biologists to piece together the patterns of gene evolution that accompanied the evolution of morphological complexity and diversity in plants and animals.

28.4 VARIATIONS ON A THEME: PLANTS VERSUS ANIMALS

Plants and animals are the best-known examples of complex multicellular organisms. The study of phylogenetic relationships makes it clear that complex multicellularity evolved independently in the two groups. In other words, they do not share a common ancestor that was multicellular. We can see the results of these separate evolutionary events by examining cell adhesion, signaling, and development in plants and animals.

Both plants and animals have evolved sophisticated systems that cause adjacent cells to adhere to each other and that promote the targeted movement of signaling molecules between cells. Plant and animal mechanisms, however, must differ because plant cells have cell walls and animal cells do not. Likewise, plants and animals have evolved similar genetic logic to govern development, but use mostly distinct sets of genes. In both plants and animals, many proteins switch genes encoding other proteins on or off, so that the spatial organization of multicellular organisms arises from networks of interacting genes and their protein products. Ancestral plants and animals simply recruited distinct families of genes to populate regulatory networks.

Cell walls shape patterns of growth and development in plants.

The plant cell wall (Chapter 5), made of cellulose, provides structural support to cells. In fact, the cell wall provides the mechanical support that allows plants to stand erect (Chapter 29). The presence of cell walls has largely determined the evolutionary fate of plants. For example, because their cells cannot engulf particles or absorb organic molecules, most plants gain carbon and energy only through photosynthesis. In addition, because all plant cells except eggs and sperm are completely surrounded by cell walls, they have no pseudopodia and no flagella (in conifers and flowering plants, even sperm have lost their flagella). This being the case, plant cells cannot move. Therefore, the plant as a whole is constrained to obtain nutrients, evade predators, and cope with environmental stress without moving parts.

The inability of plant cells to move has major consequences for development. At the level of the cell, the entire program of growth and development involves cell division, cell expansion (commonly by developing large vacuoles in cell interiors), and cell differentiation. The mechanical consequence is that plant growth is confined to **meristems** (**Fig. 28.10**), populations of actively dividing cells at the tips of stems and roots (Chapter 31). The cells in meristem regions are more or less permanently undifferentiated cells that repeatedly undergo mitosis. A few millimeters from the region of active cell division in a stem or root meristem, cells stop dividing and begin to expand. Within another few millimeters, this activity is curtailed as well. There, individual cells respond to signaling molecules that induce differentiation, forming the mature cells that will function in photosynthesis, storage, bulk flow, or mechanical support. In general, then, plant development involves site-specific cell division, followed by differentiation into distinct cell types that govern the function of the plant as a whole.

Because cell walls render plants immobile, plants have evolved mechanisms to transport water and nutrients from the soil to leaves that may be tens of meters distant without the use of any moving parts or even the expenditure of ATP. And because plants are anchored in place, they are unable to move in response to unfavorable growth conditions. Instead, plants respond to

FIG. 28.10 **A meristem.** Because of their walls, plant cells can't move. For this reason, roots grow by cell division of the undifferentiated cells in the meristem region at the root tip. *Photo source: Walker Mi./Getty Images.*

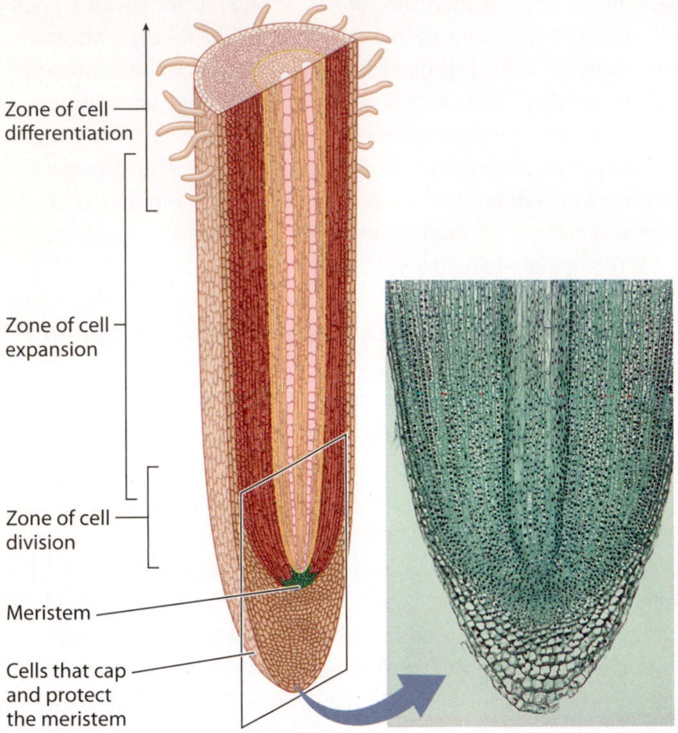

environmental signals by adjusting their meristem activity and, hence, growth pattern. Heat, drought, floods, and fire can all leave their mark in altered patterns of growth—comparing well and poorly watered tomato plants shows this effect. In addition, plants can't flee predators, so they have evolved mechanical structures (hairs and spines) and poisons to keep from being eaten. Indeed, it has been suggested that, in terms of responding to the environment, growth plays a role in plants similar to that played by behavior in animals.

The details of plant growth, development, reproduction, and function are presented in Chapters 29–33. Here, the point to bear in mind is that the properties of a pine tree or a rice plant depend fundamentally on the basic features of cell adhesion, intercellular communication, and a regulatory network to guide development, all carried out under the constraints imposed by cell walls.

Animal cells can move relative to one another.

Having outlined the growth and development of plants, we can appreciate anew the remarkable ballet that characterizes development in animal embryos (**Fig. 28.11**; Chapter 42). Fertilized eggs undergo several rounds of mitosis to form a hollow ball called a **blastula,** formed of a single layer of undifferentiated cells. Then something happens that has no parallel in plant development: Unconstrained by cell walls, animal cells can move relative to one another. Blastula cells migrate, becoming reorganized into a hollow ball that folds inward at one location to form a second layer under the first. This multilayered structure is called the **gastrula.**

Gastrula formation brings new populations of cells into direct contact with one another, inducing patterns of molecular signaling and gene regulation that begin the long process of growth and tissue specification. As cells proliferate, molecular signals are expressed in some cells and diffuse into others, generating what amounts to a three-dimensional molecular map of the organism that guides cell differentiation. Gradients in signaling molecules define top, bottom, front, back, left, and right. Plants do much the same thing, but because animal cells can move

FIG. 28.11 **Gastrulation.** Gastrulation is the movement of cells during embryogenesis that in animals transforms a blastula to a gastrula.

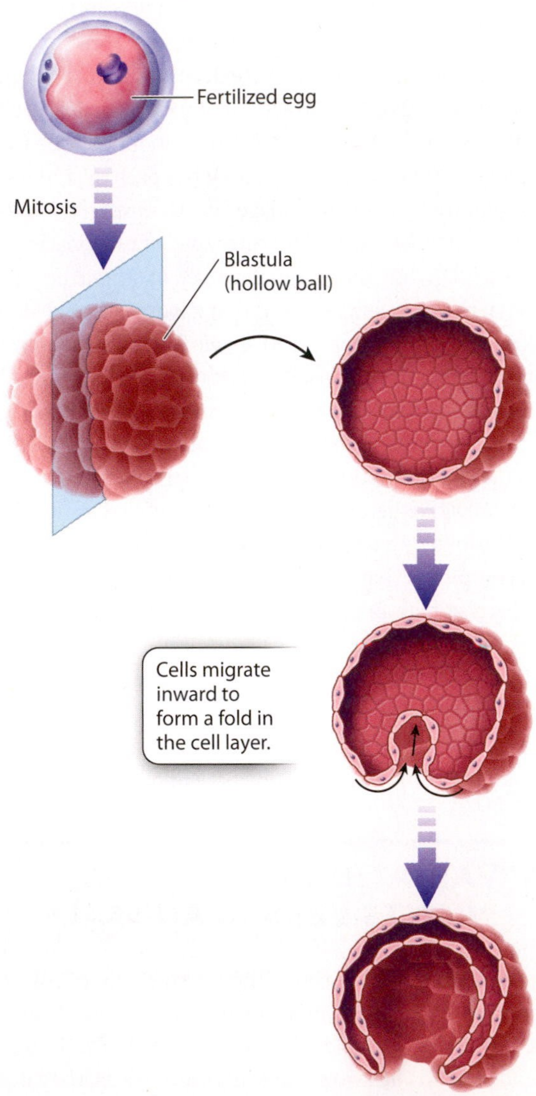

during development, animal embryos are not restricted to growth only from localized regions like meristems. Cell division and tissue differentiation occur throughout the developing animal body.

Because they are not constrained by cell walls, animals can form organs with moving parts—muscles that power active transport of food and fluids and allow movement. Thus, animals have possibilities for function that are far different from those of plants. If the environment offers a challenge, like drought or predators, animals can respond by changing their behavior—for example, they can move to a new location (Chapter 37).

Plants and animals, then, display contrasting patterns of development and function that reflect both their independent origins from different groups of protists and the constraints imposed by cell walls in plants.

→ **Quick Check 3** How do plants and animals differ in the ways their cells adhere, communicate, and differentiate during development?

28.5 THE EVOLUTION OF COMPLEX MULTICELLULARITY

The differences between an amoeba and a lobster seem vast, but research over the past decade increasingly shows how this apparent gap was bridged through a series of evolutionary innovations. Throughout this chapter, we have emphasized that complex multicellular organisms require mechanisms for cell adhesion, modes of communication between cells, and a genetic program to guide growth and development. These three requirements had to be acquired in a specific order. If the products of cell division don't stick together, there can be no complex multicellularity. Adhesion, however, is not sufficient. It must be followed by mechanisms for communication between cells. Moreover, cells must be able to send molecular messages to specific targets or regions, assisted in animals by gap junctions and in plants by plasmodesmata.

With these requirements in place, natural selection would favor the increase and diversification of genes that regulate growth and development, making possible more complex morphology and anatomy. The biological stage was finally set for the functional key to complex multicellularity: the differentiation of tissues and organs that govern the bulk flow of fluids, nutrients, signaling molecules, and oxygen through increasingly large and complex bodies. This differentiation freed organisms from the tight constraints imposed by diffusion. When all these features are placed onto phylogenies, they show both the predicted order of acquisition and the tremendous evolutionary consequences of complex multicellularity (**Fig. 28.12**).

Fossil evidence of complex multicellular organisms is first observed in rocks deposited 579–555 million years ago.

In Chapter 23, we noted that phylogenies based on living organisms make predictions about the fossil record. Characters

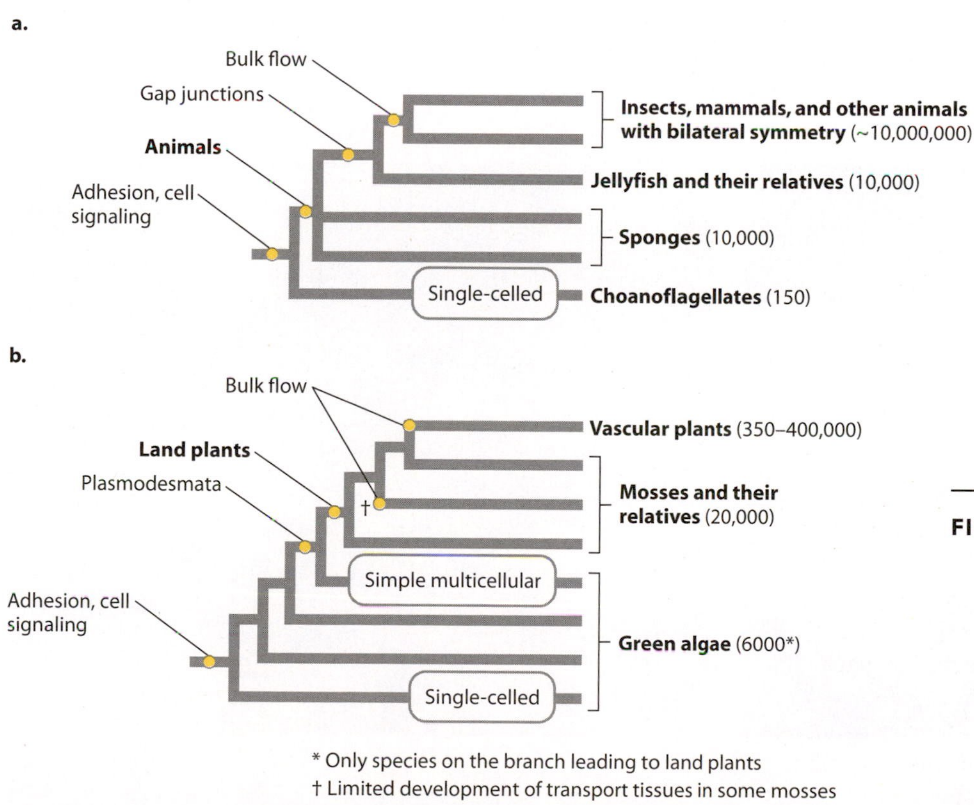

FIG. 28.11 The evolution of complex multicellularity. Phylogenetic relationships of complex (a) animals and (b) plants, with their close evolutionary relatives, show similar patterns of character accumulation and its consequences for biological diversity. Estimated species numbers are shown.

and groups associated with lower branches in phylogenies should appear as earlier fossils than those associated with later branches. Although fossils don't preserve molecular features, the record supports the phylogenetic pattern shown in Fig. 28.12.

Single-celled protists are found in sedimentary rocks as old as 1800 million years, and a number of simple multicellular forms have been discovered that are nearly as old (see Fig. 27.23). Nonetheless, the oldest fossils that record complex multicellularity are found in rocks that were deposited only 579–555 million years ago. **Fig. 28.13a** shows one of the oldest known animal fossils, a frondlike form preserved in 575-million-year-old rocks from Newfoundland, near the eastern tip of Canada. This fossil and many others found in rocks of this age are enigmatic. There is complex morphology to be sure, but it is difficult to understand these forms by comparing them to the body plans of living animals. The Newfoundland fossils show no evidence of head or tail, no limbs, and no opening that might have functioned as a mouth. They appear to be very simple organisms that gained both carbon and oxygen by diffusion. These fossils are long and wide, but they are not thick. In fact, it is likely that active metabolism occurred only along the surfaces of these structures.

Such animals lie near the base of the animal tree (Chapter 44), and probably would have developed under the guidance of regulatory genes similar to those present in modern sponges or, perhaps, jellyfish. By 560–555 million years ago, more complex animals, with distinct head and tail, top and bottom, left and right, enter the record (**Fig. 28.13b**). At the same time, we begin to see tracks and trails made by animals whose muscles enabled them to move across and through surface sediments (**Fig. 28.13c**), an indirect but complementary record of increasing animal complexity. These, then, are the first known occurrences of the types of animal that dominate ecosystems today, both in the sea and on land. These animals must have possessed a diverse array of genes to guide development, something akin to those seen today in insects or snails.

Clearly, by 579–555 million years ago, the animal tree was beginning to branch. This, however, raises a new question: Why did complex multicellular animals appear so much later than simple multicellular organisms?

Oxygen is necessary for complex multicellular life.

Earlier in the chapter, we discussed the key biological requirements for complex multicellularity. There is a critical environmental requirement as well: the presence of oxygen. There are other potential electron acceptors for respiration, like sulfate and ferric iron. However, they not only occur in much lower abundances than oxygen, but they are not gases and so do not accumulate in air. Only the oxidation of organic molecules by O_2 provides sufficient energy to support biological communities that include large and active predators like wolves and lions. And

FIG. 28.13 The earliest fossils of animals. (a) Among the oldest known macroscopic animal fossils is a frond-like form made up of tubular structures, preserved in 575-million-year-old rocks from Newfoundland. (b) The oldest known animal fossil showing bilateral symmetry is a mollusk-like form preserved in 560–555-million-year-old rocks from northern Russia. (c) The oldest tracks and trails of mobile animals are found in rocks 560 million years old and younger. *Sources: a. Guy Narbonne; b. Mikhail Fedonkin; c. Andrew Knoll, Harvard University.*

FIG. 28.14 Oxygen and the evolution of complex multicellularity.

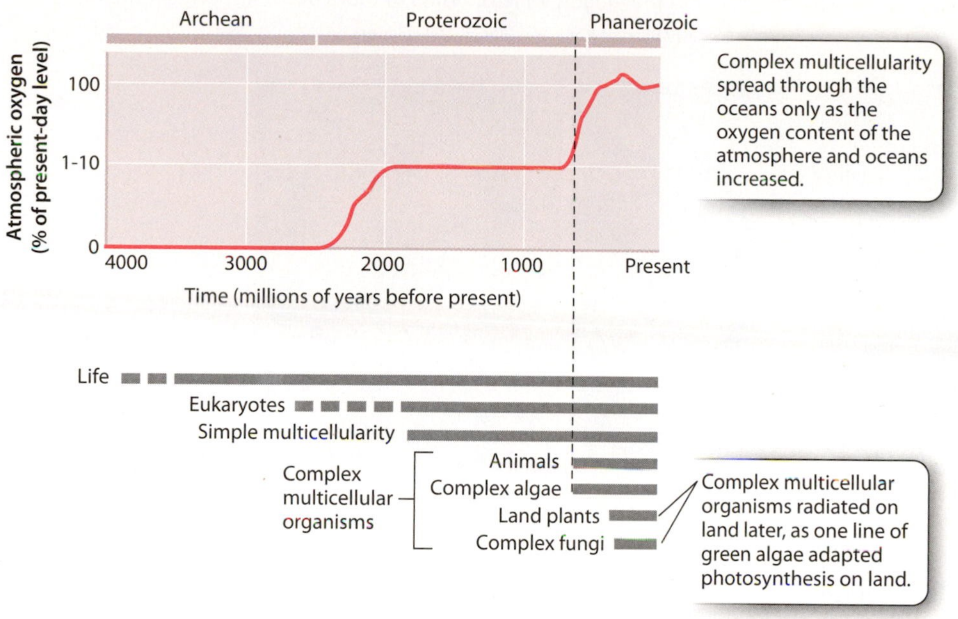

only oxygen in concentrations approaching those of the present day can diffuse into the interior cells of large, active organisms. On our planet—and, probably, on all planets—no other molecule is both sufficiently abundant and has the oxidizing power needed to support the biology of large complex organisms.

On the present-day Earth, lake or ocean bottoms with oxygen levels below about 8% of surface levels support only a limited diversity of animals, and those that they do support tend to be tiny. Large animals with energetic modes of life—carnivores, for example—live only in oxygen-rich environments. The clear restrictions on animal form and diversity observed at lower oxygen levels today have important implications for the distribution of large, active, and diverse animals through time. We noted in Chapter 26 that oxygen first began to accumulate in the atmosphere and surface oceans about 2400 million years ago, but the chemistry of sedimentary rocks deposited after that time tells us that oxygen levels remained low for a very long time. Our modern world of abundant oxygen came to exist only 580–560 million years ago, about the time when fossils first record large multicellular organisms (**Fig. 28.14**).

Scientists continue to debate the causes of this environmental transformation, but the correspondence in time between oxygen enrichment and the first appearance of large, complex animals (and algae) suggests that more oxygen permitted greater size and more energetic modes of life. Greater size in turn created opportunities for tissue differentiation, leading to transport of nutrients and signaling molecules by bulk flow. The evolution of bulk flow in turn permitted still larger size, setting up a positive feedback that eventually resulted in the complex marine organisms we see today.

With morphological complexity came new functions, including predation on other animals. Protozoan predators capture other microorganisms, but do so one cell at a time. In contrast, animals that obtain food by filtering seawater gather cells by the thousands, and larger animals can eat smaller ones, opening up endless possibilities for evolutionary specialization and, therefore, diversification. Among photosynthetic organisms, multicellular red and green algae were able to establish populations in wave-swept coastlines and other environments where simpler organisms lacked the mechanical strength to hold their position along the shoreline.

The key point is that complex multicellular organisms didn't succeed by doing the same things as simpler eukaryotes. Complex multicellularity spread through the oceans because it opened up new and unprecedented evolutionary possibilities.

Land plants evolved from green algae that could carry out photosynthesis on land.

Complex multicellular land plants originated about 465 million years ago, well after marine animals and algae. In some ways, however, the pattern of evolution of land plants parallels that of animals. Ancestral green algae evolved molecular means for cell–cell adhesion and communication between cells, and the algae that form the sister group to land plants share with them the innovation of plasmodesmata. Then, as they developed larger and more complex three-dimensional bodies, plant ancestors evolved a succession of genes and their protein products to guide cell division, expansion, and differentiation. Land plants, however, faced a different challenge from that of animals. The diversification of plants across the land surface requires that photosynthesis, evolved earlier by aquatic organisms, be carried out within tissues bathed by air. It also requires that nutrients and water be absorbed from the soil and transported throughout the plants rather than simply taken in by diffusion from the surrounding water.

Fossils indicate that by 400 million years ago, plants with specialized tissues for bulk flow of water and nutrients had begun to spread across the continents. As plant biomass built up on land, it created opportunities for additional types of complex multicellular organism. In particular, the fungi, which decompose

FIG. 28.15 **Evolution of complex fungi.** (a) This trunklike fossil, discovered in France, is actually a giant fungus that towered over early land plants 375 million years ago. (b) This image shows the interior of the fungus, which consists of tubes that transported nutrients through the large body. *Sources: Carol Hotton; b. Martha E. Cook, School of Biological Sciences, Illinois State University.*

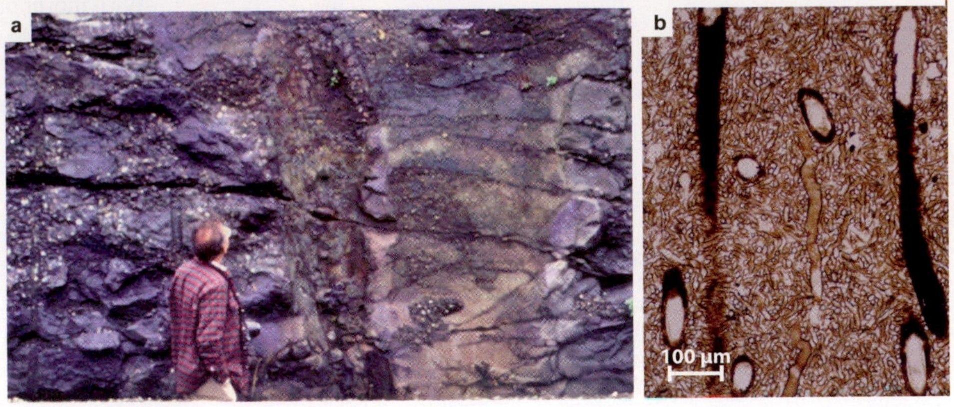

organic matter, radiated by taking advantage of this resource (**Fig. 28.15**), evolving complex multicellularity in two distinct lineages (Chapter 34).

Regulatory genes played an important role in the evolution of complex multicellular organisms.

Complex multicellular organisms account for a large proportion of all eukaryotic species. Thus, one evolutionary consequence of complex multicellularity was an increase in biological diversity. Let's compare the diversity of complex multicellular groups with their simpler phylogenetic sisters. Choanoflagellates, for example, number about 150 species; simple animals such as sponges and jellyfish about 20,000; and complex animals with circulatory systems perhaps as many as 10 million. Plants and their relatives show a comparable pattern. There are about 6000 species of green algae on the branch that includes land plants, but about 400,000 species of anatomically complex land plants capable of bulk flow. Similarly, complex fungi and red algae include more species by an order of magnitude than their simpler relatives.

How did this immense diversity arise? At one level, the answer is functional and ecological. Complex multicellular organisms can perform a range of functions that simpler organisms cannot, and so they have evolved many specific types of interaction with other organisms and the physical environment. Carnivory, for example, requires both sophisticated sensory systems and jaws or limbs to locate and capture prey, but once established, carnivorous animals radiated throughout the oceans and, eventually, on land. This explanation, however, prompts another question: What is the genetic basis for the bewildering range of sizes and shapes displayed by complex multicellular organisms?

The answer has to do with the network of genes that guide development. Since the 1950s, biologists have learned a remarkable amount about this genetic network and its host of interacting genes (Chapter 20). Many of the genes involved in development are what we might consider middle managers: A molecular signal induces the expression of a gene, and the protein product, in turn, prompts the expression or repression of another gene. It is the complex interplay of these genetic switches—turning specific genes on in one cell and off in another, depending on where those cells occur in the developing body—that results in crabs with large pincers or butterflies with patterned wings (**Fig. 28.16**).

It makes sense that if regulatory genes guide development of the body in each complex multicellular species, then mutations in regulatory genes may account for many of the differences we observe among different species. Our growing understanding of how developmental genes underpin evolutionary change has given rise to a whole new field of biology called evolutionary-developmental biology, or **evo-devo** for short.

In evo-devo research, scientists compare the genetic programs for growth and development in species found on different branches of phylogenetic trees. In addition to wanting to discover the molecular mechanisms that guide development in individual species, these scientists aim to understand the genetic differences associated with differences in form and function among species. Evo-devo is an exciting and rapidly expanding field of research because it is helping to illuminate the long-suspected relationship between the development of individuals and patterns of evolutionary relatedness among species. Continuing research promises to shed new light on the similarities and differences among plants, animals, and other complex multicellular organisms. ■

HOW DO WE KNOW?

FIG. 28.16

What controls color pattern in butterfly wings?

BACKGROUND Butterfly wings commonly show a striking pattern of color, including circular features known as eyespots. Eyespots are adaptive, for example in deterring predation by birds.

HYPOTHESIS The expression of regulatory genes during wing development governs eyespot formation on wing surfaces.

EXPERIMENT Paul Brakefield and his colleagues mapped the expression of a regulatory gene called *Distalless* in developing butterfly wings. *Distalless* genes were fused to genes for green fluorescent protein (GFP), a protein that glows vivid green when illuminated under blue light. GFP is then expressed along with *Distalless,* allowing the spatial pattern of *Distalless* expression in developing wings to be visualized.

RESULTS The expression pattern of *Distalless* (on the left in each panel of the figure) in developing wings closely resembles the pattern of eyespots on the wing (on the right in each panel). The two patterns match for both wild-type (top left) and mutant butterfly wings.

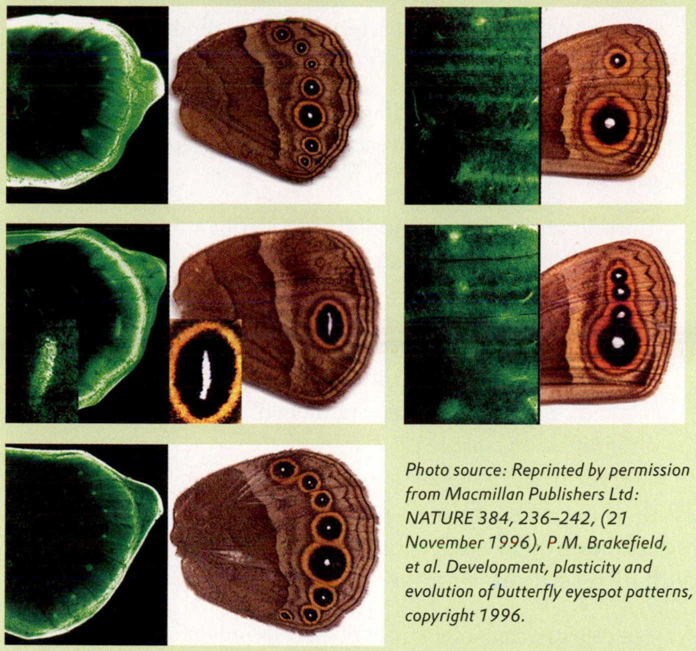

Photo source: Reprinted by permission from Macmillan Publishers Ltd: NATURE 384, 236–242, (21 November 1996), P.M. Brakefield, et al. Development, plasticity and evolution of butterfly eyespot patterns, copyright 1996.

CONCLUSION Regulatory genes play an important role in butterfly wing coloration, and mutations in these genes can account for differences in wing color patterns among species.

SOURCE Brakefield, P. M., et al. 1996. "Development, Plasticity and Evolution of Butterfly Eyespot Patterns." *Nature* 384: 236–242.

Core Concepts Summary

28.1 Complex multicellularity arose several times in evolution.

Bacteria are unicellular or form simple multicellular structures. page 578

Eukaryotes are unicellular or multicellular and exhibit both simple and complex multicellularity. page 578

Simple multicellularity involves the adhesion of cells with little cell differentiation. Complex multicellularity involves cell adhesion, cell signaling, and differentiation and specialization among cells. page 578

Complex multicellular organisms have evolved independently at least six times: in animals, vascular plants, red algae, brown algae (the kelps), and at least twice in the fungi page 580

28.2 In complex multicellular organisms, bulk flow circumvents the limitations of diffusion.

Diffusion is the random motion of molecules, with net movement occurring from regions of higher to regions of lower concentration. It generally acts over small distances, placing limits on the size of multicellular organisms. page 581

Bulk flow is an active process that allows multicellular organisms to nourish cells located far from the external environment, thereby circumventing the constraints imposed by diffusion. page 582

28.3 Complex multicellularity depends on cell adhesion, communication, and a genetic program for development.

Animal and plant cells are organized into tissues characterized by specific molecular attachments between cells. page 583

Choanoflagellates express some of the same proteins that permit cell adhesion in animals, even though choanoflagellates are unicellular. Experiments suggest that choanoflagellates use these proteins to capture bacteria, not for cell adhesion. page 583

Gap junctions in animals and plasmodesmata in plants allow cells to communicate with each other in a targeted fashion. page 585

The cells of complex multicellular organisms are genetically programmed to differentiate into multiple cell types in space. page 586

A number of gene families known to play developmental roles in multicellular organisms are also present in their unicellular relatives, where at least some of them play a role in life cycle differentiation. page 587

28.4 Plants and animals evolved complex multicellularity independently of each other and solved similar problems with different sets of genes.

The cell wall characteristic of plant cells provides structural and mechanical support, but does not allow plant cells to move. page 587

The cell wall of plants has led to distinct solutions to the problems of cell adhesion, cell communication, and development. For example, plants grow by the activity of meristems, populations of actively dividing cells at the tips of stems and roots. page 587

Animal cells do not have cell walls, allowing cell movement that is not possible in plants. For example, during animal development, cells of the embryo migrate inward to form a layered structure called a gastrula. page 588

28.5 The evolution of large and complex multicellular organisms, which required abundant oxygen, is recorded by fossils.

Oxygen may be required for the evolution of complex multicellularity because of its chemical properties and its abundance. page 590

The evolution of complex multicellularity observed in the fossil record correlates with increases in atmospheric oxygen that occurred 580–560 million years ago. page 591

Complex multicellular organisms evolved later on land, as ancestral plants evolved the capacity to photosynthesize surrounded by air rather than water. page 591

Evolutionary-developmental biology, or evo-devo, is a field of research that looks at both individual development and evolutionary patterns in an attempt to understand the developmental changes that allowed organisms to diversify and adapt to changing environments. page 592

Self-Assessment

1. Describe differences between simple and complex multicellularity.
2. Describe the phylogenetic distribution of complex multicellularity.
3. Explain how diffusion limits the size of organisms.
4. Explain how multicellular organisms get around the size limits imposed by diffusion.
5. Describe a key difference between multicellular plants and animals.
6. Describe environmental changes recorded by the sedimentary rocks that contain the oldest fossils of large active animals.

Log in to LaunchPad to check your answers to the Self-Assessment questions, and to access additional learning tools.

Agriculture

Feeding a Growing Population

You may have noticed that while your grocery store or supermarket has several types of apple for sale, there is usually only one type of banana. In fact, it's possible that every banana you have ever eaten has been the same variety. Now the popular yellow fruit may be in trouble.

Years ago, the Western world favored a banana variety known as the Gros Michel. In the first part of the twentieth century, a devastating fungal infection called Panama disease wiped out Gros Michel plantations around the world. In the 1950s, banana growers turned instead to the Cavendish, a variety that displayed some natural resistance to Panama disease. Half a century later, the Cavendish remains the top-selling banana in supermarkets in North America and beyond.

Yet the reign of the Cavendish may be coming to an end. For years, banana growers have battled black sigatoka, a fungal infection that can cause losses of 50% or more of the banana yield in infected regions. Making matters worse, Panama disease is once again posing a threat. A strain of the disease has emerged against which the Cavendish plants have no resistance. The disease has spread across Asia and Australia, and experts fear it's only a matter of time before it reaches prime banana-growing regions in Latin America.

Agriculture has transformed our planet and our species; without it, modern civilization would not be possible.

For those who enjoy bananas sliced into their breakfast cereal, the loss of bananas would be disappointing. For the millions of people in tropical countries who depend on bananas and related plantains for daily sustenance, the destruction of those crops would be much more serious. Agriculture has transformed our planet and our species; without it, modern civilization would not be possible. But as the case of the banana shows, there are challenges to overcome if we're to continue feeding a growing population.

Cultivated bananas are vulnerable to infection in a way that wild bananas are not. The bananas that we eat are sterile; they do not produce seeds. This means they must be propagated vegetatively by replanting cuttings taken from parent plants. As a result, banana plantations contain almost no genetic diversity. As Panama disease and black sigatoka are proving, that lack of diversity can have disastrous effects.

As sterile clones, bananas are particularly vulnerable to disease outbreaks. Other agricultural crops face similar threats. For example, a new, virulent strain of yellow wheat rust fungus emerged in the Middle East in 2010. In its first year, the disease wiped out as much as half of Syria's wheat crop and has since spread to several other countries in the region. Meanwhile, crop scientists have warned that Ug99, a new, virulent strain of an even more damaging fungal pathogen that first surfaced in eastern Africa, could devastate global wheat crops as it spreads.

Wheat was one of the first crops that our ancestors cultivated when agriculture emerged 10,000 years ago. The early farmers learned to choose wheat plants that were easiest to harvest and whose seeds germinated without delay. Over time, wheat evolved by this process of artificial selection. Those earliest farmers set in motion a long chain of genetic modifications that ultimately led to the high-yield wheat varieties grown today.

In the wild, natural selection favored wheat whose seeds were protected by tough barbs and separated easily from the plant, allowing efficient dispersal across the landscape. Farmers, on the other hand, preferred wheat plants whose seeds remained attached to the plant and so were easier to harvest. But the tendency for seeds to remain attached would interfere with the plant's ability to self-seed in the wild. By selecting for traits that would be disadvantageous in nature, humans created a plant dependent on humans for its survival. The codependency between domestic plants and people is a hallmark of agriculture.

In addition to selecting desirable traits in their crop plants, early agriculturalists altered the environment to be favorable for those plants. They provided water, chose

Domestication of wheat. Wild wheat (left) and modern domesticated wheat (right). The seeds of wild wheat are protected by tough barbs (the long spikes), and they fall easily from the plant, enhancing dispersal. The seeds of domesticated wheat remain attached to the plant, making the grain easier to harvest. *Sources: (left) Bob Gibbons/FLPA/Science Source; (right) Nigel Cattlin/Science Source.*

planting sites to ensure optimal sunlight, and removed nearby plants that would compete with the crops for vital resources such as water and nutrients. By changing the community structure and the availability of resources to support growth and reproduction, humans allowed domesticated crops to thrive.

By the 1940s, scientists had developed industrially fixed nitrogenous fertilizers that significantly boosted plant growth. They had also learned enough about genetics and DNA to begin applying that knowledge to plant breeding. These advances led to an amazingly productive period during the 1960s and 1970s now known as the Green Revolution. During that time, plant yields ballooned. Food production actually out-stripped population growth through the second half of the twentieth century.

Today, we are entering a new phase of agriculture, with genetic engineering being added to the traditional approaches of crossbreeding and artificial selection. Most of the corn and soy products now consumed in the United States come from plants that have been genetically engineered to resist certain pests and herbicides. Some of the genes employed come from different species altogether. For example, corn and cotton plants are often engineered to contain a gene from the bacterium *Bacillus thuringiensis* (*Bt*) that confers pest resistance.

At a glance, it would seem that we've successfully taken domesticated species under our control, and indeed, we've made great strides in increasing yields to feed an ever-growing population. But evolution never takes a break. Crop pests and pathogens continue to evolve ways to evade our savviest plant breeders.

The nature of modern agriculture only compounds the problem. Most modern farms consist largely of monocultures, single types of crop each grown over a large area. Growing only a single crop at a time makes it much easier to mechanize planting and harvesting. In a given field of corn or wheat, therefore, the individual plants are often genetically quite similar to one another. With so little genetic variation, each plant is vulnerable to pathogens in exactly the same way. This means that a pathogen that can overcome one plant's natural defenses has the potential to wipe out the entire field.

Lately, many consumers are taking a hard look at these problems and at the source of the food on their plates. Critics of modern agriculture argue that monocultures, with their heavy reliance on fungicides and pesticides, are unsustainable and unhealthy. Proponents of genetic modification and high-tech agriculture argue that we will need every tool to meet food-production demands for the expanding human population. Sometime in 2011, the human population surpassed 7 billion people. The population continues to grow, expanding at its fastest rate at any time in human history. By 2100, experts predict, our planet will be home to more than 9 billion people. One point that both sides agree on is that unless agricultural productivity continues to increase, more and more land will be needed to produce food.

In many parts of the world, agricultural productivity is limited by the availability of water or nutrients or both. Global climate change is predicted to cause regional droughts and flooding that may further weaken the global food supply. There will be no easy solutions as we move forward into civilization's next phase of agriculture. One thing that is certain, however, is that understanding how plants grow, reproduce, and protect themselves from pests will help us survive in an increasingly crowded world.

? CASE 6 QUESTIONS

Special sections in Chapters 29–34 discuss the following questions related to Case 6.

1. **How has nitrogen availability influenced agricultural productivity?** See page 616.
2. **How did scientists increase crop yields during the Green Revolution?** See page 637.
3. **What is the developmental basis for the shorter stems of high-yielding rice and wheat?** See page 649.
4. **Can modifying plants genetically protect crops from herbivores and pathogens?** See page 683.
5. **What can be done to protect the genetic diversity of crop species?** See page 708.
6. **How do fungi threaten global wheat production?** See page 734.

Frank Krahmer/Getty Images.

CHAPTER 29

Plant Structure and Function

Moving Photosynthesis onto Land

Core Concepts

29.1 The evolution of land plants from aquatic ancestors introduced a major challenge for photosynthesis: acquiring carbon dioxide without drying out.

29.2 Leaves have a waxy cuticle and stomata, both of which are important in regulating carbon dioxide gain and water loss.

29.3 Xylem allows vascular plants to replace water evaporated from leaves with water pulled from the soil.

29.4 Phloem transports carbohydrates to the non-photosynthetic portions of the plant for use in growth and respiration.

29.5 Roots expend energy to obtain nutrients from the soil.

In the ocean, where life first evolved, organisms are bathed in seawater that often matches the osmotic concentration of their cells. In contrast, organisms on land are surrounded by air and thus at risk of drying out. Excessive water loss, or **desiccation,** is a constant challenge for life on land, especially for photosynthetic organisms, which expose large surface areas to the air in order to obtain sunlight and carbon dioxide (CO_2). In Chapter 27, we saw that land plants evolved from aquatic green algae. What structural and functional innovations enabled plants to colonize the land so successfully?

The story begins about 470 million years ago, when a group of green algae began the transition from water to land. There are few fossils to show us what the first land plants looked like. We know they were small, at most only a few centimeters tall. These early colonists resembled their algal ancestors; they had no means of obtaining water from the soil and, at best, only a limited capacity to restrict water loss from cells. Therefore, their water content would have fluctuated wildly. Water is the medium for life, and metabolism is possible only when the water content of cells is high. Thus, these first land plants would have been able to carry out photosynthesis only when the surface of the land was wet.

Today, descendants of those first land plants dominate terrestrial habitats, having evolved the ability to draw water from the soil and limit water loss from their leaves. These are the vascular plants, and their evolution transformed the physical and biological environment on land. Vascular plants can carry out photosynthesis even when the soil surface is dry and can sustain the water content of photosynthetic leaves elevated as much as 100 m into the air. In many ways, the capacity to control the uptake and loss of water made possible the extraordinary evolutionary success of vascular plants.

29.1 PLANT STRUCTURE AND FUNCTION: AN EVOLUTIONARY PERSPECTIVE

We begin with a quick sketch of evolutionary relationships among land plants that provides a road map to land plant evolution. We reserve detailed discussion of plant diversity until we have discussed how plants function.

Land plants are a monophyletic group that includes vascular plants and bryophytes.

Fig. 29.1 shows that all land plants are descended from green algae; thus, land plants form a monophyletic group because they include all (and only) the descendants of a common ancestor. Land plants can be divided into two types: **vascular plants,** which can pull water from the soil, and **bryophytes,** which cannot. The first plant groups to diverge are bryophytes—namely, mosses, along with the less-familiar liverworts and hornworts.

FIG. 29.1 The phylogenetic tree of plants. Plants can be divided into two groups: the vascular plants, which actively control their hydration, and the bryophytes, which do not.

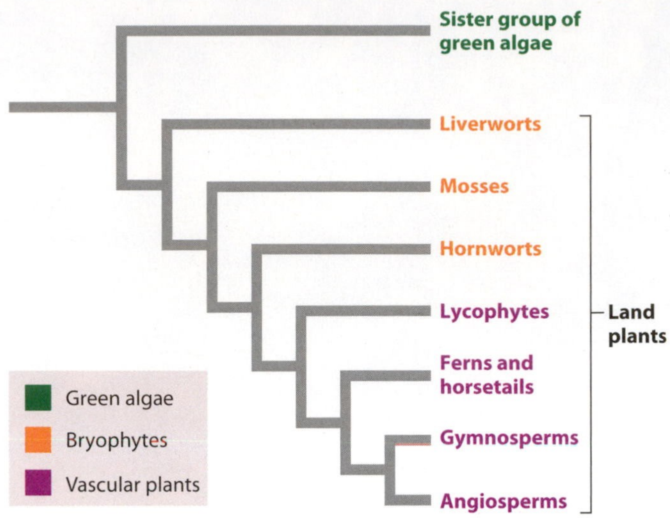

More than 95% of all land plant species found today are vascular plants, and they provide almost all the photosynthetic output of terrestrial environments. The phylogenetic tree shows that vascular plants form a monophyletic group, with four main subgroups. **Lycophytes** and **ferns and horsetails** are mostly small plants, although these groups once included tall trees. The **gymnosperms** include pine trees and other conifers, and the **angiosperms,** or flowering plants, include oak trees, grasses, and sunflowers.

The evolution of vascular plants was a game-changing event in the history of life. To appreciate why this is so, we have to understand the "game" played by the groups that preceded the vascular plants. In particular, how are bryophytes (the three groups of nonvascular plants) able to survive on land? Bryophytes do not produce roots and are totally dependent upon surface water to sustain the water content of their photosynthetic cells. As the environment dries, so do they, and all metabolic activity ceases.

All cells require a high degree of hydration to function. As bryophytes dry out, they lose the ability to carry out photosynthesis. However, many bryophytes exhibit **desiccation tolerance,** a suite of biochemical traits that allows their cells to survive extreme dehydration without damage to membranes or macromolecules (**Fig. 29.2**). Bryophytes can then resume photosynthesis when surface water is once again available. In contrast to bryophytes, vascular plants keep their photosynthetic cells hydrated with water drawn from the soil. As a result, vascular plants can photosynthesize even when the surface of the soil is dry.

Bryophytes provide insights into how the first plants may have met the challenge of carrying out photosynthesis on land.

FIG. 29.2 Desiccation tolerance in bryophytes. *Syntrichia princeps,* a moss, can withstand complete desiccation (right) but regains photosynthetic capacity rapidly when rewetted (left). *Source: Brent Mishler and Michael Hamilton.*

The ability to survive desiccation is likely to have been critical. But we must avoid the temptation to think of living bryophytes as representative of the ancestral state of all land plants. The diversity of bryophytes that we encounter today has been shaped by their long coexistence with vascular plants. Indeed, bryophytes thrive in the shady, moist understory of forests, but only in places where roots would not provide a significant advantage—on rocks and the trunks of trees, for example, and in regions too dry or too cold for roots to penetrate the soil.

We return to bryophytes when we examine plant reproduction (Chapter 30) and diversity (Chapter 33). For now, however, let's focus on the vascular plants.

29.2 THE LEAF: ACQUIRING CO_2 WHILE AVOIDING DESICCATION

Fig. 29.3 shows the general features of a vascular plant. Aboveground, we see three major types of organ—leaves, stems, and reproductive organs—which collectively form the **shoot. Roots,** which are generally hidden from view in the ground, make up the fourth major organ system. We can answer the question of how vascular plants photosynthesize on land by examining how leaves, stems, and roots work together.

The **leaf** is the principal site of photosynthesis in vascular plants. A cross section shows the leaf's three major tissues: Sheets of cells called the **epidermis** line the leaf's upper and lower surfaces; loosely packed photosynthetic cells make up the **mesophyll** (literally, "middle leaf"); and the system of vascular conduits called **veins** connects the leaf to the rest of the plant (**Fig. 29.4**). We discuss how water and carbohydrates are transported through the vascular system in sections 29.3 and 29.4. In this section, we explore how the structure of leaves allows vascular plants to photosynthesize in air without drying out.

CO_2 uptake results in water loss.

Water is the resource that most often limits a plant's ability to grow and function. We see this easily when we forget to water our houseplants or garden. And when you drive across a continent, you see that wet places look very different from dry ones. Water has such a significant effect on the growth and functioning of plants because plants use large amounts of water. Why do plants require so much water?

The answer is *not* water's role as an electron donor in photosynthesis (Chapter 8): That process accounts for less than 1% of the water required by vascular plants. Instead, most of a plant's need for water arises as a consequence of CO_2 uptake from the air.

As we saw in Chapter 25, CO_2 is a minor constituent of air, at present about 400 parts per million. This low concentration limits the rate at which CO_2 can diffuse into the leaf. Therefore, how fast a plant can take up CO_2 depends in large part on the degree to which leaves can expose their photosynthetic cells to the surrounding air. If we look inside a leaf, we see that each mesophyll cell is surrounded by air spaces. The mesophyll cells obtain the CO_2 that they need for photosynthesis directly from these air spaces. If the air spaces within the leaf were completely sealed off, photosynthesis

FIG. 29.3 Major organs of a vascular plant.

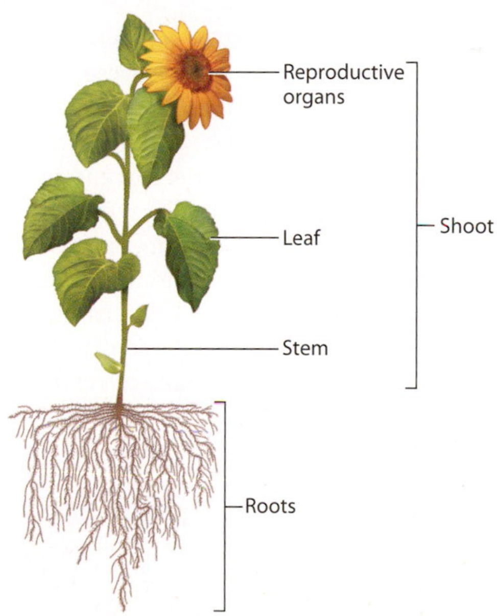

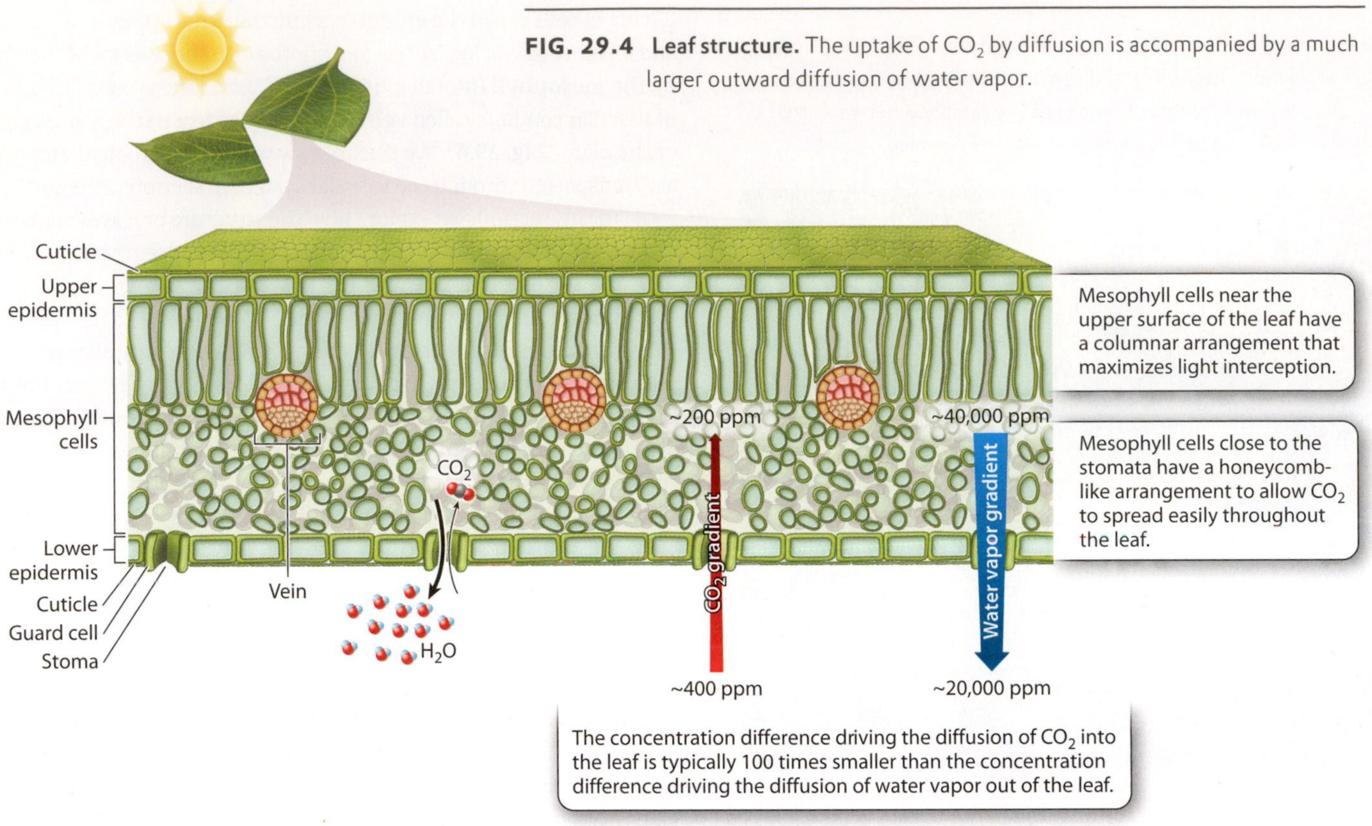

FIG. 29.4 Leaf structure. The uptake of CO_2 by diffusion is accompanied by a much larger outward diffusion of water vapor.

would quickly run out of CO_2 and come to a halt. But they're not sealed off—the leaf's air spaces are connected to the air surrounding the leaf by pores in the epidermis. As photosynthesis lowers the concentration of CO_2 molecules within the leaf's air spaces relative to the concentration of CO_2 in the outside air, the difference in concentration between the inside and the outside of the leaf causes CO_2 to diffuse into the leaf, replenishing the supply of CO_2 for photosynthesis.

The ability of leaves to draw in CO_2 comes at a price: If CO_2 can diffuse into the leaf, water vapor can diffuse out (Fig. 29.4). Furthermore, water vapor diffuses out of a leaf at a much faster rate than CO_2 diffuses inward. Molecules diffuse from regions of higher concentration to regions of lower concentration, and the rate of diffusion is proportional to the difference in concentration. On a sunny summer day, the difference in water vapor concentration between the air spaces within a leaf and the air outside can be more than 100 times greater than the difference in concentration of CO_2. Add the fact that water is lighter than CO_2, and so diffuses 1.6 times faster *for the same concentration gradient,* and it becomes clear why, on a sunny summer day, several hundred water molecules are lost for every molecule of CO_2 acquired for photosynthesis.

The loss of water vapor from leaves is referred to as **transpiration,** and the rates at which plants transpire are often quite high. A sunflower leaf, for example, can transpire an amount equal to its total water content in as little as 20 minutes. To put this rate of water loss in perspective, you would have to drink about 2 liters per minute to survive a similar rate of loss. Vascular plants can sustain such high rates of water loss because they can access the only consistently available source of water on land: the soil. Water moves from the soil, through the bodies of vascular plants, and then, as water vapor, into the atmosphere. Therefore, the challenge of keeping photosynthetic cells hydrated is met partly by the continual supply of water from the soil. As we discuss next, it is also met by limiting the rate at which water already in leaf tissues is lost to the atmosphere.

→ **Quick Check 1** Why do plants transpire?

The cuticle restricts water loss from leaves but inhibits the uptake of CO_2.

Epidermal cells secrete a waxy **cuticle** on their outer surface that limits water loss. Without a cuticle, the humidity within the internal air spaces would drop and the photosynthetic mesophyll cells would dry out. However, the cuticle prevents CO_2 from diffusing into the leaf even as it restricts water vapor from diffusing out of it.

As noted above, small pores in the epidermis allow CO_2 to diffuse into the leaf (**Fig. 29.5a**). These pores are called **stomata** (singular, **stoma**). Stomata can be numerous—there can be hundreds per square millimeter. Yet because each one is small, less than 1% to 2% of the leaf surface is actually covered by these pores. Thus, even with stomata, the epidermis is a significant barrier to the diffusion of CO_2 and water vapor.

The epidermis with its stomatal pores represents a compromise between the challenges of providing food

FIG. 29.5 **Stomata.** (a) A scanning electron microscope (SEM) image of stomata on a leaf surface; (b) guard cells of a stoma in open and closed positions. *Photo source: Dr. Jeremy Burgess/Science Source.*

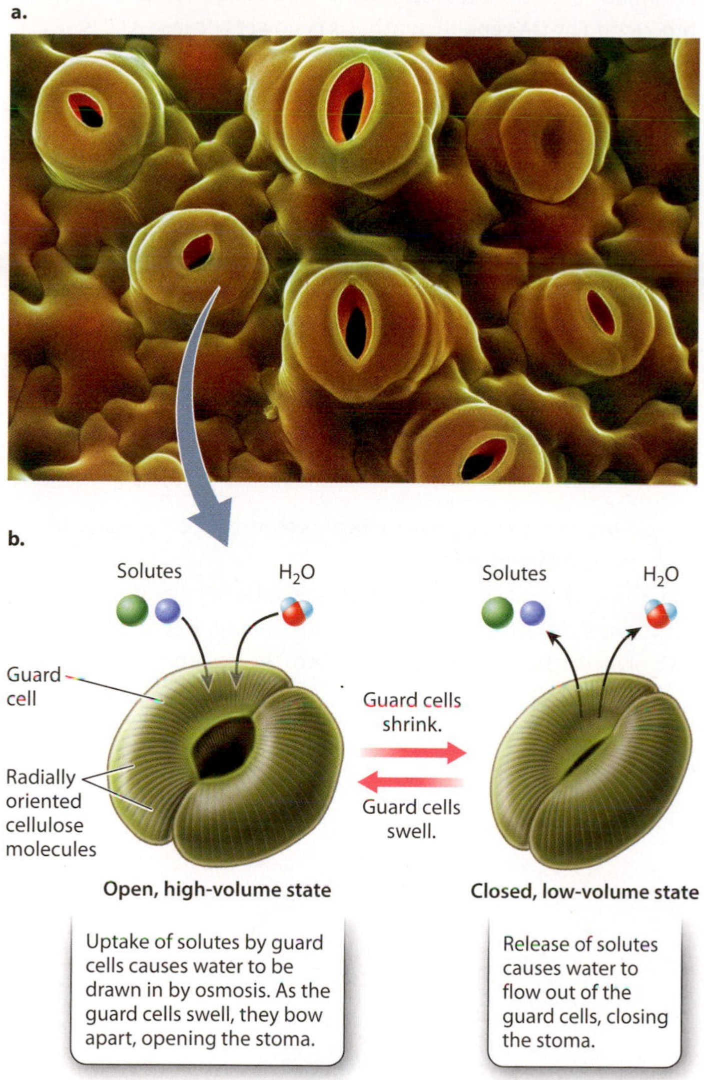

(CO_2 uptake) and preventing thirst (water loss). However, because both the dryness of the air and the wetness of the soil are variable, leaves must be able to alter the porosity of the leaf epidermis to maintain a balance between the rates of water loss to the atmosphere and water delivery from the soil. They can do this by opening and closing their stomata.

Stomata allow leaves to regulate water loss and carbon gain.

Stomata are more than just holes in the leaf epidermis; they are hydromechanical valves that can open and close. Each stoma consists of two **guard cells** surrounding a central pore. The guard cells can shrink or swell, changing the size of the pore between the guard cells (**Fig. 29.5b**). How does this valve system work?

Cellulose consists of long molecules that provide strength to plant cell walls. In guard cells, the cellulose molecules are oriented radially—that is, wrapped around the cell. This makes it easier for swelling guard cells to expand in length rather than in girth. Because the guard cells are firmly connected at their ends, this increase in length causes them to bow apart, opening the stomatal pore. Conversely, a decrease in guard cell volume causes the pore to close.

Guard cells control their volume by altering the concentration of solutes, such as potassium ions (K^+) and chloride ions (Cl^-), in their cytoplasm. ATP drives the uptake of solutes across the plasma membrane against their concentration gradient. An increase in solute concentration causes water to flow into the cell by osmosis, whereas a decrease in solute concentration causes water to flow out of the cell. Recall from Chapter 5 that osmosis is the diffusion of water molecules across a selectively permeable membrane, such as the cell's plasma membrane, from a region of higher water concentration to a region of lower water concentration. An increase in the concentration of solutes in guard cells is accompanied by a corresponding decrease in the concentration of water molecules. Thus, as guard cells add solutes, water diffuses into the cells, causing them to swell.

Stomata close by the same process, only in reverse. Solutes leave the guard cells, and as the solute concentration decreases, water diffuses out of the cells, causing them to shrink.

To function effectively, stomata must open and close in response to both CO_2 and water loss. Stomata are thus key sites for the processing of physiological information. Light stimulates stomata to open, while high levels of CO_2 inside the leaf (a signal that CO_2 is being supplied faster than photosynthesis can take it up) cause stomata to close. Guard cell volume also changes in response to signaling molecules. For example, abscisic acid, a hormone produced during drought, causes stomata to close. Thus, stomata open when the conditions for photosynthesis are favorable, and close when a water shortage endangers the hydration of the leaf.

Our own experience leads us to think of evaporation (for example, of sweat) as a means of preventing overheating. Is there any evidence that plants control the temperature of their leaves by regulating transpiration? The answer, perhaps surprisingly, is no. Although transpiration does cool leaves, plants transpire whether their leaves are too hot *or* too cold, or at their optimal temperature. When leaves are above their optimal temperature for photosynthesis, cooling through transpiration is beneficial. But when temperatures are below the optimum, transpiration results in leaf temperatures that are even less favorable. In addition, evaporative cooling is most beneficial in hot environments, yet often in these environments little water is available in the soil. For this reason, plants in arid regions tend to rely on mechanisms other than evaporative cooling, such as reflective hairs and waxes and small leaf size, to prevent their leaves from overheating.

CAM plants use nocturnal CO_2 storage to avoid water loss during the day.

In most plants, water loss and photosynthesis peak at the same time, when the sun is high overhead. What if a plant could open its stomata to capture CO_2 at night, when cool air limits rates of evaporation?

In fact, a number of plants have evolved precisely this mechanism, called **crassulacean acid metabolism,** or CAM, to help balance CO_2 gain and water loss (**Fig. 29.6**). CAM (named after the Crassulaceae family of plants, which uses this pathway) provides a system for storing CO_2 overnight by converting it into a form that will not diffuse away. The storage form of CO_2 is produced by the activity of the enzyme PEP carboxylase, which combines a dissolved form of CO_2 (bicarbonate ion, HCO_3^-) with a 3-carbon compound called phosphoenol pyruvate (PEP). The resulting product is a 4-carbon organic acid that is stored in the cell's vacuole.

When the sun comes up the next morning, the stomata close, conserving water. At the same time, the 4-carbon organic acids are retrieved from the vacuole and transferred to the chloroplasts. The 4-carbon organic acids are then decarboxylated—that is, their CO_2 is released. Because the stomata are now closed, this newly released CO_2 does not escape from the leaf. Instead, it becomes incorporated into carbohydrates by the Calvin cycle. As we saw in Chapter 8, the Calvin cycle requires a continual supply of energy in the form of ATP and NADPH produced by the light reactions of photosynthesis and so can operate only in the light. The 3-carbon PEP molecules are converted into starch and stored in the chloroplast until the sun goes down and they are again needed to capture and store CO_2 in the vacuole.

By opening stomata only at night, CAM plants greatly increase the amount of CO_2 gain per unit of water loss. Indeed, the CO_2:H_2O exchange ratio for CAM plants is in the range of 1:50, nearly 10 times higher than that of plants that open their stomata during the day. However, CAM has a drawback. The carbohydrate production by CAM photosynthesis is slower because ATP is needed to drive the uptake of organic acids into the vacuole and because only so much organic acid can accumulate overnight in the vacuole. CAM is therefore most common in habitats such as deserts, where water conservation is crucial, and among **epiphytes,** plants that grow on the branches of other plants, without contact with the soil.

Because CAM uses enzymes (for example, PEP carboxylase) and compartments (for example, vacuoles) that exist in all plant cells, it is not surprising that it has evolved multiple times. CAM occurs in all four groups of vascular plants shown in Fig. 29.1. It is most widespread in the angiosperms, occurring in 5% to 10% of species. Vanilla orchids and Spanish moss, two well-known epiphytes, as well as many cacti, are CAM plants.

C_4 plants suppress photorespiration by concentrating CO_2 in bundle-sheath cells.

Photorespiration adds another wrinkle to the challenge of acquiring CO_2 from the air. Recall from Chapter 8 that either CO_2 or O_2 can be a substrate for rubisco, the key enzyme in the

FIG. 29.6 CAM photosynthesis. CAM plants store CO_2 at night in the vacuole; leaves can then photosynthesize during the day without opening their stomata.

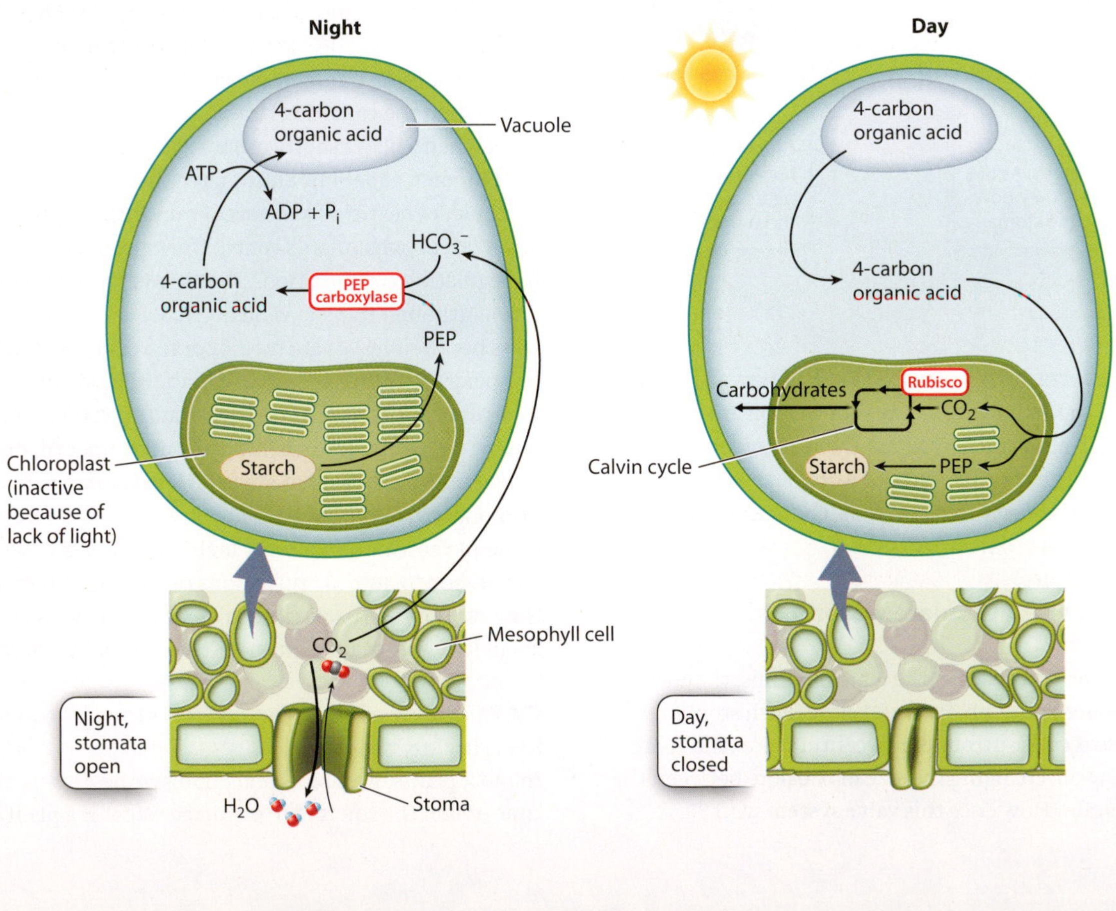

FIG. 29.7 C_4 photosynthesis. C_4 plants suppress photorespiration by concentrating CO_2 in bundle-sheath cells. *Photo source: Ken Wagner/Phototake.*

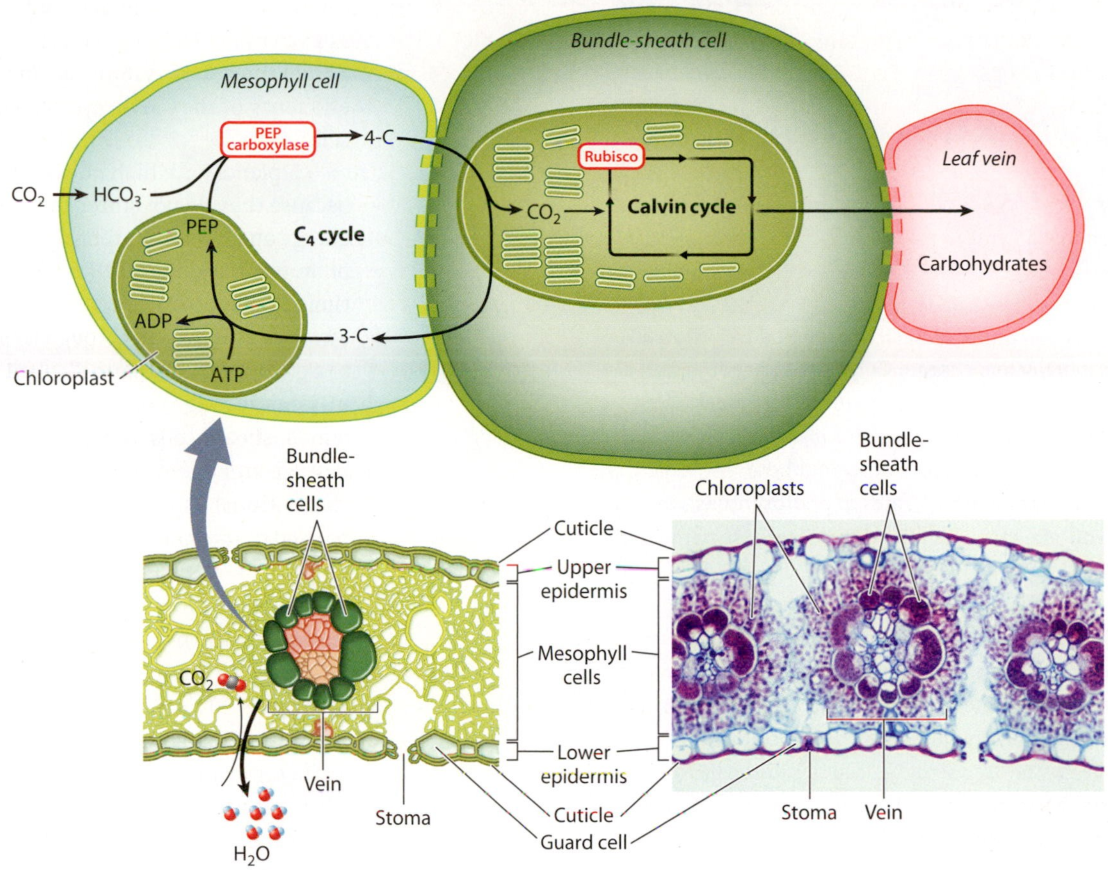

Calvin cycle. When CO_2 is the substrate, the Calvin cycle produces carbohydrates through photosynthesis. When O_2 is the substrate, there is a net loss of energy and a release of CO_2, the process called photorespiration. (Photorespiration is similar to aerobic respiration only in the sense that it uses O_2 and releases CO_2. However, plants do not gain energy; they lose it.)

Photorespiration presents a significant challenge for land plants for two reasons. The first is that air contains approximately 21% O_2 but only 0.04% CO_2. Although rubisco reacts more readily with CO_2 than with O_2, the sheer abundance of O_2 means that rubisco uses O_2 as a substrate some of the time. The second reason is that air provides much less of a thermal buffer than does water, such that organisms on land experience higher and more variable temperatures than do organisms that live in water. Temperature has a major effect on photorespiration because the selectivity of rubisco for CO_2 over O_2 is reduced as temperatures increase. At moderate leaf temperatures, O_2 is the substrate instead of CO_2 as often as 1 time out of 4. At higher temperatures, O_2 is even more likely to be the substrate for rubisco.

Some plants have evolved a way to reduce the energy and carbon losses associated with photorespiration. These are the **C_4 plants,** which suppress photorespiration by increasing the concentration of CO_2 in the immediate vicinity of rubisco. C_4 plants take their name from the fact that they, like CAM plants, use PEP carboxylase to produce 4-carbon organic acids that subsequently supply the Calvin cycle with CO_2. The Calvin cycle produces 3-carbon compounds (Chapter 8), and plants that do not use 4-carbon organic acids to supply the Calvin cycle with CO_2 are thus referred to as **C_3 plants.**

Both CAM and C_4 plants produce 4-carbon organic acids as the entry point for photosynthesis. However, in CAM plants, CO_2 capture and the Calvin cycle take place *at different times;* in C_4 plants, they take place *in different cells.*

C_4 plants initially capture CO_2 in mesophyll cells by means of PEP carboxylase, which combines a dissolved form of CO_2 (bicarbonate ion, HCO_3^-) with the 3-carbon compound PEP. This produces 4-carbon organic acids that diffuse through plasmodesmata into the **bundle sheath** (**Fig. 29.7**), a cylinder of cells that surrounds each vein. Once inside bundle-sheath cells, the 4-carbon compounds are decarboxylated, releasing CO_2 that is then incorporated into carbohydrates through the Calvin cycle (Chapter 8). The C_4 cycle is completed as the 3-carbon molecules generated during decarboxylation diffuse back to the chloroplasts in the mesophyll cells, where ATP is used to re-form PEP.

The significance of the C_4 cycle is that it operates much faster than the Calvin cycle because of the very low catalytic rate of rubisco (Chapter 8). As a result, the concentration of CO_2 within bundle-sheath cells builds up, reaching levels as much as five times

FIG. 29.8

How do we know that C_4 photosynthesis suppresses photorespiration?

BACKGROUND Studies using radioactively labeled CO_2 showed that some species initially incorporate CO_2 into 4-carbon compounds instead of the 3-carbon compounds that are the first products in the Calvin cycle. These C_4 plants also have high rates of photosynthesis. Is this a new, more efficient photosynthetic pathway? Or do C_4 plants have high rates of photosynthesis because they are able to avoid the carbon and energy losses associated with photorespiration?

HYPOTHESIS C_4 plants do not exhibit photorespiration.

METHOD "Air tests," as these experiments were first called, compared rates of photosynthesis in normal air (21% O_2) and in an experimental gas mixture in which the concentration of O_2 is only 1%. When the concentration of O_2 is low, rubisco has a low probability of using O_2 (instead of CO_2) as a substrate, and thus photorespiration does not occur.

RESULTS

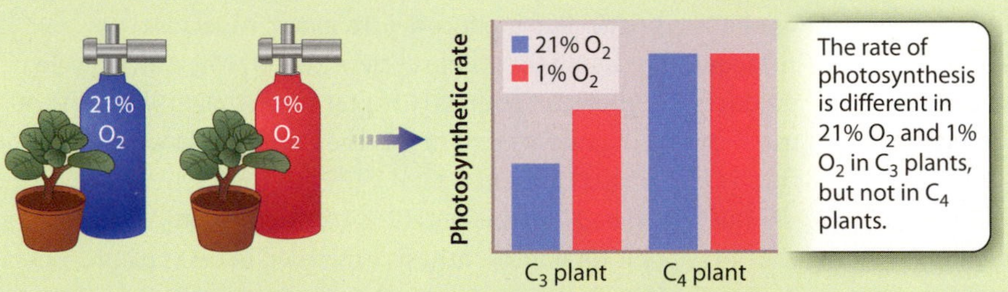

CONCLUSION Photosynthesis in C_4 plants is not affected by differences in O_4 concentration, indicating that significant photorespiration is not occurring in these plants. In contrast, the photosynthetic rate of the C_3 plants increased in the low O_4 environment, indicating that photorespiration depresses rates of photosynthesis in 21% O_4.

FOLLOW-UP WORK The higher photosynthetic efficiency of C_4 photosynthesis has prompted efforts, so far unsuccessful, to incorporate this pathway into C_3 crops such as rice.

SOURCE Bjorkman, O., and J. Berry. 1973. "High-Efficiency Photosynthesis." *Scientific American* 229:80–93.

higher than in the air surrounding the leaf. The high concentration of CO_2 in bundle-sheath cells makes it unlikely that rubisco will use O_2 as a substrate. Thus, the C_4 cycle functions like a "fuel-injection" system that increases the efficiency of the Calvin cycle in bundle sheath cells.

C_4 plants have high rates of photosynthesis because they do not suffer the losses in energy and reduced carbon associated with photorespiration (**Fig. 29.8**). At the same time, C_4 plants lose less water because they can restrict diffusion through their stomata to a greater extent than a C_3 plant while still maintaining high concentrations of CO_2 in bundle-sheath cells. However, C_4 photosynthesis has a greater energy requirement than conventional (C_3) photosynthesis, as ATP must be used to regenerate PEP in the C_4 cycle. Thus, C_4 photosynthesis confers an advantage in hot, sunny environments where rates of photorespiration and transpiration would otherwise be high. C_4 photosynthesis has evolved as many as 20 times, but is most common among tropical grasses and plants of open habitats with warm temperatures. C_4 plants include a number of important crops, including maize (corn), sugarcane, and sorghum, as well as some of the most noxious agricultural weeds.

→ **Quick Check 2** How does the formation of 4-carbon organic acids increase the efficiency of water use in both CAM and C_4 plants?

29.3 THE STEM: TRANSPORT OF WATER THROUGH XYLEM

On a summer day, a tree can transport many hundreds of liters of water from the soil to its leaves. Impressively, this feat is accomplished without any moving parts. Even more remarkably, trees and other plants transport water without any direct expenditure of energy. The structure of vascular plants allows them to use the evaporation of water from leaves to pull water from the soil.

Fig. 29.9 shows the cross section of a sunflower stem. Like a leaf, it has a surface layer of epidermal cells. This layer encloses thin-walled, undifferentiated **parenchyma** cells in the interior. Notice that the stem also contains differentiated tissues that lie in a ring near the

FIG. 29.9 Xylem and phloem. A cross section of the stem of a sunflower shows the two vascular tissues, xylem and phloem. *Photo source: Garry DeLong/Science Source.*

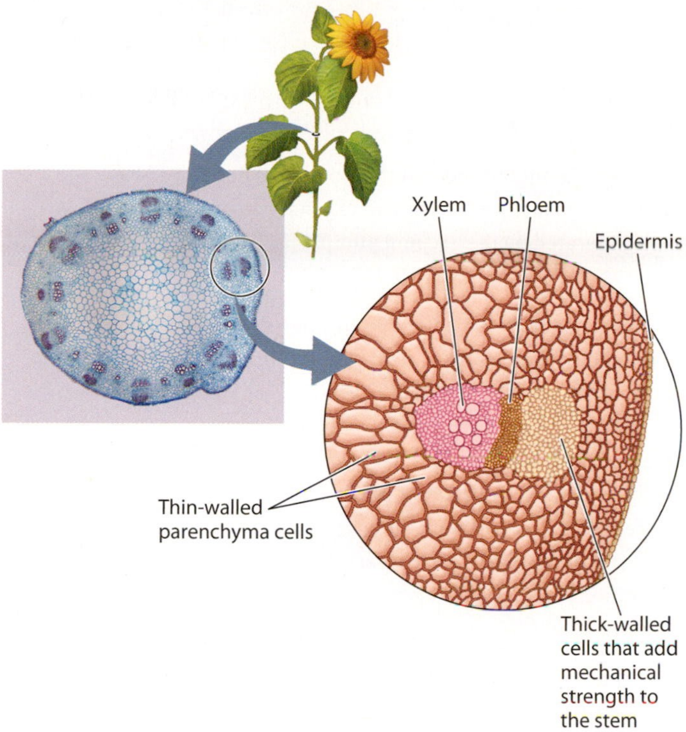

outside of the stem. These are the vascular tissues, which form a continuous pathway that extends from near the tips of the roots, through the stem, and into the network of veins within leaves. The outer tissue, called **phloem,** transports carbohydrates from leaves to the rest of the plant body. The inner tissue, called **xylem,** transports water from the roots to the leaves.

In addition, both the xylem and the phloem transport dissolved nutrients and signaling molecules, such as hormones, throughout the plant. Importantly, the xylem is the pathway by which mineral nutrients absorbed from the soil are transported to the aboveground portions of the plant. The total concentration of solutes in the xylem, however, is very low (typically less than 0.01%). We thus focus on the xylem as a means for transporting water from soil to leaves and discuss the uptake of nutrients in section 29.5.

Xylem provides a low-resistance pathway for the movement of water.

Water travels with relative ease through xylem: It would take 10 orders of magnitude greater force to drive water through living parenchyma cells than it takes to drive the same amount of water through the xylem. This ease of flow is explained by the structure of the water-transporting cells within the xylem, which have cell walls but lack any cytoplasm or membranes at maturity. Water thus flows unimpeded through the hollow centers of these cells (**Fig. 29.10**).

The walls of xylem conduits are thick and contain lignin, a chemical compound that increases mechanical strength and also prevents water from passing through the cell wall. Water enters and exits xylem conduits through **pits,** circular or ovoid regions where the lignified cell wall layer was not produced. Instead, pits contain only the thin, water-permeable cell wall that surrounded each cell as it grew. As we will see, pits play an important role in xylem because they allow the passage of water, but not air, from one conduit to another.

A xylem conduit can be formed from a single cell or from multiple cells stacked to form a hollow tube. Unicellular conduits are called **tracheids** (Fig. 29.10a), and multicellular conduits are called **vessels** (Fig. 29.10b). Because tracheids are the product of a single cell, they are typically 5–50 μm in diameter and less than 1 cm long. Vessels, which are made up of many cells called **vessel elements,** can be much wider and longer. Vessel diameters range from 10–500 μm, and lengths can be up to several meters.

Water enters a tracheid through pits, travels upward through the conduit interior, and then flows outward through other

FIG. 29.10 Xylem structure. Xylem conduits are formed by hollow cells with stiff walls and no cell content or membrane. (a) Single-celled tracheids and (b) multicellular vessels are interconnected by pits.

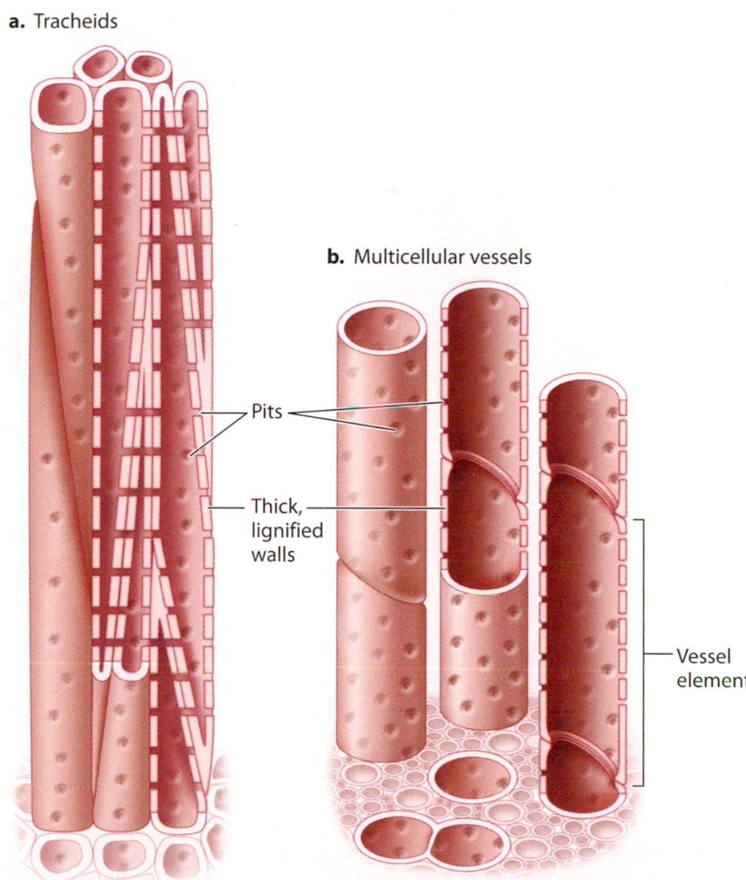

HOW DO WE KNOW?

FIG. 29.11

How large are the forces that allow leaves to pull water from the soil?

BACKGROUND The idea that water is pulled through the plant by forces generated in leaves was first suggested in 1895. However, without a way to measure these forces, there was no way to know how large they were.

HYPOTHESIS In 1912, German physiologist Otto Renner hypothesized that leaves are able to exert stronger suctions than could be generated with a vacuum pump.

EXPERIMENT Renner measured the rate at which water flowed from a reservoir into the cut tip of a branch of a transpiring plant. He next cut off about 10 cm of the branch and attached a vacuum pump in place of the transpiring plant. He then compared the flow rate through the branch when attached to the transpiring plant with the flow rate generated by the vacuum pump.

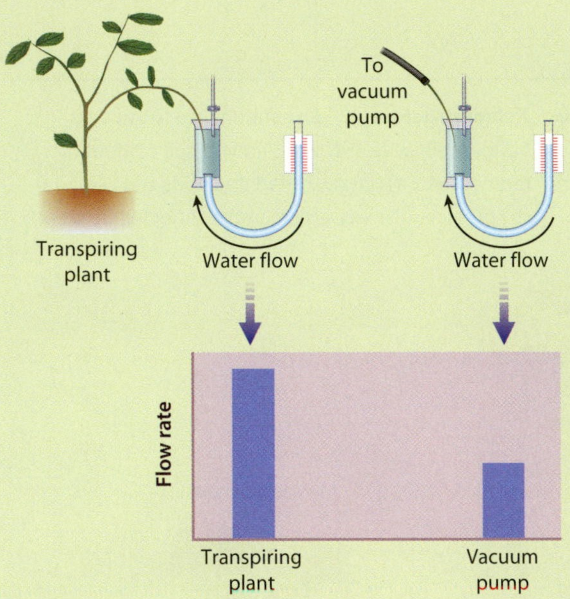

RESULTS Renner found that the flow rates generated by transpiring plants were two to nine times greater than the flow rates generated by the vacuum pump.

CONCLUSION A transpiring plant pulls water through a branch faster than a vacuum pump. Therefore, the pulling force generated by a transpiring plant must be greater than that of a vacuum pump. Because vacuum pumps create suctions by reducing the pressure in the air, the maximum suction that a vacuum pump can generate is 1 atmosphere. Thus, the maximum height that one can lift water using a vacuum pump is 10 m (33 feet). The forces that pull water through plants have no comparable limit because they are generated within the partially dehydrated cell wall. This is why plants can pull water from dry soils and why they can grow to more than 100 m in height.

FOLLOW-UP WORK In 1965, Per Fredrick Scholander developed a pressurization technique that uses reverse osmosis to measure the magnitude of the suctions exerted by leaves of intact plants. He used this technique to show that the leaves at the top of a Douglas fir tree 100 m tall exert greater pulling forces than do leaves closer to the ground.

SOURCES Renner, O. 1925. "Zum Nachweis negativer Drucke im Gefässwasser bewurzeler Holzgewachse." *Flora* 118/119:402–408.; Scholander, P. F., et al. 1965. "Sap Pressure in Vascular Plants." *Science* 148:339–346.

pits into an adjacent, partially overlapping, tracheid. Water also enters and exits a vessel through pits. Once the water is inside the vessel, little or nothing blocks the flow of water from one vessel element to the next. That is because during the development of a xylem vessel, the end walls of the vessel elements are digested away, allowing water to flow along the entire length of the vessel without having to cross any pits. At the end of a vessel, however, the water must flow through pits if it is to enter an adjacent vessel and thereby continue its journey from the soil to the leaves.

The rate at which water moves through xylem depends on both the number of conduits and their size. Water flows faster through longer conduits because the water does not need to flow so often from conduit to conduit across pits, which exert a significant resistance to flow. Like the flow through pipes, water flow through xylem conduits is also greater when the conduits are wider. The flow is proportional to the radius of the conduit raised to the fourth power, so doubling the radius increases flow sixteenfold. Because vessels are both longer and wider than tracheids, plants with vessels achieve greater rates of water transport. Tracheids are the type of xylem conduit found in lycophytes, ferns and horsetails, and most gymnosperms, whereas vessels are the principal conduit in angiosperms.

Water is pulled through xylem by an evaporative pump.

If you cut a plant's roots off under water, the leaves continue to transpire for some time. The persistence of transpiration when roots are absent demonstrates that the driving force for water transport is not generated in the roots, but instead comes from the leaves. In essence, water is pulled through xylem from above rather than being pushed from below.

The forces pulling water upward through the plant are large. Not only must these forces be able to lift water against gravity, they must also be able to pull water from the soil, which becomes increasingly difficult as the soil dries. In addition, they must be able to overcome the physical resistance associated with moving water through the plant itself. To replace water lost by transpiration with water pulled from the soil, the leaves must exert forces that are many times greater than the suctions that we can generate with a vacuum pump (**Fig. 29.11**). How do leaves exert this force?

When stomata are open, water evaporates from the walls of cells lining the intercellular air spaces of leaves. The partial dehydration of the cell walls creates a force that pulls water toward the sites of evaporation, much as water is drawn into a sponge. This force is transmitted through the xylem, beginning in the leaf veins, then down through the stem, and out through the roots to the soil (**Fig. 29.12**). Water can be pulled through xylem because of the strong hydrogen bonds that form between water molecules.

FIG. 29.12 Xylem transport from roots to leaves. Water is pulled through the plant by forces generated in the leaves, preventing them from drying out.

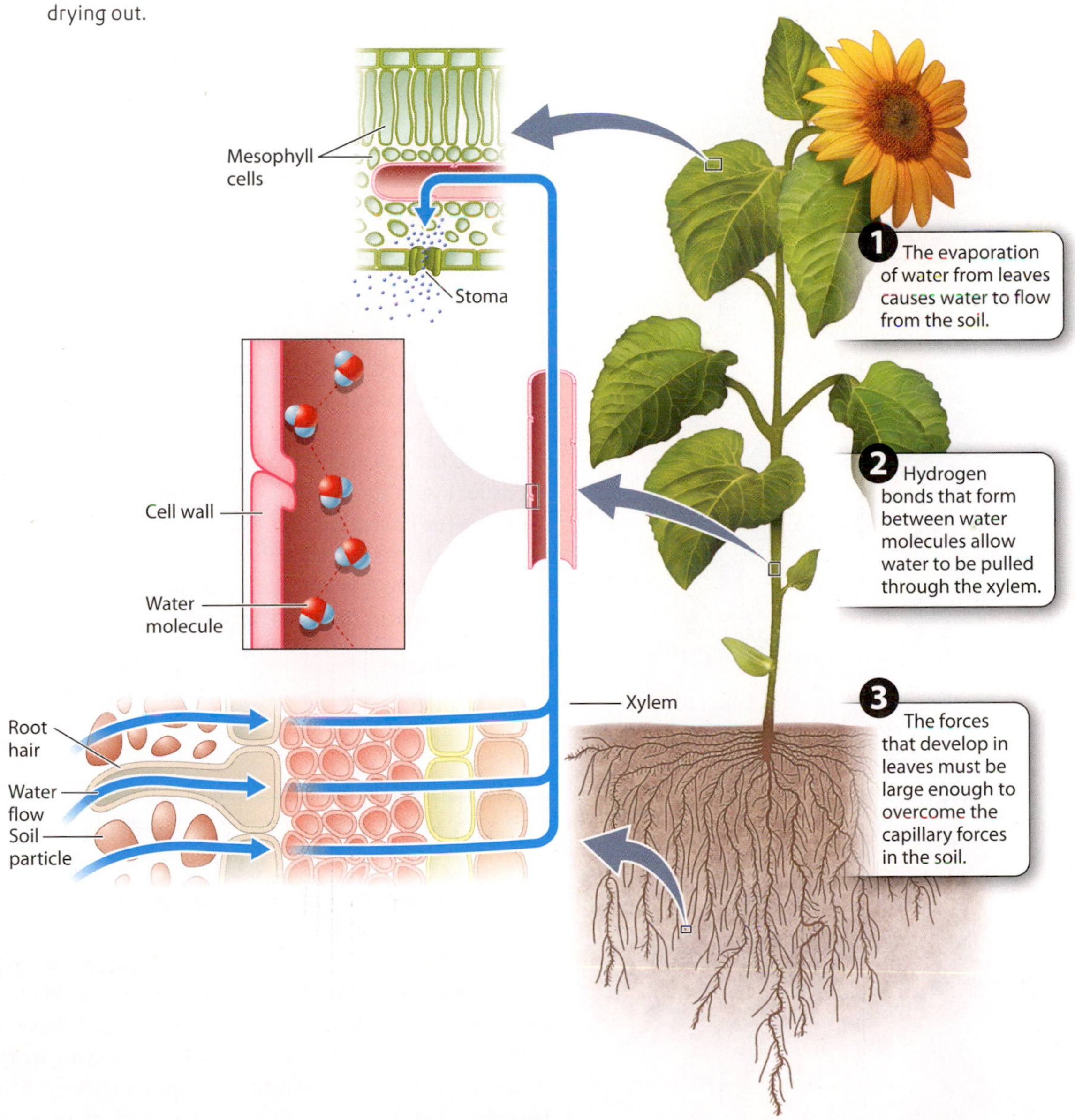

FIG. 29.13 Collapse and cavitation. The forces exerted by transpiring leaves can cause xylem conduits to collapse inward or to cavitate, in both cases blocking flow.

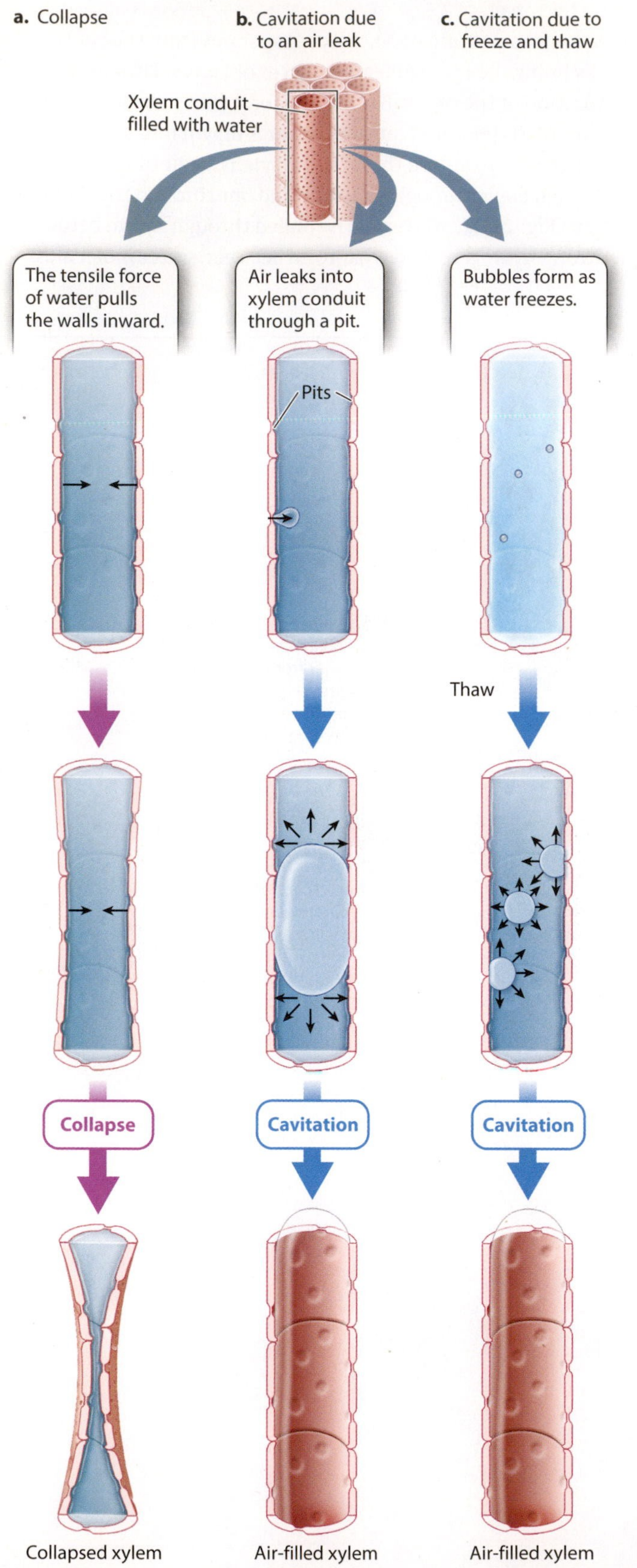

Xylem transport is at risk of conduit collapse and cavitation.

Water can be pulled through the xylem only if there is a continuous column of water that extends from the roots to the leaves. Thus, xylem conduits must be structured in such a way as to minimize two distinct risks (**Fig. 29.13**). The first is the danger of collapse. If you suck too hard on a drinking straw, it will collapse inward, blocking flow. Much the same thing can happen in xylem (Fig. 29.13a). To prevent this, xylem has thick walls that contain lignin. Although in metabolic terms lignin is more costly to produce than cellulose, lignin makes conduit walls rigid, reducing the risk of collapse.

The second danger associated with pulling water through the xylem is **cavitation,** the abrupt replacement of water by water vapor. Because cavitation disrupts the continuity of the water column, cavitated conduits can no longer transport water from the soil. Cavitation can occur in either of two ways. First, an air bubble may be pulled through a pit because of lower pressure in the water compared to the air (Fig. 29.13b). The bubble can then expand under the tensile, or pulling, force exerted by leaves. The likelihood of cavitation forming in this way increases during drought, when the tensions in the xylem are large.

The second way that cavitation occurs in xylem conduits is by gases coming out of solution during freezing (Fig. 29.13c). Gases are much less soluble in ice than in water, so as a conduit freezes, dissolved gas molecules aggregate into tiny bubbles that can cause cavitation when the conduit thaws. The wide vessels found in angiosperms are especially vulnerable to cavitation at freezing temperatures, partly explaining why boreal (that is, subarctic) forests contain few angiosperms.

The xylem is organized in ways that minimize the impact of cavitation. For example, xylem consists of many conduits in parallel, so the loss of any one conduit to cavitation does not result in a major loss of transport capacity. Similarly, as water flows from the soil to the leaves, it passes from one conduit to another, each one of finite length. The likelihood that cavitation will spread is thereby reduced because air can be pulled through pits only when the tensile forces in the xylem are large.

→ **Quick Check 3** How is the statement that water is transported through the xylem without requiring any input in energy by the plant both correct and incorrect?

29.4 THE STEM: TRANSPORT OF CARBOHYDRATES THROUGH PHLOEM

Central to the success of vascular plants is their ability to pull water from the soil. But to pull water from the soil, these plants need roots, and these roots need to be able to grow belowground. Without sunlight to power photosynthesis, growing roots depend on carbohydrates imported from aboveground. Thus, vascular plants develop a second transport tissue, the phloem, capable of moving carbohydrates efficiently across the entire length of the plant.

The phloem supports not only the growth of roots, but also the development of non-photosynthetic organs such as reproductive structures and stems.

Phloem transports carbohydrates from sources to sinks.

Whereas the direction of xylem transport is typically from the soil to the leaves, phloem moves carbohydrates from wherever they are synthesized or stored to wherever they are needed to support growth and respiration. In plants, regions that produce or store carbohydrates are referred to as **sources.** For example, leaves are sources because they produce carbohydrates by photosynthesis, and potato tubers (the part we eat) are sources because they store carbohydrates that can be supplied to the rest of the plant body when needed. **Sinks** are any portion of the plant that needs carbohydrates to fuel growth and respiration—examples are roots, young leaves, and developing fruits. Thus, the direction of phloem transport can be either up or down, depending on where the source and sink are relative to each other.

In contrast to the xylem, phloem transport takes place through living cells whose cytoplasm is rich in sugars, 10% to 30% by weight. For comparison, a can of regular Coke has a sugar content of approximately 10%. During development, the cells through which sugars will flow lose much of their intracellular structure, including the nucleus and the vacuole. At maturity, phloem transport cells retain an intact plasma membrane that encloses a modified cytoplasm containing only smooth endoplasmic reticulum and a small number of organelles, including mitochondria. Cellular functions such as protein synthesis are carried out by neighboring cells (called **companion cells** in angiosperms) to which the phloem transport cells are connected by numerous plasmodesmata.

Phloem transport cells are connected by pores, which allow sugars and other solutes to flow from one cell to another. **Phloem sap** refers to the soluble components of the cytoplasm that can pass through these pores. In angiosperms, sugar transport occurs though **sieve tubes** composed of cells that are connected by **sieve plates,** which are modified end walls with large pores (1–1.5 μm in diameter) (**Fig. 29.14**). The plasma membrane of adjacent cells is continuous through each of these pores, so each multicellular sieve tube can be considered a single cytoplasm-filled compartment.

Phloem transports carbohydrates as sucrose or larger sugars. Phloem also transports amino acids, inorganic forms of nitrogen, and ions including K^+, which are present in much lower concentrations. Finally, phloem transports informational molecules such as hormones, protein signals, and even RNA. Thus, phloem forms a multicellular highway for the movement of raw materials and signaling molecules through the entire length of the plant.

Carbohydrates are pushed through phloem by an osmotic pump.

How does phloem transport sugars from source to sink? The overall mechanism results from the buildup of sugars in sources

FIG. 29.14 **Phloem structure.** Phloem conduits are formed of living cells. (a) Sieve tubes and associated companion cells and (b) scanning electron microscope image showing sieve plates of squash (*Cucurbita maxima*). *Photo source: Michael Knoblauch.*

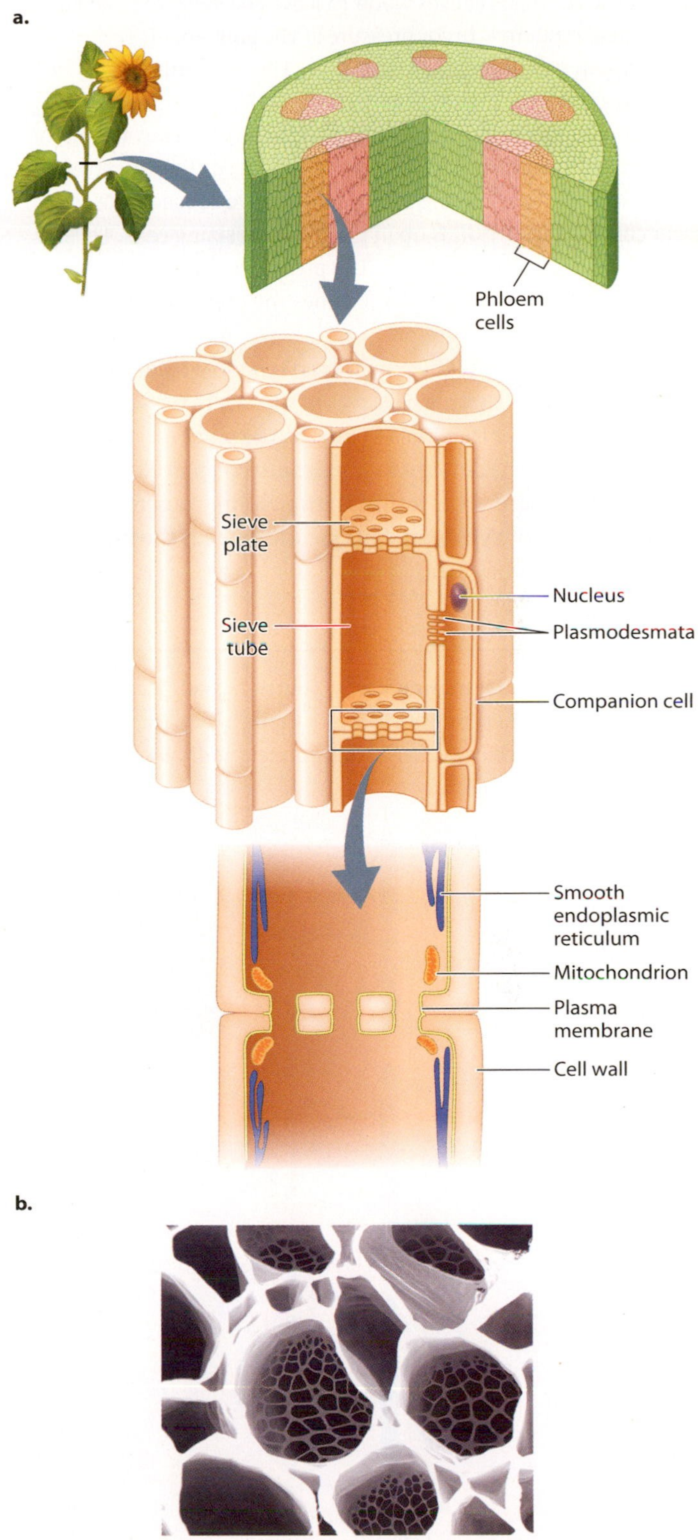

relative to a lower concentration of sugars in sinks. In sources, a high concentration of sugars in sieve tubes causes water to be drawn into the phloem by osmosis. Because the cell walls of the sieve tube resist being stretched outward, the turgor pressure (Chapter 5) at the source end increases. At sinks, sugars are transported out of the phloem into surrounding cells. This withdrawal of sugars causes water to leave the sieve tube, again by osmosis, reducing turgor pressure at the sink end. It is the difference in turgor pressure that drives the movement of phloem sap from source to sink (**Fig. 29.15**).

In some plants, sugars are actively transported into sieve tubes using energy from ATP. In others, sugars diffuse passively from the sites of photosynthesis into the sieve tubes as the sugar concentration builds up in photosynthesizing cells. These mechanisms are quite effective. The pressures that develop in sieve tubes are approximately 100 times the pressures that occur in our own circulatory system.

The water that exits the phloem can remain within the sink or it can be carried back to the leaves in the xylem (Fig. 29.15). The volume of water that moves through the phloem, however, is tiny compared to the amount that must be transported through the xylem to replace water lost by transpiration. Therefore, the number and size of xylem conduits greatly exceeds the number and size of sieve tubes.

FIG. 29.15 Phloem transport from source to sink. Sources produce carbohydrates or supply them from storage, and sinks require carbohydrates for growth and respiration.

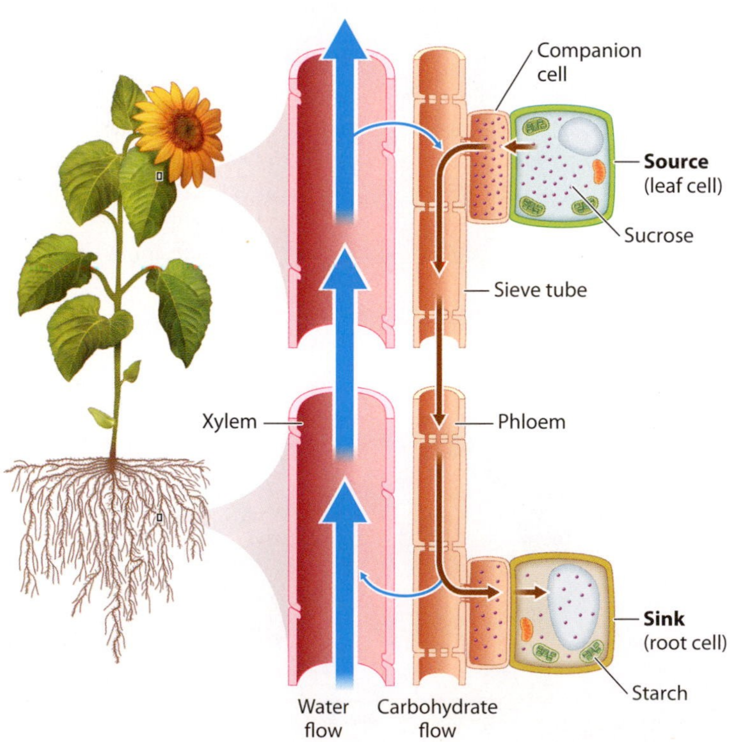

Like the xylem, the phloem is subject to risks that arise from the way the flow is generated. The contents of damaged sieve tubes can leak out, pushed by high turgor pressures in the phloem. Damage is an ever-present danger because the sugar-rich phloem is an attractive target for insects. Cell damage activates sealing mechanisms that block flow through sieve plates and thus prevent phloem sap from leaking out. In some respects, these mechanisms are comparable to the formation of blood clots in humans, except that phloem can seal itself much more rapidly, typically in less than a second.

→ **Quick Check 4** How is phloem able to transport carbohydrates from the shoot to the roots, as well as from the roots to the shoot (although not at the same time)?

Phloem feeds both the plant and the rhizosphere.

All the cells in a plant's body contain mitochondria that carry out respiration to provide a constant supply of ATP. Typically, about 50% of the carbohydrates produced by photosynthesis in one day is converted back to CO_2 by respiration within 24 hours.

Carbohydrates that are not immediately consumed in respiration can be used as raw materials for growth, or they can be stored for later use. Carbohydrates stored within roots and stems as starch, or in tubers (specialized storage organs such as potatoes), can support new growth in the spring or following a period of drought. Stored reserves can also be used to repair mechanical damage or replace leaves consumed by insects or grazing mammals.

What determines how carbohydrates become distributed within the plant? Phloem transport to reproductive organs appears to have priority over transport to stems and developing leaves, and these have priority over transport to roots. In Chapter 31, we discuss the role that hormones play in controlling the growth and development of plants and how these hormones may influence the ability of different sinks to compete successfully for resources transported in the phloem.

Phloem also supplies carbohydrates to organisms outside the plant. A fraction of the carbohydrates transported to the roots spills out into the **rhizosphere,** the soil layer that surrounds actively growing roots. This supply of carbohydrates stimulates the growth of soil microbes. As a result, the density of microbial organisms near roots is much greater than in the rest of the soil. These soil bacteria decompose soil organic matter rich in nutrients such as nitrogen and phosphorus. Thus, by releasing carbohydrates into the soil, roots are thought to acquire more nutrients from the soil.

29.5 THE ROOT: UPTAKE OF WATER AND NUTRIENTS FROM THE SOIL

As anyone who has dug a hole in a forest or a field can tell you, plants make a lot of roots (**Fig. 29.16**). Laid end to end, the roots of a single corn plant would extend over 600 km. Why do plants make such an extensive investment belowground? A major reason

FIG. 29.16 **Roots and root hairs.** Extensive branching of roots (a) and the presence of root hairs (b) create a large surface area in contact with the soil. *Sources: a. Topic Photo Agency/age fotostock; b. Photoshot/SuperStock.*

is that, with the important exception of CO_2, everything that a vascular plant needs to build and sustain its body enters through its roots.

So far, we have focused on the importance of roots for obtaining water from the soil to replace water lost to the atmosphere as stomata open to allow CO_2 to diffuse into the leaf. However, to carry out photosynthesis, plants must acquire the raw materials needed to produce the enzymes of the Calvin cycle, as well as the components of the photosynthetic electron chain. Without the substantial uptake of mineral nutrients from the soil, plants would lack the ability to photosynthesize at rapid rates. Thus, as vascular plants evolved roots to obtain water from the soil, they evolved mechanisms for acquiring nutrients as well.

Plants obtain essential mineral nutrients from the soil

On average, plants are 85% water and only 15% dry material. About 96% of the dry material, by weight, is made up of three elements: carbon, oxygen, and hydrogen. The remaining 4% is made up of **minerals,** which are elements that come from the soil. Minerals play essential roles in all metabolic and structural processes (**Table 29.1**), despite making up only a small fraction of the plant's body. A deficiency in any one of them results in a plant that grows poorly and has a low rate of photosynthesis.

TABLE 29.1 The mineral composition of plants.

ELEMENT	CONCENTRATION IN DRY MATTER (%)	FUNCTIONS
Nutrients covalently bonded with carbon compounds		
Nitrogen	1.5	Component of amino acids, nucleic acids, nucleotides, coenzymes
Phosphorus	0.2	Component of nucleic acids, nucleotides, coenzymes, phospholipids; key role in reactions involving ATP
Sulfur	0.1	Component of three amino acids, coenzyme A, and other essential organic compounds
Nutrients that remain in ionic form		
Potassium	1.0	Cofactor for many enzymes, major cation in cell osmotic balance
Calcium	0.5	Cofactor for enzymes involved in hydrolysis of ATP and phospholipids, second messenger in metabolic regulation, important structural role in cell walls
Magnesium	0.2	Enzyme cofactor, component of chlorophyll
Chlorine, manganese, sodium	each ≤0.01 %	Cofactor for enzymes involved in photosynthesis and other reactions
Nutrients involved in redox reactions		
Iron, zinc, copper, nickel, molybdenum	each ≤0.01 %	Component of enzymes that catalyze electron transfer in the electron transport chains used in cellular respiration and photosynthesis
Nutrients present in cell walls		
Silica, boron, calcium	each ≤0.01 %	Contribute to the structural integrity of cell walls and defense against herbivores

Data from Table 5.2 in Lincoln Taiz et al., *Plant Physiology and Development,* 6th ed., 2015, Sinauer Associates, Inc.; Table 8.1 in Emmanuel Epstein and Arnold J. Bloom, *Mineral Nutrition of Plants: Principles and Perspectives*, 2nd ed., 2004, Sinauer Associates, Inc.

The essential mineral nutrients can be divided up into four groups. Nitrogen, phosphorus, and sulfur are components of many organic molecules, including amino acids, nucleic acids, and phospholipids. Another group of nutrients remain in an ionic (charged) form and play important roles as enzyme cofactors and second messengers (for example, calcium) and in maintaining the osmotic balance of cells (for example, potassium). Magnesium ions are critical for photosynthesis because they occur in chlorophyll. Metal elements such as iron and zinc are primarily involved in reactions that transfer electrons from one molecule to another. The final group includes elements that contribute to the structural integrity of the cell wall, such as calcium, silica, and boron.

Nutrient uptake by roots is highly selective.

Nutrients dissolved in water from the soil enter roots and are transported from the roots to the rest of the plant in the xylem. Xylem, however, is never in direct contact with the soil. The living cells that surround the xylem allow roots to exert a high degree of selectivity over which nutrients move from the soil into the xylem. Thus, the mineral nutrient composition of the water in the xylem differs markedly from that of the water that can be easily extracted from the soil. Some elements, for example nitrogen and phosphorus, are more concentrated within the xylem, whereas others, for example sodium and calcium, are more abundant in the soil solution. Given that water travels in a continuous pathway from the soil to the leaves, how are roots able to concentrate some nutrients and exclude others?

Like leaves and stems, roots have an outer cell layer of epidermis (**Fig. 29.17**). Epidermal cells near the root tip produce slender outgrowths known as **root hairs.** Root hairs greatly increase the surface area of the root available for absorbing nutrients. Inside the epidermis is the **cortex,** composed of parenchyma cells. Depending on species, the cortex can vary in thickness from a single cell to many cells. Xylem and phloem are located at the center of the root and are surrounded by the **endodermis,** a layer of cells that acts as a gatekeeper controlling the movement of nutrients into the xylem.

Water flowing across the epidermis and the root cortex can travel through the cells, crossing plasma membranes, and can also flow through the relatively porous walls that surround each cell. At the endodermis, flow through the cell wall is blocked by the **Casparian strip,** a thin band of hydrophobic material that encircles each cell. Water can gain access to the xylem only by passing through the cytoplasm of the endodermal cells.

The Casparian strip allows the root to control what solutes enter the xylem. How does it do that? In order for solutes to pass through the plasma membrane into the cytoplasm, they must be transported by ATP-powered protein complexes in the membranes of endodermal cells. These transporters move only certain compounds through the endodermis and actively exclude others. As a result, roots can concentrate nutrients that are needed in large amounts (for example, nitrogen) in the xylem and exclude others that can damage plant cells (for example, sodium).

FIG. 29.17 Water movement through endodermal cells. As water carrying nutrients flows from the soil to the xylem of roots, it must pass through the plasma membranes of endodermal cells, which act as a selective barrier.

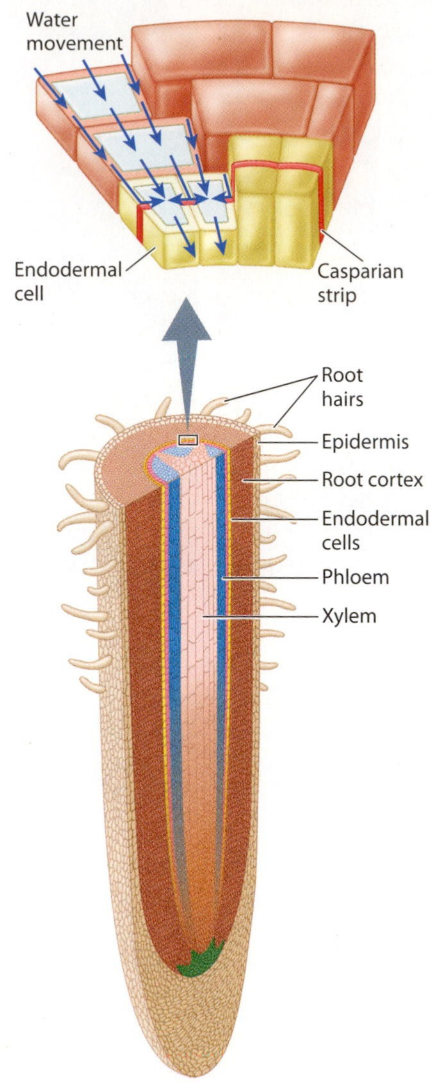

High concentrations of salts in the soil make it difficult for plants to obtain water. The reason is that to prevent salt from building up to toxic levels in leaves, plants must prevent salt from entering the xylem. As a result, salt concentrations build up in and around roots. Plants must then pull harder to overcome the osmotic effects of these salts if they are to obtain water from the soil. Poor agricultural practices can lead to salt buildup in the soil and cause crop yields to decline significantly.

Nutrient uptake requires energy.

Roots expend a great deal of energy acquiring nutrients from the soil. We have already seen that energy is required to move nutrients across the cell membranes of the endodermis on their journey to the xylem. Energy is also required to make specific nutrients in the soil accessible to the root. The mitochondria of root cells must, therefore, respire at high rates to provide energy needed for nutrient uptake. Their high respiration rates explain why plants are harmed by flooding: Waterlogged soils slow the diffusion of oxygen, limiting respiration.

To understand how roots acquire nutrients and why this requires energy, let's first look at how nutrients move through the soil. Nutrients move by diffusion through the films of water that surround individual soil particles. In addition, when transpiration rates are high, the flow of water through the soil helps to convey nutrients toward the root. The active uptake of nutrients by root cells, a process that requires an expenditure of energy in the form of ATP, decreases nutrient concentrations at the root surface, creating a concentration gradient that drives diffusion from the bulk soil to the root. However, because diffusion transports compounds efficiently only over short distances, each root is able to acquire nutrients only from a small volume of soil. This limitation explains why plants produce so many roots and so many root hairs: Both structures increase the volume of soil from which a root can obtain nutrients. It also explains why roots elongate more or less continuously, exploiting new regions of soil to obtain nutrients required for growth.

Nutrients vary dramatically in the ease and speed with which they move through the soil. For example, nitrate (NO_3^-) and sulfate (SO_4^{2-}) are extremely mobile. In contrast, interactions with clay minerals keep zinc and inorganic phosphate attached to soil particles. One way that roots gain access to highly immobile nutrients is by releasing protons or compounds that make the soil environment close to the root more acidic. This requires energy in the form of ATP, but the decrease in pH weakens the attachment of immobile nutrients to soil particles. As a result, these formerly immobile nutrients are able to move by diffusion to the root.

Soil microorganisms—both fungi and bacteria—are more adept than plants at obtaining nutrients. Fungi (Chapter 34) are particularly good at mobilizing phosphorus and at decomposing organic matter, releasing nitrogen, while some bacteria can convert gaseous nitrogen (N_2) into a chemical form that can be incorporated into proteins and nucleic acids (Chapter 26). What plants have at their disposal is a renewable source of carbohydrates. Using these carbohydrates, plants support the growth of symbiotic fungi and bacteria, in effect exchanging carbon for phosphorus or nitrogen.

Mycorrhizae enhance nutrient uptake.

Plants can increase their access to soil nutrients by forming partnerships with fungi. **Mycorrhizae** are symbioses between roots and fungi that enhance nutrient uptake: The plant provides carbohydrates produced by photosynthesis, and the fungus provides nutrients they have obtained from the soil.

Mycorrhizae are of two main types (**Fig. 29.18**). **Ectomycorrhizae** produce a thick sheath of fungal cells that surrounds the root tip. Some of the fungus extends into the interior of the root, producing a dense network of filaments that surrounds individual root cells. **Endomycorrhizae** do not form structures that are visible on the outside of the root. Instead, their networks form highly branched structures, called arbuscules, which grow inside root cells.

Why do fungi enhance the ability of roots to obtain nutrients from the soil? One reason is that fungal networks can extend more than 10 cm from the root surface, greatly increasing the absorptive surface area in contact with the soil. And because fungal filaments are so thin, they can penetrate tiny spaces between soil particles and thus access nutrients that would otherwise be unavailable to roots. Another reason is that fungi secrete enzymes that make soil

FIG. 29.18 **Mycorrhizae.** These fungus–root associations allow plants to use carbohydrates to help gain nutrients from the soil.

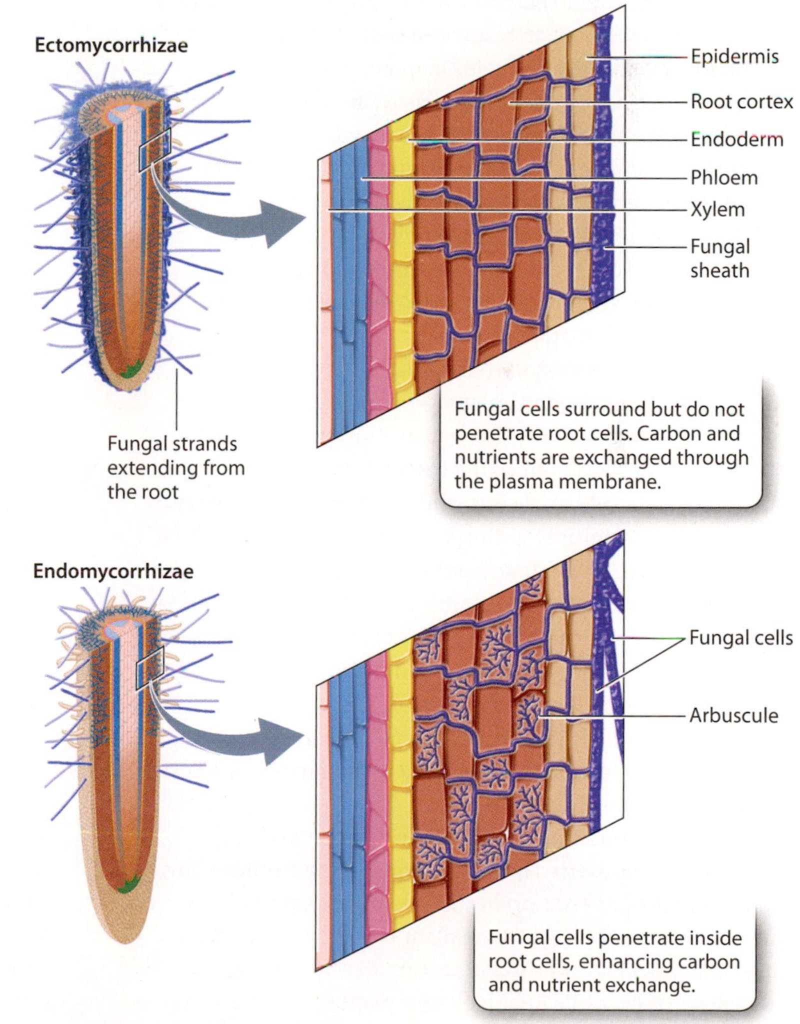

nutrients more available. Endomycorhizae are important in enhancing phosphorus uptake, while ecotomycorrhizal fungi provide host plants with access to nitrogen that would be otherwise unavailable. Both kinds of mycorrhizae may also offer other benefits to plants: For example, mycorrhizal roots may be less susceptible to infection by pathogens.

Mycorrhizal associations are costly to the plant, consuming between 4% and 20% of its total carbohydrate production. The plant has some control over whether or not the roots become infected: When nutrients are readily available, the percentage of mycorrhizal roots decreases. Nevertheless, mutualistic associations between roots and fungi are a characteristic feature of most groups of plants. Mycorrhizal roots are thus the norm rather than an exception. Fossils from more than 400 million years ago show that interactions between plants and fungi are ancient. Perhaps fungi helped plants gain a foothold on land.

Symbiotic nitrogen-fixing bacteria supply nitrogen to both plants and ecosystems.

Nitrogen is the nutrient that plants must acquire in greatest abundance from the soil, and in many ecosystems nitrogen is the nutrient that most limits growth. Nitrogen itself is not rare; 78% of Earth's atmosphere is nitrogen gas (N_2). However, as discussed in Chapter 26, plants and other eukaryotic organisms cannot use N_2 directly. Only certain bacteria and archaeons have the metabolic ability to convert N_2 into a biologically useful compound in **nitrogen fixation.** Nitrogen-fixing bacteria transform N_2 into forms such as ammonia (NH_3) that can be used to build proteins, nucleic acids, and other compounds. Some of these usable forms of nitrogen are released into the soil, where they are available to plants and other organisms.

Although many nitrogen-fixing bacteria live freely in the soil, some live in plant roots. When the level of available soil nitrogen is low, some plants form symbiotic relationships with nitrogen-fixing bacteria (**Fig. 29.19**). In response to chemical signals transmitted between the root and bacteria, bacteria multiply within the rhizosphere. At the same time, root cells begin dividing, leading to the formation of a pea-sized protrusion called a **root nodule.** Nitrogen-fixing bacteria enter the root either through root hairs or through breaks in the epidermis. They then move into the center of the nodule and begin to fix nitrogen. Even though the bacteria appear to have taken up residence within the cytoplasm of nodule cells, in fact each bacterial cell is contained within a membrane-bound vesicle produced by the host plant cell.

The plant supplies its nitrogen-fixing partners with carbohydrates transported by the phloem and carries away the products of nitrogen fixation in the xylem. As in mycorrhizal symbioses, the carbon cost to the plant is high: It is estimated that as much as 25% of the plant's total carbohydrate goes to nitrogen-fixing bacteria. Relatively few plant species have evolved symbiotic associations with nitrogen-fixing bacteria, but those that have—especially members of the legume (bean) family—provide an important way in which biologically available nitrogen enters into ecosystems. In addition, nitrogen fixation in legume roots helps to sustain fertility in many agricultural systems.

FIG. 29.19 Symbiosis between roots and nitrogen-fixing bacteria. (a) Symbiotic root nodules on alfalfa; (b) the formation of a root nodule. *Photo source: Inga Spence/Science Source.*

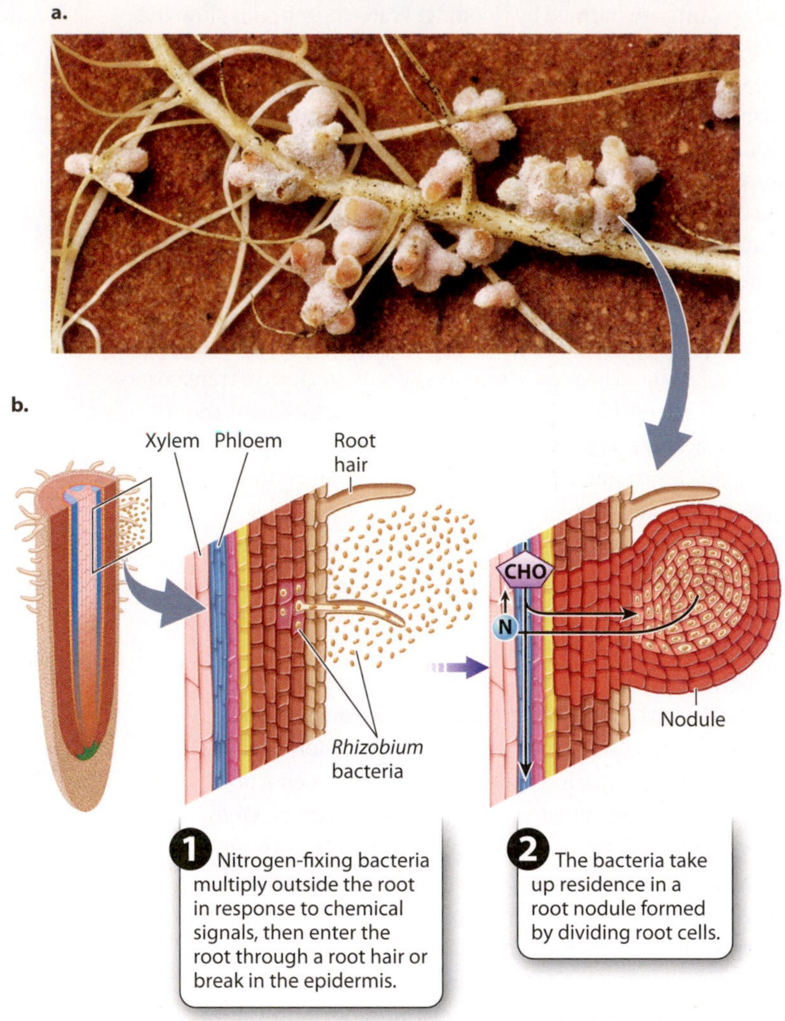

? CASE 6 AGRICULTURE: FEEDING A GROWING POPULATION

How has nitrogen availability influenced agricultural productivity?

Harvesting crops removes nutrients—including nitrogen—from an ecosystem. These nutrients must be returned if the land is to remain fertile. In the early days of agriculture, farmers left fields unplanted for long periods between crops. During these periods, nitrogen-fixing bacteria replenished soil nitrogen, either in symbiotic association with roots or as free-living bacteria in the soil.

As populations grew, fields could not be left unplanted for long periods. Organic fertilizers such as crop residues, manure, and even human feces recycled some of the lost nitrogen, but unless they are brought in from off the farm, they cannot replace all the nitrogen removed during harvest.

Crop rotation has been widely practiced throughout the history of agriculture as a means of sustaining soil fertility. Every few years, a field is planted with legumes harboring nitrogen-fixing bacteria. Typically, the legume plants are not harvested, but instead are treated as "green manure" that is allowed to decay and replenish soil nitrogen. During the eighteenth and nineteenth centuries, legume rotation contributed to the higher crop yields needed to support increasing populations. Therefore, along with the mechanization of agriculture, nitrogen-fixing symbioses contributed to the social conditions that fueled the Industrial Revolution.

By the second half of the nineteenth century, more abundant and more concentrated supplies of nitrogen fertilizers were sought. Mining of guano and sodium nitrate mineral deposits helped meet the demand for the nitrogen, but with time these natural nitrogen sources were depleted. A mechanism was needed to obtain nitrogen fertilizer from the vast storehouse of nitrogen in the atmosphere.

German chemist Fritz Haber developed an industrial method of fixing nitrogen in the years 1908–1911, and Karl Bosch subsequently adapted it for industrial production. Their method allows humans to achieve what, until then, had been accomplished only by prokaryotes. Like nitrogen fixation by bacteria, the Haber–Bosch process requires large inputs of energy; unlike nitrogen fixation by bacteria, it occurs only at high temperatures and pressures. The prospect of an essentially unlimited supply of nitrogen in a form that plants can make use of has proved irresistible. Today industrial fertilizers produced by the Haber–Bosch process account for more than 99% of fertilizer use. Industrial fixation of nitrogen is approximately equal to the entire rate of natural fixation of nitrogen, which is largely due to bacteria, although a small amount of nitrogen is fixed by lightning.

As we discuss in Chapter 49, human use of fertilizer is altering the nitrogen cycle on a massive scale, posing a threat to biodiversity and the stability of natural ecosystems. Yet abundant nitrogen fertilizers are a cornerstone of modern agriculture. The high-yielding varieties of corn, wheat, and rice introduced in the mid-twentieth century achieve their high levels of growth and grain production only when abundantly supplied with nitrogen. The fact that plants such as legumes (including beans, soybeans, alfalfa, and clover) form symbiotic relationships with nitrogen-fixing bacteria suggests a possible way around this problem. If it were possible to engineer nitrogen fixation into crops such as wheat and rice, either by inducing them to form symbiotic relationships with nitrogen-fixing bacteria or by transferring the genes that would enable them to fix nitrogen on their own, would that eliminate the need for nitrogen fertilizers? The answer is yes—but with a caveat. As we have seen, nitrogen fixation requires a large amount of energy, diverting resources that could otherwise be used to support growth and reproduction. ■

Core Concepts Summary

29.1 The evolution of land plants from aquatic ancestors introduced a major challenge for photosynthesis: acquiring carbon dioxide without drying out.

Early-branching groups of land plants, commonly grouped as bryophytes, balance CO_2 gain and water loss passively, photosynthesizing in wet conditions and tolerating desiccation in dry ones. page 600

Vascular plants maintain the hydration of their photosynthetic cells with water pulled from the soil. page 600

29.2 Leaves have a waxy cuticle and stomata, both of which are important in regulating carbon dioxide gain and water loss.

The low concentration of CO_2 in the atmosphere forces plants to expose their photosynthetic cells directly to the air. The outward diffusion of water vapor leads to a massive loss of water. page 601

The waxy cuticle on the outside of the epidermis slows rates of water loss from leaves but also slows the diffusion of CO_2 into leaves. page 602

Stomata are pores in the epidermis that open and close, allowing CO_2 to enter into the leaf and also allowing water vapor to diffuse out of the leaf. page 602

CAM plants capture CO_2 at night when evaporative rates are low. During the day, they close their stomata and use this stored CO_2 to supply the Calvin cycle, resulting in increases in the exchange ratio of CO_2 and H_2O. page 604

C_4 plants suppress photorespiration by concentrating CO_2 in bundle-sheath cells. The buildup of CO_2 in bundle-sheath cells results from the rapid production of C_4 compounds in the mesophyll cells, which then diffuse into the bundle-sheath cells and release CO_2 that can be used in the Calvin cycle. page 605

29.3 Xylem allows vascular plants to replace water evaporated from leaves with water pulled from the soil.

Water flows through xylem conduits from the soil to the leaves. page 607

Xylem is formed from cells that lose all their cell contents as they mature. page 607

Xylem conduits have thick, lignified cell walls. Water flows into xylem conduits across small thin-walled regions called pits. page 607

Xylem conduits are of two types: tracheids and vessels. Tracheids are unicellular xylem conduits; vessels are formed from many cells. page 607

The evaporation of water from leaves generates the driving force for water movement in xylem. Because of the strong hydrogen bonds between water molecules, water can be pulled from the soil and transported through xylem. page 609

Pulling water through xylem creates the risk of mechanical failure, either by the collapse inward of conduit walls or by cavitation in which a gas bubble expands to fill the entire conduit. page 610

29.4 Phloem transports carbohydrates to the non-photosynthetic portions of the plant for use in growth and respiration.

Phloem transport occurs through living cells, which lack a nucleus or vacuole at maturity. In angiosperms, these cells are connected end to end by sieve plates, forming multicellular sieve tubes. Sieve plates contain large pores that allow phloem sap to flow from one sieve tube cell to another. page 611

The active loading of solutes brings water into sieve tubes by osmosis, increasing the turgor pressure. At sites of use, the removal of solutes leads to an outflow of water and a drop in turgor pressure. page 612

The difference in turgor pressure drives the movement of phloem sap from source to sink. page 612

Carbohydrates exuded from roots provide carbon and energy sources for the rhizosphere microbial community, stimulating decomposition of organic matter and increasing nutrient availability. page 612

29.5 Roots expend energy to obtain nutrients from the soil.

The endodermis allows roots to control which materials enter the xylem. The uptake of solutes by roots is therefore highly selective. page 614

Plants exchange carbohydrates with symbiotic fungi in return for assistance obtaining nutrients from the soil. page 615

Plants supply carbohydrates to symbiotic nitrogen-fixing bacteria in exchange for available forms of nitrogen. page 616

Agricultural productivity is closely linked to nitrogen supply. page 616

Self-Assessment

1. Explain why vascular plants are better able to sustain photosynthesis than are bryophytes.
2. Explain why transpiration is best explained as a consequence of acquiring CO_2 from the atmosphere, as opposed to temperature regulation, nutrient transport, or the use of water as a substrate in photosynthesis.
3. Draw and label the features of a leaf that are necessary to maintain a well-hydrated interior while still allowing CO_2 uptake from the atmosphere.
4. Diagram how the movement of solutes results in the opening and closing of stomata.
5. Diagram how the enzyme PEP carboxylase allows CAM plants to reduce transpiration rates and C_4 plants to suppress photorespiration.
6. Contrast how transport is generated in xylem and in phloem.
7. Describe the structure of xylem conduits and sieve tubes in relation to their function.
8. Explain why phloem is necessary to produce roots.
9. List four adaptations of roots that enhance the uptake of nutrients from the soil.

Log in to LaunchPad to check your answers to the Self-Assessment questions, and to access additional learning tools.

JAPACK/amanaimagesRF/amanaimages/Corbis.

CHAPTER 30

Plant Reproduction

Finding Mates and Dispersing Young

Core Concepts

30.1 Alternation of generations evolved in plants by the addition of a diploid sporophyte generation that allows plants to disperse spores through the air.

30.2 Pollen allows the male gametophyte to be transported through the air, while resources stored in seeds support the growth and development of the embryo.

30.3 Angiosperms (flowering plants) attract and reward animal pollinators, and they provide resources for seeds only after fertilization.

30.4 Many plants also reproduce asexually.

As plants moved onto land, the challenges of carrying out photosynthesis in air were matched by the difficulties of completing their life cycle. The algal ancestors of land plants relied on water currents to carry sperm to egg and to disperse their offspring. On land, the first plants were confronted with the challenge of moving gametes and offspring through air. Air is less buoyant than water, provides a poor buffer against changes in temperature and ultraviolet radiation, and increases the risk of drying out. As plants diversified on land, structures evolved that allow gametes and offspring to survive being transported through air. For these structures to be useful, plant life cycles had to undergo radical change. The modification of land plant life cycles is a major theme in plant evolution.

The gametes and offspring of many land plants are carried passively by the movement of the air (or in a few cases, water—think coconuts). However, some plants evolved the capacity to harness animals as transport agents. In particular, the flowers and fruits of angiosperms influence animal behavior through their colors, scents, and food resources. Animals attracted by the food provided by flowers (or in a few cases, by the false promise of a mate) transport male gametes from one flower to another. Animals that eat fruits or inadvertently carry them attached to their fur are key agents of seed dispersal. As plants diversified on land, coevolution with animals emerged as a second major theme in plant evolution.

30.1 ALTERNATION OF GENERATIONS

Land plants are rooted in one place, yet must be able to disperse their offspring to minimize competition for space and resources with the parent plant. Early-diverging groups of plants evolved the capacity to disperse their offspring through the air, but they still required water for fertilization. How could these plants exist in both water and air? They couldn't—at least not at the same time. Instead, early land plants evolved a life cycle in which one generation, or phase of the life cycle, released sperm into a moist environment and the following generation dispersed offspring through the air. Thus, a pattern of alternating haploid and diploid generations, which occurs in all plants, is thought to have evolved as a means of allowing fertilization in water while enhancing the dispersal of offspring on land.

The algal sister groups of land plants have one multicellular generation in their life cycle.

Molecular sequence comparisons identify three groups of green algae as being closely related to land plants (**Fig. 30.1**). These close relatives include Coleochaetales and Charales, two groups of green algae that share a number of features with plants, including cellular structures like plasmodesmata and developmental features like apical growth (i.e, at the tip). Another group of green algae, the microscopic Zygnematales, may also be closely related to plants, but this group has gone down a distinct evolutionary path and so exhibits fewer plant-like features. The life cycles of all three groups, however, share many features. Here we use *Coleochaete* and *Chara* to gain insights into the evolution of land plant life cycles. *Coleochaete* is an inconspicuous green pincushion that grows on aquatic plants or rocks in freshwater environments, and *Chara* is an erect alga found along the margins of lakes and estuaries (**Fig. 30.2**).

In animals, the multicellular body consists of diploid ($2n$) cells. *Coleochaete* and *Chara* have a multicellular body as well, shown in Fig. 30.2, but it consists entirely of haploid ($1n$) cells. In animals,

FIG. 30.1 The phylogeny of land plants. This phylogenetic tree shows the evolutionary trajectory of the environmental setting for fertilization and dispersal.

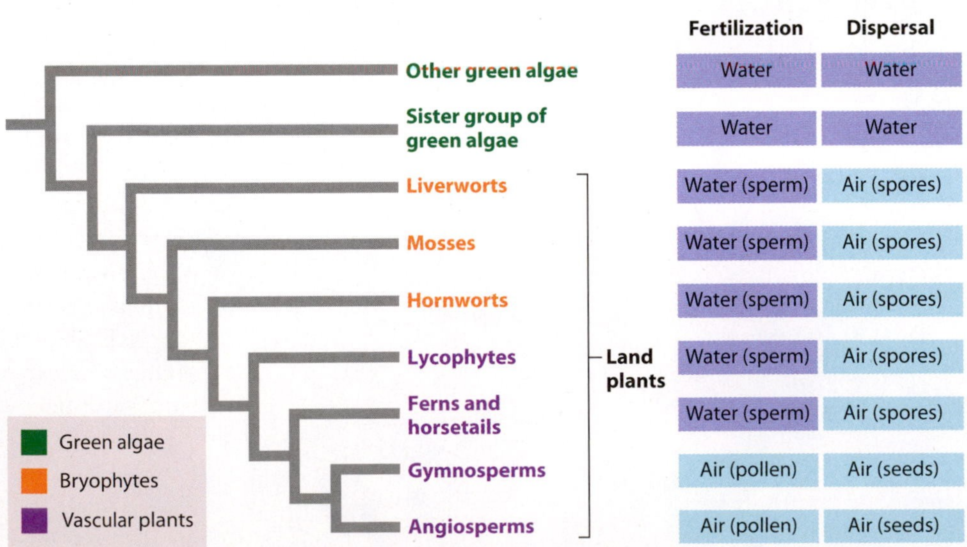

FIG. 30.2 ***Coleochaete* (left) and *Chara* (right), two green algae that are closely related to land plants.** *Sources: (left) Dr. Ralf Wagner. email<mikroskopie@dr-ralf-wagner.de>/http://www.dr-ralf-wagner.de; (right) A. Jagel/age fotostock.*

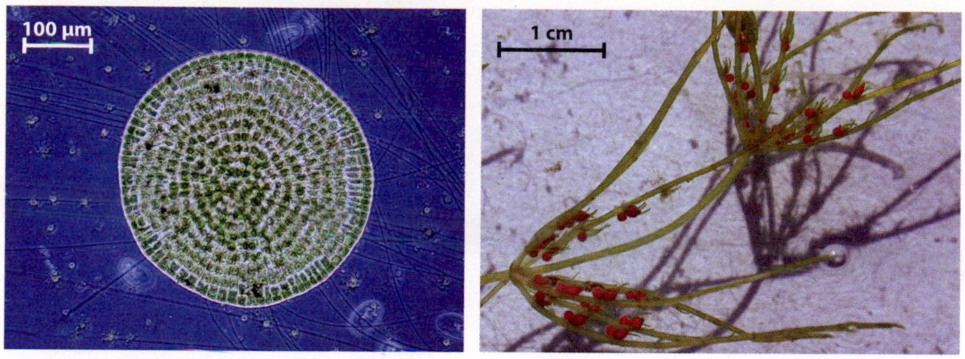

haploid gametes are formed by meiosis. In *Coleochaete* and *Chara*, haploid eggs and sperm are formed instead by mitosis. The egg is retained within the female reproductive organ and the sperm are released into the water (**Fig. 30.3**). Fertilization results when one egg and one sperm fuse to form a diploid zygote. The zygote then undergoes meiosis to produce four haploid cells. These haploid cells can develop into new multicellular (haploid) individuals, and thus they represent the beginning of the next generation. In *Coleochaete*, the haploid products of meiosis disperse by swimming through the water, propelled by flagella, before settling in a new location. In *Chara*, the zygotes are carried by water currents until they settle and undergo meiosis.

In *Chara* and *Coleochaete*, both fertilization and dispersal take place in water. Living along the margins of lakes and streams, however, *Chara* and *Coleochaete* are occasionally left high and dry by drought. Their photosynthetic bodies shrivel quickly, but these algae survive because their zygotes form a protective wall that allows the cell inside to tolerate exposure to air. When water once again covers the resting zygotes, they undergo meiosis and the resulting haploid cells escape from their protective coat. This is the condition from which the more complicated life cycles of land plants evolved.

Bryophytes illustrate how the alternation of generations allows the dispersal of spores in the air.

In Chapter 29, we saw that bryophytes—liverworts, mosses, and hornworts—make up the earliest branches on the phylogenetic tree of land plants (see Fig. 30.1). They also share many reproductive traits. Here, we use *Polytrichum commune*, a moss common in forests and the edges of fields, to illustrate how movement onto land was accompanied by the evolution of two multicellular generations: one specialized for fertilization and the other for dispersal.

Like *Chara* and *Coleochaete*, *Polytrichum* has a photosynthetic body—the familiar green tufts observed on the forest floor—that is made of haploid cells and that forms gametes by mitosis. And like *Chara* and *Coleochaete*, *Polytrichum* and other bryophytes release swimming sperm into the environment. For fertilization to occur, the sperm must swim through films of water that coat moist surfaces to reach eggs retained within reproductive organs.

Their reliance on external water for fertilization helps explain why bryophytes produce their gametes near the ground, where they are most likely to encounter a continuous film of water. It also helps explain why bryophytes tend to be small, as sperm are able to swim only a couple of centimeters. Many bryophytes release their sperm only when agitated by

FIG. 30.3 **The life cycle of the green alga *Chara corallina*.** The only diploid cell is the zygote.

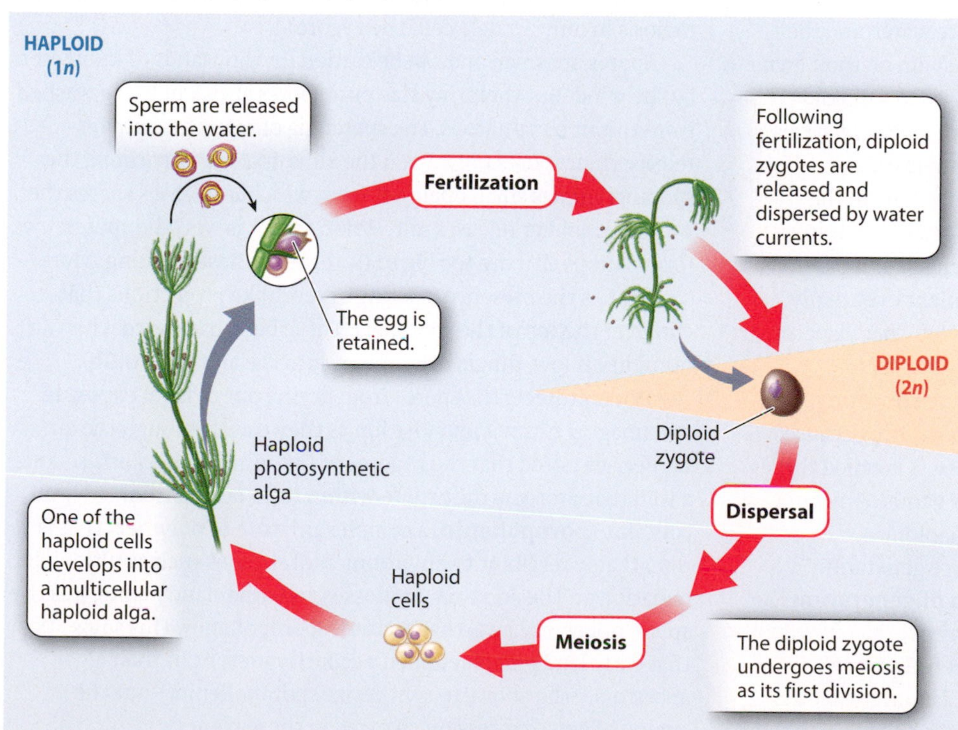

FIG. 30.4 The life cycle of the moss *Polytrichum commune*. The multicellular sporophyte generation is a major innovation of land plants. *Photo source: Malcolm Storey, 2010, www.bioimages.org.uk.*

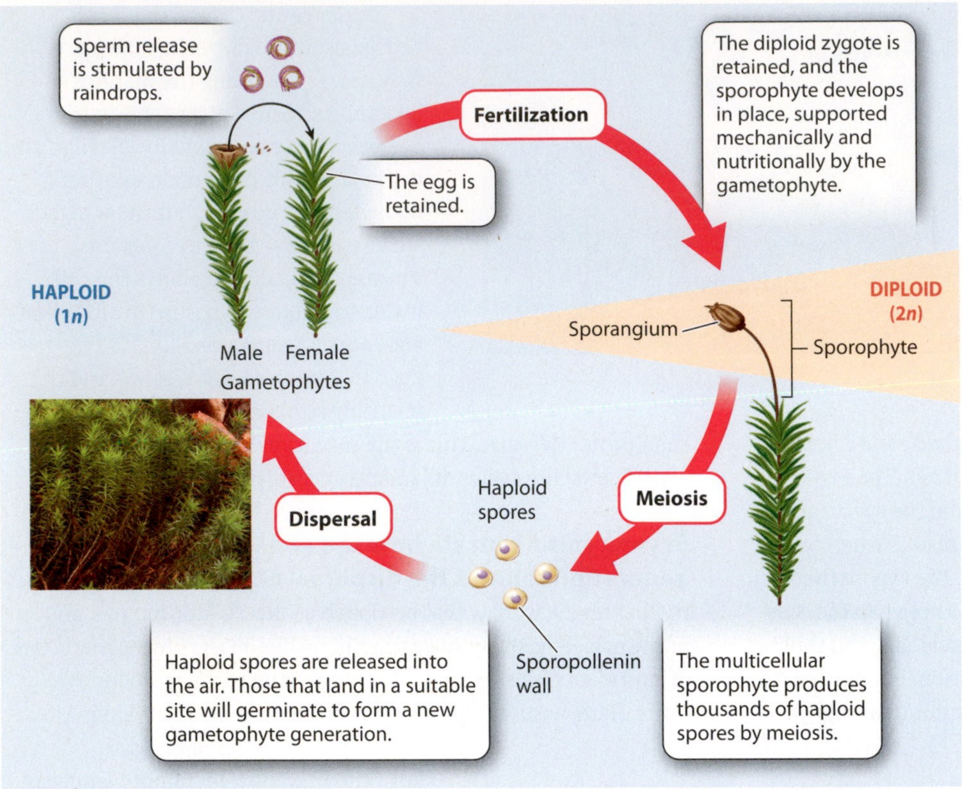

raindrops. Raindrops signal the presence of surface water, and their impact can splash sperm farther than they could swim on their own.

It is after the fusion of egg and sperm gives rise to a diploid zygote that the life cycles of plants and their green algae relatives diverge. In plants, the zygote does not undergo meiosis, and it does not disperse. Instead, the zygote is retained within the female reproductive organ, where it divides repeatedly by mitosis to produce a new multicellular plant—this one made of diploid cells (**Fig. 30.4**). Some cells within this diploid plant eventually undergo meiosis, giving rise to haploid **spores,** cells that disperse and give rise to new haploid individuals. Because the diploid multicellular plant produces spores, it is called the **sporophyte,** and this generation is called the sporophyte generation; because the haploid multicellular plant produces gametes, it is called the **gametophyte,** and that generation is called the gametophyte generation. The resulting life cycle, in which a haploid gametophyte generation and a diploid sporophyte generation follow one after the other, is called **alternation of generations,** and it describes the basic life cycle of all land plants.

Because land plants are descended from green algae with only haploid gamete-producing individuals, we can infer that the diploid sporophyte of land plants is an evolutionary novelty on this branch of the phylogenetic tree. How might this new generation have made it possible for plants to colonize land?

To answer this question, let's look at the sporophyte of *Polytrichum*. Late in the growing season, the green tufts of this moss sprout an extension—a small brown capsule at the end of a cylindrical stalk a few centimeters high (**Fig. 30.5a**). This is the sporophyte, which originated from a fertilized egg. Because the fertilized egg was retained within the female reproductive organ, the sporophyte grows directly out of the gametophyte's body. Thus, the *Polytrichum* specimen pictured in Fig. 30.5a shows both a gametophyte (green) and several sporophytes (brown). The gametophyte is photosynthetically self-sufficient, but the sporophytes must obtain water and nutrients needed for their growth from the gametophyte.

The significance of the multicellular sporophyte is that it enhances the ability of plants to disperse on land. The capsule at the top of the *Polytrichum* sporophyte is a **sporangium,** a structure in which many thousands of diploid cells undergo meiosis, producing huge numbers of haploid spores. In contrast, fertilization in *Chara* and *Coleochaete* is followed by meiosis in only a single cell (the zygote).

Spores are small and can be carried for thousands of kilometers by the wind. But their tiny size puts spores at risk of being washed from the air by raindrops. The sporangia of many bryophytes release their spores only when the air is dry. In *Polytrichum*, the sporangium has small openings from which it releases spores the way a salt shaker releases salt. When the air is wet, the spores form clumps that are too big to fit through these opening. More common is the presence of a ring of teethlike projections that curl over the top of the sporangia. These bend backward when air humidity is low, flinging the spores into the air (**Fig. 30.5b**).

What protects the spores from drying out or from exposure to damaging ultraviolet radiation as they travel through the air? Earlier, we noted that the zygotes of *Chara* and *Coleochaete* secrete a wall that protects the zygote within from desiccation. The wall contains **sporopollenin,** a complex mixture of polymers that is remarkably resistant to environmental stresses such as ultraviolet radiation and desiccation. In mosses and other land plants, it is the spore, not the zygote, that secretes sporopollenin. This suggests that early land plants retained a capacity present in their algal ancestors—the ability to synthesize sporopollenin—but the timing of gene expression changed in these plants.

FIG. 30.5 **Moss sporangia.** (a) When spores of *Polytrichum commune* are mature, a cap falls off, allowing the salt shaker–like sporangium to release the spores. (b) The sporangia of *Homalothecium sericeum* have a ring of teethlike projections that curl back as they dry, flinging spores into the air. *Sources: a. Keith Burdett/age footstock; b. Power and Syred/Science Source.*

Following dispersal, most spores germinate within a few days, but some can survive within the spore wall for months or even years. When conditions for growth are favorable, the spore wall ruptures, and the haploid cell inside can then develop into a multicellular and free-living gametophyte.

Dispersal enhances reproductive fitness in several ways.

With the evolution of the multicellular sporophyte generation, land plants were able to make many spores and disperse them across large distances on land. Dispersal is one of the most important stages in a plant's life cycle. To understand why, imagine what would happen if a plant's spores (or seeds) simply dropped straight down onto the ground below the parent plant. All the plant's offspring would then attempt to grow in a very small space that might contain enough nutrients and light for only a few to establish themselves. In contrast, offspring dispersed by wind or other means are more likely to settle in a spot where they will not be forced to compete with closely related individuals for resources.

In addition, dispersal allows offspring to avoid pathogens and parasites. Viruses and bacteria travel easily from individual to individual in a densely packed population. An individual that settles away from the parent plant is less likely to encounter a pathogen. Finally, dispersal allows offspring to colonize new habitats. The evolution of new structures to enhance dispersal on land is a major theme in plant evolution.

→ **Quick Check 1** In what ways are spores similar to gametes, and in what ways do spores and gametes differ?

Spore-dispersing vascular plants have free-living gametophytes and sporophytes.

In Chapter 29, we saw that the evolution of vascular tissues allowed plants to grow tall. Xylem and phloem are present only in the sporophyte generation because spore dispersal is enhanced by height whereas gametophytes must remain small and close to the ground to increase the chances of fertilization. A major change in the plant life cycle is that in vascular plants the sporophyte is the dominant generation both because the sporophyte is physically larger than the gametophyte and because its photosynthetic production is much higher than that of the gametophyte.

The first two groups of vascular plants—the lycophytes and the ferns and horsetails—depend on swimming sperm for fertilization and disperse by spores that are released into the air (see Fig. 30.1). In these ways, their life cycle is similar to that of the bryophytes. However, the spore-dispersing vascular plants are distinct from bryophytes in that both the gametophyte and the

FIG. 30.6 The life cycle of the bracken fern *Pteridium aquilinum*. This life cycle illustrates the alternation between a free-living gametophyte generation that remains small and a free-living vascularized sporophyte generation that grows tall. *Photo sources: (left) Biophoto Associates/Science Source; (right) Jess Merrill/Alamy.*

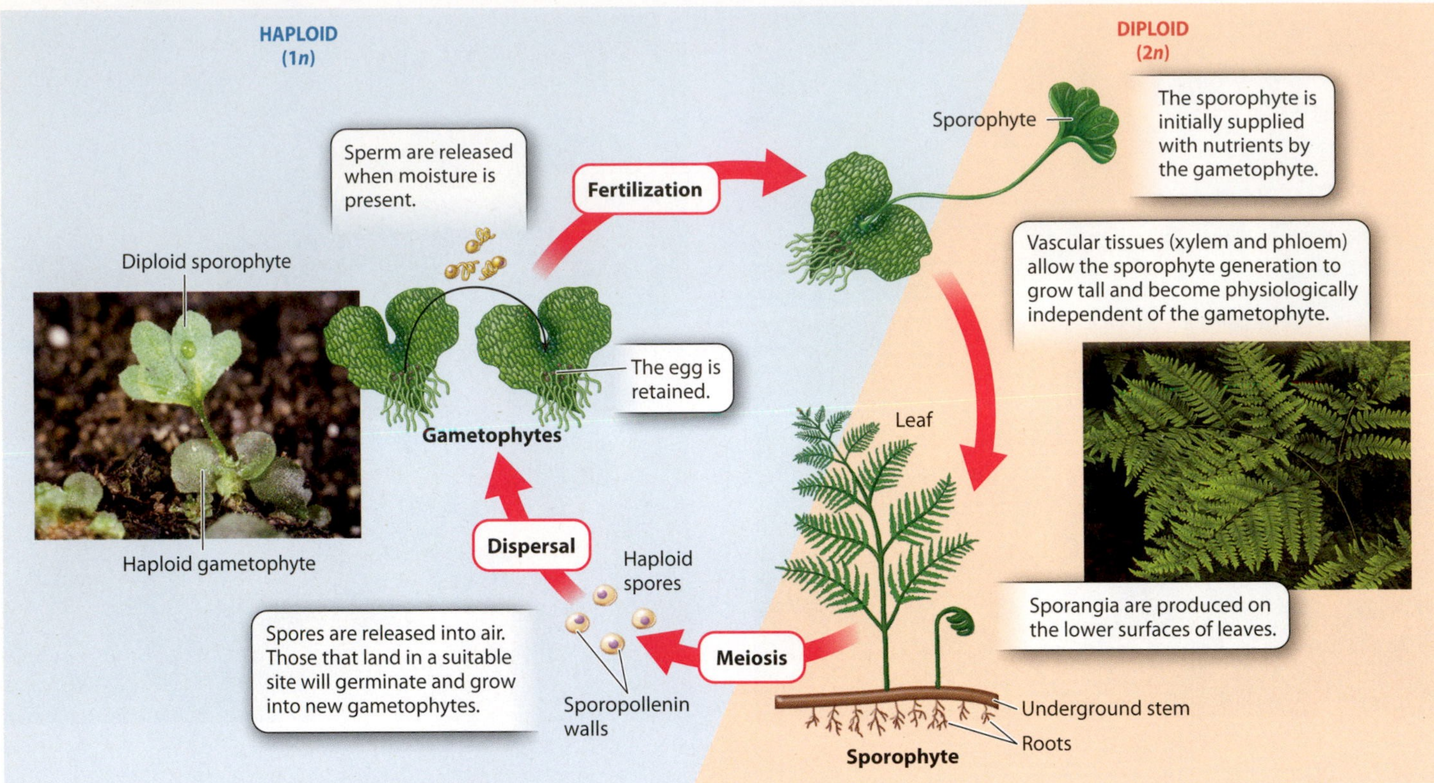

sporophyte generation are free-living, meaning that each is able to supply its own nourishment.

To illustrate this, let's examine the life cycle of the bracken fern *Pteridium aquilinum.* We begin with the largest component of the life cycle—the leafy diploid sporophyte generation (**Fig. 30.6**). If you examine a bracken leaf closely, you may find tiny brown packets along its lower margin. These are sporangia, and each contains diploid cells that undergo meiosis to generate haploid spores. The spores become covered by a thick wall containing sporopollenin, making them well suited for dispersal through the air.

As the sporangia dry out, they break open and the spores are released. Most spores travel only short distances, but a few may end up far from their parent. Bracken ferns can be found on every continent except Antarctica, as well as on isolated oceanic islands, testimony to the ability of spores to travel long distances.

Spores that land in a favorable location can germinate and grow to form the haploid gametophyte generation. You have probably never seen a fern gametophyte—or never noticed one. Typically of most ferns, the bracken gametophyte is less than 2 cm long and only one to a few cells thick (Fig. 30.6). The small size of the fern gametophyte has led many to assume that the haploid generation is a weak link in the fern life cycle. In fact, because fern gametophytes tolerate drying out, they often live longer and withstand stressful environmental conditions better than the much larger sporophytes. As discussed in Chapter 29, desiccation tolerance is important for plants that lack roots, which include both bryophytes and free-living gametophytes.

Fern gametophytes typically produce either male or female gametes. The male gametes swim to the egg on a nearby gametophyte through a film of water. The union of a male and a female gamete forms a diploid zygote, which develops in place and is supported by the gametophyte as it begins to grow. Eventually, the developing sporophyte forms leaves and roots that allow it to become a physiologically independent, diploid plant.

Ferns release swimming sperm and thus are able to reproduce only when conditions are wet. That is true of the other spore-dispersing plants (lycophytes, horsetails, and bryophytes) as well. What if a plant could eliminate the requirement for swimming sperm? Seed plants did just that.

→ **Quick Check 2** In what ways is fern reproduction similar to moss reproduction? In what ways is fern reproduction different from moss reproduction?

30.2 SEED PLANTS

Seed plants have the remarkable ability to bring male and female gametes together when conditions are dry. How is this possible? Standing beneath a pine tree on a spring day, you will see the answer to this question all around you: tiny yellow grains that cover the ground and everything else close to the tree. These are pollen grains, and they are produced by all seed plants. In fact, in seed plants we encounter two new structures: pollen and seeds. Pollen is what liberates seed plants from the need to release swimming sperm into the environment, and seeds are multicellular structures adapted for dispersing offspring. The evolution of pollen and seeds required far-reaching changes to the seed plant life cycle.

There are two groups of seed plants living today: gymnosperms, which include conifers such as pine trees, and angiosperms, which are the flowering plants (Fig. 30.1). In this section, we examine aspects of the life cycle that are common to all seed plants and use pine, a gymnosperm, to illustrate how seed plant reproduction occurs without releasing swimming sperm into the environment. In section 30.3, we examine aspects of angiosperm reproduction, such as flowers and fruits, which distinguish their life cycle from other seed plants.

The seed plant life cycle is distinguished by four major steps.

Pollen and seeds are produced only by seed plants, taking on some of the roles previously played by sperm and spores. To understand these new structures and how they relate to the life cycles we have already discussed, we consider four major steps that distinguish the life cycle of seed plants.

The first step is the formation of two types of spore, each produced in separate sporangia. One type develops into a male gametophyte and the other into a female gametophyte. Male gametophytes produce only male gametes, and female gametophytes produce only female gametes. Although this condition is not unique to seed plants (it also occurs in a small number of spore-dispersing vascular plants), it had to have evolved first or the rest of the modifications that result in the life cycle exhibited by seed plants could not have evolved.

The second step is unique to seed plants. Seed plant spores are not dispersed. Instead, gametophytes develop within their sporangium and thus remain attached to the sporophyte. The significance of this is most apparent in the development of the female gametophyte. By developing in place, the female gametophyte can draw resources from the sporophyte to produce female gametes (eggs) and to store energy and raw materials that the embryo can make use of following dispersal. Within each sporangium, only a single haploid spore develops into a female gametophyte, and this gametophyte continues growing until it fills the sporangium. The haploid female gametophyte is protected by one or more layers of sporophyte tissue that surround the sporangium wall. This entire structure, consisting of female gametophyte, sporangium and protective outer layers, is the **ovule.** Ovules, if fertilized, develop into seeds.

The male gametophyte also develops while still attached to the sporophyte. Many cells within each sporangium undergo meiosis, forming large numbers of haploid spores. Within each of these spores, a male gametophyte develops, producing a handful of cells by mitosis that all fit within the wall of the original spore. **Pollen** consists of the multicellular male gametophyte and a surrounding outer wall containing sporopollenin.

The third step begins when pollen is shed. **Pollination** is the transport of pollen, either in the air or by an animal, from the sporangium where it was produced to a location near an ovule. Because the male gametophyte produces male gametes, pollination eliminates the need to release swimming sperm into the environment. However, pollination alone is not sufficient to ensure fertilization. The male gametophyte must germinate and grow through the tissues of the ovule to deliver the male gametes to the egg. For this reason, pollen should not be thought of as flying sperm. The growth of the male gametophyte plays an essential role in bringing the male and female gametes together.

The fourth and final step is the maturation of a fertilized ovule into a **seed.** A seed consists of an embryo, stored resources, and an outer, protective coat. Seeds increase the probability that offspring will survive because their larger size allows the transport of more resources that can be used to fuel the early growth of the next generation.

Pine trees illustrate how the transport of pollen in air allows fertilization to occur in the absence of external sources of water.

To illustrate how seed plants have freed themselves from a dependence on surface moisture for fertilization, let's look at the life cycle of a pine tree (**Fig. 30.7**).

In pines, as in all vascular plants, what we think of as "the plant" is the diploid sporophyte. However, a significant part of the pine's life cycle—from the production of haploid spores to the growth of the gametophyte from a spore, to the production of gametes, and ultimately to fertilization—takes place within reproductive structures called cones. Pines develop two types of cone. **Ovule cones** typically are located on upper branches and produce spores that develop into female gametophytes, which produce female gametes. **Pollen cones** are commonly found in clusters near the tips of branches lower in the tree; they produce spores that develop into male gametophytes, which produce male gametes.

Ovule cones contain sporangia, each one surrounded by a jacket of protective tissue. Within each sporangium, a single cell undergoes meiosis, producing four haploid spores. Three of these spores degrade, leaving a single functional spore inside the sporangium. This spore is not dispersed, but instead undergoes

FIG. 30.7 **The life cycle of the loblolly pine tree, *Pinus taeda*.** *Photo source: Charles O. Cecil/Alamy.*

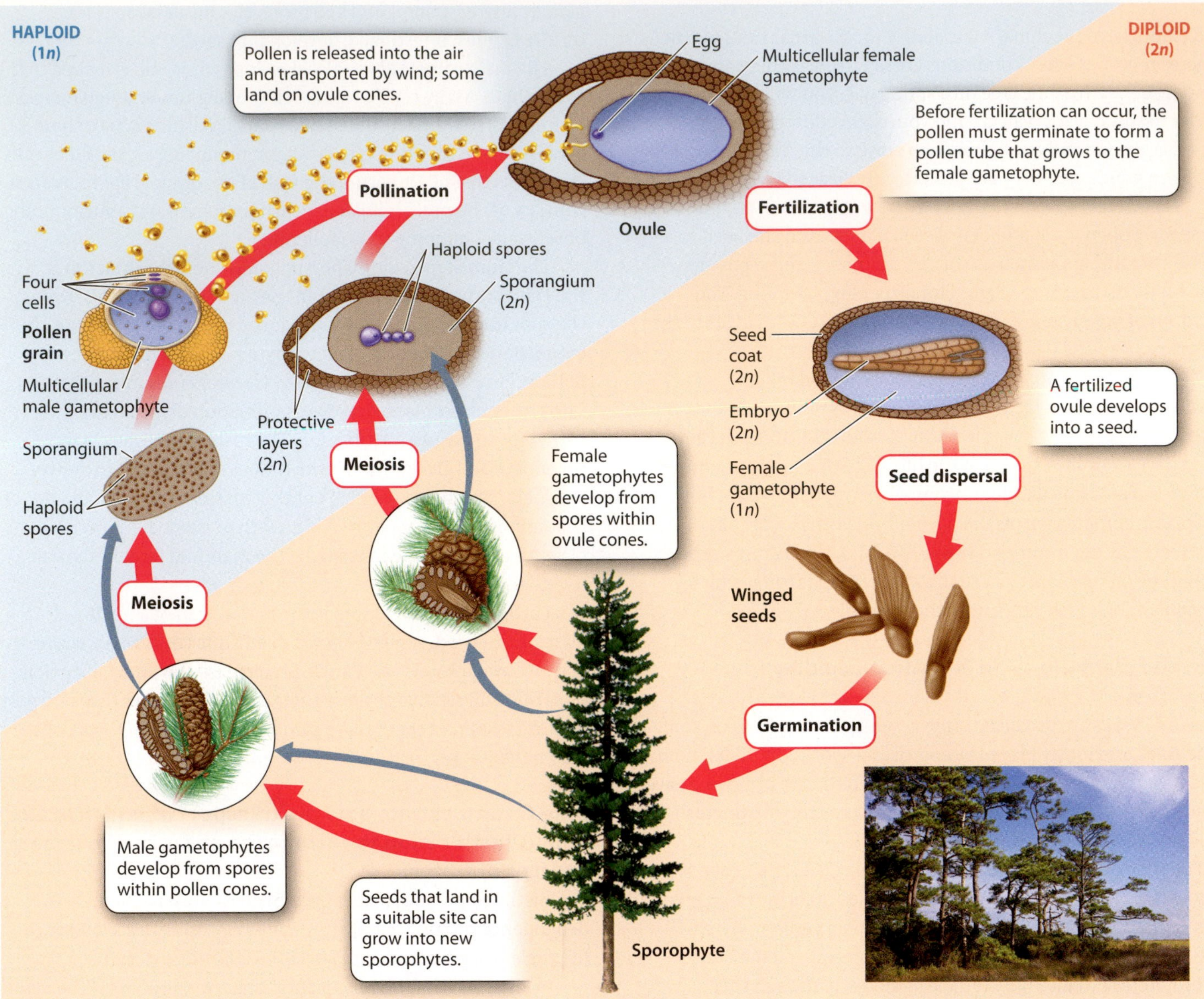

repeated mitoses to form a multicellular female gametophyte consisting of a few thousand haploid cells, one or more of which differentiate as eggs. The female gametophyte remains attached to the sporophyte as part of the ovule

Pollen cones are compact shoots with modified leaves that produce sporangia on their surface. Within the sporangia, many cells divide by meiosis to form haploid spores. The haploid spores then divide mitotically to form a multicellular male gametophyte inside the spore wall. In pine, the male gametophyte consists of only four cells at the time the pollen is shed from the parent plant.

In pine, pollen is carried to the ovule by the wind. To reach the egg, however, and penetrate the tissues of the ovule, the pollen must germinate and the male gametophyte produce a **pollen tube** that grows outward through an opening in the sporopollenin coat. In pines, pollination occurs as much as 15 months before fertilization. In fact, it is the arrival of the pollen that stimulates the development of the female gametophyte. When the female gametes are ready to be fertilized, the male gametophyte's pollen tube grows to the female gametophyte, attracted by chemical signals. Two sperm, which lack flagella, travel down the pollen tube and one of them fuses with the egg to form a zygote. Fertilization in seed plants thus takes place without the male gametes ever being exposed to the external environment.

In seed plants, the relationship between sporophyte and gametophyte has been reversed completely from that found in

FIG. 30.8 Life-cycle evolution in land plants. A major trend in the evolution of land plants is a decrease in the size and independence of the gametophyte generation and a corresponding increase in the prominence of the sporophyte generation.

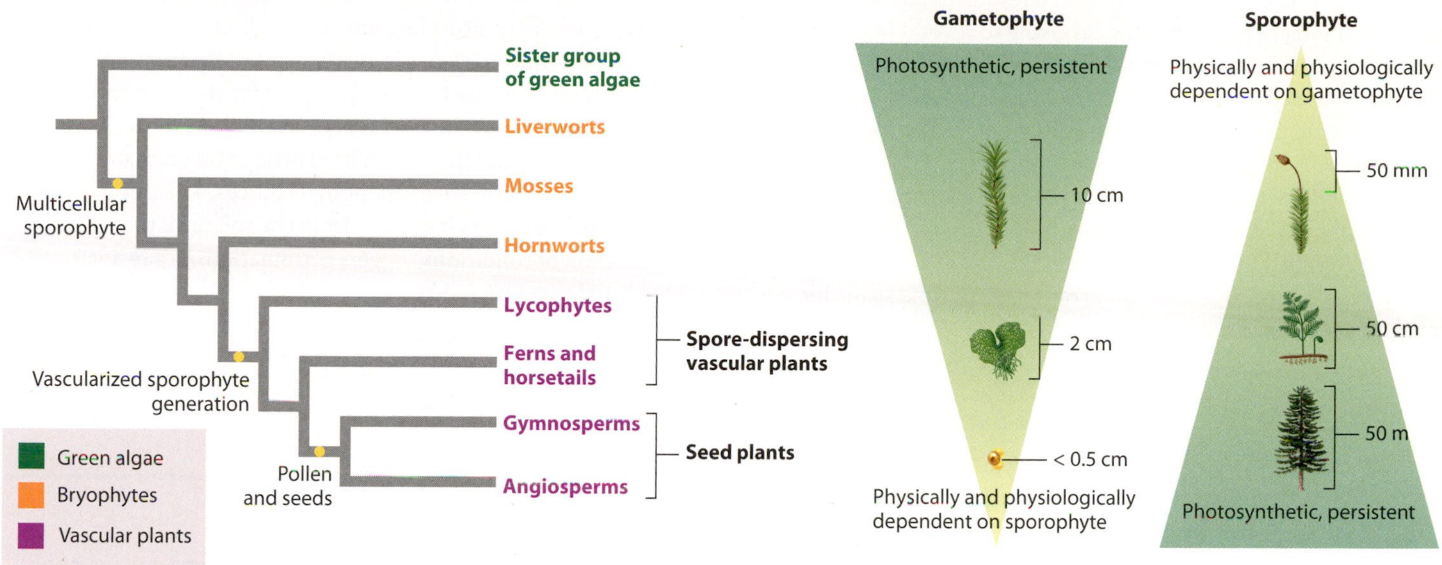

bryophytes. The gametophyte, so prominent in mosses, is reduced to a small number of cells that depend entirely on their parent sporophyte for nutrition (**Fig. 30.8**).

→ **Quick Check 3** How do the male and female gametophyte generations differ in seed plants?

Seeds enhance the establishment of the next sporophyte generation.

The union of gametes triggers the ovule to develop into a seed that can carry the new embryo away from the parent plant. Because seeds develop from fertilized ovules, they contain tissues produced by three generations (**Fig. 30.9**). On the outside is the protective **seed coat,** which is formed from tissues that surround the sporangium and thus is a product of the diploid sporophyte. At the center is the embryo, which develops from the zygote and represents the next sporophyte generation. In seed plants, the embryo develops an embryonic root at one end and an embryonic shoot with one to several embryonic leaves at the other. In gymnosperms such as pine trees, the embryo is surrounded by the haploid female gametophyte, which provides the raw materials that support the growth of the embryo. In angiosperms, the origin of this storage tissue differs, but it serves the same role of nourishing the embryo.

Compared with a unicellular spore, a seed is able to store more resources to support the growth and establishment of a new sporophyte generation. Many seeds contain high levels of protein and energy-rich oils (Chapter 2). As a result, seeds are an important food resource for animals and are the cornerstone of agricultural production. Because seeds are larger than spores, however, they are not as easily dispersed by wind. As a result, seed plants have evolved structures to enhance seed dispersal. For example, pine seeds have flattened extensions, referred to as "wings," that increase the distances over which they can be carried by a breeze. Other seed plants produce brightly colored and/or fleshy structures to attract and reward animals that may inadvertently carry seeds farther away from the parent plant than they could disperse on their own. Many plants dispersed by animals develop seed coats that allow their seeds to pass unharmed through an animal's digestive system.

FIG. 30.9 Structure of a pine seed. A pine seed contains tissues from three generations: the seed coat formed from diploid sporophyte layers, the haploid female gametophyte, and the diploid embryo, which will grow into the next sporophyte generation. *Source: Jim French.*

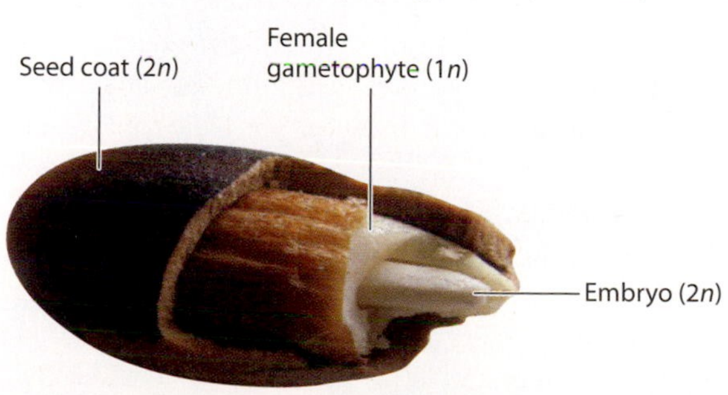

During seed maturation, most seeds lose water. And as water content falls, the seeds' metabolic activity drops to extremely low levels and the embryo ceases to grow. The combination of low metabolic activity and ample stored resources allows seeds to survive for long periods, in many cases one to several years, and sometimes much longer. Some seeds exhibit **dormancy,** meaning that they can delay germination even when conditions for growth, notably temperature and moisture, are favorable. Dormancy prevents seeds from germinating at the wrong time, for example on an uncharacteristically warm day at a time of year when the seedlings would be unable to survive. It also prevents seeds from germinating all at once, thus spreading the risk of making the transition from embryo to seedling over a larger time span and range of conditions.

FIG. 30.10 **Seed diversity.** Seeds vary tremendously in size, from (a) the dustlike orchid seeds to (b) the large seeds of the coco-de-mer. *Sources: (top) Charles Marden Fitch/age fotostock; (bottom) Kevin O'Hara/age fotostock.*

Seeds span over 11 orders of magnitude in size (**Fig. 30.10**). The 100-nanogram (ng) seeds of orchids are so small (100 ng = 0.0001 mg) that they cannot germinate successfully without obtaining nutrients from a symbiotic fungus. At the other end of the spectrum are the 30-kilogram (kg) seeds of the coco-de-mer, a palm tree found on islands in the Indian Ocean. Large seeds, being better provisioned, are well adapted for the deep shade of a forest understory. However, large seeds typically disperse shorter distances and they often have little or no dormancy. In contrast, for the same investment in resources, a plant can make many small seeds, which can persist in the soil until the right combination of conditions triggers germination. As a result, most weeds produce small seeds.

→ **Quick Check 4** Name three advantages of seeds over spores in terms of the probability that the next sporophyte generation will become successfully established.

30.3 FLOWERING PLANTS

Let's say that you want to send a postcard to a friend, but your only means of doing so is to throw the card into the air and hope the wind carries it to your friend's front door. How many cards would you have to loft in order to achieve a high probability of success? The answer, of course, is *many,* and this is precisely the reason that pines generate huge numbers of pollen grains each year. Suppose instead that you can give your card to the letter carrier whose route goes by your friend's house—your odds of success are now much greater. This analogy gives some idea of a key evolutionary innovation that appeared in the angiosperms. The "letter carriers" in angiosperm fertilization are animals, and the evolution of flowers is what allowed angiosperms to use animals to transport pollen.

Animal pollination is far more efficient than pollination by wind—that is, a higher proportion of pollen reaches an egg. Thus, a plant that relies on animal pollination can afford to produce less pollen. Flowers initially evolved to make animal pollination possible in dense tropical forests. However, as angiosperms diversified into many habitats, some species reverted to wind pollination. Today, angiosperms make use of a wide range of pollination strategies. Here, we examine how angiosperms reproduce, beginning with flowers, which when fertilized develop into fruits.

Flowers are reproductive shoots specialized for the transfer and receipt of pollen.

Angiosperms are seed plants and their life cycle is similar to that of pine trees (see Fig. 30.7), with a few important differences. One of these differences is that in angiosperms pollen and ovules are produced in a single structure—the flower (**Fig. 30.11**)— rather than in separate cones. The significance of this arrangement is that, because pollen and ovules are so close together, an animal pollinator can deliver pollen to one plant and take up pollen to carry to another in a single visit. Approximately 12% of flowering plant species produce pollen and ovules in separate flowers, but all of these species descend from ancestors whose flowers

FIG. 30.11 **Flower diversity.** Shown here are (a) a lady's-slipper orchid (*Cypripedium reginae*); (b) a magnolia (*Magnolia grandiflora*); (c) French lavender (*Lavandula stoechas*); and (d) a tropical tree (*Brownea grandiceps*). *Sources: a. Barrett & MacKay/age fotostock; b. Florida Images/Alamy; c.. Jonathan Buckley/age fotostock; d. Tim Lamen/Getty Images.*

produced both. Such "unisexual" flowers are often found in the approximately 20% of angiosperm species that have reverted to wind pollination.

Flowers are spectacularly diverse in size, color, scent, and form, but they all have the same basic organization: concentric whorls of floral organs (**Fig. 30.12;** Chapter 20). The outer whorls consist of **sepals** and **petals,** while the inner whorls are made up of pollen-producing **stamens** and ovule-producing **carpels.** Let's look at the floral organs in more detail, starting with the carpels at the center of the flower.

FIG. 30.12 **Flower organization.** The four whorls of organs in a flower are carpels, stamens, petals, and sepals. *Photo source: Cora Niele/Getty Images.*

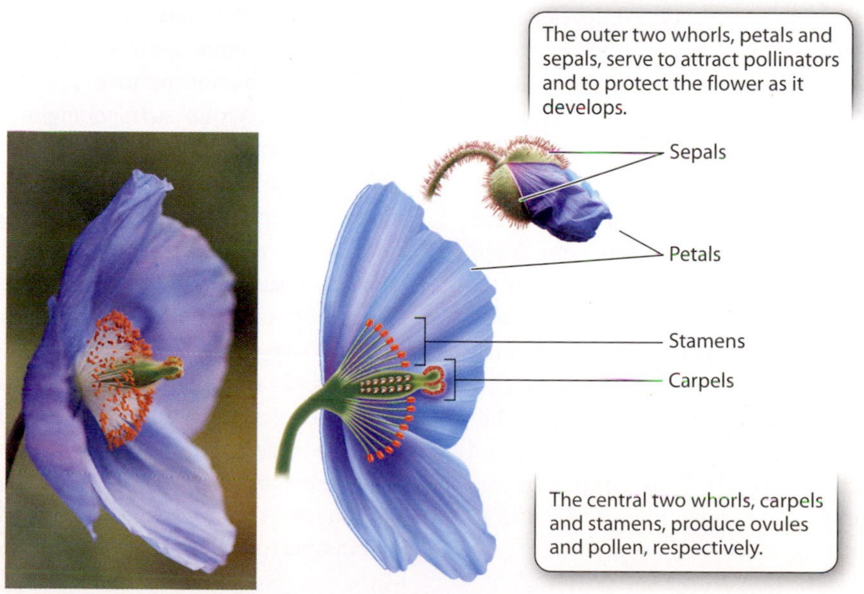

FIG. 30.13 Spore formation and gametophyte development in (a) carpels and (b) stamens.

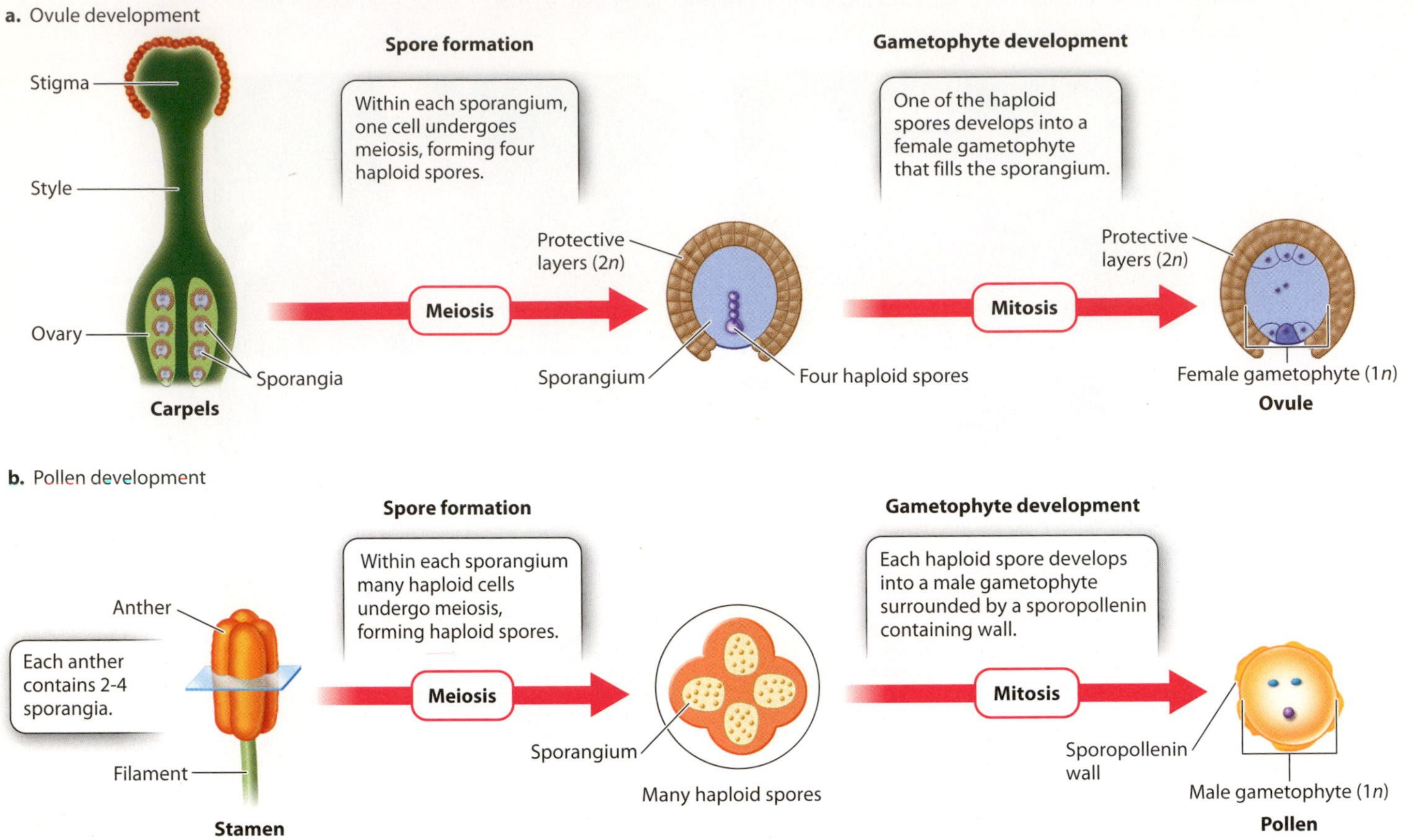

Carpels are modified leaves that have become folded over and sealed along the edges to form a hollow chamber. Sporangia are produced on the inner surface of this cavity (**Fig. 30.13a**). Within each sporangium a single cell undergoes meiosis, forming four haploid spores, one of which develops into a female gametophyte. Recall that an ovule consists of a female gametophyte surrounded by protective tissues and that ovules, when fertilized, develop into seeds. The fact that the ovules develop within the carpel is what gives rise to the name "angiosperm," which is from Greek words meaning "vessel" and "seed." In gymnosperms, the ovules are not enclosed, and their name, again from the Greek, literally means "naked seeds."

Each flower produces one or more carpels (typically 3 to 5, but sometimes more than 20), but because the carpels are often fused, there may be only a single structure at the center of the flower. At the base is the **ovary** in which one to many ovules develop. The ovary protects the ovules from being eaten or damaged by animals, but also makes it impossible for pollen to land directly on the surface of the ovule. To reach the ovules, pollen must land on the **stigma,** a sticky or feathery surface at the top of the carpel(s), and grow down through the stalklike **style.** The female gametophytes produce chemicals that guide the pollen tubes toward unfertilized ovules. Because animal pollinators can deposit many pollen grains at once, the fastest growing pollen tubes are the most likely to deliver sperm to an egg. In some plants, the style can be more than 10 cm long—corn silks, for instance, are styles—creating the opportunity for competition, and thus natural selection, between genetically distinct male gametophytes.

Immediately surrounding the carpels are the pollen-producing stamens. A stamen can have a leaflike structure bearing sporangia on its surface, but more commonly it consists of a filament that supports a structure known as the **anther,** which contains several sporangia (**Fig. 30.13b**). Within each sporangium, many cells undergo meiosis to form haploid spores. The male gametophyte then develops within the spore wall, forming pollen. In most flowers, the anther splits open, exposing the pollen grains. Once exposed, the pollen can come into contact with the body of a visiting pollinator or, in the case of a wind-pollinated species, be carried off by the wind. However, in some plants, for example tomato, small holes open at the top of anthers. To extract the pollen, bees land on the flower and vibrate at just the right frequency to shake loose the pollen inside. Orchids and milkweeds do not release individual pollen grains. Instead, they disperse their pollen all together in a package with a sticky tag that attaches to a visiting pollinator.

The outer whorls of the flower produce neither pollen nor ovules but instead contribute to reproductive success in other ways. Most

FIG. 30.14 Flowers providing food or other rewards for their pollinators. (a) A butterfly collecting pollen from flowers; (b) a hummingbird visiting a flower for nectar; (c) a bat visiting *Agave palmeri* for nectar; and (d) flies attracted to *Rhizanthes lowii* with its smell of rotting flesh. *Sources: a. Sue Kennedy/Flowerphotos/ardea.com; b. Glenn Bartley/age fotostock; c. Rolf Nussbaumer/age fotostock; d. David M. Dennis/age footstock.*

flowers have two outer whorls, of which the outermost is made up of sepals. Sepals, which are often green, encase and protect the flower during its development. In contrast, petals are frequently brightly colored and distinctively shaped. Their role is to attract and orient animal pollinators. In addition to serving as visual cues, petals in many flowers produce volatile oils. These are the source of the distinctive odors, some pleasant, some decidedly not, that many flowers use to advertise their presence to pollinators.

→ **Quick Check 5** What is the name and function of the structures in each whorl of a flower?

The diversity of floral morphology is related to modes of pollination.

The evolutionary histories of the angiosperms and their animal pollinators are closely intertwined. A rapid increase in angiosperm diversity occurred between 100 and 65 million years ago, about the same time as the diversification of bees (Hymenoptera) and butterflies (Lepidoptera). The evolution of flowers allowed animals to specialize on new food resources. At the same time, animal pollinators greatly facilitated the movement of pollen between plants within the same population.

Flowers communicate their presence through both scent and color. For pollination to be reliable, however, it must be in the interest of the animal pollinator to move repeatedly between flowers of the same species. Flowers earn fidelity from their pollinators by providing rewards, frequently in the form of food (**Fig. 30.14**). Many insects consume some of the pollen itself, while many flowers also produce nectar, a sugar-rich solution attractive to bats and birds as well as to insects.

Some species provide rewards other than food. For example, many orchids secrete chemicals that male bees need to make their own sexual pheromones. Other flowers provide an enclosed and sometimes heated chamber in which insects can aggregate. But

FIG. 30.15 ***Ophrys speculum*, a Mediterranean orchid, which mimics the shape and color of female bees.** These flowers induce males to attempt mating. Through repeated attempts on different flowers, the males transfer pollen. *Source: José Antonio Jiménez/age fotostock.*

some flowers do not provide rewards at all: Instead, they "trick" their pollinators into visiting. For example, flowers that look and smell like rotting flesh attract flies, and orchids that look like a female bee even emit similar pheromones to attract male bees (**Fig. 30.15**). Male bees find these orchids irresistible and, as they attempt to copulate with the flower, deliver and receive pollen.

Producing rewards and attractants such as nectar and petals requires resources. Thus, while animals can provide more efficient pollination than can the wind, the resources spent may be wasted if nonpollinating visitors "steal" pollen or nectar. For this reason, many flowers have mechanisms to protect their investments. For example, tubular cardinal flowers provide rewards to hummingbirds able to probe their depths, and snapdragons provide rewards only to insects that are strong enough to push apart their petals.

Many plants have evolved flowers that match rewards and attractants to the metabolic needs and sensory capabilities of their pollinators. For example, flowers pollinated by larger pollinators such as bats and birds produce copious amounts of nectar. Bat-pollinated flowers tend to be large, white or cream colored, and strongly scented, all traits appropriate to the size and nocturnal habits of their pollinators. The flowers are positioned among branches and leaves in such a way that they can be found by echolocation. In contrast, bird-pollinated flowers are often red, a color most insects cannot see, and have no scent. Because interactions between plants and pollinators affect gene flow, they can affect rates of speciation. A recent radiation in columbine occurred as the plants shifted from bumblebees to hummingbirds to hawkmoths as their primary pollinator (**Fig. 30.16**).

Angiosperms are thought to have evolved in tropical forests. Pollen cannot move freely through the air in the high density of foliage that is present year round in these forests, favoring the evolution of animal pollination. However, as angiosperms diversified and spread around the globe, some evolved wind pollination once again. The reversal to wind pollination is associated with habitats where pollinators are not reliably abundant, such as dry regions.

Wind pollination is most effective in species that are locally abundant. For example, many grasses rely on wind for pollination. Wind pollination is also common in temperate-zone trees, which produce flowers in the early spring, before leaves expand and before pollinators are active. Wind-pollinated species produce large amounts of pollen, and their petals tend to be tiny. Hay fever is an allergic reaction to pollen, virtually all of which is from wind-pollinated species.

Angiosperms have mechanisms to increase outcrossing.

When pollen is deposited onto a stigma, the pollen must germinate and form a pollen tube that grows down through the style before fertilization can occur (**Fig. 30.17**). Typically, only pollen from the same species will germinate and grow successfully,

FIG. 30.17 **Germination and growth of pollen tubes.** The pollen tubes of *Arabidopsis thaliana* can be seen here because they have been stained with a fluorescent dye. *Source: Courtesy Keun Chae.*

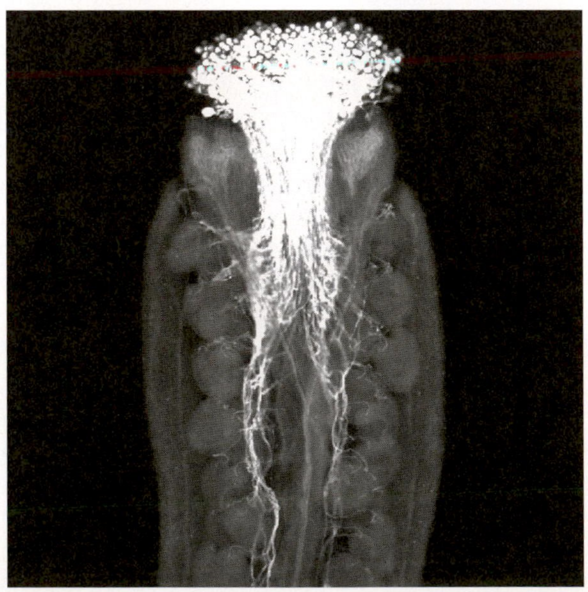

FIG. 30.16

Are pollinator shifts associated with the formation of new species?

BACKGROUND Columbines (*Aquilegia species*) are diverse and have visually striking flowers. Columbine flowers have nectar spurs, which are modified petals that form tubular outgrowths. Blue flowers with short nectar spurs are pollinated by bumblebees, red flowers with medium nectar spurs are pollinated by hummingbirds, and white or yellow flowers with long nectar spurs are pollinated by hawkmoths. The size of the nectar spur corresponds to the length of the tongue of the pollinator.

HYPOTHESIS The recent radiation of columbines corresponds with shifts from short-tongued pollinators (bumblebees) to ones with increasingly longer tongues (hummingbirds, hawkmoths).

METHOD Researchers mapped changes in flower color and nectar spur length onto a phylogenetic tree of columbines. To do this, they assigned each of 25 species examined to one of three groups based on length of nectar spur, flower color, and most common pollinator (bumblebees, hummingbirds, or hawkmoths). The group is indicated by the color circle at the tip of each branch in the phylogenetic tree in the figure. The researchers then did a statistical analysis that suggested the most likely pollinator type at each node representing an ancestor, also indicated by color circles in the figure. They used this information to infer the history of pollinator shifts during the radiation of this group.

Phylogeny of 25 North American columbine species

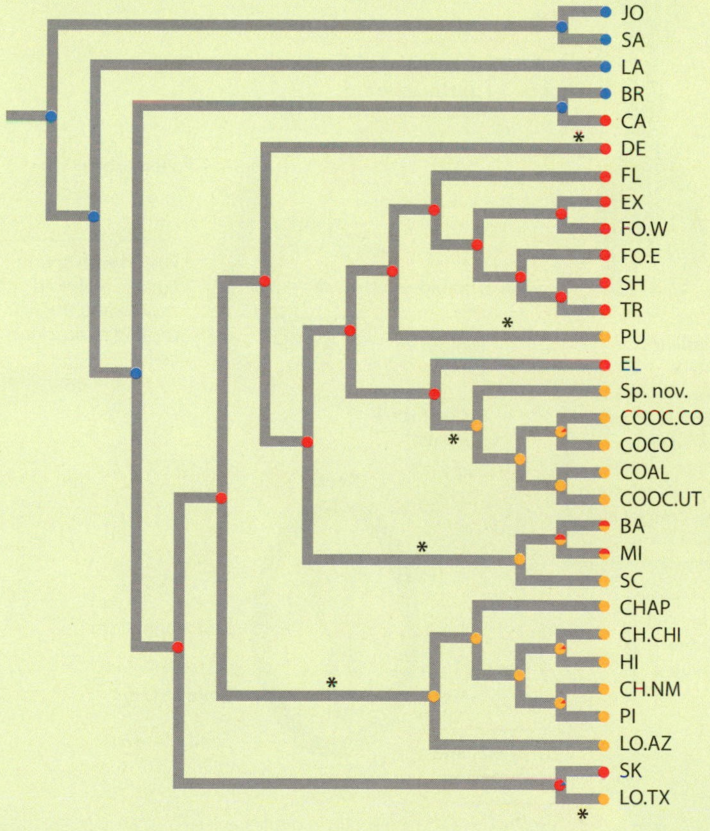

Bumblebee

Hummingbird

Hawkmoth

RESULTS Within the recent radiation of columbines, there are multiple instances of shifts from a shorter-tongued to a longer-tongued pollinator, and no reversals.

CONCLUSION The evolution of floral diversity and increasing nectar-spur length in columbines is associated with pollinator shifts. Large changes in nectar-spur length occur disproportionately at speciation events, indicating a key role in adaptive radiation.

FOLLOW-UP WORK Developmental studies show that the diversity of nectar spur length evolved as a result of changes in cell shape.

SOURCES Whittall, J. B., and S. A. Hodges. 2007. "Pollinator Shifts Drive Increasingly Long Nectar Spurs in Columbine Flowers." *Nature* 447:706–709; Puzey, J. R., et al. 2011. "Evolution of Spur Length Diversity in Aquilegia Petals Is Achieved Solely Through Cell Shape Anisotropy." *Proceedings of the Royal Society, London, Series B,* 279:1640–1645.

Photo sources: (top) Paul Oomen/Getty Images; (middle) Nancy Nehring/Getty Images; (bottom) Daniela Duncan/Getty Images

- ● Short nectar spurs, blue petals, visited by bumble bees
- ● Medium-length nectar spurs, red petals, visited by hummingbirds
- ● Long nectar spurs, white or yellow petals, visited by hawkmoths

* indicates a likely evolutionary shift in pollinator

thus preventing gametes from two different species from coming into contact.

Plants do not have the elaborate courtship rituals that many animal species use to select an appropriate mate. Nevertheless, angiosperms have a wide range of mating systems. At one extreme are **self-compatible** species in which gametes from male (pollen) and female (ovules) gametophytes produced by flowers on the same plant can form viable offspring. Self-compatible plants can reproduce even when they are physically isolated from other individuals of the same species or when pollinators are rare. Many weedy species and the majority of crops are self-compatible.

At the other extreme are species in which pollen must be transferred between different plants. Approximately half of all of angiosperm species are **self-incompatible,** meaning that pollination by the same or a closely related individual does not lead to fertilization. Self-incompatible plants must be able to recognize that a gamete comes from a closely related individual. Self-recognition is based on the proteins produced from self-incompatibility genes, or *S*-genes. If the pollen's proteins match those of the carpel, the pollen either fails to germinate or germinates but the pollen tube grows slowly and eventually stops elongating. *S*-genes may have originated as a defense against fungal invaders but subsequently evolved as a means of promoting outcrossing. Because there can be dozens of *S*-gene alleles in a population, only pollen transfers between closely related individuals are blocked. Apples are examples of self-incompatible plants, so apple orchards always contain a mixture of different varieties.

FIG. 30.18 The life cycle of an angiosperm.

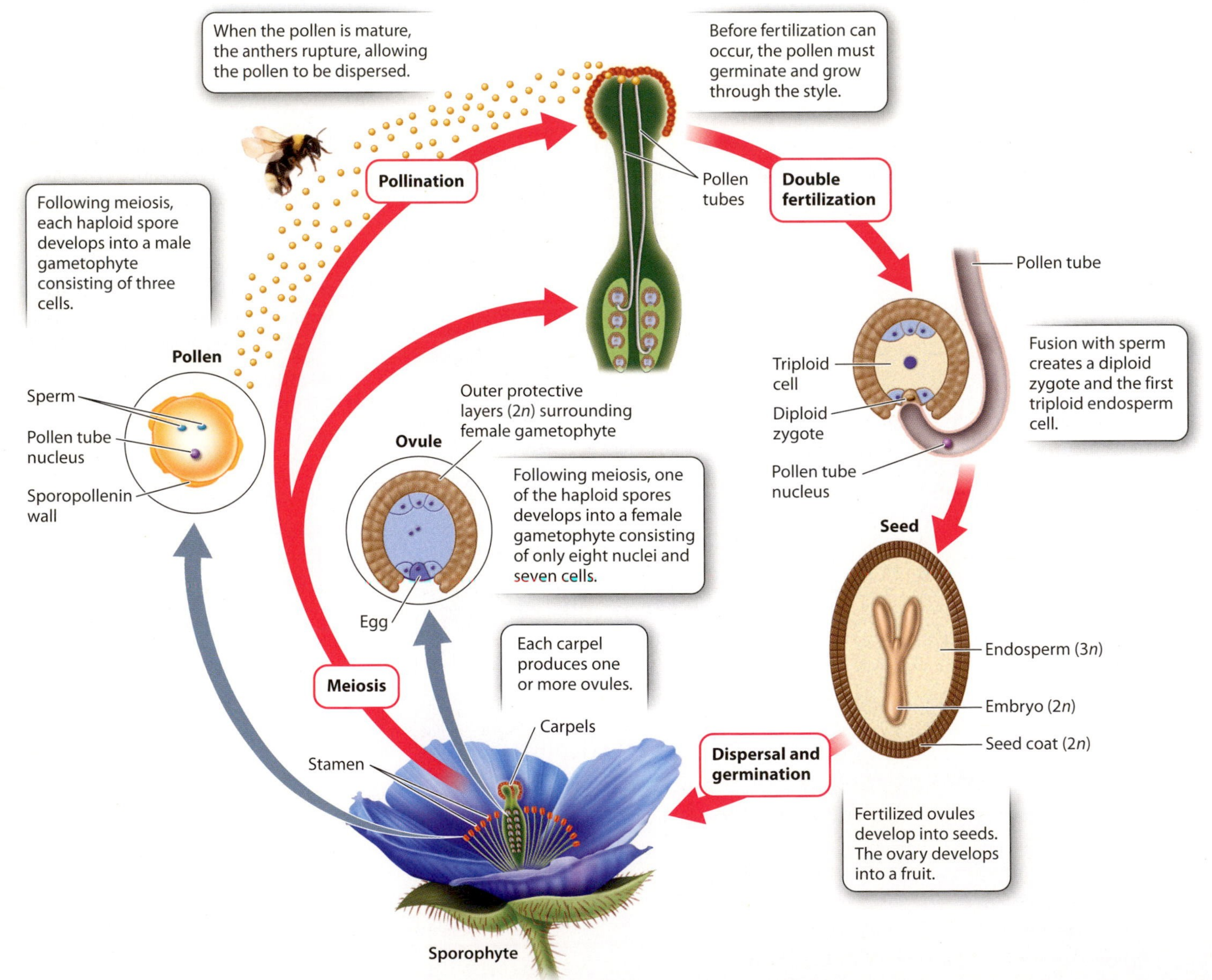

FIG. 30.19 **Double fertilization.** In double fertilization one sperm fuses with the egg to form a zygote, while a second sperm fuses with the diploid cell of the female gametophyte, leading to the formation of triploid endosperm.

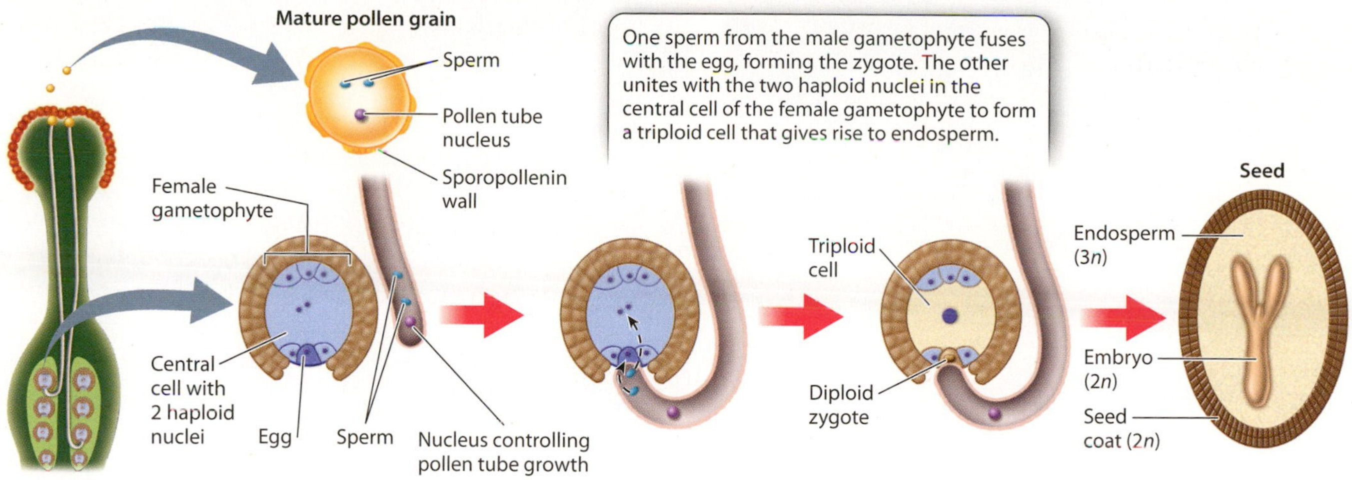

Angiosperms delay provisioning their ovules until after fertilization.

As we have seen, a major trend in the evolution of plants is the decreasing size and independence of the gametophyte generation and the increasing prominence of the sporophyte generation. In angiosperms, this trend is taken even further (**Fig. 30.18**). The male gametophyte has only three cells: One of these cells controls the growth of the pollen tube, while the other two are male gametes or sperm. In most angiosperms, the female gametophyte contains only eight nuclei, arranged as six haploid cells (one of which gives rise to the egg) plus a central cell containing two nuclei. The female gametophyte of angiosperms is thus too small to support the growth of the embryo, as it does in gymnosperms. How, then, do flowering plants provide nutrition for the embryo to grow?

This is where the central cell of the female gametophyte, which contains two haploid nuclei, comes into play. During pollination, the two sperm travel down the pollen tube and enter the ovule (**Fig. 30.19**). One of the sperm fuses with the egg to form a zygote, just as occurs in all plants. The other sperm, however, unites with the two haploid nuclei in the central cell to form a triploid ($3n$) cell. This triploid cell undergoes many mitotic divisions, forming a new tissue called **endosperm.** In angiosperm seeds, it is the endosperm that supplies nutrition to the embryo. The process in which two sperm from a single pollen tube fuse with the egg and the two haploid nuclei in the central cell is called **double fertilization.**

Double fertilization and endosperm formation offer a distinct advantage. In pines and other gymnosperms, the female gametophyte provides food for the embryo. This provisioning occurs whether or not the egg is successfully fertilized, and so is potentially a significant waste of resources. In angiosperms, the endosperm develops only when fertilization has occurred. In many flowering plants, including corn, wheat, and rice, the mature seed consists largely of endosperm (**Fig. 30.20**). The embryo itself is small. As the seeds germinate, the embryo draws resources from the endosperm to help it become established. It is no overstatement to suggest that the human population is sustained mostly by endosperm produced in the seeds of cereal crops.

In other angiosperms, the growing embryo consumes all the endosperm before the seed leaves the parent plant. For example, a

FIG. 30.20 **Endosperm.** Endosperm develops from the triploid cell produced by double fertilization. *Photo source: Greg Lawler/Alamy.*

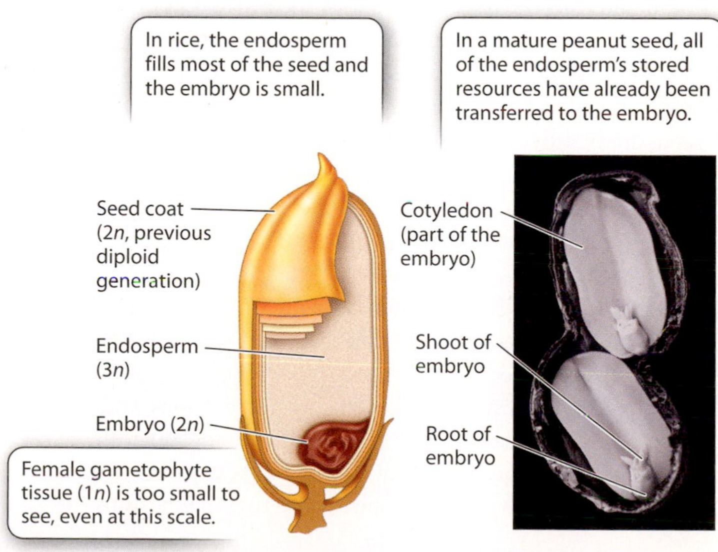

FIG. 30.21 Developmental transition from flower to fruit. Fruits, in this case a plum, develop from ovaries and provide a means of dispersing seed. *Sources: (left to right) Gert Tabak/Getty Images; Trevor Sims/age fotostock; McPHOTO/age footstock.*

peanut has virtually no endosperm left at the time it is dispersed. Almost all the resources that were previously held within the endosperm have been transferred into two large embryonic leaves, called cotyledons, which form the two halves of the part of the peanut we eat. Thus, the embryo itself contains all of the stored resources that will nourish the growth of the seedling at germination.

Fruits enhance the dispersal of seeds.

The transformation of a flower into a **fruit** is as remarkable as the metamorphosis of a caterpillar into a butterfly (**Fig. 30.21**). As the fertilized egg develops to form an embryo and the endosperm proliferates around it, the ovary wall develops into a fruit, stimulated by signaling molecules produced by the endosperm. In a tomato, for example, the fleshy fruit is a mature ovary that encloses the seeds. In some plants, other parts of the flower may also become incorporated into the fruit. For instance, the fleshy part of an apple is formed from the outer part of the ovary combined with the base of the petals and sepals. Other fruits, like pineapple, develop from multiple flowers.

Fruits serve two functions: They protect immature seeds from being preyed on by animals, and they enhance dispersal once the seeds are mature. In the case of fruits dispersed by animals, it is essential that the fruit not be consumed before the seeds are able to withstand a trip through their disperser's digestive tract. Thus, immature fleshy fruits are physically tough and their tissues highly astringent. As they ripen, the fruits are rapidly transformed in texture, palatability, and color. Ripening converts starches to sugars and loosens the connections between cell walls so that the fruit becomes softer.

In a number of species, including apples, bananas, and tomatoes, a gaseous hormone called ethylene triggers fruit ripening. It is the production of ethylene that explains why placing a ripe apple and an unripe banana together hastens the ripening of the banana. Tomatoes and bananas are usually picked in an immature state and are subsequently ripened by exposure to ethylene. The fruits can therefore be transported while they are still hard and less prone to damage. Because ethylene synthesis requires oxygen, the unripe fruits are typically stored in a low-oxygen environment. Without this precaution, the ethylene produced by the ripening of even a single fruit could set off a chain reaction in which all the stored fruit would ripen at once.

The tremendous diversity of form seen in fruits reflects the many ways in which angiosperm seeds are dispersed (**Fig. 30.22**). The winged fruit of an ash tree is carried away on the wind, and a coconut can float on ocean currents for many months, protected from exposure to salt water by its thick outer husk. The burdock fruit has small hooks that attach to fur (or clothing)—this small hitchhiker was the inspiration for Velcro. Other fruits are not dispersed, but instead open to expose their seeds. For example, as milkweed fruits mature and dry out, they split apart to release hundreds of seeds, each with a tuft of silky hairs that floats them away on even a light breeze. Still others, such as *Impatiens* (jewelweed), produce fruits that forcibly eject their seeds.

Animals are probably the most important agents of seed transport, in many cases attracted by the nutritious flesh of the fruit. In fleshy fruits, the seeds are protected by a hard seed coat and pass unharmed through the animals' digestive tract. In some species, this passage actually enhances germination rates. In many cases, however, the seed is consumed, and successful dispersal occurs only when the animals gather more seeds than they eat. Squirrels are a good example, storing many acorns that never get eaten. Humans are an important dispersal agent—we can transport seeds long distances, sometimes far outside their usual range. In some cases, accidentally introduced plant species can disrupt native plant communities.

→ **Quick Check 6** From what flower structure(s) is a fruit derived?

FIG. 30.22 Fruits and seed dispersal. (a) The seeds of the milkweed *Asclepias syriaca* are dispersed by wind; (b) a robin is eating berries of mountain ash (*Sorbus americana*); (c) cocklebur (*Xanthium strumarium*) fruits are hitching a ride on the fur of a cow; (d) a squirrel carries a fruit it will bury for later use. *Sources: a. Don Johnston/age fotostock; b. Don Johnston/age fotostock; c. blickwinkel/Alamy; d. David Norton/Alamy.*

? CASE 6 AGRICULTURE: FEEDING A GROWING POPULATION

How did scientists increase crop yields during the Green Revolution?

Human civilization and plant reproduction are closely intertwined. Not only do seeds and fruits make up a significant portion of our diet, the direct manipulation of plant reproduction plays a critical role in agriculture. Plant breeding began as the artificial selection for plants with seeds that were easy to harvest and has today become a highly quantitative field in which controlled crosses between plants are used to combine favorable traits into a single variety. Because many of the traits that affect crop productivity are the product of multiple genes, crop breeding remains a powerful tool for increasing crop yields. The spectacular increases in productivity that are often referred to as the Green Revolution resulted, in part, from controlling the movement of pollen between plants.

Norman Borlaug, an American agronomist, has been called "the father of the Green Revolution" and "the man who saved a billion lives." In the mid-twentieth century, Borlaug led a project to make Mexico self-sufficient in bread wheat. Borlaug's aim was to breed wheat that could resist infection by a devastating class of fungal pathogens known as rusts. To do this, he and his group hand-pollinated plants that exhibited resistance to rusts with pollen from plants that grew well under field conditions or had seeds that produced good flour.

Under natural conditions, wheat usually self-pollinates. To introduce new sources of pollen, Borlaug and his coworkers removed the anthers of each flower with tweezers before the flowers opened. They had to cover each flower to prevent unwanted pollen from coming into contact with the stigma. Finally, when the stigma was receptive, they dusted it with pollen collected from a plant with the desired characteristics. Under the hot sun, this is backbreaking work, but over a series of years the pollinations carried out by Borlaug and his coworkers achieved their goal of breeding wheat varieties that were disease resistant.

Yet another problem remained to be solved. To obtain high yields requires large amounts of fertilizer (Chapter 29), but under these conditions, the new wheat varieties produced so many seeds that stalks became top heavy, falling over in the wind. Borlaug knew that an answer to this problem was to cross his plants with wheat varieties that had shorter, sturdier stems. But given the selective advantage of height in natural populations, where was he to find such a plant? Japanese wheat breeders had identified and preserved a mutant dwarf plant that had arisen spontaneously, and Borlaug obtained seeds of this dwarf variety.

Much hard work and many thousands of hand pollinations lay ahead. But by 1963, more than 95% of Mexican wheat cultivation made use of Borlaug's high-yielding semidwarf wheat varieties, resulting in wheat yields six times greater than in 1944, the year he began work in Mexico. In 1965, the seeds were exported in large numbers, first to India and Pakistan and soon to the rest of the world. Used in combination with greater investments in fertilizer and irrigation, these varieties prevented the famines that had been predicted by many as a result of rising human populations. For his work in alleviating world hunger, Borlaug was awarded the Nobel Peace Prize in 1970.

30.4 ASEXUAL REPRODUCTION

Many plants can also reproduce asexually, producing new individuals without the formation and fusion of gametes. While asexual reproduction avoids the challenges associated with fertilization, asexually formed plants must still disperse if they are to avoid competition for resources with the parent plant. Sexual reproduction relies on spores and seeds; how do asexually produced individuals move across the landscape?

Asexually produced plants disperse with and without seeds.

Some species of flowering plants are able to use seeds to disperse asexually formed individuals. In these species, seeds can develop even in the absence of fertilization, a process called **apomixis.** A variety of underlying mechanisms exists that can result in "seeds without sex" —for example, an embryo can form directly from a diploid sporophyte cell. Dandelions are an example of plant species that produce seeds in the absence of sex. Apomixis would be of tremendous utility in agriculture if it could be turned on or off at will. The reason for this is that, once a desirable variety was developed, additional seeds could be produced without introducing new genetic variation. This ability to breed true would be useful in self-incompatible species such as apple and blueberry.

Most plants that reproduce asexually do so without producing seeds; instead, a new plant grows out of part of the parent plant. In these cases, dispersal of offspring occurs in different ways. In mosses and liverworts, for example, tiny plantlets form by mitosis at the base of shallow cups. Raindrops landing in this cup can dislodge the tiny plantlets, splashing some away from the parent plant.

However, most plants that reproduce asexually do so by growing to a new location and only then producing a new plant. You need look no further than a front lawn to find evidence of such **vegetative reproduction.** Grass forms horizontal stems underground from which new upright grass shoots are produced at a distance from the site where the parent plant originally germinated. Strawberries, bamboo, and spider plants (a common houseplant) also spread vegetatively by forming horizontal stems. In many cases, the connections that were needed initially to produce a new plant become severed. When this happens, genetic evidence is needed to determine if a plant is the result of sexual or asexual reproduction.

FIG. 30.23 Vegetative growth of aspen trees. Each distinct group of trees, distinguished by leaf color, is a group of genetically identical individuals produced by vegetative reproduction. *Source: Scott Smith/Corbis.*

Vegetative reproduction also occurs in woody plants. For example, when a redwood tree falls over, new upright stems can form along the now-horizontal trunk. Perhaps the most impressive example of the ability to spread vegetatively is the quaking aspen (*Populus tremuloides*), a common tree of the Rocky Mountains that produces brilliant yellow foliage in autumn (**Fig. 30.23**). Aspens reproduce vegetatively by producing upright stems from a spreading root system. Although each shoot lives for less than 200 years, a single individual can persist for tens of thousands of years by continuing to produce new stems. The current record holder is an aspen clone in Utah, consisting of nearly 50,000 stems, that is estimated to be 80,000 years old.

Plants can reproduce vegetatively because of the way they build their bodies. As we will see in the next chapter, growth and development occur throughout a plant's life, and this is the plant characteristic that allows plants to produce new upright "individuals" from roots or horizontal stems. ■

Core Concepts Summary

30.1 Alternation of generations evolved in plants by the addition of a diploid sporophyte generation that allows plants to disperse spores through the air.

The life cycle of the algal sister groups of land plants has one multicellular haploid generation. page 620

The life cycle of land plants is characterized by an alternation of haploid and diploid generations, one generation specialized for fertilization and the other for dispersal. page 621

Gametes are haploid cells produced by mitosis from cells in the multicellular haploid generation, which is referred to as the gametophyte generation. page 622

Spores are haploid cells produced by meiosis from the multicellular diploid generation, which is referred to as the sporophyte generation. page 622

Bryophytes have a free-living haploid gametophyte that makes gametes and a dependent diploid sporophyte that makes spores that are released into the air. page 622

In spore-dispersing vascular plants, both the gametophyte generation and the sporophyte generation can survive on their own. page 623

Bryophytes and spore-dispersing vascular plants rely for fertilization on swimming sperm released into the environment. page 623

30.2 Pollen allows the male gametophyte to be transported through the air, while resources stored in seeds support the growth and development of the embryo.

Pollen is the male gametophyte that develops within the spore wall, and so can be transported through the air. page 625

The female gametophyte of seed plants is retained and nourished by the sporophyte. The female gametophyte, together with the surrounding layers of protective tissues, is called an ovule. Ovules, when fertilized, develop into seeds. page 625

Pollination is the transport of pollen to the ovule, where fertilization takes place. page 625

Seeds contain the diploid embryo, a seed coat, and, in gymnosperms, nourishment in the form of the female gametophyte. page 625

The pollen and ovules of gymnosperms are produced in separate structures, whereas the pollen and ovules of angiosperms often form in one structure, the flower. page 625

A trend in the evolution of plants is the progressive reduction in size and independence of the gametophyte and the increasing role of the sporophyte. page 627

30.3 Angiosperms (flowering plants) attract and reward animal pollinators, and they provide resources for seeds only after fertilization.

Flowers consist of four whorls of organs: ovule-bearing carpels, pollen-producing stamens, petals, and sepals. page 629

Many flowers attract animals because they provide a reward, such as food, shelter, or chemicals. Animals, in turn, transfer pollen. page 631

Many flowering plants have genetic and structural mechanisms to prevent self-fertilization. page 632

Double fertilization, which is unique to angiosperms, is the formation of a diploid zygote and triploid endosperm. The endosperm nourishes the embryo that develops from the zygote. page 635

The fertilization of an ovule triggers the development of the ovary wall into a fruit, a structure that enhances seed dispersal. page 636

30.4 Many plants also reproduce asexually.

Asexual reproduction produces new individuals without the need for uniting gametes. Dispersal of offspring, however, remains important. page 638

Apomixis is the formation of seeds from diploid cells, in the absence of fertilization. Seeds formed by apomixis are thus genetically identical to the parent plant. page 638

Vegetative reproduction can occur when a plant spreads horizontally and develops new upright shoots. page 638

Self-Assessment

1. Explain how the evolution of alternation of generations is an adaptation for reproduction on land.
2. Diagram the relationship between the sporophyte and gametophyte generations in bryophytes, ferns, gymnosperms, and angiosperms. Show the relative sizes and physical interactions (if any) of the two generations.
3. Explain how the organs in the different whorls of angiosperm flowers interact to promote fertilization.
4. Contrast the investments that angiosperms and gymnosperms make and the structures that they produce to enhance pollination.
5. Diagram the structure of a mature angiosperm and a mature gymnosperm seed, indicating the ploidy ($1n$, $2n$, $3n$) of each tissue and its role in seed development and germination.
6. Describe two ways that plants can reproduce asexually and explain how each of these help to disperse offspring.

Log in to LaunchPad to check your answers to the Self-Assessment questions, and to access additional learning tools.

Dietrich Rose/Getty Images.

CHAPTER 31

Plant Growth and Development

Core Concepts

31.1 In plants, upward growth by stems occurs at shoot apical meristems, populations of totipotent cells that produce new cells for the lifetime of the plant.

31.2 Hormones are chemical signals that influence the growth and differentiation of plant cells.

31.3 Lateral meristems allow plants to grow in diameter, increasing their mechanical stability and the transport capacity of their vascular system.

31.4 The root apical meristem produces new cells that allow roots to grow downward into the soil, enabling plants to obtain water and nutrients.

31.5 Plants respond to light, gravity, and wind through changes in internode elongation and the development of leaves, roots, and branches.

31.6 Plants have sensory systems that control the timing of developmental events.

Plants build their bodies in ways that are fundamentally different from what occurs in animals. To appreciate how different, let us remind ourselves briefly how our own bodies develop from fertilized eggs (Chapter 20). Humans and other mammals grow by the repeated division of cells throughout the body. Many cells migrate from one place to another, especially early in development. The first divisions of the fertilized egg form stem cells that have the potential to give rise to many different cell types. Cells with the potential to become one of many different cell types are termed "totipotent." As development proceeds, however, cells lose this totipotency and proceed down a genetically determined developmental pathway. Once the body is mature, growth essentially stops.

In contrast to animals, cell division in plants persists in populations of totipotent cells called **meristems,** and growth commonly lasts throughout the life of the plant. A bristlecone pine in western North America may be several thousand years old, but it continues to sprout new branches, new leaves, and new cones every year. Plants respond to the world around them not by moving about, but by modifying their size and shape. Plants acquire resources by growing to them, and they compete with neighboring plants by growing beyond them. Furthermore, because development in plants can be modified in response to information about local environmental conditions, growth fulfills many of the same roles in plants that behavior does in animals (Chapter 45). And although plants can't move out of the way of danger, plants can rebuild their bodies if damaged by fire or frost, or after being struck by a falling tree. Even reproduction requires the continuing formation of new organs.

This chapter focuses on how vascular plants build their bodies. We begin by asking how the shoot system, consisting of leaves, stems, and reproductive structures, is formed. We then turn to the development of root systems, asking how plants grow belowground. Finally, we consider how plants sense the environment and how they respond to what they sense by modifying their growth and development.

31.1 SHOOT GROWTH AND DEVELOPMENT

In tropical forests, vines weave their way upward through the branches of trees. Many of these vines have long, thin stems with widely spaced leaves (**Fig 31.1a**). In contrast, a barrel cactus, living in the desert, has thick stems and leaves modified to form sharp spines (**Fig. 31.1b**). Although they look nothing alike, these two plants are constructed in the same way. Plant shoots are modular, meaning that they are formed of repeating units. Each unit consists of a **node,** the point where one or more leaves are attached, and an **internode,** the segment between two nodes (**Fig. 31.1c**). In vines the internodes are long and the leaves large, whereas in cacti the internodes are short and the leaves thin and sharp. The modular structure of plants helps us understand how the capacity for continued growth gives rise to the tremendous variation in plant form that we see around us.

Stems grow by adding new cells at their tips.

At the tip of each stem lies a tiny dome of cells called the shoot apical meristem (**Fig. 31.2**). The **shoot apical meristem** is a group of totipotent cells that, like embryonic stem cells in animals, gives rise to new tissues. As we will see, plants have meristems in several parts of their bodies, but upward growth occurs exclusively at the shoot tip. This is where cell division occurs, generating all the new cells that serve to elongate stems. The shoot apical meristem also initiates leaves and produces new meristems, which allow plants to branch.

An important characteristic of the shoot meristem is that it maintains a constant size even though it is a region of active cell division. As new cells are added near the shoot tip, cells that are farther from the shoot tip cease to divide. Cells near the shoot tip maintain their ability to divide by the expression of **meristem identity genes.** The expression of meristem identity genes is controlled by a network of chemical signals produced by cells at the very tip of the stem. Because these signals can be transmitted only

FIG. 31.1 **Shoot organization.** (a) A tropical vine and (b) barrel cacti look very different, but their shoot systems are built from (c) the same repeating units of nodes and internodes. *Sources: a. Le Do/Shutterstock; b. LianeM/Alamy.*

a.

b.

c.

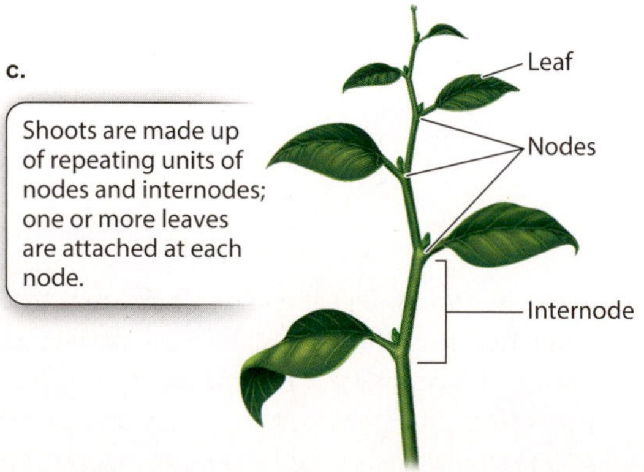

FIG. 31.2 **The shoot apical meristem.** The shoot apical meristem consists of undifferentiated cells that divide rapidly, giving rise to all the cells in stems and leaves. (a) The seedling is rowan (*Sorbus acuparia*). (b) The photo shows a longitudinal section of the shoot apical meristem of coleus (*Solenostemon scutellariodes*). (c) This scanning electron micrograph shows the shoot apical meristem of tomato (*Solanum lycopersicum*). *Sources: a. Ian Gowland/Science Source; b. M. I. Walker/Science Source; c. Siobhan Braybrook.*

a.

c.

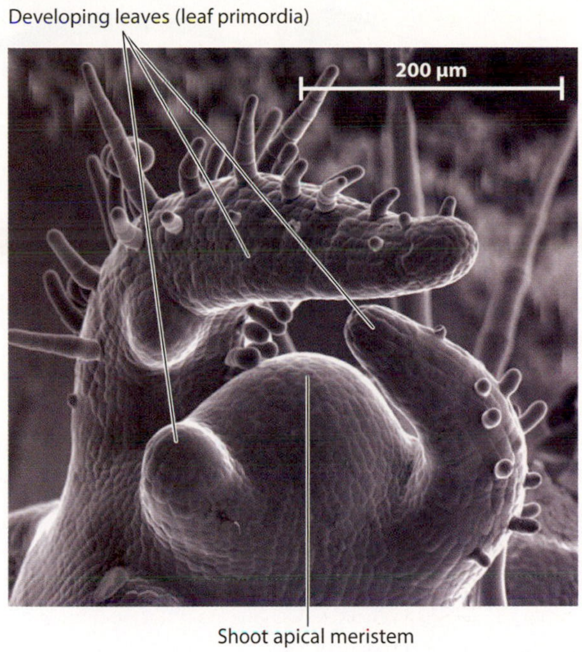

b.

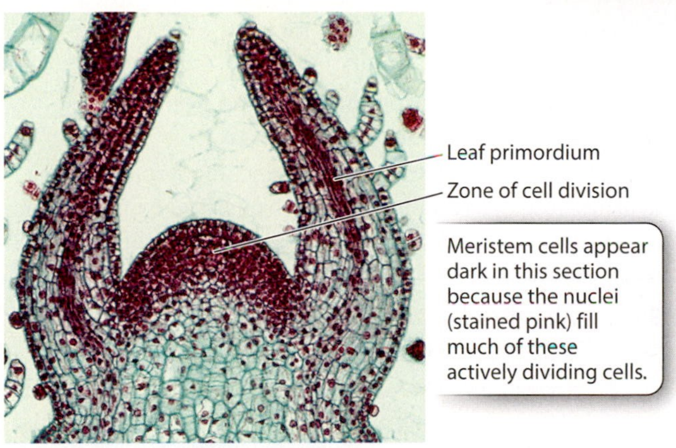

over a limited distance, only the cells close to the shoot tip express meristem identity genes.

→ **Quick Check 1** What are three functions of the shoot apical meristem?

Stem elongation occurs just below the apical meristem.

In animals, cell division and cell enlargement typically go hand in hand. In plants, most of the increase in cell size occurs after cell division is complete. Cells that are too far from the shoot tip to express meristem identity genes cease to divide, yet they continue to grow. Most of the elongation of stems takes place in the internodes located immediately beneath the shoot apical meristem (**Fig. 31.3**). The increase in cell size that occurs in this region is impressive. If all of the cells in the stem of a 100-m redwood tree remained the size they were when they left the shoot apical meristem, the tree would be less than 5 m tall.

Within an elongating internode, each cell grows many times more in length than in width. The reason is the orientation of the cellulose molecules, which are primarily wrapped around the cell perpendicular to the axis of the stem rather than parallel with it. Because cellulose is very strong, having a tensile strength similar to that of steel, cells expand more easily in length than in girth. The large central vacuole that characterizes mature plant cells forms during cell elongation. In fact, most of the increase in cell volume is due to the uptake of water and solutes that fill the vacuole. This explains in part why plant growth is markedly reduced during periods of drought.

As we move even farther from the shoot tip, cells reach their final size and complete their differentiation into mature cell types. This organization into successive zones of cell division, cell elongation, and cell maturation allows stems to grow without any predetermined limit to their length.

Once cells have matured, they no longer expand. Thus, during the upward growth of the shoot, it is the production and elongation of cells at the shoot tip that lift the meristem ever higher into the air. Imagine that, as a 10-year-old, you carved your initials into a tree. Twenty years later, your initials will be

FIG. 31.3 Stem elongation. Internodal elongation results from cells increasing many times in length, but only a small amount in width.

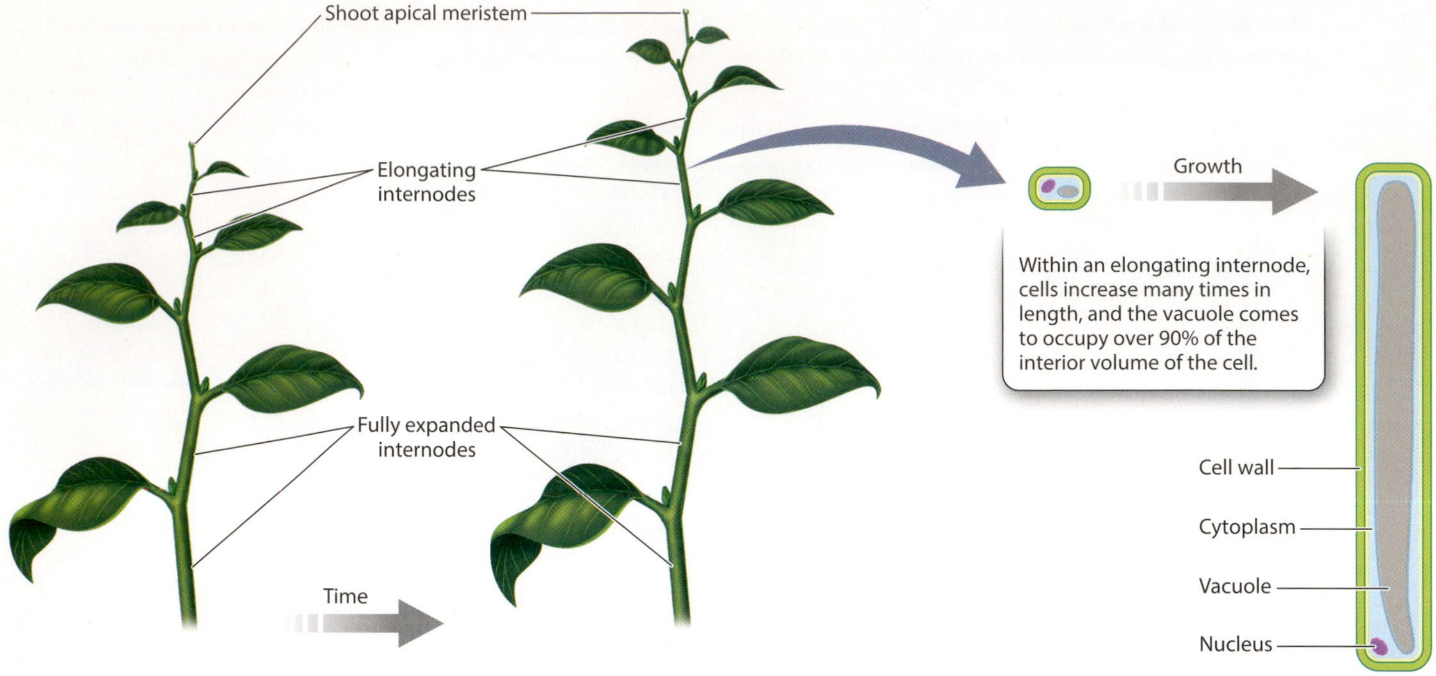

exactly the same distance from the ground as they were on the day you inscribed them. The letters, however, will be wider—a phenomenon we discuss later in this chapter.

The development of new apical meristems allows stems to branch.

Vascular plants evolved the ability to branch even before they evolved roots or leaves. Branching was important to these first plants because it allowed them to produce more sporangia. Branching allows present-day plants to support greater numbers of both reproductive structures and leaves.

In spore-dispersing vascular plants, branches form when the shoot apical meristem divides in two, giving rise to two stems, each with its own shoot apical meristem. In seed plants, branches grow out from **axillary buds,** which are meristems that form at the base of each leaf (**Fig. 31.4**). Axillary buds have the same structure and developmental potential as the apical meristem and express the same meristem identity genes. When a branch develops, the axillary bud becomes the shoot apical meristem of the new branch. Thus, most plants have many shoot apical meristems: one at the tip of each stem and branch. Axillary buds are formed at the same time as leaves, but they remain dormant until triggered to grow, and they persist even after leaves are shed. Thus, axillary buds provide seed plants with many points along their stem where new branches can form.

The shoot apical meristem controls the production and arrangement of leaves.

The earliest vascular plants were simple branching stems, and photosynthesis took place along the length of the stem. As evolution

FIG. 31.4 Axillary buds. In seed plants, branching results from axillary buds that are produced at nodes. *Photo source: Jacqueline S. Wong/ViewFinder Exis, LLC.*

FIG. 31.5 The evolution of leaves. Fossils show that leaves evolved from branches that (1) lost the ability for continued growth; (2) developed in a plane; and (3) expanded the photosynthetic surface to fill the area between veins. The photo shows *Archaeopteris,* one of the earliest plants with flattened leaves. *Photo source: Topham/The Image Works.*

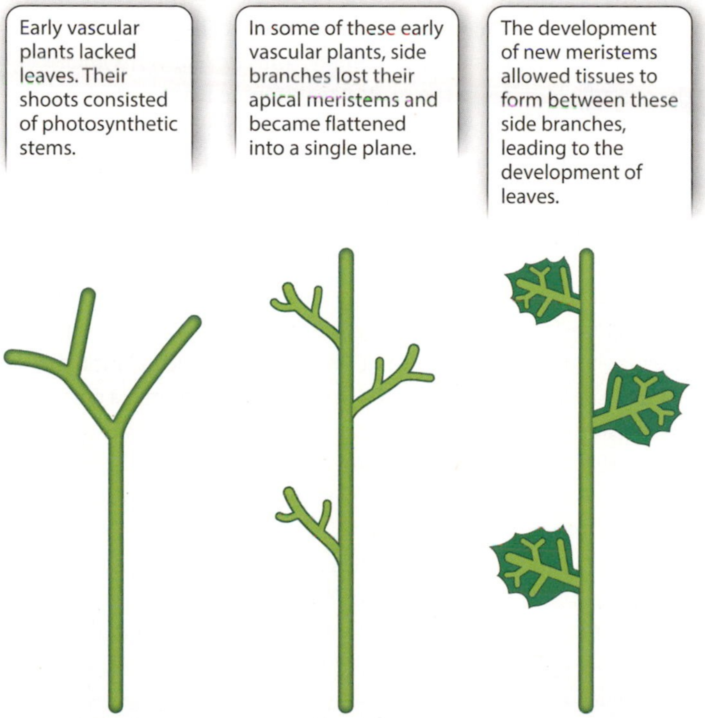

proceeded, some branch systems became flattened, increasing the capture of light. These planar branches lost the capacity for continued growth. By about 380 million years ago, the planar branches became modified into structures recognizable as leaves (**Fig. 31.5**).

The evolution of leaves required several developmental changes. Apical meristem identity genes were down-regulated and leaf identity genes up-regulated, resulting in an organ specialized for photosynthesis. New meristems evolved, enabling leaves to expand into flattened structures for capturing sunlight. In fern leaves, these regions of cell division are located along the leaf margin. In pine needles, they occur at the base of each needle. In flowering plants, meristem cells can be distributed throughout the developing leaf, making possible a diversity of leaf shapes. In contrast to cells in the shoot apical meristems, which can divide continuously throughout the lifetime of a plant, leaf meristem cells divide only for a relatively short period of time. This explains why, at a certain point, leaves stop growing.

The arrangement of the leaves along a stem has a major impact on how much sunlight each leaf receives and therefore on how much photosynthesis is carried out. Leaves begin as small bumps, called **primordia** (singular, **primordium**), which form on the sides of the shoot apical meristem in a highly regular pattern (see Fig. 31.2). Each species has a characteristic number of leaves attached at each node along the stem. Some species have only a single leaf at each node, whereas others have two or more (**Fig. 31.6**). The regular placement of leaves reduces the shading of one leaf by another and thus enhances the ability of plants to obtain sunlight.

When we think of leaves, it is the green photosynthetic ones that first come to mind. However, many plants produce leaves that

FIG. 31.6 Diversity of leaf arrangements. *Photo sources: (left to right) Biosphoto/Jonathan Buckley/GAP; Anneke Doorenbosch/Alamy; Christian Hütter/imagebroker/age fotostock.*

FIG. 31.7 Non-photosynthetic functions of leaves. (a) Protection (bud scales); (b) climbing (tendrils); (c) trapping insects (spines); and (d) attracting pollinators (color). *Sources: a. Dr. Keith Wheeler/Science Source; b. Marga Werner/age fotostock; c. J Garden/Alamy; d. Adisa/Dreamstime.com.*

are specialized for functions other than photosynthesis, including climbing, trapping insects, and attracting pollinators (**Fig. 31.7**). Plants that overwinter produce small, hard **bud scales** that protect shoot apical meristems from water loss and damage due to cold. Bud scales may not look like leaves, but they form from leaf primordia and are arranged in the same way around the stem as the green leaves produced in spring.

Young leaves develop vascular connections to the stem.

During the initial growth of leaf primordia, cells rely on diffusion to obtain the resources needed to divide and expand. As the young leaves increase in size, their demand for water and carbohydrates quickly exceeds what diffusion can supply. If they are to grow to even a fraction of their full size, leaves must establish vascular connections with the xylem and phloem in the stem. As we have seen in Chapter 29, for leaves to carry out photosynthesis requires both a supply of water through the xylem and the ability to export carbohydrates to the rest of the plant through the phloem.

When a developing leaf is still very small, files of cells within the leaf begin to elongate, forming discrete strands of **procambial cells** that extend from near the tip of the leaf to the mature vascular tissues within the stem. Procambial cells ultimately give rise to both xylem and phloem within the leaf veins and the vascular bundles in the stalk attaching the leaf blade to the stem. The common origin of xylem and phloem from procambial cells explains why these two tissues are always found in close association throughout the length of the plant.

The vascular system of an elongating stem is built from the successive addition of vascular bundles that connect each developing leaf to the vascular tissues already present within the stem. Spore-dispersing vascular plants typically have one or several vascular bundles in the center of their stems. In seed plants, the most common configuration is a ring of vascular bundles near the outside of the stem. The bundles have xylem conduits toward the center of the stem and phloem nearer the outside of the stem (**Fig. 31.8**). Typically, each leaf develops vascular connections with several of the stem's vascular bundles. The region between the epidermis and the vascular bundles is the **cortex,** and the region inside the ring of vascular bundles is the **pith.** The placement of the vascular bundles, which contain the stiff xylem conduits, near the outside of the stem increases the stem's ability to support its leaves without bending over.

Flower development terminates the growth of shoot meristems.

In vascular plants, sporangia are produced on the surfaces of leaves. In some plants, such as ferns, sporangia are located on leaves that also carry out photosynthesis. However, in conifers and some other plants, sporangia form on the surfaces of highly modified leaves and branches that form compact cones. In both cases, the arrangement of the sporangia-bearing leaves is the same as leaves that do not bear sporangia.

Flowers develop from floral meristems formed by the conversion of shoot meristems. This means that flowers can be produced at the tip of a shoot or at the base of leaves as the products of axillary buds. Floral meristems differ from shoot meristems in several ways. First, all of the cells within a floral meristem differentiate, and so floral meristems lose the capacity for continued growth. Second, although floral organs (sepals, petals, stamens, and carpels) develop from primordia and are thought to be modified leaves, their arrangement differs from the placement of leaves along the stem. Floral organs occur in whorls, with each whorl consisting of only a single organ type (see Fig. 30.12).

Flowering is triggered by florigen, a protein produced in leaves and transported through the phloem to apical meristems and

FIG. 31.8 Arrangement of vascular bundles. In most seed plants, the vascular bundles are arranged in a ring near the outside of the stem. *Photo source: Steve Gschmeissner/Science Source.*

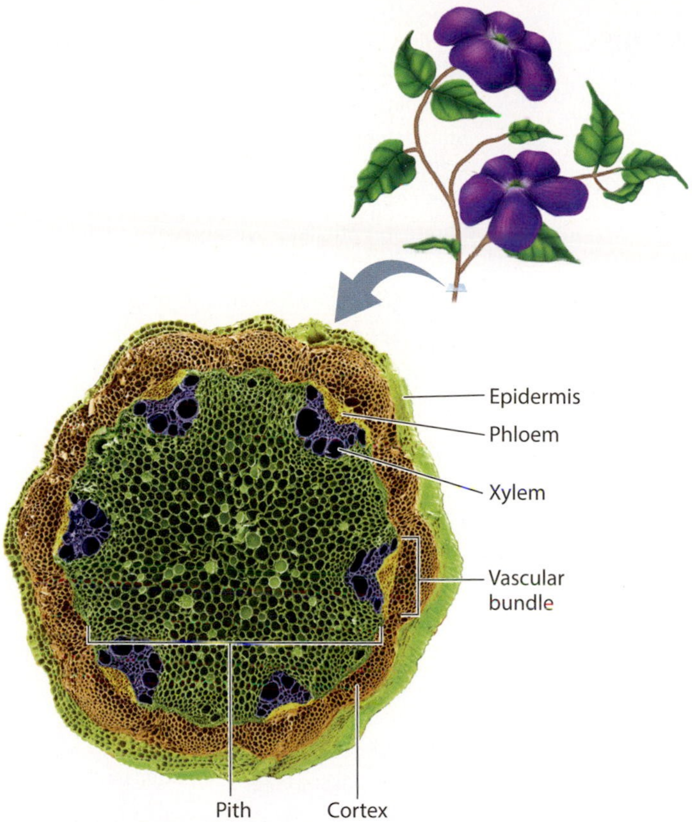

axillary buds. Florigen triggers the transition to floral meristems by initiating the down-regulation of meristem identity genes and the up-regulation of genes that govern floral identity. Once the meristem is launched along the trajectory for flower development, the identity of each whorl is controlled by the expression of homeotic genes (Chapter 20).

31.2 PLANT HORMONES

A key question in the development of multicellular organisms is what controls the identity, or "fate," of individual cells. In plants, each cell is bound to its neighbor by the cell wall and thus fixed in place. This means that cell identity must be determined by where a cell is located within the plant. For cells at the surface, the absence of neighboring cells on one side can be the signal to become an epidermal cell. For cells in the interior, chemical signals determine what sort of cell is formed. These signals provide information on both the identity of neighboring cells and the cell's position in the plant.

Some of the chemical signals that guide plant development affect only a small area. For example, in section 31.1 we mentioned that the signals that cause cells to express meristem identity genes are restricted to the region close to the shoot tip. Other signaling molecules, or their precursors, enter the vascular system and are transported the entire length of the plant. Because plants grow and develop in many places at once, signaling molecules that can move from one part of the plant to another play an important role in coordinating growth and development.

Hormones affect the growth and differentiation of plant cells.

Hormones are chemical signals that influence both growth and development. The term "growth" refers to irreversible increases in size, while "development" encompasses the formation of new tissues and organs, as well as the differentiation of specific cell types. Hormones play two roles in plant growth and development. First, they help establish and maintain basic patterns of the plant body plan: How are leaves arranged? Where do branches form? How much do internodes elongate? Second, they coordinate growth in different parts of the plant in response to both internal and external environmental factors. For example, hormones guide the growth of a stem toward sunlight and control which axillary buds will grow into branches. Hormones affect plant growth and development both by influencing physiological processes, such as the activity of membrane transport proteins, and by altering patterns of gene expression.

Traditionally, plant biologists have recognized five major plant hormones: auxin, gibberellic acid, cytokinins, ethylene, and abscisic acid (**Table 31.1**). How is a small number of hormones able to shape the continuing growth and development of plants? One characteristic of these hormones is that each triggers a wide range of developmental effects. For example, ethylene can lead to fruit ripening (Chapter 30) and also slows the elongation of roots and stems. How a cell responds to a particular hormone depends both upon its developmental stage and its position within the plant. In addition, the effect of any one hormone is often influenced by the presence of one or more other hormones.

Several chemicals with hormone-like properties have recently been discovered. Some of these have broad effects on plant development (for example, strigolactone), whereas others appear to be more specific (for example, florigen, whose only role is to stimulate flower development).

To illustrate how hormones influence plant growth and development, we consider the role of three hormones, auxin, gibberellic acid, and cytokinins, in establishing the basic patterns of shoot development. Later in the chapter, we consider the role of hormones in root development (section 31.4) and plant growth in response to light and gravity (section 31.5).

Polar transport of auxin guides the placement of leaf primordia and the development of vascular connections with the stem.

The plant hormone **auxin** was first identified when it was observed to cause shoots to elongate. As we will see in section 31.5, auxin's ability to trigger cell elongation underlies how shoots

TABLE 31.1 The Five Major Hormones

HORMONE	SYNTHESIS AND TRANSPORT	EFFECTS
Auxin 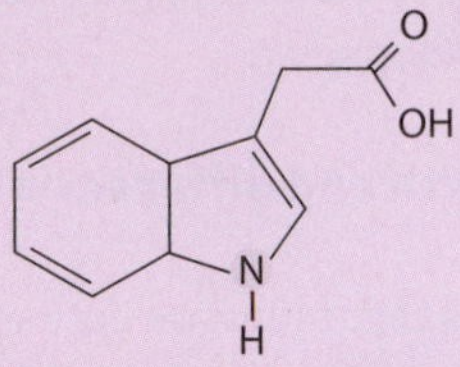 Indole-3-acetic acid	Synthesized primarily in shoot apical meristems and young leaves. Transported by polar transport and through phloem.	Induces cell wall extensibility. Inhibits outgrowth of axillary meristems. Stimulates formation of root meristems. Stimulates differentiation of procambial cells. Involved in phototropism and gravitropism. Stimulates ethylene synthesis.
Gibberellic acid 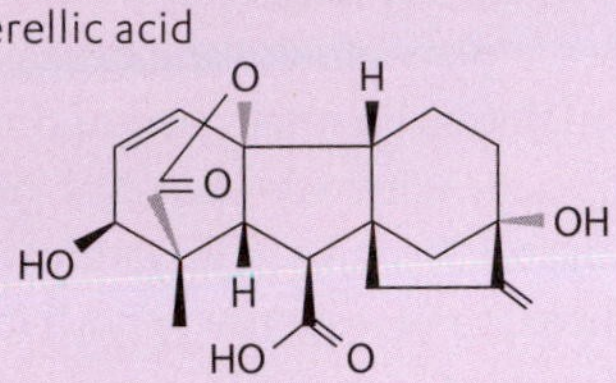	Synthesized in the growing regions of both roots and shoots.	Stimulates stem elongation and cell division. Mobilizes seed resources for developing embryo. Stimulates cytokinin synthesis.
Cytokinins 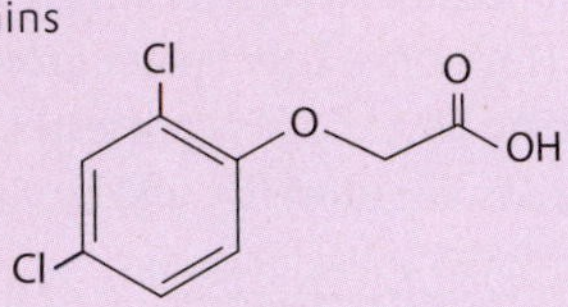 Zeatin	Synthesized in both root and shoot meristems. Transported in both xylem and phloem.	Stimulates cell division. Delays leaf senescence. Acts synergistically with auxin.
Ethylene 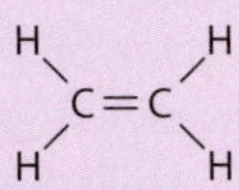	Synthesized in elongating regions of roots and stems, as well as in developing fruits. Gaseous; precursors can be transported in xylem.	Reduces cell elongation. Triggers fruit ripening. Stimulates formation of air spaces in roots.
Abscisic acid 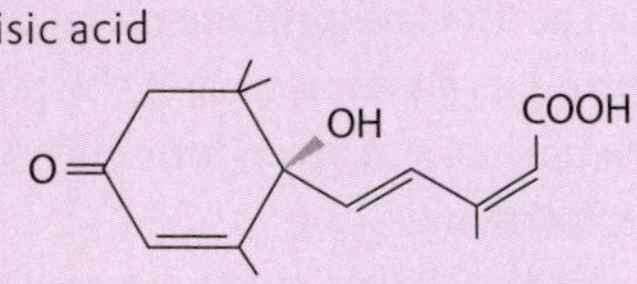	Synthesized in root cap, developing seeds, and leaves. Transported from roots to leaves in xylem.	Maintains seed dormancy. Stimulates root elongation. Triggers closing of stomata. Antagonistic with ethylene.

bend and grow toward light. Auxin also plays an important role in establishing spatial patterns of shoot development. In particular, auxin influences the spacing and placement of leaf primordia and the formation of the vascular bundles connecting young leaves to the mature vascular system of the stem.

Auxin's role in patterning results from the fact that its direction of movement across a group of cells can be controlled. The net movement of auxin in a single direction creates regions of high and low concentrations of auxin. These gradients in auxin concentration can guide the growth and development of individual cells.

Let's consider first how auxin movement influences the placement of new leaf primordia. Auxin is synthesized in rapidly dividing cells within the shoot meristem and is transported in the outermost cell layer toward the shoot tip. High levels of auxin in the outermost layer of cells trigger the growth of new leaf primordia. The movement of auxin into leaf primordia depletes auxin from surrounding regions and thus prevents new leaf primordia from forming nearby. As the shoot continues to grow, already formed leaf primordia become displaced from the shoot tip and spread farther and farther apart. This allows auxin moving upward in the outermost cell layer to accumulate to levels needed to trigger new leaf primordia close to the shoot tip.

Let's now consider how auxin transport guides the formation of vascular bundles that connect the leaf to the stem. As auxin is transported into leaf primordia, it promotes cell expansion and the young leaves grow. However, auxin does not remain in the developing leaf. Instead, auxin that has accumulated at the tip of

FIG. 31.9 Polar transport of auxin. The polar transport of auxin results in the movement of this hormone from the tip of each developing leaf to the base of the plant.

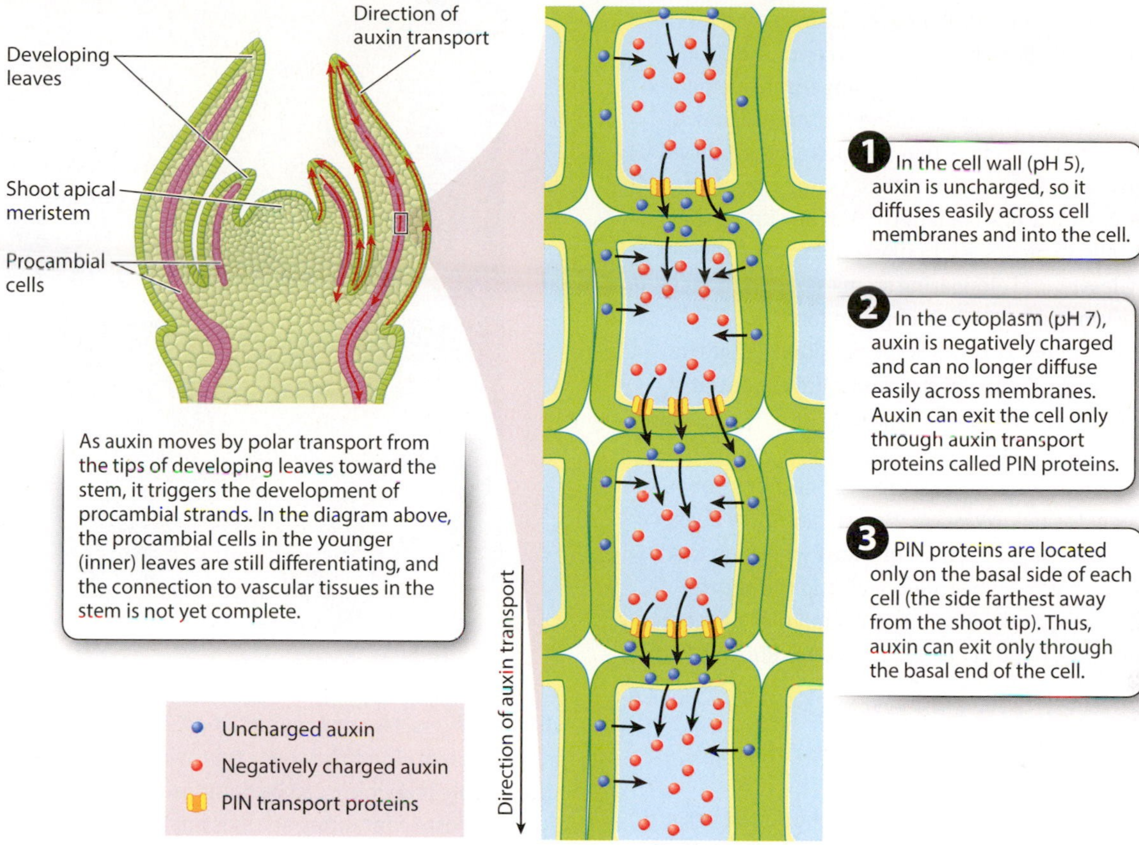

the young leaf is transported back toward the stem, but this time through cells in the interior of the leaf. As auxin moves through these cells, it causes the cells both to elongate and to produce more of the membrane transport proteins needed for auxin to exit the cell. The cells that initially, by chance, have the highest auxin levels become highly efficient at transporting auxin toward the stem. This, in turn, creates distinct "channels" through which auxin drains from the leaf. These continuous strands of elongate procambial cells eventually develop into veins and vascular bundles containing mature xylem and phloem.

These two examples illustrate how the movement of auxin in a single direction establishes spatial patterns that can guide further differentiation and development. But they leave open the question of how undifferentiated cells control the movement of auxin in the first place.

The coordinated movement of auxin across many cells in a single direction is referred to as **polar transport** (**Fig. 31.9**). Polar transport allows auxin to move independent of gravity at rates faster than the hormone could move by diffusion alone. However, it requires an input of energy.

Polar transport depends on the difference in pH between the cell wall and the cytoplasm. In the cell wall, auxin carries no net charge and thus can enter surrounding cells by diffusing across the plasma membrane. However, because the pH of the cytoplasm is higher than the pH of the cell wall, auxin loses a proton. As a result, auxin has a net negative charge when inside the cell and cannot move easily across the plasma membrane. For auxin to leave the cell, an auxin-specific plasma membrane transport protein known as PIN is required. By restricting the placement of PIN proteins to only one face of each cell, plants can control the direction of auxin movement across an otherwise undifferentiated group of cells. For example, if PIN proteins are inserted only on the side of cells farthest away from the shoot tip, auxin movement will be toward the roots. In fact, the prevailing direction of auxin transport in plants is from shoot apical meristems to the root tips.

? CASE 6 AGRICULTURE: FEEDING A GROWING POPULATION

What is the developmental basis for the shorter stems of high-yielding rice and wheat?

In natural ecosystems, competition for sunlight creates a strong selective advantage for growing tall. Compare the slender vine in Fig. 31.1, which grows in a dense forest, with the compact barrel cactus that grows where there is little danger of being shaded by a neighboring plant. Whereas the ancestors of today's crops competed for sunlight by growing taller, a key element of the high-yielding rice and wheat varieties developed during the Green Revolution is their shorter stems. These shorter-stemmed plants can support the much-larger seed mass induced by high rates of fertilizer application (Chapter 29). Moreover, by investing less in stems, these plants can allocate a greater percentage of their resources to the production of seeds. What happens inside plants of these high-yielding crop varieties to make the stems shorter?

Gibberellic acid was first identified as a chemical produced by fungal pathogens that caused the stems of young rice plants to grow to many times their normal length. We now know that it is produced naturally in plants, particularly in growing regions such as developing leaves and elongating stems. Gibberellic acid increases internode elongation by reducing the force needed to cause cell walls to expand. This raises the possibility that the high-yielding crop varieties with shorter stems have reduced production of or sensitivity to gibberellic acid.

At the time that the semidwarf varieties were developed, the underlying basis for their reduced stem growth was not known. The genes responsible for reduced stem growth have now been identified in both wheat and rice. The semidwarf wheats in fact do have reduced sensitivity to gibberellic acid, and the shorter rice varieties are deficient in the biosynthetic enzymes needed to produce this hormone. In both cases, the result is shorter but sturdier plants.

Cytokinins, in combination with other hormones, control the outgrowth of axillary buds.

Branching allows plants to support more leaves, but branches that are too close together result in leaves that overlap and shade one another. Because of the formation of axillary buds, seed plants have the potential to produce a new branch at each point where a leaf was attached. To avoid overlap, only a few axillary buds actually develop into branches. Which become branches is determined by the interaction of several hormones, including both cytokinins and auxin.

Cytokinins are plant hormones that were discovered when they were seen to stimulate cell division in tissues grown in the laboratory. Not surprisingly, cytokinins are produced in plant meristems. Although cytokinins affect a variety of physiological and developmental processes, they promote shoot growth by increasing the number of dividing cells in the apical meristem and, when applied directly to axillary buds, can stimulate them to develop into branches. This suggests that cytokinins play a role in branching in nature, but leaves open the question of what determines whether an axillary bud remains dormant or begins to grow.

Removing the shoot tip stimulates the outgrowth of axillary buds along that branch or stem (**Fig. 31.10**), suggesting that the shoot apical meristem somehow suppresses the growth of axillary buds. This suppression is referred to as **apical dominance.** Apical dominance prevents branches from being formed too close to the shoot tip, yet new branches can still be formed if the shoot meristem is damaged.

Apical dominance results because chemical signals from the shoot tip control the growth of axillary buds. One of these

FIG. 31.10 **Apical dominance.** *Source: Nigel Cattlin/Alamy.*

chemical signals is auxin, which moves by polar transport (and also in the phloem) from the shoot tip toward the root. Auxin's role in suppressing branching can be demonstrated: If the shoot tip is cut off and a paste containing auxin is applied to the cut end of the stem, the axillary buds remain dormant. If instead the shoot tip is treated with a substance that inhibits auxin transport, branching is induced.

Auxin suppresses the growth of axillary buds by inhibiting the synthesis of cytokinins, the hormones that stimulate the outgrowth of axillary buds (Table 31.1). When the apical meristem is removed, cytokinin synthesis increases in the stem nodes, triggering the growth of axillary buds.

Recent studies indicate that this top-down control is complemented by the action of another hormone, **strigolactone.** This hormone is made in roots and transported upward in the xylem. Plants that are unable to synthesize strigolactone produce many more branches than do wild-type plants. This indicates that strigolactone inhibits the outgrowth of axillary buds. One hypothesis is that strigolactones are transported to the shoot when either water or nutrients are limiting. Thus, the plant would stop growing new branches whenever the root system is unable to supply the water and nutrients they require.

31.3 SECONDARY GROWTH

Shoot apical meristems enable plants to grow in length and to branch. But a plant unable to grow in diameter would not become very tall before buckling under its own weight. And a plant unable to add vascular tissue could not supply its ever-increasing number of leaves with water. Growth in diameter thus serves two functions: It strengthens the stem, and it increases the capacity of the vascular system.

Primary growth is the increase in length made possible by apical meristems. **Secondary growth** is an increase in diameter. It results from a new type of meristem, one that forms only after elongation is complete. The fossil record shows that the ability to grow in diameter evolved millions of years after primary growth, and it did so independently in several groups, testimony to the benefits of being able to compete for sunlight by growing tall. Today, secondary growth occurs almost exclusively in seed plants, although this was not always the case in Earth's history.

Shoots produce two types of lateral meristem.

New cells must be added as a plant grows in diameter. The sources of these new cells are **lateral meristems,** which form along the length of a stem. Lateral meristems are similar to shoot apical meristems in their ability to produce new cells that grow and differentiate. However, lateral meristems differ from their apical counterparts in several ways. First, lateral meristems surround the stem, rather than occur at its tip. Second, because lateral meristems form only after elongation is complete, the new cells they produce grow in diameter but not in length. Finally, as a stem becomes thicker, the number of meristem cells needed to encircle the stem also increases. Thus, lateral meristems become larger over time.

Plants produce two distinct lateral meristems that together result in secondary growth. One of these, the **vascular cambium** (plural, **cambia**), is the source of new xylem and phloem. The xylem produced by secondary growth is familiar to you: It's wood. Plants with secondary growth are often described as woody, whereas plants that lack secondary growth are described as herbaceous. Most of the mass of a large tree consists of wood, generated from the vascular cambium.

As the plant's diameter increases, the epidermis formed during primary growth eventually ruptures. Thus, plants with secondary growth have a second lateral meristem, the **cork cambium,** which renews and maintains a protective outer layer. This outer layer protects the stem against herbivores, mechanical damage, desiccation, and, in some species, fire.

The vascular cambium produces secondary xylem and phloem.

The vascular cambium forms at the interface between xylem and phloem. In stems, the vascular cambium derives from procambial cells within each vascular bundle and parenchyma cells in between the bundles. In forming vascular cambium, both types of cell take on a new identity, becoming meristem cells.

The vascular cambium produces new cells that differentiate on both its sides. Those to the inside of the vascular cambium become **secondary xylem,** and those on the outside become **secondary phloem** (**Fig. 31.11**). The vascular cambium forms a continuous layer, one cell thick, which surrounds the xylem and runs nearly the entire length of the plant. When you peel the bark off a stem, it separates from the wood at the vascular cambium.

FIG. 31.11 **Secondary growth.** The vascular cambium is the source of new xylem and phloem, while the cork cambium maintains an outer layer of bark.

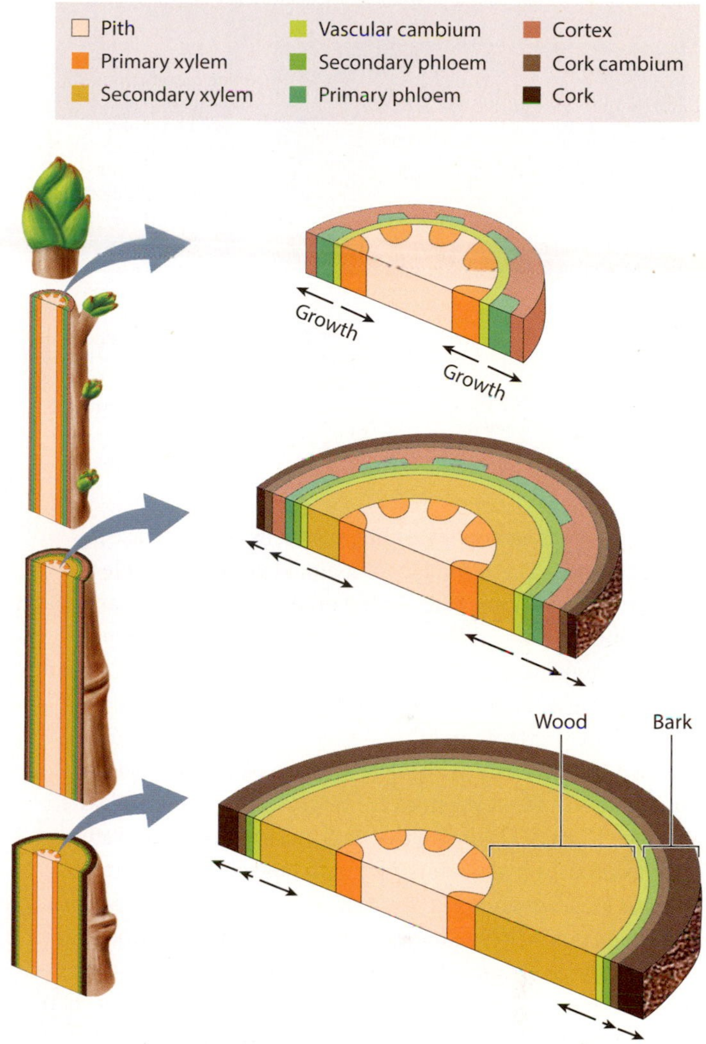

Secondary growth involves the formation of a continuous vascular cambium that produces secondary xylem toward the center of the stem and secondary phloem toward the outside. The cork cambium maintains a protective outer layer.

This is because the undifferentiated and actively dividing cells of the vascular cambium are weak and easily pulled apart. All of the material produced to the outside of the vascular cambium, including the secondary phloem, makes up the bark.

Plants must produce more secondary xylem than secondary phloem to replenish the water lost by leaves at high rates during the uptake of CO_2 for photosynthesis (Chapter 29). As the vascular cambium forms new secondary xylem cells along its inner face, the newly formed cells push the vascular cambium outward, much the way the elongation of newly formed cells pushes the shoot apical meristem upward. The vascular cambium accommodates

FIG. 31.12 **Annual growth.** In seasonal climates, tree rings provide a record of the annual accumulation of wood. The tree rings shown are from douglas-fir (*Pseudotsuga menziesii*). *Source: H.D. Grissino-Mayer.*

this outward movement by adding new cells that increase its circumference. The secondary phloem, however, is unable to produce any additional cells. As the production of secondary xylem pushes the vascular cambium outward, the phloem becomes stretched and, eventually, damaged. Consequently, the vascular cambium must continually produce new secondary phloem.

Over time, xylem conduits lose their ability to transport water. In long-lived trees, the xylem that is still active is located in a **sapwood** layer adjacent to the vascular cambium, while the **heartwood,** in the center of the stem, does not conduct water. Heartwood is often darker in color than sapwood because it contains large amounts of chemicals such as resins. For this reason, heartwood is typically more resistant to decay than sapwood.

In many plants growing in seasonal climates, secondary xylem cells produced at the end of the growing season are smaller than ones produced earlier. If you cut down a tree and inspect the top surface of the stump, the layer of smaller cells is visible as the dark edge of a **growth ring.** Counting growth rings makes it possible to determine a tree's age. Growth rings also provide a window into the past because by measuring the width of each ring, it is possible to determine how much wood was laid down in a given year (**Fig. 31.12**). A tree produces more wood when moisture and temperature levels are favorable for photosynthesis. Thus, annual growth rings provide a record of past climatic conditions, as well as events such as fire.

The cork cambium produces an outer protective layer.

The cork cambium maintains a protective layer around a stem that is actively increasing in diameter (see Fig. 31.11). The cork cambium forms initially from cortex cells that de-differentiate—that is, they regress to their earlier state—to become meristem cells. However, with time, this layer of actively dividing cells becomes increasingly distant from its source of carbohydrates—the phloem. As the cork cambium becomes cut off from its carbohydrate supply, a new cork cambium forms within the secondary phloem. Thus, each cork cambium is relatively short- lived, and in many cases, new cork cambia form in patches rather than as continuous layers. As a result, the bark of many trees has a patchy and fissured appearance (**Fig. 31.13**).

The cells produced by the cork cambium are called cork. As cork cells mature, they become coated with **suberin,** a waxy compound that protects against mechanical damage and the entry of pathogens, while also forming a barrier to water loss. However, this layer of waxy, nonliving cells impedes the inward diffusion of oxygen needed to meet the respiratory demands of living cells. The outer bark cells are less tightly packed in regions called **lenticels,** allowing oxygen to diffuse into the stem (**Fig. 31.14**).

→ **Quick Check 2** Why do plants have two types of lateral meristem?

Wood has both mechanical and transport functions.

We have seen that as a tree grows, its secondary xylem fulfills the two roles of secondary growth: to provide trees with the strength and stability to stand upright and support large crowns, and to transport water and nutrients from the roots to the leaves. In

FIG. 31.13 **Outer bark.** (a) Outside and (b and c) inside, showing how successive cork cambia are produced. The growth of many, discontinuous cork cambia results in the patchy bark of pine and walnut trees. *Sources: a. Alan Majchrowicz/age fotostock; b. Biosphoto/Dr. Keith Wheeler/Science Photo Library.*

a. Outside of tree, ponderosa pine (*Pinus ponderosa*) **b.** Cross section of bark, walnut (*Juglans regia*) **c.**

FIG. 31.14 Lenticels in (a) water birch (*Betula occidentalis*) and (b) basswood (*Tilia americana*). Lenticels allow oxygen to diffuse through the outer bark to supply the respiratory needs of the cells within the stem. *Sources: a. Robert and Jean Pollock/Science Source; b. Garry DeLong/Science Source.*

most gymnosperms, a single cell type, the tracheid, fulfills both functions (**Fig. 31.15a**). As discussed in Chapter 29, tracheids are single-celled xylem conduits that have thick lignified cell walls. They are nonliving and empty at maturity. The wood of a pine tree, for example, is relatively uniform in structure, consisting almost entirely of tracheids. This type of wood does not vary dramatically from species to species.

In contrast, the wood of angiosperms is diverse in structure because xylem cells mature into several distinct types that separate the functions of water conduction and mechanical support (**Fig. 31.15b**). **Fibers** are narrow cells with extremely thick walls and almost no lumen; their primary function is support. In contrast, **vessel elements** can be extremely wide cells; their primary function is water transport (Chapter 29). The largest vessel elements are close to 0.5 mm in diameter and thus visible to the naked eye. Vessel elements can be much wider than tracheids because the fibers provide support. Because vessel elements develop open connections that link them end to end with other vessel elements, creating multicellular xylem vessels, angiosperms can produce wood that transports water more efficiently than the wood of gymnosperms.

The ability to form xylem vessels has a number of important consequences. One is that angiosperms can have higher rates of water flow through their stems than is possible with wood that contains only tracheids. As a result, angiosperms sustain higher rates of CO_2 uptake and thus higher growth rates. In addition, because vessels can be much more efficient than tracheids at transporting water, angiosperms can devote a smaller fraction of their wood to cells for water conduction, and a larger fraction to other cell types. The wood of some angiosperms contains many fibers, in extreme cases resulting in wood that is so dense that it sinks. These dense woods are extremely strong, allowing trees to create spreading crowns with outstretched branches. Such "hardwoods" are useful as flooring materials and furniture. The wood of other angiosperms contains a high percentage of living cells instead of fibers, and this wood has a very low density. Balsa trees, whose wood is used in building model airplanes, are a good example.

FIG. 31.15 Wood structure. These scanning electron micrographs show (a) a gymnosperm, Douglas fir (*Pseudotsuga menziesii*), and (b) an angiosperm, English elm (*Ulmus procera*). *Sources: a. Scanning electron micrograph courtesy of the N.C. Brown Center for Ultrastructure Studies, State University of New York College of Environmental Science and Forestry, Syracuse, NY; b. Andrew Syred/Science Source.*

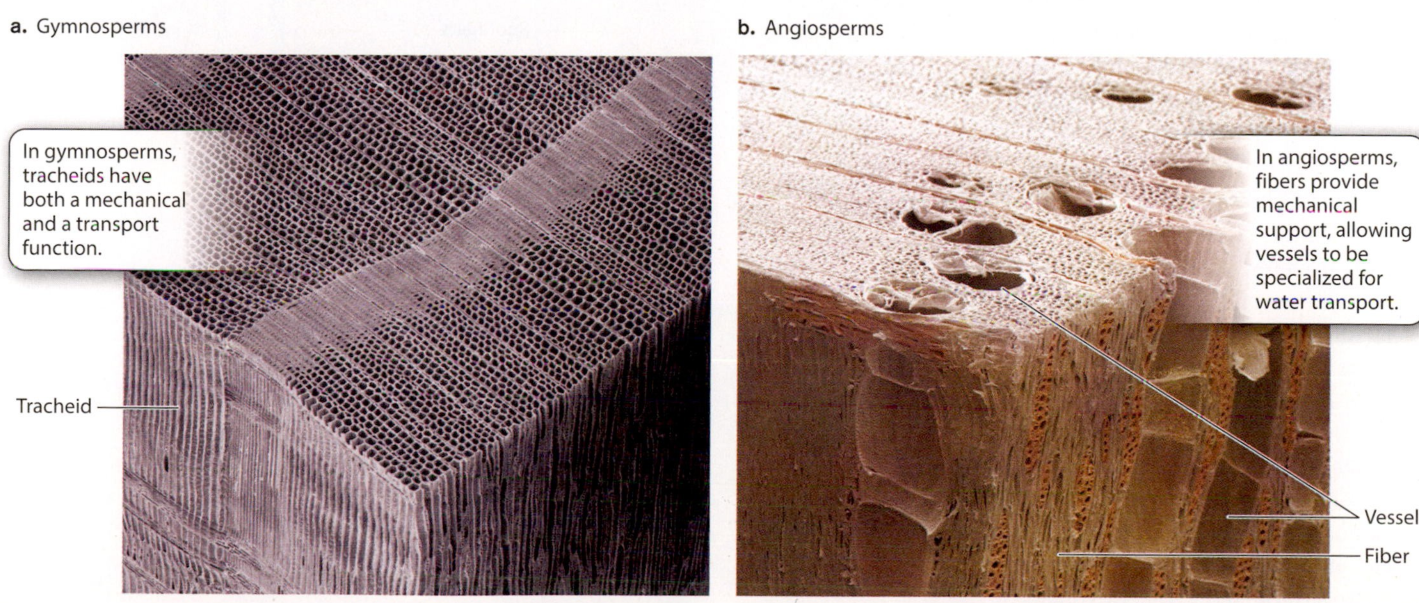

31.4 ROOT GROWTH AND DEVELOPMENT

The roots that evolved in vascular plants enabled them to obtain water and nutrients from the soil and provided a mechanical anchor for plants capable of growing tall. As in the case of leaves and wood, the fossil record shows that the first vascular plants lacked roots. These plants survived only in extremely wet environments in which they could absorb water through horizontal, and possibly submerged, stems. The evolution of roots allowed vascular plants to expand into drier habitats and ensured the stability of taller stems.

The fossil record also indicates that roots evolved at least twice. Nevertheless, the roots of all vascular plants share many features. This suggests that the physical challenges of growing through dense soil and the demands of obtaining water and nutrients from that soil place strong constraints on root structure.

Roots grow by producing new cells at their tips.

In Chapter 29, we saw that plants produce large numbers of thin roots to obtain sufficient nutrients from the soil. And to obtain water during periods without rain, plants produce roots that penetrate deep into the soil. Growing roots into the soil presents challenges different from those associated with elongating stems into the air. Although the dangers of drying out are much reduced belowground, deeper soil layers can have densities approaching that of concrete. How are the thin and permeable roots needed for water and nutrient uptake able to penetrate and grow through the soil?

In one respect, roots grow in the same way as shoots in that new cells are produced at their tip. The **root apical meristem** is the source of new cells. After cell division, these cells elongate and become differentiated. They complete their differentiation only after elongation has ceased (**Fig. 31.16**). Thus, as we move back from the tip of the elongating root, we encounter successive regions of cell division, elongation, and maturation, just as we saw in stems.

Roots, however, differ from stems in several important ways. One difference is that roots are much thinner than stems. Plants with thin roots can more efficiently produce the large surface area needed for uptake of water and nutrients from the soil. Thin roots can also more easily bend around soil particles and make use of channels made by invertebrate animals that burrow through the soil. A second difference is that the root apical meristem is covered by a **root cap** that protects the meristem as it grows through the soil. As the root elongates through the soil, root cap cells are rubbed off or damaged. Thus, the root cap is continually supplied with new cells produced by the root apical meristem. A third difference between roots and stems, and one that can be seen without a microscope, is that the root apical meristem does not produce any lateral organs, whereas the shoot apical meristem makes leaves. Root hairs, the single-celled outgrowths of root epidermal cells that greatly increase the surface area (Chapter 29), are formed only after elongation has ceased.

Root elongation and vascular development are coordinated.

A cross section taken through the zone of cell maturation shows a series of concentric tissue layers, starting with the epidermis on the outside, followed by the root cortex, and then the endodermis (Fig. 31.16). Recall that the endodermis plays a key role in the selective uptake of nutrients (Chapter 29). The Casparian strip, a waxy band surrounding each endodermal cell, forces all materials that will enter the xylem to cross the plasma membranes of these cells. At the center of the root is a single vascular bundle in which the xylem is arranged as a fluted cylinder, with phloem nestled in each groove. In cross section, the xylem is arranged in a star shape; the number of points of the star differs in different plant species.

FIG. 31.16 The structure of a typical root, including the root apical meristem and root cap.
Photo source: Walker Mi./Getty Images.

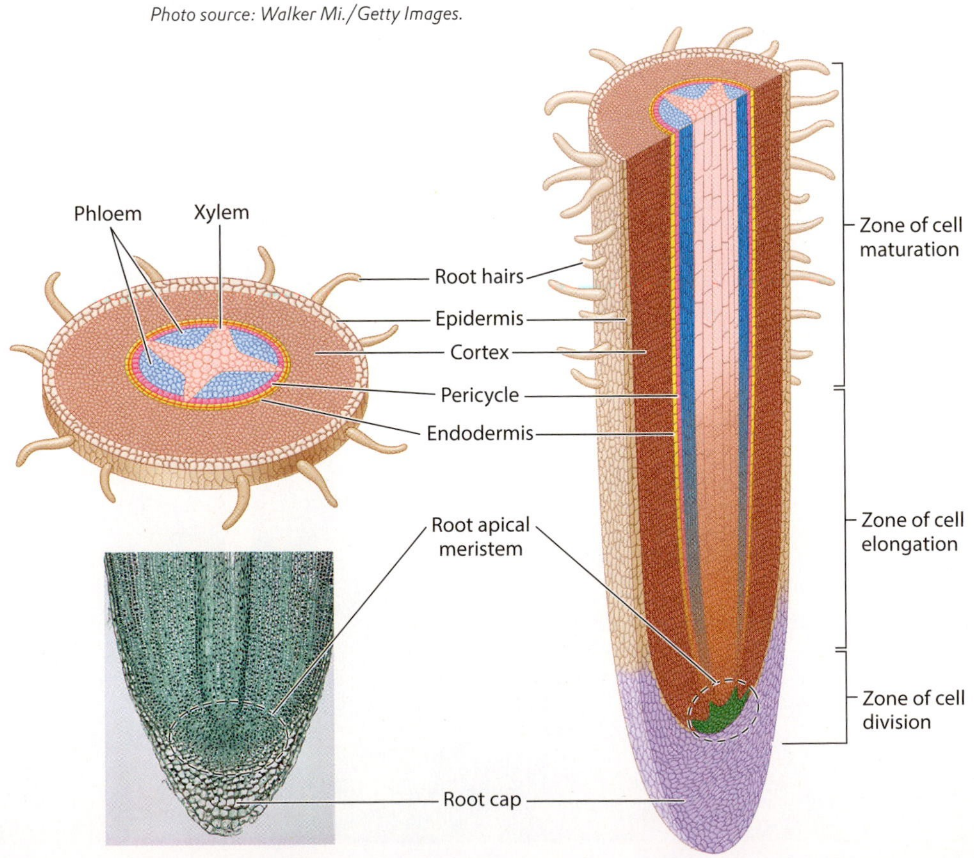

As in stems, the hormone auxin plays a central role in specifying which cells differentiate into vascular tissues. Auxin produced in shoot meristems is transported downward through the mature stem and into the roots by the phloem. At the end of the fully formed phloem, auxin moves by polar transport from cell to cell until it reaches the root apical meristem and the root cap. As auxin moves through the elongation zone, it triggers the formation of procambial cells that subsequently become phloem and xylem.

In this way, the production of auxin by developing leaves and the polar transport of auxin from cell to cell ensures that the vascular system is continuous between roots and leaves. Phloem extends closer to the root tip than do fully lignified xylem conduits. The root apical meristem does not need direct access to the xylem because it can obtain water and nutrients directly from the soil but depends entirely on the phloem for carbohydrates needed for growth.

So far, we have described only the primary growth of roots. However, roots can undergo secondary growth as well. Through secondary growth, roots increase in diameter and add new xylem and phloem. As in stems, the vascular cambium forms at the interface between the xylem and the phloem. Indeed, the vascular cambium must be continuous from the shoot to the roots to ensure that the secondary xylem and phloem provide an unbroken connection between the roots and the leaves. In older roots with extensive secondary growth, a cork cambium is also formed. As in stems, the cork cambium renews the outer layers of cells that protect the living tissues from damage and predation.

The formation of new root apical meristems allows roots to branch.

Compared to stems, roots branch extensively. Even a plant that has only a single stem aboveground will have many tens of thousands of elongating root tips. A major reason that roots branch so much is that most of the water and nutrients are taken up in a zone near the root tip where root hairs are most abundant. Thus, branching is necessary to create enough root surface area to supply the water and nutrient needs of the shoot.

For a root to branch, a new root apical meristem must be formed. In roots, new meristems develop from the **pericycle,** a single layer of cells just inside the endodermis (**Fig. 31.17**). The pericycle is immediately adjacent to the vascular bundle, so the new root is connected to the vascular system from the beginning. Because the new root meristem develops internally, the new root must grow through the endodermis, cortex, and epidermis before reaching the soil.

One consequence of initiating new root meristems from the pericycle is that new roots can form anywhere along the root. As a result, plants have tremendous flexibility in the development of their root systems and roots can proliferate locally in response to nutrient abundance. For example, the presence of nitrate, the most common form of nitrogen in soil, triggers the formation of lateral roots. Only a relatively small number of a plant's roots, however, continue to elongate and branch indefinitely. Most new roots are active for only a short time, after which they die and are shed. This turnover of fine roots allows the root system to respond to changes in the availability of water and nutrients but represents a major expenditure of carbon and energy.

Up to this point, we have described how new roots form from existing roots. New roots can also be produced by stems. Many plants that lack secondary growth produce horizontal stems that grow on or beneath the soil surface. In these plants, new roots form at each successive node, so the uptake capacity of the root system is directly coupled to the elongation of the stem.

When the stem of a plant is severed, new roots often form at the cut end. The formation of new root meristems is stimulated by auxin, which is produced by the young leaves and accumulates by polar transport at the cut end. Because their cut stems can form new roots, plants are able to survive damage and in some cases even proliferate afterward. Many commercially important plants are propagated through cut stems.

→ **Quick Check 3** How is root development similar to and how is it different from stem development?

The structures and functions of root systems are diverse.

How a plant's roots are distributed in the soil has a tremendous impact on the ability of the root system to supply the shoot with

FIG. 31.17 Branching in roots. New root meristems develop from the pericycle, and the developing lateral root grows through the root cortex before reaching the soil. The light micrographs show willow (*Salix*) roots. *Source: Science Source.*

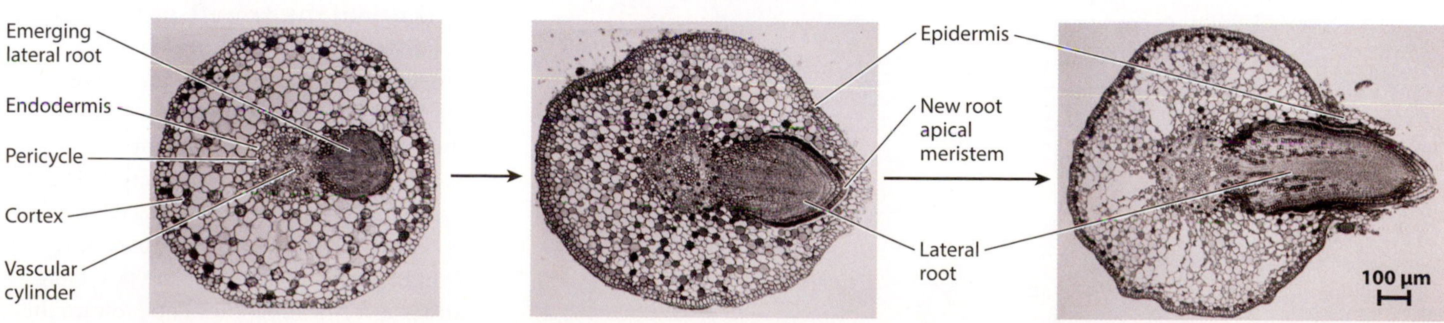

water. Some plants produce very deep roots that provide access to groundwater that is independent of the variability of rainfall. For example, roots have been observed in mine shafts and caves more than 50 m below the soil surface. Producing and maintaining these non-photosynthetic structures is costly for the plant.

At the other extreme are plants that produce shallow and spreading root systems. For example, barrel cacti produce roots that penetrate less than 10 cm into the soil but extend out several meters from the base of the plant. These roots cast a wide but shallow net to capture as much water as possible from intermittent rain showers.

Some plants produce distinctive roots whose principal role is *not* the absorption of water and nutrients from the soil (**Fig. 31.18**). For example, climbing plants such as poison ivy produce roots along their stems that allow them to adhere to the sides of trees (Fig. 31.18a). The conspicuous prop roots produced by some tropical trees provide mechanical stability that allows these plants to grow tall despite their slender stems (Fig. 31.18b).

Many plants produce swollen roots that store resources such as starch (Fig. 31.18c). Plants with belowground storage can produce new leaves and shoots quickly following either damage or a period of drought or cold that has killed the shoots. Cassava and sweet potato are examples of economically important species with storage roots.

Perhaps the most unusual rooting structures are the "breathing" roots produced by trees that grow in water. For example, black mangroves produce pencil-sized roots that extend vertically upward out of the sandy soil (Fig. 31.18d). Breathing roots do not actually breathe; instead, they contain internal air spaces that provide an easier pathway for oxygen to diffuse into the roots than through the waterlogged soil. For this reason, breathing roots must extend above the surface of the water.

FIG. 31.18 **Root diversity.** (a) Climbing roots of poison ivy; (b) prop roots of a *Pandanus* tree; (c) storage roots of yams; and (d) breathing roots of black mangrove (shown at low tide). *Sources: a. Doug Wechsler/age fotostock; b. A Jagel/age fotostock; c. PhilipYb/Shutterstock; d. Patrick Lynch/Alamy.*

31.5 THE ENVIRONMENTAL CONTEXT OF GROWTH AND DEVELOPMENT

The capacity for continued growth allows plants to respond to changes in their environment by growing more in one direction than another. To respond in a manner that enhances fitness, however, plants must first gain information about the world around them. Plants rely on three types of sensory receptors: Photoreceptors sense the availability of light needed to drive photosynthesis; mechanical receptors sense physical influences such as gravity and wind; and chemical receptors detect the presence of specific chemicals, as well as chemical gradients.

We saw in Chapter 9 how a signal can be conveyed to a cell by a change in the shape of a receptor molecule on the cell surface or within the cell. Sensory receptors in plants work in a similar manner. For example, the absorption of light by a photoreceptor changes the chemical properties of the photoreceptor, similar to the photoreceptors in your eye. When a plant's sensory receptor is activated, it produces a signal that triggers changes in the cell's metabolism or alters patterns of gene expression. Hormones play a key role in translating information gained by the plant's sensory receptors into an appropriate developmental response.

Plants orient the growth of their stems and roots by light and gravity.

If you lay a plant on its side, its stems will turn upward and its roots will bend downward. Similarly, if you place a houseplant near a window, the shoot will grow toward the light. **Tropism** is the bending or turning of an organism in response to an external signal such as light or gravity. Plant stems are positively **phototropic,** meaning they bend toward the

HOW DO WE KNOW?

FIG. 31.19

How do plants grow toward light?

BACKGROUND In 1926, Frits Went, a Dutch graduate student, followed up on the experiments of Charles and Francis Darwin. In particular, Went sought to understand how the perception of light by the tip of the plant led to a growth response farther down the stem. He reasoned that if the signal from the shoot tip was a chemical, he could isolate it by cutting the tip off of a growing plant and putting it on a block of gelatin. The gelatin would act like a trap, capturing any chemical that moved out of the shoot tip.

HYPOTHESIS A chemical signal links the perception of light by the shoot tip with the growth of the stem toward light.

EXPERIMENT Went cut off the tips of young oat plants (*Avena sativa*) and placed them on gelatin blocks for periods of 1 to 4 hours. Removing the tip of the plant caused the plants to stop elongating. When the gelatin block was placed on top of the cut plant (Experiment 1), growth resumed. This experiment demonstrated that a growth-inducing chemical was exported from the tip of the young oat plants.

Interestingly, Went found that if the block of gelatin was not centered on the tip but instead was a bit off center, the plant bent away from the side of the gelatin block (Experiment 2). In this case, the side of the plant exposed to the gelatin block grew faster, causing bending, because it received more of the chemical signal.

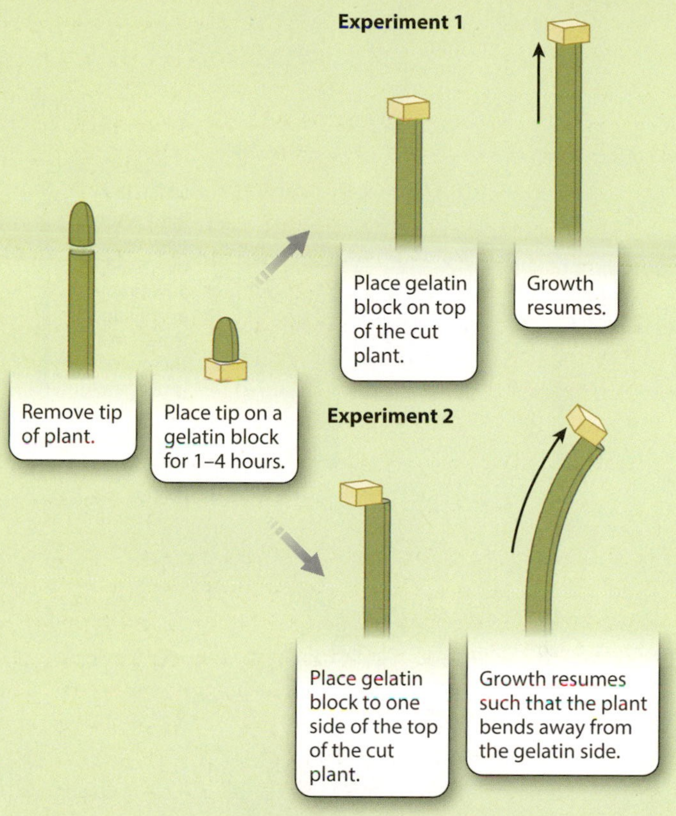

CONCLUSION A signal is produced in the tip of the plant and travels downward through the stem, toward the roots. This signal stimulates growth. Light causes the signal to move to the opposite side, stimulating growth on that side, which in turn causes the plant to bend toward the light. This signal is now known to be a plant hormone, called auxin from the Latin word *augere*, "to increase."

FOLLOW-UP WORK Went followed up his experiments by determining how the level of auxin relates to the degree of curvature. This assay became known as the Avena curvature test, after the experimental species. Other experiments showed that auxin moves in a directed fashion by polar transport. Later, auxin was shown to be indole-3-acetic acid (IAA).

SOURCES Darwin, C., and F. Darwin. 1880. *The Power of Movement in Plants*. London: John Murray; Went, F. W. 1926. "On Growth-Accelerating Substances in the Coleoptile of *Avena sativa*." *Proceedings of the Royal Academy of Amsterdam* 30:10–19.

light. Plant stems are also negatively **gravitropic,** meaning they grow upward against the force of gravity. In contrast, plant roots grow down and away from the light. Thus, roots are positively gravitropic and negatively phototropic.

Observing that grass seedlings bend toward light, Charles Darwin and his son Francis conducted experiments to determine what part of the plant detects light. When they covered just the tip of the young plants, the plants no longer grew toward the light, but when they buried them in fine black sand so that only their tips were exposed, they did. This experiment established the shoot tip as the site where the light is perceived but left open the question of what kind of signal caused the plants to bend (**Fig. 31.19**).

Photoreceptors allow plants to detect and respond to light. If a plant is exposed to light from only one side, only the photoreceptors on the illuminated side are activated. Phototropism in plants involves a photoreceptor that absorbs blue wavelengths of light. The absorption of blue light by the photoreceptor is thought to trigger a change in the placement of

FIG. 31.20 Phototropism in shoots, controlled by the lateral redistribution of the hormone auxin.

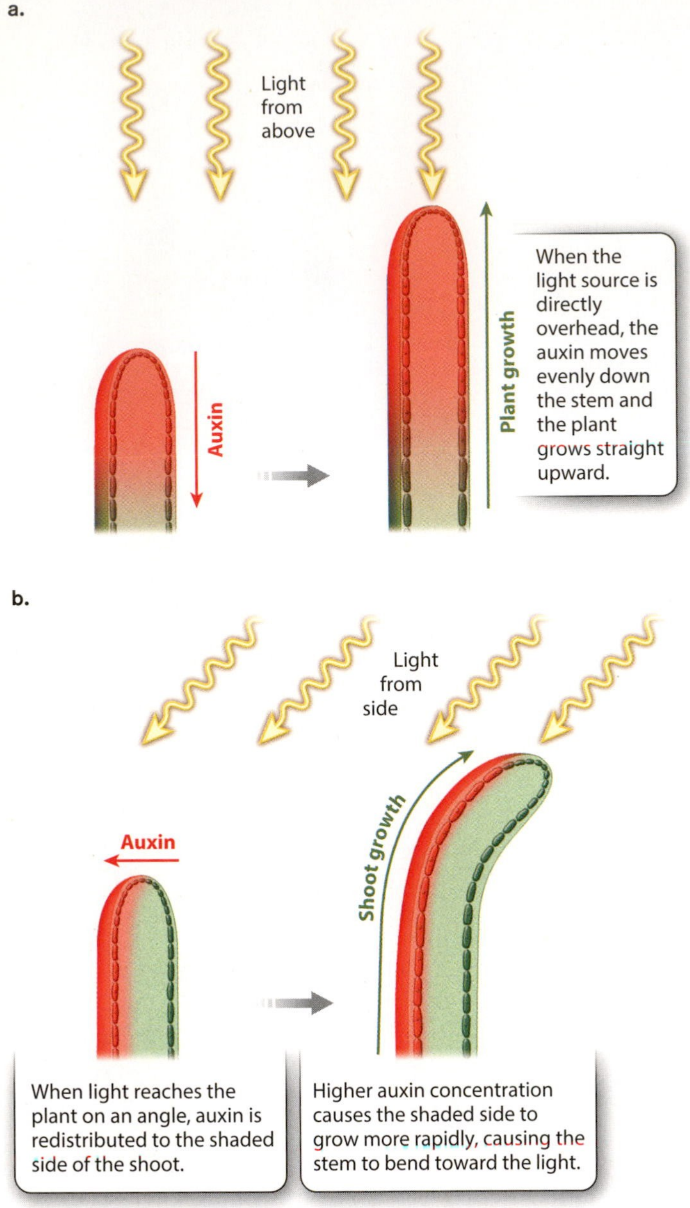

PIN proteins so that auxin is transported to the shaded side of the stem. As a result, concentrations of auxin are higher on the shaded side and lower on the illuminated side. Because auxin stimulates cell expansion in stems, the shoot grows faster on the shaded side than it does on the illuminated side, causing the plant to bend toward the light (**Fig. 31.20**).

Phototropism in roots is similar, in that illuminating a root from the side causes auxin to be redistributed toward the shaded side. In this case, auxin is transported to the shaded side within the root cap, and then moves upward in the outermost layer of cells into the zone of cell elongation. However, higher auxin concentrations *decrease* the rate of cell elongation in roots, causing roots to bend *away* from light. The reason that auxin redistribution leads to such a different response in roots compared with shoots is that roots and stems differ in their sensitivity to auxin and that in roots high concentrations of auxin trigger the production of the hormone ethylene, which reduces cell elongation.

A similar mechanism orients plants with respect to gravity. Moving a plant to a horizontal position results in the transport of auxin to the lower side. The accumulation of auxin on the lower surface causes stems to bend up and roots to turn down. How is the plant able to detect which way is up and which way is down? Recall that plants store carbohydrates as starch. Starch is a large molecule that has a density greater than water. Starch-filled organelles

FIG. 31.21 Root orientation with respect to gravity. Starch-filled statoliths settle to the bottom of cells, triggering a redistribution of auxin to the lower side of the root, where it slows elongation, causing the root to bend downward. *Photo source: Courtesy Dr. Christophe Jourdan.*

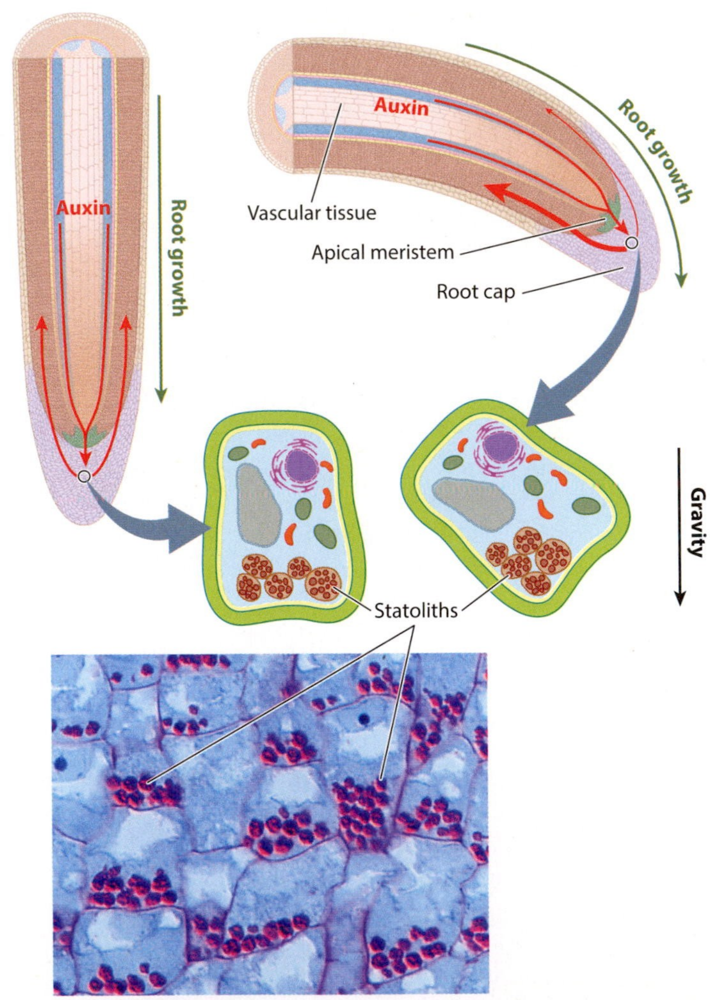

HOW DO WE KNOW?

FIG. 31.22

How do seeds detect the presence of plants growing overhead?

BACKGROUND To study the effect of light on seed germination, scientists exposed lettuce seeds that had been kept continually in the dark to different wavelengths of light and then counted what fraction of the seeds germinated. Red light had the greatest ability to stimulate germination, but surprisingly, far-red light inhibited germination such that fewer seeds germinated than in the control seeds, which were kept in darkness. In the rush to conduct more experiments, petri dishes with the light-treated seeds piled up by the sink until someone noticed that the seeds that had been experimentally treated with far-red light were now germinating.

HYPOTHESIS This observation suggested that the inhibitory effect of far-red light can be overcome by a subsequent exposure to red light.

EXPERIMENT The scientists exposed lettuce seeds to red and far-red light in an alternating pattern, ending with either red light or far-red light. They then placed the seeds in the dark for two days and afterward counted the number of seeds that had germinated.

RESULTS When the lettuce seeds were exposed to red light last, nearly 100% of the seeds germinated. By contrast, when the last exposure of the seeds was to far-red light, the percentage of seeds that germinated was dramatically reduced.

Lettuce Seeds Germinating After Exposure to Red and Far-Red Light in Sequence

SEQUENCE OF LIGHT EXPOSURE	SUBSTRATE-LEVEL GERMINATION (%)
Dark (control)	8.5
Red	98
Red, far-red	54
Red, far-red, red	100
Red, far-red, red, far-red	43
Red, far-red, red, far-red, red	99
Red, far-red, red, far-red, red, far-red	54
Red, far-red, red, far-red, red, far-red, red	98
Time →	

CONCLUSION Seed germination in lettuce is triggered by exposure to red light and is inhibited by exposure to far-red light in a reversible fashion. As a result, plants are able to track changes in the relative amount of red and far-red light, which provides information on the presence or absence of plants overhead (see Fig. 30.23).

FOLLOW-UP WORK Additional studies showed that this result was due to a single pigment in a photoreceptor, now known as phytochrome, that is converted into an active form by red light and reversibly converted into an inactive form by far-red light.

SOURCE Borthwick, H. A., et al. 1952. "A Reversible Photoreaction Controlling Seed Germination." *Proceedings of the National Academy of Sciences* 38: 662–666.

thus sink to the bottom of the cell. The weight of these organelles pressing on the cytoskeleton or membranes is thought to be the signal that causes auxin to move toward the lower side of the plant. In roots, gravity is a critical source of information regarding orientation. Specialized gravity-sensing cells in the root cap contain large starch-filled organelles known as **statoliths** (**Fig. 31.21**).

Seeds can delay germination if they detect the presence of plants overhead.

For a small seed that has few stored reserves, germinating in the shade of another plant could be fatal. But how can a seed detect the presence of plants overhead?

As sunlight passes through leaves, the red wavelengths are absorbed by chlorophyll but the far-red wavelengths are not. The ratio of red to far-red light is thus much higher under an open sky than it is in the shade of another plant. When scientists first began to study the effects of light on seed germination, they found that red light is particularly good at stimulating germination, while far-red light inhibits germination. A simple "flip-flop" experiment by H. A. Borthwick and colleagues showed that the inhibitory effect of far-red light could be overcome by a subsequent exposure to red light (**Fig. 31.22**). Ultimately, it was shown that only a single photoreceptor was needed to detect both far-red and red light.

Phytochrome is a photoreceptor that switches back and forth between two stable forms depending on its exposure to light (**Fig. 31.23a**). Red light causes phytochrome to change into a form that absorbs primarily far-red light (P_{fr}); far-red light causes phytochrome to change back into the form that absorbs red light (P_r). Because red light stimulates seed germination, we know that P_{fr} is the active form of phytochrome. That is, red light causes phytochrome to change into its P_{fr} form and P_{fr} sends a signal that triggers seed germination.

Phytochrome allows dormant seeds to detect the presence of plants overhead (**Fig. 31.23b**). Because it responds to the ratio of red to far-red light, the proportion of phytochrome that is in the active P_{fr} form is much greater in open environments than in the forest understory. Furthermore, because the ratio of red to far-red light is independent of light intensity, it provides a reliable signal that leaves are overhead even for seeds covered by a thin layer of soil. Independent of whether light is present or not, P_{fr} slowly converts back into the inactive P_r form. The light-independent conversion of P_{fr} to P_r prevents seeds that are too deep in the soil, yet still receive some light, from germinating.

FIG. 31.23 Detection of shading. Phytochrome detects the depletion of red light under the canopy.

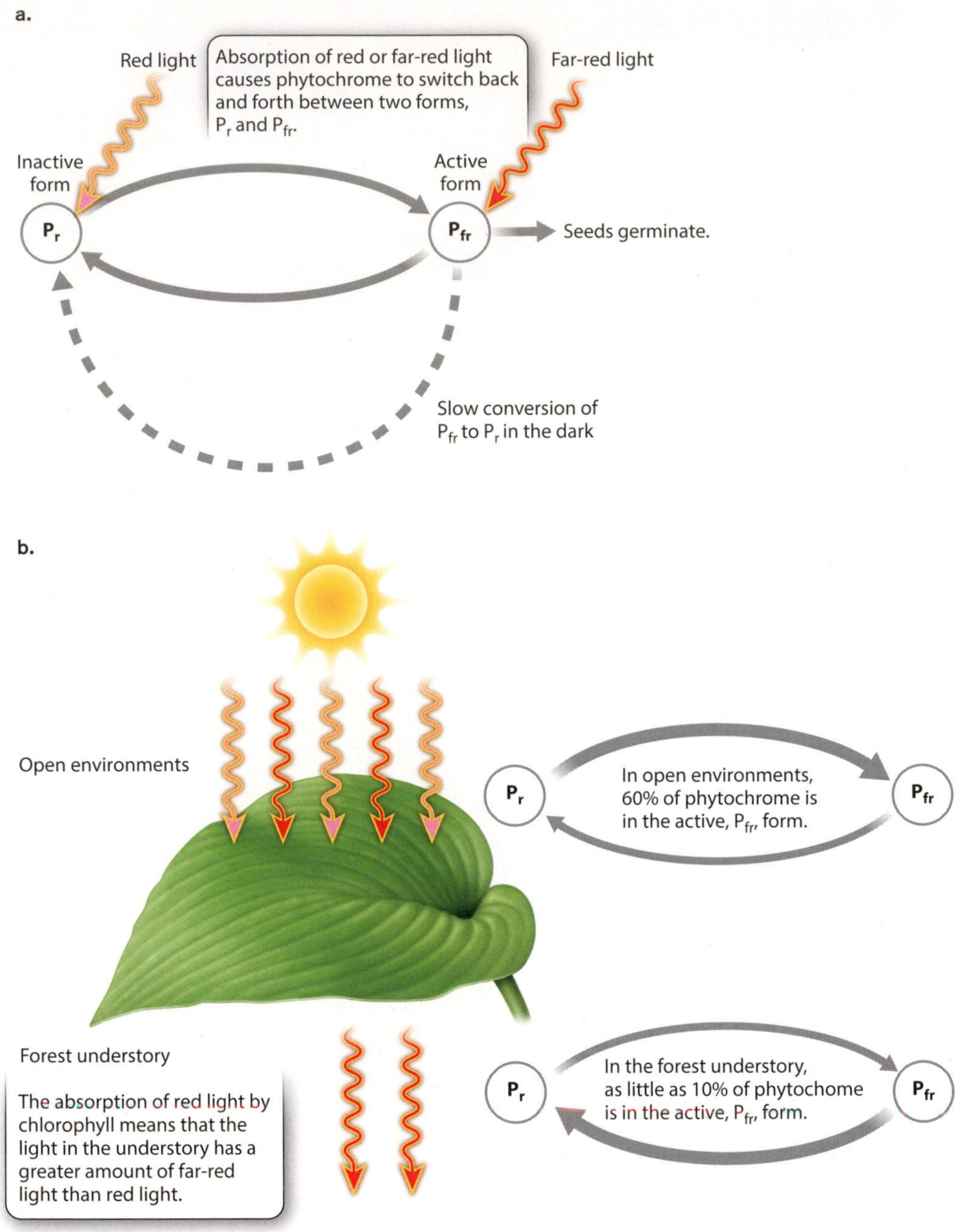

Plants grow taller and branch less when growing in the shade of other plants.

A newly germinated seedling must reach the soil surface before its stored resources are used up or it will die. Seeds that germinate in the dark produce completely white seedlings with thin, highly elongated internodes and small leaves (**Fig. 31.24**). By putting all their resources into elongation, seedlings grown in the dark maximize their chances of reaching the light. If and when they do, internodal elongation slows markedly and green leaves expand. These dramatic changes are triggered by photoreceptors that signal the presence of light. In its transformation from a dark-grown to a light-exposed seedling, the seedling receives signals from several types of photoreceptor, including one that absorbs blue wavelengths and one, phytochrome, that absorbs both red and far-red wavelengths.

Even after a plant emerges above the soil, its ability to sense and respond to light can enhance its chances for survival. Some species are adapted for shade, producing thin leaves that efficiently use low levels of light; others grow best in direct sunlight. When such sun-adapted species are grown in the shade of an into growing tall. These plant species produce stems that have long, thin internodes and fewer branches, and they invest less in root growth.

Interestingly, the effect is much less if the shade is produced by anything other than another plant. Experiments have shown that the taller, thinner stems are the result of a plant's sensing the presence of its neighbors. The reason a plant is able to sense neighboring plants is that chlorophyll absorbs red, but not far-red, wavelengths of light. Thus, as light passes through leaves, it becomes depleted in red wavelengths. As a result, phytochrome is converted from the P_{fr} to the P_r form. Furthermore, light reflected off the leaves of a plant's neighbors is also depleted in red wavelengths, so phytochrome allows a plant to detect not only when it is underneath another plant, but also when it is surrounded

FIG. 31.24 Seedlings grown in the light and the dark (left) or in full sun and shade (right). *Photo source: Nigel Cattlin/Science Source.*

by other plants that could grow to shade it in the future. Thus, phytochrome provides an early-warning system that competitors are nearby.

Roots elongate more and branch less when water is scarce.

The developmental sensitivity of plant shoots to their environment is mirrored underground. When a plant experiences drought, it produces more roots, and these roots penetrate farther into the soil, in part because they produce fewer lateral branches. By producing deeper roots, a plant increases its chances of reaching moister soil.

The root cap appears to play an important role in allowing roots to sense and respond to the amount of water in the soil. Because the root cap is not well connected to the plant's vascular system, the water status of its cells provides a good indication of the moisture content of the soil. As soils dry, root cap cells produce a hormone called **abscisic acid,** abbreviated as ABA (Table 31.1). Abscisic acid stimulates root elongation, leading to roots that penetrate deeper into the soil. It also triggers stomata to close, reducing the demand for water.

Abscisic acid stimulates root elongation by suppressing ethylene synthesis. Ethylene slows root elongation by influencing the orientation of cellulose in the cell wall. Cells treated with ethylene expand in diameter rather than in length. Thus, by inhibiting the production of ethylene, the production of ABA in drought-exposed roots leads to roots that penetrate deeper into the soil.

Exposure to wind results in shorter and stronger stems.

In the 1980s, scientists studied the model plant *Arabidopsis thaliana* to learn which genes are expressed in response to different hormones. The experimental treatment involved spraying plants with different solutions containing different hormones; control plants were sprayed with water. The hormone treatments were successful in that several genes were strongly up-regulated. However, the same genes were also turned on in the plants sprayed with water. Further experiments showed that what the plants were responding to was neither the hormones nor the water, but the fact that their stems were bent back and forth when the plants were sprayed. This led to the discovery of touch-sensitive genes in plants, which are activated by mechanical perturbation.

This discovery confirmed what foresters and horticulturists had long known. Plants exposed to wind produce stems that are shorter and wider than ones grown in more protected sites. In fact, commercial greenhouses often install fans, in part so that their plants will produce stems robust enough that the plants can thrive when moved outdoors. Flexing a stem back and forth triggers an increase in the synthesis of ethylene. As noted in section 31.4, cells treated with ethylene expand more in diameter and less in length, resulting in shorter and thicker stems.

31.6 TIMING OF DEVELOPMENTAL EVENTS

As plants moved onto land, they encountered environments with distinct seasons. In temperate regions, temperatures vary from summer to winter; in most tropical habitats, wet seasons alternate with dry ones. Thus, plants evolved mechanisms to control the timing of developmental transitions such as the production of flowers and other reproductive organs, the germination of spores or seeds, or the formation of overwintering buds. To accomplish this, plants need reliable information that tells them what season it is. While temperature tends to be unreliable—think of unusually warm or cold summers—day length is a dependable source of information about the seasons that works everywhere except close to the equator.

Flowering time is affected by day length.

In the 1910s, Wightman Garner and Henry Allard, scientists at the U.S. Department of Agriculture, set out to understand why Maryland Mammoth, a variety of tobacco, does not produce flowers during the summer like other tobacco plants. Instead, these plants continue to grow throughout the summer, reaching heights several times that of a typical tobacco plant and producing many more leaves. Only in autumn does the Maryland Mammoth tobacco plant finally produce flowers, too late in the year for seeds to mature.

One possible explanation for the observed delay in flowering time is that Maryland Mammoth must achieve a greater size before it produces flowers. However, when Garner and Allard grew Maryland Mammoth plants in a heated greenhouse during the winter, the plants flowered at the same time and size as other tobacco plants. Therefore, size does not determine when Maryland Mammoth flowers. Because the heated greenhouse was set to the same temperatures as those found during the summer, temperature could also be eliminated as a factor. Garner and Allard were left with only one environmental variable that was different between the summer and the winter: the length of the day.

In a series of experiments in the early part of the last century, Garner and Allen artificially shortened the day length during summer and extended it during winter. Their results showed that in Maryland Mammoth, as well as in many other plants, flowering is controlled by day length. They coined the term **photoperiodism** to describe the effect of the "photo period," or day length, on flowering (**Fig. 31.25**). **Short-day plants,** like Maryland Mammoth, flower only when the day length is less than a critical value. When the light period exceeds this threshold, a short-day plant continues to produce new leaves, but no flowers are formed. In contrast, **long-day plants,** which include radish, lettuce, and some varieties of wheat, flower only when the light period exceeds a certain length. Flowering of **day-neutral plants** is independent of any change in day length.

Sensitivity to day length ensures that plants flower only when they are large enough to support the development of a large number of seeds *and* there is enough time for those seeds to mature before the onset of winter. A short-day plant that germinates in the spring will not flower until late summer, when the decreasing day length falls below the critical value for that species (**Fig. 31.26a**). This flowering strategy allows short-day plants to grow as large as possible and switch from producing leaves to producing flowers in time to complete seed development before winter. In contrast, many long-day plants produce flowers during midsummer. These plants germinate in spring and are triggered to flower when the day length exceeds a critical value (**Fig. 31.26b**).

FIG. 31.25 **Control of flowering by daylength.** Short-day plants flower only when hours of daylight are below a certain number. Long-day plants flower only when hours of daylight are above a certain number. *Photo sources: (top left) Darlyne A. Murawski/Getty Images; (bottom left) Liz Van Steenburgh/Shutterstock; (top right) Dr. Jeremy Burgess/Science Source; (bottom right) Biosphoto/Dr. Jeremy Burgess/Science Photo Library.*

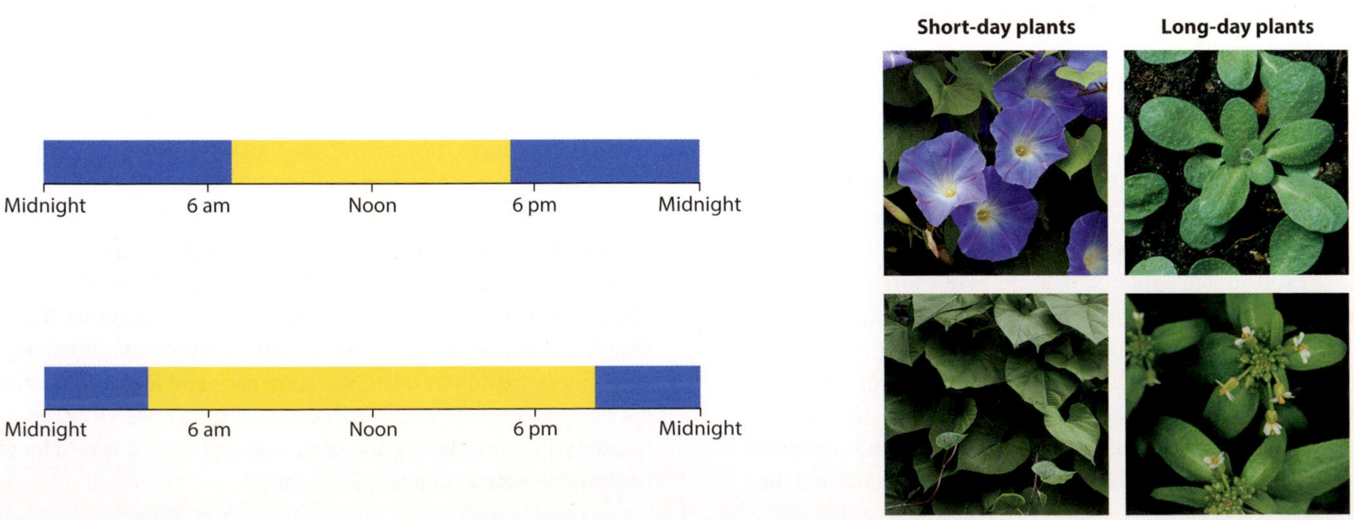

FIG. 31.26 Flowering strategies for short-day and long-day plants. (a) A short-day plant that germinates in the spring flowers in the late summer when day length falls below a critical value. (b) A long-day plant flowers in the summer after a critical day length is reached. (c) A long-day plant that requires vernalization does not flower until the critical day length is reached in the second summer.

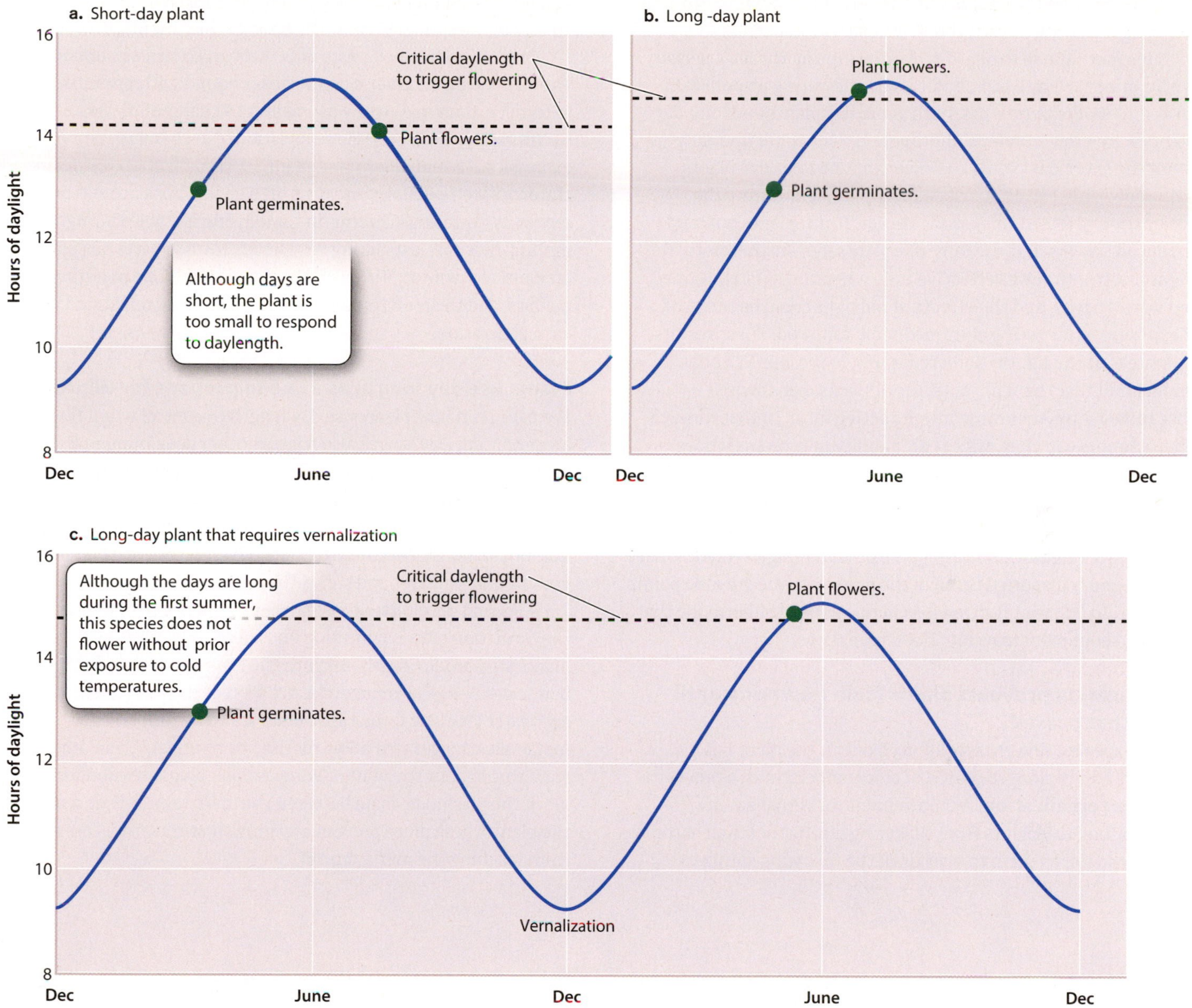

Plants use their internal circadian clock and photoreceptors to determine day length.

To determine day length, one first needs to know what time it is, specifically how long it has been since the sun came up, and then whether it is light or dark outside. Plants, like all organisms, have an internal circadian clock. Circadian clocks are biochemical mechanisms that oscillate with a 24-hour period and are coordinated with the day–night cycle. Phytochrome is the photoreceptor responsible for synchronizing the plant's internal oscillator with the first light of dawn. The circadian clock tells the plant how long it has been since the sun came up, while phytochrome tells the plant whether it is light or dark during a specific phase of the circadian cycle.

The actual mechanism used by plants to determine day length relies on the fact that the plant's circadian clock affects the transcription of many genes. For example, in *Arabidopsis thaliana*, a long-day plant, a gene whose protein product is critical for triggering flowering is transcribed at increasingly higher rates throughout the day, peaking ~16 hours after dawn. After that, the rate of transcription declines. The protein product of this gene is targeted for rapid degradation unless light is present. Thus, whether the protein builds up to a level sufficient to trigger flowering depends on day length. During long days, the protein reaches levels needed to trigger flowering because peak production occurs during daylight, when the protein is not targeted for destruction. During short days, the protein product does not reach levels needed to trigger flowering because the gene is transcribed at high rates only after the sun has gone down, so the protein is rapidly degraded.

This mechanism explains why interrupting the dark period with a brief exposure to light alters the flowering response. In an early set of experiments, both short-day plants and long-day plants were grown under a photoperiod that would normally cause the short-day plants to flower and the long-day plants to produce only leaves. A brief exposure to light during the dark period reversed this result. The short-day plants no longer produced flowers, but the long-day plants did. Furthermore, the night interruption was effective only when the light contained red wavelengths, and the effects of red light could be reversed if exposure to far-red light immediately followed. These results indicate that the light is detected by phytochrome. We now understand that the night interruption acts like a switch that converts phytochrome into its active form. In that form, phytochrome can then affect the stability of proteins whose synthesis follows a circadian rhythm.

Other experiments demonstrated that day length is sensed by the leaves and not by the buds where the flowers will form. When conditions are right for flowering, the protein florigen is synthesized in leaves and transported through the phloem to the growing points of a plant. In this way, florigen is a chemical signal or hormone that converts shoot meristems into floral meristems.

Vernalization prevents plants from flowering until winter has passed.

In some species, flowering is induced only if the plant has experienced a prolonged period of cold temperatures, a process known as **vernalization.** A requirement for vernalization prevents long-day plants from flowering during their first summer, and instead forces them to wait until the following spring to flower, when the chances for successful pollination and seed development are greater (**Fig. 31.26c**). Even parts of the plant that form long after the cold stimulus is past seem to "remember" that they received the required cold treatment. This "memory" of winter is important because many plants that overwinter grow rapidly in the spring. What is the basis of this "memory"?

One common mechanism acts through chromatin remodeling. The DNA of all eukaryotic organisms is bound with proteins to form chromatin. We saw in Chapter 19 that modifications to chromatin, such as DNA methylation, can change gene expression. In *Arabidopsis*, for example, vernalization results in chromatin remodeling that turns off a gene whose protein product represses flowering. Chromatin remodeling is stable through mitotic divisions, explaining why newly formed parts of a plant "remember" winter. However, the slate is wiped clean during meiosis, and the requirement for vernalization is reinstated with each generation.

Plants use day length as a cue to prepare for winter.

We have seen how plants use day length to control when flowers are produced. Day length also triggers other developmental events, particularly ones that allow plants that persist for more than a year to prepare for winter. One such developmental change is the formation of storage organs, for example in roots. Carbohydrates that accumulate in these structures can support the growth of new leaves and stems the following spring.

A second developmental change in response to day length is the formation of overwintering buds. As the days begin to shorten, plants stop producing photosynthetic leaves and begin forming bud scales. Bud scales surround and protect the meristem from ice and water loss. The formation of bud scales accompanies a series of metabolic changes that allow meristems to remain in a dormant state throughout the winter. For example, plugs are produced that block the plasmodesmata between the meristem and the rest of the plant. These plugs prevent any growth-stimulating compounds from reaching the meristem. ■

Core Concepts Summary

31.1 In plants, upward growth by stems occurs at shoot apical meristems, populations of totipotent cells that produce new cells for the lifetime of the plant.

Stems are built from repeating modules consisting of nodes, where one or more leaves are attached, and internodes, which are the regions of stem between the nodes. page 642

The shoot apical meristem, located at the tip of each stem, is the source of new cells from which the stem and the leaves are formed. page 642

Stems contain zones of cell division, elongation, and maturation. Most of the increase in size occurs in the zone of cell elongation. page 643

In seed plants, branching occurs at axillary buds at the base of each leaf. page 644

Leaves differ from stems in that they differentiate fully and thus have no capacity for continued growth. page 645

Leaves develop from leaf primordia produced by the shoot apical meristem. page 645

In most seed plants, the stem's vascular bundles are organized as a ring. page 646

Shoot apical meristems and axillary buds can be transformed into floral meristems, at which point they lose their capacity for continued growth. page 646

31.2 Hormones are chemical signals that influence the growth and differentiation of plant cells.

Traditionally, five plant hormones have been identified: auxin, gibberellic acid, cytokinins, ethylene, and abscisic acid. page 647

Xylem and phloem differentiate from procambial cells. page 649

The polar transport of auxin through PIN proteins guides the formation of leaf primordia and the development of vascular channels (xylem and phloem) that connect leaves and stems. page 649

Gibberellic acid causes internodal elongation of shoots. page 650

Apical dominance is the suppression of the outgrowth of axillary buds caused by hormones produced by roots (strigolactone) and shoots (auxin). page 650

31.3 Lateral meristems allow plants to grow in diameter, increasing their mechanical stability and the transport capacity of their vascular system.

The formation of two lateral meristems—the vascular cambium and the cork cambium—allows plants to grow in diameter. page 651

The vascular cambium produces secondary xylem toward the center and phloem toward the outside. page 651

The cork cambium produces a protective outer layer. page 652

Gymnosperm tracheids function in both mechanical support and water transport. Angiosperm xylem produces fibers that influence the strength of wood and vessel elements that function in water transport. page 653

31.4 The root apical meristem produces new cells that allow roots to grow downward into the soil, enabling plants to obtain water and nutrients.

The root apical meristem is the source of new cells that allow roots to elongate. page 654

The root apical meristem has a cap that protects it from damage as the root grows through the soil. page 654

The root's vascular tissues are located in the center of the root, surrounded by the pericycle, the endodermis, the cortex, and the epidermis. page 654

Lateral roots develop from new root apical meristems that form in the pericycle. page 655

Roots alter their development in response to conditions in the soil. page 655

Some roots are involved in adhesion, storage, and gas diffusion. page 656

31.5 Plants respond to light, gravity, and wind through changes in internode elongation and the development of leaves, roots, and branches.

Plant stems grow toward the light and away from gravity as a result of redistribution of auxin. page 656

Roots grow away from light and toward gravity as a result of redistribution of auxin. page 656

Starch-filled statoliths allow roots to sense and respond to gravity. page 659

Phytochrome is a photoreceptor that allows plants to determine the amount of red versus far-red light. Because light filtered through leaves has lower levels of red wavelengths, phytochrome allows plants to detect the presence of plants overhead. page 659

Seedlings germinating in the dark produce elongated internodes and suppress leaf expansion to increase their ability to reach the soil surface before their stored resources are depleted. page 660

Some plants respond to the presence of neighboring plants by elongating more rapidly and suppressing branching, which allow them to maximize height growth. page 660

Mechanical stress, such as the bending of stems in the wind, results in stems that are short and wide. page 661

31.6 Plants have sensory systems that control the timing of developmental events.

Many plants are photoperiodic, meaning that their flowering is controlled by the day length. page 662

Plants measure day length through an interaction between photoreceptors and their internal circadian clock. page 663

Vernalization is a prolonged period of exposure to cold temperatures necessary in some temperate-zone species before flowering can be induced. page 664

Decreases in photoperiod trigger plants to prepare their meristems for winter by producing bud scales and reducing internode elongation. page 664

Self-Assessment

1. Diagram the zones of cell division, elongation, and maturation, and explain why this organization allows stems to grow without a predetermined limit to their length.

2. Name one role of the plant hormone auxin and describe how auxin is transported within a plant.
3. Explain why a plant that has a vascular cambium also has a cork cambium.
4. Explain why the vascular cambium forms a continuous sheath that runs from near the tips of the branches to near the tips of the roots, whereas the cork cambium is discontinuous in both space and time.
5. List three structural differences between roots and shoots that allow roots to grow through the soil. Explain how you would tell whether an isolated piece of a plant came from the shoot or from the root.
6. Discuss the similarities and differences in the ways stems and roots respond to gravity and light.
7. Diagram how the amount of red versus far-red light allows plants to determine if they are in the open or in the forest understory.
8. Give an example of how a plant's ability to sense its environment improves the plant's chances for survival and reproduction.
9. Explain why a short-day plant that germinates in the spring will not flower until late summer and why a long-day plant that germinates at the end of summer will not flower until late the following spring.
10. Describe how vernalization can have an effect in cells that were not formed at the time of the cold treatment.

Log in to LaunchPad to check your answers to the Self-Assessment questions, and to access additional learning tools.

Don Johnston/Getty Images.

CHAPTER 32

Plant Defense

Keeping the World Green

Core Concepts

32.1 Plants have evolved mechanisms to protect themselves from infection by pathogens, which include viruses, bacteria, fungi, worms, and even parasitic plants.

32.2 Plants use chemical, mechanical, and ecological defenses to protect themselves from being eaten by herbivores.

32.3 The production of defenses is costly, resulting in trade-offs between protection and growth.

32.4 Interactions among plants, pathogens, and herbivores contribute to the origin and maintenance of plant diversity.

Plants have complicated interactions with animals. As we saw in Chapter 30, many plants rely on animals for pollination or seed dispersal. In return, plant seeds provide a rich source of food for animals, and leaves furnish a vast, but lower-quality, food resource. The plant body also provides habitats for an array of viruses, bacteria, and protists, not to mention other plants, which may grow on or within their tissues. How do plants defend themselves against the animals that would eat them? How do they defend against pathogens? The answers to these questions help us understand how plant populations are distributed in nature, and they help us solve practical problems as well. In the United States alone, crop damage by herbivores and pathogens costs farmers more than $2 billion every year.

Plants can't run away and hide as animals can. Nonetheless, they can sense danger and they are able to deploy a diverse and powerful defensive arsenal that includes physical and chemical deterrents as well as aerial surveillance, armed guards, and sticky traps.

Plants produce a wide array of chemicals that alter the metabolism and influence the behavior of animals that seek to consume them. These molecules also interact with our own biochemistry, and so compounds derived from plants have been used since antiquity as sources of medicine. Our early ancestors turned to the leaves and bark of willow trees in much the same way that we rely on aspirin—and for good reason. Willows produce a compound that deters herbivores but in humans relieves aches and pain; this compound in fact provides the chemical basis of aspirin. Quinine, derived from the tropical cinchona tree, is used in the treatment of malaria, while belladonna, a relative of the tomato, produces toxins that the ancient Romans used to eliminate rivals but in very low doses relieve gastrointestinal disorders. In fact, nearly 30% of all medicines are based on chemicals that plants produce to defend themselves.

FIG. 32.1 A potato plant infected with *Phytophthora infestans*, an oomycete that causes blight. *Source: Nigel Cattlin/Science Source.*

32.1 PROTECTION AGAINST PATHOGENS

In September 1845, potato plants across Ireland began to die. Their leaves turned black and the entire plant began to rot. When the potatoes were dug up, those that were not already rotten soon shriveled and decayed into a foul-smelling and inedible mush. The biological agent that had destroyed the potato crop looked like a fungus, but in fact was an oomycete called *Phytophthora infestans,* a type of protist (Chapter 27).

Phytophthora infestans is one example of the diverse pathogens that infect plants. Plant pathogens include viruses, bacteria, fungi, nematode worms, and even other plants, but what they all have in common is the ability to grow on and in the tissues of plants, extracting resources to support their own growth and causing an array of diseases. The outcome for the plant is often death (**Fig. 32.1**). The fact that most plants appear disease free is testimony to their ability to defend themselves. Plants have an immune system that allows them to recognize and respond to pathogens that would otherwise exploit them. In the Irish potato famine, which lasted from 1845 to 1852, these defenses were breached, allowing *P. infestans* to spread and grow with devastating effects on the potato crop and, in consequence, the people of Ireland.

Plant pathogens infect and exploit host plants by a variety of mechanisms.

The epidermis, with its thick walls and waxy cuticle, is a plant's first line of defense against pathogens. Viruses and bacteria cannot penetrate the cuticle and so commonly enter through wounds or on the piercing mouthparts of insects and nematodes. Stomata form a natural entry point for bacteria, oomycetes, and fungi (**Fig. 32.2**). Some bacteria and fungi even secrete chemicals that prevent stomata from closing. Other pathogens, notably parasitic plants and some species of fungi, secrete enzymes that weaken epidermal cell walls. These invaders then force their way through the weakened epidermis. Mechanical cultivation can damage crop plants, thus also providing entry points for pathogens.

Once inside the plant, fungi can spread by growing through plant cell walls. In contrast, viruses move from one plant cell to another through plasmodesmata. Because the genomes of most plant viruses are single strands of RNA, they can travel by the plant's natural mechanism for moving messenger RNA between cells. The plant vascular system provides a highway that allows

FIG. 32.2 Entry points for pathogens. A scanning electron micrograph shows *Phytophthora infestans* gaining access to the leaf interior by growing through stomata. *Source: Clouds Hill Imaging Ltd/Science Source.*

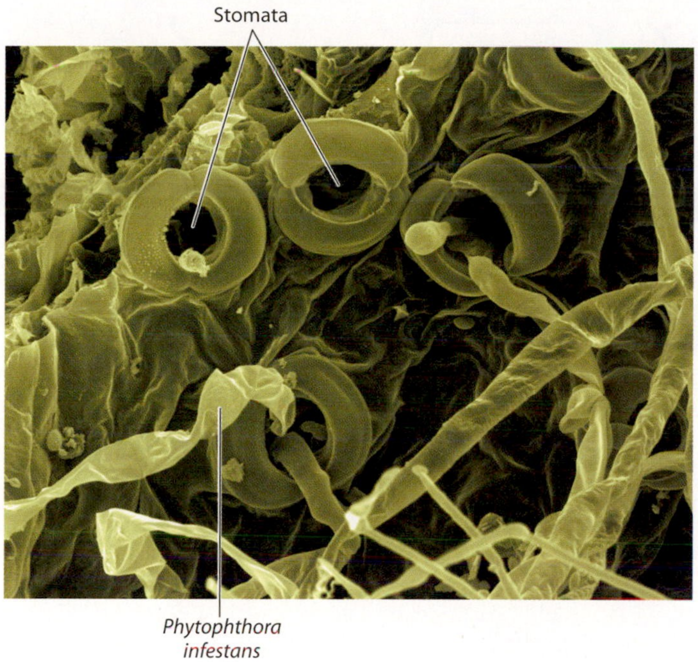

pathogens to move over long distances. Viruses are transported within the phloem, whereas bacteria and fungi move through the xylem. During the 1950s, commercial banana production in Mesoamerica was nearly wiped out by *Fusarium oxysporum,* a fungal pathogen whose spores are carried in the xylem. More recently, the California wine industry has been threatened by Pierce's disease, which results from infection by *Xylella fastidiosa,* a bacterium that also moves through the xylem.

Biotrophic pathogens obtain resources from living cells, whereas **necrotrophic pathogens** kill cells before exploiting them. Most biotrophic pathogens do not enter into the host plant's cells. Instead, they produce chemicals that stimulate the host cells to secrete compounds that the pathogen can use as food. Viruses, however, are biotrophic pathogens that do enter the cell: They must hijack a living cell's machinery to reproduce. Necrotrophic pathogens produce toxins that kill cells; the pathogens then feed on compounds that leak from the dead and dying cells. Bacteria and fungal pathogens can be either biotrophic or necrotrophic. For example, wheat leaf rust (*Puccinia* species) is a biotrophic fungus that significantly reduces wheat yields (Chapter 34). Rice blast (*Magnaporthe grisea*) is a necrotrophic fungus that each year reduces rice production by an amount that would feed 60 million people.

Parasitic plants obtain resources by infecting other plants (**Fig. 32.3**). Parasitic plants produce structures that penetrate into the stems or roots of other plants, eventually tapping into the host plant's vascular system. Some, like some mistletoes (Fig. 32.3a), are xylem parasites: They obtain water and nutrients from the xylem of the host plant without expending the costs of producing roots themselves. Other plant pathogens target both the xylem and the phloem. In the most extreme cases, the parasite receives all its carbohydrates from its host and carries out no photosynthesis of its own. *Rafflesia* (Fig. 32.3b), famous for producing the largest flower in the world, grows entirely within the tissues of its host, emerging only when it produces its

FIG. 32.3 Parasitic plants. (a) Mistletoe carries out photosynthesis, but relies on its host plant for water and nutrients. (b) *Rafflesia,* which produces the world's largest flower, also obtains everything it needs from its host. (c) *Cuscuta,* or dodder, parasitizes the stems of many plants. It does not photosynthesize and instead receives everything it needs from its host. *Sources: a. © Mark Boulton/NHPA/Photoshot; b. Paul Kennedy/Gettty Images; c. Clive Varlack/clivevbugs/AGPix.*

FIG. 32.4 The plant immune system. (a) In basal resistance, plasma membrane receptors recognize molecules produced by broad classes of pathogens. (b) Specific resistance depends on *R* genes that allow plant cells to identify and deactivate AVR proteins produced by specific pathogens.

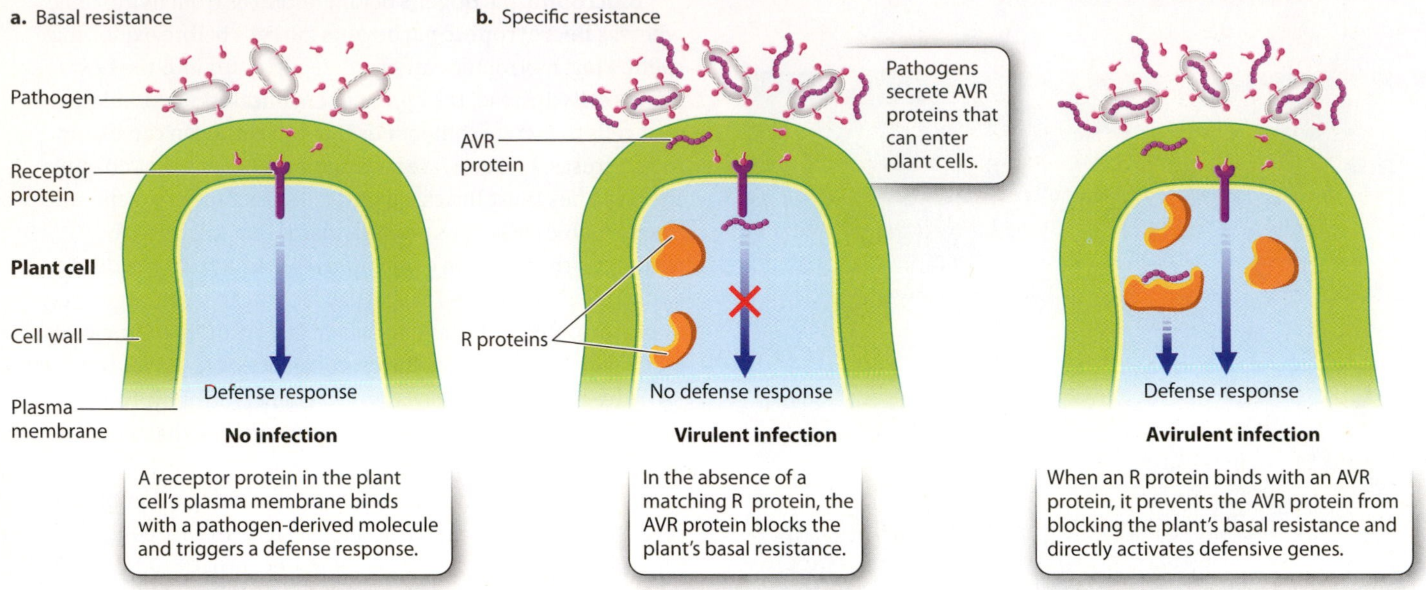

giant flowers. The ability to parasitize other plants has evolved independently multiple times. Currently, more than 4000 species of parasitic plants have been identified.

How do parasitic plants find an appropriate host? Mistletoe fruits, which are eaten and dispersed by birds, have a sticky layer that adheres to the stems and branches of trees that the plant will parasitize. Dodder, a parasitic vine (Fig. 32.3c), orients its growth toward a potential host plant, guided by volatile organic compounds produced by the species it infects. One of the most destructive plant pathogens is witchweed (*Striga* species), which infects the roots of a number of crop species, including corn, sorghum, and sugarcane. *Striga* seeds remain dormant in the soil until their germination is triggered by chemicals released from the roots of host species. *Striga* infects more than 40% of all agricultural areas in Africa, driving down crop yields, in some cases to zero.

Plants are able to detect and respond to pathogens.

Most plants are at risk of infection by only a small subset of the many known plant pathogens. **Host plants** of a particular pathogen are the species that can be infected by that pathogen. Not all infections significantly damage the host plant, however. **Virulent** pathogens can overcome the host plant's defenses and lead to disease. In contrast, **avirulent** pathogens damage only a small part of the plant because the host plant is able to contain the infection. Whether an interaction is virulent or avirulent depends on the genotypes of both the pathogen and the host plant.

In your own body, the innate immune system provides a first line of defense against pathogens (Chapter 43). Plants, too, have an immune system that allows them to detect pathogens and mount an appropriate response. A key feature of the immune system is that plants, like animals, can recognize a cell or virus as foreign. How do they accomplish this?

Plant cells have protein receptors (Chapter 9) that bind to molecules produced by pathogens and recognize them as foreign to the plant body. Binding changes the conformation of the receptor, triggering a cascade of reactions that enhance the plant's ability to resist infection.

The plant immune system has two components (**Fig 32.4**). The first consists of receptors located on the plasma membrane (Fig 32.4a). These receptors recognize highly conserved molecules generated by broad classes of pathogens. Examples of these molecules are flagellins, present in the flagella of bacteria, and chitin, a component of the fungal cell wall. When one of these molecules binds to the plant's receptor, an array of defense mechanisms is triggered that help protects the plant from infection. Protection from pathogens conferred by this component of the immune system is called basal resistance.

The second component of the plant immune system allows plants to resist pathogens that have evolved means of overcoming basal resistance. Pathogens produce proteins called AVR proteins that enter into plant cells and facilitate infection ("AVR" stands for "avirulence"). Some AVR proteins block defense responses triggered by the basal resistance of plants, while others cause the cell to secrete molecules that the pathogen needs to support its own growth. The second component of the plant immune system targets these AVR proteins.

Like basal resistance, this component of the immune system also consists of receptors, but in this case they are located inside the cell rather than on the plasma membrane (Fig. 32.4b). Plants harbor hundreds of these receptors, called **R proteins,** each expressed by a different gene—these genes are collectively called **R genes** (the "R" in both cases stands for "resistance"). Each R protein recognizes a specific AVR protein. When an R protein binds with a pathogen-derived AVR protein, it both prevents the AVR protein from blocking the plant's defenses and activates additional defenses. Because this component of the immune system depends on interactions between specific plant and pathogen genes, it is commonly called the gene-for-gene model of plant immunity. Protection from pathogens conferred by this component of the immune system is called specific resistance.

The large number of *R* genes is the result of an evolutionary arms race that occurs as pathogens evolve ways to evade their host's detection systems. Pathogens improve their success rate when mutation results in an AVR protein that does not bind with any of the R proteins produced by its host plant. For example, *P. infestans,* the pathogen responsible for the Irish potato famine, produces many more AVR proteins than related *Phytophthora* species. The large number of AVR proteins found in *P. infestans* increases the probability that at least one of these proteins is not detected by the infected plant. Conversely, when pathogens are successful, natural selection favors plant populations that produce new R proteins capable of binding to the AVR proteins of the invader.

→ **Quick Check 1** Can the specific component of the plant immune system (the one using *R* genes) defend against both biotrophic and necrotrophic pathogens?

Plants respond to infections by isolating infected regions.

Once a plant has detected a pathogen, how does it protect itself? One way is by reinforcing its natural barriers against pathogens. For example, plants commonly defend themselves by strengthening their cell walls, closing their stomata, and plugging their xylem. Or the plant can counterattack. Plants produce a variety of antimicrobial compounds, including ones that attack bacterial or fungal cell walls. A third type of defense is known as the **hypersensitive response.** In this case, uninfected cells surrounding the site of infection rapidly produce large numbers of reactive oxygen species, which trigger cell wall reinforcement and cause the cells to die (**Fig. 32.5**). The death of surrounding cells creates a barrier of dead tissue that prevents the spread of biotrophic pathogens.

Plants can defeat pathogens by isolating infected regions to a far greater degree than is possible in most animals. For example, plants respond to the presence of xylem-transported pathogens by plugging xylem conduits. If the infection is caught in time, the pathogen is prevented from spreading throughout the plant. But if containment fails and the pathogen spreads, the plant may contribute to its own demise by blocking its entire vascular system. Virulent pathogens that move through the xylem cause a variety of disorders collectively referred to as vascular wilt diseases (**Fig. 32.6**). The widespread death of native elm trees in both Europe and North America is the result of a vascular wilt disease caused by a fungus introduced from Asia but first identified in the Netherlands and therefore referred to as Dutch elm disease.

FIG. 32.5 **Hypersensitive response.** A tobacco plant infected with tobacco mosaic virus responds by producing a region of dead tissue that surrounds the site of infection and prevents the virus from spreading to other parts of the plant. *Source: Norm Thomas/Science Source.*

Mobile signals trigger defenses in uninfected tissues.

In 1961, American plant pathologist A. F. Ross reported that when individual tobacco leaves were infected with the tobacco mosaic virus (TMV), they developed necrotic patches characteristic of a hypersensitive response. However, when uninfected leaves of the same plant were subsequently exposed to the virus, they suffered little or no damage (**Fig. 32.7**). Ross called this ability to resist future infections **systemic acquired resistance (SAR).** Initially reported for viral infections, SAR occurs in response to a wide range of pathogens, including bacteria and fungi.

How do uninfected leaves acquire resistance to pathogens? One hypothesis is that a chemical signal is transported from the infected region through the phloem. This chemical signal then triggers the expression of genes encoding many of the same proteins that defend against the pathogen in infected cells. The identity of the mobile signal has proved elusive, but experiments show that salicylic acid—the chemical basis of aspirin—is required for SAR. The methylated form of salicylic acid, known familiarly as oil of wintergreen, is also a potent inducer of defense responses.

FIG. 32.6 Vascular wilt diseases. Pathogens that are transmitted through the xylem can kill their host if plant defenses also block water transport to the leaves. (a) Panama disease, which affects bananas, is caused by a fungal pathogen. (b) Pierce's disease, which affects grapes, is caused by a pathogenic bacterium. *Sources: a. Guy Blomme/Bioversity International; b. Bruce Fleming/Cephas Picture Library.*

HOW DO WE KNOW?

FIG. 32.7

Can plants develop immunity to specific pathogens?

BACKGROUND Tobacco is susceptible to an infectious disease that causes its leaves to turn a mottled yellow, hence the name of the disease—tobacco mosaic disease. In the 1890s, experiments with filters showed that the infectious agent was smaller than a bacterium, a finding that led to the discovery of viruses. Some plants succumb to the disease, but others do not. Are the plants that survive the initial infection immune from further attacks?

HYPOTHESIS Plants that survive infection with tobacco mosaic virus (TMV) have acquired immunity and therefore resist further attack.

EXPERIMENT American plant pathologist A. F. Ross infected one leaf on a tobacco plant with TMV. One week later, he exposed another leaf on the same plant to the virus.

RESULTS Plants that had been previously exposed to TMV showed no signs of infection when exposed to the virus a second time.

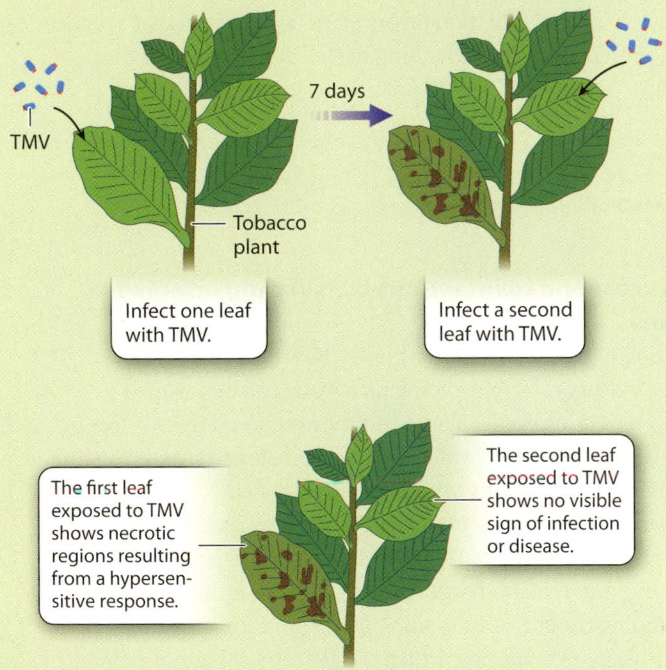

CONCLUSION Tobacco leaves that have never been exposed to TMV can acquire resistance to the pathogen if another leaf on the same plant has been previously exposed. This result indicates that a signal has been transmitted from the originally infected leaf to the undamaged parts of the plant, and that the transmitted signal subsequently triggers the development of an immune response that protects the plant from further infection.

FOLLOW-UP WORK Plants are now known to acquire immunity to a wide range of pathogens in addition to viruses.

SOURCE Ross, A. F. 1961. "Systemic Acquired Resistance Induced by Localized Virus Infections in Plants." *Virology* 14:340–358.

FIG. 32.8 Plant defense by small interfering RNAs (siRNAs). siRNAs target complementary single-stranded RNA for destruction.

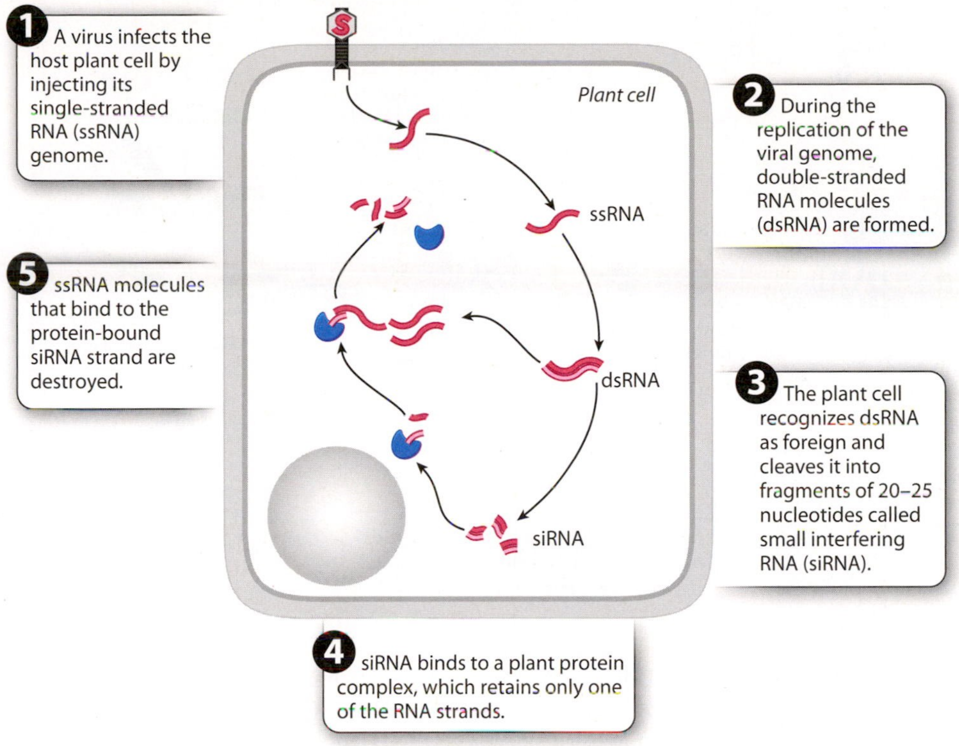

Because methyl salicylic acid vaporizes readily, the transport of this compound through the air may play a role in signaling the presence of pathogens. Airborne signals may be particularly important for leaves: Because leaves export carbohydrates rather than receive them, they are not able to receive signaling molecules transported in the phloem.

Up-regulation of salicylic acid occurs when infection by a biotrophic pathogen triggers a hypersensitive response. Plants respond to necrotrophic pathogens by producing different signaling molecules, notably jasmonic acid and the plant hormone ethylene. A general feature of these responses is that they activate defenses effective against broad classes of pathogens. Only in the case of viral infections, discussed next, is the response targeted to an individual pathogen.

→ **Quick Check 2** How does the plant hypersensitive response differ from systemic acquired resistance (SAR)?

Plants defend against viral infections by producing siRNA.

Viruses can be potent infectious agents, but plants have evolved responses to viral infection—including a hypersensitive response—that are much like those mounted against other pathogens. In addition, plants have a form of defense that targets the virus itself (**Fig. 32.8**). Most plant viruses have genomes made of single-stranded RNA (ssRNA). During the replication of the viral genome inside the plant cell, double-stranded RNA molecules (dsRNA) are formed. Because plant cells do not normally make double-stranded RNA, the replicating viral genomes are identified as foreign. Enzymes produced by the plant cell cleave the double-stranded RNA molecules into small pieces of 20 to 25 nucleotides, forming fragments called **small interfering RNA,** or **siRNA** (Chapter 3). These fragments bind with specific protein complexes in the cell. The RNA–protein complexes then play a role in targeting and destroying single-stranded RNA molecules that have a complementary sequence—that is, the viral genome (Chapter 19). The virus is unable to replicate and thus cannot spread. In this way, the plant cell uses RNA chemistry to eliminate viral infections.

When a cell is attacked by a virus, the siRNA molecules that are produced in response to the initial infection can also spread, allowing the plant to acquire immunity against specific viruses. The siRNA molecules move through plasmodesmata, enter the phloem, and from there spread throughout the plant. Thus, the systemic response to viral infection consists of both general defense responses triggered by salicylic acid and the transport of highly specific molecules that can target and destroy viral genomes throughout the plant.

A pathogenic bacterium provides a way to modify plant genomes.

Some pathogenic bacteria alter their host plant's biology by inserting their own genes into the host's genome. One such pathogen, *Rhizobium radiobacter* (formerly known as *Agrobacterium tumifaciens*), is useful to humans, despite the fact that it can cause widespread damage to crops such as apples and grapes. The reason? This bacterium's naturally evolved ability to incorporate its DNA into the genome of its host plant provides a surprisingly easy way to genetically modify plants.

Rhizobium radiobacter is a soil bacterium with virulent strains that infect the roots and stems of many plants. Virulent bacteria enter a plant through wounds and move through the plant's cell walls, propelled by flagella. A tumor, or gall, forms at the point of infection, often where the stem emerges from the soil, and for this reason *R. radiobacter* infection is called crown gall disease. *R. radiobacter* alters the growth and metabolism of infected cells by inserting some of its own genes into the chromosomal DNA of the host plant (**Fig. 32.9**).

Virulent strains of *R. radiobacter* contain a plasmid, called the **Ti plasmid,** which is a small (~200 kb), circular DNA molecule

FIG. 32.9 The formation of a crown gall tumor. *Rhizobium radiobacter,* a soil bacterium, can insert genes into the plant genome, altering the growth and metabolism of the plant in ways that enhance the growth of *R. radiobacter* bacteria.

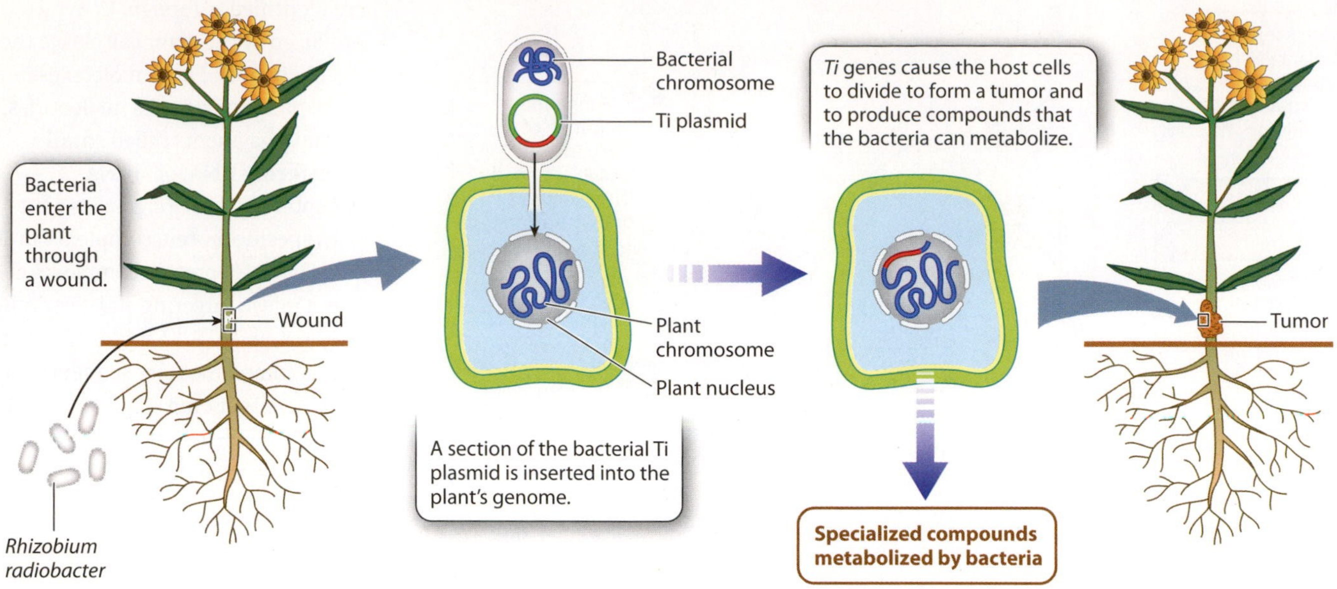

containing the genes that are integrated into the host cell's genome, as well as all the genes needed to make this transfer. Some of the transferred genes—including ones that result in the synthesis of the two plant hormones auxin and cytokinin (Chapter 31)—cause infected cells to proliferate and form a gall. Other transferred genes induce cells in the gall to produce specialized compounds that the bacteria use as sources of carbon and nitrogen for growth. Note that because plant cells do not move (in contrast to animal cells), there is no possibility that the tumor-forming cells will spread to other parts of the plant.

Biologists interested in genetically modifying crops and other plants make use of *R. radiobacter*'s ability to insert genes into the host genome. The first step is to replace the genes in the Ti plasmid that are usually inserted into the plant genome. Genes for gall formation are removed, and genes of interest are inserted. Examples of genes that have been transferred into plants in this way include genes whose products increase the plant's nutritional value or confer resistance against disease. In addition, researchers need a way to identify which cells have taken up the genes, and so a marker gene, for example one conferring antibiotic resistance, is typically included. In this way, cells that survive an application of the antibiotic are known to have successfully incorporated the Ti plasmid into their genome.

32.2 DEFENSE AGAINST HERBIVORES

The first land plants were followed closely in time by the ancestors of spiders and scorpions. Fossils show that these earliest land animals soon began to feed on plant fluids and tissues. As plants diversified, so did herbivores, and the pressure on plants to defend themselves against organisms from caterpillars to cows only increased. Given that they are unable to run away, plants would seem to be the ideal food. But plants have evolved a diversity of mechanisms that deter would-be consumers. Because these mechanisms deplete resources that could otherwise be directed toward growth and reproduction, plants have also evolved means of deploying their defenses in a cost-effective manner. Here, we explore the mechanical, chemical, developmental, and even ecological means by which plants protect themselves in a world teeming with hungry herbivores.

Plants use mechanical and chemical defenses to avoid being eaten.

Milkweeds (*Asclepias* species) are commonly found in open fields and roadsides across North America. They illustrate well how plants protect themselves from herbivorous animals. In fact, most generalist herbivores, which eat a diversity of plants, do not use milkweeds as a food source. Only specialists, such as caterpillars of the monarch butterfly (*Danaus plexippus*), can feed on these well-defended plants.

In part, milkweed's defenses are mechanical. The leaves of many milkweed species are covered with dense hairs. An insect crawling on a leaf must make its way through as many as 3000 hairs per square centimeter, a journey likely more daunting to an insect than wading into a blackberry thicket is to a human. Monarch caterpillars may spend as much as an hour mowing off these hairs before they begin to feed. This long handling time represents a significant cost. To understand why monarch caterpillars make this investment, let's look at milkweed's other defenses.

A network of extracellular canals filled with a white, sticky liquid called **latex** runs through the milkweed leaf, both within

FIG. 32.10 How monarch caterpillars feed on milkweed. Young monarch caterpillars disarm milkweed defenses by digging trenches, while older, larger caterpillars sever major veins. Both methods prevent the sticky latex from flowing into the feeding area.

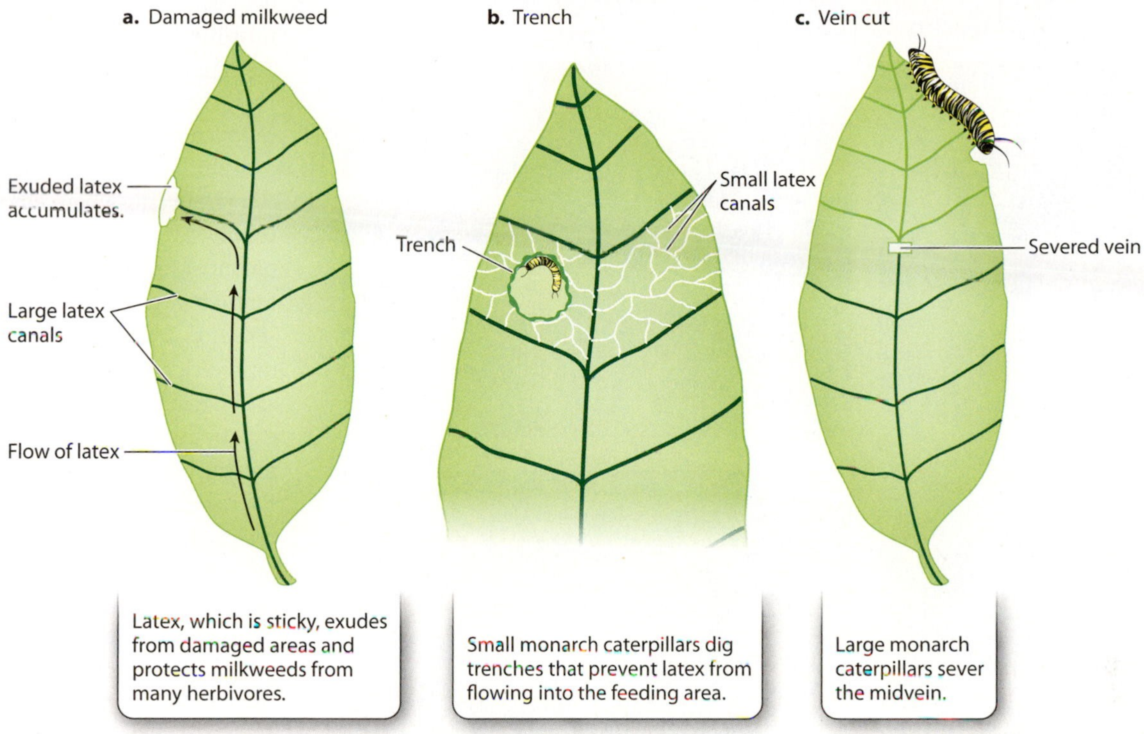

the major veins and extending into the regions between veins (**Fig. 32.10**). Latex canals are under pressure in the intact leaf. Thus, when the leaf is damaged, the latex flows out, sticking to everything it comes into contact with and rapidly congealing on exposure to air. A feeding insect is in danger of having its tiny mouthparts glued together, with potentially fatal consequences. Small monarch caterpillars cut trenches into the leaf that prevent latex from flowing onto the region where they are feeding. Larger caterpillars sever the midvein, relieving the pressure within the latex canals downstream of the cut. The caterpillars are then able to feed without risk on the part of the leaf that lies beyond the severed midvein.

Milkweed latex is not only sticky but also toxic. In particular, milkweed latex contains high concentrations of cardenolides, steroid compounds that cause heart arrest in animals. (In small doses, these compounds are used to treat irregular heart rhythms in humans.) Monarch caterpillars do not metabolize the cardenolides that they consume, but instead sequester them within specific regions of their bodies as a chemical defense against their own predators. Because of the presence of these cardenolides, monarch caterpillars and even the adult butterflies are highly toxic to most birds and other animals that would otherwise eat them. The protection conferred by these chemicals helps explain why monarch caterpillars have evolved to feed on leaves that are both metabolically difficult and time-consuming to eat.

Most plant species have mechanical or chemical defenses against herbivores, or even both. Mechanical defenses are varied. Hairs are common on leaves, and, as anyone who has ever rubbed against a stinging nettle knows, these hairs are sometimes armed with chemical irritants. Grasses and some other plants have a hard, mineral defense consisting of silica (SiO_2) plates formed within epidermal cells. Silica wears down insect mouthparts, so the insects feed less efficiently and grow more slowly. On a larger scale, some tropical trees have prickles, spines, or thorns (**Fig. 32.11**). Even something as simple as how tough a leaf is can have a large impact on the probability of the leaf's being eaten. In fact, leaves are most likely to be eaten while they are still growing and their tissues not fully hardened. Nonetheless, the most spectacular diversity in plant defenses lies in the realm of chemistry.

Diverse chemical compounds deter herbivores.

The cardenolides produced by milkweeds are only one example from the vast chemical arsenal available to plants for protection. Some plants produce **alkaloids,** nitrogen-bearing compounds that damage the nervous system of animals. Commonly bitter tasting, alkaloids include such well-known compounds as nicotine, caffeine, morphine, theobromine (found in chocolate), quinine (a treatment for malaria), strychnine, and atropine. Alkaloids are a costly defense because they are rich in nitrogen, an essential and often limiting element that plants need to build proteins for photosynthesis.

FIG. 32.11 **Mechanical defenses.** (a) Spines on the beach palm make it hard for animals to climb or penetrate its stem, and hairs on the leaves of (b) nettle and (c) *Arabidopsis* deter much smaller herbivores. *Sources: a. Fletcher & Baylis/Science Source; b. Dr. Keith Wheeler/Science Source; c. Dr. John Runions/Science Source.*

However, because alkaloids affect specific aspects of metabolism, even very small concentrations can deter herbivores. One of the first chemotherapy drugs (**Table 32.1**), vincristine, is an alkaloid extracted from the Madagascar periwinkle. It inhibits microtubule polymerization and thus prevents cell division (Chapter 11). Vincristine does not harm the plant's own cells because it is stored within an extracellular system of latex-containing canals.

A second group of defensive compounds, the **terpenes,** do not contain nitrogen. As a result, plants can produce them for defense without having to make use of nitrogen that could otherwise be used in protein synthesis. Small terpenes are volatile, vaporizing easily, and so make up many of the essential oils associated with plants. The distinctive smells of lemon peel, mint, sage, menthol, pine resins, and geranium leaves are all due to terpenes. While we find these odors pleasant, they are feeding deterrents to mammals such as squirrels and moose. In addition, terpenes obstruct the growth and metabolism of both fungi and insects. As a result, pyrethrin, a terpene derivative extracted from chrysanthemums, is marketed commercially as an insecticide. Other compounds derived from terpenes interfere with insect development. For example, exposure to compounds that are replicas of insect hormones cause insects to molt prematurely and thereby keep their populations in check. Taxol, which is used in chemotherapy, is a terpene that was first extracted from the bark of the Pacific yew tree.

Phenols form the third main class of defensive compounds, illustrated by the **tannins** found widely in plant tissues. Tannins bind with proteins, reducing their digestibility. Plants store tannins in cell vacuoles, and so these compounds come into contact with the protein-rich cytoplasm only when cells are damaged. Herbivores attempting to feed on tannin-producing plants obtain a poor reward for their efforts. For this reason, natural selection favors individuals that avoid tannin-rich plants. Many unripe fruits are high in tannins, and the unpleasant experience of biting into an unripe banana illustrates how tannins deter consumers. For thousands of years, humans have taken advantage of the protein-binding properties of tannins, using extracts from tree bark to process animal skins by "tanning" to produce leather.

Some of the chemical defenses used by plants are stored in a separate compartment from the enzymes that activate them. When you bite into a plant belonging to the cabbage family, the chemicals stored in the vacuole comingle with enzymes in the cytosol. The enzymes then catalyze the production of mustard oils from these compounds. Mustard oils give cabbage and its relatives their distinctive smell. They serve as feeding deterrents because they interfere with insect growth.

Cassava roots, an important food crop in Africa and South America, release the toxin hydrogen cyanide when their cells are damaged. Before cassava roots can be safely consumed, the roots must be ground and the cyanide-producing chemicals removed in running water. Although some cassava varieties, so-called sweet cassava, lack these chemicals and can be eaten without any special preparation, the bitter varieties are preferred crops in many cases because they often have higher yields.

Some chemical defenses found in plants are protein based. Plants and animals use the same 20 amino acids to construct proteins, but some plants produce additional amino acids as well. Plants do not incorporate these additional amino acids into their proteins, but herbivores that ingest them do. The resulting proteins can no longer fulfill their function. Thus, insect herbivores that consume nonprotein amino acids grow slowly and often die early. A second protein-based defense is the production of antidigestive proteins called **protease inhibitors.** These proteins bind to the active site of enzymes that break down proteins in the herbivore's digestive system. This prevents proteins from being broken down into their individual amino acids and therefore reduces the nutritional value of

TABLE 32.1 A SAMPLING OF PLANT COMPOUNDS USED IN MEDICINE

CLASS	EXAMPLE	PLANT SOURCE	MEDICAL USE
Alkaloids	Atropine	Deadly nightshade	Dilate pupils; treat extremely low heart rate
	Vincristine	Madagascar periwinkle	Treat cancer
	Codeine	Opium poppy	Cough suppressant
	Morphine	Opium poppy	Painkiller
	Quinine	Cinchona tree	Treat malaria
	Ephedrine	*Ephedra* (Mormon tea)	Asthma relief
Morphine	Turocurare (curare)	*Strychnos toxifera*	Muscle relaxant
	Digitoxin	Foxglove	Treat congestive heart failure and arrhythmia
	Berberine	*Berberis* (barberry)	Antibiotic, anti-inflammatory
Terpenoids 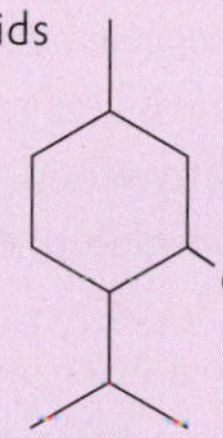	Menthol	Peppermint	Topical analgesic; mouthwash
	Camphor	Camphor laurel	Anti-itch creams
	Eucalyptol	*Eucalyptus* species	Mouthwash; cough suppressant
	Taxol	Pacific yew	Treat cancer
	Artemisinin	*Artemisia annua*	Treat malaria
Menthol			
Phenols 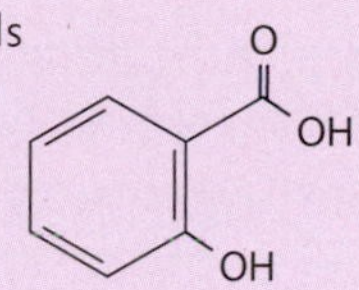 Salicylic acid	Salicylic acid	Willow tree	Treat aches and pains

the plant tissue. Insects that feed on plants that produce protease inhibitors have reduced growth rates.

Some plants provide food and shelter for ants, which actively defend them.

While many plants have mechanical defenses against herbivores and nearly all have at least one form of chemical defense, a smaller number of species have evolved ecological defenses against herbivory. These plants "employ" animals as bodyguards in exchange for shelter or nourishment, much the way that many plants provide food or other rewards to animals that transfer pollen and disperse seeds (Chapter 30).

Nectar is typically associated with flowers. However, many plants produce nectar in glands located on their leaves (**Fig. 32.12**). These extrafloral nectaries attract ants, which move actively throughout the plant in search of nectar. When the ants encounter the eggs or larvae of other insects, they consume these as well. The value of this relationship to the plant is easily demonstrated: When ants are prevented experimentally from patrolling certain branches, those branches suffer higher rates of herbivore damage.

FIG. 32.12 **Ants attracted by an extrafloral nectary.** These ants receive a food reward and also consume any insect eggs or larvae they encounter on the plant. *Source: Alex Wild/alexanderwild.com.*

FIG. 32.13 The symbiosis between ants and the bullhorn acacia. (a) Ants live in the plant's thorns, and (b) the plant provides food (the yellow food bodies). The ants vigorously defend their home against animals and plants alike. *Source: Dan L. Perlman/EcoLibrary.org.*

A small number of plants, several hundred in total, have evolved a much closer relationship with ants. These so-called ant-plants provide both food and shelter for an entire colony of ants, which then defend their host. The bullhorn acacia (*Acacia cornigera*), which grows in dry areas of Mexico and Central America, produces hollow spines at the base of each leaf, as well as protein- and lipid-rich food bodies on the tips of its expanding leaves. Symbiotic ants (*Pseudomyrmex ferruginea*) live in the spines and feed on the food bodies (**Fig. 32.13**). To protect this source of food and shelter, the ants actively defend the plant. When the ants detect an invader, they produce an alarm pheromone that causes the entire colony to swarm over the plant, attacking any insects or vertebrates (including scientists) that they encounter. The ants will attack the growing tip of a vine that might climb up the host and even destroy plants growing in the soil beneath the host *Acacia*.

This system works well until it comes time to reproduce. Without some mechanism to control the hypervigilant ants, any visiting pollinator would be attacked as though it were a truly unwanted guest. So that pollinators may visit flowers unharmed, flowers of the bullhorn acacia release a chemical that repels the resident ants.

The interactions between ant-plants and their ants are highly coevolved. The ant species are found only in association with their host plants. And while ant-plants can be grown without ants, herbivores would rapidly consume an undefended plant in nature because ant-plants lack the chemical defenses present in related species that do not host ants.

Grasses can regrow quickly following grazing by mammals.

By far, the most important vertebrate herbivores are the bison, zebra, antelope, and their relatives that populate the great plains of East Africa and North America. Studies of herbivore damage in an East African savanna indicate that vertebrates annually consume approximately the same amount of leaf biomass as do insects.

Grasses are well adapted to cope with grazing mammals (**Fig. 32.14**). The vertically oriented leaves grow from stems with limited internodal elongation. This allows the shoot apical meristem to remain close to the ground, where it is less likely to be damaged. Grasses have persistent zones of cell division and elongation at the base of each leaf. Because grass leaves elongate from their base, grasses can quickly replace leaf tissues removed by a grazing mammal, fire, or a lawn mower. Only when grasses produce their flowering stalks does the apical meristem emerge above the leaves.

Grasses are thought to have evolved in forest environments. In fact, grass-dominated ecosystems have become widespread

FIG. 32.14 Adaptation for grazing. The ability of grass leaves to elongate from the base allows grasses to regrow following grazing.

only during the past 25 million years. The fossil record shows that as grasslands expanded, grazing mammals evolved along with them. Horses, for example, originated as small browsers that fed on the leaves of small trees and shrubs. As grasslands expanded, larger horse species with teeth adapted to eating grass evolved. Why would specialized teeth be advantageous? The small plates of silica in grass cell walls increases their strength and may provide mechanical defense against herbivores. Horses and other grazers that feed on these grasses wear down their tooth enamel grinding against the silica, so teeth with a thick enamel layer are favored by natural selection. It is not clear whether grasses evolved in response to grazing or to disturbances such as fire, but without question, their distinctive pattern of meristem activity minimizes the damage caused by grazing mammals.

32.3 ALLOCATING RESOURCES TO DEFENSE

By now, it should be clear that plants invest substantial resources in defense, resources that might otherwise be used for growth and reproduction. Given the considerable cost of plant defenses, their benefits must be correspondingly large. If a commonly encountered pathogen or herbivore is lethal, the resources invested in defense provide a clear reproductive advantage. But what if the pathogen or herbivore usually damages only half of a plant's leaves? Or 10%? Or encounters the plant only infrequently?

Some defenses are always present, whereas others are turned on in response to a threat.

In nature, a balance between cost and benefit is achieved in several different ways. When a threat is common, plants produce constitutive defenses—defenses that are always present. When a threat is uncommon, however, there is an advantage to mounting a defense only when the threat is encountered. A defense that is activated only when the plant senses the threat is an inducible defense. An inducible defense can be either an increase in the level of an already present (constitutive) defense or the production of a new type of defense. The ability to sense and respond to the presence of herbivores allows plants to use resources efficiently.

When a plant is attacked by herbivores, it begins to express a new set of genes. Many of these genes increase the production of chemical defenses, while others stimulate the mechanical reinforcement of cell walls. In addition, herbivore damage triggers the synthesis of jasmonic acid, a signal that is transmitted through the phloem. Exposure to jasmonic acid induces the transcription of defensive genes even in parts of the plant untouched by herbivores. A version of the compound called methyl-jasmonic acid is highly volatile and can move through the air to reach portions of the plant that are not actively importing carbohydrates through the phloem.

→ **Quick Check3** What would happen to a plant in which jasmonic acid synthesis is blocked?

Plants can sense and respond to herbivores

For more than 15 years, biologists have studied how coyote tobacco (*Nicotiana attenuata*), a relative of tobacco native to dry open habitats of the American West (**Fig. 32.15**), defends itself against herbivores. These studies provide a fascinating window into how plants allocate resources toward defense under natural conditions.

FIG. 32.15 ***Nicotiana attenuata*, the coyote tobacco.** This species has been developed as a model organism for studying how plants protect themselves against herbivores. (a) *Nicotiana attenuata* in a Utah desert; (b) *Manduca sexta*, a specialist herbivore, feeding on *N. attenuata*.
Sources: a. Danny Kessler, Max Planck Institute for Chemical Ecology, Jena, Germany; b. Celia Diezel, Max Planck Institute for Chemical Ecology, Jena, Germany.

HOW DO WE KNOW?

FIG. 32.16

Can plants communicate?

BACKGROUND Plants respond to herbivore damage by producing chemicals that make their tissues less palatable. Field experiments in which caterpillars were added to target plants suggested that undamaged neighboring plants increased their production of defensive chemicals, whereas plants farther away did not.

HYPOTHESIS Volatile chemicals released from plants that have been attacked by herbivores elicit a defensive response in undamaged plants.

EXPERIMENT One set of plants was placed downwind of other plants in which two leaves were torn, simulating herbivore damage, and a set of control plants was placed downwind of an empty chamber. The concentrations of defensive compounds in all sets of plants were measured and compared.

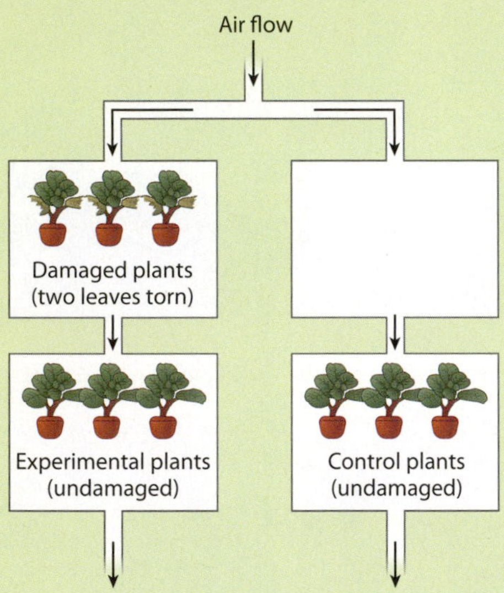

RESULTS When measured 52 hours after the initial damage, both the damaged plants and the undamaged experimental plants exhibited elevated levels of defensive compounds in their leaves compared to control plants.

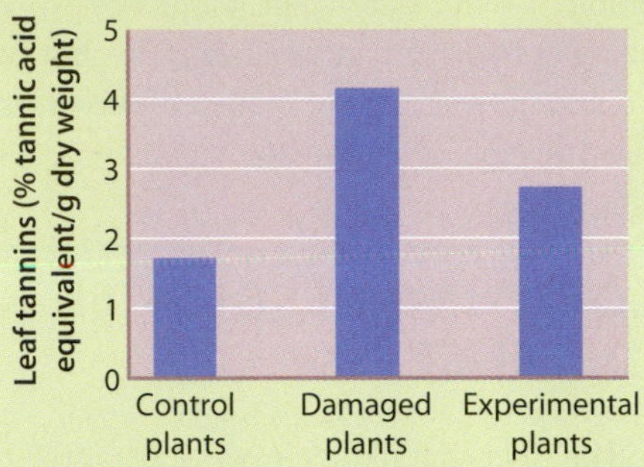

CONCLUSION Volatile chemicals released from damaged plant tissues can trigger the production of defensive chemicals in undamaged plants.

FOLLOW-UP WORK The volatile chemicals released from damaged plants have been identified and shown to attract predatory animals that feed on the herbivores or parasitize their eggs or both. Thus, when plants respond to damaged neighbors, they are "eavesdropping" on signals that likely evolved to attract predatory insects, rather than to warn neighboring plants.

SOURCE Baldwin, I. T., and J. C. Schulze. 1983. "Rapid Changes in Tree Leaf Chemistry Induced by Damage: Evidence for Communication Between Plants." *Science* 221:277–279.

Nicotiana attenuata is attacked by a wide variety of herbivores, both generalists and specialists. A major component of *N. attenuata's* defense against being eaten is the production of the alkaloid nicotine. Nicotine is an effective defense against many generalist herbivores. However, nicotine is also an expensive form of chemical defense because it contains nitrogen, a nutrient whose availability can limit growth. Thus, *N. attenuata* plants that have not been exposed to herbivores produce only low levels of nicotine as a constitutive defense. When *N. attenuata* is attacked by a generalist herbivore, nicotine concentrations in leaves increase dramatically—typically fourfold, but in some cases as much as tenfold. Because nicotine is synthesized in roots and transported to leaves in the xylem, the activation of this inducible defense requires the transport of signaling molecules in the phloem from the sites of insect damage to the roots.

Nicotiana attenuata is also attacked by a specialist herbivore, the tobacco hornworm caterpillar, *Manduca sexta.* These caterpillars, which can reach 7 cm in length, have evolved the ability to sequester and secrete nicotine. By this means, these caterpillars are able to consume leaves with high levels of nicotine. If nicotine is not an effective defense against this voracious herbivore, what can *N. attenuata* do to protect itself?

It turns out that *N. attenuata* can recognize specific chemicals in the saliva of *M. sexta.* When damaged by most herbivores—or a pair of scissors—*N. attenuata* responds by producing nicotine. But when attacked by *M. sexta, N. attenuata* does not waste time and resources synthesizing nicotine. Instead, it protects itself in an entirely different way, by attracting other insects that will attack its herbivores, as we discuss next.

Plants produce volatile signals that attract insects that prey upon herbivores.

How does *N. attenuata* attract its insect allies? A hint comes from the smell of a newly mown lawn. This distinctive smell originates from chemicals released from the cut blades of grass. Some of the chemicals released when plants are damaged represent nothing more than the exposure of cell interiors to the air. Others, however, are produced specifically in response to damage. When *N. attenuata* is attacked by *M. sexta* caterpillars, it produces volatile signals that attract insects that prey on the eggs and larvae of *M. sexta*. The adage "The enemy of my enemy is my friend" seems to hold true when it comes to protecting plants from herbivore damage.

Studies first conducted in the 1980s showed that undamaged plants increase their synthesis of chemical defenses when neighboring plants are attacked by herbivores (**Fig. 32.16**). Although these plants were described in the popular press as "talking trees," the plants under attack are not altruistically warning their neighbors. Instead, subsequent studies suggest that the undamaged plants are detecting the signals released by their neighbors that are directed at other insects, then ramping up their own defenses. Recent work demonstrates that even the sound of a neighboring plant being chewed on by insects can trigger defensive responses in a nearby plant.

Nutrient-rich environments select for plants that allocate more resources to growth than to defense.

Some plant species allocate far more resources to defense than others. Because defenses are costly to produce, you might think that plants growing in nutrient-rich habitats should be the best defended. However, consider the counterargument: Plants growing on nutrient-poor soils might invest heavily in defense because they cannot afford to replace tissues lost to herbivores. Such plants would grow relatively slowly because their resources are being used to build defenses rather than to produce new leaves and roots. By contrast, plants from nutrient-rich habitats would favor growth over defense because they can more readily replace lost or damaged tissues.

This situation is an example of a **trade-off:** Something is gained and, at the same time, something is lost. You can't have your cake and eat it, too—you have to make a choice. With plants, there can be a trade-off between growth and defense in which allocating resources for one means that those resources cannot be allocated for the other.

A recent study in the Amazon rain forest has focused on a possible trade-off between growth and defense. Researchers compared species found on nutrient-rich clay soils with closely related species found on nutrient-poor sandy soils. They found that species from the nutrient-rich clay soils invested less heavily in defenses than did species from the sandy sites. Because the clay-soil species divert fewer resources to defense, they grow faster in either soil type compared to sandy-soil species—as long as they are grown under a net that excludes herbivores (**Fig. 32.17**). However, when grown out in the open, each species grows best in the soil type on which it is naturally found.

The clay-soil species grow poorly on sandy soils when not protected by a net because, without ample soil nutrients available, they are unable to replace tissues consumed by herbivores. Conversely, when grown on the nutrient-rich soils, species from

FIG. 32.17 A trade-off between growth and defense. Plants from nutrient-rich clay habitats can overcome herbivory by growing fast, whereas plants from low-nutrient sandy habitats grow slowly and invest more in defense. The photos show that, without defenses, clay-soil plants are prone to being eaten by herbivores. *Source: Paul Fine*

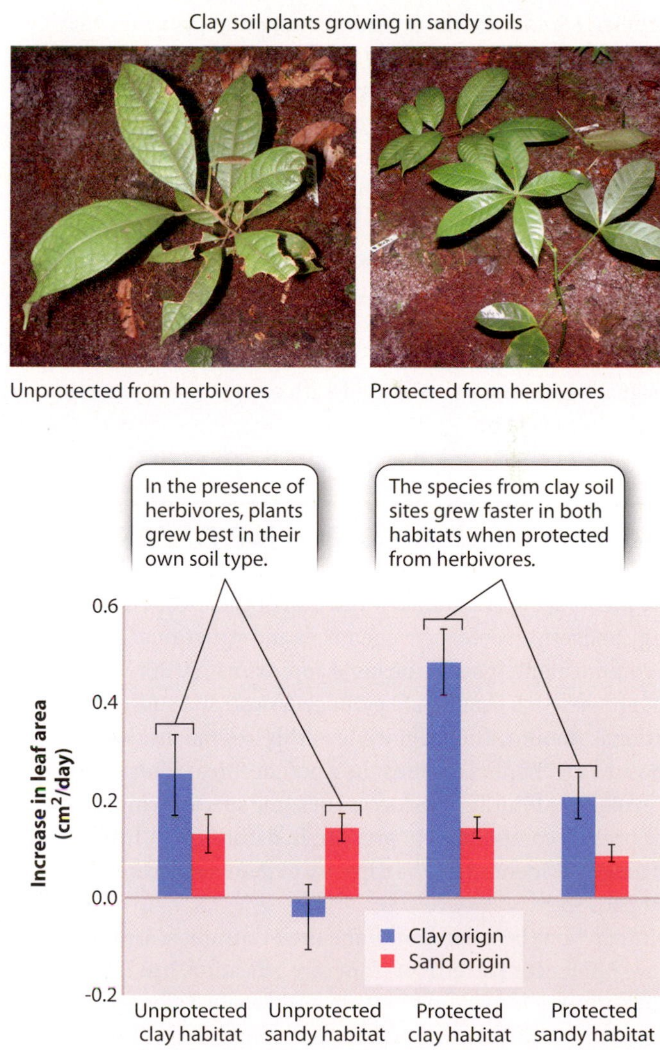

the sandy habitats are poor competitors because they allocate resources that could have been used for growth to defensive compounds and structures. What we see is a classic example of a trade-off.

→ **Quick Check 4** Does the trade-off between growth and defense favor a single species that dominates in all soil types or different species specialized for each habitat?

Exposure to multiple threats can lead to trade-offs.

In the real world, plants are routinely confronted with more than one threat. A plant may be attacked simultaneously by both a pathogen and an herbivore, or it might be in danger of being shaded by a fast-growing neighbor at the same time that it must defend its leaves from being eaten. Given that the resources a plant can draw on are limited, how do plants respond to multiple threats?

Often, the response to multiple threats suggests a trade-off. For example, a plant exposed to a pathogen may be less able to respond to a later herbivore attack. If tobacco plants are exposed to tobacco mosaic virus and systemic acquired resistance is activated, they become more susceptible to the hornworm caterpillar, *Manduca sexta.* Studies indicate that plants do not defend themselves as vigorously against subsequent herbivore attacks because of crosstalk between the signaling pathways for responding to biotrophic pathogens and those for responding to herbivores.

There can also be trade-offs between the threat of competition with neighboring plants and the threat of attack by herbivores. For example, when the phytochrome receptor in plants detects the presence of neighboring plants (Chapter 31), plants allocate more resources to growing tall, but they also produce fewer defensive chemicals in response to herbivore damage. Interactions among the signaling molecules associated with each of these processes underlie this trade-off.

32.4 DEFENSE AND PLANT DIVERSITY

Many plants spend a significant fraction of their resource and energy budgets on defensive chemicals and structures. Despite this commitment, insects consume approximately 20% of all new plant growth each year, while pathogens destroy a substantial additional amount. The magnitude of this loss means that pathogens and herbivores must be a potent force in both ecology and evolution, limiting the success of some species while allowing some of their competitors to prosper. In nature, plant fitness may be strongly influenced by the capacity to deter herbivores and resist pathogens.

Interactions between plants and their consumers are thought to have influenced patterns of plant diversification. Just like pollinating insects, plant-eating insects diversified along with angiosperms (the flowering plants). The coincidence of timing is consistent with an important role for plant–insect coevolution in generating diversity. Further evidence that defense has influenced plant diversification comes from plant genes and secondary compounds. For example, the *R* genes that are instrumental in pathogen recognition form one of the largest gene families in plants, while chemical defenses against herbivores include about 6000 alkaloids and more than 10,000 terpenoid compounds. Such observations suggest that plants are locked in an evolutionary arms race with herbivores and pathogens. This arms race is very much in evidence in the efforts by farmers to limit crop losses by reducing the numbers of the pathogens and herbivores in fields and orchards.

The evolution of new defenses may allow plants to diversify.

American scientists Paul Ehrlich and Peter Raven observed that closely related species of plants were fed upon by closely related species of butterflies. In 1964, they cited this observation as evidence that plants and their herbivores are caught up in an evolutionary arms race: Plants evolve new forms of defense, and herbivores evolve mechanisms to overcome these defenses. In particular, Ehrlich and Raven hypothesized that the evolution of novel defenses has been an important force in the diversification of both plants and herbivorous insects.

A novel form of defense may allow a plant population to expand into new areas. If the population in a new area remains separated from the original population, it may evolve into a new species. Similarly, a novel means of overcoming plant defenses may allow an insect or pathogen population access to new plant resources. If these insects or pathogens became separated from their original population, then they might also evolve into new species. Ehrlich and Raven's "escape and radiate" hypothesis thus predicts a burst of diversification following the evolution of novel defenses.

This hypothesis makes intuitive sense, but it could only be tested with the advent of phylogenetic trees based on DNA sequence comparisons (Chapter 23). Biologists knew that latex or resin canals evolved independently in more than 40 different groups of plants. Reasoning that these features represent a novel defense, they asked whether the groups that evolved latex or resin canals were more diverse than closely related groups that lacked these forms of defense. In 14 of the 16 groups examined, the groups with the protective canals were significantly more diverse than their closest relatives that lacked latex or resin canals. This pattern supports the hypothesis that the evolution of latex and resin canals provided plants with the freedom to expand into new habitats.

Escalation of defenses, however, is not the only possible evolutionary outcome. Phylogenetic research shows that, as milkweeds diversified, many of the newly evolved species produced *fewer* chemical defenses such as latex canals and cardenolides. Instead, these species are found in resource-rich habitats and replace tissues by growing quickly following damage. The presence of specialized herbivores such as monarch caterpillars that can disarm milkweeds' defenses may help explain

FIG. 32.18 Density-dependent mortality promotes tree diversity. Seedlings of rare species have a greater probability of surviving because they are more likely to escape infection by species-specific pathogens.

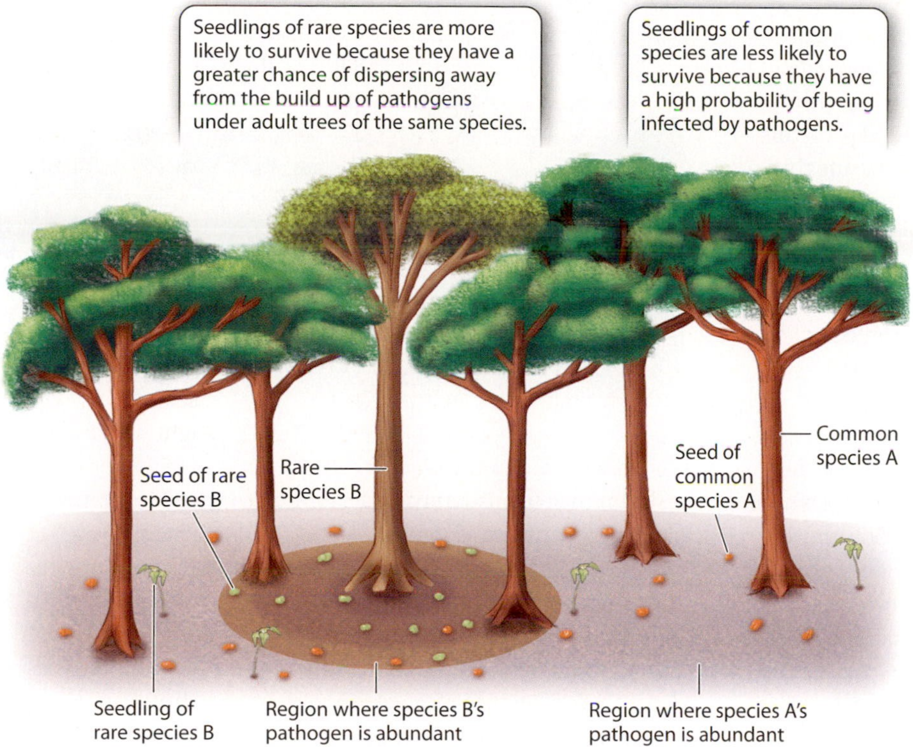

why these more recent plant species reduce investments in defense in favor of more rapid growth.

Pathogens, herbivores, and seed predators can increase plant diversity.

The lowland rain forests of eastern Ecuador have the highest diversity of tree species recorded anywhere: More than 1000 species of trees were identified in a 500 m × 500 m plot. Because trees compete for sunlight, water, and nutrients, the species best able to acquire resources would be expected, over time, to outcompete all others. What allows so many tree species to coexist is one of the great questions of modern biology. In the 1970s, two ecologists, Daniel Janzen and Joseph Connell, independently proposed that interactions with host-specific pathogens, herbivores, or seed predators play an important role.

The Janzen–Connell hypothesis, as it is now known, is based on the observation that seeds or seedlings often suffer high rates of mortality when growing directly beneath an adult tree of the same species (**Fig. 32.18**). Seed predators often gather in higher densities near trees with mature fruit, so seeds that disperse far from their maternal plant may escape notice by seed predators. By a similar argument, pathogens are common in the soil underneath an adult tree, so seedlings that grow farther away have a better chance of escaping these pathogens. In both cases, high-density populations of seeds and seedlings are more at risk.

Such density-dependent mortality provides rare species with an advantage over more common species. In other words, the offspring of a rare species is more likely to survive because its seeds are more likely to be dispersed into a location where there are no neighboring adult plants of the same species. By favoring the survival of less common species, density-dependent mortality prevents one species from outcompeting all others.

In the tropical rain forests of Ecuador and elsewhere, only a small number of tree species are common; the majority of tree species are rare. These rare species are able to escape herbivore and pathogen damage, but they need reliable ways of getting pollen from one plant to another. Thus, animal pollination, herbivory, and infection by pathogens are thought to play complementary roles in promoting angiosperm diversity.

? CASE 6 AGRICULTURE: FEEDING A GROWING POPULATION

Can modifying plants genetically protect crops from herbivores and pathogens?

The density-dependence of herbivore and pathogen damage is a particular challenge for agriculture as practiced in most countries. When acre after acre are covered by a single crop species, successful invaders are not slowed by low target density. On the contrary, pest populations build up to high levels, and the tendency to plant genetically uniform varieties increases the losses in yield. Not surprisingly, then, herbivores and pathogens exact a large toll on agricultural crops, and protecting crops from damage is a high priority for farmers.

The production of pesticides and herbicides increased dramatically in the second half of the twentieth century. Together with irrigation and fertilizers, these compounds are important contributors to the increased yields of the Green Revolution. However, large-scale application of pesticides is not without risk. In addition to the potential for toxic effects, widespread chemical treatments provide a strong selective force for the evolution of resistance. As pests become resistant to pesticides, one response is to apply more or stronger chemicals. The consequences of this approach, however, can spiral out of control as chemical applications indiscriminately kill off the natural predators of the pests themselves.

During the 1980s, farmers applied increasing amounts of pesticides to rice paddies in South Asia, yet crop losses continued to increase. This dangerous course was reversed by the introduction of a program first developed in the 1950s. Farmers in this program carefully determine what pests are actually present, whether or not they are at damaging levels, and whether or not there are natural predators of the pests. Only after understanding the ecology of the farm do farmers remove pests, starting with mechanical means, then biological control using natural predators, and, only as a last resort, industrial pesticides. Farmers using this approach, called integrated pest management, can minimize routine pesticide use and therefore prevent pest species from evolving resistance.

Another approach is to improve the ability of crops to defend themselves against herbivores and pathogens through breeding and genetic modification. Crop breeders must emulate—on a faster timescale—the role played by evolution in selecting for resistance to newly evolving pests. For example, Norman Borlaug's original task was to develop wheat varieties for Mexico that were resistant to fungal pathogens. To do this, he obtained seeds of wheat varieties from around the world that showed evidence of pathogen resistance. He then crossed these disease-resistant varieties with varieties adapted to the local conditions. Although it took thousands of hand-pollinations, Borlaug was successful at introducing resistance genes into wheat varieties that grew well in Mexico.

Crop breeding is a powerful way to alter the genetic makeup of cultivated species. However, because it relies on sexual reproduction, crop breeding is limited to the genetic diversity that exists within varieties that can interbreed. In contrast, *Rhizobium radiobacter* allows scientists to introduce specific genes found in distantly related organisms into cultivated species. One of the first commercial uses of this technology was to introduce chemical defenses from a bacterium into crop plants.

Bacillus thuringiensis (*Bt*) is a soil-dwelling bacterium that produces proteins toxic to insects. It has been used to control insect outbreaks in agriculture since the 1920s. At first, farmers applied spores containing the toxins or the toxins themselves directly to plants. In 1985, the genes that encode the toxins from *B. thuringiensis* were inserted into tobacco, where their expression conferred substantial protection against insects. Today, *Bt*-modified crops are some of the most widely planted genetically modified plants.

The introduction of *Bt* crops markedly reduced the application of pesticides, to the benefit of both farmworkers and the environment. However, the danger of planting *Bt*-expressing crops widely is that constant exposure to *Bt* toxins increases the probability that pest populations will evolve resistance. Organic farmers are particularly concerned because they continue to use traditional application of *B. thuringiensis* spores to control insect outbreaks. To prevent the evolution of resistance, U.S. farmers are required by law to plant a fraction of their fields, known as a refuge, with non–*Bt*-expressing plants (**Fig. 32.19**). Pests that are not resistant to *Bt* toxins survive by feeding on the non–*Bt*-expressing plants. These non-resistant pests interbreed with individuals feeding on the *Bt*-expressing plants, thus slowing the evolution of *Bt*-resistance. The efficacy of this practice, however, is called into question by reports of *Bt*-resistant pests and the need to apply increasing amounts of industrial pesticides to *Bt* crops.

FIG. 32.19 ***Bt*-crops co-planted with non-*Bt* varieties to prevent the development of resistance.** *Source: John L. Obermeyer, IPM Specialist Department of Entomology, Purdue University.*

If there is one practical lesson to be learned from our experience with *Bt*, it is that pathogens and herbivores will continue to evolve new ways of circumventing the defenses of crop plants. Thus, agriculture will continue to require new forms of crop protection. While the use of chemical pesticides will probably remain widespread, crop protection will increasingly be based on the application of genetics, through both crop breeding and biotechnology. In either case, success will depend upon the availability of existing genes honed by natural selection. Without access to genes involved in pathogen recognition (*R* genes) or in the production of chemical and mechanical defenses, we will have the tools, but not the raw materials, to protect our crops.

In some cases, crop protection will rely on genes from distantly related organisms, as illustrated by *Bt*. Commonly, however, defense strategies for crops will rely on genes from other plants, often ones that are closely related. For example, the specific *R* genes used for pathogen recognition typically function only within a single species or closely related species. Thus, to develop potatoes that are resistant to the protist *Phytophthora infestans* that causes potato blight, scientists are transferring *R* genes from a wild relative of potato. The continuation of such efforts requires a commitment to safeguarding the genetic diversity found in natural populations of crop species and their close relatives. Only by tapping into these natural genetic resources can breeders produce crops that stay one step ahead of the constantly evolving threats of pathogens and herbivores. ■

Core Concepts Summary

32.1 Plants have evolved mechanisms to protect themselves from infection by pathogens, which include viruses, bacteria, fungi, worms, and even parasitic plants.

Pathogens enter plants through damaged tissue or stomata. page 668

Plant pathogens spread by growing or moving through the plant vascular system. Viruses move in the phloem, whereas bacteria and fungi move in the xylem. page 669

Biotrophic pathogens gain resources from living cells; necrotrophic pathogens kill cells before exploiting them. page 669

Plants have an immune system that allows them to detect and respond to pathogens. page 670

One part of the plant immune system acts generally and recognizes highly conserved molecules produced by broad classes of pathogens. page 670

A second part of the plant immune system acts specifically and consists of "resistance" or *R* genes that encode R proteins, which recognize pathogen-derived AVR proteins page 670

Plants respond to pathogens by actively killing the cells surrounding the infection (the hypersensitive response) and by sending a signal to uninfected tissues so that they can mount a defense (systemic acquired resistance). page 671

The bacterium *Rhizobium radiobacter* infects plants by inserting some of its genes into the plant's genome, resulting in the formation of a tumor and providing a way to genetically engineer plants. page 673

32.2 Plants use chemical, mechanical, and ecological defenses to protect themselves from being eaten by herbivores.

Plant defenses against herbivory include dense hairs, latex, and chemicals. page 675

Alkaloids are nitrogen-bearing compounds that interact with the nervous system of animals. page 675

Terpenes, volatile compounds that give rise to many of the essential oils we associate with plants, deter herbivores. page 676

Tannins bind with proteins, reducing their digestibility. page 676

Ant-plants, such as the bullhorn acacia, provide food and shelter for ants, which defend their host plant. page 677

32.3 The production of defenses is costly, resulting in trade-offs between protection and growth.

Plants produce both constitutive defenses, which are produced whether or not a threat is present, as well as inducible defenses, which are triggered when a plant detects that it is being attacked. page 679

Plants produce volatile signals that attract insects that prey on herbivores. page 681

A trade-off is sometimes observed between plant growth and defense, in which allocation to defense results in slower growth rates. page 681

In nutrient-rich environments, species that allocate more resources to growth and less to defense can outcompete plants that invest more resources to defense and less to growth; the opposite is true on nutrient-poor soils. page 681

32.4 Interactions among plants, pathogens, and herbivores contribute to the origin and maintenance of plant diversity.

The "escape and radiate" pattern of plant evolution suggests that plants undergo a burst of diversification following the evolution of a new form of defense against a pathogen or an herbivore. page 682

The Janzen–Connell hypothesis proposes that interactions with pathogens and herbivores increase plant diversity. page 683

Herbivores and pathogens are a major concern for agriculture. page 683

Crop protection includes the use of chemical pesticides, integrated pest management, application of spores or toxins from *Bacillus thuringiensis* to plants, and inserting genes that encode for toxins from *B. thuringiensis* to make *Bt*-modified plants. page 684

Self-Assessment

1. Describe how pathogens enter and move within the plant body.
2. Distinguish between biotrophic and necrotrophic plant pathogens.
3. Name and describe the two components of the plant immune system.
4. Describe and contrast the hypersensitive response and systemic acquired resistance.

5. Describe a feature of Rhizobium radiobacter that makes it a useful tool in biotechnology.
6. Describe three ways that plants protect themselves from being eaten by herbivores.
7. Explain why there are often trade-offs between plant growth and plant defense.
8. Explain how pathogens and herbivores can increase plant diversity.
9. Draw a phylogenetic tree that illustrates an "escape and radiate" pattern of diversification for plants that evolve novel defenses.
10. Describe one benefit and one disadvantage of herbicide and pesticide use in agriculture.

Log in to LaunchPad to check your answers to the Self-Assessment questions, and to access additional learning tools.

Gaby Wojciech / age fotostock.

CHAPTER 33

Plant Diversity

Core Concepts

33.1 Angiosperms make up approximately 90% of all plant species found today.

33.2 Bryophytes form persistent, photosynthetic gametophytes and small, unbranched sporophytes; today they grow in environments where the ability to pull water from the soil does not provide an advantage.

33.3 Spore-dispersing vascular plants today are primarily small plants that grow in moist environments, but in the past included tall trees.

33.4 Gymnosperms produce seeds and woody stems and are most common in seasonally cool or dry regions.

33.5 Angiosperms are distinguished by flowers, fruits, double fertilization, and xylem vessels; their diversity is the result of traits that increase the efficiency of completing their life cycle and building their bodies.

Five hundred million years ago, the land surface supported little more than a green crust made up of a mixture of photosynthetic and non-photosynthetic microbes. Then, through the process of natural selection, a group of freshwater algae began the transition to survival on land. Fast-forward 465 million years—to the present time—and plants are everywhere. Much of what makes up a plant, including roots, leaves, tree trunks, and seeds, evolved quickly. Within the first quarter of the time since these algal ancestors began to colonize the land, all of these features were present. But one thing was missing: flowers. Flowering plants, also known as angiosperms, do not appear on the scene until the last quarter of this history, and yet today they make up a remarkable 90% of all plant species. Many of the plant groups that diverged before angiosperms appeared are still present, but where they live and grow has been shaped by the evolution of flowering plants.

33.1 PLANT DIVERSITY: AN EVOLUTIONARY OVERVIEW

The closest living relatives of plants are green algae that grow in streams and ponds (**Fig. 33.1**). Some features found in plants are also present in the close algal relatives, including cellular structures such as plasmodesmata and the enzymes used to reduce CO_2 loss during photorespiration. But for the most part, plants are quite distinct from their aquatic progenitors. The reason is that, when plants moved onto land, they had to survive in a new medium, air. Algae can rely on water for hydration, support, and protection from ultraviolet radiation and temperature fluctuations, and they can use water currents to transport sperm and disperse offspring. Surviving on land required the evolution of new ways of growing and reproducing. Here, we highlight key adaptations that allowed plants to photosynthesize and reproduce on land, explored in Chapters 29–31, and discuss how plant diversity changed as new ways of growing and reproducing evolved.

Four major transformations in life cycle and structure characterize the evolutionary history of plants.

The first major transformation was the evolution of alternation of generations (Fig. 33.1; Chapter 30). To understand this change, let's first consider the presumed ancestral life cycle. The algal relatives of plants produce multicellular bodies made up entirely of haploid cells. These algae form gametes by mitosis, and fertilization results in a zygote, which is the only diploid cell. The algal ancestor of land plants is thought to have released male gametes (sperm) into the surrounding water and relied on water currents to carry the zygote away from the parent plant. The zygote then underwent meiosis, producing haploid cells that developed into new multicellular algae.

Like their green algal relatives, the first land plants would have produced a multicellular body composed entirely of haploid cells and that produces gametes. During periods of rain, the male gametes could swim to female gametes in surface water layers. But zygotes dispersed this way would not be able to travel very far from the parent plant. To enhance dispersal, land plants evolved alternation of generations: The multicellular haploid generation alternates with a multicellular generation composed of diploid

FIG. 33.1 **Phylogenetic tree of land plants.** The number of species is shown in parentheses. Note that 90% of all land plant species are angiosperms.

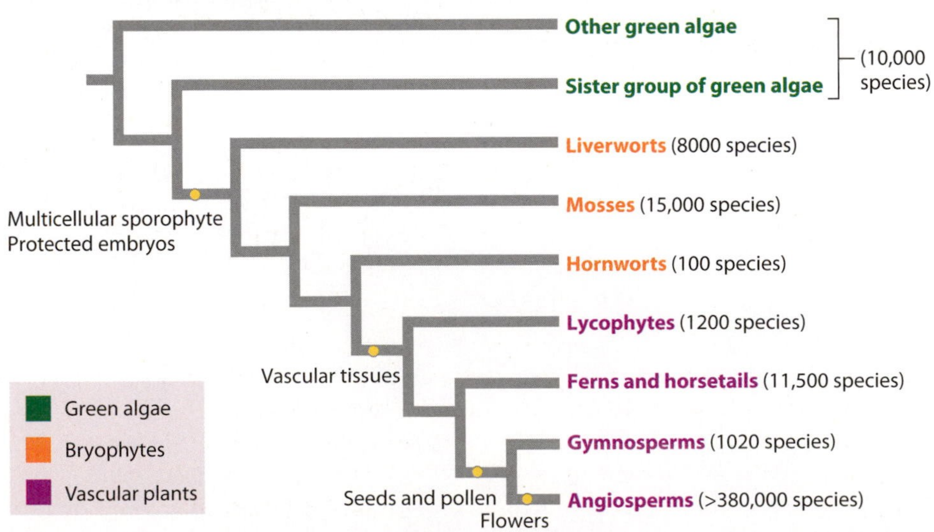

cells. Specifically, the diploid zygote develops by mitosis into a multicellular spore-producing plant, called a sporophyte, while still attached to and supported by the haploid gamete-producing plant, called a gametophyte. In the first diverging groups of plants, the multicellular diploid generation produces many haploid spores by meiosis and, because it is typically erect, can release these spores into the air where they can be carried off by a breeze. Furthermore, because the walls of the spores contain sporopollenin, a polymer that is highly resistant to decay, the spores can survive prolonged exposure to air. Thus, the sporophyte generation and the formation of spores represent evolutionary innovations that enhance dispersal on land.

The first plants lacked roots and thus would have depended entirely on surface water both for fertilization and to maintain the hydration of their cells. As a result, these plants would have been small, with their photosynthetic tissues and gamete-producing structures remaining in close contact with the ground. Photosynthesis would have taken place only when conditions were wet enough to keep cells well supplied with water. Only the sporophyte generation is likely to have extended above the surface, to increase the chances of spores being carried off by a breeze.

This situation radically changed with the second major event in the evolutionary history of plants: the evolution of vascular plants. These plants produce elongate cells for the internal transport of water and other materials. Xylem cells transport water and dissolved nutrients, and phloem cells transport carbohydrates produced by photosynthesis (Chapter 29). The cell walls of the xylem conduits contain lignin, a chemical that greatly strengthens the cellulose wall. Conduits made rigid by lignin allow vascular plants to pull water from the soil and to transport it efficiently through their stems. As a result, vascular plants are taller and able to photosynthesize over a much wider range of conditions than plants dependent solely on surface moisture. The phloem conduits that transport sugars allow roots to grow into the soil, where there is no sunlight to power photosynthesis.

In many vascular plants, xylem and phloem are formed only as stems and roots elongate; they cannot be added to an already formed stem. However, some vascular plants evolved the ability to produce additional or "secondary" xylem and phloem through the formation of a vascular cambium, a layer of actively dividing and differentiating cells that surrounds stems and allows them to increase in diameter (Chapter 31). A second layer of dividing cells, the cork cambium, maintains an intact layer of protective outer bark. Vascular and cork cambia provide the support and water transport capacity needed for plants to grow tall and to support increasing numbers of leaves. Thus, the evolution of secondary growth opened the way to the development of trees and forests.

As plants moved onto land, they retained their ancestral pattern of releasing swimming sperm into the environment. As a result, the gamete-producing generation remained small, constrained by the requirement for surface moisture for fertilization, even as the evolution of vascular tissues allowed the sporophyte to become large. These plants could only reproduce where and when surfaces were sufficiently wet. Eliminating this dependence—through the evolution of pollen and seeds—is the third major evolutionary event in the history of plant diversification.

The life cycle of seed plants differs from that of spore-dispersing vascular plants in a number of important ways (Chapter 30): (1) Spores are not dispersed; instead, they germinate and develop into morphologically distinct male and female gametophytes while still attached to the sporophyte. (2) Male and female gametes are brought together by the transport of the male gametophyte, which is so small that is fits within the spore wall, forming a pollen grain. (3) Following fertilization, the embryo and surrounding tissues develop into a seed, a multicellular structure that replaces unicellular spores as the dispersal unit. In seed plants, alternation of generations persists, but the dominant phase of the life cycle is the sporophyte, and pollen and seeds carry out tasks previously accomplished by swimming sperm and spores. The effect of this transformation is a life cycle that is freed from dependence on surface moisture for fertilization and in which embryos are dispersed along with resources that they can draw upon during germination.

The fourth evolutionary event is the evolution of the flowering plants, also referred to as the angiosperms. Angiosperms are seed plants and thus their life cycle contains all the traits just described. In addition, four new reproductive features are thought to have contributed to angiosperm diversity and success. The first is the flower, a reproductive structure that attracts animal pollinators, increasing the efficiency of pollen transfer compared to wind pollination. The second is the carpel—the closed "vessel" in which seeds develop and that gives angiosperms their name. Because pollen must grow through the carpel to reach the female gametophyte, interactions between pollen and carpel genotypes can affect the probability of fertilization. A third feature is double fertilization, in which one sperm nucleus from the male gametophyte fuses with the egg, and a second sperm nucleus fuses with two haploid nuclei from the female gametophyte. The first union results in the diploid zygote, and the second gives rise to endosperm, which forms the nutritive tissue within the seeds of angiosperms. A consequence of double fertilization is that angiosperms do not expend resources for the next generation until the egg is fertilized. A fourth feature is the development of fruits, structures that surround seeds and attract animals to enhance seed dispersal.

Angiosperms are also distinguished by the presence of wood containing xylem vessels (Chapter 29). Most other seed plants produce only tracheids, which both support the stem and transport water. In contrast, angiosperms have thick-walled, elongate fibers that support the stem and xylem vessels that transport water. Because xylem vessels are not a stem's sole source of support, they can be wide and long, allowing angiosperms to transport water through their stems at higher rates than plants with tracheids. Their uptake of CO_2 for photosynthesis increases as well because CO_2 uptake is directly linked to water lost to the atmosphere. As a result, angiosperms compete well for light and space and are the dominant plants in most terrestrial ecosystems.

Plant diversity has changed over time.

If we tabulate the numbers of species within each major branch on the phylogenetic tree shown in Fig. 33.1, one fact stands out: Of the nearly 400,000 species of plants present today, approximately 90% are angiosperms. Furthermore, angiosperms are relative newcomers (**Fig. 33.2**). They appeared in the fossil record about 140 million years ago. The oldest known evidence of land plants is found in rocks approximately 465 million years old. Thus, for more than 300 million years, terrestrial vegetation was made up of plants other than angiosperms. During this period, lycophytes, ferns and horsetails, and gymnosperms, as well as many groups of now-extinct plants, dominated the land, forming forests and landscapes unlike those that surround us today.

Once the flowering plants gained an ecological foothold, however, the number of angiosperm species increased at an unprecedented rate. As the number of angiosperm species rose, the number of species in other groups fell. Yet the total number of plant species increased dramatically. Thus, angiosperms' hold on plant diversity is the result of both squeezing out other groups and increasing the number of species that can coexist.

One way that angiosperm evolution may have increased the overall number of plant species on Earth was by transforming the environments into which they diversified. Climate models suggest that angiosperms may have been necessary for the formation of tropical rain forests as we know them today. Without the higher rates of transpiration exhibited by angiosperms, many tropical regions would have higher temperatures and lower rainfall. The warmer, drier conditions would have been less conducive to the luxuriant plant growth found in tropical rain forests. The new tropical forests, with their dense shade and humid understories, provided new habitats into which both angiosperms and non-angiosperms could evolve. Thus, although early angiosperms outcompeted many species for space and light, their evolution also stimulated radiations of new species in at least some of the groups that preceded them.

FIG. 33.2 The diversity of vascular plant groups through time. First appearing about 140 million years ago, angiosperms have come to dominate the fossil record. *Source: Adapted from A. H. Knoll and K. J. Niklas, 1987, "Adaptation, Plant Evolution, and the Fossil Record," Review of Palaeobotany and Palynology, 50:127–149.*

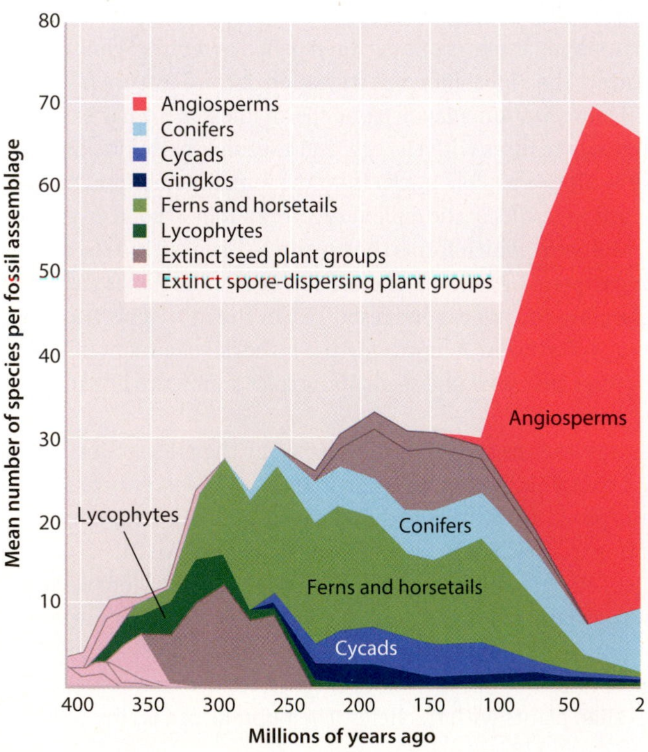

33.2 BRYOPHYTES

Liverworts, mosses, and hornworts diverged before the evolution of xylem and phloem (Fig. 33.1). The hypothesis shown in Fig. 33.1 implies that liverworts, mosses, and hornworts evolved independently for hundreds of millions of years and that they do not form a monophyletic group. Recall from Chapter 23 that a monophyletic group includes a common ancestor and *all* of its descendants, whereas a paraphyletic group includes a common ancestor and *some* of its descendants. However, it is important to recognize that the phylogenetic relationship between liverworts, mosses, and hornworts remains uncertain. Because these three groups share many features, we refer to them collectively as bryophytes.

As the living representatives of the first plant groups to diverge after plants moved onto land, bryophytes provide us with insights into how plants gained a foothold in the terrestrial environment. The insights from bryophytes are particularly welcome because the first plants are poorly represented in the fossil record. Only tiny spores and fragments of a cuticle-like covering record these early events. Of course, we must not lose sight of the fact that bryophytes have continued to evolve as the conditions for life on land have changed over the past 400+ million years. Where bryophytes grow today and what they look like have been shaped by their long coexistence with vascular plants.

→ **Quick Check 1** Look at Fig. 33.1 and determine whether green algae, bryophytes, and vascular plants are monophyletic or paraphyletic.

Bryophytes are small, simple, and tough.

Mosses are the most widely distributed of the bryophyte groups and the most diverse, with about 15,000 species. You may have

FIG. 33.3 Bryophyte diversity. Bryophytes include (a) mosses (*Polytricum commune* is shown); (b) liverworts (*Pellia ephiphylla,* a thalloid liverwort, is shown); and (c) hornworts (*Anthoceros* species). *Sources: a. NHPA/Photoshot; b. ARCO/Reinhard H/age fotostock; c. Daniel Vega/age fotostock.*

seen moss growing on a shady log or on top of rocks by a stream. However, mosses grow in all terrestrial environments, from deserts to tropical rain forests. Liverworts (about 8000 species) and hornworts (about 100 species) are less widespread and also less diverse. In considering the diversity of these three groups, let's start by reviewing some of their features that allow them to carry out photosynthesis on land.

Bryophytes have small, simple bodies (**Fig. 33.3**). Some produce only a flattened photosynthetic structure called a **thallus.** Others consist of slender stalks and have a leafy appearance. However, these leaflike structures are quite different from the leaves of vascular plants in that they are only one to several cells thick and lack internal air spaces. Both thalloid and leafy species are found in the liverworts, whereas all mosses are of the leafy type and all hornworts are of the thalloid type.

Because bryophytes do not produce lignified xylem conduits, they cannot pull water from the soil. Instead, they absorb water, nutrients, and CO_2 through their surfaces. For this reason, bryophytes have little or no waxy cuticle to protect their photosynthetic tissues. Bryophytes can absorb enough water to remain metabolically active when the environment is wet, but they must be able to tolerate desiccation when the environment is dry. Thus, while the small size and lack of differentiated structures might suggest that bryophytes are delicate, nothing could be further from the truth. Bryophytes can be found from the equator to both the latitudinal and altitudinal limits of vegetation, and from swamps to deserts.

Before the evolution of vascular plants, bryophytes likely covered much of the land surface. However, because bryophytes are small and they can only photosynthesize when surface water is present, they are poor competitors for light and space with vascular plants. Today, bryophytes thrive in local environments where roots do not provide an advantage—for example, on the surfaces of rocks. Many bryophytes live on the branches and trunks of trees rather than on the ground. Plants that grow on other plants are called **epiphytes** (from the Greek words *epi,* "on," and *phyton,* "plant"). Bryophytes are well suited to growing on other plants because they do not depend on the soil as a source of water.

The small gametophytes and unbranched sporophytes of bryophytes are adaptations for reproducing on land.

The easily visible and persistently photosynthetic bryophyte bodies described above are those of the haploid, gamete-producing generation. Because bryophytes release swimming sperm into the environment, their mode of fertilization also limits their size. Bryophyte sperm must swim through films of surface water or be transported by the splash of a raindrop. Sperm can travel in this manner only a relatively limited distance and only if water is present. A few mosses grow to be more than half a meter tall, but these giants of the bryophyte world are found only in the understory of very wet forests. In most bryophytes, the gametophyte grows to only a couple of centimeters in height.

Following fertilization, the diploid sporophyte develops while remaining physically attached to and, in varying degrees, nutritionally dependent upon the gametophyte. Recall that the diploid, spore-producing generation represents a new component of the life cycle that evolved as plants moved onto land. This sporophyte generation addresses the challenges plants face in dispersing offspring through the air. In most bryophytes, the unbranched sporophyte extends several centimeters above the gametophyte, greatly increasing the chances of a gust of wind dispersing spores through the air.

In mosses and liverworts, the sporophyte is short-lived, drying out after the spores are dispersed. However, in hornworts, the

FIG. 33.4 Moss specializations. (a) The yellow moose-dung moss, *Splachnum luteum,* attracts insects that then transport its spores. (b) *Dawsonia superba* has internal transport tissues that allow it to grow more than 40 cm tall. *Sources: a. Biopix: A Neumann; b. blickwinkel/Alamy.*

sporophyte can live nearly as long as the gametophyte because it can produce new cells at its base, similar to the way grass blades elongate (Chapter 32). Thus, bryophytes illustrate an important trend in the evolution of plants, an increase in the persistence of the spore-producing generation.

Bryophytes exhibit several cases of convergent evolution with the vascular plants.

Bryophytes are interesting partly because they are so different from the more familiar vascular plants. However, given that they have long evolved in parallel with vascular plants, it is not surprising that bryophytes have evolved similar solutions to environmental challenges, a process referred to as convergent evolution (Chapter 23). For example, some mosses depend on insects to transport their spores (**Fig. 33.4a**). The brightly colored sporangia of these mosses emit volatile chemicals that mimic compounds released by rotting flesh or herbivore dung. Attracted by these chemicals, insects land on the sporangia, and their legs become covered with spores. When the insect subsequently lands on a real pile of dung, some of the spores fall off into what is, for them, an ideal site for supporting the growth of a new gametophyte.

Another example of convergent evolution is the presence of cells specialized for the transport of water and carbohydrates.in some of the largest bryophytes (**Fig 33.4b**). The structure of these cells indicates that they evolved independently from the xylem and phloem of vascular plants. In particular, the water-conducting cells in bryophytes do not have lignified cell walls and thus are not sufficiently rigid to pull water from the soil. Nevertheless, because of these elongated cells, water and carbohydrates can move from one part of the plant to the other efficiently.

Stomata are present on the sporophytes of some mosses and hornworts, particularly in regions where a waxy cuticle also develops. As discussed in Chapter 29, stomata are pores in the epidermis of vascular plants that open and close, providing an active means of controlling water loss. Bryophyte stomata may play an additional role: A sporophyte with open stomata will dry out quickly, so its spores are more easily released into the air. Further study is needed to determine if these stomata are the result of convergent evolution.

→ **Quick Check 2** If the stomata of bryophytes and of vascular plants are not an example of convergent evolution, what is an alternative explanation for their appearance in these groups?

Sphagnum moss plays an important role in the global carbon cycle.

In most ecosystems, bryophytes make only a small contribution to the total biomass. The one exception is **peat bogs,** wetlands in which dead organic matter accumulates rather than decomposes. A major component of peat bogs is sphagnum moss, any of the several hundred species of the genus *Sphagnum*. These mosses play a key role in creating wet and acidic conditions that slow rates of decomposition. They have specialized cells that hold onto water, much like a sponge, and they secrete protons that acidify the surrounding water (**Fig. 33.5**). Unlike many plants with roots, sphagnum moss thrives under wet, acidic conditions. In addition, sphagnum mosses produce phenols, organic compounds that slow decomposition under waterlogged conditions.

Peat bogs occupy only 2% to 3% of the total land surface, but they store large amounts of organic carbon—on the order of 65 times the amount released each year from the combustion of fossil fuels. Thus, any increase in the decomposition rate of these naturally formed peats has the potential to increase significantly the CO_2 content of Earth's atmosphere.

What might cause an increase in the decomposition rate? Peat bogs are vulnerable to changes in the water table. As water levels fall, the peat becomes exposed to the air, allowing it to be broken down by microbes. Most peat bogs are located in northern latitudes, where the effects of climate change are predicted to be greatest. If rising summer temperatures result in higher rates of transpiration from surrounding forests, the result could be a lowering of the water table and a relatively rapid release of CO_2 from this moss-dominated ecosystem. Thus, peat bogs have the potential to accelerate rates of climate change over the next century.

FIG. 33.5 Sphagnum moss. Peat bogs have low rates of organic decomposition due to the wet and acidic conditions created by sphagnum moss. *Sources: (top) Duncan Shaw/Science Source; (middle) All Canada Photos/Alamy; (bottom) blickwinkel/Alamy.*

33.3 SPORE-DISPERSING VASCULAR PLANTS

Vascular plants first appear in the fossil record approximately 425 million years ago. The evolution of xylem and phloem gave vascular plants advantages in size and hydration. They were able to outcompete bryophytes for light and other resources and to become the dominant plants on land.

Vascular plants can be divided into two groups according to how they complete their life cycle. Lycophytes, as well as ferns and horsetails, disperse by spores and rely on swimming sperm for fertilization, just like bryophytes. Gymnosperms and angiosperms are seed plants: They disperse seeds and rely on the aerial transport of pollen for fertilization. In this section, we focus on the spore-dispersing vascular plants: lycophytes, and ferns and horsetails. In both groups, the sporophyte dominates in physical size; the gametophyte is only a few centimeters across, thalloid in structure, and typically tolerant of desiccation. Thus, our discussion emphasizes the sporophytes of these plants.

Rhynie cherts provide a window into the early evolution of vascular plants.

Fossils recovered from a single site in Scotland document key stages in the evolution of vascular plants. Four hundred million years ago, near the present-day village of Rhynie, a wetland community thrived in a volcanic landscape. Mineral-laden springs similar to those found today in Yellowstone Park encased the newly deposited remains of local populations in chert, a mineral made of silica (SiO_2). These fossils preserve plants, fungi, protists, and small animals in remarkable detail.

Rhynie cherts provide our best view of early vascular plants. These plants were small (up to about 15 cm tall) photosynthetic structures that branched repeatedly, forming sporangia at the tips of short side branches (**Fig. 33.6**). They had no leaves, and their only root structures were small hairlike extensions on the lower parts of stems that ran along the ground. The stems had a cuticle with stomata, and a tiny cylinder of vascular tissue ran through the

FIG. 33.6 Rhynia, a fossil vascular plant from the 400-million-year-old Rhynie chert. All vascular plants are descended from simple plants such as this. *Photo sources: (top) Courtesy of Hans Kerp © Palaeobotanical Research Group, University of Münster; (middle & bottom) Hans Steur, The Netherlands.*

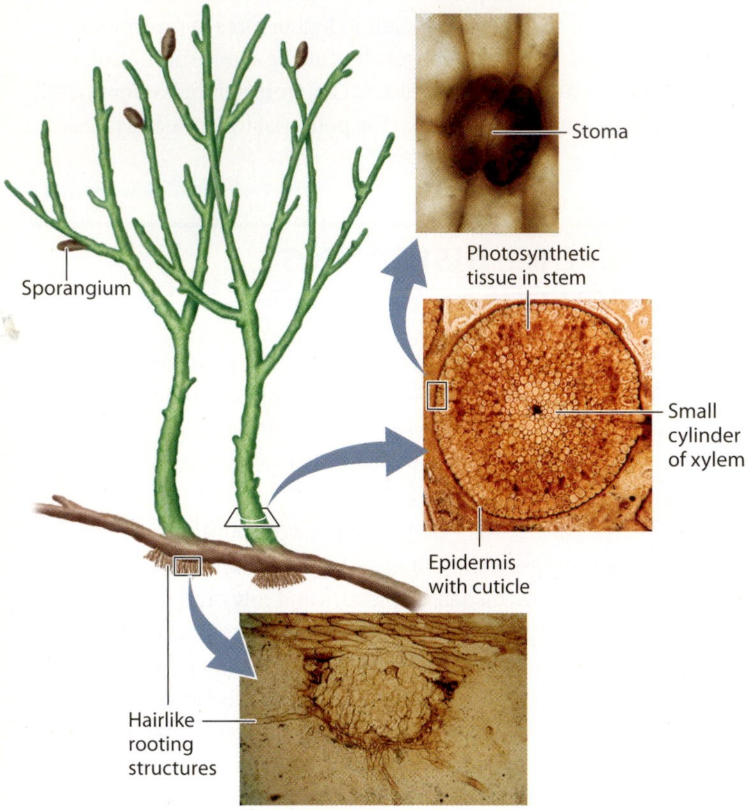

center. Many of the gametophytes also had erect cylindrical stems, not at all like the reduced gametophytes of living vascular plants.

Rhynie cherts provide another important glimpse of early plant evolution: Some fossils have a branched sporophyte that includes cell walls that are reinforced with lignin, but there is no evidence of vascular tissue. Evidently, plants evolved the upright stature we associate with living vascular plants before they evolved the differentiated tissues of the vascular system (xylem and phloem). Once vascular tissues had evolved, however, they played a key role in shaping subsequent evolution of the plant sporophyte.

Lycophytes are the sister group of all other vascular plants.

One of the fossil populations in Rhynie cherts is notably larger and more complex than the others. These fossils show leaflike structures arranged on the stem in a spiral pattern. Through the stem runs a thick-lobed cylinder of xylem (**Fig. 33.7**). These are among the earliest fossils of lycophytes, a group of vascular plants that can be found from tropical rain forests to Arctic tundra. Because lycophytes are the earliest branching group of vascular plants, they are the sister group to all other vascular plants (see Fig. 33.1).

Fossils indicate that the leaves and roots of lycophytes evolved independently from those of ferns, horsetails, and seed plants. A distinctive feature of lycophyte leaves is that they have only a single vein, in contrast to the more complex leaf venation found in other vascular plants. About 1200 species of lycophytes can be found today. Some grow low to the ground and produce branching stems covered by small leaves (**Fig. 33.8a**) or leaves in rows held in a single plane (**Fig. 33.8b**), while others grow as epiphytes in tropical forests. The most distinctive living lycophytes are the quillworts (genus *Isoetes*), which are small plants that live along the margins of lakes and slow moving streams (**Fig. 33.8c**). A few species are desiccation tolerant and can live in ponds that dry up in summer. Uniquely among living lycophytes, they have a vascular cambium. Seeing a quillwort today, you would never guess that its relatives once included enormous trees.

FIG. 33.7 An early lycophyte preserved in the 400-million-year-old Rhynie chert. The fossil has helically arranged leaflike structures, adventitious roots (that is, roots that arise from the shoot and not from the root system), and a central strand of xylem, features much like those of living lycophytes. *Sources: (top) Photo by William Chaloner, Crown Copyright; (bottom) Photo by John Hall © Botanical Society of America.*

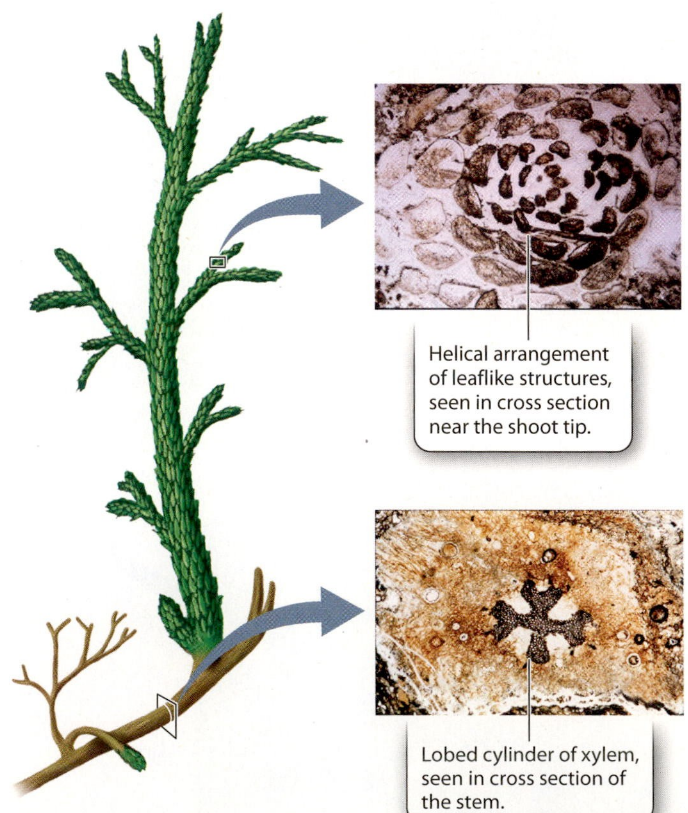

FIG. 33.8 **Lycophyte diversity.** Living lycophytes include about 1200 species, in three major groups: (a) *Lycopodium annotinum,* showing yellow sporangia-bearing leaves at the shoot tips of this 5-to-30-cm tall plant; (b) *Selaginella wildenowii,* in a close-up showing the flattened arrangement of its 3-mm-long leaves; and (c) *Isoetes lacustris,* an aquatic lycophyte that grows from 8 to 20 cm tall. *Photo sources: a. Philippe Clement/Nature Picture Library; b. blickwinkel/Alamy; c. Biopix: JC Schou.*

Ancient lycophytes included giant trees that dominated coal swamps about 320 million years ago.

Fossils show that ancient lycophytes evolved additional features convergently with seed plants, including a vascular cambium and cork cambium that enabled them to form trees up to 40 m tall (**Fig. 33.9**). Swamps that formed widely about 320 million years ago were dominated by tree-sized lycophytes. Unlike the vascular cambium of seed plants, the vascular cambium of these lycophytes (**Fig. 33.10**) produced relatively little secondary xylem and no secondary phloem. Instead, the giant lycophytes relied on thick bark for mechanical support. Thus, tree-sized lycophytes were markedly different from the trees familiar to us today.

These forest giants were not outcompeted by seed plants, but persisted for millions of years in swampy environments alongside early seed plants. It was environmental change that spelled their doom. When changing climate dried out the swamps, the large lycophytes disappeared, along with their habitat. Little *Isoetes* was left as the only living reminder of a once-dominant group of plants.

These ancient forests left one other legacy: the coal deposits we mine today. As trees died and fell over into the swamp, their bodies decomposed slowly. Over time, as material accumulated

FIG. 33.9 **Giant lycophytes.** (a) Artist's depiction of giant lycophytes. (b) A fossil grove of lycophyte trees showing rooting structures and thick stems. *Sources: a. John Gurche; b. Biophoto Associates/Science Source.*

HOW DO WE KNOW?

FIG. 33.10

Did woody plants evolve more than once?

OBSERVATIONS Today, the vascular cambium occurs almost exclusively among seed plants, with only limited development of secondary xylem in some adder's tongue ferns and quillwort lycophytes (*Isoetes* species). The fossil record, however, shows that other plants, now extinct, also had a vascular cambium.

HYPOTHESIS The vascular cambium evolved convergently in several different groups of vascular plants.

EXPERIMENT AND RESULTS The vascular cambium is recorded in fossils by xylem cells in rows oriented radially in the stem or root. Thus, the giant tree lycophytes of Carboniferous coal swamps (the top photo) had a vascular cambium, as did extinct tree-sized relatives of the horsetails (the bottom photo) and other extinct horsetail relatives. Phylogenetic trees generated from morphological features preserved in fossils unambiguously show that woody lycophytes, woody horsetail relatives, and the group of seed plants and progymnosperms did not share a common ancestor that had a vascular cambium. Moreover, anatomical research shows that the vascular cambium of extinct woody lycophytes and the giant horsetails *Archaecalamites* and *Calamites* generated secondary xylem but not secondary phloem, unlike the vascular cambium of seed plants. Living horsetails do not have a vascular cambium, but fossils show that they are descended from ancestors that did make secondary xylem and have lost this trait through evolution.

CONCLUSION Fossils and phylogeny support the hypothesis that the vascular cambium and, hence, wood evolved more than once, reflecting a strong and persistent selection for tall sporophytes among vascular plants.

SOURCE Taylor, T. N., E. L. Taylor, and M. Krings. 2009. *Paleobotany.* Amsterdam: Elsevier.

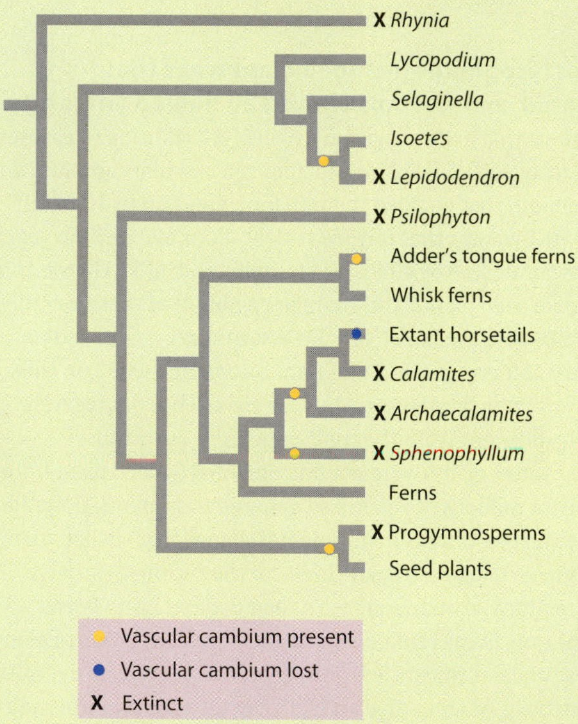

A small branch of *Lepidodendron*, an ancient giant lycophyte

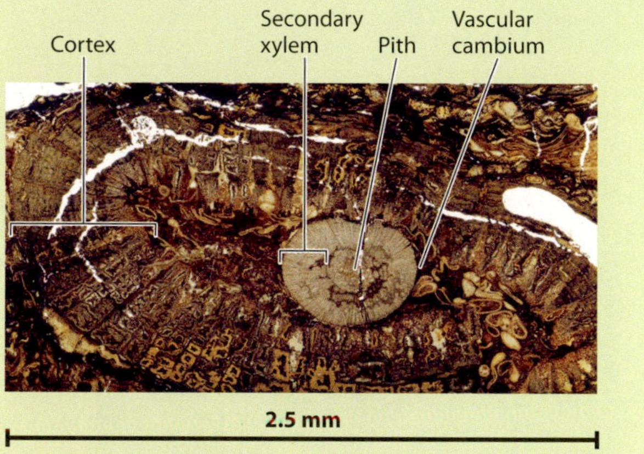

A small stem of *Calamites*, an ancient giant horsetail

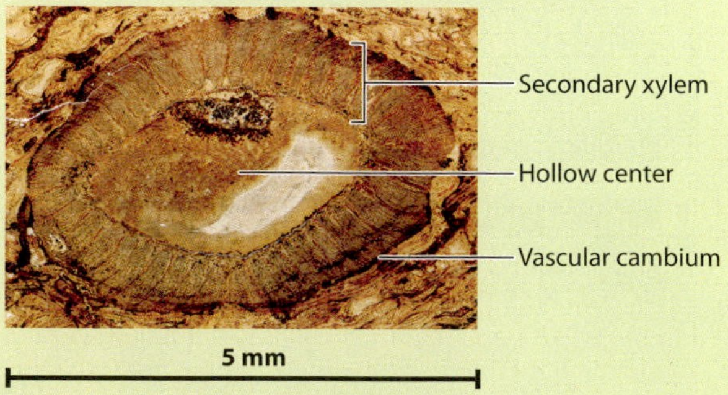

Photos source: Andrew Knoll, Harvard University.

FIG. 33.11 Diversity of ferns and horsetails. (a) The adder's tongue fern *Ophioglossum*, (b) the whisk fern *Psilotum*, (c) the horsetail *Equisetum*, (d) a marattioid fern, (e) an aquatic fern (*Salvinia*), and (f) a polypod fern (in the middle of the photo is *Athyrium felix-femina*, the common lady-fern). *Photo sources: a. Jonathan Buckley/age fotostock; b. Biophoto Associates/Science Source; c. CuboImages srl/Alamy; d. Courtesy Gary Higgins; e. FLPA/Chris Mattison/age fotostock; f. Ron Watts/First Light/Corbis.*

and became buried, the combined action of high temperature and pressure converted the dead organic matter first into peat and then into the carbon-rich material we call coal. Thus, a major source of energy we use today derives originally from photosynthesis carried out by lycophyte trees.

Ferns and horsetails are morphologically and ecologically diverse.

Ferns and horsetails form a monophyletic group that is the sister group to the seed plants (**Fig. 33.11**). The majority of species in this group are ferns. You can recognize ferns by their distinctive leaves that uncoil during development from tightly wound "fiddleheads" (**Fig. 33.12**). This group also includes horsetails and whisk ferns, which traditionally were considered distinct groups because of their unique body organizations. Whisk ferns have photosynthetic stems without leaves and roots (see Fig. 33.11b). Thus, they look similar to fossils of early vascular plants. Molecular sequence comparisons, however, support the hypothesis that whisk ferns are the simplified descendants of plants that produced both leaves and roots.

The horsetails, represented by 15 living species of the genus *Equisetum*, produce tiny leaves arranged in whorls, giving the stem a jointed appearance (see Fig. 33.11c). Horsetail stems are hollow, and their cells accumulate high levels of silica, earning them the name "scouring rush." Today, horsetails are small plants, most less than a meter tall. However, tree-sized versions grew side by side with the ancient lycophyte trees (see Fig. 33.9).

FIG. 33.12 A fern leaf uncoiling from a "fiddlehead." *Source: Andre Nantel/Shutterstock.*

FIG. 33.13 A tree fern and an aquatic fern. (a) Silver tree ferns (*Cyathea meullaris*) grow 10 m tall or more. (b) *Azolla* ferns floating on the surface of still freshwater. Each fern is about 10 mm in diameter. *Sources: a. AA World Travel Library/Alamy; b. Huetter, C/age fotostock.*

In contrast, many ferns produce large leaves—in some cases up to 5 m long, although typically the photosynthetic surfaces are divided into smaller units called **pinnae.** Yet, because ferns are not capable of secondary growth, their stems remain slender and frequently grow underground, with only the leaves emerging into the air. Many ferns exhibit little stem elongation and produce their leaves in tight clumps. However, some species, such as bracken fern, produce spreading stems with widely spaced leaves. Because bracken is toxic to cattle and other livestock, its tendency to invade pastures is a serious problem.

Despite the absence of secondary growth, some ferns can grow quite tall. Tree ferns, which can grow more than 10 m tall, produce thick roots near the base of each leaf that descend parallel to the stem to the ground. These roots increase the mechanical stability of the plant and also allow it to transport more water (**Fig. 31.13a**). Other ferns grow tall by producing leaves that twine around the stems of other plants, using the other plant for support. At the other extreme are tiny aquatic ferns such as *Salvinia* (see Fig. 33.11e) and *Azolla* (**Fig. 33.13b**) that float on the surface of the water.

Fern diversity has been strongly affected by the evolution of angiosperms.

The fossil record shows that ferns originated more than 360 million years ago, and between 300 and 100 million years ago they were the most diverse group of plants (Fig. 33.2). Then, as angiosperms evolved, their diversity markedly decreased. It is not clear why angiosperm evolution should have so affected ferns and horsetails, which had coexisted with a variety of seed plant groups for more than 200 million years.

However, the most diverse group of ferns today evolved after the rise of the angiosperms. These are the polypod ferns (about 9300 species; see Fig. 33.11f). Thus, many of the fern species present today are likely to have evolved to occupy habitats newly created by the development of angiosperm forests. Approximately 40% of living ferns are epiphytes, many of them growing in tropical rain forests dominated by angiosperm trees.

Polypod ferns may owe some of their evolutionary success to an enhanced capacity for spore dispersal. Horsetails, whisk ferns, and a few other fern groups make large sporangia similar to those produced by early vascular plants. Most ferns, however, have a distinctive sporangium (called a leptosporangium) whose wall is only a single cell thick. Polypod ferns have leptosporangia with a line of thick-walled cells that run along the sporangium surface. When the spores mature, the sporangium dries out and these cells contract, forcibly ejecting spores into the air.

An aquatic fern contributes to rice production.

Azolla is a free-floating aquatic fern that has been used for centuries in Asia as a biofertilizer of rice paddies (see Fig. 33.13b). Its success as a fertilizer is due to its symbiotic association with nitrogen-fixing cyanobacteria. *Azolla* is common worldwide on still water such as ponds and quiet streams. Its leaves contain an internal cavity in which colonies of nitrogen-fixing cyanobacteria become established. The cyanobacteria provide *Azolla* with a source of nitrogen, while the fern provides shelter and carbohydrates. The cyanobacteria are reported to have higher rates of nitrogen fixation when they are housed within an *Azolla* leaf than when they live freely in the environment.

When added to newly planted rice fields, this fast-growing fern suppresses weed growth by forming a dense, floating layer. At the same time, it provides organic nitrogen that eventually finds its way into the developing rice plants. The floating fertilizer factories provided by *Azolla* may have made possible the long history of rice cultivation in Asia.

33.4 GYMNOSPERMS

Seed plants have come to dominate terrestrial environments since their first appearance about 365 million years ago. While the coal-swamps of 300 million years ago were populated by spore-producing giants, the understory of these strange forests included plants that produced pollen and seeds. By the time of the dinosaurs, seed plants had become the largest and most abundant plants on Earth.

What factors have contributed to the success of seed plants? One is that they do not require external water for fertilization. Instead, the male gametophyte is transported through the air in pollen. As a result, seed plants are able to reproduce successfully in both rain forest and desert. Pollination, if successful, leads to fertilization and the formation of a seed. Seeds in turn enhance dispersal. The nutrient-rich tissue in the seed makes it more likely that the embryo will grow into a photosynthetically self-sufficient plant. Before these innovations, the ancestors of seed plants evolved the ability to form large woody stems through the formation of both a vascular and a cork cambium. Therefore, woody stems and roots are the ancestral condition in seed plants.

Although the fossil record shows that there were once many different groups of seed plants, today there are only two: gymnosperms and angiosperms. Gymnosperms total only a bit more than 1000 species, whereas angiosperms total 380,000 species or more. Despite their much lower diversity, in cool or dry parts of the world, gymnosperms such as pine trees can make up more of the forest canopy than do angiosperms. In this section, we examine the evolutionary history of the four groups of living gymnosperms (**Fig. 33.14**) and explore how their diversity and distribution has been affected by the rise of the angiosperms.

Cycads and ginkgos are the earliest diverging groups of living gymnosperms.

It's easy to confuse cycads with palm trees because both have unbranched stems and large leaves (**Fig. 33.15**). However, the presence of cones rather than flowers clearly marks cycads as gymnosperms. Cones are modified branch tips inside which reproductive structures develop, with pollen and ovules produced in separate cones. In cycads, the cones are large and woody, and thus provide protection for the developing pollen and seeds.

Cycads have low photosynthetic rates and they grow slowly. Although they have large stems, their vascular cambium produces little additional xylem, and most of their bulk is made up of unlignified (nonwoody) cells in the center of the stem (pith) and surrounding the vascular tissues (cortex). Cycads are able to grow in nutrient-poor environments by forming symbiotic associations with nitrogen-fixing cyanobacteria. Cycads are also able to survive wildfires because their apical meristem is protected by a dense layer of bud scales.

The approximately 300 species of cycads occur most commonly in tropical and subtropical regions. Most species have a limited and often fragmented distribution, suggesting that they were much more widespread in the past. Most cycad populations are small and often vulnerable to extinction. Cycads may owe their persistence in part to the fact that they rely on insects for pollination, as well as to their ability to live in nutrient-poor and fire-prone environments. Insects transfer pollen more efficiently than wind, and therefore insect-pollinated species with small populations are able to reproduce successfully. Many cycads are pollinated by beetles, which are attracted by chemical signals produced by the cones. The large seed cones provide shelter for the insects, and the pollen cones provide them with food.

FIG. 33.14 Seed plant diversity. The phylogenetic tree shows a hypothesis of evolutionary relationships among gymnosperms. *Photo sources: (top to bottom) Robert Davis/age fotostock; Michael P. Gadomski/Science Source; Pete Ryan/Getty Images; Patti Murray/age fotostock.*

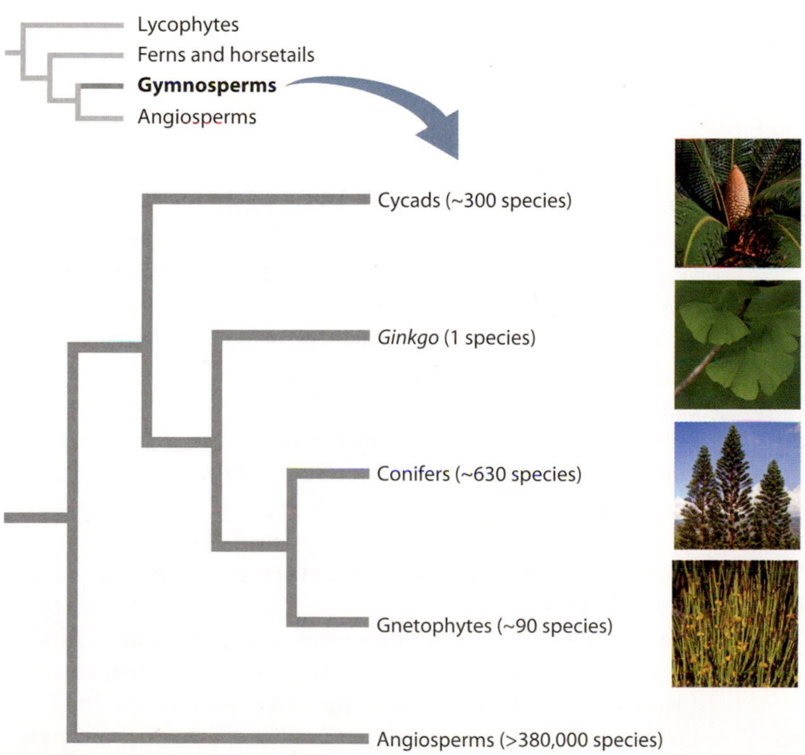

FIG. 33.15 **Cycads.** (a) Most living cycads have thick, unbranched stems that support large leaves and cones. Shown here is *Encephalartos friderici-guilielmi.* (b) Many cycads, such as *Encephalartos transvenosus,* have large cones. (c) Mature seeds are large (here are shown those of *Cycas circinalis*) and contain a relatively large female gametophyte that provides nutrition for the embryo and germinating sporophyte. *Sources: a. A. Laule/age fotostock; b. & c. Andrew Knoll, Harvard University.*

Following fertilization, brightly colored, fleshy seed coats attract a variety of birds and mammals that serve as dispersal agents. Because of their small numbers and fragmented distribution, two-thirds of all cycads are on the International Union for the Conservation of Nature's "red list" of threatened species, the highest percentage of any group of plants.

Cycads are an ancient group, with fossils at least 280 million years old. During the time of the dinosaurs, they were among the most common plants. Cycad diversity and abundance declined markedly during the Cretaceous Period, at the same time the angiosperms rose to ecological prominence. A recent study shows that present-day cycad species date from a burst of speciation that took place approximately 12 million years ago. At that time, the positions of the continents and ocean circulation patterns were changing in such a way that climates became increasingly cooler and more seasonal. This climate change provided new habitats for emerging species while dooming older species to extinction.

A second group of gymnosperms has only a single living representative: *Ginkgo biloba.* This species forms tall, branched trees, with fan-shaped leaves that turn brilliant yellow in autumn (**Fig. 33.16**). *Ginkgo* is wind pollinated, but, like cycads, develops fleshy seeds, perhaps a feature inherited from their common seed plant ancestor. Ginkgos date back about 270 million years, and during the time of the dinosaurs, they were common trees in temperate forests. Like cycads, ginkgos declined in abundance, diversity, and geographic distribution over the past 100 million years, as a result of both changing climates and angiosperm diversification.

Ginkgos have long been cultivated in China, especially near temples, and it is unclear whether any truly natural populations exist in the wild. Today, however, *Ginkgo* once again is found all over the world, as its ability to tolerate the pollution and other stresses of urban environments has made it a tree of choice for street plantings.

Conifers are woody plants that thrive in dry and cold climates.

The approximately 630 species of conifers are, for the most part, trees. Conifers include the tallest (over 100 m) and oldest (more than 5000 years) trees on Earth (**Fig. 33.17**). Well-known conifers include pines, junipers, and redwoods. Most conifers have strong apical dominance, and thus develop a single straight stem from

FIG. 33.16 ***Ginkgo biloba.*** (a) The fan-shaped leaves of *Ginkgo* turn yellow in autumn. (b) The fleshy seeds enclose a large female gametophyte, as in cycads. *Sources: a. JTB/Photoshot; b. Kathy Merrifield/Science Source.*

which extend much smaller horizontal branches. Conifers have traditionally served as masts for sailing ships, and today are used for telephone poles. Conifer xylem consists almost entirely of tracheids, and so its wood is strong for its weight, making it well suited as a building material. Conifers supply much of the world's timber, as well as the raw material for producing paper.

Most conifers are evergreen, meaning that they retain their often needle-like leaves throughout the year. In some species, the leaves are remarkably long lived, remaining green and photosynthetic for decades. Many conifers produce resin canals in their wood, bark, and leaves that deter insects and fungi. Conifer resins are harvested and distilled to produce turpentine and other solvents. In addition, conifer resins are the source of a number of important chemicals, including the drug taxol used in the treatment of cancer. Amber, prized for its beauty, is fossilized resin.

Conifer reproduction tends to be slow: After pollination, it can take more than a year before the seeds are ready to be dispersed. Pollen is produced in small cones, and seeds in larger cones. The seed cones become woody, providing protection, as the fertilized ovules develop slowly into seeds. Conifers are wind pollinated, and most species also rely on wind for seed dispersal. However, some conifers, such as junipers and yews, produce seeds surrounded by fleshy and often brightly colored tissues that attract birds and other animals.

Conifers dominate the vast boreal forests of Canada, Alaska, Siberia, and northern Europe. Conifers tend to become more abundant as elevation increases, and they are common in dry areas, such as the western parts of North America and much of Australia. Only a small number of conifer species are found in tropical latitudes, most commonly at higher elevations. Before the rise of the angiosperms, however, conifers were more evenly distributed across the globe, including in the lowland tropics. One of the great questions in the evolutionary history of plants is why conifers were displaced from tropical latitudes by the angiosperms but continued to thrive in cold and dry environments.

An important difference between angiosperms and conifers is that angiosperms transport water in the xylem vessels, whereas conifers transport water in tracheids. In wet and warm environments, the xylem vessels of angiosperms can be both wide and long, increasing the efficiency of water transport through their stems. Thus, one hypothesis is that the evolution of xylem vessels gave angiosperms a competitive advantage in moist tropical environments. With more water flowing through their stems, angiosperms could have higher rates of photosynthesis and

FIG. 33.17 **Conifers.** Giant sequoia (*Sequoiadendron giganteum*) can reach 85 m in height and have stems that are 8 m in diameter. *Source: Charles J. Smith.*

FIG. 33.18 Gnetophytes. The three genera in this group include some of the most distinctive seed plants: (a) *Welwitschia mirabilis;* (b) *Gnetum* species; and (c) *Ephedra* species. *Sources: a. M. Philip Kahl/Science Source; b. Ahmad Fuad Morad [Flickr page] 072112; c. Bob Gibbons/Science Source.*

grow larger leaves. Once conifer populations began to dwindle, they could have disappeared completely because wind pollination requires large populations. In colder or drier regions, the dangers of cavitation due to drought or freeze followed by thaw (Chapter 29) constrain the size of angiosperm vessels. This helps level the playing field and explains the greater coexistence of conifers and angiosperms in cold or dry climates.

Gnetophytes are gymnosperms that have independently evolved xylem vessels and double fertilization.

Gnetophytes are a small group of gymnosperms (about 90 species) made up of three genera that look completely different from one another (**Fig. 33.18**). *Welwitschia mirabilis* is found only in the deserts of southwestern Africa. It produces only two straplike leaves that elongate continuously from their base. *Gnetum* grow in tropical rain forests, where they form woody vines and small trees with large broad leaves. *Ephedra,* which is native to arid regions, produces shrubby plants with photosynthetic stems and tiny leaves.

Gnetophytes have long attracted the attention of plant biologists because they exhibit a number of traits that are typical of angiosperms, notably the formation of xylem vessels and double fertilization. However, double fertilization in gnetophytes does not result in the formation of a triploid tissue, as it does in angiosperms. Moreover, the water transport capacity of gnetophyte xylem, while higher than other gymnosperms, is less than that of the angiosperms sharing their habitats.

For years, many botanists viewed gnetophytes as the sister group of angiosperms. DNA analysis, however, indicates that gnetophytes are not closely related to the angiosperms but are, instead, more closely related to conifers. Thus, both xylem vessels and double fertilization evolved independently in this group.

The fossil record shows that gnetophytes originated as much as 200 million years ago, reaching their greatest diversity at approximately the same time as the angiosperms began to diversify. The subsequent decline in gnetophyte diversity may result from competition with angiosperms.

➔ **Quick Check 3** Double fertilization evolved independently in gnetophytes and in flowering plants. Name two other features usually associated with angiosperms that evolved independently in gymnosperms.

33.5 ANGIOSPERMS

The evolution of angiosperms transformed life on land. The total number of plant species increased significantly as angiosperms diversified. Some non-angiosperm plant groups thrived as they expanded into new niches created by angiosperms, while others declined as many of their species went extinct. Terrestrial ecosystems became more productive as the more efficient angiosperm xylem allowed higher rates of photosynthesis. As a result, more food became available for insects, birds, and mammals. All of these animal groups diversified along with the flowering plants.

Angiosperms may have originated in the shady understory of tropical forests.

The last common ancestor of angiosperms and living gymnosperms is thought to have lived more than 300 million years ago. Yet the first unambiguous fossils of angiosperms appear in rocks about 140 million years ago. This 160-million-year gap means that there is no closely related sister group to help us understand

FIG. 33.19 **Angiosperm phylogeny.** Most angiosperm species belong to the monocots and eudicots. *Photo sources: (top to bottom) Kenneth Setzer; Tim Gainey/Alamy; Liz Cole/age fotostock; Jo Whitworth/age fotostock; Tropicals JR Mau/PhotoResourceHawaii.com; Image Source/Getty Images.*

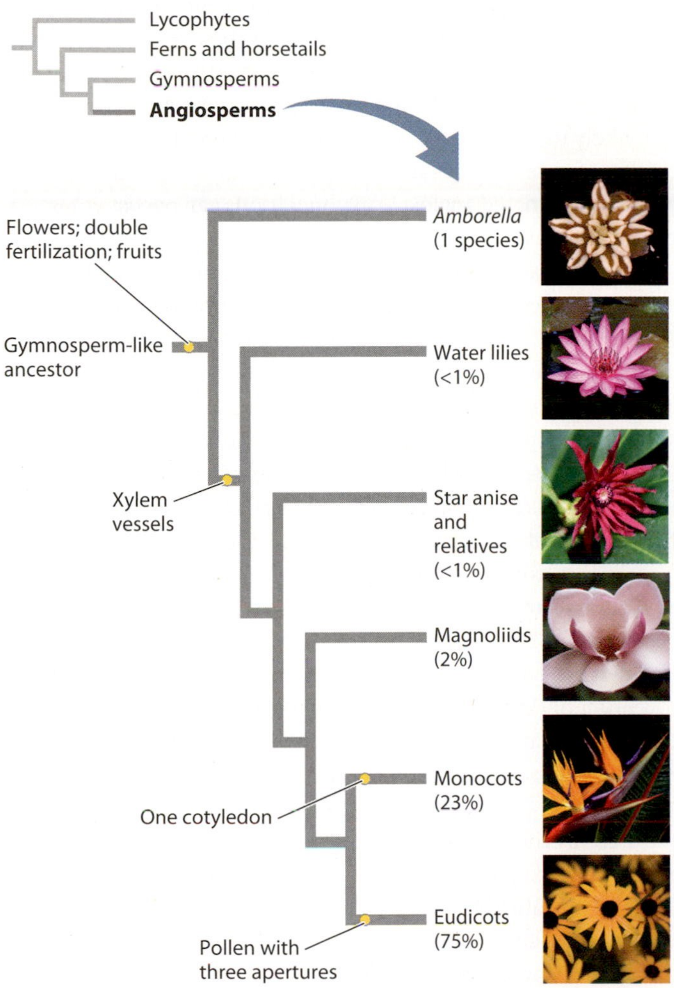

what type of environment the early angiosperms evolved in. But we can look at where living plants from the earliest-branching groups of angiosperms are found today (**Fig. 33.19**). Most grow in the understory of tropical rain forests; water lilies, which grow in shallow ponds, are one notable exception. This observation suggests that angiosperms may have evolved in wet and shady habitats. How could life in the understory of a tropical forest dominated by conifers have contributed to the evolution of flowers and xylem vessels?

Recall from Chapter 30 that flowers are a plant's way of enlisting animals to help them reproduce. Brightly colored petals and volatile scents advertise the presence of flowers, while rewards such as nectar encourage pollinators to return repeatedly to individuals of the same species. Because many flowers produce pollen and ovules close to each other, the movement of animals from one flower to another provides an efficient mode of pollen transfer.

Insect pollination may have allowed the first angiosperms to exploit understory habitats of tropical rain forests. Wind provides an effective means of pollen transfer in open habitats or in plants that can produce reproductive structures at the top of the canopy. In the dense understory of a tropical forest, the likelihood that air movements alone would bring pollen to an ovule is close to zero. All of the living members of the earliest angiosperm groups produce flowers that are visited by insects. **Fig. 33.20** shows the flower of Japanese star anise, which is visited by beetles. The fruit of a close relative (*Illicium verum*) is the source of a popular seasoning used in Asian cooking.

Let's turn now to a second important feature of angiosperms, the formation of wood containing xylem vessels. In the humid understory of a tropical forest, where transpiration rates are extremely low, there would seem to be little advantage in producing wood with a greater capacity for water transport. In fact, the sole existing species of the earliest-diverging angiosperm group—*Amborella trichopoda*—produces only tracheids. Furthermore, early-diverging angiosperm groups that do produce xylem vessels do not conduct water at higher rates than those that do not. Thus, xylem vessels in angiosperms may not have evolved, at least initially, as a means of enhancing water transport.

FIG. 33.20 **Japanese star anise, an early-branching angiosperm.** The flowers of *Illicium anisatum* are pollinated by beetles. *Source: Biosphoto/Richard Bloom/GAP.*

Instead, the formation of separate cell types for water transport and mechanical support may have given the early angiosperms greater flexibility of form. Vines and sprawling shrubs are two examples of growth forms exhibited by early diverging angiosperms. Both growth forms allow plants to grow toward small patches of sunlight that filter through the forest canopy. In the heavily shaded tropical rain forest understory, this ability may have provided a distinct advantage.

Angiosperm diversity results from flowers and xylem vessels, among other traits, as well as coevolutionary interactions with animals and other organisms.

The fossil record provides evidence that for the first 30 to 40 million years of their evolution, angiosperms were neither diverse nor ecologically dominant. In fact, fossil pollen indicates that, between about 140 and 100 million years ago, gnetophytes diversified just as much as angiosperms. Later in the Cretaceous Period, beginning about 100 million years ago, angiosperm diversity began to increase at a much higher rate. Trees belonging to the magnoliids, the branch of the angiosperm tree that today includes magnolias, black pepper, and avocado, emerged as ecologically important members of forest canopies. More important, however, was the divergence of the two groups that would come to dominate both angiosperm diversity and the ecology of many terrestrial environments: the monocots and the eudicots.

Before looking at the diversity of these two groups, let's consider how characteristics of angiosperms may have promoted their diversification. While many factors likely have contributed, we'll focus on two themes. The first is that angiosperms are more efficient at building their bodies and completing their life cycle than are other plants. Recall that the xylem of angiosperms has vessels supplying water required for photosynthesis and fibers providing mechanical support. Because water transport and mechanical support are separated, angiosperms are able to grow efficiently into a wide variety of forms, ranging from short-lived herbaceous plants to trees with large spreading crowns. This diversity of shapes and sizes reduces competition for light and space.

Angiosperms complete their life cycle as efficiently as they build their bodies. Not only are animal-pollinated plants able to reduce their production of pollen, but also double fertilization reduces the costs of reproduction by allowing angiosperms to delay provisioning their offspring until after fertilization. Angiosperms can thus reproduce quickly and with a minimum of resources. One possible consequence is that angiosperms can make use of habitats and resources that are only fleetingly available, such as ephemeral pools or open areas that appear following a fire.

A second theme focuses on the ways in which angiosperms interact with other types of organisms. The interaction of flowers and animal pollinators contributes to angiosperm diversity. Plants whose flowers attract different pollinators tend to more readily diverge to form new species than wind-pollinated species. In addition, the ability of carpels to block the growth of pollen from different species and from closely related individuals (including self) suggests that these genetic recognition systems contribute to the reproductive isolation required for speciation (Chapter 22).

It is important to recognize that diversity results from both high rates of species formation and low rates of species loss. The fossil record suggests that the persistence of species may play a particularly important role in angiosperm diversity. Wind-pollinated plants can reproduce only when their populations have a relatively high density. Since no wind-carried pollen is likely to fall on isolated individuals, species with low population density are more likely to go extinct. In contrast, animal pollinators actively searching for rare species are much more likely to find them, and so animal-pollinated angiosperm populations can persist at low population densities.

Another reason rare species may be more persistent is that they are less likely to encounter soil pathogens and seed predators. These threats to survival tend to gather near an adult of the same species. The seeds of rare species are more likely to land away from an individual of the same species and thus suffer less from this density-dependent mortality than do more common species. However, this ecological strategy works only if rare isolated individuals can remain in contact with a large enough population of potential mates, underlining again the key role of flowers and animal pollinators in persistence.

If animal pollination has so many advantages, why did approximately 20% of angiosperm species subsequently evolve a dependence on wind for pollination? Wind-pollinated species grow primarily in seasonal environments, where fewer animal pollinators are available than in tropical forests, and the pressures resulting from soil pathogens and seed predators may be much less.

That angiosperm diversity is the result of multiple factors is reinforced by the observation that some of these "angiosperm" features are also present in other groups of plants. For example, cycads rely on animal pollination, whereas gnetophytes produce xylem vessels, and yet neither group is particularly diverse. The initially slow but increasingly rapid buildup of angiosperm diversity may reflect the compounding effect of coevolutionary interactions with pollinators and dispersers, as well as adaptive radiations triggered by chemical arms races with pathogens and herbivores (Chapter 32). This is what is meant by the adage "Diversity begets diversity," a phenomenon further explored in Chapter 47.

Monocots are diverse in shape and size despite not forming a vascular cambium.

Monocots or monocotyledons make up nearly one-quarter of all angiosperms. Monocots come in all shapes and sizes and are found in virtually every terrestrial habitat on Earth (**Fig. 33.21**). Coconut palms, hanging Spanish moss (a relative of pineapple), and tiny floating duckweeds, less than 2 mm across, are all monocots, as are the sea grasses of tropical lagoons and the lilies and daffodils in the garden. Some monocots grow in arid regions—including

FIG. 33.21 Monocot diversity and phylogeny. This diverse group of angiosperms includes (a) *Agave shawii,* a desert succulent; (b) *Leucojum vernum,* a spring wildflower; (c) bamboo, a forest grass; and (d) *Costus* species, which grow in the rain forest understory. (e) Monocot phylogeny indicates that orchids (~22,000 species) and grasses (~11,500) make up much of monocot diversity. *Photo sources: a. Biosphoto/François Gohier; b. Kerstin Hinze/naturepl.com/NaturePL; c. Axle71/Dreamstime.com; d. Tropicals JR Mau/PhotoResourceHawaii.com.*

agave, which is used to make tequila. Others, such as orchids, grow primarily as epiphytes in tropical forests. A few monocots form vines—for example, the rattan palms, whose stems are used to produce rattan furniture. Monocots are the most important group of plants in terms of what we eat. Grasses, for example corn, rice, wheat, and sugarcane, are staples in the diets of most people. Other monocots that make their way to our dinner table include banana, yams, ginger, asparagus, pineapple, and vanilla.

Monocots take their name from the fact that they have one embryonic seed leaf, or cotyledon, whereas all other angiosperms have two. However, monocots are distinct in form in so many other ways that one rarely has to count the number of cotyledons to identify a member of this large, monophyletic group. Monocots represent a major evolutionary departure in the way seed plants build their bodies, in that a vascular cambium is never formed. How then has this group been so successful?

In monocot leaves, the major veins are typically parallel and the base of the leaf surrounds the stem, forming a continuous sheath. This type of leaf base means that only one leaf can be attached at any node, consistent with the formation of a single cotyledon. Although monocots do not form a vascular cambium, they can still produce stems that are quite large. For example, both corn and coconut palms are monocots. In monocots, all of the increase in stem diameter occurs in a narrow zone immediately below the apical meristem. As a result of this lateral expansion, the vascular bundles become distributed throughout the stem instead of being arranged in a ring as in all other seed plants. Monocots that have slender stems—for example, some species of climbing yams—have their vascular bundles in a ring.

The lack of a vascular cambium has a profound impact on the way monocots form roots. Because individual roots cannot increase their vascular capacity, monocots continuously initiate new roots from their stems. Thus, the root systems of monocots are more similar to those found in ferns and lycophytes than in other seed plants. Finally, monocot flowers typically produce organs in multiples of 3 (for example, having 3, 6, or 9 stamens), whereas eudicot flowers typically produce organs in multiples of 4 or 5.

Monocots have a relatively poor fossil record, in part because they do not produce wood. What factors might have led to such a radical revision in how they build their bodies? This question requires that we look at where and how living monocots grow. One hypothesis is that monocots evolved from ancestors that

HOW DO WE KNOW?

FIG. 33.22

When did grasslands expand over the land surface?

BACKGROUND Today, prairies, steppes, and other grasslands occur widely in the interiors of continents. Grasses, however, do not fossilize readily. How, then, can we understand how grasslands developed through time?

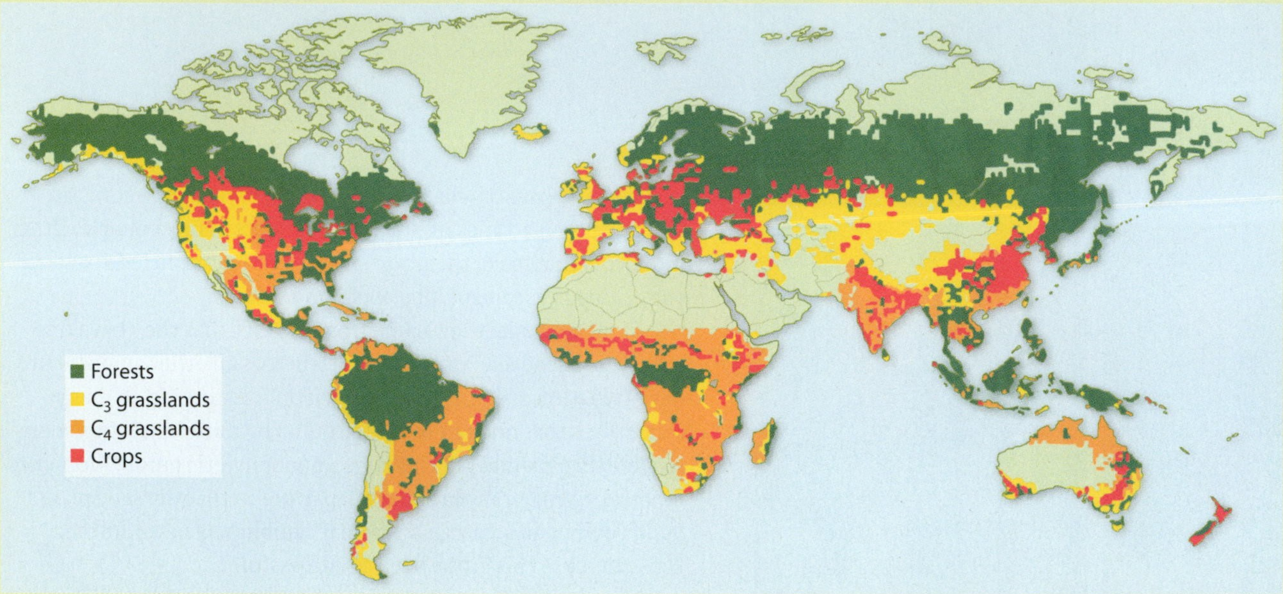

Map is from Edwards, E. J., C. O. Osborne, C. A. E. Stromberg, S. A. Smith, and the C4 Grasses Consortium. 2010. "The Origins of C_4 Grasslands: Integrating Evolutionary and Ecosystem Science." Science 328:587–591

HYPOTHESIS Grasslands expanded as climate changed over the past 50 million years.

OBSERVATIONS AND EXPERIMENTS Grasses commonly make phytoliths, small structures of silica (SiO_2) in their cells. These preserve well, providing a direct record of grass expansion. Moreover, mammals that feed on grasses evolved high-crowned teeth, which also preserve well, giving us an indirect record of grassland history. Finally, C_4 grasses, which are adapted to hot sunny environments with limited rainfall, have a distinctive carbon isotopic composition imparted by the initial fixation of CO_2 by PEP carboxylase (Chapter 29). Measurements of ^{13}C and ^{12}C in mammal teeth, soil carbonate minerals, and more recently, tiny amounts of organic matter incorporated into phytoliths, allow scientists to track C_4 grasslands through time.

produced creeping, horizontal stems as they grew along the shores of lakes and other wetlands. Many monocots today, including the earliest diverging groups of monocots, grow in such habitats, and many of the features of the monocot body plan are well adapted to environments with loose substrates, flowing water, and fluctuating water levels. For example, their leaf base provides a firm attachment that prevents leaves from being pulled off by flowing water. Furthermore, many monocots produce strap-shaped leaves that elongate from a persistent zone of cell division and expansion located at the base of the leaf blade. By continually elongating from the base, monocot leaves can extend above fluctuating water levels. Finally, it is easier to imagine evolutionary changes that affected the vascular system occurring within an environment that makes only modest demands for water transport.

We may never know for sure in what environment the monocots first evolved. We thus turn our attention to one of the most diverse groups within the monocots: the grasses. Many grasses produce stems that grow horizontally and branch, allowing them to cover large areas. Grasses have linear leaves that elongate from the base, allowing them to survive grazing as well as fire and drought. Many grasses have evolved the ability to tolerate dry environments by producing roots that extend deep into the soil. In addition, C_4 photosynthesis has evolved within the grasses multiple times. As described in Chapter 29, C_4 photosynthesis allows plants to avoid photorespiration and thus photosynthesize

RESULTS Studies of phytoliths, mammal tooth structure, and carbon isotopic composition of teeth enamel from the North American midcontinent clearly show that grasslands expanded 20 to 15 million years ago, and C_4 grasslands expanded later, about 8 to 6 million years ago.

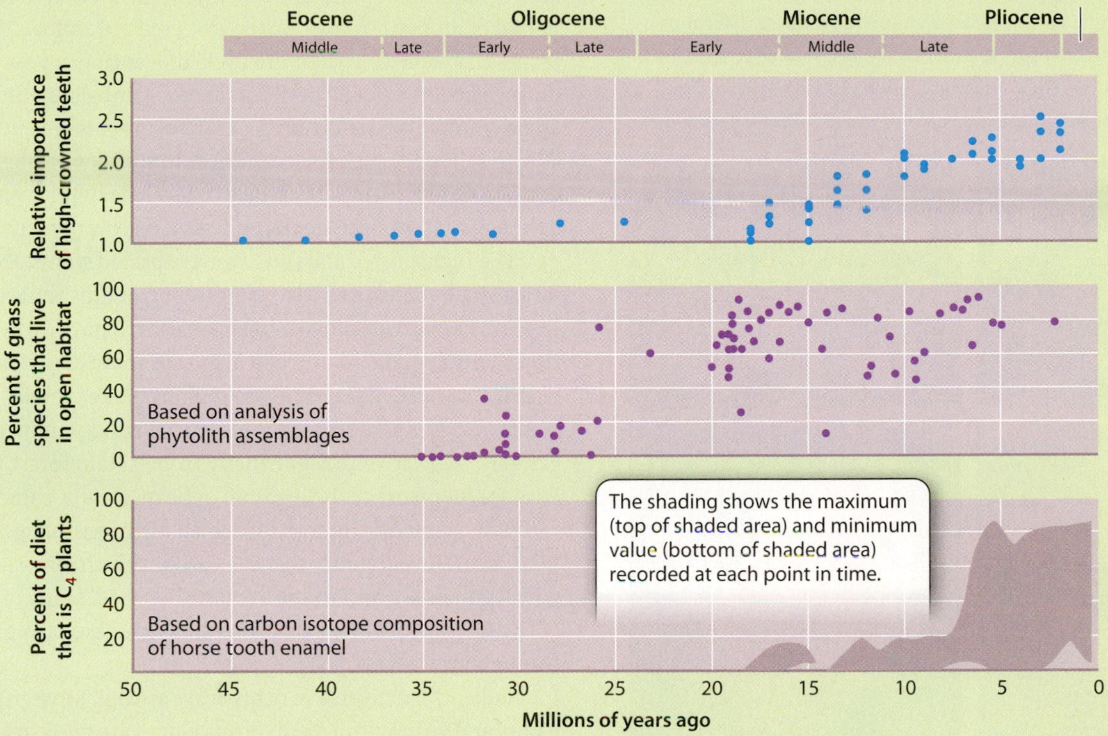

CONCLUSION In North America, grasslands expanded as atmospheric CO_2 levels declined and climates became drier. Other continents show evidence of a similar linkage of grassland expansion to climate change.

FOLLOW-UP WORK At present, atmospheric CO_2 levels are increasing rapidly, which may affect the competitive abilities of C_3 and C_4 grasses. Scientists are working to understand how global change will influence grasslands and other vegetation.

SOURCE Stömberg, C. A. E. 2011. "Evolution of Grasses and Grassland Ecosystems." *Annual Review of Earth and Planetary Sciences* 39:517–544.

with greater efficiency. Grasses are among the most successful group of plants, becoming widespread within the past 20 million years as climates changed (**Fig. 33.22**). Today, nearly 30% of terrestrial environments are grasslands.

→ **Quick Check 4** What are some of the distinctive features of monocots?

Eudicots are the most diverse group of angiosperms.

Eudicots first appear in the fossil record about 125 million years ago and, by 90 to 80 million years ago, most of the major groups we see today were present. Today, there are estimated to be approximately 160,000 species of eudicots, nearly three-quarters of all angiosperm species (**Fig. 33.23**). Eudicots are well represented in the fossil record, in part because their pollen is easily distinguished. Each eudicot pollen grain has three openings from which the pollen tube can grow, whereas pollen from all other seed plants has only a single opening. Eudicots take their name from the fact that they produce two cotyledons, whereas monocots produce one. But, because the magnoliids and early angiosperm groups also have two cotyledons, this largest of all angiosperm groups is referred to as the "eu-" or "true" dicots.

Many eudicots produce highly conductive xylem. High rates of water transport, and thus high rates of photosynthesis, may explain why eudicot trees were able to replace magnoliid trees as the most ecologically important members of forest canopies. In

FIG. 33.23 Eudicot diversity and phylogeny. Eudicots include (a) red oak trees (*Quercus rubra*), (b) tropical passionflower vines (*Passiflora caerulea*), (c) *Banksia* shrubs, native to Australia, and (d) beavertail cactus (*Opuntia basilaris*). (e) The tree shows a phylogeny of eudicots, with well-known species indicated. Groups with <3000 species are not shown. *Photo sources: a. Michael P Gadomski/Getty Images; b. Sergio Hayashi/Dreamstime.com; c. Joanne Harris/Dreamstime.com; d. Danita Delimont/Getty Images.*

addition, many eudicot trees lack strong apical dominance and produce crowns with many spreading branches. Today, tropical rain forest trees are an important component of the diversity of eudicots. Important eudicot trees in temperate regions include oaks, willows, and eucalyptus.

At the other end of the size spectrum are the herbaceous eudicots, which make up a substantial fraction of eudicot diversity. Herbaceous plants do not form woody stems. Instead, the aboveground shoot dies back each year rather than withstand a period of drought or cold. At the extreme are annuals, herbaceous plants that complete their life cycle in less than a year, persisting during the unfavorable period as seeds. Annuals are unique to angiosperms and the majority of these annuals are eudicots. Why might eudicots be successful as both trees and herbs? One possibility is that the ability to produce highly conductive xylem may allow herbaceous eudicots to grow quickly and to produce inexpensive and thus easily replaced stems. Examples of herbaceous eudicots include violets, buttercups, and sunflowers.

The herbaceous growth form appears to have evolved many times within eudicots as tropical groups represented by woody plants expanded into temperate regions. For example, the common pea is an herbaceous plant that resides in many vegetable gardens; its relatives include many tropical rain forest trees. Peas are members of the legume, or bean, family, some of which can form symbiotic interactions with nitrogen-fixing bacteria (Chapter 29). The ability to trade carbon for nitrogen may explain why legumes are important in many habitats. Many of the trees of the African savannas, as well as trees in seasonally dry regions throughout the world, are legumes.

Eudicots are diverse in other ways as well. Most parasitic plants and virtually all carnivorous plants are eudicots, as are water-storing cacti that grow in deserts. Some, such as roses and blueberry, are woody shrubs. Others grow as epiphytes in the canopy of rain forests, and still others, such as grapes and honeysuckle, are vines. Perhaps the most unusual are the strangler fig trees. These begin as epiphytes and then produce roots that descend to the forest floor and fuse to form a solid cylinder that "strangles" their host tree. Finally, eudicots make a significant contribution to the diversity of the dinner table. Apples, carrots, pumpkins, and potatoes are all eudicots, as are coffee, cacao (the source of chocolate), and tea. Many plants with oil-rich seeds, such as olives, walnuts, soybean, and canola, are eudicots, as are buckwheat and quinoa.

The reproduction, growth, and physiology of angiosperms are summarized in **Fig. 33.25** on pages 710 and 711.

? CASE 6 AGRICULTURE: FEEDING A GROWING POPULATION

What can be done to protect the genetic diversity of crop species?

Of the more than 400,000 species of plants on Earth today, we eat surprisingly few. Of those we do eat, nearly all are angiosperms. The only exceptions are the female gametophyte of a small number of gymnosperms (pine nuts and ginkgos), as well as the young leaves of a few ferns. Most of the plants we eat come from species that are grown in cultivation, and many of these cultivated species have, as a consequence of artificial selection,

FIG. 33.24 "Centers of origin" where different crop species were domesticated.

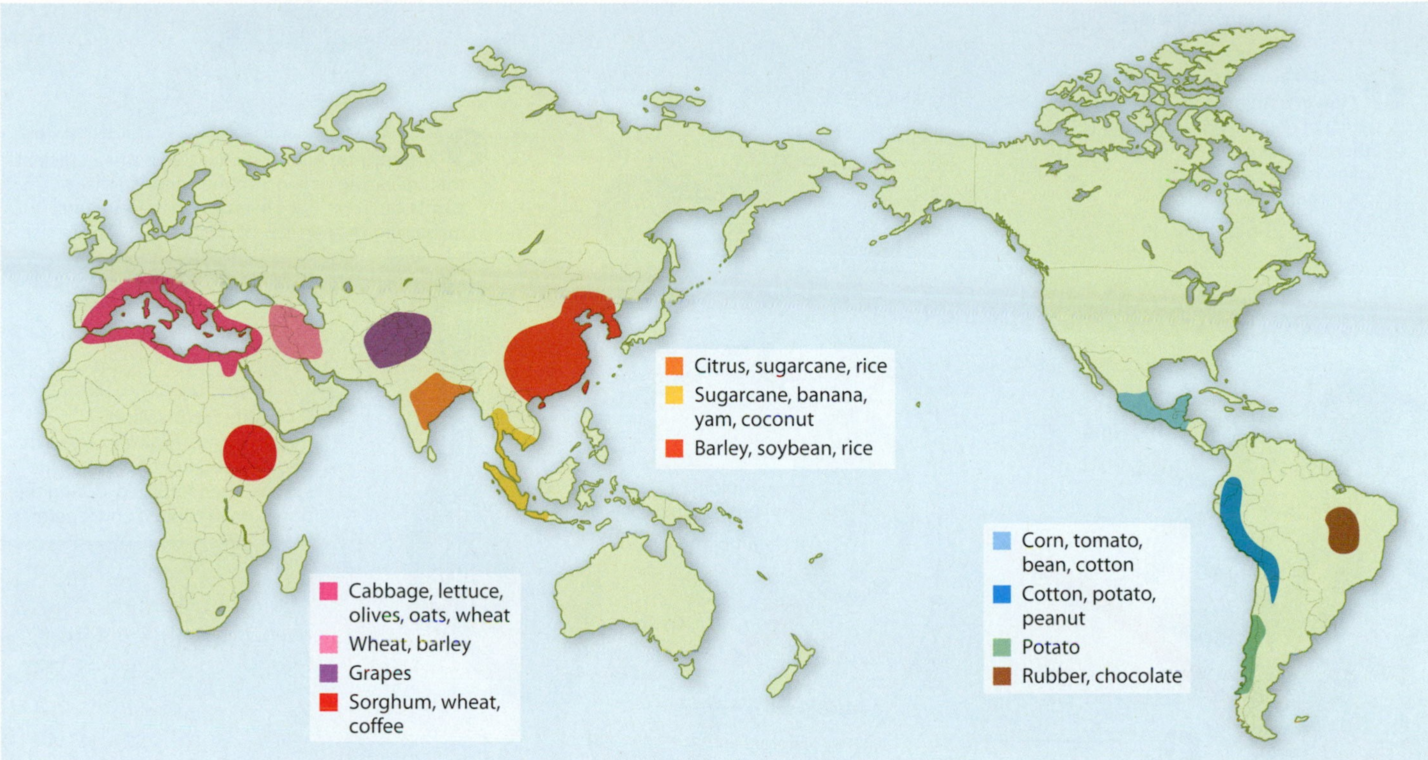

lost the ability to survive and reproduce on their own. Of the approximately 200 such domesticated species, 12 account for over 80% of human caloric intake. Just three—wheat, rice, and corn—make up more than two-thirds.

Crop breeders select for varieties that grow well under cultivation and are easy to harvest, but a consequence of selective breeding is a decrease in genetic diversity. Thus, not only do we depend upon a small number of species for food, agriculture itself contributes to a narrowing of plant diversity. Because pathogens and pests continue to evolve, any loss in genetic diversity of crop species creates substantial risks. For example, in 1970, the fungal pathogen *Bipolaris maydis*, a disease-causing organism that had previously destroyed less than 1% of the U.S. corn crop annually, destroyed more than 15% of the corn crop in a single year. The severity of the epidemic was due to the genetic uniformity of the corn varieties planted at that time. In natural plant populations, genetic variation for pathogen resistance makes it highly unlikely that a newly evolved pathogen strain will be able to infect every plant.

Nicolai Vavilov, a Russian botanist and geneticist, was one of the first to recognize the importance of safeguarding the genetic diversity of both crop species and their wild relatives. In the early twentieth century, he mounted a series of expeditions to collect seeds from around the globe. Vavilov's observations of cultivated plants and their relatives led him to hypothesize that plants had been domesticated in specific locations from which they had subsequently expanded as a result of human migration and commerce (**Fig. 33.24**). Vavilov called these regions "centers of origin," and he believed that they coincided with the centers of diversity for both crop species and their wild relatives. In World War II, during the siege of Leningrad in which more than 700,000 people perished, scientists at the Vavilov Institute sought to protect what was then the world's largest collection of seeds. Their dedication—at least one of the self-appointed caretakers died of starvation, despite being surrounded by vast quantities of edible seeds—illustrates the priceless nature of the genetic diversity on which our food supply rests.

Today, seed banks help preserve the genetic diversity of crop species and their wild relatives. However, seed banks can store only a fraction of the genetic diversity present in nature. To help make up the difference, some have suggested establishing protected areas that coincide with Vavilov's centers of origin. As we discuss in Chapter 49, new threats lie just over the horizon. Meeting these threats will require that every genetic resource be brought to bear. ■

VISUAL SYNTHESIS

FIG. 33.25

Angiosperms: Structure and Function

Integrating concepts from Chapters 29–31, and 33

Reproduction

Many angiosperms rely on animals for pollination and dispersal.

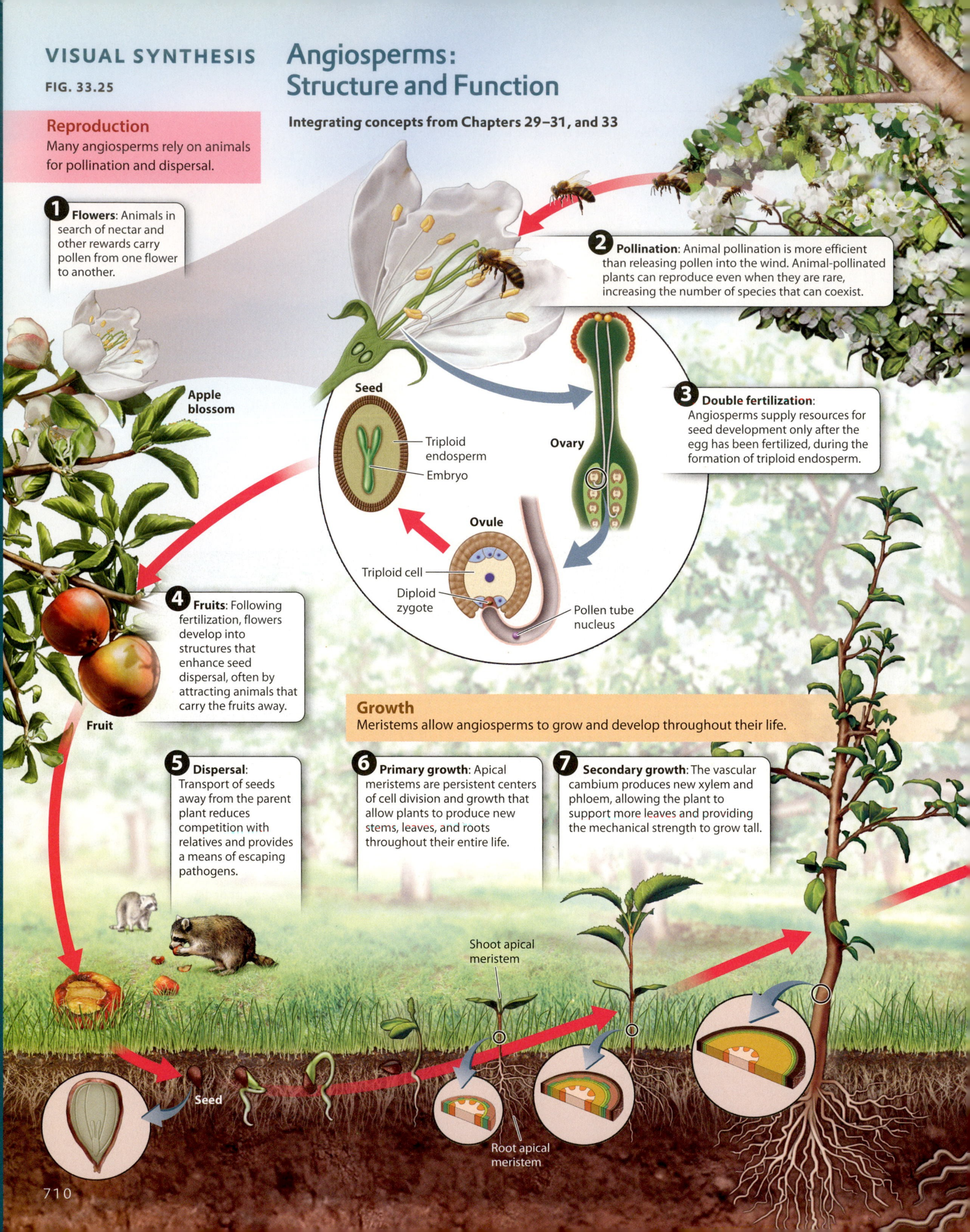

1 Flowers: Animals in search of nectar and other rewards carry pollen from one flower to another.

2 Pollination: Animal pollination is more efficient than releasing pollen into the wind. Animal-pollinated plants can reproduce even when they are rare, increasing the number of species that can coexist.

3 Double fertilization: Angiosperms supply resources for seed development only after the egg has been fertilized, during the formation of triploid endosperm.

4 Fruits: Following fertilization, flowers develop into structures that enhance seed dispersal, often by attracting animals that carry the fruits away.

5 Dispersal: Transport of seeds away from the parent plant reduces competition with relatives and provides a means of escaping pathogens.

Growth

Meristems allow angiosperms to grow and develop throughout their life.

6 Primary growth: Apical meristems are persistent centers of cell division and growth that allow plants to produce new stems, leaves, and roots throughout their entire life.

7 Secondary growth: The vascular cambium produces new xylem and phloem, allowing the plant to support more leaves and providing the mechanical strength to grow tall.

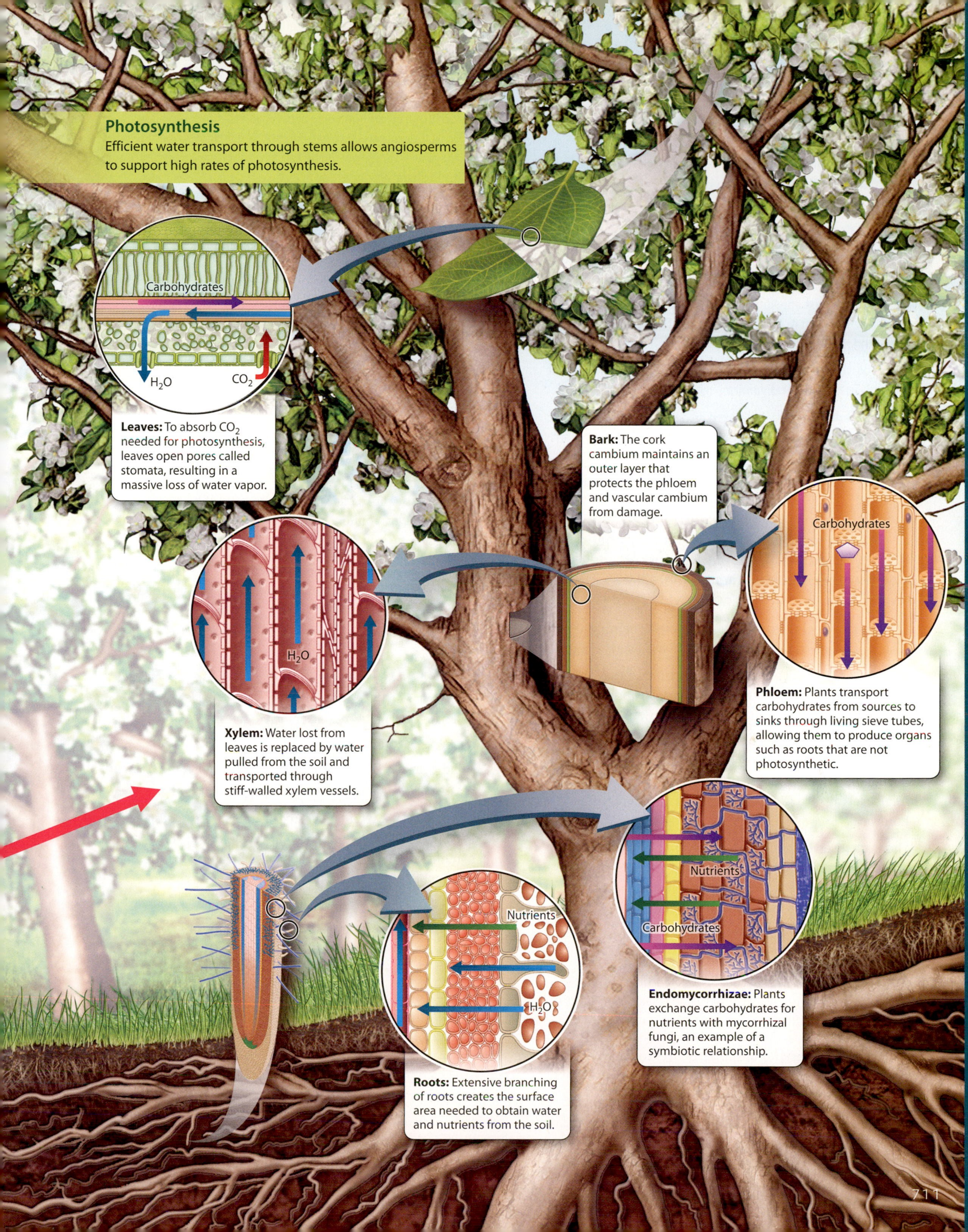

Photosynthesis
Efficient water transport through stems allows angiosperms to support high rates of photosynthesis.
Carbohydrates
H_2O
CO_2
Leaves: To absorb CO_2 needed for photosynthesis, leaves open pores called stomata, resulting in a massive loss of water vapor.
Bark: The cork cambium maintains an outer layer that protects the phloem and vascular cambium from damage.
Carbohydrates
H_2O
Phloem: Plants transport carbohydrates from sources to sinks through living sieve tubes, allowing them to produce organs such as roots that are not photosynthetic.
Xylem: Water lost from leaves is replaced by water pulled from the soil and transported through stiff-walled xylem vessels.
Nutrients
Carbohydrates
Nutrients
H_2O
Endomycorrhizae: Plants exchange carbohydrates for nutrients with mycorrhizal fungi, an example of a symbiotic relationship.
Roots: Extensive branching of roots creates the surface area needed to obtain water and nutrients from the soil.

Core Concepts Summary

33.1 Angiosperms make up approximately 90% of all plant species found today.

There are thought to be nearly 400,000 species of plants living today. Of these, approximately 90% are angiosperms. page 690

The other 10% of plant species is distributed among the other six major groups of plants. page 690

Angiosperms first appear in the fossil record about 140 million years ago, more than 300 million years after plants first moved onto land. page 690

The evolution of angiosperms resulted in a rapid and dramatic increase in total plant diversity. page 690

As angiosperm diversity increased, other plant groups, such as gymnosperms and ferns, declined in diversity. page 690

Moist tropical rain forests dominated by angiosperms provided new types of habitat into which other plants could evolve. page 690

33.2 Bryophytes form persistent, photosynthetic gametophytes and small, unbranched sporophytes; today, they grow in environments where the ability to pull water from the soil does not provide an advantage.

Bryophytes constitute a paraphyletic group consisting of three groups of plants: mosses, liverworts, and hornworts. page 690

Bryophytes are small plants that produce one of two morphological types: a flattened thallus or an upright leafy type. They do not form roots, but instead absorb water through their surfaces. page 691

The persistent component of the life cycle is the gametophyte. Sporophytes range from tiny and non-photosynthetic in some liverworts to relatively long-lived and photosynthetic in hornworts. page 691

There are several examples of convergent evolution between bryophytes and vascular plants, including insect dispersal of spores in some mosses and the evolution of internal transport cells in some mosses and liverworts. page 692

Sphagnum moss is the dominant plant of peat bogs. page 692

Sphagnum moss produces water-holding cells that allow it to soak up water, and it acidifies the environment. Both characteristics help slow decomposition, so large amounts of organic carbon build up year after year. page 692

33.3 Spore-dispersing vascular plants today are primarily small plants that grow in moist environments, but in the past included tall trees.

Only two groups of seedless vascular plants have living relatives today: lycophytes, and ferns and horsetails. page 693

Three hundred million years ago, lycophytes included large trees that dominated swamp forests. Today, lycophytes are small plants that either grow in the forest understory as epiphytes or occur in shallow ponds. page 694

Ferns and horsetails are morphologically diverse. Ferns produce large leaves that uncoil as they grow; horsetails have tiny leaves; and whisk ferns with no leaves at all. page 697

Although ferns have a long history, most present-day species are the result of a radiation that occurred after the rise of the angiosperms. page 698

33.4 Gymnosperms produce seeds and woody stems and are most common in seasonally cool or dry regions.

Of the many groups of seed plants that have evolved, only two can be found today: the gymnosperms, with fewer than 1000 species, and the angiosperms, with more than 380,000 species. page 699

Gymnosperms are composed of four groups of woody plants: cycads, ginkgos, conifers, and gnetophytes. page 699

Cycads produce large leaves on stout, unbranched stems. Although they once were widely distributed, they now occur in small, fragmented populations, primarily in the tropics and subtropics. Many cycads are insect pollinated, and all form symbiotic associations with nitrogen-fixing bacteria. page 699

Ginkgo is the single living species of a group that was distributed globally before the evolution of the angiosperms. *Ginkgo* is wind pollinated and produces tall, branched trees. page 700

Conifers include the tallest and longest-lived trees on Earth. Wind-pollinated and largely evergreen, they are found primarily in cool to cold environments. page 700

Before the angiosperms appeared, conifers were widespread. Their almost complete absence in the tropics and persistence in temperate regions may be due to their dependence upon wind pollination and having xylem formed entirely of tracheids. page 701

The gnetophytes are a small group, containing only three genera and few species, that has independently evolved xylem vessels. page 702

33.5 Angiosperms are distinguished by flowers, fruits, double fertilization, and xylem vessels; their diversity is the result of traits that increase the efficiency of completing their life cycle and building their bodies.

Xylem vessels make it possible for angiosperms to have a diversity of form and to grow toward light. page 704

Angiosperms reproduce quickly and with less use of resources as a result of pollination and double fertilization, allowing them to diversify into ephemeral habitats. page 704

Angiosperm diversity may result as much from low rates of extinction as from high rates of speciation. page 704

Plants having flowers can reproduce even if they are far apart, allowing rare species to persist and reproduce. page 704

The angiosperm phylogeny indicates a major split between two diverse groups, monocots and eudicots. page 704

Monocots include grasses, coconut palms, bananas, ginger, and orchids. page 704

Monocots have a single cotyledon, or embryonic seed leaf, and they do not form a vascular cambium. page 705

Eudicots are characterized by pollen grains with three openings through which the pollen tube can grow. page 707

Eudicots are diverse, and include many familiar plants such as legumes, roses, cabbage, pumpkin, coffee, tea, and cacao. page 708

Modern agriculture is based on just a few plant species with low genetic diversity, making crops vulnerable to pests and pathogens. page 709

Self-Assessment

1. Using information in Fig. 33.2, draw a pie chart that depicts proportional plant diversity today.
2. Imagine a world in which mosses, liverworts, and hornworts form a monophyletic group. How would your ability to infer what the first land plants looked like be affected?
3. Describe three environments that allow bryophytes to coexist with vascular plants.
4. Describe the habitats in which lycophytes are found today.
5. List three ways that ferns, which lack secondary growth, are able to elevate their leaves and thus access more sunlight.
6. Describe how fern diversity has been affected by the evolution of the angiosperms.
7. Contrast the ways in which the evolution of angiosperms has affected the distribution of cycads and of conifers.
8. Explain how xylem produced by conifers differs from that of angiosperms and how that difference may have influenced the present-day distribution of conifers.
9. Compare the movement of pollen from an animal-pollinated angiosperm and a wind-pollinated conifer, noting what features of angiosperm reproduction increase the efficiency (or lower the costs) of pollen transfer.
10. Name two features that have contributed to the diversity and success of angiosperms and discuss possible advantages of these features.

Log in to LaunchPad to check your answers to the Self-Assessment questions, and to access additional learning tools.

Chris Ted/Getty Images.

CHAPTER 34

Fungi

Structure, Function, and Diversity

Core Concepts

34.1 Fungi are heterotrophic eukaryotes that feed by absorption.

34.2 Fungi reproduce both sexually and asexually, and disperse by spores.

34.3 Other than animals, fungi are the most diverse group of eukaryotic organisms.

In tropical rain forests, vascular plants make up most of the biomass. Animals eat the plants, obtaining food for growth and reproduction, but they feed selectively: They consume leaves, fruits, and seeds, but largely avoid a much more abundant tissue—wood. In fact, most leaves also escape grazing and eventually fall, dead, onto the forest floor. Animals, too, may be eaten by predators, but some die of other causes and contribute their remains to the soil. Given this constant rain of biological materials, we might expect wood, leaves, and animal carcasses to accumulate in great piles, but a walk through the rain forest reveals that the forest floor is remarkably clean. What happened to all the dead plant and animal tissues?

On land, the organisms principally responsible for the decomposition of plant and animal tissues are **fungi,** one of the most abundant and diverse groups of eukaryotic organisms. Many of us have seen mushrooms and toadstools in a meadow or woodland, but most fungal biomass and most of the metabolic work that fungi do occurs within the soil, out of sight. Fungi play a critical role in cycling carbon because of their ability to locate and break down the complex molecules and bulky tissues in plant and animal bodies. Moreover, because they form intimate relationships with living plants and animals, fungi affect the growth and reproduction of many other organisms. For example, fungi associated with plant roots dramatically increase plant productivity by enhancing the uptake of mineral nutrients from the soil (Chapter 29). Other fungi are major agricultural pests (Chapter 32). Fungi enable us to turn wheat into bread, barley into beer, and milk into cheese, but still other fungi cause athlete's foot, yeast infections, and, especially in patients with compromised immune systems, overwhelming infections that can lead to death.

34.1 GROWTH AND NUTRITION

Fungi are heterotrophs, meaning that they depend on preformed organic molecules for both carbon and energy. Unlike animals, fungi do not have organs that enable them to ingest food and break it down in a digestive cavity (Chapter 40). Instead, fungi absorb organic molecules directly through their cell walls. This mode of feeding presents two major problems. First, although simple molecules like amino acids and sugars pass readily through the cell wall, more complicated molecules do not. Fungi secrete a diversity of enzymes that break down complex organic molecules like starch or cellulose into simpler compounds that can be absorbed. Fungi, then, digest their food first and take it into the body afterward.

A second challenge is to find food in the environment. Other heterotrophic organisms such as bacteria, protists, and animals commonly move through their habitat, actively searching for food. Fungi, however, have no means of locomotion, and so these organisms use the process of growth itself to find nourishment.

Hyphae permit fungi to explore their environment for food resources.

Most fungi consist of highly branched, multicellular filaments called **hyphae** (**Fig. 34.1**). The hyphae are slender, typically 10 to 50 times thinner than a human hair. The numerous long, thin hyphae provide fungi with a large surface area for absorbing nutrients.

Hyphae maintain their slender form by growing only at their tips. Elongating hyphae thus penetrate ever farther into their environment, encountering new food resources as they grow. Where resources are low, growth is slow or may stop entirely. On the other hand, when fungi encounter a rich food resource, they grow rapidly and branch repeatedly, forming a network of branching hyphae called a **mycelium** (plural, **mycelia**). Mycelia can grow to be quite large—the largest known individual of the fungus *Armillaria ostoyae* covers over 2000 acres in the Blue Mountains of Oregon and weighs many hundreds of tons. We may be impressed by blue whales and redwoods, but in fact some fungi are the largest living organisms on Earth.

A strong but flexible cell wall is key to hyphal growth and nutrient transport. In fungi, cell walls are made of **chitin,** the same compound found in the exoskeletons of insects. Chitin is a modified polysaccharide that contains nitrogen. Because of their chemical makeup, fungal cell walls are thinner than plant cell walls. Fungal cell walls prevent hyphae from swelling as water flows into the cytoplasm by osmosis. Thus, the wall prevents cells from rupturing when exposed to dilute solutions such as those in

FIG. 34.1 **Fungal hyphae.** These thin filaments have enormous surface area that increases absorption of nutrients. Shown here are hyphae of *Trichophyton,* a fungus that infects the skin of humans, as seen under a scanning electron microscope. *Source: Susumu Nishinaga/Science Source.*

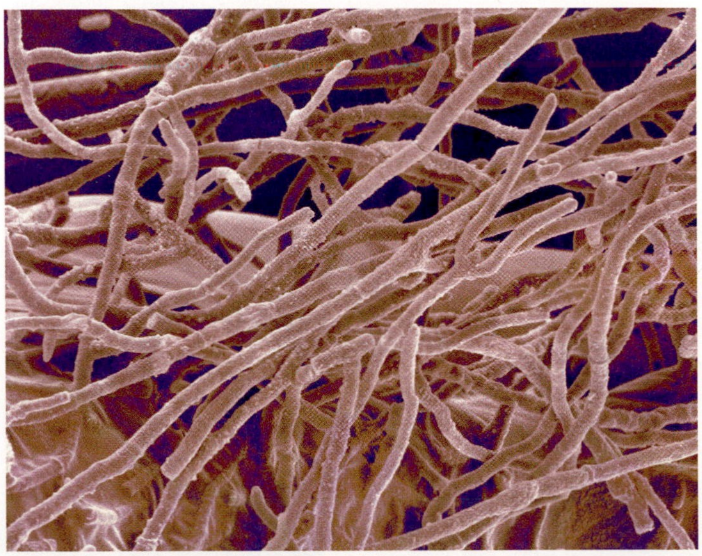

FIG. 34.2 Septa separating individual cells in fungi. Septa, partitions between cells, have pores that allow transport of solutes along the hypha.

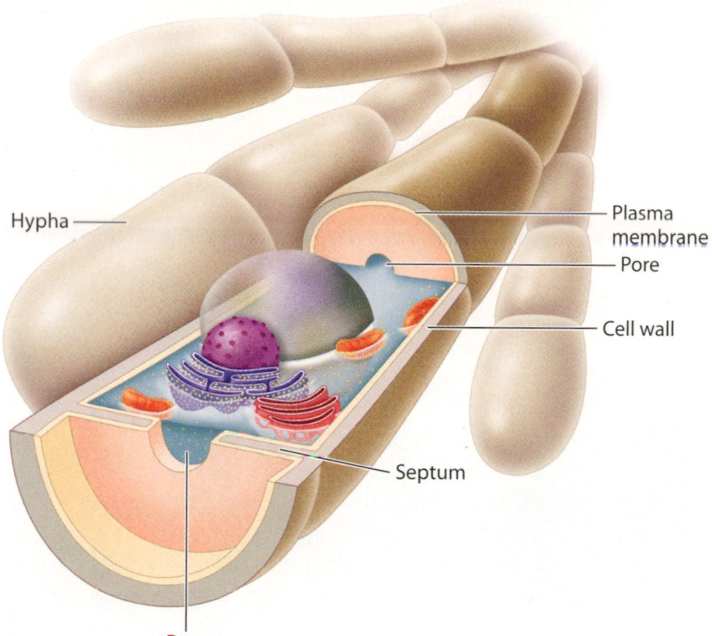

freshwater environments and in the soil. Moreover, the inflow of water by osmosis provides the force that enables fungi to explore their surroundings: the inflow of water generates positive turgor pressure (Chapters 5 and 29), pushing the hyphal tip ever deeper into the local environment.

Fungi transport materials within their hyphae.

Fungi can transport food and signaling molecules across long distances in mycelia. As a result, they are able to grow between resource patches and to produce reproductive structures such as mushrooms that rise up above the ground. Molecules taken up actively from the environment drive water into the cell by osmosis and thus increase turgor pressure. At the same time, growth and respiration consume these molecules, resulting in a decrease in turgor pressure. Such differences in turgor pressure along hyphae drives bulk flow in a manner similar to phloem transport in plants (Chapter 29). Bulk flow carries materials obtained in a nutrient-rich location so that they can fuel hyphal elongation across nutrient-poor locations. Transport by bulk flow also allows fungi to build relatively large reproductive structures aboveground. All the raw materials used to build a mushroom must be transported from hyphae in contact with a source of nutrients, such as the soil or a rotting log.

A continuous stream of cytoplasm, not divided by barriers, is essential for the long-distance movement of materials within mycelia. In early-diverging groups, the hyphae have many nuclei but no cell walls to separate them. In later-diverging groups, nuclear divisions are accompanied by the formation of **septa** (singular, **septum**), walls that partially divide the cytoplasm into separate cells (**Fig. 34.2**). Each septum contains one or more pores that allow water and solutes to move freely between cells. Septa play an important role when hyphae are damaged. Injury activates sealing mechanisms that plug pores in the septa, preventing the loss of pressurized cytoplasm.

→ **Quick Check 1** Name two roles of hyphae.

Not all fungi produce hyphae.

Yeasts are single-celled fungi found in moist, nutrient-rich environments. In these circumstances, the search for food does not require hyphae. All yeasts are descended from ancestors that developed hyphae, but hyphal development has been lost independently several times. Thus, the lack of hyphae is an example of convergent evolution, in which the same trait develops independently in different lineages in response to similar selective pressures (Chapter 23).

Most yeasts divide by budding. A small outgrowth increases in size and eventually breaks off to form a new cell (**Fig. 34.3**). The localized outgrowth that results in budding is similar to growth by elongation at hyphal tips. In fact, some yeasts can form hypha-like structures under certain conditions. For example, in *Candida albicans,* a yeast that infects humans, hyphae are produced during the transition of the fungus from a harmless to a pathogenic form.

FIG. 34.3 The yeast *Saccharomyces cerevisiae.* Yeasts are unicellular fungi that often divide by budding smaller cells off larger ones. *Source: Andrew Syred/Science Source.*

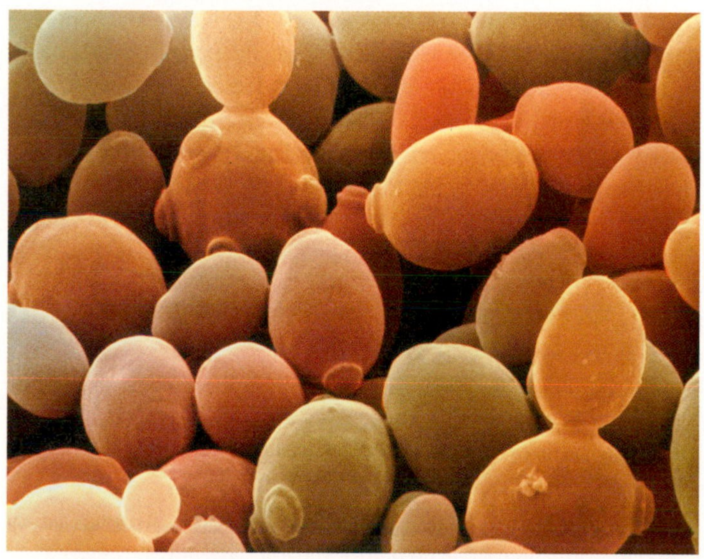

Yeasts are common on the surfaces of plants, and to a lesser extent on the surfaces and in the guts of animals. Humans have long used the yeast *Saccharomyces cerevisiae* to ferment plant carbohydrates and thereby to produce leavened bread and alcoholic beverages. *Saccharomyces cerevisiae* is also an important model organism used in laboratory studies of genetics.

Fungi are principal decomposers of plant tissues.

Most fungi use dead organic matter as their source of energy and raw materials. As noted in Chapter 25, on land the organic matter in dead tissues on and within soils far exceeds the amount in the living biomass. Thus, the ground beneath our feet provides tremendous resources for heterotrophic growth. Bacteria and some protists can use this resource, but for the most part it is the fungi that convert dead organic matter back to carbon dioxide and water. This helps to keep the biological carbon cycle in balance (Chapter 25) and returns nutrients to the soil, where they will be available for new plant growth.

A brief consideration shows why fungi have the advantage on a forest floor or in soil. Like fungi, bacteria secrete enzymes that break down organic molecules in their immediate environment, but in the soil bacteria have only a limited capacity to move from place to place. Flagella can propel them a few micrometers through a thin film of water, but when local resources become depleted, bacteria must wait for new food to appear. In contrast, the growing hyphae of fungi can search actively through the soil for new food resources. Furthermore, because fungi can penetrate their food, they are particularly well suited for breaking down the bodies of multicellular organisms, such as tree stems.

Fungi can gain nutrition from dead animal, protozoan, or even bacterial cells, but the most abundant biomolecules on and within soils are cellulose and lignin, the principal components of plant cell walls (Chapter 31). Cellulose is a polymer of the sugar glucose, so it is a particularly rich source of carbon and energy. Cellulose is difficult to degrade because individual cellulose polymers bind tightly to one another. In addition, in wood, cellulose and lignin are closely associated, which makes the cellulose even harder to get at. Lignin is difficult for enzymes to break apart, in part because it lacks a regular chemical structure. A few bacteria are known to degrade lignin, but in nature, fungi account for most of the decomposition of wood (**Fig. 34.4**).

To break down lignified plant cell walls, fungi require oxygen. However, oxygen concentrations are typically low in water-logged environments, and in that case decay proceeds slowly if at all. For this reason, woody biomass, once submerged in water, tends to accumulate, leading to the buildup of partially decayed plant material as peat. Over geologically long time intervals, peat becomes coal, an energy staple of the industrial world that owes its existence to the inhibition of fungal decay.

Fungi are important plant and animal pathogens.

Fungi not only feed on dead plants and animals, but they are also well adapted to infect living tissues (**Fig. 34.5**). Plants, for example, are vulnerable to a diverse array of fungal pathogens—rusts, smuts, and molds—that cause huge losses in agricultural production, as we discuss in section 34.3.

Invasion is the key to success for plant pathogens (Chapter 32). Successful pathogens must be able to get past a plant's physical or chemical defenses. This explains why fungal pathogens exhibit host specificity, meaning that they are able to infect only one or a small number of host species. In many cases, fungi infect plants through wounds, which provide a route through a plant's outer defenses. Some fungi enter through stomata, while others penetrate epidermal cells directly, degrading the wall with enzymes and then using turgor pressure to push their hyphae into the plant interior. The extent of infection usually depends on the ability of hyphae to expand into new tissues. Vascular wilt fungi (Chapter 32) send their

FIG. 34.4 Decomposition of wood by fungi. (a) A dead tree being decomposed by brown rot fungi. (b) These scanning electron microscope images show (left) normal cedar wood and (right) rotted cedar wood from the coffin in the tomb of King Midas in Gordion, Turkey. *Sources: a. Science Photo Library/Alamy; b. Timothy R. Filley, Robert A. Blanchette, Elizabeth Simpson, and Marilyn L. Fogel. Nitrogen cycling by wood decomposing soft-rot fungi in the "King Midas tomb," Gordion, Turkey PNAS 2001 98 (23) 13346-13350. Copyright 2001 National Academy of Sciences, U.S.A. Photos courtesy of Robert A. Blanchette, University of Minnesota.*

FIG. 34.5 Fungal infection of living tissue. (a) This stained section of heart tissue (pink) is infected with the fungus *Candida albicans* (the purple threadlike structures). (b) A scanning electron micrograph shows hyphae of *Ceratocystis ulmi*, a vascular wilt fungus, growing in xylem vessels of an elm tree. *Sources: a. Science Source; b. Biophoto Associates/Science Source.*

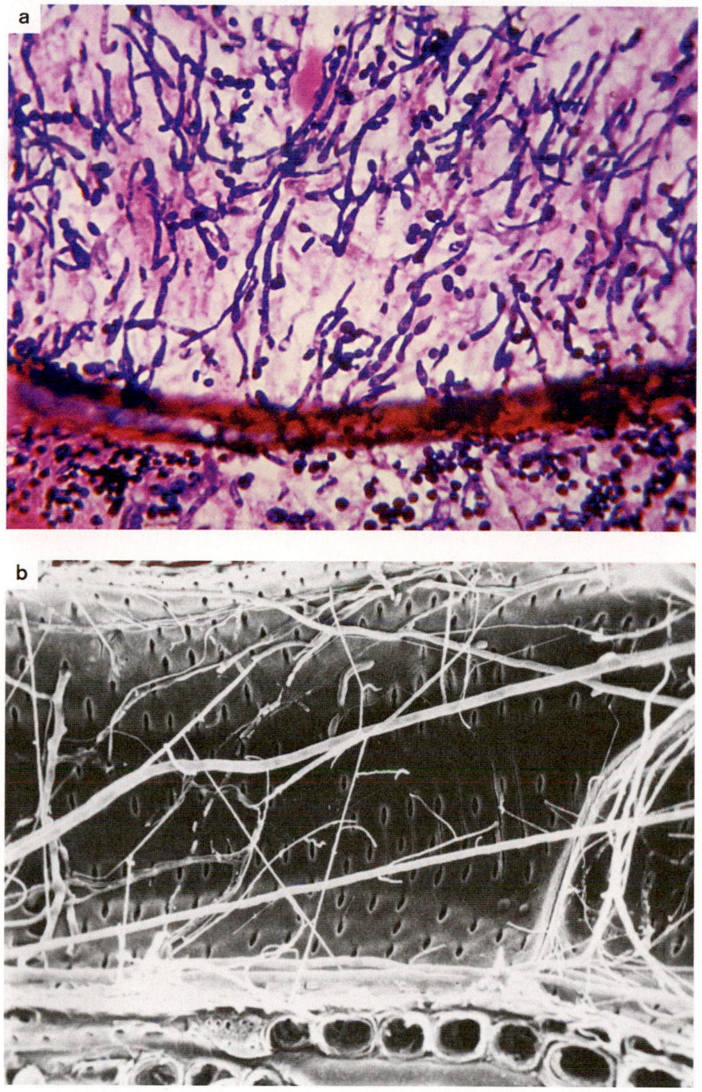

hyphae through xylem vessels, enabling them to spread throughout the entire plant. Old photographs of towns in the eastern United States show streets shaded by American elm trees, but today nearly all those trees are dead, victims of a vascular wilt fungus introduced into the United States early in the twentieth century.

Aboveground, plant infections are usually transmitted by fungal spores, carried either by the wind or on the bodies of insects. For example, many fungi that attack fruits are transmitted by insects also seeking those fruits; yeasts that feed on nectar within flowers hitch rides on pollinators. Some fungi secrete sugar-rich droplets that attract insects, which then transport their spores to other sites.

Belowground, infection is typically transmitted by hyphae that penetrate the root. Hyphae can feed on organic matter in the soil until they encounter the root of a suitable host plant. The persistence of fungal hyphae in soils means that populations of pathogenic fungi can build up in the soil near a host plant. This is why seedlings germinating close to an adult tree of the same species are at risk of infection, whereas seedlings of distantly related species may be spared. By favoring the establishment of seedlings of less common species, fungal pathogens are thought to play an important role in maintaining high levels of species diversity in tropical rain forests (Chapter 32).

Insects and other invertebrates are also susceptible to fungal pathogens. Certain fungi exploit living animals in a unique way: by trapping them. Some fungal predators form sticky traps with their hyphae, whereas others actually lasso their prey (**Fig. 34.6**). The hyphae of these fungi form rings that can inflate within seconds, trapping anything within the ring. Nematodes, tiny worms common in soils, inadvertently pass through the rings, setting off hyphal inflation. The nematodes are then trapped, providing food for the fungus.

Fungal infection is relatively rare in vertebrates. Severe infections are more frequent among fish and amphibians than in mammals, perhaps because fungi grow poorly at mammalian body temperatures. An apparent exception is the fungal infection that has caused dramatic declines in North American bat populations. But these fungi infect the bats during their winter hibernation, when the bats have lowered body temperatures that conserve energy. In humans, most fungal infections are annoying rather

FIG. 34.6 A fungal lasso. Hyphae of *Drechslerella doedycoides* form inflated rings that capture nematode worms moving through the soil. *Source: Reprinted from Current Opinion in Neurobiology, 22, Jennifer K Pirri, Mark J Alkema, The neuroethology of C. elegansescape Fig. 3, Copyright 2011, with permission from Elsevier. Reproduced with permission from Mark Alkema.*

than life threatening—athlete's foot and yeast infections are prime examples. However, in individuals with immune systems compromised by HIV or illness, fungal infection can be fatal.

Many fungi form symbiotic associations with plants and animals.

While some interactions between species benefit the fungus at the expense of its host, others benefit both partners. In Chapter 29, we saw that mycorrhizal fungi supply plant roots with nutrients such as phosphorus and nitrogen from the soil and in return receive carbohydrates from their host (**Fig. 34.7**). There are two main types of mycorrhizae. The hyphae of **ectomycorrhizal** fungi surround, but do not penetrate, root cells. In contrast, the hyphae of **endomycorrhizal** fungi penetrate into root cells, where they produce highly branched structures that provide a large surface area for nutrient exchange. As we discuss in section 34.3, the endomycorrhizal fungi constitute a monophyletic group dating back over 400 million years. In contrast, ectomycorrhizae are present in multiple groups and appear to have evolved more than once.

Other fungi, called **endophytes,** live within leaves. While endophytes are much less studied than mycorrhizae, they may be as widespread. Endophytic hyphae grow within cell walls and in the spaces between cells. Long thought to be harmless, endophytes are now recognized as beneficial. They may help the host plant by producing chemicals that deter pathogens and herbivorous insects.

Mutually beneficial associations between fungi and animals are much less common, but a few examples are known, none more striking than insects that actually grow fungi for food. The fungi benefit because the insects provide shelter, food, and protection from predators and pathogens. Insect–fungal agriculture has evolved at least three times: in leaf-cutter ants that live in tropical forests, in a group of African termites, and in some wood-boring beetles. These beetles maintain fungal gardens within tunnels that they excavate in damaged or dead trees.

FIG. 34.7 Mutualistic relationship between fungi and plants. Mycorrhizal fungi (white) surround roots (brown) and produce extensively branched hyphae that extend into soil. *Source: Dr. Jeremy Burgess/Science Source.*

Lichens are symbioses between a fungus and a green alga or a cyanobacterium.

Lichens are familiar sights in many environments, often forming colorful growths on rocks or tree trunks (**Fig. 34.8**). **Lichens** look, function, and even reproduce as single organisms, but they are actually stable associations between a fungus and a photosynthetic microorganism, usually a green alga but

FIG. 34.8 Lichens. A lichen is a mutualistic association between a fungus and a photosynthetic microorganism. *Sources: (left to right) Wallace Garrison/Getty Images; Stephen Sharnoff; Stephen Sharnoff.*

FIG. 34.9 Lichen anatomy. The fungus provides structure, and photosynthetic algae form a thin layer under the surface. The scanning electron micrograph shows a section through *Xanthoria flammea*. *Photo source: Eye of Science/Science Source.*

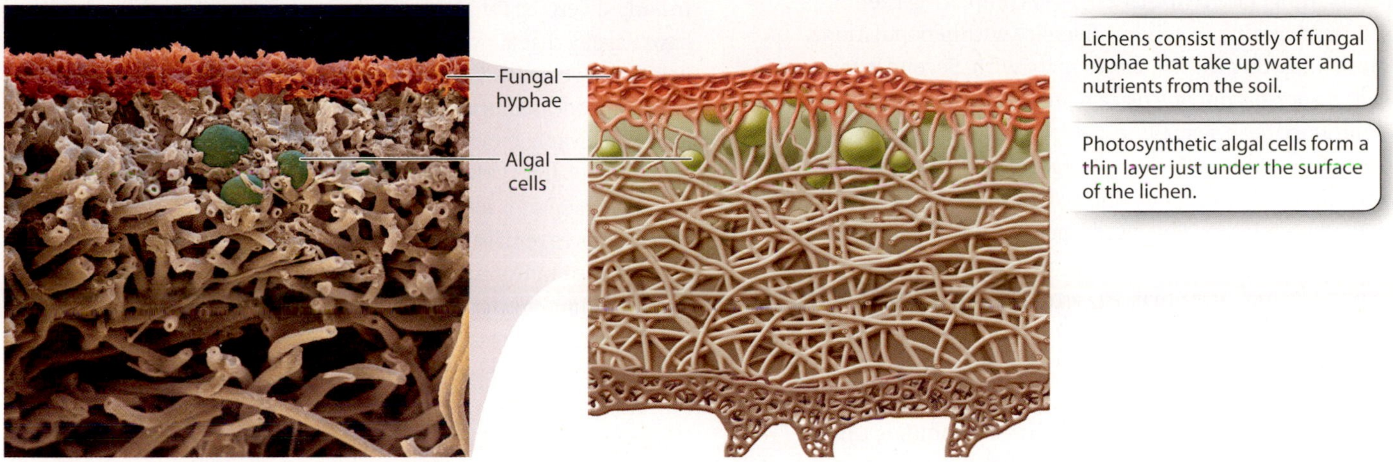

sometimes a cyanobacterium. Nearly 15% of all known fungal species grow as lichens.

The dual nature of lichens was first proposed by a Swiss botanist, Simon Schwendener, in 1867, but his hypothesis found little favor at the time. The existence of a composite "organism" challenged the idea that life could be divided into discrete categories such as animals and plants and raised the question of how an association of distinct species could function as an integrated whole. Given what we now understand about the widespread nature of mutualisms, the opposition to Schwendener's proposal may seem surprising. Yet it was only in 1939, when Eugen Thomas showed that lichens could be separated into their individual parts and then reassembled, that their dual nature became widely accepted.

Lichens consist mostly of fungal hyphae, with the photosynthetic algae or cyanobacteria forming a thin layer just under the surface (**Fig. 34.9**). The hyphae anchor the lichen to a rock or tree, aid in the uptake and retention of water and nutrients, and produce chemicals that protect against excess light and herbivorous animals. In turn, the photosynthetic partners provide a source of reduced carbon, such as carbohydrates. Cyanobacteria living in lichens are capable of nitrogen fixation (Chapter 26) and thus can use atmospheric N_2 as a source of nitrogen. The two partners exchange nutrients through fungal hyphae that tightly encircle or even penetrate the walls of the photosynthetic cells. As a result, lichens are able to thrive where neither partner could exist on its own.

It remains an open question whether either partner can exist independently in nature. In cases where the fungal partner can be grown in the laboratory in culture, it produces a relatively undifferentiated hyphal mass. Thus, while the bulk of the lichen structure comes from the fungus, chemical signals from the photosynthetic partner influence the form and shape of the fungus.

Surprisingly, these associations are not species specific. There are approximately 13,500 known lichens, but only about 100 participating photosynthetic species. This means that different lichens can have the same algae or cyanobacteria. Conversely, a given fungal partner may associate with more than one photosynthetic species without affecting the lichen's outward form. Because of the intimate association between lichen morphology and fungal species, the lichen and fungus are assigned the same scientific name.

Lichens spread asexually by fragmentation or through the formation of dispersal units consisting of a single photosynthetic cell surrounded by hyphae. Sexual reproduction, at least by the fungi, is common, and the photosynthetic cells reproduce asexually by mitotic cell division. Whether sexual or asexual reproduction is more important in allowing lichens to establish in new habitats is not known.

Lichens are remarkable for their ability to grow on the surfaces of rocks and tree trunks. They are among the first colonizers of lava flows and the barren land left after glacial retreat. In these habitats, lichens obtain nutrients from rainfall or by secreting organic acids that help release some nutrients from even rocky surfaces. Not surprisingly, lichens in these harsh environments grow slowly. Some lichens grow on soil, for example reindeer "moss," which grows in the Arctic tundra and is eaten by caribou. Reindeer moss is thought to compete successfully for space by secreting chemicals that hinder the growth of plants.

Given their small size and exposure to the environment, lichens must be able to tolerate drying out from time to time. All lichens have a high tolerance for desiccation, and can tolerate wide fluctuations in temperature and light. In contrast, lichens are quite sensitive to air pollution, particularly sulfur dioxide (SO_2). For this reason, lichen growth is sometimes used as an indicator of industrial pollution.

34.2 REPRODUCTION

Like plants, fungi face two challenges in completing their life cycles. First, to maintain genetic diversity within populations, they must find other individuals to mate with. Second, they must be able to disperse from one place to another. The majority of fungi reproduce both sexually and asexually. Sexual reproduction in many fungi is characterized by a feature of the life cycle that distinguishes it from all other eukaryotes, and some asexual fungi have a unique way to generate genetic diversity. Fungal adaptations for dispersal broadly resemble those of plants. Fungi rely on wind, water, or animals to carry spores through the environment. Recall from Chapter 30 that spores are specialized cells well adapted for dispersal and long-term survival.

We most often come into contact with fungi through their reproductive structures. Fungal spores commonly cause respiratory illness, and mushrooms attract our attention because they are delicious or poisonous. Yet many aspects of fungal reproduction are not easily observed. In this section, we emphasize the most general principles and patterns.

Fungi proliferate and disperse using spores.

The tissues, living or dead, that support fungal growth often have a patchy distribution. For this reason, fungi must be able to disperse from one food source to another. The extensive networks of hyphae within soils or host organisms can spread locally but cannot disperse over great distances, so fungi produce spores that can be carried by the wind (**Fig. 34.10a**), in water, or attached to (or within) animals. In early-diverging fungi that live in aquatic environments, spores have flagella that allow them to swim. The great majority of fungi, however, live on land, and their spores lack flagella. Instead, the spores are encased in a thick wall that protects them as they are dispersed over habitats unsuitable for growth.

The probability that any given spore will come to rest in a favorable habitat is low, so fungi produce large numbers of spores. Fungal spores remain viable, able to grow if provided with an appropriate environment, for periods ranging from only a few hours in some species to many years in others. Thus, spores allow fungi to use resources that are patchy in time as well as in space. In fact, a shortage of resources is one of the cues that triggers spore formation.

Spores can form by meiotic cell division as part of sexual reproduction, and they can also form asexually. Asexual spores are formed by mitotic cell division and therefore are genetically identical to their parent. In many species, asexual spores are produced within sporangia that form at the ends of erect hyphae, allowing the release of the spores into the air. A close look at a moldy piece of bread reveals that the surface is covered with hyphae carrying sporangia containing asexual spores (**Fig. 34.10b**).

Fungal species vary in the extent to which they rely on asexual as opposed to sexual spore production. In some groups,

FIG. 34.10 Fungal spores dispersed by wind. *Sources: a. Biosphoto/Yves Lanceau; b. (top) Gregory G. Dimijian/Science Source; b. (bottom) Garry DeLong/Getty Images.*

FIG. 34.11 Fruiting bodies, complex multicellular structures built from hyphae. Fruiting bodies of *Hygrophorus miniatus,* known commonly as the vermillion waxcap. *Photo source: Matt Meadow/Getty Images.*

such as the group that includes the most common mushrooms, sexual reproduction is the dominant means of spore production. On the other hand, in a small number of fungi that includes the endomycorrhizal species, spores produced by meiotic cell division have never been observed. Although sexual reproduction appears to be absent in some fungal species, asexual fungi have other mechanisms for producing genetic diversity, as discussed later in this section.

Multicellular fruiting bodies facilitate the dispersal of sexually produced spores.

Fungi employ an astonishing array of mechanisms to enhance spore dispersal. Particularly conspicuous are the multicellular **fruiting bodies** produced by some fungi, which are structures that allow the dispersal of sexually produced spores. (Fungal fruiting bodies should not be confused with the fruits of flowering plants, which are unrelated and very different structures.) Mushrooms, stinkhorns, puffballs, bracket fungi, truffles, and many other well-known structures are fungal fruiting bodies.

Fruiting bodies are highly ordered and compact structures compared with the mycelia from which they grow, yet they are constructed entirely of hyphae (**Fig. 34.11**). In fact, in many cases their mechanisms of spore dispersal demand a high degree of structural precision. Yet we know little of the developmental processes that allow tip-growing hyphae to produce such complex and regular structures.

The fruiting bodies of many fungi rise above the ground or grow from the trunks of dead trees, so the sexually produced spores are released high above the ground. But elevation by itself is not enough to ensure dispersal. Thus, many fungi forcibly eject their spores, achieving velocities of more than 1 m/s that allow the tiny spores to penetrate and travel beyond the layer of unstirred air that surrounds the fruiting body. Once airborne, the spores are transported efficiently through air (**Fig. 34.12**). Other fungi rely on external agents such as raindrops or animals to move their spores around. In section 34.3, we will encounter examples of the diverse dispersal mechanisms found in the fungi.

The fungal life cycle often includes a stage in which haploid cells fuse, but nuclei do not.

Like other sexually reproducing eukaryotes, fungi have life cycles that include haploid ($1n$) and diploid ($2n$) stages. The nuclei in fungal hyphae are haploid, and the fungal life cycle is therefore similar to the life cycles of eukaryotic organisms that exist normally in the haploid stage (see Fig. 27.3a). In these haploid-dominant organisms, asexual reproduction takes place through the production of haploid spores by mitosis, while sexual reproduction takes place as two haploid cells (often male and female gametes) fuse to form a diploid zygote, which undergoes meiosis as its first division. However, sexual reproduction in fungi differs from all other haploid-dominant organisms in one important respect: In fungi, the fusion of haploid cells is not immediately followed by the fusion of their nuclei.

In most fungi, the sexual phase of the life cycle takes place through the fusion of hyphal tips rather than specialized reproductive cells, or gametes. For mating to occur, two hyphae grow together and release enzymes that digest their cell walls at the point of contact. Once the dividing wall is gone, the contents of the two hyphal cells merge, forming a single cell with two haploid nuclei. Eventually, the two haploid nuclei merge, forming

HOW DO WE KNOW?

FIG. 34.12

What determines the shape of fungal spores that are ejected into the air?

BACKGROUND Many fungi disperse their spores by ejecting them forcibly into the air. For spores to be picked up by the wind, however, they must escape a layer of still air called the boundary layer that lies close to the ground. Escape from the boundary layer is easier if spores have a size and shape that minimize drag, the resistance of air to the movement of an object.

HYPOTHESIS In fungi that eject their spores into the atmosphere, natural selection favors spore shapes that minimize drag.

EXPERIMENT Working from computer models for minimizing drag on airplane wings, scientists modeled spore shapes that minimize drag. These models took into account spore size as well as the physical characteristics of the air through which spores travel. The scientists then measured spore shape for more than 100 species of spore-ejecting fungi.

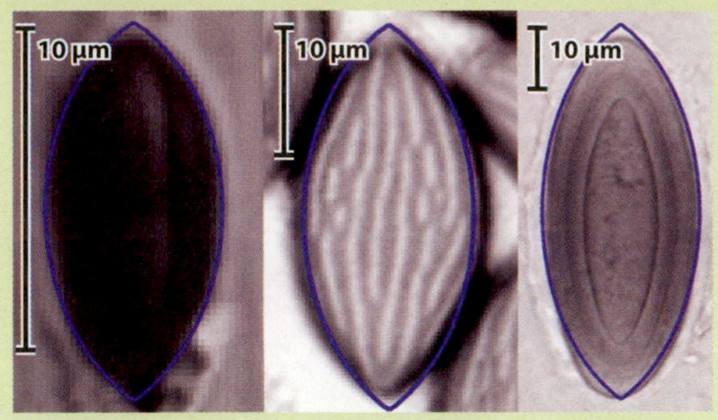

Photo source: (left) Reprinted from Mycological Research, 102/11, R. Dulymamode, P. F. Cannon, A. Peerally, Fungi from Mauritius Anthostomella Species of Pandanus, 1998, with permission from Elsevier.

Fungal spores with outlines (blue) of computer-modeled shapes that minimize drag.

RESULTS Nearly three-quarters of the examined spores had shapes that came within 1% of the shape calculated to minimize drag (shown in the figure as blue outlines). Related species that do not eject spores forcibly were less likely to have drag-minimizing shapes.

CONCLUSION Natural selection has acted on fungi to facilitate spore dispersal by wind.

SOURCE Roper, M., et al. 2008. "Explosively Launched Spores of Ascomycete Fungi Have Drag-Minimizing Shapes." *Proceedings of the National Academy of Sciences, USA* 105:20583–20588.

a diploid zygote. The zygote divides by meiotic cell division, giving rise to sexually produced haploid spores.

In most sexually reproducing organisms, when two gametes merge, their nuclei fuse almost instantly to form a diploid zygote. In fungi, however, the cytoplasmic union of two cells (**plasmogamy**) is not always followed immediately by the fusion of their nuclei (**karyogamy**). Instead, the haploid nuclei retain their independent identities, resulting in what is referred to as a **heterokaryotic** ("different nuclei") stage (**Fig. 34.13**). In the heterokaryotic stage, a cell contains nuclei from two parental hyphae, but the nuclei remain distinct. The heterokaryotic stage ends with pairs of nuclei fusing, which leads to the formation of a diploid zygote.

In some groups, the heterokaryotic stage consists of a single cell with many haploid nuclei. In other groups, plasmogamy is followed by mitosis, which produces hyphae in which each cell contains two haploid nuclei, one from each parent. Heterokaryotic cells with two genetically distinct haploid nuclei are referred to as **dikaryotic,** or $\boldsymbol{n + n}$**,** cells. In fungi with dikaryotic cells, called dikaryotic fungi, there may be a small number of dikaryotic cells or extensively developed hyphae made up of dikaryotic cells. Some edible mushrooms found on market shelves consist entirely of dikaryotic hyphae.

Dikaryotic fungi account for more than 98% of all known fungal species, suggesting that the separation of plasmogamy and karyogamy in time and space may confer an evolutionary advantage. One such advantage can be seen when we think about the environments where fungi live. Mating takes place principally within the soil or inside the trunks of rotting trees, sites where hyphae are most likely to come into contact. Dispersal, however, is most effective when spores can be released into the air. The separation of plasmogamy and karyogamy allows mating and spore production to occur where each is most effective. In some dikaryotic fungi, however, the separation of plasmogamy and karyogamy is neither distant in space nor long in time. In these fungi, the highly branched dikaryotic hyphae may serve primarily as a means of increasing the number of cells in which karyogamy will eventually take place, therefore leading to the production of more sexually produced spores.

At present, we really don't know why fungi proliferate $n + n$ cells rather than forming a multicellular diploid phase, and there

FIG. 34.13 Generalized fungal life cycle. In many fungi, plasmogamy and karyogamy are separated by a heterokaryotic stage.

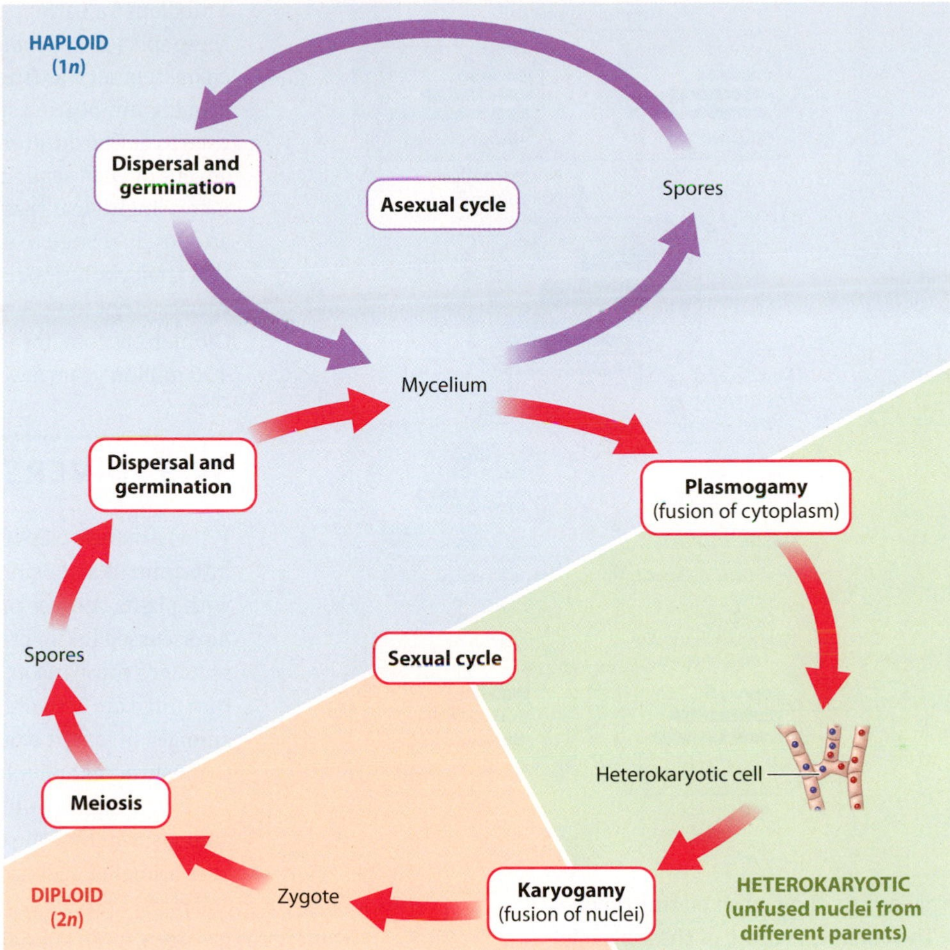

may be no single answer to the question of why the dikaryotic fungi have this unique cell type.

→ **Quick Check 2** What is the difference between a diploid cell and a dikaryotic cell?

Genetically distinct mating types promote outcrossing.

Under most conditions found in nature, genetically diverse populations persist better than those lacking diversity. Indeed, sexual reproduction is widely viewed as a means of promoting genetic diversity for long-term ecological success in variable environments (Chapters 11 and 42). With the exception of early-diverging aquatic groups, fungi do not produce male and female gametes. What prevents an individual from mating with itself?

Fungi have different **mating types** that are genetically determined and prevent self-fertilization. The mating type of an individual is determined by a mating-type gene. Fertilization can take place only between individuals that have different alleles at the mating-type gene. In some species, there are only two mating-type alleles. In this case, mating patterns are identical to those for species with male and female sexes. If the two mating-type alleles are spread evenly throughout the population, an individual should be able to mate with 50% of the general population.

Some fungi have more than two mating-type alleles. These fungi have a greater likelihood of encountering a compatible genotype in the general population. That likelihood is even higher for fungi in which mating type is determined by two different mating-type genes that each have multiple alleles. Fungi in the group that includes the common mushrooms can have as many as 20,000 different mating-type alleles. In this case, the odds of finding a compatible mate are close to 100%.

Genetic compatibility is a prerequisite for mating. How do fungi go about finding a suitable mate? The answer is that they

FIG. 34.14 **Parasexuality.** Crossing over during mitosis is thought to provide genetic diversity in asexual fungi.

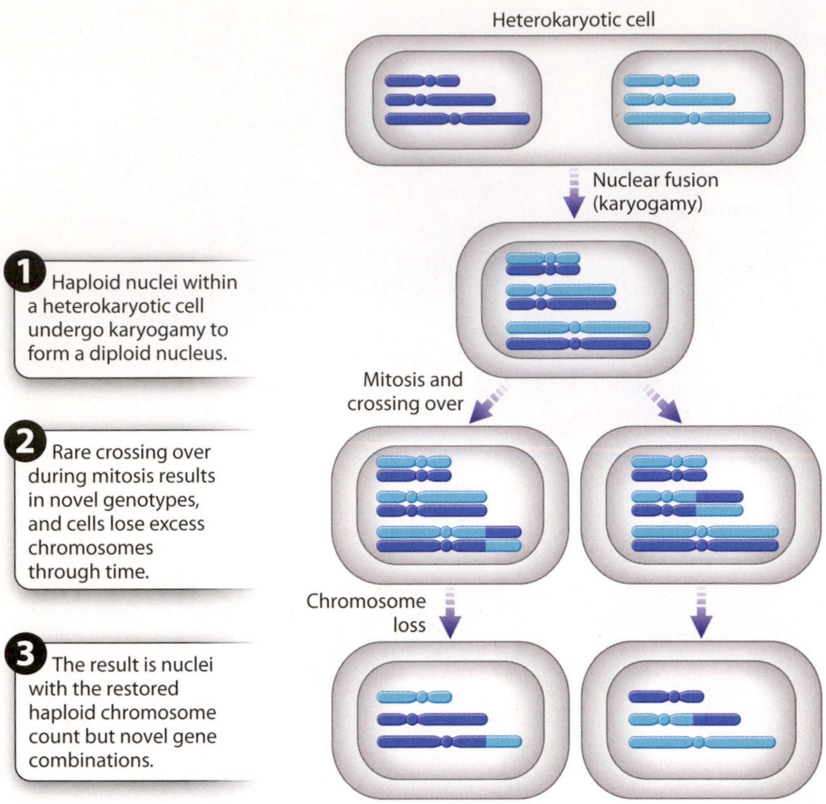

secrete chemical signals that attract fungi with mating types different from their own. The chemical nature of these signals differs among groups.

About 20% of fungi appear to lack sexual reproduction altogether. Among these are such well-known groups as *Penicillium* (the source of the antibiotic penicillin), *Aspergillus* (the major industrial source of vitamin C), and all the endomycorrhizal fungi. Spores formed by mitosis (asexual reproduction) allow these fungi to proliferate and disperse.

How are asexual species able to persist without a mechanism for generating genetic diversity? The short answer is, they don't. Fungal species that lack an observable sexual cycle are hypothesized to have another mechanism of generating genetic diversity: the crossing over of DNA during mitosis (**Fig. 34.14**). Such species are described as **parasexual.** In parasexual fungi, haploid nuclei within a heterokaryotic cell undergo karyogamy to form a diploid nucleus—so far, a familiar pattern. Next, however, during mitosis, a rare crossing over may occur between the two sets of chromosomes, resulting in novel genotypes. Cells lose excess chromosomes over time, such that the haploid state is restored. Like sexual reproduction, these processes produce novel gene combinations. The key difference between parasexuality and sexual reproduction is that parasexual species do not undergo meiosis.

Entirely asexual species of fungi posed a problem for early taxonomists because these species lack many of the morphological characters such as fruiting bodies or sporangium-bearing stalks previously used to define different groups. With the advent of DNA sequence data, it has become increasingly clear that many asexual species are closely related to sexual forms, indicating that their evolutionary history as asexual species is not long. A major exception are the Glomeromycota, for which fossils as old as 450 million years have been reported.

34.3 DIVERSITY

When Aristotle classified the living world into animals and plants, he grouped fungi with plants because of their lack of motility. As discussed in Chapter 27, molecular sequence comparisons now make it clear that fungi are actually related more closely to animals. In fact, it would not be amiss to refer to mushrooms and molds as our "cousins."

Fungi share a number of features with animals: Their motile cells, when present, have a single flagellum attached to their posterior ends; fungi and many animals synthesize chitin; and fungi store energy as glycogen, as do animals. Nevertheless, the last common ancestor of fungi and animals was a single-celled microorganism that lived in aquatic environments about 1 billion years ago. Multicellularity evolved independently in fungi and animals (Chapter 28), and so the fungi have many features that are unique in the biological world.

Fungi are highly diverse.

By any measure, fungi are diverse, although just how diverse is not known. About 75,000 species have been formally described, but estimates of true fungal diversity run as high as 5 million species. Among eukaryotes, the only organisms more diverse than fungi are the animals. The availability of DNA sequence data has greatly advanced our understanding of phylogenetic relationships within the fungi, which are shown in **Fig. 34.15.** The tree clearly shows how the characters present in familiar mushrooms accumulated through the course of evolution: first cell walls of chitin, then hyphae, then regularly placed septa and, finally, the complex multicellular reproductive bodies we call mushrooms. Motile single cells are found only in the most ancient groups, while dikaryotic cells and complexity of form are found only on the most recent branches.

FIG. 34.15 A phylogeny of the fungi. Approximate numbers of described species are given in parentheses.

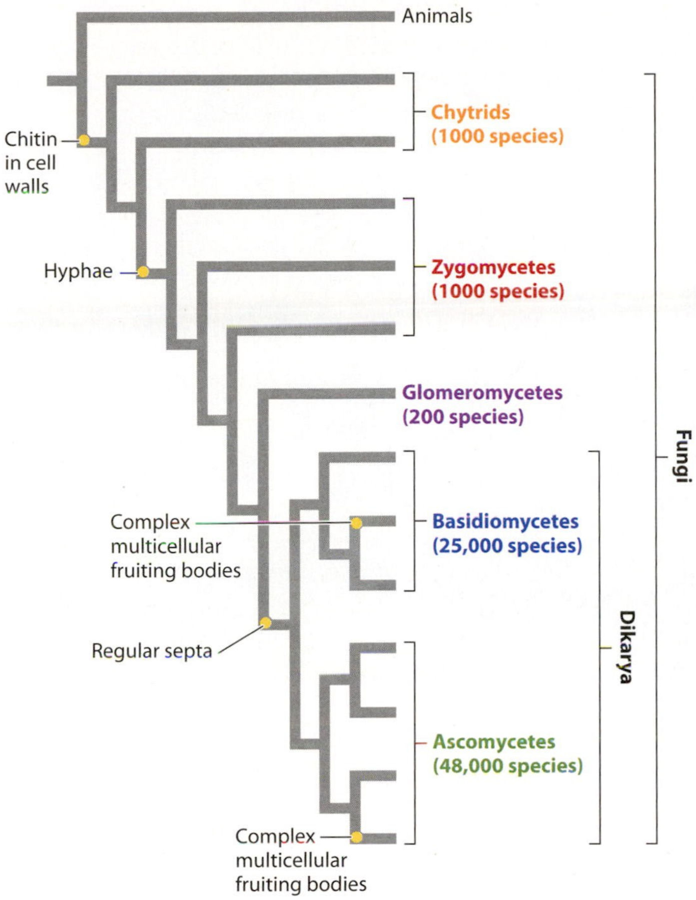

The evolutionary history of fungi can be viewed as the transition from a flagellated ancestral form that lived in aquatic environments to soil-dwelling forms in which hyphae divided by septa develop complex morphologies visible to the naked eye. Moreover, in the earliest diverging fungal group, karyogamy follows quickly on the heels of plasmogamy and there is no heterokaryotic stage. Thus, another evolutionary trend is the progressive separation in time between plasmogamy and karyogamy in fungal life cycles. Many of the distinguishing features of fungi evolved as adaptations to the challenges faced by nonmotile heterotrophs on land that absorb nutrients directly from their substrate.

We also see that the numbers of species are not spread evenly across the phylogeny. Instead, more ancient groups include less than 2% of known species, while the two dikaryotic groups include more than 98% of known fungi. Clearly, dikaryotic fungi are well adapted to many different habitats, and these include other organisms, both living and dead.

→ **Quick Check 3** From the phylogenetic tree shown in Fig. 34.15, would you say that complex multicellular fruiting bodies evolved once or twice (independently) in fungi?

Fungi evolved from aquatic, unicellular, and flagellated ancestors.

Fungi are opisthokonts, members of the eukaryotic superkingdom that includes animals (Chapter 27). Among fungi, the Chytridiomycota (commonly referred to as **chytrids**) were the first groups to diverge. Molecular clock analysis estimates that chytrids began to diversify 800 to 700 million years ago. Today, there are about 1000 species of chytrids found in aquatic or moist environments (**Fig. 34.16**). Many chytrids are single cells with walls of chitin. They may also form short multinucleate structures, but they lack the well-defined hyphae characteristic of other fungi. Chytrids, therefore, do not form a true mycelium, although in some species elongated cellular outgrowths called rhizoids penetrate into organic substrates. Rhizoids anchor the organism in place and absorb food molecules.

Chytrids generally disperse by flagellated spores. Sexual reproduction appears to be rare, but there is substantial life-cycle diversity within this group. Chytrids lack a heterokaryotic stage but form flagellated gametes that swim through their aqueous environment.

Most chytrids are decomposers, and a few of them are pathogenic. Notably, infection by the chytrid *Batrachochytrium dendrobatidis* may be a cause of a widespread decline in amphibian populations (Chapter 49).

FIG. 34.16 Chytrids. Chytrids are aquatic fungi that attach to decomposing organic matter by rhizoids. *Source: John Hardy, University of Wisconsin–Stevens Point Department of Biology.*

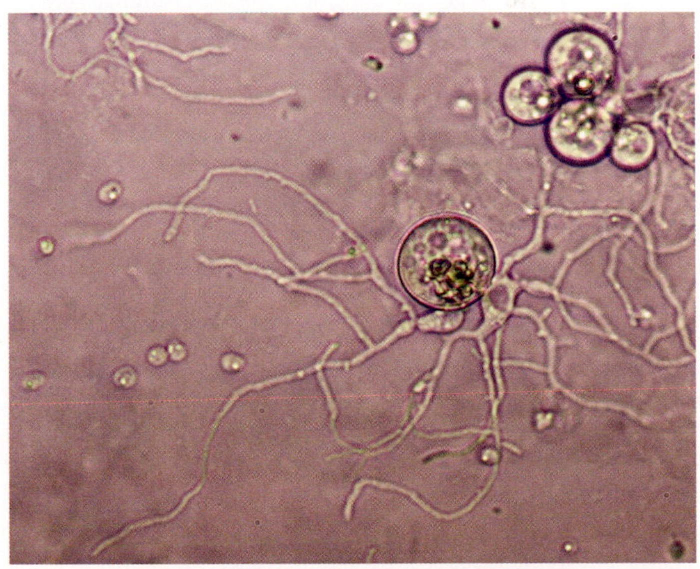

Zygomycetes produce hyphae undivided by septa.

Moving through the phylogenetic tree, we next encounter several groups traditionally united as the Zygomycota, or **zygomycetes** (see Fig. 34.15). These groups share a number of traits.

Collectively, zygomycetes make up less than 1% of the known fungal species. Some are decomposers, specializing on dead leaves, animal feces, and food. There are probably zygomycetes in your kitchen. Others live on and in plants, animals, and even other fungi. Zygomycetes have many typical fungal traits, including the growth of a mycelium and the production of aerial spores. These traits may have evolved as adaptations for finding food and dispersing on land. The loss of flagellated spores may also reflect adaptation for life on land. Unlike more complex fungi, zygomycetes do not form regular septa along their hyphae, nor do they produce multicellular fruiting bodies.

Earlier, we introduced the black bread mold *Rhizopus,* which is a zygomycete (see Fig. 34.10b). *Rhizopus* and its relatives are specialists on substrates containing abundant, easy-to-digest carbon compounds, such as bread, ripe fruits, and the dung of herbivorous animals. These fungi consume their substrates rapidly, and once their meal is finished, they release large numbers of aerial spores to locate another food source.

Sexual reproduction occurs when two compatible hyphal tips fuse to form a thick-walled structure containing many nuclei of each mating type. Karyogamy and meiosis are followed by germination of the haploid cell to form an elevated stalk. Each stalk develops a sporangium that contains spores produced asexually by mitotic cell division. If you look closely at a bread mold, you can see the white filamentous hyphae and small black sporangia on slender stalks. Each sporangium can produce as many as 100,000 spores that are dispersed by the wind, explaining how these organisms seem to get everywhere.

Although zygomycete fungi do not form multicellular fruiting bodies, some have evolved truly spectacular means of dispersal. *Pilobilus,* which consumes the dung of herbivorous animals, has a life cycle similar to *Rhizopus* except that, instead of releasing individual spores, it forcibly ejects the entire sporangium (**Fig. 34.17**). Turgor pressure generated in the supporting stalk propels the sporangia as far as 2 m. Light-sensitive pigments in the stalk's hyphae control the orientation of this water cannon, ensuring that the spores have the best chance of escaping the dung pile and landing on vegetation that is attractive to grazing herbivores. Feeding herbivores disperse the spores further, while supplying them with fresh dung for food. The remarkable dispersal of *Pilobilus* sporangia benefits other species, as well: Parasitic nematode worms hitch a ride on the sporangia to find a new host.

FIG. 34.17 **Spore dispersal in a zygomycete.** In *Pilobilus,* light-sensing pigments orient the turgor-propelled ejection of the sporangia. *Source: Carolina Biological Supply Company/ Phototake.*

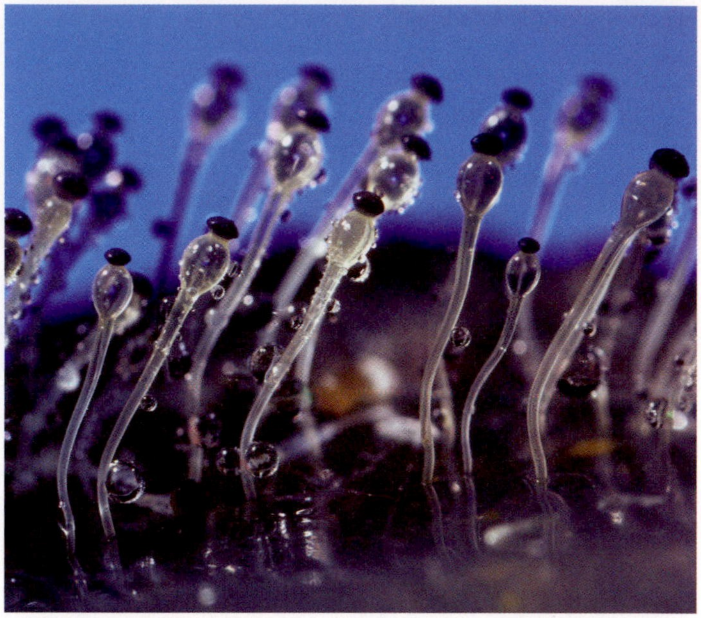

Glomeromycetes form endomycorrhizae.

The Glomeromycota, or **glomeromycetes,** are a monophyletic group of apparently low diversity (there are about 200 described species) but tremendous ecological importance. All known glomeromycetes occur in association with plant roots. Most famous are those that form endomycorrhizae. Molecular clock estimates suggest that glomeromycetes diverged from other fungi 600 to 500 million years ago, before the evolution of vascular plants. Establishment of glomerophyte–root symbiosis changed the course of evolution for both participants. Today, most vascular plant species harbor endomycorrhizae, dramatically increasing nutrient uptake from the soil.

Glomeromycetes are hard to study because they cannot be grown independently of their plant partners, and so many unanswered questions remain. For example, no evidence of sexual reproduction has ever been found in this group. Yet genetic studies of glomeromycete populations show some genetic diversity, which is thought to result from parasexual processes. Glomerocytes produce extremely large (0.1 to 0.5 mm diameter) multinucleate spores asexually. These large spores can persist in soils until they come into contact with an uninfected root.

The Dikarya produce regular septa during mitosis.

The final major node, or branching point, in the fungal phylogeny marks the origin of the **Dikarya,** a group that includes about 98% of all described fungal species. A key feature of the dikaryotic fungi is that every mitotic division is accompanied by the formation of a new septum. This innovation allows these fungi to control the number of nuclei within each cell and thus to proliferate dikaryotic cells. The Dikarya include all the edible mushrooms, the yeast species used in the production of beer, bread, and cheese, the major wood-rotting fungi, and pathogens of both crops and humans.

FIG. 34.18 **Two groups of dikaryotic fungi.** (a) The morel (*Morchella esculenta*) is an ascomycete, whereas (b) the shiitake mushroom (*Lentinula edodes*) is a basidiomycete. *Sources: a. Matt Meadows/Getty Images; b. Topic Photo Agency/age fotostock.*

The Dikarya contain two monophyletic subgroups (see Fig. 34.15), each named for the shape of the cell in which the key reproductive processes of nuclear fusion (karyogamy) and meiosis take place. The Ascomycota, or **ascomycetes,** are sometimes called sac fungi because nuclear fusion and meiosis take place in an elongated saclike cell called an ascus (*askos* is Greek for "leather bag"). The Basidiomycota, or **basidiomycetes,** are popularly known as club fungi because the nuclear fusion and meiosis take place in a club-shaped cell called a basidium. ("Basidium" is derived from the Greek *basis,* which means "base" or "pedestal"; the reference is to the position of sexually produced spores at the tips of this club-shaped body.)

Although ascomycetes and basidiomycetes have the same basic life cycle, they diverged from each other as much as 600 to 500 million years ago and evolved complex multicellular structures independently of each other. The morel in the supermarket is an ascomycete, whereas the shiitake mushroom next to it is a basidiomycete (**Fig. 34.18**). In addition to asci and basidia, the two groups differ in several other features. For example, the septal pores that connect adjacent cells within hyphae are distinct in the two groups, as are the ways in which cells can plug the pores to minimize damage to a mycelium. Also, when ascomycetes reproduce, mating cells form and their nuclei fuse in relatively close succession. Thus, the heterokaryotic stage is brief. In contrast, in the basidiomycetes, the nuclei of two fused cells may remain separate for a long time. In this case, hyphae containing both types of nuclei (dikaryotic hyphae) can continue to absorb nutrients from the substrate for a while before nuclear fusion occurs.

Ascomycetes are the most diverse group of fungi.

Ascomycetes make up 64% of all known fungal species. They include important wood-rotting fungi, many ectomychorrizal species, and significant pathogens of both animals and plants. Ascomycetes form the fungal partner in most lichens—approximately 40% of all ascomycete species occur in lichens. Ascomycetes loom large in human history, contributing the baker's and brewer's yeasts used to make bread and beer, the antibiotic penicillin, and model systems employed in laboratory investigations of eukaryotic cell biology and genetics. Ascomycetes are used to produce soy sauce, sake, rice vinegar, and miso, and to transform milk into Brie, Camembert, and Roquefort cheeses. Athlete's foot and other skin infections are caused by ascomycetes, as are many more serious fungal diseases.

Ascomycetes include both species that form multicellular fruiting bodies and ones that form unicellular yeasts. A remarkable feature of ascomycete fruiting bodies is that they contain both dikaryotic and haploid cells. The proliferation of dikaryotic cells leads to the production of many asci, while the growth of haploid cells contributes to the bulk of the fruiting body. Two ascomycete lineages consist largely of yeasts. The earliest branching members in these groups, however, produce hyphae. Thus, the unicellular nature of yeasts is a derived feature evolved through loss of hyphae and is not an ancestral condition.

FIG. 34.19 The sexual cycle of a common ascomycete, the brown cup fungus (*Peziza* species). *Photo source: Ed Reschke/Getty Images.*

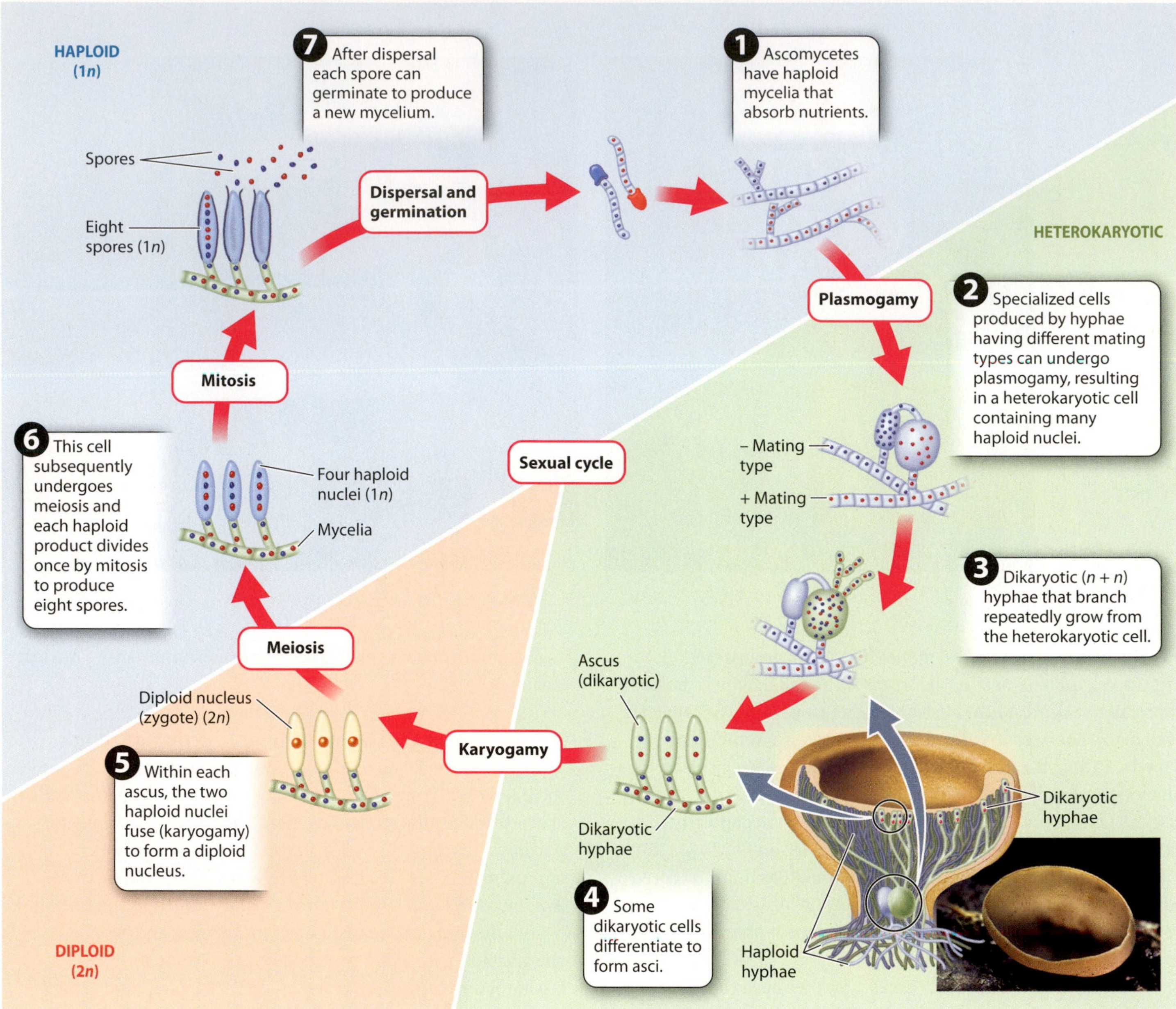

Fig. 34.19 illustrates the life cycle of a common ascomycete, the brown cup fungus. In ascomycetes, meiosis is followed by a single round of mitosis, resulting in asci that contain eight haploid spores. When mature, the spores are ejected from the top of the asci, expelled by turgor pressure. In many ascomycetes, fruiting bodies elevate the asci on one or more cup-shaped surfaces, from where the spores are easily caught and carried away by wind.

In the fruiting bodies of some ascomycetes, the asci are completely enclosed by a layer of tissue and thus must be dispersed by other organisms rather than by the wind. The truffles prized in cooking provide an example. The edible truffle is the fruiting body of an ascomycete that grows as an ectomycorrhizal fungus on tree roots. Not only do truffles encase their spores in protective tissues, but they also develop underground. How, then, can their spores be dispersed? Developing truffles release androstenol, a hormone also produced by boars before mating. The hormone attracts female pigs, which unearth and consume the fruiting body. The spores pass through the pig's digestive tract without damage and are released into the environment in feces, thereby dispersing the spores.

Another ascomycete is thought to have played a role in the events that unfolded in Salem, Massachusetts, in 1692. That year, several girls came down with an unknown illness manifested by delirium, hallucinations, convulsions, and a crawling sensation on

FIG. 34.20

Can a fungus influence the behavior of an ant?

BACKGROUND A curious death ritual unfolds in the rain forest of Thailand. Spores of the ascomycete fungus *Ophiocordyceps* infect *Camponotus leonardi* ants, growing hyphae inside their bodies. The fungus eventually kills the ant, and fruiting bodies emerge from the dead ant's head to disperse spores that begin the life cycle anew. *Camponotus leonardi* nests and forages for food high in the forest canopy. Infected ants, however, undergo convulsions that cause them to fall to the forest floor. The infected ants wander erratically and are unable to climb more than a few meters above the ground before convulsions make them fall again. In their final act, the ants bite into the undersides of leaves and die. The fungus is then able to complete its life cycle within the humid forest understory. Is the ants' behavior, including both the convulsions and leaf biting, induced by the fungi?

An infected ant attached to a leaf by having bitten into it in a "death grip."

HYPOTHESIS Parasitic fungi manipulate ant behavior to complete their own reproductive cycle.

OBSERVATIONS Infected and uninfected ants were observed for many hours and behavioral events were recorded.

RESULTS Infected ants have repeated convulsions (indicated by vertical bars in the graph), but uninfected ants show no such behavior (not shown). Transitions from erratic wandering to a "death grip," in which the ant bites into a leaf (indicated by red triangles in the graph and shown in the photograph), occurred at about the same time of day. Dissections showed that the ants' death grip results from jaw-muscle wasting caused by the fungus.

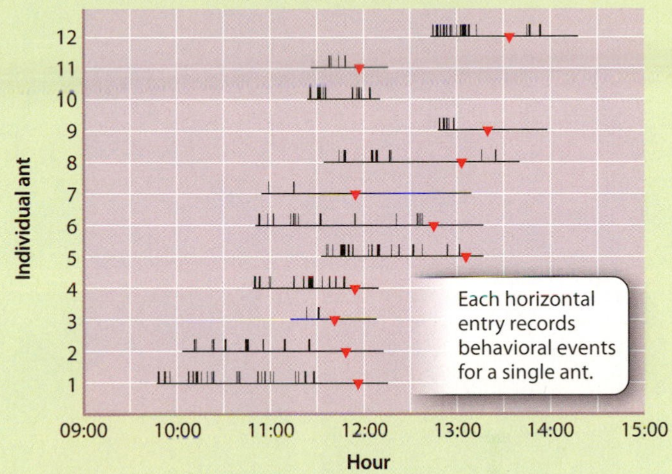

Observations of numerous infected ants (vertical bars indicate convulsions; red triangles indicate the time of ant biting).

CONCLUSION The fungi change the behavior of infected ants, inducing a stereotypical behavior that facilitates completion of the fungus's life cycle. The molecular basis of this manipulation is not yet known.

FOLLOW-UP WORK The zombie-like behavior induced by the fungus is an example of what has been called the extended phenotype: the idea that a phenotype should include the effects that an organism has on its environment—in this case, its effects on host behavior.

SOURCE Hughes, D. P., et al. 2011. "Behavioral Mechanisms and Morphological Symptoms of Zombie Ants Dying from Fungal Infections." *BMC Ecology* 11:13–22.

the skin. The girls were thought to be bewitched, and subsequent accusations led to the infamous Salem witch trials. They were executed for witchcraft, but today many scholars prefer a simpler diagnosis for the illness: poisoning by the ascomycete fungus ergot. Ergot is a common pathogen of rye and related grasses, and ingestion of contaminated plants can lead to the symptoms reported in Salem. Like many defensive compounds produced by plants, alkaloid molecules produced by ergot and their derivatives are used in medicine in low doses, such as in the treatment of migraines.

Ascomycetes are even known for producing so-called zombie ants, a topic explored in **Fig. 34.20.**

Basidiomycetes include smuts, rusts, and mushrooms.

The basidiomycetes, which make up 34% of all described fungal species, include three major groups (**Fig. 34.21**). Two of these—the smuts and rusts—consist primarily of plant pathogens. Smut fungi, which infect the reproductive tissues of grasses and related plants, take their name from the black sooty spores that they produce. *Ustilago maydis,* the corn smut, turns developing corn kernels into soft gray masses that are a culinary delicacy in Mexico. Because smuts infect seeds, their spread is magnified by the harvest, storage, and eventual sowing of crop seeds. Before the introduction of chemical seed treatments in the 1930s, infection by *Tilletia* (the stinking smut) commonly caused the loss of up to

FIG. 34.21 Basidiomycetes. (a) Corn smut (*Ustilago maydis*); (b) pseudoflowers that form when plants are infected by the rust *Puccinia monoica*; (c) a toadstool fungus—the highly toxic *Amanita*. *Sources: a. Inga Spence/Science Source; b. Bitty Roy, University of Oregon; c. David Clapp/Getty Images.*

50% of the wheat harvest. Smuts remain a problem in parts of the world where untreated seeds are still planted.

Rusts can also have profound impacts on plant function and development. For example, the rust *Puccinia monoica* alters leaf development in its host plant *Arabis*, resulting in a pseudoflower that attracts insects by visual and olfactory cues as well as nectar rewards. These pseudopollinators transport spores, thus enhancing outcrossing as well as dispersal of the fungi, but provide no service to the plant. We discuss the biology of rusts in greater detail in the next section when we examine Ug99, a highly virulent wheat rust first reported in Uganda in 1999.

The third basidiomycete group is characterized by the formation of multicellular fruiting bodies, including the iconic toadstools with their central stalk and umbrella-like cap. However, this group also includes the diverse shapes seen in stinkhorns, puffballs, and bracket fungi (**Fig. 34.22**).

Fig. 34.23 shows the life cycle of a typical basidiomycete mushroom. The life cycle is similar to that of ascomycetes (see Fig. 34.19), except that the fruiting body is made up entirely of dikaryotic hyphae instead of a combination of dikaryotic and haploid hyphae as in ascomycetes. In addition, nuclear fusion (karyogamy) takes place in a specialized cell called a basidium rather than in an ascus. Finally, the haploid products of meiosis (the spores) do not undergo mitosis, so four spores are produced from each basidium rather than eight spores from each ascus.

→ **Quick Check 4** When you eat a mushroom, what stage of its life cycle are you consuming?

Most species in this group live by forming ectomycorrhizal associations or by decomposing wood and other substrates. The fruiting bodies that we see elevated above a rotting log or the soil are connected to extensive mycelia that provide resources for fruiting-body development.

FIG. 34.22 Diverse fruiting bodies of basidiomycetes. (a) Stinkhorn (*Dictyophora indusiata*); (b) puffball (*Calvatia gigantea*); (c) bracket fungi (*Ganoderma australe*). *Sources: a. Bill Gozansky/age fotostock; b. Wally Eberhart/Getty Images; c. Bob Gibbons/Science Source.*

FIG. 34.23 The sexual cycle of a basidiomycete that forms multicellular fruiting bodies.

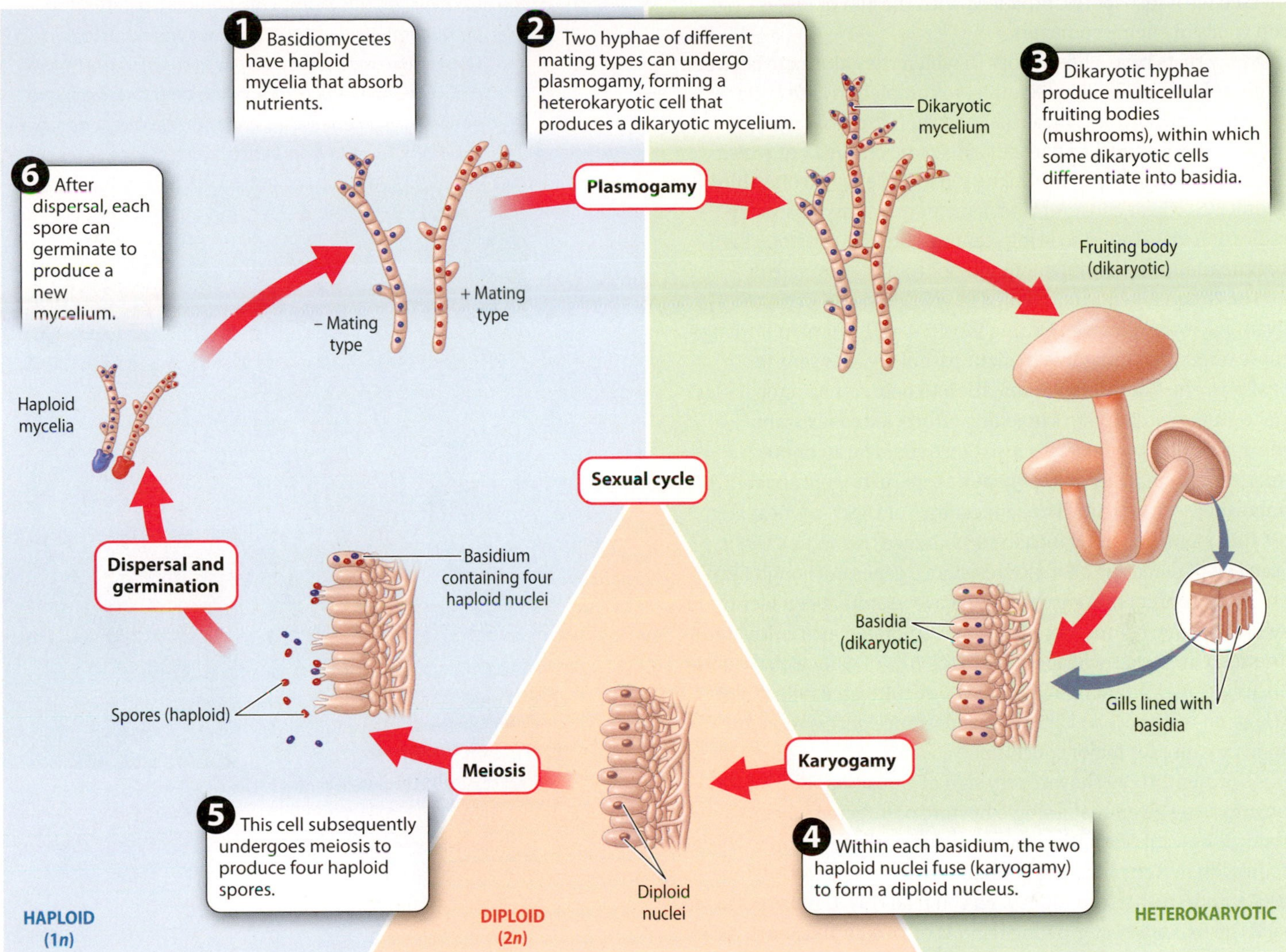

Elevation alone does not ensure dispersal because the spores must penetrate a boundary layer of air that surrounds the fruiting body. Many basidiomycetes depend on surface tension to catapult their spores through the air. The placement of their four spores on the ends of short stalks is the key to this mechanism. Both the spores and the supporting basidial cell actively secrete solutes that cause water to condense on their surfaces. At first, the droplets are independent, but as they grow they eventually come into contact with one another, and at this point the high surface tension of water causes them to merge into a single, smooth droplet. This change in shape shifts the center of mass of the water with such force that it catapults the spore into the air.

Basidiomycetes that form conspicuous fruiting bodies have diverse mechanisms for dispersing their spores. For example, puffballs function like bellows: The impact of raindrops forces the loose, dry spores out of a small hole at the tip of the ball (see Fig. 34.10a). Bird's nest fungi function like splash cups: In this case, raindrops displace groups of spores (the "eggs" in the nest) onto nearby vegetation. Herbivores consume the spores when they feed on the vegetation and transport them to a new dung pile. Stinkhorns produce spores in a stinky, sticky mass that attracts insects, which then carry off spores that stick to their legs and body.

? CASE 6 AGRICULTURE: FEEDING A GROWING POPULATION

How do fungi threaten global wheat production?

Throughout much of human history, crops have been vulnerable to infection by *Puccinia graminis*, the black stem rust of wheat. *Puccinia graminis* infects wheat leaves and stems through their stomata and then extends within the plant's living tissues to fuel its own growth. Rust-colored pustules erupt along the stem and then release a vast number of spores, leaving the host plant to wither and die. Crop losses from *P. graminis* can be devastating. In the past, both the United States and the Soviet Union tried to develop black stem rust as a biological weapon. The U.S. version was cluster bombs (since destroyed) containing turkey feathers smeared with rust spores, produced with the aim of disrupting crop yields in enemy countries.

Many rusts have a life history in which they alternate between different host species. For example, to complete its life cycle, *P. graminis* must infect wheat first and then barberry bushes (*Berberis* species). After a particularly severe outbreak of wheat rust during World War I, in which crop losses equaled one-third of total U.S. wheat consumption, a major program of barberry eradication was started to bring stem rusts under control. This program reduced the threat, but did not totally eliminate it.

The Green Revolution seemed to give growers a decisive advantage over stem rust. During World War II, Norman Borlaug, whose original training was in plant pathology, was sent to Mexico by the Rockefeller Foundation to help combat crop losses due to stem rust. Borlaug's breeding efforts led to resistant wheat varieties. Much of this resistance could be attributed to variation in a small number of genes. Stem rusts were apparently vanquished—that is, until the appearance of Ug99, a wheat stem rust first identified in Uganda in 1999 (hence the name). Ug99 is capable of defeating all major resistance genes. Although new genes that confer resistance to Ug99 have recently been identified, it will take time before these can be incorporated into cultivated varieties. Given that wheat accounts for 20% of daily global calorie consumption by humans and that 90% of cultivated wheat has little or no resistance to Ug99, the appearance of this resistant fungus is cause for major concern.

At present, Ug99 shows every sign of spreading (**Fig. 34.24**)—not surprising given that a single hectare of infected wheat produces upward of a billion spores. It currently devastates wheat production in Kenya and has been found in Yemen (2006), Iran (2007), and South Africa (2009). As you read this, Ug99 is poised to reach the major wheat growing regions of Turkey and South Asia. Even more rapid spread is possible if spores accidentally lodge on cargo or airline passengers.

Just before his death in 2009, Norman Borlaug urged the world to take this threat seriously. Today, plant breeders are actively searching ancestral wheat varieties for genetic sources of resistance to Ug99, while other scientists are trying to understand how plants defend themselves against rusts and what makes Ug99 so virulent. The resurgence of *P. graminis* as a serious threat to world food production underscores the importance of safeguarding the genetic diversity of agricultural species so that the inevitable evolutionary battles with pathogenic species may be waged successfully. ■

FIG. 34.24 **The spread of Ug99.** Ug99 is a highly resistant stem rust of wheat and a major threat to global food production. *Sources: After "The Spread of Wheat StemRust UG99 Lineage," Food and Agriculture Organization of the United Nations, JPEG image, 2010, http://www.fao.org/agriculture/crops/rust/stem/rust-report/stem-ug99racettksk/en/; (photos) Yue Jin, USDA-ARS, Cereal Disease Laboratory.*

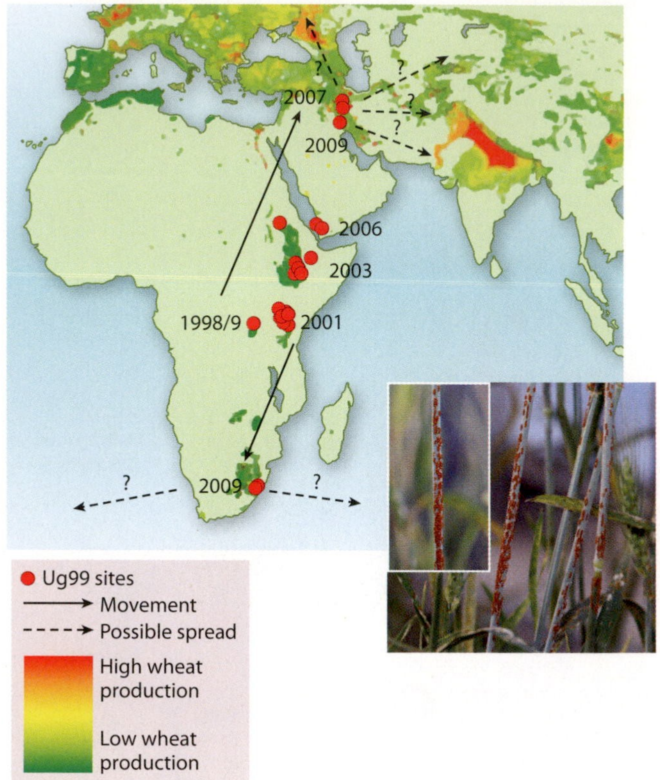

Core Concepts Summary

34.1 Fungi are heterotrophic eukaryotes that feed by absorption.

Fungi are heterotrophs that depend on preformed organic molecules for both carbon and energy. page 716

Fungi break down their food and then absorb it. page 716

Fungi use the process of growth to find and obtain food in their environment. page 716

Most fungi have numerous branched filaments called hyphae, which absorb nutrients. page 716

When fungi encounter a rich source of food, their hyphae form a network called a mycelium. page 716

Fungal cell walls are made of chitin, the same compound found in the exoskeletons of insects. page 716

In early fungal groups, the hyphae form a continuous compartment, with many nuclei but with no cell walls, but in later evolving groups, nuclear divisions are accompanied by the formation of septa that divide the cytoplasm into separate cells. page 717

Some fungi, such as yeasts, do not produce hyphae. page 717

Most fungi feed on dead organic matter. page 718

Fungi are critical elements of the carbon cycle, converting dead organic matter back into carbon dioxide and water. page 718

Many fungi feed on living animals and plants, causing diseases. page 718

Fungi have repeatedly evolved mutually beneficial relationships with plants and animals. page 720

Lichens are stable associations between a fungus and a photosynthetic microorganism that look, function, and even reproduce as single organisms page 720

34.2 Fungi reproduce both sexually and asexually, and disperse by spores.

Fungi produce haploid spores that are dispersed by wind, water, or animals. page 722

Spores can be produced asexually (by mitosis) or sexually (by cell fusion and meiosis). page 722

During sexual reproduction, many fungi produce spores within multicellular fruiting bodies that enhance spore dispersal. page 723

The fungal life cycle is similar to haploid-dominant life cycles found in many eukaryotic organisms, except that plasmogamy (cell fusion) and karyogamy (nuclear fusion) are often separated, resulting in a stage (called a heterokaryotic stage) in which there are genetically distinct haploid nuclei in one cell. page 724

Genetically determined mating types prevent self-fertilization in many fungi. page 725

Asexual fungal species have mechanisms for generating genetic diversity; these are thought to involve fusion of haploid nuclei, followed by crossing over during mitosis and chromosome loss. page 726

34.3 Other than animals, fungi are the most diverse group of eukaryotic organisms.

About 75,000 fungal species have been formally described, but total diversity may be as high as 5 million species. page 726

Chytrids are the first diverging groups of fungi. They occur in aquatic or moist habitats and only about 1000 species have been identified. page 727

Zygomycetes produce hyphae without septa and make up less than 1% of known fungal species. page 728

The glomeromycetes are a group of fungi of low diversity but tremendous ecological importance because of their association with plant roots. page 728

Most fungi belong to the Dikarya (ascomycetes and basidiomycetes), which form septa along hyphae. During the heterokaryotic stage, these fungi produce cells that contain two haploid nuclei, one from each parent. page 728

Ascomycetes include wood-rotting fungi, many ectomychorrizal species, plant and animal pathogens, the fungal partner in most lichens, and baker's and brewer's yeasts. page 729

Basidiomycetes include familiar toadstools and puffballs, as well as plant pathogens page 732

Self-Assessment

1. Describe two ways in which fungi differ from other heterotrophic organisms in how they obtain and digest their food.
2. Explain how hyphae and cell walls made of chitin allow fungi to obtain nutrients from their environment.
3. Describe how fungi contribute to the terrestrial carbon cycle.
4. Name the two organisms that make up a lichen and describe how each partner benefits from the association.
5. Describe how fungi disperse.
6. Draw the life cycle of an ascomycete, and indicate the heterokaryotic stage.
7. Draw the life cycle of a basidiomycete, and indicate the heterokaryotic stage.
8. Name and describe several key innovations in the evolutionary history of fungi that allowed them to move from water to land.
9. Explain how the evolution of vascular plants has provided opportunities for fungal diversification.

CASE 7

Predator–Prey

A Game of Life and Death

In his 1850 poem *In Memoriam A.H.H.*, Alfred, Lord Tennyson wrote of "Nature, red in tooth and claw." The phrase has endured as a powerful description of the brutality of wild nature. Thanks to nature documentaries and our human fascination with wild animals, most of us have witnessed plenty of images of this red-toothed nature: a lion chasing down a zebra, a hawk scooping a mouse into its talons, a shark sinking its teeth into a sea lion.

Predator–prey interactions are important in every ecosystem on Earth. After all, every living animal can be classified as either "predator" or "prey"—and often, both labels apply. To a fly, a toad is a fearsome predator. To a snake, that same toad may be choice prey.

Very few species enjoy the luxury of having no natural predators; these are called top predators. For the vast majority of animals, the threat of predation is simply a fact of life. Not surprisingly, that threat has been a powerful evolutionary force. Predator–prey dynamics have influenced both the evolution of individual organisms and the shape of the ecosystems in which they live.

On Isle Royale, an island in Lake Superior, ecologists have been studying moose and wolves since 1958 in the longest-running study of a single predator–prey system in the world. The moose arrived on the island about 100 years ago, presumably by swimming about 15 miles from the nearest shoreline. They arrived to find a predator-free paradise, and the moose population quickly exploded. Then, about 1950, a pair of wolves crossed an ice bridge to Isle Royale. As the wolf population grew, the moose population declined, until a delicate balance was established.

As a predator population grows, the prey population shrinks—a seemingly logical relationship. But predator–prey interactions are complex and sometimes even counter-intuitive. Removing a top predator can actually reduce biodiversity. Sea stars, for instance, prey on mussels in the rocky intertidal zone of the Pacific Northwest. When sea star numbers fall, mussel numbers increase. The mussels crowd out other species, such as barnacles and seaweed, which compete for space on the rocks. The result is a drop in the overall number of species in that habitat.

The same patterns have been found among birds of prey in the Italian Alps. These raptors—including the goshawk and four types of owl—are the top predators of their food chains. Spanish researchers found that locations where the birds were present had a greater diversity of trees, birds, and butterflies than did comparable control sites.

Predator–prey dynamics have influenced both the evolution of individual organisms and the shape of the ecosystems in which they live.

One reason that the predator and prey populations eventually achieve a delicate balance is that both are well adapted to their roles. Predation has exerted powerful evolutionary pressure that has influenced anatomy and physiology over the long term. That pressure has shaped predators by giving them claws, teeth, venom, and powerful muscles for hunting prey.

Of course, predation pressure has also shaped the form and function of prey. Some animals, including some insects and frogs, have evolved toxins to deter would-be predators. These toxic species often exhibit warning colors that tell the predators to steer clear. Other prey animals have evolved camouflage colors and forms to help them hide from hungry carnivores. Still others have developed protective behaviors, such as living together in herds for security against predation.

Often, predator and prey evolve in lockstep, driving each other's adaptations. Clams may have evolved thicker shells to protect themselves from hungry crabs. In turn, the crabs evolved larger, stronger claws for cracking clamshells.

Wolves and moose on Isle Royale. The populations of this predator and prey have been studied for more than 50 years, and the two have achieved a delicate balance over time. *Sources: (wolf) Samuel R. Maglione/Science Source; (moose) Ron Erwin/age fotostock; (graph) Data from J. A. Vucetich and R. O. Peterson, 2012, "The population biology of Isle Royale wolves and moose: an overview," www.isleroyalewolf.org.*

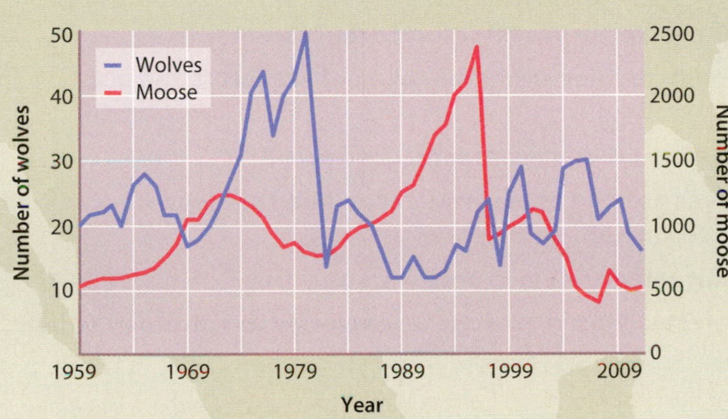

This pattern of back-and-forth change is often described as an evolutionary arms race. And it has happened time and time again.

The interactions between eaters and eaten have also influenced the evolution of physiological systems. The skills that a predator needs to hunt its prey and that its prey needs to escape depend in large part on adaptations of their sensory systems, musculoskeletal systems, nervous systems, and even their circulatory and respiratory systems.

Consider sensory systems. Animals rely on visual, auditory, tactile, and chemical stimuli to warn them of approaching predators—or to guide them to suitable prey. A bat relies on sonar to locate insects. A spider responds to the flutter of silk when an insect becomes ensnared in its web. Gazelles are always attentive, watching and listening for signs of a cheetah or lion in the grass.

When predators are nearby, prey species experience fear and anxiety. Stress hormones produced by prey animals can influence an animal's physiology in a number of ways. In snowshoe hares, levels of the stress hormone cortisol increase when predators such as lynx and coyote are plentiful. The hormones trigger behaviors, such as alertness and fearfulness, which help the hares avoid becoming a lynx's lunch. But the behavioral benefit comes at a cost: Research has shown that stressed hares give birth to fewer and smaller offspring.

In short, predation has left a physical imprint on both predator and prey, from nose to tail—their body shapes, their behaviors, their physiology, even their muscle fibers have been influenced over time by predator–prey interactions.

In the modern world, our own species, *Homo sapiens,* occupies a unique position at the top of the food web. While humans are occasionally killed by animals such as bears or sharks, we are, for the most part, no longer constrained by the fear of predation by other species. But it wasn't always so. Some scientists have proposed that fear of predation among early humans led to the evolution of cooperative social behavior and large brains. Others have theorized that it was our hunting of other animals that led us to evolve big brains and the ability to work together. Perhaps both factors played a role. Either way, the importance of the predator–prey system can't be ignored. To understand our roots, it seems, we must take a good look at nature, red in tooth and claw.

? CASE 7 QUESTIONS

Special sections in Chapters 35–41 discuss the following questions related to Case 7.

1. **What body features arose as adaptations for successful predation?** See page 742.
2. **How have sensory systems evolved in predators and prey?** See page 771.
3. **How do different types of muscle fiber affect the speed of predators and prey?** See page 797.
4. **How does the endocrine system influence predators and prey?** See page 821.
5. **How do hormones and nerves provide homeostatic regulation of blood flow as well as allow an animal to respond to stress?** See page 843.
6. **Does body temperature limit activity level in predators and prey?** See page 854.
7. **Can the loss of water and electrolytes in exercise be exploited as a strategy to hunt prey?** See page 880.

CHAPTER 35

Animal Nervous Systems

Core Concepts

35.1 Animal nervous systems allow organisms to sense and respond to the environment, coordinate movement, and regulate internal functions of the body.

35.2 The basic functional unit of the nervous system is the neuron, which has dendrites that receive information and axons that transmit information.

35.3 The electrical properties of neurons allow them to communicate rapidly with one another.

35.4 Animal nervous systems can be organized into central and peripheral components.

Gazelle grazing on an African plain are alert to the threat of predators nearby. Members of the herd scan the distant landscape for visual cues that might reveal the stealthy approach of a cheetah (**Fig. 35.1**). Olfactory (smell) and auditory (sound) cues from the environment can also warn of the cheetah's approach. Herd members communicate the presence of the cheetah to one another by the directed attention of a dominant male or the pawing of a hoof. At what point is the threat real? When does each animal decide to stop grazing and turn to run? What direction should the escape take to keep the herd together as a group?

The cheetah scans the horizon, its keen vision spotting the grazing gazelle. The cheetah's ability to recognize wind direction enables it to move downwind of the gazelle to minimize auditory and olfactory cues that might otherwise be carried by the air toward suspecting prey. At what point does the cheetah decide to sprint in pursuit of its quick-running prey?

FIG. 35.1 Predator and prey. Interactions of predator and prey, such as this cheetah and gazelle, are mediated by the nervous system. *Source: Winfried Wisniewski/Corbis.*

The perceptions and reactions of these two animals depend on the activity of their nervous systems. In this and the following chapter, we explore the organization of the nervous system, including the brain and the nerve cells that compose it. The nervous system has the remarkable ability to process all kinds of information. It takes in information from the environment through the animal's sensory organs and processes that information through the activity of networks of nerve cells. The results of that processing guide an animal's motor responses. Nervous system processes underlie the perceptions and interaction of cheetah and gazelle as predator and prey, as well as the complex emotional and cognitive responses triggered, for example, by seeing a friend or relative after a long separation.

35.1 NERVOUS SYSTEM FUNCTION AND EVOLUTION

Most multicellular animals possess a **nervous system,** a network of many interconnected nerve cells. This network allows an animal to sense and respond to the environment, coordinate the action of muscles, and control the internal function of its body. Nerve cells, or **neurons,** are the basic functional units of nervous systems.

Nervous systems first evolved in simple animals. These early nervous systems could sense basic features of the environment, such as light, temperature, chemical odors, and physical forces such as touch or pressure. Guided by these cues, early nervous systems could trigger movements useful for obtaining food, finding a mate, or choosing a suitable habitat. As animals evolved more complex bodies with increasingly specialized organ systems for respiration, circulation, digestion, and reproduction, their nervous systems also became more complex. These more complex nervous systems could regulate internal body functions in response to sensory cues received from the environment. They also made possible more sophisticated behaviors that relied on improved decision making, learning, and memory. These abilities, possessed by many invertebrate and vertebrate animals, increased their fitness by improving their ability to survive, find and select a mate, and reproduce.

Animal nervous systems have three types of nerve cell.

All animal nervous systems are made up of three types of nerve cell: **sensory neurons, interneurons,** and **motor neurons.** Each type of neuron has a different function. Sensory neurons receive and transmit information about an animal's environment or its internal physiological state. These neurons respond to physical features such as temperature, light, and touch or to chemical signals such as odor and taste. Interneurons process the information received by sensory neurons and transmit it to different body regions, communicating with motor neurons at the end of the pathway to produce suitable responses. For example, a motor neuron may stimulate a muscle to contract to produce movement. Other motor neurons may adjust an animal's internal physiology, constricting blood vessels to adjust blood flow or causing wavelike contractions of the gut to aid digestion. As a result, nervous system function is fundamental to **homeostasis,** the ability of animals, organs, and cells to actively regulate and maintain a stable internal state (Chapter 5).

Most nerve cells have fiberlike extensions, some of which receive information and others that transit information. The result is a network of interconnected nerve cells that form a circuit. These circuits allow information to be received, processed, and delivered. In most animals, interneurons form specialized circuits that may consist of a diverse array of nerve cells having different shapes, sizes, and chemical properties.

The ability to process information first evolved with the formation of **ganglia** (singular, **ganglion**). Ganglia are groups of nerve cell bodies that process sensory information received from a local, nearby region, resulting in a signal to motor neurons that control some physiological function of the animal. They can be thought of as relay stations or processing points in nerve cell circuits. Eventually, the centralized concentration of neurons that we call a **brain** evolved, and this organ was able to process increasingly complex sensory stimuli received from the environment or anywhere in the body. Neurons in the brain form complex circuits, allowing the brain to regulate a broad array of behaviors, as well as to learn and retain memories of past experiences.

Nervous systems range from simple to complex.

Even single-celled animals have cell-surface receptors that enable them to sense and respond to their environment. By contrast, most multicellular animals have a nervous system. In general, the organization and complexity of an animal's nervous system reflects its lifestyle. Sessile, or immobile, animals, like sea anemones and clams, have relatively simple sense organs and nervous systems. In contrast, active animals such as arthropods (which include insects, spiders, and crustaceans), squid, and vertebrates have more sophisticated sense organs linked to a brain and more complex nervous systems. The nervous systems of these animals guide homeostatic regulation of their bodies and coordinate more complex behaviors that require decision making, forming memories, and learning.

Sponges are the only multicellular animals that lack a nervous system (**Fig. 35.2**). Sponges are marine animals that you may recognize from aquariums or photographs of coral reefs. They have tissues and organs, but lack any form of body symmetry. Many small pores along the side and a larger opening at the top draw nutrient-laden water into a central cavity. The cells lining the cavity capture food particles and dissolved organic matter by endocytosis, then metabolize these nutrients in the cell interior. The functions of different body regions of sponges are coordinated in the absence of nerve cells or a nervous system. Instead, groups of cells specialized for functions such as contracting or taking in water respond to local chemical and physical cues. For example, a pore-like cell that allows water to enter the sponge closes when touched.

Among multicellular organisms, the simplest nervous systems are found in cnidarians, which are radially symmetric animals such as jellyfish and sea anemones (Fig. 35.2). At one end of a cnidarian's body is a mouth, surrounded by tentacles, that opens into a central body cavity where food is digested. The nervous system of these animals has relatively few neurons arranged like a net, but no ganglia or central brain (**Fig. 35.3a**). The nerve net of a sea anemone is organized around its body cavity. Sensory neurons sensitive to touch signal when one of the anemone's tentacles has captured food prey, and then motor neurons react to coordinate the tentacle's movement toward the mouth for feeding. Sensory neurons responding to a noxious stimulus signal the sea anemone to retract or detach from its substrate.

FIG. 35.2 A simplified phylogenetic tree of multicellular animals. All multicellular organisms, except sponges, have a nervous system. *Photo sources: (top) Andrew J. Martinez/Science Source; (middle) UMI NO KAZE/a.collectionRF/Getty Images; (bottom) Christof Wermter/age fotostock.*

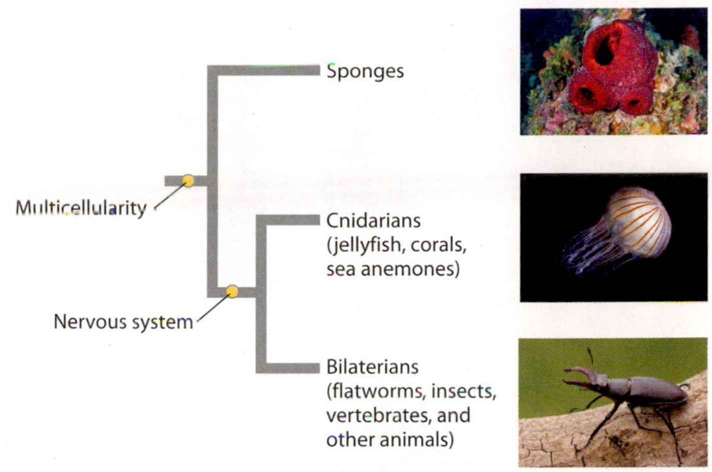

More complex nerve connections develop in the head end of flatworms, segmented annelid worms, insects, and cephalopod mollusks, such as squid and octopus. These animals are all bilaterians, animals like us with symmetrical right and left sides and a distinct front and back end (see Fig. 35.2; Chapter 44). Flatworms are representatives of one of the earliest branching groups of bilaterians, and their bodies are well adapted for forward locomotion. Flatworms are elongated, flattened organisms that glide along a streambed, trapping small animals under their bodies and sucking them in through a muscular opening in their underside. Sensory neurons are most numerous in the animal's front end, so the flatworm can perceive the environment ahead of it as it moves forward. The flatworm's nervous system can compare inputs from different sensory neurons and then send signals to the appropriate muscles to move toward a target. Paired ganglia (one on each side of the animal) in the head end of flatworms handle this processing task (**Fig. 35.3b**).

In animals with an organized nervous system, information is transmitted to different regions along the length of the animal's body by distinct **nerve cords** and **nerves.** These are bundles of the long fiberlike extensions from multiple nerve cells.

FIG. 35.3 Animal nervous systems. (a) Sea anemones have a nerve net; (b) flatworms have paired ganglia; (c) earthworms, (d) insects, (e) squid, and (f) frogs have a brain.

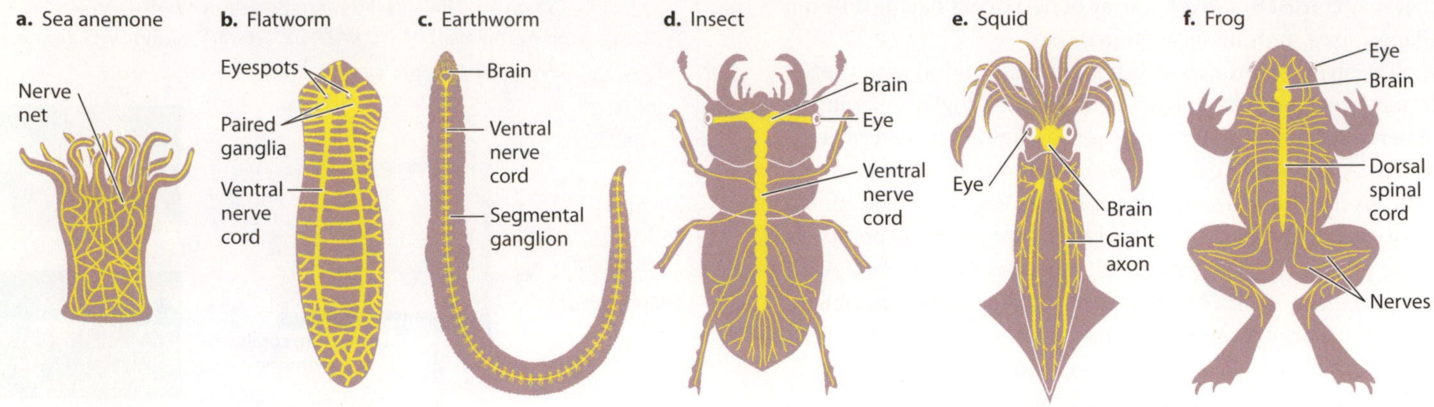

In earthworms, paired ganglia, each made up of many interneurons, are located in each of the body's segments (**Fig. 35.3c**). These segmental ganglia help to regulate the muscles within each body segment that move the earthworm's body. Paired ganglia near the head regulate the activity of motor neurons that control movements of the mouth for feeding. Ganglia serve to regulate key processes in local regions and organs of the animal's body. For example, local ganglia assist in processing information from the eyes (Chapter 36) or control the digestive state of an animal's gut (Chapter 40). The homeostatic regulation of internal physiological processes is a fundamental role of nervous systems.

Centralized collections of neurons forming a brain are present in earthworms and other annelid worms (Fig. 35.3c), insects (**Fig. 35.3d**), and mollusks (**Fig. 35.3e**). Sense organs also developed in the head region, from eyespots in flatworms that respond to light to more sophisticated image-forming eyes in octopus and squid. The evolution of a brain in these invertebrate animals enabled them to learn and perform complex behaviors. Many of these nervous system capabilities are shared with vertebrate animals.

In common with certain invertebrates, such as arthropods and cephalopod mollusks, the evolution of a brain is a key feature of vertebrate nervous systems (**Fig. 35.3f**). It enabled these organisms to evolve complex behaviors that rely on learning and memory and, in certain vertebrates, the ability to reason. At the same time, animal brains became increasingly complex in the number of nerve cells they contain (fruit fly, about 0.25 million; cockroach, about 1 million; mouse, about 75 million; humans, about 100 billion), in the number of connections between neurons, and in the variety of shapes and sizes of nerve cells and specialized connections between them.

Although animal nervous systems differ in organization and complexity, their nerve cells have fundamentally similar molecular and cellular features and use fundamentally the same mechanisms to communicate with neighboring cells. These mechanisms have been retained over the course of evolution to code and reliably transmit information. The ability to receive sensory information from the environment, and transmit and process this information within a nervous system, is therefore shared across a wide diversity of animal life, contributing to the success of multicellular animals.

? CASE 7 PREDATOR–PREY: A GAME OF LIFE AND DEATH

What body features arose as adaptations for successful predation?

A notable feature of most animal nervous systems is that nervous system tissue, including specialized sense organs such as eyes, becomes concentrated at one end of the body. For example, the paired ganglia and eyespot of the flatworm are located at one end of its body, as are the brain and sense organs of the earthworm, squid, and insect, as shown in Fig. 35.3. The concentration of nervous system components at one end of the body, defined as the "front," is referred to as **cephalization.** Cephalization is a key feature of the body plan of most multicellular animals, including all vertebrates.

Cephalization evolved independently multiple times in different animal groups and is therefore thought to confer certain advantages. Cephalization is thought be an adaptation for forward locomotion because it allows animals to take in sensory information from the environment ahead of them as they move forward. In addition, the nearness of the sensory organs to central ganglia or the brain makes it possible to process this information quickly to enable a suitable behavioral response. As the quality and amount of sensory information taken in increased, brain size and complexity increased. Cephalization is also considered to be an adaptation for predation, allowing animals to better detect and capture prey.

Although cephalization is a feature of many animals, it has been particularly well studied in vertebrates. In vertebrates, the brain, many sense organs, and mouth are all located in the head. Vertebrates also evolved several novel features, including a jaw, teeth, and tongue. These are all thought to be adaptations for predation, or more generally the acquisition and processing of food.

As a result of evolutionary selection for enhanced sensory perception and the ability to respond to important cues in the environment, the brain, sensory organs, and nervous system of many animals are complex in their organization. These are linked to more sophisticated abilities that allow for a broad range of behaviors. These abilities are critical to the success of both predators and prey and underlie the complex interactions that occur among members of a species when they mate, reproduce and disperse, and care for their young.

35.2 NEURON STRUCTURE

We have seen that nerve cells can be classified as sensory neurons, interneurons, or motor neurons. Yet all three receive and transmit information, and all three share the same basic organization. Neurons have extensions that receive information and relay it to the cell body. They also have other extensions that transmit information away from the cell body. Although all neurons share this basic plan, they differ remarkably in size and number of extensions, according to their specific function.

Neurons share a common organization.

Neurons all share some basic features. These include a cell body from which emerge two kinds of fiberlike extension, **dendrites** and **axons** (**Fig. 35.4**). These extensions are the input and output ends of the nerve cell. Both types of cellular extension can be highly branched, enabling neurons to communicate over large distances with many other cells. A neuron's dendrites receive signals from other nerve cells or, in the case of sensory nerves, from specialized sensory endings. These signals travel along the dendrites to the neuron's cell body. At the junction of the cell body and its axon, the **axon hillock,** the signals are summed. If the sum of the signals is high enough, the neuron fires an **action potential,** or nerve impulse, that travels down the axon. An action potential is a brief electrical signal transmitted from the cell body along one or more axon branches.

Axons generally transmit signals away from the nerve's cell body. The end of each axon forms a swelling called the axon terminal. An axon terminal communicates with a neighboring cell through a junction called a **synapse.** A space, the **synaptic cleft,** separates the end of the axon of the presynaptic cell and the neighboring postsynaptic cell. The synaptic cleft is commonly only about 10 to 20 nm wide.

How does a signal cross the synaptic cleft? Molecules called **neurotransmitters** convey the signal from the end of the axon to the postsynaptic target cell. The arrival of a nerve signal at the axon terminal triggers the release of neurotransmitter molecules from vesicles located in the terminal. The vesicles fuse with the axon's membrane, releasing neurotransmitter molecules into the synaptic cleft. The molecules diffuse across the synapse and bind to receptors on the plasma membrane of the target cell. The binding of neurotransmitters to these receptors causes a change in the electrical charge across the membrane of the receiving postsynaptic cell, continuing the signal in the second cell. Most neurons communicate by the hundreds to thousands of synapses that they form with other cells. The massive number of interconnections enables the formation of complex nerve cell circuits and the transmission of information from one circuit to others in the nervous system.

FIG. 35.4 Neuron organization. Nerve cells typically have dendrites that receive signals and axons that transmit signals.

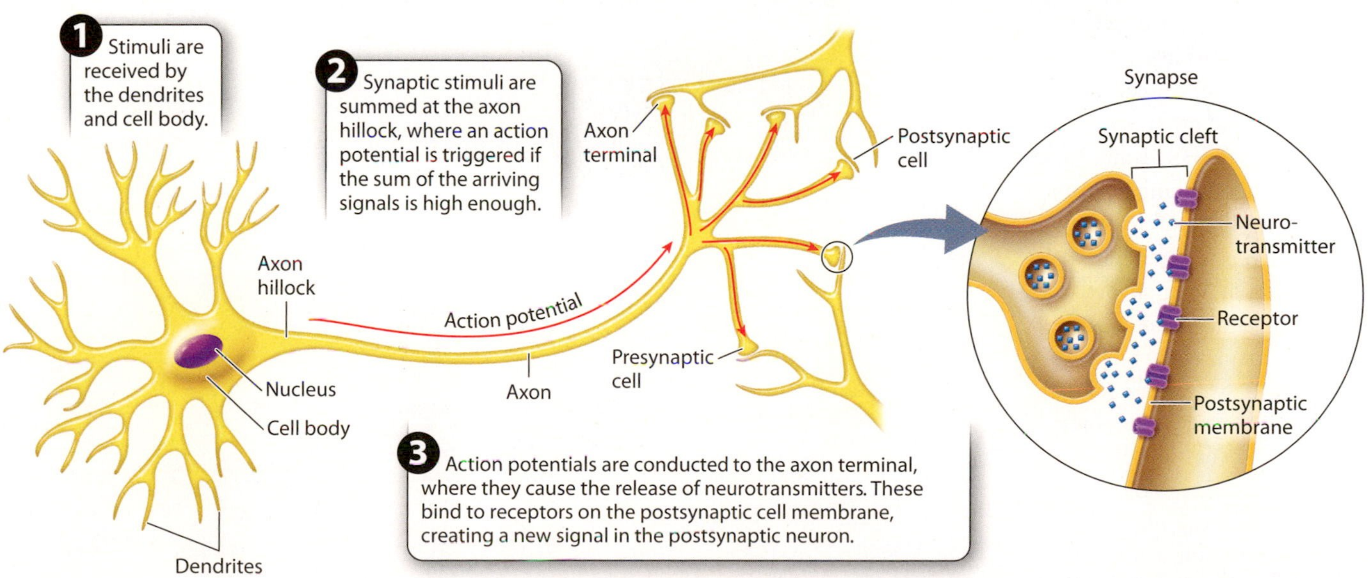

FIG. 35.5 Variation in neuron shape. Neurons display a diversity of shapes, reflecting their different functions.

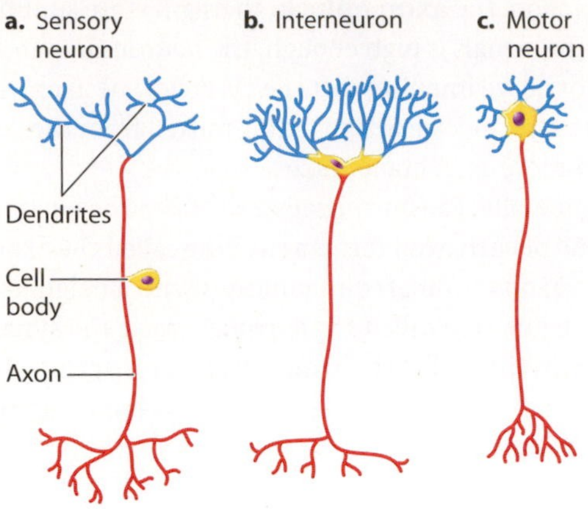

Some neurons do not synapse with other neurons, but instead with other types of cell that produce some physiological response in the animal. Examples are muscle cells, which contract in response to nerve signals, and secretory cells in glands, which release hormones in response to nerve signals.

→ **Quick Check 1** How are signals transmitted from one end of a nerve cell to the other, and from one neuron to another?

Neurons differ in size and shape.

In spite of their common organization, neurons show a remarkable diversity in size and shape. Some neurons, such as interneurons, are very short, whereas others, like the motor neuron that extends from your spinal cord to your big toe, are several feet long. The number of extensions coming off the cell body can also vary in different neurons (**Fig. 35.5**). Some sensory neurons involved in smell, sight, and taste have just two extensions coming off the cell body (Fig. 35.5a), whereas other neurons have many more (Fig. 35.5b and 35.5c).

The cell bodies of some neurons have distinctive shapes. For example, pyramidal neurons, as their name suggests, have triangular-shaped cell bodies (**Fig. 35.6**). These cells, located in the mammalian brain, were elegantly drawn by the Spanish neuroanatomist Santiago Ramón y Cajal, who shared the 1906 Nobel Prize in Medicine or Physiology with the Italian scientist Camillo Golgi "in recognition of their work on the structure of the nervous system."

The degree of branching of a neuron's extensions can vary from one neuron to the next. A neuron's dendrites form a finely branched field that receives inputs from many other nerve cells. A neuron with more dendritic branches receives inputs from more sources than one with fewer branches. Some neurons receive information only from nearby regions through a few dendrites that branch close to the cell body. Other neurons receive information from many branching dendrites. Axons typically have fewer branches than dendrites.

Signals from all the dendrites, converging on the neuron cell body, determine the strength, or frequency, and timing of signals carried by the neuron's axon. The process of bringing together information gathered from different sources is referred to as the integration of information. The degree of branching and the number of synapses that a neuron makes with neighboring nerve cells, therefore, reflect how information is processed and integrated by the cell as part of a larger circuit.

Neurons are supported by other types of cell.

Neurons in many body regions, and in particular within the brain, are supported by other types of cell that do not themselves transmit electrical signals. **Glial cells** are a major class of supporting cell. Indeed, the human brain has more glial cells than neurons.

Glial cells have many functions. They surround neurons and provide them with nutrition and physical support. Some glial cells, called **astrocytes,** support endothelial cells that make up blood vessels in the brain. These endothelial cells are linked by tight junctions (Chapter 10) to form a selective barrier, called the blood–brain barrier, which limits the types of compounds that can diffuse

FIG. 35.6 Pyramidal cell. (a) These neurons, drawn by Santiago Ramón y Cajal (1852–1934), show the triangular cell body, highly branched dendrites, and single axon. (b) The triangular shape of the cell body is clear in the photomicrograph at right. *Sources: a. Science Source; b. Bob Jacobs, Ph.D., Laboratory of Quantitative Neuromorphology, Department of Psychology, Colorado College.*

a

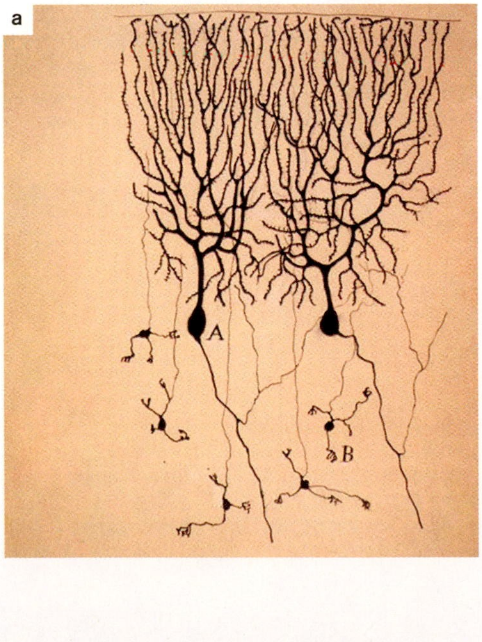

b

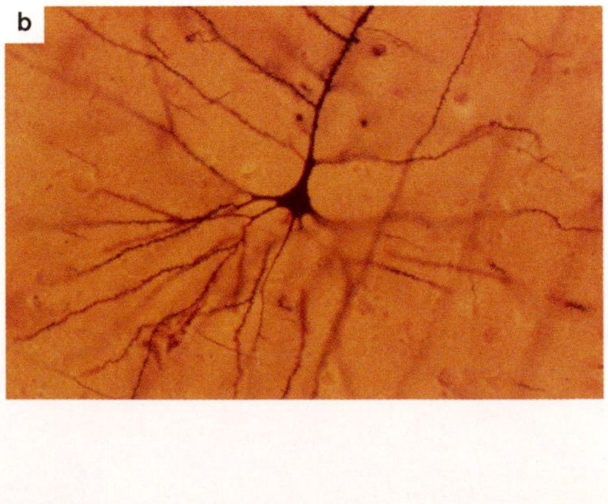

from the blood into the brain. This barrier prevents pathogens and toxic compounds in the blood from entering the brain. Because nerve cells have limited capacity to regenerate after damage, protection of the brain and spinal cord is critically important. Nevertheless, lipid-soluble compounds such as alcohol and certain anesthetics readily diffuse across the blood–brain barrier, and thus they can affect the functioning of the brain and an animal's mental state. They can also damage and even destroy neurons.

During development, glial cells help orient neurons as they develop their connections. They also provide electrical insulation to vertebrate neurons that allows nerve signals to be transmitted rapidly, a topic we discuss next.

35.3 NEURON FUNCTION

We have seen that neurons send signals from one neuron to the next by releasing neurotransmitters that diffuse across the synaptic cleft and bind to receptors. In Chapter 9, we discussed that this is a general mechanism of cell communication: A signaling cell releases a chemical signal that binds to a receptor on a receiving cell. However, once the receptor has bound the neurotransmitter and is activated, in neurons the signal is not propagated by a chemical signal diffusing through the cell, which would be much too slow. Instead, as we have seen, neurons send signals electrically from one end of the cell to the other.

All neurons are electrically excitable cells that transmit information in the form of electrical signals. Nerve signals travel at high speeds, up to 200 m/s (450 mph). Although the speed of nerve signals is fast by biological standards, it is slow by comparison with the speed of electrical signals transmitted by a wire. In this section, we discuss the electrical properties of neurons and how signals are propagated along neurons. We also describe special features of some neurons that enable even faster signal transmission and more rapid responses by animals to environmental cues.

The resting membrane potential is negative and results in part from the movement of potassium ions.

The inside surface of a neuron carries an electrical charge that is due to the presence of charged ions, and the same is true of the outside surface. However, the electrical charge on the inside and the electrical charge on the outside are not the same because unequal numbers of charged ions are located inside and outside the cell. When no signal is present, there are more negative ions inside the cell, and the inside is thus negatively charged relative to the outside.

The charge difference between the inside and the outside of a neuron due to differences in charged ions is the cell's **membrane potential,** measured in volts. Other cells also have membrane potentials, but only nerve cells and muscle cells respond to changes in their membrane potential. Therefore, these cells are considered electrically excitable.

The key to producing an electrical signal in a nerve cell is the movement of positively and negatively charged ions across the cell membrane. This movement creates a change in electrical charge across the membrane that constitutes the electrical signal.

Let's first consider the membrane potential when the neuron is at rest and no signal is being received or sent. Under these conditions, the cell's membrane voltage is negative on its inside relative to its outside (**Fig. 35.7**). The resting membrane potential of the cell is said to be **polarized.** This means that there is a

FIG. 35.7 Neuron resting membrane potential. The resting potential of a neuron is negative and results primarily from the movement of K^+ ions out of the cell.

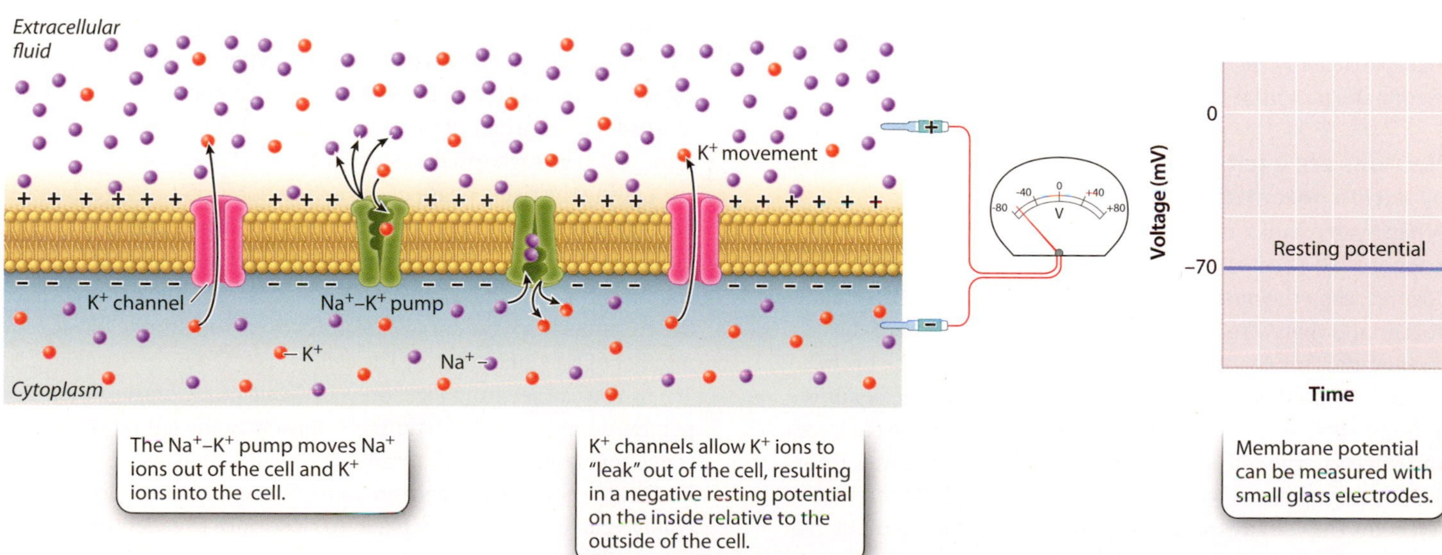

buildup of negatively charged ions on the inside surface of the cell's plasma membrane and of positively charged ions on its outer surface. The negative voltage across the membrane at rest is referred to as the cell's **resting membrane potential.** The resting membrane potential ranges from −40 to −85 millivolts (mV) depending on the type of nerve cell and most commonly is about −65 to −70 mV. The voltage of the cell's interior can be measured with respect to its outside voltage by small glass electrodes on the inside and outside of the cell.

At rest, nerve (and muscle) cells have a greater concentration of sodium (Na^+) ions outside the cell than inside, and a greater concentration of potassium (K^+) ions inside the cell than outside. This distribution of ions results in part from the action of the sodium-potassium pump, discussed in Chapter 5. The sodium-potassium pump uses the energy of ATP to move three Na^+ ions outside the cell for every two K^+ ions moved in (Fig. 35.7). The action of the pump makes the inside of the cell less positive, and therefore more negative, than the outside of the cell.

The exact value of the resting membrane potential, however, depends on the movement of K^+ ions back out of the cell by passive diffusion through potassium ion channels (Fig. 35.7). As discussed in Chapter 5, ion channels are protein pores embedded in the cell membrane. The most important ions that move across the cell membrane are sodium (Na^+), potassium (K^+), chloride (Cl^-), and calcium (Ca^{2+}). When a nerve cell is at rest, more K^+ ion channels are open, giving K^+ greater permeability compared with all other ions. As a result, K^+ ions move out of the cell. The movement of K^+ ions from the inside to the outside causes positive ions to build up on the outside of the cell, whereas negatively charged ions (largely proteins) remain on the inside of the cell, making the inside of the nerve cell more negative than its outside.

The relative proportion of ions does not by itself determine the resting membrane potential. This is because the number of charged ions that build up at the cell's surface is a tiny fraction of the total number of charged ions and proteins located inside and outside the cell. It is the *movement* of K^+ ions relative to other ions, particularly Na^+ ions, that largely determines the resting membrane potential.

Neurons are excitable cells that transmit information by action potentials.

When a nerve cell is excited, its membrane potential becomes less negative—that is, the inside becomes less negative, or more positive, than the outside of the cell. The increase in membrane potential is therefore referred to as a **depolarization** of the membrane. This depolarization does not occur over the entire cell at once. Rather, it starts at the terminal end of the dendrite, in response to neurotransmitter binding to membrane receptors, and travels to the cell body, losing strength along the way. If the depolarization is still strong enough at the axon hillock of the cell body, the cell fires an action potential that carries the signal from the cell body to the terminal ends of the axon.

How is the action potential generated? At the axon hillock, the summed membrane depolarization of the cell's dendrites causes voltage-gated sodium channels to open, allowing Na^+ ions to enter the cell. **Voltage-gated channels** open and close in response to changes in membrane potential (**Fig. 35.8**; Chapter 9). As Fig. 35.8 shows, they play a key role in how neurons fire action potentials.

The influx of Na^+ causes a flow of positive charge, like a current, along the inside of the cell, toward more negatively charged nearby regions. This flow of charge is reduced at longer distances and quickly dissipates so that the membrane potential returns to a resting state unless additional excitatory stimuli further depolarize the membrane.

If the excitatory signal is strong enough to depolarize the membrane of the nerve cell body to a voltage of approximately 15 mV above the resting membrane potential (about −55 mV), the nerve fires an action potential at the axon hillock. An action potential is a rapid, short-lasting rise and fall in membrane potential, as shown in Fig. 35.8. The critical depolarization voltage of −55 mV required for an action potential is the cell's **threshold potential.** Nerve cells in different animals have slightly different resting and threshold potentials, but all operate similarly with respect to firing an action potential. When the threshold potential is exceeded, the nerve fires an action potential in an all-or-nothing fashion. This means that once the threshold potential has been exceeded, the magnitude of the action potential is always the same and is independent of the strength of the stimulating input. At this point, a rapid spike in voltage to approximately +40 mV occurs over a very brief time period of 1 to 2 msec.

An action potential results from dramatic changes in the state of voltage-gated Na^+ and voltage-gated K^+ channels in the axon membrane. As the cell membrane crosses its threshold potential, a large number of voltage-gated Na^+ channels suddenly open, allowing Na^+ ions to rush inside the cell. The rise in voltage that results causes additional voltage-gated Na^+ channels to open. This is an example of **positive feedback,** in which a signal (depolarization) causes a response (open voltage-gated Na^+ channels) that leads to an enhancement of the signal (more depolarization) that leads to an even larger response (more open voltage-gated Na^+ channels).

During the rising phase of the action potential, voltage-gated K^+ channels also begin to open, but these channels are slow to activate relative to the rapidly opening voltage-gated Na^+ channels. As a result, relatively few K^+ ions leave the cell and the cell membrane potential continues to rise due to the large number of Na^+ ions entering the cell.

As you can see from Fig. 35.8, the rising phase of the action potential (rapid depolarization) is followed by the falling phase (rapid repolarization). What causes this sudden reversal of the membrane potential? There are two important factors. First, voltage-gated Na^+ channels begin to close once the membrane potential becomes positive. Second, voltage-gated K^+ channels

FIG. 35.8 An action potential. An action potential results from the opening and closing of voltage-gated ion channels and allows the membrane potential briefly to be positively charged.

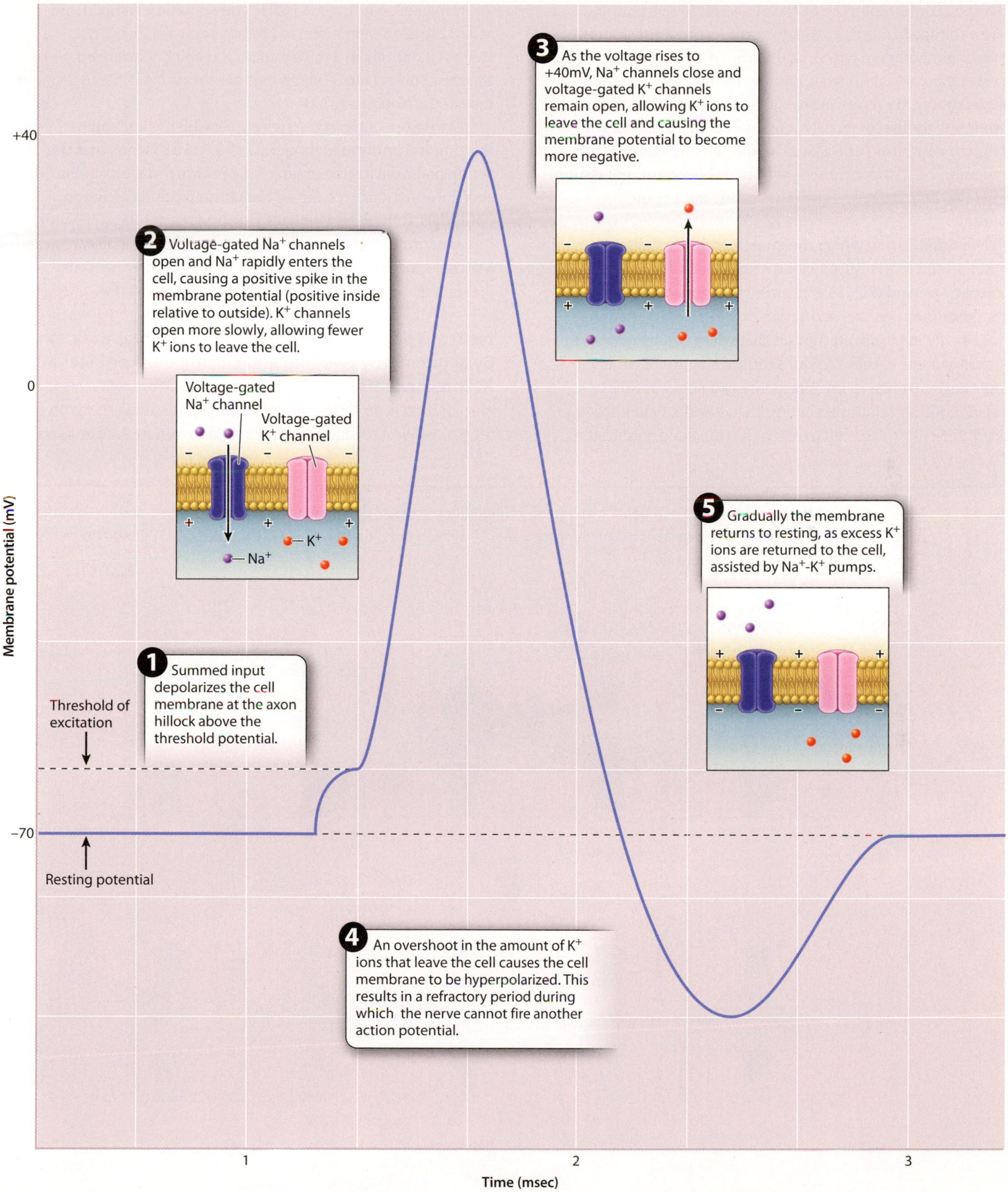

continue to open in response to the change in voltage. Because they are slower to respond than the voltage-gated Na^+ channels, the membrane voltage peaks and then falls as K^+ ions diffuse out of the axon through the open voltage-gated K^+ channels.

The voltage inside the axon does not return immediately to the resting membrane potential. Instead, it briefly falls below the resting potential (in what is known as hyperpolarization or undershoot), then returns to the resting potential after another few milliseconds as K^+ channels close to restore the resting concentration of Na^+ and K^+ ions on either side of the cell membrane. The continuous action of the sodium-potassium pumps also helps to reestablish the resting membrane potential.

The period during which the inner membrane voltage falls below and then returns to the resting potential is the **refractory period** (Fig. 35.8). During the refractory period, a neuron cannot fire a second action potential. The refractory period results in part from the fact that when voltage-gated Na^+ channels close, they require a certain amount of time before they will open again in response to a new wave of depolarization. In addition, open voltage-gated K^+ channels make it difficult for the cell to reach the threshold potential. The duration of the refractory period varies for different kinds of nerve cell but limits a nerve cell's fastest firing frequency to less than 200 times per second. Most nerve cells fire at lower rates.

It might seem that many Na^+ and K^+ ions need to diffuse across the axon membrane to produce an action potential. However, only about 1 in 10 million Na^+ and K^+ ions need to cross the membrane. Thus, enough ions are always available to generate repeated action potentials.

The nature of the action potential, which is both all-or-nothing and uniform in its electrical features, means that the action potential itself contains little information. Most neurons code information by changing the rate and timing of action potentials. Generally, a higher firing frequency codes for a more intense stimulus, such as a brighter light or louder noise, or a stronger signal transmitted by the nerve cell to other cells it contacts.

Neurons propagate action potentials along their axons by sequentially opening and closing adjacent Na^+ and K^+ ion channels.

Once initiated, an action potential propagates along the axon (**Fig. 35.9**). The local membrane depolarization initiated at the axon

FIG. 35.9 Propagation of action potentials. Local depolarization of the membrane triggers the opening of nearby voltage-gated Na^+ channels, producing an action potential that spreads along the membrane. The action potential moves in only one direction (shown here as left to right) because of the refractory period.

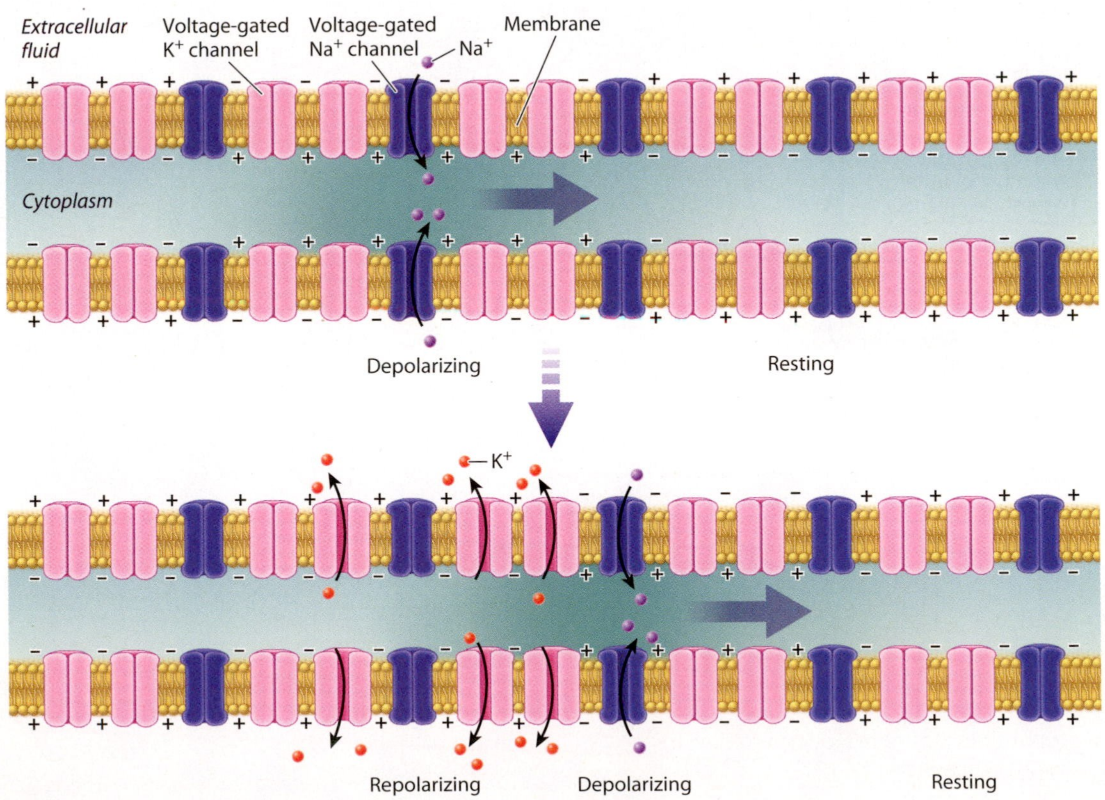

hillock triggers the opening of neighboring voltage-gated Na^+ channels farther along the axon. The inward sodium current depolarizes the membrane above threshold, triggering the opening of nearby voltage-gated Na^+ channels still farther along the axon. By this means, the depolarization—the location of the action potential spike—spreads down the axon. Neighboring voltage-gated K^+ channels subsequently also open and close to reestablish a resting membrane potential after an action potential has fired. Action potentials are thus **self-propagating.** Action potentials propagate only in one direction, normally from the cell body at the axon hillock to the axon terminal. The refractory period following an action potential prevents the membrane from reaching threshold and firing an action potential in the reverse direction.

The conduction speed of action potentials is limited by the neuron's membrane properties and the width of its axon. Yet fast transmission speeds are essential for animals to respond quickly to stimuli. For example, when you mistakenly touch something very hot, your fingers sense and relay this information rapidly to ensure that you withdraw your hand quickly and avoid worse injury.

The neurons of vertebrate animals have an adaptation that lets them respond rapidly to stimuli in their environment. Glial cells form lipid-rich layers or sheaths called **myelin** that wrap around the axons of these neurons (**Fig. 35.10**). Myelin gives many nerves their glistening white appearance. The myelin sheath insulates the axon's membrane, spreading the charge from a local action potential over a much greater distance along the axon's length than would be possible in the absence of myelin.

FIG. 35.10 Saltatory conduction. Myelination insulates axons, enabling faster conduction of action potentials. *Photo source: Don W. Fawcett/Science Source.*

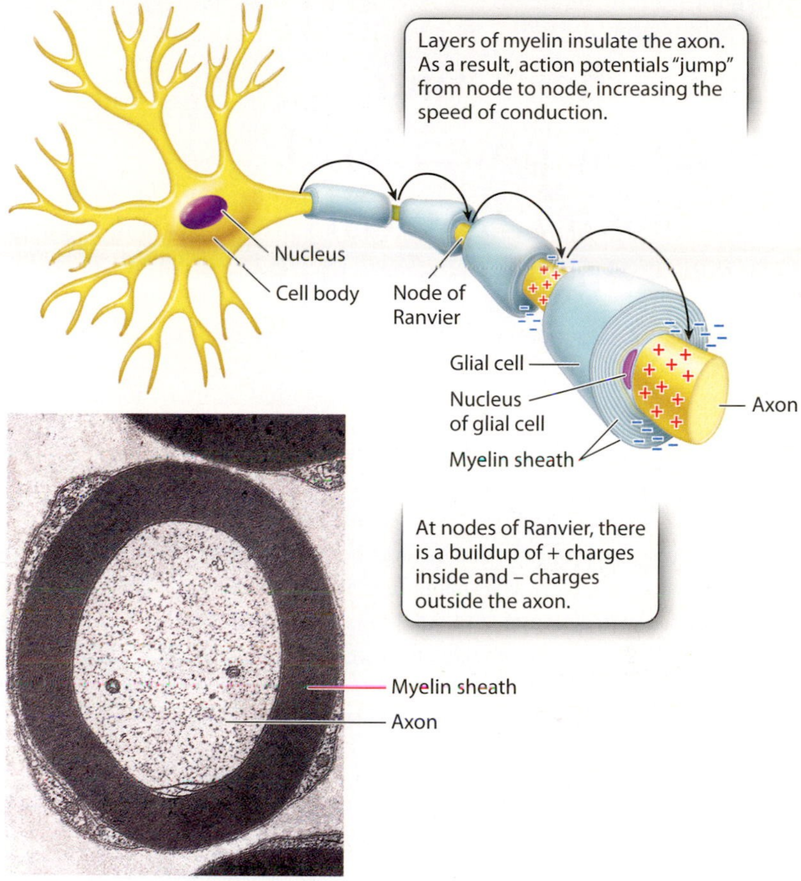

At regular intervals, the axon membrane is exposed at sites called **nodes of Ranvier** that lie between adjacent segments wrapped with myelin (Fig. 35.10). Voltage-gated Na^+ and K^+ channels are concentrated at these nodes. As a result, rather than being conducted in a continuous fashion, as is the case for unmyelinated axons, the action potentials in myelinated axons "jump" from node to node. This **saltatory propagation** (from the Latin *saltus*, "jump") of action potentials by myelinated axons greatly increases the speed of signal transmission.

→ **Quick Check 2** Multiple sclerosis is a disease in which the immune system attacks and destroys the myelin sheath surrounding nerve cells. What effect would you expect the loss of myelin to have on the speed of nerve impulses?

The British neurophysiologists Alan Hodgkin and Andrew Huxley first worked out the electrical properties of the axon plasma membrane by studying the giant axon in squid that stimulates contractions of their body muscles for swimming (**Fig. 35.11**). For this work, they shared the 1963 Nobel Prize in Physiology or Medicine along with John Eccles, who worked out the membrane properties of synapses, which we explore next.

Neurons communicate at synapses.

We have seen that nerve cells communicate at specialized junctions called synapses. There are two types of synapse, electrical and chemical. Electrical synapses provide direct electrical communication through gap junctions that form between neighboring cells (Chapters 10 and 28). They enable rapid communication but limit the ability to process and integrate information. Electrical synapses are found in the giant axons of squid, as well as in large motor nerves in fish used for escape swimming. Electrical synapses can also be found in the mammalian brain. They likely help to speed up information processing, but their role in the brain has not been well studied.

HOW DO WE KNOW?

FIG. 35.11

What is the resting membrane potential? How does electrical activity change during an action potential?

BACKGROUND In order to record the voltage of the inside of a nerve cell relative to the outside, a small electrical recording device called a microelectrode can be inserted into a neuron. The technique is more easily performed on large cells, such as the squid giant axon. This axon, as its name suggests, is quite large, measuring 0.5 mm in diameter (Fig. 35.11a). The large diameter of the axon allows electrical signals to be propagated to the muscles very quickly. Vertebrate neurons are much smaller; they rely on myelin sheaths rather than large size for rapid conduction of electrical signals.

EXPERIMENT In 1939, two British neurophysiologists, Alan Hodgkin and Andrew Huxley, inserted a microelectrode into a squid giant axon and placed a reference electrode on the outside (Fig. 35.11b). They then used a separate set of electrodes (not shown) to depolarize the cell to threshold, triggering an action potential.

RESULTS Fig. 35.11c is a trace from Hodgkin and Huxley's 1939 paper showing the resting membrane potential and the course of an action potential recorded by the electrode inside a giant squid axon. Note that the resting potential (–45 mV in squid) is negative on the inside of the axon relative to the outside, and that the action potential is a rapid spike in potential, with the inside of the cell quickly becoming positive, then negative again. The large size of the spike was a surprise.

FOLLOW-UP WORK This work was performed before anyone had identified the ion channels responsible for the changes in current across the membrane. Subsequent work focused on identifying these channels and their roles in generating the action potential.

a. Squid, *Loligo forbesi*

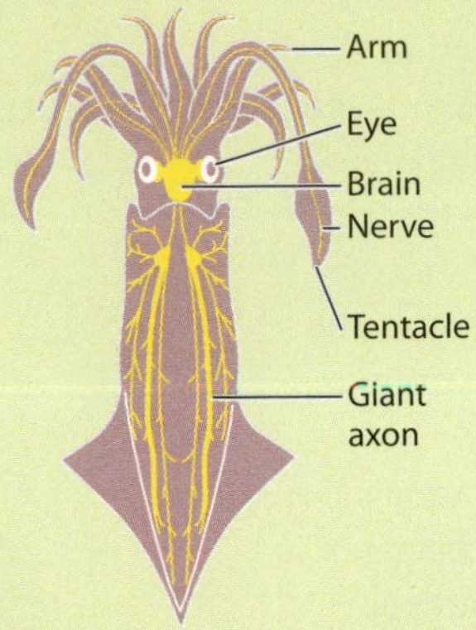

b.

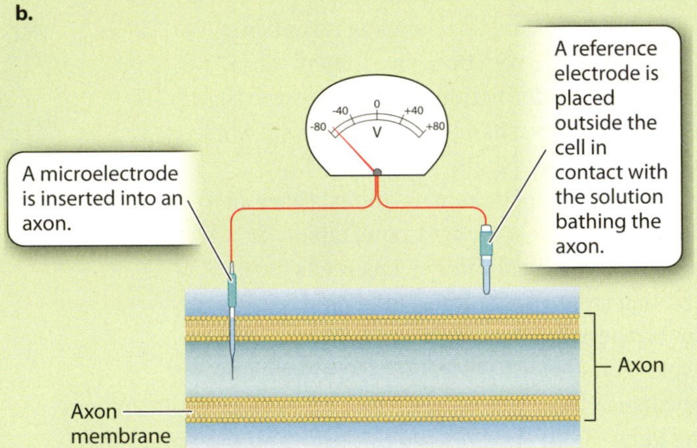

c.

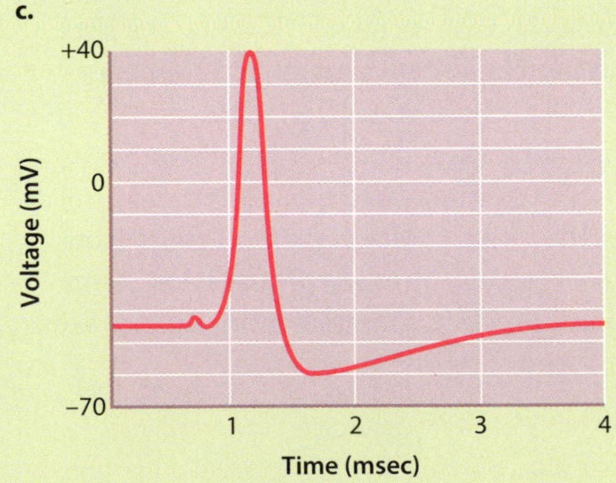

SOURCES Hodgkin, A. L., and A. F. Huxley. 1939. "Action Potentials Recorded from Inside a Nerve Fibre." *Nature* 144:710–711; R. D. Keynes. 1989. "The Role of Giant Axons in Studies of the Nerve Impulse." *Bioessays* 10:90–94.

FIG. 35.12 A chemical synapse. Neurons communicate with other neurons or with muscle cells by releasing neurotransmitters at a synapse.

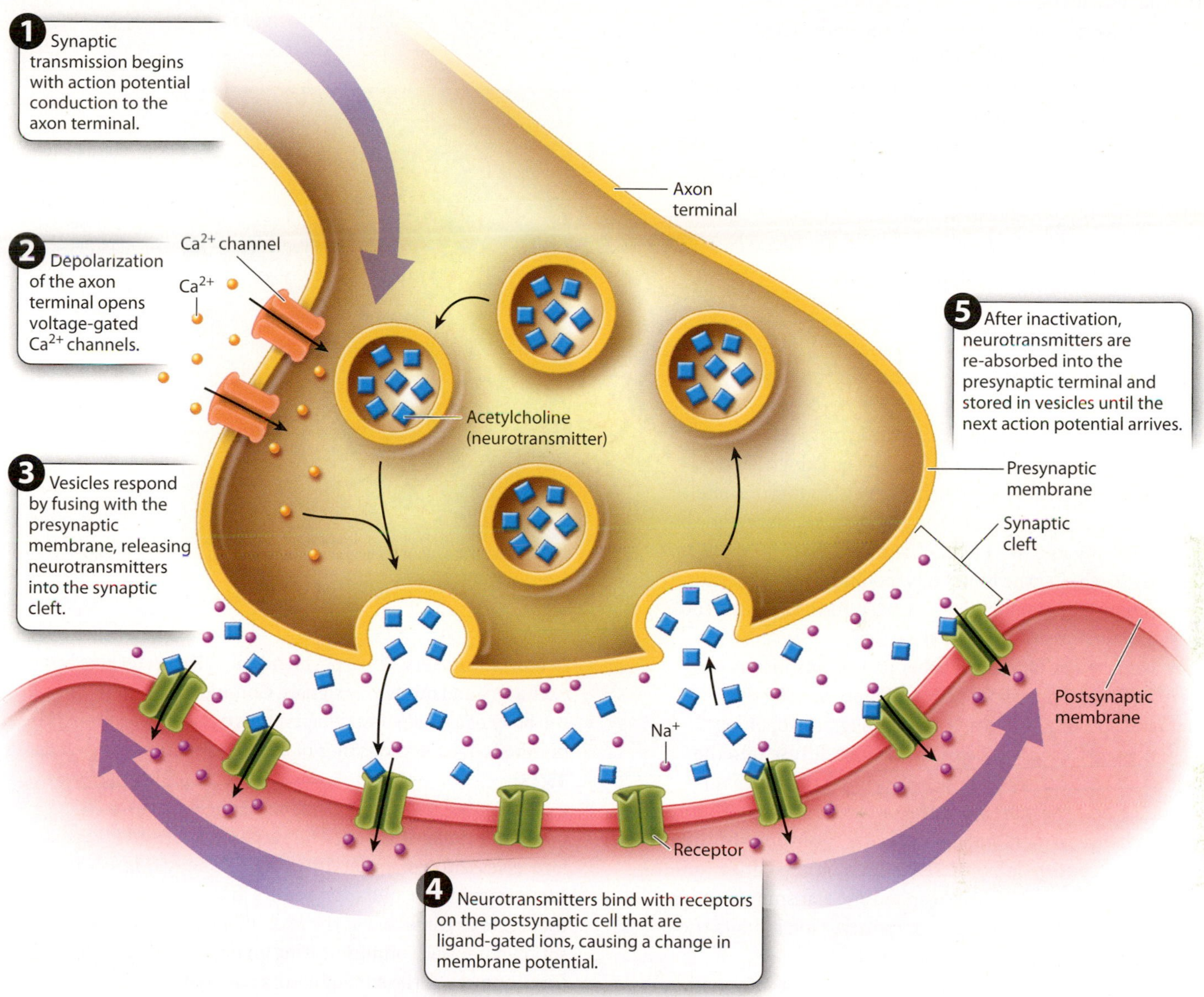

Chemical synapses are by far the more common type of synapse in animal nervous systems. The signals conveyed at chemical synapses are neurotransmitters contained within small vesicles in the axon terminal. When an action potential reaches the end of an axon (**Fig. 35.12**), the resulting depolarization induces voltage-gated Ca^{2+} ion channels to open. These channels are found only in the axon terminal membrane. Because of their higher concentration outside the cell, Ca^{2+} ions diffuse through these channels into the axon terminal. In response to the rise in Ca^{2+} concentration, the vesicles fuse with the presynaptic membrane and release neurotransmitter molecules into the synaptic cleft by exocytosis. The neurotransmitters diffuse rapidly across the cleft and bind to postsynaptic membrane receptors of the neighboring cell. The binding of neurotransmitters opens or closes ion channels, causing a change in the postsynaptic cell membrane potential that allows the signal to propagate along the next neuron.

These postsynaptic membrane receptors are called **ligand-gated ion channels** because when the neurotransmitter binds as a ligand it causes the ion channel to open. This allows specific ions to enter the postsynaptic cell, changing its membrane potential. Ligand-gated ion channels underlie the signal transmission that occurs at synapses, regulating the activity of nerve and muscle cells.

The neurotransmitter's effect on the postsynaptic membrane ends shortly after binding. Neurotransmitters become unbound from the receptor, causing the ligand-gated ion channels to close, and neurotransmitters in the synaptic cleft are either broken down by deactivating enzymes or taken up again by the presynaptic cell and reassembled in newly formed vesicles.

→ **Quick Check 3** How do the structural and chemical properties of the synapse ensure that communication proceeds in a single direction between nerve cells?

Signals between neurons can be excitatory or inhibitory.

Neurotransmitters bound to postsynaptic membrane receptors can either stimulate or inhibit the firing of action potentials in the postsynaptic neuron. The binding of some neurotransmitters depolarizes the postsynaptic membrane by opening ligand-gated Na^+ channels. As a result, Na^+ ions diffuse into the cell, making the membrane potential more positive. The positive change in membrane potential is termed an **excitatory postsynaptic potential,** or **EPSP.**

In contrast, the binding of other neurotransmitters can hyperpolarize the postsynaptic membrane, making the membrane potential more negative through the opening of ligand-gated Cl^- or K^+ channels. Cl^- ions diffuse into the cell or K^+ ions diffuse out of the cell, causing the membrane potential to become more negative, or hyperpolarized. The negative change in membrane potential is called an **inhibitory postsynaptic potential,** or **IPSP.** Membrane depolarization makes a neuron more likely to respond and transmit an action potential, whereas membrane hyperpolarization inhibits the neuron from sending an action potential.

Excitatory synapses tend to transmit relevant information between nerve cells, and inhibitory synapses often serve to filter out unimportant information. For example, when a grazing gazelle senses a predator, its nervous system must transmit the key sensory information about the presence of a predator but filter out irrelevant information such as the flies buzzing over its back and ears. Inhibitory synapses are also important for controlling the timing of muscle activity required for coordinated movement, discussed in Chapter 37.

There are a number of different neurotransmitters, but a single nerve cell releases only one type of neurotransmitter. Consequently, a particular nerve cell exerts either an excitatory or an inhibitory effect on neighboring cells, but not both. In contrast, postsynaptic membranes typically contain multiple types of membrane receptor that bind different neurotransmitters. Thus, postsynaptic cells can receive both excitatory and inhibitory signals. Commonly, the dendrites or cell body of a single postsynaptic nerve cell make connections to hundreds or even thousands of axons from other nerve cells (**Fig. 35.13**).

FIG. 35.13 **Integration of information.** Multiple synapses—often hundreds of them—from communicating neurons enable a neuron to integrate diverse sources of information. *Source: Prof. Mary B. Kennedy, Division of Biology and Biological Engineering, Caltech.*

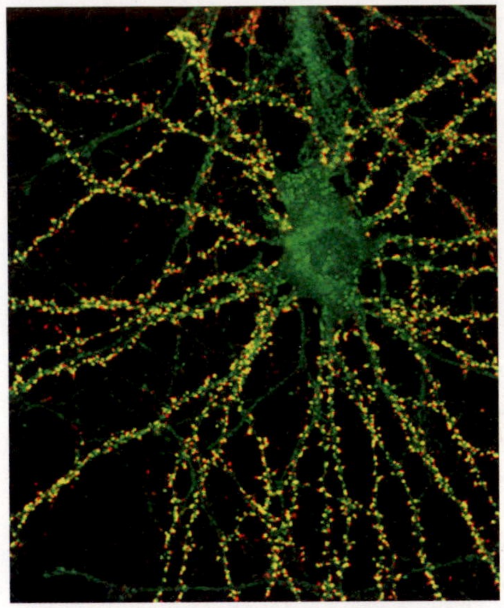

The postsynaptic nerve cell sums the excitatory and inhibitory synaptic inputs (the summed EPSPs and IPSPs) that it receives through its dendrites and cell body (**Fig. 35.14**). If the sum of synaptic stimulation results in a membrane potential that exceeds threshold at the axon hillock, the neuron fires an action potential, communicating an impulse to cells that it in turn contacts. If the sum of inputs does not exceed threshold, no action potential fires (Fig. 35.14a).

EPSPs and IPSPs can be summed over time and space. When summed over time, the frequency of synaptic stimuli determines whether the postsynaptic cell fires an action potential (Fig. 35.14b). This is referred to as **temporal summation.** When summed over space, the number of synaptic stimuli received from different regions of the postsynaptic cell's dendrites determines if the cell fires an action potential (Fig. 35.14c). This is referred to as **spatial summation.** Sometimes, excitatory and inhibitory signals may cancel each other out (Fig. 35.14d). Temporal and spatial summation of EPSPs and IPSPs are the fundamental forms of information processing carried out by the nervous system. They provide mechanisms for determining whether a particular sensory stimulus is responded to or ignored.

FIG. 35.14 Summation of excitatory and inhibitory postsynaptic potentials. (a) Widely spaced EPSPs do not sum. Other EPSPs and IPSPs can be summed (b) in time or (c) space, or even (d) cancel each other out.

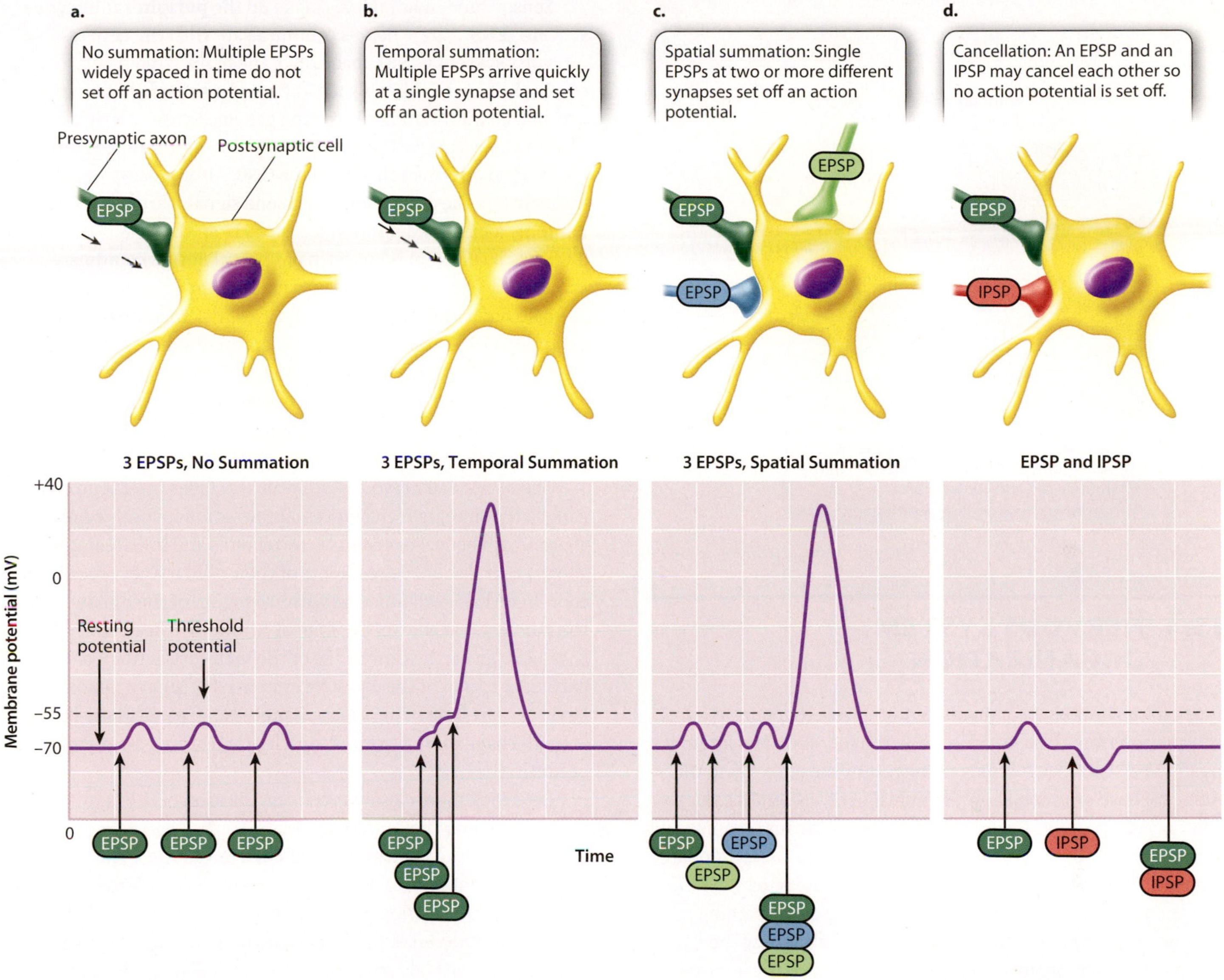

The role of temporal and spatial summation is explored further in Chapter 36.

More than 25 neurotransmitters are now recognized, and more are likely to be discovered. The amino acids glutamate (excitatory), glycine (inhibitory), and GABA (inhibitory) are key neurotransmitters that operate at synapses in the brain. The related amino acid derivatives dopamine, norepinephrine, and serotonin are also neurotransmitters acting in the brain. Simple peptides serve as neurotransmitters for sensory neurons. More recently, two gases, nitrous oxide and carbon monoxide, have been discovered to function as messengers between some nerve cells even though they do not behave as standard neurotransmitters that bind to membrane receptors.

Acetylcholine is one of the key neurotransmitters produced in a variety of nerve cells. It is the excitatory neurotransmitter released by motor neurons to stimulate muscle fibers (**Fig. 35.15**). All vertebrate motor synapses rely on acetylcholine and are therefore excitatory only. The release of acetylcholine produces a depolarization of the muscle cell membrane by opening ligand-gated Na^+ channels, causing the muscle to contract (Chapter 37).

→ **Quick Check 4** Acetylcholinesterase is the enzyme that breaks down and inactivates acetylcholine. Some nerve gases used as chemical weapons block acetylcholinesterase. What effect would such nerve gases have on muscle contraction?

FIG. 35.15 Vertebrate motor endplate. The motor endplate is an excitatory synapse between a motor nerve and a muscle cell; acetylcholine is the neurotransmitter. *Source: David B. Fankhauser, Ph.D., Professor of Biology and Chemistry, University of Cincinnati, Clermont College.*

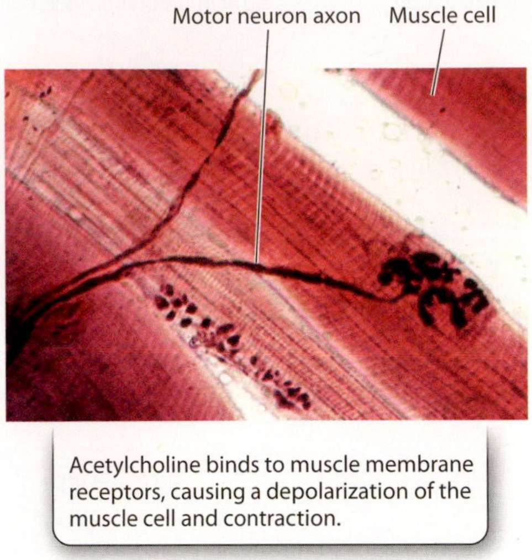

35.4 NERVOUS SYSTEM ORGANIZATION

The brain is the body's command center: It receives sensory information from the eyes, ears, nose, tongue, skin, and internal organs, and sends instructions to the rest of the body. Up to this point, we have examined how individual neurons send signals to other nerve and muscle cells. How are these neurons organized in the body to sense stimuli, process them, and issue an appropriate response? How does the brain coordinate the movement of limbs necessary to walk or run, and adjust heart rate and breathing in response to exercise? In this section, we look at how the nervous system is organized to allow the brain and the body to communicate with each other.

Nervous systems are organized into peripheral and central components.

As animals evolved the ability to sense and coordinate responses to increasingly complex stimuli, their nervous systems became organized into peripheral and central components (**Fig. 35.16**). Your eyes, receptors for the sense of touch, and other sensory organs are located on the surface of the body, where they can receive signals from the environment. In contrast, the brain, centrally located ganglia, and a main nerve cord are located in the interior.

Nerves form the lines of communication between these nervous system structures. In general, nerve cell bodies are grouped together in sensory organs, ganglia, and a main nerve cord that extends from the brain. Nerves are composed mainly of axons from many different nerve cells. For example, the optic nerve contains axons that travel from nerve cells in the eye to the brain.

Sensory and motor nerves make up the **peripheral nervous system (PNS).** These nerves communicate with the **central nervous system (CNS),** made up of the brain, a main nerve cord, or—in the case of animals such as flatworms that lack a brain—centralized information-processing ganglia. Sensory organs at or near the body surface transmit information by afferent neurons, which send information *toward* the CNS. After processing the information, the ganglia and brain send signals through efferent neurons, which send information *away* from the CNS. The signals delivered by efferent neurons in peripheral nerves coordinate the activity of muscles and glands in different regions of the animal's body. This organization ensures that an animal achieves a coordinated behavioral and physiological response to the stimuli received from its environment, which is fundamental to its ability to maintain homeostasis. Among multicellular animals with nervous systems, only the diffuse nerve nets of cnidarians lack defined central and peripheral components.

The peripheral nervous system also includes interneurons and ganglia that integrate and process information in local regions of the animal's body. For example, vertebrates and invertebrates have ganglia that lie outside segments of their primary nerve cord that coordinate localized function within a region of their body, such as the muscles that control the limb of a running insect, the muscles that allow an earthworm to dig, or the secretory activity of a particular region of the digestive system of a fish or a mammal. However, the bulk of information processing, particularly in animals that exhibit a greater capacity for learning and memory, occurs in the CNS.

Let's look more closely at the organization of the vertebrate nervous system. The central nervous system of vertebrates includes both the brain and **spinal cord** (Fig. 35.16). The spinal cord is a central tract of neurons that passes through the vertebrae to transmit information between the brain and the periphery of the body. The vertebrate spinal cord is divided into segments, each controlling body movement in a particular region along the animal's length. Each spinal cord segment contains axons from peripheral sensory neurons, a set of interneurons, and a set of motor neuron cell bodies. These are distinct from, but often associated with, the ganglia mentioned above that lie outside the spinal cord.

In humans and other vertebrate animals, the peripheral nervous system is organized into left and right sets of **cranial nerves** located within the head and **spinal nerves** running from the spinal cord to the periphery. Most of the cranial nerves and all of the spinal nerves contain axons of both sensory and motor neurons. Cranial nerves link specialized sensory organs (eyes, ears, tongue) to the brain. Cranial nerves also control eye movement, facial expression, speech, and feeding. Some cranial nerves, such as the olfactory and optic nerves, contain only sensory axons.

FIG. 35.16 Human central (yellow) and peripheral (blue) nervous systems.

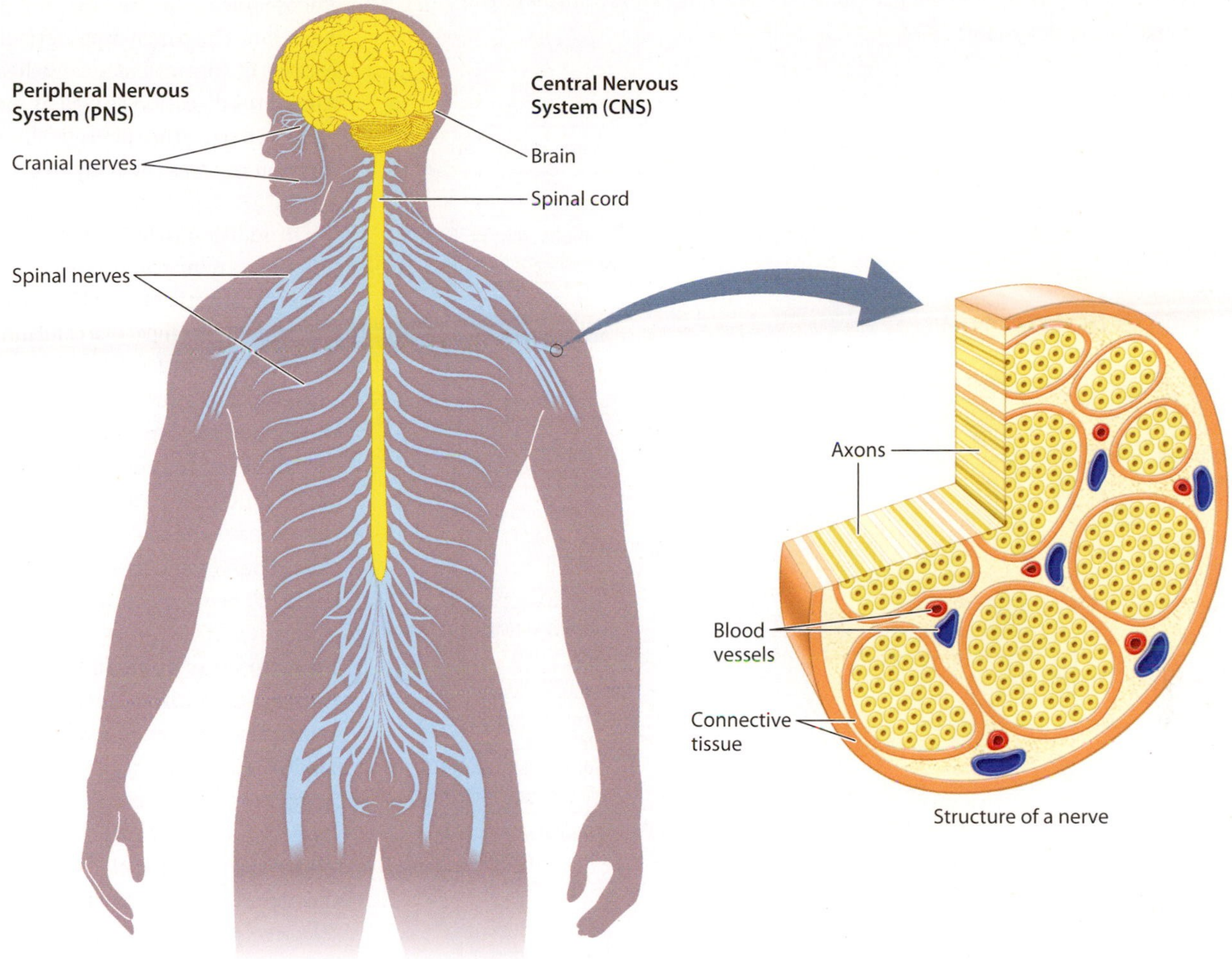

Spinal nerves exit from the spinal cord through openings located between adjacent vertebrae to thread through the trunk and limbs of an animal's body. These nerves receive sensory information from receptors in nearby body regions along the length of the body and send motor signals from the spinal cord back to those regions (Fig. 35.16).

Peripheral nervous systems have voluntary and involuntary components.

As bodies with distinct internal organ systems evolved, two separate components of the nervous system emerged. When a gazelle senses a predator, some nerve circuits send a signal to run, an action that is under conscious control, while others signal the heart to beat faster and blood vessels supplying muscles to dilate, actions that occur unconsciously. Conscious reactions are under the control of the **voluntary component** of the nervous system, and unconscious ones are under the control of the **involuntary component.** Voluntary components mainly handle sensing and responding to external stimuli, whereas involuntary components typically regulate internal bodily functions. Both nervous system components are found in invertebrate and vertebrate animals.

In insects and crustaceans, for example, nerve circuits from the involuntary component regulate the animal's digestive system. Nerve circuits connecting sensory structures such as the antennae and eyes to the animal's brain and muscles are part of the voluntary system.

In vertebrates, the peripheral nervous system is divided into **somatic** (voluntary) and **autonomic** (involuntary or visceral) components. The somatic nervous system is made up of sensory neurons that respond to external stimuli and motor neurons that synapse with voluntary muscles. This system is considered voluntary because it is under conscious control. However, many reflexes are controlled lower in the spinal cord (discussed in the following section) or by the brainstem, independent of conscious control by the central nervous system.

FIG. 35.17 The human involuntary (autonomic) nervous system. The autonomic system, which controls involuntary functions, is divided into sympathetic and parasympathetic systems.

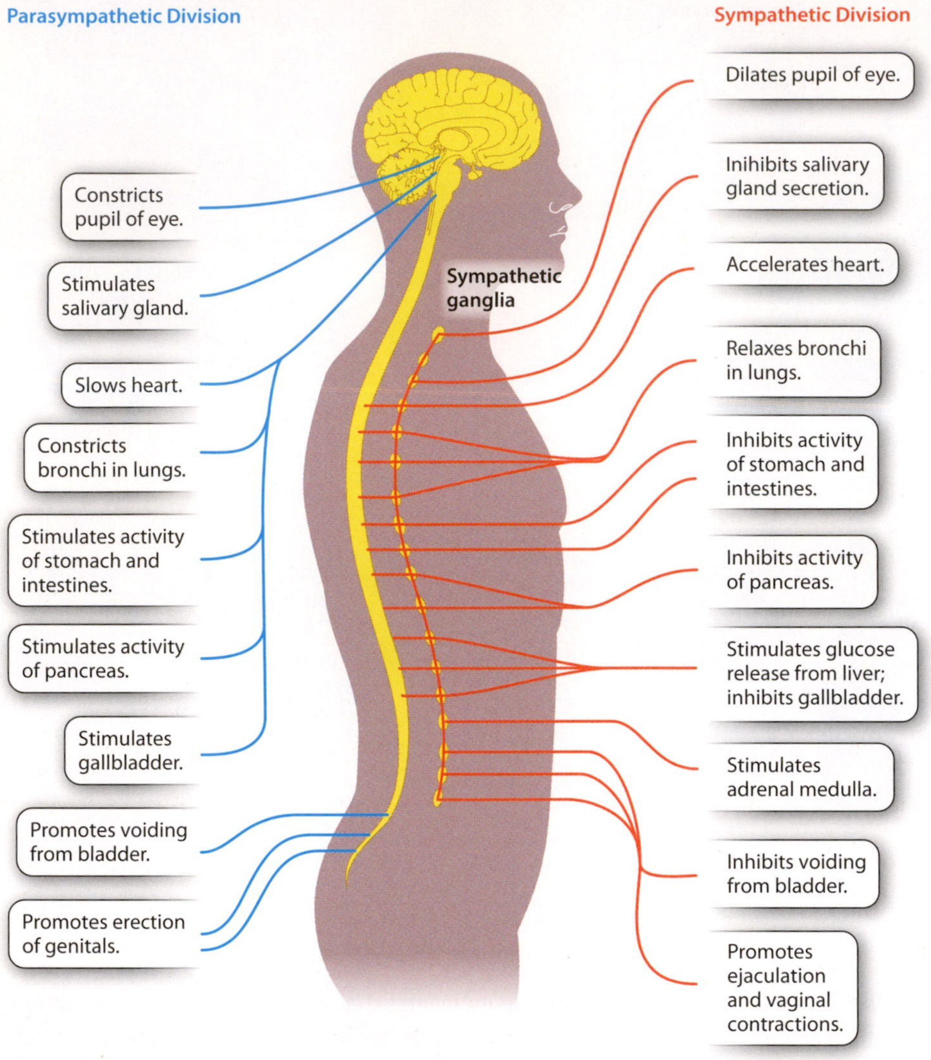

The autonomic nervous system (**Fig. 35.17**) controls internal functions of the body such as heart rate, blood flow, digestion, excretion, and temperature. It includes both sensory and motor components, which usually act without our conscious awareness. The autonomic nervous system, in turn, is divided into two major subdivisions, a **sympathetic division** and a **parasympathetic division.** Both divisions continuously monitor and regulate internal functions of the body. Generally, sympathetic and parasympathetic nerves have opposite effects, but not always. For example, sympathetic neurons stimulate the heart to beat faster, whereas parasympathetic neurons cause the heart to beat slower.

The sympathetic nervous system generally results in arousal and increased activity. It is the pathway activated when animals are exposed to threatening conditions, resulting in what is referred to as the "fight or flight" response. As part of this response, the sympathetic nervous system increases heart rate and breathing rate, stimulates glucose release by the liver, and inhibits digestion. The parasympathetic nervous system, in contrast, slows the heart and stimulates digestion as well as metabolic processes that store energy—in other words, it enables the body to "rest and digest."

In addition to having different functions, sympathetic and parasympathetic nerves also have different anatomical distributions, as shown in Fig. 35.17. Sympathetic nerves leave the CNS from the middle (thoracic and lumbar) region of the spinal cord, forming ganglia along much of the length of the spinal cord. Parasympathetic nerves leave the CNS from the brain by cranial nerves and from lower (sacral) levels of the spinal cord.

The nervous system helps to maintain homeostasis.

The autonomic nervous system plays a key role in the ability to maintain homeostasis, or a steady physiological state, in the face of changing environmental conditions. In Chapter 5, we discussed how the selectively permeable cell membrane helps maintain homeostasis within individual cells. For example, it keeps ion concentrations within narrow ranges for normal cell function. Similarly, the firing of action potentials by neurons requires particular ion concentrations on either side of the membrane. Temperature and pH are other parameters that do not vary much because enzymes often work effectively only in narrow temperature and pH ranges. Recent evidence suggests that nerve cell firing rates are maintained at a steady level in the brain of rats, regardless of sensory stimulus or deprivation, or whether the animals are awake or asleep. A shift in the homeostatic set point of nerve cell firing rates may therefore underlie disorders of low brain activity, such as autism.

Homeostasis is also maintained for the whole body. Many physiological parameters are maintained in a narrow range of conditions throughout the body, including pH, temperature, and ion concentrations. Similarly, the water content of the body as a whole is kept stable through the careful regulation of ions and other solutes, as discussed in Chapter 41.

The concept of homeostasis was first described as regulation of the body's "interior milieu" in the late 1800s by the French

FIG. 35.18 Temperature regulation by negative feedback in (a) a house and (b) a mammal. In negative feedback, a response (such as heat) opposes the stimulus (cold), leading to a stable state (a steady temperature).

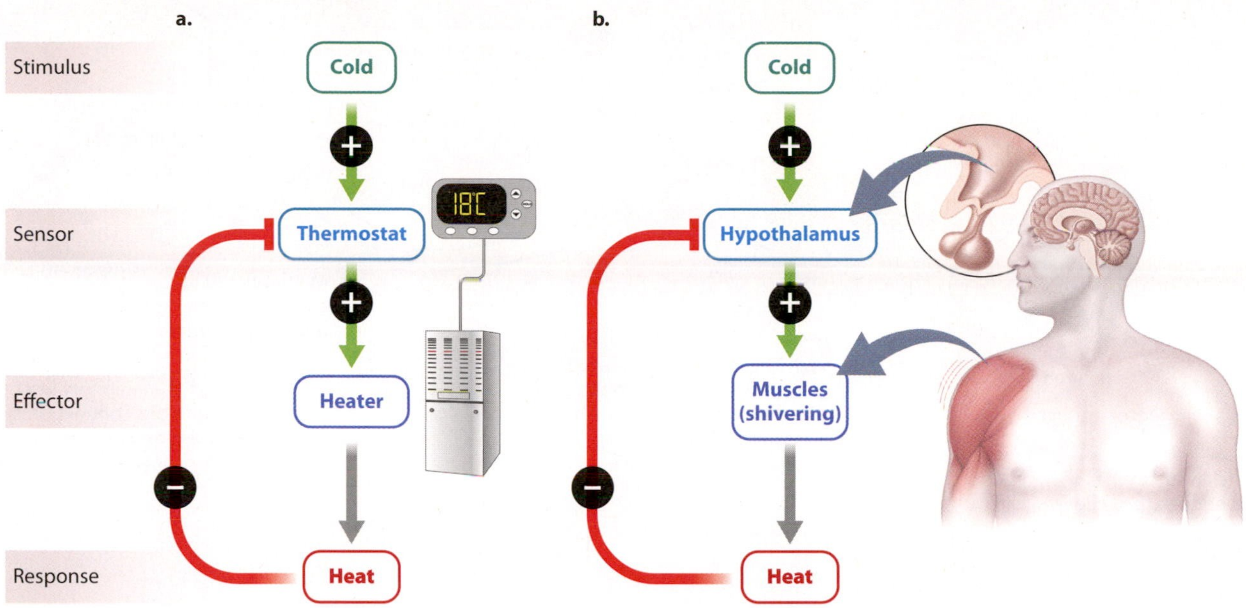

physiologist Claude Bernard, who is often credited with bringing the scientific method to the field of medicine. The term "homeostasis" was coined by the American physiologist Walter Cannon, whose book *The Wisdom of the Body* (first published in 1932) popularized the concept.

Maintaining steady and stable conditions takes work in the face of changing environmental conditions. That is, a cell (or organism) actively maintains homeostasis. For example, many animals are often faced with long periods of drought that challenge their ability to remain hydrated and maintain a stable water and ion balance. These animals must rapidly respond by changing the permeability of their skin and respiratory organs so that they can retain as much water as possible.

How does the body, and in particular the nervous system, maintain homeostasis? Homeostatic regulation often depends on **negative feedback** (**Fig. 35.18**). In negative feedback, a stimulus acts on a sensor that communicates with an effector, which produces a response that opposes the initial stimulus. For example, negative feedback is used to maintain a constant temperature in a house. Cool temperature (the stimulus) is detected by a thermostat (the sensor). The thermostat sends a signal to the heater (the effector), producing heat (the response). The response (heat) opposes the initial stimulus (cool temperature), and no further heat will be produced until the temperature drops below the temperature setting of the thermostat. In this way, a stable temperature is maintained (Fig. 35.18a).

In a similar way, humans and other mammals maintain a steady body temperature even as the temperature outside fluctuates. Nerve cells in the hypothalamus (located in the base of the brain) act as the body's thermostat (Fig. 35.18b). When a decrease in the temperature in the environment causes a drop in body temperature, the lowered body temperature signals the hypothalamus to activate the somatic nervous system to induce shivering and the production of metabolic heat, as discussed in Chapter 40. At the same time, the hypothalamus activates the autonomic nervous system, causing peripheral blood vessels to constrict. The reduction in blood flow near the body's surface reduces heat loss to the surrounding air. By contrast, an increase in temperature signals sweat glands to secrete moisture and peripheral blood vessels to dilate to aid heat loss from the skin.

Homeostatic regulation, therefore, relies on negative feedback to maintain a set point, which in this case represents an animal's preferred body temperature. The ability to maintain a constant body temperature is known as thermoregulation, and it is just one of many physiological set points that the body actively maintains, as we discuss in subsequent chapters.

Simple reflex circuits provide rapid responses to stimuli.

An animal that has perceived a predator has an advantage if it can move quickly. Fast responses are made possible by simple reflex circuits that bypass the brain. These circuits connect sensory neurons (detecting the presence of the predator) directly with motor neurons (providing the quick movement necessary to escape the predator). Reflex circuits are common in both the somatic and autonomic components of the nervous system.

FIG. 35.19 The knee-extension reflex. The knee-extension reflex involves a sensory neuron, a single synapse, and a motor neuron.

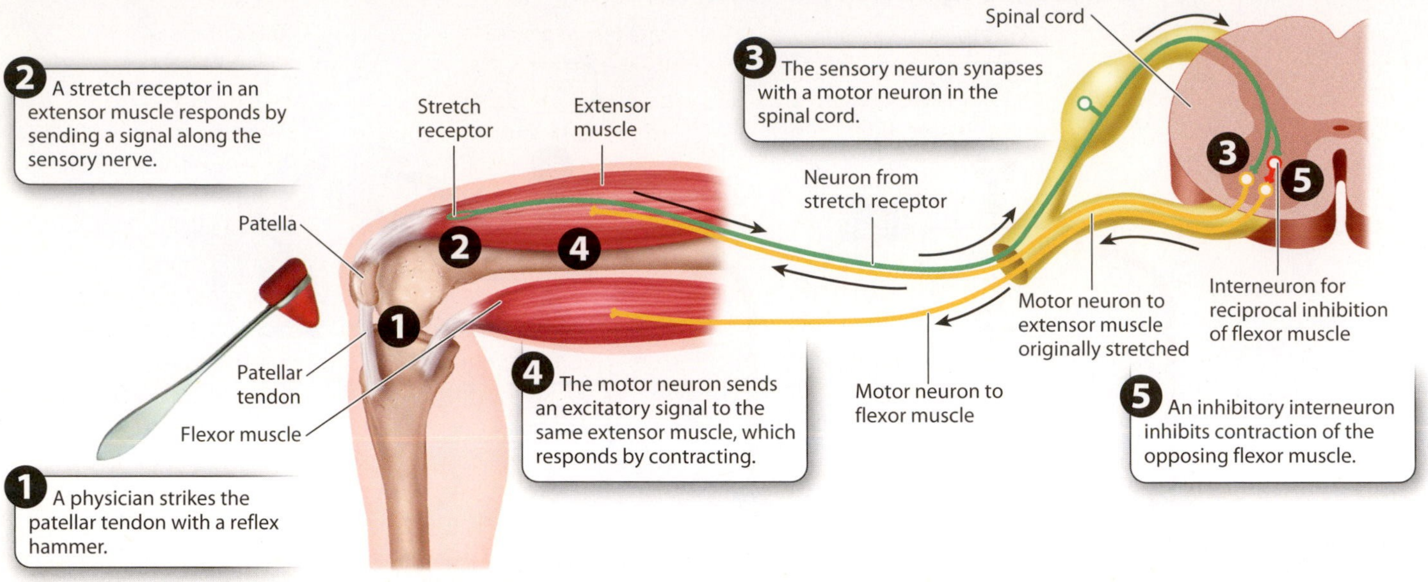

Simple spinal reflex circuits connect sensory and motor neurons directly in the spinal cord of vertebrates. The large Mauthner neurons of fish are an example of such a circuit. Sensory neurons respond to threatening visual, tactile, or vibratory cues from the environment and transmit signals through the spinal cord to the hindbrain of a fish, activating a large Mauthner neuron on the side opposite the stimulus. The Mauthner cell in turn rapidly activates motor neurons along the body, causing the fish to bend away from the threatening stimulus to initiate a rapid escape. At the same time, the Mauthner cell and motor nerves on the other side of the fish are inhibited. The giant axons of squid are part of a similar reflex circuit that allows squid to swim quickly away from a predator.

The **knee-extension reflex** in humans is an example of a simple nerve circuit that includes only a single synapse between two neurons—a sensory neuron and a motor neuron (**Fig. 35.19**). Physicians commonly use this reflex to evaluate the performance of the peripheral nervous and muscular systems. The strike of the physician's hammer below the kneecap activates the extensor muscles on the front of the thigh, including the quadriceps muscle, causing the bent leg to straighten.

Let's follow this reflex pathway. It starts with specialized "stretch" receptors located on the dendrites of a sensory neuron in the extensor muscles of the leg. These dendrites extend to cell bodies in ganglia alongside the spinal cord. These cell bodies have axons that extend into the spinal cord. The stretch receptors sense the stretch of the muscle that occurs during movement or in response to a physician's strike of a reflex hammer. In response to a stretch, a signal is sent from the stretch receptor, through dendrite and cell body to the axon. In the spinal cord, the axon of the sensory neuron forms synapses with motor neurons that travel from the spinal cord back to the muscle where the stretch originated. The signal from the muscle stretch receptor stimulates the motor neurons to increase the activation of the muscle: the muscle contracts, and the leg extends at the knee.

This reflex arc does not include an interneuron: It is composed of just two neurons and one synapse. Transmission is delayed by communication at synapses, so the muscle contracts quickly because only one synapse is required to relay the sensory information back to the muscle.

As well as being a useful medical test, the knee-extension reflex has a normal physiological role. In running or landing from a jump, the knee joint first flexes (bends), stretching the quadriceps muscles. This action sends the same sensory signal to the spinal cord as when a physician's hammer taps the muscle's tendon. As a result, after flexing, the knee reflexively extends (straightens). Knee flexion followed by extension allows rapid adjustments in muscle force and knee position that help stabilize the body during running and jumping.

Because muscles can only contract, opposing movements at a joint such as flexion and extension require flexor and extensor muscles on either side of the joint (Chapter 37). These are often activated out of phase (at different times) so that, when one set of muscles is activated (contracting), the other is inhibited (relaxed). The alternating extension and flexion of a leg during walking and running provide an example. This pattern of joint and limb movement is achieved by **reciprocal inhibition:** When stretch receptors of the knee extensor muscles are activated to stimulate these muscles to extend the knee, they also inhibit the activity of opposing muscles that flex the knee (Fig. 35.19).

Reciprocal inhibition of opposing sets of muscles occurs in the spinal cord. Axons of the stretch receptor neurons not only synapse with motor neurons of extensor muscles, but also synapse with inhibitory interneurons that inhibit motor neuron stimulation of the opposing flexor muscles. This inhibitory reflex pathway contains two synapses, one between the sensory neuron and interneuron and the second between the interneuron and the motor neuron to the flexor muscle.

→ **Quick Check 5** Why is an interneuron needed to provide reciprocal inhibition of the flexor muscle when an extensor muscle is activated?

Reciprocal inhibition also operates between the right and left sides of the body. For example, reciprocal inhibition helps to control the movement of the right and left limbs. In this case, interneurons cross the spinal cord to control the timing of activity by flexor and extensor muscles of the opposite limb. Reciprocal inhibition is also involved in the Mauthner cell circuit of fish to ensure that only one side of the fish's body bends away from the stimulus to escape. The local spinal circuits that provide reciprocal inhibition are therefore fundamental to the alternating motion of the body and limbs that characterizes the movements of most animals.

In other cases, more complex circuits in the brain may act to coordinate sensory input and motor output. These circuits integrate other sources of sensory information, such as vision and balance, with conscious commands from the brain to voluntarily control motor behavior. In the next chapter, we look more closely at these sensory circuits. ■

Core Concepts Summary

35.1 Animal nervous systems allow organisms to sense and respond to the environment, coordinate movement, and regulate internal functions of the body.

Nerve cells, or neurons, receive and send signals and are the functional unit of the nervous system. page 740

Animal nervous systems include three types of neuron: sensory neurons that respond to signals, interneurons that integrate and process sensory information, and motor neurons that produce a response from muscle. page 740

Ganglia are localized collections of nerve cell bodies that integrate and process information. page 741

Simply organized animals, such as cnidarians, have a nerve net to coordinate sensory and motor function. page 741

Forward locomotion led to the evolution of specialized sense organs in the head, along with concentrated groupings of nerve cells to form ganglia and a brain. page 741

35.2 The basic functional unit of the nervous system is the neuron, which has dendrites that receive information and axons that transmit information.

Neurons share a common organization: They have dendrites that receive inputs, a cell body that receives and sums the inputs, and axons that transmit signals to other nerve cells. page 743

Most neurons communicate by chemical synapses formed between an axon terminal and a neighboring nerve or muscle cell. page 743

Glial cells provide nutritional and physical support for neurons. page 744

35.3 The electrical properties of neurons allow them to communicate rapidly with one another.

Neurons have electrically excitable membranes that code information by changes in membrane voltage and transmit information in the form of electrical signals called action potentials. page 745

Ion channels open and close in response to changes in membrane voltage, underlying the production of action potentials in nerve cells. page 746

Action potentials fire in an all-or-nothing fashion; a brief refractory period follows. page 746

In vertebrates, glial cells also produce the myelin sheath that insulates axons, increasing the speed of nerve impulses. page 749

Action potentials are conducted in a saltatory fashion in myelinated axons, firing at nodes of Ranvier, where the axon membrane is exposed and not insulated by myelin. page 749

Communication across the synapse occurs when an arriving action potential triggers the release of neurotransmitters from vesicles within the axon terminal. page 751

Neurotransmitters released from presynaptic vesicles bind to receptors in the postsynaptic membrane, causing either an excitatory or an inhibitory stimulus. page 752

Excitatory stimuli depolarize the membrane, producing an excitatory postsynaptic potential (EPSP), whereas inhibitory stimuli hyperpolarize the membrane, producing an inhibitory postsynaptic potential (IPSP). page 752

35.4 Animal nervous systems can be organized into central and peripheral components.

Animal nervous systems are organized into central and peripheral components called the central nervous system (CNS) and peripheral nervous system (PNS). page 754

The central nervous system includes the brain and one or more main trunks of nerve cells, such as the spinal cord. The peripheral nervous system is distributed throughout the animal's body and is composed of sensory and motor nerve cells. page 754

In many invertebrates and vertebrates, the peripheral nervous system is divided into voluntary and involuntary components. page 755

In vertebrates, the voluntary component is referred to as the somatic nervous system, and the involuntary component is referred to as the autonomic nervous system. page 755

The autonomic system regulates body functions through opposing actions of the sympathetic and parasympathetic divisions. page 756

The nervous system helps to regulate physiological functions to actively maintain stable conditions inside a cell or an organism, a process known as homeostasis. page 756

Homeostasis is often achieved by negative feedback, in which the response inhibits the stimulus. page 757

Simple reflex circuits can involve as few as two neurons: a sensory neuron from the periphery that synapses with a motor neuron in the spinal cord that sends a signal to a muscle. page 757

Self-Assessment

1. Diagram a simple nervous system of an animal that lacks cephalization and compare that system with the general organizational features of a nervous system that exhibits cephalization.
2. Name the three basic types of neuron and describe their functions.
3. Diagram and label the basic features of a neuron, indicating where information is received and where it is sent.
4. Graph an action potential, showing the change in electrical potential on the *y*-axis and time on the *x*-axis. Indicate on the graph the phases when voltage-gated Na^+ and K^+ ion channels are opened and when they are closed.
5. Explain what is meant by saying action potentials are "all-or-nothing."
6. Explain why action potentials propagate along an axon only in a single direction.
7. Create a table that lists the sequence of events that create an action potential, identifying the membrane potential at each stage, which channels are opening or closing, and in which direction ions are moving across the cell membrane.
8. Briefly describe how myelinated axons increase the speed of signal transmission.
9. Diagram a chemical synapse, labeling the vesicles that contain neurotransmitter molecules and the receptors that bind the neurotransmitter to produce either an inhibitory or an excitatory stimulus in the postsynaptic cell.
10. Describe how neurotransmitter binding to receptors on a postsynaptic cell causes inhibition or excitation.
11. Describe how temporal summation of EPSPs from two presynaptic neurons results in a larger depolarization of the postsynaptic cell compared with how an inhibitory presynaptic neuron causes an IPSP that negates the EPSP of an excitatory neuron, and how spatial summation of three excitatory presynaptic neurons produces an even larger depolarization of the postsynaptic neuron.
12. List which functions of an animal are controlled by voluntary and by involuntary components of the nervous system.
13. Diagram a simple circuit that includes a sensory neuron that synapses with a motor neuron to produce a reflex. Indicate where in the nervous system this synapse is found.

Log in to LaunchPad to check your answers to the Self-Assessment questions, and to access additional learning tools.

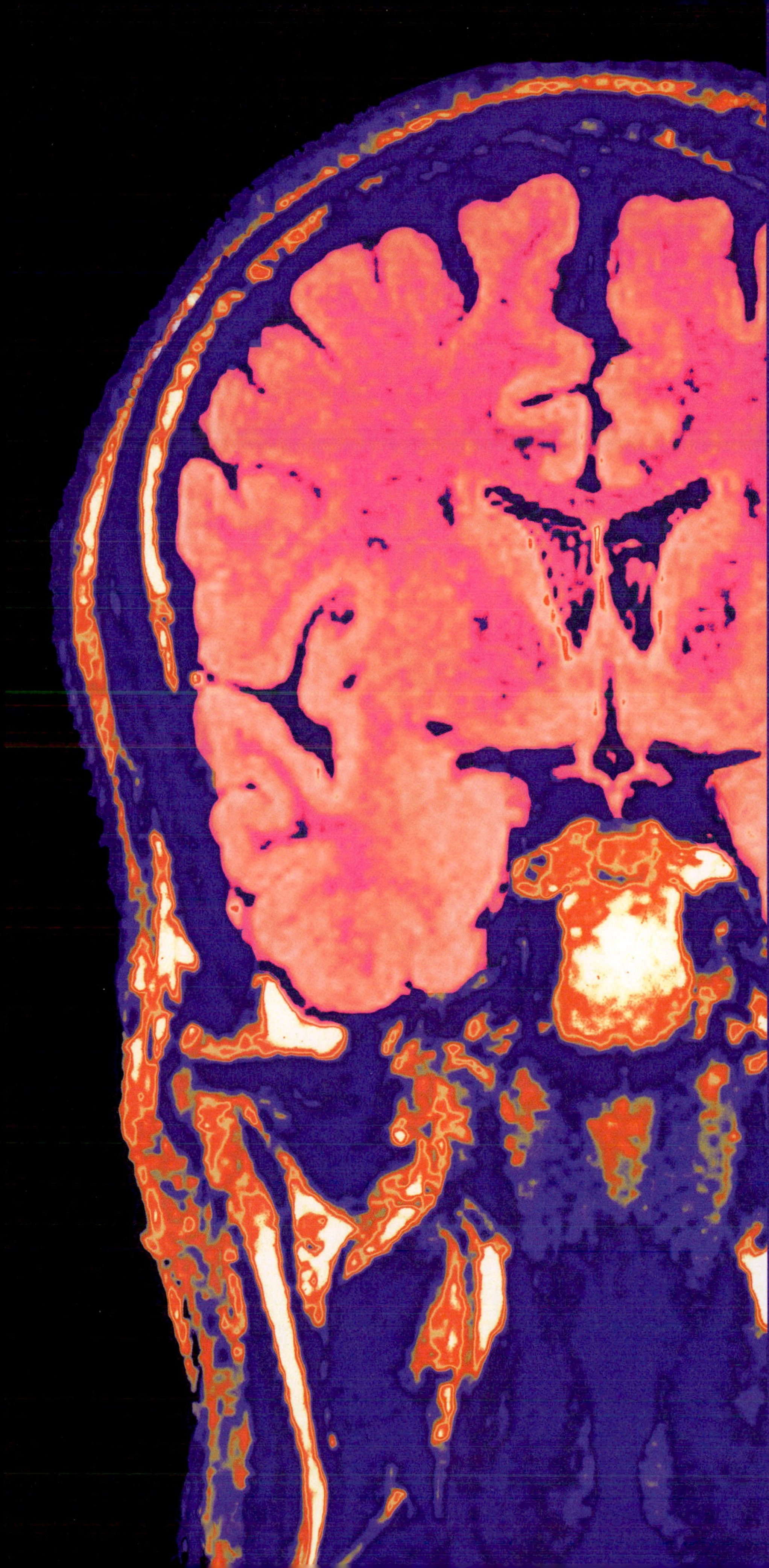

CHAPTER 36

Animal Sensory Systems and Brain Function

Core Concepts

36.1 Animal sensory receptors detect physical and chemical stimuli by changes in membrane potential.

36.2 Specialized chemoreceptors relay information about smell and taste.

36.3 Hair cells convey information about gravity, movement, and sound.

36.4 The ability to sense light and form images depends on photosensitive cells with light-absorbing proteins.

36.5 The brain processes and integrates information from multiple sensory systems, with tactile, visual, and auditory stimuli mapped topographically in the cerebral cortex.

36.6 Cognition is the ability of the brain to process and integrate complex information, remember and interpret past events, solve problems, reason, and form ideas.

To see, your eyes detect light, which is a form of electromagnetic radiation. To hear, your ears detect sound waves. To smell, your nose detects odor molecules present in the air. In all these cases, your sense organs are detecting a physical or chemical stimulus in the environment. The first half of this chapter explores how specialized sensory cells detect these signals and code them as information that can be transmitted and processed by the nervous system. How is a dog able to distinguish thousands of different odors? How is a hawk able to see a small rodent far below? How are humans able to detect small differences in the pitch of a sound?

The second half of the chapter follows nervous system pathways from these sensory cells to the brain, on the way exploring basic principles of brain function. Processing of sensory information in the brain takes place in regions specialized for each sense, and within many of these regions information is represented in the form of topographic maps.

36.1 ANIMAL SENSORY SYSTEMS

Animals can sense the physical properties of their environment, including light, chemicals, temperature, pressure, and sound, that are useful in finding mates and food and avoiding predators and noxious environments. Early in evolutionary history, organisms evolved specialized protein receptors located in the cell membrane that were able to detect these critical features in their environment. For example, before the evolution of nerve cells and a nervous system, bacteria evolved membrane receptors that sensed osmotic pressures that might otherwise rupture their cell membrane. Bacteria and sponges also evolved membrane receptors that detect chemicals—including nutrients—in their environment.

The senses of smell, taste, and sight in multicellular organisms with a nervous system also rely on membrane receptors. These receptors are embedded in specialized **sensory receptor cells.** These cells either communicate with neurons, as for taste and sight, or are themselves neurons, as for smell. Note that the shorthand term "sensory receptor" refers to the entire sensory cell or sensory neuron, not just the membrane protein.

Sensory receptors represent a key cellular unit of the sensory component of animal nervous systems. In most multicellular animals, the sensory receptors are organized into specialized **sensory organs** that convert particular physical and chemical stimuli into signals that are communicated to the brain. Cnidarians (including jellyfish, corals, and anemones) and roundworms (including the laboratory organism *Caenorhabditis elegans*) evolved sensory receptors that sensed physical contact, and cnidarians and flatworms were among the first multicellular animals to evolve simple light-sensing organs.

The conversion of physical or chemical stimuli into nerve impulses is called **sensory transduction.** For example, receptors located in the ear convert the energy of sound waves into nerve impulses that allow an animal to distinguish loud versus soft sounds and high-pitched versus low-pitched sounds. Although the sense organs of different animals share many similar properties, differences have also evolved. Consequently, different animals perceive the world differently. Many insects, for instance, are sensitive to ultraviolet light; nocturnal snakes can see at night by sensing infrared radiation; and dogs can distinguish odor compounds as much as 100 million times lower in concentration than humans can detect (**Fig. 36.1**).

Specialized sensory receptors detect diverse stimuli.

Animals have evolved a diverse array of sensory receptors that respond to different stimuli, among them touch, light, and the oscillations in air pressure we call sound waves. How is a physical phenomenon in the world outside an organism transformed into a nerve impulse? The initial transformation takes place inside the sensory receptor (**Fig. 36.2**). A physical or chemical stimulus

FIG. 36.1 Diverse sensory receptors. (a) Many insects can detect ultraviolet light. (b) Rattlesnakes can see at night by detecting infrared radiation. (c) Dogs have a keen sense of smell. *Sources: a. Russell Burden/Getty Images; b. Andy Hunger/age fotostock; c. Monika Ondrušová/Dreamstime.com.*

FIG. 36.2 Sensory transduction, which converts an external stimulus into a change in membrane potential. (a) Chemoreceptors are located in the antennae of the luna moth; (b) mechanoreceptors below the cuticle of a roundworm; and (c) photoreceptors in the eyes of a vertebrate. *Photo sources: a. Rolf Nussbaumer Photography/Alamy; b. Sinclair Stammers/Science Source; c. Eyal Nahmias/Alamy.*

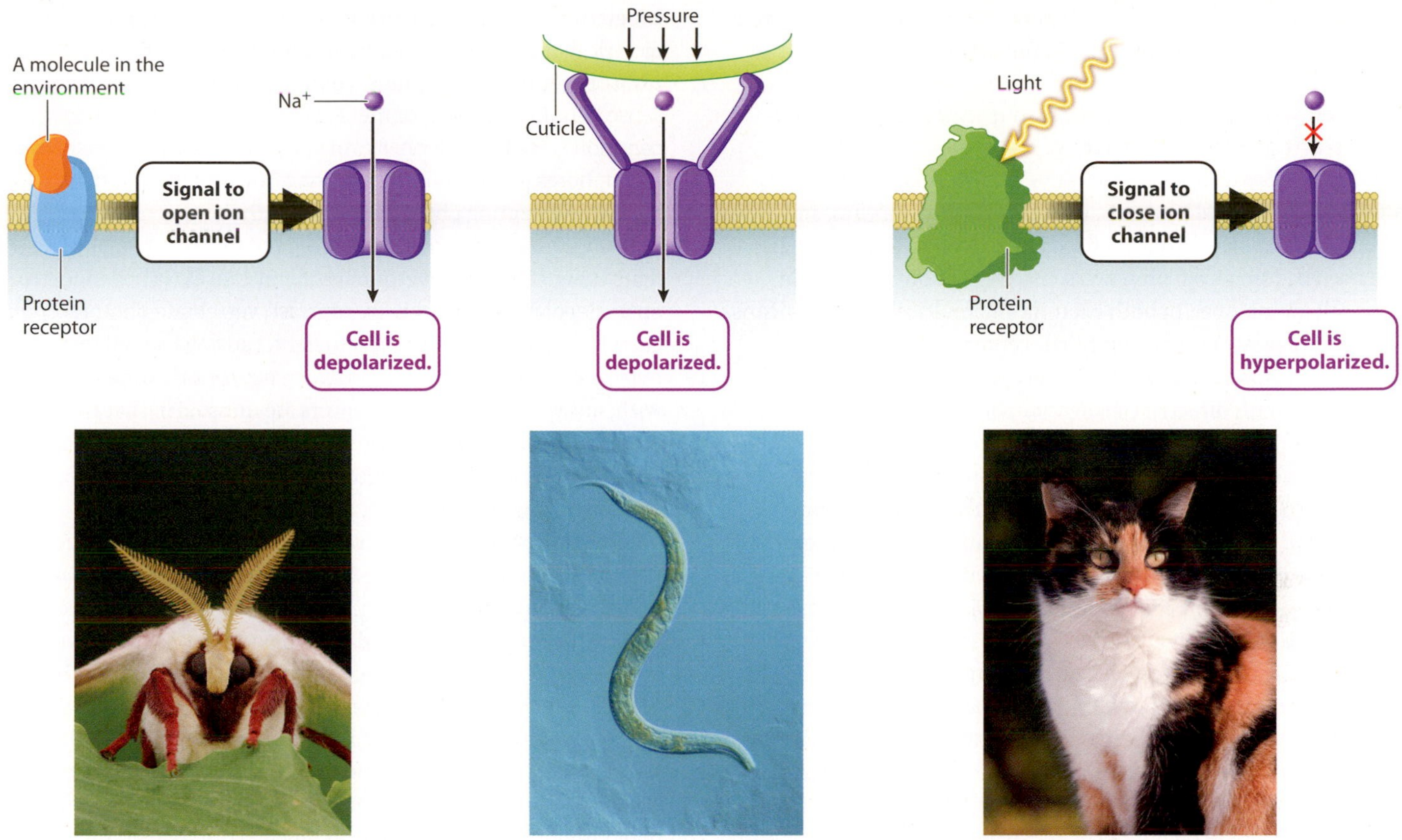

is converted into a change in the sensory receptor's membrane potential. Recall from Chapter 35 that a cell's membrane potential is the electrical charge difference between the inside and the outside of the cell membrane, and that a depolarization in membrane potential is the first step in firing an action potential. Sensory receptors either fire action potentials themselves or synapse with neurons that fire action potentials, which are then transmitted to the central nervous system.

In many cases, the sound wave, touch, or other stimulus causes ion channels in the cell's plasma membrane to open. The influx of ions changes the membrane potential by altering the distribution of charged ions on either side of the membrane. Let's look at examples of how three different types of stimuli from the environment—molecules, touch, and light—can be transformed into a nerve signal.

Chemoreceptors are universally present in animals.

The most ancient type of sensory detection is chemoreception. All organisms respond to chemical cues in their environment. Even the earliest branching groups of Bacteria and Archaea have protein receptors in their cell membranes that respond to molecules in the environment.

In animals, **chemoreceptors** are sensory receptor cells that respond to molecules that bind to specific protein receptors on the cell membrane (Fig. 36.2a). Many animals detect food in their environment by sensing key molecules such as oxygen (O_2), carbon dioxide (CO_2), glucose, and amino acids. Female mosquitoes track CO_2 levels to detect prey for blood meals, and coral polyps respond to simple amino acids in the water, extending their bodies and tentacles toward areas of greater concentration to feed. Other arthropods, such as flies and crabs, have chemosensory hairs on their feet. These animals taste potential food sources by walking on them. Salmon rely on chemoreception to detect chemical traces of the home waters of the river where they hatched, and where they will return to mate and spawn.

Chemoreception underlies the sense of smell and taste. Consider taste as an example. In most cases, the binding of molecules to a protein receptor on a taste receptor causes the protein receptor to change conformation. That change in conformation in turn triggers the opening of Na^+ channels through G protein signal transduction pathways (Chapter 9). The influx of Na^+ ions depolarizes the receptor cell. In the case of taste receptors, no action potential fires, but the depolarization travels far enough down the receptor's short axon to trigger the release of neurotransmitters. Other chemoreceptors, such as those involved in the sense of smell or located on antennae to detect chemicals in the air, are sensory neurons that do fire action potentials.

Mechanoreceptors are a second general class of ancient sensory receptors.

Mechanoreceptors respond to physical deformations of their membrane produced by touch, stretch, pressure, motion, and sound. Mechanoreceptors are found in all multicellular animals, and even bacteria have pressure-sensitive protein receptors in their cell membrane. In both bacteria and multicellular organisms, evidence suggests that the protein receptor is also a sodium ion channel. Thus, deformation of the receptor membrane opens sodium channels directly, causing a depolarization of the endings of the cell's dendrites (Fig. 36.2b). Early pressure-sensitive protein receptors in bacteria sensed internal cell pressure: When water moving into the cell by osmosis increases the risk of bursting (Chapter 5), the stretching of the membrane opens channels that let water leave the cell. Other mechanoreceptors in roundworms and anemones are linked to externally projecting cilia that sense forces at the animal's body surface.

One well-studied example of a mechanoreceptor is a touch receptor in the roundworm *Caenorhabditis elegans.* Roundworms, also called nematodes, are one of the most diverse and widespread groups of animals on Earth. Deformation of the surface of the worm—its cuticle—exerts pressure on proteins connected to an ion channel. The mechanical force changes the shape of the ion channel, causing it to open and produce a change in membrane potential.

A mechanoreceptor in humans and other mammals is the sensory receptor found in the skin that senses touch and pressure. The cell bodies of these neurons are located in ganglia near the spinal cord, with extensions going in two directions: to the skin and to the spinal cord. In the skin, the neuron has branched tips containing ion channels sensitive to deformations of the membrane. These branched tips are the initial sensors of touch and pressure. If the stimulus is strong enough, local depolarization leads to the firing of an action potential that travels all the way to the spinal cord. Thus, in contrast to taste chemoreceptors, this type of mechanoreceptor transmits an action potential. The axon is much too long for simple depolarization to spread to its other end.

Mechanoreceptors are also found at the base of whiskers that rodents, cats, dogs, and other mammals use to sense touch with their snouts. Stretch receptors found in muscles are also mechanoreceptors that influence a muscle's motor activation, helping to control its length and force. A very different group of specialized mechanoreceptors called hair cells are the sensory receptors for balance, gravity sensing, and hearing, discussed in section 36.3.

Electromagnetic receptors sense light, thermoreceptors sense temperature, and nociceptors sense pain.

Electromagnetic receptors are sensory cells that respond to electrical, magnetic, and light stimuli. Of these, light-detecting **photoreceptors** are the most common and diverse. Most animals sense light in their environment. Photoreceptors have been common since the first ones evolved in cnidarians (jellyfish) and ctenophores (comb jellies) more than 500 million years ago.

Light-detecting photoreceptors are the sensory receptors in eyes (Fig. 36.2c). Photoreceptors respond to individual photons of light energy. In most invertebrates, light causes Na^+ channels to open, depolarizing the cell. By contrast, vertebrate photoreceptors close Na^+ channels in response to light, causing the cell to become hyperpolarized. Most receptors excite neurons that they synapse with, but vertebrate photoreceptors are unusual in that they can either excite or inhibit neurons in the eye. Photoreception in the vertebrate eye is discussed in section 36.4.

Some fish, such as catfish, contain specialized **electroreceptors** arranged in a lateral line along their bodies. Electroreceptors enable these fish to detect weak electrical signals emitted by all organisms. They likely evolved as an adaptation for locating prey or potential predators in poorly lit habitats where vision was less useful. Some specialized "weakly electric" fish actually generate an electromagnetic field by emitting pulses from an electric organ located in the tail. Disturbances in the field detected by electroreceptors of the lateral line system signal the location of nearby prey. These fish also inhabit rivers with poor visibility. The bill of the duckbilled platypus also contains electroreceptors that locate prey in dimly lit water.

Finally, specialized receptors with dendritic branches in the skin can respond to heat and cold (**thermoreceptors**) or to pain (**nociceptors**). Thermoreceptors help to control an animal's metabolism, and they also regulate body temperature by controlling patterns of blood flow that in turn alter rates of heat gain and loss (Chapter 40). As a result, they help to maintain homeostasis. Pain receptors send action potentials to the brain or spinal cord when exposed to excessive heat, force, or chemical damage. A quick withdrawal from the painful stimulus is the typical response.

Stimuli are transmitted by changes in the firing rate of action potentials.

Sensory reception depends on converting the energy of a physical or chemical stimulus into an action potential, either in the sensory receptor itself or in the neuron it synapses with. How do these nerve impulses convey information to the brain? In effect, action potentials can be considered a code that the brain deciphers.

To convert information from the environment into a code, the nerve impulses conveyed by sensory organs carry out the following functions:

FIG. 36.3 Sensory receptor signals. The firing rate of action potentials increases with (a) the intensity of the stimulus and (b) a novel compared to a continuous stimulus.

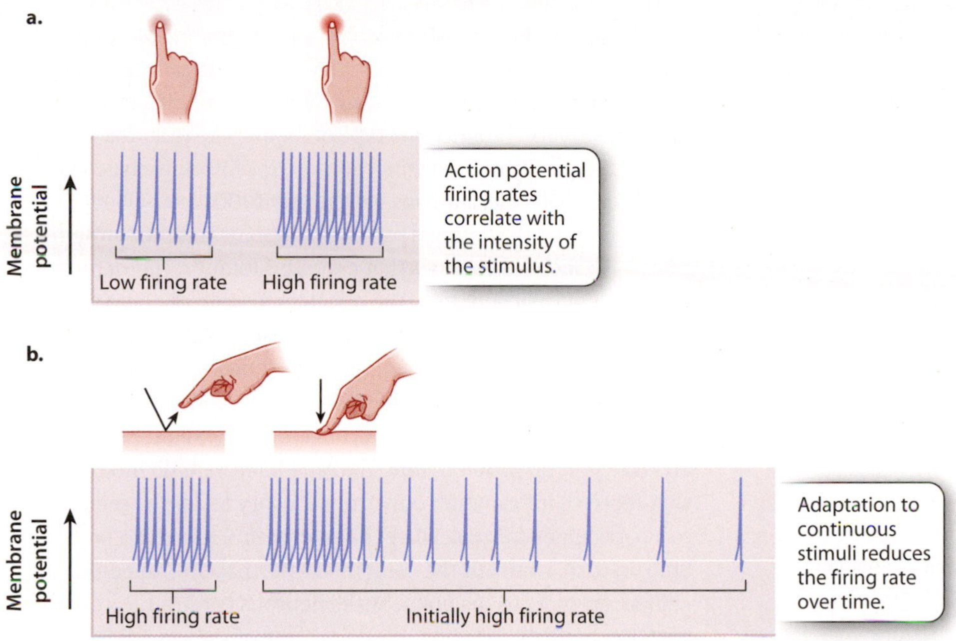

1. Convey the strength of the incoming signal—for example, the brightness of light, the loudness of a sound, or the intensity of an odor.
2. Carry information about even weak signals when necessary.
3. Convey the location of a signal's source—for example, where the body is touched.
4. Filter out unimportant background signals.

Because nerve cells transmit action potentials in an all-or-nothing fashion, information about the strength of the signal is coded by the number of nerve cells involved and the nerve cell's **firing rate,** which is the number of action potentials fired over a given period of time. Most commonly, weak signals produce low or background firing rates and strong signals produce high firing rates (**Fig. 36.3a**). For example, the firing rate of mechanoreceptors increases as the tactile pressure they sense increases.

Several mechanisms increase a nerve cell's sensitivity to even weak signals. Following sensory transduction, the signal typically undergoes initial processing. As discussed in Chapter 35, sensory stimuli may be summed over space or time. Signals from multiple sensory receptors often converge onto a neighboring neuron, which increases its firing rate proportionally to the number of signals received. This is an example of **spatial summation.** Combining the synaptic input increases the receptor organ's sensitivity to weak external signals.

Another way to increase sensitivity to external signals is through **temporal summation,** summing sensory stimuli that are received repeatedly over time by the same sensory cell. When a sensory cell is stimulated more frequently, the excitatory postsynaptic potentials (EPSPs) sum, and the cell is more likely to become depolarized above threshold and fire an action potential.

Sensory receptors also distinguish between discrete (intermittent) and continuous stimuli, providing a way to ignore background noise. Novel stimuli are generally the most important for an animal. Consequently, sensory receptors initially respond most strongly when a stimulus is first received (**Fig. 36.3b**). If the stimulus continues over a longer time period, sensory receptors typically reduce their firing rate through a process called **adaptation** (this is *not* the adaptation that results from natural selection). Adaptation of firing rate enables an animal's sensory receptors to focus on novel stimuli, which are those likely to be most important. You can experience adaptation by touching your skin lightly at a particular location. Initially, skin mechanoreceptors fire in response to the touch, but after a while their firing rate diminishes and the sensation of a touch becomes less noticeable. Adaptation also ensures that the clothes touching your skin do not cause an irritating and distracting long-lasting sensation.

Sensory receptors locate a signal's source by **lateral inhibition** (**Fig. 36.4**). Lateral inhibition enhances the strength of a sensory signal locally but diminishes it peripherally. Sensory receptors have multiple extensions. Some make contact with nearby interneurons, while others make contact with interneurons farther away, in adjacent regions. In lateral inhibition, receptors send inhibitory signals to adjacent regions of the sensory organ, and send excitatory signals to interneurons within their local region. This increases the contrast in signal strength between adjacent regions. For example, lateral inhibition in the retina of the eye enhances edge detection and sharpens image contrast. Lateral inhibition in the skin enables the sense of touch to be localized to a small region of the body surface. The inhibitory and excitatory signals are transmitted to the interneurons by synapses that have different neurotransmitters.

We have not yet described all the information in the code. How do action potentials convey pitch or color or a particular odor? This information is conveyed by the identity of the particular sensory receptors that are activated. Sensory organs are constructed so that, for example, tone A activates "tone A" sensory neurons in the ear, red light activates "red" sensory neurons in the eye, and the fragrance of vanilla activates "vanilla" sensory neurons in the nose. The way in which the "correct" sensory receptor is

FIG. 36.4 **Lateral inhibition.** The firing rate of action potentials is higher in the center of the sensory receptive field than in surrounding regions.

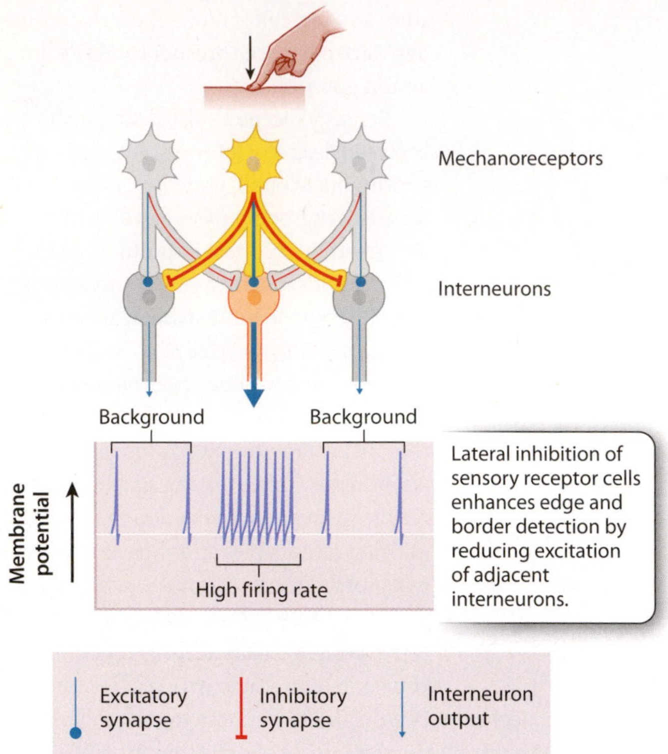

excited varies for each of the senses. In the next three sections, we discuss how this happens for smell and taste, hearing, and vision.

→ **Quick Check 1** Aspirin and ibuprofen reduce pain by inhibiting the synthesis of chemicals called prostaglandins. What effect would you predict prostaglandins to have on the firing rate of nociceptors?

36.2 SMELL AND TASTE

In many animals, specialized sensory organs for smell (**olfaction**) and taste (**gustation**) have evolved that detect a variety of chemical compounds. These organs contain distinct types of protein membrane receptor, and each type binds to a specific class of molecule. A large number of genes code for different chemical odor receptors, and the presence of so many receptors enables the detection of broad categories of smell and taste.

In contrast to the senses of hearing and sight, the sensory receptors of smell and taste do the "work" of detecting an odor or taste by themselves, without other structures being necessary. All that is needed is exposure to the environment and protection from damage. Both are achieved by the placement of chemoreceptors in the nasal and mouth cavities.

Smell and taste depend on chemoreception of molecules carried in the environment and in food.

The sense of smell in mammals depends on specialized sensory neurons that line the nasal cavity (**Fig. 36.5**). These neurons have long, hairlike extensions containing membrane receptors that project into the nasal mucus. Odor molecules are captured by the mucus during inhalation and bind to membrane receptors. Each sensory receptor has one type of membrane receptor, but there are typically many distinct sensory receptors, each detecting a different molecule. Humans have nearly 1000 genes (about 5% of our genome) that express particular odorant receptor proteins. Mice have about 1500 genes that express odorant receptor proteins (about 7% of their genome). The olfactory system can detect specific odor molecules in quite low concentrations.

When bound to odor molecules, the sensory receptors produce excitatory postsynaptic potentials (EPSPs). If enough odor molecules bind to the receptor cell, the EPSPs are summed and transmit an action potential to an olfactory bulb interneuron that receives information only from sensory receptors sensitive to a particular odor molecule. The interneurons in the olfactory bulb, in turn, transmit the combined information from different chemoreceptors to the brain. These neurons communicate directly with the brain by the olfactory nerve, one of the cranial nerves (Chapter 35).

Taste is achieved by clusters of chemosensory receptor cells located in specialized **taste buds,** the sensory organs for taste (Fig. 36.5). Some fish have taste buds in their skin that sense amino acids in the water, helping them to locate food. The taste buds of terrestrial vertebrates are in the mouth; most of the taste buds in mammals are on the tongue. In humans, the tongue has about 10,000 taste buds, which are contained within raised structures called papillae, giving the tongue its rough surface. Each taste bud has a pore through which extend fingerlike projections called microvilli that contact food items. The microvilli provide a large surface area that is rich in membrane receptors. These receptors bind to specific chemical compounds that give the food its taste. Taste bud sensory cells synapse with afferent neurons, which fire action potentials when a sufficient number of sensory cells are depolarized by a particular food item. Compared with olfaction, taste generally requires higher concentrations of food chemicals coming into contact with taste bud receptors to stimulate receptor depolarization.

Human taste can be divided into five categories—sweet, sour, bitter, salty, and savory. The tongue contains only five types of taste receptor, each specialized for one of these categories. Variation in taste is achieved by combining the signals received by receptors in the five categories. Your sense of taste is substantially enhanced by airborne odors that emanate from a food or beverage and are sensed by chemoreceptors in the nose. Signals from taste bud and olfactory receptors are combined in the brain, eliciting the subtle sensation of a particular flavor.

→ **Quick Check 2** Why is your sense of smell and taste diminished when you have a cold?

FIG. 36.5 The senses of smell and taste. Olfactory receptors respond to chemical odors in the nasal passages and communicate directly with the brain. Taste cells are chemosensory receptors located in taste buds.

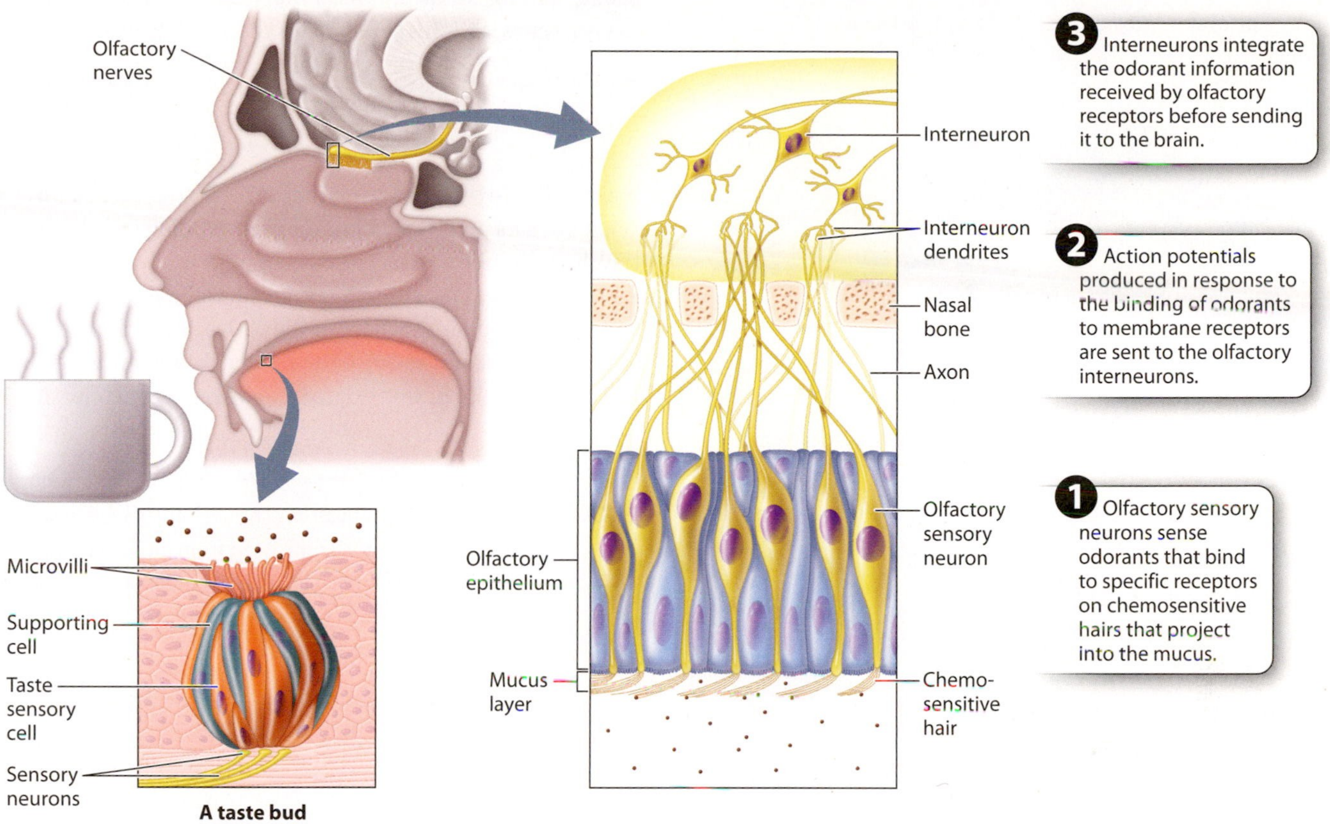

36.3 GRAVITY, MOVEMENT, AND SOUND

The ability to orient with respect to gravity, to detect motion, and to hear all depend on specialized mechanoreceptors called **hair cells** that sense movement and vibration. They are found in fishes and amphibians, in which they detect movement of the surrounding water, and in many invertebrates, in which they sense gravity and other forces acting on the animal. They are also found in the ears of terrestrial vertebrates, in which they sense sound, body orientation, and motion.

In all these cases, hair cells sense mechanical vibrations. These vibrations move small, nonmotile, hairlike projections from the surface of the hair cell called **stereocilia.** In turn, the motion of stereocilia causes a depolarization of the cell's membrane by opening or closing ion channels. Despite their name, stereocilia are more similar to microvilli than to cilia since cilia can move on their own but stereocilia cannot. Hair cells themselves do not fire action potentials but instead, when depolarized, release neurotransmitters that alter the firing rate of adjacent neurons.

Unlike chemoreceptors, hair cells are generally similar in structure; there is not a specific hair cell for a particular pitch of sound or body orientation. Thus, for hair cells to provide information about mechanical stimuli, they must interact with other structures within the sensory organ in which they reside. These more elaborate structures within the sensory organ influence how individual hair cells respond to physical motion, influencing the information they provide.

Hair cells sense gravity and motion.

Hair cells detect gravity and motion in many animals. Hair cells contained in the **lateral line system** of fishes and sharks sense water vibrations that indicate nearby threats or prey, as well as their own motion (**Fig. 36.6a**). The lateral line is a sensory organ along both sides of the body that uses hair cells to detect movement of the surrounding water.

A sense of gravity helps animals to orient their body within the environment, providing a sense of "up" and "down." Even buoyant aquatic animals depend on a sense of gravity to keep oriented to the water's surface or within the water column. Gravity-sensing organs called **statocysts** are found in most invertebrates (**Fig. 36.6b**). The statocyst is an internal chamber lined by hair cells with stereocilia that project into the chamber. Small granules of sand or other material form a dense particle called a **statolith** that is free to move within the statocyst organ. By pressing down on hair

FIG. 36.6 **Motion and gravity sensation.** (a) Hair cells within the lateral line system of fish sense water vibrations for motion and prey detection. (b) Gravity-sensing organs (statocysts) in lobsters provide cues about body orientation.

a. Lateral line

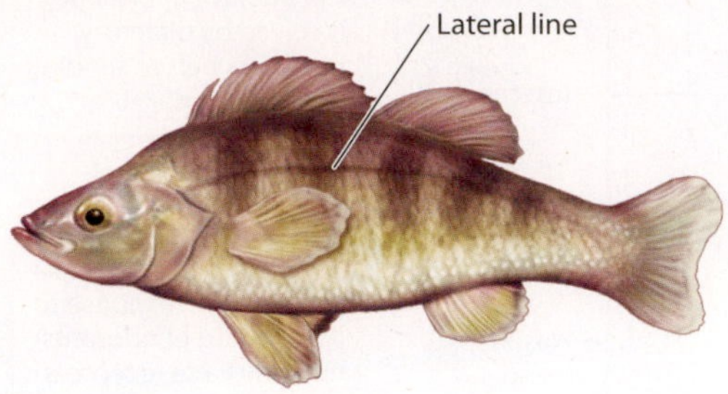

b. Statocyst

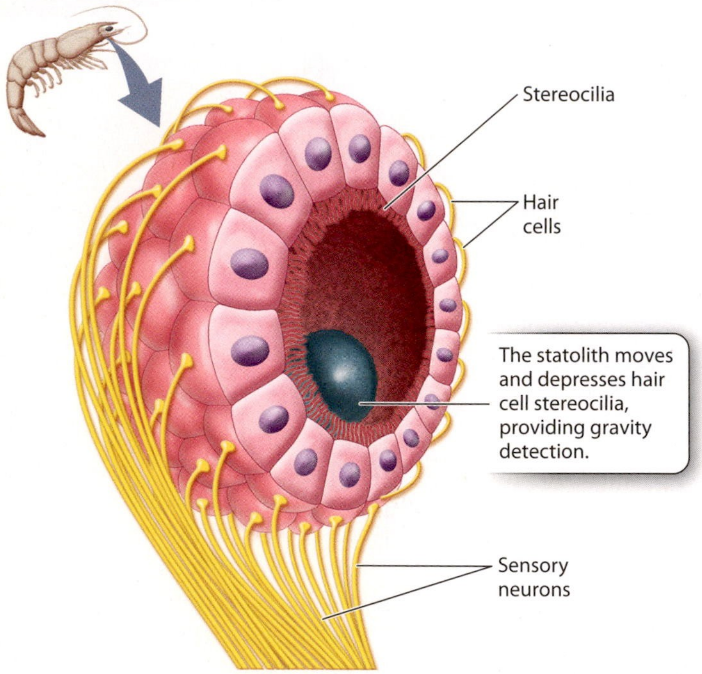

cells at the "bottom" of the chamber, the statolith activates those cells. Statoliths help anemones and jellyfish orient themselves in the water and direct their tentacles. Statoliths of lobster and crayfish can be replaced by magnetic particles and subjected to an experimental magnetic field that repositions the statolith to the top or sides of the chamber. In this case, the lobsters and crayfish respond by swimming upside down or on their sides.

The mammalian inner ear contains organs that sense motions of the head and its orientation with respect to gravity. These organs make up the **vestibular system** (**Fig. 36.7**), which consists of two statocyst chambers and three **semicircular canals.** These statocyst chambers are similar to those of invertebrates, providing a sense of gravity and body orientation with respect to motion. Hair cells located within the semicircular canals sense angular motions of the head in three perpendicular planes, providing a sense of balance. When the head rotates, gelatinous fluid in the semicircular canals is accelerated, deflecting the stereocilia of the hair cells to activate sensory neurons. The brain interprets differences in the motion of the hair cells among the three semicircular canals to resolve angular motions of the head. As a result, mammals and birds react much faster to head motion than to visual cues, and this ability helps them to stabilize their gaze.

→ **Quick Check 3** Why are you unstable when you try to walk after you spin in place or take a ride on a merry-go-round?

Hair cells detect the physical vibrations of sound.

What our ears interpret as sound are alternating waves of pressure traveling through the air. The frequency of the waves—the number of waves per second—determines the pitch of a sound, and the amplitude, or height, of the waves determines its loudness. Sound waves cause the stereocilia of hair cells to bend, and the hair cells become excited, depolarizing and releasing

FIG. 36.7 **The vestibular system.** The vestibular system senses angular rotations of the head from movements of fluid within three semicircular canals that deflect hair cells. It detects gravity and posture by means of statoliths in two statocyst chambers.

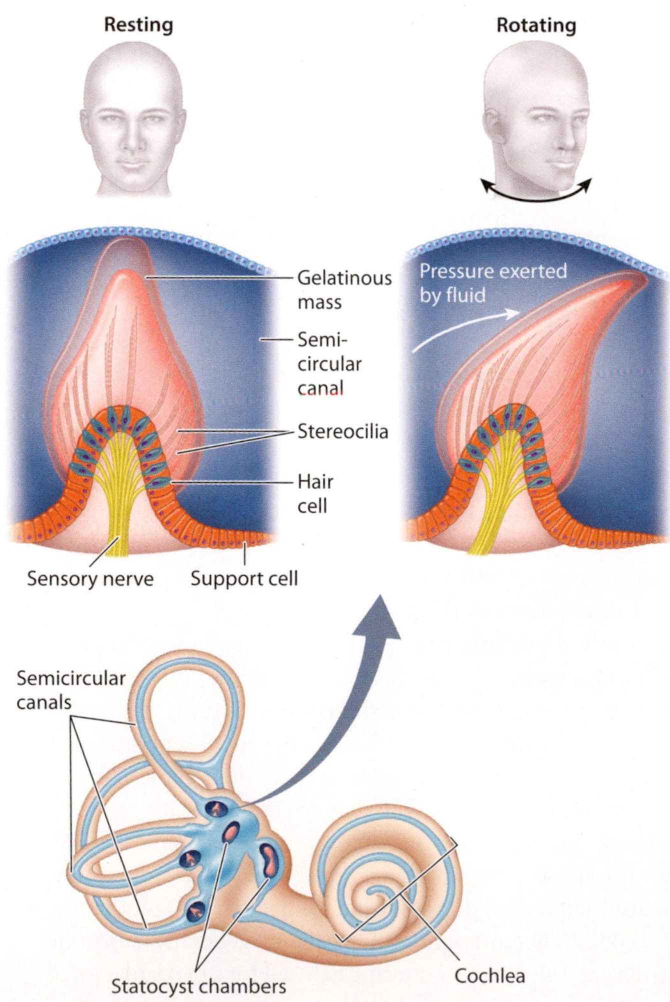

FIG. 36.8 **The human ear.** (a) The tympanic membrane and bones of the middle ear transmit external sound vibrations to the oval window of the inner ear. (b) Hair cells in the cochlea respond to the vibrations, sending signals to the brain that distinguish sound amplitude and pitch. (c) A scanning electron micrograph shows the V-shaped arrangement of stereocilia on hair cells. *Photo source: SPL/Science Source.*

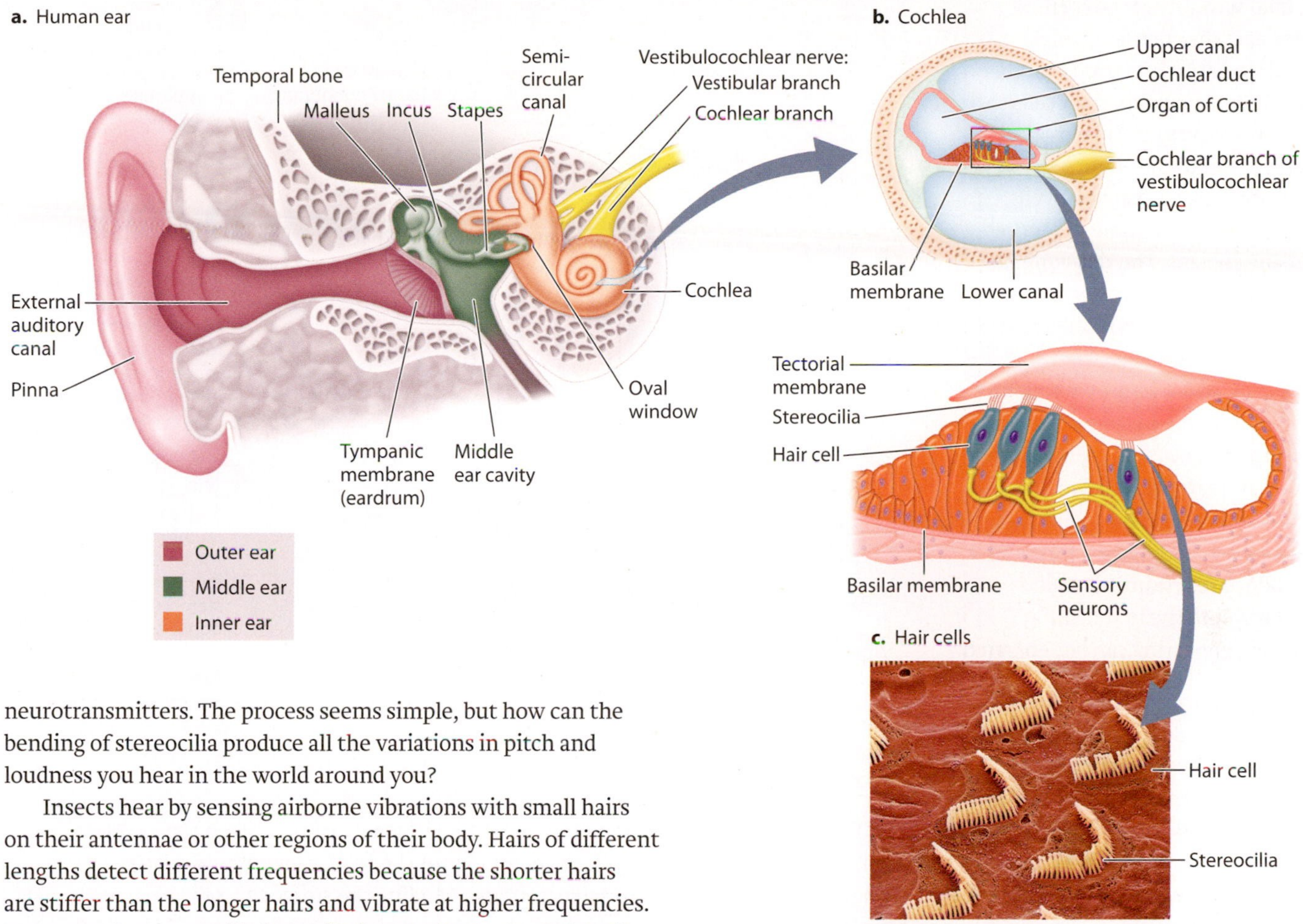

neurotransmitters. The process seems simple, but how can the bending of stereocilia produce all the variations in pitch and loudness you hear in the world around you?

Insects hear by sensing airborne vibrations with small hairs on their antennae or other regions of their body. Hairs of different lengths detect different frequencies because the shorter hairs are stiffer than the longer hairs and vibrate at higher frequencies. Frequency is measured in hertz (Hz), or cycles per second. Male mosquitoes have small antennal hairs that vibrate in response to the 500-Hz hum of a female mosquito's flight, which attracts them to the female to mate.

Many insects also have specialized ears that sense sound by means of a **tympanic membrane.** The tympanic membrane is a thin sheet of tissue at the surface of the ear that vibrates in response to sound waves, amplifying airborne vibrations. The vibration excites hair cells attached to the inside of the tympanic membrane, and these hair cells in turn excite other neurons to fire action potentials. Crickets and other singing insects use their ears to sense the chirping or buzzing of potential mates.

Hearing is most widespread and developed in terrestrial vertebrates. The ears of amphibians, reptiles, and birds have an external tympanic membrane that vibrates when sound waves strike its surface. This membrane is similar to the tympanic membrane of insects but evolved independently and therefore is an example of convergent evolution. Terrestrial vertebrate ears can detect a broad range of sound frequencies. For humans, the range of audible frequencies is from 20 to 20,000 Hz. For dogs, the audible range is 70 to 44,000 Hz, and some rodents can hear up to 80,000 Hz. Birds generally hear in the same frequency range as mammals, which is one reason bird songs are pleasing to our ear. Frogs and reptiles generally hear at lower frequencies than mammals and birds. Vertebrate ears can also make fine distinctions between close frequencies and can detect softer sounds because external structures amplify the sounds before they reach the hair cells.

The ears of mammals have an external structure, the **pinna,** which enhances the reception of sound waves contacting the ear. Notice the heightened sensitivity that this structure provides for hearing when you cup your hands behind your ears. It is no coincidence that dogs and rabbits perk up their ears when alerted by a sound: This action orients them to the direction of the sound's source.

Fig. 36.8 shows the structure of the human ear. The pinna is part of the **outer ear,** which also includes the ear canal and tympanic membrane, or mammalian **eardrum,** which transmits airborne sounds into the ear. The **middle ear** contains three small bones, or ossicles, called the **malleus, incus,** and **stapes,** which amplify the waves that strike the tympanic membrane. The stapes connects to a thin membrane called the **oval window** of the cochlea in the **inner ear.** The **cochlea** (which means "snail" in Latin) is a coiled chamber within the skull that contains hair cells that convert pressure waves into an electrical impulse that is sent to the brain.

The mammalian eardrum evolved from the tympanic membrane of reptiles and earlier amphibians. The three middle ear bones are also evolutionarily related to three bones in reptiles, but only the stapes functions in reptiles to transmit vibrations of the tympanic membrane to the inner ear. In reptiles, the malleus and incus help support jaw movements during feeding. As mammals evolved from their reptilian ancestors, these two bones were incorporated with the stapes to form the middle ear bones that transmit sounds from the eardrum to the inner ear.

The process of hearing can be separated into three stages:

FIG. 36.9 Activation of hair cells. Upward movement of the basilar membrane pushes hair cells into the tectorial membrane, causing bending of the stereocilia.

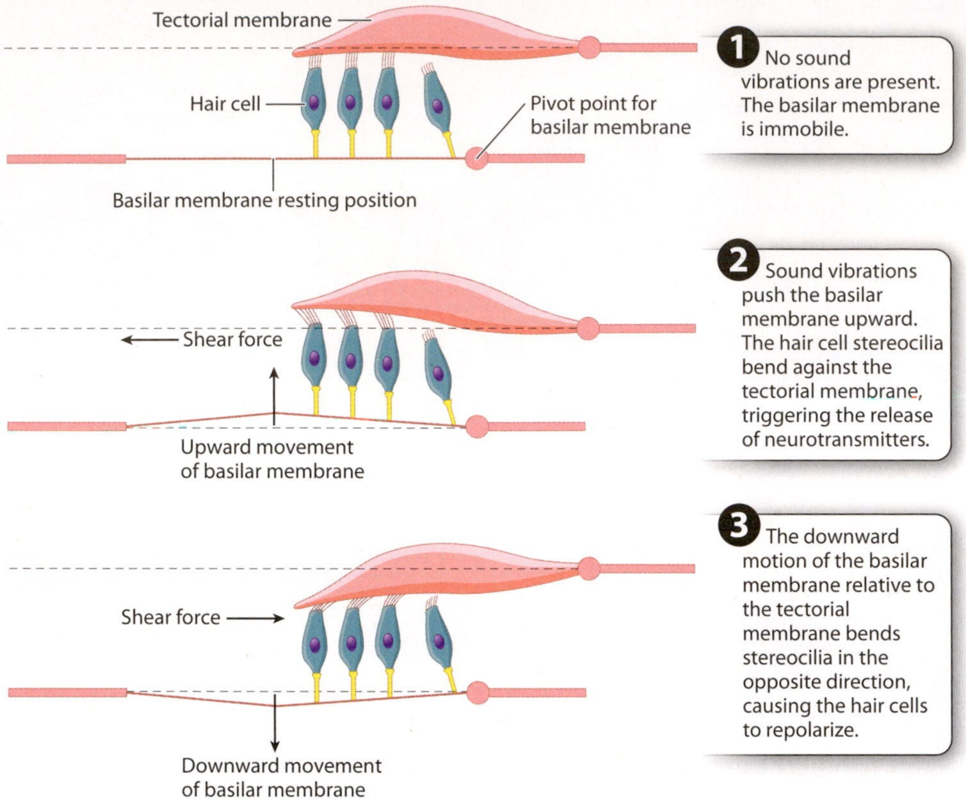

1. *Amplification.* Sound vibrations received by the outer ear are transmitted by the eardrum and amplified by the three bones in the middle ear (Fig. 36.8a). Through piston-like actions, the stapes transmits the energy of its movements to vibrations of the oval window of the cochlea in the inner ear. Vibrations transmitted from the eardrum through the middle ear to the oval window are amplified more than 30 times because of the larger size of the eardrum compared to the oval window and the lever-like action of the three bones.

2. *Transfer of sound vibration to fluid pressure waves.* The cochlea contains fluid and an upper and a lower canal separated by a **basilar membrane** and **cochlear duct** (Fig. 36.8b). Vibrations of the oval window cause fluid pressure waves in both canals at nearly the same time.

3. *Mechanoreception by hair cells within the cochlea.* The cochlear duct houses the **organ of Corti,** which contains specialized hair cells with stereocilia. These hair cells are supported by the basilar membrane. The stereocilia of mammalian hair cells form a "V" and project into a rigid **tectorial membrane** that does not move (Fig. 36.8c).

So far, we have seen pressure waves in the air transformed into vibrations of solid structures in the ear and then into waves of fluid. In the cochlea, motion is finally transduced into a nerve signal. The fluid vibrations within the cochlear canals induce motions of the basilar membrane relative to the tectorial membrane, bending the stereocilia of the hair cells back and forth in localized regions of the cochlea (**Fig. 36.9**). Bending of the stereocilia stimulates them to release excitatory neurotransmitters that cause postsynaptic neurons to fire action potentials. These are sensed as sound by neuronal networks in the **auditory cortex,** the area of the brain that processes sound.

Sound is characterized by amplitude (loudness) and frequency (pitch). Louder sounds produce larger fluid vibrations that cause the stereocilia to bend more, increasing their release of excitatory neurotransmitters. The release of more neurotransmitters increases the firing rate of the postsynaptic sensory neuron. The firing rate indicates intensity, or, in the case of hearing, loudness.

Many people can distinguish pure tones that differ by only a fraction of a percentage point in frequency. The ability to discriminate different sound frequencies is largely the result of differences in the mechanical properties of the basilar membrane along its length. At the apex, the basilar membrane is widest but thinnest and most flexible. At the base, the basilar membrane is narrow but thick and stiff. Thus, the basilar membrane is mechanically tuned to respond in different regions to different frequencies. Low-frequency vibrations excite hair cells in the apex of the basilar membrane, mid-frequency vibrations excite hair cells in the middle region of the basilar membrane, and high-frequency vibrations excite hair cells close to the base.

Sound amplitude and frequency, together with vestibular sensing of gravity and head motion, are transmitted by the vestibulocochlear nerve (another cranial nerve) to the brain.

? CASE 7 PREDATOR–PREY: A GAME OF LIFE AND DEATH

How have sensory systems evolved in predators and prey?

Many insect-eating bats rely on **echolocation** to find and apprehend their flying prey (**Fig. 36.10**). As it flies, a bat emits short bursts of high-frequency sound. These sound pulses bounce off surrounding objects and are reflected back to the bat. The echoes are detected by the ears and processed by the brain to locate the prey. Bats typically increase their call rate as they approach the prey to locate it more accurately and judge its flight trajectory and speed. The sharpness of an image is measured by its resolution, the smallest distance between two features or objects that can be perceived. Bats can resolve prey and other objects at less than 1 mm, a resolution that exceeds that of the most sophisticated sonar developed by human engineers.

Together with mosquitoes and other flying insects, nocturnal moths are one of the bat's primary food sources. In response to the evolution of echolocation by bats, several moths have evolved the ability to emit sounds that interfere with the bat's sonar signal. These moths are more likely to escape capture. The evolution of sophisticated sensory systems in both groups of animals, and the brain organization that goes with it, is the result of a kind of evolutionary arms race. It illustrates how animal sensory systems and brains can achieve impressive detection and information-processing performance that improves an animal's chance of survival.

FIG. 36.10 **A bat about to catch a moth by high-frequency auditory echolocation.** *Source: Daniel Tuitert/age fotostock.*

36.4 VISION

Most multicellular animals have photoreceptor cells that respond to light. Animals with simple forms of photosensitivity can move toward or away from light sources. The photoreceptors of other animals are arranged to form eyes that produce an image. The eyes of some of these animals, such as those of annelid worms and arthropods, are adapted to detecting the motion of shapes and simple patterns of light. Other animals, such as vertebrates and cephalopod mollusks like squid and octopus, perceive sharper images.

Although animals have evolved a variety of light-sensing organs, all rely on the same light-sensitive protein, called **opsin,** to convert light energy into electrical signals in the receptor cell. Opsins are ancient molecules. They evolved early in the tree of animal life as receptor proteins sensitive to many different stimuli, including light. They are G protein-coupled receptors. As we saw in Chapter 9, G protein-coupled receptors activate G proteins, leading to a cellular response. In this case, the cellular response is a change in membrane potential. The observation that different types of animal eye use a common light-sensitive protein suggests a common evolutionary origin for animal eyes about 500 million years ago in the Cambrian Period.

Animals see the world through different types of eyes.

Although opsins are the universal photoreceptor protein among animals, the way the visual world is perceived—what it looks like to the organism—depends on the structure of the eye in which the receptors are embedded. Here, we look at the three main types of eye structure shown in **Fig. 36.11: eyecups, compound eyes,** and **single-lens eyes.** Each has very different capabilities.

In flatworms, two simple eyecups on the back surface of the head detect the direction and intensity of light sources (Fig. 36.11a). Each eyecup contains photoreceptors that point up and to the left or right. A pigmented epithelium blocks light from behind, so the photoreceptors receive light only from above and in front of the animal. In response to light, flatworms turn to move directly away from the light source, seeking a dark region that hides them from potential predators. To find a dark region, the flatworm nervous system compares the light intensity received by photoreceptors of the two eyecups, and then moves in the direction of lower light intensity. Many invertebrates use simple light-sensitive photoreceptors in a similar way.

A key innovation for detecting images was the evolution of a lens. Two major types of image-forming eye evolved in other invertebrates: compound eyes and single-lens eyes. Insects and crustaceans have compound eyes that consist of individual light-focusing elements, each having a lens, called **ommatidia** (singular, **ommatidium**) (Fig. 36.11b). The number of ommatidia in the eye determines the resolution, or sharpness, of the image. The compound eyes of ants have only a few ommatidia, and those of fruit flies have about 800. In contrast, the eyes of dragonflies have

FIG. 36.11 Three types of eye. *Photo sources: a. David M. Dennis/age fotostock; b. Eye of Science/Science Source; c. Danté Fenolio/Science Source.*

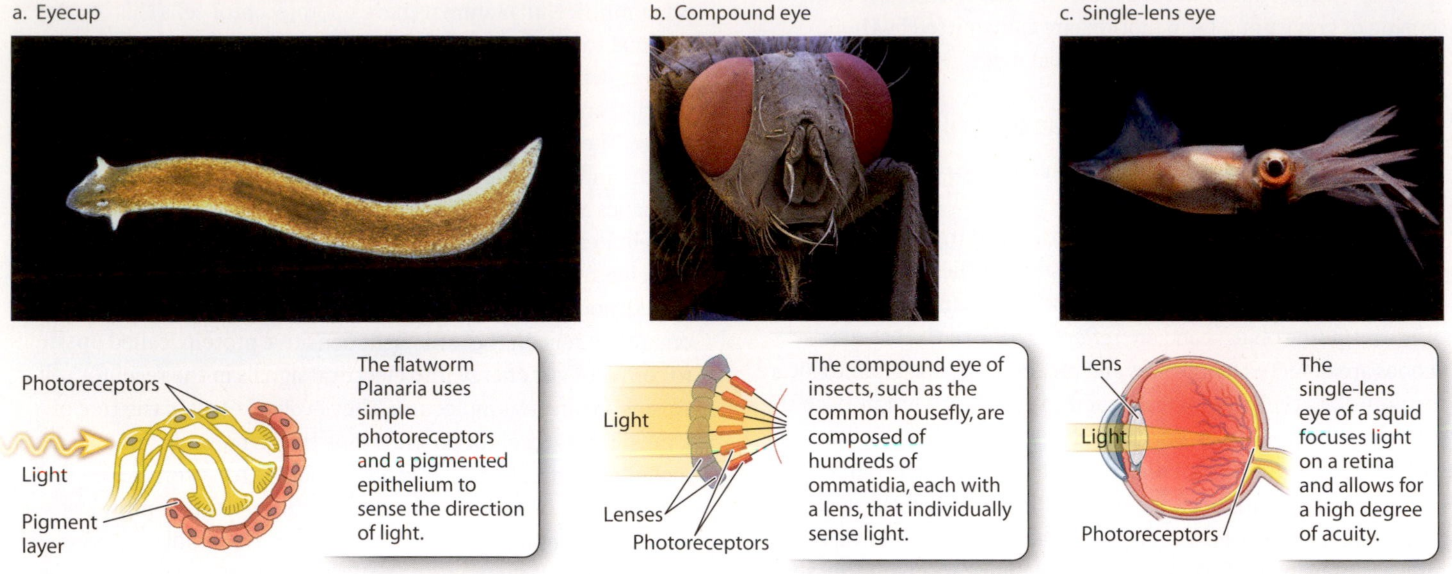

10,000 or more. Dragonflies are predators, and their high number of ommatidia is an adaptation for visually tracking their prey.

Within a single ommatidium, light is focused through a lens onto a central region formed from multiple overlapping photoreceptors. Each ommatidium is sensitive to a narrow angle of light (about 1° to 2°) in the animal's visual field. Compound eyes of insects and crustaceans provide a mosaic image because individual light regions are sensed by separate ommatidia. It is likely that the image is sharpened within the brain, but the resolution of images produced by compound eyes is not nearly as good as that produced by single-lens eyes. Nevertheless, compound eyes are extremely good at detecting motion and rapid flashes of light, more than 300 per second compared with 50 per second for a human eye. Insect eyes also have good color vision, and many insects can perceive ultraviolet light. Many pollinating insects sense UV light to locate the flowers they prefer to visit for nectar (**Fig. 36.12**).

The single-lens eyes of vertebrates and cephalopod mollusks like squid and octopus work like a camera to produce a sharply defined image of the animal's visual field (see Fig. 36.11c). In spite of their similar outward appearance, there is clear evidence that single-lens eyes evolved independently in vertebrates and in some mollusks (**Fig. 36.13**). Other invertebrates, such as some spiders and annelid worms, also evolved single-lens eyes independently. An advantage of single-lens eyes compared to compound eyes is that the single lens can focus light rays on a particular region of photoreceptors, improving both image quality and light sensitivity. As a result, animals with single-lens eyes can detect prey effectively by their motion and shape.

FIG. 36.12 The same flower seen with visual and UV light. Many pollinating insects see ultraviolet light, which can dramatically alter the image that attracts them to a particular flower. *Source: Bjorn Rorslett/Science Source.*

FIG. 36.13 Convergent evolution of single-lens eyes. Single-lens eyes evolved independently in vertebrates (fish and mammals) and in some mollusks (octopus and squid).

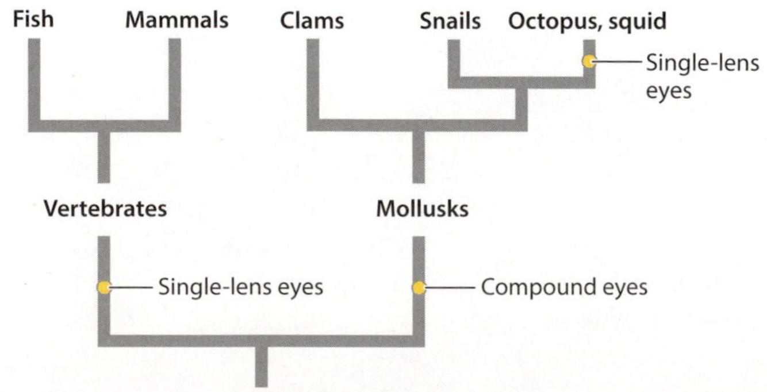

FIG. 36.14 **Anatomy of the human eye.** The eyes of other mammals and of birds have a similar organization.

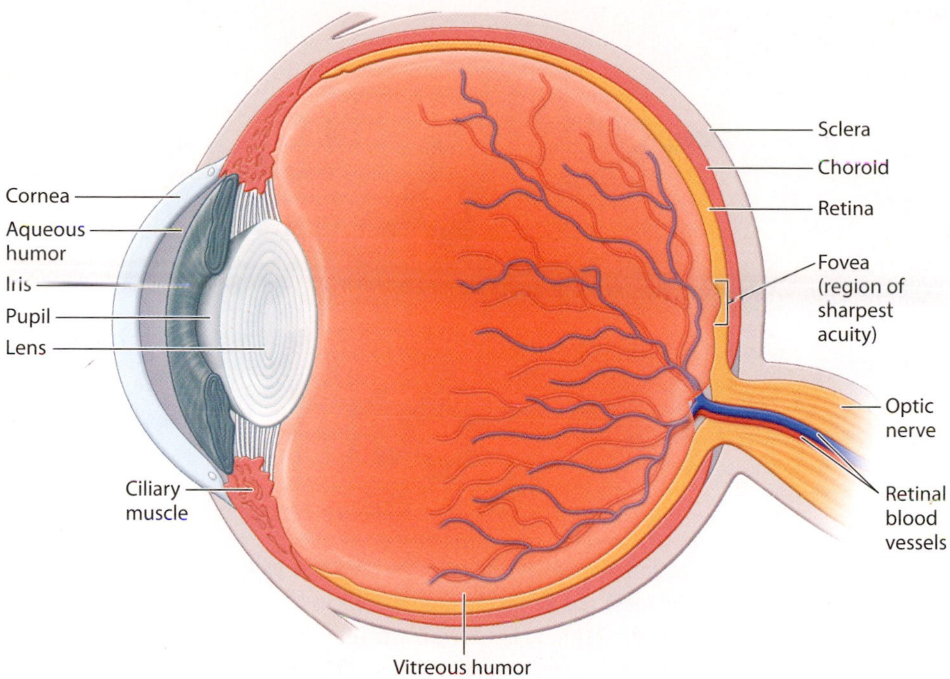

We focus much of our subsequent discussion on the human eye, but many of its organizational features and properties apply to vertebrates more generally.

The structure and function of the vertebrate eye underlie image processing.

Vertebrate eyes are protected within the eye socket of the skull. The anatomy of the eye is shown in **Fig. 36.14.** The eye is surrounded by the **sclera,** a tough, white outer layer. Below the sclera, a thinner layer, called the choroid, carries blood vessels to nourish the eye. Mucus secretions produced over the sclera keep the eye moist. The portion of the sclera in the front of the eye is the **cornea,** which is transparent. Light passes through the cornea, enters the eye through an opening called the **pupil,** and then passes through a convex **lens.** Small muscles called ciliary muscles attach between the sclera and lens. Contraction or relaxation of these muscles adjusts the shape of the lens to focus light images.

At the front of the eye, surrounding the pupil, the choroid forms a donut-shaped **iris** that opens and closes to adjust the amount of light that enters through the pupil, in the manner of the aperture of a camera. In bright light, the iris constricts to make the pupil small. In dim light, the iris dilates to allow more light to enter the pupil. The iris also gives the eye its color.

The focusing muscles and lens separate the eye into two regions. The interior region in front of the lens is filled with a clear watery liquid, the **aqueous humor.** This liquid is produced continuously and drained by small ducts at the base of the eye. The large cavity behind the lens is filled with a gel-like substance, the **vitreous humor,** which makes up most of the eye's volume.

To form an image, the cornea and lens bend incoming light rays, focusing them on the **retina** (Fig. 36.14). In doing so, the light rays cross, inverting the image on the retina and changing up to down and left to right. The retina is a thin tissue located in the posterior of the eye. It contains the photoreceptors and nerve cells that sense and initially process light stimuli. Humans, other mammals, and birds focus light on the retina by changing the shape of the lens (**Fig. 36.15**). When we focus on objects at close range, as when we read a book, the ciliary muscles contract, making the lens more rounded to bend the light rays more. When we focus on distant objects, the ciliary muscles relax, allowing the lens to flatten, reducing the bending of light. In contrast, the eyes of octopus and squid focus by moving the lens back and forth within the eye, as a camera does, rather than by changing its shape.

FIG. 36.15 **Focusing an image.** The eyes of humans and other mammals and of birds focus the light of an image on the retina by changing the shape of the lens.

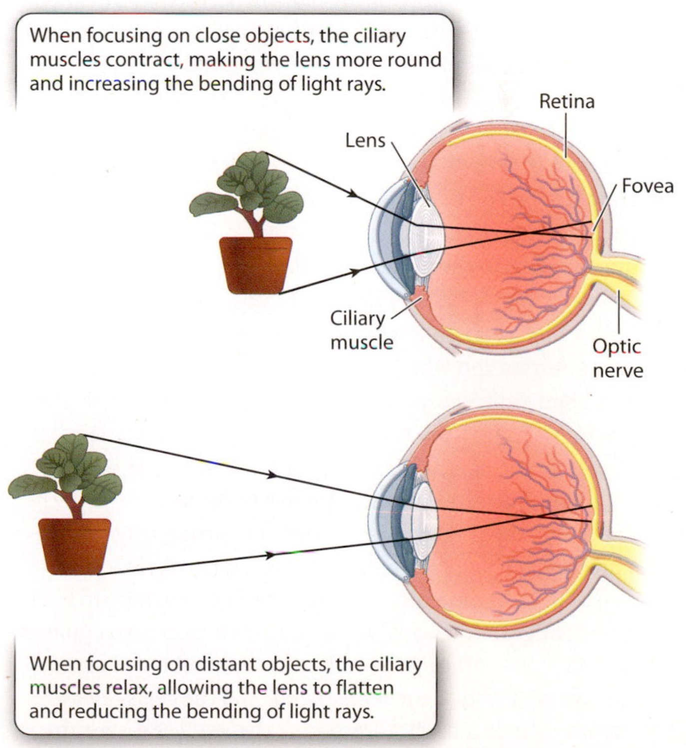

FIG. 36.16 Opsin bound to the light-absorbing pigment retinal. Retinal changes conformation from cis to trans when it absorbs a photon of light. *Rendered by Dale Muzzey.*

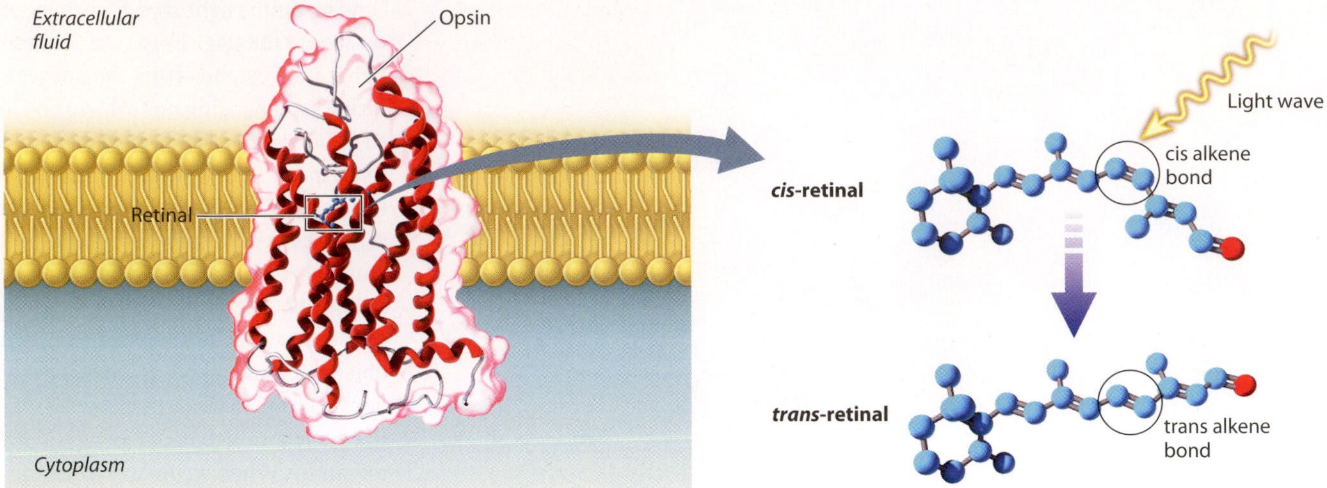

Why do so many animals have two eyes? One reason is that two eyes provide a wider field of vision. Close one eye and you will see the visual field of the other eye. With both eyes open, the visual field is wider than the visual field of either eye alone. You'll also notice considerable overlap in the visual field of both eyes. This overlap exists in humans and other primates, as well as in other mammals and birds of prey, and makes possible binocular vision, the ability to combine images from both eyes to produce a single visual image. Binocular vision allows for depth and distance perception.

Prey animals, such as some birds, antelope, and rabbits, have eyes positioned more to the side, expanding their visual field up to 360°, but limiting or preventing binocular vision. Animals that require greater visual sharpness, such as predatory hawks, cats, and snakes, have eyes pointing forward, increasing their binocular field of view.

Vertebrate photoreceptors are unusual because they hyperpolarize in response to light.

In photoreceptors of the retina, opsin molecules are arranged in cylindrical groups in the plasma membrane of most photosensitive cells. Each opsin protein contains a light-absorbing pigment molecule, called **retinal.** Retinal is a derivative of vitamin A that changes conformation when it absorbs a photon (**Fig. 36.16**).

Vertebrate photoreceptor cells have leaky Na^+ channels that let some Na^+ ions into the cell even at rest. As a result, their resting membrane potential (in the dark) is less negative (about −35 mV) than that of nerve cells. When a photoreceptor is in the dark and its resting potential is −35 mV, it continuously releases the neurotransmitter glutamate. However, when retinal absorbs a photon of light, it undergoes a conformational change from a cis to a trans configuration, causing Na^+ channels to close completely. Without Na^+ ions entering the cell, the cell membrane becomes hyperpolarized (rather than depolarized) and the cell's release of glutamate is reduced. Photoreceptors themselves do not fire action potentials, but when stimulated by light they increase or decrease the firing rate of neurons in the retina of the eye. That is, the reduction in neurotransmitter triggers ESPNs in the dendrites of some connecting neurons, and ISPNs in others. Differences in the firing rates of these neurons provide information about the intensity and location of light.

Color vision detects different wavelengths of light.

Color vision is crucial for many invertebrate and vertebrate animals. It is achieved by photoreceptor cells called **cone cells** that contain opsins sensitive to different wavelengths of light (**Fig. 36.17**). The more numerous **rod cells** are also sensitive to light, and most sensitive to blue-green light. Because all rod cells have the same opsin (called rhodopsin), the brain interprets

FIG. 36.17 Vertebrate photoreceptors: rods and cones. *Photo source: Science Source.*

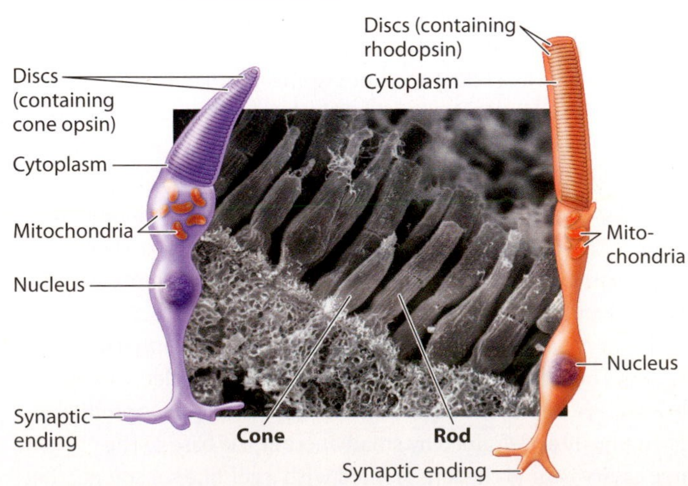

FIG. 36.18 Absorption spectra of the three opsins in human cone cells. Each cone cell expresses just one of the three opsins.

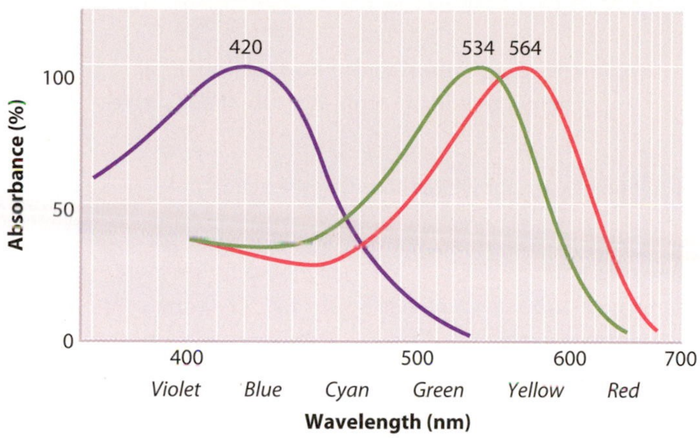

the light detected by rod cells as shades of gray, not as blue-green. The human retina contains about 6 million cone cells and about 125 million rod cells. Because of their greater number and sensitivity to light, rod cells enable animals to see in low light. Together, rod and cone cells constitute approximately 70% of all sensory receptor cells in the human body, highlighting the importance of vision in perceiving our environment.

Vertebrate cone cells likely evolved from a rod cell precursor. Most non-mammalian vertebrates have four types of cone cell, each with a different opsin, whereas most mammals have only two types of cone cell, and two opsins. It is likely that two opsins were lost early in mammal evolution, when mammals were small, nocturnal, and burrowing. Old World primates, apes, and humans, as well as some New World primates, regained trichromatic (three-color) vision by means of gene duplication, likely in response to selection for better ability to locate fruit, a key part of their diet. Each human cone cell therefore has one of three opsins, which absorb light at blue, green, or red wavelengths (**Fig. 36.18**). Stimulation of cone cells in varying combinations of these three opsins allows humans and other primates to see a full range of color (violet to red). Fish, amphibians, reptiles, and birds also have good color vision, including for many the ability to see ultraviolet.

Cone cells require higher levels of light than rod cells to become stimulated. The low sensitivity of cone cells to light makes it hard to detect color at night. Cone cells are most concentrated within the **fovea** of the retina, the center of the visual field of most vertebrates (see Fig. 36.15). Cone cells provide the sharpest vision. Animals with particularly sharp vision, such as hawks and other birds of prey, have an extremely concentrated number of cone cells in the fovea, approaching 1 million/mm^2 (compared with about 150,000/mm^2 in the human fovea). Birds of prey can see small prey on the ground from high in the air. Many birds also have eyes with two foveae, one projecting forward for binocular vision and one projecting to the side to enhance the sharpness of their side vision.

In contrast, rod cells are absent in the fovea but predominate in the periphery, making it easier to detect motion in the periphery, particularly at night. The eyes of many nocturnal predatory mammals, such as foxes and cats, contain mainly rod cells, enhancing nighttime vision. These animals also have a pigment located behind the retina that reflects light past the photoreceptors, enhancing the ability to sense light under low-light conditions. The reflection of light by this pigment creates the "eyeshine" that you see when a flashlight or a car's headlights shines on a nocturnal predator's eyes at night. Great white sharks are able to hunt at night because they also have a reflective pigment within the retina.

→ **Quick Check 4** What is the primary role of rod cells, and what are the two roles of cone cells in the retina?

Local sensory processing of light determines basic features of shape and movement.

The rods and cones detect the intensity, color, and pattern of light entering the eye through the lens. However, it is their interaction with a highly ordered array of neurons in the retina that begins the first steps of visual sensory processing (**Fig. 36.19**). The retina consists of five layers of cells that form a signal-processing network just in front of a pigmented epithelium. The rods and

FIG. 36.19 Cellular organization of the human retina. The retina is made up of five layers of cells: rods and cones, bipolar cells, ganglion cells, horizontal cells, and amacrine cells.

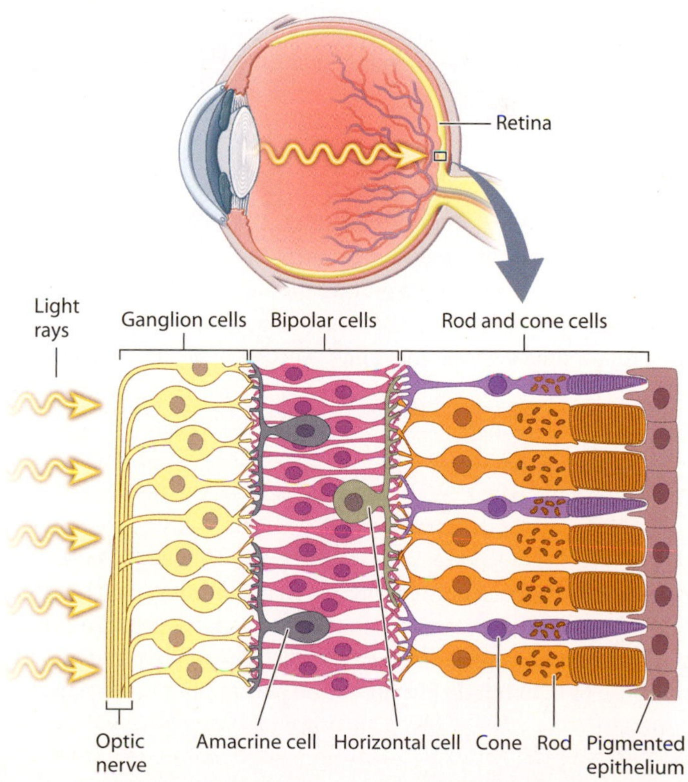

HOW DO WE KNOW?

FIG. 36.20

How does the retina process visual information?

BACKGROUND In the 1950s, the American neurophysiologist Stephen Kuffler was interested in understanding how the retina helps to process light information before it is sent to the brain. He focused on the activity of ganglion cells in the retina because they receive input from the photoreceptors and bipolar cells.

EXPERIMENT Kuffler stimulated different regions of a cat's retina with localized points of light while recording the action potentials produced by ganglion cells.

RESULTS Kuffler found that there are two types of ganglion cell: on-center and off-center cells. On-center ganglion cells fire more action potentials when light shines on the center of the cell's receptive field compared to the surrounding region, and off-center cells fire more when light is shown in the periphery and less on the center. These patterns are explained by lateral inhibition of input by the photoreceptors and varying excitation or inhibition of bipolar cells to the ganglion cells in the retina.

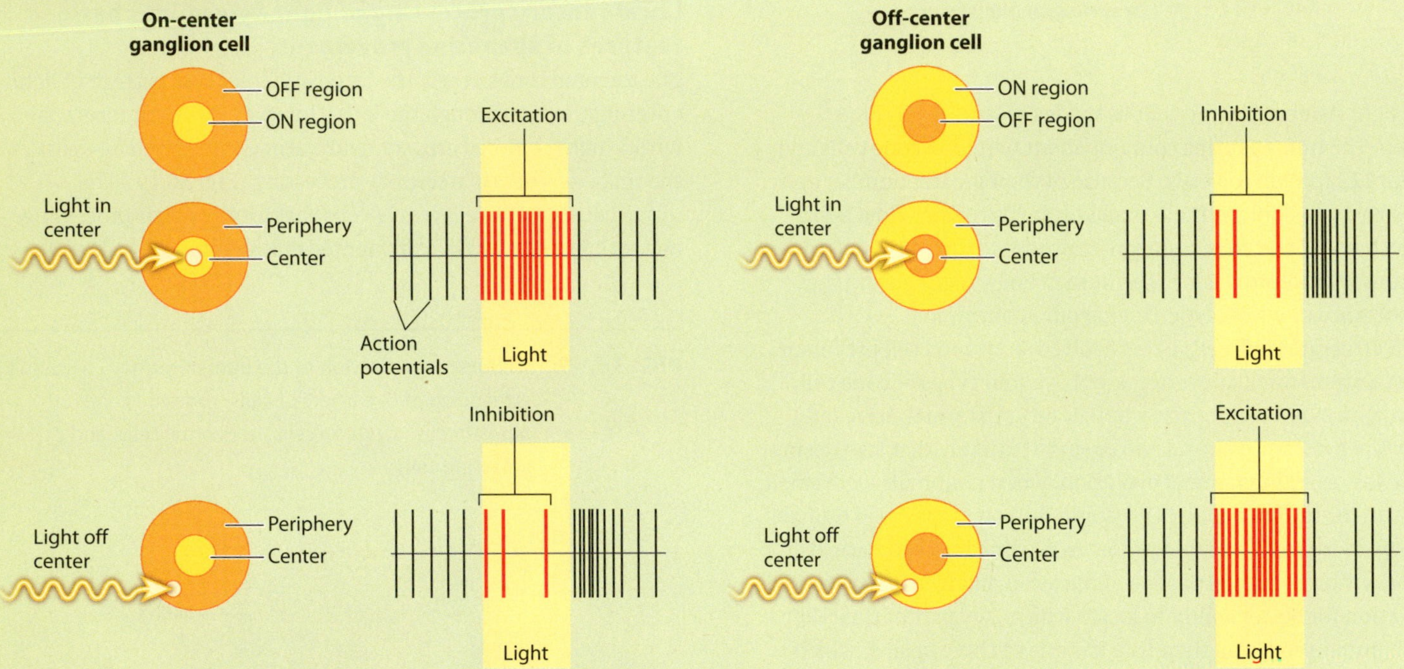

FOLLOW-UP WORK In the 1960s, Hubel and Wiesel found similar center–surround neural receptive fields, though with enhanced opposition, in part of the thalamus and in the visual cortex of the brain. Cells with these fields enable cats and other mammals to detect shapes of a given orientation moving through their visual field. Similar center–surround receptive fields have also been found in the somatosensory and auditory cortex, highlighting the use of lateral inhibition to enhance sensory acuity and edge detection. Other studies have found similar center–surround sensory processing in invertebrates and other vertebrates.

SOURCES Kuffler, S. W. 1953. "Discharge Patterns and Functional Organization of Mammalian Retina." *Journal of Neurophysiology* 16:37–68; Hubel, D. H. 1963. "The Visual Cortex of the Brain." *Scientific American* 209:54–62.

cones are at the back of the retina, so that light must pass through several layers before reaching these light-sensitive cells. Because they are photoreceptors, rod and cone cells hyperpolarize in response to light but do not fire action potentials. They synapse on to **bipolar cells,** which also do not fire action potentials, but rather adjust their release of neurotransmitter in response to the input from multiple rod and cone cells. Depending on their receptors, some bipolar cells are inhibited when rod and cone cells reduce their release of glutamate, whereas other bipolar cells are excited by the reduction in glutamate release.

The bipolar cells, in turn, synapse on to **ganglion cells** located on the front of the retina. If activated, ganglion cells transmit action potentials by the **optic nerve** (a cranial nerve) to the **visual cortex** in the brain, the part of the brain that processes images. The optic nerve begins at the front of the retina and exits at the back, creating an area without light-sensitive cells: This is the blind spot. In octopus and squid, photoreceptors are in the front of the retina, not in the back as they are in vertebrates. As a result, octopus and squid don't have a blind spot.

In addition to these three primary layers of neurons, two additional sets of nerve cells combine information laterally across the retina. **Horizontal cells** communicate between neighboring groups of photoreceptors and bipolar cells, enhancing contrast and adjusting photoreceptor sensitivity to light levels. Although their function is less well understood, we know that **amacrine cells** communicate between neighboring bipolar cells and ganglion cells, enhancing motion detection and likely reinforcing the adjustment of photoreceptor light sensitivity by horizontal cells.

Photoreceptor cells, bipolar cells, ganglion cells, horizontal cells, and amacrine cells form a neural network that processes visual information before it is sent to the brain for further processing and interpretation. Visual signals from approximately 131 million photoreceptors converge to approximately 1 million ganglion cell neurons that communicate with the brain. Properties of this network were first studied in the mammalian retina by the American neurophysiologists Stephen Kuffler, David Hubel, and Torsten Wiesel in the 1950s and 1960s (**Fig. 36.20**). Hubel and Wiesel shared the 1981 Nobel Prize in Physiology or Medicine for their later work on how visual information is processed in the brain.

36.5 BRAIN ORGANIZATION AND FUNCTION

Up to this point, we have looked at how sensory information, such as smell, taste, hearing, and vision, is detected by sensory receptors. In each case, the sensory information is sent to the brain, where it is processed and integrated. Much of what we know about how the brain functions has been accumulated over many years by studies of patients who suffered brain injuries.

A well-known example is Phineas Gage, a railroad worker who in 1848 survived severe damage to his brain's frontal lobe. A tamping iron he was using to set a dynamite charge caused the dynamite accidentally to explode, sending the iron through his skull and the frontal cortex of his brain (**Fig. 36.21**). Gage had been hard working, responsible, and well liked; after the accident, he was given to fitful arguments, lacked self-control, and seemed less intelligent. These changes provided some of the first evidence that particular brain regions control particular aspects of a person's behavior and cognitive function.

FIG. 36.21 The skull of Phineas Gage. Phineas Gage survived an accident in which a tamping iron passed through his skull, but his personality changed and his intellectual capacity diminished. *Source: The National Library of Medicine.*

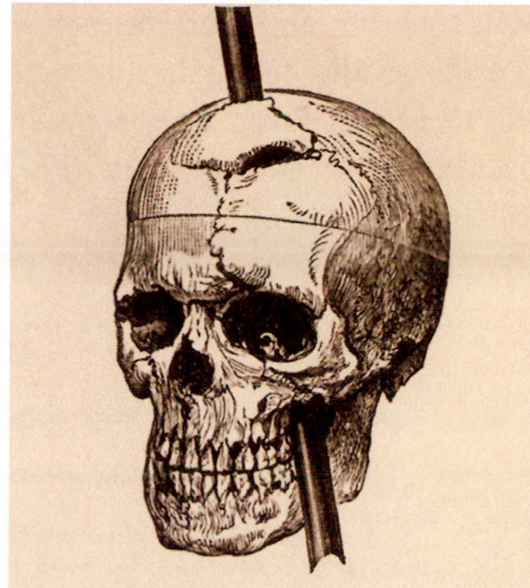

The remainder of this chapter explores the functional organization of the vertebrate brain. We discuss examples that show how the brain organizes sensory information, and how memories of sights and sounds, for example, are formed.

The brain processes and integrates information received from different sensory systems.

With the evolution of greater cognitive ability, animal brains became larger and more complex. Different brain regions became specialized to carry out different functions. The vertebrate brain is organized into a **hindbrain, midbrain,** and **forebrain,** with a **cerebral cortex** formed from a portion of the forebrain (**Fig. 36.22a**). The hindbrain and midbrain control basic body functions and behaviors, and the forebrain, particularly the cerebral cortex, governs more advanced cognitive functions. The cerebral cortex of mammals—and particularly of primates and humans—is greatly expanded relative to that of other vertebrates.

In adult vertebrates, the hindbrain develops into the **cerebellum** and a portion of the **brainstem** (Fig. 36.22a). The remainder of the brainstem develops from the midbrain. The cerebellum coordinates complex motor tasks, such as catching a ball or writing. It integrates both motor and sensory information. The brainstem, consisting of the **medulla, pons,** and midbrain, initiates and regulates motor functions such as walking and controlling posture, and coordinates breathing and swallowing. The brainstem activates the forebrain by relaying information from lower spinal levels. High levels of activity within the

FIG. 36.22 Human brain development. (a) The brain develops from the forebrain, midbrain, and hindbrain. (b) Part of the forebrain develops into the limbic system, which controls drives and emotions.

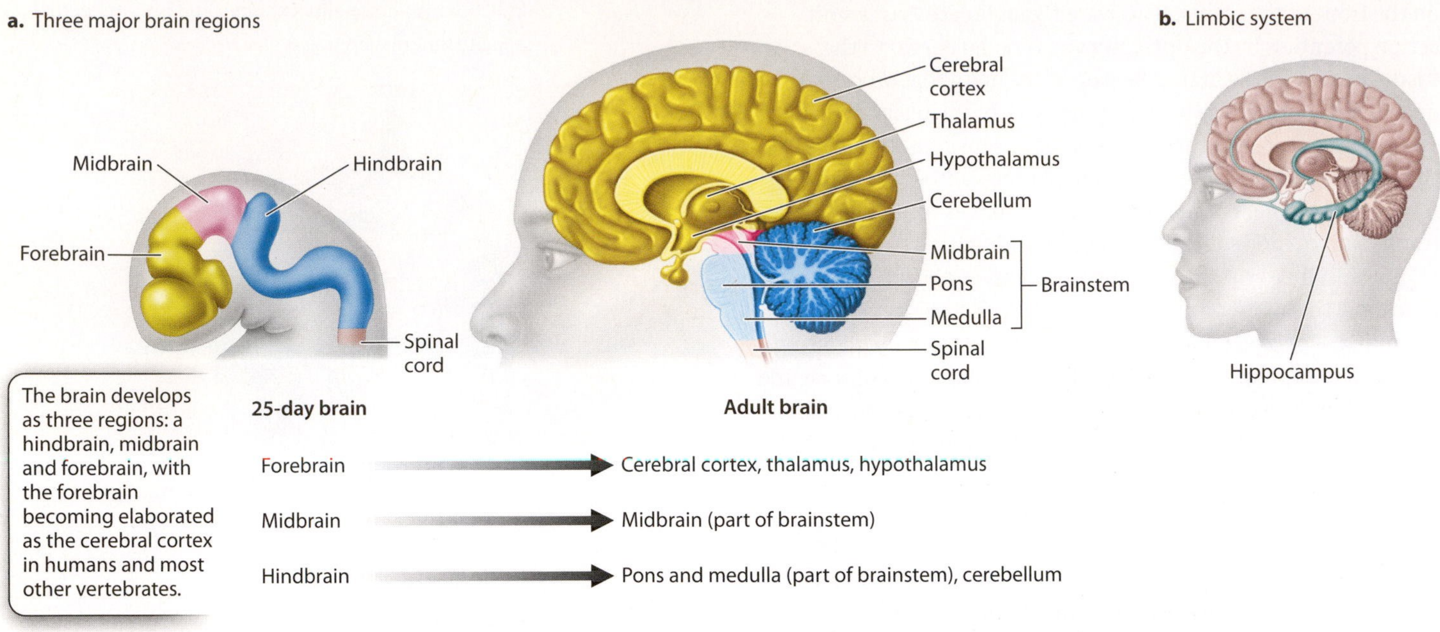

brainstem maintain a wakeful state; low levels enable sleep. If the midbrain is damaged, loss of consciousness and then coma result.

The forebrain consists of an inner brain region that forms the **thalamus** and the underlying **hypothalamus,** and a more anterior region that develops into the **cerebrum,** the outer left and right hemispheres of the cerebral cortex (Fig. 36.22a). The thalamus is a central relay station for sensory information sent to higher brain centers of the cerebrum. The hypothalamus interacts closely with the autonomic and endocrine systems to regulate the general physiological state of the body. In humans and most other primates, the cerebral cortex is the largest part of the brain, overseeing sensory perception, memory, and learning. Whereas the forebrain is essential to normal behavior in mammals, it appears less important for other vertebrates.

Other inner components of the forebrain constitute the **limbic system,** which controls physiological drives, instincts, emotions, and motivation and, through interactions with midbrain regions, the sense of reward (**Fig. 36.22b**). Stimulation of the limbic system can induce strong sensations of pleasure, pain, or rage. A posterior region of the limbic system, the **hippocampus,** is involved in long-term memory formation, discussed in the next section.

Sensory information reaches the cerebral cortex from the cranial nerves and nerves passing through the spinal cord. This information passes through the brainstem, and then through the thalamus, the central relay station for sensory information. From the thalamus, information for each of the senses goes to a different region of the brain specialized to further process that information, in a manner discussed next.

The brain is divided into lobes with specialized functions.

The cerebral hemispheres are the largest structures of the mammalian brain (**Fig. 36.23**). They consist of a highly folded outer layer of **gray matter** about 4 mm thick that forms the cerebral cortex, which is made up of densely packed neuron cell bodies and their dendrites (Fig. 36.23a). The folds greatly increase the surface area of the cortex and are the result of selection for an increased number of cortical neurons within the limited volume of the skull. Deep inside the cerebral cortex is the **white matter,** which contains the axons of cortical neurons. The fatty myelin produced by glial cells surrounding the axon makes this region of the brain white. These axons are the means by which neurons in different regions of the gray matter communicate with neurons in other regions of the cortex and in deeper regions of the forebrain, midbrain, and hindbrain.

Major regions of the cerebral hemispheres are defined by clearly visible anatomical lobes separated in most cases by deep crevices called **sulci** (singular, **sulcus**) (Fig. 36.23b). The **frontal lobe** is located in the anterior region of the cerebral cortex (behind your forehead). It is important in decision making and planning. The **parietal lobe** is located posterior to the frontal lobe and is separated from the frontal lobe by the central sulcus. The parietal lobe controls body awareness and the ability to perform complex tasks, such as dressing. The **temporal lobe** lies below the parietal lobe. It is involved in processing sound, as well as performing other functions discussed below. Located behind the parietal lobe, at the back of the brain, is the **occipital lobe,** which processes visual information.

FIG. 36.23 Human brain organization. (a) The folding of the brain and (b) the lobes of the cerebrum.

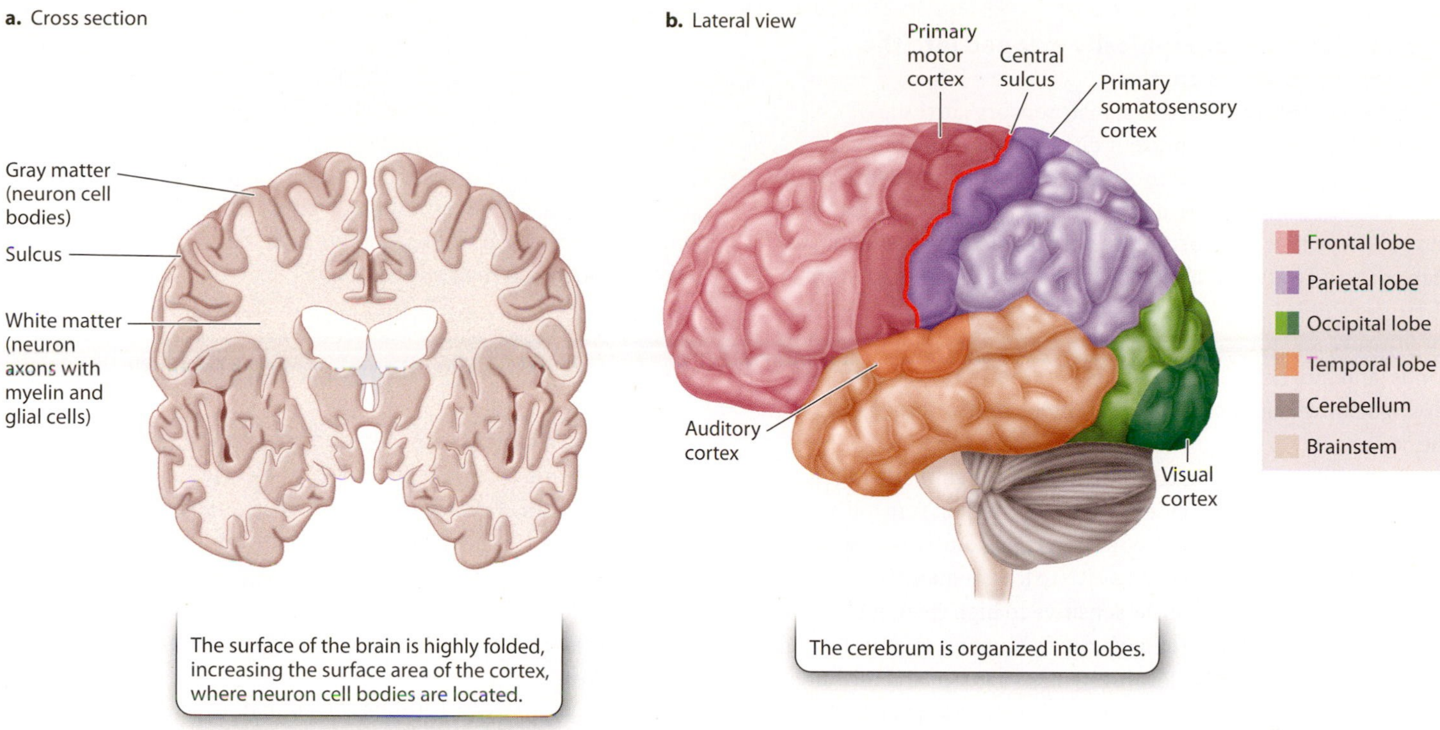

The central sulcus separates the **primary motor cortex** of the frontal lobe from the **primary somatosensory cortex** of the parietal lobe (**Fig. 36.24**). "Command" neurons in the primary motor cortex produce complex coordinated behaviors by controlling skeletal muscle movements. The primary somatosensory cortex integrates tactile information from specific body regions and relays this information to the motor cortex. Both cortices make connections to the opposite side of the body by sending axons that cross over in the brainstem and spinal cord. As a result, the right cortex controls the left

FIG. 36.24 Somatosensory topographic map.

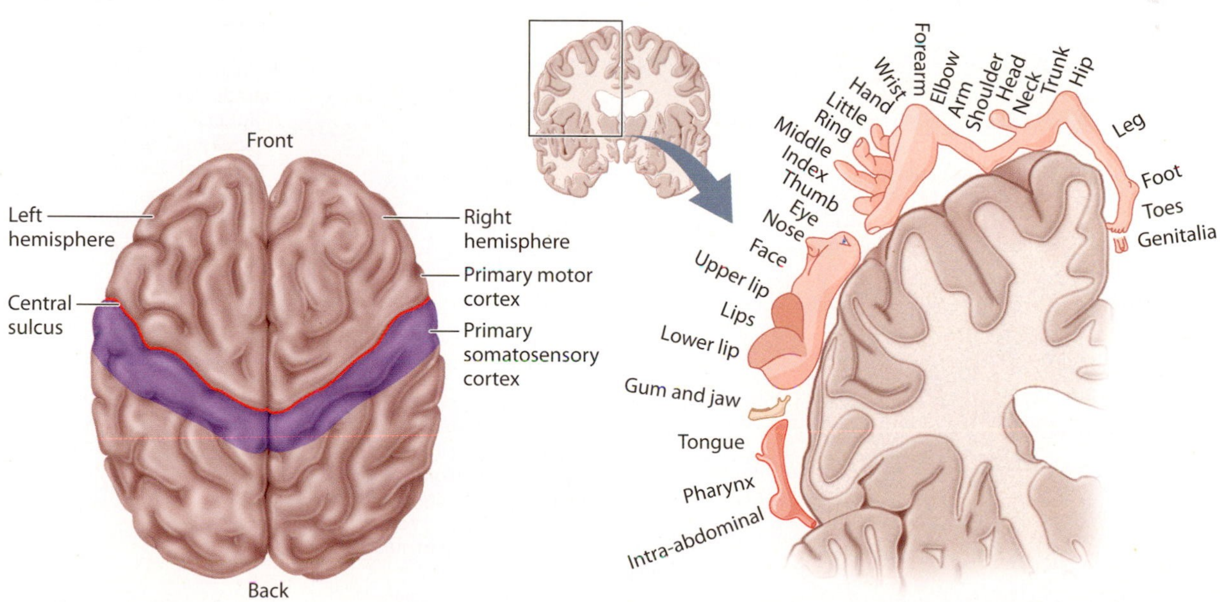

side of the body and the left cortex controls the right side of the body.

Information is topographically mapped into the vertebrate cerebral cortex.

The primary motor cortex and somatosensory cortex are organized as maps that represent different body regions, as shown in Fig. 36.24. The neurons that control movements of the foot and lower limb are located near the mid-axis of the motor cortex. Those that control motions of the face, jaws, and lips map to the side and farther down in the primary motor cortex. There is a similar topographic map in the somatosensory cortex for pressure and touch sensation. Regions such as the fingers, hands, and face that are involved in fine motor movements and distinguishing fine sensations are represented by larger areas of the cortex.

The auditory cortex in the temporal lobe is similarly organized as a map. This region processes sound information transmitted from the cochlea of the ear. Neurons in the auditory cortex are organized by pitch: Neurons sensitive to low frequencies are located at one end and neurons sensitive to high frequencies at the other. With the evolution of language in humans, regions farther back in the temporal lobe developed into language and reading centers. These centers are linked by pathways to a specialized motor center controlling speech, located at the base of the primary motor cortex of the frontal lobe. Functions of the temporal lobe include object identification and naming. An individual with damage to the temporal lobe may lose the ability to recognize faces, even though he or she may still recognize others by voice or other body features.

The occipital lobe topographically maps visual information received from the optic nerves. The topographic mapping of information from retinal ganglion cells enables the brain to detect patterns and motion that are precisely related to different regions of the visual field. As a result, highly visual animals, such as humans, primates, and birds, are able to detect complex patterns and movements in their environment.

36.6 MEMORY AND COGNITION

The ability of the brain to process and integrate complex sources of information, interpret and remember past events, solve problems, reason, and form ideas is broadly referred to as **cognition.** In this section, we consider memory and learning, as examples of cognitive abilities.

The brain serves an important role in memory and learning.

Memory is the basis for learning. An animal that can remember its encounter with a predator or noxious plant will avoid that danger in the future. A memory is formed by changes to neural circuits

FIG. 36.25 Synaptic plasticity. Memory and learning depend on long-term potentiation (LTP) of synapses within neural circuits.

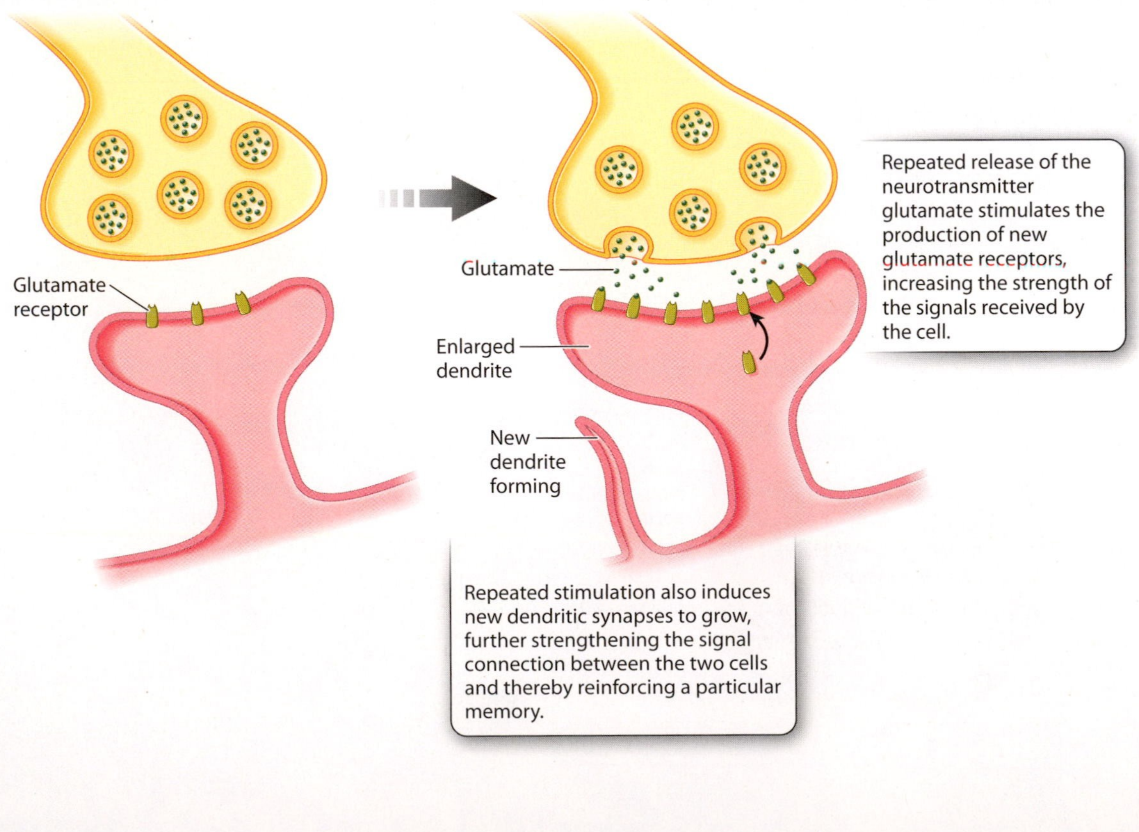

within specific regions of the brain, specifically, by changes in the synaptic connections between neurons in a neural circuit. These changes enable an animal, for example, to recognize other individuals in its social group or to recognize the threat of a predator. In humans and other primates, the hippocampus, a brain structure in the limbic region, plays a special role in memory formation. We can remember much of the information we receive for several minutes. However, unless reinforced by repetition or by paying particular attention, the information is typically forgotten. The hippocampus transforms reinforced short-term memories into long-term memories. Long-term memory is essential for learning.

The hippocampus forms long-term memories by repetitively relaying information to regions of the cerebral cortex. Although the details remain uncertain, memory and learning depend on establishing particular neural circuits in the hippocampus and cerebral cortex. These circuits can be activated to recall a memory that is triggered by a particular sound, or sight, or other relevant stimulus. Establishing a neural circuit requires establishing a particular set of connections between neurons. Thus, the formation of a memory circuit requires changes in synapses. Some synapses are weakened or removed altogether, and others are formed or strengthened.

The ability to adjust synaptic connections between neurons is called **synaptic plasticity** (**Fig. 36.25**). Long-term potentiation (LTP) is an example of synaptic plasticity that is believed to underlie memory and learning. In response to repeated excitation, neurons in excited circuits within the hippocampus release the neurotransmitter glutamate, opening Na^+ and Ca^{2+} channels. The increased Ca^{2+} stimulates signaling pathways leading to protein synthesis. New receptors are placed in the postsynaptic cell membrane and new dendrites are formed, strengthening synaptic signaling between the two cells. These changes make the cells more responsive to subsequent stimulation.

Through repeated activation of neurons within the hippocampus and associated regions of the cerebral cortex, neural circuits are established that may be reinforced by LTP to create a memory that can be retrieved over long periods of time. Although memories may be associated with certain regions of the cortex, they also rely on the interaction of nerve cells and networks within multiple brain regions.

Although the brain regions involved in memory and learning vary among different animals, the basic features of how memories are constructed by means of synaptic plasticity within neural circuits are likely broadly shared across animals capable of forming memories. For example, the ability of an octopus to learn and remember how to perform motor tasks rivals that of certain vertebrates. Even the roundworm *C. elegans* has been found to be capable of memory and learning: Individual roundworms learn to avoid pathogenic bacteria in favor of bacteria that are nutritious.

→ **Quick Check 5** Which three areas of the human brain are important for remembering and recognizing a face?

Cognition involves brain information processing and decision making.

Cognitive brain function ultimately gives rise to a state of consciousness—an awareness of oneself and one's actions in relation to others. Consciousness allows judgments to be made about events and experiences in one's environment. When you suddenly see someone you recognize from long ago, rapid neural associations within the cortex identify that person by name, retrieve multiple memories of past events you shared with that person, and elicit conscious thoughts, including your opinion of that person. At the same time, extraneous sensory information is filtered and other thoughts recede, allowing you to focus your attention on the person you have just seen and recognized.

Mental processes are studied by relating changes in brain activity to changes in an animal's behavioral state. The study of the brain is challenging because of the complexity of the circuits involved. Nevertheless, considerable progress is being made through non-invasive neuroimaging techniques (**Fig. 36.26**). These techniques allow the interactions of different brain regions within human and other animal subjects to be investigated while the subjects perform a particular mental or motor task. The results of neuroimaging studies can be linked to studies of the molecular and cellular properties of nerve cells, to

FIG. 36.26 **Non-invasive imaging of brain function.** PET scan imaging reveals regions of metabolic brain cell activity. The reddish areas show the locations of (upper left) the visual area activated by sight, (upper right) the auditory area activated by hearing, (lower left) the tactile area activated by touching braille, and (lower right) areas of the frontal cortex activated by generating words. *Source: WDCN/University College London/Science Source.*

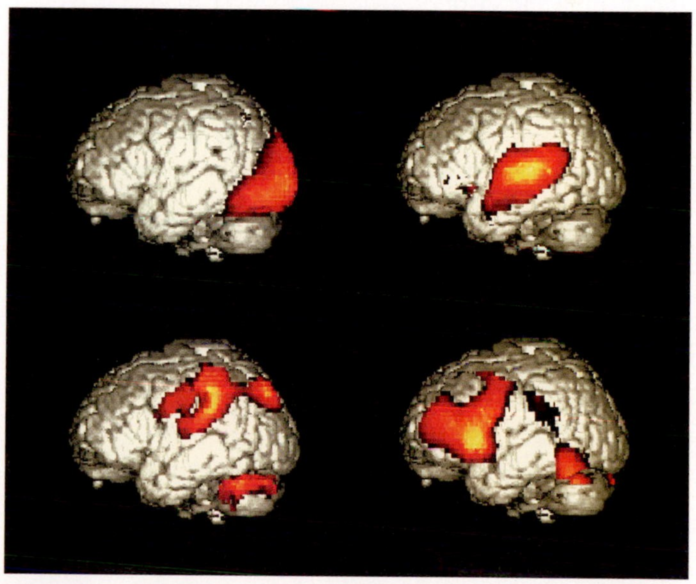

advance our understanding of how nerve cells form and maintain neural circuits among different brain regions, and of how these circuits are linked to particular behaviors and cognitive processes.

Do animals other than humans have a conscious state of awareness? Although the conscious awareness of other animals is difficult to study, it seems clear that many animals exhibit conscious states of awareness that influence their experience and behavior. Donald Griffin, known for his codiscovery as an undergraduate of bat echolocation, advocated the view not only that animals are conscious, but also that animal consciousness is amenable to scientific study. Behavioral biologist Marion Dawkins summarized her review of studies on animal cognition and consciousness as follows:

> Our near-certainty about [human] shared experiences is based, amongst other things, on a mixture of the complexity of their behavior, their ability to "think" intelligently, and on their being able to demonstrate to us that they have a point of view in which what happens to them matters to them. We now know that these three attributes—complexity, thinking, and minding about the world—are also present in other species. The conclusion that they, too, are consciously aware is therefore compelling. ■

Core Concepts Summary

36.1 Animal sensory receptors detect physical and chemical stimuli by changes in membrane potential.

Sensory receptors include chemoreceptors, mechanoreceptors, thermoreceptors, pain receptors, and electromagnetic receptors. page 764

Action potential firing rate correlates with the strength of a stimulus. Generally, strong stimuli induce high firing rates and weak stimuli induce low firing rates. page 764

Animal sensory receptors increase their sensitivity to stimuli by temporal and spatial summation, enhance their acuity by lateral inhibition, and adapt to continuous stimuli. page 765

36.2 Specialized chemoreceptors relay information about smell and taste.

Odorant molecules bind to membrane receptors on the surface of olfactory chemoreceptors. page 766

Olfactory sensory cells are neurons that fire action potentials when sending odor information to the brain. page 766

In humans, five taste receptors—for sweet, bitter, sour, salty, and savory—are activated in combination to determine a specific taste. page 766

Taste receptors are sensory cells that do not fire action potentials but communicate by synapses with afferent neurons. page 766

36.3 Hair cells convey information about gravity, movement, and sound.

Hair cells are specialized mechanoreceptors that detect motion and vibration. page 767

The statocysts of invertebrates are sensitive to gravity, orienting an organism to "up" and "down." page 767

The vestibular system of the vertebrate ear provides a sense of balance and gravity. page 768

The tympanic membrane in the outer ear transmits airborne sound waves to the middle ear, where the signal is amplified. The amplified sound waves are transmitted to the cochlea in the inner ear, where sound waves are converted to fluid pressure waves. These waves are then sensed by hair cells that convert the signal to an electrical impulse. page 769

36.4 The ability to sense light and form images depends on photosensitive cells with light-absorbing proteins.

Opsin is a G protein-coupled receptor and the universal photoreceptor protein in all animal eyes. page 771

The eyecups of flatworms detect light and dark. page 771

The compound eyes of arthropods sense light using individual light-focusing elements called ommatidia, providing low acuity but rapid motion detection. page 771

The single-lens eyes of vertebrates and cephalopod mollusks focus images on a retina, which contains photoreceptor cells and interneurons that process the light stimuli. page 773

The protein opsin present in photoreceptor cells converts light energy into chemical signals, altering the firing rate of neurons. page 774

Photoreceptor cells of the retina include rod cells, which detect light intensity, and three types of cone cells, which detect different wavelengths of light and allow color vision. page 774

Photoreceptor cells, bipolar cells, ganglion cells, horizontal cells, and amacrine cells form a network that processes visual information in the retina. page 775

36.5 The brain processes and integrates information from multiple sensory systems, with tactile, visual, and auditory stimuli mapped topographically in the cerebral cortex.

The vertebrate brain is organized into a hindbrain, midbrain, and forebrain, with the forebrain elaborated into the cerebral cortex in birds and mammals. page 777

The cerebrum is divided into frontal, parietal, temporal, and occipital lobes, which are specialized for different functions. page 778

Somatosensory, motor, auditory, and visual information is topographically mapped to specific areas in the cerebral cortex. page 780

36.6 Cognition is the ability of the brain to process and integrate complex information, remember and interpret past events, solve problems, reason, and form ideas.

The cerebral cortex integrates and processes information from diverse sources, giving rise to problem solving, reasoning, and decision making. page 780

The brain stores memories of past experiences and enables learning through the creation of more long-lasting neural circuits. page 781

Consciousness is an awareness of one's own identity in relation to others. page 781

Self-Assessment

1. Give an example of a chemosensory, mechanosensory, and electromagnetic sensory cell.
2. Diagram a generalized sensory receptor cell and show how it changes its firing rate in response to detected stimuli.
3. State a hypothesis that explains why all animals have chemoreceptors.
4. Describe the three main stages by which the mammalian ear detects and codes sound.
5. Compare and contrast the roles of rod cells and cone cells in the retina.
6. Describe the role of the cornea and lens in vertebrate eyes.
7. Describe three different types of eye in animals.
8. Draw the brain, label the lobes, and describe their primary functions.
9. Describe the importance of topographic mapping of sensory input to the cortex, using the primary somatosensory cortex as an example.
10. Explain how brain function can be understood by studying patients with brain injuries.

Log in to LaunchPad to check your answers to the Self-Assessment questions, and to access additional learning tools.

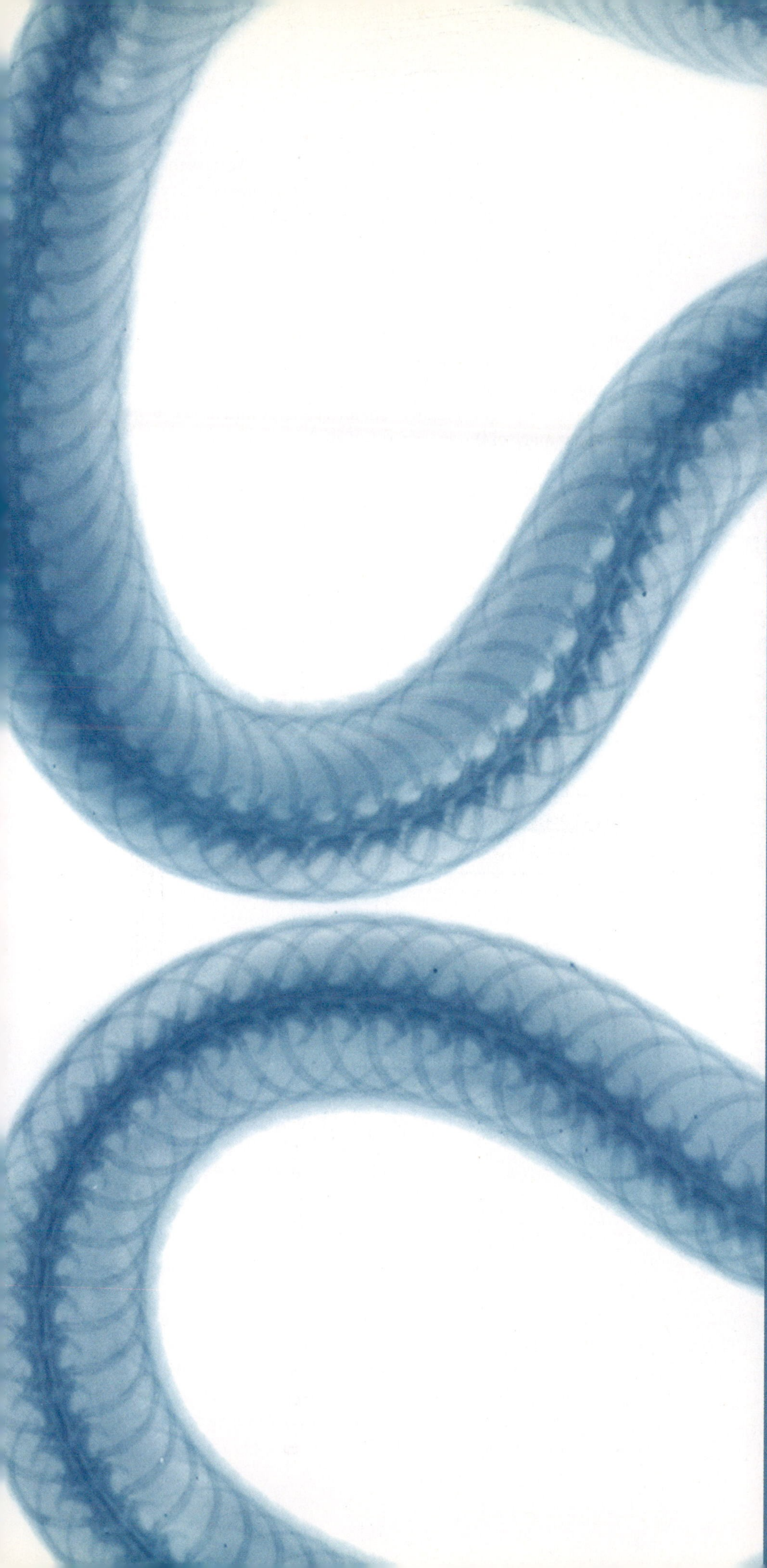

Gustoimages / Science Source.

CHAPTER 37

Animal Movement

Muscles and Skeletons

Core Concepts

37.1 Muscles are biological motors composed of actin and myosin that generate force and produce movement for support, locomotion, and control of internal physiological functionsl.

37.2 The force generated by muscle depends on muscle size, degree of actin–myosin overlap, shortening velocity, and stimulation rate.

37.3 Hydrostatic skeletons, exoskeletons, and endoskeletons provide animals with mechanical support and protection

37.4 Vertebrate endoskeletons allow for growth and repair and transmit muscle forces across joints.

The ability to move is a defining feature of animal life, allowing animals to explore new environments and avoid inhospitable ones, escape predators, mate, feed, and play. An animal's motor and nervous systems (Chapters 35 and 36) work together to enable it to sense and respond to its environment. How are movements produced, and how are they controlled? What determines an athlete's performance? How do muscles and the skeleton work together to provide movement and support? And how do muscles convert the chemical energy of ATP into force and movement during a contraction? This chapter explores the organization and function of the muscles and skeletons that power the movements of larger multicellular animals and provide mechanical support of their bodies (**Fig. 37.1**).

37.1 MUSCLES: BIOLOGICAL MOTORS THAT GENERATE FORCE AND PRODUCE MOVEMENT

We rely on our muscles for all kinds of movement. Muscles function as biological motors within the body because they generate force and produce movement. Some of these movements are obvious—running, climbing stairs, raising your hand, playing the piano. We are less aware of others—moving food through the digestive tract, breathing in and out, pumping blood through the body. A muscle's ability to produce movement depends on electrically excitable muscle cells containing proteins that can be activated by the nervous system.

The contractile machinery of muscles is ancient. The proteins underlying muscle contraction are found in eukaryotes that lived more than 1 billion years ago, and the first muscle fibers common to all animals are found in cnidarians (jellyfish and sea anemones), which evolved at least 600 million years ago. Thus, basic features of muscle organization and function are conserved across the vast diversity of eukaryotes. As we will discuss, the geometry and organization of these proteins largely determines how muscles contract to produce force and movement.

Muscles use chemical energy to produce force and movement.

Muscles are composed of elongated cells called muscle **fibers.** Muscle fibers use ATP generated through cellular respiration (Chapter 6) to generate force and change length during a contraction. This conversion of chemical energy into muscle work underlies much of the energy cost of animal movement discussed in Chapter 40.

A **force** is a push or pull by one object interacting with another object. The forces generated by skeletal muscles, for example, act as pulling forces at specific sites of the skeleton. The work performed by a muscle is equal to force times length change. For example, work is increased when either the force produced or the distance the skeleton is moved increases. Similarly, a muscle that produces a pulling force twice as much as another muscle but moves the skeletal segment only half the distance does the same work as that other muscle.

Because muscles only exert pulling forces, pairs of muscles are arranged to produce movements in two opposing directions at specific joints of the skeleton. For example, muscles pull on the bones of your leg and foot to swing them forward when you kick a ball. Opposing sets of muscles contract to swing your leg back and flex your knee before and after the kick.

Muscles can shorten very quickly, and they can produce large forces for their weight. The forces they produce can be many times an animal's body weight, allowing animals to move quickly. Some small muscles contract extremely quickly: The muscles of many flying insects contract 200 to 500 times per second. Even in vertebrate animals, muscles can contract as many as 50 to 100 times per second. Other muscles, such as the closing muscle of a clam, contract slowly, but can do so for prolonged periods without tiring. To see how muscles produce forces and contract to change length, we now consider how muscle fibers are organized at the molecular level.

FIG. 37.1 Muscles and bones. Muscles and bones work together to produce movement.
Photo source: Tom Brakefield/Corbis.

Muscles can be striated or smooth.

All muscles contain the same contractile proteins that enable them to shorten and produce force—actin and myosin (Chapter 10). These proteins are organized into bundles called **filaments** that interact with one another to cause muscles to shorten.

FIG. 37.2 Light micrographs of (a) skeletal muscle, (b) cardiac muscle, and (c) smooth muscle. Skeletal and cardiac muscles have a striated, or striped, appearance; smooth muscle does not. *Sources: a. Innerspace Imaging/Science Source; b. Manfred Kage/Science Source; c. SPL/Science Source.*

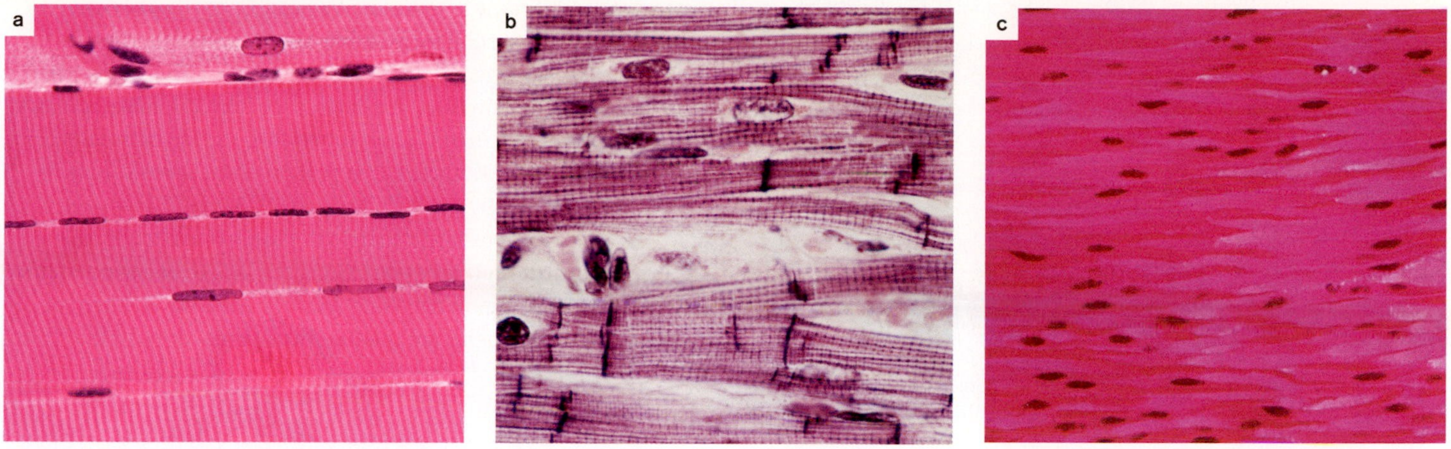

Although all muscle fibers have filaments of actin and myosin, the filaments are arranged differently in different types of muscle. Muscles are divided into two broad groups on the basis of their function and appearance under a light microscope (**Fig. 37.2**). As their name suggests, **striated muscles** appear striped under a light microscope because the actin and myosin filaments are arranged in a regularly repeating pattern. Striated muscles include **skeletal muscles** (Fig. 37.2a), which connect to the body skeleton to move the animal's limbs and torso, and **cardiac muscle** (Fig. 37.2b), which contracts to pump blood. Skeletal muscle cells are elongated and have many nuclei in each cell, whereas cardiac muscle cells are typically less elongated than skeletal muscle cells, often branched, and contain one or more nuclei per cell.

In contrast to skeletal and cardiac muscle, **smooth muscles** appear unstriated under the light microscope because the organization of actin and myosin filaments in smooth muscles is irregular (Fig. 37.2c). Smooth muscles are found in the walls of vertebrate and certain invertebrate (squid and octopus) arteries to regulate blood flow, in the respiratory system to control airflow, and in the digestive and excretory systems to help transport food and waste products. Smooth muscles contract slowly compared with cardiac and skeletal muscles. Consequently, bivalve mollusks rely on smooth muscle fibers to keep their shells closed for long periods without having to expend much energy.

Skeletal and cardiac muscle fibers are organized into repeating contractile units called sarcomeres.

The visible bands in skeletal and cardiac muscles are an important clue to the mechanism that produces muscle contractions. Whole muscles are made up of parallel bundles of individual muscle fibers (**Fig. 37.3**). Recall that a muscle fiber is a muscle cell. Each muscle fiber in turn contains hundreds of long, rodlike structures called **myofibrils.** Myofibrils contain parallel arrays of actin and myosin filaments that cause a muscle to contract.

FIG. 37.3 Skeletal muscle organization. Muscles are made up of bundles of muscle fibers (cells), each of which contains myofibrils.

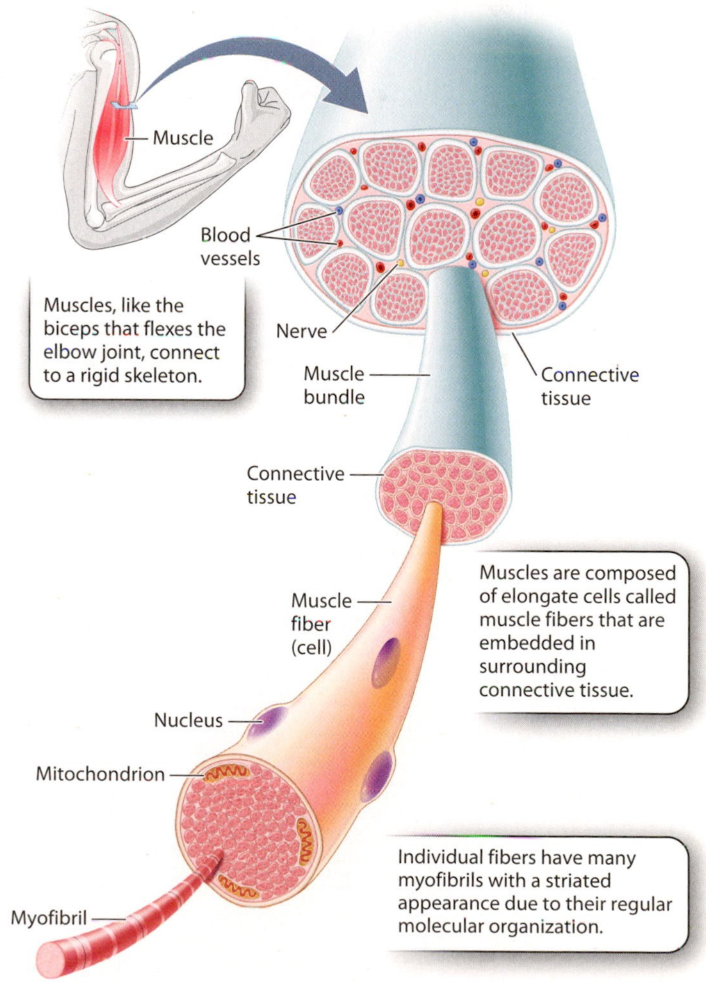

FIG. 37.4 **Thin and thick filaments.** A thin filament is made up of actin subunits arranged in a double helix. A thick filament is made up of numerous myosin molecules arranged in parallel.

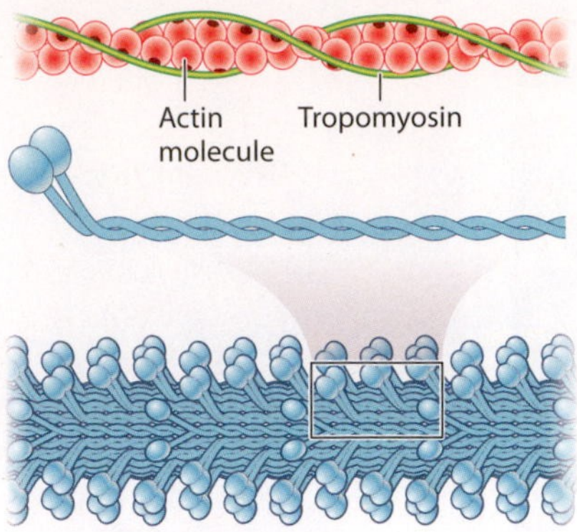

Let's consider the structural organization of myosin and actin filaments in skeletal muscles in more detail (**Fig. 37.4**). Each myosin molecule consists of two long polypeptide chains coiled together, each ending with a globular head. Consequently, the myosin molecule resembles a double-headed golf club. The myosin molecules are arranged in parallel to form a **thick filament,** with the numerous myosin heads extending out from their flexible necks along the myosin filament. Actin subunits are arranged as a double helix to form a **thin filament.** The protein **tropomyosin** runs in the grooves formed by the actin helices.

The thin filaments are attached to protein backbones called **Z discs** that are regularly spaced along the length of the myofibril (**Fig. 37.5a**). The region from one Z disc to the next is called the **sarcomere.** Sarcomeres are the functional units of muscles. In

FIG. 37.5 **Muscle sarcomere.** (a) Longitudinal and (b) cross-sectional views of a sarcomere, including two transmission electron micrographs. The sarcomere is the region between Z discs. It is the contractile unit of a muscle. *Photo sources: a. Don W. Fawcett/Science Source; b. Biophoto Associates/Science Source.*

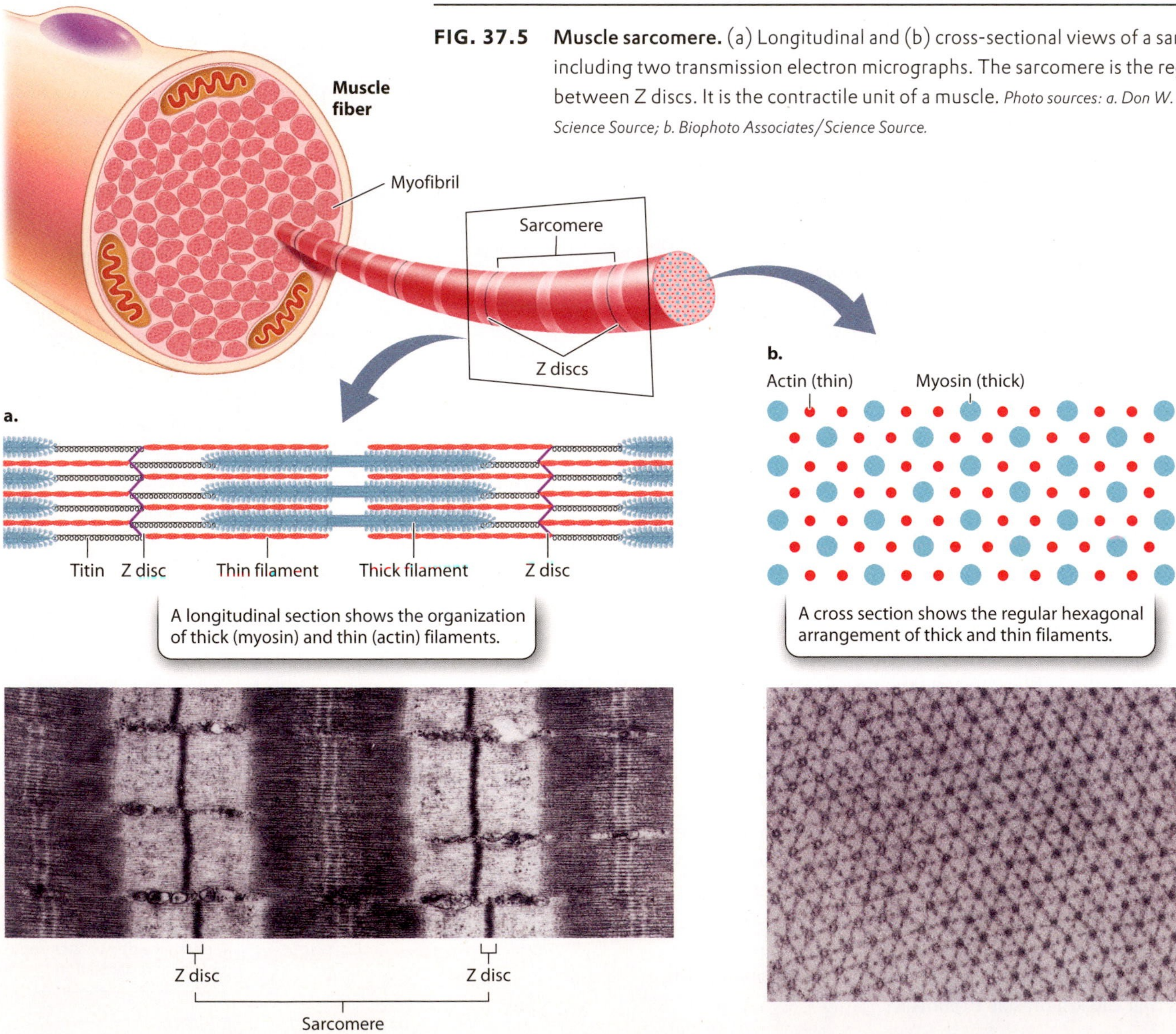

FIG. 37.6 Sliding filament model. Muscle contraction, or shortening, results from the sliding of actin thin filaments relative to myosin thick filaments. Filament sliding changes the sarcomere banding pattern between Z discs. *Photo source: Reprinted by permission from Macmillan Publishers Ltd: NATURE 173, H. Huxley and J. Hanson, Changes in the Cross-Striations of Muscle during Contraction and Stretch and their Structural Interpretation, Copyright 1954.*

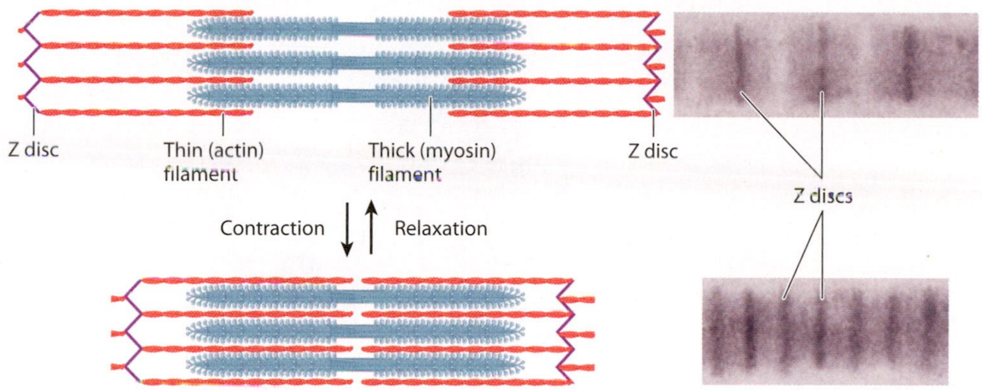

when rabbit or frog myofibrils were stimulated to contract to different lengths. Their results led them to hypothesize that muscles produce force and change length by the sliding of actin filaments relative to myosin filaments. Their theory is known as the **sliding filament model** of muscle contraction.

The banding patterns of sarcomeres revealed the extent of overlap between actin and myosin filaments. When myofibrils contracted to short lengths, the sarcomeres were observed to have increased actin–myosin overlap. When myofibrils were stretched to longer lengths, actin–myosin overlap decreased (**Fig. 37.6**). However, the lengths of the myosin and actin filaments never varied. Consequently, nearly all of the length change during a muscle contraction results from the sliding of actin filaments with respect to myosin filaments within individual sarcomeres. The length change of the whole muscle fiber is therefore a sum of the fractions by which each sarcomere shortens along the fiber's length.

A muscle's ability to generate force and change length is largely determined by the properties of its sarcomeres. Whereas sarcomere length is quite uniform in vertebrate animals (about 2.3 μm at rest), it is variable in invertebrate animals (ranging from as short as 1.3 μm to as long as 43 μm). Longer sarcomeres allow a greater degree of shortening. Thus, length changes are more limited in vertebrate muscles compared to those in some invertebrate muscles with long sarcomeres. For example, when the sarcomeres in an octopus tentacle contract, they can produce large motions of the tentacle. However, because vertebrates have shorter sarcomeres than an octopus, vertebrate muscles can shorten to produce movements more quickly.

Interactions between the myosin and actin filaments are what cause a muscle fiber to shorten and produce force. Along the myosin filament, one of the two heads of a myosin molecule at a given point in time binds to actin at a specific site to form **cross-bridges** between the myosin and actin filaments. The myosin filaments pull the actin filaments toward each other by means of these cross-bridges. The key to making the filaments slide relative to each other is the ability of the myosin head to undergo a conformational change and therefore pivot back and forth. The myosin head attaches to the actin filament and pivots forward, sliding the actin filament toward the middle of the sarcomere. The myosin head then detaches from the actin filament, pivots back, reattaches, and the cycle repeats. The movement of the myosin head is powered by ATP. Together, these events describe the **cross-bridge cycle.**

other words, the shortening (contraction) of a muscle is ultimately the result of the shortening of thousands of sarcomeres along a myofibril.

Sarcomeres are arranged in series along the length of the myofibril. Sets of actin thin filaments extend from both sides of the Z discs toward the midline of the sarcomere (Fig. 37.5a). In the middle of the sarcomere, not directly contacting the Z discs, are myosin thick filaments. The thin filaments overlap with the myosin thick filaments, forming two regions of overlap within a sarcomere. Toward either end of the sarcomere, as well as in the middle, are regions where the actin and myosin filaments do not overlap. The regions that overlap appear dark under an electron microscope, while the regions of non-overlap appear lighter in color. A third large protein, called titin, links the myosin filaments to the Z discs at the ends of the sarcomere. Titin is believed to help with assembly and protect the sarcomeres from being overstretched, thus contributing to muscle elasticity.

As we have seen, the regular pattern of actin and myosin filaments within sarcomeres along the length of the fiber gives skeletal muscles their striated appearance (Fig. 37.5a). In cross section, thick and thin filaments are arranged in a hexagonal lattice. This arrangement allows each thick filament to interact with six adjacent thin filaments (**Fig. 37.5b**). The precise geometry of thin and thick filaments is critical to their interaction, as we discuss next.

Muscles contract by the sliding of myosin and actin filaments.

Using high-resolution light microscopy, two independent teams (physiologists Hugh Huxley and Jean Hanson in the United States and Andrew Huxley and Rolf Niedergerke in England) were able to quantify the changing banding pattern of sarcomeres

FIG. 37.7 **Cross-bridge cycle.** The myosin head binds to actin and uses the energy of ATP to pull on the actin filament, causing muscle shortening.

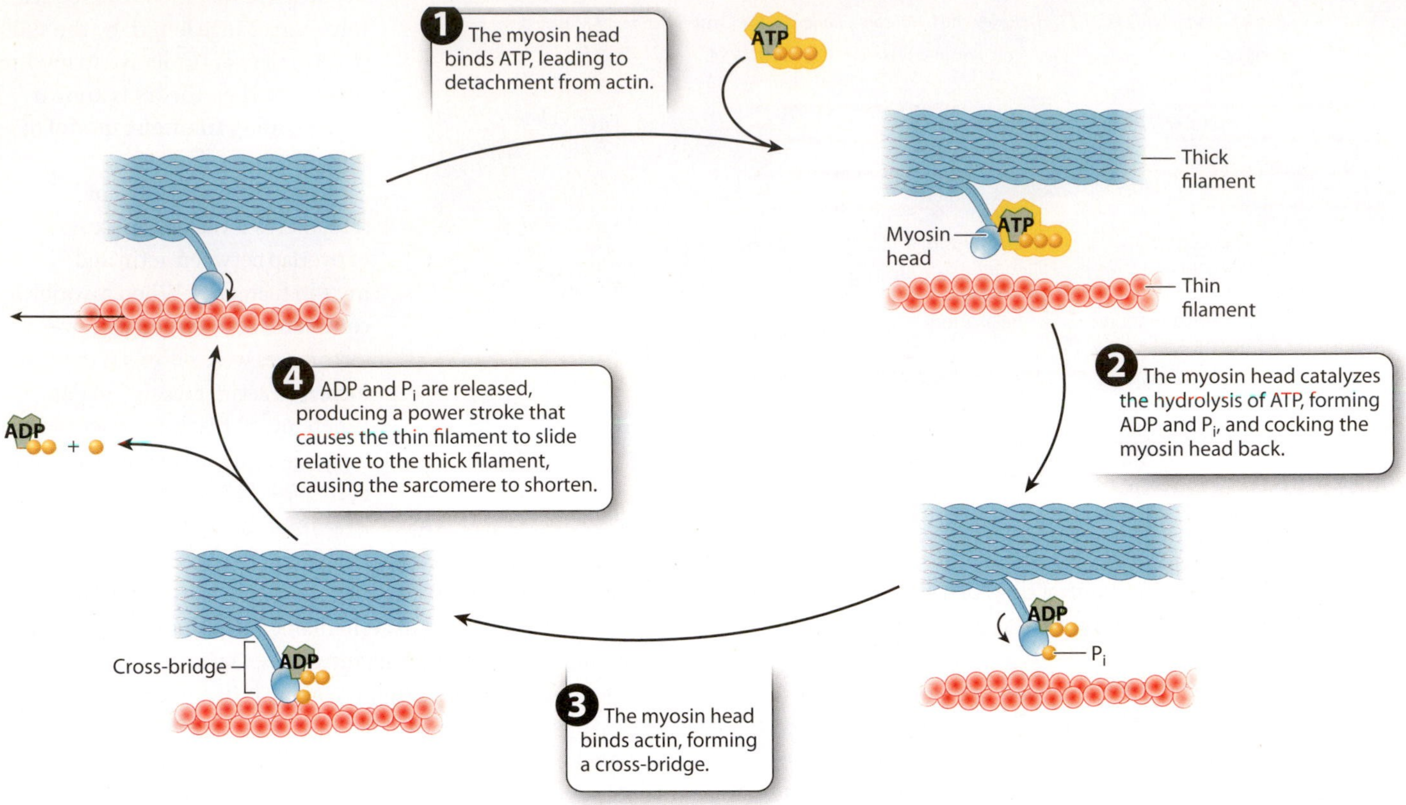

The cross-bridge cycle is just that—a cycle (**Fig. 37.7**). Let's look at it in more detail:

1. The myosin head binds ATP. Binding of ATP allows the myosin head to detach from actin and readies it for attachment to actin.
2. The myosin head hydrolyzes ATP to ADP and inorganic phosphate (P_i). Hydrolysis of ATP results in a conformational change in which the myosin head is cocked back. ADP and P_i remain bound to the myosin head. Because ADP and P_i are bound rather than released, the myosin head is in a high-energy state.
3. The myosin head then binds actin, forming a cross-bridge.
4. When the myosin head binds actin, the myosin head releases ADP and P_i. The result is another conformational change in the myosin head, called the **power stroke.** During the power stroke, the myosin head pivots forward and generates a force, causing the myosin and actin filaments to slide relative to each other over a distance of approximately 7 nm. The power stroke pulls the actin filaments toward the sarcomere midline.

The myosin head then binds a new ATP molecule (step 1 again), allowing it to detach from actin, and the cycle repeats. After detachment, the myosin head again hydrolyzes ATP, allowing it to return to its original conformation and bind to a new site farther along the actin filament. With the release of ADP and P_i, the myosin head undergoes another cycle of force generation and movement.

Individual muscle contractions are the result of many successive cycles of cross-bridge formation and detachment. During these cycles, muscle cells convert the chemical energy released by ATP into force and the kinetic energy of movement. Muscle fibers that contract especially quickly, such as those that power insect wing flapping or produce the sound of cicadas and rattlesnake tails, express myosin molecules that have high rates of ATP hydrolysis, allowing faster rates of cross-bridge cycling, force development, and shortening. Thus, myosin functions as both a structural protein and an enzyme.

Because each thick filament can interact with as many as six actin filaments, at any one time numerous cross-bridges anchor the lattice of myosin filaments, while other myosin heads are detached to find new binding sites. When summed over millions of cross-bridges, these molecular events generate the force that shortens the whole muscle.

→ **Quick Check 1** When an animal dies, its limbs and body become stiff because its muscles go into rigor mortis (literally, *rigor mortis* means "stiffness of death"). Why would the loss of ATP following death cause this to happen?

Calcium regulates actin–myosin interaction through excitation–contraction coupling.

We have discussed *what* makes muscles contract. We next consider *when* they contract. Skeletal and smooth muscle fibers are both activated by the nervous system. Whereas vertebrate skeletal muscles are innervated by the somatic nervous system, smooth muscles are innervated by the autonomic nervous system, as are cardiac muscles (Chapter 35).

Like nerve cells, muscle fibers are electrically excitable. Skeletal muscle fibers are activated by impulses transmitted by motor nerves to synaptic junctions (**Fig. 37.8**). Motor neuron axons have branches that allow them to synapse with multiple muscle fibers. When action potentials traveling down a motor neuron arrive at the neuromuscular junction, the neurotransmitter acetylcholine is released into the synaptic cleft. The neurotransmitter binds with postsynaptic receptors on the muscle cell at a region called the **motor endplate,** triggering the opening of Na^+ channels. The resulting influx of Na^+ ions in turn initiates a wave of depolarization that passes from the neuromuscular junction toward both ends of the muscle fiber (Chapter 35).

How does membrane depolarization lead to the cross-bridge cycle? Actin and myosin filaments can form cross-bridges only when the myosin-binding sites on actin are exposed. At rest, these binding sites are blocked by the protein tropomyosin. The wave of depolarization in the muscle cell initiates a chain of events that moves tropomyosin away from these binding sites, allowing cross-bridges between actin and myosin to form and the muscle to contract (Fig. 37.8).

Let's look at the chain of events that concludes with the exposure of the myosin-binding sites in more detail. You saw in Chapter 5 that eukaryotic cells contain several types of membrane-bound internal organelle. The myofibrils of muscle cells are surrounded by a highly branched membrane-bound organelle called the **sarcoplasmic reticulum (SR),** a modified form of the endoplasmic reticulum (Fig. 37.8). Depolarization initiated in the plasma membrane is conducted to the SR through infoldings of the plasma membrane (Fig. 37.8). When the muscle is at rest, the SR contains a large internal concentration of calcium (Ca^{2+}) ions transported in by calcium pumps in its membranes.

A muscular contraction is initiated when depolarization of the muscle fiber causes the SR to release Ca^{2+}. The Ca^{2+} diffuses into the myofibrils and binds to a protein called **troponin,** causing the troponin molecule to change shape. This conformational change of troponin, in turn, causes tropomyosin to move, exposing myosin-binding sites along the actin filament. Now myosin cross-bridges can form with actin, producing a contraction. Note that at this stage in the cross-bridge cycle, the myosin head has already

FIG. 37.8 Excitation–contraction coupling. Depolarization (excitation) leads to shortening (contraction) of the muscle. *Photo source: Don W. Fawcett/Science Source.*

a. Scanning electron micrograph of a motor nerve endplate

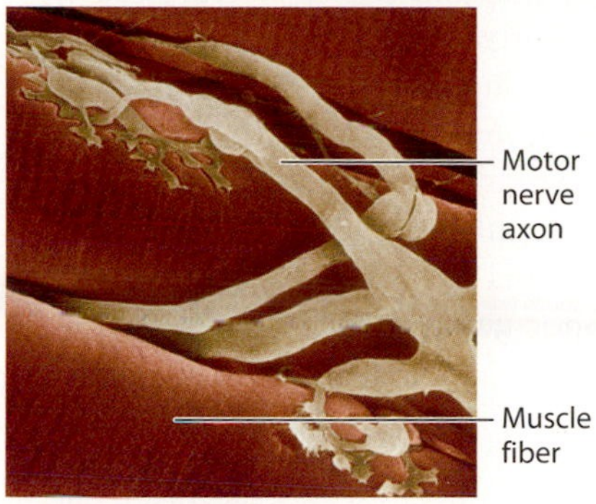

b. Muscle excitation–contraction coupling

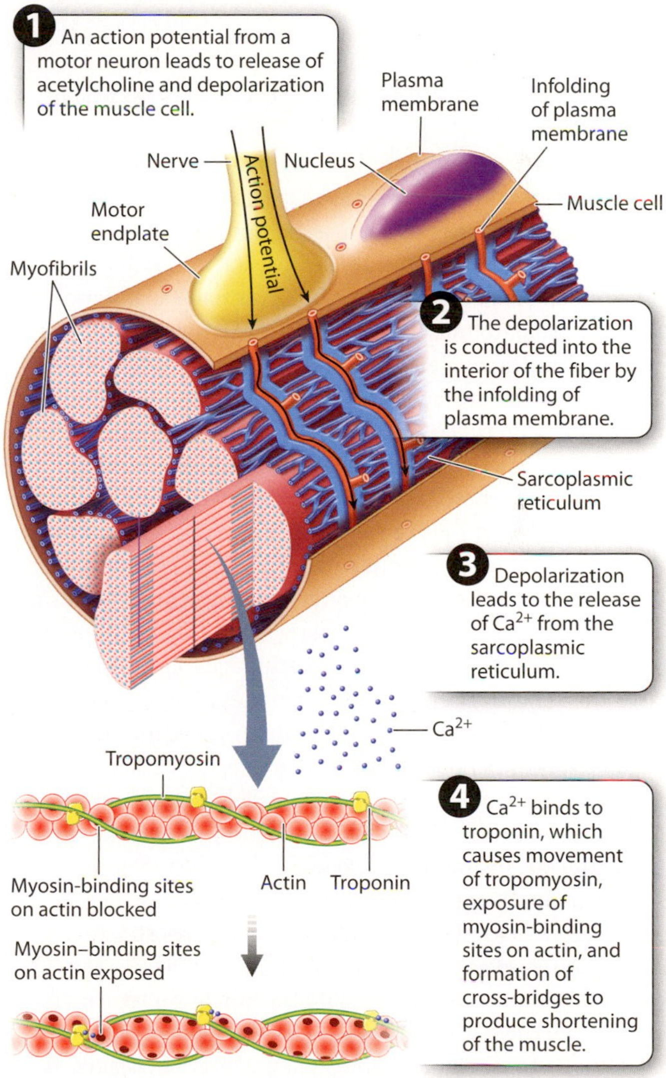

hydrolyzed ATP, is bound to ADP and P_i, and is in the cocked-back "ready" position. Binding to actin then allows the power stroke to occur. In this way, contraction is initiated immediately following depolarization and release of Ca^{2+}.

The process by which membrane depolarization leads to Ca^{2+} release from the SR and the formation of myosin-actin cross-bridges is called **excitation–contraction coupling** because excitation of the muscle cell is coupled to contraction of the muscle, producing force and movement. Together, these events are the molecular "switch" that causes a muscle to contract. The muscle relaxes when neural stimulation ends. Acetylcholine is broken down or reabsorbed, and Ca^{2+} is actively transported back into the sarcoplasmic reticulum, allowing tropomyosin molecules to once again block myosin-binding sites along the actin filaments.

→ **Quick Check 2** Curare is a paralyzing compound that blocks the action of the neurotransmitter acetylcholine at the muscle fiber's motor endplate. What effect do you think curare has on the release of calcium ions from the sarcoplasmic reticulum of the muscle cell?

Calmodulin regulates Ca^{2+} activation and relaxation of smooth muscle.

How is the contraction of smooth muscle regulated? The smooth muscle that controls many internal organs can be activated by the autonomic nervous system and also responds to stretch of the muscle, local hormones, and other local factors such as pH, oxygen, carbon dioxide, and nitric oxide by intracellular signaling. For example, smooth muscle in the walls of arteries contracts when stimulated by the release of epinephrine to reduce blood flow to a body region (Chapter 39), and smooth muscle within the mammalian uterus contracts when oxytocin is released to stimulate contractions (Chapter 42). As in skeletal muscle, these initiators trigger the release of Ca^{2+} from the SR. However, in smooth muscle, Ca^{2+} also enters through voltage-gated and stretch-receptor calcium channels in the cell's plasma membrane.

Smooth muscle lacks the troponin–tropomyosin mechanism for regulating contraction. Instead, activation of smooth muscle cells results when the protein **calmodulin** binds with Ca^{2+} released from the SR or entering through the cell's membrane. The calmodulin–Ca^{2+} complex activates a myosin kinase that phosphorylates the smooth muscle myosin heads, causing them to bind actin and begin the cross-bridge cycle. A second enzyme dephosphorylates the myosin heads, disrupting their ability to bind to actin, and the muscle relaxes.

Smooth muscle contracts slowly compared with skeletal muscle, using less ATP per unit time. The SR of smooth muscle cells is less extensive and has many fewer calcium pumps than the SR of skeletal muscle cells. As a result, calcium is returned more gradually from the myofibrils back to the SR, or pumped back out through the cell's membrane, slowing the relaxation of smooth muscle. Using the slow contractions of smooth muscle, clams hold their shells closed for long periods, arteries contract and restrict blood flow to particular body regions, and the gut moves digestive contents through the gastrointestinal tract.

37.2 MUSCLE CONTRACTILE PROPERTIES

Actin and myosin are evolutionarily conserved across all animal life, so all muscles have similar force-generating properties. The huge muscles of a weight lifter are evidence that larger muscles can exert more force. Why is that? The summed force produced by individual myofibrils of activated muscle fibers determines the force the whole muscle can exert. This means that larger muscles with more fibers produce greater forces than smaller muscles with fewer fibers. However, the force that a muscle produces also depends on its contraction length and the speed of contraction, as we discuss next.

Muscle length affects actin–myosin overlap and generation of force.

Why is it difficult to jump while standing on your toes? A muscle's ability to generate force depends in part on how much it is stretched before contraction begins. The sliding filament model for muscle contraction, like any good model, makes specific predictions. In this case, it predicts that the amount of overlap between actin and myosin filaments within the sarcomere determines the number of cross-bridges that can form. The more overlap, the more force that can be produced. Conversely, when a muscle fiber is pulled to long lengths and then contracts, it generates less force because the overlap between myosin and actin filaments is reduced. When a muscle fiber contracts at short lengths, it also produces less force because myosin filaments begin to run into the Z disc at each end of the sarcomere, disrupting the geometry of the myosin filament lattice and hindering cross-bridge formation. At intermediate lengths, the muscle fiber generates maximum force because actin–myosin filament overlap is greatest. These predictions were tested by experiments (**Fig. 37.9**).

This influence of muscle fiber length on force generated is the reason that you can't jump well if you are on your toes and your ankles are fully extended: This position shortens your calf muscle fibers. Conversely, if your ankles are overly flexed, your calf muscles are lengthened, and again jumping force is limited. Through training, athletes and dancers learn techniques, such as bending the knees before jumping, so that their leg muscles contract at intermediate lengths to maximize muscle-force output.

The relationship between force and length of striated muscle also explains Starling's Law for the function of the heart as a pump (discussed in Chapter 39), which ensures that the heart contracts more strongly when it is filled by larger amounts of blood returning from the veins.

→ **Quick Check 3** Why can you lift a larger load when your elbow is slightly flexed and your biceps muscle is at an intermediate length than you can when your elbow is fully extended?

FIG. 37.9

How does filament overlap affect force generation in muscles?

BACKGROUND From the 1930s through the 1960s, British muscle physiologists studied the properties of striated frog muscle fibers. In the first set of experiments, A. V. Hill examined how a muscle's shortening velocity affects its force production. In the second set of studies, American physiologists Albert Gordon and Fred Julian, together with British physiologist Andrew Huxley, studied how a muscle's length when it contracts affects the muscle's force-generating ability.

HYPOTHESIS Gordon, Huxley, and Julian hypothesized that changes in overlap between myosin and actin affect the number of cross-bridges and thus the force the muscle can produce.

EXPERIMENT The three scientists isolated a single muscle fiber and used optical microscopy to measure sarcomere length. They kept this length constant while the fiber was stimulated to produce force. Measurements of force were then obtained at different sarcomere lengths. The length at which the muscle fiber generated maximum force is indicated on the graphs as 100%. Other lengths are expressed relative to this length.

RESULTS The top graph shows the data obtained by Gordon, Huxley, and Julian, and the bottom graph reinterprets their data in terms of force and percent of muscle length. Their experiments showed that the muscle fiber produced maximal force at an intermediate sarcomere length (approximately 2.3 μm). At this length, the greatest number of cross-bridges were hypothesized to form, generating maximal force. As the sarcomere length was decreased or stretched, and the sarcomere then stimulated to contract, force declined. These results were consistent with a decrease in the number of cross-bridges that could form as a result of decreased actin–myosin overlap.

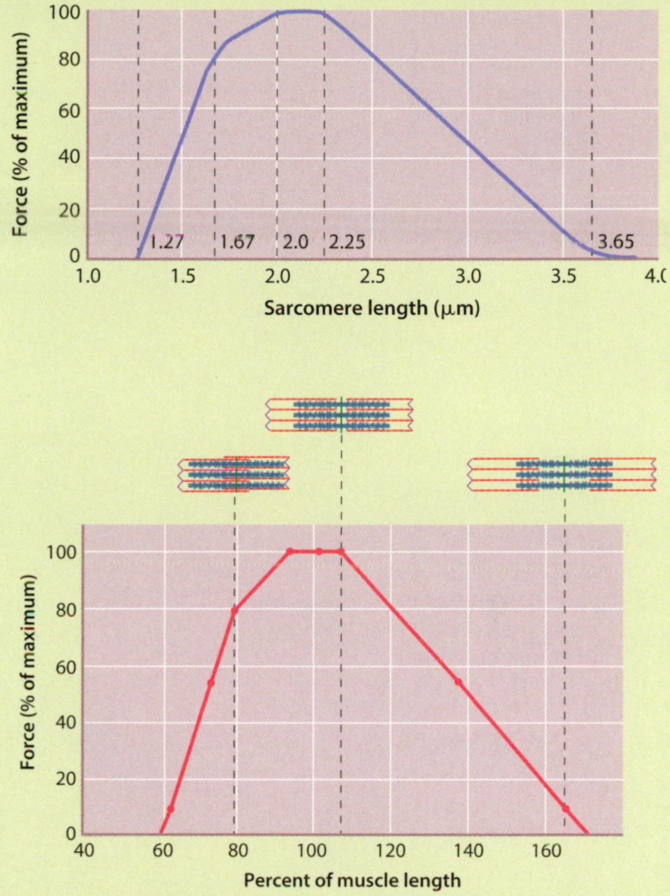

FOLLOW-UP WORK Other studies confirmed that changes in myosin cross-bridge formation are linked to changes in actin and myosin filament overlap at different sarcomere lengths, supporting the sliding filament model of muscle contraction.

SOURCE Gordon, A. M., A. F. Huxley, and F. J. Julian. 1966. "The Variation in Isometric Tension with Sarcomere Length in Vertebrate Muscle Fibres." *Journal of Physiology* 184:170–192.

Muscle force and shortening velocity are inversely related.

In experiments carried out in the 1930s, A. V. Hill (who won the 1922 Nobel Prize in Physiology or Medicine for his work on muscle energy use) observed that a muscle shortens fastest when producing low forces (**Fig. 37.10**). To produce larger forces, the muscle must shorten at progressively slower velocities. For example, your arm moves faster when you throw a light ball than when you throw a heavy one. The reason is that the faster a muscle fiber shortens, the fewer cross-bridges can form within it, reducing the force that each fiber and the muscle as a whole can produce.

The term "muscle contraction" suggests that muscles shorten when stimulated to generate force, but in fact muscles may also exert force and remain a uniform length or even be lengthened.

FIG. 37.10 Force–velocity relationship of muscle. Skeletal muscles exert greater force when they shorten more slowly and produce even greater force when they are stretched relative to being isometric.

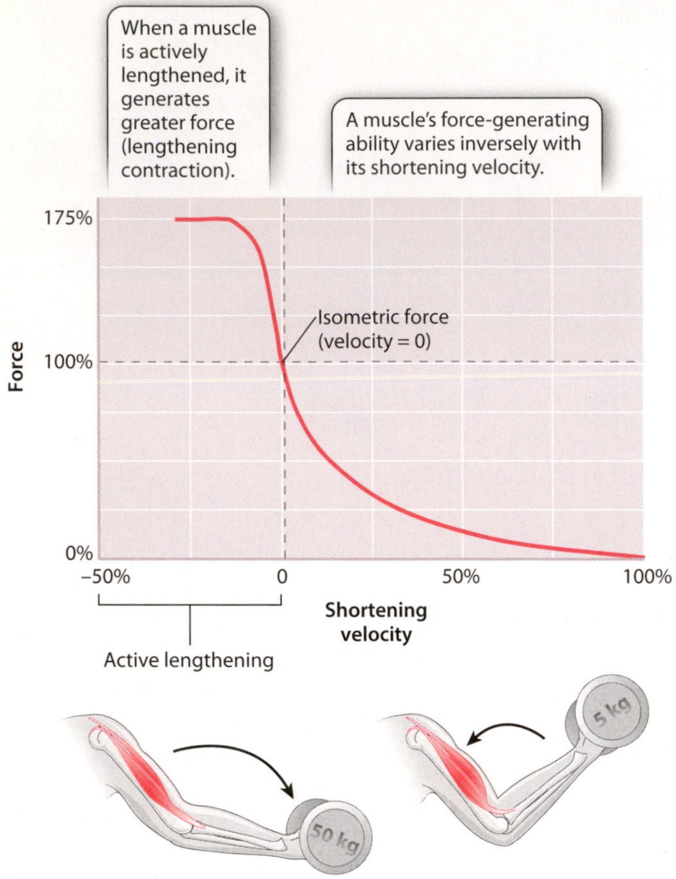

Consequently, the term "contraction" needs to be broadened beyond its familiar reference to muscle shortening to include those instances when a muscle exerts force while remaining the same length or being actively stretched.

When a muscle contracts but cannot shorten (for example, when you carry a heavy bucket of water), the muscle's shortening velocity is zero (Fig. 37.10), and it generates an **isometric** force (*iso-* means "same" and *metric* refers to "measure"). That is, the muscle still generates force, even though it stays the same length and does not shorten. Isometric exercises use this type of muscle contraction: Holding a weight in a fixed position or exerting force against an immovable object helps maintain muscle strength and may be beneficial in rehabilitation after injury by avoiding joint motion.

Muscles can also be stretched when the external load against which they contract exceeds the force the muscle produces. This action is known as an active **lengthening contraction.** During a lengthening contraction, the muscle generates more force (up to 75% more, Fig. 37.10) than when it remains isometric or shortens. This kind of contraction occurs when you lower a box from a shelf and your biceps muscle lengthens as it exerts force, or when your calf and quadriceps muscles are stretched to support your body in landing from a jump. It also happens during every stride you take during a run as your leg joints flex when a leg contacts the ground, stretching contracting leg muscles. Active muscle lengthening reduces the muscle's energy consumption because fewer fibers are recruited to generate a given level of force.

Injury can result if a muscle is stretched too rapidly or repetitively. Rapid and repetitive stretching often occurs during prolonged downhill hiking, causing muscle damage and soreness, particularly in the quadriceps muscles in the front of the thigh. Suitable training and warm-up exercises help muscles to stretch without damage.

Antagonist pairs of muscles produce reciprocal motions at a joint.

Muscles can generate force only by pulling, not pushing, on the skeleton. Therefore, to produce reciprocal joint and limb movements, they are arranged as pairs of **antagonist muscles** that pull in opposite directions. **Flexion** is the joint motion that moves bone segments closer together, and **extension** is the joint motion that moves them farther apart. For example, muscles at the elbow joint are organized as flexors (the biceps muscle) that contract to bring the arm toward the shoulder, and extensors (the triceps muscle) that contract to extend the arm away from the shoulder (**Fig. 37.11**). Additional pairs of antagonist muscles are present for movements in other directions. Muscles in the body wall of cnidarians (jellyfish and sea anemones) and segmented annelid worms, as well as the smooth muscles in the walls of intestines, are organized in longitudinal (lengthwise) and

FIG. 37.11 Antagonist muscles. Skeletal muscles can produce force only by pulling on a skeletal attachment, so they are arranged as antagonists to produce opposing motions at a joint.

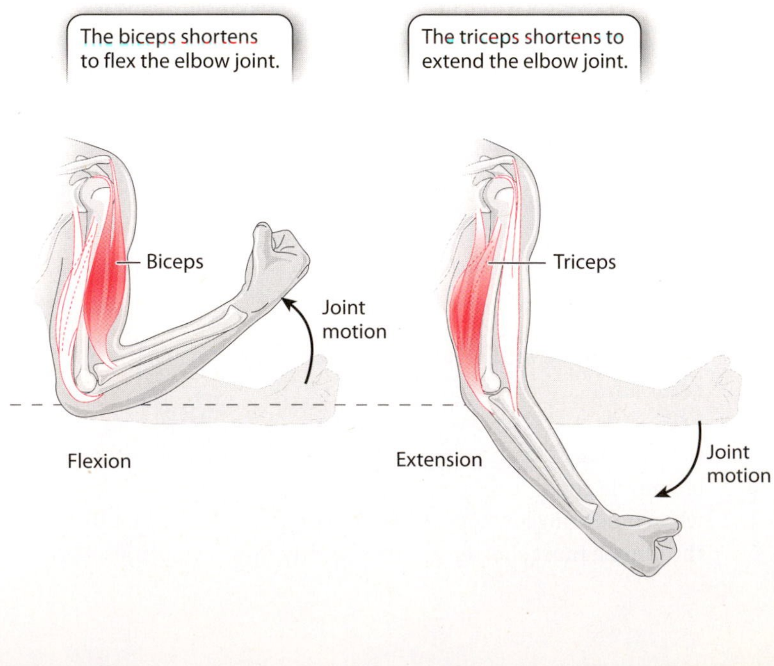

FIG. 37.12 **(a) Twitch and (b) fused tetanus muscle contractions.** *Source: Adapted from E. P. Widmaier, H. Raff, and K. T. Strang, 2011, Vander's Human Physiology, New York: McGraw-Hill.*

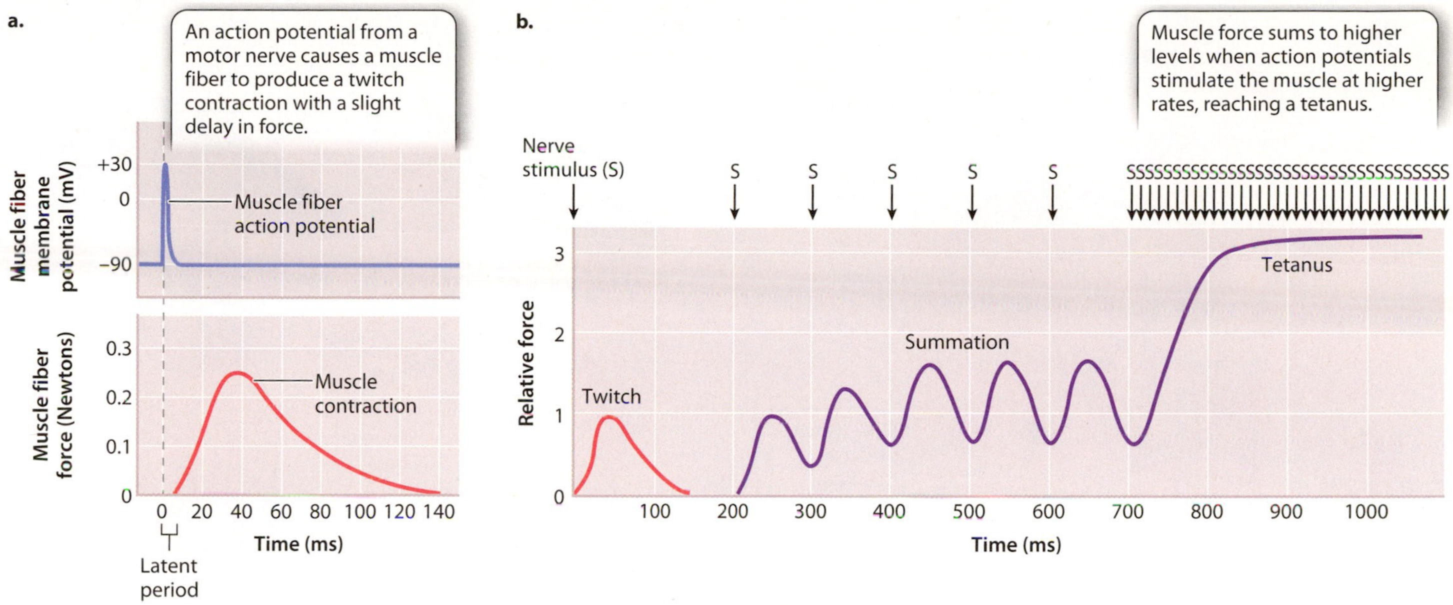

circumferential (circular) layers, allowing them to function as muscle antagonists that control movement and shape.

In contrast, when muscles combine to produce similar motions, they are termed muscle **agonists.** Muscles arranged as agonists increase the strength and improve the control of joint motion. For example, the three heads of the triceps are agonists of each other (Fig. 37.11).

Muscle force is summed by an increase in stimulation frequency and the recruitment of motor units.

How do animals vary the amount of force exerted by a muscle? As we have seen, skeletal muscles are stimulated by motor nerves, which conduct action potentials to the neuromuscular junction. The force exerted by a muscle depends on the frequency of stimulation by the motor nerve. As action potentials become more frequent, the amount of calcium that is released to activate cross-bridge formation in the muscle fibers also increases.

A single action potential results in a **twitch** contraction of a certain force (**Fig. 37.12a**). If sufficient time is allowed for the muscle to pump Ca^{2+} back into the sarcoplasmic reticulum, the contraction ends, force falls to zero, and a second stimulus can elicit a twitch contraction of the same force. However, if a second action potential arrives before the muscle has relaxed, a greater force is produced because Ca^{2+} remains within the myofibrils to activate more cross-bridges. With increasing frequency of stimulation, muscle force increases (**Fig. 37.12b**). This property of muscle contraction is called force summation.

At a sufficiently high stimulation frequency, muscle force reaches a plateau and remains steady as stimulation continues. This muscle contraction of sustained force is called a **tetanus.**

Measurements in animals indicate that skeletal muscles commonly produce tetanic contractions, which generate steady and large forces. Tetanus is a normal physiological response of a muscle, but it can also be induced by a toxin from the bacterium *Clostridium tetani,* so named because it causes severe muscle spasms, particularly of the jaw, chest, back, and abdomen.

A motor neuron and the population of muscle fibers that it innervates are collectively termed a **motor unit** (**Fig. 37.13**).

FIG. 37.13 **Motor units.** A motor unit consists of the motor neuron and the population of cells (fibers) it innervates.

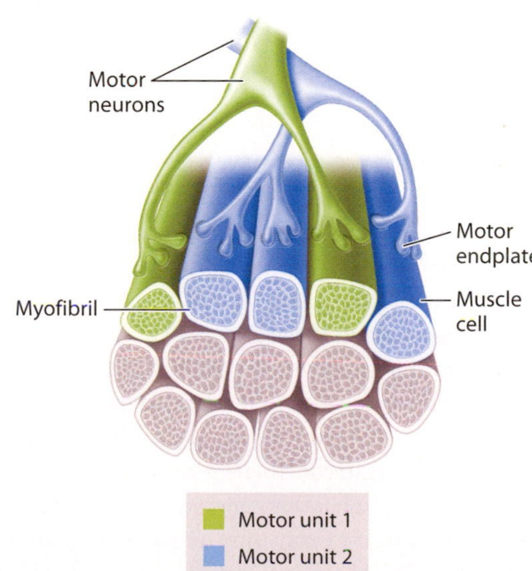

FIG. 37.14 Slow-twitch (darkly stained) and fast-twitch (lightly stained) skeletal muscle fibers. These two types of fiber provide differences in endurance and the speed of contraction. (a) Mammalian limb muscle; (b) muscles of two different fish species. *Photo source: Biophoto Associates/Science Source.*

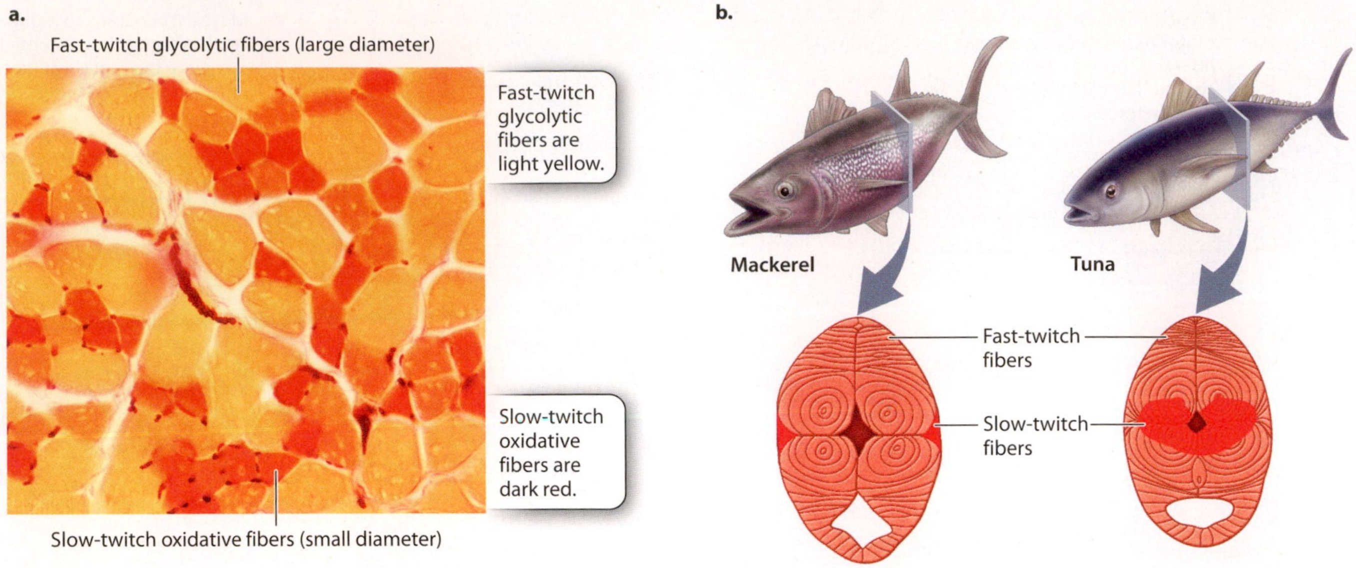

Motor units can include relatively few to several hundred muscle fibers. The number of muscle fibers innervated by a given motor nerve, and hence the size of the motor unit, affects how finely a vertebrate muscle's force can be controlled. For example, finger muscles have small motor units allowing fine adjustments in muscle force, whereas leg muscles have large motor units providing larger adjustments in force. Together with stimulation frequency, a vertebrate muscle's force output depends on the number of motor units (and therefore the number of muscle fibers) that are activated. Within each motor unit, muscle fibers generally share similar contractile and metabolic properties. Differences in these properties can give muscles very different capabilities, as discussed in the next section.

→ **Quick Check 4** What are the two mechanisms by which the force of contraction can be increased in vertebrate muscles?

Skeletal muscles have slow-twitch and fast-twitch fibers.

The difference between a sprinter and a marathoner is not just training, but also the properties of their muscles. There are different types of muscle fiber (**Fig. 37.14**), allowing variation in the way muscles contract. In vertebrates, these are most simply classified as red **slow-twitch** fibers and white **fast-twitch** fibers (Fig. 37.14a). Slow-twitch fibers are found in muscles that contract slowly and consume ATP more slowly to produce force. An animal uses slow-twitch fibers to control its posture or move slowly and economically. Fast-twitch fibers generate force quickly, producing rapid movements, but they consume ATP more quickly. An animal depends on fast-twitch fibers to sprint or lunge at its prey. A key difference between the two types of fiber is that slow-twitch fibers obtain their energy through oxidative phosphorylation (aerobic respiration), and fast-twitch fibers obtain their energy mainly through glycolysis (Chapter 7).

Although slow-twitch fibers develop force more slowly, they have greater resistance to fatigue in response to repetitive stimulation—that is, their loss of force over time is less. Their greater endurance is explained by the ability of their mitochondria to supply ATP to muscle fibers by aerobic respiration. Slow-twitch fibers have many mitochondria and are well supplied by capillaries. They contain an abundance of **myoglobin,** an oxygen-binding protein related to hemoglobin that facilitates oxygen delivery to the mitochondria (Chapter 39). The iron in myoglobin gives muscles with these fibers their red appearance. Slow-twitch muscle fibers express a chemically "slow" form of myosin with a relatively low rate of ATP hydrolysis, limiting their speed of cross-bridge cycling and force development.

White fast-twitch muscle fibers have fewer mitochondria and capillaries, contain little myoglobin, and rely heavily on glycolysis to produce ATP. Fast-twitch fibers express a chemically "fast" form of myosin with a high rate of ATP hydrolysis, favoring rapid force development and movement. However, fast-twitch fibers also fatigue quickly. Because white fibers are larger than red fibers, they generate more force. In general, most skeletal muscles are composed of mixed populations of red and white fiber types, providing a broad range of contractile function.

Athletes who have more oxidative slow-twitch fibers excel at endurance and long-distance competitions. In contrast, athletes who have larger numbers of fast-twitch fibers perform better in sprint races and weight-lifting competitions. Because the distribution and properties of muscle fiber types are strongly inherited, it is largely true that champions are born and not made. Nevertheless, aerobic training can significantly increase the energy output through oxidative phosphorylation of an individual's muscle fibers, just as weight lifting and speed training can enhance an individual's strength and sprinting ability. Contrary to popular belief that weight lifting builds muscle by straining the muscle to build new fibers, it actually builds muscle by increasing the size of existing muscle fibers through the synthesis of additional myosin and actin filaments. This synthesis increases the cross-sectional area of the fiber and thus its force-generating capacity.

? CASE 7 PREDATOR–PREY: A GAME OF LIFE AND DEATH

How do different types of muscle fiber affect the speed of predators and prey?

We have just seen that slow-twitch muscle fibers provide endurance for sustained activity, whereas larger fast-twitch fibers produce greater speed and strength but fatigue quickly. The relative distribution of these fiber types in the muscles of different animals is influenced mainly by genetic heritage. This affects the locomotive behavior and capacities of both predator and prey animals. Whereas cheetahs have exceptional speed, their muscles tire quickly. As their prey, antelope have muscles that enable endurance long-distance running and contract rapidly for maneuvering.

Highly aerobic animals such as dogs and antelope have large fractions of slow-twitch fibers *and* specialized fast-twitch muscle fibers that have a high oxidative capacity and use glycolysis for rapid ATP synthesis. These muscle fibers allow the animals to move for extended periods of time at relatively fast speeds. In contrast, cats have mainly fast-twitch fibers, enabling them to sprint and pounce quickly. Nevertheless, cats and other animals use particular muscles with high concentrations of slow-twitch fibers (these are the dark-staining fibers shown in Fig. 37.14a) for slower postural movements and stealth. Animals such as sloths and lorises that move slowly have mostly oxidative slow-twitch fibers and few glycolytic fast-twitch fibers.

More generally, animals with high body temperatures and high metabolic rates (birds and mammals) have larger numbers of oxidative fibers compared with animals that have lower body temperatures and lower metabolic rates (reptiles). As a result, lizards use glycolytic fast-twitch fibers to sprint for brief periods but must then recover from the acid buildup produced by their muscles' anaerobic activity in longer periods of rest (Chapter 40). In contrast, birds and mammals can sustain activity over longer time periods because they have oxidative muscle fibers supported by aerobic ATP supply.

Most fish have a narrow band of slow-twitch red muscle fibers that runs beneath their skin, which they use for slow steady swimming. Tuna and some sharks have evolved deeper regions of red muscle that they can keep warm (Fig. 37.14b), enabling their muscles to contract longer and faster to enhance their swimming. When escaping from a predator or attacking prey, fish recruit their larger fast-twitch white trunk musculature for rapid swimming.

The evolutionary arms race between predators and prey has resulted in selection for particular muscle fiber types and other musculoskeletal specializations for rapid movement and maneuvering. Features of the skeleton that contribute to these specializations are discussed next.

37.3 ANIMAL SKELETONS

Most animal skeletons provide a rigid set of elements that meet at joints (articulate) to transmit muscle forces for movement and body support. They also enable many animals to manipulate their environment by digging burrows, building nests, or constructing webs to catch food. Animals have evolved three types of skeletal system: **hydrostatic skeletons, exoskeletons,** and **endoskeletons.** All three types of skeleton have rigid elements that resist the pull of antagonist sets of muscles.

Hydrostatic skeletons support animals by muscles that act on a fluid-filled cavity.

Hydrostatic skeletons evolved early in multicellular animals with the first cnidarians (jellyfish and anemones) approximately 600 million years ago. They are found in nearly all multicellular animals as well as in many vascular plants. In animals that depend on a hydrostatic skeleton, fluid contained within a body cavity serves as the supportive component of the skeleton. Muscles exert pressure against the fluid to produce movement.

Two sets of opposing muscles surround the fluid, controlling the width and length of the body cavity and, in many cases, of the whole animal (**Fig. 37.15a**). Circular muscles reduce the diameter of the body cavity, and longitudinal muscles reduce its length. A sea anemone can bend its body in various directions to feed or to resist water currents by closing its mouth to keep its fluid volume constant and contracting its circular and longitudinal muscles. Sea anemones extend themselves into the water column by contracting their circular muscles and relaxing their longitudinal muscles. Controlling the amount of fluid in their body cavity also allows sea anemones to adjust their shape. When threatened, a sea anemone rapidly retracts by contracting its longitudinal muscles and allowing water to escape from its body cavity.

Similarly, earthworms burrow through the soil by moving a series of hydrostatic segments along their body (**Fig. 37.15b**). By contracting longitudinal muscles in a few segments, they shorten those segments and also widen them to anchor those segments against the soil. They then extend intervening segments between anchor points by contracting circular muscles, causing the segments to lengthen and move the animal's body through the soil.

Because of their simple but effective organization, hydrostatic skeletons are well adapted for many uses. In animals such as squid

FIG. 37.15 Diverse examples of hydrostatic skeletal support. (a) In sea anemone; (b) in earthworm; (c) intervertebral discs; (d) articular cartilage. *Photo sources: a. Carole Valkenier/Getty Images; b. Nigel Cattlin/Alamy.*

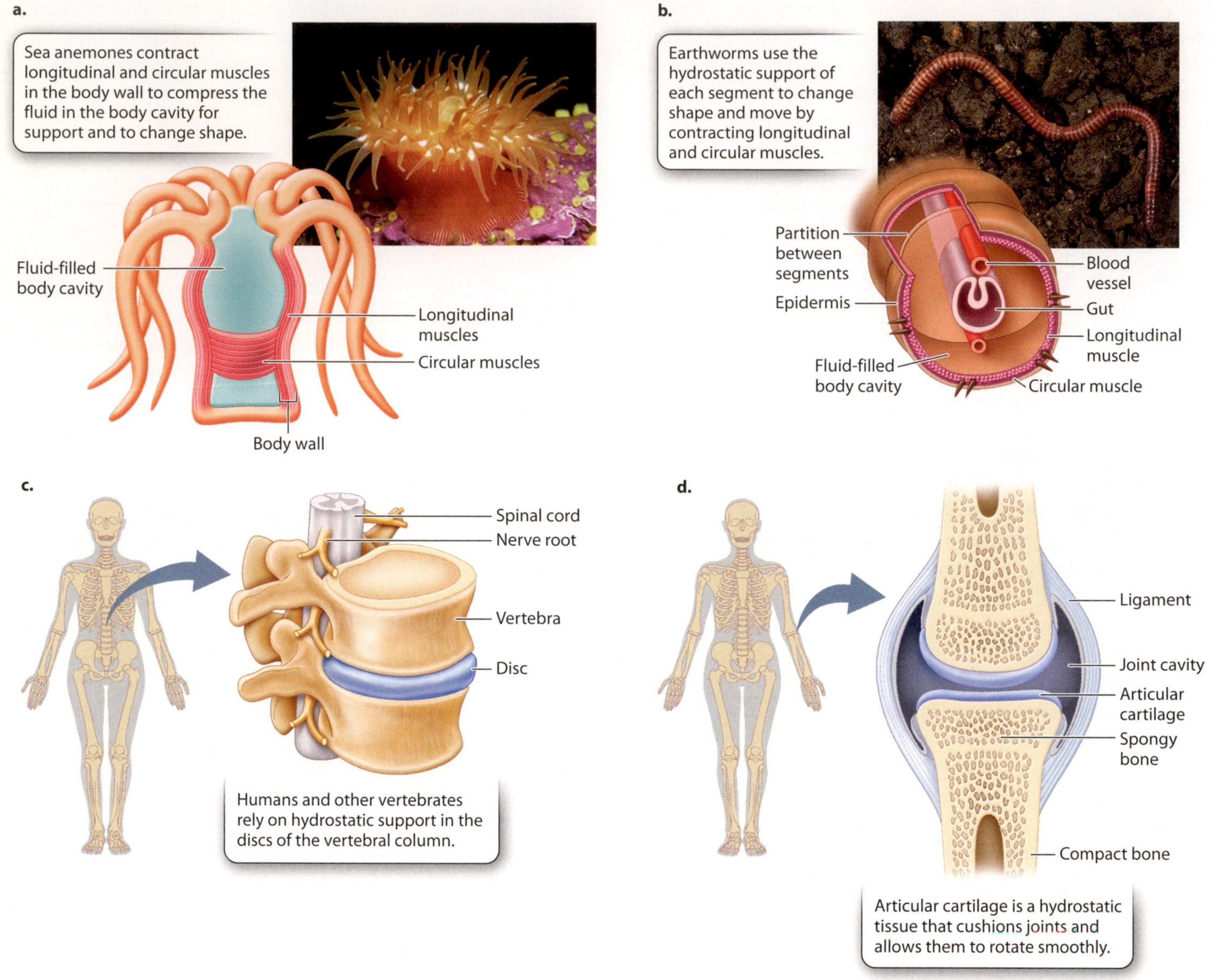

and jellyfish, the hydrostatic skeleton has become specialized for jet-propelled locomotion. Large circular muscles contract to eject fluid from the body cavity, propelling the animal in the opposite direction. The muscles of other mollusks also exert pressure against a hydrostatic skeleton. By this means, for example, a clam uses its muscular "foot" to burrow into the sediment, and an octopus adeptly manipulates its flexible tentacles. (Clams also have shells that form an exoskeleton to protect the body.) These same principles allow an elephant to control its trunk, and our tongue to manipulate food and assist in speech.

Even vertebrate animals with rigid endoskeletons have hydrostatic elements that provide flexibility and cushion loads transmitted by the skeleton. These include **intervertebral discs** and **cartilage.** Intervertebral discs are sandwiched between the bony vertebrae of the backbone (**Fig. 37.15c**). Each disc has a wall reinforced with connective tissue that surrounds a jelly-like fluid. Intervertebral discs enable the backbone to twist and bend. Disc walls damaged by repeated stress can rupture, causing significant lower back pain. Articular cartilage, another fluid-filled tissue, forms the joint surfaces between adjacent bones (**Fig. 37.15d**). The fluid within the cartilage resists the pressure between the ends of articulating bones, cushioning forces transmitted across the joint.

Exoskeletons provide hard external support and protection.

An exoskeleton is a rigid skeleton that lies outside the animal's soft tissues. The first mineralized skeletons arose with sponges about 650 million years ago, but the more common exoskeletons

FIG. 37.16 **Examples of exoskeletons.** The shells of (a) a bivalve mollusk (b) and nautilus, made of calcium carbonate, surround and protect internal soft body parts. The exoskeleton of arthropods, such as (c) a grasshopper, is made of stiff cuticle that surrounds internal muscles and tendons and other internal body parts. *Sources: a. Rita van den Broek/age fotostock; b. Reinhard Dirscherl/Getty Images; c. Jim Simmen/Getty Images.*

subsequently evolved in other invertebrate groups adapted for life in aquatic and terrestrial environments (**Fig. 37.16**). Because they are on the exterior of the animal, exoskeletons provide hard external support and protection, allowing muscles to attach from the inside. However, a main disadvantage is that exoskeletons limit growth.

Different invertebrates have different kinds of exoskeleton, but all types have relatively few cells for their volume of tissue. The shells of bivalve mollusks, such as clams and mussels, or the cephalopod mollusk *Nautilus* that has existed largely unchanged for nearly 500 million years, form a hard, mineralized calcium carbonate exoskeleton (Figs. 37.16a and 37.16b). The calcium carbonate is reinforced by proteins and is therefore an example of a composite material, one that combines substances with different properties. Because of its composite nature, the shell is less brittle and more difficult to break than if it were made of just one substance. The shell expands as the organism grows and epidermal cells deposit new layers of mineralized protein in regular geometric patterns. You can see the growth rings if you examine a mollusk shell. Although some marine bivalves achieve large size over many years, growth is limited by the rate at which new skeletal material can be deposited on the outside of the shell.

Arthropods have more complex exoskeletons. More than half of all known animal species, including insects and crustaceans, are arthropods. Arthropods are defined in part by their jointed legs, and another important feature is their exoskeleton formed of a **cuticle** that covers their entire body (Fig. 37.16c). Arthropod exoskeletons, which first evolved in aquatic crustaceans, protect animals from desiccation and physical insults. Consequently, exoskeletons were key to the success and diversification of insects in terrestrial environments about 450 million years ago.

The cuticle of insects is composed mainly of **chitin,** a nitrogen-containing polysaccharide. When initially formed, the cuticle of arthropods is soft, so the animal can grow before the cuticle hardens. Mature cuticle consists of two layers. A thin outer, waxy layer minimizes water loss. Water loss is especially dangerous for terrestrial arthropods, given their small size. The much thicker inner layer, whether flexible or stiff, is tough and hard to break. The cuticle remains flexible at joints, allowing motion between body segments. Because a rigid exoskeleton restricts growth, arthropods shed their cuticle at intervals, a process called **molting.** Molting allows arthropods to expand and grow before forming a new rigid exoskeleton.

Marine crustaceans, such as crabs and lobsters, incorporate calcium carbonate in their cuticle. Calcium carbonate makes the exoskeleton hard and stiff, much like the hardened cuticle of terrestrial insects, helping to protect these crustaceans from predators.

While offering several protective benefits, exoskeletons pose risks as well. Animals are vulnerable when the newly formed exoskeleton has not yet hardened. Exoskeletons are also hard to repair. If a skeleton is damaged while growing, the animal must produce an entirely new one. Because the exoskeleton is a thin-walled structure, it is prone to breaking if its surface area is very large. Consequently, the imperviousness of the monstrous insects depicted in science-fiction films is improbable—they would easily break from a blow to their exoskeleton.

The rigid bones of vertebrate endoskeletons are jointed for motion and can be repaired if damaged.

Endoskeletons first evolved about 350–500 million years ago when the first vertebrates evolved cartilaginous endoskeletons. Lampreys are descendants of these early vertebrates. Subsequently, bony fishes evolved a more rigid mineralized internal skeleton, a defining feature of this diverse group of vertebrates. Sharks and other elasmobranchs (the cartilaginous fishes) evolved unusual skeletons of calcified cartilage, losing the earlier mineralized bony skeletons of their bony fish ancestors.

A bony or cartilaginous endoskeleton lies internal to most of an animal's soft tissues. In contrast to exoskeletons, the endoskeletons of vertebrate animals can grow extensively and, when broken, can be repaired. Endoskeletons also provide protection for certain key internal organs, such as the brain, lungs, and heart.

The bones of vertebrate skeletons consist of a variety of tubular, rodlike, and platelike elements that form a scaffold to which the muscles attach. Because bones are rigid and hard, there are joints between adjacent bones to enable movement. Muscles attach to the skeleton by connective tissue and specialized **tendons** made of collagen. Tendons transmit muscle forces, allowing the forces to be redirected and transmitted over a wide range of joint motion. Tendons, like a spring, also store and recover elastic energy.

FIG. 37.17 The vertebrate skeleton, showing axial and appendicular regions.

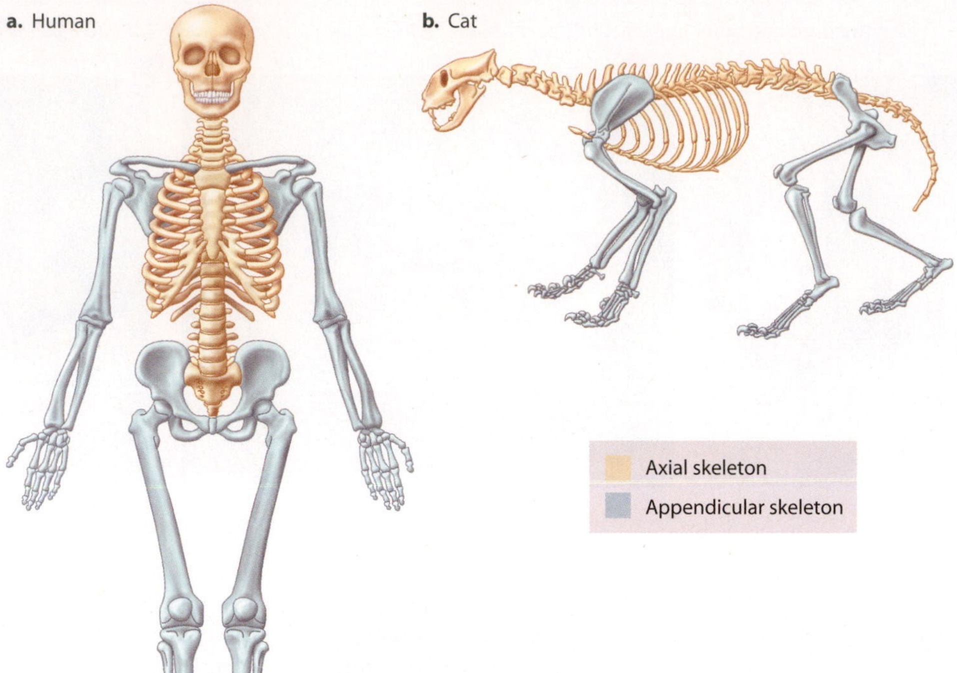

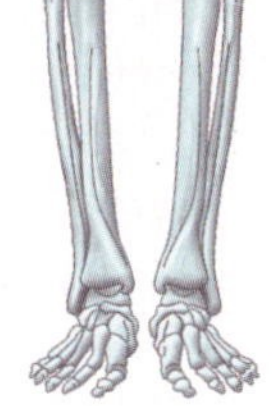

The vertebrate skeleton can be divided into **axial** and **appendicular** regions (**Fig. 37.17**). The axial skeleton consists of the skull and jaws of the head, the vertebrae of the spinal column, and the ribs. Bones of the skull protect the brain and sense organs of the head. The appendicular skeleton consists of the bones of the limbs, including the shoulder and pelvis. The axial and appendicular regions reflect the evolutionary ancestry of vertebrates. The axial skeleton formed first, to protect the head and provide support and movement of the animal's body by bending its body axis. As a result, the axial skeleton is the main skeletal component of fishes. The appendicular skeleton includes the pectoral and pelvic fins of fish, which evolved into limbs when vertebrate animals first moved onto land. In a special group of lobed-fin fish, the fin bones became adapted for an amphibious lifestyle, evolving into the limb bones that terrestrial vertebrates use for support against gravity and for movement on land.

Skeletal tissues have relatively few cells for their volume because they consist mostly of an extensive **extracellular matrix** that is secreted by specialized cells, forming connective tissue external to the cells (Chapter 10). In addition to tendon, three main extracellular tissues are produced in the formation of the vertebrate skeleton: bone, tooth enamel, and cartilage.

Bone tissue–forming cells called **osteoblasts** synthesize and secrete calcium phosphate as **hydroxyapatite** mineral crystals in close association with the protein **collagen.** Generally, bone consists of two-thirds hydroxyapatite mineral and one-third type I collagen protein. The combination of mineral and protein makes bone a composite material, like exoskeletons. Bone is rigid (because of the properties of the mineral), but also hard to break because it absorbs much energy before fracture (because of the properties of the collagen). Bone gradually becomes more mineralized and brittle as a normal process of aging. The genetic bone disorder osteogenesis imperfecta leads to bones that are brittle and fragile because the osteoblasts produce insufficient collagen in the extracellular matrix.

Articular cartilage located at the joint surfaces of a bone forms a cushion between bones meeting at a joint. The cartilage is a gel-like matrix that resists fluid being squeezed out when forces press on the cartilage during movement. Articular cartilage consists of 70% water, 15% large, branched molecules called proteoglycans, and the remaining 15% type II collagen. The collagen fibers reinforce the cartilage, allowing the cartilage to cushion and distribute loads as the joint moves. Cartilage may degenerate for a variety of reasons, including rheumatoid arthritis and repetitive physical injury, but in general it lasts over an individual's lifetime.

➔ **Quick Check 5** Glass is strong and rigid but can be easily broken. In contrast, a bone is strong and rigid but hard to break. How can this be?

37.4 VERTEBRATE SKELETONS

Although both vertebrate skeletons and invertebrate exoskeletons are rigid and have flexible joints, certain features distinguish them.

FIG. 37.18 Bone formation from a cartilage model. (a) Most bones other than the skull are formed as a cartilage first, followed by vascularization and the transformation into bone tissue by osteoblasts. (b) Cartilage model of a human fetal hand, showing initial sites of bone tissue formation at about 9 weeks of embryonic development. *Photo source: Science Photo Library/Science Source.*

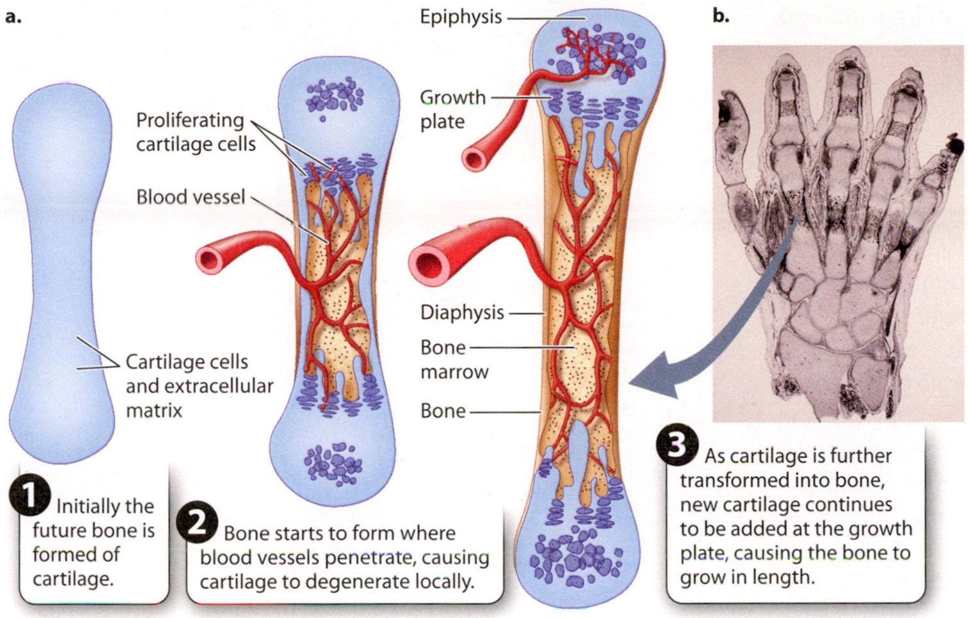

Here we explore the biology of vertebrate skeletons, focusing on how bones develop and grow and how they are repaired and remodeled.

Vertebrate bones form directly or by forming a cartilage model first.

Vertebrate bones form and develop by two distinct processes. Bones of the skull and the ribs form when precursor cells differentiate into osteoblasts that immediately begin producing bone as a skeletal sheet or membrane. These bones are often referred to as membranous bones.

This method of forming bone is slow, however, and, if most bones in the body formed this way, the fetus would take a very long time to develop. Most other bones of the body—the vertebrae, shoulder, pelvis, and limb bones—are formed as cartilage first (**Fig. 37.18**). The precursor cells initially become cartilage-producing cells, known as chondroblasts, which in the embryo grow into cartilage models of most bones of the developing body (Fig. 37.18a). For example, at approximately 9 weeks of development, a human fetal hand consists mainly of this cartilage model (Fig. 37.18b). This embryonic cartilage is the same as the articular cartilage that remains at the ends of bones to form the joint surfaces. Because cartilage is a pliable, fluid-filled matrix, it grows by expansion within the structure as well as at the surface, enabling the fetal skeleton to grow rapidly. As the fetus grows and its skeleton matures, the cartilage model of the bone is invaded by blood vessels. The presence of blood vessels triggers the transformation of cartilage into hard, mineralized bone.

Two main types of bone are compact and spongy bone.

The cartilage template of a bone is replaced by one of two main types of bone (**Fig. 37.19**). Both types contribute to the overall form of many bones in the body. **Compact bone** forms the walls of the bone's shaft. It consists mainly of dense, mineralized bone tissue containing bone cells derived from osteoblasts and the network of blood vessels supplying them with oxygen and nutrients.

A second type of bone, called **spongy bone,** consists of small plates and rods known as **trabeculae** with spaces between them. Spongy bone is found in the ends of limb bones and within the vertebrae. It reduces the weight of bone in these regions, but it

FIG. 37.19 Compact and spongy bone. *Source: Dave King/Getty Images.*

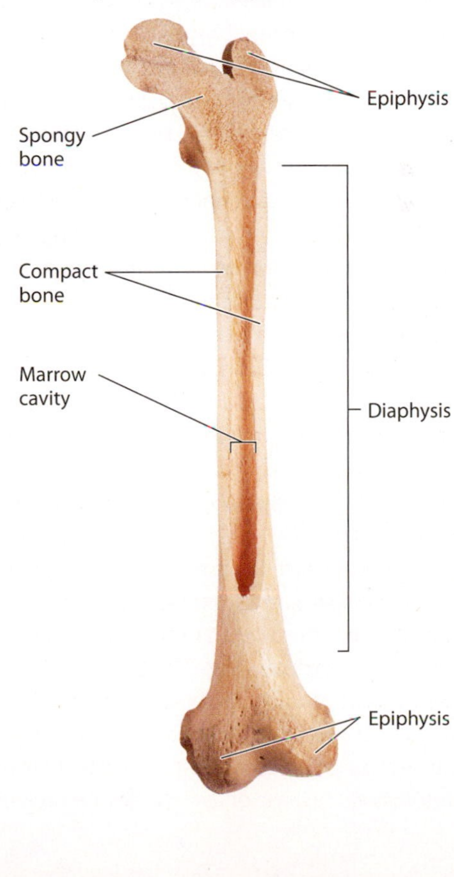

is also stiff so that forces are transmitted effectively from the joint to the bone shaft. **Bone marrow,** a fatty tissue found between trabeculae and also within the bone's central cavity, contains many important cells, including blood-forming cells, other stem cells (Chapter 20), immune system cells (Chapter 43), and fat cells.

Bones grow in length and width, and can be repaired.

Bones can grow in length, at least until an animal reaches maturity. Growth in length occurs at a **growth plate** (see Fig. 37.18), a region of cartilage between the middle region, called the **diaphysis,** and the end, called the **epiphysis.** The growth plate remains in place when the rest of the cartilage is transformed into bone. The growth plate adds new cartilage toward the bone's diaphysis, enabling bone length to continue to increase after birth (see Fig. 37.18). At maturity, each growth plate fuses as the remaining cartilage is replaced with bone, preventing further growth in length. The fusion of growth plates is typical of most mammals and birds. In amphibians and reptiles, the growth plates often do not fuse and growth may continue at a slower pace over much of their lifetime.

Limb bones grow in diameter when osteoblasts deposit new bone on the bone's external surface. At the same time, bone is removed from the inner surface, expanding the marrow cavity. Bone removal is slower than bone growth, thickening the walls during growth. Bone is removed from the marrow cavity by a group of cells called **osteoclasts** that secrete digestive enzymes and acid to dissolve the calcium mineral and collagen. These dissolved compounds are reabsorbed and recycled for bone formation in other regions of the skeleton. Through this process, the vertebrate skeleton serves as an important store of calcium and phosphate ions. For example, in female birds, calcium removed from the skeleton is regularly used to form the eggshell.

A great advantage of endoskeletons compared with exoskeletons is that a damaged endoskeleton can be repaired by osteoblasts and osteoclasts forming and removing mineralized tissue in particular regions of the bone. Generally, physically active younger adults have thicker bones than less active younger adults because physical loading stimulates osteoblasts to produce more bone during growth. However, in older adults, bone tissue is gradually lost as osteoclasts remove more bone than is produced by osteoblasts. Bone loss is particularly severe in women after menopause, in part because of hormone shifts as well as reduced physical activity, and can lead to osteoporosis, a condition that significantly increases the risk of bone fracture.

Joint shape determines range of motion and skeletal muscle organization.

The shapes of the bone surfaces that meet at a joint determine the range of motion at that joint. Joints range from simple **hinge joints** that allow one axis of rotation to **ball-and-socket joints** that allow rotation in three axes (**Fig. 37.20**). The human elbow joint and the ankle joint of a dog are examples of hinge joints. The shoulder and hip joints are examples of ball-and-socket joints, which allow the widest range of motion, as when you throw or kick a ball. The joints at the base of each finger are intermediate: They allow you to flex and extend your fingers as well as spread them laterally or move them together when making a fist or grasping objects.

FIG. 37.20 Hinge and ball-and-socket joints. The elbow and knee are examples of hinge joints. The shoulder and hip are examples of ball-and-socket joints.

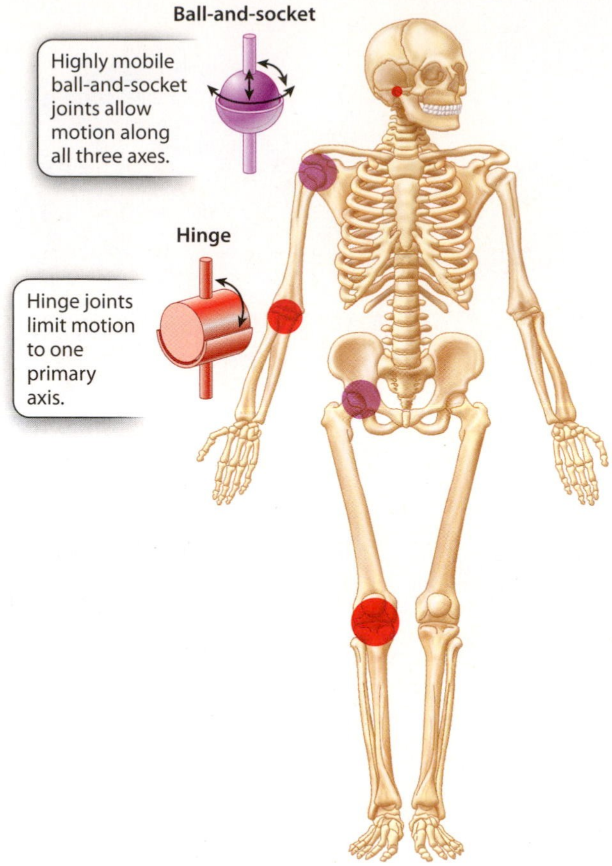

Joints with a broader range of motion are generally less stable. The shoulder is the most mobile joint in the human body, but it is also the most often dislocated or injured. In contrast, the ankle joint of dogs, horses, and other animals is a stable hinge joint that is unlikely to be dislocated, but its range of motion is limited to flexion and extension. Because muscles are arranged as paired sets of antagonists to move a joint in opposing directions, hinge joints are controlled by as few as two antagonist muscles (generally referred to as a flexor and an extensor). In contrast, ball-and-socket joints have at least three sets of muscle antagonists to control motion in three different planes. As a result, a more complex organization of muscles is needed to control the movements of the arm at the shoulder joint or the leg at the hip joint.

Muscles exert forces by skeletal levers to produce joint motion.

By serving as a rigid set of levers, the skeleton enables muscles to transmit forces that cause joint rotation. Muscles that attach

FIG. 37.21 **Torque.** Torque or rotational strength ($F \times r$) depends on the muscle's force (F) and the distance the force acts from the joint's axis of rotation (r).

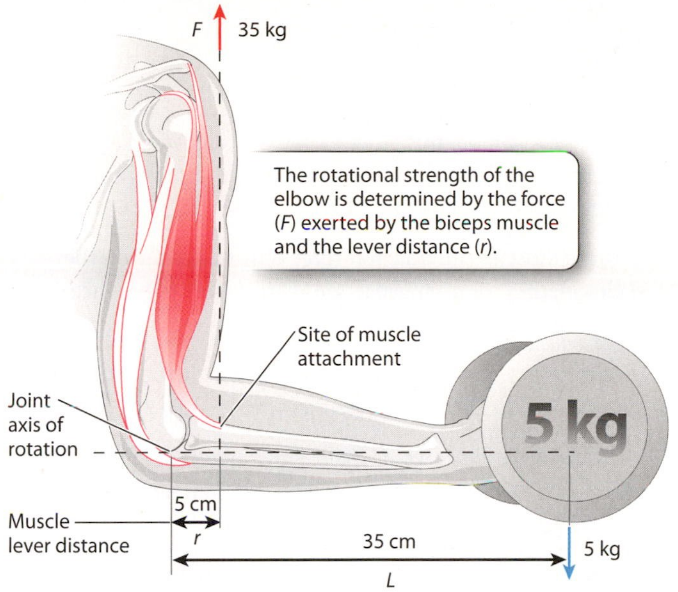

farther from a joint's axis of rotation produce stronger but slower movements, whereas muscles that attach closer to a joint produce weaker but faster movements. This is why you open a door by pulling on a doorknob located opposite from its hinges. The doorknob's location increases the lever distance, making it easier to pull the door open than if it were placed next to the hinge.

The strength of rotational movement, or torque, is determined by the product of a muscle's force (F) and lever distance (r) from the axis of joint rotation: $F \times r$. Therefore, muscles produce stronger rotational movements either by exerting a larger force or by having a longer lever distance between the joint and muscle attachment site. Let's consider a simple example: a weight carried by your hand (**Fig. 37.21**). In this case, the axis of rotation is the elbow, and contraction of the biceps muscle causes the arm to flex at the elbow. The strength of elbow rotation is determined by the biceps muscle's force (F) and the lever distance (r) between the elbow joint and the site of muscle attachment, where the muscle pulls on the arm bone.

In addition to the muscle's force (F) and lever distance (r), the distance from the joint to the site where the force is exerted (L) also affects movement (Fig. 37.21, **Fig. 37.22**). Animal joints located a long distance from the muscle attachment site (large r) and a short

FIG. 37.22 **Functional anatomy.** The armadillo forelimb skeleton is adapted for digging, whereas the horse forelimb is adapted for speed and weight support. *Sources: Based on M. J. Smith and R. J. G. Savage, 1956, "Some Locomotory Adaptations in Mammals," Zoological Journal of the Linnean Society 42:603–622. Photos (left to right): Morales/age footstock; J. L. Klein & M. L. Hubert / Science Source.*

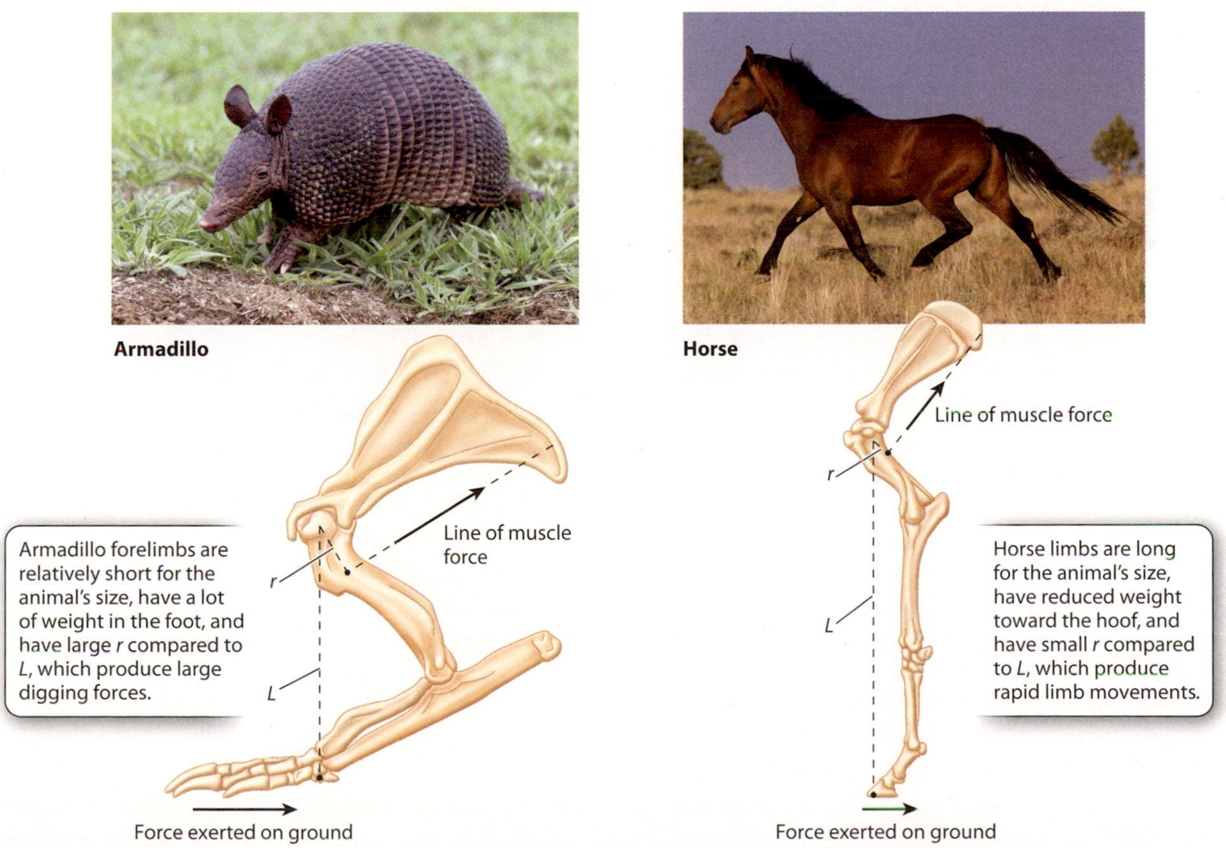

distance from where the output force is exerted (small *L*) produce strong but slow movements. Moles, armadillos, and spade-foot toads are examples of digging animals that have this type of muscle–joint arrangement. In contrast, joints located a short distance from the muscle attachment site (small *r*) and a long distance from where the output force is exerted (large *L*) produce faster movements but with less force. These are found in animals adapted for high-speed running, such as antelope, cheetahs, and horses. ■

Core Concepts Summary

37.1 Muscles are biological motors composed of actin and myosin that generate force and produce movement for support, locomotion, and control of internal physiological functions.

There are two main types of muscle: striated (skeletal and cardiac) and smooth. page 787

Skeletal and cardiac muscle fibers appear striated when viewed under the microscope because of the regular spacing of sarcomeres along their length. page 787

Smooth muscle fibers, which regulate airflow for breathing, blood flow through arteries, and the passage of food through the gut, lack regular sarcomere organization and appear smooth when viewed under the microscope. page 787

Skeletal muscle fibers are long thin cells composed of parallel sets of myofibrils built up from smaller parallel arrays of actin and myosin filaments. page 787

The sarcomere is the basic contractile unit of a skeletal muscle. Sarcomeres are arranged in series along the length of a myofibril. page 788

Muscles change length and produce force by the formation of actin–myosin cross-bridges, causing myosin and actin filaments to slide relative to each other. page 789

Muscles are stimulated by motor neurons at the fiber's motor endplate, leading to depolarization of the muscle cell that triggers the release of Ca^{2+} ions from the sarcoplasmic reticulum. Ca^{2+} binds troponin, which moves tropomyosin off the myosin-binding sites on actin, allowing myosin heads to form cross-bridges with actin. page 791

Excitation–contraction coupling is the process by which depolarization of the muscle cell leads to its shortening. page 792

37.2 The force generated by muscle depends on muscle size, degree of actin–myosin overlap, shortening velocity, and stimulation rate.

Muscles exert their greatest force when actin–myosin filaments have maximal overlap, allowing the most cross-bridges to form. page 792

Muscles exert more force when they contract at slow velocities compared with when they contract at high velocities. page 793

Because muscles can transmit force only by pulling on the skeleton, they are arranged as antagonist pairs to produce reciprocal motions of a joint or limb. page 794

In vertebrates, a motor unit consists of a single motor neuron and the muscle fibers (cells) it innervates. page 795

Muscle force is increased by increasing the motor neuron firing rate and, in vertebrates, by increasing the number of motor units that are activated. page 796

Vertebrate muscles have two types of fiber: red slow-twitch fibers that contract slowly over longer time periods and white fast-twitch fibers that contract rapidly but fatigue quickly. page 796

37.3 Hydrostatic skeletons, exoskeletons, and endoskeletons provide animals with mechanical support and protection.

The rigid element of a hydrostatic skeleton is an incompressible fluid within a body cavity, used to support the body and change shape. page 797

Invertebrate exoskeletons form an external rigid support system, which protects the animal and limits water loss, but also limits growth and repair. page 798

Vertebrates have a bony endoskeleton that provides rigid support and protection of body organs, and that can grow and be repaired. page 799

The vertebrate endoskeleton is organized into axial (central) and appendicular (limb) components. page 800

Vertebrate endoskeletons consist of bone and cartilage. page 800

Bone is a composite tissue that consists of calcium phosphate mineral and type I collagen and that is rigid and hard to break. page 800

Cartilage is a fluid-based gel reinforced by type II collagen and proteoglycans, providing cushioning support at joint surfaces page 800

37.4 Vertebrate endoskeletons allow for growth and repair and transmit muscle forces across joints.

Vertebrate bones develop by one of two processes: either directly or by way of a cartilage model. page 801

Cartilage forms much of the embryonic skeleton and remains as growth plates within the bone to provide rapid growth after birth, as well as forming the bone's joint (articular) surfaces. page 801

Bone has two basic structures: Compact bone forms the solid walls of a bone's shaft, and spongy bone forms a mesh that supports the cartilage at the bone's ends. page 801

Bone formation by osteoblasts and removal by osteoclasts is a continuing process of growth, shape change, and repair. page 802

The shape of a bone's joint surfaces largely determines the range of motion and stability of a joint. page 802

Muscles produce joint movements by transmitting forces by means of rigid skeletal levers. page 802

Self-Assessment

1. Diagram a sarcomere, showing the basic organization of thin (actin) and thick (myosin) filaments. Indicate on your diagram the regions where myosin cross-bridges form with actin.
2. Describe the sequence of events that occurs when an action potential arrives at a muscle fiber's motor endplate, causing the muscle to be depolarized.
3. Describe the sequence of events involved in myosin cross-bridge attachment and detachment. Relate these events to how muscles generate force and shorten when they are stimulated to contract, identifying the role that ATP plays.
4. Draw a graph of the isometric force–length relationship of striated muscle, indicating where maximal overlap between actin and myosin filaments occurs.
5. Draw a graph of the force–shortening velocity relationship of striated muscle.
6. Compare the force produced by a muscle over time when it is stimulated by a single twitch stimulus with the force produced by multiple stimuli at low frequency, and then with the force produced when the stimulation frequency is increased.
7. Name the two basic structural elements common to all animal musculoskeletal systems.
8. Compare and contrast three features of an exoskeleton and an endoskeleton.
9. Identify the two primary components of bone tissue and explain how each of these contributes to a bone's strength, stiffness, and resistance to fracture.
10. Diagram a limb bone, such as the tibia, showing the regions of articular cartilage, spongy trabecular bone, compact bone tissue, and the marrow cavity.

Log in to LaunchPad to check your answers to the Self-Assessment questions, and to access additional learning tools.

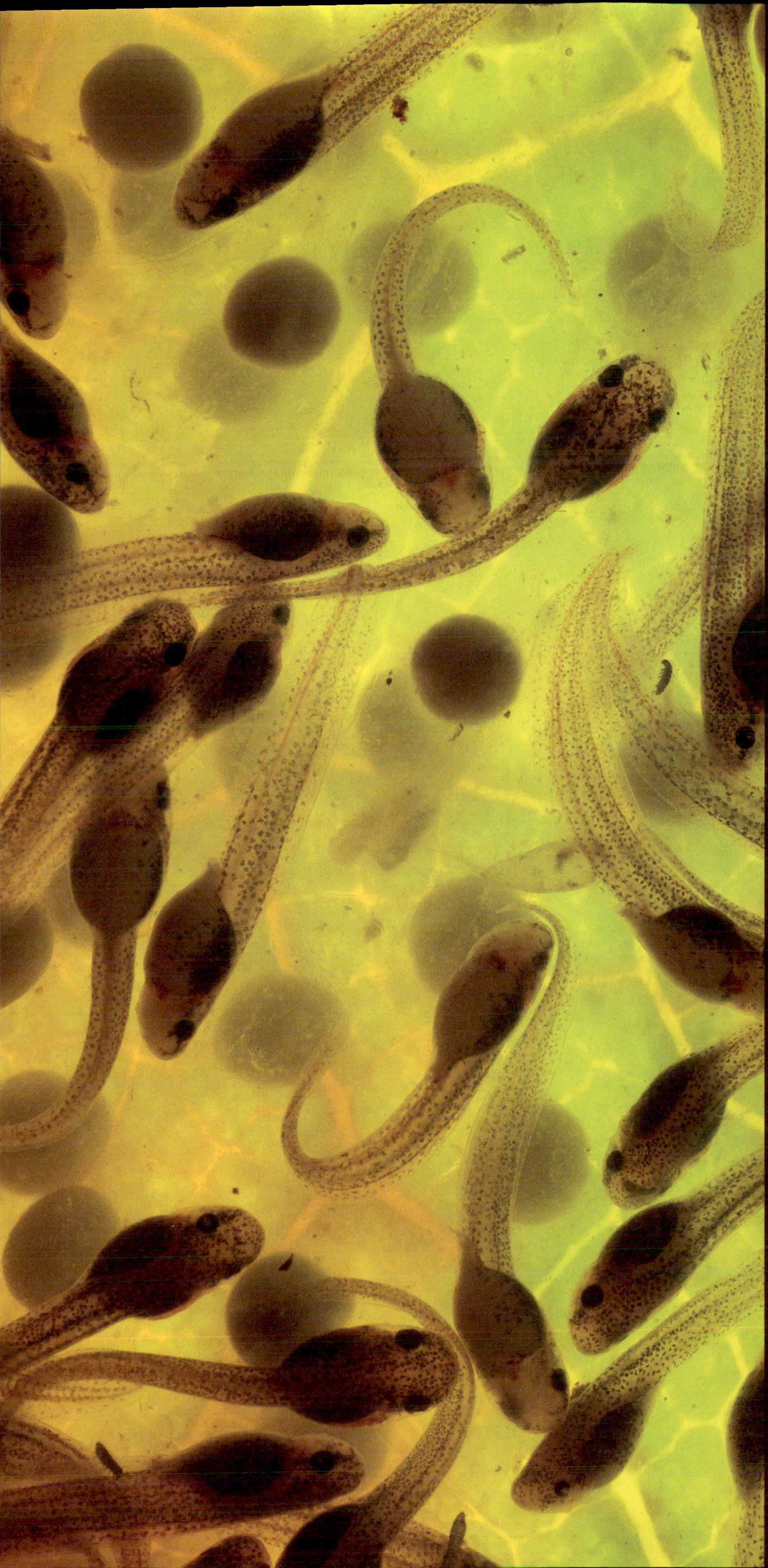

CHAPTER 38

Animal Endocrine Systems

Core Concepts

38.1 Animal endocrine systems release chemical signals called hormones into the bloodstream that respond to environmental cues, regulate growth and development, and maintain homeostasis.

38.2 Hormones bind to receptors on or inside target cells, and their signals are amplified to exert strong effects on target cells.

38.3 In vertebrate animals, the hypothalamus and the pituitary gland control and integrate diverse bodily functions and behaviors.

38.4 Chemical communication can also occur locally between neighboring cells or, through the release of pheromones, between individuals, coordinating social interactions.

FLPA/Chris Mattison/age fotostock.

We have seen how the nervous system allows an animal to sense and respond to its environment, coordinating the animal's movement and behavior (Chapters 35–37). The nervous system also works closely with the **endocrine system** to regulate an animal's internal physiological functions. This coordination is accomplished by signals sent from the nervous system to the endocrine system, which distributes chemical signals throughout the body. The endocrine system is made up of an interacting set of secretory glands and organs, located in different regions of the body, which respond to nervous system signals as well as to blood-borne signals from other organs to regulate internal bodily functions.

The signals communicated by the nervous and endocrine systems differ greatly in their modes of transmission and the times over which they act. The nervous system sends signals rapidly by action potentials running along nerve axons, and communication occurs between adjacent nerve cells by means of neurotransmitters. The endocrine system, by contrast, relies on cells and glands that secrete chemical signals called **hormones,** which influence the actions of other cells in the body. Whereas some hormones act locally, many others are released into the bloodstream, allowing them to circulate and influence distant target cells throughout the body. As a result, endocrine communication is generally slower and more prolonged than the rapid and brief signals transmitted by nerve cells. Hormones exert specific effects on particular target cells within the body by binding to receptors on or in target cells. The downstream influence of hormones is amplified in a series of steps so that a small amount of hormone can have dramatic effects in the body (Chapter 9).

In this chapter, we explore the properties and actions of hormones that help to coordinate and maintain a broad set of physiological functions of the organism. In particular, we examine the mechanisms by which hormones trigger cellular responses in their target organs. Because most hormones are evolutionarily conserved, their chemical structure is similar across a diverse array of animals. However, their functions often evolve rapidly, enabling hormones to serve new and broader roles.

38.1 AN OVERVIEW OF ENDOCRINE FUNCTION

The word "hormone" often popularly connotes teenagers and the changes that occur to their bodies as they grow and mature. Hormones, as we will see, do play a key role in animal growth and development—but that is only one of their roles. Hormones play diverse physiological roles in the body, in particular regulating an organism's response to the environment and helping to maintain stable physiological conditions within cells or within the animal as a whole. We start by highlighting these functions before addressing the molecular mechanisms of how hormones work in section 38.2.

The endocrine system helps to regulate an organism's response to its environment.

Organisms face constant changes in their environment. These changes often present physiological challenges or stresses that the organism must respond to by altering the functional state of its body. Changes in light or temperature, the threat of a predator, or the presence of a potential mate stimulate responses of an animal's nervous system and endocrine system. Endocrine

FIG. 38.1 Growth and development in (a) moths and caterpillars and in (b) grasshoppers.

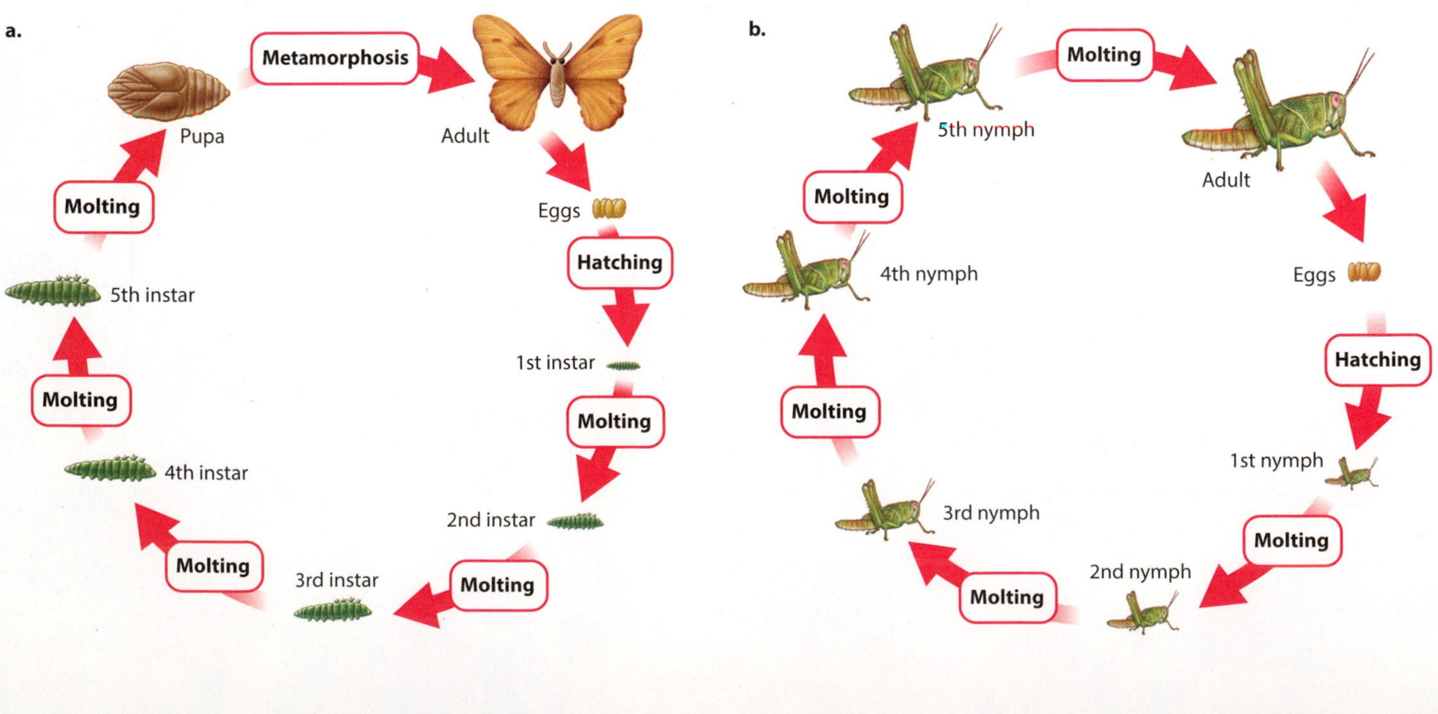

responses are commonly triggered by sensory signals received by the nervous system that are relayed to the endocrine system. The endocrine system, with its slower and more prolonged signaling, reinforces physiological changes in the animal's body that better suit the environmental cues received by its nervous system.

When a gazelle sees or smells a predator, its endocrine system helps to ready its body for rapid escape. When a female cardinal sees a bright red, singing male early in spring, endocrine signals released from within the female cardinal's brain initiate changes in its reproductive organs and increase its behavioral responsiveness, encouraging it to select the male as its mate. And when particular smells stimulate a sense of hunger and digestive function in animals such as dogs and humans, the animals are attracted to food. In each of these cases, the endocrine system helps an organism respond appropriately to environmental cues.

The endocrine system regulates growth and development.

Growth and development require broad changes in many different organ systems. The release of circulating hormones from the endocrine system accomplishes these changes, regulating how animals develop and grow. For example, we discuss in Chapter 42 the importance of the sex hormones estrogen and testosterone in determining female or male sexual characteristics both in the embryo and during puberty.

As another example, let's consider how the well-studied endocrine system of insects regulates their growth and development. Insects produce a rigid exoskeleton (Chapter 37), which must be periodically shed and replaced by a larger new one to enable the insect to grow (**Fig. 38.1**). Shedding of the exoskeleton is referred to as **molting.** Each molting produces a new larval stage. In some insects, such as moths, the animal's body also changes dramatically, undergoing **metamorphosis** at a key stage in development (Fig. 38.1a). In other insects, such as grasshoppers and crickets, the animal grows larger after each molting, but with little change in body form (Fig. 38.1b).

Molting and metamorphosis are regulated by hormones released from tissues in the insect's head. British physiologist Vincent Wigglesworth studied their effects on the blood-sucking insect *Rhodnius.* Normally, *Rhodnius* goes through five juvenile stages before developing into its final adult body form. Like grasshoppers and crickets, it does not undergo metamorphosis. Each of the developmental stages is triggered by a blood meal. Wigglesworth showed that he could prevent molting if the head of the insect were removed shortly after a blood meal (**Fig. 38.2**). *Rhodnius* can survive for long periods despite having its head removed. If the insect takes a blood meal and the head is removed a week later, the animal's body undergoes a molt and grows into its adult form. Wigglesworth hypothesized that a diffusing substance from the animal's head controls its molt. In the first case, this substance does not have time to reach the body to trigger a molt, but in the second it does.

The German endocrinologist Alfred Kuhn found a similar result in experiments performed on moth caterpillars. Tying the caterpillar's head off from the rest of the body prevented the body from molting. Wigglesworth and Kuhn considered the diffusing substance to be a brain hormone. Subsequent work showed that cells in a specialized region of the insect brain secrete the peptide now known as prothoracicotropic hormone (PTTH) (**Table 38.1**). In *Rhodnius,* PTTH is released after a blood meal and acts to trigger

TABLE 38.1 Major Invertebrate Hormones

SECRETING TISSUE OR GLAND	HORMONE	TARGET GLAND OR ORGAN	ACTION
Brain	Brain hormone (PTTH) (peptide)	Prothoracic gland	Stimulates release of ecdysone to trigger molt and metamorphosis
Corpora allata	Juvenile hormone (peptide)	All body tissues	Inhibits metamorphosis to adult stages; stimulates retention of juvenile characteristics
Prothoracic gland	Ecdysone (steroid)	All body tissues	Stimulates molt and, in absence or low levels of juvenile hormone, stimulates metamorphosis
Nervous system	Ecdysone (steroid)	All body tissues	Stimulates growth and regeneration in sea anemones, flatworms, nematodes, annelids, snails, and sea stars
Brain	Melanocyte-stimulating hormone (peptide)	Chromatophores	Stimulates pigmentation changes in cephalopods (octopus, squid, and cuttlefish)
Brain and eye stalks	Chromatotropins (peptides)	Chromatophores	Stimulates pigmentation changes in crustaceans
Reproductive gland	Androgen (peptide)	Reproductive tract	Regulates development of testes and male secondary sexual characteristics of crustaceans

FIG. 38.2

How are growth and development controlled in insects?

BACKGROUND During the 1930s, British physiologist Vincent Wigglesworth studied how a blood meal taken by the bug *Rhodnius* triggered its molting and growth. This work pioneered the discovery of hormonal substances that stimulated its growth. *Rhodnius* goes through five successive larval stages (called nymphs) before becoming a winged adult, as shown in Figure 38.2a. Each developmental step is triggered by a blood meal.

HYPOTHESIS Wigglesworth hypothesized that a substance (specifically, a hormone) that diffuses from the head triggers molt in *Rhodnius.*

EXPERIMENT Wigglesworth decapitated juvenile bugs at different intervals of time after a blood meal and observed whether or not molting occurred. (Decapitation does not kill the insect.)

RESULTS Wigglesworth showed that if a bug is decapitated less than an hour after a blood meal, it fails to molt. If a bug is decapitated 1 week after the blood meal, it molts (Fig. 38.2b, Experiment 1). He also found that if he used a fluid tube to join the body of a bug decapitated immediately after a blood meal with the body of one that wasn't decapitated until a week after feeding, the bodies of both bugs molted (Fig. 38.2b, Experiment 2). This experiment demonstrated that the diffusing hormone could trigger molting in the bug that lacked the hormone because of immediate decapitations.

CONCLUSION A substance diffuses from the head in *Rhodnius* and triggers the molting process.

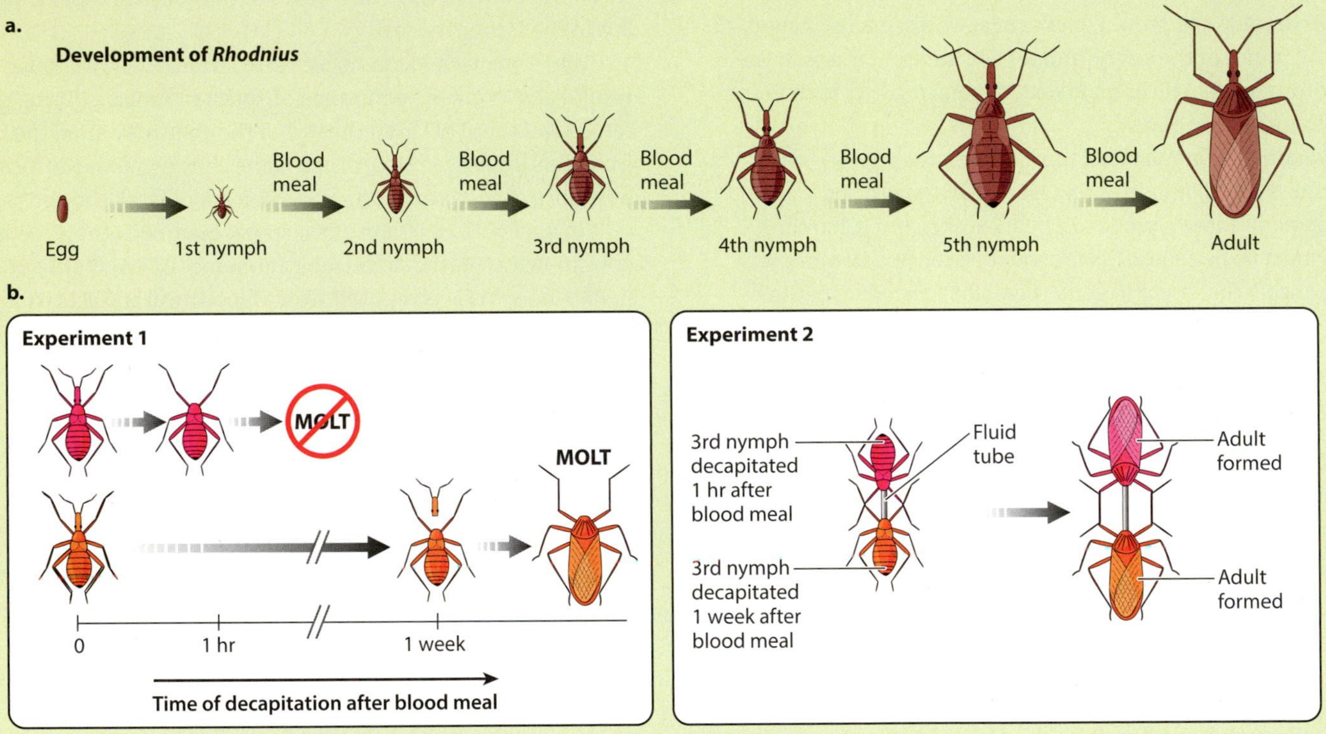

FOLLOW-UP WORK Subsequent studies by the German endocrinologist Alfred Kuhn confirmed the result in experiments performed on moth caterpillars. More recent molecular analyses performed by Japanese molecular endocrinologists Nagasawa and colleagues (1984) and Kawakami and colleagues (1990) isolated and sequenced the structure of the hormone from silkworm moths. It is now known as prothoracicotropic hormone, or PTTH, because it acts on the prothoracic gland of the insect to release the molting hormone, ecdysone.

SOURCES Wigglesworth, V. B. 1934. "The Physiology of Ecdysis in *Rhodnius prolixus* (Hemiptera). II. Factors Controlling Moulting and 'Metamorphosis.'" *Quarterly Journal of Microscopical Sciences* 77:191–223; Nagasawa, H., et al. 1984. "Amino-Terminal Amino Acid Sequence of the Silkworm Prothoracicotropic Hormone: Homology with Insulin." *Science* 226:1344–1345; Kawakami, A., et al. 1990. "Molecular Cloning of the *Bombyx mori* Prothoracicotropic Hormone." *Science* 247:1333–1335.

the animal's molt by stimulating the release of a second hormone from the prothoracic gland.

Two decades later, the German endocrinologist Peter Karlson isolated and purified this second hormone, termed ecdysone. PTTH triggers molting by stimulating the release of ecdysone, a steroid hormone that coordinates the growth and reorganization of body tissues during a molt. The action of ecdysone is a good example of how a hormone can precisely coordinate broad changes in body organization and function, in contrast to the specific regulation that nerves provide. Karlson was able to purify a small amount of ecdysone from large numbers of silk moth *Bombyx* caterpillars. He ultimately isolated a mere 25-mg sample from 500 kg of moth larvae! This tiny amount highlights the general principle that relatively few hormone molecules can have a large effect on an organism.

In his studies of *Rhodnius,* Wigglesworth also noted that, whatever the larval stage, the decapitated bug always molted into an adult after its blood meal. By removing just the front brain region of the head, Wigglesworth was able to show that a region in the head close to the brain releases a hormone that normally prevents the earlier larval stages from molting into the adult form (Table 38.1). This hormone, called juvenile hormone, is released in decreasing amounts during each successive larval molt. After the fifth and final stage, the level of juvenile hormone is so low that it no longer blocks maturation, and at the final molt the insect undergoes its final growth into the adult form (**Fig. 38.3**).

The brains of insects, similar to the nervous systems of other invertebrate and vertebrate animals, contain **neurosecretory cells.** These cells are neurons that release hormones, which act on endocrine glands or other targets within the insect instead of secreting neurotransmitters that bind to another neuron or to muscle. In the case of insects, peptide hormones released from neurosecretory brain cells act on the corpora allata (paired endocrine glands in insects) and the prothoracic gland (Fig. 38.3).Neurosecretory cells stimulate the corpora allata to produce juvenile hormone so that molting larvae retain juvenile characteristics. At appropriate times, other neurosecretory brain cells secrete the peptide hormone PTTH, which stimulates the release of ecdysone from the prothoracic gland, triggering the transition from larval to pupal and adult forms.

These studies reveal how small amounts of hormones released from key glands within the animal's body regulate major stages of growth and changes in body form during metamorphosis. When an animal's body requires coordinated changes in multiple organ systems, hormonal regulation by the endocrine system plays a critical role.

Hormones also regulate growth in humans and other vertebrate animals. Growth hormone controls the growth of the skeleton and many other tissues in the human body. Growth hormone is produced by the **pituitary gland,** which is located beneath the brain. Tumors of the pituitary that cause an overproduction of growth hormone lead to gigantism, and tumors that result in too little growth hormone cause pituitary dwarfism. The role of the pituitary gland in regulating body function is described later in the chapter.

The endocrine system underlies homeostasis.

Endocrine control of internal body functions is also central to **homeostasis,** the maintenance of a steady physiological state within a cell or an organism (Chapters 5 and 35). Without some way to maintain a stable internal environment, changing environmental conditions would lead to dangerous shifts in an animal's physiological function. For example, an animal's body weight depends on the regulation of its energy intake relative to energy expenditure. Disruption of hormones that regulate appetite and food intake can lead to obesity on the one hand or to weakness and lethargy on the other. Similarly, hormones that regulate the concentration of key ions in the body, such as Na^+ and K^+, are fundamental to healthy nerve and muscle function (Chapters 35 and 36) and fluid balance within the body (Chapter 41).

How does the body, and in particular the endocrine system, maintain homeostasis? Maintaining homeostasis depends on

FIG. 38.3 Hormonal control of growth and development in *Rhodnius.*

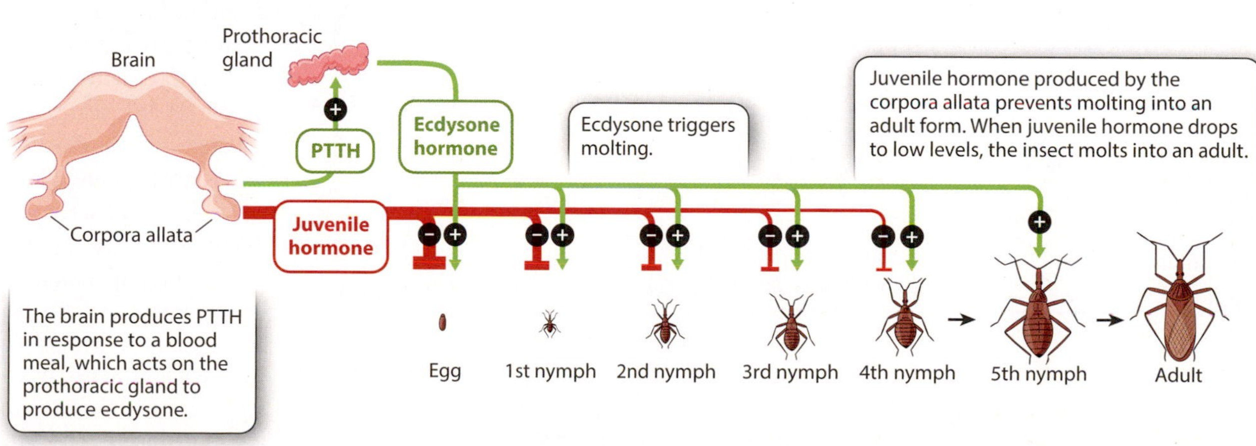

FIG. 38.4 **Negative feedback.** Negative feedback results in steady conditions, or homeostasis.

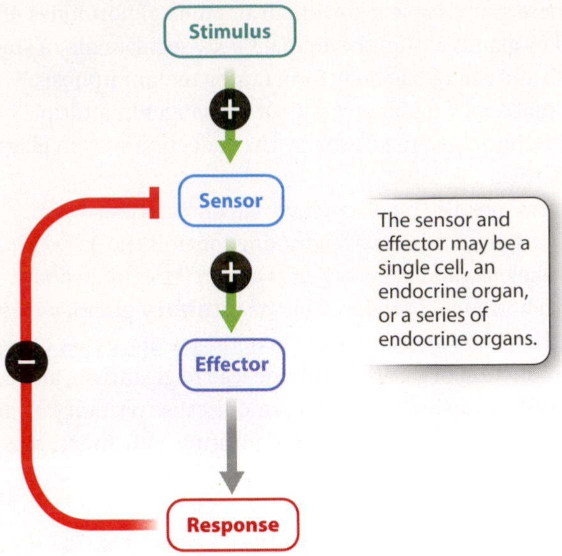

feedback from the target organ to the endocrine gland that secretes the hormone. Because hormones are transmitted through the bloodstream, this feedback can occur over varying distances within the body, coordinating the functions of several organs at any one time. In response to this feedback, the endocrine gland modifies its own subsequent production of hormone, either increasing or decreasing it. There are two general types of feedback, **negative feedback** and **positive feedback.**

Homeostasis typically depends on negative feedback. In Chapter 35, we saw how both the temperature in a house and the core body temperature of an animal are maintained at a constant level by this type of control.

Many physiological parameters—such as blood-glucose and calcium levels—are maintained at relatively steady levels by negative feedback. For example, a change in glucose or calcium level causes a response that brings the level back to the starting point, called the set point. In the case of the endocrine system, a change in level (the stimulus) is detected by a sensor in the endocrine organ (**Fig. 38.4**). The endocrine organ (the effector) releases a hormone, and the hormone causes a response in the body that opposes the initial stimulus such that the glucose or calcium level moves back toward the set point. The reversal of these levels causes secretion of the hormone to be decreased, and there is a limit to further change. Constant feedback between the response and sensor maintains a set point. That is, the maintenance of homeostasis is an active process.

Let's examine an example in more detail. Like core body temperature, the amount of glucose in the blood of animals is maintained at a steady level (**Fig. 38.5**). If glucose levels are too low, cells of the body do not have a ready source of energy. If glucose levels are too high for too long, they can damage organs.

Maintaining steady blood-glucose levels is challenging because glucose levels rise immediately after a meal when glucose is absorbed into the bloodstream from the intestine, then fall as glucose is taken up by cells to meet their energy needs. How then are constant levels maintained in the body? After a meal, when blood glucose rises, β (beta) cells of the pancreas secrete the hormone insulin, which circulates in the blood (Fig. 38.5a). In response to insulin, muscle and liver cells take up glucose from the blood and either use it or convert it to a storage form called glycogen (Chapter 7). In this way, insulin guards against high levels of glucose in the blood.

What happens if blood-glucose levels fall too low several hours after a meal? In this case, a different population of cells in the pancreas, called α (alpha) cells, secretes the hormone glucagon, which has effects roughly opposite to those of insulin (Fig. 38.5b). Glucagon stimulates the breakdown of glycogen into

FIG. 38.5 Control of glucose levels in the blood by negative feedback.

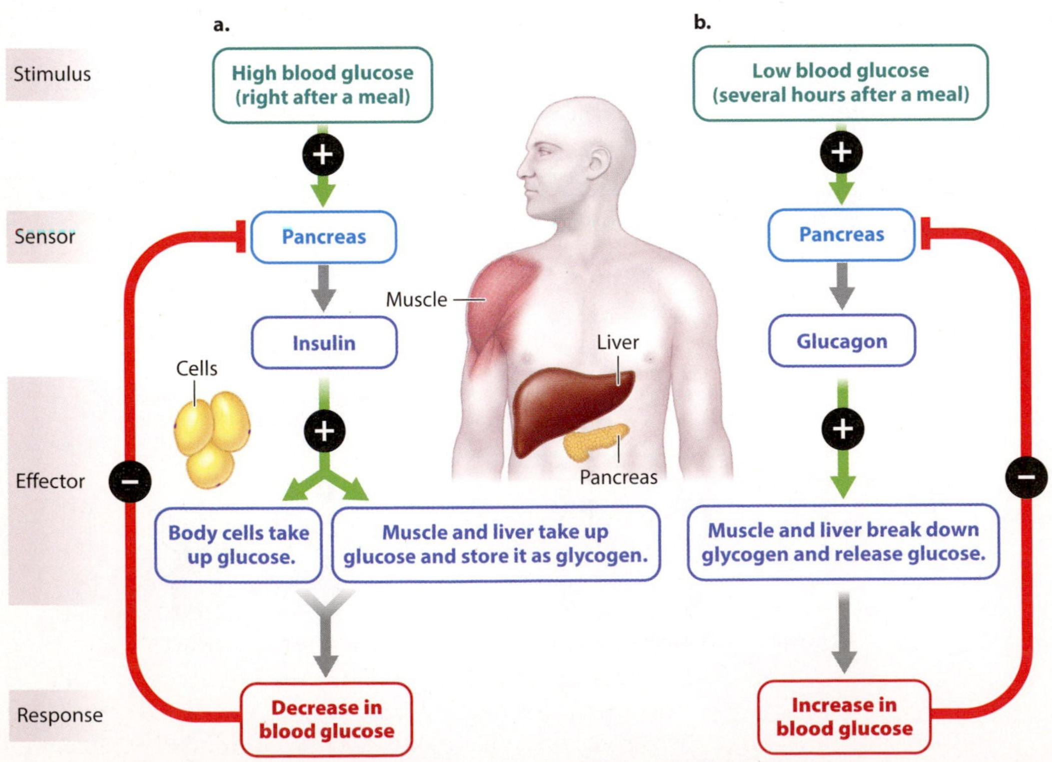

glucose and its release from muscle and liver cells. The result is that blood-glucose levels rise. In both cases, the stimulus (either high or low blood-glucose levels) is sensed by cells of the pancreas (β cells or α cells) and triggers a response (secretion of insulin or of glucagon, bringing blood-glucose levels back to the set point). Note that in each case the response feeds back to the secreting cells to reduce further hormone secretion.

When the control of blood-glucose levels by insulin fails, a disease called **diabetes mellitus** results. When untreated, diabetic individuals excrete excess glucose in their urine because blood-glucose levels are too high. Diabetes causes cardiovascular and neurological damage, including loss of sensation in the extremities, particularly the feet.

➔ **Quick Check 1** Diabetes mellitus is a disease characterized by high blood-glucose levels. What two different physiological conditions can produce diabetes?

In some instances, it is necessary to accelerate the response of target cells for a period of time. Positive feedback provides this enhancement (**Fig. 38.6**). In positive feedback, a stimulus causes a response in the same direction as the initial stimulus, which leads to a further response, and so on. In positive feedback in the endocrine system, a stimulus leads to secretion of a hormone that causes a response, and the response causes the release of more hormone. The result is an escalation of the response. A positive feedback loop reinforces itself until it is interrupted or broken by some sort of external signal outside the feedback loop but from within the body.

Positive feedback occurs in mammals during birth (Chapter 42). In response to uterine contractions, the hormone oxytocin is released from the pituitary gland. The release of oxytocin (stimulus) by the pituitary gland causes the uterine muscles (effector) to contract more forcefully. The uterine contractions in turn stimulate (positive feedback) the pituitary gland to secrete more oxytocin, causing the uterine muscles to contract more forcefully and more frequently.

FIG. 38.6 **Positive feedback.** In positive feedback, a stimulus causes a response, and that response causes a further response in the same direction.

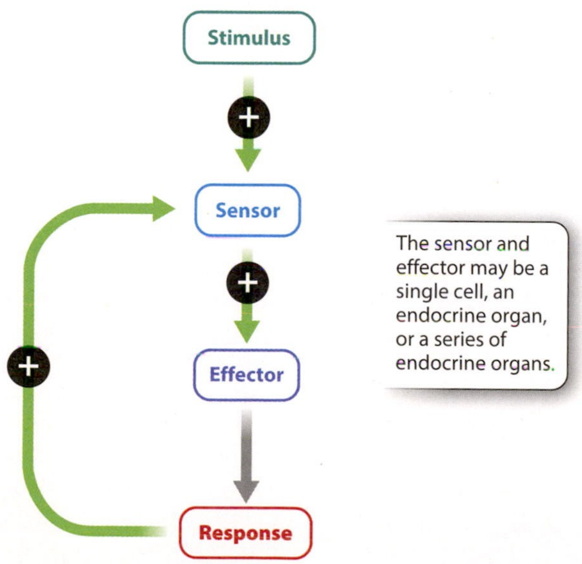

38.2 PROPERTIES OF HORMONES

Because hormones are released into the bloodstream, they have the potential to affect many organ systems. How do hormones circulating in the blood target specific tissues? How can one hormone exert its effect on one cell type and another hormone exert its effect on a different cell type? The answer lies in the presence of receptors on cells within target organs that bind specific hormones.

Hormones act specifically on cells that bind the hormone.

Hormones circulating in the bloodstream bind to receptors located either on the surface of or inside target cells. Therefore, it is the presence or absence of a receptor for a given hormone that determines which cells respond and which ones do not. For example, when the hormone oxytocin is released into the bloodstream of mammals, it affects only cells that express a receptor on the cell surface capable of binding oxytocin as it flows by in the bloodstream. These cells are uterine muscle cells and secretory cells in breast tissue. When oxytocin binds these cell-surface receptors, uterine muscle cells are stimulated to contract, and secretory cells in breast tissue release milk during breastfeeding. The cells of other organs do *not* express the receptor, so the hormone can exert its effect only on these specific tissues and no others.

The binding of a hormone to its receptor triggers changes in the target cell, resulting in a cellular response (Chapter 9). The specific action of a hormone depends on the kind of response that it triggers in the target cell. For example, the binding of a hormone can alter ion flow across the cell membrane, activate an intracellular signal transduction cascade that leads to changes in the biochemical activity of a target cell, or initiate more substantial changes in gene and protein expression.

➔ **Quick Check 2** What general features make a chemical compound a hormone, and how do hormones achieve specificity for certain kinds of target cell?

Two main classes of hormone are peptide and amines, and steroid hormones.

Some hormone receptors are located on the cell surface, others are located inside the cell. Where a receptor is located depends on characteristics of the hormone molecules themselves. Hormones can be grouped into two general classes: hydrophilic molecules, which cannot diffuse across the plasma membrane, and hydrophobic molecules, which can. The hydrophilic hormones are

peptide hormones and **amine hormones,** and the hydrophobic hormones are **steroid hormones,** examples of which are illustrated in **Fig. 38.7.**

Peptide hormones and amine hormones are both derived from amino acids (Figs. 38.7a and 38.7b). Peptide hormones are short chains of amino acids, whereas amine hormones are derived from a single aromatic amino acid, such as tyrosine or phenylalanine (Chapter 4). All steroid hormones are derived from cholesterol (Fig. 38.7c). Whereas peptide hormones (which are sometimes large enough to be considered proteins) can evolve through changes in their amino acid sequence, steroid hormones cannot. Evolutionary changes in steroid hormone function depend instead on changes in their synthetic enzymes or changes in the receptors they bind and the cellular responses they trigger.

Peptide and amine hormones are more abundant than steroid hormones and are more diverse in their actions. Because most peptide and amine hormones are hydrophilic and cannot diffuse across the plasma membrane (exceptions are thyroid hormones), nearly all these hormones bind to membrane receptors on the surface of the cell (**Fig. 38.8a**). Peptide and amine hormones alter the biochemical activity of the target by initiating signaling cascades within the target cell, as discussed in Chapter 9: The bound receptor activates an enzyme or other molecule within the cell, and it in turn activates another and so on. Typically, the activated enzymes are protein kinases, which phosphorylate other proteins. These signaling cascades can lead to changes in gene expression or they can alter metabolism by turning on or off metabolic enzymes. For example, a peptide or amine hormone could trigger a cell to grow, divide, change shape, or to release another hormone. Peptide and amine hormones act on timescales of minutes to hours.

There are many examples of peptide hormones. In vertebrates, for instance, growth hormone stimulates protein synthesis and the growth of many body tissues, particularly the musculoskeletal system (**Table 38.2**, on pages 816–817). As we saw, insulin and glucagon are peptide hormones released by the pancreas that regulate glucose metabolism by the liver, muscles, and other tissues. Gastrin and cholecystokinin are peptide hormones that regulate mammalian digestive function (Chapter 40). In addition to serving as a neurotransmitter released by the sympathetic nervous system (Chapter 35), epinephrine and norepinephrine (also known as adrenaline and noradrenaline) are amine hormones, supporting an animal's fight-or-flight response.

Because steroid hormones are hydrophobic, they diffuse freely across the cell membrane to bind with receptors in the cytoplasm or nucleus, forming a steroid hormone–receptor complex (**Fig. 38.8b**). Hormone–receptor complexes that form in the cytoplasm are transported into the nucleus of the cell. These complexes most commonly act as transcription factors: They stimulate or repress gene expression and thereby alter the proteins produced by the target cell. Consequently, steroid hormones typically have profound and long-lasting effects on the cells and tissues they target, on timescales of days to months.

The sex hormones estrogen, progesterone, and testosterone (Table 38.2) are steroid hormones that regulate the development of the vertebrate reproductive organs. Cortisol is a steroid hormone that regulates a vertebrate animal's response to stress and inhibits inflammation. Steroid hormones similar to cortisol are used medically to reduce the symptoms of certain inflammatory

FIG. 38.7 **(a) Peptide, (b) amine, and (c) steroid hormones.**

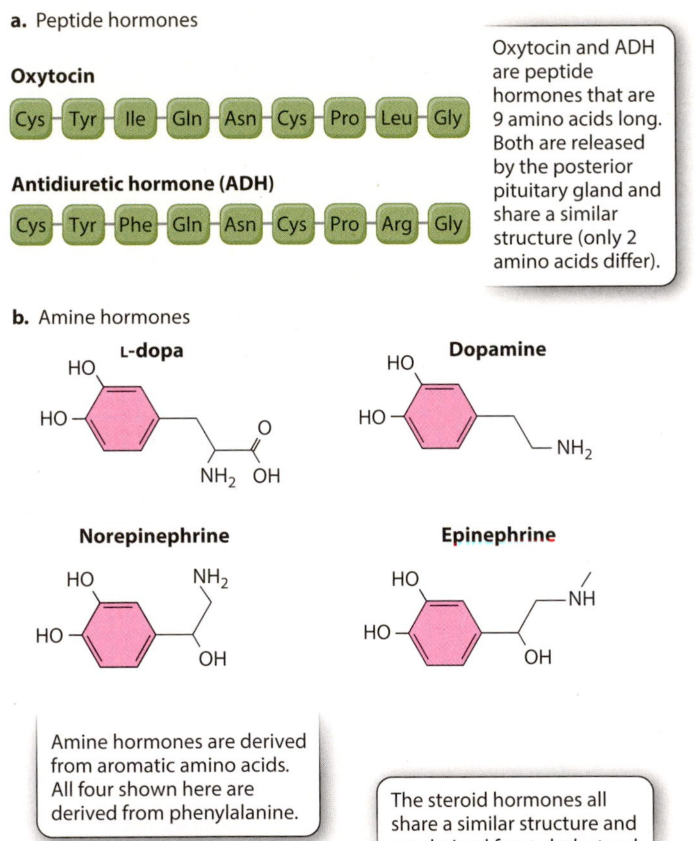

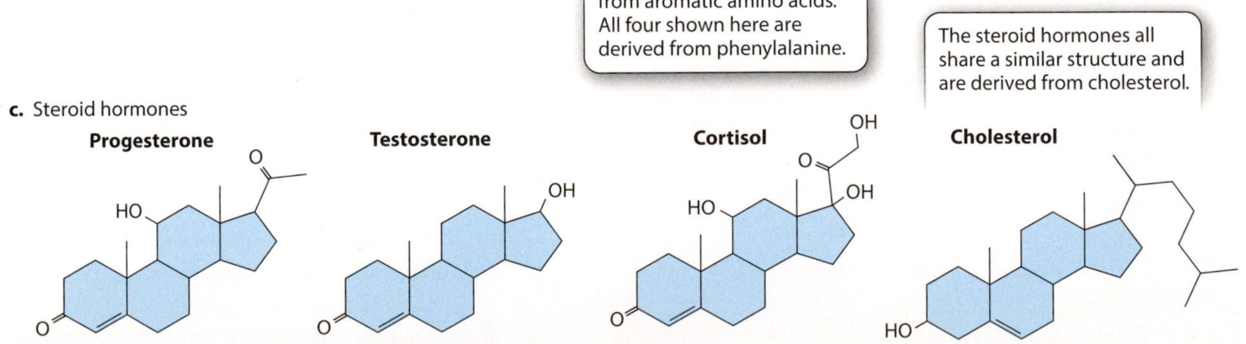

FIG. 38.8 Mechanism of action of (a) peptide and (b) steroid hormones.

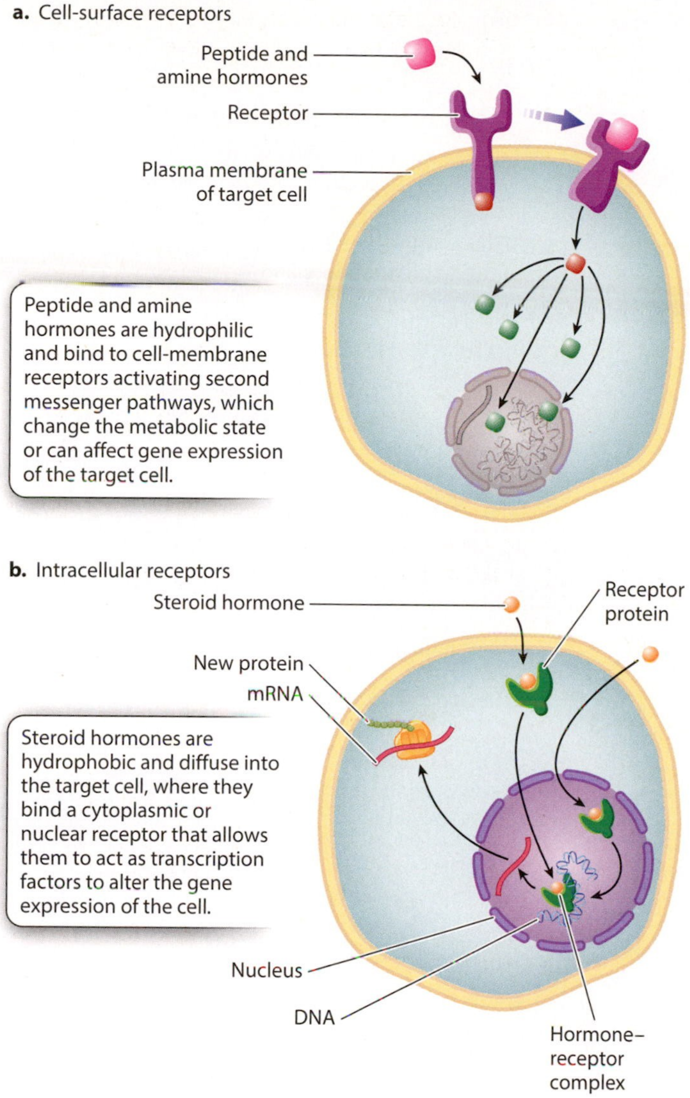

disorders. However, they carry risks because they also suppress the immune system, increasing the chance of infection and diminishing wound healing.

→ **Quick Check 3** Why do peptide and steroid hormones bind different kinds of receptor, and how does this difference affect the resulting signaling pathways in the target cell?

Hormonal signals are amplified to strengthen their effect.

Hormones are typically released in small amounts. So how is it possible for them to have large effects on the overall physiology of an organism? The answer is that hormone signals are amplified (Chapter 9). Amplification occurs during a series of signaling steps, both before the hormone binds to a cell receptor and inside the cell afterward. The first set of signaling steps is the passing of a signal from one endocrine gland to the next in a hormonal pathway, and the second set is signal transduction in the target cell. Signals are amplified at each step of the pathway, resulting in a large effect on the target cell or organ.

Hormonal signaling pathways between endocrine glands and tissues are often referred to as endocrine axes. In vertebrates, hormonal signals are amplified along a pathway called the hypothalamic–pituitary axis that goes from the hypothalamus to the anterior pituitary gland, and then from the anterior pituitary gland to target glands or tissues in the body. The hypothalamus initially releases trace amounts of peptide hormones called **releasing factors** (**Fig. 38.9**). The releasing factors bind to receptors on cells in the anterior pituitary gland, leading that organ to release a much larger amount of associated hormones. For example, the hypothalamus secretes a corticotropin releasing factor, which stimulates the anterior pituitary gland to release a larger amount of adrenocorticotropic hormone (ACTH).

FIG. 38.9 Amplification of a hormonal signal.

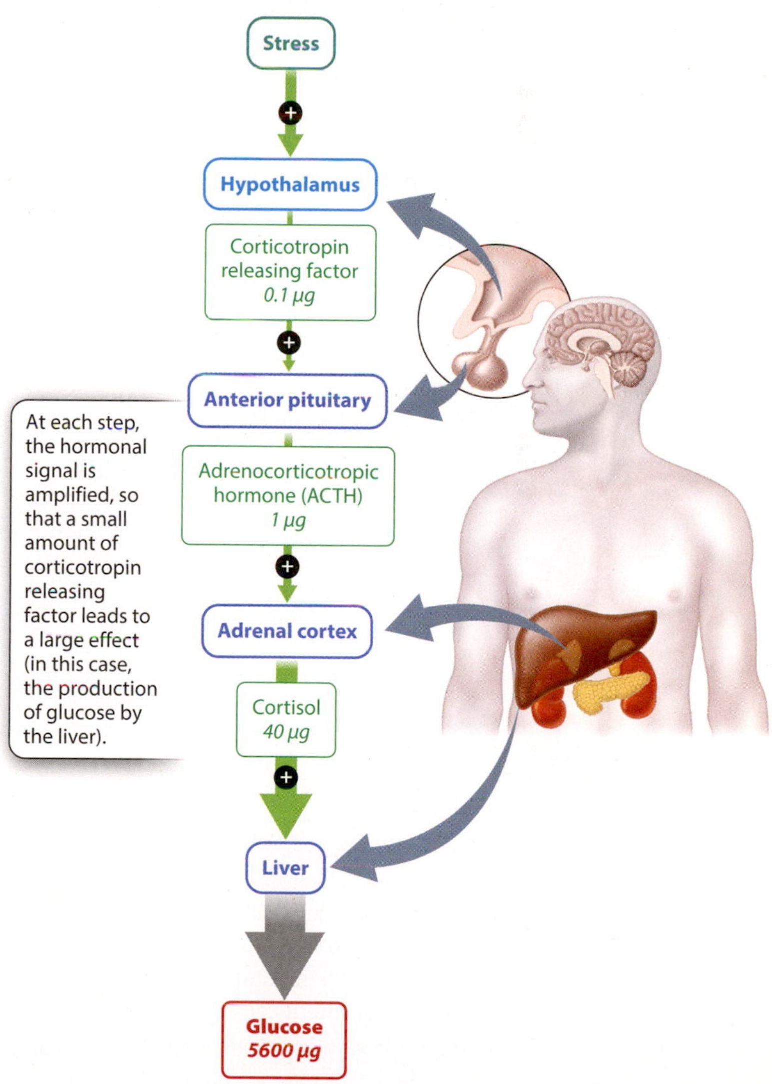

TABLE 38.2 Major Vertebrate Hormones

SECRETING TISSUE OR GLAND	HORMONE	TARGET GLAND OR ORGAN	ACTION
Hypothalamus 	Releasing factors (peptides)	Anterior pituitary gland	Stimulate secretion of anterior pituitary hormones
Anterior pituitary gland 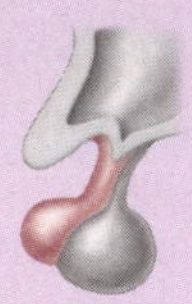	Thyroid-stimulating hormone (TSH) (glycoprotein)	Thyroid gland	Stimulates synthesis and secretion of thyroid hormones by the thyroid gland
	Follicle-stimulating hormone (FSH) (glycoprotein)	Gonads	Stimulates maturation of eggs in females; stimulates sperm production in males
	Luteinizing hormone (LH) (glycoprotein)	Gonads	Stimulates production and secretion of sex hormones in ovaries (estrogen and progesterone) and testes (testosterone)
	Adrenocorticotropic hormone (ACTH) (peptide)	Adrenal glands	Stimulates production and release of cortisol
	Growth hormone (GH) (protein)	Bones, muscles, liver	Stimulates protein synthesis and body growth
	Prolactin (protein)	Mammary glands	Stimulates milk production
	Melanocyte-stimulating hormone (peptide)	Melanocytes	Regulates skin (and scale) pigmentation
Posterior pituitary gland 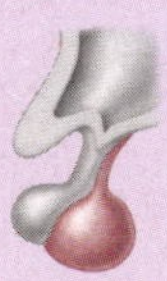	Oxytocin (peptide)	Uterus, breast, brain	Stimulates uterine contraction and milk let-down; influences social behavior
	Antidiuretic hormone (ADH; vasopressin) (peptide)	Kidneys, brain	Stimulates uptake of water from the kidneys; involved in pair bonding
Thyroid gland 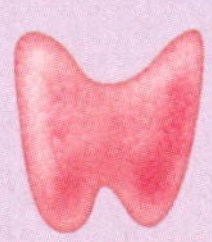	Thyroid hormones (peptides)	Many tissues	Stimulate and maintain metabolism for development and growth
	Calcitonin (peptide)	Bone	Stimulates bone formation by osteoblasts
Ovaries 	Estrogen (steroid)	Uterus, breast, other tissues	Stimulates development of female secondary sexual characteristics and regulates reproductive behavior
	Progesterone (steroid)	Uterus	Maintains female secondary sexual characteristics and sustains pregnancy

Hormones released by the anterior pituitary gland in turn bind cell receptors in the target organ. In this case, ACTH acts on cells of the adrenal cortex, stimulating their secretion of the hormone cortisol. Cortisol acts on many different cells and tissues in the body, causing what is known as an acute stress response. Among its effects, it causes the liver to convert glycogen and amino acids to glucose. This action yields 56,000 times more glucose than the initial weight of releasing factors secreted by the hypothalamus.

We can also see an example of amplification in invertebrates in the regulation of insect molting and metamorphosis (see Fig. 38.3).

TABLE 38.2 Major Vertebrate Hormones *(continued)*

SECRETING TISSUE OR GLAND	HORMONE	TARGET GLAND OR ORGAN	ACTION
Testes 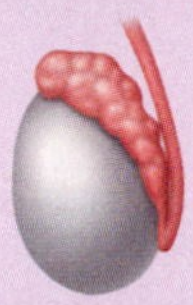	Testosterone (steroid)	Various tissues	Stimulates development of male secondary sexual characteristics; regulates male reproductive behavior and stimulates sperm production
Adrenal cortex 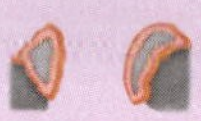	Cortisol (steroid)	Liver, muscles, immune system	Regulates response to stress by increasing blood glucose levels and reduces inflammation
Adrenal medulla 	(Nor-) Epinephrine (peptide)	Heart, blood vessels, liver	Stimulates heart rate, blood flow to muscles, and elevation of blood glucose level as part of fight-or-flight response
Parathyroid glands 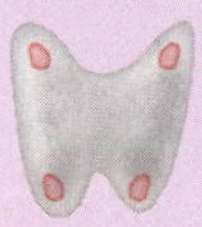	Parathyroid hormone (PTH) (protein)	Bone	Stimulates bone resorption by osteoclasts to increase blood Ca^{2+} levels
Pancreas 	Insulin (protein)	Liver, muscles, fat, other tissues	Stimulates uptake of blood glucose and storage as glycogen
	Glucagon (protein)	Liver	Stimulates breakdown of glycogen and glucose release into blood
	Somatostatin (peptide)	Digestive tract	Inhibits insulin and glucagon release; decreases digestive activity (secretion, absorption, and motility)
Stomach 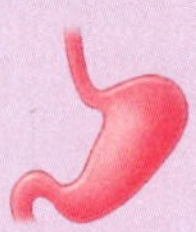	Gastrin (peptide)	Stomach	Stimulates protein digestion by secretion of digestive enzymes and acid; stimulates gastric motility
Small intestine 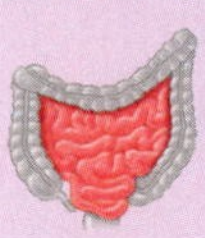	Cholecystokinin (peptide)	Pancreas, liver, gallbladder	Stimulates secretion of digestive enzymes and products from liver and gallbladder
	Secretin (peptide)	Pancreas	Stimulates bicarbonate secretion from pancreas
Pineal gland 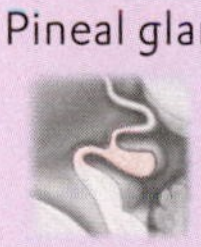	Melatonin (peptide)	Brain, various organs	Regulates circadian rhythms

This pathway has two signaling steps in which an endocrine gland releases a hormone. First, the brain releases the hormone PTTH. That hormone signals the prothoracic gland, leading to the release of a much larger amount of ecdysone, the hormone that regulates growth and metamorphosis of the insect's body. Ecdysone in turn binds to cell receptors in the target organ. Signaling cascades within these cells further amplify the hormone signal. Thus, amplification applies both to signal transduction cascades within a cell and to the chemical signals that are transmitted as endocrine hormones between glands and tissues.

Hormones are evolutionarily conserved molecules with diverse functions.

Most hormones have an ancient evolutionary history. The structures of many hormones are evolutionarily conserved. For example, some vertebrate hormones can also be found in many invertebrates. Since the first vertebrate animals diverged from invertebrates over 500 million years ago, some animal hormones are even older. In many cases, their roles in many invertebrate animals have yet to be discovered.

Typically, the same hormone serves different functions in vertebrates and in invertebrates, and may even serve different functions within distinct groups of vertebrates. Thyroid-stimulating hormone (TSH), a hormone released by the anterior pituitary gland that targets the thyroid gland, regulates metabolism in mammals but triggers metamorphosis in amphibians and feather molt in birds. It has even been found in snails and other invertebrates that lack a thyroid gland. Its function in snails appears to be stimulation of the number of sperm or eggs produced.

Recent genomic analysis has shown that the receptors for many hormones evolved well before the hormones with which they now interact. Even though the structure of a hormone or its receptor is often largely unchanged across diverse groups of organisms, hormones and their receptors can readily be selected to take on new roles as organisms evolve new behaviors and exploit new environments. An example is the shift in function of TSH to regulate metabolism in mammals and feather molt in birds from its role stimulating sperm or egg production in snails.

Another intriguing finding is that many peptides originally identified as hormones in various tissues have also been found to function as neurotransmitters in the nervous system. For example, oxytocin, which stimulates uterine contraction and the release of milk, also serves as a neurotransmitter in the brain and is believed to influence social behavior, as well as stimulate sexual arousal in mammals. Similarly, antidiuretic hormone, a peptide hormone that regulates water uptake in the kidneys (Chapter 41), also functions as a neurotransmitter in the brain influencing mammalian mating and pair-bonding behavior. Hence, the same compound, expressed in two different cell types, may have entirely different functions. The roles of hormones as chemical messengers are varied within an organism and easily changed over the course of evolution.

38.3 THE VERTEBRATE ENDOCRINE SYSTEM

The vertebrate endocrine system regulates changes in the animal's physiological and behavioral states in response to sensory cues received by its nervous system both from the environment and from internal organs. Whereas certain sensory signals may communicate directly to endocrine glands or tissues in the body, many are processed within the brain and transmitted to the endocrine system by the hypothalamus, which is located in the forebrain (**Fig. 38.10**). The hypothalamus, in turn, relays these signals to the pituitary gland, which lies just below it. The pituitary gland is a central regulating gland of the vertebrate endocrine system: It releases several hormones that coordinate the action of many other endocrine glands and tissues. These glands, illustrated in Fig. 38.10, control the growth and maturation of the body, regulate

FIG. 38.10 The hypothalamus–pituitary axis. The hypothalamus sends signals to the pituitary gland, which in turn sends signals to diverse cells and organs throughout the body.

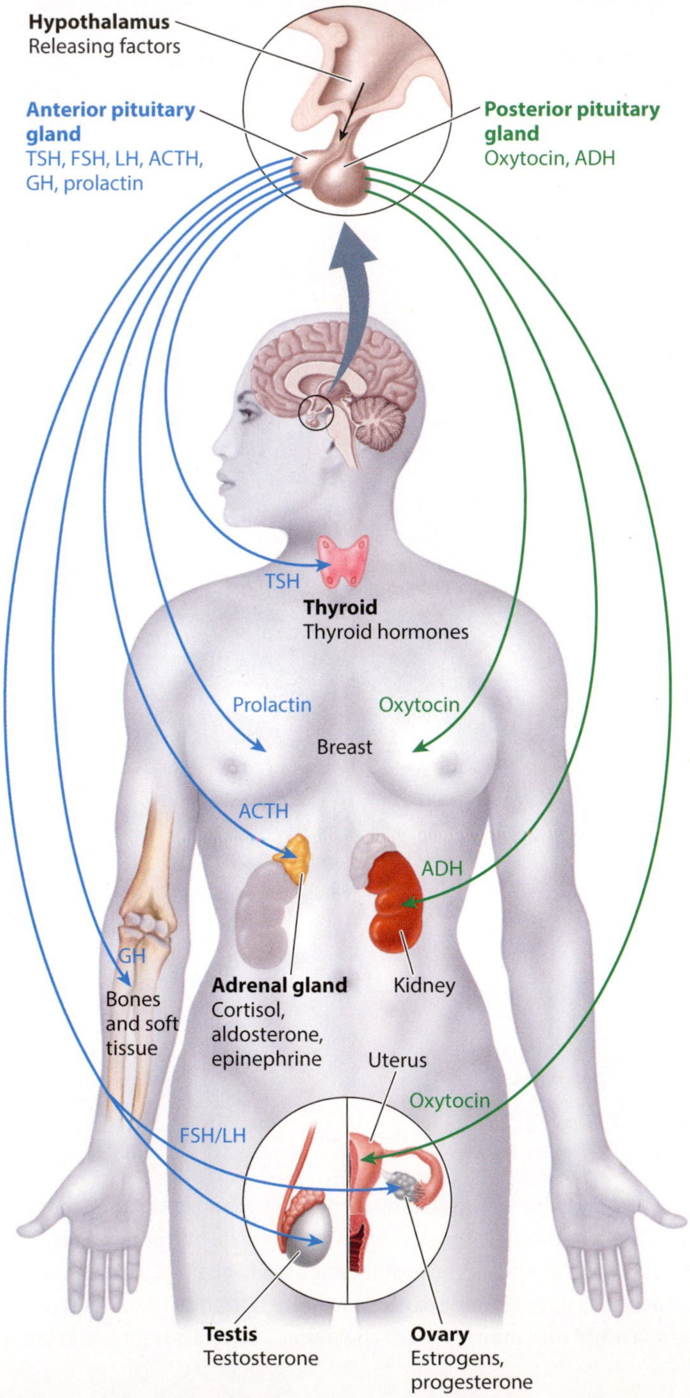

reproductive development and metabolism, and control of water balance. These effects reflect the broad range of roles that the endocrine and nervous systems perform in regulating body maturation and maintaining homeostasis.

The pituitary gland communicates with many cells, tissues, and organs of the body. Some of these, like the thyroid and adrenal glands, secrete hormones in response to signals from the pituitary gland and therefore have exclusively endocrine functions. Others, like the lungs, kidneys, and digestive tract, harbor endocrine cells that secrete hormones and also have other physiological roles in the body. Consequently, the vertebrate endocrine system is not localized in one part of the body, but is present throughout.

The pituitary gland integrates diverse bodily functions by secreting hormones in response to signals from the hypothalamus.

The hypothalamus is the main route by which nervous system signals are transmitted to the vertebrate endocrine system. The function of the hypothalamus is to transmit these signals to the pituitary gland, the endocrine gland that acts as a control center for many other endocrine glands in the body.

The pituitary gland is divided into anterior and posterior regions (**Fig. 38.11**). This division is not arbitrary—the two regions have distinct functions, organizations, and embryonic origins. The **anterior pituitary gland** forms from epithelial cells that develop and push up from the roof of the mouth, whereas the **posterior pituitary gland** develops from neural tissue at the base of the brain. Both sets of cells form pouches that develop into glands, which come to lie adjacent to each other. The anterior and posterior pituitary glands should therefore be thought of as two distinct glands, not one.

Because of these developmental differences, the anterior and posterior pituitary glands receive input from the hypothalamus in different ways. As does the brain of arthropods, the vertebrate hypothalamus contains neurosecretory cells, which are part of the brain and therefore are neurons. As with invertebrate neurosecretory cells, these cells release hormones into the bloodstream. Some of these neurosecretory cells communicate with the anterior pituitary gland. In this case, they secrete hormones called releasing factors into small blood vessels that travel to and supply the anterior pituitary gland (Fig. 38.11). In response, cells of the anterior pituitary gland release hormones into the bloodstream. These hormones circulate throughout the body and bind to receptors on target cells, tissues, and organs.

In contrast, communication between the hypothalamus and the posterior pituitary gland does not involve hypothalamic releasing factors. Instead, the posterior pituitary gland contains the axons of neurosecretory cells whose cell bodies are located in the hypothalamus. These axons release hormones directly into the bloodstream, and these hormones are transported in the blood to distant sites (Fig. 38.11). Consequently, the posterior pituitary is part of the nervous system itself.

In response to signals from the hypothalamus, distinct hormones are secreted by the anterior and posterior pituitary glands (Table 38.2). The hormones released by the anterior pituitary hormone act on an endocrine gland to cause release of other hormones. Hormones that control

FIG. 38.11 **The anterior and posterior pituitary glands.** The hypothalamus communicates with the anterior pituitary gland by secreting releasing factors into the blood. The hypothalamus communicates with the posterior pituitary gland by extending axons of neurosecretory cells into it.

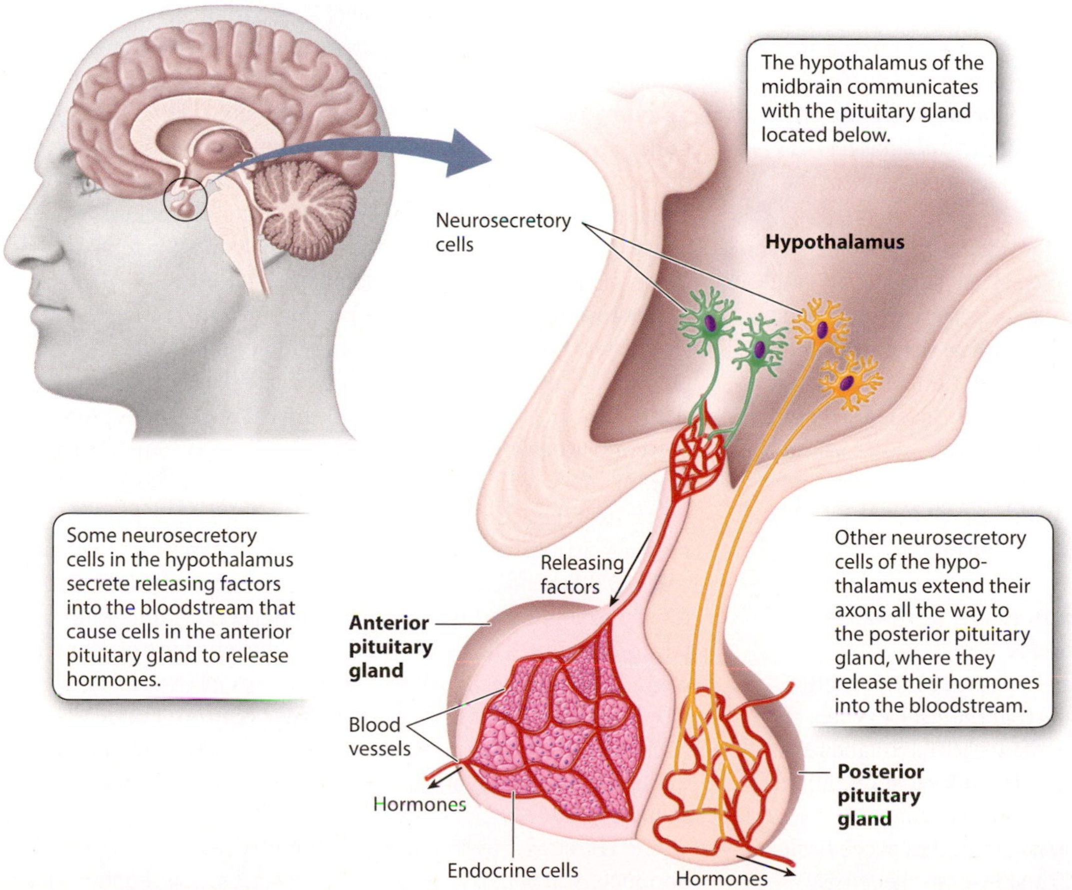

the release of other hormones are called **tropic hormones.** In response to thyroid-stimulating hormone (TSH), the thyroid gland releases thyroid hormones that regulate the metabolic state of the body. In response to follicle-stimulating hormone (FSH) and luteinizing hormone (LH), the ovaries release estrogen and progesterone and the testes release testosterone. In response to adrenocorticotropic hormone (ACTH), the adrenal glands release cortisol, which has diverse effects that include stimulating glucose release into the bloodstream, maintaining blood pressure, and suppressing the immune system.

The anterior pituitary gland also secretes the hormones growth hormone (GH) and prolactin. Growth hormone acts generally on the muscles, bones, and other body tissues to stimulate their growth, and prolactin stimulates milk production in the breasts of female mammals in response to an infant's suckling at the mother's nipple.

The hormones released by the posterior pituitary gland include **oxytocin** and **antidiuretic hormone** (**ADH;** also called **vasopressin**). These are two evolutionarily related peptide hormones of similar structure. As we discussed earlier, oxytocin plays several roles related to female reproduction: It causes uterine contraction during labor and stimulates the release of milk during breastfeeding. Antidiuretic hormone acts on the kidneys (Chapter 41), regulating the concentration of urine that an animal excretes, which is critical to maintaining water and solute balance in the body.

In addition to their roles in reproduction and kidney function, oxytocin and antidiuretic hormone are released by cells in the brain and may play roles in social behaviors. For instance, recent evidence indicates that oxytocin may be important in regulating maternal behavior toward infants. Recent research suggests that oxytocin may also have a role in behaviors such as trust as a prelude to mating in vertebrate animals. These functions reflect the long evolutionary history of oxytocin, as it plays a role in the social behavior of insects.

Recent evidence also suggests that antidiuretic hormone plays a parallel role in regulating male social behavior and parental behavior in mammals. This is not surprising because oxytocin and antidiuretic hormone have very similar chemical structures (see Fig. 38.7). The integration of reproductive function with parental and social behavior for the care of young is likely favored strongly by natural selection. Receptors for these hormones are expressed within the brain. Thus, in addition to acting on reproductive and renal organs of the body, oxytocin and antidiuretic hormone affect mammalian behavior by acting directly on the brain.

Many targets of pituitary hormones are endocrine tissues that also secrete hormones.

As we have seen, some of the hormones released by the anterior pituitary gland act on endocrine organs that themselves then release hormones. These tropic hormones are TSH, FSH and LH, and ACTH (see Fig. 38.10; Table 38.2). Here, we look at their target organs in more detail.

TSH acts on the **thyroid gland,** which is located in the front of the neck (see Fig. 38.10; Table 38.2) and leads to the release of two peptide hormones, thyroxine (T4) and triiodothyronine (T3). These peptide hormones are sufficiently hydrophobic that they enter cells, binding to intracellular receptors to increase the resting metabolic rate of cells throughout the body. As levels of T3 and T4 rise, these hormones act as a brake on the release of TSH by the anterior pituitary gland. This regulation of TSH by negative feedback is critical to maintaining stable TSH levels and a stable metabolic state of the body as a whole. Overproduction of thyroid hormones (hyperthyroidism) creates symptoms that reflect an overly active metabolic state (increased appetite and weight loss), whereas thyroid hormone deficiency (hypothyroidism) creates symptoms that reflect a metabolic state that is too low (fatigue and sluggishness). Both conditions are diagnosed by blood tests that monitor circulating levels of thyroid hormones, and both can be treated by medication.

Because the thyroid hormones T3 and T4 require iodine for their formation, individuals who do not acquire enough iodine from their diets produce too little thyroid hormone. Additionally, in cases of iodine insufficiency, the anterior pituitary gland increases its production of TSH because it is no longer receiving negative feedback. Over time, in response to TSH the thyroid gland enlarges to form a goiter, which is observed as an enlargement of the throat. The introduction of iodized salt has eliminated goiter formation and hypothyroidism in many countries, but in many underdeveloped areas these remain significant public health problems.

The gonadotropic hormones are FSH and LH. They target the female and male gonads—the **ovaries** and **testes** (see Fig. 38.10; Table 38.2). In response to FSH and LH, the ovaries and testes each secrete sex hormones that regulate their own development as well as the development of secondary sexual characteristics. Female sex hormones include the steroids estrogen and progesterone. The principal male sex hormone is the androgen testosterone. These sex hormones are common to a vast majority of vertebrates, regulating sexual differentiation, gonadal maturation, and reproductive behavior. The role of these sex hormones in regulating reproductive function is discussed in greater detail in Chapter 42.

Testosterone is a naturally occurring anabolic steroid that stimulates the synthesis in the testes of proteins needed for sperm production and the development of male sexual features and body tissue growth, particularly in muscles. Anabolic steroids promote protein synthesis to build body tissues and anabolic metabolism to store energy within cells. A variety of synthetic anabolic steroids that mimic the action of testosterone have been developed to treat muscle wasting due to loss of appetite and diseases such as cancer and AIDS. Anabolic steroids are used illegally by some male and female athletes to promote muscle strengthening, undermining fair competition. The risks to health of continued use of anabolic steroids are considerable: liver damage, heart disease, and high blood pressure. Tissues become damaged because elevated circulating levels of testosterone produced by anabolic steroids inhibits the normal synthesis of tropic hormones by the anterior pituitary gland.

ACTH released by the anterior pituitary gland is also a tropic hormone. As noted in section 38.1 in explaining how amplification strengthens the action of downstream hormones in a signaling pathway, ACTH acts on the cortex (the outer portion) of the paired **adrenal glands** (see Fig. 38.10; Table 38.2). The adrenal glands are located adjacent to the kidneys (*ad-* meaning "near" and "renal" referring to the kidneys). In humans, each small adrenal gland lies just above each kidney. During times of stress, such as starvation, fear, or intense physical exertion, ACTH stimulates adrenal cortex cells to secrete cortisol. The release of cortisol in response to stress affects a broad range of bodily functions, including blood-glucose levels (Chapter 41), immune function (Chapter 43), and blood pressure (Chapter 39). Whereas acute (short-term) responses that lead to the release of ACTH and elevated cortisol by the adrenal cortex are healthy physiological responses that enable animals to respond to stressful situations, chronic (long-term) stress often leads to a broad range of unhealthy conditions that include impaired cognitive and immune function, sleep disruption, and fatigue.

Other endocrine organs have diverse functions.

Although many endocrine organs receive signals from the hypothalamic–pituitary axis, some respond instead to internal physiological states of the body or to external environmental cues. One of these is the **parathyroid gland,** which is located on the thyroid gland. The parathyroid gland secretes parathyroid hormone (PTH). Working together with calcitonin, secreted by the thyroid gland, PTH controls the levels of calcium in the blood (Table 38.2). The two hormones act by regulating the actions of bone cells (osteoblasts and osteoclasts; Chapter 37) that control bone formation and bone removal.

When circulating levels of calcium fall too low, the parathyroid gland releases PTH, which stimulates osteoclasts to reabsorb bone mineral, halting bone formation and releasing calcium into the bloodstream. When calcium levels are too high, the production of PTH is inhibited and calcitonin is released to shift bone metabolism toward net bone formation, building bone that stores calcium in the skeleton. Note that, as with blood-glucose levels, calcium levels are maintained in a narrow range by negative feedback: The response to the hormone (increasing or decreasing levels of calcium in the blood) is the opposite of the stimulus (low or high levels of calcium in the blood) so that a stable set point is maintained.

Several organs of the digestive system, including the pancreas, stomach, and duodenum, produce hormones that regulate appetite and digestion. Three of these hormones are discussed in Chapter 40.

The **pineal gland** is located in the thalamic region of the brain. In response to darkness, the pineal gland secretes melatonin, a hormone that helps control an animal's state of wakefulness (Table 38.2). Melatonin secretion is inhibited when environmental light cues sensed by the retina are conveyed to the gland by the autonomic nervous system. In other vertebrates, such as lampreys, some fishes, and reptiles, the pineal gland responds to light sensed by a "third eye," photosensitive cells located at the skull's surface between the animal's actual eyes. In both cases, melatonin levels rise at night. In diurnal animals (those that are active during the day), the rise in melatonin levels at night causes the animals to sleep. In nocturnal animals (those that are active during the night), the rise in melatonin levels causes them to become active. The release of melatonin in response to daily (and seasonal) light cycles helps to maintain circadian rhythms of the body. Circadian rhythms are cycles of about 24 hours in which an animal's biochemical, physiological, and behavioral state shifts in response to changes in daily and seasonal environmental conditions. To overcome jet lag, travelers are advised to seek sunlight and often to take melatonin before sleeping to help reset their 24-hour circadian rhythm.

In several animals, including ground squirrels, bats, bears, and rattlesnakes, the pineal gland regulates hibernation, controlling the animal's metabolic state over longer time periods. In many animals, release of melatonin also regulates seasonal breeding cycles.

? CASE 7 PREDATOR–PREY: A GAME OF LIFE AND DEATH

How does the endocrine system influence predators and prey?

The endocrine system works closely with the nervous system to enable animals to respond to external cues in the environment. In Chapters 35 and 36, we discussed how prey animals, such as gazelles, use visual and olfactory cues processed by the nervous system to detect the distant movement of a predator, such as a lion (**Fig. 38.12**). We saw how the sympathetic nervous system,

FIG. 38.12 **Predator and prey.** Animals rely on rapid integration of sensory information with coordination of bodily functions to escape a predator or to catch their prey. *Source: Federico Veronesi/Getty Images.*

in the fight-or-flight response, can trigger broad physiological changes, such as increased heart and breathing rates and changes in metabolism that ready the predator to attack or the prey to respond.

These actions are coordinated by the endocrine system. The sympathetic nervous system sends axons to the **adrenal medulla** (see Figure 38.10), the inner part of the adrenal gland. In response to stimulation by the sympathetic nervous system, cells of the adrenal medulla secrete two hormones, epinephrine and norepinephrine (also known as adrenaline and noradrenaline). These are hormones released by the adrenal gland with targets throughout the body, leading to the physiological changes of the fight-or-flight response. Interestingly, epinephrine and norepinephrine also act as neurotransmitters in the brain. In fact, cells of the adrenal medulla are modified nerve cells that have lost their axons and dendrites.

The same set of responses occurs when, walking along a deserted sidewalk, we note a shadowy movement or hear a footstep behind us. The perceived threat sets in motion a coordinated physiological reaction, mediated by the combined action of the nervous and endocrine systems. These changes make us alert and ready for action to avoid or resist a perceived threat. They also inhibit digestion and eliminate a sense of appetite.

Earlier, we considered the importance of homeostasis in maintaining a variety of parameters, such as temperature and glucose levels, at a steady level. In this case, there is an adjustment of the set points of different organs, which is critical in enabling an animal to respond to external cues and adjust its physiological response appropriately. When the real or perceived threat is gone, the body is able to return to its prior state by reestablishing earlier set points.

Set points are also changed when an animal acclimatizes to a new environment, such as when it moves from fresh water to a marine environment or to a new altitude. An adaptive change in body function to a new environment is referred to as **acclimatization.** Examples of acclimatization are the increase in the number of red blood cells and their capacity to carry more oxygen at higher altitude (Chapter 39) and seasonal changes in the fur of mammals that achieve the insulation required to maintain a stable body temperature (Chapter 40). These physiological changes result from establishing new set points in a process coordinated by the endocrine system.

38.4 OTHER FORMS OF CHEMICAL COMMUNICATION

As we have seen, hormones are chemical compounds secreted by glands that enter the circulation and act on distant cells, those having receptors that bind the hormones. However, other chemical signals act over short distances, and still others are released into the environment to signal other individuals of the same species. Here, we explore additional modes of chemical communication and the compounds that produce them.

FIG. 38.13 Modes of chemical signaling. Chemical signaling can occur (a) over long distances (endocrine), or (b) locally (paracrine and synaptic).

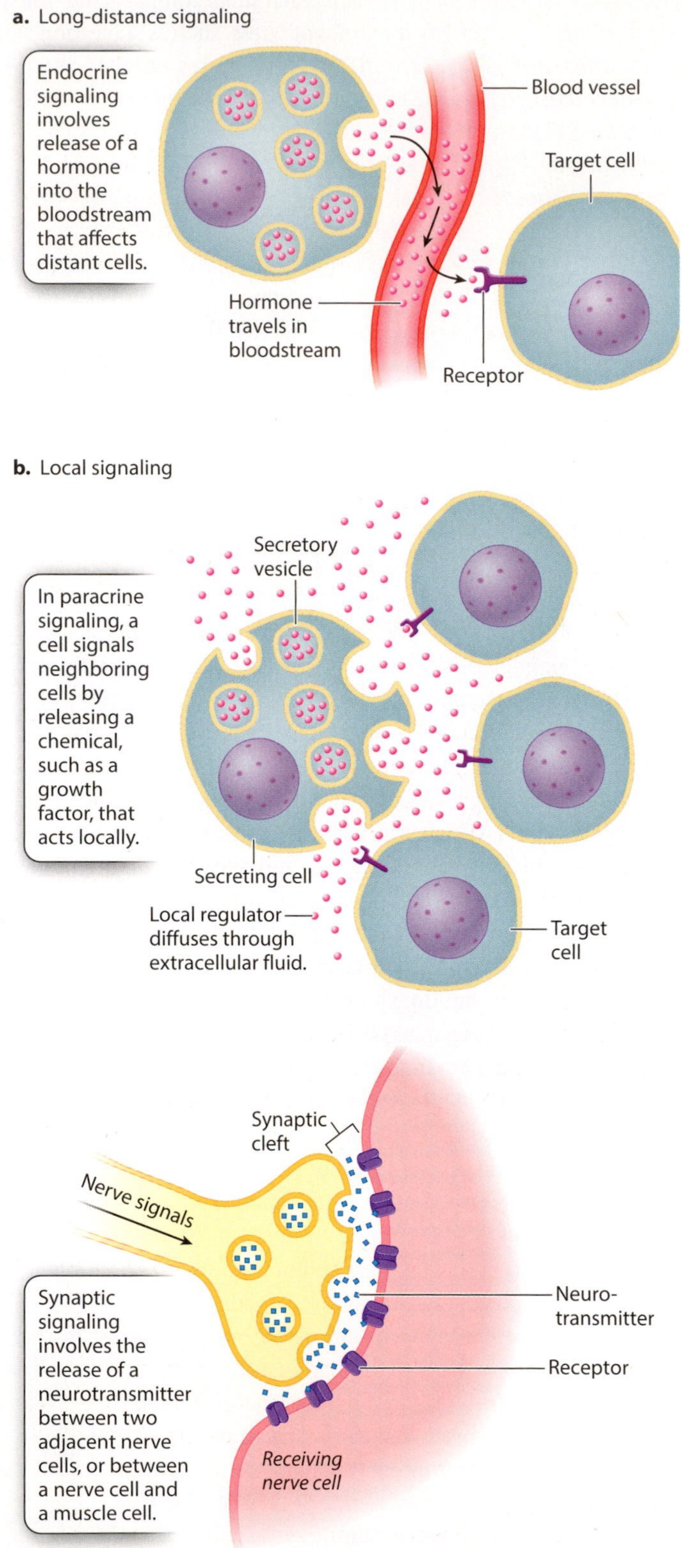

FIG. 38.14 **Pheromone action.** (a) Pheromones are means of attracting mates, as in these ladybugs; (b) wolves secrete pheromones in their urine that mark their territory; (c) honeybees swarm to defend their nest after a bee that stings a threatening animal releases an alarm pheromone. *Sources: a. Krzysztof Bisztyga/age footstock; b. Jorg & Petra Werner/Animals Animals/Earth Scenes; c. Steven Smith/Alamy.*

Local chemical signals regulate neighboring target cells.

Whereas hormones enter the bloodstream to be transmitted to more distant target cells (**Fig. 38.13a**), other chemical compounds act locally on neighboring cells (**Fig. 38.13b**). In order to take up the chemical signal, these cells must have receptors for the compound to bind to before it degrades or diffuses into the bloodstream.

Chemical compounds that act locally are said to have paracrine function (Fig. 38.13b). If the compounds act on the secreting cell itself, they are said to have autocrine function—that is, the chemical signals may stimulate or inhibit their own secretion. Growth factors (also referred to as cytokines and distinct from the growth hormone released by the anterior pituitary gland) and histamine are examples of paracrine chemical agents released in small amounts that act locally on neighboring cells.

Growth factors (Chapter 9) enhance the differentiation and growth of particular kinds of tissue. For example, bone morphogenetic proteins (BMPs) promote the formation of bone in skeletal growth, and fibroblast growth factors (FGFs) stimulate the formation of connective tissue. Similarly, nerve growth factors (NGFs) promote the survival and growth of nerve cells. When tissues are damaged, specialized cells nearby release histamine. This chemical signal triggers the dilation of blood vessels, allowing blood proteins and white blood cells to move into the region to fight infection and induce repair (Chapter 43). The release of histamine causes the swelling, redness, and warmth that commonly surround a wound.

Another form of short-range chemical signaling is the release of neurotransmitters into the synapse between directly communicating nerve cells or between nerve cells and muscles at neuromuscular junctions (Fig. 38.13b). Synaptic signaling can be extremely rapid and brief compared to the other modes of chemical communication.

Pheromones are chemical compounds released into the environment to signal physiological and behavioral changes in other species members.

Many animals release small chemical compounds into the environment to signal and influence the physiology and behavior of other members of their species. These water- or airborne compounds are called **pheromones.** For example, bees and wasps release pheromones to find a nest, locate food, and regulate the development and activities of other colony members. Even plants, such as orchids, release pheromones. Some plant pheromones attract pollinating bees or wasps by mimicking the chemical signals released by the females of these species. There are many different pheromones, which serve very different functions, as shown in **Fig. 38.14.**

Female silk moths use pheromones to attract males over long distances. The sex pheromone bombykol, released from a gland in the female's abdomen, signals her location to males looking for a mate. Sex pheromones play a key role in the mating behavior of many invertebrate and vertebrate species. In most mammals, amphibians, and reptiles, pheromones are detected by a vomeronasal organ with chemosensory neurons in the nasal region of the skull. The release of sex pheromones to attract a mate and trigger ritualistic social and mating behaviors is one of the most common types of pheromone signaling (Chapter 45). Humans and nonhuman primates lack a vomeronasal organ, and most studies indicate that they do not produce pheromones.

Other pheromones signal territorial boundaries or predatory threats. Dogs and other canids (this family includes wolves and coyotes), as well as cats, secrete pheromones in their urine that

FIG. 38.15 **Trail pheromones in ants.** Here, a chemical trail has been laid down by worker ants signaling a food source for the colony. *Source: LatitudeStock–Patrick Ford/Getty Images.*

FIG. 38.16 **A brightly colored male cichlid fish.** The behavior and markings of this fish are under hormonal control and triggered by social context. *Source: blickwinkel/Alamy.*

mark their territories. Social seabirds, such as cormorants and boobies, mark their nesting sites by pheromones.

Ants and other social insects deploy alarm pheromones to warn of an attack by an advancing army from another colony. The release of alarm pheromones from a stinging bee causes other bees to swarm and join the attack to defend their nest. In many other species, an animal under attack releases an alarm pheromone, warning its neighbors. Similarly, social mammals such as deer warn of an approaching predator by means of alarm pheromones.

Social insects, such as ants, deploy trail pheromones that provide chemical cues that other worker ants can follow to a food source (**Fig. 38.15**). An ant returning to the colony from a food source releases trail pheromones from its abdomen onto the soil. Other worker ants detect the trail with their antennae and track the concentration of trail pheromone, guiding them along the trail. If you have watched ants marching in line across a sidewalk, a trail, or into and out of your house, you have seen worker ants using pheromone cues that signpost the way to food.

Aquatic species also depend on pheromone signaling. Fish, salamanders, and tadpoles release alarm pheromones into the water to warn of a threatening predator. Prey species that share the same habitat may even sense and respond to the release of another species' alarm pheromone.

Fish also release pheromones to coordinate mating and to regulate social interactions. In dense populations of cichlid fish living in Lake Tanganyika in central Africa, the ratio of brightly colored breeding males (**Fig. 38.16**) to females is highly regulated. Only about 10% of males are brightly colored and defend their territory to attract and mate with females. This restriction on the number of dominant breeding males is regulated by behavioral cues and chemical pheromones. When a dominant male dies, subordinate males fight for the vacated territory. When a new male assumes dominance, it becomes brightly colored to attract females and defend its newly won territory. The dominant breeding male reinforces its dominance by releasing pheromones in its urine that signal females and subordinate males in its breeding area. The color change is mediated by hormones released from the male fish's pituitary gland, which act on pigment cells called melanocytes that are located in the epidermis of its scales. Thus, the fish makes use of both endocrine and pheromone signals to change its behavior and the behavior of other members of its species.

→ **Quick Check 4** What distinguishes paracrine and pheromone signals from endocrine signals? ■

Core Concepts Summary

38.1 Animal endocrine systems release chemical signals called hormones into the bloodstream that respond to environmental cues, regulate growth and development, and maintain homeostasis.

The nervous system responds rapidly to sensory information, providing immediate physiological responses, whereas the endocrine system acts more slowly and over prolonged periods to effect broad changes in the physiological and behavioral state of an animal. page 808

The endocrine system helps an animal respond to its environment. page 808

The endocrine system is involved in growth and development. page 809

The nervous and endocrine systems regulate physiological functions to maintain homeostasis. page 811

Homeostasis is often achieved by negative feedback, in which the response opposes the stimulus. page 812

A stimulus can be amplified by positive feedback, in which the response amplifies the initial stimulus. page 813

38.2 Hormones bind to receptors on or inside target cells, and their signals are amplified to exert strong effects on target cells.

Hormones are chemical compounds that are secreted and transported by the circulatory system, often to distant targets. page 808

Hormones bind to specific receptors located on the surface or inside the target cells that receive their signals. page 813

Two main classes of hormone are peptides and amines, and steroid hormones. page 813

Peptide and amine hormones are hydrophilic and bind to cell-membrane receptors, stimulating second messenger pathways to alter the biochemical function of target cells. page 814

Steroids are hydrophobic and bind to intracellular or nuclear receptors, which function as transcription factors to regulate gene expression and protein synthesis. page 814

Hormones communicate by signaling cascades to amplify the strength of their downstream effect on target cells. page 815

The hypothalamic–pituitary axis forms the central endocrine pathway by which vertebrates regulate and integrate diverse bodily functions. page 815

Hormones are evolutionarily conserved molecules common to diverse groups of animals, but which have evolved novel functions. page 818

38.3 In vertebrate animals, the hypothalamus and the pituitary gland control and integrate diverse bodily functions and behaviors.

The anterior pituitary gland secretes hormones in response to releasing factors produced and secreted by the hypothalamus. page 819

Hormones released by the anterior pituitary gland include thyroid-stimulating hormone (TSH), the gonadotropic hormones follicle-stimulating hormone (FSH) and luteinizing hormone (LH), adrenocorticotropic hormone (ACTH), growth hormone (GH), and prolactin. page 820

Tropic hormones are hormones that act on endocrine glands to produce additional hormones. TSH causes the thyroid gland to release thyroid hormones; the gonadotropic hormones FSH and LH act on the female ovaries to produce estrogen and progesterone and the male testes to produce testosterone; and ACTH acts on the adrenal gland to release cortisol. page 820

The posterior pituitary gland contains the axons of neurosecretory cells that release the peptide hormones oxytocin and antidiuretic hormone into the bloodstream. page 820

Some endocrine glands do not respond to signals from the pituitary gland. These include the parathyroid gland, which releases parathyroid hormone that helps to control calcium levels; the pancreas, which secretes insulin and glucagon to regulate blood glucose levels; and the pineal gland, which secretes melatonin in response to light and dark cycles. page 821

38.4 Chemical communication can also occur locally between neighboring cells or, through the release of pheromones, between individuals, coordinating social interactions.

Paracrine signals act locally over short times to affect neighboring cells, compared with the more distant and longer-lasting signals transmitted by blood-borne hormones. page 823

Pheromones are chemical signals released into the air or water, signaling behavioral cues between individuals of a species. page 823

Sex pheromones attract mates and coordinate ritualized mating behavior; alarm pheromones signal predatory threats; trail pheromones signal the location of food sources; and territorial pheromones signal a male's dominance and breeding area. page 823

Self-Assessment

1. Describe three general functions of the endocrine system and provide an example of each.
2. Diagram an endocrine pathway, such as parathyroid regulation of blood-calcium levels, which maintains homeostasis of a bodily function by means of negative feedback.
3. Diagram an endocrine pathway, such as oxytocin regulation of female labor during birth, which relies on positive feedback.
4. Identify and provide an example of each of the three general classes of hormone.
5. Explain how hormones act specifically on only some cells or organs although they are transported by the bloodstream throughout the body.
6. Describe how hormones exert substantial and broad effects on their target cells although they are released in small amounts by the initial signaling gland.
7. Explain how peptide hormones affect the function of a target cell.
8. Explain how steroid hormones affect the function of a target cell.
9. Explain the difference between the way in which the hypothalamus controls the anterior pituitary gland and the way it controls the posterior pituitary gland.
10. Explain the difference in action between a hormone, a paracrine signal, a neurotransmitter, and a pheromone.

Log in to LaunchPad to check your answers to the Self-Assessment questions, and to access additional learning tools.

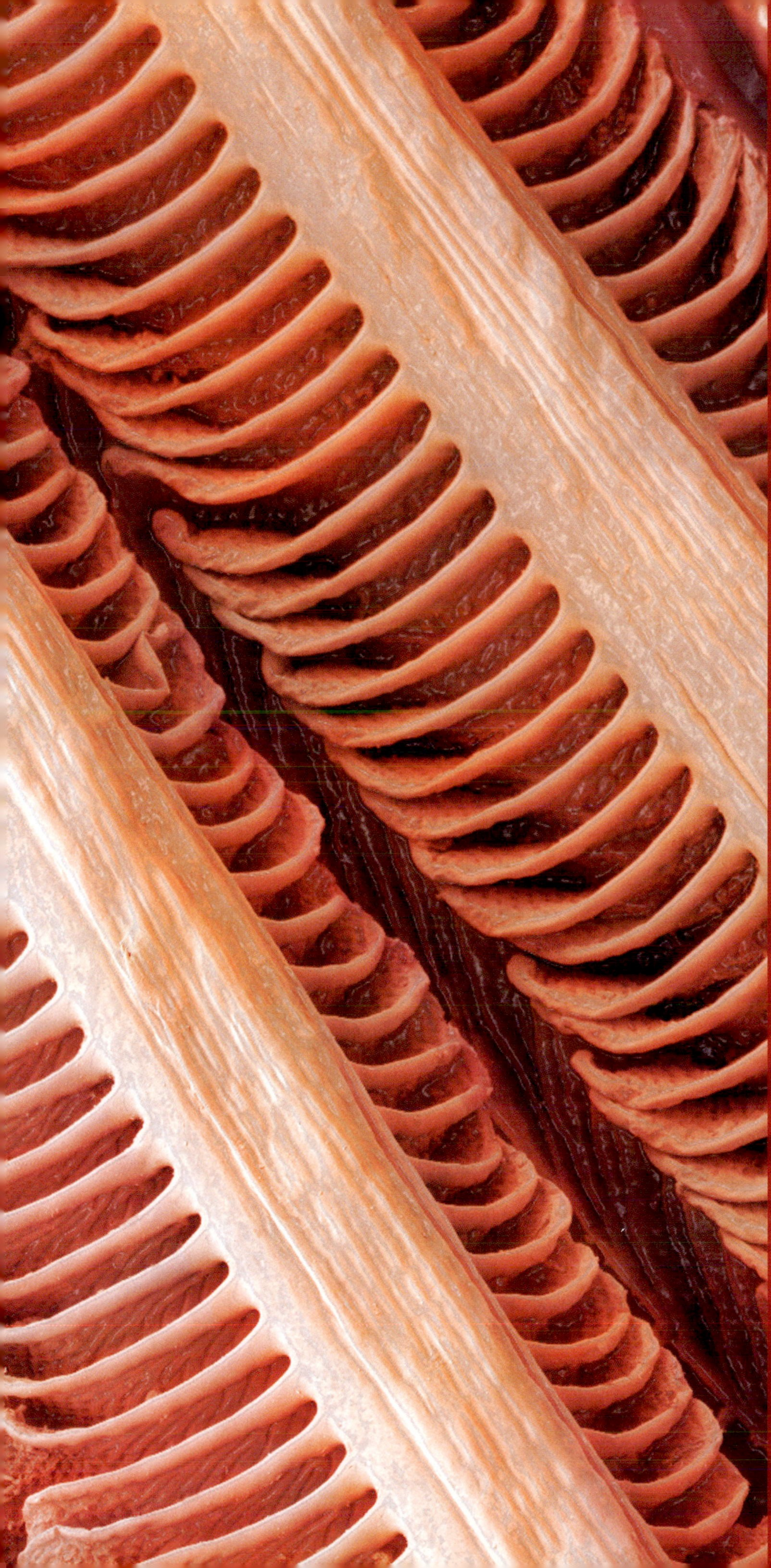

Source: Power and Syred/Science Source.

CHAPTER 39

Animal Cardiovascular and Respiratory Systems

Core Concepts

39.1 Respiration and circulation depend on diffusion over short distances and bulk flow over long distances.

39.2 Respiration provides oxygen and eliminates carbon dioxide in support of cellular metabolism.

39.3 Red blood cells produce hemoglobin, greatly increasing the amount of oxygen transported by the blood.

39.4 Circulatory systems have vessels of different sizes that facilitate bulk flow and diffusion.

39.5 The evolution of animal hearts reflects selection for a high metabolic rate, achieved by increasing the delivery of oxygen to metabolically active cells.

As animals evolved to be larger and more complex, they faced the challenge of obtaining an adequate supply of oxygen (O_2) and nutrients and, at the same time, eliminating carbon dioxide (CO_2) and other waste products. In this chapter, we explore how animal respiratory and circulatory systems evolved to meet the metabolic needs of an animal's cells. In particular, we focus on the uptake of O_2 and the release of CO_2, key gases in cellular respiration (Chapter 7).

Gases, nutrients, and wastes are continually being exchanged between an organism and its environment. Because of their small size or thin bodies, single-celled organisms and simple multicellular animals exchange compounds with the environment by diffusion. Because diffusion works effectively only over very short distances, larger and more complex animals cannot rely on diffusion as the sole means of exchanging compounds (Chapter 28). These animals evolved respiratory structures at the body's surface for taking up O_2 and releasing CO_2, and internal circulatory systems for long-distance transport to different body regions.

Because similar physical laws govern gas and liquid movement, the respiratory and circulatory systems of different animals have many features in common. However, fish obtain O_2 from water, whereas land animals breathe air. Water and air have very different physical properties. Thus, aquatic and terrestrial environments pose distinct challenges to the animal in obtaining O_2. Because life first began in the water, the evolution of life on land required substantial changes in how animals exchange gases, take up nutrients, and eliminate wastes.

39.1 DELIVERY OF OXYGEN AND ELIMINATION OF CARBON DIOXIDE

Eukaryotic cells use O_2 to burn organic fuels—carbohydrate, lipid, and protein—for ATP production (Chapter 7). The increase in atmospheric O_2 with the evolution of photosynthesis was key to the evolution of O_2-based cellular respiration, and that in turn contributed to the diversification of multicellular animal life during the Cambrian Period, about 500 million years ago. In the process of burning fuel by aerobic respiration, CO_2 is produced. Animals have evolved mechanisms to acquire O_2 from their environment and eliminate excess CO_2 from their bodies to the environment. These mechanisms make up the process known as animal respiration—or "breathing." The transport of O_2 and CO_2 between an animal and its environment is referred to as **gas exchange,** and is fundamental to all eukaryotic animals, as well as to photosynthetic plants.

Diffusion governs gas exchange over short distances.

Single-celled organisms and simple animals and plants exchange gases by diffusion. **Diffusion** is the random movement of individual molecules (Chapter 5). If there is a difference in concentration of a substance between two regions, diffusion results in the net movement of molecules from regions of higher concentration to regions of lower concentration until the molecules are evenly distributed.

Given a large enough surface, diffusion is an extremely effective way to exchange gases and other substances over short distances between two compartments, for example between the lungs and the blood vessels supplying the lungs. The rate of diffusion is directly proportional to the surface area over which exchange occurs and to the concentration difference, and inversely proportional to the distance over which the molecules move (or the thickness of the barrier). Diffusion is generally ineffective over distances exceeding about 100 μm, or 0.1 mm. Consequently, respiratory and circulatory structures have large surface areas and thin interfaces.

Because effective transport by diffusion is limited to short distances, some of the earliest and simplest invertebrate animals are composed of thin sheets of cells with a large surface area. For example, sponges and sea anemones have a thin body with a central cavity through which water circulates, and flatworms have a flattened body surface (**Fig. 39.1**).

Net movement of a substance results from diffusion when there is a concentration difference between two regions. When considering the diffusion of a gas, however, we usually refer to its partial pressure rather than its concentration. The **partial pressure (*p*)** of a gas is defined as its fractional concentration relative to other gases present, multiplied by the overall (that is, the atmospheric)

FIG. 39.1 **The thin body of a flatworm.** Gases can be exchanged by diffusion in flatworms because their bodies are only a few cell layers thick. *Photo source: Amar & Isabelle Guillen/SeaPics.com.*

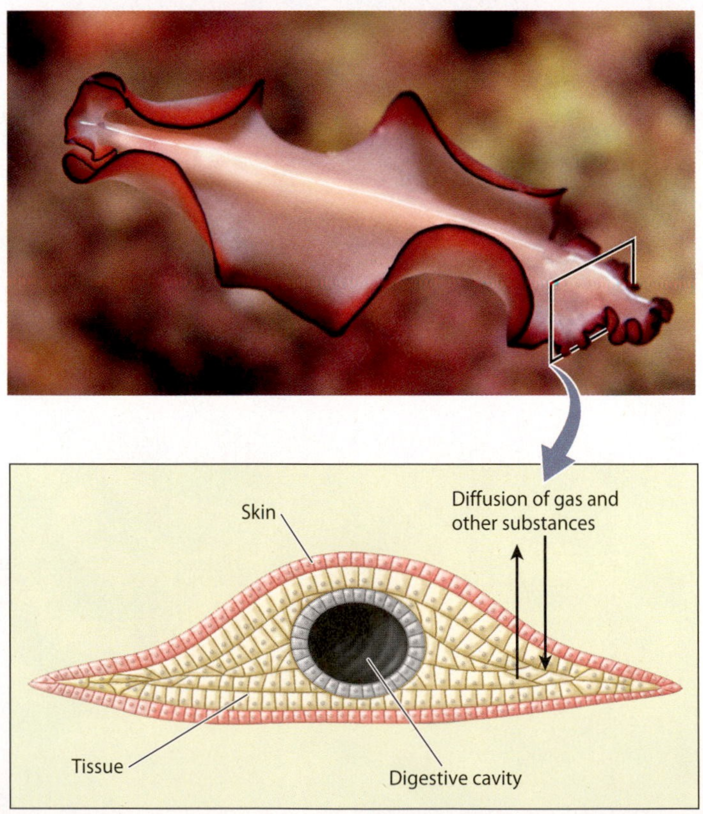

FIG. 39.2 Partial pressure. The partial pressure of a gas is its fractional concentration multiplied by the total atmospheric pressure.

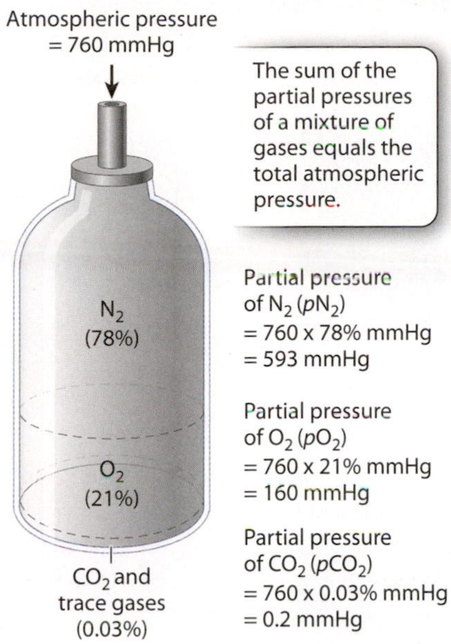

pressure (**Fig. 39.2**). Pressure is measured in units of millimeters of mercury (mmHg). Oxygen makes up approximately 21% of the gases in air (nitrogen contributes about 78%, carbon dioxide about 0.03%, and the remainder are trace gases). At sea level, atmospheric pressure is approximately 760 mmHg, which means that O_2 in the atmosphere exerts a partial pressure (pO_2) of about 160 mmHg (0.21×760). For O_2 to diffuse from the air into cells, the partial pressure of O_2 inside the cells must be lower than the partial pressure of O_2 in the atmosphere.

→ **Quick Check 1** What two structural features favor diffusion?

Bulk flow moves fluid over long distances.

Larger, more complex animals transport O_2 and CO_2 longer distances to cells within their bodies. These animals rely on **bulk flow** in addition to diffusion. Bulk flow is the physical movement of fluid (that is, liquids and gases), and the compounds carried by the fluid, over a given distance. Specialized pumps, like the heart, generate the pressure that is required to move the fluid.

Bulk flow occurs in two steps to meet the gas exchange needs of cells in larger animals. The first is **ventilation.** Ventilation is the movement of the animal's respiratory medium—water or air—past a specialized respiratory surface. The second is **circulation.** Circulation is the movement of a specialized body fluid that carries O_2 and CO_2. The circulatory fluid is called **hemolymph** in invertebrates and **blood** in vertebrates. This fluid delivers O_2 to cells within different regions of the body and carries CO_2 back to the respiratory exchange surface.

Ventilation and circulation each require a pump to produce a pressure (P) that drives flow (Q) against the resistance to flow (R). The rate of flow is governed by the following simple equation:

$$Q = P/R$$

Resistance to flow measures the difficulty of pumping the fluid through a network of chambers and vessels located within the respiratory and circulatory systems. If the resistance to flow doubles, the flow rate is halved. The longer the network of vessels and the narrower the vessels themselves, the greater the vessels' resistance to the fluid (whether air, water, or blood) moving through them. Consider how much harder it is to blow fluid out of a long thin tube than a short wide one.

In summary, we can view gas exchange in complex multicellular organisms as four steps that are linked in series (**Fig. 39.3**). To deliver O_2 to the mitochondria in an animal's cells, (1) fresh air or water moves past the respiratory exchange surface in the process of ventilation (in bulk flow). Ventilation maximizes the concentration of O_2 in the air or water on the outside of the

FIG. 39.3 The four steps of gas transport and exchange.

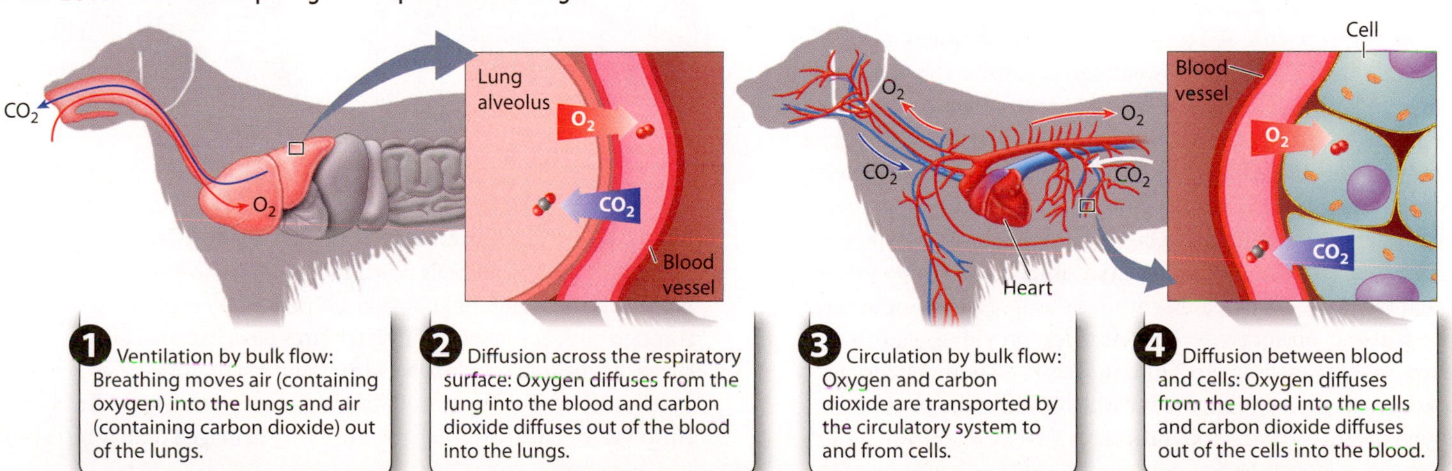

respiratory surface. (2) The buildup of O_2 favors the diffusion of O_2 into the animal across its respiratory surface. (3) Following diffusion into the blood, O_2 is transported by the circulation (in bulk flow) to the tissues. Internal circulation again serves to maximize the concentration of O_2 outside cells. (4) Oxygen then diffuses from the blood across the cell membrane and into the mitochondria, where it burns fuels for ATP production.

The same four steps occur in reverse to remove CO_2 from the body: CO_2 moves from cells to the blood by diffusion, then is carried in the circulation to the respiratory surface by bulk flow, then moves across the respiratory surface by diffusion, and finally moves out of the animal by bulk flow. Because diffusion and bulk flow are coupled, it is important that each step of transport (whether for a gas, nutrient, or waste) has a similar capacity—that is, the same amount of O_2 enters the blood through diffusion as is transported by the circulation.

→ **Quick Check 2** Those who suffer an asthma attack have difficulty breathing. From the discussion of bulk flow, what do you think happens to these individuals' airways that makes it difficult for them to breathe?

39.2 RESPIRATORY GAS EXCHANGE

All animals obtain O_2 from the surrounding air or water. Aquatic animals like crabs, aquatic salamanders, and fish that take in O_2 from water breathe through **gills,** highly folded delicate structures that facilitate gas exchange with the surrounding water (**Fig. 39.4a**). Some animals, such as aquatic frogs and terrestrial adult salamanders, breathe through their skin. In contrast, many terrestrial animals, such as reptiles, certain terrestrial salamanders, birds, and mammals, have internal **lungs** for gas exchange (**Fig. 39.4b**). Internal respiratory exchange surfaces are adaptations that protect the lung surfaces from drying out in a terrestrial environment. Instead of lungs, terrestrial insects evolved a system of air tubes called **tracheae** that branch from openings along their abdominal surface into smaller airways. These smaller airways, termed tracheoles, supply air directly to the cells within their bodies.

For both aquatic and terrestrial animals, ventilation supplies O_2 to gas exchange surfaces, and then O_2 crosses the surface through diffusion. Diffusion is much slower over longer distances than the bulk flow of ventilation. How can the rate of diffusion through the gas exchange surface in lungs or gills keep up with the rate of O_2 supplied by ventilation? The gas exchange surface must have a large surface area and be extremely thin. The respiratory surfaces of more complex and active animals are highly folded, creating a large surface area within a small space. These surfaces are also only one or two cell layers thick, providing a diffusion distance of as little as 1 to 2 μm. In fact, this short distance is all that separates the air you breathe into your lungs from the bloodstream into which O_2 diffuses.

FIG. 39.4 Animal gas exchange organs. Gas exchange organs are shaped and folded to provide large surfaces for gas exchange. *Photo sources: (top) Reinhard Dirscherl/age fotostock; (bottom) John Cancalosi/Alamy.*

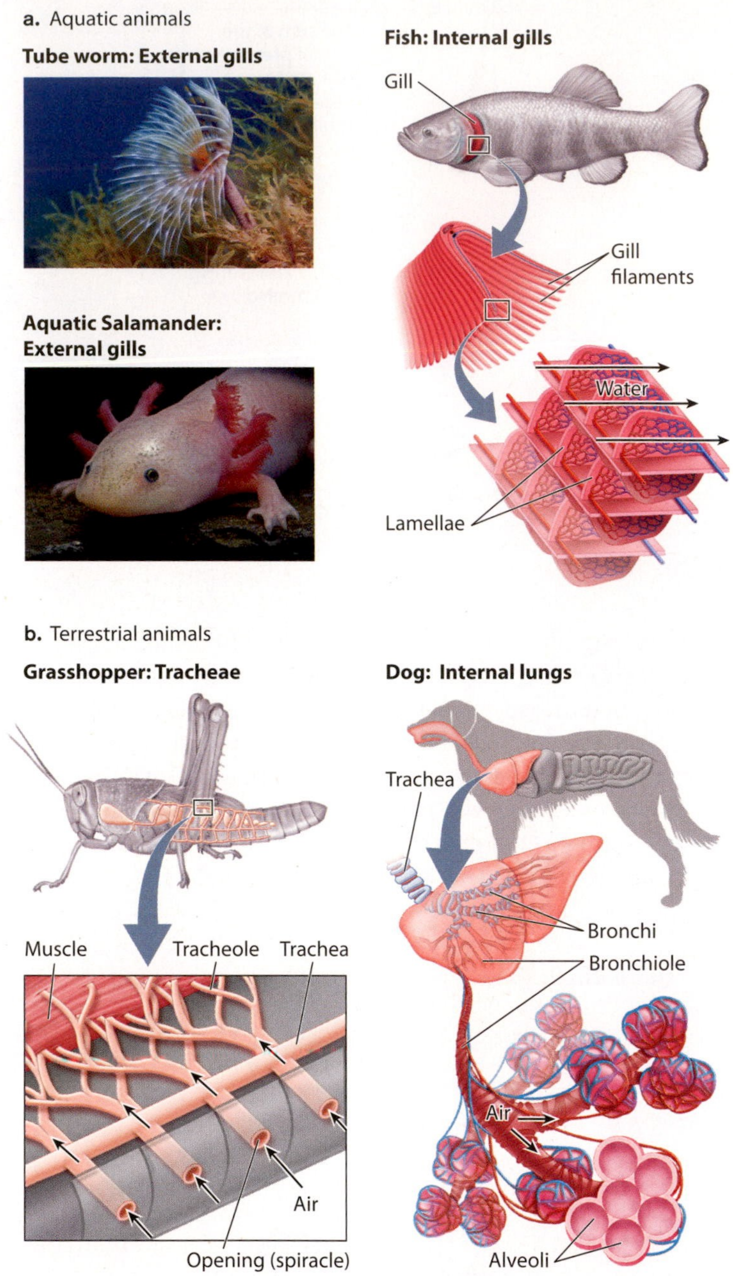

Many aquatic animals breathe through gills.

Most large aquatic animals (with the notable exception of marine mammals like dolphins and whales) breathe O_2-containing water through gills. Some sessile invertebrates have external gills; these organisms include tube worms and sea urchins, as well as nudibranch mollusks and aquatic amphibians (see Fig. 39.4a). However, because external gills are easily damaged or may be eaten

FIG. 39.5 Respiration by fish gills. Fish gills extract O_2 from water efficiently through a unidirectional flow of water and countercurrent flow.

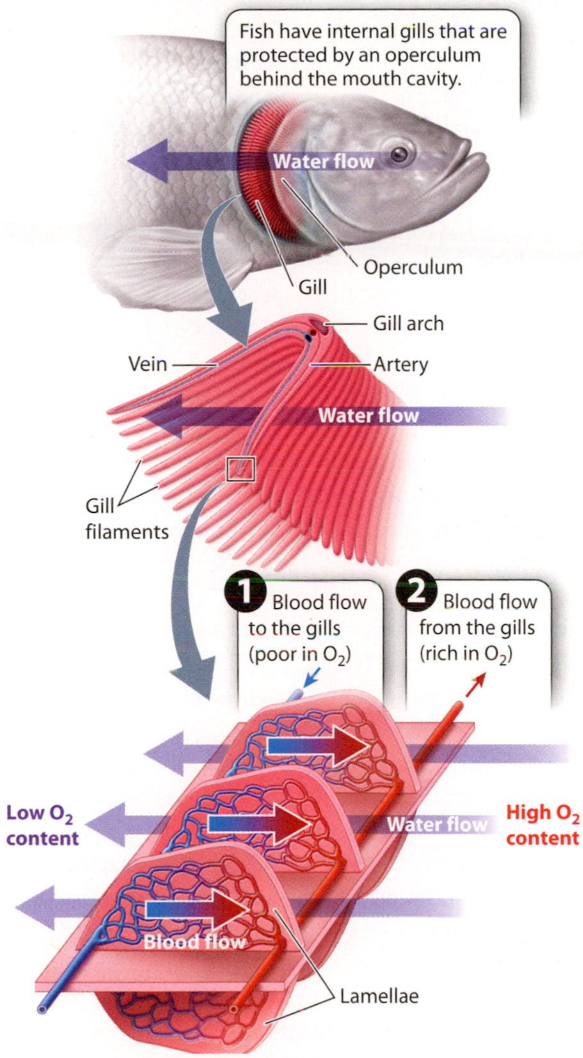

by predators, most mollusks, crustaceans, and fishes have evolved internal gills that, although still formed as outfoldings of the body surface, are contained within protective cavities. Whereas the external gills of invertebrates often rely on the natural motions of the water to move water past the gills, invertebrates with internal gills often have cilia that direct water over the gill's surface.

Fish need sufficient O_2 to meet the energy demands of swimming. As a result, they actively pump water through their mouth and over the gills, which are located in a chamber behind the mouth cavity (**Fig. 39.5**). The effort required to ventilate the gills could be high because of the density and viscosity of water, but fish greatly reduce the energy cost of moving water by maintaining a continuous, unidirectional flow of water past their gills. In bony fishes, a protective flap overlying the gills, called the operculum, expands laterally to draw water over the gills while the mouth is refilling before its next pumping cycle.

The gills of fish typically consist of a series of gill arches located on either side of the animal behind the mouth cavity and, in bony fishes, beneath the operculum. Each gill arch consists of two stacked rows of flat leaf-shaped structures called gill filaments. Numerous **lamellae** (singular, **lamella**), thin sheetlike structures, are evenly but tightly spaced along the length of each gill filament and extend upward from the filament's surface. A series of blood vessels brings O_2-poor blood from the heart to the lamellar surfaces. The lamellae are composed of flattened epithelial cells and are extremely thin, so the water passing outside and blood passing inside the lamellae are separated by only about 1 to 2 μm. The lamellae give the gills an enormous surface area for their size. For this reason, O_2 in the water passing over the gill can readily diffuse into the small blood vessels of the lamellae.

The lamellae are oriented so that the blood flowing through them in a capillary network moves in a direction opposite to the flow of water past the gills (Fig. 39.5). This type of organization, in which fluids with different properties move in opposite directions, is an efficient way to exchange properties between the two fluids. The property can be a chemical property, such as O_2 concentration, or a physical property, such as heat. The movement of two fluids with different properties in opposite directions is called countercurrent flow.

Consider for a moment two tubes right next to each other, one carrying hot water and the other carrying cold water (**Fig. 39.6**). If we want to make the hot water cold and the cold water hot, the efficiency of heat transfer between the two tubes depends on the direction of flow. When the water in the tubes flows in the same direction (concurrent flow), the hot water gets colder and the cold water gets hotter since heat is transferred between the two tubes (Fig. 39.6a). Given enough distance, the water will be at an intermediate (warm) temperature at the end of the tubes, reaching the same average temperature in both tubes.

Now consider what happens when the water in the two tubes flows in opposite directions (countercurrent flow). In this case, the cold water encounters increasingly hot temperatures and, given enough distance, becomes nearly as hot as the hot water entering the other tube (Fig 39.6b). Similarly, the hot water encounters increasingly cold temperatures, and becomes nearly as cold as the cold water entering the other tube. With countercurrent flow, the two essentially exchange properties (heat, in this example).

This mechanism, known as **countercurrent exchange,** is used widely in nature. As a result of countercurrent exchange, fish gills can extract nearly all the O_2 in the water that passes over them (Fig. 39.6c). The O_2 in the oxygen-rich water moves down its concentration gradient into the oxygen-poor blood of the gills along the entire length of the lamellae. At the same time, CO_2 readily diffuses out of the blood vessels and into the water that leaves the gill chamber. A set of blood vessels carries the O_2-rich blood away from the gills to supply the fish's body.

FIG. 39.6 Concurrent and countercurrent exchange. Countercurrent exchange is used widely in nature, as in fish gills, in cases where two fluids exchange properties (such as O_2, CO_2, or heat) efficiently.

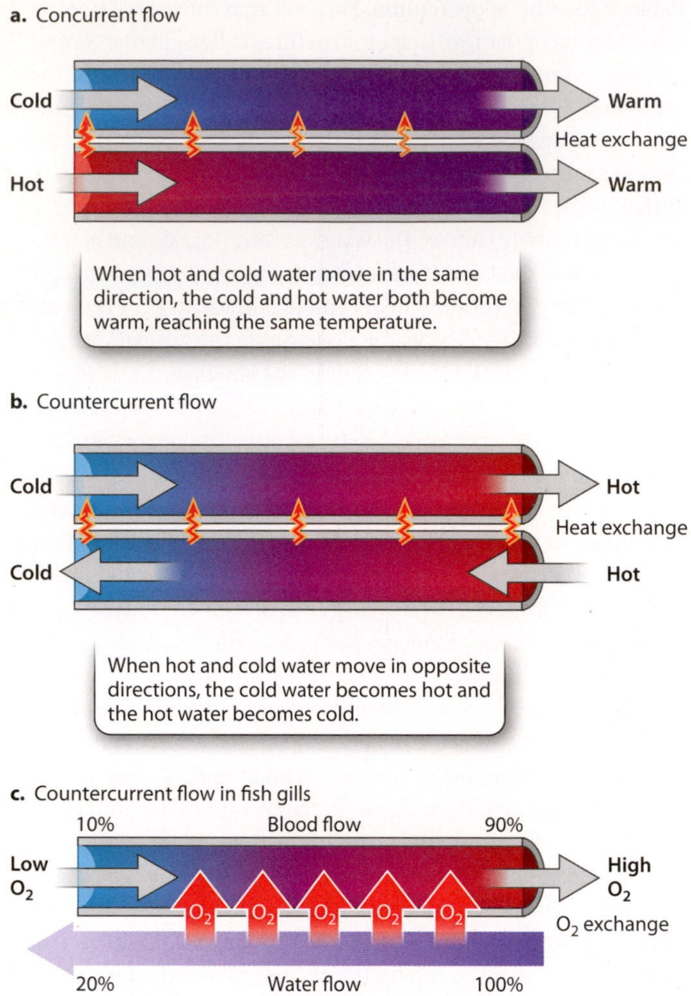

Insects breathe air through tracheae.

With the evolution of life on land, animals that breathe air were able to increase their uptake of O_2 and achieve higher rates of metabolism. Oxygen uptake increased for the following reasons:

1. The O_2 content of air is typically 50 times greater than that of a similar volume of water. Fresh air contains approximately 210 mL O_2 per liter of air (21%), whereas the concentration of O_2 in well-mixed freshwater or salt water rarely exceeds 4 mL O_2 per liter of water (0.4%).

2. Oxygen diffuses about 8000 times faster in air than in water. For this reason, the O_2 content can be close to zero a few centimeters beneath the surface of a stagnant pond.

3. Water is 800 times denser and 50 times more viscous than air. Thus, it requires more energy to pump water than to pump air past a gas exchange surface.

Because of their small size, insects can use a direct pathway of air transport that gets air right to their tissues (see Fig. 39.4b). In contrast to terrestrial vertebrates, insects rely on a two-step process of ventilation and diffusion to supply their cells with O_2 and eliminate CO_2. First, air enters an insect through openings, called **spiracles,** along either side of its abdomen. The spiracles can be opened or closed to limit water loss and regulate O_2 delivery, much like the stomata of leaves in plants (Chapter 29). Inside the insect body, air is ventilated through a branching series of air tubes—the tracheae and tracheoles—directly to the cells. Second, diffusion occurs at the cell: O_2 supplied by the fine airways diffuses into the cells, and CO_2 diffuses out and is eliminated through the insect's tracheae.

The mitochondria of the flight muscles and other metabolically active tissues of insects are located within a few micrometers of tracheole airways. Because O_2 and CO_2 diffuse rapidly in air, the tracheal system of insects delivers O_2 to cellular mitochondria at high rates. Active insects also have an air sac system connected to their tracheae. Powered by contracting abdominal muscles, the air sacs act like bellows to pump air through the tracheae, speeding the movement of air to tissues so that gas exchange is faster.

Most terrestrial vertebrates breathe by tidal ventilation of internal lungs.

Unlike fish gills, the lungs of most land vertebrates inflate and deflate to move fresh air with high levels of O_2 into the lungs and expire stale air with high levels of CO_2 out of the lungs. The low density and viscosity of air enables these animals to breathe by **tidal ventilation** without expending too much energy. In tidal respiration, air is drawn into the lungs during **inhalation** and then moved out during **exhalation** (**Fig. 39.7a**).

Mammals and reptiles expand their thoracic cavity to draw air inside their lungs on inhalation. The expansion of the lungs causes the air pressure inside the lungs to become lower than the air pressure outside the lungs. The resulting negative pressure draws air into the lungs. By contrast, amphibians inflate their lungs by pressure produced by their mouth cavity. In most animals, exhalation is passively driven by elastic recoil of tissues that were previously stretched during inhalation. Elastic recoil compresses the lungs, causing the air pressure inside the lungs to become higher than the air pressure outside the lungs. The resulting positive pressure forces air out of the lungs.

In mammals, inhalation during normal, relaxed breathing is driven by contraction of the **diaphragm,** a domed sheet of muscle located at the base of the lungs that separates the thoracic and abdominal cavities (Fig. 39.7a). Exhalation occurs passively by elastic recoil of the lungs and chest wall. During exercise, other muscles come into play to assist with inhalation and exhalation. For example, **intercostal muscles,** which are attached to adjacent pairs of ribs, assist the diaphragm by elevating the ribs on inhalation and depressing them during exhalation. The action of the intercostal muscles helps to produce larger changes in the volume of the thoracic cavity, increasing the negative pressure that draws air into

FIG. 39.7 **Tidal ventilation of the lungs.** (a) The movement of the diaphragm causes lung volume to increase and decrease during breathing, drawing in air and then expelling it. (b) The change in lung volume is typically about 0.5 L when breathing at rest.

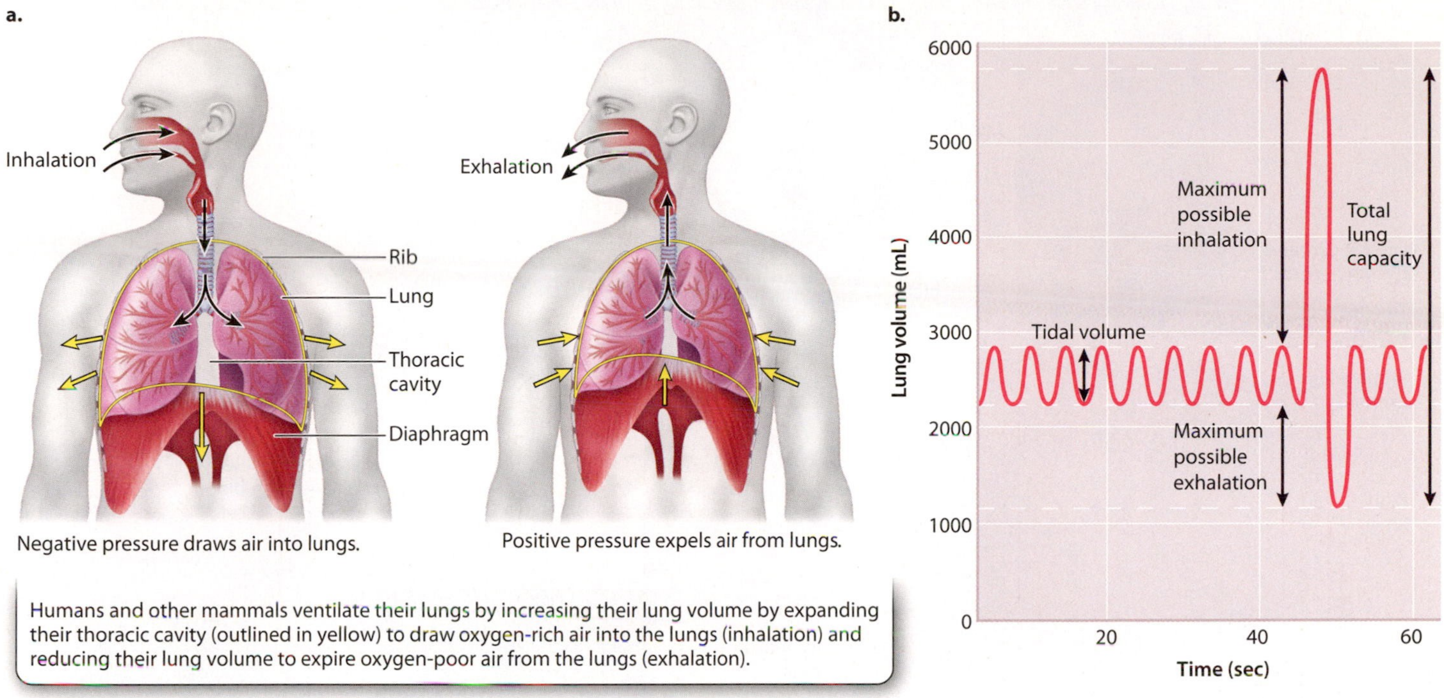

the lungs during inhalation, and assisting elastic recoil of the lungs and chest wall to pump air out of the lungs during exhalation.

At rest, we inhale and exhale about 0.5 L of air every cycle (**Fig. 39.7b**). This amount represents the **tidal volume** of the lungs. With a breathing frequency of 12 breaths per minute, the ventilation rate (breathing frequency × tidal volume) is 6 L per minute. When more O_2 is needed during exercise, both breathing frequency and tidal volume increase to elevate ventilation rate.

The fresh air inhaled during tidal breathing mixes with O_2-depleted stale air that remains in the airways after exhalation. As a result, the pO_2 in the lungs (approximately 100 mmHg in humans) is lower than the pO_2 of freshly inhaled air (approximately 160 mmHg at sea level). Consequently, the fraction of O_2 that can be extracted is lower than the fraction that can be extracted by the countercurrent flow of fish gills, which can reach 90% or more. Typically, mammalian lungs extract less than 25% of the O_2 in the air, and reptile and amphibian lungs extract even less. Nevertheless, the disadvantages of tidal respiration are offset by the ease of ventilating the lung at high rates and the high O_2 content of air.

→ **Quick Check 3** What properties of air make it possible for humans and other mammals to breathe by tidal ventilation?

Mammalian lungs are well adapted for gas exchange.

Despite the limitations of tidal respiration, mammalian lungs are well adapted for breathing air. They have an enormous surface area and a short diffusion distance for gas exchange. Consequently, the lungs of mammals supply O_2 quickly enough to support high metabolic rates.

Except in the case of birds (discussed below), the lungs of air-breathing vertebrates are blind-ended sacs located within the thoracic cavity. Air is taken in through the mouth and nasal passages and then passes through the **larynx,** an organ in the throat made of muscle and cartilage that helps to isolate breathing from swallowing and within which the **vocal cords** are located. (The vocal cords enable speech, song, and sound production.) Air then enters the **trachea,** the central airway leading to the lungs (**Fig. 39.8**). The trachea divides into two airways, called the **primary bronchi,** one of which supplies each lung. The trachea and bronchi are supported by rings of cartilage that prevent them from collapsing during respiration, ensuring that airflow meets with minimal resistance. You can feel some of these cartilage rings at the front of your throat below your larynx (the part of the larynx you can feel is your Adam's apple). The primary bronchi divide into smaller secondary bronchi and again into finer **bronchioles.** This branching continues until the terminal bronchioles have a diameter of less than 1 mm. The very fine bronchioles end with clusters of tiny thin-walled sacs, the **alveoli** (singular, **alveolus**), where gas exchange by diffusion takes place. As a result of this branching pattern, each lung consists of several million alveoli, providing a large surface area. Together, human lungs consist of 300 to 500 million alveoli with a combined surface area as large as a tennis court—about 100 m^2!

FIG. 39.8 **Human lung anatomy.** (a) Millions of alveoli provide a vast surface for gas exchange. (b) A scanning electron micrograph of the lung alveolar surface. (c) A transmission electron micrograph showing a cross section of the alveolar wall and lung capillary. *Source: Reprinted from Respiratory Physiology & Neurobiology, 148, E. Weibel, B. Sapoval, M. Filoche, Design of peripheral airways for efficient gas exchange, 3–21. Copyright 2005, with permission from Elsevier.*

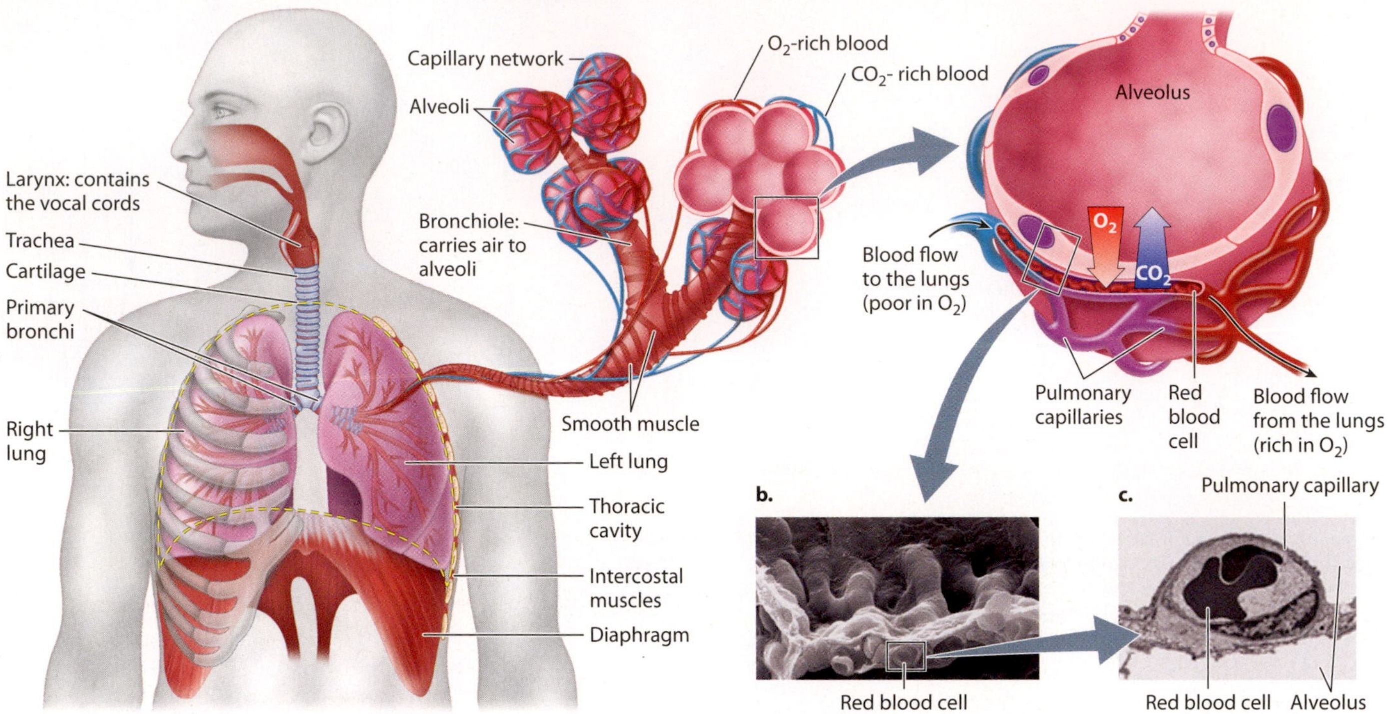

Small blood vessels, the **pulmonary capillaries,** supply the alveolar wall. Each alveolus is lined with thin epithelial cells in intimate contact with the endothelium of these small blood vessels (Fig. 39.8). As a result, the diffusion distance from the alveolus into the capillary is extremely short (about 2 μm), similar to that of gill lamellae and the terminal airways of bird lungs.

Keeping the surfaces of the alveoli moist is critical because moisture helps move O_2 molecules from the air into solution and thus helps them diffuse across the alveolar wall. Certain alveolar epithelial cells produce a **surfactant,** a compound that acts to reduce the surface tension of the fluid film (soap, for example, is a surfactant). Surface tension is a cohesive force that holds the molecules of a liquid together, causing the surface to act like an elastic membrane. Because of surface tension, it requires greater pressure to inflate a small balloon than a large one. The surfactant allows the lungs to be inflated easily at low volumes when the alveoli are partially collapsed and small. It also ensures that alveoli of different sizes inflate with similar ease, enabling more nearly uniform ventilation of the lung.

Mucus-secreting cells line the airways of the lung not only to keep them moist, but also to trap and remove foreign particles and microorganisms that an animal may breathe in with air. Beating cilia on the surface of these cells move the mucus and foreign debris out of the lungs and into the throat, where it is swallowed and digested. Smoking even one cigarette can stop the cilia's beating for several hours or even destroy them. This is one reason why smokers cough—they must clear the mucus that builds up.

The structure of bird lungs allows unidirectional airflow for increased oxygen uptake.

Birds extract O_2 efficiently to supply the considerable energy needed for flight. An efficient O_2 supply is particularly important for migratory birds that fly at high altitude where the air's partial pressure of O_2 is relatively low. The lungs of birds are unique in that they are rigid and do not inflate and deflate. As a consequence, birds do not use tidal breathing. Instead, they benefit from unidirectional airflow that enhances gas exchange by maintaining larger concentration gradients for diffusion.

A bird's respiratory system consists of two sets of air sacs: one located posterior to the lungs (behind them) and one located anterior to the lungs (in front of them). Air is pumped through these air sacs continuously in a bellows-like action (**Fig. 39.9**). As the set of posterior air sacs expand, air is drawn from the bird's mouth into those sacs. The sacs are then compressed by surrounding muscles acting on the skeleton, pumping air through the bird's rigid lungs. The lungs consist of a series of larger air channels that branch into fine air capillaries having a large surface area. The air capillaries direct the air so that it moves **crosscurrent** (that is, at 90°) to the path of blood flow through the walls of the air capillaries. This crosscurrent flow is not as efficient as the

FIG. 39.9 Respiration by bird lungs. Anterior and posterior air sacs move air unidirectionally through a rigid lung, increasing the uptake of O_2 at low pO_2.

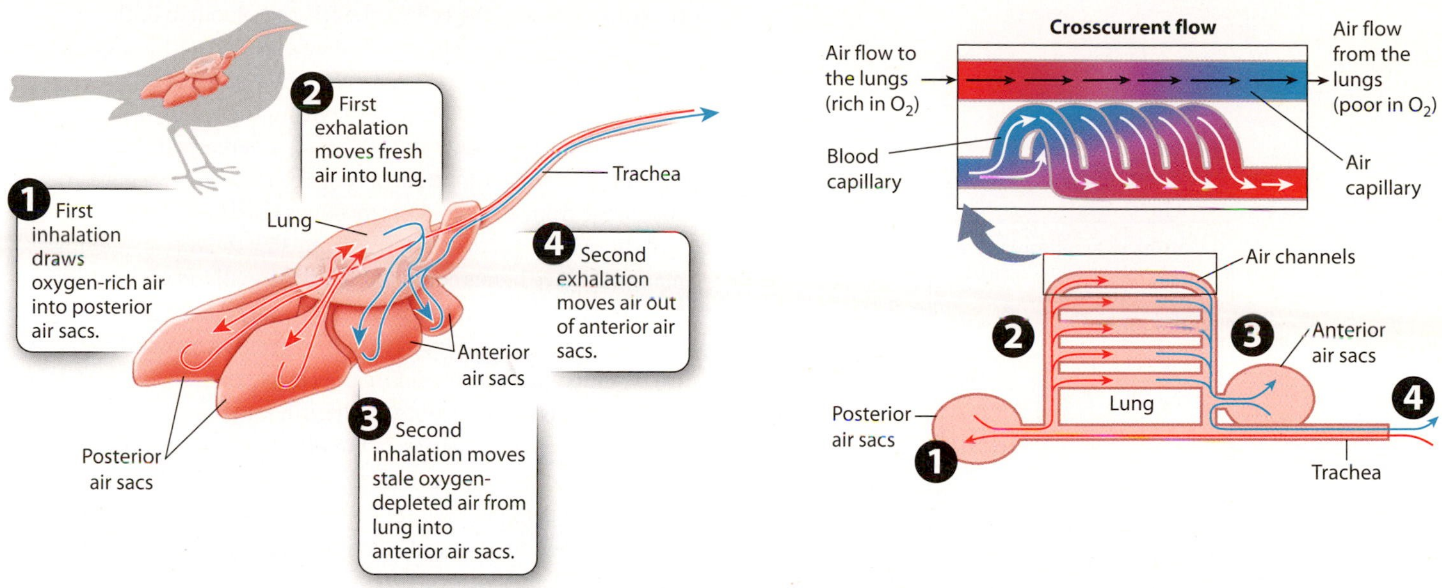

countercurrent flow of fish gills, but it enables birds to extract more O_2 from the air than mammals can.

After losing O_2 and gaining CO_2, the air leaves the lungs and enters the set of anterior air sacs. Fresh air is again inhaled into the posterior air sacs, and then the stale air in the anterior air sacs is exhaled out of the bird's trachea and mouth. As a result, birds achieve a continuous supply of fresh air into the lung during both inhalation *and* exhalation. In contrast to other air-breathing vertebrates, therefore, birds move air through their respiratory system in two cycles of ventilation.

Voluntary and involuntary mechanisms control breathing.

Because an animal's need for O_2 varies with activity level, animals adjust their respiratory rate to meet their cells' changing demand for O_2. Respiration is a unique physiological process in that it is controlled by both the voluntary and involuntary components of the nervous system (Chapter 35). In sleep, breathing is maintained at a resting rate by the involuntary part of the nervous system. Indeed, in most circumstances, breathing is controlled unconsciously.

The regulation of blood O_2 levels is a key example of homeostasis, like core body temperature (Chapter 35) and blood-glucose levels (Chapter 38). Recall that homeostasis often depends on sensors that monitor the levels of the chemical being regulated. In the case of breathing, the sensors are chemoreceptors located within the brainstem and in sensory structures called the **carotid** and **aortic bodies** that are located in the neck and near the heart (**Fig. 39.10**). The carotid bodies sense O_2 and proton (H^+) concentrations of the blood going to the brain, and the aortic bodies monitor their levels in blood moving to the body. In contrast, chemoreceptors in the brainstem sense CO_2 and H^+ concentrations. The most important factor in the control of breathing is the amount of CO_2 in the blood. If the concentration of CO_2 in the blood is too high, chemoreceptors in the brainstem stimulate motor neurons that activate the respiratory muscles to contract more strongly or more frequently. Stronger or faster breathing rids the blood of excess CO_2 and increases the supply of O_2 to the body.

Breathing can also be controlled voluntarily. It is a simple matter to choose to hold your breath. Holding the breath makes it possible for humans and marine mammals to dive under water; it is also critical to the production of speech, song, and sound for communication. Sound is produced by voluntarily adjusting the magnitude and rate of airflow over the vocal cords of mammals, the syrinx of songbirds, or the glottal folds of calling amphibians, such as some toads and frogs.

39.3 OXYGEN TRANSPORT BY HEMOGLOBIN

Once O_2 is extracted from the environment and diffuses into the animal, it is transported to respiring cells. Oxygen is transported in the circulation by specialized body fluids—the blood of vertebrates or the hemolymph of invertebrates.

Blood is composed of fluid and several types of cell.

In addition to its role in oxygen and carbon dioxide transport, which is our focus in this chapter, blood serves many other functions. These include heat transport for temperature regulation and nutrient transport to metabolically active tissues (Chapter 40), hormone transport (Chapter 38), waste transport to excretory organs (Chapter 41), and transport of immune cells

FIG. 39.10 Homeostatic control of breathing. Breathing becomes stronger or faster when CO_2 levels increase or O_2 levels decrease.

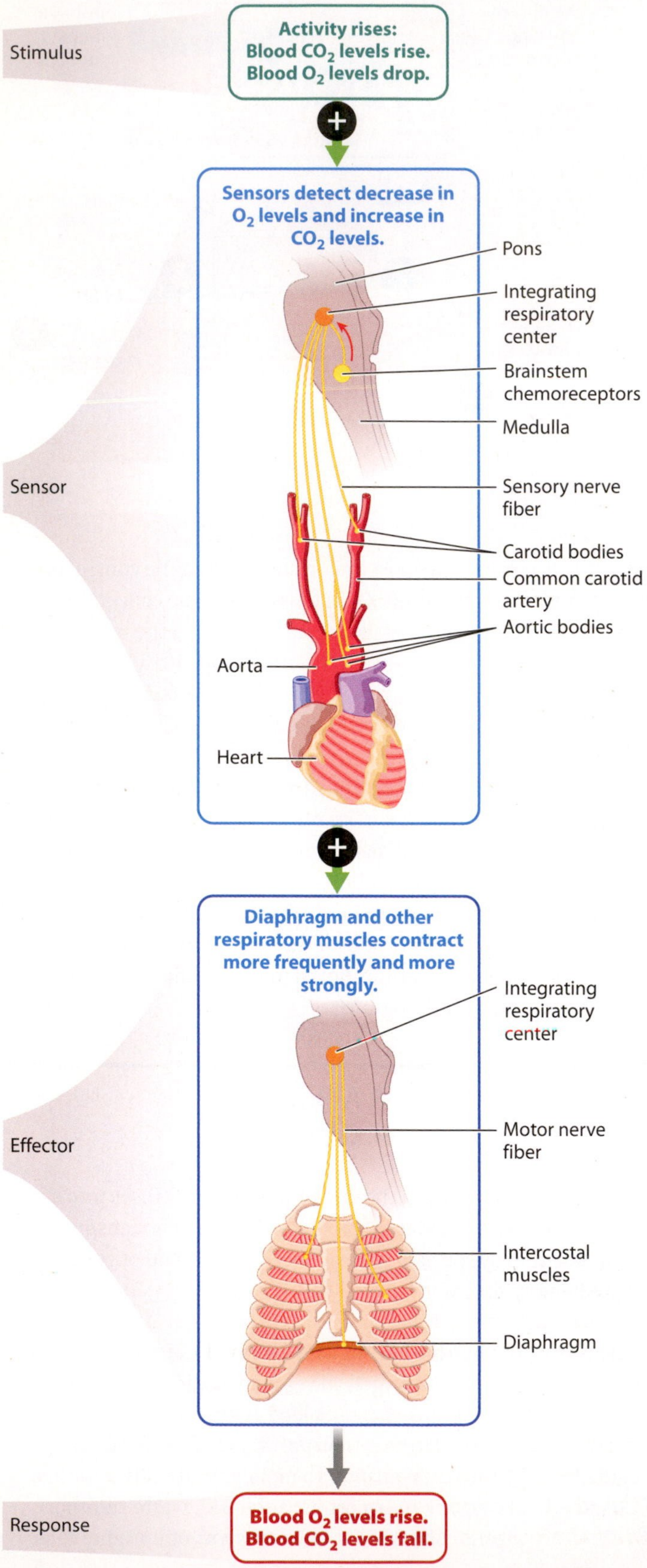

involved in fighting pathogens (Chapter 43), as well as wound healing. Blood can be separated into a cellular fraction and a fluid fraction, or blood plasma. Red blood cells that carry **hemoglobin** constitute most of the cellular fraction of blood. In both vertebrates and invertebrates, hemoglobin evolved as a specialized iron-containing, or heme, molecule for O_2 transport.

The fraction of red blood cells within the blood of vertebrates is defined as the **hematocrit.** Whereas a hematocrit of 45% is considered normal for humans (**Fig. 39.11**), fish have hematocrits that range from 20% to 35%, crocodiles have hematocrits ~20%, birds have hematocrits that vary from 35% to 58%, and diving mammals have hematocrits as high as 65%. Blood with a lower hematocrit flows with less resistance (that is, has a lower viscosity), but carries less oxygen. Humans with low hematocrits are considered to be iron deficient because their blood has fewer red blood cells and therefore too little hemoglobin.

A much smaller fraction (~1%) of cells are white blood cells, which help to defend the body against pathogens. Their role is discussed in detail in Chapter 43. The remaining cells are platelets, which respond to damaged blood vessels by helping to form a clot to prevent blood loss.

When blood is spun in a centrifuge, the red blood cells, the densest component of blood, form a layer in the bottom of a centrifuge tube, and a thin "buffy coat" layer of white blood cells and platelets forms at the top of the cellular fraction. The remaining contents are the blood plasma and solutes, which rise to the top of the centrifuge tube (Fig. 39.11).

Blood plasma can hold only as much O_2 or CO_2 as can be dissolved in solution. How much O_2 goes into solution at a given partial pressure is a measure of its **solubility.** Because O_2 is about 30 times less soluble than CO_2, only about 0.2 mL of O_2 can be

FIG. 39.11 Blood composition in (a) mammals and (b) fish. Blood consists of a cellular fraction, which includes red blood cells, white blood cells, and platelets, and a fluid fraction called plasma.

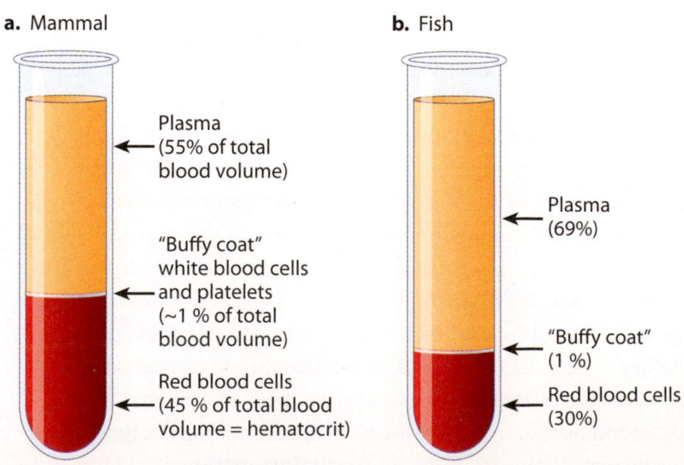

FIG. 39.12

What is the molecular structure of hemoglobin and myoglobin?

BACKGROUND In the 1950s, the Austrian Max Perutz worked with John Kendrew in the Cavendish Laboratory of the University of Cambridge to determine the molecular structure of globular proteins. Perutz and Kendrew were interested in understanding how the structures of hemoglobin and myoglobin enabled binding and transport of O_2.

EXPERIMENT Perutz and Kendrew developed and applied the new technique of X-ray crystallography to determine the three-dimensional molecular structures of hemoglobin and myoglobin.

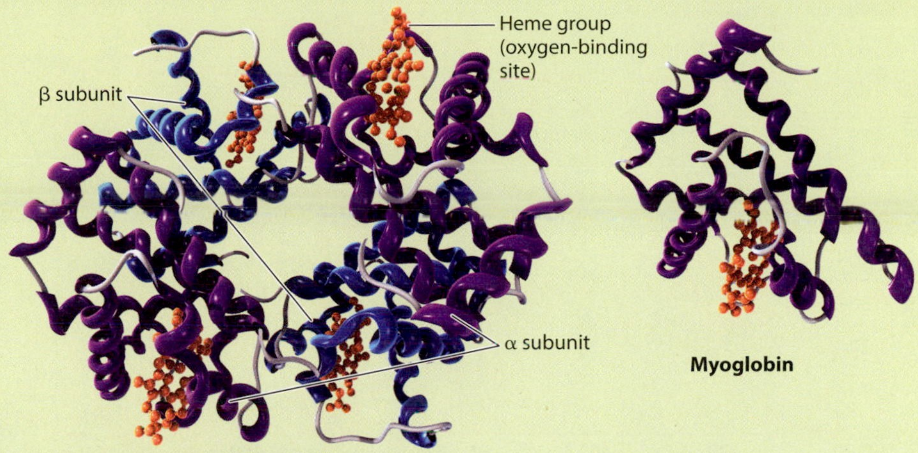

Figure rendered by Dale Muzzey.

RESULTS Perutz and Kendrew showed that adult hemoglobin consists of four subunits, two α (alpha) and two β (beta) subunits. Each subunit contains a heme group that contains iron, which is the site of O_2 binding. By contrast, myoglobin consists of only a single subunit with one heme group. These differences in molecular structure underlie the O_2-binding and dissociation properties of the two O_2 transport proteins.

FOLLOW-UP WORK Work by Perutz in the 1960s showed how the molecular structure of hemoglobin changes when it binds O_2. Perutz also supervised Francis Crick as a doctoral student (Crick later worked with James Watson to resolve the structure of DNA). Perutz and Kendrew received the 1962 Nobel Prize in Chemistry for their work.

SOURCES Kendrew J. C., and M. F. Perutz. 1948. "A Comparative X-Ray Study of Foetal and Adult Sheep Haemoglobins." *Proceedings of the Royal Society, London, Series A* 194:375–98; Perutz, M. F. 1964. "The Hemoglobin Molecule." *Scientific American* 211:64–76; Perutz, M. F. 1978. "Hemoglobin Structure and Respiratory Ttransport." *Scientific American* 239:92–125.

carried in 100 mL of blood. By binding O_2 and removing it from solution, hemoglobin increases the amount of O_2 in the blood a hundredfold. Because of its greater solubility, CO_2 is transported in solution within the blood. Rather than remaining dissolved in solution as a gas, most of the CO_2 (about 95%) is converted to carbonic acid, which dissociates to form bicarbonate ions (HCO_3^-) and protons (Chapter 6).

Whereas some invertebrate hemoglobins exist in solution in the hemolymph and others exist within cells carried by the hemolymph, all vertebrate hemoglobins are produced by and reside in red blood cells. Hemoglobin gives the cells and the blood their red appearance.

Hemoglobin is an ancient molecule with diverse roles related to oxygen binding and transport.

Hemoglobin is an ancient molecule that has been found in organisms representing all kingdoms of life. In bacteria, archaeons, and fungi, similar forms of hemoglobin evolved as a terminal electron acceptor in cellular respiration and may have also served a role in scavenging O_2 to avoid damage due to oxidative stress.

Plant hemoglobins are known that bind and transport O_2 within the cell. In soybeans, hemoglobin binds O_2 in root nodule cells to keep concentrations low enough to avoid inhibiting nitrogenase, the enzyme used by symbiotic rhizobial bacteria in the root nodules for nitrogen fixation (Chapter 29).

Invertebrate hemoglobins found in roundworms, annelid worms, and arthropods bind O_2 in the circulatory fluid, as discussed above. They share considerable gene and amino acid sequence similarity with vertebrate hemoglobins, indicating that a common ancestor to invertebrate and vertebrate animals evolved a shared form of hemoglobin more than 670 million years ago.

Hemoglobin reversibly binds oxygen.

Hemoglobin exists in large concentrations within red blood cells. It is a globular protein that consists of four polypeptide units (**Fig. 39.12**). Each of these four units surrounds a heme group that contains iron, which reversibly binds one O_2 molecule. After

FIG. 39.13 (a) Hemoglobin and (b) myoglobin dissociation curves.

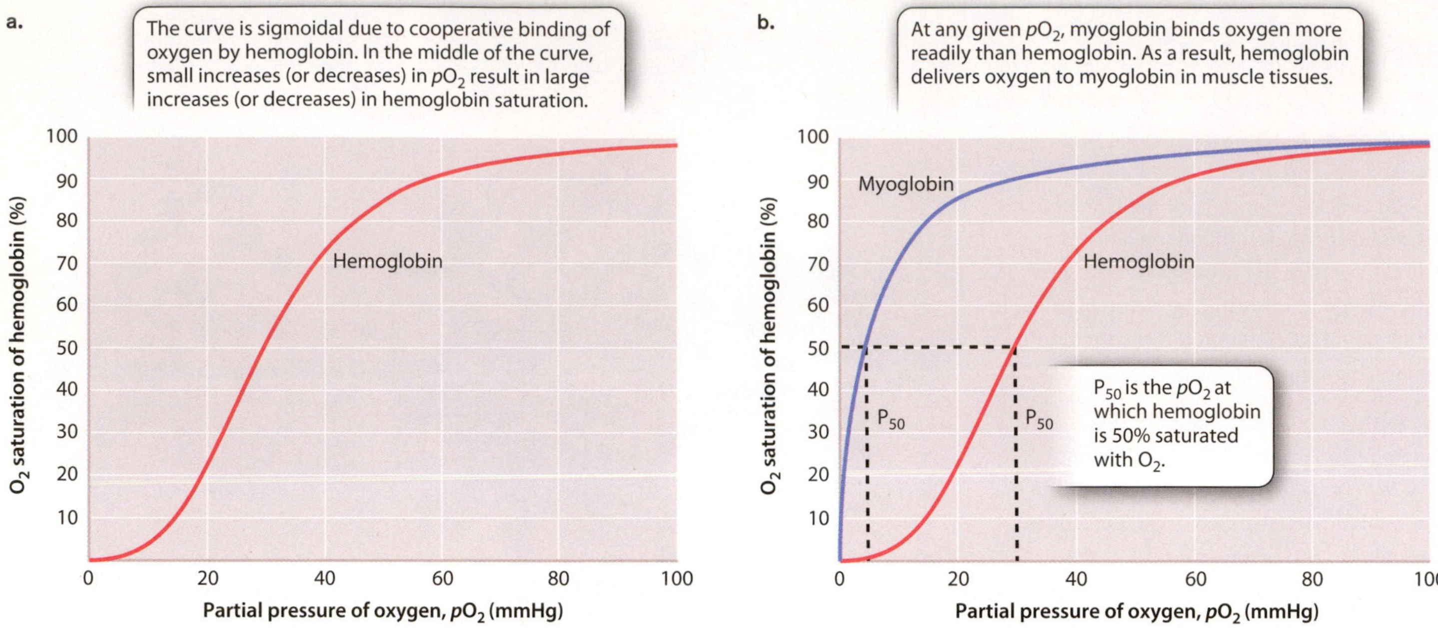

O_2 diffuses into the blood, it diffuses into the red blood cells and binds to the heme groups in hemoglobin. Hemoglobin's binding of O_2 removes O_2 from solution, keeping the pO_2 of the red blood cell below that of the blood plasma, so O_2 continues to diffuse into the cell. The removal of O_2 from the plasma, in turn, keeps the pO_2 of the plasma below that of the lung alveolus, so O_2 continues to diffuse from the lungs into the blood.

When more O_2 is present in blood plasma, we expect more O_2 to become bound to hemoglobin, and that is what happens. But there is an interesting wrinkle that enables hemoglobin to readily bind O_2 leaving the lung. If we plot blood pO_2 against the percentage of O_2 bound to hemoglobin (the relation defined as "hemoglobin saturation"), we get the curve shown in **Fig. 39.13a.** As blood pO_2 increases, hemoglobin saturation rises slowly at first, then more steeply, and then more slowly again until it levels out. At 100% saturation, all hemoglobin molecules bind four O_2 molecules. The curve is called the hemoglobin's **oxygen dissociation curve,** and it has a sigmoidal shape—that is, a shape like an "S".

The shape of this curve can be explained as a result of changes in the ability of hemoglobin to bind O_2 (a property called its binding affinity) at different O_2 partial pressures. At 25% saturation, each hemoglobin molecule binds on average one O_2 molecule. As pO_2 rises, hemoglobin binds O_2 with increasing binding affinity as a result of the interaction of adjacent heme groups. After one heme group binds the first O_2 molecule, hemoglobin undergoes a conformational change that increases the binding affinity of the remaining heme groups for additional O_2. The increase in binding affinity with additional binding of O_2 is called **cooperative binding,** and it gives the O_2 dissociation curve for hemoglobin its sigmoidal shape.

Cooperative binding has important physiological consequences. In the middle part of the O_2 dissociation curve, small increases in O_2 concentration lead to large increases in hemoglobin saturation. Therefore, in normal circumstances, hemoglobin in the blood leaving the lung is fully saturated with O_2—each hemoglobin molecule is bound to four O_2 molecules.

The shape of the O_2 dissociation curve also helps to explain how O_2 is delivered to respiring cells. When the hemoglobin in red blood cells reaches tissues needing O_2 to supply their mitochondria, the O_2 is released from the hemoglobin. As active cells consume O_2, they reduce the local pO_2 of the cell and surrounding tissues to 40 mmHg or less. At these lower pO_2 values, the hemoglobin's O_2 dissociation curve has a steep slope (Fig. 39.13a). The steepness of the slope indicates that for a relatively small decrease in pO_2, large amounts of O_2 can be released from hemoglobin to diffuse into the cell.

Myoglobin stores oxygen, enhancing delivery to muscle mitochondria.

Myoglobin is a specialized O_2 carrier within the cells of vertebrate muscles. In contrast to hemoglobin, myoglobin is a monomer that contains only a single heme group. Because there are no interacting subunits, the O_2 dissociation curve for myoglobin has a different shape from that of hemoglobin (**Fig. 39.13b**). In fact, over the physiological range of pO_2, myoglobin has a greater affinity for O_2 than hemoglobin does and binds O_2 more tightly. As a result, hemoglobin releases O_2 to exercising muscles. As the exercising muscles consume the available O_2, the intracellular pO_2 drops, and the myoglobin releases its bound O_2 to the mitochondria.

Red muscle cells that depend mainly on aerobic respiration to produce ATP (Chapter 37) store large amounts of myoglobin. The myoglobin in these cells can release O_2 quickly at the onset of activity, before respiratory and circulatory systems have had time to increase the supply of O_2. Diving marine mammals, such as whales and seals, carry large amounts of myoglobin within their muscles. The myoglobin loads up with O_2 when the animals breathe at the surface before a dive. The O_2 bound to the myoglobin is then used to supply ATP during the dive, when the animal cannot breathe. These marine mammals can stay under water for 30 minutes or longer and dive to considerable depths (**Fig. 39.14**).

Many factors affect hemoglobin–oxygen binding.

Obtaining enough O_2 is a special challenge in some circumstances. How does a mammalian fetus in its mother's uterus take up O_2 from its mother's blood? How do animals that live at high altitude obtain O_2 under conditions of low pO_2? In each case, the species adapts by evolving forms of hemoglobin with higher binding affinity.

In mammals, the mother's respiratory and cardiovascular systems must provide the fetus with O_2 and remove CO_2, as well as supply nutrients and remove waste. Maternal and fetal blood do not mix. Rather, they exchange gases across the placenta. Yet the O_2 concentration gradient at this exchange point does not strongly favor the movement of O_2 from the mother's blood to the fetal blood. Fetal mammals solve the problem of extracting O_2 from their mother's hemoglobin by producing a form of hemoglobin before birth that has a higher affinity for O_2 than their mother's hemoglobin. The O_2 dissociation curve for fetal hemoglobin is shifted to the left of the curve for maternal hemoglobin (**Fig. 39.15a**), allowing the fetal hemoglobin to extract O_2 from its mother's circulation. At birth, when the newborn begins to breathe, its red blood cells rapidly shift to producing the adult form of hemoglobin, maintaining this form throughout life.

Many animals live at high altitude. For example, birds have been observed flying over the highest peaks (taller than 9000 m) despite the fact that air at this altitude has less than one-third the O_2 content of air at sea level. Similarly, llamas inhabit the Andes at altitudes up to 5000 m, where at a pO_2 of approximately 80 mmHg the air has about one-half the O_2 content of air at sea level. Consequently, the lung pO_2 of a llama is only about 50 mmHg. How do high-flying birds and llamas obtain enough O_2? Their hemoglobin has a higher affinity for O_2 and binds O_2 more readily than the hemoglobin of mammals living at sea level. Consequently, their O_2 dissociation curve is shifted to the left, favoring the binding of more O_2 at a given pO_2.

Tissue and blood pH also have an important physiological effect on the O_2 affinity of hemoglobin. When pH falls during exercise (that is, when H^+ concentration increases), the affinity of hemoglobin for O_2 *decreases*, resulting in a rightward shift in the O_2 dissociation curve (**Fig. 39.15b**). This phenomenon is called the Bohr effect after Christian Bohr, the Danish physiologist who first

FIG. 39.14 Myoglobin as an O_2 store for long-lasting dives. The graph plots the dives of a beaked whale (*Ziphius cavirostris*) fitted with a tracking device. *Source: Based on Fig. 3A in P. L. Tyack, M. Johnson, N. Aguilar de Soto, A. Sturlese, and P. T. Madsen, 2006, "Extreme Diving Behaviour of Beaked Whale Species Known to Strand in Conjunction with Use of Military Sonars, Journal of Experimental Biology 209:4238–4253. Photo: Bill Curtsinger/Getty Images.*

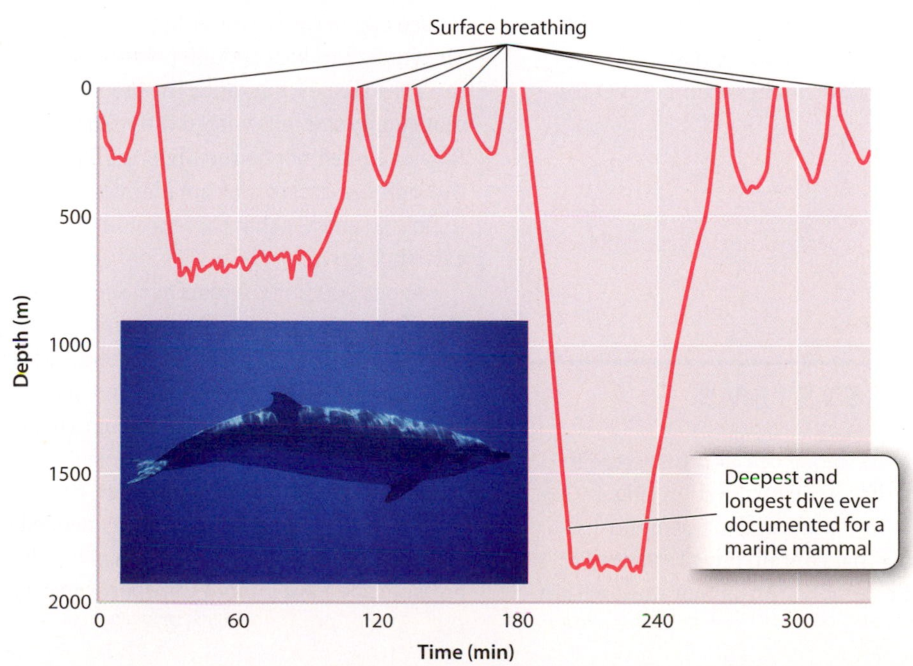

FIG. 39.15 Changes in hemoglobin's O_2 dissociation curve.

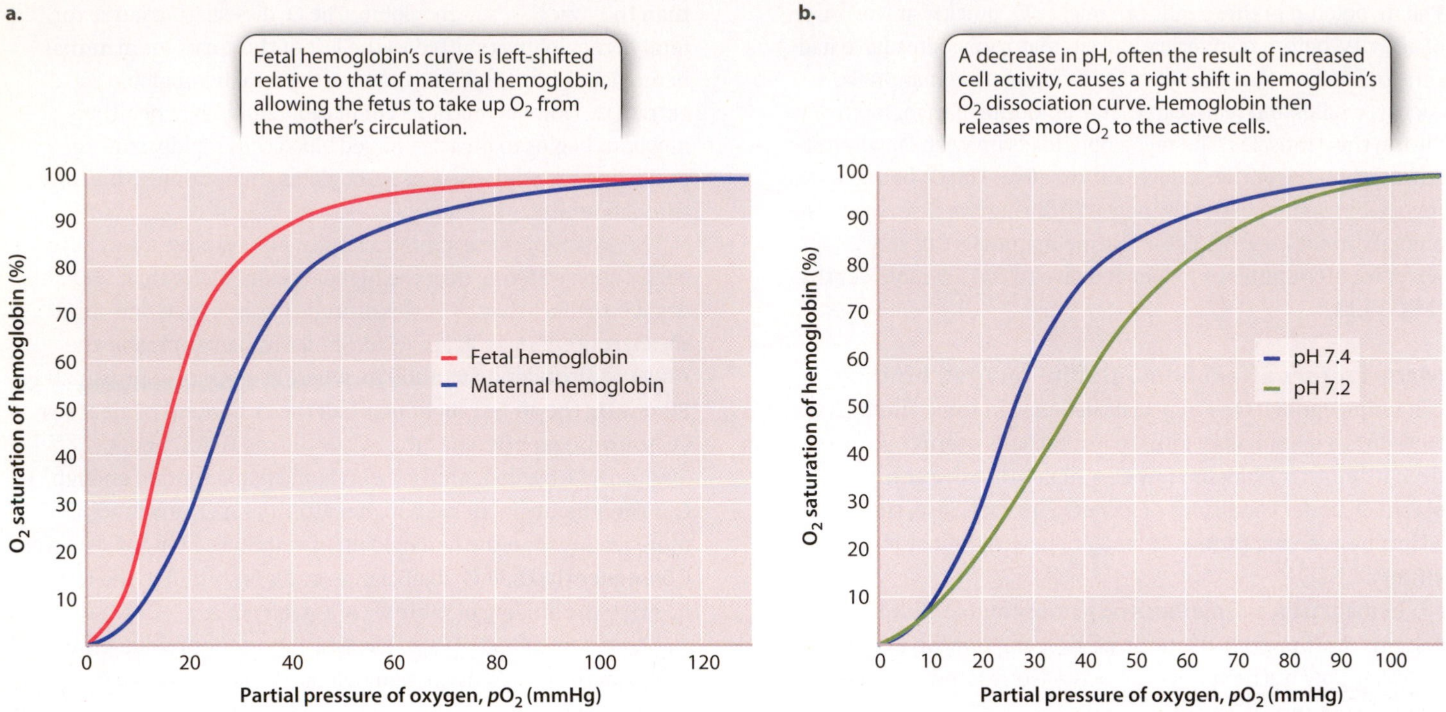

described it in 1904. Decreases in pH occur when CO_2 is released from metabolizing cells, or when an inadequate supply of O_2 leads to the production of lactic acid (Chapter 7). Because hemoglobin's affinity for O_2 is reduced, more O_2 is released and supplied to the cells for aerobic ATP synthesis.

Carbon dioxide also reacts with the amine (NH_2) groups of hemoglobin, reducing hemoglobin's affinity for O_2. Thus, when released from respiring tissues, CO_2 promotes increased O_2 delivery both through its direct effect on hemoglobin and its contribution to a decrease in blood pH by the Bohr effect. While the production of CO_2 promotes O_2 release at the tissues, its elimination at the lung increases hemoglobin's affinity for O_2, enhancing O_2 uptake.

→ **Quick Check 4** Sickle-cell anemia is a genetic disorder of individuals homozygous for a mutation of hemoglobin that causes their red blood cells to be sickle shaped and stiff under conditions of low pO_2. Why is this disease life threatening?

39.4 CIRCULATORY SYSTEMS

The O_2 carried by hemoglobin in red blood cells is transported to tissues throughout the body by the circulatory system (**Fig. 39.16**). Generally, smaller animals, such as insects and many mollusks, have **open circulatory systems** that contain few blood vessels: Most of the circulating fluid, the hemolymph, is contained within the animal's body cavity. The hemolymph bathes the animal's tissues and organs (Fig. 39.16a). Open circulatory systems have limited control of where the fluid moves. Muscles that are active in locomotion can assist circulation of the hemolymph, and some invertebrates have simple hearts with openings that help pump fluid between different regions of the animal's body cavity.

The **closed circulatory systems** of larger animals are made up of a set of internal vessels and a pump—the **heart**—to transport the blood to different regions of the body (Fig. 39.16b). Closed circulatory systems have two conflicting requirements: They must produce enough pressure to carry the circulating blood to all the tissues, but, once the blood reaches smaller vessels that supply the cells within the tissues, the blood pressure and flow rate must not be too high. High pressure would push fluid through the walls of the smaller vessels and cause the blood to flow so quickly that there would not be enough time to exchange gases.

About 500 million years ago, segmented worms first evolved closed circulatory systems that allow them to expand individual body segments to burrow underground. About 480 million years ago, the first cephalopods (squid and octopus) evolved closed circulatory systems that enabled them to be early successful predators in Paleozoic oceans. A closed circulatory system delivers O_2 at high rates to exercising tissues, necessary to enable their mitochondria to provide the energy needed to chase prey. In contrast, open circulatory systems generally operate under low pressure and have limited transport capacity. As a result, animals

FIG. 39.16 (a) Open and (b) closed invertebrate circulatory systems.

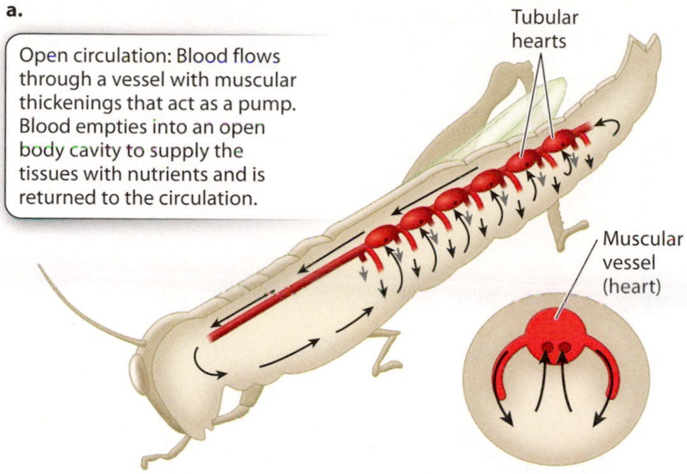

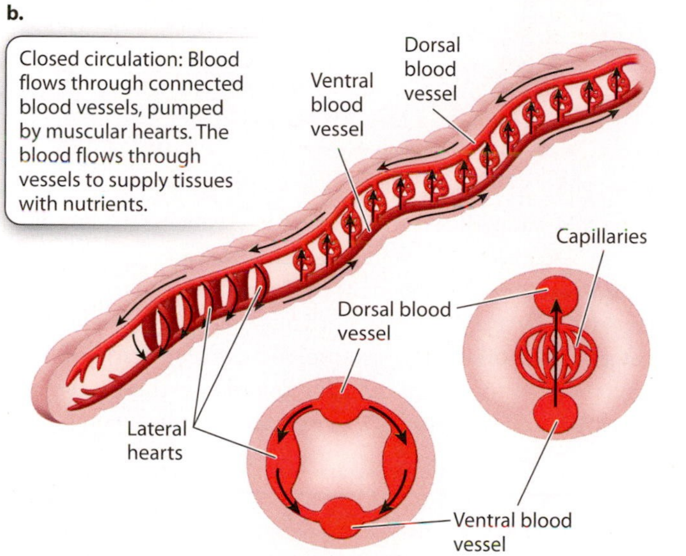

with open circulatory systems are less active. Their ability to control the delivery of respiratory gases and metabolites to specific tissues and regions is also limited. However, insects can be very active despite having an open circulatory system because they obtain O_2 by their tracheal system independently of the circulatory system.

Closed circulatory systems can control blood flow to specific regions of the body by varying the resistance to flow. For example, wading birds reduce blood flow to their legs when they are in cold water to reduce heat loss, and vertebrates can increase blood flow to their muscles to deliver O_2 and nutrients during exercise. In order to pump blood through a set of closed interconnected vessels, a muscular heart is needed to produce sufficient pressure to overcome the flow resistance of the vessels.

Circulatory systems have vessels of different sizes to facilitate bulk flow and diffusion.

For a fluid like blood to flow through a set of pipes, pressure (P) is required to overcome the resistance to flow (R). The pressure developed by a heart when it contracts establishes a rate of blood flow that is governed by P/R. That is, the rate of blood flow increases with an increase in pressure and decreases with an increase in resistance. The resistance to flow is determined in part by the fluid's viscosity and the vessel's length. More viscous fluid and longer vessels of a given diameter have greater resistance. However, the main factor governing resistance to flow is the vessel's radius (r). Resistance is proportional to $1/r^4$, which means that, if a vessel's radius is reduced by half, its resistance to flow increases 16 times.

The high resistance of narrow vessels presents a challenge. To keep diffusion distances short, gas exchange between cells and blood requires narrow vessels. How does the circulatory system overcome the dramatic increase in resistance with narrower diameters?

The organization of animal circulatory systems provides an answer (**Fig. 39.17**). Animal circulatory systems are organized to have blood flow over longer distances in a relatively few large-diameter vessels with low resistance to flow (Fig. 39.17a). **Arteries** are the large, high-pressure vessels that move blood flow away from the heart to the tissues. **Veins** are the large, low-pressure vessels that return blood to the heart. Arteries branch into blood vessels of progressively smaller diameter called **arterioles,** and these arterioles ultimately connect to finely branched networks of very small blood vessels called **capillaries** that are only one cell in diameter or less. Indeed, the diameter of a capillary is so small that red blood cells pass through one at a time, and must be flexible to do even that.

It is at the capillaries that gases are exchanged by diffusion with the surrounding tissues. The number of smaller-diameter vessels at each branching point in the circulatory system greatly exceeds the number of larger-diameter vessels. The reverse organization is found on the return side of circulation: Numerous capillaries drain into vessels of progressively larger diameter called **venules,** and the venules drain into a few larger veins that return blood to the heart.

This organization has two advantages. First, it maintains the same volume of blood flow at all levels within the circulatory system: The increased resistance to flow in the capillaries is offset by the large increase in the number of capillaries. Second, it enables the blood to flow more slowly in smaller vessels (Fig. 39.17b), providing time for gases and metabolites to diffuse into and out of the neighboring cells.

Arteries are muscular, elastic vessels that carry blood away from the heart under high pressure.

When animals become active after a period of rest and feeding, blood flow must be increased to their muscles and reduced to

FIG. 39.17 Flow and vessel size. (a) Vessel hierarchy of circulatory systems; (b) total vessel cross-sectional area versus flow velocity.

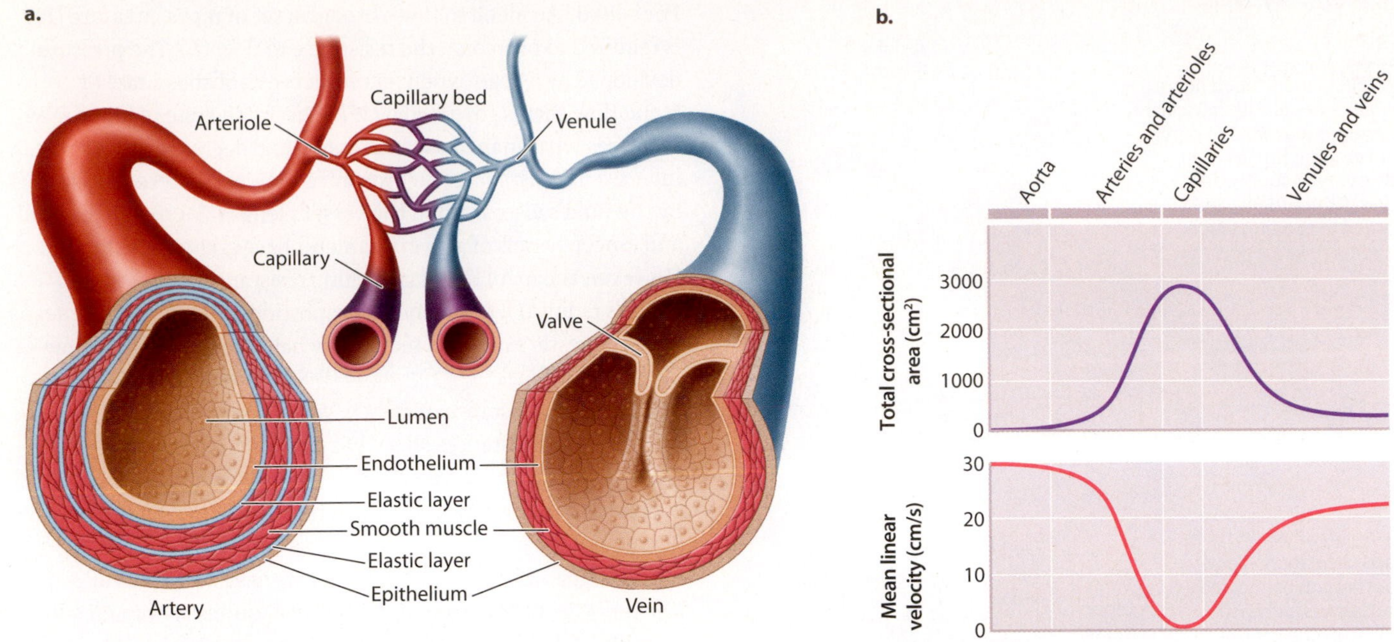

their digestive organs (**Fig. 39.18**). Animals meet this challenge by changing the radius of vessels. Recall that resistance to flow is strongly affected by the vessel radius. Arterioles supplying regions of the body that need less blood become narrower by contracting the circular smooth muscle fibers in their walls. This change dramatically increases the resistance to flow and thereby reduces the rate of blood flow. For example, when you put your hands or feet into cold water, the arterioles constrict, reducing the loss of heat from blood flowing to your hands and feet.

Arterioles supplying regions of the body that need more blood become wider by relaxing the smooth muscles lining their walls. The larger vessels offer less resistance, so more blood can flow through them. As the rate of cellular respiration in these cells increases, they release more CO_2 as well as other metabolites,

FIG. 39.18 Changes in blood flow from resting to exercise as measured in humans. Numbers in parentheses indicate percent of total blood flow. *Source: Data from C. B. Chapman and J. H. Mitchell, 1965, "The Physiology of Exercise," Scientific American 212:88–96.*

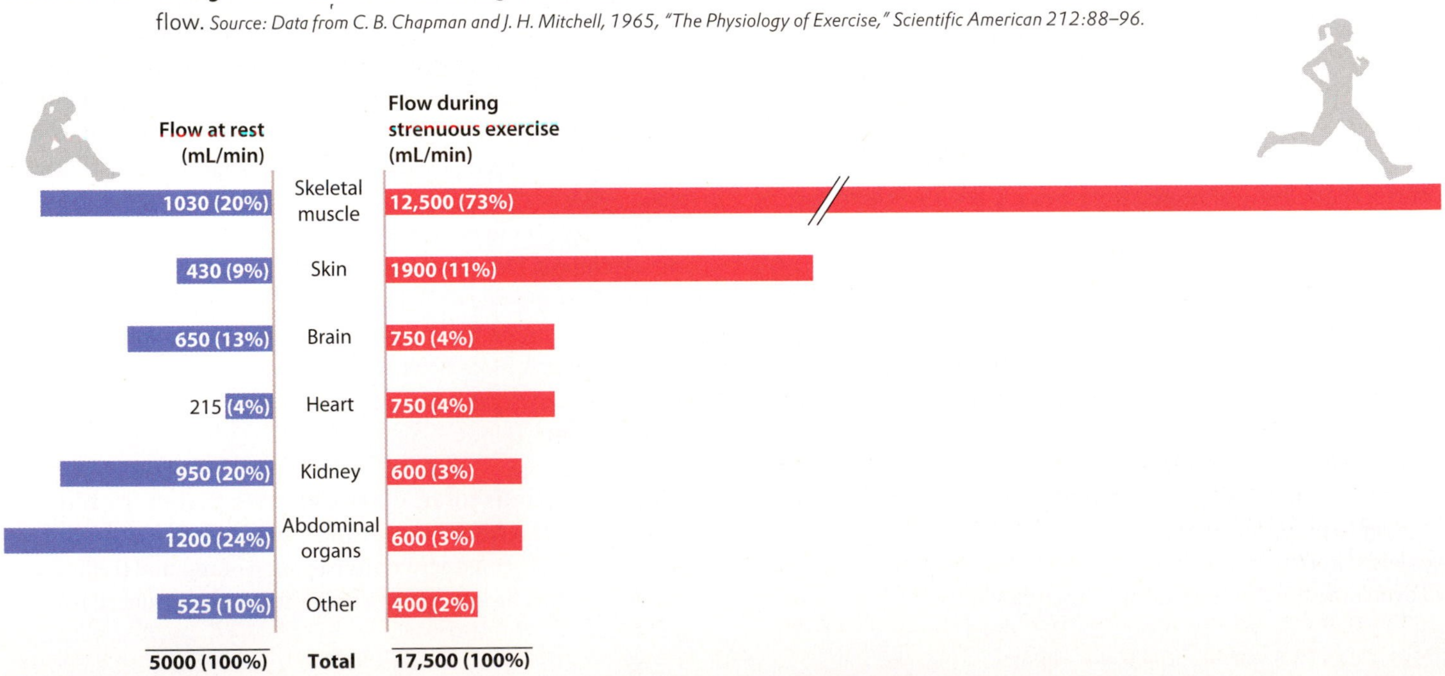

such as lactate and hydrogen ions. The increase of these molecules in the bloodstream acts as a signal to the smooth muscle in the arteriole walls to relax, allowing the arterioles to increase in diameter. In this way, blood flow increases to the capillaries to match O_2 and nutrient delivery to the metabolic need of the cells.

A strong pulse of flow pushes against the arterial walls about every second in a human with each heartbeat that pumps blood through the arteries. This pulse of flow causes a momentary expansion of the arteries. Artery walls can withstand these repeated pressure pulses because they contain multiple elastic layers composed of two proteins: **collagen** and **elastin.** Collagen fibers are strong and resist the expansion of the arterial wall during these pressure pulses. Together with elastin fibers, the collagen provides an elastic rebound of the arterial wall once the pulse has passed, returning energy to help smooth out blood flow. The collagen and elastin resist overexpansion of the arterial wall. However, if an artery wall deteriorates, the collagen and elastin can become so thin that the artery bulges outward. This outward bulge, called an aneurysm, can lead to a life-threatening rupture.

→ **Quick Check 5** What change in vessel shape most affects resistance to blood flow?

Veins are thin-walled vessels that return blood to the heart under low pressure.

Blood collected from local capillary networks returns to the heart through progressively larger veins that ultimately drain into the two largest veins, the **venae cavae** (singular, **vena cava**). The venae cavae drain blood from the head and body into the heart. There is little pressure available to push the blood forward in the veins because pressure has been lost to the resistance of the arterioles and capillaries. Consequently, veins are thin walled and have little smooth muscle or elastic connective tissue. Because of the low pressure, blood tends to accumulate within the veins: As much as 80% of your total blood volume resides in the venous side of your circulation at any one time. When you stand or sit for a long time, blood may pool within the veins of your limbs.

Veins located in the limbs and in the body below the heart have one-way valves that help prevent blood from pooling owing to gravity (see Fig. 39.17a). However, the most important mechanism of returning blood to the heart is the voluntary muscle contractions that occur during walking and exercise, which exert pressure on the veins. Above the level of the heart, the return of blood to the heart through the veins is assisted by gravity.

Compounds and fluid move across capillary walls by diffusion, filtration, and osmosis.

We have seen that O_2 and CO_2 move between capillaries and tissues by diffusion. In addition, water, certain ions, and other small molecules, but not proteins and blood cells, move from capillaries into the surrounding interstitial fluid (the extracellular fluid surrounding vessels) forced by blood pressure. The blood is thus filtered as it passes through the capillary wall.

Why doesn't the blood plasma lose all its water and ions over time? As water is filtered out of the capillaries, some ions, cells, proteins, and other large compounds remain inside, increasing in concentration (**Fig. 39.19**). As a result, there is a tendency of water to flow back into the capillaries by osmosis (Chapter 5). At the same time, blood pressure decreases because of the flow resistance imposed by the capillaries. When the pressure is no longer high enough to counter osmosis, water moves back into the capillaries.

In vertebrates, excess interstitial fluid is also returned to the bloodstream by means of the **lymphatic system,** a network of vessels distributed throughout the body. The fluid that enters the lymphatic system is called **lymph.** Small lymphatic vessels merge with progressively larger thin-walled vessels, draining the lymph into lymphatic ducts, which empty into the venous system and then the heart. Lymphatic vessels have one-way valves, similar to those in the veins, which assist the return of lymph to the circulatory system. Local muscle contractions also help to pump the lymph toward the thoracic duct, the final lymphatic vessel through which lymph returns to the circulatory system of vertebrates. In addition to its role in returning plasma to the bloodstream, the lymphatic system has important functions in the immune system (Chapter 43).

? CASE 7 PREDATOR–PREY: A GAME OF LIFE AND DEATH

How do hormones and nerves provide homeostatic regulation of blood flow as well as allow an animal to respond to stress?

An animal that is dehydrated or has lost blood following an injury may experience a drop in blood pressure. In such cases, the posterior pituitary gland releases antidiuretic hormone (ADH), also known as vasopressin (Chapter 38), into the circulation. ADH causes the

FIG. 39.19 Pressures pushing fluid out of blood vessels (blood pressure) and into them (osmotic pressure).

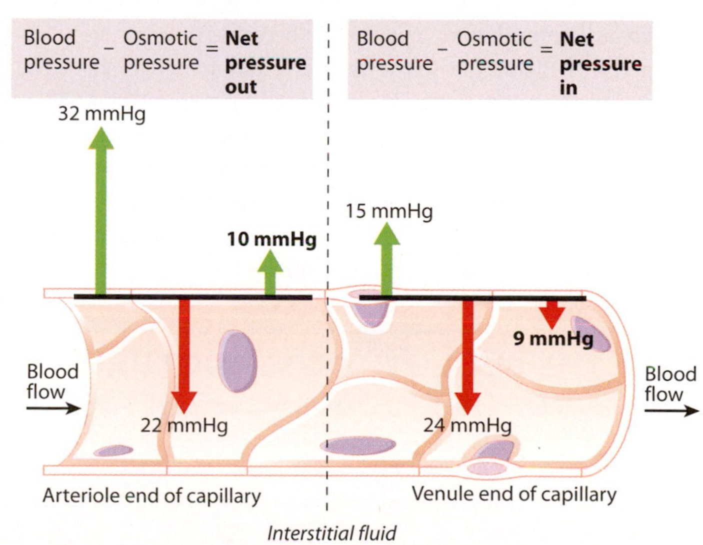

arteries to constrict, increasing their resistance to blood flow. The higher resistance elevates blood pressure throughout the body.

Even under less stressful conditions, blood pressure is continually being adjusted to remain within normal physiological limits, another example of homeostasis. The autonomic nervous system plays an important role in keeping blood pressure at healthy levels (Chapter 35). If blood pressure in the general circulation drops, animals reduce the supply of blood to the limbs by constricting arterioles that supply the limbs. This response helps maintain blood pressure to the heart, brain, and kidneys. In this case, sympathetic neurons synapsing on the smooth muscles of arterioles in the limbs stimulate these muscles to contract. In contrast, when blood pressure is too high, sympathetic neurons that synapse on the smooth muscles are inhibited. The smooth muscles relax, reducing resistance in the arterioles and increasing blood flow. These two responses are called **vasoconstriction** and **vasodilation,** respectively.

In some circumstances, it may be advantageous to an animal to have blood pressure rise higher than normal levels. One example is exhibited during the fight-or-flight response of animals faced with a stressful situation, as when prey are alerted to the presence of predators, and when predators locate a potential prey. The sympathetic nervous system stimulates the adrenal gland to release the hormone **adrenaline** (also called **epinephrine**). The surge of adrenaline into the bloodstream—the "adrenaline rush"—causes an animal to become alert and aroused. At the same time, the sympathetic nervous system acts to increase heart rate and respiratory rate and inhibit digestive processes. Arterioles supplying limb muscles dilate to enhance blood flow to exercising muscles, while arterioles supplying organs within the digestive system constrict. The increase in heart and breathing rates increases the rate of O_2 delivery to the exercising muscles to meet their increased demand for ATP. These physiological changes, along with a heightening of sensory acuity, prepare an animal for the more intensive activity needed to chase or flee.

In contrast, when an animal has recently eaten a meal, its sympathetic nervous system is inhibited and its parasympathetic nervous system is stimulated. In response, digestive secretions increase and smooth muscle in the gut becomes more active (Chapter 40). Blood flow to the digestive system increases as well. In this way, the regulation of blood flow to different organ systems is coordinated with more general feedback from the nervous and endocrine systems, providing regulation of blood pressure linked to the physiological state of the animal. These shifts in blood flow are readily controlled by vasoconstriction or dilation of the arterioles supplying the different organs.

39.5 THE EVOLUTION, STRUCTURE, AND FUNCTION OF THE HEART

The pressure required to drive blood through blood vessels is generated by the heart. Animal hearts are pumps made of muscle that rhythmically contracts, producing this pressure. Early hearts were as simple as a muscular thickening of a small region of a blood vessel. Such hearts are still found in insects (see Fig. 39.16a). More complex hearts have chambers that expand and fill with deoxygenated blood returning from the animal's tissues. After filling, the heart muscle contracts, pumping deoxygenated blood to lungs or gills and the newly oxygenated blood onward to the animal's tissues. The closed circulatory system of cephalopod mollusks has three hearts—two that pump deoxygenated blood through the gills and another that pumps oxygenated blood to the body. In contrast, vertebrates have a single heart with at least two chambers, one for receiving blood and the other for pumping blood to the body during each heartbeat cycle. Hearts also have one-way valves to ensure that blood does not flow backward when the heart contracts.

A theme runs through the evolution of vertebrate circulatory systems: the progressive separation of circulation to the gas exchange organ from circulation to the rest of the body. In fish, deoxygenated blood is first pumped to the gills to gain O_2; then the freshly oxygenated blood flows to the rest of the body and

FIG. 39.20 Fish heart and circulatory system. (a) Fish have a two-chambered heart. (b) Blood flows in a single circuit from the heart, through gills, to the tissues for gas exchange, and back to the heart.

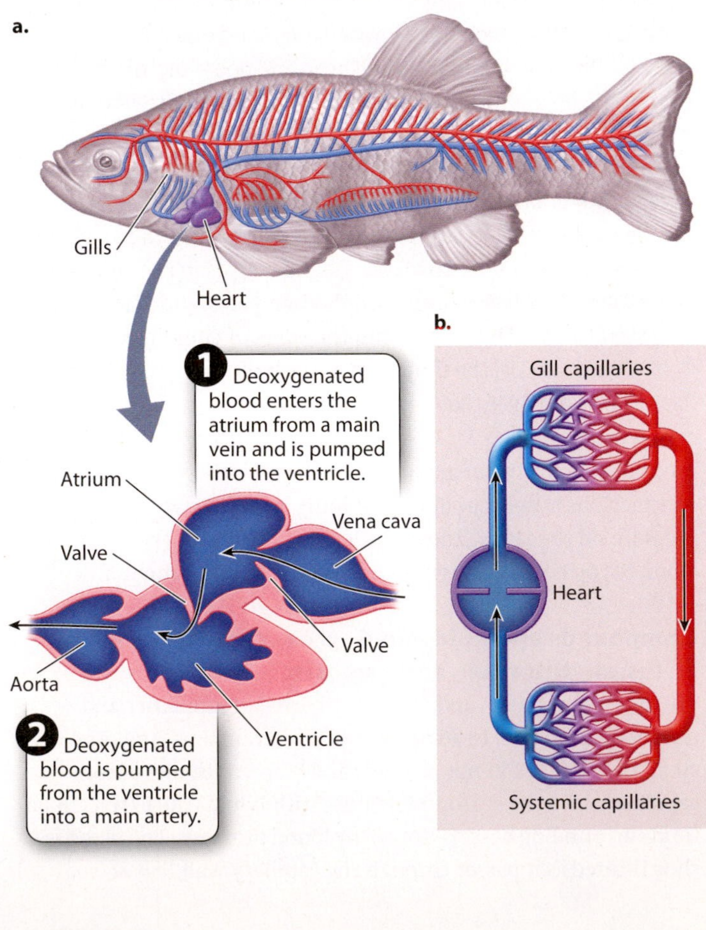

FIG. 39.21 Amphibian heart and circulatory system. (a) The frog heart has three chambers: two atria and one ventricle. (b) Mixed blood leaving the common ventricle goes either to the lung and back to the heart after being re-oxygenated, or to the tissues and back to the heart.

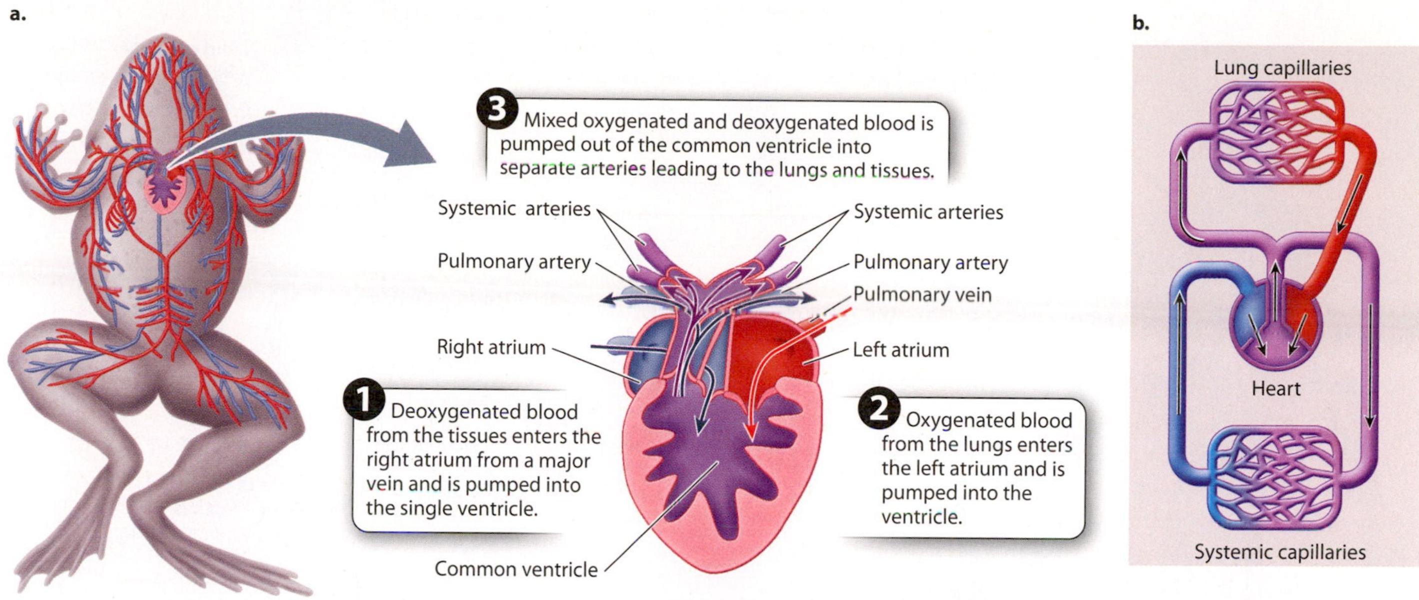

returns to the heart. This single circulation path limits the rate of blood flow to metabolically active body tissues. By contrast, birds and mammals evolved a separate **pulmonary circulation** to the lungs and **systemic circulation** to the rest of the body. This double circulation was made possible by the evolution of a four-chambered heart. This organization has two advantages: It increases the supply of oxygenated blood to active tissues, and it increases the uptake of O_2 at the gas exchange surface.

Fish have two-chambered hearts and a single circulatory system.

Fish hearts have two chambers (**Fig. 39.20**), an **atrium** (plural, **atria**) and a **ventricle.** Deoxygenated blood returning from the fish's tissues enters the atrium, which fills and then contracts to move the blood into a thicker-walled ventricle. The muscular ventricle pumps the blood through a main artery to the gills for uptake of O_2 and elimination of CO_2. Oxygenated blood collected from the gills travels to the tissues through a large artery called the **aorta.**

The small gill capillaries impose a large resistance to flow. As a result, much of the blood pressure is lost in moving blood through the gills. This loss of pressure limits the flow of oxygenated blood to body tissues.

Amphibians and reptiles have three-chambered hearts and partially divided circulations.

As animals moved onto land, the transition from breathing water to breathing air had important consequences for the organization of vertebrate circulatory systems. Land vertebrates evolved hearts that separated the circulation of deoxygenated blood pumped to their gas exchange organs from circulation of oxygenated blood delivered to their body tissues. Gas exchange became more efficient and O_2 delivery increased. Metabolic rates rose, and animals became capable of greater activity.

Reflecting their aquatic and terrestrial lifestyle, amphibians evolved a variety of ways to breathe. Some retain gills, while others use lungs or their skin, or some combination of these, to breathe. Amphibian hearts evolved an additional atrium, becoming divided into two separate atria and a single ventricle (**Fig. 39.21**). This arrangement partially separates the pulmonary and systemic circulations, so freshly oxygenated blood can be pumped under higher pressure to the body than is the case in fishes. However, because amphibians have a single ventricle, oxygenated blood returning to the heart from the gills or lungs mixes with deoxygenated blood returning from the animal's body before being pumped from the ventricle. Anatomical features of the ventricle help to limit this mixing, but nonetheless mixing still hinders the supply of oxygenated blood to an amphibian's body.

Most reptiles also have a three-chambered heart, but their single ventricle has internal ridges to improve the separation of deoxygenated and oxygenated blood. As a result, reptiles are able to deliver more O_2 to their tissues than amphibians can. At the same time, the incompletely divided ventricle provides reptiles with versatile circulatory responses during exercise and when diving underwater. For example, when they dive, the blood of turtles and sea snakes bypasses their lungs, which they cannot use

FIG. 39.22 The mammalian heart and circulatory system. (a) Mammals have a fully divided four-chambered heart with two atria and two ventricles. (b) Mammals have separate pulmonary and systemic circulations.

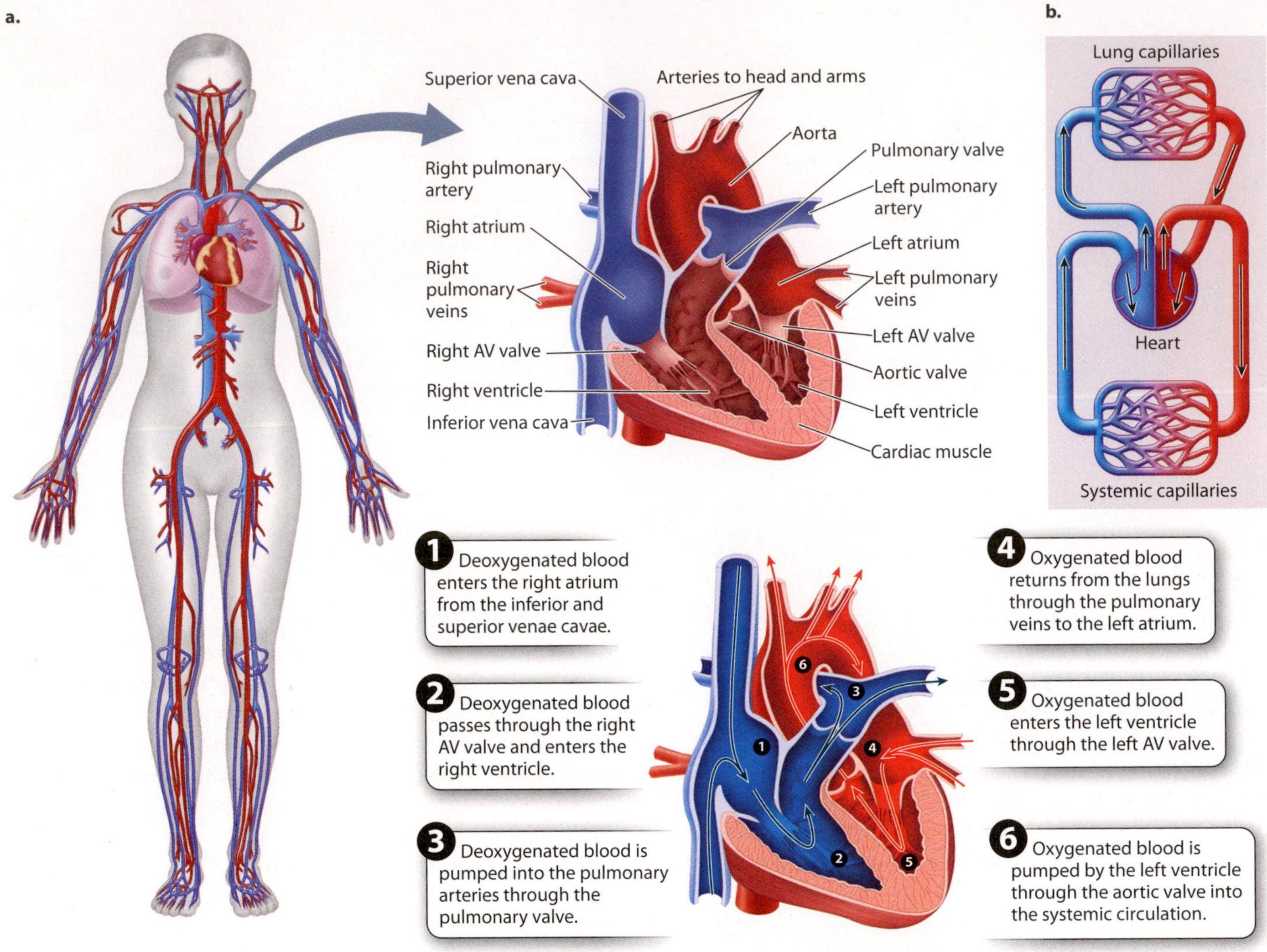

underwater. Notably, crocodiles and alligators have evolved four-chambered hearts, similar to those of mammals and birds, which are discussed next.

Mammals and birds have four-chambered hearts and fully divided pulmonary and systemic circulations.

Mammals and birds evolved four-chambered hearts with separate atria and separate ventricles (**Fig. 39.22**). This arrangement completely separates blood flow to the lungs from blood flow to the tissues. As a result, these animals can pump blood to their lungs under lower pressure, allowing increased uptake of O_2, and at the same time supplying blood gases and nutrients to their tissues at high pressure. Whereas the systemic blood pressure of fish is only a few mmHg, in a bird or mammal it is 100 mmHg or higher. Moreover, blood flowing to the tissues is fully oxygenated at the outset, delivering more O_2 to body tissues at a higher rate. This change in heart structure and circulatory pattern was linked to an increase in the metabolic rates and the evolution of endothermic lifestyles in birds and mammals (Chapter 40).

Following the pattern established in amphibians and reptiles, deoxygenated blood enters the right atrium from the venae cavae and, when the atrium contracts, moves through an **atrioventricular (AV) valve** into the right ventricle (Fig. 39.22). When the right ventricle contracts, blood is pumped through the **pulmonary valve** into the pulmonary trunk, which divides into the left and right **pulmonary arteries,** and then to the lungs for oxygenation. The atrioventricular and pulmonary valves ensure that blood flows in one direction into and then out of the right ventricle. The walls of the right ventricle are thinner than those of the left ventricle, and its weaker contractions eject blood at a

FIG. 39.23 The cardiac cycle.

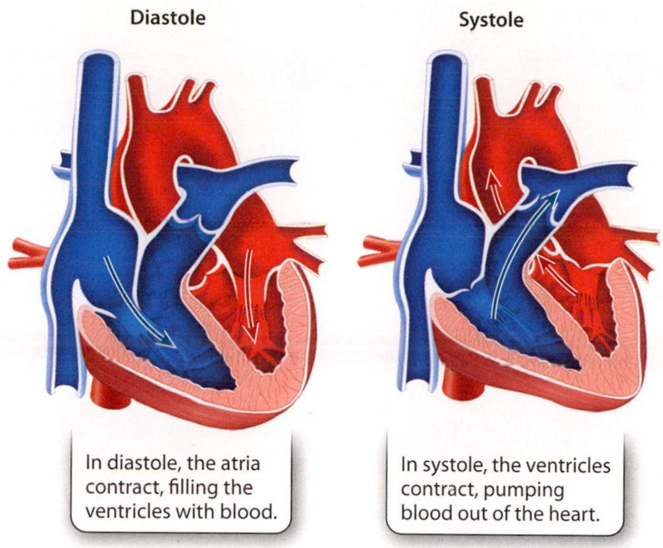

lower pressure. As a result, the blood of the pulmonary circulation moves at a slower rate, allowing greater time for gases to diffuse into and out of the lungs.

The oxygenated blood returns from the lungs through the **pulmonary veins** and enters the left atrium of the heart. When the left atrium contracts, blood is pumped through a second atrioventricular valve into the left ventricle. The thick muscular walls of the left ventricle eject the blood under high pressure to the body. Oxygenated blood leaving the left ventricle passes through the **aortic valve** and then flows to the head and the rest of the body through the large artery called the aorta. The one-way aortic valve ensures that blood leaving the left ventricle does not flow back into the left ventricle as it relaxes.

The contraction of the two atria followed by the contraction of the two ventricles makes up the **cardiac cycle** (**Fig. 39.23**). The cardiac cycle is divided into two main phases. **Diastole** is the relaxation of the ventricles, and **systole** is the contraction of the ventricles. Systole, therefore, is the phase of the cardiac cycle in which blood is pumped from the heart into the pulmonary and systemic circulations, and diastole is the phase in which the atria contract and the ventricles fill with blood.

→ **Quick Check 6** What pressure difference between the ventricle and the atrium causes the atrioventricular valve to close?

Cardiac muscle cells are electrically connected to contract in synchrony.

For the heart to function effectively as a pump, the **cardiac muscle** cells that make up the walls of the atria and ventricles must contract in a coordinated fashion. Atrial muscle cells must be activated to contract in synchrony during diastole to fill the ventricles, and ventricular muscle cells must contract in synchrony during systole to eject the blood from the heart. For cardiac muscle cells to pump in unison, they must be depolarized in unison by an action potential.

Cardiac muscle cells have contractile properties similar to those of skeletal muscle cells (Chapter 37). However, cardiac muscle cells are distinct in two ways. First, specialized cardiac muscle cells can generate action potentials on their own, independently of the nervous system. Second, cardiac muscle cells are in electrical continuity with one another through gap junctions (Chapter 10), meaning that they can pass their action potentials to adjacent cells. These two features of cardiac muscle cells ensure that all muscle cells in a region surrounding a heart chamber are activated and contract in unison.

The specialized cardiac muscle cells capable of generating action potentials independently function as a **pacemaker** that causes the heart to beat with a basic rhythm. These cells stimulate neighboring cells to contract in synchrony. They are found in two specialized regions of the heart—the **sinoatrial (SA)** and **atrioventricular (AV) nodes** (**Fig. 39.24**).

Whereas most nerve cells maintain their resting membrane potential until they receive a signal from another cell, the critical feature of pacemaker cells is that their resting membrane potential gradually becomes less negative on its own until it reaches threshold and the cell fires an action potential. The tendency of their membrane potential to become less negative is due to slow leakage of sodium ions into the cell and decreased flow of potassium out of the cell. When the cell reaches threshold, voltage-gated calcium ion channels open, causing more rapid depolarization. Like nerve cells, after firing an action potential, the pacemaker cells repolarize to reach their resting potential, keeping the heart relaxed between contractions. Also like nerve cells, pacemaker cells repolarize by opening voltage-gated potassium channels to allow potassium ions to leave the cell.

The heartbeat is initiated at the sinoatrial node, located at the junction of the vena cava and right atrium (Fig. 39.24). Because adjacent cardiac muscle cells are in electrical contact, action potentials initiated at the sinoatrial node spread rapidly from one cell to the next. Thus, when the pacemaker cells in the SA node fire an action potential, their depolarization spreads electrically throughout the right and left atria, causing them to contract in unison. Because there is no electrical contact between the atria and ventricles, the ventricles do not contract. Instead, the depolarization in the atria reaches a second set of pacemaker cells, located in the atrioventricular (AV) node.

Activation of the AV node transmits the action potential to the ventricles. A modified set of cardiac muscle fibers transmits the action potential from the AV node to the base of the ventricles. From there, the depolarization spreads throughout the ventricle walls, causing the ventricles to contract in unison. The delay in

FIG. 39.24 Control of heart rate and rhythm. The pacemaker cells of the sinoatrial and atrioventricular nodes keep the heart beating with a basic rhythm. This rhythm can be assessed by recording the electrocardiogram (EKG) of the heart.

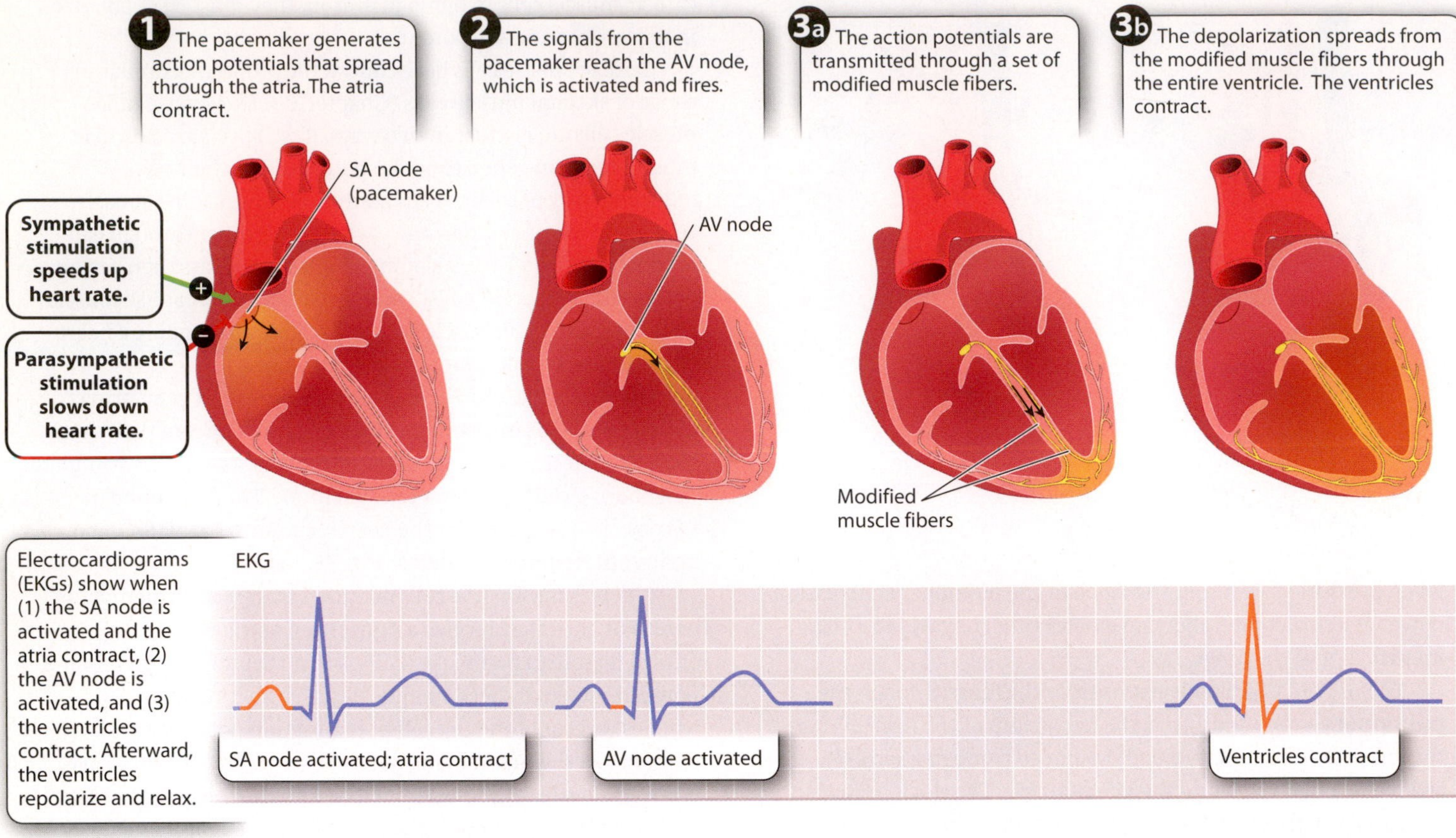

transmission from the AV node and through the conducting fibers ensures that the ventricles do not contract until they are fully filled with blood from the atria.

Electrodes placed over the surface of the chest and other body regions can record the electrical currents produced by the depolarization of the heart while it beats. This type of recording is called an electrocardiogram (abbreviated "EKG" because "kardia" is Greek for heart, although "ECG" is also used; Figure 39.24). Cardiologists use EKGs to diagnose heart problems or to determine what region of the heart has been damaged following a heart attack.

Heart rate and cardiac output are regulated by the autonomic nervous system.

The volume of blood pumped by the heart over a given interval of time is its **cardiac output (CO).** This quantity is a key measure of heart function. Cardiac output is determined by calculating the product of **heart rate (HR)** and the volume of blood pumped during each beat, the **stroke volume (SV):** $CO = HR \times SV$. Animals can therefore increase their cardiac output by increasing heart rate or stroke volume or both. Whereas humans and other mammals rely most heavily on increases in heart rate to increase cardiac output, fish rely more on increases in stroke volume.

Cardiac output rises or falls in response to the metabolic demand for O_2, which in turn depends on an animal's state of activity. When an animal sleeps, cardiac output is minimal, but when it runs, cardiac output increases to meet the demand for O_2 of its active muscles.

The nervous system controls the heart rate. Nerve signals influence the rate at which the cardiac pacemaker cells depolarize in between action potentials. Stimulation by sympathetic nerves causes the pacemaker cells of the sinoatrial node to depolarize more rapidly and the heart rate to speed up. Stimulation by parasympathetic nerves causes these cells to depolarize more slowly and the heart rate to slow down. Adrenaline released by the adrenal gland into the circulation also increases the heart rate for more prolonged periods of time. This hormone is a key component of the fight-or-flight response of many vertebrate animals.

Stroke volume is adjusted directly in response to changes in blood flow triggered by a change in activity. During exercise, when an animal's limb muscles regularly contract, the return of blood to the heart is increased. The influx of additional blood stretches the walls of the atria and ventricles to a greater extent, causing them to contract more forcefully. As a result, more blood is ejected during each heartbeat. The adjustment of stroke volume to changes in activity ensures that heart emptying is matched to heart filling (that is, to venous return). The relationship between the volume of blood filling the heart and stroke volume is described by **Starling's Law:** A greater volume of blood returned to the heart increases stretching of the muscle, which leads to more forceful ejection of blood and a greater stroke volume.

Increased sympathetic stimulation to the heart and circulating adrenaline also increase the contractile strength of the heart muscle, further increasing the heart's stroke volume. These changes in heart rate and stroke volume underlie the increase in cardiac output that occurs when an animal changes from resting to activity (see Fig. 39.18).

As we have noted, circulation is linked to several other functions in addition to the transport of respiratory gases, nutrients, and wastes. The distribution of blood flow, controlled by changes in arterial resistance through vasodilation or vasoconstriction, is an important factor controlling heat loss or gain between an animal's body and its environment. Thus, the circulatory system plays a central role in the temperature regulation of many animals (Chapters 35 and 40). The circulatory system also provides the route by which hormones (Chapter 38) and immune cells (Chapter 43) are distributed throughout the body, enabling the endocrine system to maintain homeostatic control of diverse bodily functions and the immune system to defend the body against pathogens. In doing so, the circulatory system serves as a general pathway by which hormonal and immune communication occurs to regulate, integrate, and protect the functional state of the whole organism. ■

Core Concepts Summary

39.1 Respiration and circulation depend on diffusion over short distances and bulk flow over long distances.

Diffusion is effective only over short distances and requires large exchange surface areas with thin barriers. page 828

Ventilation and circulation provide bulk flow over long distances. page 829

Oxygen is delivered to tissues in four steps: (1) bulk flow of water or air past the respiratory surface (gills and lungs); (2) diffusion of O_2 across the respiratory surface into the circulatory system; (3) bulk flow through the circulatory system; and (4) diffusion of O_2 into tissues and cells. page 829

39.2 Respiration provides oxygen and eliminates carbon dioxide in support of cellular metabolism.

Many aquatic animals exchange respiratory gases with water through gills. page 830

Countercurrent flow of water relative to blood in gills enhances O_2 extraction from water. page 831

Terrestrial animals breathe air by means of internal tracheae or lungs. page 832

Terrestrial vertebrates inflate and deflate their lungs by bidirectional tidal ventilation driven by changes in pressure. page 832

Unidirectional flow of air maximizes O_2 uptake by bird lungs. page 834

Both the voluntary and involuntary nervous systems control breathing. page 835

39.3 Red blood cells produce hemoglobin, greatly increasing the amount of oxygen transported by the blood.

Hemoglobin greatly increases the capacity of the blood to transport O_2. page 836

Red blood cells contain hemoglobin with iron-containing heme groups that reversibly bind and release O_2. page 837

A sigmoidal O_2-dissociation curve describes the change in the binding affinity of hemoglobin for O_2 with changes in partial pressure. The shape of the curve results from cooperative binding of O_2 by hemoglobin. page 838

Myoglobin binds and stores O_2 in muscle cells, increasing the delivery of O_2 to muscle mitochondria for activity in general and for diving in marine mammals. page 838

Fetal hemoglobin is expressed by the mammalian fetus to allow O_2 uptake from the mother's blood. page 839

The affinity of hemoglobin for O_2 is reduced by a decrease in pH and an increase in CO_2 levels. page 839

39.4 Circulatory systems have vessels of different sizes that facilitate bulk flow and diffusion.

The fluid in an open circulatory system moves through only a few vessels and is mostly contained within the animal's body cavity. page 840

The fluid in a closed circulatory system travels through a set of internal vessels, moved by a pump, the heart. page 840

Resistance to blood flow is most strongly controlled by vessel radius. page 841

Arteries are thick-walled vessels that contain collagen and elastin fibers for support against pressure and elastic rebound. page 841

Capillaries are small-diameter, thin-walled vessels that facilitate diffusion of O_2 and CO_2. page 841

Arterioles control blood flow within the body by changing their resistance through contraction and relaxation of the smooth muscle in their walls. page 842

Veins are thin-walled vessels that return blood to the heart under low pressure. page 843

Liquids, gases, and other compounds move across capillary walls by diffusion, filtration, and osmosis. In vertebrates, some fluid is returned to the bloodstream by the lymphatic system. page 843

39.5 The evolution of animal hearts reflects selection for a high metabolic rate, achieved by increasing the delivery of oxygen to metabolically active cells.

Fish have two-chambered hearts and a single circulatory path. page 845

Amphibians and reptiles have three-chambered hearts and a partially divided circulatory path. page 845

Birds and mammals have four-chambered hears and a double circulatory path, with a pulmonary circulation, from the heart to the lungs, and systemic circulation, from the heart to the rest of the body, to enhance O_2 uptake and delivery to metabolizing tissues. page 846

One-way valves control blood flow through the heart. page 847

The cardiac cycle has discrete ventricular filling (diastole) and ventricular emptying (systole) phases. page 847

Cardiac muscle cells are electrically connected, allowing them to contract in synchrony as a unified heartbeat. page 847

Contraction of the heart is controlled by signals from pacemaker cells of the sinoatrial and atrioventricular nodes, transmitted by electrical conducting fibers to the ventricles. page 847

The vertebrate autonomic nervous system regulates the heart's cardiac output by changes in heart rate and stroke volume. page 848

Self-Assessment

1. Diagram the four basic steps of O_2 transport from an animal's respiratory medium (air or water) to its cells.
2. Describe two features of the gills of aquatic animals that favor diffusion of O_2 and CO_2.
3. Name two differences in the physical properties of water and of air that affect gas exchange in aquatic versus terrestrial animals.
4. Explain how the tracheal respiratory system of insects enables high metabolic rates.
5. Explain how gill respiration is facilitated by unidirectional water flow, whereas lung ventilation can depend on bidirectional airflow.
6. Describe the pressure changes that are needed to draw air into the lungs and to expel air out of the lungs.
7. Explain how cooperative binding by hemoglobin facilitates O_2 uptake into blood.
8. Explain how the branching of larger arteries into many smaller vessels affects the rate of and resistance to blood flow in the smaller vessels.
9. Diagram the path of blood flow through the heart, lungs, and body of a fish and of a mammal or bird.
10. Describe two mechanisms by which the cardiac output of an animal's heart can be adjusted and indicate which is more important for increasing cardiac output in mammals and which is more important in fish.

Log in to LaunchPad to check your answers to the Self-Assessment questions, and to access additional learning tools.

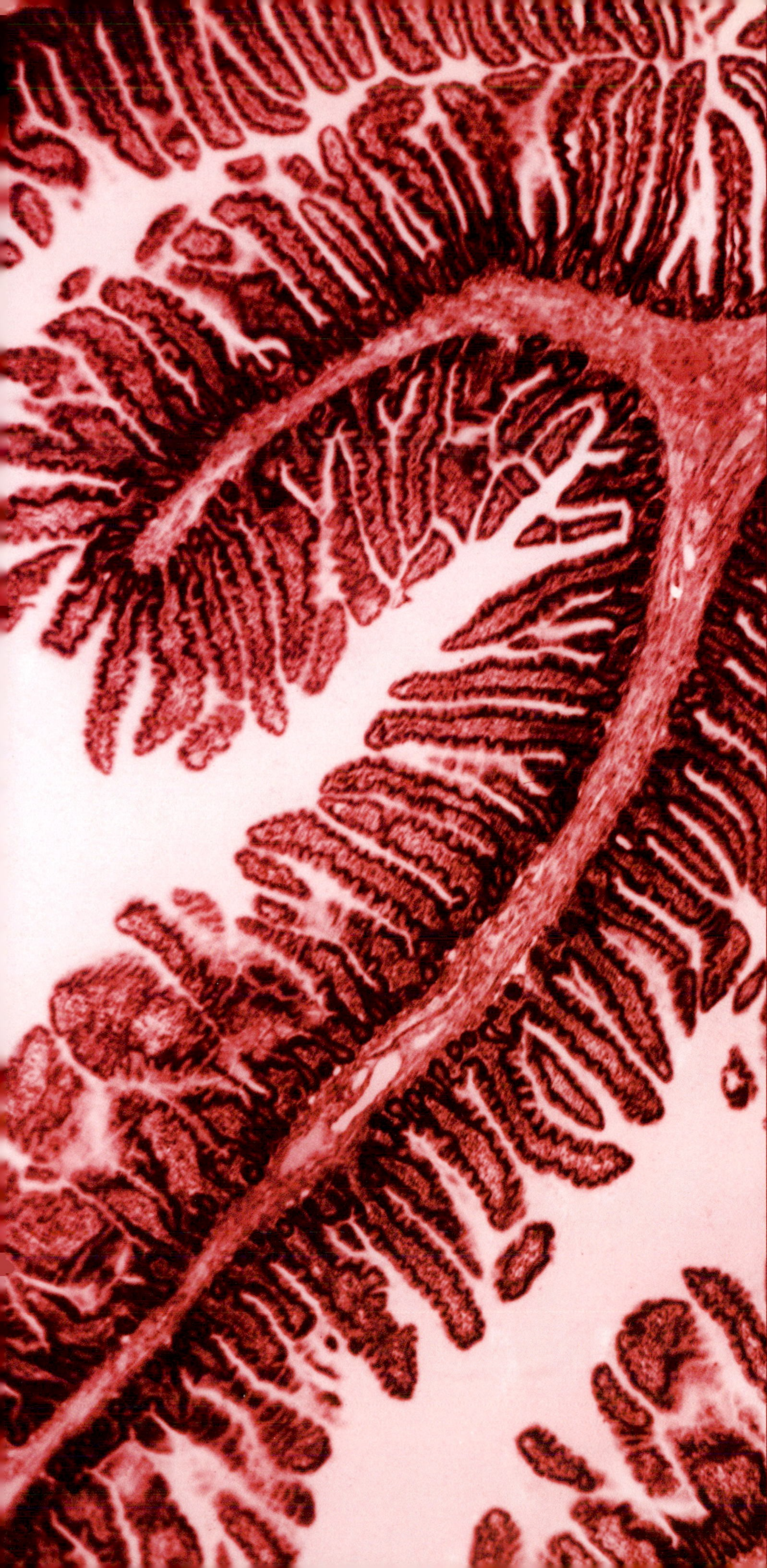

Science Source.

CHAPTER 40

Animal Metabolism, Nutrition, and Digestion

Core Concepts

40.1 Metabolic rate depends on activity level, body size, and body temperature.

40.2 An animal's diet supplies the energy the animal needs for homeostasis and essential nutrients it cannot synthesize on its own.

40.3 Animals have different adaptations for feeding.

40.4 The digestive tract has regions specialized for different functions, including digestion, absorption, storage, and elimination.

It is literally true that "you are what you eat." Not only do animals build their bodies and obtain energy from the food they eat, but also their very appearance often reflects what they eat. Consider a cat. We know it's a carnivore—an organism that mainly eats animals—just from looking at it. It has sharp teeth and powerful jaws specialized for tearing meat; strong legs, sharp claws, and keen eyesight for hunting prey; and a short gut because breaking down animal protein is easier than breaking down plant material. Humans are omnivores, adapted for eating both other animals and plants, and this is reflected in our anatomy. Our teeth include sharp canines for tearing meat and flat molars for grinding plant material. Finally, consider herbivores, such as cows, that eat plants. Cows have evolved a variety of adaptations, including a four-chambered stomach, that allow them to digest cellulose, which is tough but rich in energy.

Animals are heterotrophs, obtaining food from other organisms (Chapter 6). The energy animals obtain from food is essential for building bodies, moving, surviving, and reproducing. Much of an animal's activity therefore centers on obtaining food. In turn, the biology of an animal is in large part shaped by the type of food the animal consumes. Furthermore, an animal's metabolism, the chemical reactions by which it breaks down its food, is closely linked to the energy and nutrients contained in that food.

The food that an animal eats also affects its place in an ecosystem. Primary producers, such as plants and some algae, are food for a great diversity of animals. Animals that feed on plants tend to exist in large numbers. Large animals that feed on other, smaller animals exist in smaller numbers and commonly represent fewer species.

This chapter examines how animals obtain and break down the food they eat. We begin at the cellular level, building on the biochemistry of metabolism discussed earlier (Chapter 7). Then, we examine the dietary needs and the organization and function of animal digestive systems: how an animal eats, digests, and absorbs food to supply its cells with the energy required for its function. The relationship between an animal's diet and its place in an ecosystem is reserved until Chapter 47.

40.1 PATTERNS OF ANIMAL METABOLISM

As discussed in Chapter 7, the biochemical pathways that make up animal metabolism are remarkably conserved among diverse groups of organisms. These pathways evolved early in the history of life and have been retained throughout the evolution and diversification of life. In Chapter 7, we examined the pathway of glucose breakdown by cellular respiration, which provides ATP to fuel the work of a cell. Here, we put this and related pathways in an organismal context. We explore how carbohydrates, proteins, and fats are metabolized for building body structures and supplying energy needs and how metabolic rate varies with an animal's level of activity, internal body temperature, and size.

Animals rely on anaerobic and aerobic metabolism.

Nearly all animals depend on three main classes of molecule as sources of energy and building blocks for growth and development: carbohydrates, fats, and proteins. These molecules are broken down, and in the process the energy in their chemical bonds is transferred to biologically useful forms, typically **adenosine triphosphate (ATP).** While other nucleic acid triphosphates also serve as usable forms of energy, ATP is by far the one most commonly used by animals. The breakdown of carbohydrates, fats, and proteins to produce ATP takes place through a series of chemical reactions. Reactions that break down food sources to fuel the energy needs of a cell are termed **catabolic.** In contrast, reactions that result in net energy storage within cells and the organism are **anabolic.**

Let's first review the breakdown of carbohydrates, including sugars (such as glucose) and starches, food sources that are chemically similar but clearly different in taste and texture. Glucose can be partially broken down in the absence of oxygen by glycolysis (**Fig. 40.1**). Glycolysis occurs in the cytosol of the cell, and the breakdown of each glucose molecule results in the production of 2 molecules of pyruvate, 2 molecules of ATP, and 2 molecules of NADH. In animals, if oxygen is not present, or is present only in small amounts, pyruvate is converted by fermentation into lactic acid. The production of ATP by **anaerobic** ("absence of oxygen") **metabolism** provides rapid but short-term energy to the cell and organism.

In contrast, **aerobic** ("of oxygen") **metabolism** (specifically, cellular respiration), carried out within the mitochondria of eukaryotic animals, provides a steady supply of ATP for longer-term, sustainable activity. When enough oxygen diffuses into the mitochondria, pyruvate can be further broken down by the **citric acid cycle** rather than converted to lactic acid (Fig. 40.1). In a series of oxidation–reduction reactions, energy-rich NADH and $FADH_2$ molecules are produced. These molecules then enter the electron transport chain in the inner mitochondrial membrane. The **electron transport chain** couples the transfer of electrons to the pumping of protons across the inner mitochondrial membrane, resulting in a proton electrochemical gradient that drives the synthesis of ATP by oxidative phosphorylation. Because oxygen is reduced to water in the process, oxygen is consumed and water is produced as a by-product.

Overall, aerobic respiration results in the formation of a total of 32 molecules of ATP when 1 molecule of glucose is broken down, much more than the 2 molecules of ATP from 1 molecule of glucose produced anaerobically. We can define metabolic efficiency as the amount of energy captured in a usable form (at least 234 kcal of useful ATP energy) per amount of energy in the starting molecule (686 kcal in 1 mol of glucose). The metabolic efficiency of aerobic respiration is therefore at least 34% (or $234/686 \times 100$). The remaining 66% of this energy is converted to heat and either lost to the environment or used to warm the animal. Glucose that is not metabolized is stored as glycogen, primarily in the liver and muscles.

FIG. 40.1 Cellular metabolism, which includes anaerobic and aerobic pathways. Animals obtain energy to fuel their activities and development from the breakdown of carbohydrates, fats, and proteins.

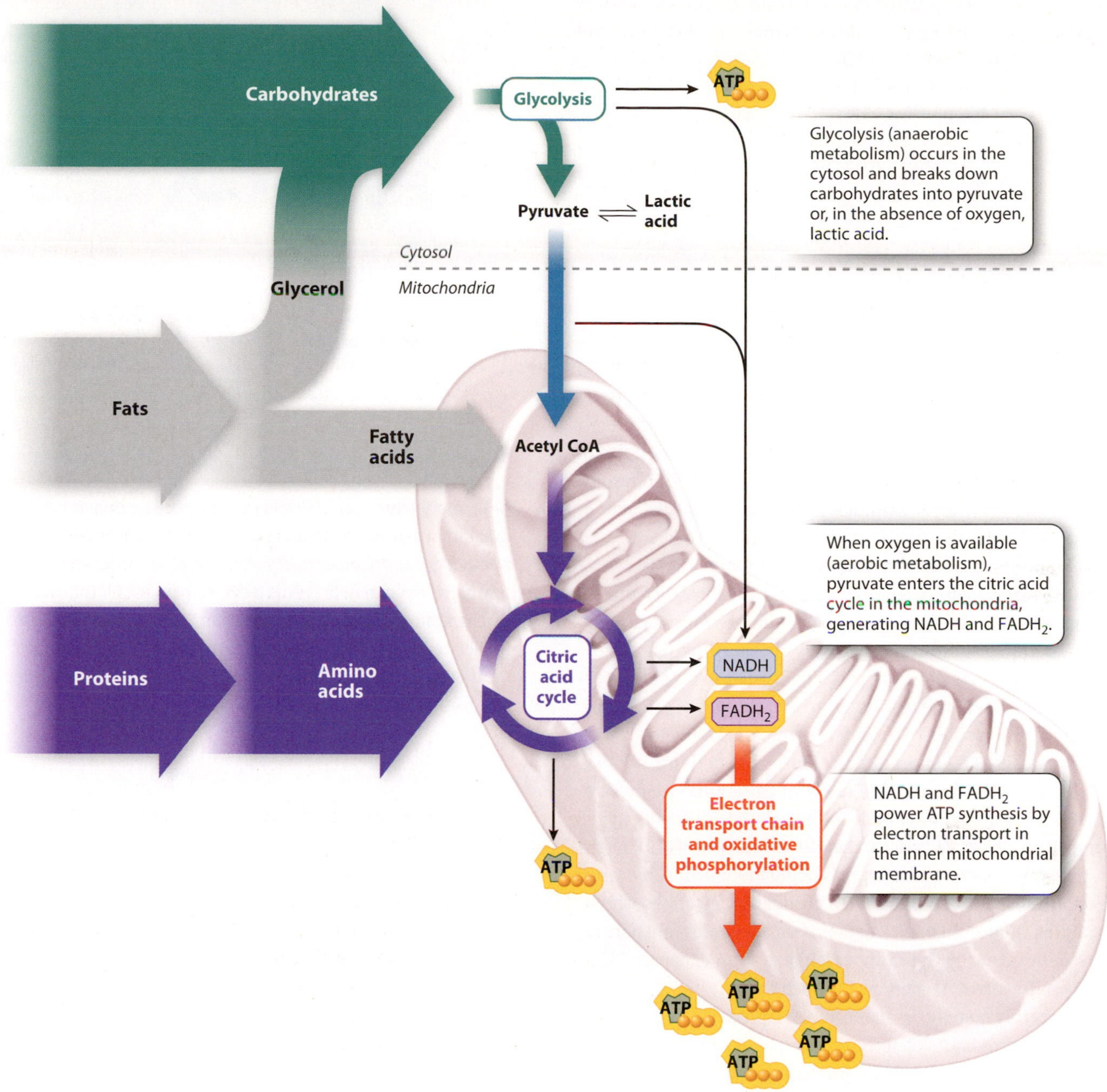

Lipids are another important energy source for most animals. Lipids, such as triacylglycerol, consumed in the diet are broken down to glycerol and free fatty acids, which enter glycolysis or the citric acid cycle to yield ATP by oxidative phosphorylation (Chapter 7). Alternatively, they can be stored as fat (adipose tissue). Storing energy as fat is a particularly efficient way to store energy because it yields more than twice the energy supply per unit weight (9.4 kcal/g) as glycogen (4.2 kcal/g). Like lipids consumed in the diet, stored fats are also broken down into glycerol and free fatty acids, which yield ATP by the same pathways as lipids from the diet.

Proteins consumed in the diet can also be a useful energy source for animals. Carnivores such as cats get most of their energy from the protein in the meat they eat. Proteins, which are needed for building and maintaining the body, constitute the enzymes and structural elements of cells and tissues. Only following prolonged food deprivation, when fat and carbohydrate reserves are depleted, do animals break down protein reserves (mainly from their muscles) to form ATP. Animals preferentially rely on fats for long-term energy supply. Well-nourished humans have enough fat reserves to support their metabolic needs for 5–6 days before they start to use protein reserves. Because nerve cells depend primarily on glucose to produce ATP, carbohydrates are spared for the brain and nervous system when animals face starvation.

→ **Quick Check 1** Muscle protein, fat, and glycogen are all reservoirs of energy. In what order are they used during a prolonged fast?

Metabolic rate varies with activity level.

An animal's overall rate of energy use is its **metabolic rate.** Metabolic rate can be measured by the animal's rate of oxygen consumption, which reflects the rate of aerobic production of ATP.

Metabolic rate is affected by many factors, one of which is the activity level of an organism. When an animal shifts from rest to activity, its metabolic rate and oxygen consumption rise to meet its increased demand for ATP. The onset of activity requires immediate energy, which in animals is provided by specialized energy stores in their tissues. In vertebrate muscle cells, for example, phosphocreatine is a ready source of high-energy phosphates that is hydrolyzed to synthesize ATP directly from ADP at the onset of activity. This reaction is followed by the relatively rapid production of ATP by anaerobic glycolysis. Although glycolysis produces relatively few ATP molecules per molecule of glucose broken down, the reactions are extremely fast, providing a rapid short-term supply of energy for animals. Animals rely on anaerobic glycolysis for short bursts of intensive activity.

With longer activity, the ATP needs of the animal are met by aerobic respiration in mitochondria. Therefore, the rate of oxygen consumption initially increases, then levels off (**Fig. 40.2**). At this point, the animal's need for energy is being met entirely by aerobic respiration. Once mitochondrial ATP production increases, the steady state production of ATP depends on adequate transport of oxygen to the tissue by the cardiovascular system (Chapter 39) as well as by diffusion of oxygen into mitochondria.

Sprinters rely heavily on anaerobic ATP production for intense short-term bouts of activity, but longer races are run at progressively slower speeds for increased endurance as the distance runner relies more heavily on aerobic metabolism to produce ATP. The world record for the 100-m sprint averages just above 10 m/s, but for the 1500-m race it is about 6.6 m/s.

When activity ends, the animal's oxygen consumption rate declines but does not immediately return to resting levels, as shown in Fig. 40.2. The elevated consumption of oxygen following activity is the animal's **recovery metabolism.** It represents the continued metabolic energy required to reestablish the resting metabolic state of the cells. During recovery metabolism, cells resynthesize depleted ATP stores and metabolize the end products

FIG. 40.2 Oxygen consumption during physical activity.

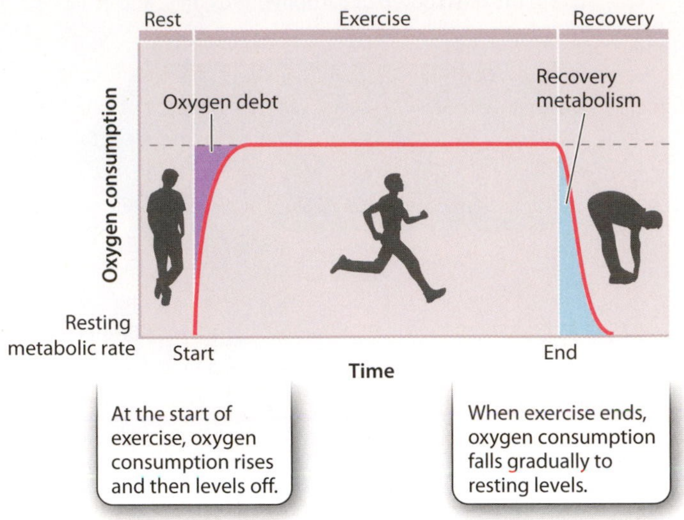

of fermentation, particularly lactic acid. The difference between an animal's immediate energy need at the onset of activity and energy supplied by aerobic metabolism is referred to as the animal's "oxygen debt." This debt is "paid back" following exercise by the animal's recovery metabolism. Recovery metabolism is associated with the elevated breathing and heart rate that you (and other animals) experience when resting after moderate to intense exercise.

? CASE 7 PREDATOR–PREY: A GAME OF LIFE AND DEATH

Does body temperature limit activity level in predators and prey?

We just saw that metabolic rate increases with activity level. A by-product of metabolism is the generation of heat, which is a necessary consequence of the second law of thermodynamics (Chapters 1 and 6). Animals have several mechanisms for dissipating excess heat, including sweating, changes in blood flow, and (in dogs, for example) panting.

Cheetahs, as top predators, are one of the fastest land animals, capable of speeds up to 70 mph (114 km/h). Such extreme speed generates a lot of heat. Interestingly, instead of releasing the excess heat as it builds up, the cheetah stores it, dissipating it only after a chase.

The amount of heat that cheetahs store during a sprint can be considerable, on the order of 60 times the heat produced at rest. Heat storage raises the cheetah's body temperature. Cheetahs can therefore reach remarkable speeds but can sustain them only for short durations and must rest for long periods of time between sprints. Because their bodies can support only so much heat storage, cheetahs do not run when the temperature of their environment reaches approximately 105°F (41°C). In other words, the amount of time a cheetah runs and the distance it covers are limited by its body temperature.

Gazelles, common prey of cheetahs, have similar heat storage mechanisms and are similarly known for their speed and agility. Goats, by contrast, do not store heat as they increase their activity levels, instead dissipating it through evaporative mechanisms. But goats are not known for their speed. Heat storage therefore seems to enable quick bursts of speed, and has evolved as an adaptation among both predators and prey to achieve short, rapid sprints.

Metabolic rate is affected by body size.

In addition to activity level, an animal's size influences its metabolic rate. At rest, larger animals consume more energy and have higher metabolic rates than smaller ones. However, resting metabolic rate does not increase linearly with an animal's mass. Instead, measurements of resting metabolic rate in a wide range of organisms show that metabolic rate increases with animal mass raised to the ¾ power (**Fig. 40.3**). This means that the average rate at which each gram of body tissue consumes energy is *less* in larger animals compared to smaller ones. Or, put another way, the larger the organism, the *lower* the metabolic rate *per gram* of body tissue. This scaling pattern of cellular energy metabolism is remarkable because it holds across a diverse size range of unicellular and multicellular organisms, as can be seen in Fig. 40.3.

In addition to comparing resting metabolic rate in different-sized animals, we can also examine how metabolic rate varies with respect to level of activity in different-sized animals. In this case, for example, we are looking at how energy is used to allow animals to move at different speeds. The American comparative physiologist C. Richard Taylor has extensively studied how metabolic rate varies with running speed in different-sized terrestrial animals (**Fig. 40.4**). He found that metabolic rate increases linearly with speed, and that larger animals expend less energy per unit mass than smaller ones.

→ **Quick Check 2** Does a dog that is twice as heavy as a cat have twice the resting metabolic rate?

FIG. 40.3 Resting metabolic rate and body size in diverse organisms. The relationship plotted here shows that larger animals have lower metabolic rates per gram of tissue than smaller animals. *Source: Adapted from A. M. Hemingsen, 1960, "Energy metabolism as related to body size and respiratory surfaces, and its evolution." Re. Steno Memorial Hospital Nordisk Insulinlaboratorium 9:1–110.*

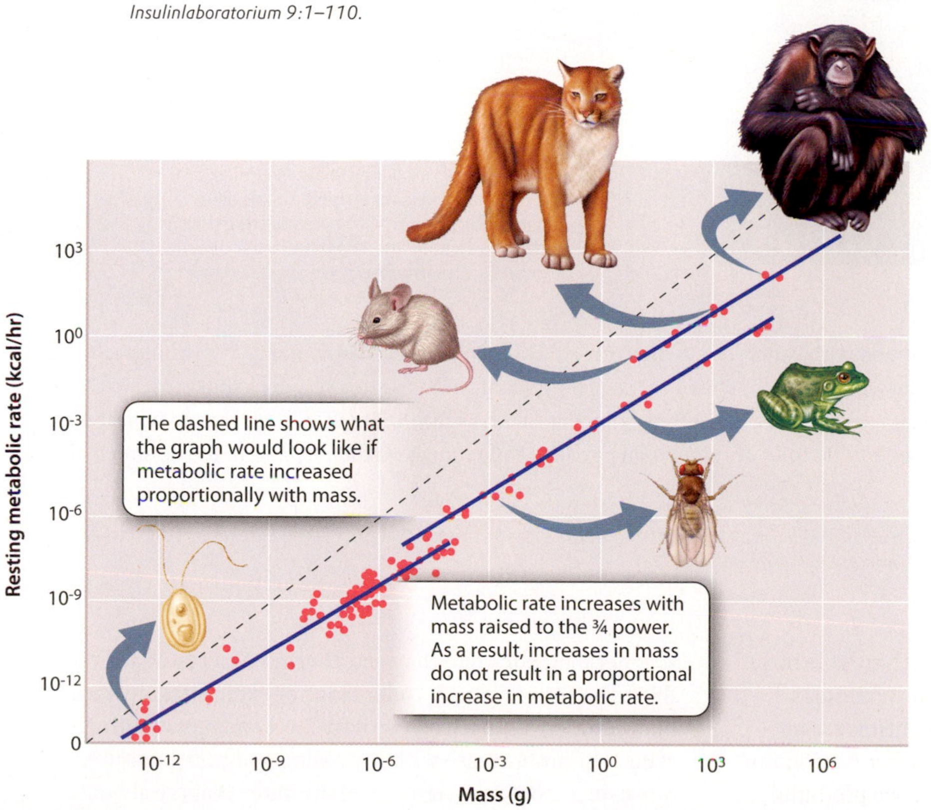

Metabolic rate is linked to body temperature.

Metabolic rate is also affected by an animal's internal body temperature. Temperature affects the rates of chemical reactions, which in turn determine how fast fuel molecules can be mobilized and broken down to supply energy for a cell. Animals can be categorized by the sources of most of their heat. Animals that produce most of their own heat as by-products of metabolic reactions, including the breakdown of food, are **endotherms.** These animals usually, but not always, maintain a constant body temperature that is higher than that of their environment. As we saw earlier (Chapter 35), the maintenance of a constant core body temperature is a form of homeostasis and requires balancing heat production and heat loss. Mammals, including humans, and birds are endotherms.

By contrast, animals that obtain most of their heat from the environment are **ectotherms.** Ectotherms often regulate their body temperatures by changing their behavior—moving into or out of the sun, for example. Think of a turtle or some other reptile sunning itself on a rock: It is raising its body temperature. The temperatures of these animals usually fluctuate with the outside environment. Most fish, amphibians, reptiles, and invertebrates are ectotherms.

Endotherms are sometimes referred to as "warm-blooded" and ectotherms as "cold-blooded," but these terms are misleading because, depending on

HOW DO WE KNOW?

FIG. 40.4

How is metabolic rate affected by running speed and body size?

BACKGROUND In the 1970s and 1980s, the American physiologist C. Richard Taylor and his colleagues performed studies on the relationship between metabolic rate and running speed in mammals of different sizes. They were interested in understanding the energetic costs of running in different animals.

EXPERIMENT Taylor and colleagues measured oxygen consumption during running for a wide range of organisms. They trained each animal to run on a treadmill at different speeds and measured oxygen consumption using either a face mask or enclosure. Oxygen consumption was used as a measure of metabolic rate.

RESULTS The researchers found that there is a linear increase in metabolic rate with speed in different-sized animals and that larger animals expend less energy per unit body mass to move a given distance compared to smaller ones, as shown on the graph.

Photo: Cary Wolinsky/Getty Images.

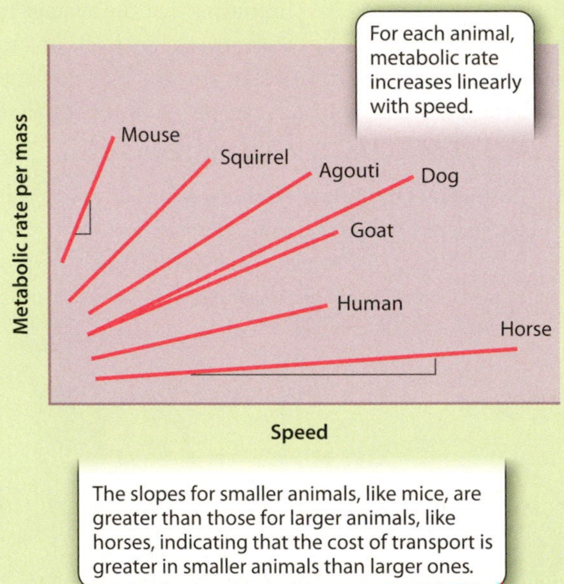

FOLLOW-UP WORK Similar studies were performed with kangaroos, which move by hopping rather than running. Interestingly, it was found that it is "cheaper" (that is, it requires less energy) to hop than to run. This finding perhaps explains why kangaroos and other hopping animals survived, while many animals that run on four legs became extinct in Australia 30,000 to 20,000 years ago with the arrival of early humans. Studies have also compared the energetic costs of running on two legs and running on four legs, with the finding that the cost is the same.

SOURCES Dawson, T. J., and C. R. Taylor. 1973. "Energetic Cost of Locomotion in Kangaroos." *Nature* 246:313–314; Taylor, C. R., N. C. Heglund, and G. M. O. Maloiy. 1982. "Energetics and Mechanics of Terrestrial Locomotion." *Journal of Experimental Biology* 97:1–21.

environmental conditions, cold-blooded animals can have core body temperatures higher than warm-blooded animals. It should also be kept in mind that these thermoregulatory mechanisms are not completely distinct, but instead represent extremes of a continuum.

Nevertheless, thermoregulatory mechanisms have profound implications for an animal's metabolic rate. Endotherms have a higher metabolic rate than ectotherms. As a result, they are able to be active over a broader range of external temperatures than ectotherms. Both the activity level and metabolic rate of ectotherms increase with increasing body temperature. However, ectotherms have metabolic rates that are about 25% of endotherms of similar body mass and at similar body temperatures. Although they can achieve activity levels similar to those of endotherms when their body temperatures are similar, ectotherms cannot sustain prolonged activity. Therefore, ectotherms such as lizards or insects rely on brief bouts of activity, followed by longer periods of inactivity compared to endotherms.

Endotherms benefit from having an active lifestyle. However, their high metabolic rate requires long periods of foraging to acquire enough food to keep warm and be active. By contrast, although ectotherms cannot sustain activity for long periods, they can survive much longer periods without food.

As we have discussed, the control of core body temperature, or thermoregulation, is a form of homeostasis and requires the coordinated activities of the nervous, muscular, endocrine, circulatory, and digestive systems. We summarize how these systems work together in endotherms and ectotherms in **Fig. 40.5** on the following pages.

40.2 ANIMAL NUTRITION AND DIET

Running, jumping, moving, growing, reproducing, and all other activities performed by animals require energy. Autotrophs like plants and certain microbes are able to capture energy from the sun or from inorganic compounds (Chapters 8 and 26), but animals must acquire food in the form of plants or other animals. Food does not consist just of carbohydrates, fats, and proteins, however. To enable various cell processes, animals must also obtain in their diet certain minerals and chemical compounds that they cannot synthesize on their own. In this section, we explore what animals need to obtain from their food, and how much of it, in order to sustain their cells and tissues.

Energy balance is a form of homeostasis.

Like core body temperature (Fig. 40.5), blood glucose levels (Chapter 38), and blood pressure (Chapter 39), the **energy balance** of an organism is often maintained at a constant level. An animal in energy balance takes in over time the same amount of calories of energy from food that it uses over time to meet its metabolic needs. Energy balance can be thought of as a form of homeostasis. We consider sources of energy, or **energy intake,** and ways in which energy is expended, or **energy use.**

For animals, the source of energy is the diet. In turn, energy is used to do work—maintain tissues, grow, move about, and the like. Energy required for basic life processes accounts for the majority (about 70%) of energy use. Interestingly, digestion and absorption of food themselves require energy, and this use is included in the 70% figure. The balance (about 30%) can be used for physical activity, with higher levels of activity using more energy, as we saw.

When energy intake does not equal energy used, there is an energy imbalance. If an animal eats more food than it requires, energy stores such as fat deposits grow over time. The result is that the body shifts its metabolism mostly to anabolic processes that build energy stores. Many animals achieve a net positive energy balance during the late summer and fall when food is plentiful, before it becomes scarce in winter. Other animals maintain a constant energy intake throughout the year by migrating to areas with more abundant food and avoiding colder temperatures that require increased energy expenditure to remain active and warm. Still other animals hibernate, or become less active, to conserve their energy use over the winter.

Animals that cannot acquire enough food are in negative energy balance and become undernourished. During prolonged periods of an inadequate food supply or starvation, an animal consumes its own internal fuel reserves. Starvation forces animals to deplete their glycogen and fat reserves first, and then, if no food is found, to resort to protein stores, primarily in muscle tissue. This ultimately leads to muscle wasting. Undernourishment and starvation are particularly serious human health problems in many developing countries, especially those ravaged by war and political instability.

Humans, like most other animals, store excess food calories as fat. We do so because, over much of our evolutionary history, food was less abundant and more unpredictable in its availability than it is today. With the development of agriculture and domestication of livestock, food supplies rapidly increased, and many human populations were able to grow and consume increasing amounts of food. With the rise of mechanized agricultural food production and the availability of highly processed foods developed by the modern food industry, excessive intake of food calories has led to an increasing and now critical public health problem: obesity.

Obesity is now an epidemic in many industrialized nations. In the United States, about 36% of the adult population is considered obese. Obesity is a major public health concern because it increases the risk of diabetes, heart disease, and stroke, and contributes to a shorter life-span. For most animals, however, acquiring food and storing its products efficiently in the body allow them to have a fuel reserve to meet seasonal energy requirements. This rationing of stored energy remains an essential part of their metabolic and digestive physiology.

An animal's diet must supply nutrients that it cannot synthesize.

Through their metabolic pathways, animals are able to obtain energy from the environment as well as to synthesize many of the compounds needed to sustain life. However, many nutrients necessary for life cannot be synthesized by an animal's metabolism and therefore must be acquired in the food that they eat.

Recall that amino acids are the basic units of proteins (Chapter 4). There are 20 different amino acids that are typically found in proteins (see Fig. 4.2). Although all of them are necessary for life, an **essential amino acid** is defined as one that cannot be synthesized by cellular biochemical pathways and instead must be ingested. Most animals can synthesize about half of their amino acids.

VISUAL SYNTHESIS

FIG. 40.5

Homeostasis and Thermoregulation

Integrating concepts from Chapters 35–40

Thermoregulation in an endotherm

Endotherms regulate their body temperature primarily by physiological means, such as increasing or decreasing their metabolic activity, which generates heat as a by-product.

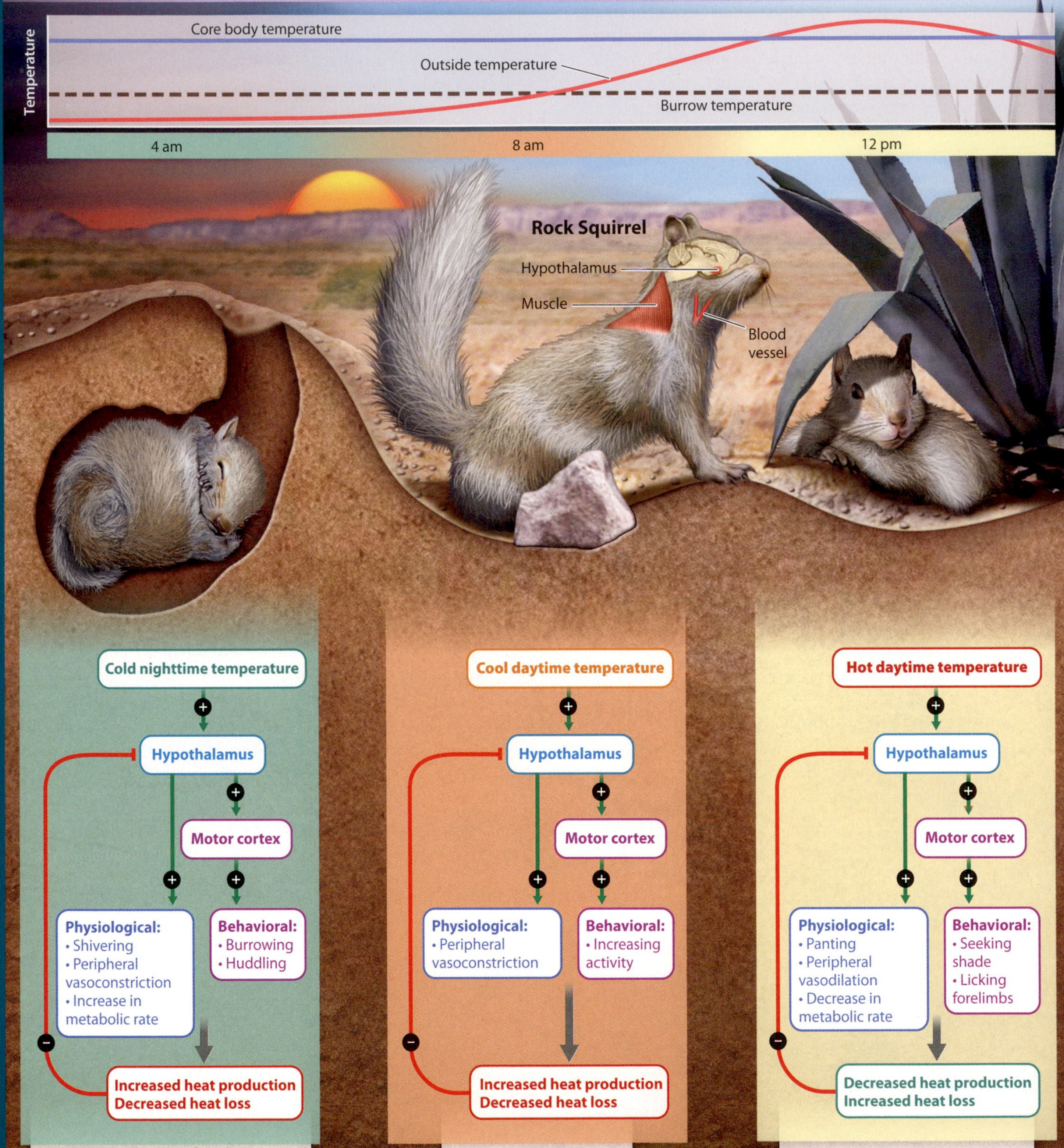

At night when it is cold, the rock squirrel enters its burrow, and physiological mechanisms generate heat to maintain a constant core body temperature.

The squirrel is most active in the early morning and late afternoon, when it is cool.

To avoid overheating during the day, the squirrel decreases its exposure to the sun, and physiological mechanisms cool the body.

Thermoregulation in an ectotherm

Ectotherms gain heat from or lose heat to their environment, and so their primary means of regulating their body temperature is behavioral, increasing or decreasing their exposure to the sun and hot surfaces.

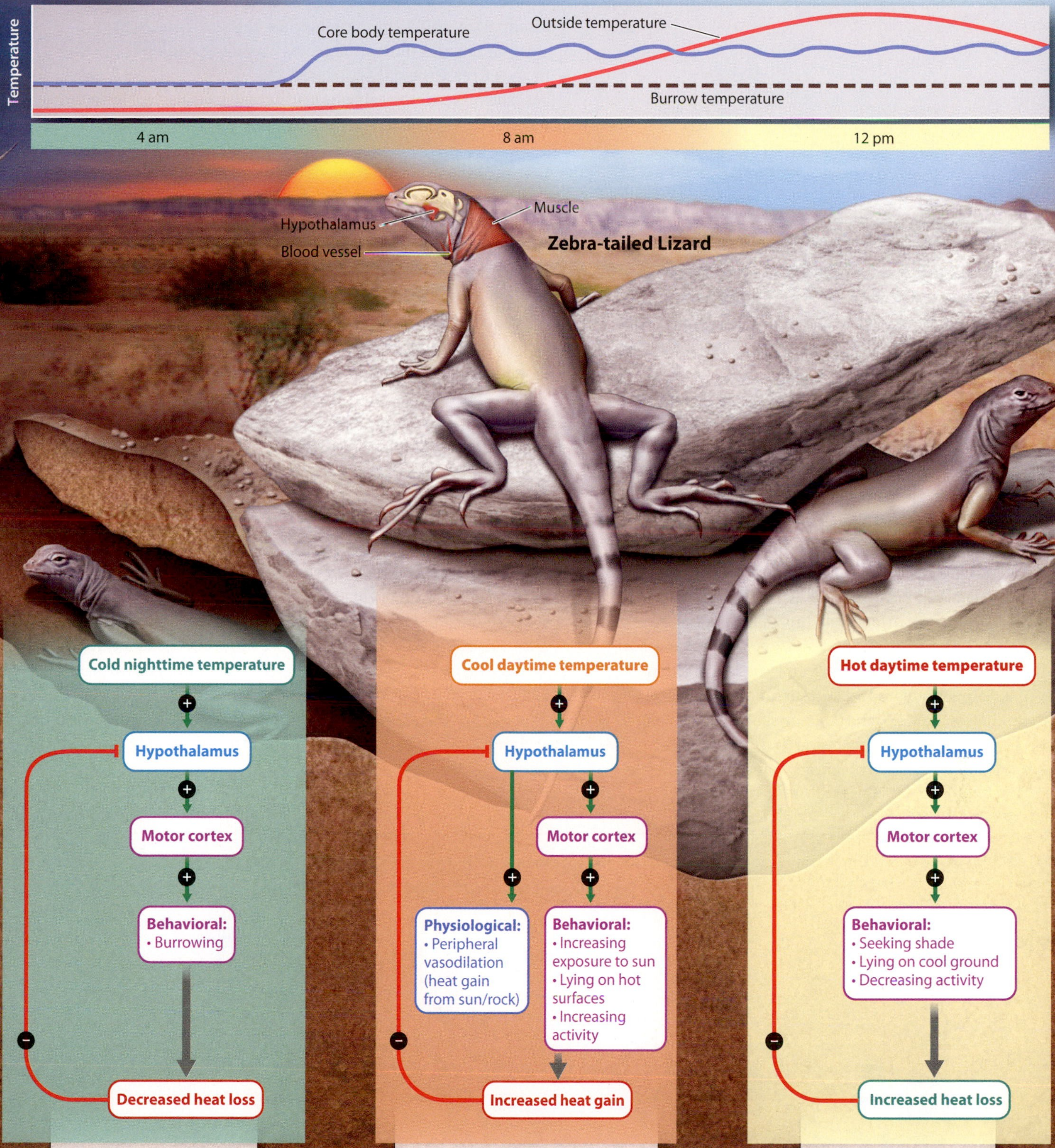

At night, lizards retreat to burrows in order to maintain a constant, intermediate temperature.

In the morning, the lizard basks in the heat of the sun. In the evening, it warms itself by heat radiated from rocks heated by the daytime sun.

During the hottest parts of day, the lizard regulates its temperature by seeking shade.

TABLE 40.1 Essential Amino Acids for Humans

ESSENTIAL AMINO ACID	COMMON SOURCES
Isoleucine	Meat, dairy, lentils, wheat
Leucine	Meat, cottage cheese, lentils, peanuts
Lysine	Meat, beans, lentils, spinach
Methionine	Meat, dairy, whole grains, corn
Phenylalanine	Meat, dairy, peanuts, seeds
Threonine	Meat, dairy, nuts, seeds
Tryptophan	Meat, dairy, oats
Valine	Meat, dairy, grains

Humans are unable to synthesize 8 of the 20 amino acids (**Table 40.1**). We have to obtain these eight essential amino acids in our diets. Infants require additional amino acids. The most reliable source of all eight essential amino acids is meat. This is not to say that only omnivorous humans can lead healthy lives. While most plant proteins lack at least one of the eight amino acids essential to humans, and diets heavy in a particular food, such as corn (which lacks lysine), can lead to protein deficiency, vegetarians can achieve a healthy intake of all essential amino acids by combining plant foods. For example, beans supply the lysine that corn lacks, and corn supplies the methionine that beans lack.

Dietary minerals are chemical elements other than carbon, hydrogen, oxygen, and nitrogen that are required in cells and must be obtained in the food that an animal eats (shown in red in **Fig. 40.6**). They include magnesium (Mg), phosphorus (P), potassium (K), calcium (Ca), iron (Fe), and zinc (Zn). Magnesium and zinc are sometimes associated with enzymes as cofactors (Chapter 6); calcium is required for neuron function (Chapter 35), muscle function (Chapter 37), and building skeletons (Chapter 37); and iron in hemoglobin binds oxygen and transports it in the blood (Chapter 39). Humans typically obtain the minerals sodium (Na) and chloride (Cl) through common table salt and other foods. Many animals seek exposed rock that they lick to obtain minerals and salts.

In addition to amino acids and minerals, animals must ingest essential **vitamins,** organic molecules that are required in very small amounts in the diet. Vitamins have diverse roles: Some bind to and increase the activity of particular enzymes; others act as antioxidants or chemical signals. Different animals have different vitamin requirements, so a molecule that functions as a vitamin for one may not be a vitamin for another. Thirteen essential vitamins have been identified for humans (**Table 40.2**). Knowing which vitamins are required for an animal's health is critical to proper medical and veterinary care. Animals have evolved diets that ensure that their need for vitamins, as well as for essential amino acids and minerals, is met through the combination of foods that they eat.

In humans as well as in other animals, vitamin deficiency can have serious consequences. Whereas most mammals can synthesize ascorbic acid (vitamin C), which is necessary for building connective tissue, primates (including humans) cannot. Without ingesting sufficient vitamin C, humans develop scurvy, a disease characterized by bleeding gums, loss of teeth, and slow wound healing. Until it was discovered that green vegetables and fruit supply vitamin C, scurvy was a common disease that could lead to death for sailors who were deprived of fresh produce for long periods of time while at sea.

Deficiencies of vitamins B_1, B_6, and B_{12} can cause nervous system disorders and various forms of anemia (evidenced by low levels of red blood cells or hemoglobin in the blood). Vitamin D is essential for the absorption of calcium in the diet and thus for skeletal growth and health. With adequate exposure to ultraviolet solar radiation, skin cells synthesize enough vitamin D to sustain a growing body. However, people inhabiting more northern regions of the world do not get a lot of sunlight. As a result, they produce less vitamin D and require more of it in their diet. Lack of adequate vitamin D can lead to rickets, a disease often seen in malnourished children in which

FIG. 40.6 **Commonly recognized minerals required by humans.** Dietary minerals, shown in red, are chemical elements other than carbon, hydrogen, oxygen, and nitrogen that are required in the diet.

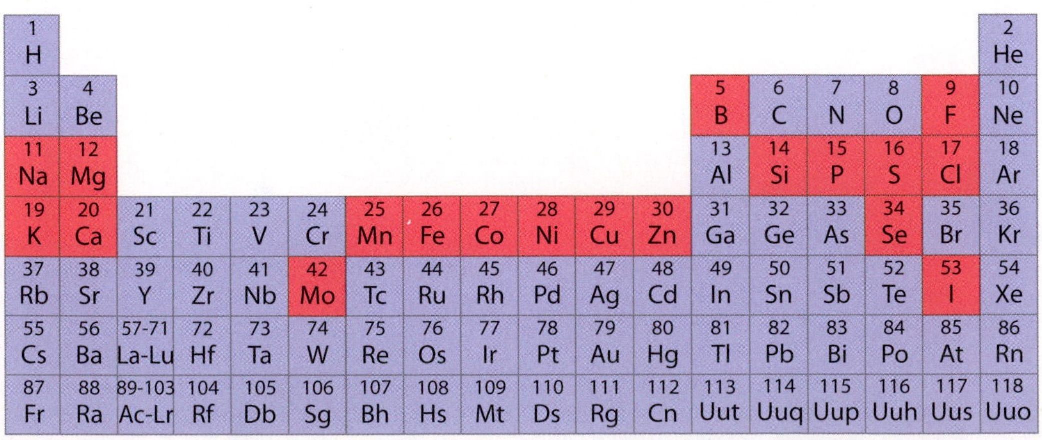

TABLE 40.2 Essential Vitamins for Humans

VITAMIN DESIGNATION	NAME	COMMON SOURCES
Vitamin A	Retinol	Green and orange vegetables, liver, milk
Vitamin B_1	Thiamine	Pork, grains
Vitamin B_2	Riboflavin	Dairy products, eggs, leafy green vegetables
Vitamin B_3	Niacin	Meat, dairy products, eggs, vegetables
Vitamin B_5	Pantothenic acid	Meat, grains
Vitamin B_6	Pyridoxine	Meat, grains
Vitamin B_7	Biotin	Meat, eggs
Vitamin B_9	Folic acid	Leafy green vegetables, eggs
Vitamin B_{12}	Cobalamin	Liver, eggs
Vitamin C	Ascorbic acid	Fruit, vegetables
Vitamin D	Calciferol	Fish, eggs
Vitamin E	Tocopherol	Fruit, vegetable oils
Vitamin K	Phylloquinone	Leafy green vegetables

bones do not mineralize properly. A clear role for vitamin E remains uncertain, but its absence is sometimes linked to anemia.

→ **Quick Check 3** In the context of nutrition, what does "essential" mean?

40.3 ADAPTATIONS FOR FEEDING

Animals have evolved a variety of adaptations to enhance their ability to acquire energy and nutrients from the environment. Animals eat other organisms, whether they are other animals, plants, or fungi. Herbivores are adapted to eat plants; carnivores are adapted to eat animals; and omnivores are adapted to eat both. To acquire energy and nutrients effectively from the food that they eat, animals have evolved a diversity of ways to capture food and mechanisms for breaking down that food before its digestion in the gut.

Suspension filter feeding is common in many aquatic animals.

The most common form of food capture by animals is **suspension filter feeding,** in which water with food suspended in it is passed through a sievelike structure (**Fig. 40.7**). Suspension filter feeding is possible only in aquatic environments, where it is common in a great diversity of animals found in oceans and freshwater bodies. This form of feeding evolved multiple times independently on many different branches of the tree of life, and so these occurrences of filter feeding are examples of convergent evolution (Chapter 23).

Many worms and bivalve mollusks, such as scallops, clams, and oysters, pump water over their gills to trap food particles suspended in the water (Fig. 40.7a). Cells of the gills produce mucus that acts as an adhesive to trap food particles, and the cells' beating cilia sweep food trapped in the mucus to the mouth and gut.

Other aquatic organisms move water with food particles suspended in it through filters in their oral cavity to separate the food from the water, and then convey the food into the gut. Large baleen whales have long comblike blades of baleen (which is made of keratin, the same material that makes up hair) in their mouth to capture small shrimp called krill that are plentiful in ocean waters (Fig. 40.7b).

FIG. 40.7 Suspension filter feeding by (a) a scallop and (b) a baleen whale. *Sources: a. Andrew J. Martinez/SeaPics.com; b. Cornforth Images/Alamy.*

FIG. 40.8 Suction feeding. This form of feeding involves a rapid expansion of the fish's mouth that draws in water and prey. *Source: Marevision/Getty Images.*

Large aquatic animals apprehend prey by suction feeding and active swimming.

Aquatic animals that do not rely on suspension filter feeding capture their prey in other ways. Many fish feed by suction. A rapid expansion of the fish's mouth cavity draws water and the desired prey into the mouth (**Fig. 40.8**). After the fish closes its mouth, the water is pumped out of the mouth past the gills. The prey is trapped inside the mouth, moves into the pharynx (part of the throat), and is broken up by specialized pharyngeal jaws (a second set of jaws in the throat) before being swallowed. Suction feeding is the most common form of prey capture among fish, as it allows them to be "sit-and-wait" predators, hiding within a coral reef or under a rock before rapidly striking to capture prey moving in front of them. Suction feeding is effectively used by a wide variety of fish, as well as by aquatic salamanders, and has contributed to their evolutionary diversification and success.

Many insects that bite to obtain a blood meal rely on suction to draw the blood of their prey into their digestive system. Young mammals feed by suckling milk from their mother's breast, using their tongue and also generating suction as fish do.

As top predators in marine and freshwater environments, larger fish and marine mammals like sharks, whales, and dolphins actively swim to capture their prey (**Fig. 40.9**). Their larger size and speed enables them to catch smaller or similar-sized prey. As discussed below, the evolution of jaws and teeth considerably enhanced the ability of these and other vertebrate animals to capture new kinds of food, allowing them to lead a more active and energetically demanding lifestyle.

Jaws and teeth provide specialized food capture and mechanical breakdown of food.

Jaws and teeth were an important evolutionary innovation for active predators. Some of the first vertebrates, the jawed fishes, became dominant in their aquatic environment through the ability to swim and bite forcefully to obtain their food. **Fig. 40.10** shows a phylogenetic tree of vertebrates indicating the appearance of jaws. Jawed fish evolved from jawless ancestors, and jaws can

FIG. 40.9 Feeding by active swimming. Large marine predators, like this great white shark, can use their size and speed to capture prey. *Source: C & M Fallows/SeaPics.com.*

FIG. 40.10 A phylogeny of vertebrates, showing the appearance of jaws. Jaws, which evolved more than 400 million years ago from a group of jawless ancestors, contributed to the success of many groups of animals.

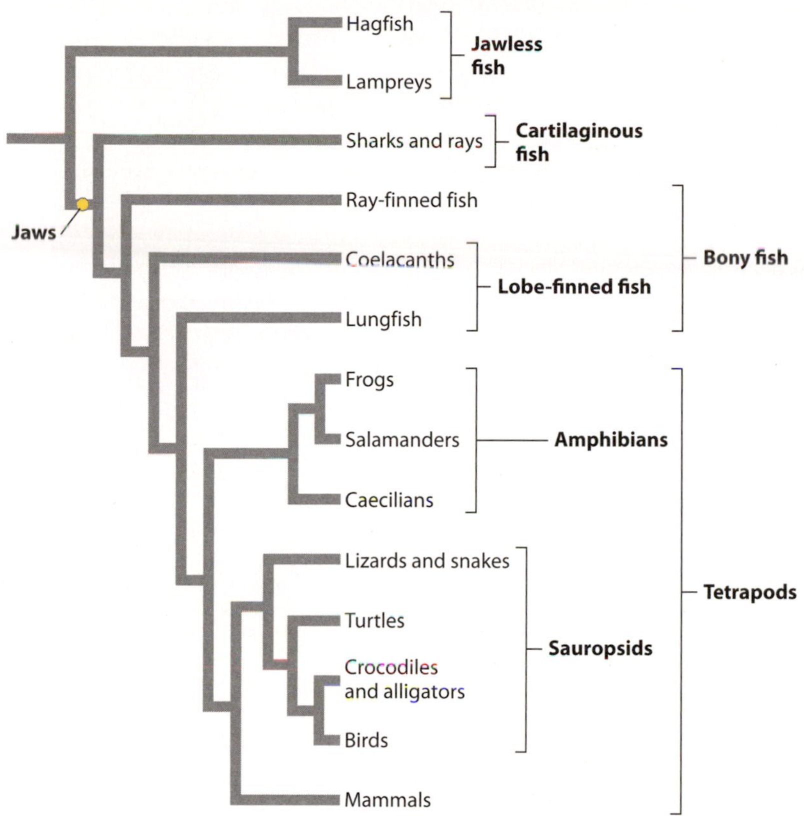

be found in present-day fish, amphibians, reptiles, and mammals. As you can see from the large number of jawed groups, jaws are key to the evolutionary success of vertebrates. Jaws evolved from cartilage that supported the gills, providing a spectacular example of an organ adapted for one function (gill support) changing over time to become adapted for an entirely different function (predation).

Among vertebrates, mammals evolved a specialized jaw joint, the **temporomandibular joint,** as well as a great diversity of specialized forms of teeth. These teeth have cutting and crushing surfaces, enabling mammals to break down a variety of foods mechanically before swallowing. **Fig. 40.11** illustrates the arrangement of specialized teeth in mammals with different diets. Teeth in the front of the mouth, the **incisors,** are specialized for biting. Others, such as the **canines** of dogs, cats, and other carnivores, are specialized for piercing the body of their prey. Saber-toothed cats, which lived as recently as 10,000 years ago, had remarkably long canines (up to 16 inches or 40 cm!) specialized for killing large prey. **Premolars** and **molars** are teeth in the back of the mouth that are well adapted for crushing and shredding tougher foods, such as meat and fibrous plant material.

Herbivorous animals, such as cattle, sheep, and horses, have specialized premolars and molars with prominent surface ridges. These teeth can shred tough plant material before it is swallowed and digested.

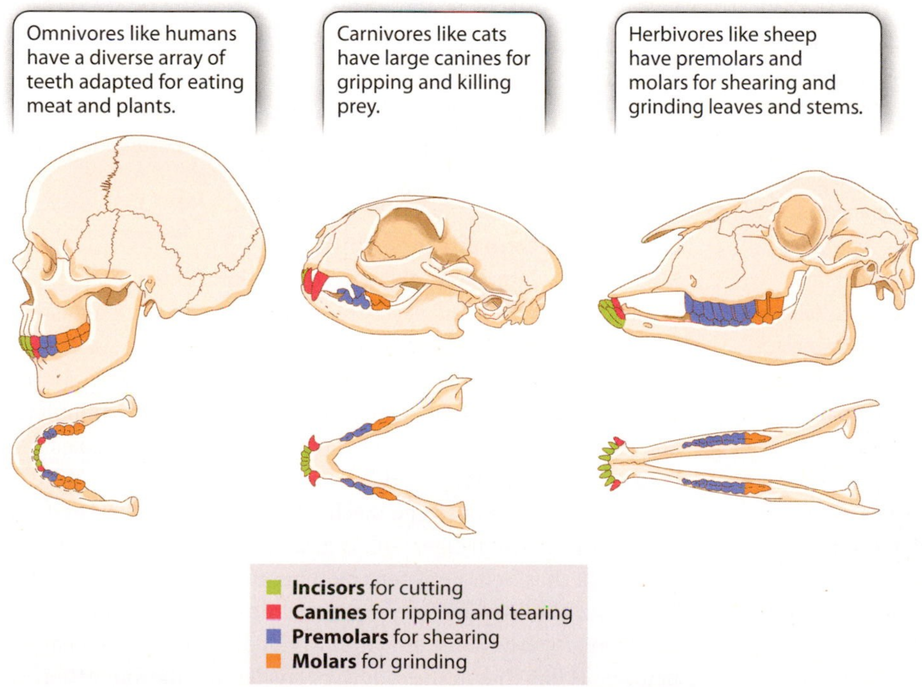

FIG. 40.11 Jaws and teeth of carnivorous, herbivorous, and omnivorous mammals. Mammals evolved a diverse arrangement of teeth for diverse diets.

Mammalian herbivores use their front incisors and canines to bite grasses and leaves, and then move the food to the back of the mouth, where it is ground and crushed between their ridged premolars and molars.

40.4 DIGESTION AND ABSORPTION OF FOOD

After animals obtain food, they break it down in the process of digestion. The food is isolated in a specialized organelle or compartment so that it can be broken down chemically without damaging other organelles or structures of the body. Organisms on different branches of the tree of life have evolved very different means of isolating and digesting food. The process in single-celled protists, which obtain food particles by phagocytosis, is **intracellular digestion,** in which food is broken down within cells. Lysosomes containing hydrolytic enzymes fuse with vacuoles containing food particles, so the enzymes are mixed with the food (Chapter 5). Following their chemical breakdown, the food products of intracellular digestion are available for use by the cell. In most animals, however, the process is **extracellular digestion,** in which food is isolated and broken down outside a cell in a body compartment. Following digestion, the breakdown products are taken up into the bloodstream in a process called **absorption.**

FIG. 40.12 Organization of animal digestive tracts: foregut, midgut, and hindgut.

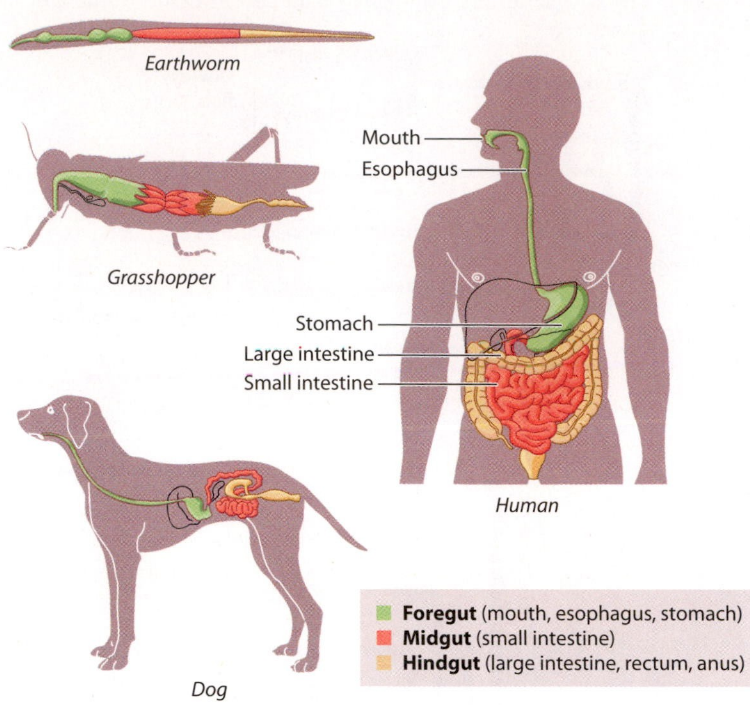

The digestive tract has regional specializations.

In extracellular digestion, food is broken down in a specialized cavity or body compartment. Some animals, like sea anemones, carry out digestion in an internal cavity. Food flows directly into this cavity through the mouth. Cells lining the cavity secrete digestive enzymes that break down the food. These very same cells then absorb the smaller chemical compounds that are produced.

Other animals have evolved more elaborate digestive systems that allow the transport of food by a digestive tube that runs from an animal's mouth to its anus. Collectively, the passages that connect the mouth, digestive organs, and anus constitute an animal's **gut** or **digestive tract.** Because the food is moved in a single direction through a tubelike gut, particular regions of the digestive tract are specialized for different functions. These functions include the mechanical and chemical breakdown of food, absorption of released nutrients, and storage and elimination of waste products.

These various functions take place in the three main parts of the digestive tract: the **foregut, midgut,** and **hindgut** (**Fig. 40.12**). Most animals' digestive systems have the same basic plan, though some animals have also evolved specialized structures. The foregut includes the **mouth, esophagus,** and **stomach** or **crop,** which serves as an initial storage and digestive chamber. Next is the midgut, which includes the **small intestine,** where the remainder of digestion and most nutrient absorption takes place. Here, specialized organs secrete enzymes and other chemicals that aid in the breakdown of particular macromolecules, such as fats and carbohydrates. Finally, the digested material reaches the hindgut, which includes the **large intestine** and **rectum.** In the hindgut, water and inorganic molecules are absorbed, leaving the waste products, or feces, which are stored in the rectum until being eliminated from the body.

Food and its breakdown products move through the gastrointestinal tract by **peristalsis,** waves of smooth muscle contraction and relaxation. These waves keep substances moving from one end of the tract to the other, and help to prevent backward movement.

Digestion begins in the mouth.

Most animals jump-start digestion by breaking down food mechanically, which increases the surface area exposed to taste buds in the tongue and available for enzyme action. As we saw in the previous section, many animals break down food with the aid of jaws and teeth. Insects, which lack jaws and teeth, manipulate and break down their food with mouthparts called mandibles before it passes into their gut.

After being ingested and mechanically broken down, the food is ready for chemical digestion. Digestion is carried out by enzymes that break the bonds between the subunits of large molecules. These enzymes require the proper chemical environment to function properly: Some enzymes work best at neutral or near neutral pH like that in the mouth or small intestine, but other

TABLE 40.3 Major Sites of Chemical Digestion of Macromolecules

MACROMOLECULE	SITE OF DIGESTION	ENZYME
Protein	Stomach	Pepsin
	Small intestine	Trypsin
Carbohydrate	Mouth	Salivary amylase
	Small intestine	Pancreatic amylase, lactase
Lipid	Mouth	Lingual (tongue) lipase
	Stomachs	Gastric (stomach) lipase
	Small intestine	Pancreatic lipase

enzymes work best in acidic environments like that of the stomach. This need for different environments is met by carrying out digestion in different compartments. In some animals, such as mammals, chemical digestion begins in the mouth. Food entering the mouth is mixed with salivary secretions that contain **amylase,** an enzyme that breaks down carbohydrates, and secretions from the tongue that contain **lipase,** an enzyme that breaks down lipids (**Table 40.3**).

The muscular tongue of mammals and other land vertebrates manipulates food and transports it within the mouth cavity. The tongue moves the food into position within the mouth for effective cutting and maceration into smaller pieces by the teeth. When food has been chewed and is ready to be swallowed, the tongue moves it to the rear of the mouth cavity.

Swallowing is a complex set of motor reflexes carried out by several muscles and structures in the rear of the mouth and in the **pharynx,** the region of the throat that connects the nasal and mouth cavities (**Fig. 40.13**). Once initiated, swallowing reflexes are involuntary, under the control of the autonomic nervous system. During swallowing, food is pushed through the pharynx, over the **epiglottis,** a flap of tissue that protects food from entering the trachea and lungs, and into the esophagus, a tube that connects the pharynx to the stomach. In the esophagus, peristalsis moves the food toward the stomach.

Birds, alligators, crocodiles, and earthworms break down food into smaller pieces farther along their digestive tracts in the **gizzard,** a compartment with thick muscular walls. Birds and earthworms often ingest small rocks or sediment that help grind the food into smaller pieces within the muscular gizzard. Breaking food into smaller sizes aids the chemical digestion that follows.

→ **Quick Check 4** The rocks and sediments in a gizzard that break apart food mechanically play a role similar to what structures in vertebrates?

Further digestion and storage of nutrients take place in the stomach.

Digestion, begun in the mouth, continues in the stomach. The stomach is one of the main sites of protein and lipid breakdown. The acidic environment of the stomach facilitates digestion. In

FIG. 40.13 **Swallowing.** Swallowing moves food from the mouth to the stomach.

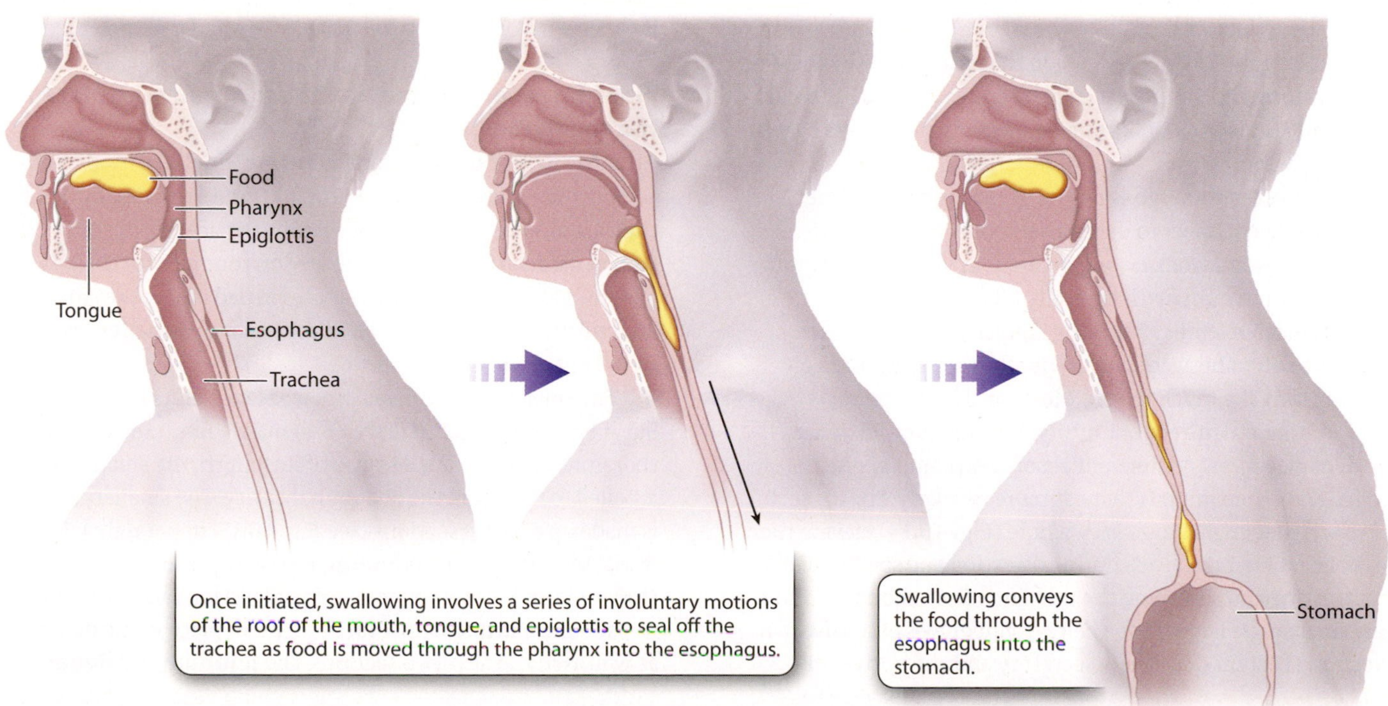

addition, several enzymes have evolved to act at a very low pH of 1–3, that found in the stomach. Most enzymes, by contrast, are denatured (that is, unfolded and thus made inactive) in acidic environments.

The stomach is maintained at a low pH through the secretion of hydrochloric acid (HCl) by specific cells lining the stomach. A low pH could damage the stomach when no food is present, so the amount of HCl is carefully regulated. The sight, smell, and taste of food send signals to the brain that stimulate a sense of appetite, and to the stomach that stimulate the secretion of HCl and digestive enzymes that break down proteins and lipids.

When food arrives in the stomach and digestion there begins, cells lining the stomach secrete a peptide hormone called **gastrin.** Gastrin stimulates the cells lining the stomach to increase their production of HCl still further. If the pH of the stomach becomes too low, gastrin secretion is inhibited, an example of a negative feedback loop (Chapter 38).

The stomach has several mechanisms to protect itself from the acidic environment and presence of digestive enzymes. For example, glands in the lining of the stomach secrete mucus to protect the stomach wall. In addition, cells in the stomach secrete some digestive enzymes in an inactive form; otherwise, the cells themselves would be digested. After secretion, the inactive enzymes are activated by a change in their chemical structure. The primary digestive enzyme produced in the stomach is **pepsin,** an enzyme that breaks down proteins into amino acids (Table 40.3). Cells lining the stomach release pepsin in an inactive form called pepsinogen that is activated by the low pH of the stomach. The stomach also secretes lipases that break down lipids (Table 40.3). Like pepsin, gastric lipase works best in an acidic environment.

Animals often eat large amounts of food in a short time, but it takes much longer to digest that food. Therefore, the stomach serves as a storage compartment as well as a digestion compartment. In humans, it typically takes about four hours for the stomach to empty, allowing digestion and absorption of nutrients to occur between meals. In other animals, stomach emptying can take much longer. Most carnivorous animals consume large and infrequent meals that are followed by long periods of digestion and nutrient absorption compared with humans and other animals that eat more regularly. Snakes such as pythons that engulf large prey whole (**Fig. 40.14**) elevate their metabolic rate to high levels during digestion and spend several days digesting and absorbing the nutrients from their meal. To digest their single large meal, their gut undergoes extensive remodeling that produces new secretory and absorptive cell surfaces. These extra cells are not retained between meals because of the high energy cost of maintaining them.

FIG. 40.14 **Python and prey.** Because the prey is eaten whole, the python's metabolic rate is high during digestion, which takes place over several days. *Source: Biosphoto/Ken Griffiths/Photoshot.*

The stomach walls contract to mix the contents of the stomach, aiding their digestion. Waves of muscular contraction also move the food toward the base of the stomach. There, the **pyloric sphincter,** a band of muscle, opens and allows small amounts of digested food to enter the small intestine. Opening and closing of the pyloric sphincter regulates the rate at which the stomach empties, allowing time for the food products released into the small intestine to be further digested and then absorbed.

Final digestion and nutrient absorption take place in the small intestine.

After the food enters the small intestine, final digestion of protein, carbohydrates, and fats occurs. The nutrients released by digestion then are absorbed by the body. Finally, the acid produced by the stomach is neutralized to create an environment for the intestinal enzymes to act. These functions are carried out by the small intestine with the help of two other organs, the **pancreas** and **liver** (**Fig. 40.15**).

The small intestine is so named because of its small diameter, but it is a long (about 20 feet or 6 m in an adult human) and therefore a large organ. Because of its length, the small intestine is coiled to fit within the abdominal cavity. The small intestine is made up of three sections (see Fig. 40.15, upper right). The initial section is the **duodenum,** into which the food enters from the stomach. Most of the remaining digestion takes place in the duodenum, and nutrient absorption takes place in the duodenum as well as in the next two sections, the **jejunum** and **ileum.**

FIG. 40.15 **The small intestine and associated organs.** Digestion in the small intestine is aided by secretions from the liver, gallbladder, and pancreas. The small intestine is the main site of nutrient absorption.

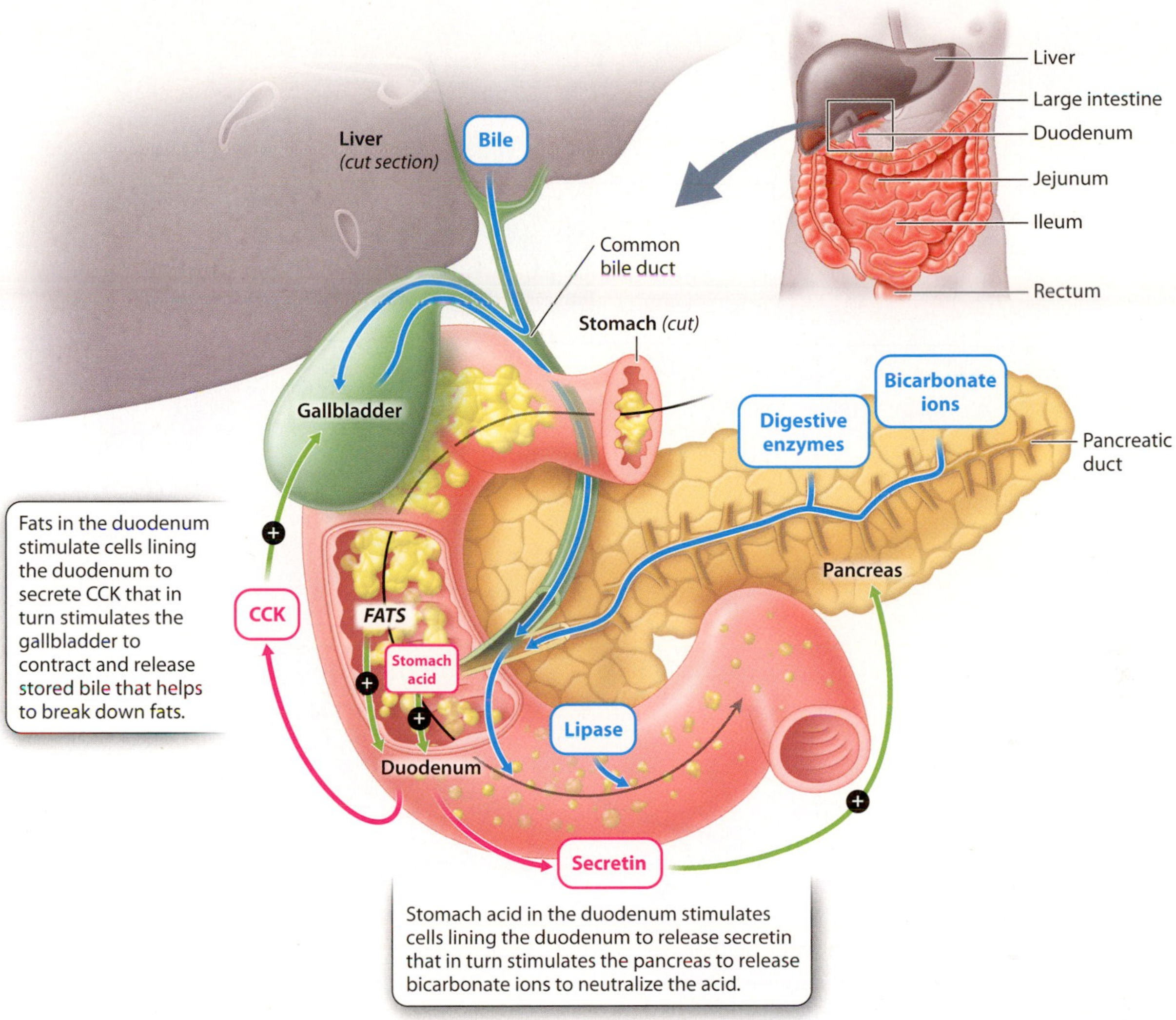

Some digestive enzymes are produced by the small intestine. For example, cells of the small intestine produce lactase, an enzyme that breaks down the sugar lactose into glucose and galactose (Table 40.3). Lactose is found in milk, and so lactase is important in newborn mammals during suckling. In humans, the gene that encodes lactase has an interesting evolutionary history. Some people cannot drink milk as adults, a condition called lactose intolerance. In these individuals, the gene that encodes lactase is down-regulated and therefore not expressed after weaning. Other people can continue to drink milk as adults. In these individuals, the gene for lactase continues to be expressed. Researchers have found that the continued expression of lactase in these individuals is the result of mutations that occurred and then were selected for when cattle were domesticated and milk became part of the diet 10,000–5000 years ago.

Many of the digestive enzymes that act in the small intestine are not produced by the small intestine, but instead by the pancreas. The pancreas is an organ that lies below the stomach, behind the duodenum (Fig. 40.15). It secretes hormones directly into the blood, as discussed in Chapter 38, and secretes other substances into ducts that connect to the duodenum. The pancreas produces a variety of digestive enzymes, including lipase, which breaks down fats; amylase, which breaks down carbohydrates; and **trypsin,** which breaks down proteins (Table 40.3). Like pepsin produced by the stomach, the pancreas produces trypsin in an inactive form called trypsinogen to avoid digesting itself. Trypsinogen is converted to trypsin by an enzyme produced by intestinal cells.

The pancreas also secretes bicarbonate ions, which neutralize the acid produced by the stomach. The pancreas is stimulated

FIG. 40.16 **Intestinal villi and microvilli.** These structures increase the surface area for absorption of nutrients. *Photo source: Biophoto Associates/ Science Source.*

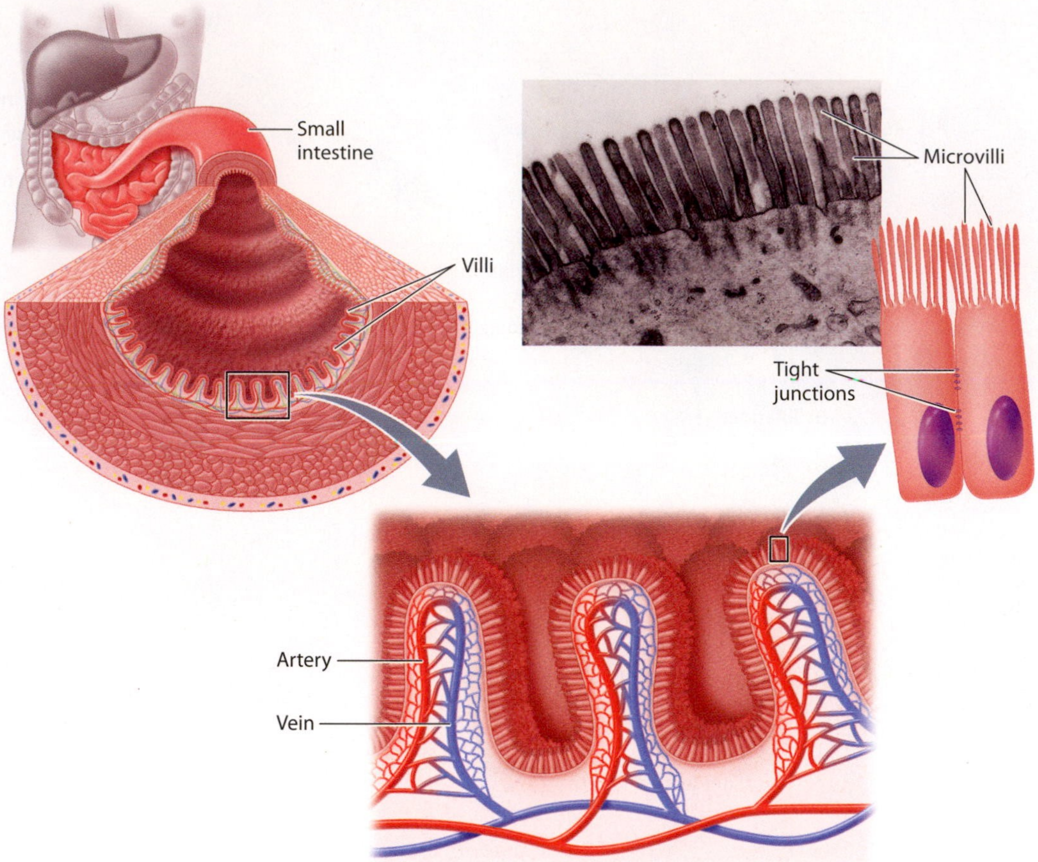

to secrete bicarbonate ions by the hormone **secretin** (the first hormone to be identified). Cells lining the duodenum secrete the hormone in response to the acidic pH of the stomach contents entering the small intestine.

The liver participates in digestion by producing **bile,** which is composed of bile salts, bile acids, and bicarbonate ions. Bile aids in fat digestion by breaking large clusters of fats into smaller lipid droplets, a process called emulsification. Bile produced by the liver is stored in the **gallbladder** (Fig. 40.15). When fats enter the duodenum, cells lining the duodenum release a peptide hormone called **cholecystokinin (CCK),** which causes the gallbladder to contract, thus releasing the bile into the duodenum. By breaking up the fats into smaller droplets, bile allows lipases to digest the lipids more effectively.

When large macromolecules are broken down by digestive enzymes, they are ready to be absorbed, another function of the small intestine. The small intestine has highly folded inner surfaces, called **villi,** and the cells constituting the villi themselves have highly folded surfaces, called **microvilli** (**Fig. 40.16**). Together, villi and microvilli greatly increase the surface area for the absorption of nutrients. Cells that line the small intestine are connected by tight junctions (Fig. 40.16; Chapter 10). These junctions force the products of digestion to be absorbed across the microvilli surfaces, rather than leaking between the cells. Membrane transporters can then control their movement through the cell and into the bloodstream.

Nutrient molecules, such as glucose and amino acids, are often co-transported into cells lining the intestine with sodium ions from gut contents. A sodium ion binds to a transmembrane protein on the surface of cells lining the intestine that also binds the nutrient molecule. Absorption of the nutrient molecule into the cell is driven by the movement of the sodium ion down its concentration gradient (**Fig. 40.17**). The concentration of sodium ions inside the cell is kept low by the action of the sodium-potassium pump (Chapter 5). Molecules such as glucose are then moved from the intestinal cell into the blood by a pump on the basal side of the cell. Because the products of fat digestion—fatty acids and glycerol—are lipid soluble, they do not require a carrier protein for transport. Instead, they readily diffuse across the plasma membrane, entering the mucosal cell, and are eventually transported to the bloodstream. There, lipids bind molecules that allow them to be soluble and therefore transported in the aqueous (water-based) blood.

FIG. 40.17 Glucose absorption by cells of the small intestine. Absorption of glucose into the cell is driven by the movement of the sodium ions down a concentration gradient.

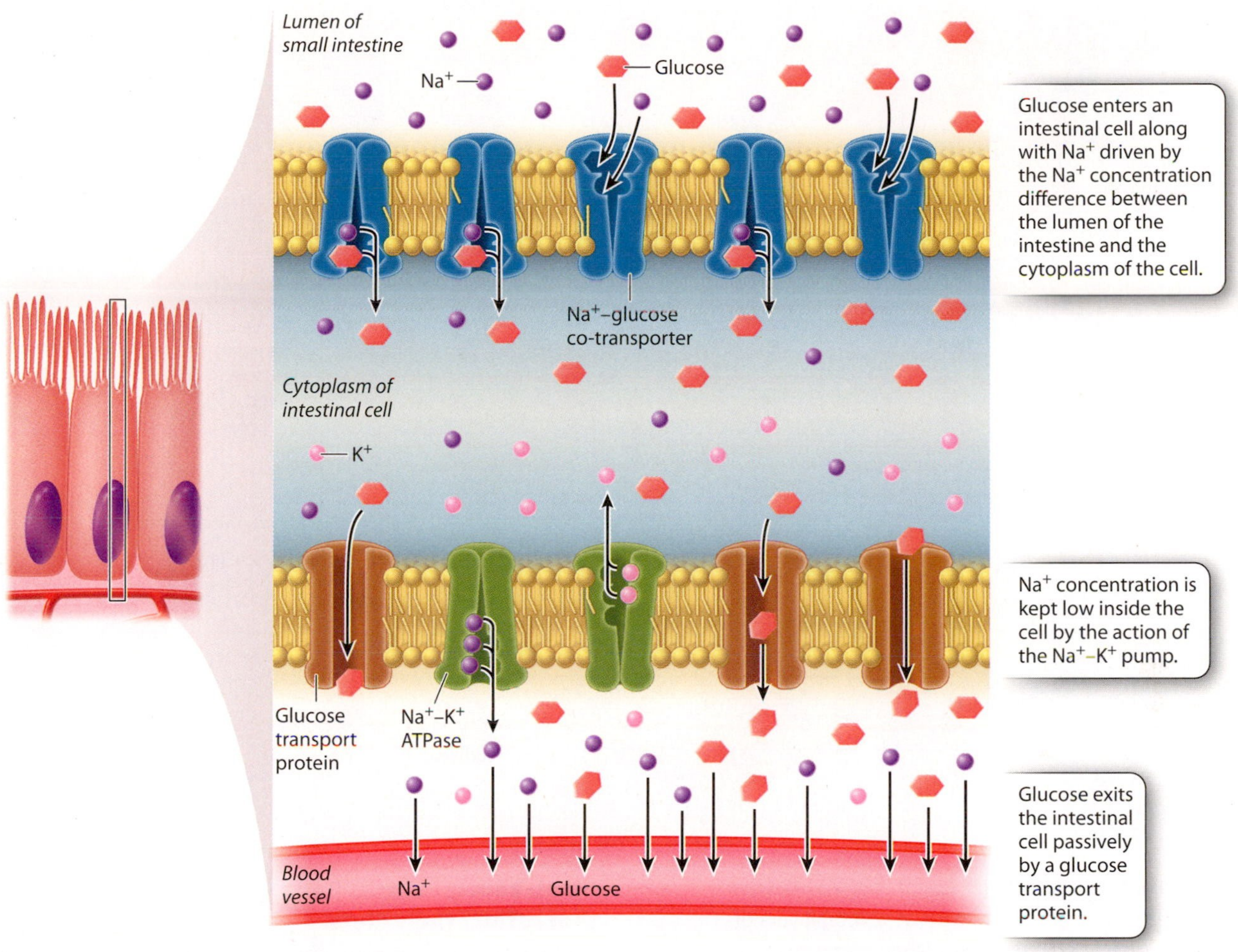

The large intestine absorbs water and stores waste.

After the absorption of nutrients in the small intestine, water and inorganic ions are absorbed in the large intestine, or **colon,** of the hindgut (see Fig. 40.12). Peristalsis moves the contents of the small intestine into the colon at a pace that allows time for nutrient absorption in the small intestine. By the time the gut contents reach the large intestine, the nutrients have been absorbed into the body, but water and inorganic ions remain. These are absorbed in the large intestine until the contents form semisolid feces. Excess water absorption can cause constipation, whereas too little water absorption results in diarrhea. Diarrhea and constipation also can be caused by toxins released by microorganisms, such as the bacterium *Vibrio cholera,* the cause of cholera, in the food an animal eats. Feces are stored in the final segment of the colon, the rectum, until they are periodically eliminated through the anus.

As we saw in Case 5: The Human Microbiome, digestion is not a solo endeavor. Large populations of bacteria reside in the small and large intestines and help extract nutrients that the animal's body cannot extract itself. The principle gut resident is *Escherichia coli,* the bacterium commonly used for research in molecular biology laboratories. In the process of nourishing themselves by aiding in the digestion of the host's gut contents, the bacteria provide nutrients and certain vitamins, such as biotin and vitamin K, that the animal cannot produce itself. Vitamin deficiency therefore can sometimes result from prolonged antibiotic medication that kills large numbers of gut bacteria. Another by-product of bacterial action in the gut is gases such as methane and hydrogen sulfide. These are expelled when foods that, like beans, contain compounds that the bacteria can digest but that the host animal cannot. The mutual benefits to the host animal and the bacteria ensure the success of their symbiotic relationship.

The lining of the digestive tract is composed of distinct layers.

As we have seen, the digestive tract is not a passive tube. It secretes enzymes and other chemical compounds for digestion, absorbs nutrients, and actively moves food through the body.

FIG. 40.18 Cross section of the mammalian intestine.

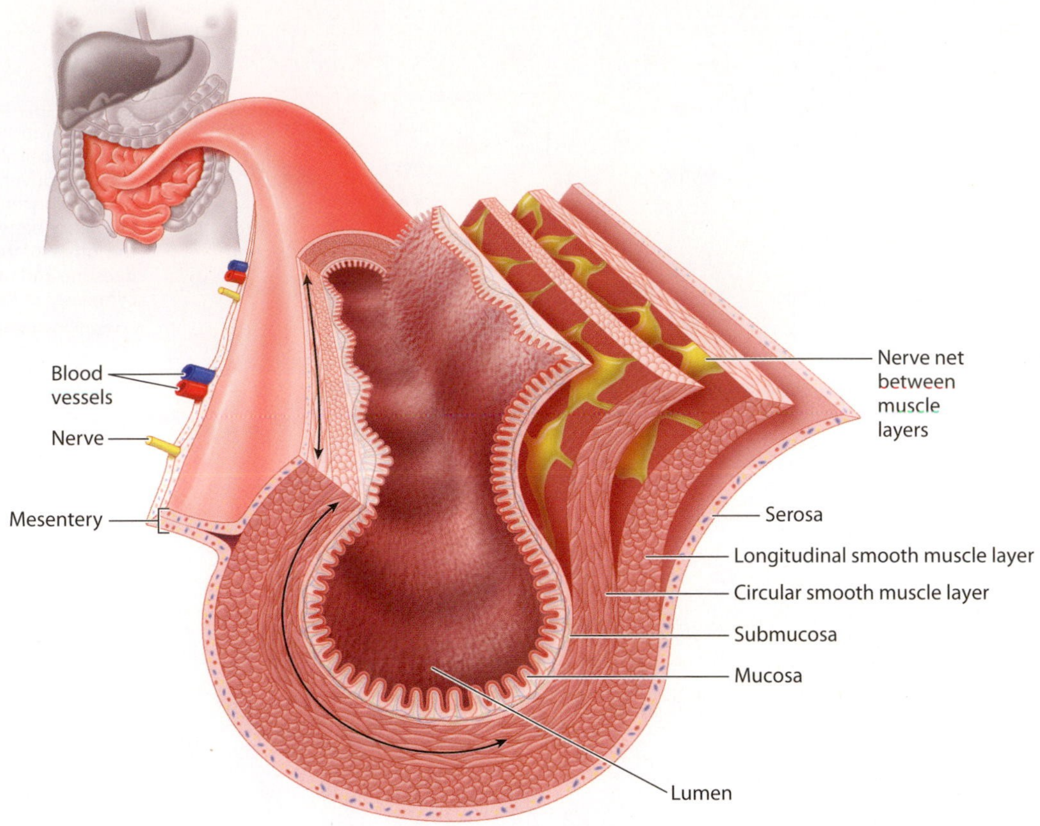

How does a series of tubes accomplish all of this? As shown in **Fig. 40.18,** the digestive tract is made up of several layers of tissue, each with a specialized function. The central space, or **lumen,** through which the gut contents travel is surrounded by an inner tissue layer, the **mucosa,** which both secretes enzymes and absorbs nutrients. The cells of the mucosa secrete mucus to protect the gut wall from digestive enzymes, and, in the stomach, hydrochloric acid. Surrounding the mucosa is the **submucosa,** a layer containing blood vessels, lymph vessels, and nerves.

Outside these layers are two smooth muscle layers (Fig. 40.18). An inner **circular muscle** layer contracts to reduce the size of the lumen. An outer **longitudinal muscle** layer contracts to shorten small sections of the gut. These two muscle layers contract alternately to mix gut contents, producing a traveling wave during peristalsis that moves the contents along the digestive tract from compartment to compartment. Between the two muscle layers are autonomic nerves that control the contractions of the two sets of smooth muscle.

An outer layer of cells and connective tissue called the **serosa** covers and protects the gut (Fig. 40.18). The gut is supported in the abdominal cavity by a membrane called the **mesentery,** through which blood vessels, nerves, and lymph travel to supply the gut.

Plant-eating animals have specialized digestive tracts that reflect their diets.

Herbivorous (that is, plant-eating) animals evolved specialized digestive tracts that enhance their ability to digest cellulose and other plant compounds. Animals, including herbivores, lack **cellulase,** the enzyme that breaks down cellulose. Instead, they have specialized compartments in their digestive tract that contain large bacterial populations, which do produce cellulase. As a result, the bacteria are able to digest the cellulose. This association of herbivores and their gut microbes is another example of a symbiotic relationship.

Rather than the single stomach of other mammals, ruminants (cattle, sheep, and goats) have a four-chambered stomach that is highly specialized to enhance the ability of their gut bacteria to digest plants (**Fig. 40.19a**). The first two chambers—the **rumen** and the **reticulum**—harbor large populations of anaerobic bacteria that break down cellulose. To facilitate breakdown, the rumen contents are frequently regurgitated (in a process called rumination) for further chewing in the mouth. This repeated chewing decreases the size of the plant fibers, increasing the surface area available for bacterial digestion in the rumen. The bacteria break down cellulose by fermentation, and some of the products are used as nutrients

FIG. 40.19 Digestive systems of (a) a foregut fermenter and (b) a hindgut fermenter.

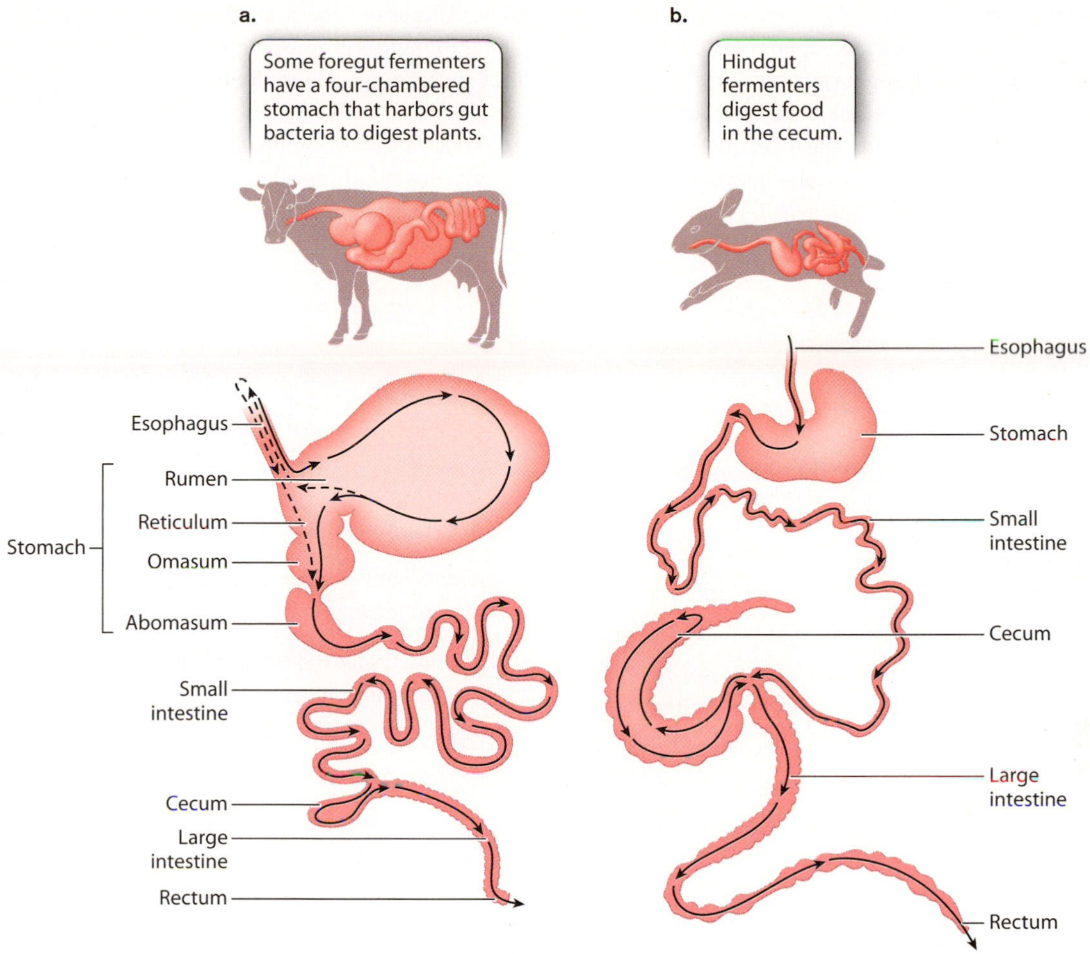

by the host. (The site of fermentation provides the name "foregut fermenters" for these mammals.)

Carbon dioxide and methane gas are also produced as a result of bacterial fermentation. The amount of methane produced by domesticated ruminants is second only to the amount produced industrially. Since methane is a greenhouse gas that traps heat from the sun, methane production by cattle, sheep, and other animals contributes to atmospheric increases in methane and global climate change (Chapter 25).

After leaving the reticulum, the mixture of food and bacteria passes into a third chamber, the **omasum,** where water is absorbed. It then enters the **abomasum,** the last of the four chambers, where protein digestion begins as it does in other mammals. The acid and protease secretions kill the bacteria, and the nutrients are absorbed in the small intestine. The bacteria, too, are an important protein source for the ruminant, yielding as much as 100 g of protein each day for the cow. This is important because plants are a poor source of protein but contain considerable nitrogen that the bacteria use to synthesize their own amino acids. The bacteria lost to digestion are balanced by the continual reproduction of bacteria in the rumen.

In contrast to the foregut fermentation of cows and sheep, other mammalian herbivores, such as horses, rabbits, and koalas, digest the plant material they eat by hindgut fermentation (**Fig. 40.19b**). Hindgut fermentation occurs in the colon and in the **cecum,** a chamber that branches off the large intestine. Because the fermentation products released from the cecum have already passed through the small intestine, the main site of nutrient absorption, hindgut fermentation is a less efficient nutrient extractor than foregut fermentation.

In humans and other animals such as lemurs and squirrels, the cecum is present but small and includes the **appendix,** a narrow, tubelike structure that extends from the cecum. The appendix is an example of a **vestigial structure,** a structure that has lost its original function over time and is now much reduced in size. Although the appendix has no clear function in nonherbivorous animals, evidence suggests it may play a role in the immune system (Chapter 43). The appendix can become infected when partially digested material enters or blocks it. Unless the infected appendix is removed, it can burst, releasing gut contents into the abdominal cavity, a condition that can be life threatening.

Charles Darwin put forth a hypothesis explaining how the appendix might have evolved to take on its current form in humans. He suggested that in the ancestors of modern humans the cecum harbored bacteria that aided in the digestion of plant material such as cellulose. However, as their diet became less reliant on plant material, the cecum became smaller and lost its function over time, becoming what we see today as the appendix. Vestigial structures, such as the appendix in humans, remnants of hind limbs in whales, small eyes in cave-dwelling animals, and wings in flightless birds, provide an illuminating window into the past and strong evidence for evolution.

→ **Quick Check 5** Some animals, such as rabbits, whose digestive systems include hindgut fermentation, re-ingest the products of digestion. What function do you think this serves? ■

Core Concepts Summary

40.1 Metabolic rate depends on activity level, body size, and body temperature.

Animals acquire carbohydrates, fats, and protein from their diet for energy to fuel cellular processes and activities. page 852

Carbohydrates are broken down rapidly in the cytosol by anaerobic glycolysis to produce a small amount of ATP in the absence of oxygen. page 852

Carbohydrates and fats produce large amounts of ATP by aerobic respiration through the citric acid cycle and oxidative phosphorylation. page 852

An animal's metabolic rate is its overall rate of energy use. page 854

Metabolic rate increases with increased physical activity. page 854

Larger animals consume more energy and have higher metabolic rates than smaller ones, but the increase in metabolic rate is less than proportional to mass. page 855

Endothermic animals maintain elevated body temperature by metabolic heat production and have high metabolic rates. page 855

Ectothermic animals depend on external heat sources to warm their bodies. They have lower metabolic rates compared to endothermic animals. page 855

40.2 An animal's diet supplies the energy the animal needs for homeostasis and essential nutrients it cannot synthesize on its own.

An animal is in energy balance when the amount of energy that it takes in is equal to the amount it uses to sustain life and perform its functions. page 857

Animals must ingest essential amino acids, minerals, and vitamins, those that cannot be synthesized in their cells. page 857

40.3 Animals have different adaptations for feeding.

Aquatic animals take in food by suspension filter feeding, suction feeding, and active swimming. page 861

Vertebrates have specialized jaws and teeth for biting and mechanically processing food. page 862

Carnivorous animals have enlarged canines and slicing molars for catching prey and eating meat. page 863

Herbivorous animals have molars and premolars with surface ridges that enable them to grind plant matter. page 863

40.4 The digestive tract has regions specialized for different functions, including digestion, absorption, storage, and elimination.

The animal digestive tract is a long tubelike structure that allows different regions to have different functions. page 864

Peristalsis mixes and moves food through the gut by means of smooth muscle contraction and relaxation. page 864

Most animals break down their food into small pieces in the mouth, making it easier to digest. page 864

Initial chemical digestion begins in the mouth. page 865

The stomach stores food and initiates protein digestion through secretions of hydrochloric acid and digestive enzymes, including pepsin. page 865

Proteins, carbohydrates, and lipids are further broken down in the duodenum of the small intestine, facilitated by the secretion of digestive enzymes by the pancreas and of bile, bile salts, and bicarbonate by the liver. page 866

Nutrient absorption takes place in the small intestine. page 866

Small intestine nutrient absorption is enhanced by an enlarged surface area of villi and smaller microvilli on the surface of cells. page 868

The large intestine absorbs water and inorganic molecules before waste products are eliminated as semisolid feces from the rectum through the anus. page 869

Herbivorous animals have specialized digestive chambers (the rumen and cecum) that enable bacterial fermentation to break down plant cellulose. page 870

Self-Assessment

1. Discuss how metabolic rate changes with levels of activity, body-temperature regulation, and size.
2. Describe how ATP is produced for short and rapid activities versus long and sustained activities.
3. Explain how you would measure the metabolic rate of an animal.
4. Explain why the world record speed for a 10-m race is faster than that for a marathon.
5. Describe what happens in an endotherm and an ectotherm when outside temperature gets cold, and what happens when it gets hot.
6. Describe the order in which energy reserves are used in negative energy balance, such as starvation.
7. Name four ways in which animals capture prey and, for each, name an organism that uses it.
8. Draw the path of food through the vertebrate digestive tract, naming and describing the major function of each part.
9. Name the principal sites of digestion of proteins, carbohydrates, and fats, and indicate the principal enzymes involved in breaking down each.
10. Compare and contrast foregut and hindgut fermentation.

Log in to LaunchPad to check your answers to the Self-Assessment questions, and to access additional learning tools.

CHAPTER 41

Animal Renal Systems

Water and Waste

Core Concepts

41.1 All animals regulate the water and electrolyte levels within their cells.

41.2 Excretory organs eliminate nitrogenous wastes and regulate water and electrolyte levels.

41.3 The mammalian kidney can produce urine that is more concentrated than blood as an adaptation for living on land.

Source: Steve Allen/Getty Images.

In the 1960s, researchers at the University of Florida developed a new sports drink for players on the football team, the Gators. In addition to water and sugar, the drink contained electrolytes, ions such as sodium (Na^+) and potassium (K^+) that give the solution electrical properties. The researchers who came up with the formula for the drink reasoned that it was important to replace both water and electrolytes lost in sweat. The drink—called Gatorade—is now well known.

Water and electrolytes are two key substances that allow cells and organisms to maintain their size and shape, as well as to carry out essential functions. Life originated and diversified in a watery environment. Consequently, the chemical functions of cells and organisms crucially depend on the properties of water (Chapter 2) and the relative amounts of water and electrolytes inside a cell. In general, the internal environment of a cell, including the concentration of electrolytes, is maintained within a narrow window of conditions. The active maintenance of a stable internal environment in the face of a changing external environment is known as **homeostasis** (Chapter 5).

Animals must also maintain homeostasis in terms of the molecules produced by normal metabolic activity. Metabolism generates wastes that are not needed and, in many cases, can be toxic to the cell. These must be eliminated. In most organisms, the elimination of wastes is closely tied to the maintenance of water and electrolyte balance.

In this chapter, we explore the physiological and cellular mechanisms that underlie water and electrolyte balance, as well as waste elimination. We examine how animals meet the demands for water and electrolyte balance in different environments. We also consider how water and electrolyte balance is linked to waste elimination and explore the diversity of excretory organs animals have evolved to regulate these processes. The organization and function of mammalian excretory organs, the kidneys, is covered in the last section of the chapter.

41.1 WATER AND ELECTROLYTE BALANCE

A healthy person can live several weeks without food but cannot survive much more than a few days without water. A loss of more than 10% of body water threatens survival. Homeostasis of water and electrolyte concentration is critical to normal physiological function, not just in humans, but in all organisms.

Osmosis governs the movement of water across cell membranes.

Water, like all molecules, moves by diffusion, or the random motion of molecules (Chapter 5). When there is a difference in water concentration between two regions, there is a net movement of water molecules from the region of higher *water* concentration to the region of lower *water* concentration. Water often contains other dissolved molecules, such as electrolytes, amino acids, and sugars, which are collectively called **solutes.** Because we typically describe solutions in terms of the concentration of dissolved solutes, it is sometimes easier to think about water moving from a region of lower *solute* concentration (higher water concentration) to a region of higher *solute* concentration (lower water concentration). Whether considered in terms of water concentration or solute concentration, the direction of net water movement is the same.

A barrier can sometimes separate two solutions. When a barrier allows water or solutes to diffuse freely, it is said to be permeable. When it blocks the diffusion of water or solutes entirely, it is impermeable. When it allows the movement of some molecules but not others, it is **selectively permeable.**

In organisms, the plasma membrane can serve as a barrier between solutions, as we saw in Chapter 5. The plasma membrane is a lipid bilayer with embedded proteins. It is selectively permeable: Water and smaller solutes can diffuse freely through the membrane, but larger solutes, like proteins, cannot. Water is able to move directly through lipid bilayers by diffusion, or through channels called **aquaporins** by facilitated diffusion.

The process by which water moves from regions of higher water concentration to lower water concentration across a selectively permeable membrane is called **osmosis** (**Fig. 41.1**). Consider a situation in which there is a difference in solute concentration on the two sides of a selectively permeable membrane that is freely permeable to water but impermeable to the solute (Fig. 41.1a). Water moves by osmosis from regions of higher *water* concentration (lower *solute* concentration) to regions of lower *water* concentration (higher *solute* concentration) (Fig. 41.1b).

→ **Quick Check 1** What happens when a cell is placed in a hypotonic solution (one with a lower solute concentration than that of the cell)? What happens when a cell is placed in a hypertonic solution (one with a higher solute concentration than that of a cell)?

The pressure needed to prevent water from moving across a selectively permeable membrane is referred to as **osmotic pressure.** Osmotic pressure is the tendency of water to move from one solution into another by osmosis. The higher the solute concentration, the higher the osmotic pressure of that solution (that is, there is a greater tendency for water to move *into* that solution). As water diffuses across a selectively permeable membrane into a region with a higher solute concentration (higher osmotic pressure), the hydrostatic pressure of the more concentrated solution increases. Hydrostatic pressure can result from gravity (if the solutions have different levels) or from the stiffness of the container walls. When osmotic pressure equals hydrostatic pressure, the net movement of water stops and equilibrium is reached (Fig. 41.1c). At equilibrium, water molecules still move across the membrane in both directions by diffusion, but there is no net movement of water molecules.

FIG. 41.1 Osmosis and osmotic pressure.

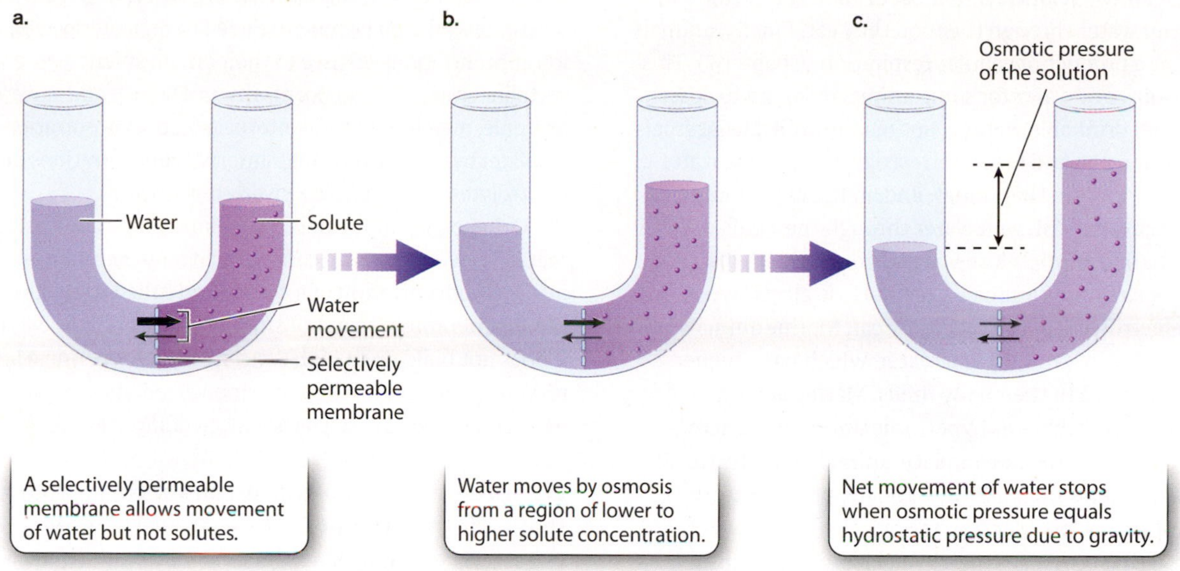

Whether water moves by osmosis into or out of a cell depends on the solute concentration of the inside of the cell relative to the outside. Similarly, whether water moves by osmosis into or out of an organism depends on solute concentration inside an animal relative to its environment, as well as the properties of the sheets of cells that make up the skin and other surfaces exposed to the environment.

Note that water movement is driven by differences in *total* solute concentration between the two sides of a cell membrane. That is, if a particular electrolyte such as potassium has a high concentration inside the cell and low concentration outside the cell, but another electrolyte, such as sodium, has exactly the opposite distribution, there will be no net water movement because the total concentrations of the solutes on the two sides of the membrane are the same.

Also note that water moves across selectively permeable membranes in response to concentration differences of *any* solute. That is, the solute can be an electrolyte, such as sodium or potassium, or another dissolved molecule, such as glucose or an amino acid or even waste, as long as the membrane is impermeable to it.

→ **Quick Check 2** A solution of water and dissolved ions is separated by a selectively permeable membrane that allows the passage of water but not small ions. The concentration of sodium is higher on one side of the membrane than on the other side. However, there is no net movement of water across the membrane. How is this possible?

Osmoregulation is the control of osmotic pressure inside cells and organisms.

Animal life on Earth can exist in a wide range of environments by adopting different means of **osmoregulation,** which is the regulation of osmotic pressure. Osmoregulation can be thought of as the regulation of water content, keeping internal fluids from becoming too concentrated (high osmotic pressure) or too dilute (low osmotic pressure). Osmoregulation, like energy balance (Chapter 40), is a form of homeostasis. High osmotic pressures inside a cell can damage the cell, sometimes even causing it to burst and thus disrupt some of the animal's functions. Low osmotic pressure can lead to dehydration. Although cells and tissues can tolerate dehydration, often as high as 50% or more, excessive dehydration impairs a cell's metabolic function because many chemical reactions, such as hydrolysis reactions that break down proteins and nucleic acids into individual subunits, depend on the presence of water.

How do cells control their internal osmotic pressure? We have seen that water follows solutes across selectively permeable cell membranes. To regulate water and solute levels, and hence the osmotic pressure inside cells, the cell controls the solute concentration of the inside of the cell relative to the solute concentration outside of the cell. At the level of the cell, then, osmoregulation is achieved by the movement of solutes, particularly electrolytes.

At the level of the organism, osmoregulation is achieved by balancing input and output of water and electrolytes. Let's first consider how animals gain and lose water. Animals gain water by drinking water that is less concentrated in solutes (hypotonic) than their body fluids. Humans and many other vertebrates, both freshwater and terrestrial, gain most of their water this way. Animals also gain water through the food they eat. Finally, animals produce water as a product of cellular respiration (Chapter 7). This is the primary source of water for some animals adapted to deserts and oceans, where drinkable water is not readily available. Animals lose water in their urine and feces. Terrestrial animals lose water through evaporation from their lungs, and, in the case of humans, by sweating. Freshwater fish gain water through their gills, whereas most marine animals lose water through their gills.

Now let's consider how animals gain and lose electrolytes. Animals gain electrolytes in the food they eat. Marine animals can gain electrolytes by drinking salt water, which has a higher solute concentration than their body fluids. Marine aquatic animals also gain electrolytes as hypertonic water moves across their gills. In contrast, freshwater aquatic animals lose electrolytes through diffusion across their gills into the hypotonic watery environment. Humans lose electrolytes as well as water when they sweat. Drinking beverages high in electrolytes is helpful before demanding physical activities—including sports—because sweating during these activities can result in substantial water and electrolyte loss. Electrolytes are also lost by specialized glands (discussed below) and in urine and feces.

Osmoconformers match their internal solute concentration to that of the environment.

There are two basic ways in which animals maintain homeostasis with respect to water and electrolytes. Some animals match their internal osmotic pressure to that of their external environment: These animals are **osmoconformers.** They keep their internal fluids at the same osmotic pressure as the surrounding environment, which reduces the movement of water and solutes into or out of their bodies. Because the solute concentrations inside and outside osmoconformers are similar, osmoconformers don't have to spend a lot of energy regulating osmotic pressure. However, they have to adapt to the solute concentration of their external environment. Osmoconformers tend to live in environments like sea water that have stable solute concentrations because maintaining stable internal solute concentrations is easier in stable environments.

Although osmoconformers generally match the *overall* concentration of their tissues with their external environment, they often expend energy to regulate the concentrations of *particular* ions such as sodium, potassium, and chloride and of other solutes such as amino acids and glucose. For example, whereas the intracellular space of nearly all multicellular animals has a relatively high concentration of potassium ions and low concentration of sodium ions, the extracellular space has concentrations that are just the opposite—low in potassium and high in sodium. This means that these cells are actively pumping sodium out and potassium in (Chapter 5).

Most marine invertebrates, such as sea stars, mussels, lobsters, and scallops, are osmoconformers, matching their total intracellular solute concentration to the solute concentration of seawater. These organisms maintain high concentrations of sodium and chloride in their cells to achieve an overall solute concentration close to that of seawater. Consequently, they have few specializations for osmoregulation beyond the need to regulate specific internal ion concentrations relative to their environment. Some marine vertebrates are also osmoconformers. Hagfish and lampreys, for example, match their total internal solute concentration to that of seawater by maintaining high internal concentrations of particular electrolytes, just as marine invertebrates do.

Other osmoconformers, like sharks, rays, and coelacanths, match seawater's solute concentration by maintaining a high internal concentration of a compound called **urea.** Urea is a waste product of protein metabolism that many animals excrete, so it does not build up to high concentrations (section 41.2). By retaining this solute, these marine vertebrates are able to achieve osmotic equilibrium with the surrounding seawater.

Osmoregulators have internal solute concentrations that differ from that of their environment.

The second way in which animals achieve water and electrolyte homeostasis is by maintaining internal solute concentrations different from that of the environment. These animals are **osmoregulators.** Osmoregulators expend considerable energy pumping ions across cell membranes in order to regulate the movement of water into or out of their bodies. Some freshwater and marine fish that are osmoregulators are estimated to expend 50% of their resting metabolic energy on osmoregulation. However, the ability to regulate osmotic pressure allows osmoregulators to live in diverse environments—in salt water, in fresh water, and on land.

Teleosts (or bony fish, the largest group of marine vertebrates) are osmoregulators. So are all freshwater and terrestrial animals. Marine bony fish maintain concentrations of electrolytes much lower than the surrounding sea water, whereas freshwater fish and amphibians maintain concentrations that are higher than the surrounding fresh water.

Let's consider how osmoregulators achieve osmotic balance in salt water, in fresh water, and on land. In salt water, marine organisms live in an environment that is hypertonic relative to their internal fluids, like the cytoplasm of the cell, blood, and other internal fluids. These organisms maintain internal electrolyte concentrations lower than the surrounding salt water. They are thus faced with the problem of water loss (across their gills, as well as in their urine and feces) and electrolyte gain (across their gills and in seawater that they drink). As a result, they need to take in as much water as possible and eliminate excess electrolytes. They obtain water primarily through the food they eat and cellular respiration. They avoid water loss by producing concentrated urine (discussed in section 41.2).

Bony fish have evolved specialized gills that allow them to excrete excess electrolytes as well as to breathe. The thin gill filaments, with their extensive surface area, are useful sites for electrolyte transport out of the body in addition to being well adapted for gas exchange (**Fig. 41.2**). Specialized cells in the gills

FIG. 41.2 Osmoregulators. Osmoregulators actively maintain osmotic pressure different from that of their environment.

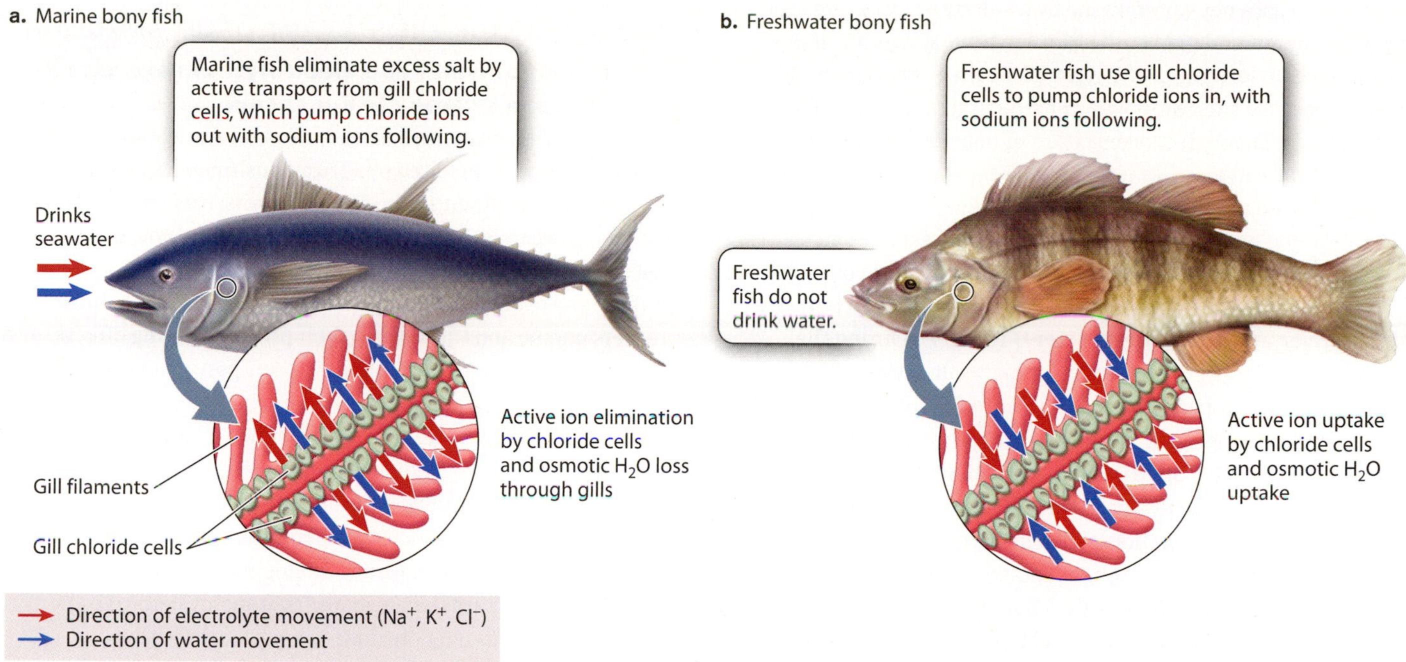

called **chloride cells** pump chloride ions into the surrounding seawater. By pumping negatively charged chloride ions out of the body, chloride cells create an electrical gradient that is balanced by positively charged sodium ions moving out of the body as well (Fig. 41.2a).

In fresh water, the environment is hypotonic relative to an organism's internal fluids. In this case, the problem is just the opposite as that faced by marine organisms: water gain and salt loss. As a result, freshwater organisms tend to minimize their water intake and maximize water elimination. For example, in contrast to marine fish, they do not drink water. In addition, freshwater fish produce very dilute urine (discussed in section 41.2).

To maintain electrolyte concentration higher than the surrounding water, freshwater fish have gill chloride cells with reversed polarity relative to marine bony fish. Their chloride cells pump chloride ions (with sodium ions following) into the body to counter their ongoing loss from the gills into the surrounding fresh water (Fig. 41.2b).

Nearly all amphibians also live in freshwater environments and therefore face challenges similar to those facing freshwater fish. They take up electrolytes across their skin and produce dilute urine.

On land, the challenge is water loss. Terrestrial animals lose water in their urine and feces, through evaporation from their lungs, and, in humans, by sweating. To counter these losses, they drink water that is hypotonic relative to their body fluids. They also acquire water by eating food and by producing water during cellular respiration. They produce concentrated urine to minimize water loss.

Many marine birds have evolved specialized nasal salt glands to rid themselves of the high salt content of their diet (**Fig. 41.3**). These salt glands actively pump ions from the blood into the

FIG. 41.3 Salt excretion from glands. Many marine birds have glands on the top of the skull that actively excrete salt.
Photo source: David Wall/Alamy.

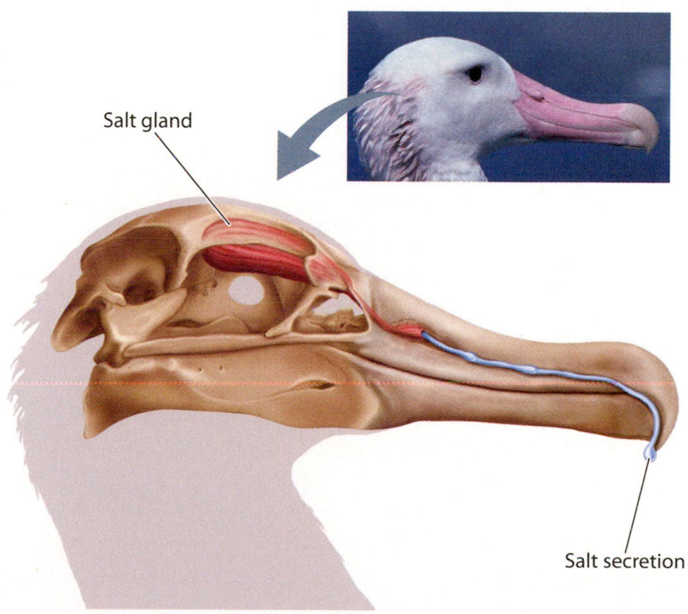

cells that make up the gland, and then excrete salt from the body through the gland and the nasal cavity. Their salt glands allow these birds to gain net water intake by drinking seawater. In contrast, humans and many other non–marine-adapted animals are incapable of deriving useful net water gain by drinking sea water because they cannot eliminate salt in high enough concentration. The high concentration of magnesium sulfate in sea water results in diarrhea if too much sea water is consumed, exacerbating the net loss of water in the feces.

An extreme example of the ability of an animal to regulate water and electrolyte levels is provided by the life cycle of salmon. Mature salmon swim in the salt water of the open ocean but return upriver to spawn in fresh water (**Fig. 41.4**). During the course of their return, they swim from high-salt ocean water through water of intermediate saltiness into freshwater rivers and streams that have little salt. After spawning, the young feed and grow in fresh water for up to 3 years, eventually returning to the open ocean, where they live until they become reproductively mature and return to their home river system to spawn again.

These changes in aquatic habitat demand that salmon make dramatic adjustments in the way they handle water and solutes. When in the ocean, salmon must avoid loss of water and uptake of electrolytes. However, when mature salmon migrate upriver to spawn and as young salmon continue to grow in freshwater streams and rivers, the challenge is reversed: They must avoid gaining too much water and losing critical electrolytes across their gills. In salmon, radical changes in osmoregulation are required over the course of an animal's lifetime to enable it to reproduce in freshwater streams and return to the sea for further growth and maturation. Next, we explore how gills and kidneys meet these challenges.

? CASE 7 PREDATOR–PREY: A GAME OF LIFE AND DEATH

Can the loss of water and electrolytes in exercise be exploited as a strategy to hunt prey?

As we have seen, exercise can lead to loss of water and electrolytes, a response that is exploited by African bushmen as they hunt large prey. African bushmen are hunter-gatherers; their tribes live in hot, dry grasslands. Equipped only with wood spears, bows, and poisoned arrows, they have historically hunted kudus, antelope, buffalo, and other large grazing herbivores (**Fig. 41.5**). Bushmen hunters pursue and track down their prey, exploiting effects of the physiological stress that occur when their prey run during the heat of the day.

Before hunting, bushmen drink as much water as possible, hydrating their bodies. They select a large prey animal, which they chase from a shaded refuge into the open sun when temperatures exceed the animal's core body temperature. The bushmen can deal with the high temperatures, cooling their bodies through sweating, but because they hydrated thoroughly before the hunt, their sweating does not dehydrate them. Not so for the prey. The bushmen hunters track and chase their prey until the prey becomes dehydrated and hyperthermic, thus rendered unable to escape.

Even if the bushmen hunters strike the animal successfully with a poisoned arrow, they still must track it for long distances and over long periods of time. Hunts typically last several hours, but sometimes as much as 1 or 2 days, and cover distances of 25 km

FIG. 41.4 **The salmon life cycle.** Salmon live in a range of environments at different stages of their life cycle, which require different osmoregulatory strategies.

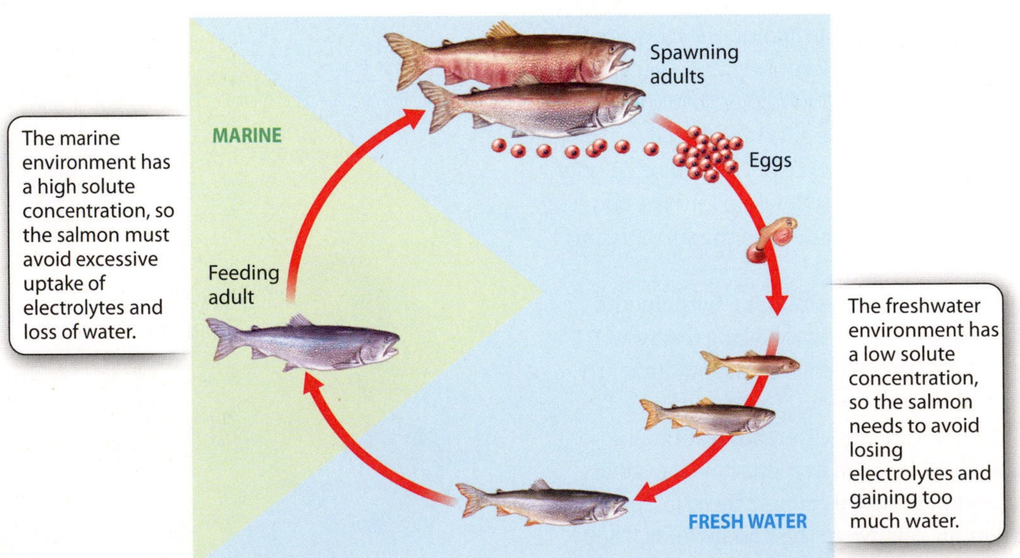

FIG. 41.5 Exploiting an animal's physiological stress. (a) A kudu will be forced to run in the heat of the day by (b) African bushmen hunters. The animal will lose water and electrolytes, and will eventually become too dehydrated to escape. *Sources: a. Peter Betts/iStockphoto; b. Roger De La Harpe, Gallo Images/Corbis.*

(about 16 miles) or more. The loss of water in intensive heat disrupts the animal's water and electrolyte balance and thus contributes to its inability to move and defend itself.

41.2 EXCRETION OF WASTES

In most organisms, the regulation of water and solute levels is closely tied to waste **excretion.** Excretion is the elimination of waste products and toxic compounds from the body. Some of these wastes are generated as by-products of metabolism, particularly protein and nucleic acid breakdown. Others were originally ingested, like the undigested remains of food, and must be eliminated from the organism.

Most multicellular animals have evolved specialized excretory organs, such as the vertebrate kidney, to isolate, store, and eliminate toxic compounds and metabolic waste products. Because these compounds are eliminated as solutes that are dissolved and transported within the blood or other body fluids, their elimination from the body is intimately tied to water and electrolyte balance. The excretory organs of most animals therefore have two functions: They maintain water and electrolyte balance and they eliminate waste products.

Both the gastrointestinal tract (Chapter 40) and excretory organs like the vertebrate kidneys eliminate wastes from the body. However, the gastrointestinal tract is continuous with the outside world, and therefore its role in the elimination of wastes is quite different from that of excretory organs that separate toxic compounds from essential nutrients, electrolytes, and cells circulating in the blood.

The excretion of nitrogenous wastes is linked to an animal's habitat and evolutionary history.

When proteins and nucleic acids are broken down, one of the by-products is ammonia (NH_3). Ammonia contains nitrogen and is toxic to organisms, so it is a form of **nitrogenous waste (Fig. 41.6)**. By far, most ammonia is produced from the breakdown of proteins. Ammonia can disturb the pH balance of cells and damage neurons. At high levels, it is lethal to organisms. It is either eliminated from the body or converted into a less toxic form.

Some organisms, such as fish, are able to excrete ammonia directly into the surrounding water, primarily through their gills. As a result, ammonia does not accumulate to toxic levels. This process is not energetically expensive, but it requires high volumes of water in the surrounding environment to dilute the highly toxic ammonia.

When animals moved from water to land, they faced the challenge of how to eliminate waste. No longer could they rely on simple diffusion of ammonia into the surrounding water. Instead, they evolved biochemical pathways to convert ammonia to less toxic forms. One of these forms is urea (Fig. 41.6). Because urea is less toxic than ammonia, it can be stored in a fairly concentrated form before being excreted. Mammals (including humans), many amphibians, sharks, and some bony fish excrete nitrogenous waste

FIG. 41.6 Nitrogenous waste. Animals excrete nitrogenous waste in three major forms—ammonia, urea, and uric acid—that vary in their toxicity and cost to produce.

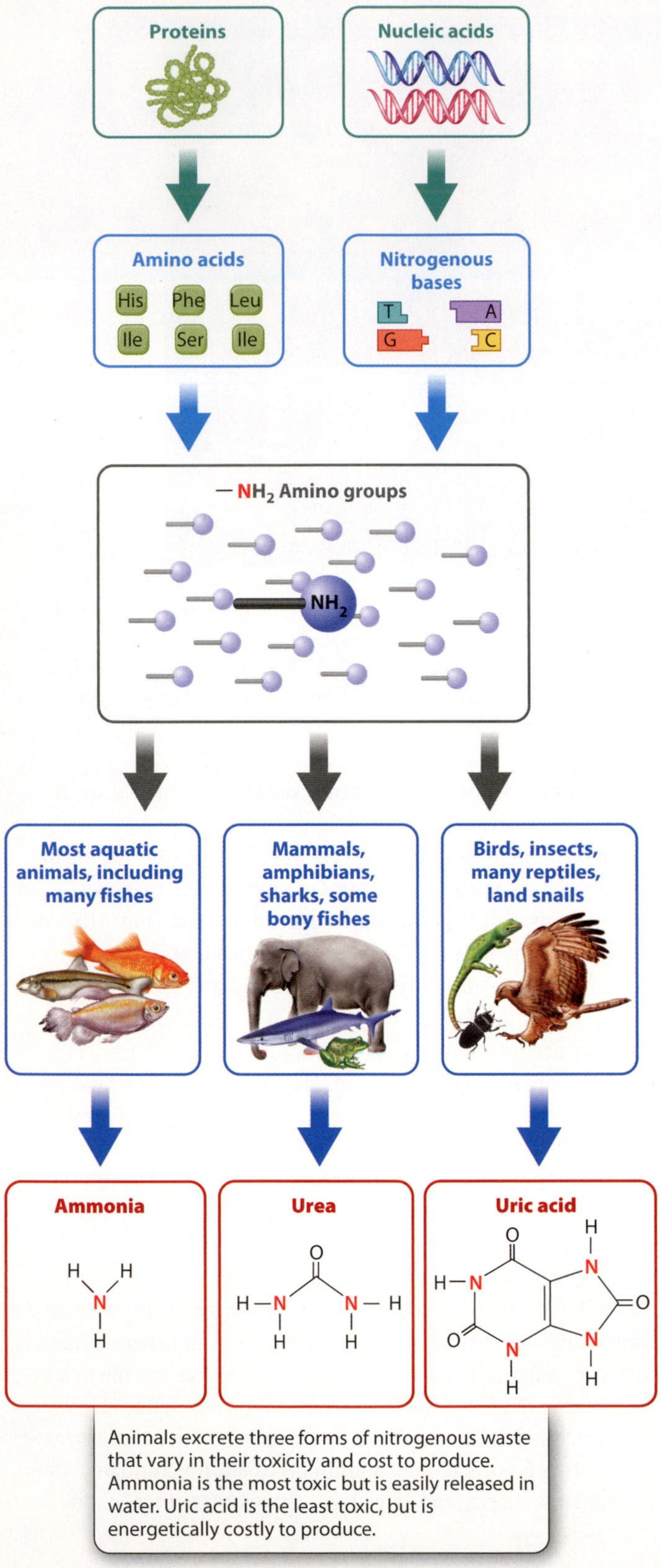

in the form of urea. In mammals, urea is produced in the liver and is then carried by the blood to the kidneys, from which it is eliminated. Urea is less toxic than ammonia but requires energy to produce and water to eliminate.

Ammonia can also be converted into uric acid (Fig. 41.6). Uric acid is the least toxic of the three forms of nitrogenous waste, so animals can store it at a higher concentration than urea. Birds and many reptiles, arthropods (including insects), and land snails excrete nitrogenous waste in the form of uric acid. These groups evolved the ability to convert ammonia to uric acid independently of one another as an adaptation to living on land, providing a good example of convergent evolution (Chapter 23). Uric acid precipitates from solution and forms a semisolid paste (familiar as bird droppings). Because uric acid is not dissolved in water, it does not exert osmotic pressure and is eliminated with minimal water loss. By limiting water loss, many reptiles and insects are able to live in extremely hot and dry environments. However, uric acid is energetically expensive to produce.

Many amphibians, like frogs, excrete ammonia early in their development, when they live in water, but then excrete urea when they move onto land after metamorphosis. Alligators and turtles also adjust what they excrete: ammonia when in the water and uric acid when on land.

Ammonia, urea, and uric acid are the three major forms of nitrogenous wastes. However, they are not the only forms. Animals excrete nitrogen as trimethylamine oxide, creatine, creatinine, and even amino acids. The form of nitrogenous waste that an animal excretes is ultimately linked to its environment and evolutionary history.

→ **Quick Check 3** Why do mammals convert ammonia to urea rather than simply excreting it, as fish do?

Excretory organs work by filtration, reabsorption and secretion.

There are two ways that organisms eliminate waste and other compounds out of the blood: **filtration** and **secretion.** Filtration is the process by which the blood is filtered into an extracellular space, allowing some substances to pass through but retaining others. Secretion is the active transport of molecules from the blood into the extracellular space. Filtration is energetically inexpensive and allows removal of novel compounds, but also removes compounds that the body needs, such as water, electrolytes, and essential nutrients. Secretion is specific but energetically expensive. So both are used. Initially, filtration is used to isolate most waste, but because it is nonspecific, **reabsorption** is required. Reabsorption is the process by which essential molecules are transported back into the blood. Reabsorption and secretion together help to balance water and electrolyte levels, producing dilute or concentrated urine as required by the animal.

The first step of excretion is to isolate the waste into an extracellular space. In the case of unicellular organisms such as

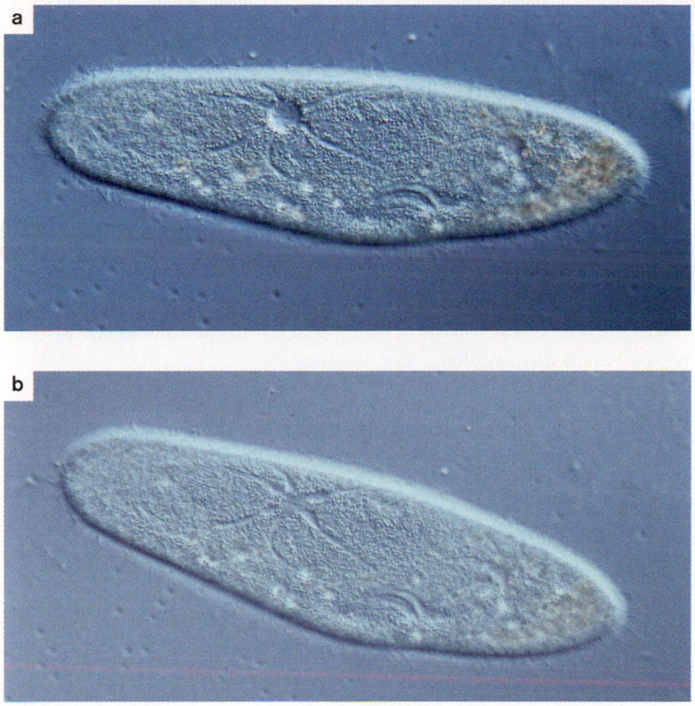

FIG. 41.7 **Contractile vacuole of a Paramecium when (a) full and (b) after emptying.** *Source: Michael Abbey/Science Source.*

Paramecium and simple multicellular animals such as sponges and cnidarians (jellyfish and sea anemones), wastes are isolated in a cellular compartment called a **contractile vacuole** (**Fig. 41.7**). Fusion of the contractile vacuole with the cell membrane eliminates the waste contents by exocytosis from the cell (Chapter 5).

Multicellular animals with pressurized circulatory systems (annelid worms, mollusks, and vertebrates) isolate wastes from the blood by filtration (**Fig. 41.8**). Circulatory pressure pushes fluid containing the wastes through specialized filters into an extracellular space. The filtered fluid, called the filtrate, contains waste products along with water, electrolytes, and other solutes. The filtrate drains into **excretory tubules** that connect to the outside. The excretory filter, like a cell's plasma membrane, is selectively permeable. It allows smaller ions and solutes to pass into the tubule along with water, but retains larger solutes such as proteins and cells.

Filtration is followed by reabsorption of key ions and solutes from the filtrate into the blood (Fig. 41.8). Reabsorption can be an active process, requiring active transporters for some molecules, or a passive process driven by diffusion for others. In both cases, substances that are needed by the animal, such as electrolytes, amino acids, vitamins, and simple sugars, are taken up by cells of the excretory tubule and returned to the bloodstream. Water is reabsorbed into the cells by osmosis, driven by the osmotic pressure established by the active transport of solutes into the cell.

The final step in waste removal is the secretion of additional toxic compounds and excess ions (Fig. 41.8). Secretion is an active process, involving protein transporters in the plasma membranes of cells lining the tubules. These transporters use the energy of ATP to pump molecules from the blood into the excretory tubules. Secretion eliminates substances that were not filtered from the blood, as well as helps to fine tune electrolyte levels in the blood.

Filtration, reabsorption, and secretion are the three steps that almost all animal excretory systems perform to eliminate toxic wastes, retain essential nutrients, and regulate water and electrolyte levels.

Animals have diverse excretory organs.

As we have seen, the processes of filtration, reabsorption, and secretion serve two important and complementary functions: They allow wastes to be isolated and subsequently removed from the body, and they allow the organism to adjust the amounts of water and electrolytes required to meet its osmoregulatory needs. These two functions are carried out by diverse excretory organs.

For example, consider waste excretion in flatworms, small flattened animals that glide along the streambed floor. In flatworms, waste is excreted without being filtered: The fluid from the body cavity is isolated in excretory organs called

FIG. 41.8 **Filtration, reabsorption, and secretion.** These three processes eliminate wastes and provide osmoregulation in multicellular animals with pressurized circulatory systems.

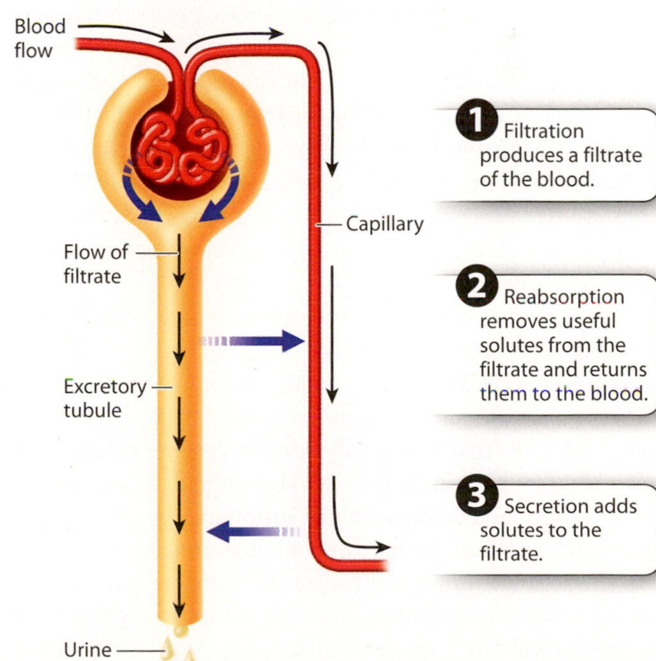

FIG. 41.9 Excretory organ of a flatworm.

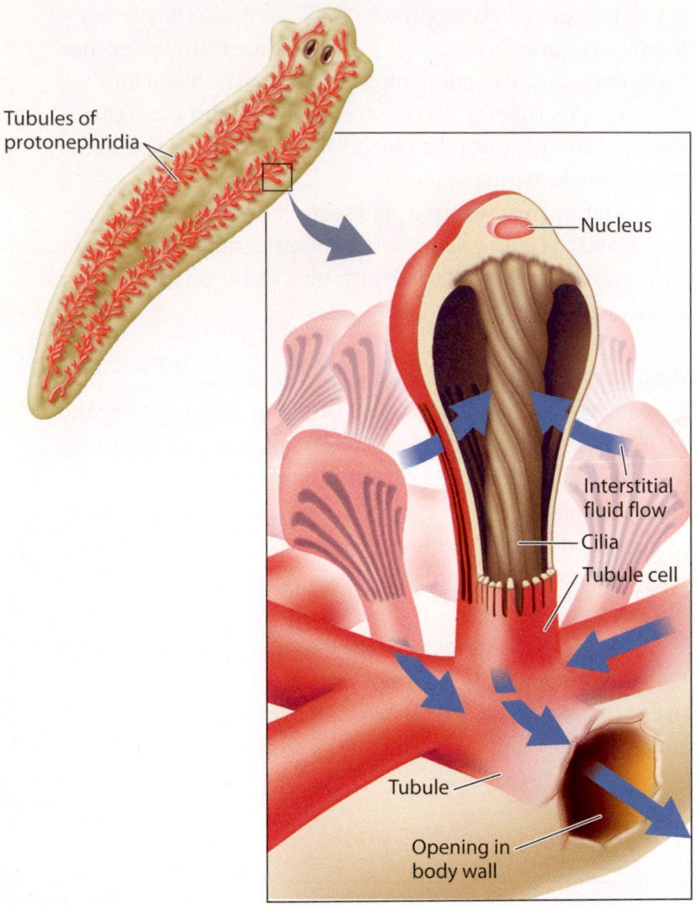

protonephridia (**Fig. 41.9**). Cells with beating cilia move the fluid down the lumen of a tubule to an excretory pore. Freshwater flatworms, such as *Planaria,* have an extensive network of tubules that eliminate excess fluid along with waste products. As the fluid passes down the tubule, cells lining the tubule modify the fluid's contents through reabsorption and secretion. The urine leaving a freshwater flatworm contains nitrogenous waste and is less concentrated (that is, more dilute) than its internal body fluids to balance the water it ingests as it feeds, as is typical of freshwater animals, as we saw earlier.

In segmented annelid worms such as earthworms, the blood is filtered through small capillaries into a pair of excretory organs called **metanephridia** within each body segment (**Fig. 41.10**). The fluid enters a metanephridium at a funnel-shaped opening surrounded by cilia that drains body fluid into a series of tubes in the next body segment and then out of the animal through an excretory pore. Blood vessels and cells surrounding the tubules reabsorb and secrete additional compounds as urine is formed. Like freshwater flatworms, earthworms reabsorb electrolytes and eliminate dilute urine containing dissolved nitrogenous wastes.

In insects and other terrestrial arthropods, body fluid called hemolymph passes from the main body cavity into a series of tubes called **Malpighian tubules** that empty into the hindgut (**Fig. 41.11**). The cells lining the tubules actively secrete uric acid and excess electrolytes into the lumen of the tubules while retaining key solutes. Muscle fibers in the walls of the tubules help push the fluid to the hindgut. In the hindgut, the fluid becomes increasingly acidic because of the buildup of uric acid, causing uric acid to precipitate from solution when it enters the rectum. Because uric acid is removed from the fluid, it does not influence the movement of water. Water is therefore free to diffuse back into the tissues as electrolytes are reabsorbed from the rectum. The uric acid and remaining dried digestive wastes are then eliminated. Malpighian tubules represent one of the most successful adaptations for excreting nitrogenous wastes with minimal water loss, enabling insects to live in diverse and arid terrestrial environments.

Vertebrates filter blood under pressure through paired kidneys.

The excretory organs of vertebrates are the **kidneys.** Like all excretory organs, they not only serve to eliminate wastes, but also play an important role in maintaining water and electrolyte

FIG. 41.10 Excretory organ of an earthworm.

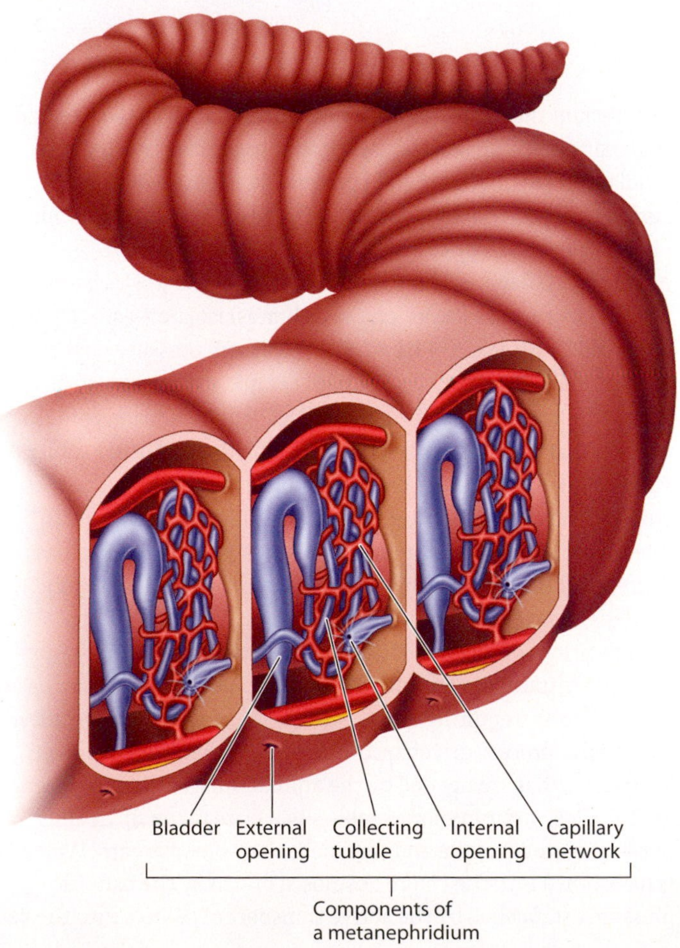

FIG. 41.11 Excretory organ of an insect.

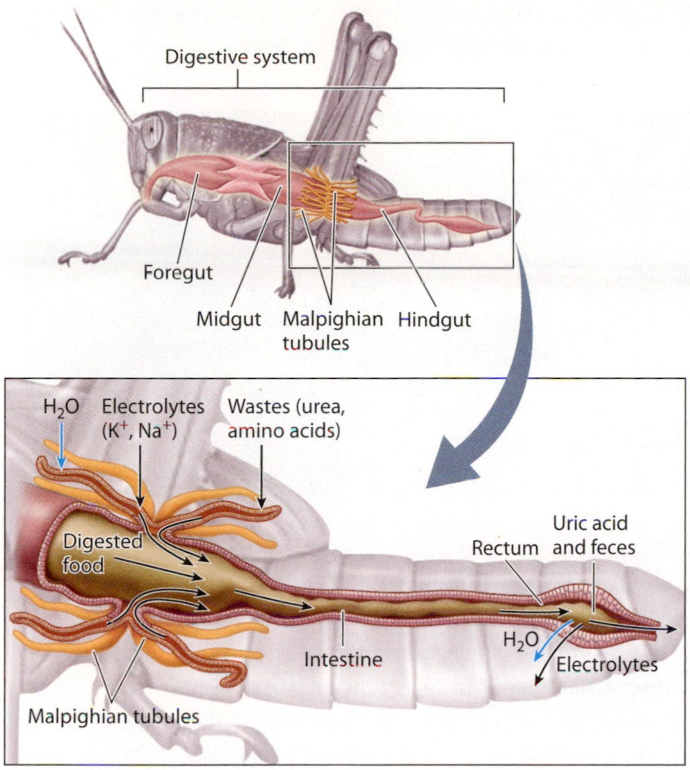

balance. Vertebrate kidneys also reflect the evolutionary transition of vertebrates from an aquatic to a terrestrial environment, where there is a need to minimize water loss. Mammals and birds, for example, can excrete urine that is more concentrated in solute than their blood and tissues.

The kidneys lie posterior to the abdominal cavity. The kidneys of humans and other mammals are bean-shaped, whereas the kidneys of birds and other vertebrate animals are elongated and segmented (**Fig. 41.12**). Paired vertebrate kidneys receive blood supply from the heart by renal arteries that branch from the abdominal aorta.

Vertebrates filter their blood through specialized capillaries that have openings in their walls. These porous capillaries form a tufted loop called a **glomerulus** (**Fig. 41.13**). There are thousands to millions of glomeruli in each kidney. Driven by circulatory pressure, nitrogenous wastes, electrolytes, other small solutes, and water move through small holes in the capillary walls into an extracellular space surrounded by a capsule (called Bowman's capsule after the surgeon William Bowman, who identified it in 1842). The filtrate then moves through a series of **renal tubules,** which further process the filtrate by reabsorption and secretion before it enters the **collecting ducts** as urine. The collecting ducts converge on a larger tube called the **ureter,** which brings urine from the kidneys to a hollow organ called the **bladder** for storage and elimination from the body.

FIG. 41.12 Location and shape of kidneys in different vertebrates.

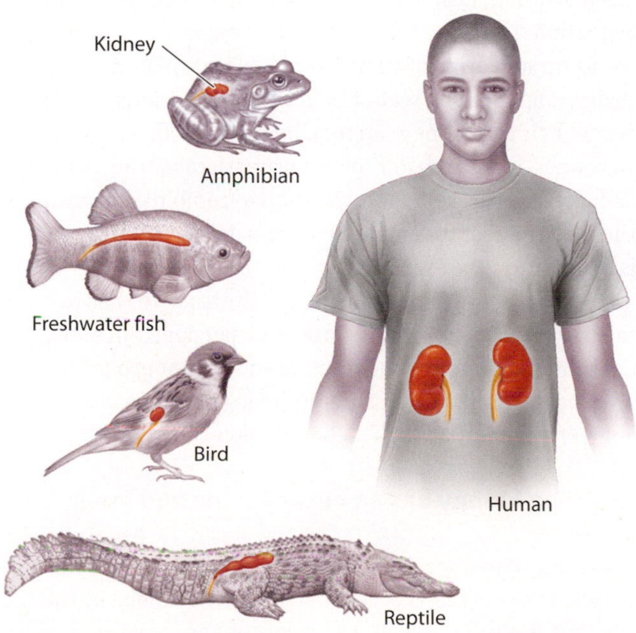

FIG. 41.13 Vertebrate kidney and nephron.

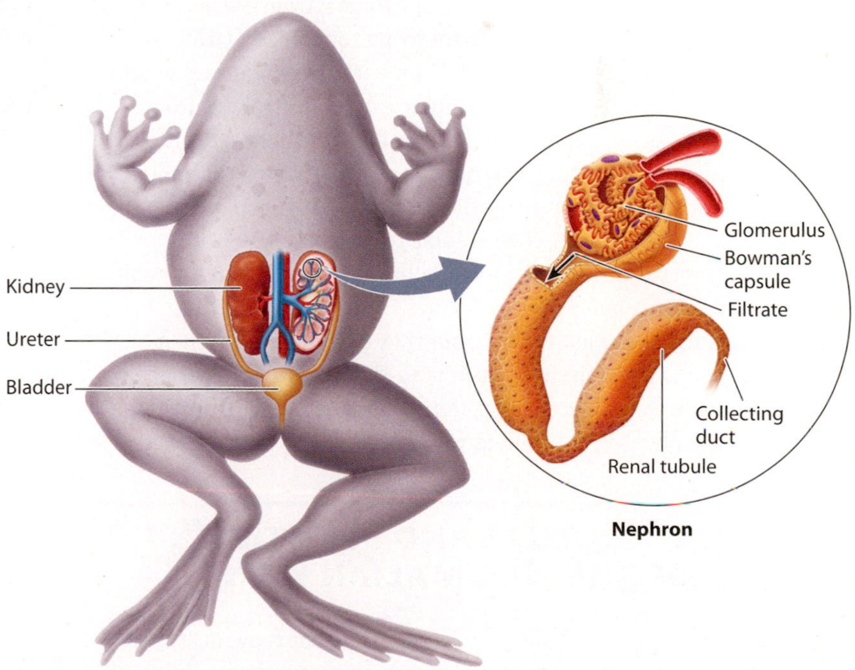

The glomerulus, capsule, renal tubules, and collecting ducts make up a **nephron,** the functional unit of the kidney. Nephrons perform the three basic steps of excretion and osmoregulation: filtration of blood passing through the glomerulus, reabsorption from the renal tubule back to the bloodstream of key electrolytes and solutes, and secretion of additional wastes and electrolytes by the renal tubules.

In saltwater and freshwater fish, the kidneys run along the length of the body, with nephrons arranged segmentally. Because fish excrete nitrogenous waste in the form of ammonia through their gills, the kidney is primarily an organ of water and electrolyte balance rather than one of waste elimination. Recall that saltwater and freshwater fish face different challenges in achieving water balance. Saltwater fish must eliminate excess electrolytes that enter through their gills and in the sea water that they drink, whereas freshwater fish must eliminate excess water that enters through their gills. In saltwater fish, the renal tubules are readily permeable to water, allowing water to diffuse out of the tubule by osmosis, and producing relatively concentrated urine (high concentration of electrolytes and low concentration of water). By contrast, the renal tubules in freshwater fish are relatively impermeable to water while allowing reabsorption of electrolytes, sugars, and amino acids. This enables freshwater fish to excrete very dilute urine, eliminating the excess water that continually enters through their gills.

The kidneys of amphibians are similar to those of freshwater fish. Amphibians take in excess water across their gills and skin as they breathe. Amphibian kidneys therefore produce extremely dilute urine, enabling them to maintain an electrolyte concentration in their body that exceeds that of their freshwater environment.

Reptiles are found in a variety of habitats, from oceans to deserts, and are well adapted to their environments. Marine iguanas, for example, ingest salt water with their food, and so rely on specialized salt glands to eliminate excess electrolytes from their body. Terrestrial reptiles living in desert regions face the challenge of minimizing water loss in an environment of limited water availability. Because reptiles excrete nitrogenous waste in the form of uric acid, they are able to reabsorb most of the water from the urine. This process allows many reptiles to inhabit extremely dry environments.

Mammals also live on land, but they excrete most of their nitrogenous waste in the form of urea dissolved in water. Their kidneys are well adapted for excreting waste and balancing their water and electrolyte levels, as we discuss now.

41.3 STRUCTURE AND FUNCTION OF THE MAMMALIAN KIDNEY

Mammals excrete urea dissolved in solution, but they must be able to do so without losing too much water, an adaptation to living on land. To minimize the loss of water, they remove water from the waste-containing fluid by osmosis as it moves through the kidney.

FIG. 41.14 **A giant kangaroo rat (*Dipodomys ingens*).** Giant kangaroo rats do not drink. Instead, they obtain water through cellular respiration and from the seeds they eat, which contain small amounts of water. *Source: Tom McHugh / Science Source.*

In this way, the urine is concentrated without the kidney ever actively transporting water itself.

The cells of the kidneys are surrounded by a solution called the interstitial fluid. Mammals concentrate their urine by generating concentrated (hypertonic) interstitial fluid deep in the kidney, through which the collecting ducts pass. Then, as the filtrate passes through the collecting ducts, water moves out of collecting ducts and into the interstitial fluid by osmosis, leaving concentrated urine in the collecting ducts.

This process is quite efficient, and allows animals to thrive in diverse environments. For example, the giant kangaroo rat shown in **Fig. 41.14,** a desert mammal, reabsorbs so much water that it does not need to drink: It can obtain enough water from seeds and cellular respiration. The giant kangaroo rat is thought to produce the most concentrated urine of all mammals. Marine mammals, such as whales, dolphins, and seals, also have an exceptional ability to concentrate their urine. They can remain at sea for long periods of time but cannot drink seawater for net water uptake. They, too, rely on water produced by cellular respiration in combination with water in the food they eat to remain in water balance.

In this final section, we examine the mammalian kidney in more detail. The kidneys of birds share many features with those of mammals. However, these features arose independently in birds and mammals and are therefore the result of convergent evolution, presumably as adaptations for efficient processing of wastes in endothermic vertebrates with high metabolic rates (Chapter 40).

The mammalian kidney has an outer cortex and inner medulla.

The paired kidneys of mammals are organized into two layers, an outer layer called the **cortex** and a deeper layer, surrounded by the cortex, called the **medulla** (**Fig. 41.15**). Nephrons are organized into wedge-shaped renal pyramids (Fig. 41.15). Each nephron's glomerulus is located within the renal cortex or outer layer of

FIG. 41.15 The organization of the mammalian kidney and nephron.

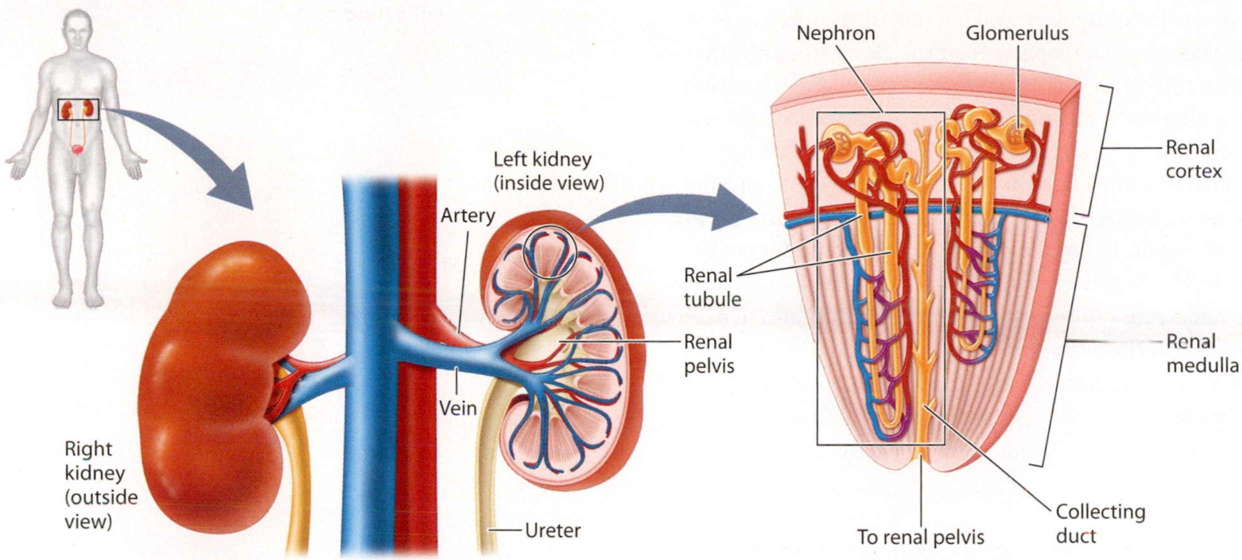

the medulla, and feeds into a renal tubule that starts in the renal cortex before dipping into the medulla at the base of the renal pyramid and then looping back up to the cortex. The collecting ducts of the renal tubules also descend into the pyramid's base, emptying into the renal pelvis and draining through the ureter into the bladder.

Glomerular filtration isolates wastes carried by the blood along with water and small solutes.

The glomerulus is a tuft of capillaries, with blood entering by an afferent ("toward") arteriole and leaving by an efferent ("away") arteriole (**Fig. 41.16**). The tuft of capillaries is encased in a membranous sac called **Bowman's capsule,** and the space enclosed

FIG. 41.16 **The mammalian glomerulus and Bowman's capsule.** Three layers create the filter: endothelial cells, basal lamina, and podocytes.

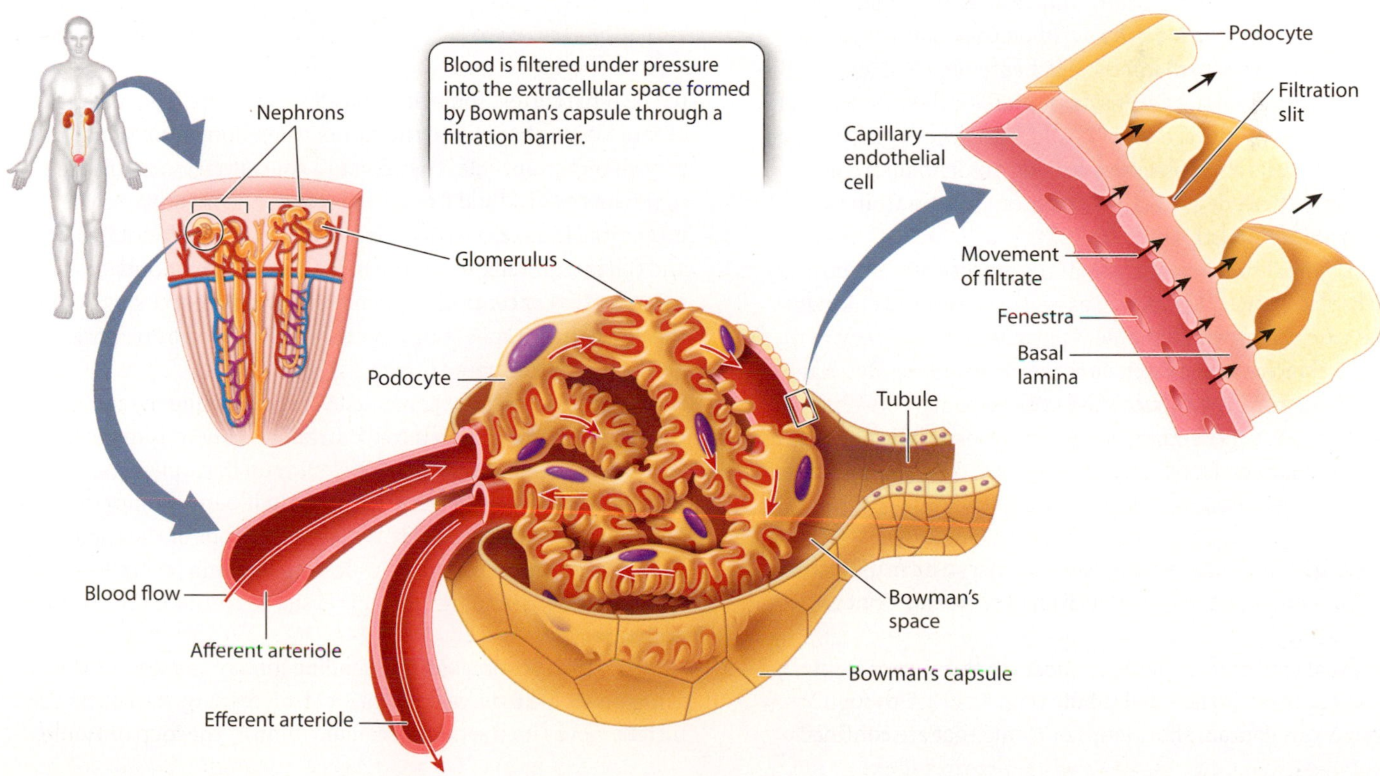

by the capsule is called Bowman's space. Blood is filtered as it passes through the tuft of capillaries. Water, wastes, and solutes from the blood pass through the capillary wall into Bowman's space.

To filter the blood, the capillaries of the glomerulus have a different structure from that of capillaries elsewhere in the body. The cells that line the capillaries, known as endothelial cells, have pores called fenestrae (which means "windows") that allow fluid to pass through. The filtration barrier of the glomerulus is made up of three layers (Fig. 41.16): the endothelial cells, a basal lamina, and a layer of cells with footlike projections called **podocytes,** which loosely interlock to create slits in the cell layer. These three layers together create a filter that allows small molecules to pass through but blocks the passage of large proteins and cells, leaving them in the blood. As a result, wastes such as urea, as well as water and solutes such as electrolytes, glucose, and amino acids, all end up in the filtrate. After passing into Bowman's space, the filtrate enters the renal tubule.

The proximal convoluted tubule reabsorbs solutes by active transport.

The renal tubule of the mammalian kidney is divided into three sections—the **proximal convoluted tubule,** the **loop of Henle,** and the **distal convoluted tubule**—as shown in **Fig. 41.17.** Although the renal tubule is one long tube, these three sections are specialized for different functions because of differences in permeability to different substances.

The filtrate initially moves into the first portion of the renal tubule, the proximal convoluted tubule (Fig. 41.17). Here, electrolytes and other nutrients that the animal requires are reabsorbed into the blood. To maximize absorption, the proximal convoluted tubule has thick walls composed of epithelial cells with fingerlike projections called **microvilli** on their surfaces, similar to those found in the small intestine (Chapter 40). The microvilli form a brush border along the inside (lumen) of the tubule, greatly increasing the cells' surface area for solute and water reabsorption. The epithelial cells also possess numerous mitochondria that produce the ATP needed to actively reabsorb solutes from the filtrate into the bloodstream.

The proximal tubule reabsorbs all the glucose and amino acids filtered by the glomerulus, as well as most sodium and chloride ions. Because the proximal tubule is permeable to water, water diffuses by osmosis in the same direction as the electrolytes and solutes. By the time the filtrate leaves the proximal convoluted tubule, about 75% of the water, along with most of the electrolytes and solutes, has been reabsorbed into the bloodstream. What's left is primarily water, urea, and electrolytes.

The loop of Henle acts as a countercurrent multiplier to create a concentration gradient from the cortex to the medulla.

The proximal convoluted tubule connects to the loop of Henle, the second portion of the renal tubule (Fig. 41.17). Although many nephrons contain short loops of Henle that are confined

FIG. 41.17 **The organization of the mammalian renal tubules and collecting duct.**

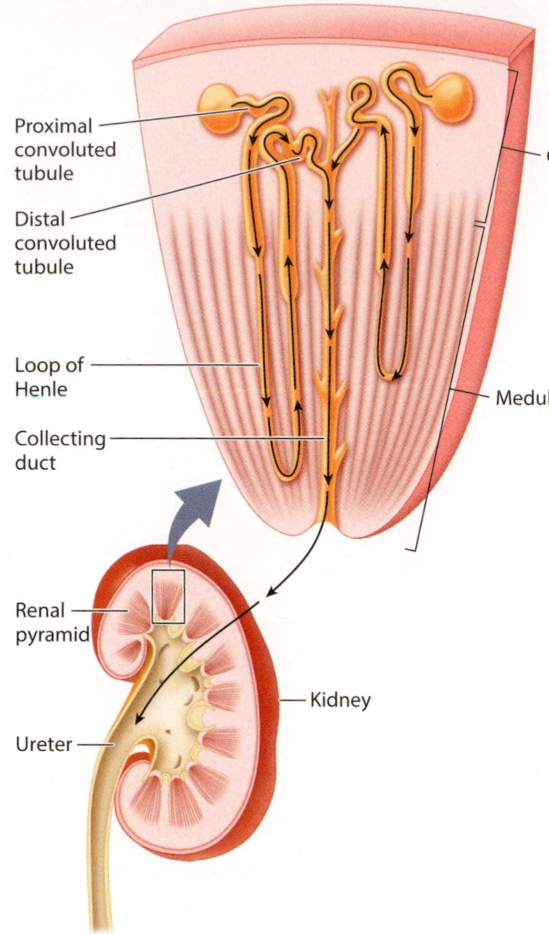

to the renal cortex, some descend all the way into the medulla before looping back up to the cortex. These longer loops of Henle play an important role: They create a concentration gradient in the interstitial fluid from the cortex to the medulla, so the interstitial fluid of the cortex is less concentrated in solutes and the interstitial fluid of the medulla is more concentrated in solutes. This concentration gradient allows water passing through the downstream collecting duct to be reabsorbed by osmosis, producing concentrated urine.

How is the gradient generated? The key to understanding how the loop of Henle creates a concentration gradient is to recognize that the two limbs of the loop of Henle run in parallel but opposite directions, and they differ in their permeability to water and ability to actively transport electrolytes out of the tubule. Although the filtrate enters the descending limb before looping around to the ascending limb, it is easier to consider the ascending limb first.

The thick portion of the ascending limb of the loop of Henle is impermeable to water but not electrolytes (**Fig. 41.18**). As the filtrate moves up the thick ascending limb of the loop of Henle,

FIG. 41.18 **The mammalian nephron.** The loop of Henle creates a concentration gradient, with higher concentration in the medulla and lower concentration in the cortex.

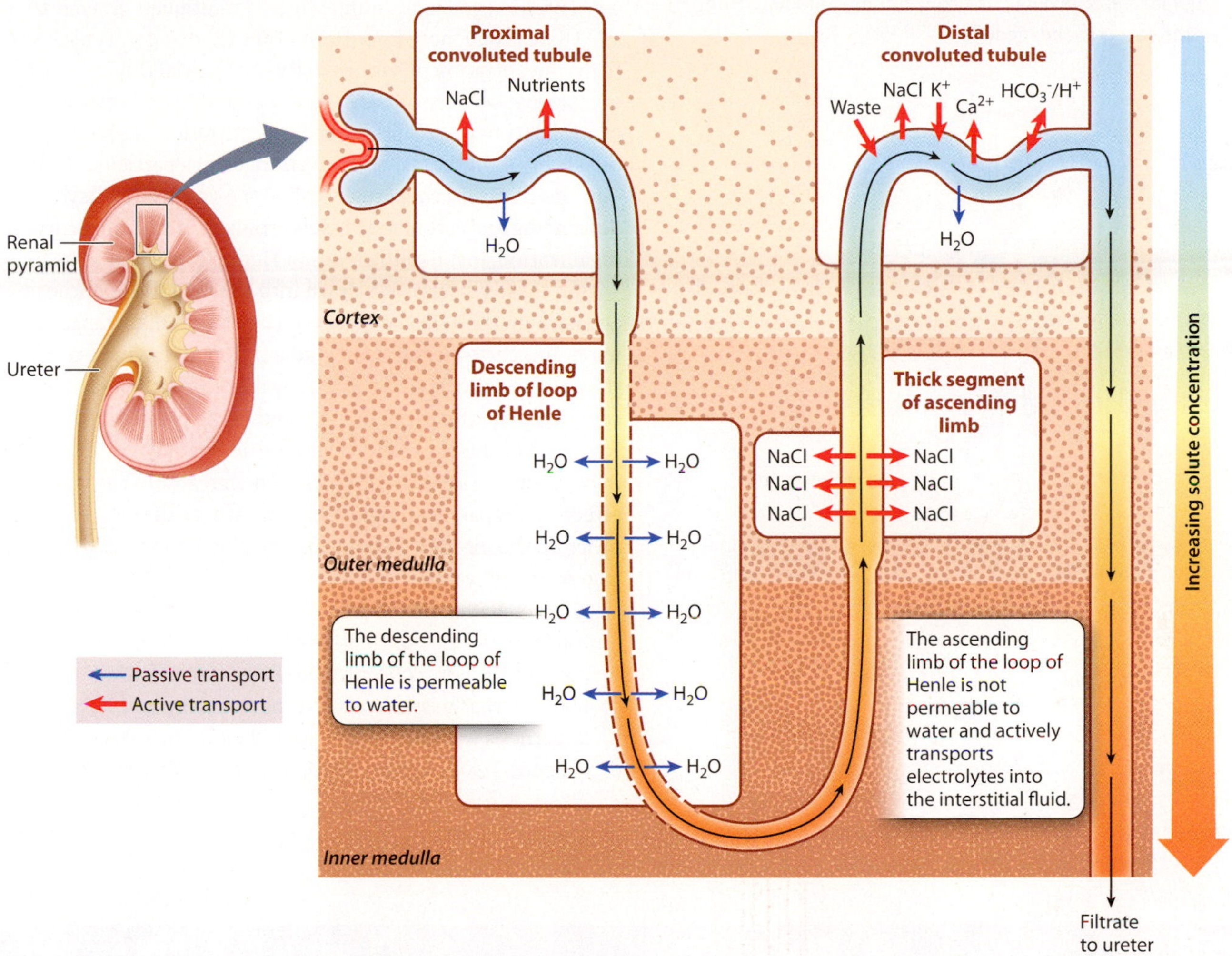

electrolytes are actively transported out of it into the surrounding interstitial fluid. As a result, the interstitial fluid becomes more concentrated, and the filtrate passing through the ascending limb becomes less concentrated.

In contrast to the ascending limb, the descending limb is permeable to water. Therefore, water moves passively out of the descending limb, driven by osmosis due to the electrolytes pumped into the interstitial fluid by the ascending limb. As water moves out of the descending limb, the filtrate becomes more and more concentrated, matching the concentration in the surrounding interstitial space. We have already seen that, as electrolytes are pumped out of the thick ascending limb, the filtrate becomes less and less concentrated. Thus, the filtrate enters and leaves the loop of Henle at about the same concentration. Although filtrate concentration is unchanged, the trip around the loop creates a concentration gradient in the interstitial fluid surrounding the loop of Henle, with fluid in the inner medulla much more concentrated than that in the outer cortex.

The gradient results from the combination of active transport of electrolytes in the ascending limb, osmosis of water in the descending limb, and the movement of the filtrate through the loop of Henle. The filtrate is most concentrated at the base of the loop, and the concentration of solutes inside and outside the loop is about the same, leaving the medulla more concentrated than the cortex.

In Chapter 39, we saw how the movement of water and blood in opposite directions in fish gills provides an efficient way to extract oxygen from the water and move it into the blood. This mechanism, called countercurrent exchange, is an efficient mechanism by which two fluids can exchange properties such as amounts of oxygen. In the loop of Henle, active transport of electrolytes creates a concentration gradient, which is then multiplied because

FIG. 41.19 **Vasa recta.** The blood in the vasa recta moves in a countercurrent fashion, just like the filtrate in the loop of Henle, to maintain the concentration gradient from the cortex to the medulla.

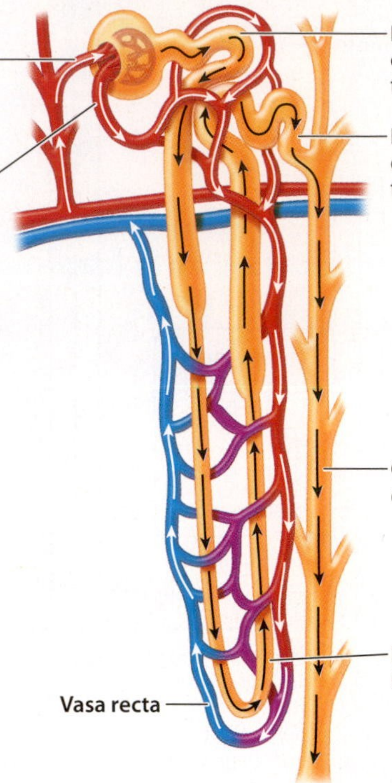

the descending and ascending limbs move in parallel but opposite directions. This system is therefore known as a **countercurrent multiplier.** While countercurrent exchangers *maintain* a concentration gradient, countercurrent multipliers *generate* them.

There is one more important point to consider. Typically, a high concentration of solutes in the interstitial fluid would be quickly lost because of water flowing out of blood vessels by osmosis. To avoid dissipating the concentration gradient, the blood vessels in the kidneys are arranged in a particular fashion. The blood vessels, collectively called the **vasa recta,** have descending and ascending vessels arranged in a countercurrent organization, just like the loops of Henle (**Fig. 41.19**). The blood within these vessels flows down into the medulla in one arm and back up through the cortex in the other. This arrangement maintains the concentration gradient. Blood flowing from the cortex to the base of the kidney encounters increasingly concentrated interstitial fluid, allowing the passive diffusion of water out of the blood. However, the descending vessels pass in proximity to ascending vessels, which reabsorb the water. In effect, water passes straight from one arm of the vasa recta to the other, so the interstitial fluid and the bloodstream have no net gain or loss of water.

The generation of a concentration gradient by a countercurrent multiplier in the loop of Henle is a key step in the production of urine that is more concentrated than the blood. It was first proposed by the Swiss chemist Werner Kuhn in the 1950s. Although his hypothesis was initially met with skepticism, subsequent experimental work confirmed its predictions (**Fig. 41.20**).

HOW DO WE KNOW?

FIG. 41.20

How does the mammalian kidney produce concentrated urine?

BACKGROUND The mammalian kidney produces urine that is more concentrated than the blood. How it concentrates urine was one of the most perplexing problems in renal physiology.

HYPOTHESIS One hypothesis is that the kidney actively transports water out of the collecting ducts as the filtrate leaves the kidney. Another, first suggested by the Swiss chemist Werner Kuhn in the early 1950s, is that the loop of Henle creates a concentration gradient from the cortex to the medulla by a countercurrent multiplier mechanism. According to this model, the solute concentration is low in the cortex and high in the medulla, so water leaves the collecting ducts passively as it moves from the cortex to the medulla.

PREDICTION A prediction of Kuhn's hypothesis is that the solute concentrations of fluid taken at the same level from the medulla, loop of Henle, and collecting ducts should be similar to one another, and higher than that of the blood. If, instead, water is simply pumped out of the collecting ducts, then the fluid in the collecting ducts should be more concentrated than that in the loop of Henle.

EXPERIMENT American physiologists Carl W. Gottschalk and Margaret Mylle tested Kuhn's hypothesis. They used innovative micropuncture techniques in hamsters to measure the solute concentration of fluid taken from the bend of the loop of Henle and the collecting ducts at the same level.

→ **Quick Check 4** The filtrate is dilute at the start and end of the loop of Henle. What, then, is the function of the loop of Henle?

The distal convoluted tubule secretes additional wastes.

The distal convoluted tubule is located in the cortex. The filtrate entering the tube is dilute, having lost nearly all of its electrolytes in the ascending limb of the loop of Henle, as we have seen. Urea is the principal solute in the filtrate.

The distal convoluted tubule is one of the main sites of secretion. Additional wastes not filtered by the glomerulus are actively secreted into the distal convoluted tubule from the bloodstream. In addition, the distal convoluted tubule participates in the regulation of key electrolytes, such as potassium, sodium, and calcium. Potassium is secreted from the blood into the tubule, whereas sodium and calcium are reabsorbed from the tubule into the blood in response to hormones. Finally, the distal convoluted tubule helps to regulate blood pH: bicarbonate (HCO_3^-) and protons (H^+) are secreted into the tubule or reabsorbed out of it, as needed.

As a result, the filtrate leaving the distal convoluted tubule and entering the collecting ducts is dilute and contains mainly urea and other wastes.

The final concentration of urine is determined in the collecting ducts and is under hormonal control.

The final segment of the nephron is the collecting duct. The collecting duct is where water levels are adjusted to meet the osmoregulatory needs of the organism and therefore maintain homeostasis. The concentration gradient from cortex to medulla established by the loop of Henle allows the collecting ducts to retain or lose water as needed depending on the water levels of the organism.

The dilute filtrate enters the collecting ducts, passing from the cortex to the medulla. These ducts drain into the renal pelvis at the base of the kidney. In response to the concentration gradient from the cortex to the medulla, water moves out of the collecting ducts into the interstitial fluid by osmosis as long as the collecting ducts are permeable to water.

The permeability of the collecting ducts to water can be adjusted to determine how dilute or concentrated the urine is. The water permeability of the collecting ducts is controlled by the peptide hormone **antidiuretic hormone (ADH),** or **vasopressin.** ADH is secreted by the posterior pituitary gland (Chapter 38) in response to signals from the hypothalamus, where receptors monitor the concentration of solutes in the blood. When a person is dehydrated, solute concentration is high and ADH is secreted. In the presence of ADH, the collecting duct walls become permeable to water (**Fig. 41.21a**) as aquaporins insert into cell membranes. Water in the collecting duct diffuses into the interstitial fluid through the aquaporins and the urine becomes more concentrated. When a person is dehydrated, the urine leaving the collecting ducts can become as concentrated as the interstitial fluid at the base of the kidney.

When a person is well hydrated, no ADH is released. In the absence of ADH, the collecting ducts are impermeable to water,

Solute concentration (milliosmoles per kilogram of water)

HAMSTER NO.	LOOP OF HENLE	COLLECTING DUCT	PLASMA
1	1391	1402	308
2	725	720	336
3	1270	1206	325
4	453	453	

RESULTS Their results, shown in the table, are consistent with the predictions of Kuhn's hypothesis. That is, the solute concentrations of the fluid in the loop of Henle and in the collecting ducts are similar to each other and higher than that of the blood.

CONCLUSION These results are consistent with a model in which concentrated urine is produced by first generating a concentration gradient between the cortex and the medulla, and then allowing water to move out of the collecting ducts by osmosis.

FOLLOW-UP WORK Similar measurements in other areas of the kidney and in other mammals and birds supported the model.

SOURCES Gottschalk, C. W., and M. Mylle. 1958. "Evidence That the Mammalian Nephron Functions as a Countercurrent Multiplier System." *Science* 128:594; Gottschalk, C. W., and M. Mylle. 1959. "Micropuncture Study of the Mammalian Urinary Concentrating Mechanism: Evidence for the Countercurrent Hypothesis." *American Journal of Physiology* 196 (4):927–936.

FIG. 41.21 Hormonal control of urine concentration by ADH. (a) In the presence of ADH, the collecting duct is more permeable to water than (b) in the absence of ADH.

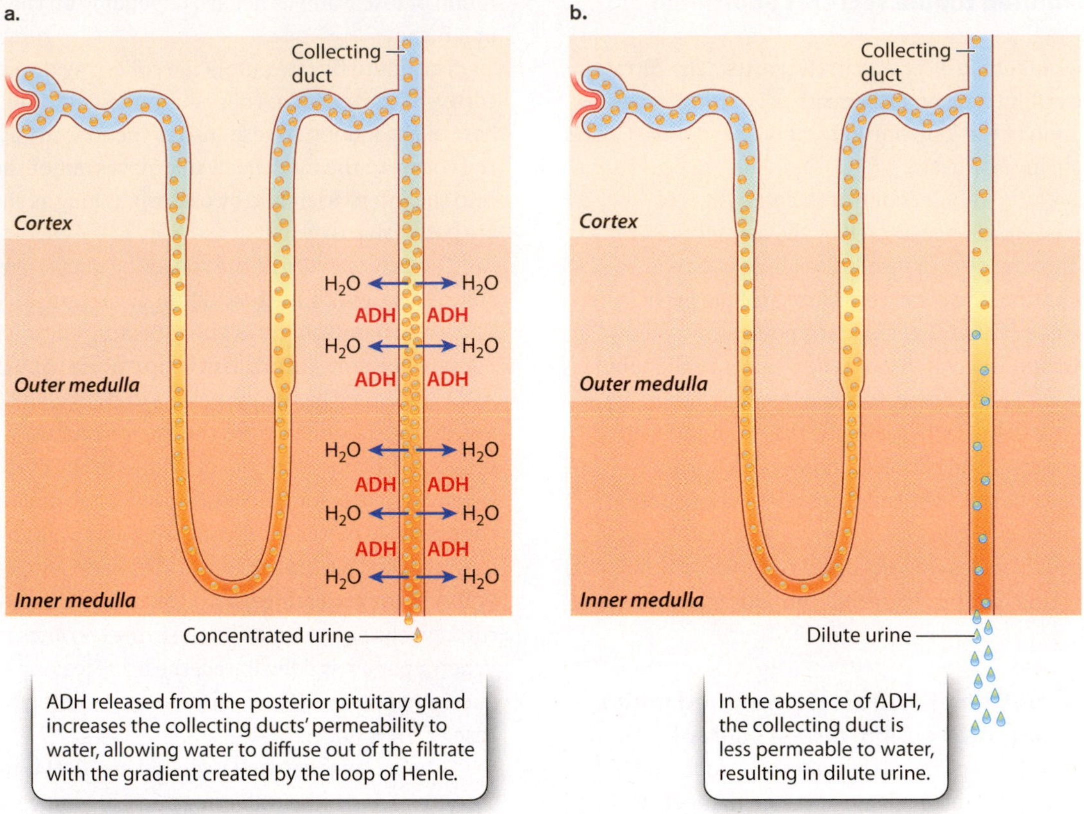

preventing the water from being reabsorbed (**Fig. 41.21b**). The result is dilute urine. Some diuretic substances ("diuretic" is from a Greek word meaning "to flow through"), such as alcohol, promote the dilution of urine by interfering with ADH production and therefore making the collecting ducts impermeable to water. When dilute urine is produced, there is net water loss. Consequently, diuretic drinks often cause dehydration.

The regulation of water and solute reabsorption by the kidney is remarkably precise. The kidneys of a human filter 180 L of blood each day (125 mL/min). The total blood volume of a human is about 5 L (of which 3 L is plasma), which means that the blood is filtered about 60 times per day. Urine production in humans averages 0.5 L per day, so under normal conditions, the kidneys reabsorb 99.9% of the water entering the renal tubules. In addition, the kidneys reabsorb all the glucose and amino acids that are filtered out of the blood, together with 99.9% of the electrolytes in the blood.

The exceptional ability of the kidneys to filter and reabsorb electrolytes, solutes, and water also explains why kidney damage can have such serious consequences. People with kidney disease must have their blood filtered by dialysis treatment as frequently as three times per week, each treatment lasting several hours. Otherwise, toxic wastes build up to dangerous levels in the body. Kidney disease remains a substantial health problem in the world today, costing more than $34 billion in the United States alone.

The kidneys help regulate blood pressure and blood volume.

Because the kidneys receive a large fraction of the blood leaving the heart, they are well suited for monitoring changes in blood pressure and maintaining it at relatively constant levels, another example of homeostasis. Specialized cells of the efferent arteriole leaving the glomerulus of each nephron form the **juxtaglomerular apparatus** (**Fig. 41.22**). In response to a drop in blood pressure, these cells secrete the hormone **renin.** Renin converts angiotensinogen, which is produced in the liver and circulates in the blood, into angiotensin I. Angiotensin I in turn is converted by an enzyme present in the lungs and elsewhere to its active form, **angiotensin II.** Angiotensin II acts on the smooth muscles of arterioles throughout the body, causing them to constrict, thus increasing blood pressure and directing more blood back to the heart.

Angiotensin II also stimulates the release of the hormone **aldosterone** from the adrenal glands (Fig. 41.22). Aldosterone stimulates the distal convoluted tubule and collecting ducts to increase the reabsorption of electrolytes and water back into

FIG. 41.22 **Control of blood volume and blood pressure by the kidney.** Cells of the efferent arteriole of the glomerulus sense blood pressure and start a hormone cascade when blood pressure gets too low.

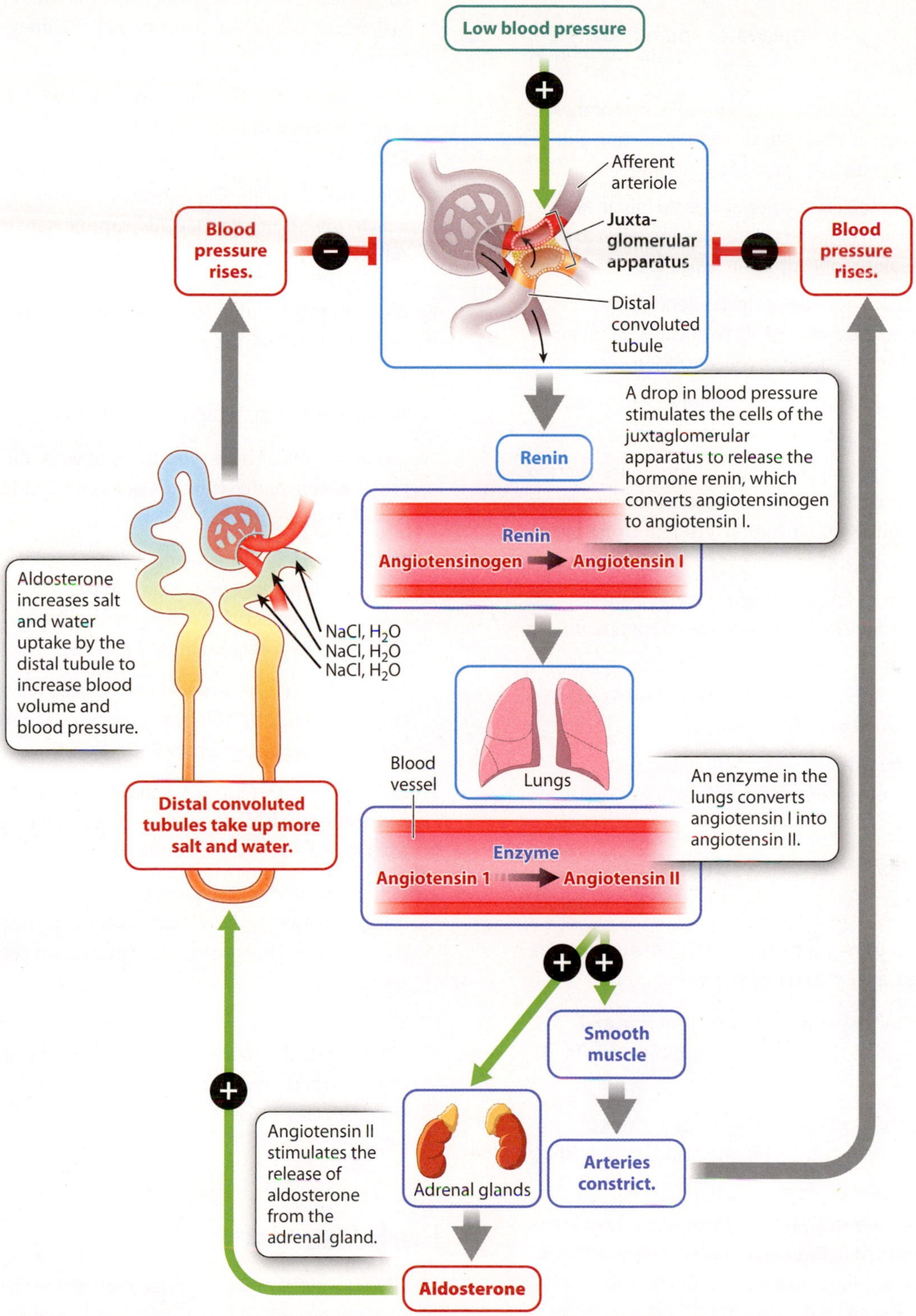

the blood. Blood volume increases and therefore blood pressure increases, too. The kidneys, therefore, play many roles. In addition to eliminating wastes and other toxic compounds, they maintain homeostasis in several ways, including through the maintenance of water and electrolyte balance and the regulation of blood volume and blood pressure. ■

Core Concepts Summary

41.1 All animals regulate the water and electrolyte levels within their cells.

Cell membranes act as selectively permeable membranes, allowing the passage of water but restricting or controlling the movement of many solutes. page 876

Water moves across selectively permeable membranes by osmosis from a region of lower solute concentration to a region of higher solute concentration. page 876

Osmotic pressure is a measure of the tendency for water to move by osmosis across a selectively permeable membrane into a solution with higher solute concentration. page 876

Osmoregulation is the regulation of osmotic pressure inside cells or organisms. page 877

Osmoconformers maintain an internal solute concentration similar to that of the environment, whereas osmoregulators have an internal solute concentration different from that of the environment. page 878

Both osmoconformers and osmoregulators regulate the levels of particular solutes, especially sodium, potassium, and chloride. page 878

Osmoregulators that live in high-salt environments excrete excess electrolytes and minimize water loss. page 878

Osmoregulators that live in low-salt or freshwater environments excrete excess water and minimize electrolyte loss. page 879

Osmoregulators that live on land minimize water loss. page 879

41.2 Excretory organs eliminate nitrogenous wastes and regulate water and electrolyte levels.

Osmoregulation and excretion are closely coordinated processes controlled by the excretory organs of animals. page 881

Ammonia, urea, and uric acid are three major forms of nitrogenous waste that animals excrete, depending on their evolutionary history and habitat. page 881

Ammonia is toxic but readily diffuses into water; urea is less toxic and can be stored before it is eliminated; uric acid is a semisolid and can be eliminated with minimal water loss as an adaptation for living on land. page 881

All animal excretory systems first isolate or filter fluid into an extracellular space. Key electrolytes, solutes, and water are then reabsorbed from the filtrate. Additional wastes are then secreted into the remaining filtrate before it is eliminated from the body. page 882

Examples of excretory organs are the protonephridia of flatworms, the metanephridia of segmented annelid worms, the Malpighian tubules of insects, and the kidneys of vertebrates. page 884

The urine produced by the kidneys and other excretory organs is stored in a bladder until it is eliminated from the body. page 885

The functional unit of the vertebrate kidney is the nephron, which consists of a glomerulus, capsule, renal tubules, and collecting duct. page 885

41.3 The mammalian kidney can produce urine that is more concentrated than blood as an adaptation for living on land.

The kidney has an outer cortex and an inner medulla. page 886

The renal tubules have specialized regions, including the proximal convoluted tubule, the loop of Henle, and the distal convoluted tubule. page 888

The loop of Henle of some nephrons extends all the way into the medulla before looping back to the cortex. These nephrons create a concentration gradient from the outer cortex to the deeper medulla of the kidney. page 888

The ability to create a concentration gradient from the cortex to the medulla results from a countercurrent multiplier mechanism of the loops of Henle. page 890

The final concentration of urine is regulated by adjusting water permeability within the walls of the collecting ducts, which pass from the cortex to the medulla. page 891

Release of antidiuretic hormone by the posterior pituitary gland increases the water permeability of the collecting ducts, concentrating the urine that empties from vertebrate kidneys. page 891

The kidneys help to regulate blood volume and pressure by secreting the hormone renin, leading to the production of angiotensin II, which constricts blood vessels, and aldosterone, which increases reabsorption of salt and water by the kidneys. page 892

Self-Assessment

1. Given a selectively permeable membrane that is permeable to water but not to a particular solute and given that there are different solute concentrations on the two sides of the membrane, show the direction of water movement and label the side with the higher osmotic pressure.
2. Describe how animals gain and lose water and electrolytes.

3. Name two animals that are osmoconformers and two that are osmoregulators. Explain the difference between the two types of animal.
4. List three forms of nitrogenous waste and describe how each is an adaptation for the environment in which the animal lives.
5. Describe the three steps in which organisms excrete wastes.
6. Draw a mammalian nephron, label and describe the primary function of each part, and show the direction of water and electrolytes in each part.
7. Explain how the loop of Henle creates a concentration gradient from the cortex to the medulla.
8. Describe the role of ADH in the regulation of urine concentration.
9. Explain how the kidneys help to regulate blood volume and blood pressure.

Log in to LaunchPad to check your answers to the Self-Assessment questions, and to access additional learning tools.

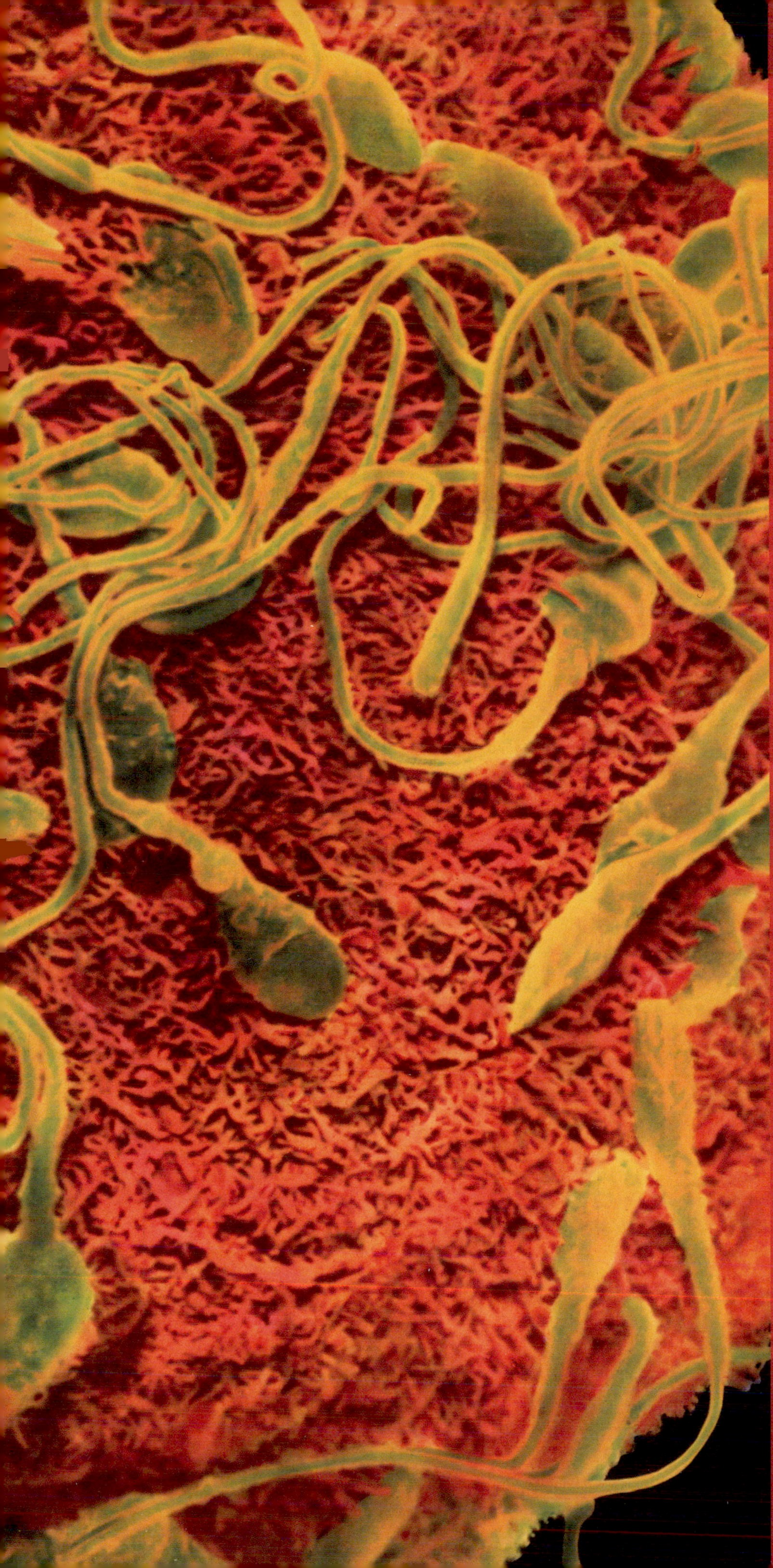

SCIEPRO/SPL/age fotostock.

CHAPTER 42

Animal Reproduction and Development

Core Concepts

42.1 Reproduction is a basic feature of living organisms and occurs both asexually and sexually.

42.2 The movement of vertebrates from water to land involved changes in reproduction, including internal fertilization and the amniotic egg.

42.3 The male reproductive system is adapted for the production and delivery of sperm, and the female reproductive system is adapted for the production of eggs and, in some cases, support of the developing fetus.

42.4 Human reproduction involves the formation of gametes, fertilization, and growth and development.

Children are sometimes gently introduced to the subject of human reproduction by reference to "the birds and the bees." Since birds and bees reproduce, it is assumed that they can substitute for humans in discussions about reproduction. But how relevant is bird and bee reproduction to human reproduction? It might come as a surprise that most male birds don't have penises, and that male bees don't even have fathers. Perhaps, then, the lesson of "the birds and the bees" is not so much in what it teaches us about human reproduction, but, as we explore in this chapter, in showing us that reproductive strategies among organisms are spectacularly diverse.

Reproduction is a striking and conspicuous feature of the natural world. Flowers burst into bloom each spring to attract pollinators; male fireflies light up brilliantly on warm summer nights as they signal to females in the grass below; birds sing and display ornate plumage to attract mates. Most living organisms have the ability to reproduce—that is, to make new organisms like themselves. This remarkable ability is not unique to living organisms—after all, computer viruses and self-replicating molecules are capable of reproduction but are not themselves living. However, reproduction is a key attribute of life. All living things came about by reproduction, and most are capable in turn of reproduction. Of course, this statement does not address the question of how life originated in the first place, a topic considered in Case 1: The First Cell.

This chapter focuses on reproduction and development. We begin by considering reproduction broadly among diverse organisms, then turn our attention to animal, and specifically vertebrate, reproduction and the surprise that even among vertebrates there is quite an array of reproductive strategies. In the last two sections, we focus on human reproduction, starting with reproductive anatomy and physiology, and then following reproduction from gamete formation to fertilization, pregnancy, and birth.

42.1 THE EVOLUTIONARY HISTORY OF REPRODUCTION

While the ability to reproduce is shared across all domains of life, the particular way in which animals reproduce varies among species, from the familiar to what might strike us as truly bizarre. To take just one of many examples: Soon after it is born, a male anglerfish finds a female anglerfish, bites her skin, and fuses with her. His body atrophies, leaving the male reproductive organs physically attached to the female and allowing her eggs to be fertilized when she is ready. To begin to make sense of this and other reproductive strategies, we start by considering how different forms of reproduction evolved, what organisms use them, and their adaptive significance.

Asexual reproduction produces clones.

Organisms have two potential ways to reproduce. One is the production of genetically identical cells or individuals called **clones.** This form of reproduction is called **asexual reproduction,** and it occurs in a wide variety of organisms (**Fig. 42.1**).

For example, bacteria (Fig. 42.1a) and archaeons reproduce by binary fission, a form of cell division in which the genome replicates and then the cell divides in two (Chapter 11). The result is two cells that are either identical to the parent cell or nearly identical, differing only because of chance mutations.

Prokaryotes can nevertheless increase genetic variation by horizontal gene transfer, as described in Chapter 26. For example, they can transfer DNA from one individual to another, usually of the same species, and much more rarely to an individual of another species, during conjugation. They can also obtain DNA directly from the environment and by viral infection, incorporating genes that might have evolved in even distantly related species. These processes allow prokaryotes to acquire novel genetic sequences that can affect fitness.

Eukaryotic cells, on the other hand, divide by mitotic cell division (Chapter 11), a process that evolved early and was present in the common ancestor of all living eukaryotes. Mitotic cell division is a form of growth and development in all eukaryotes, and is the basis of asexual reproduction in many eukaryotes.

One form of asexual reproduction in fungi, plants, and some animals is **budding,** in which a bud, or protrusion, forms on an organism and eventually breaks off to form a new organism that is smaller than its parent. The budding yeast, *Saccharomyces cerevisiae,* which is used to make bread and beer, is so named because of its ability to produce new individuals by budding (Fig. 42.1b). Budding can also occur in animals such as the water-dwelling hydra, which can grow and subsequently break off part of itself to form a new, free-living individual. Whether in unicellular or in multicellular organisms, budding involves an unequal division between a mother and daughter cell or organism. Budding is the result of mitosis and therefore, like binary fission, results in an offspring that is genetically identical to the parent.

Another form of asexual reproduction is **fragmentation,** in which new individuals arise asexually by the splitting of one organism into pieces, each of which develops into a new individual. Certain molds, algae, worms, sea stars, and corals can reproduce by fragmentation. Fragmentation occurs naturally, but humans can use this mode of reproduction to propagate corals in the restoration of coral reefs (Fig. 42.1c). Once again, the new individual is a clone of the parent.

A final example of asexual reproduction is **parthenogenesis,** literally "virgin birth." In this form of asexual reproduction, females produce eggs that are not fertilized by males but divide by mitosis and develop into new individuals. Parthenogenesis explains why male bees don't have fathers. It occurs in invertebrates, including some insects and crustaceans, as well as in vertebrates, including some fish and reptiles (Fig. 42.1d). The details of parthenogenesis differ depending on the organism. In most sexually reproducing animals, egg

FIG. 42.1 Asexually reproducing organisms. (a) Bacteria divide by binary fission. (b) The yeast *Saccharomyces cerevisiae* reproduces by budding. (c) Corals are propagated by fragmentation. (d) Komodo dragons can reproduce by parthenogenesis. *Sources: a. CNRI/Science Source; b. SPL/Science Source; c. WaterFrame/Alamy; d. Mike Lane/Alamy.*

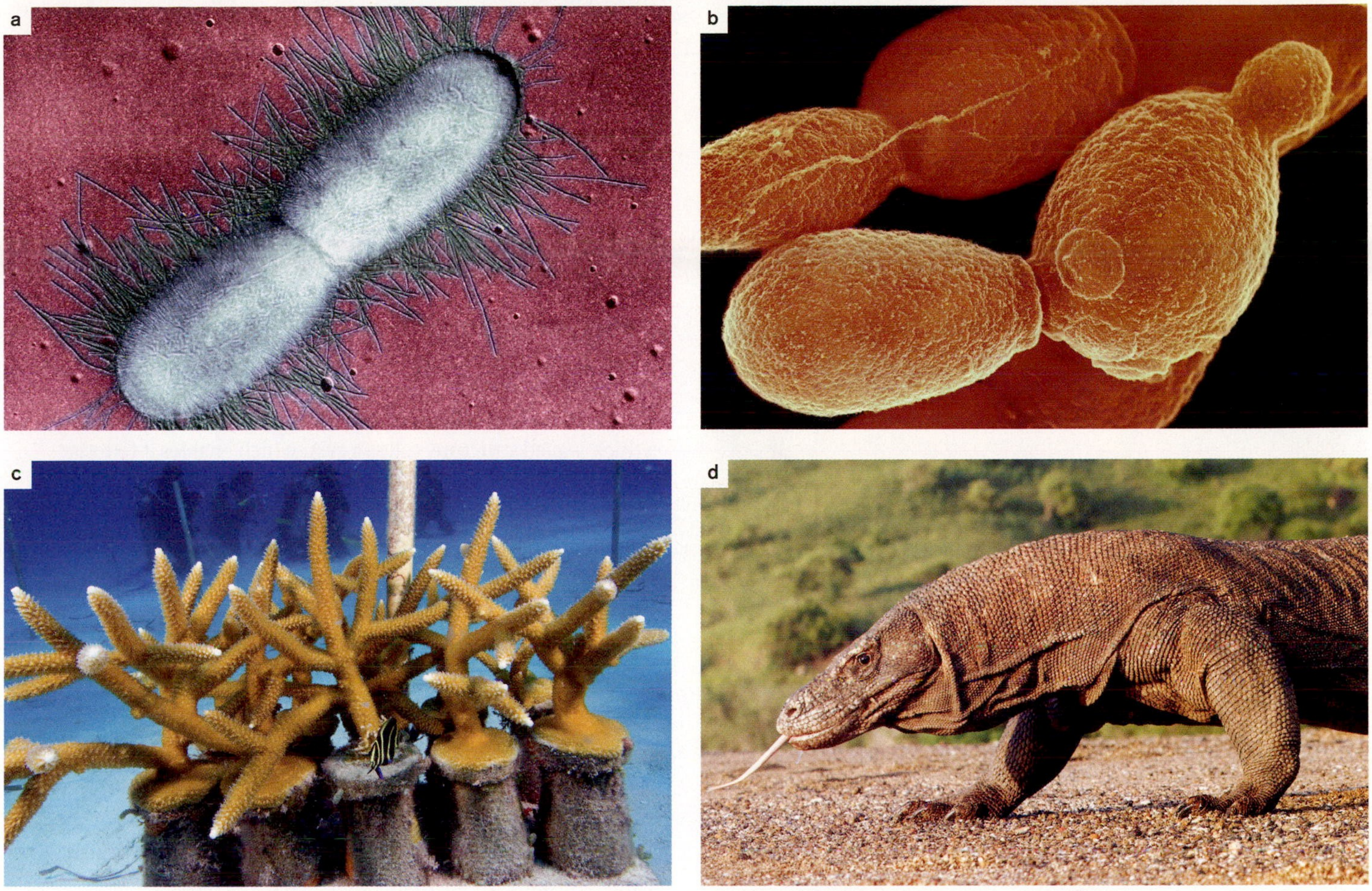

cells are usually the product of meiosis and are haploid, having one set of chromosomes. In some parthenogenic organisms, this haploid egg cell divides repeatedly by mitosis to become a haploid adult; in others, the DNA content doubles during development so the adult is diploid, having two sets of chromosomes; and in still others, the egg cell is not the product of meiosis (it is diploid) and divides by mitosis to become a diploid adult. In all cases, parthenogenesis involves a single maternal parent.

Sexual reproduction involves the formation and fusion of gametes.

The second way that organisms reproduce is by combining complete sets of genetic information from two individuals to make a new genetically unique individual. This form of reproduction is called **sexual reproduction.**

At its core are two basic biological processes—**meiotic cell division** and **fertilization** (**Fig. 42.2**). Meiotic cell division halves the number of chromosomes; fertilization involves fusion of the products of meiotic cell division to restore the original chromosomal content. Meiotic cell division evolved in the ancestors of modern eukaryotes, and so sexual reproduction occurs only in eukaryotes.

As we saw in Chapter 11, meiotic cell division consists of one round of chromosome replication followed by two rounds of cell division, resulting in four cells each with half the number of chromosomes of the parent cell. So, in a diploid species with $2n$ chromosomes, the result of meiosis is four cells with n chromosomes. The resulting cells are haploid and are called **gametes** or **spores.** Human individuals, for example, have 23 pairs of chromosomes ($2n = 46$ chromosomes) and human gametes have half this number ($n = 23$ chromosomes).

Many species produce two types of gamete that differ in shape and size. The smaller, male gametes are called **spermatozoa** or **sperm,** and the larger, female gametes are called **ova** or **eggs.** For other organisms, including green algae, diatoms, slime molds, and

FIG. 42.2 Sexual reproduction. This form of reproduction involves the production of gametes by meiosis and their fusion by fertilization.

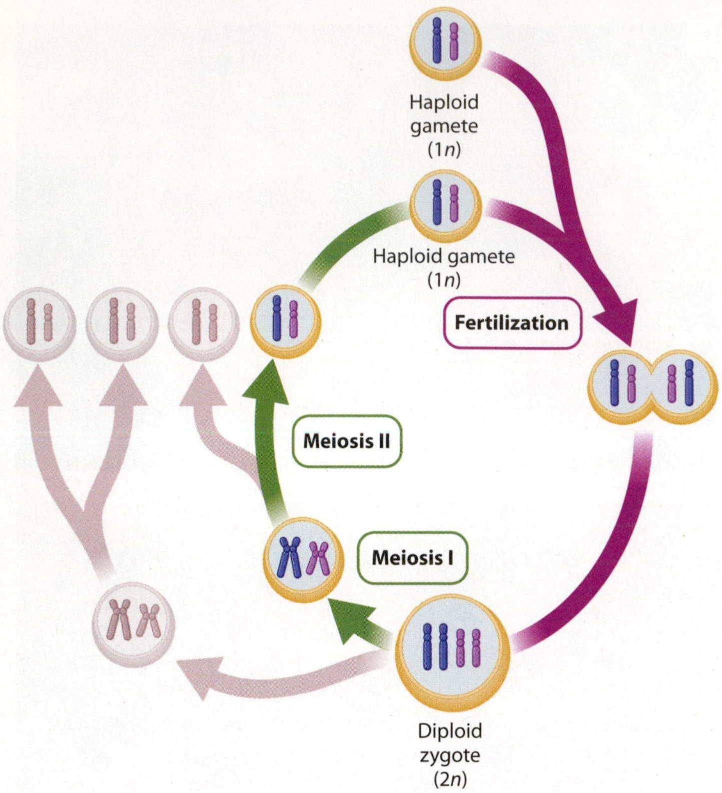

most fungi, the two types of gamete are the same size. In spite of their similar size and appearance, however, they come in two (or more) distinct mating types and cannot reproduce sexually with individuals of the same type.

Fertilization involves the fusion of two gametes, one from each parent (Fig. 42.2). The result is a **zygote.** If each gamete has *n* chromosomes, the resulting zygote has 2*n* chromosomes. For diploid multicellular organisms, fertilization is the direct basis for reproduction. The zygote divides by mitosis and develops into an **embryo,** an early stage of multicellular development.

Meiotic cell division and fertilization have important genetic consequences. During meiotic cell division, pairs of homologous chromosomes undergo recombination, shuffling the combination of alleles between the chromosomes. In addition, meiotic cell division randomly sorts homologous chromosomes into different gametes. As a result, the products of meiotic cell division, the haploid gametes, are each genetically unique and represent a mixture of the parental genetic makeup. Then, during fertilization, two of these unique haploid gametes fuse, creating a new genetic combination and a unique diploid individual.

→ **Quick Check 1** Sexual reproduction results in offspring that are genetically different from one another and from their parents. Describe four mechanisms that produce this genetic variation. Do any of these mechanisms occur in asexual reproduction?

Many species reproduce both sexually and asexually.

Most organisms that reproduce asexually are also capable of reproducing sexually. For example, hydras reproduce asexually by budding, but their ability to reproduce asexually does not preclude their ability to reproduce sexually. Even budding yeast are capable of sexual reproduction. Most aphids, fish, and reptiles capable of parthenogenesis can also reproduce sexually.

What determines when an organism reproduces sexually and when it reproduces asexually? This question has been examined in the freshwater crustacean *Daphnia. Daphnia* reproduces asexually by parthenogenesis in the spring, when food is plentiful, perhaps because asexual reproduction is a rapid form of reproduction that allows the organism to take advantage of abundant resources (**Fig. 42.3**). Later in the season, when conditions are less favorable because of increased crowding and decreased food and temperature, *Daphnia* reproduces sexually, mating to produce a zygote that starts to form an embryo. At an early stage, however, development is arrested and metabolism slows dramatically. The embryo forms a thick-walled structure, called a cyst, which enables it to resist freezing and drying, allowing it to make it through the winter.

This link between environmental conditions and mode of reproduction has been observed in many other species, from algae to slime molds. The life cycle of *Daphnia* hints that asexual reproduction and sexual reproduction have costs and benefits, a subject we turn to now.

FIG. 42.3 Sexual and asexual reproduction in *Daphnia.* *Daphnia* can reproduce sexually or asexually, depending on environmental conditions.

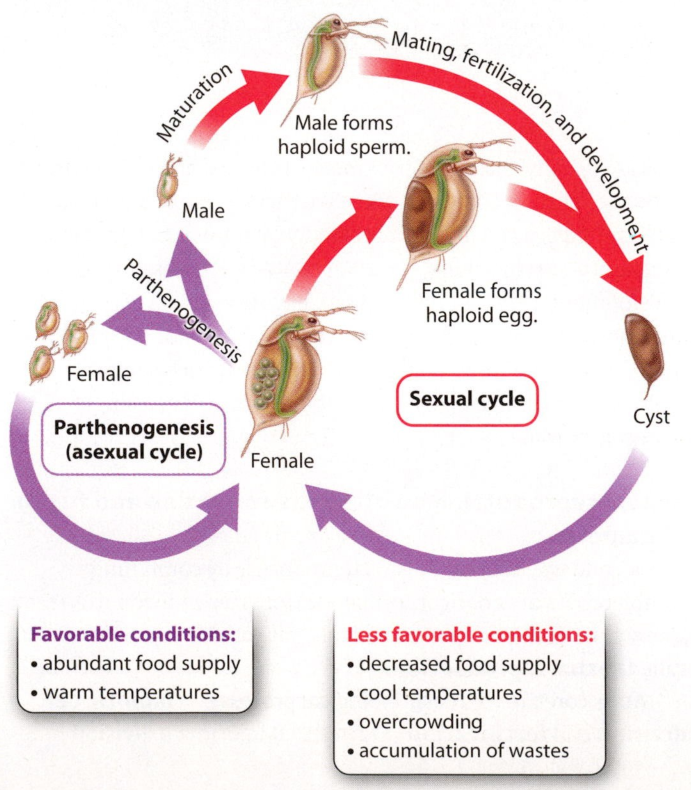

FIG. 42.4 Time and energy costs of sexual reproduction. (a) A male bird of paradise has colorful plumage to attract a mate. (b) A bowerbird builds an elaborate hut to entice a female. *Source: Tim Laman Photography.*

Exclusive asexuality is often an evolutionary dead end.

Asexual reproduction has some advantages compared to sexual reproduction. It does not involve finding and attracting a mate, which takes time and energy. Certain birds, for example, go to elaborate lengths to attract mates. Male birds of paradise invest a significant amount of energy in growing colorful plumage, and male bowerbirds take time to build elaborate homes for potential mates (**Fig. 42.4**). For other organisms, finding a mate can be difficult because of very low population density—this is the case for anglerfish, and may help to explain their unusual reproductive strategy, described earlier.

In addition, asexual reproduction is rapid, allowing organisms to increase their numbers quickly. Consider rates of population growth for organisms that reproduce asexually and those that reproduce sexually. In asexual organisms, all offspring can produce more offspring, allowing for exponential population growth if conditions are favorable. By contrast, in sexually reproducing organisms, it takes two individuals to produce new offspring. The consequence of this difference is that an asexual group can have a much greater rate of population growth compared with that of a sexually reproducing one. The British evolutionary biologist John Maynard Smith called this effect the **twofold cost of sex** (**Fig. 42.5**), and it is one of the key disadvantages of sexual reproduction.

In spite of the advantages of asexual reproduction, a survey of the natural world reveals a striking pattern: Most organisms reproduce sexually at least part of the time. And even prokaryotes, as we have seen, have mechanisms for genetic exchange between individuals. The observation that the vast majority of organisms reproduce sexually at some point suggests that sexual reproduction offers advantages over asexual reproduction. Asexual reproduction is quick, but it produces offspring that are genetically identical to one another and to the parent. The only

FIG. 42.5 The twofold cost of sex. Population size increases more rapidly in asexually reproducing organisms compared to sexually reproducing organisms.

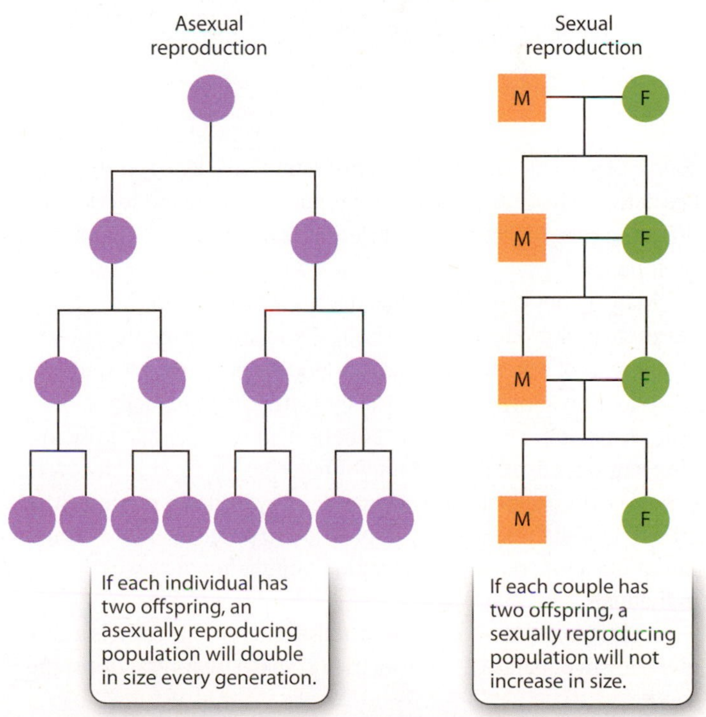

HOW DO WE KNOW?

FIG. 42.6

Do bdelloid rotifers reproduce only asexually?

BACKGROUND Many groups of animals reproduce asexually, but most of these organisms reproduce sexually at least some of the time. The bdelloid rotifer, one of several species of microscopic aquatic animals called rotifers, may be an exception.

OBSERVATION All bdelloid rotifers found to date are females. No males or hermaphrodites (organisms with both male and female reproductive organs) have ever been observed. Meiosis has never been documented.

HYPOTHESIS Bdelloid rotifers are asexual and reproduce solely by parthenogenesis, in which unfertilized eggs develop into adult females.

PREDICTION In the absence of meiosis, the two copies (alleles) of a gene do not recombine, and mutations that accumulate are not reshuffled between them. As a result, two alleles of a gene in an asexually reproducing organism will show a much higher degree of sequence difference compared with two alleles of a gene in a sexually reproducing organism.

RESULTS At right is the sequence from the two copies of the *hsp82* gene in a sexually reproducing rotifer species (Fig. 42.6a) and an asexually reproducing rotifer species (Fig. 42.6b). As can be seen, the two copies are similar in Fig. 42.6a but very different in Fig. 42.6b.

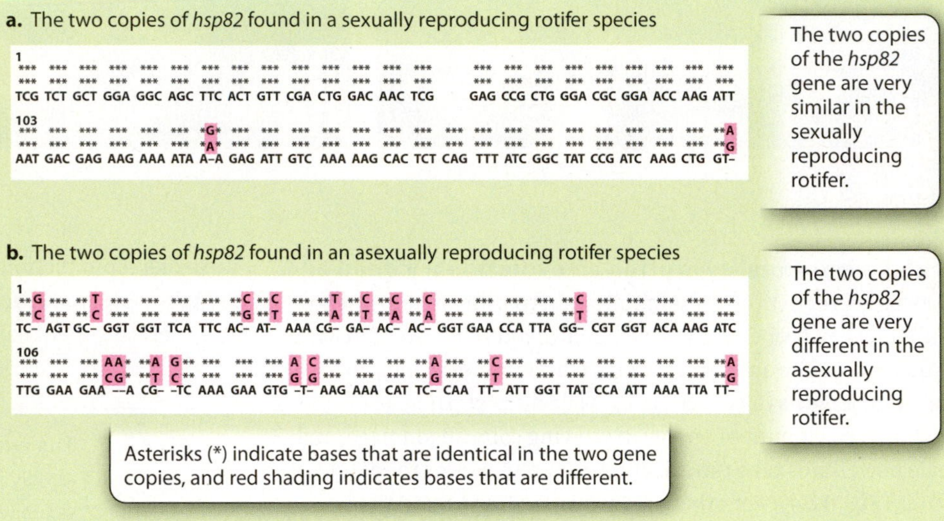

CONCLUSION The evidence from this experiment suggests that bdelloid rotifers are exclusively asexual.

SOURCE Welch, M. D., and M. S. Meselson, 2000. "Evidence for the Evolution of Bdelloid Rotifers Without Sexual Reproduction or Genetic Exchange." *Science* 288 (5469):1211–1215.

source of genetic variation is chance mutations. By contrast, one of the defining features of sexual reproduction is the production of offspring that are genetically different from one another and from their parents.

Many hypotheses have been developed to explain the advantages of producing genetically distinct offspring in spite of the costs of sexual reproduction. One proposes that sexual reproduction allows species to adapt faster than asexually reproducing ones because rare beneficial mutations that arise in different organisms can be brought together, increasing the overall fitness of the population. Another set of hypotheses proposes that sexual reproduction allows a population to purge itself of harmful mutations more quickly than could a population of asexual individuals.

A third hypothesis proposes that sexual reproduction is a mechanism of parasite defense. In host–parasite interactions, the host evolves defenses against the parasite, and the parasite evolves new ways to infect the host, so each evolves rapidly simply to maintain its ecological position. Called the Red Queen hypothesis, this idea takes its name from Lewis Carroll's *Through the Looking-Glass,* in which the Red Queen tells Alice that she needs to keep running to stay in the same place. This kind of rapid evolution requires lots of genetic variation on which natural selection can act, and sexual reproduction increases genetic variation among individuals more than asexual reproduction does.

These different hypotheses for the prevalence of sexual reproduction are not mutually exclusive. It may be that several mechanisms work together, or that different ones are important for different organisms.

It is instructive to consider groups of eukaryotes that have lost the ability to reproduce sexually, and those that have lost the ability to reproduce asexually. While there are organisms such as the New Mexico whiptail lizard (*Cnemidophorus neomexicanus*) that reproduce only asexually, these species are not very old in evolutionary terms. In other words, there are very few, if any, *ancient* asexual groups. It appears that asexual reproduction on its own rarely allows a species to persist for very long in evolutionary terms.

A possible exception to this rule may be bdelloid rotifers, a group of microscopic freshwater invertebrates that are thought to have reproduced asexually for at least 35 to 40 million years (**Fig. 42.6**). If indeed bdelloid rotifers have persisted for tens of millions of years without sexual reproduction, we would predict that they have evolved some process other than sexual reproduction for increasing genetic variation. In fact, evidence suggests that they are able to take up DNA from their environment, similar to what bacteria do through transformation (Chapter 26). The bdelloid rotifers seem to be fantastic "collectors" of DNA from other organisms, including bacteria, plants, and fungi, totaling over 500 different species. However, comparison of DNA sequences among bdelloids has recently demonstrated that different individuals can share genes that are identical or almost identical in sequence, which is not possible without at least occasional sexual reproduction. More research is needed to really understand reproduction in this group.

In contrast to organisms that lose the ability to reproduce sexually, mammals are unusual in the opposite way: They have lost the ability to reproduce asexually. Parthenogenesis, as we have seen, occurs in some vertebrates, but does not occur in mammals. Although parthenogenesis may seem odd from our point of view, scientists have wondered just the opposite—why can't female mammals reproduce without a father? Asked another way, why do mammalian embryos need both a maternal and a paternal genome? Wouldn't two maternal genomes from a single parent do the trick?

Evidently not. Scientists have known for some time that mammalian maternal and paternal genomes are not quite the same. There are some genes, called imprinted genes, that are expressed differently depending on their parent of origin (Chapter 19). Imprinting ensures that mammals develop from embryos with one maternal and one paternal genome. When researchers manipulated key imprinted genes in a maternal genome, they were able to breed a parthenogenetic mouse—one with two copies of a maternal genome from a single parent. We can hypothesize that early mammals gained fitness advantages in the evolution of imprinted genes, but lost the ability to reproduce parthenogenetically.

42.2 MOVEMENT ONTO LAND AND REPRODUCTIVE ADAPTATIONS

We have considered the two modes of reproduction observed across all living organisms. In this section, we narrow our focus to vertebrates: fish, amphibians, reptiles, birds, and mammals. Interestingly, even among vertebrates, there are many different reproductive strategies. Understanding these strategies not only gives us a picture of how these organisms reproduce but also sheds light on how they live and behave.

How vertebrates reproduce is also intimately connected to the transition of vertebrates from water to land 400 to 375 million years ago. Moving from an aquatic environment to a terrestrial one involved many adaptations. Among these were changes in the way that reproduction took place, so it is helpful to consider reproductive strategies in an evolutionary context.

Fertilization can take place externally or internally.

To survive, eggs and sperm require a wet environment. In aquatic organisms like fish and amphibians, eggs and sperm can be released directly into the water by **external fertilization.** External fertilization, as you might imagine, is a chance affair: Somehow, sperm must meet egg in a large environment. Animals that release gametes externally therefore have developed strategies to increase the probability of fertilization. Some species release large numbers of gametes. Many female fish, including salmon and perch, lay eggs on or near a substrate, such as a rock, and males release their sperm onto the eggs to increase the probability of fertilization. Externally fertilizing animals may also come close together physically to improve chances that the sperm will fertilize the eggs. Male frogs grasp females, even though fertilization occurs externally, releasing sperm as the female lays her eggs (**Fig. 42.7**).

FIG. 42.7 External fertilization. Tree frogs reproduce by external fertilization, with the male frog holding on to the female and fertilizing her eggs as she lays them. *Source: George Grall/National Geographic Society/Corbis.*

A challenge for land-dwelling organisms is keeping gametes from drying out. **Internal fertilization,** in which fertilization takes place inside the body of the female, is an adaptation for living on land. Reptiles, birds, and mammals all use internal fertilization. However, this mode of fertilization is not exclusive to land dwellers. Marine mammals such as whales and dolphins use internal fertilization, as did the land-dwelling ancestors from which they evolved. Even some fish, such as guppies, reproduce by internal fertilization.

r-strategists and *K*-strategists differ in number of offspring and parental care.

While the terms "external" and "internal" fertilization focus our attention on where fertilization takes place, the two modes of reproduction are associated with a host of other behavioral and reproductive differences. One key difference is the number of offspring produced.

External fertilization is generally associated with the release of large numbers of gametes and the production of large numbers of offspring, each of which has a low probability of survival in part because the offspring typically receive little, if any, parental care. External fertilization, in other words, is a game of numbers. For internal fertilization, however, there is much more control over fertilization and subsequent development of the embryo because both fertilization and embryonic development take place inside the female. As a result, animals that use internal fertilization typically produce far fewer offspring and invest considerable time and energy into raising those offspring.

These two strategies represent two ends of a continuum of reproductive strategies that was first described by the American ecologists Robert H. MacArthur and E. O. Wilson in 1967. Organisms that produce large numbers of offspring without a lot of parental investment are ***r*-strategists,** and those that produce few offspring but put in a lot of parental investment are ***K*-strategists** (**Fig. 42.8;** Chapter 46). Fish are *r*-strategists; humans are *K*-strategists.

According to this model, the environment places selective pressures on organisms to drive them in one or the other direction. In general, *r*-strategists evolve in unstable, changing, and unpredictable environments. In such an environment, there is an advantage to reproducing quickly and producing many offspring when conditions are favorable. By contrast, *K*-strategists often evolve in stable, unchanging, and predictable environments, where there tends to be more crowding and larger populations. Here, there is intense competition for limited resources, so traits such as increased parental care and few offspring are favored.

These two strategies represent the extremes of a spectrum, and there are notable exceptions. An octopus, for example, can lay up to 150,000 eggs, typical of *r*-strategists, but the female guards and takes care of them for a month or more, typical of *K*-strategists. However, the two strategies highlight different reproductive patterns across animals and suggest important links between number of offspring, parental care, and the environment.

Animals either lay eggs or give birth to live young.

When we speak of parental care, it is natural to think of parents tending the young. However, parental care actually begins much earlier in development, in the provisioning of nutrients for the developing embryo. The eggs of animals that are externally fertilized have a substance called **yolk** that provides all the nutrients that the developing embryo needs until it hatches. Water and oxygen are obtained directly from the aquatic environment and metabolic wastes easily diffuse out.

The movement onto land was associated with important changes in the anatomy of the egg. No longer could the developing embryo depend simply on diffusion of key substances into and out of a watery environment. A key innovation is the **amnion,** a membrane surrounding a fluid-filled cavity that allows the embryo to develop in a watery environment. Reptiles, birds, and mammals are known as amniotes to reflect the presence of this key membrane (**Fig. 42.9**). A second membrane, called the **allantois,** encloses a space where metabolic wastes collect. Finally,

FIG. 42.8 ***r*-strategists and *K*-strategists.** *r*-strategists, like fish, produce many offspring with low probability of survival, whereas *K*-strategists, like primates, produce few offspring and put a lot of care into raising their young.

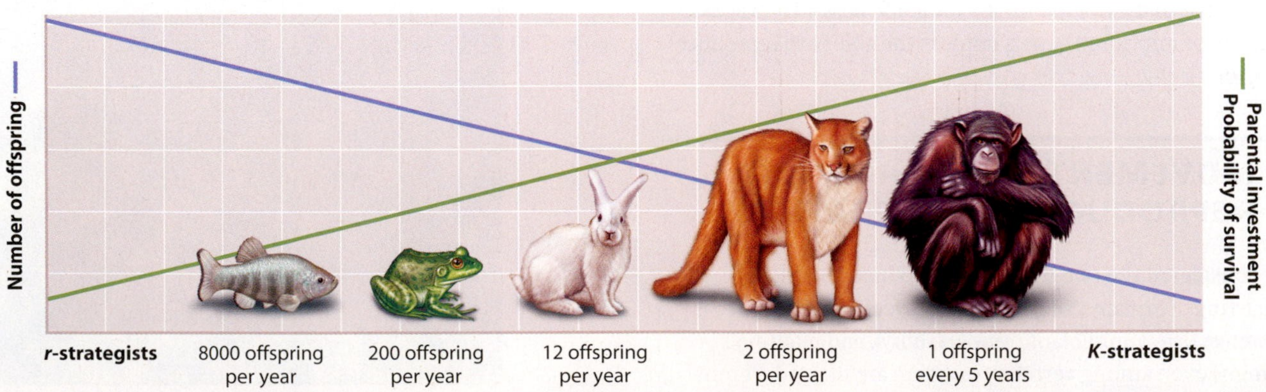

FIG. 42.9 The evolution of the amnion. The amnion is a membrane enclosing a fluid-filled cavity surrounding the developing embryo that evolved during the transition from water to land.

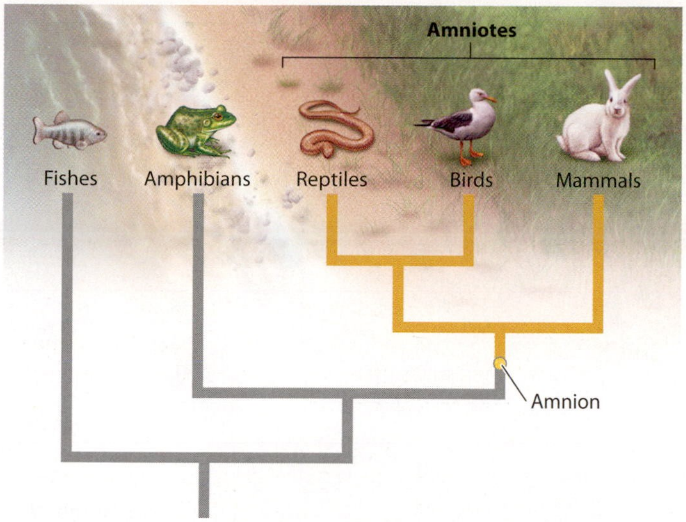

FIG. 42.10 The amniotic egg. The amniotic egg includes four extraembryonic membranes: the yolk sac, amnion, allantois, and chorion. *Photo source: Thanunkorn Klypaksi/iStockphoto.*

a third membrane, called the **chorion,** surrounds the entire embryo along with its yolk and allantoic sac. The embryo is further protected in many species by a hard shell just outside the chorion. Together, the chorion and shell limit water loss, and the chorion, shell, and allantois allow gas exchange with the environment. The yolk sac, amnion, allantois, and chorion are all **extraembryonic membranes,** sheets of cells that extend out from the developing embryo (**Fig. 42.10**).

Animals that are internally fertilized can either lay eggs or give birth to live young. Animals that lay eggs are oviparous (**oviparity** means "egg birth"). Oviparity is used by most insects, fish, amphibians, and reptiles, as well as all birds. Even some mammals—the monotremes, which include the spiny anteater and platypus—lay eggs. For oviparous animals, there is very little, if any, embryonic development inside the mother, and all the nutrients for the developing embryo come from the yolk.

Other animals give birth to live young. These animals differ in where they get their nourishment during embryonic development. In some animals, embryos develop inside eggs that are retained inside the mother until they hatch. These embryos receive nutritional support from the yolk, not from the mother. Such animals are ovoviviparous (**ovoviviparity** means "egg live birth"). Some sharks and snakes are ovoviviparous.

In other animals, the embryo develops inside the mother and receives all its nourishment from the mother. These animals are viviparous (**viviparity** means "live birth"). Most mammals and some sharks and snakes are viviparous. In fact, viviparity has evolved many times independently. In placental mammals, such as humans, the chorion and allantois fuse to form the **placenta,** an organ that allows nutrients to be obtained directly from the mother.

42.3 HUMAN REPRODUCTIVE ANATOMY AND PHYSIOLOGY

In this and the next section, we continue to narrow our focus in order to consider reproduction in more detail. Here, we discuss human reproduction and development. Many features of human reproduction are seen in other mammals: Humans are amniotes, reproduce sexually, use internal fertilization, and are viviparous. Other aspects, however, are unique. Our focus on human reproduction allows for a detailed treatment of the topic in one species and also gives us a chance to explore ways we intervene in the process to promote, prevent, and follow pregnancy.

The male reproductive system is specialized for the production and delivery of sperm.

Let's start by considering the human male reproductive system. Male gametes, called spermatozoa or sperm, are produced in the two male **gonads,** or **testes** (singular, **testis; Fig. 42.11**). Over the course of his life, a man is capable of making hundreds of billions of sperm. In fact, a single ejaculate typically contains several hundred million sperm. If it takes just one sperm to fertilize an egg, why do males make so many? A likely explanation involves sperm competition. In species like humans, where females can mate with more than one male, a male's reproductive success is at least partially correlated with the number of sperm—the more sperm, the better chance of fertilizing the egg. A second explanation relates to the female reproductive tract, which is hostile to sperm; many sperm are needed to ensure that at least one makes it to the egg. Of the several hundred million sperm released by a male during a single ejaculation, only several hundred sperm make it to the egg.

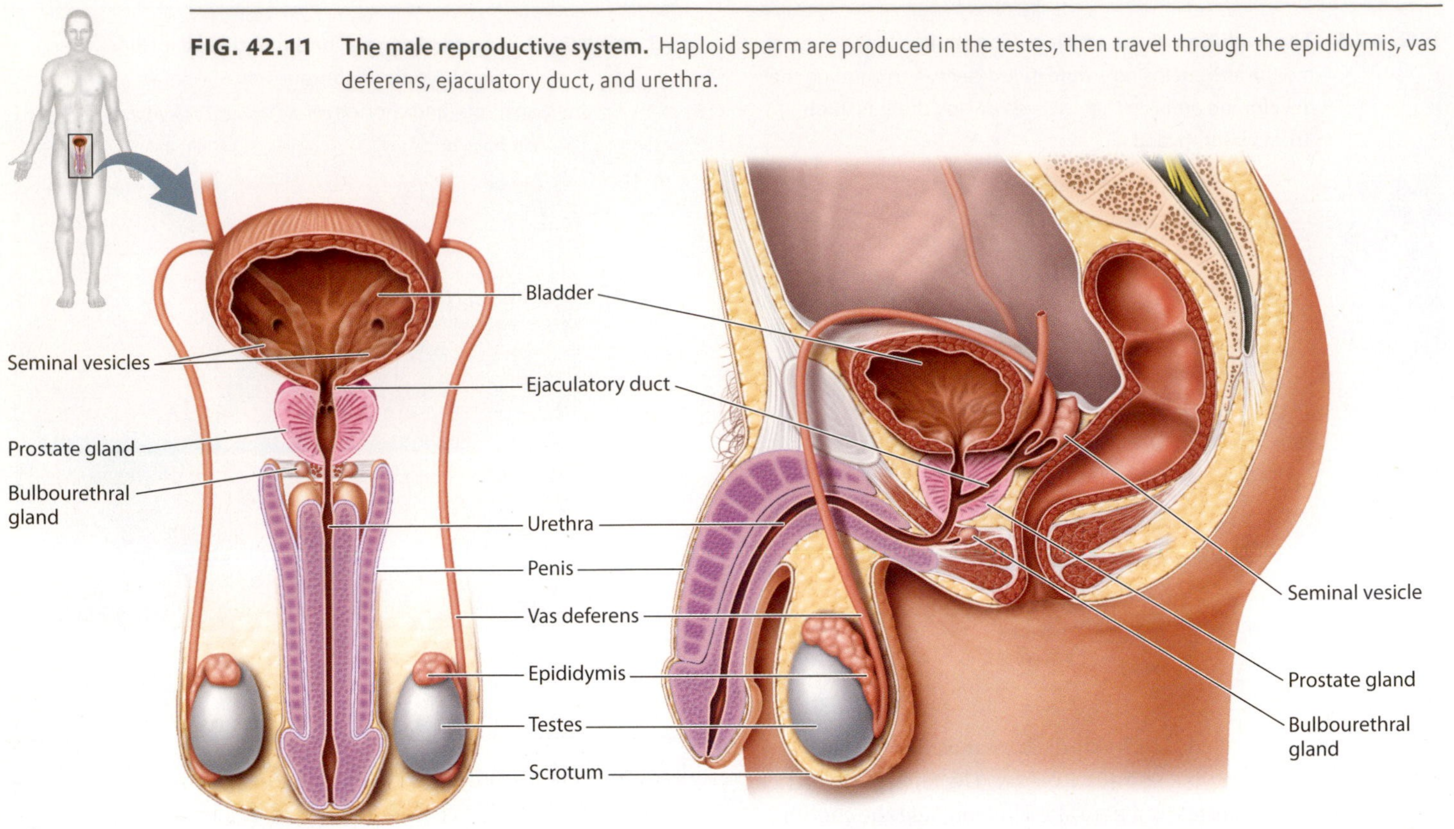

FIG. 42.11 The male reproductive system. Haploid sperm are produced in the testes, then travel through the epididymis, vas deferens, ejaculatory duct, and urethra.

Sperm are haploid cells with a unique shape and properties adapted for their function of delivering genetic material to the egg (**Fig. 42.12**). They have a small head containing minimal cytoplasm and a very densely packed nucleus. The head is surrounded by a specialized organelle, the **acrosome,** which contains enzymes that enable the sperm to transverse the outer coating of the egg. Sperm have a long tail, or **flagellum,** which moves the cell by a whipping motion powered by the sperm's mitochondria.

The testes are located outside the abdominal cavity in a sac called the **scrotum** (see Fig. 42.11). This location is critical for the production of sperm, which require cooler temperatures than are found in the abdominal cavity of humans. However, the testes of some mammals, such as elephants, whales, and bats, are located inside the abdominal cavity, indicating that there are species-specific differences in sperm temperature requirements.

Sperm are produced within the testes in a series of tubes called the **seminiferous** ("seed-bearing") **tubules** (**Fig. 42.13**). This is where diploid cells undergo meiosis to produce haploid sperm. When sperm are first produced, they are not fully mature: They are unable to move on their own and incapable of fertilizing an egg. They mature as they move through the male reproductive system.

From the seminiferous tubules, sperm travel to the **epididymis,** where they become motile (Fig. 42.13). This is also where sperm are stored prior to ejaculation. From there, sperm enter the **vas deferens,** a long, muscular tube that follows a circuitous path, starting off in the scrotum of each testis, passing into the abdominal cavity, running along the bladder, and then connecting with an **ejaculatory duct** (see Fig. 42.11). The two ejaculatory ducts merge at the **urethra,** just below the bladder. The urethra is used by both the urinary and reproductive systems in males.

Along this path, there are several exocrine glands that produce components of **semen,** the fluid that nourishes and sustains sperm as they travel in the male and then female reproductive tracts. The first is the **prostate gland,** located just beneath the bladder (see Fig. 42.11). The urethra and ejaculatory ducts pass directly through this gland. The prostate gland produces a thin, slightly alkaline fluid that helps maintain sperm motility and counteracts the acidity of the female reproductive tract. Two **seminal vesicles,** located at the junction of the vas deferens and the prostate gland, secrete a protein- and sugar-rich fluid that makes up most of the semen and provides energy for sperm motility. Finally, the **bulbourethral glands,** below the prostate gland, produce a clear fluid that lubricates the urethra for passage of sperm.

The urethra is enclosed in the **penis,** the male copulatory organ (see Fig. 42.11). When a male is sexually aroused, the penis becomes erect as a result of changes in blood flow—the arteries become dilated and the veins constricted. During ejaculation, semen is expelled from the penis.

The male reproductive system is well adapted for the production, storage, and delivery of sperm. Therefore, male contraceptive methods focus on blocking sperm. For example, a condom is a sheath of rubber or latex that covers the penis and prevents sperm from entering the female during sexual intercourse.

FIG. 42.12 **Sperm.** Sperm are specialized cells with minimal cytoplasm, a densely packed nucleus, an acrosome with digestive enzymes, and a flagellum. *Photo source: Roland Birke/Getty Images.*

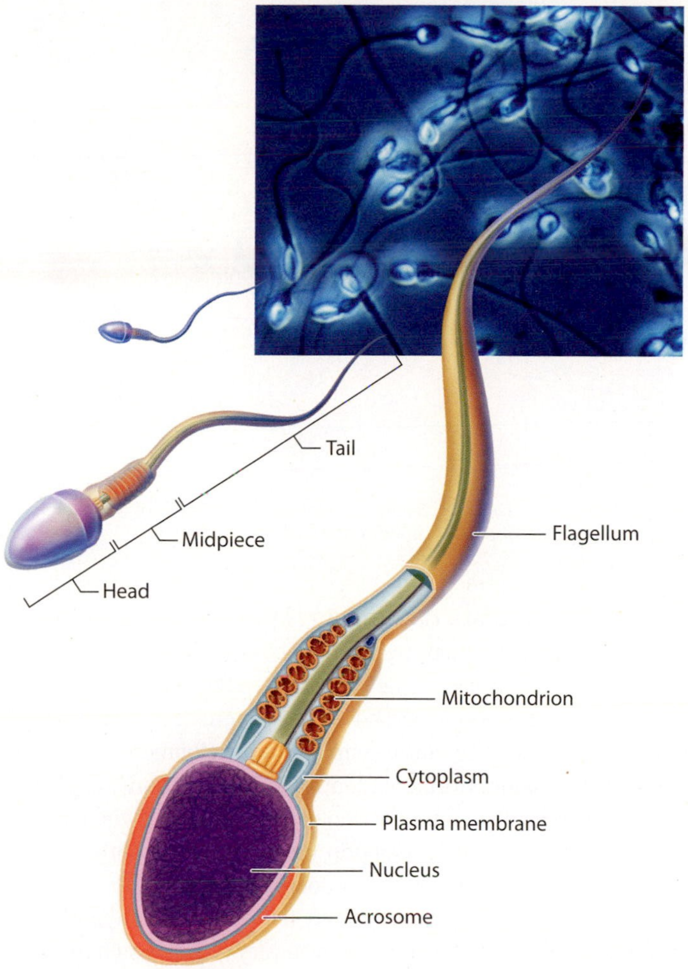

The human penis is a relatively simple organ, consisting of a shaft and a head called the **glans penis.** In other animals, the penis can be quite elaborate, with spikes, hooks, and grooves whose functions are to stimulate the female, grip the female, and in some cases even remove sperm from a previous copulation. In marsupial mammals like kangaroos and koalas, males have forked penises, and in snakes and lizards, males have two penises, called hemipenes. In insects, the external genitalia are often distinct enough to be used in species identification. By contrast, most species of birds do not have a penis.

→ **Quick Check 2** What effect would cutting the vas deferens on both sides and tying off the ends have on male fertility?

The female reproductive system produces eggs and supports the developing embryo.

The female reproductive system, like the male reproductive system, produces gametes, but is also specialized to support the developing embryo. Developing female gametes, called **oocytes,** which mature into ova, are produced in the two female gonads, called **ovaries** (**Fig. 42.14**), and are released monthly in response to hormones in the process of menstruation, discussed in the next section.

The oocyte is a remarkable and unique cell (**Fig. 42.15**). In contrast to sperm, which are small and motile, the oocyte is the largest cell by volume in the human body and is visible to the naked eye; it is about the size of the period at the end of this sentence. Its large size reflects the presence of large amounts of cytoplasm containing molecules that direct the early cell divisions following fertilization and food supply in the form of yolk. Whereas a male produces millions of sperm, a female usually produces just one oocyte each month, as she will nourish and support the developing embryo if the oocyte is fertilized.

Following release from the ovary, the oocyte travels through the **fallopian tube,** or **oviduct** (see Fig. 42.14). There are two fallopian tubes, one for each ovary. The fallopian tubes have featherlike projections on their ends that help channel the egg into the tube and not into the abdominal cavity.

The oocyte travels from the fallopian tube to the **uterus,** or womb (see Fig. 42.14). The uterus is a hollow organ with thick, muscular walls that is adapted to support the developing embryo if fertilization occurs and to deliver the baby during birth. In humans, the uterus is usually a single pear-shaped structure, but it can sometimes have two horns, which can interfere with pregnancy. Other mammals differ in the shape of the uterus, ranging from pear-shaped (humans and other primates), to two horns (pigs, horses), to more completely divided but still connected at the base (cats, dogs), to completely separate (rodents, rabbits).

FIG. 42.13 **Seminiferous tubules, epididymis, and vas deferens.**

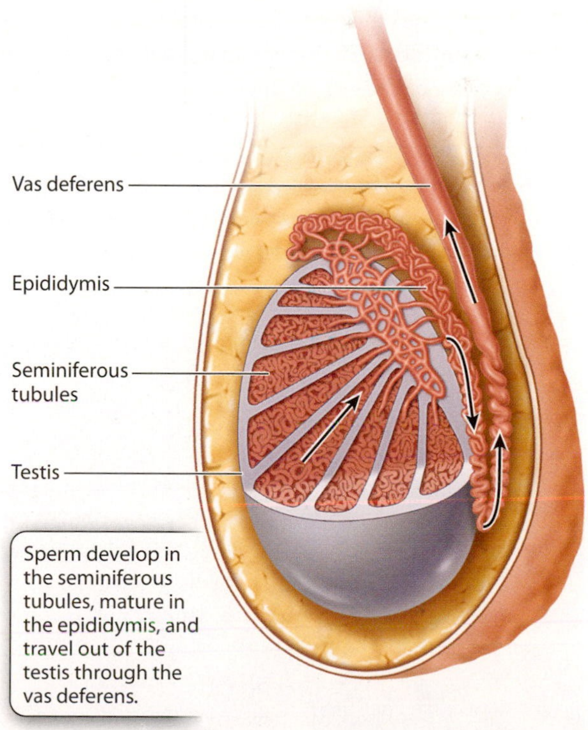

FIG. 42.14 The female reproductive system. Oocytes are produced in the ovaries and then travel through the fallopian tubes to the uterus, where they implant if they are fertilized; if they are not fertilized, they are shed during menstruation.

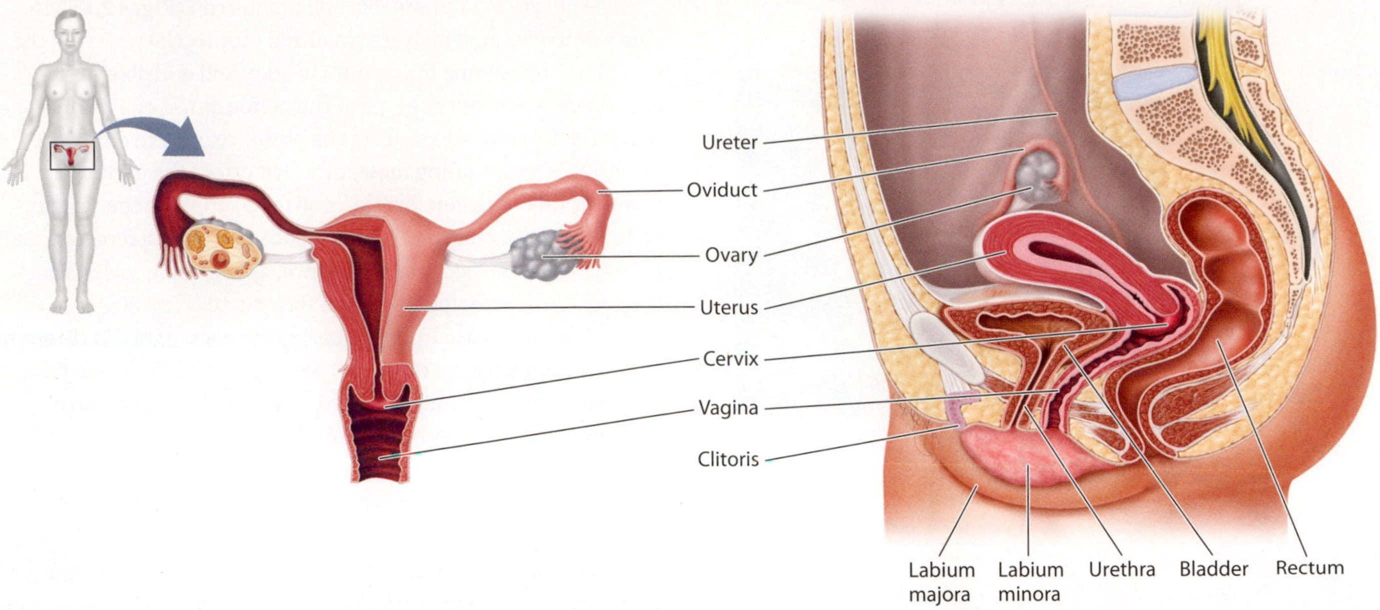

The end of the uterus is called the **cervix,** or neck, of the uterus (see Fig. 42.14). Under the influence of hormones, the cervix produces different kinds of mucus capable of either blocking or guiding sperm through the cervix.

The uterus is continuous with the **vagina,** or birth canal (see Fig. 42.14). The vagina is a tubular channel connecting the uterus to the exterior of the body. It is highly elastic, allowing it to stretch during sexual intercourse and birth.

The external genitalia of the female are collectively called the **vulva** (**Fig. 42.16**). The vulva includes two folds of skin, the **labia majora** and **labia minora.** The labia minora meet at the **clitoris,** the female homolog of the glans penis.

FIG. 42.15 A female gamete. Oocytes are produced in the ovaries, mature into ova (eggs), and are by volume the largest cells in the human body. *Source: Science Source.*

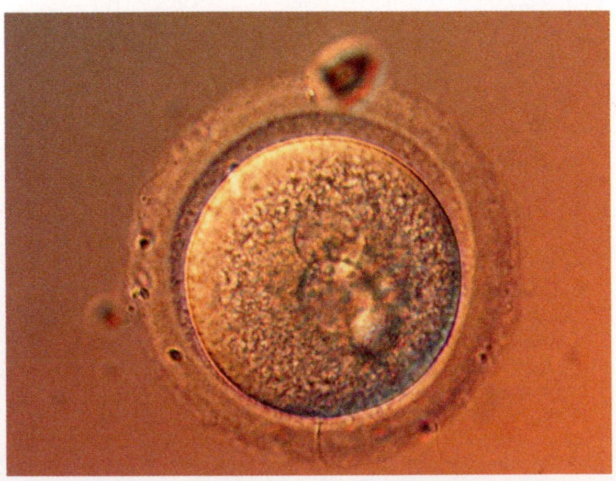

Hormones regulate the human reproductive system.

The male testes and female ovaries are not only the site of gamete production, but also part of the endocrine system (Chapter 38). That is, they respond to hormones and in turn secrete hormones. These hormones play key roles during embryonic development and puberty, and also in the adult. Hormonal control of reproduction is critically important, as it allows reproduction to take place only when, for example, a woman is able to support the developing embryo.

The master regulator of the endocrine system is the hypothalamus, located in the brain (Chapter 38). The hypothalamus secretes hormones that act on the pituitary gland, which in turn releases hormones that act on target organs (**Fig. 42.17**). In the case of the reproductive system, the hypothalamus releases **gonadotropin-releasing hormone (GnRH),** which stimulates the anterior pituitary gland to secrete **luteinizing hormone (LH)** and **follicle-stimulating hormone (FSH).** These hormones act on the male and female gonads. In males, these hormones are secreted almost continually, whereas in females they are secreted cyclically. In response, the testes secrete the hormone **testosterone** and the ovaries secrete the hormones **estrogen** and **progesterone.**

In males, LH acts on cells in the testes called **Leydig cells,** which secrete testosterone. Testosterone is a steroid hormone derived from cholesterol that plays key roles in male growth, development, and reproduction. Testosterone is important during embryogenesis to direct development of the male reproductive organs. At puberty, levels of testosterone increase, leading to the development of male **secondary sexual characteristics,** traits that characterize and differentiate the two sexes but that do not relate directly to reproduction. For males, these include a deep voice; facial, body, and pubic hair; and increased muscle mass. Testosterone together with FSH act on another group of cells in the testes,

FIG. 42.16 The female external genitalia.

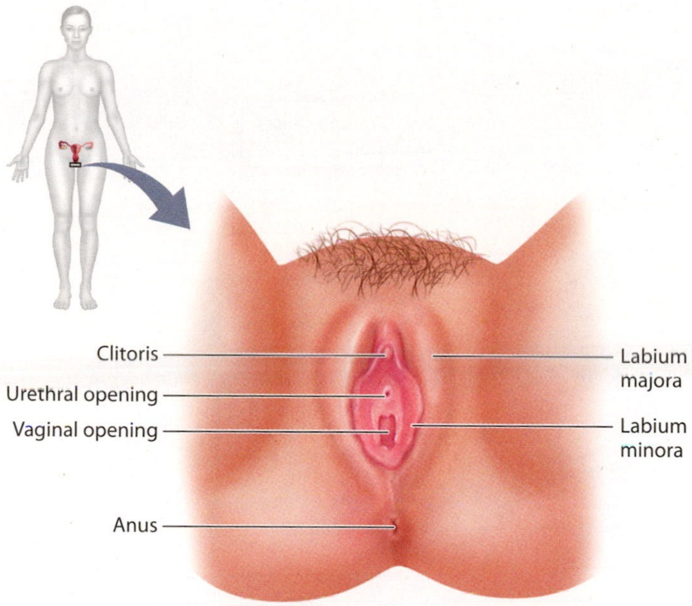

called **Sertoli cells,** to stimulate sperm production. Sertoli cells, located in the seminiferous tubules, support the process of sperm production. In this way, hormones help to regulate the timing of sexual maturity.

In females, an increase in the level of estrogen at puberty leads to the development of female secondary sexual characteristics, such as enlargement of the breasts, growth of body and pubic hair, and changes in the distribution of muscle and fat. Also at puberty, females begin a monthly cycle, the **menstrual cycle,** in which an oocyte matures and is released from the ovary under the influence of hormones (**Fig. 42.18**). The menstrual cycle consists of cyclical and coordinated changes in the ovaries and uterus, timing the release of an oocyte from the ovary with the growth of the uterine lining so it can support the developing embryo if fertilization occurs.

The menstrual cycle has two phases, the **follicular phase** and **luteal phase.** In the follicular phase, an oocyte develops and is eventually released from the ovary. The oocyte is located in a supporting shell of cells called a **follicle.** At the start of each menstrual cycle, FSH stimulates specific follicle cells called granulosa cells to secrete a form of estrogen called estradiol. In response to estradiol, oocytes in several follicles begin to mature each month, but usually only one becomes fully mature and the others die off. Estradiol also acts on the lining of the uterus, causing it to thicken. Finally, estradiol acts on the hypothalamus and anterior pituitary gland, where it reduces the secretion of GnRH, FSH, and LH by negative feedback (Fig. 42.18).

At the end of the follicular phase, this negative feedback of estradiol changes to positive feedback (Fig. 42.18). The mechanism for this change is not well understood, but may be due in part to high levels of estradiol. Positive feedback of estradiol on the hypothalamus and anterior pituitary causes a rapid increase followed by a sharp decrease in the level of LH produced by the anterior pituitary gland. This LH surge causes ovulation, the release of the oocyte from the follicle in the ovary.

Ovulation marks the end of the follicular phase and beginning of the luteal phase. The follicle, now devoid of the oocyte, is converted to a structure known as the **corpus luteum,** literally "yellow body," so called because of its appearance. The corpus luteum is a temporary endocrine structure that secretes the hormone progesterone, which maintains the thickened and vascularized uterine lining, and inhibits the secretion of GnRH, FSH, and LH (Fig. 42.18).

The oocyte, meanwhile, is swept into the fallopian tube and travels to the uterus. If it is fertilized, the developing embryo implants into the uterine lining. In this case, the corpus luteum

FIG. 42.17 Hormonal control of the reproductive system.

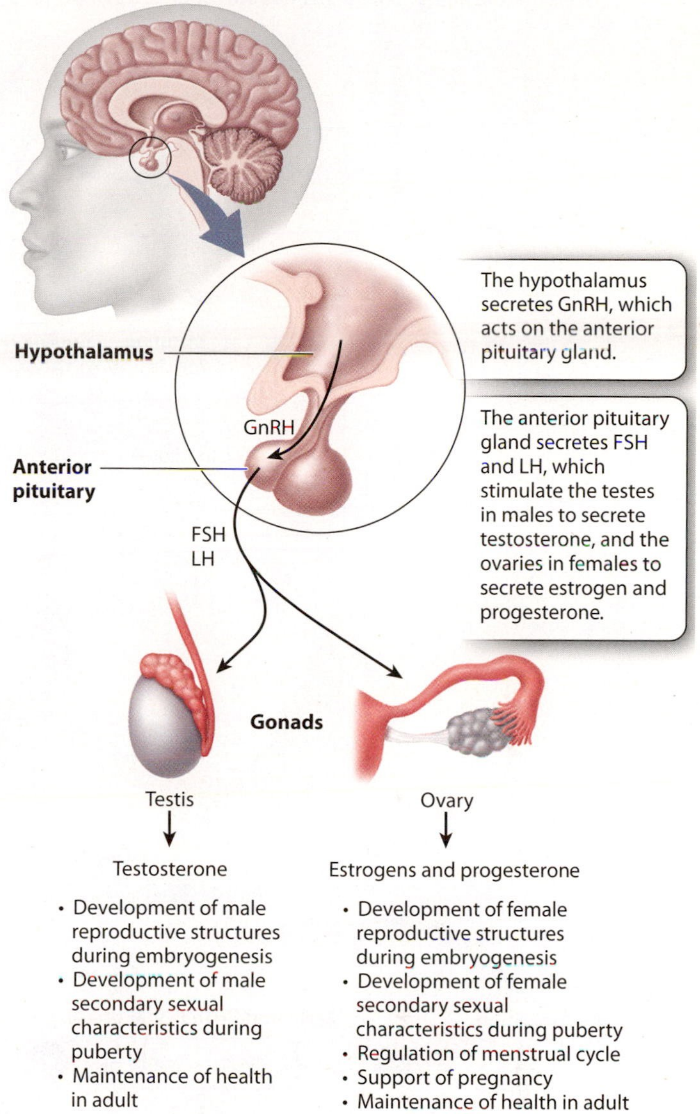

FIG. 42.18 **The menstrual cycle.** Note the cyclical and coordinated changes in the anterior pituitary hormones, ovary, ovarian hormones, and uterine lining. These changes are regulated by hormonal feedback loops.

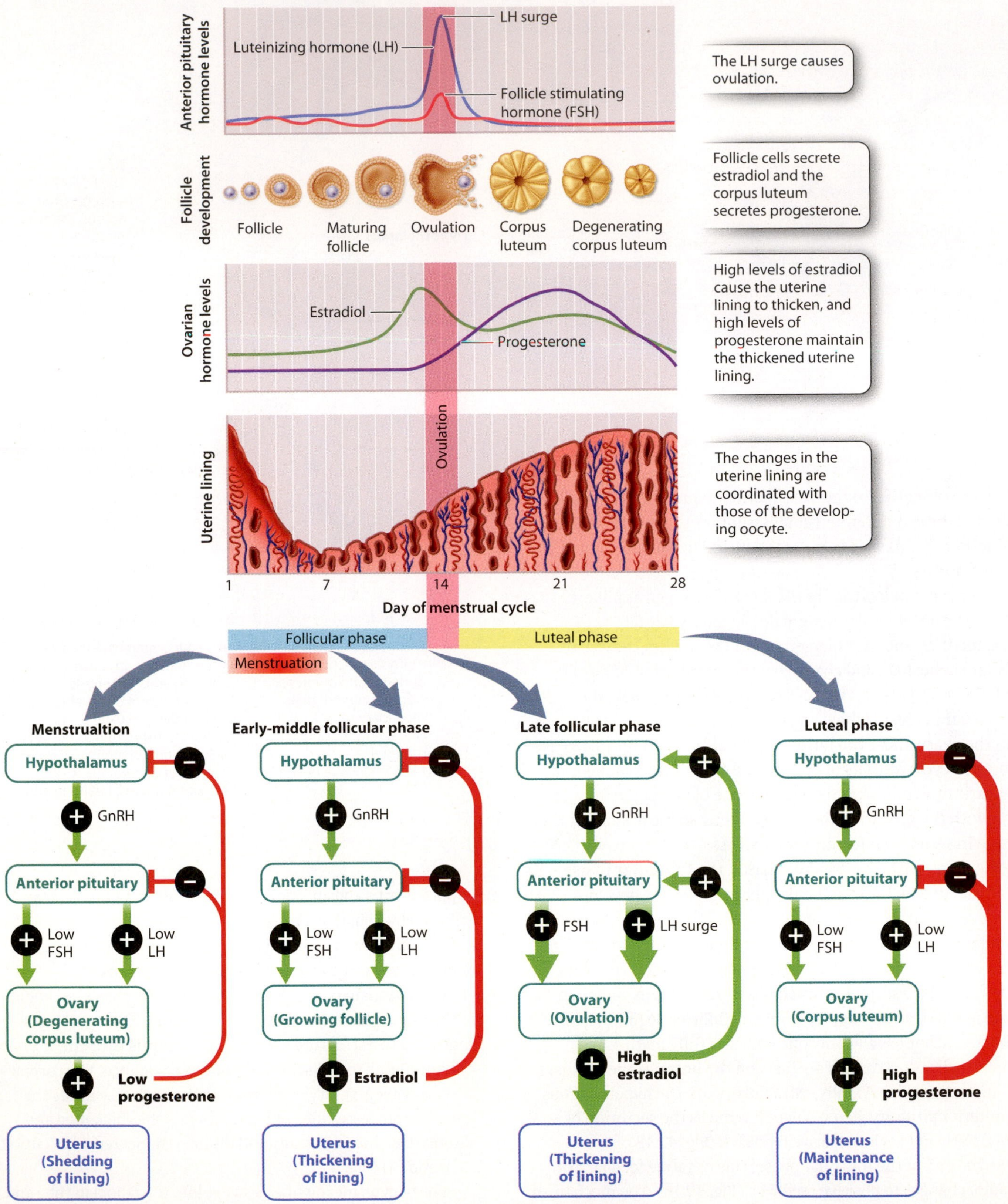

continues to secrete progesterone, and is first maintained by LH and then by the hormone **human chorionic gonadotropin (hCG),** which is released by the developing embryo. Eventually, the placenta takes over estrogen and progesterone production to maintain the uterine lining and stimulate growth of the uterus. High levels of estrogen and progesterone during pregnancy also block ovulation because they suppress the release of GnRH, FSH, and LH.

If the oocyte is not fertilized, the corpus luteum degenerates, estrogen and progesterone levels drop, and the uterine lining is shed. The monthly shedding of the uterine lining is known as **menstruation.** Menstrual cycles usually start occurring around age 12 in the United States and continue until approximately age 45 to 55. The cessation of menstrual cycles, called **menopause,** results from decreasing production of estradiol and progesterone by the ovaries.

As we have seen, the brain, anterior pituitary, ovaries, uterus, and placenta pass hormonal messages back and forth to make sure that behavior, gamete release, the uterine lining, and embryonic development are coordinated (Fig. 42.18). If hormone levels are low or absent, as occurs normally before puberty or as the result of malnutrition or some diseases, ovulation does not occur. Similarly, too much hormone at the wrong time can interfere with fertility, which is the basis for oral contraceptive pills ("the pill"). Oral contraceptive pills contain various combinations of synthetic estrogens or progesterones or both. Estrogens and/or progesterones in the pill suppress GnRH, FSH, and LH and therefore block oocyte development and ovulation.

Humans and chimpanzees both have menstrual cycles. Other placental mammals have an **estrus cycle,** which is characterized by phases in which females are sexually receptive. In these mammals, the uterine lining is reabsorbed instead of shed if fertilization does not occur.

→ **Quick Check 3** A monthly course of an oral contraceptive pill is usually three weeks of hormone pills followed by one week of placebo (a sugar pill with no hormone). Why do you think the placebos are sometimes used?

42.4 GAMETE FORMATION TO BIRTH IN HUMANS

With an understanding of male and female reproductive anatomy and endocrinology, we are ready to follow human reproduction and development from the formation of sperm and eggs, to fertilization, to subsequent development of the embryo and fetus, and finally to birth. Along the way, we consider how these processes vary in different organisms. Our understanding of these processes is in large part due to studies in model organisms, including the sea urchin (*Strongylocentrotus purpuratus*), roundworm (*Caenorhabditis elegans*), fruit fly (*Drosophila melanogaster*), African clawed frog (*Xenopus laevis*), and house mouse (*Mus musculus*). These organisms have particular reproductive and developmental features that make them amenable to laboratory study.

Male and female gametogenesis have both shared and distinct features.

The formation of gametes is called **gametogenesis.** Gametogenesis follows the same steps in males and in females, but there are striking differences in timing (**Fig. 42.19**). **Spermatogenesis** is the formation of sperm. It occurs in the seminiferous tubules of the testes beginning at puberty and continuing throughout life. The process starts with spermatogonia, diploid stem cells that divide by mitosis to ensure a continuous population of precursor cells. During puberty and throughout adulthood, these cells differentiate into diploid **primary spermatocytes,** which undergo the first meiotic division to produce two **secondary spermatocytes** and the second meiotic division to produce four haploid spermatids joined by cytoplasmic bridges. They subsequently separate from each other to become distinct haploid cells. The maturation process, in which spermatids become spermatozoa or sperm, begins in the testes and continues in the epididymis, where they acquire motility. Spermatogenesis takes about 2 to 3 months.

Oogenesis is the formation of ova or eggs. The diploid precursor cells are called oogonia. These cells form during embryonic development and divide by mitosis. Some oogonia differentiate into diploid **primary oocytes** during fetal development. These enter the first meiotic division but arrest immediately in prophase I, the first stage of meiosis. All these steps take place before the female is even born. At birth, a female has about 1 to 2 million primary oocytes, all arrested in prophase I of meiosis. They do not resume the first meiotic division until a menstrual cycle. Therefore, remarkably, they all remain arrested for at least 12 years (from birth to the age of menarche) and some as long as 50 years (to the age of menopause).

In the first phase of the menstrual cycle, approximately 10 to 20 follicles, each containing an oocyte, begin to mature in response to rising FSH levels. One of these follicles becomes dominant, allowing the primary oocyte to complete the first meiotic division. This division is asymmetric, leading to the formation of a large **secondary oocyte** and a much smaller **polar body.** The unequal division allows most of the cytoplasm to be partitioned into the secondary oocyte. The secondary oocyte is released from the ovary and immediately enters the second meiotic division, but again arrests, this time in metaphase II. The secondary oocyte remains arrested in metaphase II until fertilization takes place. Upon fertilization, the second meiotic division continues, producing an ootid and another polar body, again allowing most of the cytoplasm to end up in the developing egg. The ootid quickly develops into a mature ovum or egg.

→ **Quick Check 4** When are primary oocytes and spermatocytes produced, and how many gametes result from one diploid cell in females and males?

FIG. 42.19 Spermatogenesis in the male and oogenesis in the female.

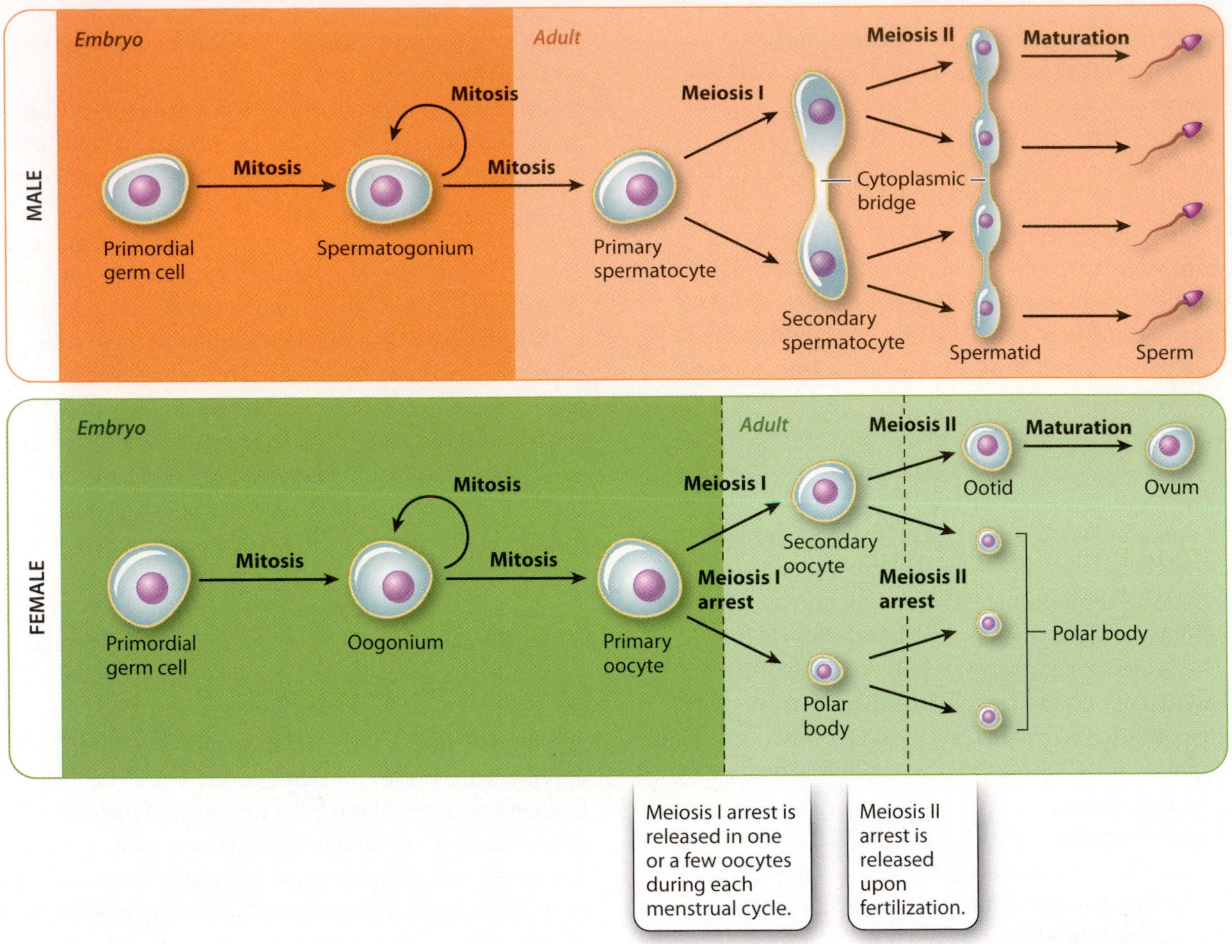

Fertilization occurs when a sperm fuses with an oocyte.

Fertilization involves the fusion of gametes, thus restoring the diploid chromosome content (**Fig. 42.20**). In humans, male and female gametes are usually brought together by sexual intercourse. Sexual arousal leads to a host of physiological changes, including erection of the penis and erection of the clitoris and lubrication of the vagina. The erect penis is inserted into the vagina until orgasm occurs. Orgasm is characterized by feelings of pleasure and contractions of the pelvic muscles. In males, it is usually accompanied by ejaculation, in which semen, containing sperm, is delivered into the vagina.

Sperm travel through the cervix and into the uterus and fallopian tubes. Fertilization usually occurs in one of the fallopian tubes. As a result, many of the details of how sperm travel through the female reproductive tract are poorly understood. Sperm are motile, but from studies of the time it takes sperm to reach the fallopian tubes, it appears that the female reproductive tract plays an active role in moving sperm to the oocyte, for example through muscular contractions.

In the female reproductive tract, sperm undergo **capacitation,** a key developmental step that is unique to mammals. Capacitation is a series of physiological changes that allow the sperm to fertilize the egg. These changes include alterations in the fluidity of the plasma membrane, loss of some surface membrane proteins, and changes in the charge across the membrane, called the membrane potential (Chapter 35). As a result, sperm show increased motility.

In many species, the oocyte releases soluble molecules, creating a concentration gradient that guides the sperm to the oocyte. On reaching the oocyte, the sperm passes through two layers—an outer layer of follicle cells called the corona radiata and an inner matrix of glycoproteins called the zona pellucida—before fusing with the plasma membrane of the oocyte (Fig. 42.20). In sea urchins, by contrast, there is a single vitelline membrane surrounding the egg. To facilitate its passage through these various outer layers, the head of the sperm releases enzymes from the acrosome. In addition, surface proteins on the gametes allow them to interact in a species-specific manner. For organisms with external fertilization, such as sea urchins, these proteins are especially important.

After passing through the layers surrounding the ooctye, the plasma membranes of the sperm and oocyte fuse (Fig. 42.20). The oocyte then completes meiosis II and undergoes changes that prevent **polyspermy,** or fertilization by more than one sperm.

FIG. 42.20 Fertilization. Many sperm cluster around an oocyte, but usually just one fuses with it to form a zygote. *Photo source: David M. Phillips/ Science Source.*

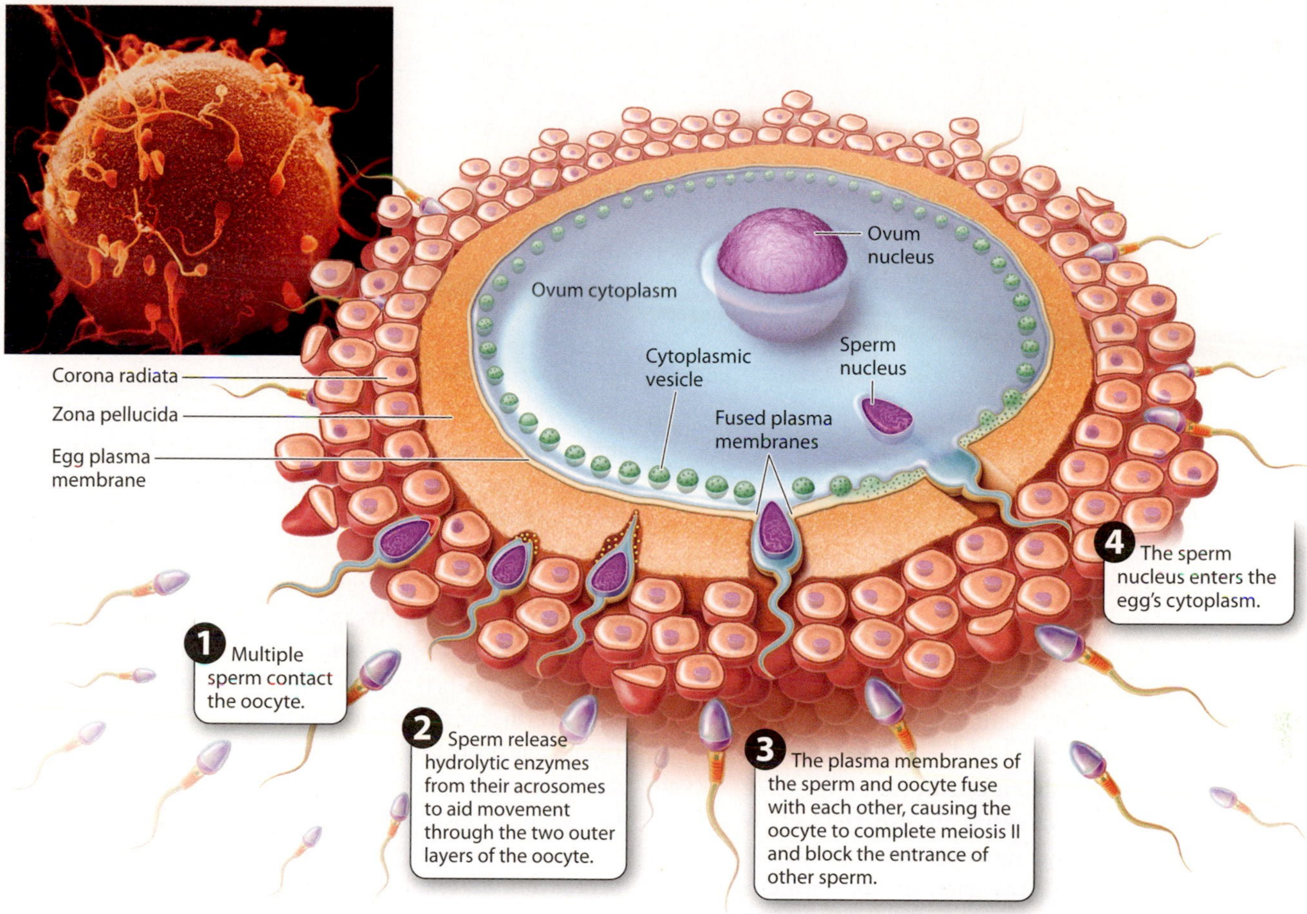

These mechanisms are classified according to how quickly they act and have been extensively studied in sea urchins. The first line of defense, or fast block to polyspermy, is caused by changes in the membrane potential of the egg, which is usually negative but quickly becomes positive on fertilization. The fast block is quick but temporary and does not occur in mammals.

There is also a second, more stable defense, or slow block to polyspermy. In this process, vesicles in the oocyte fuse with the plasma membrane and release their contents, leading to modification of the structure of the zona pellucida in mammals and the vitelline membrane in sea urchins. The fertilized, diploid egg is a zygote—the first cell in the development of a new organism.

Although sexual intercourse is one way that gametes are brought together, there are now a variety of other methods to achieve fertilization and pregnancy in humans. These are collectively called assisted reproductive technologies and are typically used to treat infertility. For example, **in vitro fertilization (IVF)** is a process in which eggs and sperm are brought together in a petri dish, where fertilization and cell divisions occur. The developing embryos are then inserted into the uterus. The first IVF or "test tube" baby was born in 1978, and since then several million babies have been conceived in this way.

Assisted reproductive technologies such as IVF have opened the door to pre-implantation genetic diagnosis, in which oocytes or embryos are screened for genetic diseases before implantation. This technique raises ethical concerns, as it can be used to determine and select the gender of the embryo and other traits not associated with disease.

The first trimester includes cleavage, gastrulation, and organogenesis.

How does a single cell become a complex, multicellular organism? Development depends on a precisely orchestrated set of cell divisions, movements, and specializations coordinated by both maternal and zygotic genetic instructions (Chapter 20). In this section, we focus on three key processes in early embryonic development: cleavage, gastrulation, and organogenesis.

After fertilization, the single-celled zygote divides by mitosis in a process called **cleavage** (**Fig. 42.21**). In cleavage, the single large egg divides into many smaller cells. In most organisms, these mitotic divisions are rapid, occur at the same time, and are not accompanied by an overall increase in size of the embryo. Furthermore, they occur in the absence of expression of the zygote's genes. Instead, molecules in the egg's cytoplasm, including

FIG. 42.21 Cleavage divisions. *Photos source: Anatomical Travelogue/Science Source.*

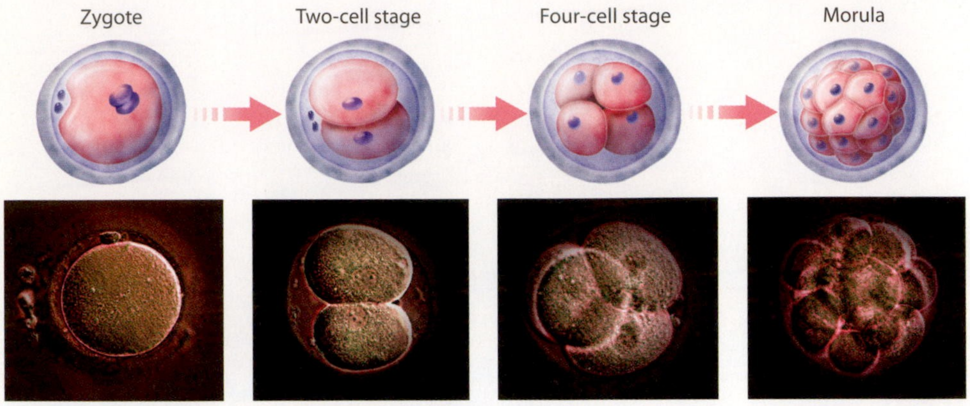

mRNA and proteins encoded by maternal genes (Chapter 20) and stored in the egg, direct these early cell divisions.

Cleavage in mammals is unique in several respects. It is slow compared with cleavage in other organisms. In addition, in mammals the early divisions do not occur at the same time, so it is not unusual to find an odd number of cells in the developing embryo. Finally, zygotic transcription begins very early, as early as the first mitotic division, or two-cell stage.

The pattern of cleavage is different for different organisms. It is determined in part by the amount of yolk present in the egg because the presence of yolk inhibits cell division (**Fig. 42.22**). In animals with lots of yolk, such as insects, fish, and birds, cell division does not extend through the entire egg (Fig. 42.22a). In animals with very little yolk, such as mammals and sea urchins, cell division extends completely through the egg (Fig. 42.22b). In the absence of sufficient yolk, the developing embryo needs another source of nutrition, which in humans and other placental mammals is provided by the placenta.

Cleavage results in a solid ball of cells called a **morula** (see Fig. 42.21). Further cell divisions result in a fluid-filled ball of cells called a **blastula** (or a blastocyst in humans and other mammals). It is around this time, about 5 days after fertilization, that the developing embryo implants into the uterine lining, which is thickened and highly vascularized under the influence of progesterone from the corpus luteum.

Implantation marks the beginning of **pregnancy,** the carrying of one or more embryos in the uterus. It lasts approximately 38 weeks from fertilization. Because the time of fertilization is often not known, human pregnancy is typically measured not from implantation but as 40 weeks from the first day of the last menstrual period. Pregnancy is divided into three periods called **trimesters,** each lasting about 3 months. The divisions between the three trimesters are arbitrary, but they provide a convenient way to follow the major changes that occur during pregnancy.

The first trimester is characterized by spectacular changes in the developing embryo under the influence of genes and developmental pathways that are largely conserved across vertebrates (Chapter 20). During this time, the embryo develops all its major organs and begins to be recognizably human. Toward the end of the first trimester, a human embryo is known as a **fetus.**

In mammals, such as humans, the blastocyst has an inner cell mass and an outer layer of cells (**Fig. 42.23a**). The inner cell mass becomes the embryo itself, and the outer cell layer forms part of the placenta. The placenta is composed of both maternal and embryonic cells and is the site of gas exchange, nutrient uptake, and waste excretion between maternal and fetal blood. The placenta also secretes the hormones estrogen and progesterone, which maintain the uterine lining during pregnancy. Meanwhile, the inner cell mass differentiates into a two-layered (bilaminar) structure consisting of the epiblast and the hypoblast (**Fig. 42.23b**).

Gastrulation is the second key developmental process in embryogenesis. It is a highly coordinated set of cell movements that leads to a fundamental reorganization of the embryo. In humans, cells of the epiblast divide and migrate inward (**Fig. 42.23c**). These migrating epiblast cells form three layers or sheets of cells called **germ layers,** which include the outer **ectoderm,** intermediate **mesoderm,** and inner **endoderm** (**Fig. 42.23d**). All of the adult tissues and organs are made from the three germ layers. At this stage, the embryo is known as a trilaminar (three-layered) embryo to reflect the three germ layers.

As with cleavage, the details of gastrulation vary in different organisms. One notable difference is the fate of the opening through which cells migrate during gastrulation. In protostomes ("mouth first"), such as worms, mollusks, and arthropods, it becomes the mouth, whereas in deuterostomes ("mouth second"), such as vertebrates and sea urchins, it becomes the anus and the mouth forms later (Chapter 44).

Organogenesis, the third key developmental process, is the transformation of the three germ layers into all the organ systems of the body. The ectoderm becomes the outer layer of the skin and nervous system, the mesoderm makes up the circulatory system, muscle, and bone, and the endoderm becomes the lining of the digestive tract and lungs (Fig. 42.24d). Because all the basic structures and organ systems develop during the first trimester of pregnancy, it is during this time that the developing fetus is particularly vulnerable to toxins, drugs, and infections.

It is also during the first trimester that the embryo begins to develop male or female characteristics. In humans, **sex determination** is achieved by the presence or absence of the Y chromosome. The Y chromosome contains a gene, *SRY,* which encodes a protein called testis-determining factor (TDF). TDF acts on the developing gonad to cause it to differentiate into a testis. In the absence of TDF in females, the immature gonad becomes an ovary. Once sex is determined, sexual differentiation follows under the influence of hormones produced by the gonads. In humans, then, sex determination is genetic, as is the case in other mammals

FIG. 42.22 Cleavage patterns. (a) Partial cleavage, shown in this zebrafish embryo. (b) Full cleavage, shown in this sea urchin embryo. *Photo sources: a. © 2002 Steve Baskauf; b. Carolina Biological Supply Co./Phototake.*

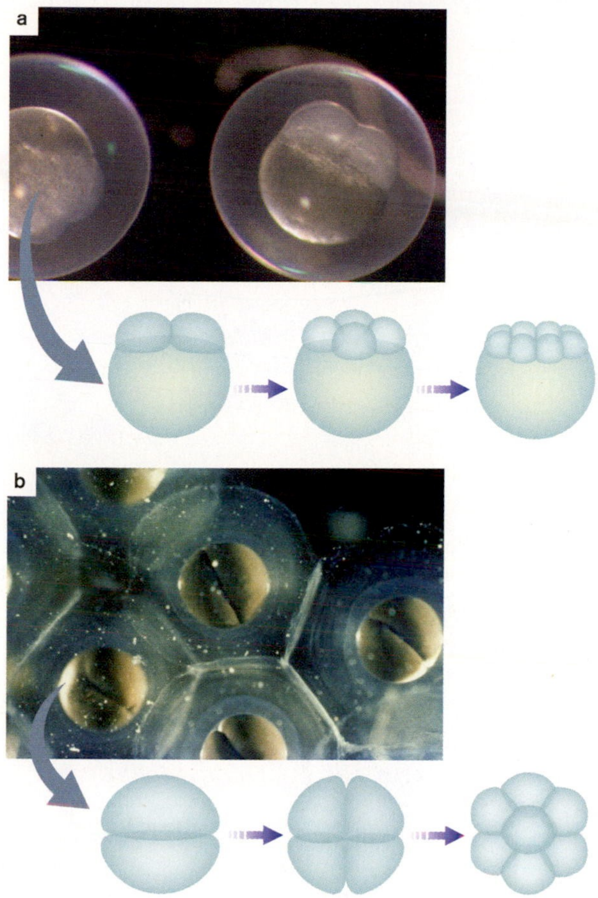

FIG. 42.23 Early human embryonic development.

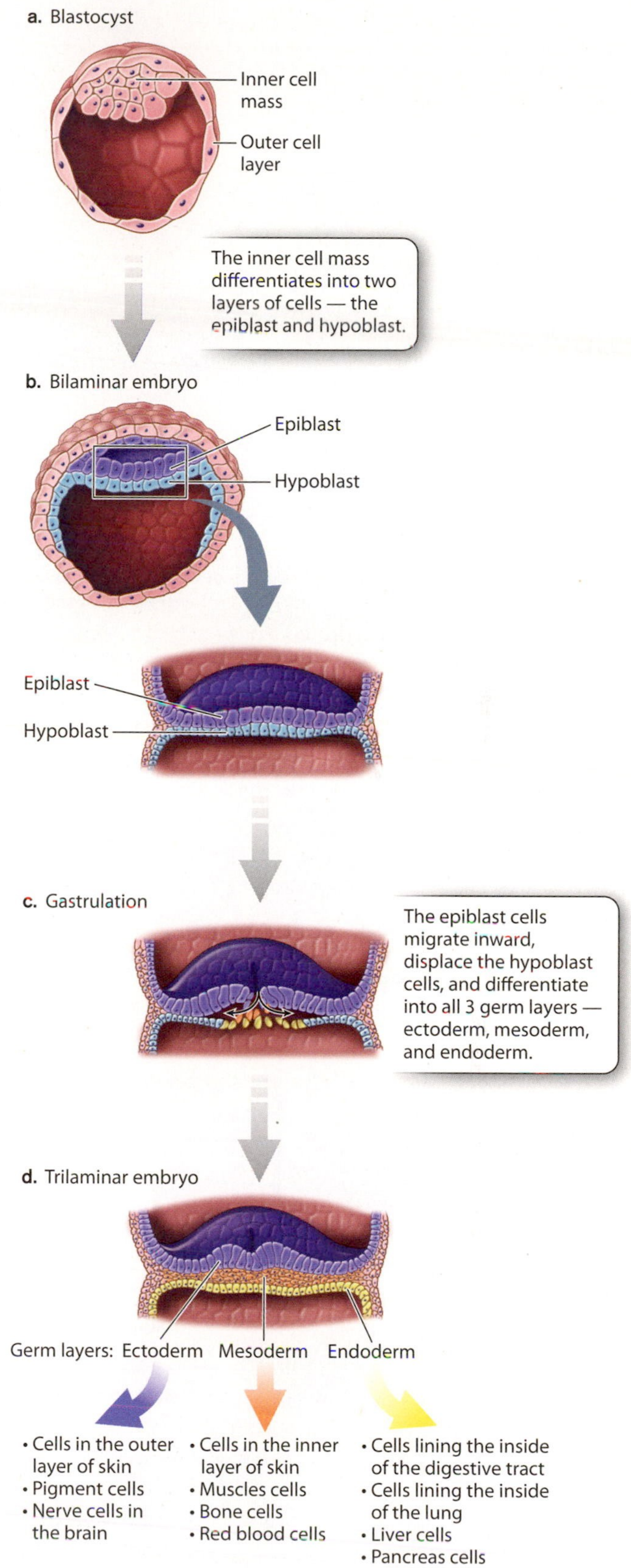

and in birds. For other organisms, such as some reptiles and fish, sex determination depends instead on environmental factors, including cues such as temperature and crowding.

The second and third trimesters are characterized by fetal growth.

The second trimester, from around 13 to 27 weeks, is characterized by continued development and growth of the fetus. This is the time when fetal movements can first be felt by the mother. Because of improvements in medical care, babies born in the last few weeks of the second trimester are often able to survive.

The third trimester, from 28 to 40 weeks, is characterized by rapid growth, with the fetus doubling in size during the last 2 months. Weight gain is critical, as birth weight is a major factor in infant mortality. The fetus often moves less frequently during this trimester as it becomes increasingly compressed by limited space. Although 40 weeks (from the last menstrual period) is the average length of pregnancy, a baby born between 37 and 42 weeks is considered term. Human development from fertilization to the end of pregnancy is summarized in **Fig. 42.24.**

VISUAL SYNTHESIS

FIG. 42.24

Reproduction and Development

Integrating concepts from Chapters 10, 20, and 42

Early development

The fertilized egg is a totipotent stem cell because it can give rise to a complete organism. During development, cells become increasingly restricted in their developmental potential as a result of changes in gene expression. During gastrulation, cells further differentiate and move relative to one another until the embryo is organized in three germ layers.

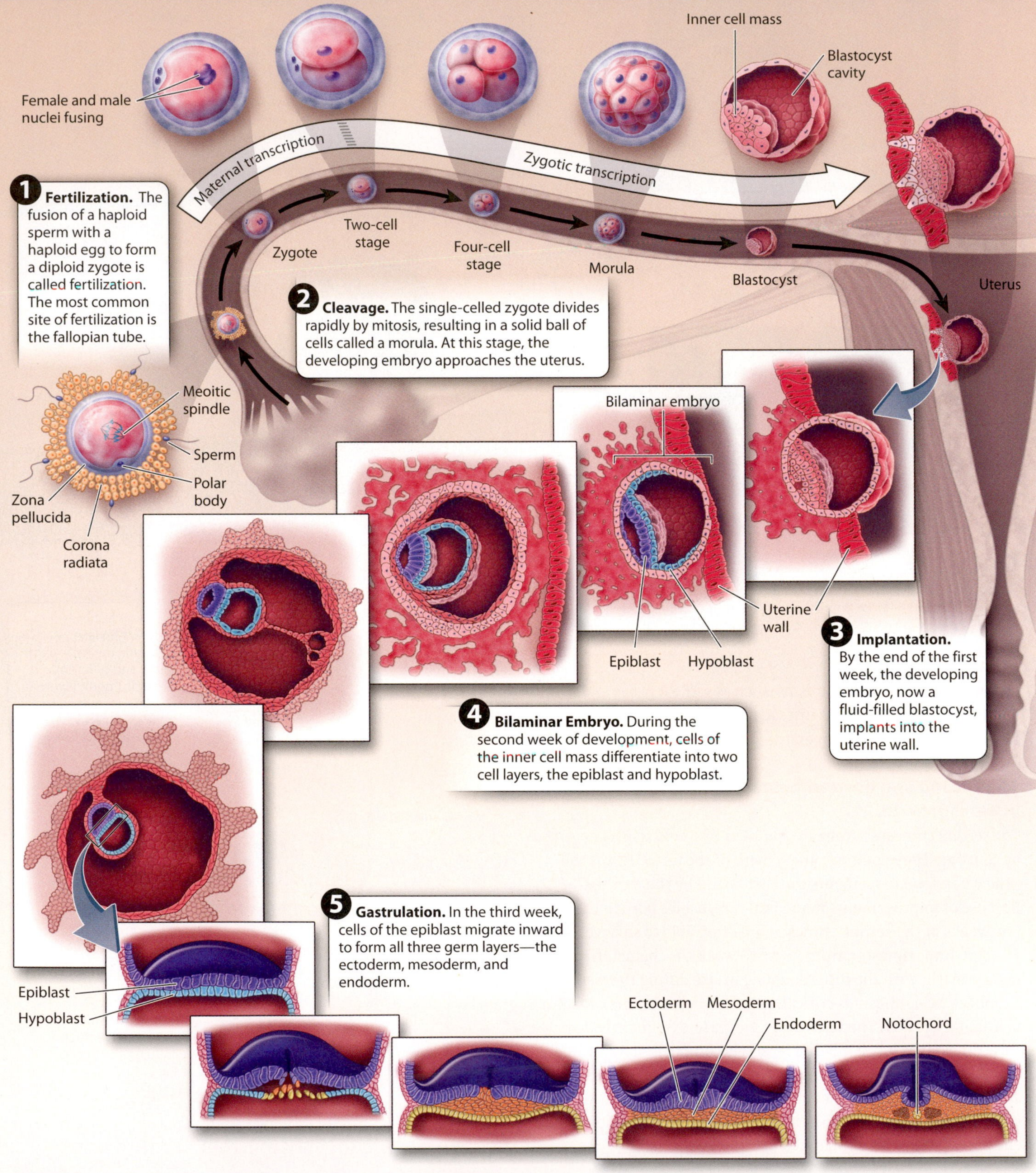

Late development

The three germ layers form all of the organs of the fetus by organogenesis. The organs, such as the heart, lungs, and kidneys, are specialized for different functions but work in a coordinated fashion.

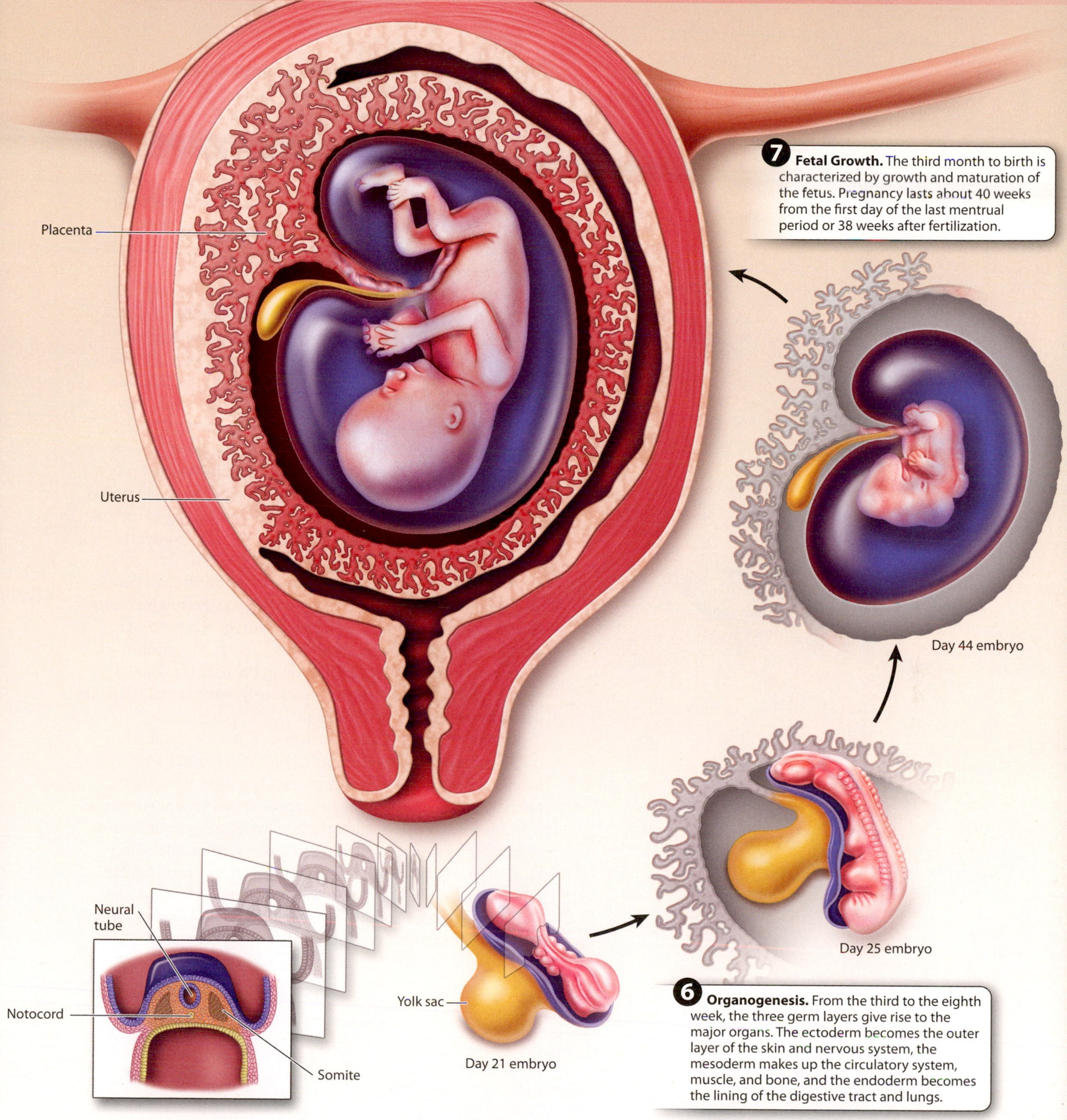

7 Fetal Growth. The third month to birth is characterized by growth and maturation of the fetus. Pregnancy lasts about 40 weeks from the first day of the last mentrual period or 38 weeks after fertilization.

6 Organogenesis. From the third to the eighth week, the three germ layers give rise to the major organs. The ectoderm becomes the outer layer of the skin and nervous system, the mesoderm makes up the circulatory system, muscle, and bone, and the endoderm becomes the lining of the digestive tract and lungs.

FIG. 42.25 Childbirth. Childbirth involves changes to the cervix, delivery of the baby, and delivery of the placenta.

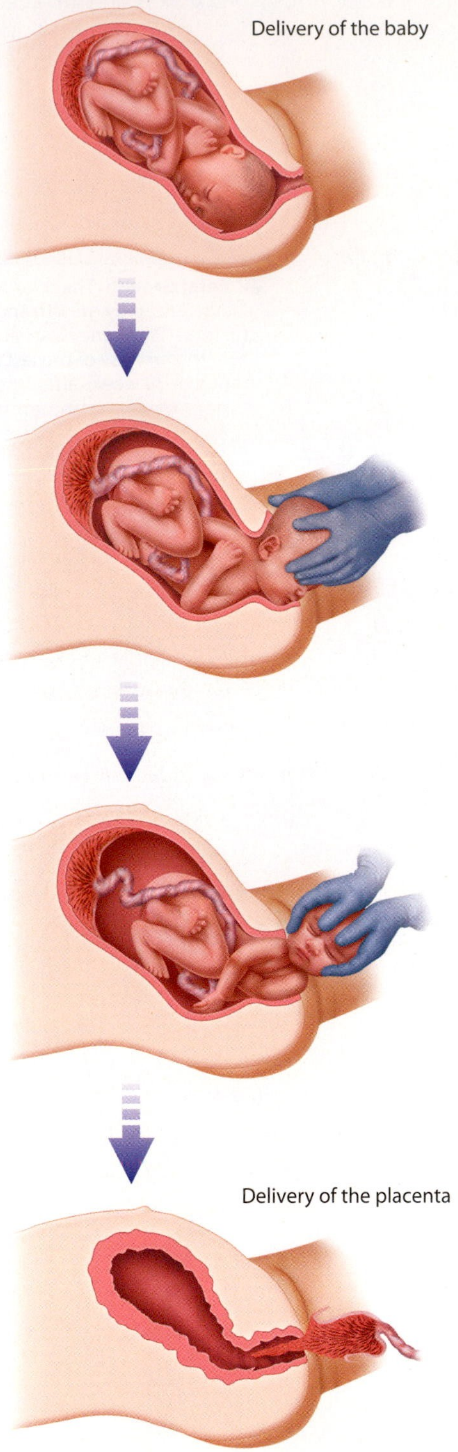

→ **Quick Check 5** A multicellular organism starts off as a single cell, which divides by mitosis to produce many cells. Therefore, for the most part, all the cells in a multicellular organism are genetically identical—they are clones—yet in time different groups of cells look different and function quite differently from one another. How can this be?

Childbirth is initiated by hormonal changes.

Childbirth marks the end of pregnancy. Although we tend to think of childbirth as the delivery of the baby, it actually consists of three events—changes in the cervix to allow passage of the baby, descent and delivery of the baby, and delivery of the placenta. These three events result from strong muscular contractions of the uterus mediated by the hormone oxytocin from the posterior pituitary gland. Oxytocin causes uterine contractions, stimulating more oxytocin release, which in turn leads to further contractions, and so on, providing a good example of positive feedback.

During pregnancy, the cervix is a tube about 4 cm long closed with a mucus plug. To allow passage of the infant, the cervix shortens and widens. These changes are not enough to accommodate passage of the baby during delivery. Humans have a narrow pelvis relative to that of other primates as an adaptation for bipedalism. To facilitate its passage through the pelvis, the baby goes through a series of movements (**Fig. 42.25**). Babies are usually born head first with the face pointing down (toward the mother's back) when passing through the birth canal. Other positions make vaginal birth more difficult, sometimes impossible.

The third and final stage of childbirth is the delivery of the placenta, which usually occurs about 15 to 30 minutes after the baby is born.

Although birth normally occurs vaginally, in some cases, such as when the baby is too large or positioned feet first, it is necessary to remove the baby by making an incision in the abdomen and uterus in a surgical procedure known as a Caesarian section. Interventions such as Caesarian sections, the use of antibiotics, and pre- and postnatal care have greatly reduced maternal and fetal mortality associated with childbirth.

Whichever way the baby is delivered, when the umbilical cord is cut and the baby takes its first breath, it begins its life independent of the mother—anatomically independent, that is. Otherwise, human newborns are completely dependent on others. Humans are unusual among animals in having an extended period of immaturity while their brains and bodies grow and develop. This length of time means that parents usually put considerable effort into raising and caring for their offspring, and it allows for extended periods of learning, the development of strong social bonds, and the transmission of culture to the next generation. ■

Core Concepts Summary

42.1 Reproduction is a basic feature of living organisms and occurs both asexually and sexually.

Asexual reproduction involves a single parent. Offspring are produced by binary fission or mitotic cell division and are clones of the parent. page 898

Examples of asexual reproduction include budding, fragmentation, and parthenogenesis. page 898

Sexual reproduction involves the production of haploid gametes by meiotic cell division and the fusion of gametes to make a diploid zygote. page 899

Many organisms reproduce both sexually and asexually. For some organisms, environmental conditions favor one mode or the other. page 900

Asexual reproduction allows rapid population growth but produces clones and results in genetic uniformity. page 901

Sexual reproduction produces genetically unique offspring, a feature that is thought to explain why most eukaryotes reproduce sexually at least some of the time. page 901

42.2 The movement of vertebrates from water to land involved changes in reproduction, including internal fertilization and the amniotic egg.

Fertilization can occur outside or inside the body of the female. External fertilization can occur only in an aquatic environment. Internal fertilization is an adaptation to terrestrial living. page 903

External fertilization is usually associated with the production of many offspring, little or no parental care, and high mortality. These are characteristics of *r*-strategists. page 904

Internal fertilization is often associated with the production of fewer offspring, increased parental care, and low mortality. These are characteristics of *K*-strategists. page 904

A watertight sac, called the amnion, is an adaptation for reproduction on land. page 904

Oviparous animals lay eggs; ovoviviparous and viviparous animals give birth to live young, but ovoviviparous animals retain eggs until they hatch and commonly get nutritional support from the yolk, whereas viviparous animals get nutritional support from the mother. page 905

42.3 The male reproductive system is adapted for the production and delivery of sperm, and the female reproductive system is adapted for the production of eggs and, in some cases, support of the developing fetus.

Male gametes, called sperm, develop in the seminiferous tubules of the testes. page 905

Sperm travel from the seminiferous tubules to the epididymis to the vas deferens to the ejaculatory duct and finally to the urethra before being released during ejaculation. page 905

Semen is a mixture of sperm and fluid from the prostate gland, seminal vesicles, and bulbourethral glands. page 906

Developing female gametes, called oocytes, are produced in the ovaries. Mature oocytes are called eggs, or ova. page 907

An oocyte is released from the ovary, travels through the fallopian tube, and implants in the uterus if it is fertilized. page 907

The reproductive system responds to and produces hormones. The hypothalamus releases GnRH, which stimulates the anterior pituitary gland to release FSH and LH, which in turn stimulate the ovaries to release estrogen and progesterone or the testes to secrete testosterone. page 908

In males, LH stimulates Leydig cells to secrete testosterone, and FSH and testosterone act on Sertoli cells to support sperm production. page 908

In females, monthly menstrual cycles are driven by the interplay of hypothalamic, anterior pituitary, and ovarian hormones, leading to ovulation, which is the maturation and release of an oocyte from the ovary, and to changes to the uterine lining. page 909

42.4 Human reproduction involves the formation of gametes, fertilization, and growth and development.

Gametogenesis is called spermatogenesis in males and oogenesis in females. page 911

In males, spermatogonia differentiate into primary spermatocytes that undergo meiosis to make immature haploid spermatids and mature spermatozoa or sperm. page 911

Spermatogenesis begins at puberty and continues throughout life. page 911

In females, oogonia undergo differentiation into primary oocytes during fetal development, which begin meiosis but arrest in prophase I. During a menstrual cycle, primary oocytes are released from arrest, finish meiosis I, but then arrest in metaphase II. They remain arrested in metaphase II until they are fertilized. page 911

Fertilization is the fusion of the oocyte and sperm plasma membranes. It usually occurs in the fallopian tube and results in the formation of a diploid zygote. page 912

Cleavage is the division of the zygote by mitotic cell division into smaller cells. At the blastocyst stage, the developing embryo implants in the uterus. page 913

Pregnancy lasts about 38 weeks from fertilization until birth and is divided into three trimesters. page 914

Gastrulation results in the formation of the three germ layers—ectoderm, mesoderm, and endoderm. page 914

Organogenesis is the formation of organs. page 914

Childbirth, mediated by the posterior pituitary hormone oxytocin, involves changes in the cervix, delivery of the baby, and delivery of the placenta. page 918

Self-Assessment

1. List similarities and differences between asexual and sexual reproduction.
2. Explain the roles of meiotic cell division and fertilization in sexual reproduction.
3. Describe the costs and benefits of asexual and of sexual reproduction.
4. Provide an explanation for the observation that there are very few, if any, ancient asexual organisms.
5. Name three adaptations that allow reproduction to take place on land.
6. Describe the pathway of sperm from its site of production to the urethra.
7. Describe the pathway of an oocyte (and the egg it develops into) from its site of production to the uterus.
8. Explain the relationships among changes in levels of anterior pituitary and ovarian hormones, oocyte development, and changes in the uterine lining during a menstrual cycle.
9. Describe similarities and differences between male and female gametogenesis.
10. Name and describe three key developmental steps in the development of a single-celled zygote to a multicellular individual.

Log in to LaunchPad to check your answers to the Self-Assessment questions, and to access additional learning tools.

Source: Power and Syred/Science Source.

CHAPTER 43

Animal Immune Systems

Core Concepts

43.1 The immune system protects organisms from pathogens.

43.2 The innate immune system acts generally against a diversity of pathogens.

43.3 The adaptive immune system includes B cells that produce antibodies against specific pathogens that reside outside cells in tissues.

43.4 The adaptive immune system also includes helper and cytotoxic T cells that attack infected, diseased, and foreign cells.

43.5 Some pathogens have evolved mechanisms that enable them to evade the immune system.

All organisms have mechanisms that protect them against harmful pathogens. In turn, pathogens evolve strategies to evade these mechanisms. And organisms in turn evolve new responses. In animals, the result of this evolutionary back and forth is an elaborate and sophisticated immune system. The immune system is remarkable in its ability, in most cases, to protect organisms from harmful pathogens. Consider how well your body is able to defend itself from the common cold and bacterial infections from cuts and scrapes without taking antibiotics. In fact, we don't even notice most of the pathogens we encounter on a daily basis because of the effectiveness of our immune system. A key aspect of the vertebrate immune system is that it has evolved both defenses that act against a broad range of pathogens and defenses that target specific pathogens.

We begin with an overview of the immune system and then discuss general defenses that protect against a diverse array of pathogens. We then turn to a powerful part of the immune system that is tailored to attack specific pathogens and that has the ability to learn and remember, attributes we usually associate with the nervous system. Finally, we consider three infections that illustrate how pathogens have evolved ways to evade these defenses.

43.1 AN OVERVIEW OF THE IMMUNE SYSTEM

The immune system is different from other systems we have considered up to this point. Most organ systems are made up of discrete tissues or organs. Consider the cardiovascular system, which includes the heart and associated blood vessels. Or think of the gastrointestinal tract, which in vertebrates consists of a series of connected tissues and spaces, each with a specialized function for digesting and absorbing nutrients from food.

By contrast, the immune system consists of organs, cells, proteins, and even biochemical reactions that are spread throughout the body. This makes sense, in that the immune system must be able to interact with pathogens wherever they might be found. There are some organs that are dedicated to the immune system, such as the thymus in many vertebrates (see below), but otherwise it's a diffuse system of cells, molecules, and processes. In fact, the phrase "immune system" was not even used until sometime in the late 1960s. It was introduced to draw attention to the fact that it's truly a *system* made up of many different components. These components work by different mechanisms, but all share a common function: to protect an organism from pathogens.

Pathogens cause disease.

Pathogens are organisms and other agents that cause disease. For a bacterium, pathogens include viruses that infect bacteria, as we have seen (Chapter 19). These viruses, called bacteriophages (literally "bacteria eaters"), can cause the bacterial cell to lyse, or burst, as viral progeny are produced inside the bacterium. In response, bacteria protect themselves with general mechanisms such as the cell wall or specific mechanisms such as CRISPR, which allows bacteria to "remember" past viral infections and mount a response on subsequent infections (Chapter 12).

For a fungus, pathogens include bacteria and viruses. In turn, fungi have evolved molecules that kill or inhibit the growth of bacteria. We have isolated many of these substances and learned how to use them as antibiotics. Penicillin is a famous example. It was discovered in 1928 by Alexander Fleming when he noticed that a fungus in the genus *Penicillium* interfered with the growth of bacteria in a petri dish.

For mammals and other vertebrates, the focus of this chapter, pathogens include viruses, bacteria, fungi, single-celled eukaryotes commonly called protists, and worms (**Fig. 43.1**). They cause disease in diverse ways. In some cases, they infect a particular type of cell. For example, HIV destroys a specific type of cell involved in the immune system itself, making the body susceptible to other pathogens that the body would usually be able to easily defend itself against. In other cases, pathogens proliferate in an organ and interfere with its function. Pneumonia, for example, can result when bacteria or fungi proliferate in lung tissue and limit gas exchange. Sometimes, pathogens produce toxins that cause disease. Cholera, for instance, results from infection with a toxin-producing bacterium called *Vibrio cholera.* The toxin interferes with membrane transporters in the small intestine and causes diarrhea, resulting in dehydration and sometimes death.

The immune system distinguishes self from nonself.

Critical to the functioning of the immune system is the ability to recognize pathogens as foreign. That is, it must be able to distinguish the molecules and cells of its own organism, described as host or **self,** from molecules and cells of other organisms, described as foreign or **nonself,** and then respond only to those that are nonself.

The ability of the immune system to distinguish self from nonself depends largely on proteins that normally reside on the surface of cells. For example, in organ transplants, the immune system sometimes recognizes proteins on the surface of transplanted organs as nonself. The transplanted organs are then attacked by the immune system, and they are rejected.

Sometimes, the immune system mistakes self and nonself. If the immune system reacts inappropriately to self, it can end up attacking its own cells or tissues. This occurs in **autoimmune diseases,** such as rheumatoid arthritis, type 1 diabetes, and multiple sclerosis. Alternatively, the immune system sometimes recognizes as nonself harmless substances such as pollen or dust, leading to allergic reactions. Or the immune system may not react when it should and therefore does not adequately protect an organism. A weak immune system is most likely to occur at the extremes of age or when infections or diseases compromise its function, a condition called **immunodeficiency.**

FIG. 43.1 Human pathogens. (a) Human immunodeficiency virus (HIV); (b) cholera bacteria; (c) a malaria parasite (green) inside a swollen red blood cell; (d) ringworm (a fungus); (e) schistosome (a worm). *Sources: a. James Cavallini/Science Source; b. Juergen Berger/Science Source; c. Dr. Tony Brain/Science Source; d. David Scharf/Science Source; e. Scott Camazine/Getty.*

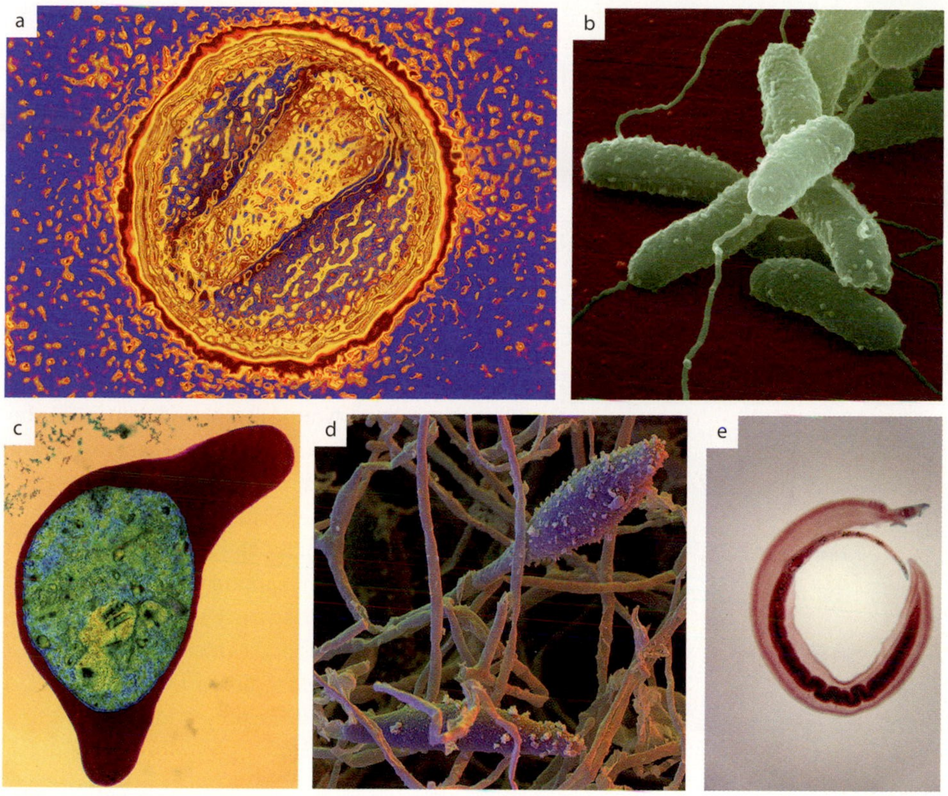

As a result, the activity of the immune system is carefully regulated. One way this regulation occurs is by keeping immune cells and biochemical processes in an inactive state that is activated only when they come into contact with a foreign molecule or cell. In this way, the immune system reacts only when it is needed to attack a pathogen.

TABLE 43.1 Comparison of the Innate and Adaptive Immune Systems

	INNATE IMMUNE SYSTEM	ADAPTIVE IMMUNE SYSTEM
Targets	Diverse pathogens	Diverse pathogens
Ability to distinguish self from nonself?	Yes	Yes
Specificity?	No	Yes
Memory?	No	Yes
Organisms	All organisms	Vertebrates

The distinction between self and nonself is a useful way to think about how the immune system functions, but it has limitations. For example, as we will see, the immune system is able to recognize and attack diseased or infected self cells, but it also tolerates bacteria and other nonself microorganisms that make up our microbiome (Case 5: The Human Microbiome). French immunologist Polly Matzinger suggested that another way to think about the immune system is that it recognizes and attacks agents that cause injury or damage, not necessarily those that are foreign. Her idea suggests that there is more work to do to fully understand how the immune system functions.

The immune system consists of innate and adaptive immunity.

The immune system consists of two parts that work in different ways and interact with each other to protect against infection. Each part is made up of different cell types and pathways. The first, called **innate immunity** (or natural immunity), provides protection against all kinds of infection in a nonspecific manner. This form of immunity is present in plants, fungi, and animals and is an evolutionarily early form of immunity. It is considered *innate* because it does not depend on prior exposure to a pathogen.

By contrast, the second form of immunity, called **adaptive immunity** (or acquired immunity), is specific to a given pathogen. This form of immunity "remembers" past infections, meaning that subsequent encounters with the same pathogen generate a stronger response. In other words, the immune system *adapts* over time, and immunity is *acquired* after an initial exposure. Adaptive immunity evolved later than innate immunity and is unique to vertebrates. **Table 43.1** summarizes the major differences between these two forms of immunity.

43.2 INNATE IMMUNITY

The innate immune system provides general, nonspecific defenses against a broad array of pathogens. In animals, it includes tissues, such as the skin, that act as a barrier to infection, cells that act as sentries against pathogens, and proteins and other molecules circulating in the blood that signal the presence of infection. All these defenses are nonspecific, act immediately, and do not depend on past exposure to elicit a response.

FIG. 43.2 Physical barriers against infection. (a) Tree bark; (b) leaf cuticle; (c) insect exoskeleton; (d) egg shell; (e) skin. *Sources: a. Christian Beier/age fotostock; b. Michael P. Gadomski/Science Source; c. Charles J. Smith; d. Inga Spence/Alamy; e. Janaka Dharmasena/Shutterstocs.*

The skin and mucous membranes provide the first line of defense against infection.

Organisms have a barrier that surrounds them, which can at times be breached. Among its other functions, the barrier surrounding an organism ordinarily protects against infection. Examples include the cell wall of bacteria, the bark of trees, the cuticle of leaves, the exoskeleton of insects, the scales of fish, the shells of eggs, and the skin of mammals (**Fig. 43.2**). These structures act as physical barriers against infection.

To understand the importance of the barrier, consider what happens when it is breached. Burn victims who lose the outer layer of their skin are particularly vulnerable to infection. The bite of a dog or the sting of an insect can insert viruses, bacteria, or parasites directly into the bloodstream, bypassing the protective layer of skin. Even a small puncture wound can provide entry for the bacterium *Clostridium tetani,* the cause of tetanus.

Any surface that is in contact with the external environment is a potential entry site for pathogens. Like the skin, the mucous membranes of the respiratory, gastrointestinal, and genitourinary tracts act as physical barriers. However, they have additional protective mechanisms. For example, parts of the respiratory tract have **cilia,** hairlike projections that

FIG. 43.3 White blood cells. There are many different types of white blood cell that function in the innate and adaptive immune systems. They develop from stem cells in the bone marrow.

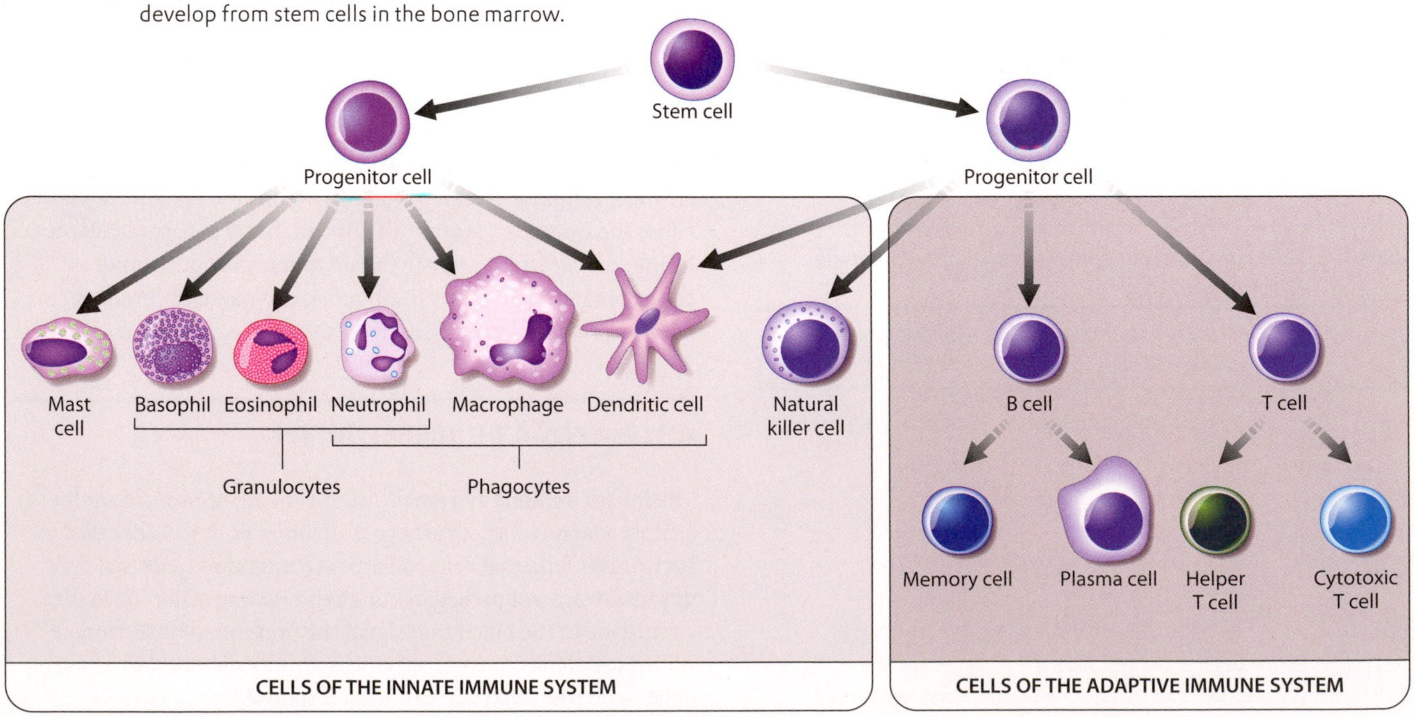

sweep mucus containing trapped particles and microorganisms out of the body. Saliva and tears contain enzymes that help to protect against bacteria. The stomach, with its low pH and enzymes that break down food, is inhospitable to many microorganisms.

In addition to acting as a physical and chemical barrier, the skin and mucous membranes provide a home to many nonpathogenic microorganisms. These microorganisms in some cases provide protection against harmful pathogens by competing with them for food and space. They may also maintain health in other ways, such as by aiding digestion. In this way, we live in a mutualistic relationship with these microorganisms (Case 5: The Human Microbiome).

→ **Quick Check 1** Certain diseases, as well as cigarette smoke, can damage cilia lining the respiratory tract. What effects would you expect from damage to respiratory cilia?

FIG. 43.4 **Phagocytosis.** A phagocyte recognizes, engulfs, and digests foreign particles and cells. The scanning electron micrograph shows a macrophage engulfing tuberculosis bacteria. *Photo source: SPL/Science Source.*

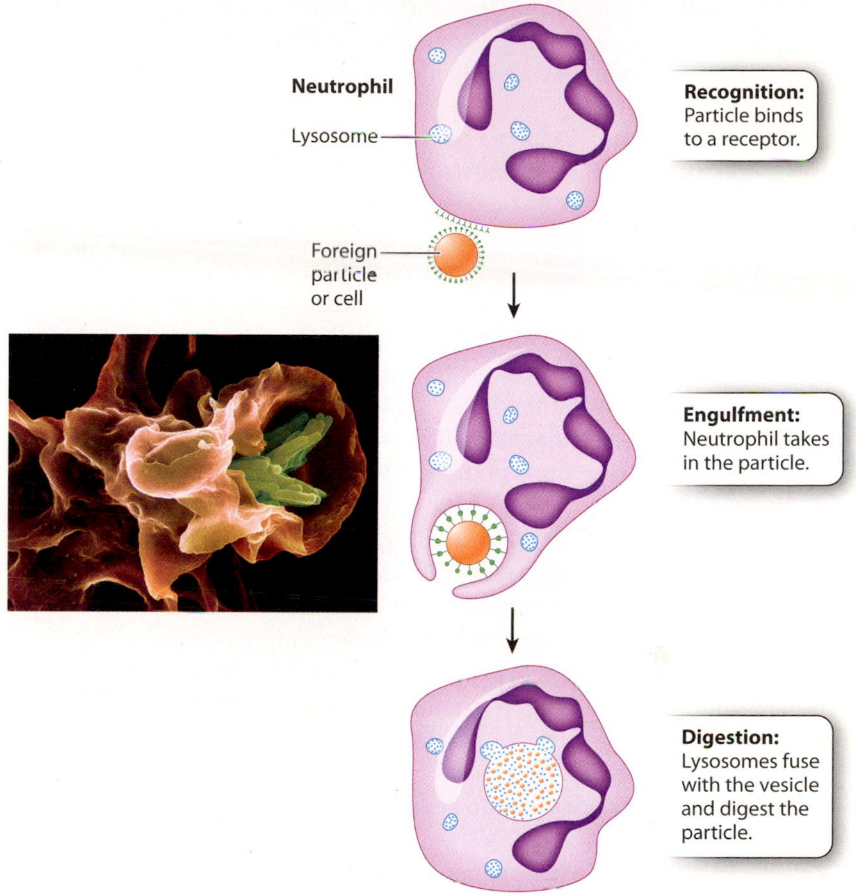

White blood cells provide a second line of defense against pathogens.

If a pathogen makes it past the skin and mucous membranes of an animal, there is a second line of defense: **white blood cells** (leukocytes). There are many different types of white blood cell, which arise by differentiation from stem cells in the bone marrow (**Fig. 43.3**). Just as the skin and mucous membranes are an external defense against pathogens, white blood cells are an internal defense. Some white blood cells mount an early response to a pathogen; these are nonspecific innate immune cells. If they are unable to clear the pathogen, they alert the more specialized white blood cells of the adaptive immune system. Here, we focus on the role of white blood cells in innate immunity.

Phagocytes (literally, "eating cells") are white blood cells that engulf and destroy foreign cells or particles. The engulfing of a cell or particle by another cell is called **phagocytosis.** The Russian microbiologist Ilya Mechnikov first described this remarkable process, for which he shared the Nobel Prize in Physiology or Medicine with the German scientist Paul Ehrlich in 1908.

The process begins when a phagocyte encounters a cell or particle that it recognizes as foreign and binds to it. The phagocyte then extends its plasma membrane completely around the cell or particle until it is contained in a compartment or vesicle inside the phagocyte (**Fig. 43.4**). In some cases, the vesicle merges with a lysosome, and enzymes within the lysosome digest the foreign cell or particle. Phagocytosis can also trigger a respiratory burst, a process that generates reactive oxygen species, such as superoxide radicals and hydrogen peroxide, and reactive nitrogen species, such as nitrogen oxide and nitrogen dioxide. These molecules have antimicrobial properties that directly damage pathogens.

There are three major types of phagocytic cell: **macrophages, dendritic cells,** and **neutrophils** (see Fig. 43.3). In each case, their names give clues to their shape and size. Macrophages are large cells that patrol the body (*macro* means "large"). Dendritic cells have long cellular projections that resemble the dendrites of neurons (Chapter 35). These cells are typically part of the natural defenses found in the skin and mucous membranes. Neutrophils are members of a group of cells called **granulocytes** because of the presence of granules in their cytoplasm. The granules of neutrophils take up both acidic and basic dyes, so their staining pattern is considered neutral. Neutrophils are abundant in the blood and are often the first cells to respond to infection.

The innate immune system includes several other white blood cell types in addition to phagocytic cells (see Fig. 43.3). Two types of granulocyte, **eosinophils** and **basophils,** defend against parasitic infections, but they also contribute to allergies. **Mast cells** release **histamine,** an important contributor to allergic reactions and inflammation. Finally, **natural killer cells** do not recognize foreign cells, but instead recognize and kill host cells that are infected by a virus or have become cancerous or otherwise abnormal.

Phagocytes recognize foreign molecules and send signals to other cells.

Phagocytes attack foreign cells that they encounter but leave host cells alone. How is a phagocyte able to distinguish a foreign cell from a host cell? All cells have on their surface an array of proteins and other molecules, as we saw in Chapters 5 and 9. Phagocytes detect surface molecules that are common to pathogens, but not to the host.

Toll-like receptors (TLRs) are proteins on the surface of phagocytes that recognize and bind to molecules on pathogens. First discovered in the fruit fly *Drosophila*, TLRs are found on cells in vertebrates and invertebrates, and similar molecules are found in plants and bacteria, indicating that they are an early, evolutionarily conserved part of the innate immune system. They provide one of the earliest signals that an infection is present. TLRs recognize evolutionarily conserved surface molecules on a wide range of microorganisms. For example, one type of TLR recognizes a glycolipid on the cell wall of certain bacteria. Another recognizes a protein that is part of the bacterial flagellum.

Binding of TLRs to surface molecules on the pathogen is a signal to the phagocyte to engulf and destroy its target. Following binding of a foreign molecule to the TLR, phagocytes send a signal to the rest of the immune system that the body is under attack by releasing chemical messengers called **cytokines** that recruit other immune cells to the site of injury or infection. Cytokines, like hormones, provide long-distance communication between cells (Chapters 9 and 38).

Inflammation is a coordinated response to tissue injury.

While there are generally some immune cells already present at the site of tissue injury or infection, many more must be recruited to fight infection successfully. That recruitment happens as part of a process called inflammation. The redness and swelling you see following an injury, a break in the skin, or an infection are signs of inflammation. More precisely, **inflammation** is a physiological response of the body to injury that removes the inciting agent if present and begins the healing process, such as cell division to heal a wound. It works through the coordinated responses of the local tissue, the vascular system, and the innate immune system. Inflammation is typically beneficial, but if left unchecked or allowed to continue for a long time, it can be debilitating.

Inflammation is characterized by four signs, described by their Latin names: *rubor* (redness), *calor* (heat), *dolor* (pain), and *tumor* (swelling). The process begins following tissue injury, such as a cut, or infection, as shown in **Fig. 43.5.** Certain cells in the tissue, such as dendritic cells and mast cells, recognize a pathogen as foreign and release cytokines and other chemical messengers that recruit more white blood cells to the site or make it easier for them to reach the site quickly. For example, histamine, released by mast cells and basophils, acts directly on blood vessels to cause vasodilation, increasing blood flow to the site of infection or injury. Vasodilation causes the characteristic redness and heat of inflammation. In addition, histamine increases the permeability of the blood vessel wall. Fluid leaks out of the blood vessel, carrying white blood cells into the damaged tissue. The increased fluid in the tissue surrounding the blood vessels is visible as swelling. Some chemical messengers act directly on nerve fibers, causing pain.

FIG. 43.5 Inflammation. Inflammation involves increased blood flow, leaky blood vessels, and activation of the immune system.

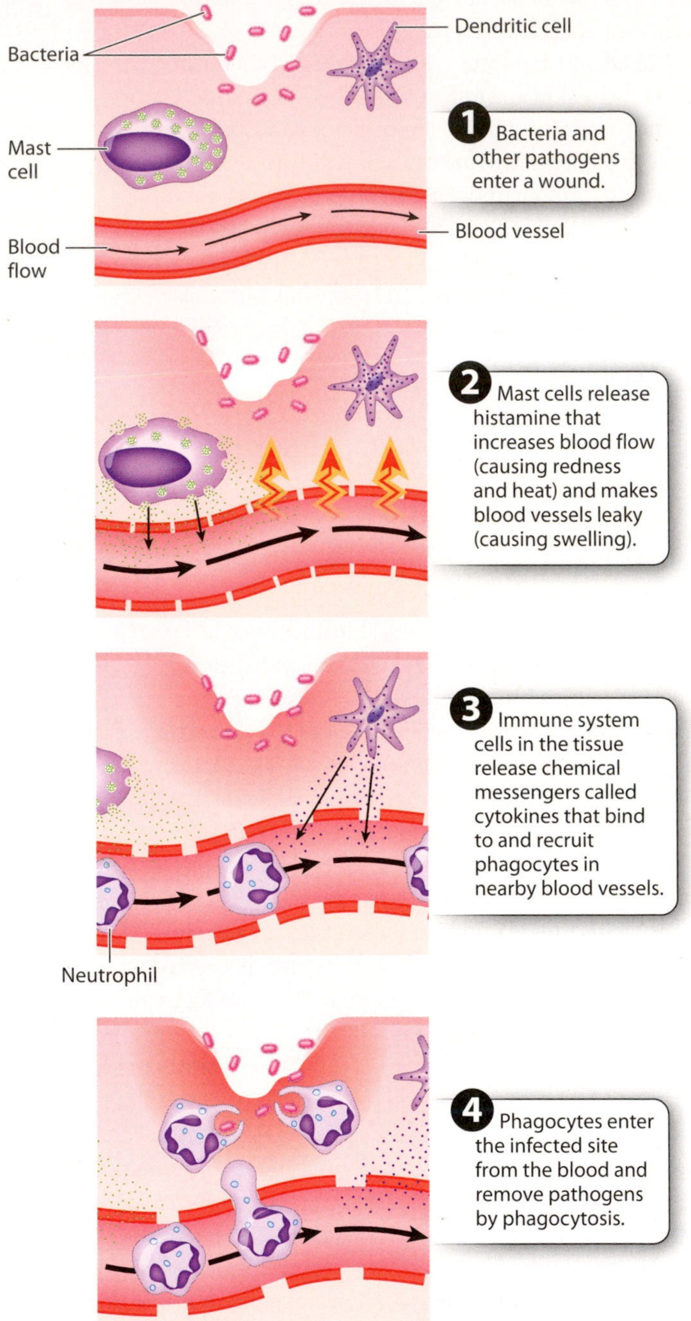

FIG. 43.6 Extravasation. In extravasation, a phagocyte moves from a blood vessel to the surrounding tissue.

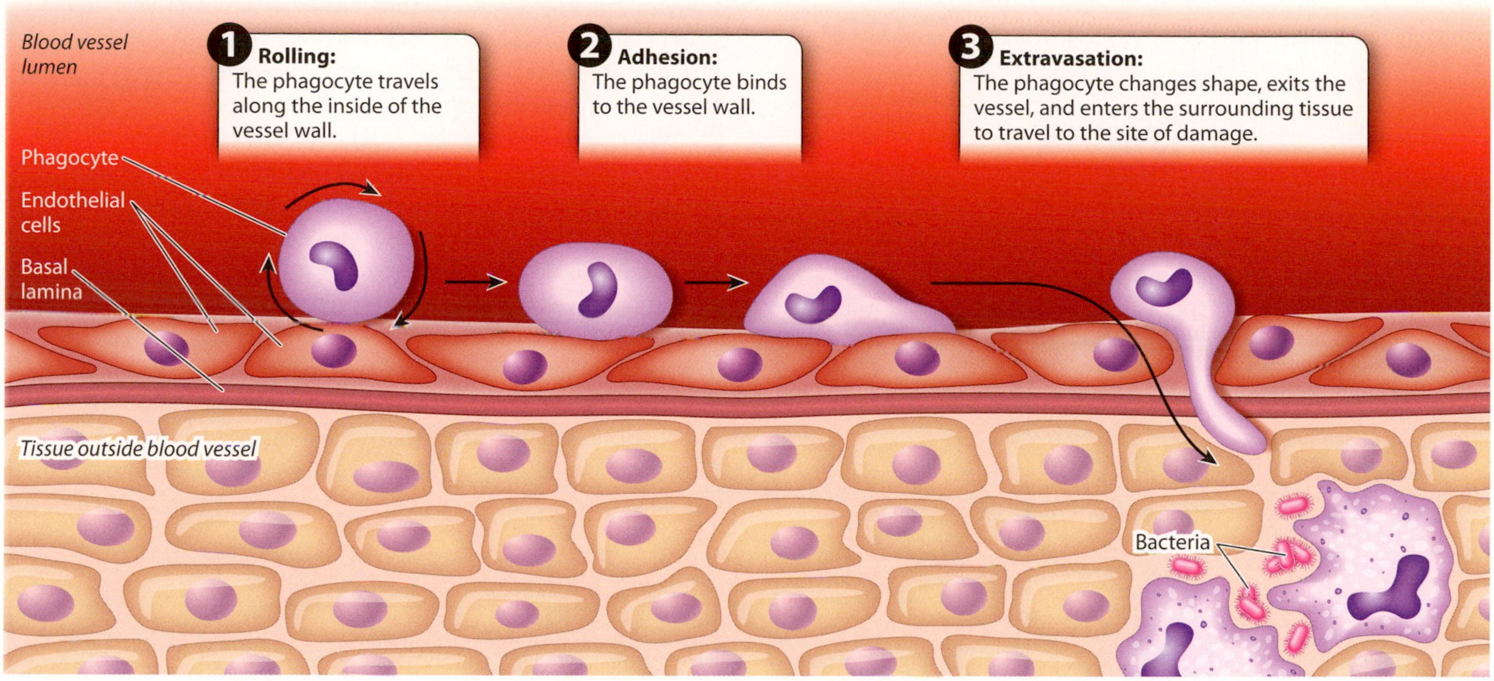

During inflammation, phagocytes in the blood often move from a blood vessel to the site of infection by **extravasation** (**Fig. 43.6**). In this process, the phagocyte travels along the vessel wall in a rolling motion, grabbing weakly onto the wall as glycoproteins on the phagocyte surface bind transiently to proteins on the endothelial cells. Stronger interactions put a brake on this rolling motion and allow the phagocyte to adhere more firmly to the vessel wall. As it nears the site of infection, the phagocyte encounters and binds to cytokines. The phagocyte then changes shape and moves in between cells lining the blood vessel and into the surrounding tissue.

The complement system participates in the innate and adaptive immune systems.

We have seen that many white blood cells participate in innate immunity, providing a general defense against pathogens. Other white blood cells participate in adaptive immunity, and they can be activated by a set of proteins that circulate in the blood. Collectively, these proteins make up the **complement system,** and they are also considered a part of the innate immune system. First described by Paul Ehrlich in the late 1800s, this system is so named because it supplements or *complements* other parts of the immune system.

The complement system consists of more than 25 types of protein that circulate in the blood in an inactive form. The system is activated when these proteins bind to molecules found only on invading microorganisms or to antibodies (discussed below). Activation sets off a biochemical cascade in which the product of one reaction is an enzyme that catalyzes the next. This kind of sequential activation leads to amplification of the response, as we saw earlier in cell signaling (Chapter 9) and in the endocrine system (Chapter 38), where a small stimulus can have a dramatic response.

Activation of the complement system has three effects (**Fig. 43.7**). The most dramatic occurs when complement proteins circulating in the blood are activated in such a way that they assemble to form a pore, called a **membrane attack complex (MAC),** in bacterial cells. The MAC causes the bacterial cell to break open, or lyse. Second, the complement system coats bacteria with a protein that phagocytes recognize, targeting them for phagocytosis. The binding of a protein to a pathogen to facilitate phagocytosis is called **opsonization.** Finally, activation of the complement system leads to the production of activated proteins that attract other components of the immune system. For example, some of these proteins stimulate mast cells, which release histamine and initiate the inflammatory response.

The importance of the complement system can be appreciated by considering what occurs in individuals who lack key components. Complement deficiencies are a form of immunodeficiency, a condition in which the immune system does not function properly. For example, one type of complement deficiency makes individuals susceptible to bacterial infections because fewer pathogens undergo cell lysis and opsonization. The intensity and type of symptoms can vary depending on which of the components of the complement system is lacking, from very mild, to recurrent infections, to other immunological disorders.

FIG. 43.7 The complement system. The proteins of the complement system contribute to the defense against pathogens by cell lysis, phagocytosis, and inflammation.

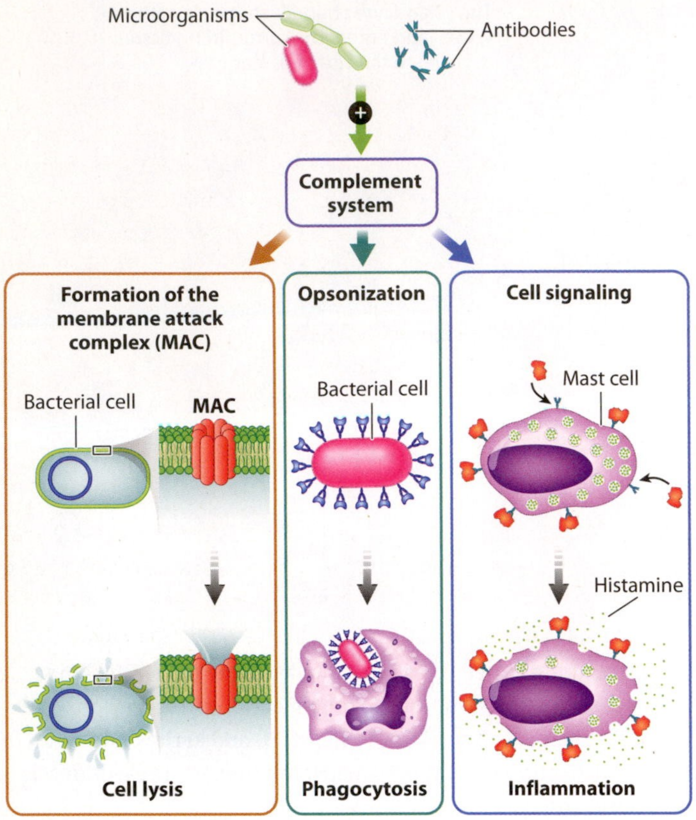

Because antibodies are an important signal that activates the complement system, and because the presence of complement proteins on a pathogen helps to activate the adaptive immune system, the complement system provides an important bridge between the innate, nonspecific part of the immune system and the adaptive, specific part, a topic we consider next.

43.3 ADAPTIVE IMMUNITY: B CELLS AND ANTIBODIES

The innate immune system can react to diverse pathogens and has the capacity to distinguish self from nonself. The adaptive immune system shares these features, and has two additional features not present in the innate immune system. The first is specificity: The adaptive immune system produces an array of molecules, each of which has the potential to target a specific pathogen it has never before encountered. As a result, it is often more effective at eliminating a pathogen than the innate immune system. The second is memory: The adaptive immune system remembers past infections and mounts a stronger response on re-exposure.

Consider for a moment the number of pathogens that we might encounter—on the order of millions—and remember that they, too, are evolving all the time. How can the adaptive immune system tailor its response to each one of these pathogens? And how does it remember past infections? We explore answers to these questions in the next two sections.

While many cells and tissues play a role in adaptive immunity, two white blood cell types are particularly important: **B lymphocytes** or **B cells,** and **T lymphocytes** or **T cells** (see Fig. 43.3). B and T cells are named for their sites of maturation. In birds, B cells mature in the bursa of Fabricius, an organ located off the digestive tract (**Fig. 43.8**). In mammals, B cells mature in the bone marrow. In most vertebrates, T cells mature in the thymus, an organ of the immune system located just behind the sternum, or breastbone. B and T cells circulate in blood vessels and lymphatic vessels, which are thin-walled vessels that absorb excess fluid from tissues (Chapter 39). B and T cells also travel through the spleen, liver, and lymph nodes, where they sample the body

FIG. 43.8 Organs that participate in the immune system.

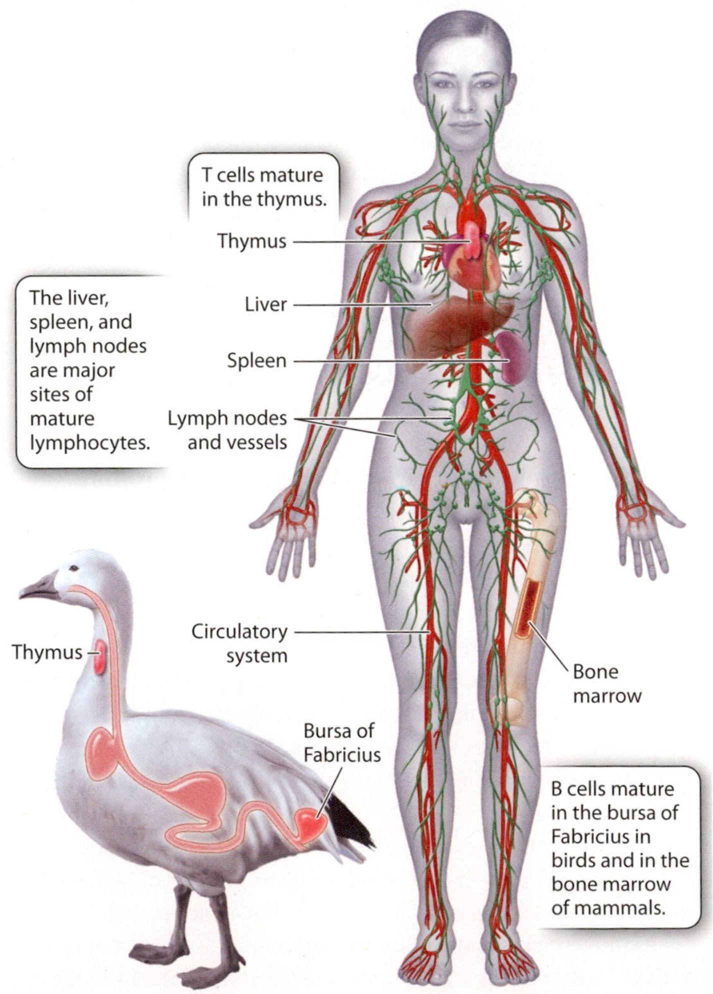

FIG. 43.9 Structure of one type of antibody. An antibody has variable regions that distinguish it from the billions of other antibodies and allow it to bind a specific antigen.

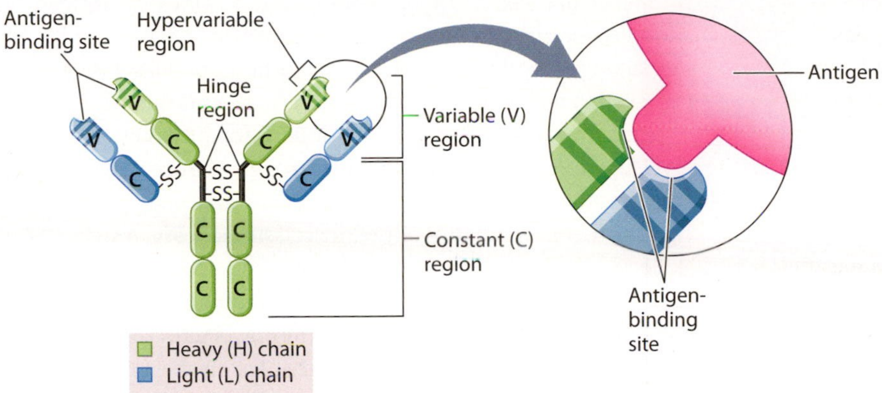

for pathogens. We begin with a discussion of the role of B cells in adaptive immunity.

B cells produce antibodies.

One of the hallmarks of adaptive immunity is the ability to target specific novel pathogens. But there are many different viruses, bacteria, single-celled eukaryotes, fungi, and worms that can lead to infection. How does the adaptive immune system recognize each one? The specificity of the adaptive immune system is in part the result of antibodies produced by B cells.

An **antibody** is a large protein found on the surface of B cells or free in the blood or tissues. An antibody binds to molecules on the surface of foreign cells, called **antigens,** which, by definition, are molecules that lead to the production of antibodies. (The word "antigen" is a contraction of "antibody" and "generator.") Any molecule that naturally occurs on or in a microorganism can be an antigen. The great diversity of antigens on different microorganisms is matched by a great diversity of antibodies produced by an individual. It is estimated that humans produce approximately 10 billion different antibodies, each with the ability to recognize one or a few antigens on microorganisms.

Antibodies have a distinctive structure that reflects their specificity for a given antigen. The simplest antibody molecule, shown in **Fig. 43.9,** is a Y-shaped protein made up of four polypeptide chains—two identical **light (L) chains** and two identical **heavy (H) chains,** so called because of their relative molecular weights. Covalent disulfide bonds hold the four chains together.

The light and heavy chains are both subdivided into **variable (V)** and **constant (C)** regions (Fig. 43.9). The constant region, as the name implies, is the same in all antibodies of a given class. The variable region is unique to each antigen that the antibody targets. Within the variable regions of the L and H chains are regions that are even more variable, called hypervariable regions. These hypervariable regions interact in a specific manner with just a portion of the antigen.

Binding of antibodies to antigens is the first step in recognition and removal of pathogens. Binding alone is sometimes enough to disable a pathogen. More often, the pathogen is killed by a different component of the immune system, for example by a phagocyte or the complement system. In this case, the function of an antibody is to recognize the pathogen, then recruit other cells of the immune system or activate the complement system.

Mammals produce five classes of antibody with different functions.

In mammals, there are five classes of antibody, called **IgG, IgM, IgA, IgD,** and **IgE,** each with a different function (**Fig. 43.10**). IgG is the Y-shaped antibody depicted in Figs. 43.9 and 43.10. IgG circulates in the blood and is particularly effective against bacteria and viruses.

IgM is a pentamer in mammals and a tetramer in fish. (A pentamer is a molecule made of five monomers; a tetramer is a molecule made of four monomers.) The individual units—the monomers—are linked by a joining chain. IgM also exists as a monomer on the surface of B cells. IgM is important in the early response to infection.

IgA is usually a dimer consisting of two antibody molecules linked by a joining chain. It is the major antibody on mucosal surfaces, such as those of the respiratory, gastrointestinal, and genitourinary tracts. It is also present in secretions, including tears, saliva, and breast milk.

FIG. 43.10 Five classes of antibody.

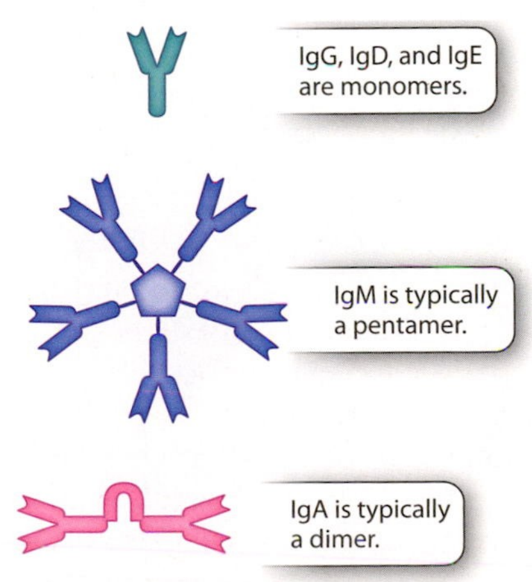

IgD and IgE are monomers. IgD is typically found on the surface of B cells. This immunoglobulin helps initiate inflammation. IgE plays a central role in allergies, asthma, and other **immediate hypersensitivity reactions,** which are characterized by a heightened or inappropriate immune response to common antigens. Immediate hypersensitivity reactions result from the binding of IgE to receptors on the surface of mast cells and basophils. Binding leads to the release of histamine and cytokines, resulting in inflammation that can be life threatening.

Clonal selection is the basis for antibody specificity.

A central question in immunology is how an individual can generate a great diversity of antibodies, each of which is specific for a given antigen. There are seemingly limitless types of antigen that an organism might come across. How can B cells generate antibodies that interact only with specific antigens among this tremendous diversity?

Two alternative hypotheses were suggested to explain antibody specificity. In 1900, Paul Ehrlich suggested that there exists a large pool of B cells, each with a different antibody on its surface. The antigen binds to one or a few of these antibodies, stimulating the B cell to divide. In this model, specificity exists before any exposure to the antigen. Ehrlich's model drew on concepts from chemistry such as the binding of an enzyme to a substrate in the manner of a lock and key (Chapter 6).

Later, in the early 1900s, a different hypothesis was suggested. This hypothesis proposed that the antigen instructs the antibody to fold in a particular way so that the two interact in a specific manner. Here, the antigen plays a more active role in the process, and antibody specificity is determined only after interaction with the antibody.

Which of these two hypotheses is correct? Does the antigen *select* a B cell with a preexisting surface antibody with which it interacts specifically, as in the first hypothesis, or does it *instruct* the B cell to make a specific antibody, as in the second?

Experimental data on the genetic basis for antibody diversity (described below) provided strong evidence for the first hypothesis. Every individual has a very large pool of B cells, each with a different antibody on its cell surface (**Fig. 43.11**). Collectively, these antibodies recognize a great diversity of antigens, but each individual antibody can recognize only one antigen or a few antigens. An antigen interacts with the antibody on the B cell, or a small set of B cells, that has the best fit for the antigen. Binding leads the B cell to divide and differentiate into two types of cell. **Plasma cells** are daughters of the B cell that secrete antibodies that bind to the antigen and elicit an immune response against the pathogen. **Memory cells** are long-lived daughters of the B cell that contain membrane-bound antibody having the same antigen specificity as the parent cell. Since the resulting population of B cells arose from one B cell, they are all clones of one another.

The process by which antigen binding generates a clone of B cells is called **clonal selection,** and it is one of the central principles of immunology. Almost counterintuitively, antibody

FIG. 43.11 Clonal selection. Those B cells that recognize a foreign antigen are selected to divide and differentiate into plasma cells and memory cells.

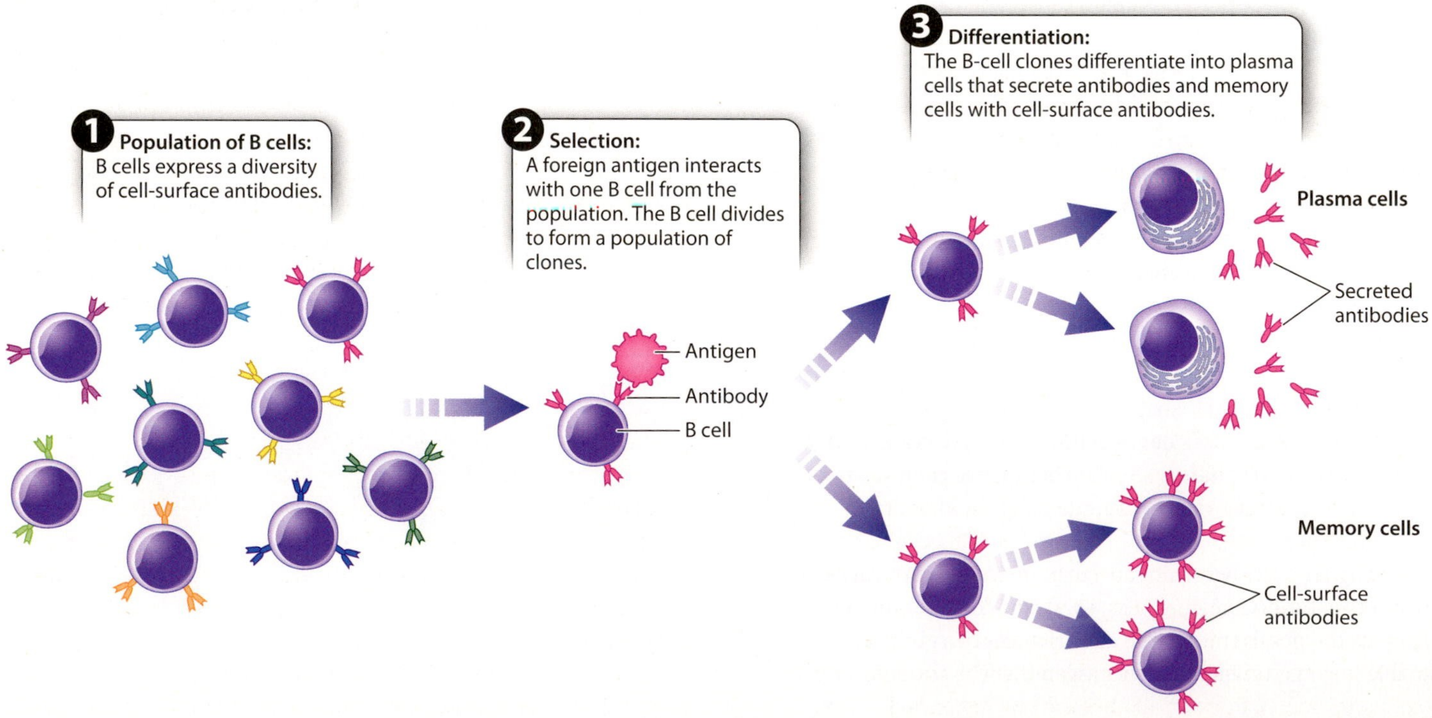

FIG. 43.12 Primary and secondary responses. The immune system responds more strongly to the second exposure to an antigen than to the first exposure because of the presence of memory B cells.

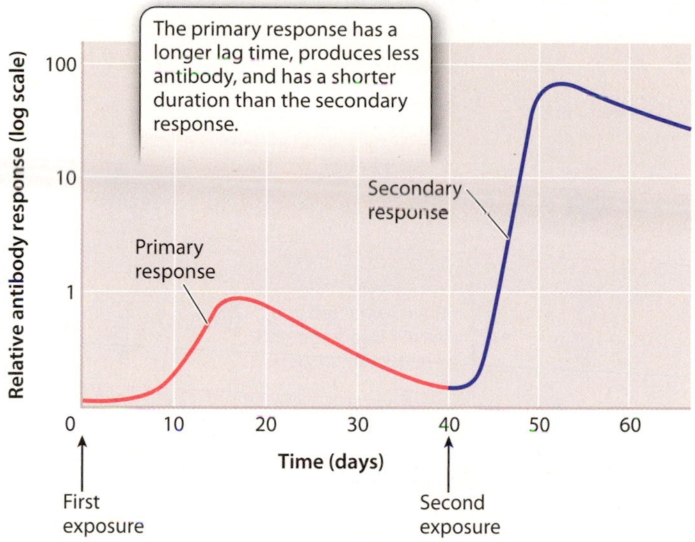

specificity is achieved before any exposure to the antigen. In this way, clonal selection resembles natural selection (Chapter 21). In natural selection, there is genetic variation in a population and the environment selects among this variation, allowing some individuals to survive and reproduce more than others. Similarly, in clonal selection, there is variation in a population of B cells, and the antigen selects among this variation, allowing some cells to proliferate more than others.

Clonal selection also explains immunological memory.

Clonal selection explains antibody specificity. It also explains the second feature of the adaptive immune system: the ability to remember past infections—that is, to react more vigorously on re-exposure. The first encounter with an antigen leads to a **primary response** (**Fig. 43.12**). In this response, there is a short lag before antibody is produced. The lag is the time required for B cells to divide and form plasma cells. The plasma cells secrete antibodies, typically IgM. The level of IgM in the blood increases, peaks, and then declines.

On re-exposure to the same antigen, even months or years after the first exposure, there is a **secondary response,** which is quicker, stronger, and lasts longer than the primary response (Fig. 43.12). During this response, memory B cells differentiate into plasma cells that release IgG antibodies. These memory B cells respond more quickly to antigen exposure than naïve B cells (B cells that haven't been previously exposed to antigen). In addition, the pool of memory B cells is larger than the pool of naïve B cells for a given antigen, explaining why the secondary response produces more antibodies for a longer period of time.

The presence of long-lived memory cells produced following the primary response provides long-lasting immunity after an infection such as chicken pox. We have learned to make use of immunological memory in **vaccination.** Vaccination involves deliberately giving a patient an antigen from a pathogen to induce a primary response but not the disease, thereby providing future protection from infection by the same pathogen.

Vaccination was discovered by a combination of observation and experiment. Until the early 1700s, it was common practice to inoculate people with smallpox to induce a mild disease and prevent a more severe, even lethal, infection. It was also well known that milkmaids, exposed to the relatively benign cowpox virus from milking cows, were immune to the related but more deadly smallpox virus. In 1796, the English scientist Edward Jenner specifically tested the consequences of exposure to cowpox by inoculating a young boy with cowpox and demonstrating that he became immune to smallpox. In fact, the word "vaccine" is derived from the Latin *vacca,* which means "cow."

Today, vaccines take many forms: They can be a protein or a part of a protein from the pathogen, a live but weakened form of the pathogen, or a killed pathogen. They are among the most effective public health measures ever developed.

→ **Quick Check 2** Look again at Fig. 43.12. Imagine that a different novel antigen is added on day 40 rather than a second injection of the same antigen. Would you predict that there would be a primary or a secondary response to this second antigen?

Genomic rearrangement generates antibody diversity.

We pointed out earlier that antibody specificity is determined before exposure to the antigen. Clonal selection therefore requires a mechanism for producing a great diversity of antibodies. How is this diversity achieved? Diversity would result if each antibody were encoded by a separate gene. However, this cannot be the case, as the number of antibodies produced by an organism far exceeds the total number of genes in the genome. For example, the human genome consists of about 25,000 protein-coding genes, but humans can produce about 10 billion different antibodies.

In 1965, American biologists William Dreyer and J. Claude Bennett suggested a novel hypothesis to explain antibody diversity. As they put it, this hypothesis was "radically different from anything found in modern molecular genetics." They proposed that a single antibody is made by separate gene segments that are brought together by recombination. According to this model, many copies of each gene segment are present, each slightly different from the others. Only one copy of each gene segment ends up in the final, recombined antibody gene (and hence in the antibody protein). Diversity is therefore achieved by combining different copies of each segment in different B cells as they mature.

More than 10 years after Dreyer and Bennett proposed their model, direct experimental evidence showed that their hypothesis

HOW DO WE KNOW?

FIG. 43.13

How is antibody diversity generated?

BACKGROUND Humans produce an estimated 10 billion different antibodies. Antibodies are proteins, encoded by genes. However, there are only about 25,000 genes in the human genome. How then can humans (and other vertebrates) produce such a diversity of antibodies from a limited set of genes? This was one of the central problems in immunology until the 1960s and 1970s.

HYPOTHESIS In 1965, American biologists William Dreyer and J. Claude Bennett proposed a model in which there are many copies of two gene segments, which they called the V (variable) and C (constant) segments. An antibody gene is assembled by recombining a single copy of each of these segments. At the time, there was no direct experimental evidence to support this hypothesis.

EXPERIMENT In 1976, Japanese immunologists Nobumichi Hozumi and Susumu Tonegawa tested the Dreyer and Bennett hypothesis. Using mice as a model system, they isolated DNA from embryonic cells (before recombination was thought to occur) and from adult B-cell tumor lines that produce one type of antibody (after recombination was thought to occur). The procedure they followed was slightly different from what is described here but conceptually the same. DNA is cut by restriction enzymes, the fragments are separated on a gel and transferred to a filter paper, and light chain DNA (containing both V and C regions) is used as a radioactive probe to determine the size and positions of the V and C regions (Southern blot, Chapter 12).

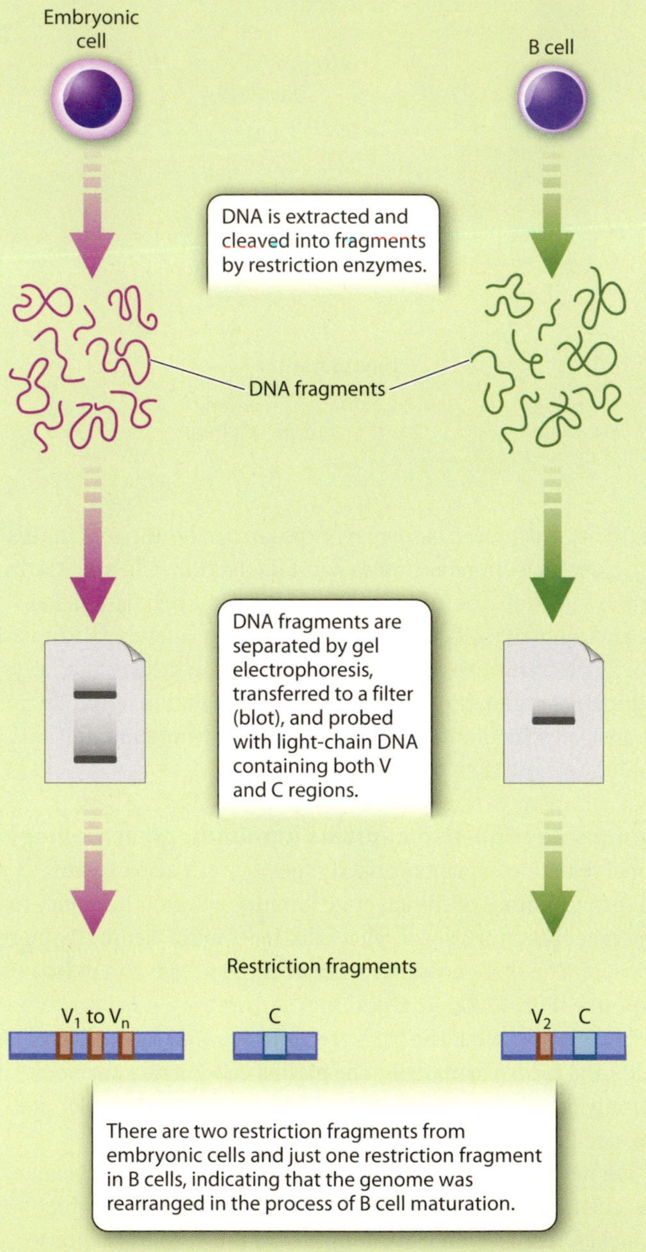

RESULTS Hozumi and Tonegawa found that the V and C regions were far apart in the DNA of embryonic cells and close together in adult cells, suggesting that the gene segments were brought together during B cell differentiation.

CONCLUSION Antibody genes are assembled by recombination of individual gene segments, providing a mechanism for generating antibody diversity.

FOLLOW-UP WORK Further studies showed that there are multiple copies of several different gene segments, now called the V, D (diversity), J (joining), and C regions for heavy chains, and V, J, and C regions for light chains. These are recombined in unique patterns in different B cells, so each B cell produces one antibody and a population of B cells produces a diversity of antibodies.

SOURCES Dreyer, W. J., and J. C. Bennett. 1965. "The Molecular Basis of Antibody Formation: A Paradox." *Proceedings of the National Academy of Sciences* 54:864–869; Hozumi, N., and S. Tonegawa. 1976. "Evidence for Somatic Rearrangement of Immunoglobulin Genes Coding for Variable and Constant Regions." *Proceedings of the National Academy of Sciences* 73:3628–3632.

is correct (**Fig. 43.13**). As a B cell differentiates, different gene segments are joined in a process called **genomic rearrangement** that produces a specific antibody.

Genes for H chains are composed of multiple different V (variable), D (diversity), J (joining), and C (constant) gene segments (**Fig. 43.14**). Genes for mice H chains, for example, can be composed of any of 100 different V segments, 30 D segments, 6 J segments, and 8 C segments. During B cell differentiation, the DNA for all of these gene segments undergoes recombination so that just one of each type of segment is present in the DNA, and the intervening DNA is deleted. The assembled VDJ segment encodes the variable portion, and the C segment encodes the constant region. L chain genes also contain multiple copies of gene segments, but they do not have a D gene segment, so the VJ segment encodes the variable portion and the C segment encodes the constant region. The result of this process is that each B cell encodes a single H chain and a single L chain, and each B cell expresses a unique antibody.

It is worth pausing here to consider the significance of this mechanism. We generally think of DNA as being a stable blueprint present in identical copies in all of the somatic cells in our bodies. B cells are an exception. As a result of genomic rearrangement, the DNA in each mature B cell is different from the DNA in every other mature B cell and different from the DNA in other cells in the body.

Genomic rearrangement is the primary mechanism that creates antibody diversity, but other mechanisms contribute as well. One is simply the association of different L and H chains to make a functional antibody. In addition, recombination is sometimes imprecise, creating different sequences at the junctions between the different gene segments. There are also mechanisms for adding or replacing nucleotides within VJ and VDJ segments. Finally, alternative splicing of mRNA (Chapter 3) allows a single B cell to express both membrane-bound and secreted antibodies.

We have seen that B cells as a group produce an enormous diversity of antibodies, whereas a given B cell makes one and only one antibody. But B cells are diploid, containing maternal and paternal copies of the genes that encode for antibodies. Maternal and paternal alleles of the antibody genes are similar, but not identical. Therefore, a B cell able to use both alleles would make two different antibodies, yet B cells always make one. Remarkably, once one allele undergoes genomic rearrangement, the other allele is prevented from doing so. As a result, B cells function as if they have a single allele for each antibody gene—they express the maternal or the paternal gene, but not both.

The mechanisms we have considered pertain to humans and mice. Whereas B cells mature in the bone marrow in humans and mice, they mature in gut-associated lymphoid tissue (GALT) in many other vertebrates. Furthermore, genomic rearrangement

FIG. 43.14 **Genomic rearrangement in mice.** In genomic rearrangement, different gene segments recombine to generate a functional gene encoding the parts of an antibody.

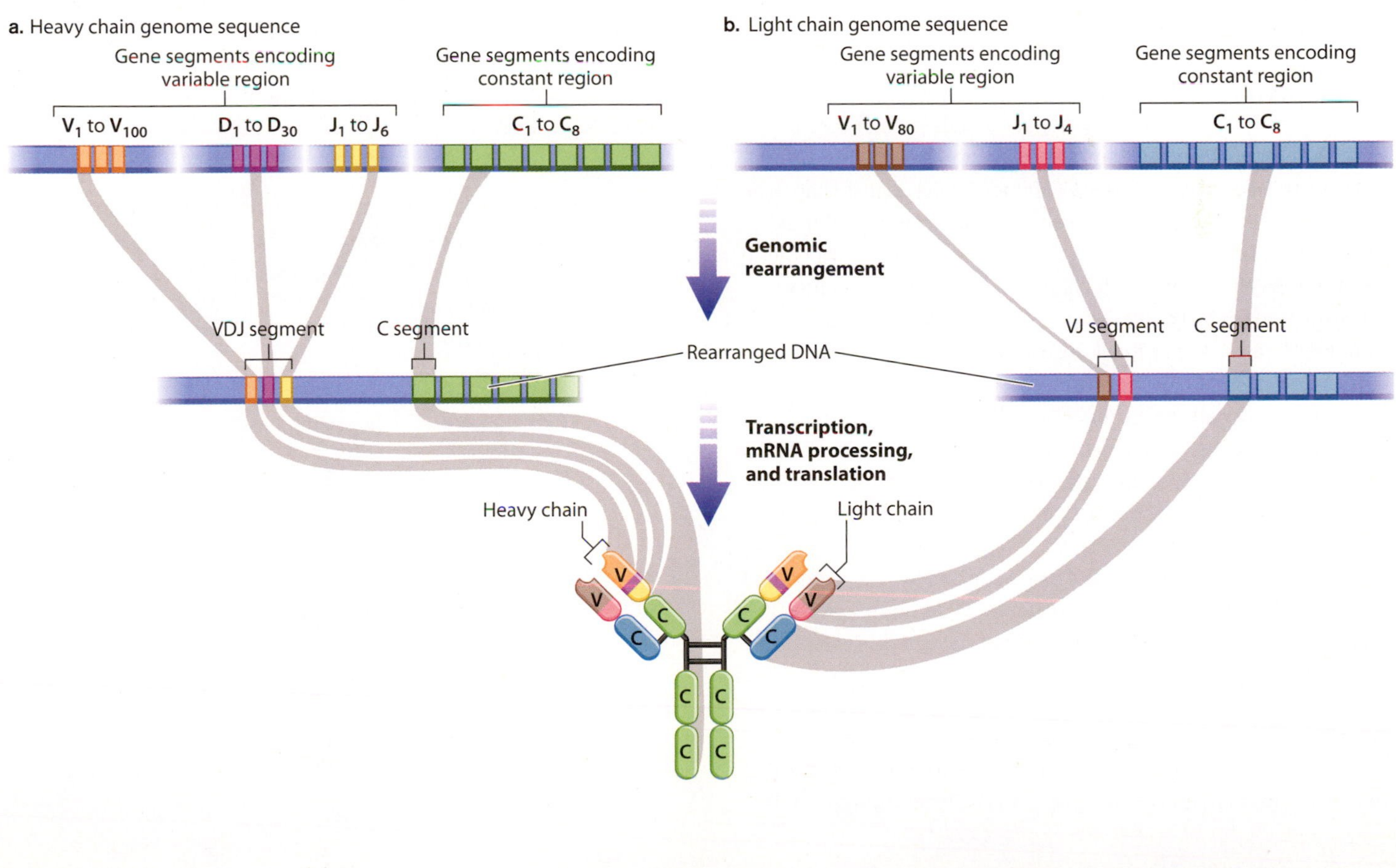

is less important in many vertebrates. Instead, other mechanisms appear to play a more important role.

→ **Quick Check 3** If a given B cell produces only one type of antibody, how do organisms produce a great diversity of antibodies?

43.4 ADAPTIVE IMMUNITY: T CELLS AND CELL-MEDIATED IMMUNITY

Antibodies are remarkable for their diversity and specificity. They provide protection against many kinds of bacteria, especially those enclosed in a polysaccharide capsule, such as bacteria that cause pneumonia and meningitis. However, B cells and their antibodies have limitations. For example, B cells on their own can make antibodies against only some antigens; for other antigens, they require the assistance of other cells. Furthermore, they are sometimes ineffective against pathogens that take up residence inside a host cell, as do the bacteria that cause tuberculosis.

To handle these and other kinds of pathogens, there is another part of the adaptive immune system that depends on the second type of lymphocyte: the T cell. T cells do not secrete antibodies. Instead, they participate in **cell-mediated immunity,** so named because cells, not antibodies, recognize and act against pathogens. T cells target infected or diseased cells. Their importance is most dramatically demonstrated by individuals taking immunosuppressive drugs or infected by HIV, both of which suppress the activity of T cells. The result can be overwhelming infections and even tumors.

T cells include helper and cytotoxic cells.

There are two major subpopulations of T cells with different names, functions, and cell-surface markers—**helper T cells** and **cytotoxic T cells** (see Fig. 43.3). Helper T cells do just that—they *help* other cells of the immune system by secreting cytokines. One of their key roles is to activate B cells to secrete antibodies. Although B cells can work on their own, particularly against encapsulated bacteria, in most cases they require the participation of helper T cells. Helper T cells also activate macrophages, cytotoxic T cells, and other cells of the immune system.

Cytotoxic T cells *kill* other cells—that's what "cytotoxic" means. Like B cells, they are activated by cytokines released from helper T cells. They are particularly effective against altered host cells, such as those infected with a virus or that have become cancerous.

These two types of T cell are distinguished by the presence of different glycoproteins on their surface—CD4 on helper T cells and CD8 on cytotoxic T cells. The ratio of these two classes of T cell can be used to assess immune function. In healthy individuals, the CD4 : CD8 ratio is usually about 2:1. A decreased ratio, indicating the presence of fewer helper T cells relative to cytotoxic T cells, is typical of individuals infected with HIV. HIV infects and kills helper T cells, disarming a key player in the immune system. In addition to helping distinguish the two types of T cell, CD4 and CD8 assist receptors on T cells carry out their functions, a topic we discuss next.

T cells have T cell receptors on their surface that recognize an antigen in association with MHC proteins.

All T cells, like B cells, originate in the mammalian bone marrow. However, unlike B cells, T cells mature in the thymus. A mature T cell is characterized by the presence on the plasma membrane of a **T cell receptor (TCR),** a protein receptor that recognizes and binds to the antigen.

In many ways, TCRs are similar to antibodies on the surface of B cells. For example, TCRs recognize antigens with a specific structure. In addition, there is a great diversity of TCRs that differ from one another, but each T cell has just one type of TCR on its surface. Binding of TCR to an antigen triggers the T cell to divide into clones, resulting in a pool of T cells that are each specific for a given antigen. Finally, like antibodies, the diversity of TCRs among different T cells results from genomic rearrangement of V, D, J, and C gene segments.

However, TCRs are different from antibodies in important ways. First, they are composed of two, rather than four, polypeptide chains (**Fig. 43.15**). Second, they are not secreted like antibodies, but are always membrane-bound on the T cell surface. Third, the TCR does not recognize an antigen by itself. Instead, it recognizes an antigen in association with proteins that appear on the surface of most mammalian cells and that are encoded by the **major histocompatibility complex,** or **MHC.** MHC proteins were first discovered in transplantation biology because it is their presence that leads to acceptance or rejection of transplanted tissues.

The MHC is a cluster of genes present in all mammals that encode proteins on the surface of cells. The MHC is composed

FIG. 43.15 **The T cell receptor (TCR).** The TCR is a protein on the surface of T cells that binds antigens in association with MHC proteins.

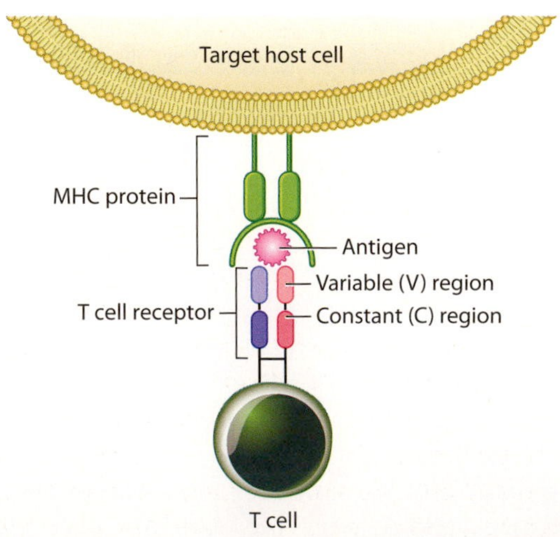

FIG. 43.16 Three-way interactions among the T cell receptor, MHC protein, and processed antigen. Helper T cells recognize processed antigen in association with class II MHC protein, and cytotoxic cells recognize processed antigen in association with class I MHC protein.

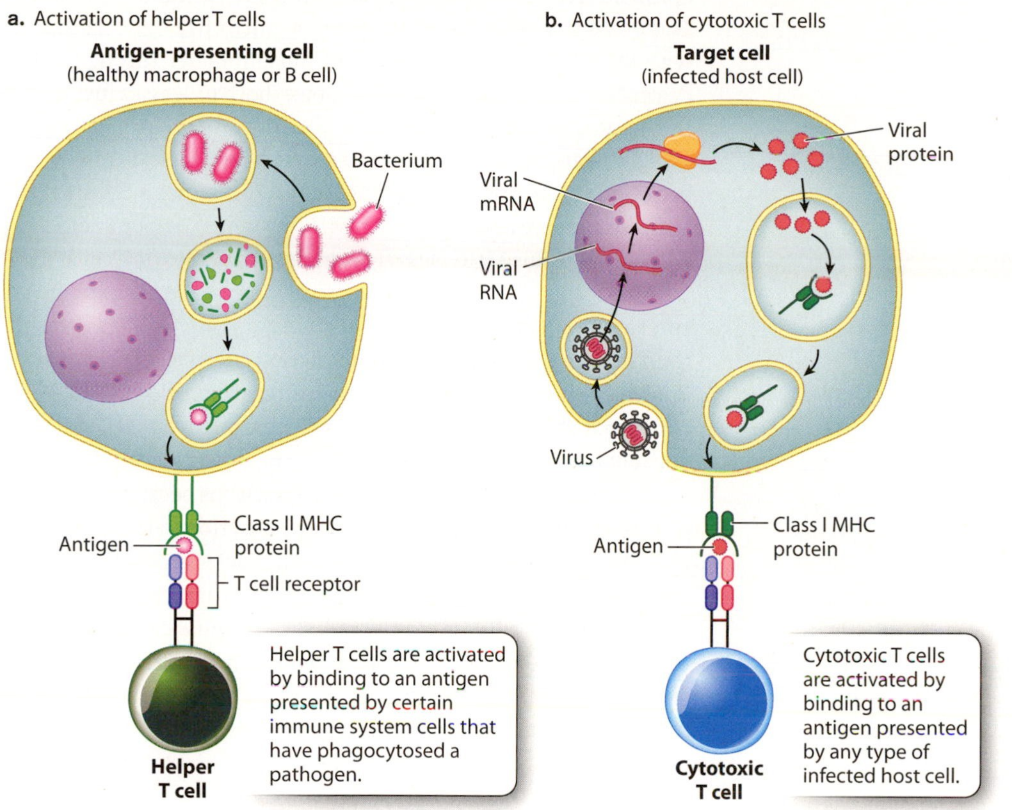

of many genes with a high rate of polymorphism, meaning that there is a lot of variation in the gene sequence (and consequently in the protein sequence) among different individuals. In humans and mice, the genes are divided into three classes: **Class I** genes are expressed on the surface of all nucleated cells; **class II** genes are expressed on the surface of macrophages, dendritic cells, and B cells; and **class III** genes encode several proteins of the complement system and proteins involved in inflammation.

How do helper and cytotoxic T cells recognize and respond to antigens? Let's consider the activation of helper T cells first (**Fig. 43.16**). When an antigen enters the body, it may be recognized by an antibody directly or be taken up by **antigen-presenting cells.** These cells, which include macrophages, dendritic cells, and B cells, take up the antigen and return portions of it to the cell surface bound to MHC class II proteins. Helper T cells recognize processed antigen along with MHC class II proteins by their T cell receptors. When TCR binds to antigen and MHC class II proteins, the helper T cells release cytokines that activate other parts of the immune system, including macrophages, B cells, and cytotoxic T cells.

Cytotoxic T cells also recognize antigen displayed by host cells, but only antigens that are associated with MHC class I proteins (Fig. 43.16). Because class I proteins are present on virtually all cells, cytotoxic T cells recognize and kill any host cell that becomes abnormal in some way. For example, a virus-infected cell often expresses viral antigens with MHC class I proteins on its surface. Cytotoxic T cells recognize the antigen and MHC class I proteins and kill the cell. Tumor cells express novel antigens along with MHC class I proteins, and in some cases may be eliminated by cytotoxic T cells. **Table 43.2** summarizes the major features of helper and cytotoxic T cells.

Once activated, T cells divide and form clones of helper or cytotoxic T cells. Some cells of each type are memory cells that provide long-lasting immunity following an initial infection, as in the case of B cells. Also like B cells, T cells can sometimes be activated too strongly. Earlier, we saw how IgE bound to mast cells and basophils can lead to immediate hypersensitivity reactions characteristic of allergies and asthma. The counterpart in T cells is a **delayed hypersensitivity reaction,** which, as its name suggests, does not begin right away. For example, if you touch poison ivy, your skin will turn red and start itching only after a delay of several hours or days. Delayed hypersensitivity reactions are initiated by helper T cells, which release cytokines that attract macrophages to the site of exposure, which is typically the skin.

→ **Quick Check 4** How does T cell activation differ from B cell activation?

TABLE 43.2 Comparison of Helper and Cytotoxic T Cells

	HELPER T CELL	CYTOTOXIC T CELL
Functions	Activation of macrophages Activation of B cells Activation of cytotoxic T cells Secretion of cytokines	Cell killing
Surface molecule	CD4	CD8
MHC protein recognized by the T cell	Class II	Class I

FIG. 43.17 **Positive and negative selection.** Positive selection ensures that T cells react with self MHC proteins, and negative selection ensures that T cells do not react with other self antigens.

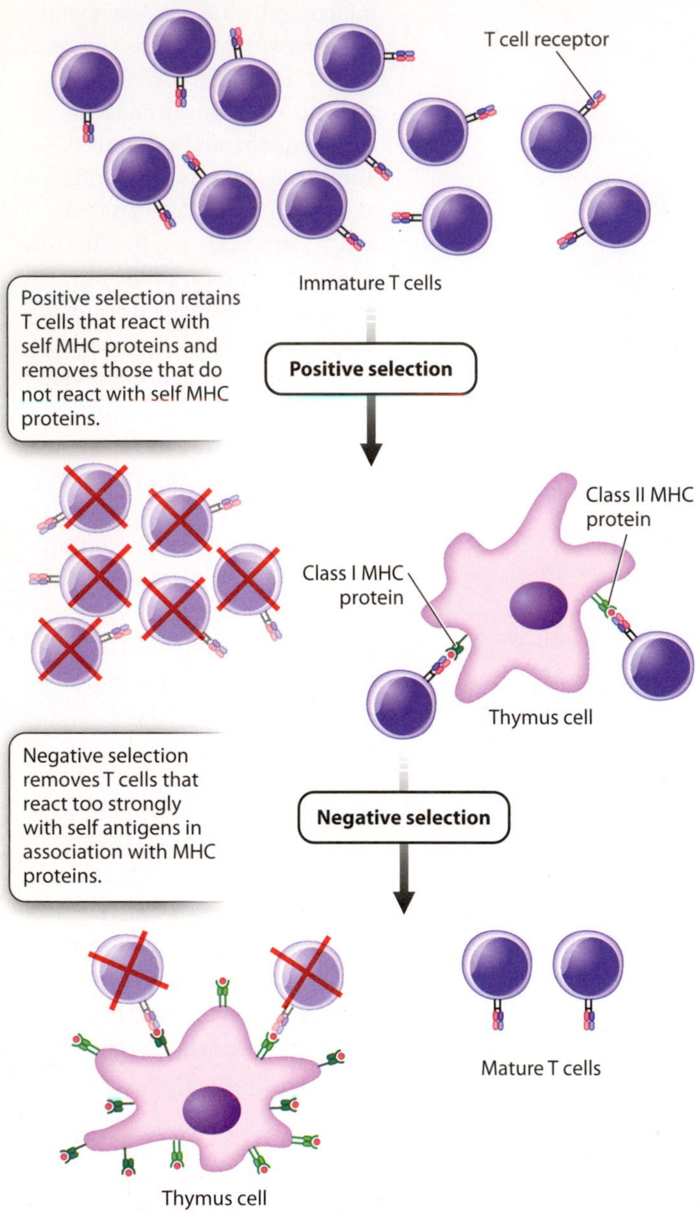

The ability to distinguish between self and nonself is acquired during T cell maturation.

Of all the possible T cell receptors that are generated by genomic rearrangement, only some are useful. Useful T cell receptors must be able to react with the host's own MHC proteins. In addition, useful T cell receptors must not react with other molecules that are normally present in or on cells of the host. In other words, T cells must respond only to self MHC proteins in association with nonself antigens, but not to other self molecules of an organism's own cells.

A sorting process is therefore necessary so that only some T cells mature and others are eliminated (**Fig. 43.17**). As T cells mature in the thymus, they interact with cells of the epithelium of the thymus. The T cells that recognize self MHC proteins on epithelium cells are **positively selected** and continue to mature. The T cells that react too strongly to self antigens in association with MHC are **negatively selected** and eliminated through cell death. Note that, in spite of their names, both processes involve the elimination of some T cells.

The result of this sorting process is twofold. First, T cells become MHC restricted, meaning that they interact only with antigens that are associated with MHC proteins (helper T cells with antigen plus MHC class II proteins, and cytotoxic T cells with antigen plus MHC class I proteins). Second, T cells exhibit **tolerance**—in other words, they do not respond to self antigens, even though the immune system functions normally otherwise. Those antigens present as T cells mature are identified as self and do not elicit a response; those not present are nonself and, if encountered, do elicit a response.

B cells also exhibit tolerance to self antigens because they go through a similar process of negative selection in the bone marrow. During that process, B cells that react strongly to self antigens are eliminated through cell death. Although most self-reactive T and B cells are eliminated, some T and B cells that react to self antigens escape this developmental check and end up in the circulation. Additional mechanisms are in place to eliminate these self-reactive cells, either through the activity of T cells or through cell death.

The ability to distinguish self from nonself is critical. Failure leads to autoimmune disease, in which tolerance is lost and the immune system becomes active against antigens of the host. Autoimmune diseases can be debilitating as T cells or antibodies attack cells and organs of the host. In rheumatoid arthritis, self-reactive T cells attack the joints, causing chronic inflammation and tissue damage; in type 1 diabetes, attack is directed against insulin-producing cells of the pancreas; and in multiple sclerosis, the target is the myelin sheath surrounding nerves of the central nervous system. It is not fully understood how these self-reactive T cells evade checks on their development.

43.5 THREE PATHOGENS: A VIRUS, BACTERIUM, AND EUKARYOTE

Although we discussed the different components of the immune system one at a time, they in fact work together to provide a coordinated response to pathogens. We have seen how B cells often require T cell help, how the complement system can be activated by antibodies produced by B cells, and how macrophages present antigens in association with MHC proteins to T cells. In turn, pathogens have evolved means of evading the immune system. In this section, we consider the flu virus, the bacterium that causes tuberculosis, and the malaria parasite. These three pathogens illustrate how pathogens evolve and adapt in response to the immune system, and how the immune system in turn evolves and adapts in response to pathogens.

The flu virus evades the immune system by antigenic drift and shift.

Everyone knows the symptoms: fever, chills, sore throat, cough, weakness, and fatigue, perhaps a headache and muscle pain. Many viruses cause these common flu-like symptoms. However, the common cold, stomach flu, and 24-hour flu are caused by unrelated viruses, not by the flu virus.

The flu—its full name is influenza—is caused by an RNA virus that infects mammals and birds. It is notable for causing seasonal outbreaks, or epidemics. Four times in the last century—in the Spanish flu of 1918, the Asian flu of 1957–58, the Hong Kong flu of 1968–69, and the swine flu of 2009—the virus spread more widely and infected more people, and the disease reached pandemic levels. The flu is more than just a nuisance: During pandemics, it causes hundreds of thousands of deaths worldwide.

Part of the success of the virus lies in its ability to spread easily. It can spread in the air by small droplets produced when an infected person coughs or sneezes. It can also spread by direct contact or by touching a contaminated surface, such as a doorknob. Hand washing, therefore, is one of the most effective means of protection.

Another characteristic of the flu is its ability to evade the immune system. The flu virus can change its outer coat so memory cells, produced from earlier infections, no longer recognize it. There are three major types of flu virus—called A, B, and C—and many different strains. For example, the swine flu pandemic of 2009 was caused by a type A flu virus of the strain H1N1. These strains differ in several respects, including the structure of a cell-surface glycoprotein called hemagglutinin (HA). HA binds to epithelial cells and controls viral entry into these cells. It can bind only to cells that display a complementary cell-surface protein. As a result, hemagglutinin determines in part which organisms the flu virus infects—humans, pigs, or birds—and which part of the body it infects—the nose, throat, or lungs.

Cells infected by viruses secrete cytokines that bring macrophages, T cells, and B cells to the site of infection, as we saw earlier. These cytokines, produced in abundance, lead to many of the symptoms commonly associated with the flu. For example, the cytokine interleukin-1 causes fever.

Once in the cell, the virus replicates its genome and makes more virus particles (Chapter 19). However, viral replication is prone to error, so there is a high rate of mutation that leads to changes in the amino acid sequences of antigens present on the viral surface, including HA. This process is called **antigenic drift** (**Fig. 43.18a**). It allows a population of viruses to evolve over time and evade memory B and T cells that remember past infections.

Antigenic drift leads to a gradual change of the virus over time. The flu virus is also capable of sudden changes by a process known as **antigenic shift** (**Fig. 43.18b**). The viral genome consists of eight

FIG. 43.18 **(a) Antigenic drift and (b) antigenic shift.** Virus populations evolve over time and evade the immune system's memory cells.

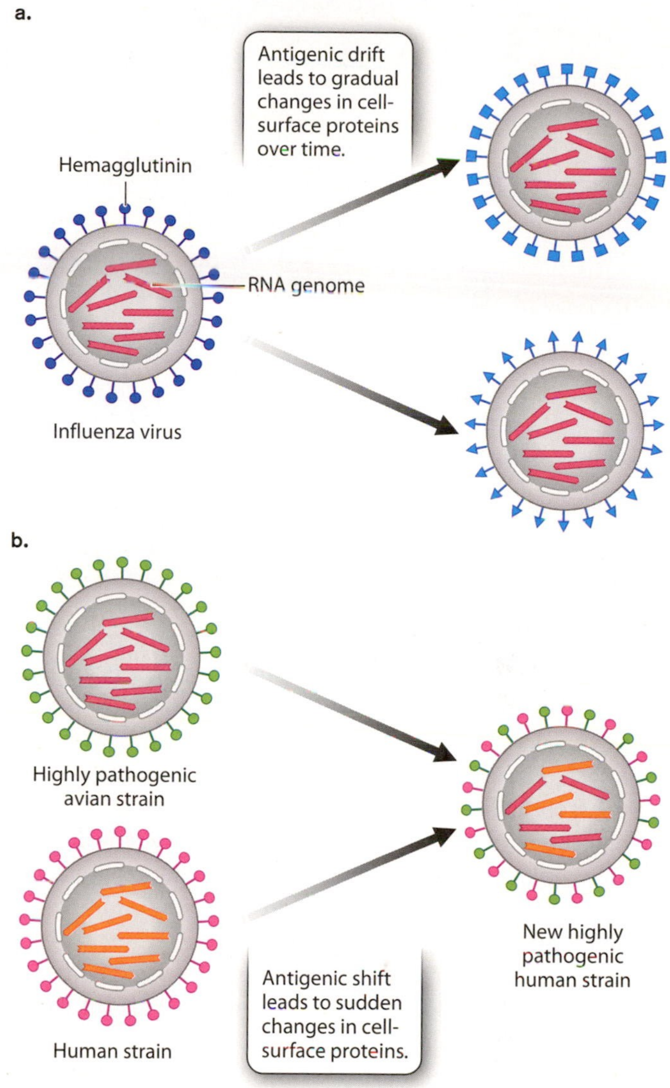

linear RNA strands. If a single cell is infected with two or more different flu strains at one time, the RNA strands can reassort to generate a new strain. H1N1, for example, has genetic elements from human, pig, and bird flu viruses. Antigenic drift and shift make it difficult to predict from year to year which strains will be most prevalent and therefore what vaccine will be most effective.

Tuberculosis is caused by a slow-growing, intracellular bacterium.

Our second example is tuberculosis, also known as TB. Like the flu, it is spread through small droplets dispersed in the air when a person with active disease coughs or sneezes. Also like the flu, the TB pathogen replicates inside host cells. But unlike the flu, TB is caused by a bacterium, not a virus. TB results from

FIG. 43.19 A granuloma in a tubercular lung. *Photo source: Biophoto Associates/Science Source.*

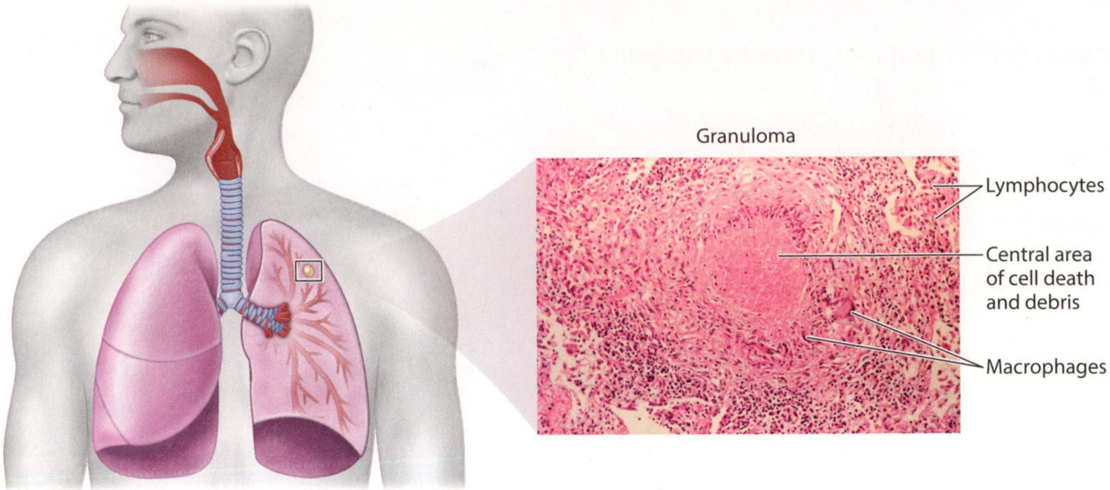

infection by *Mycobacterium tuberculosis,* a bacterium with several unusual properties. It is very small with a compact genome, lacks a plasma membrane outside the cell wall, and replicates exceedingly slowly. *M. tuberculosis* takes about 16–20 hours to divide, compared to 20 minutes for the common intestinal bacterium *E. coli.*

Tuberculosis is very common: It is estimated that about a third of the world's population has TB. Symptoms are absent in most cases because the bacteria are in a dormant state. About 10% of cases are active, characterized by chronic cough, fever, night sweats, and the weight loss that gives TB its other name: consumption.

Tuberculosis primarily affects the lungs. The bacteria infect macrophages in the alveoli of the lungs, where they replicate. In response, infected macrophages release cytokines, recruiting T and B cells to the site of infection. These lymphocytes surround the infected macrophages, forming a structure called a granuloma (**Fig. 43.19**). The granuloma helps to prevent the spread of the infection and aids in killing infected cells.

Many TB cases remain asymptomatic. However, the bacteria can sometimes overcome host defenses, especially if the immune system is compromised or suppressed, as in the case of co-infection with HIV. Treatment for active TB requires a long course of multiple antibiotics. In many areas of the world, however, TB has evolved antibiotic resistance, a growing and alarming problem.

Diagnosis poses challenges, especially in the case of asymptomatic infections. The tuberculin skin test is performed by injecting protein from the bacteria under the skin and observing whether or not a reaction, in the form of a hard, raised area, occurs. A positive skin test is an example of a delayed hypersensitivity reaction caused by the recruitment of memory T cells to the site of inoculation, indicating prior infection with *M. tuberculosis.* Unfortunately, the test is not always accurate, so newer tests are being developed.

Developing an effective vaccine is challenging. The only one currently available is BCG (Bacillus Calmette-Guérin, named for Albert Calmette, a French bacteriologist, and Camille Guérin, a French veterinarian, who developed it early in the twentieth century). BCG is a live but weakened strain of a related bacterium that infects cows. This vaccine was developed following the success of using cowpox to immunize against smallpox. However, BCG is only sometimes effective, and a public health disaster resulting from contaminated vaccine occurred in 1930 in Germany. Consequently, it is not used in the United States.

The malaria parasite changes surface molecules by antigenic variation.

"Bad air." That's the literal translation of "malaria," a contraction of the Italian *mala* ("bad") and *aria* ("air"). But malaria is not caused by the air. Nor is it caused by a virus or bacterium. Malaria is caused by a single-celled eukaryote of the genus *Plasmodium* (Case 4: Malaria). Several species infect humans, but *P. falciparum* is the most common and most virulent.

Malaria is both devastating and common: There are about 500 million cases and 2 million deaths every year, mostly in sub-Saharan Africa, and mostly of children. Malaria causes cyclical fevers and chills and can lead to coma and death.

The malaria parasite is transmitted to humans by the bite of an infected mosquito (**Fig. 43.20a**). Using mosquitoes as a vector, the malaria parasite deftly bypasses the natural protective barrier provided by the skin. The malaria parasite spreads to the liver and then to red blood cells, where it completes its life cycle. The progeny can be taken up again by mosquitoes.

Malaria is an ancient parasite that has coevolved with humans. We have already seen how altered hemoglobin provides some protection against infection (Chapter 21). But the malaria parasite, too, has evolved—it is able to evade our immune system.

The malaria parasite resides in liver cells and red blood cells. Infected red blood cells can be removed by the spleen, but the malaria parasite has evolved a way to avoid this filtering action. It expresses an adhesive protein, PfEMP1 (*Plasmodium falciparum* erythrocyte membrane protein 1), that inserts itself on the surface

FIG. 43.20 **Malaria.** (a) The malaria parasite, a single-celled eukaryote, is transmitted by the bite of a mosquito. (b) Following infection, red blood cells adhere to blood vessel walls by interactions with a malaria-encoded cell-surface protein and receptors on the blood vessel wall. *Photo source: a. Science Source.*

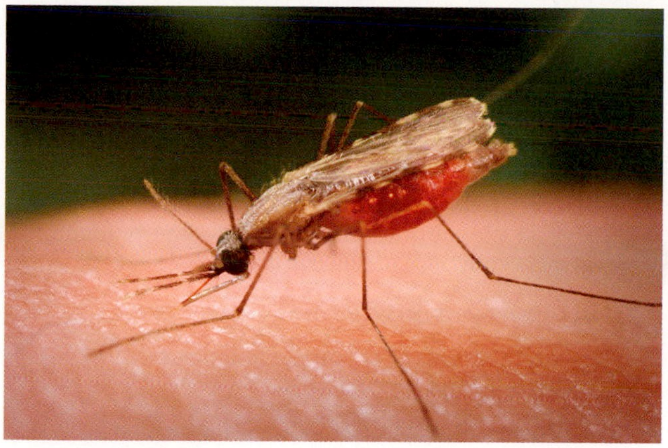

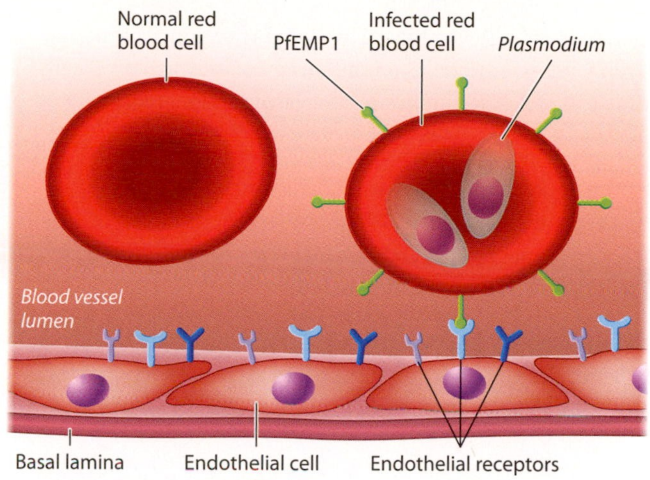

of red blood cells. This protein interacts with proteins on the surface of cells lining blood vessels, helping infected cells stick to blood vessel walls and keeping them in the circulatory system and out of the spleen (**Fig. 43.20b**). PfEMP1 could conceivably be a target for antibodies or TCRs. However, it is encoded by one of about 60 genes, each slightly different from the others. The parasite expresses just one of these genes at a time and can change which one is expressed, in a process called **antigenic variation.** The great and ever-changing diversity of this protein, both in a single parasite and in the population as a whole, makes it a moving target for the immune system.

Despite continuing research, there is no vaccine. Researchers hope that the sequence of the malaria genome, completed in 2002, will provide new targets for vaccine research. In the meantime, efforts to control malaria focus on measures to control mosquitoes, such as nets and insecticides, and on treatment with antimalarial medications. The evolution of resistance, however, is an all too familiar problem. ■

Core Concepts Summary

43.1 The immune system protects organisms from pathogens.

Pathogens are agents or organisms that cause disease. page 922

The immune system is able to distinguish host molecules and cells, called self, from foreign molecules and cells, called nonself. page 922

The immune system consists of two parts: The innate immune system acts nonspecifically and the adaptive immune system acts specifically against pathogens. page 923

43.2 The innate immune system acts generally against a diversity of pathogens.

The skin is a physical barrier and contains chemicals, cells, and microorganisms that provide protection against pathogens. page 924

Phagocytes engulf other cells and debris. page 925

Eosinophils, basophils, mast cells, and natural killer cells participate in the innate immune system. page 925

Toll-like receptors on the surface of phagocytes play a key role in recognition of foreign molecules. page 926

Inflammation is the body's response to infection or injury. It involves an increase in blood flow and blood vessel permeability, recruitment of phagocytes, and production of cytokines. page 926

The complement system consists of circulating proteins that activate one another. Activation results in the formation of a membrane attack complex that lyses bacteria, production of proteins that enhance phagocytosis, and generation of cytokines. page 927

43.3 The adaptive immune system includes B cells that produce antibodies against specific pathogens that reside outside cells in tissues.

B cells produce antibodies that bind to antigens. page 929

Antibodies have two light chains and two heavy chains joined by covalent disulfide bonds. page 929

Light and heavy chains have constant and variable regions. page 929

Mammals produce five types of antibody: IgG, IgM, IgA, IgD, and IgE. page 929

Antibody specificity is achieved by clonal selection, in which an antigen binds to antibodies on the surface of one or a few B cells, leading to B cell differentiation into plasma and memory cells. page 930

The primary immune response is characterized by a long lag period and a short response. The secondary immune response is quicker and more robust, and provides the basis for immunity following infection or vaccination. page 931

Antibody diversity is generated by genomic rearrangement, in which individual gene segments are joined to make a functional gene. page 931

43.4 The adaptive immune system includes helper and cytotoxic T cells that attack infected, diseased, and foreign cells.

Two types of T cell are helper T cells and cytotoxic T cells. Helper T cells activate other cells of the immune system, including macrophages, B cells, and cytotoxic T cells. Cytotoxic T cells kill altered host cells. page 934

T cells are activated by the binding of a T cell receptor to an antigen in association with an MHC protein. page 934

As T cells mature in the thymus, they undergo positive and negative selection, so only T cells that recognize MHC, but not self antigens, survive. page 936

The ability to distinguish self from nonself antigens is critical for proper immune function; autoimmune diseases result when the immune system attacks cells of the host. page 936

43.5 Some pathogens have evolved mechanisms that enable them to evade the immune system.

The flu is caused by an RNA virus that can change its surface proteins by antigenic drift and antigenic shift. page 937

The bacterium that causes tuberculosis grows slowly and resides inside cells, particularly macrophages. page 937

Malaria is caused by a single-celled eukaryote with a complex life cycle having stages in mosquitoes and in humans. The malaria parasite has evolved several mechanisms, including antigenic variation, that help it reproduce in humans without being eliminated by the immune system. page 938

Self-Assessment

1. Compare and contrast the innate and adaptive immune systems.
2. List components of the innate immune system.
3. Describe how cells of the innate immune system recognize pathogens.
4. Explain how the four signs of inflammation relate to the changes that occur in the underlying tissues.
5. Draw a picture of an antibody molecule, labeling each of the components and the places where antigens bind.
6. Describe how B cells produce a great variety of different antibodies, each specific for a particular antigen.
7. Explain how immunological memory is achieved.
8. Compare and contrast the activation of B cells and of T cells.
9. Describe how positive and negative selection leads to a population of T cells that react to MHC molecules in association with nonself antigens, but not self antigens, and what happens when tolerance to self antigens is lost.
10. Describe three mechanisms that allow pathogens to evade the immune system.

Log in to LaunchPad to check your answers to the Self-Assessment questions, and to access additional learning tools.

CASE 8

Biodiversity Hotspots

Rain Forests and Coral Reefs

When we think about mass extinctions, the demise of the dinosaurs is often the first example to spring to mind. But many conservationists believe we are in the midst of a mass extinction today. The International Union for the Conservation of Nature has estimated that 800 plant and animal species in habitats around the world have died out in the last 500 years, and more than 16,000 are currently at risk of extinction.

The cause of this die-off isn't an asteroid impact or a cataclysmic volcanic event. Instead, human activity is dramatically altering the planet and its biological diversity.

It's clearly not possible to save every single species threatened with extinction. Conservationists must make difficult choices, prioritizing which species to protect. To get the most bang for every conservation buck, ecologists have identified 25 priority areas, or biodiversity hotspots.

These hotspots are areas with disproportionately large concentrations of endemic species—that is, species found nowhere else. For example, scientists estimate that as many as 44% of all vascular plants and 35% of vertebrate species are confined to these 25 hotspots. Those figures are impressive, considering that the identified hotspots make up just 1.4% of Earth's land surface.

The islands of the Caribbean—the West Indies—constitute one such biodiversity hotspot. The Caribbean Sea contains more than 7000 islands rich in biological diversity, including an estimated 6550 native plant species. All the Caribbean islands are flush with flora and fauna, but one island in particular epitomizes the concept of a biodiversity hotspot. Hispaniola—the second largest of the Caribbean islands—was the site of the first European colonies in the New World, founded by Columbus. Today, Hispaniola is shared by two countries, Haiti to the west and the Dominican Republic to the east.

Hispaniola is only about the size of Ireland, yet it contains a striking assortment of habitats, including desert, savanna, rain forests, dry forests, pine forests, coastal estuaries, and interior mountains. With such a rich abundance of habitats, Hispaniola is, in some ways, a microcosm of the planet's many ecosystems. And like the Earth's other natural places, Hispaniola faces many challenges to its native biological diversity.

Of all the habitats on Hispaniola, the forests have the greatest variety of plants, animals, and other organisms. Ferns, orchids, and bromeliads—spiky relatives of the pineapple—grow in treetops far above the forest floor. These plants, known as epiphytes, absorb water from dew, rainwater, and even the air. Their roots never reach the soil.

The forest is alive with activity. Ants march silently. Crickets and cicadas buzz and whine in the treetops. Vertebrate vocalists such as frogs and birds fill the air with chirps and whistles, making themselves known to both potential mates and rivals. Bats swoop through the darkened skies, using ultrasonic calls to locate insect prey. About 30,000 species of multicellular organisms are found on Hispaniola, and a third of those are endemic—they are found nowhere else in the world. One example is the Hispaniolan Solendon, a long-nosed mammal that resembles an overgrown shrew and secretes toxic saliva. Listed as endangered, the solenodon has flirted with extinction for decades.

Conservationists must make difficult choices, prioritizing which species to protect. To get the most bang for every conservation buck, ecologists have identified 25 priority areas, or biodiversity hotspots.

How did all these species come to live on the island? Over tens of millions of years, organisms traveled to Hispaniola by air and by sea. Geographically separated from their ancestors on the mainland, many of these organisms evolved to become distinct species. As they adapted to fill vacant niches on the island, animals such as *Anolis* lizards, frogs, and crickets evolved an impressive diversity of forms.

In some cases, scientists have been able to explore Hispaniola's evolutionary history directly. They have found remains of insects and other species entombed in fossilized tree resin, or amber, 30–23 million years ago. Many of these preserved specimens resemble species still living on Hispaniola. Others are unlike any creatures found on the island today.

The stunning diversity of Hispaniola doesn't exist just on land. Conservationists consider the warm, shallow reefs of the Caribbean one of the world's marine biodiversity hotspots. Coral reefs in general are among the world's most biologically diverse ecosystems. Species from all recognized phyla of animals are found on reefs.

Around the world, reefs support many thousands of species, including snails, corals, lobsters, and fish. Many, but not all, of these creatures spend their entire life in this habitat. Reefs also act as vital nursery grounds for animals that live in other parts of the ocean, from sea turtles to sharks. Reef biodiversity is critical to maintaining fisheries that provide food and income for millions of people.

Hispaniola's rain forests and reefs—and indeed all the biodiversity hotspots identified by conservationists—aren't just notable for their biological richness. They have also been given priority status because they are facing serious threats.

Reefs around the world are being damaged by human activities. Overfishing and pollution take their toll on reef ecosystems. Climate change is increasing seawater temperature and increasing CO_2 is altering ocean chemistry. Those changes are threatening the coral organisms that form the backbone of the reefs.

Ecosystems on land, too, face significant challenges. The Caribbean islands retain only about 11% of their original vegetation. Humans have cleared great swaths of land for agriculture and chopped trees to use as fuel. Humans have also introduced non-native, or invasive, species to the islands, both intentionally and accidentally. On Hispaniola, Europeans introduced the Javanese mongoose to control rats in sugarcane fields. The mongoose has devastated local populations of reptiles, amphibians, and birds.

If Hispaniola serves as a microcosm of the planet's ecosystems, it also serves as a cautionary tale. Two divergent paths have emerged on the single island. Haiti has historically suffered greater poverty than the Dominican Republic, and has much greater population density. Those factors put more pressure on the environment in Haiti, and have contributed to much greater deforestation on the western half of the island.

The case of Hispaniola illustrates the complex challenges facing conservationists. Solutions to environmental

Biodiversity hotspots. The areas highlighted in orange have a particularly high diversity of species found nowhere else. *Source: Reprinted by permission from Macmillan Publishers Ltd: NATURE 403, 853–858 (24 February 2000), N. Meyers, et al. Biodiversity hotspots for conservation priorities. Copyright 2000.*

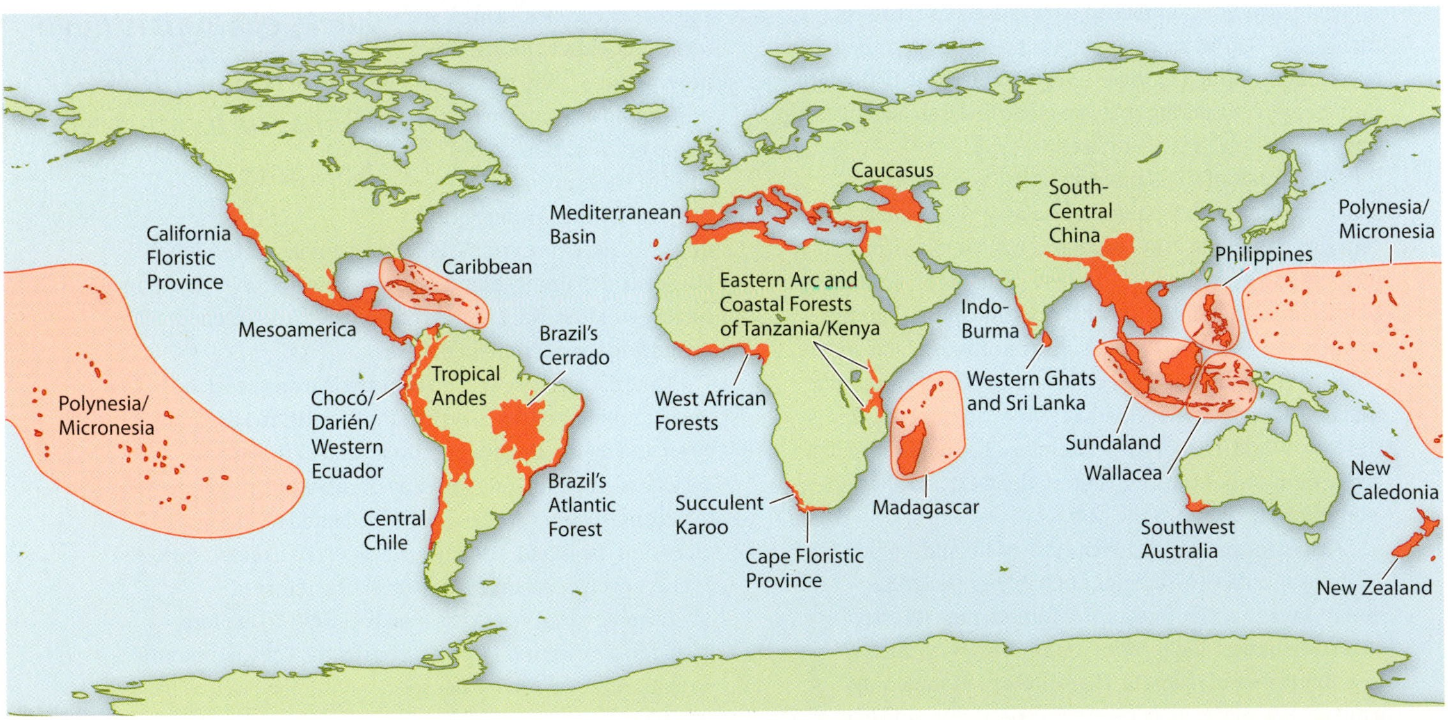

Coral reefs. Coral reefs support many species but are threatened by ocean warming and acidification. *Source: Jeff Hunter/Getty Images.*

challenges must take into consideration the effects of poverty, culture, and other social pressures. Balancing the needs of humans with the needs of natural ecosystems is no easy task. But as a relatively small island, Hispaniola may in fact be in a better position to turn things around than larger, more politically complex countries. The world's biodiversity hotspots might yet lead the charge in finding ways for humans and the rest of the planet's inhabitants to coexist peacefully.

? CASE 8 QUESTIONS

Special sections in Chapters 44–49 discuss the following questions related to Case 8.

1. **How have reefs changed through time?** See page 975.
2. **How do islands promote species diversification?** See page 1019.
3. **Can competition drive species diversification?** See page 1025.
4. **How is biodiversity measured?** See page 1032.
5. **Why does biodiversity decrease from the equator toward the poles?** See page 1066.
6. **How has global environmental change affected coral reefs around the world?** See page 1077.
7. **What are our conservation priorities?** See page 1091.

Source: Brian Farrell.

CHAPTER 44

Animal Diversity

Core Concepts

44.1 Phylogeny charts the evolutionary relationships among more than a million animal species.

44.2 The earliest branches of the animal tree include sponges, cnidarians, ctenophores, and placozoans.

44.3 Bilaterians, including protostomes and deuterostomes, have bilateral symmetry and develop from three germ layers.

44.4 Most vertebrates have a bony cranium and vertebral column; vertebrates include fish, amphibians, reptiles, birds, and mammals.

44.5 Animals first evolved more than 600 million years ago in the oceans, and by 500 million years ago the major structural and functional body plans of animal phyla were in place.

In previous chapters, we saw how the organ systems of animals enable them to move, feed, and behave in different ways. Indeed, animals have evolved a remarkable variety of structures to fulfill these and other functions. Take, for example, the interactive system of muscles and skeleton that governs locomotion (Chapter 37). Sponges and placozoans have no muscle cells at all. In jellyfish, muscle fibers contract to squeeze a fluid-filled cavity, powering movement by jet propulsion. Mammals have limbs for locomotion, powered by the coordinated actions of muscles attached to an internal skeleton. Insects do much the same, but their skeleton is external. Other organ systems show comparable variation among animals. The wide range of functions made possible by these anatomical and physiological variations has permitted animals to diversify to a degree unmatched by plants, fungi, algae, or protozoans.

Biologists have described about 1.8 million species of eukaryotic organisms from the world's forests, deserts, grasslands, and oceans. Of these, about 1.3 million species are animals. There is reason to believe that animal diversification began in the oceans, but today most animal species are found on land, and a majority of all animal species are insects. How can we understand the evolutionary relationships among animals, and how, in turn, can we understand the biological traits that have given rise to such great diversity on a few branches of the animal tree of life?

44.1 A TREE OF LIFE FOR MORE THAN A MILLION ANIMAL SPECIES

As discussed in Chapter 23, it is relatively easy to understand how the visible features of humans, chimpanzees, and gorillas show them to be more closely related to one another than any of them is to other animal species. Furthermore, it isn't hard to see that humans, chimpanzees, and gorillas are more closely related to monkeys than they are to lemurs, and that humans, chimpanzees, gorillas, and monkeys are more closely related to lemurs than to horses. Anatomy and morphology reveal key evolutionary relationships among vertebrate animals, but how do we come to understand the place of vertebrates in a broader tree that includes all animals? More generally, how can we construct an animal phylogeny that includes organisms with body plans as different as those of sponges, sea stars, earthworms, and mussels?

Phylogenetic trees propose an evolutionary history of animals.

Before considering the evolutionary tree of animals, it is useful to begin by asking what an animal is. What features differentiate animals from all other organisms? And, more specifically, what features differentiate animals from their nearest relatives?

As introduced in Chapter 27, the organisms most closely related to animals are choanoflagellates, a type of protist. This relationship was first proposed in the nineteenth century on the basis of cell shape and was confirmed in the twenty-first century by molecular sequence comparisons. Multicellularity differentiates animals from most choanoflagellates, which are unicellular, but multicellularity is not a unique feature of animals. Plants and fungi are multicellular as well. Animals are heterotrophs, gaining energy and carbon from preformed organic molecules (Chapter 6). This character differentiates animals from plants, but not from fungi, which are also heterotrophs. Animals follow a pattern of early embryological development that includes gastrulation (Chapter 42), a character not found in fungi or other multicellular groups.

Animals can therefore be described as multicellular heterotrophic eukaryotes that form a gastrula during development. Other features, both molecular and biochemical, also distinguish animals from other organisms, but basic features of nutrition and development set the stage for considerations of animal diversity.

In Chapter 23, we saw that the fundamental mechanism by which biological diversity increases is speciation, the divergence of two populations from a common ancestor. Played out repeatedly through time, speciation gives rise to a treelike pattern of evolutionary relatedness, with more closely related taxa branching from points close to the tips of the tree and more distantly related taxa diverging from branch points nearer its base. Distinctive features of organisms, called characters, evolve throughout evolutionary history, and the time when they arose can be estimated from their shared presence in the descendants of the population in which the character first evolved.

With these points in mind, we can read **Fig. 44.1** as showing that animals are closely related to choanoflagellates, but differ from them in the presence (among other features) of persistent multicellularity, the formation of a gastrula during early development, and the synthesis of the structural protein collagen. Cnidarians (the phylum that contains jellyfish and sea anemones) and bilaterians (most other animals, including humans, insects, and snails) are differentiated from sponges by the presence of well-defined tissues.

Note that this figure does not tell us that sponges are older than more complex animals. Instead, it says that sponges and more complex animals diverged from a common ancestor, but it doesn't tell us what that common ancestor looked like. Similarly, cnidarians and more complex animals with bilateral symmetry diverged from their last common ancestor at a single point in time and so are equally old, but structurally distinct.

Morphology and development provide clues to animal phylogeny.

Early in the history of biology, taxonomists such as Carolus Linnaeus recognized that for all their diversity, animals can be sorted into a limited number of distinctive body plans, each group today formally called a phylum. The problem for biologists has long been how to understand the evolutionary relationships among animal phyla, still an active field of research.

Nineteenth-century scientists recognized that animals vary in their structural complexity. Sponges, for example, have only a few cell types, and those cells are not organized into tightly coordinated tissues or organs. Jellyfish and sea anemones have

FIG. 44.1 A simple phylogenetic tree of animals. The tree shows increasing complexity from the closest protistan relatives of animals, the choanoflagellates, to bilaterian animals such as insects and mammals. *Photo sources: (top to bottom) Mark Dayel (http://www.dayel.com/choanoflagellates) and Nicole King (http://kinglab.berkeley.edu); James D. Watt/SeaPics.com; UMI NO KAZE/a.collectionRF/Getty Images; British Antarctic Survey/Science Source.*

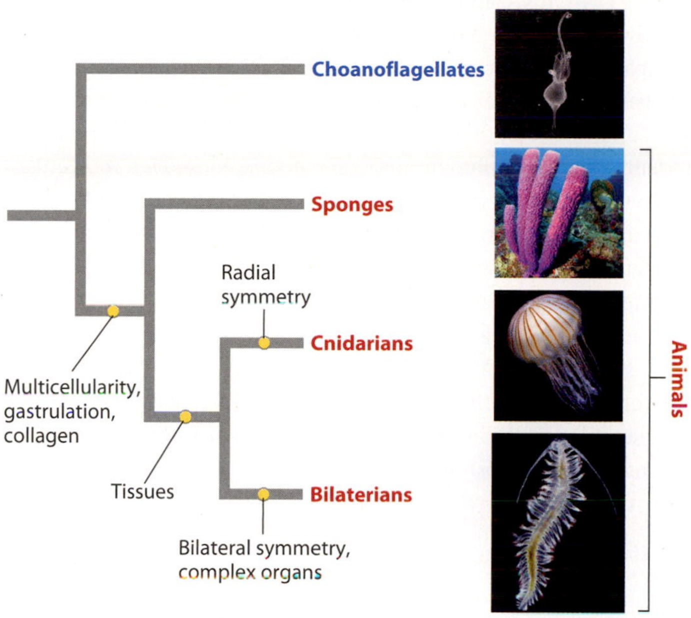

simple tissues but do not have complex organs, whereas crabs and alligators have complex organs specialized for movement, behavior, digestion, and gas exchange.

Early biologists could also see that animals differ in the symmetry of their bodies (**Fig. 44.2**). Sponges are often irregular in form, with no plane of symmetry. In contrast, jellyfish and sea anemones display radial symmetry, meaning that their bodies have an axis that runs from mouth to base with many planes of symmetry through this axis (Fig. 44.2a). This organization allows jellyfish to move up and down in the water column by flexing muscles around their bell-like bodies, and permits sea anemones and corals to wave their ring of food-gathering tentacles in all directions at once. Most animals, however, show bilateral symmetry: Their bodies have a distinct head and tail, front and back, and right and left, with a single plane of symmetry running between right and left at the midline (Fig. 44.2b). Bilateral symmetry enables animals to move in one horizontal direction to capture prey, find shelter, and escape from enemies. It also allows the development of specialized sensory organs at the front end for guidance (Chapter 36), and specialized appendages along both sides for locomotion, grasping, or defense.

Within the bilaterian group, however, understanding evolutionary relationships proved more difficult to early taxonomists. Some biologists pointed to the presence or absence of a cavity, or **coelom,** surrounding the gut, dividing bilaterians into three groups: those without a body cavity (acoelomates) and those with a body cavity (coelomates and pseudocoelomates,

FIG. 44.2 Symmetry in animal form. (a) Radial or (b) bilateral symmetry distinguishes cnidarians and bilaterian animals.

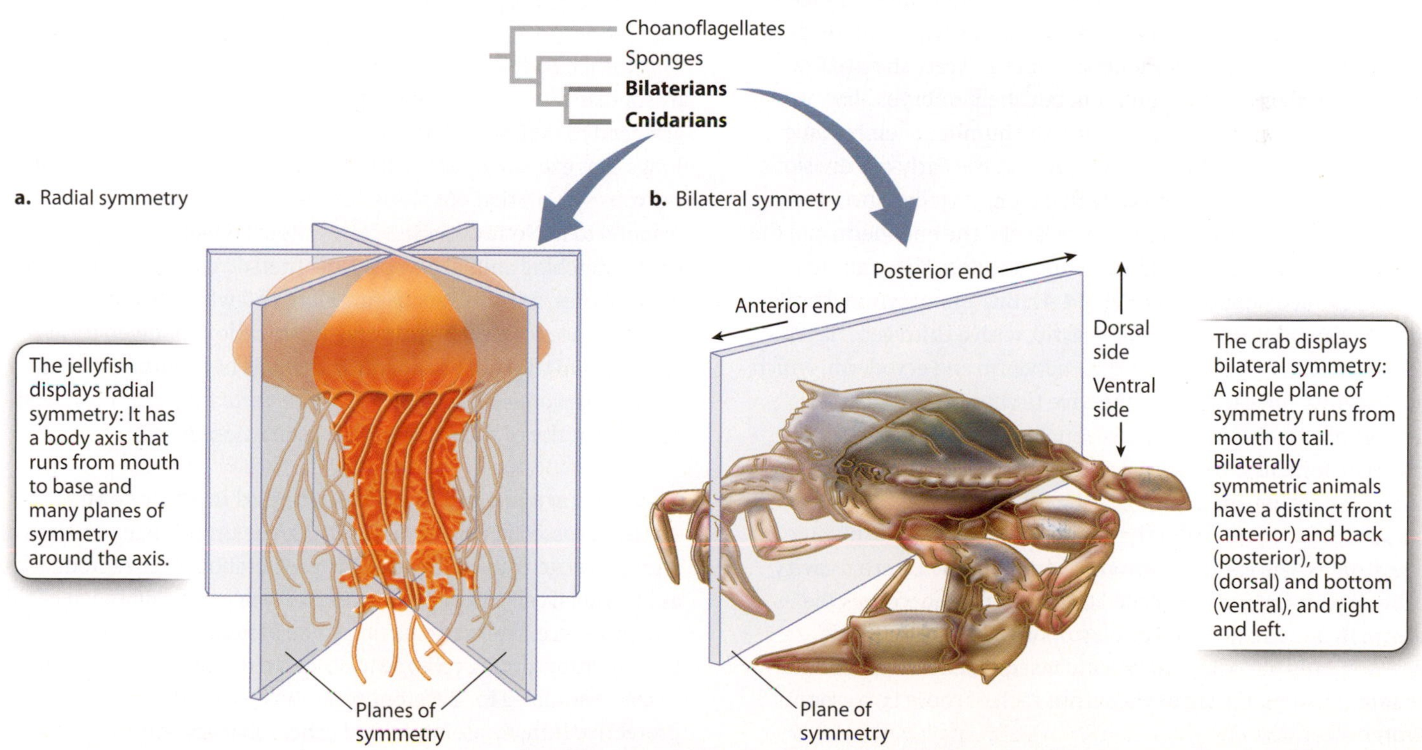

FIG. 44.3 Major types of anatomical organization of bilaterian animals. Bilaterians vary in the presence, absence, or structure of a body cavity.

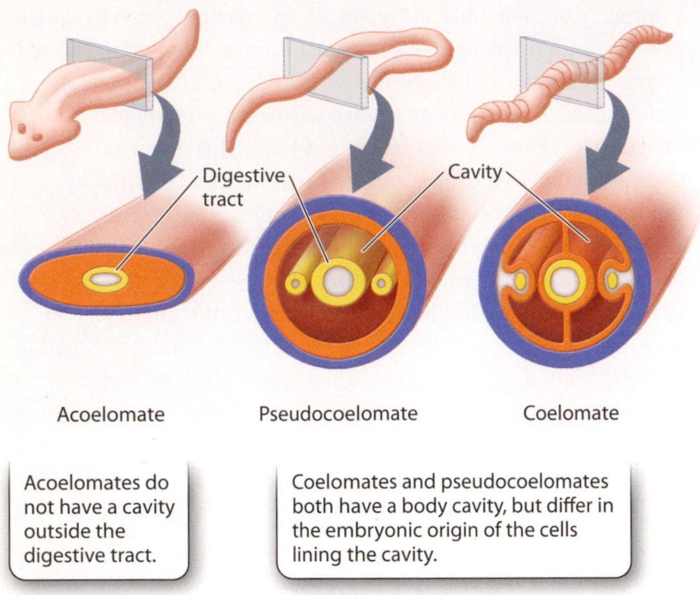

which differ in the embryonic origin of the cells lining the cavity). These body plans are shown in **Fig. 44.3.** A body cavity cushions the internal organs against hard blows to the body and enables the body to turn without twisting these organs. It also allows internal organs like the stomach to expand, enhancing digestive function.

New insights became possible with the advent of microscopes that enabled the direct study of development from its earliest stages. One important observation was that some animals that look very different as adults share patterns of early embryological development. For example, adult sea stars and catfish have few morphological features in common, but their embryos show a number of key similarities, including the number of embryonic tissue layers they develop and the way that the early cell divisions occur. In the phylum **Cnidaria,** which includes jellyfish and sea anemones, the embryo has two germ layers, the endoderm and the ectoderm, from which the adult tissues develop. These animals are therefore called diploblastic (**Fig. 44.4**). Bilaterally symmetrical animals, the **Bilateria,** are triploblastic, with a third germ layer, the mesoderm, lying between the endoderm and ectoderm, which develops into muscles and connective tissues (Fig. 44.4).

Comparative embryology also enabled biologists to divide bilaterian animals into the two groups shown in **Fig. 44.5,** which are called **protostomes** (from the Greek for "first mouth") and **deuterostomes** (from the Greek for "second mouth"). In protostomes, the earliest-forming opening to the internal cavity of the developing embryo, called the blastopore, becomes the mouth. In deuterostomes, the blastopore becomes the anus. However, only in the age of molecular sequence comparisons have the relationships among phyla within each of these two groups become clear.

Molecular sequence comparisons have confirmed some relationships and raised new questions.

Anatomical and embryological observations by early biologists laid the groundwork for understanding animal phylogeny, but, as was true for phylogenetic relationships in Bacteria, Archaea, and Eukarya as a whole, a modern understanding of evolutionary relationships among animals had to wait for the revolution in molecular sequencing. Over the past two decades, comparisons among DNA, RNA, and amino acid sequences have greatly improved our understanding of animal phylogeny. Molecular comparisons support many of the conclusions reached earlier on the basis of comparative anatomy and embryonic development, including the early divergence of sponges and the separation of radially and bilaterally symmetrical animals. Moreover, molecular sequence comparisons confirm the view that choanoflagellates are the closest protistan relatives of animals.

Other hypotheses, however, have been rejected. For example, the traditional phylogenetic division of bilaterians into acoelomate, coelomate, and pseudocoelomate groups gets little support from molecular studies. Similarly, molecular sequence comparisons do not support the once widespread view that segmented bodies indicate a close relationship between earthworms and lobsters.

→ **Quick Check 1** Which animals are more closely related to sponges, Cnidaria or Bilateria?

44.2 THE SIMPLEST ANIMALS: SPONGES, CNIDARIANS, CTENOPHORES, AND PLACOZOANS

The human body is complex, our diverse functions made possible by sophisticated, interacting organ systems: from muscle pairs attached to skeletons that allow us to run, jump, and dance to lungs that exchange gases with the atmosphere and a brain and nervous system that enable us to sense our environment and respond to it. None of these organ systems is well developed in early-branching animal groups. How then do sponges, cnidarians, ctenophores, and placozoans function, and what are the consequences of their anatomical organization for diversity and ecology? On the other hand, these animals have survived a half billion years of Earth history. What can we learn from the diversity and continuity of their remarkable adaptations?

Sponges are simple and widespread in the oceans.

As we discussed in Chapters 27 and 28, the unicellular ancestors of animals carried out all functions from metabolism to reproduction as individual cells. The cells making up sponges retain much of this independence, while also having some benefits of multicellularity. Coordination among cells means that there can be distinct kinds of cell specialized for different functions—cells that form a protective skin, for example, and others that secrete enzymes for

FIG. 44.4 Variation in embryonic development. Early development of germ tissues in the embryo separates the diploblastic cnidarians from the triploblastic bilaterians.

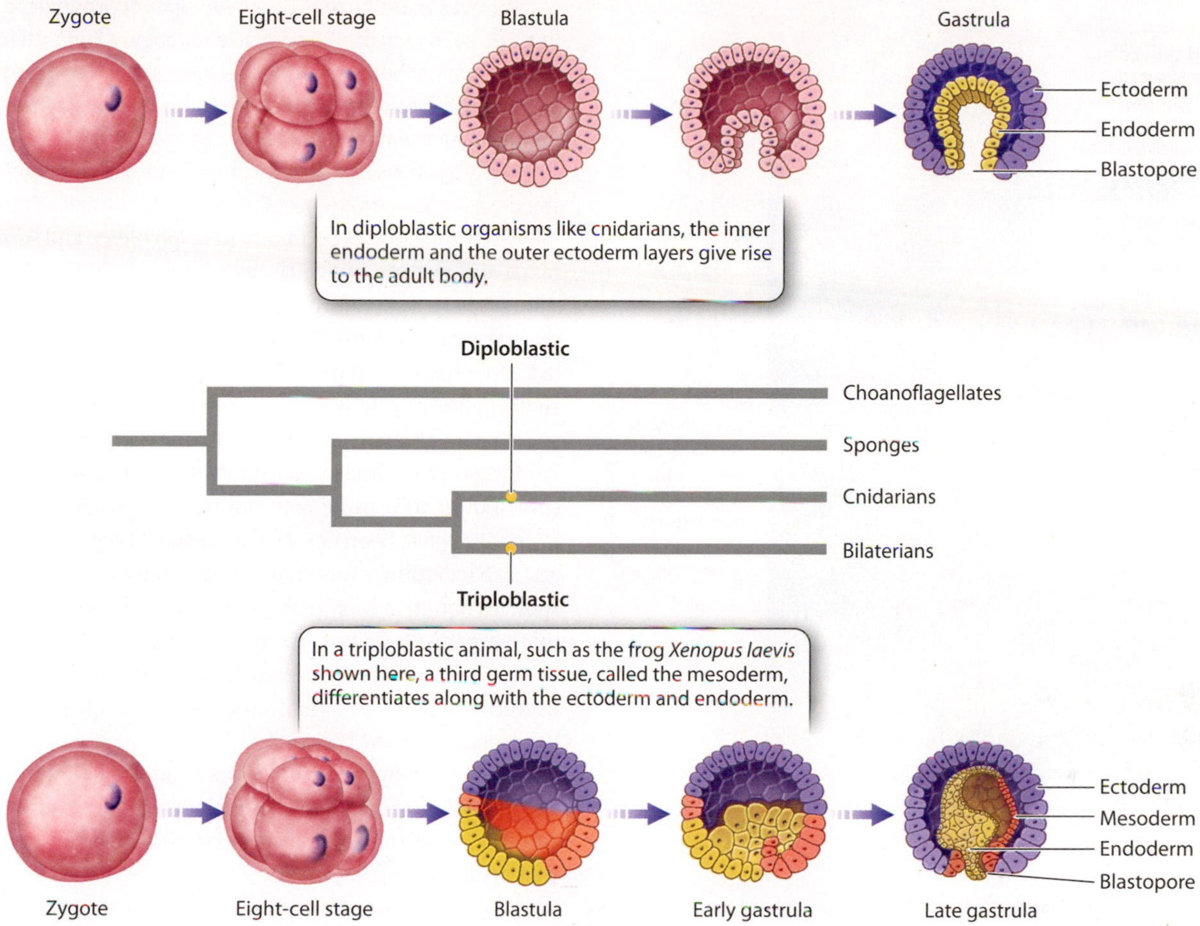

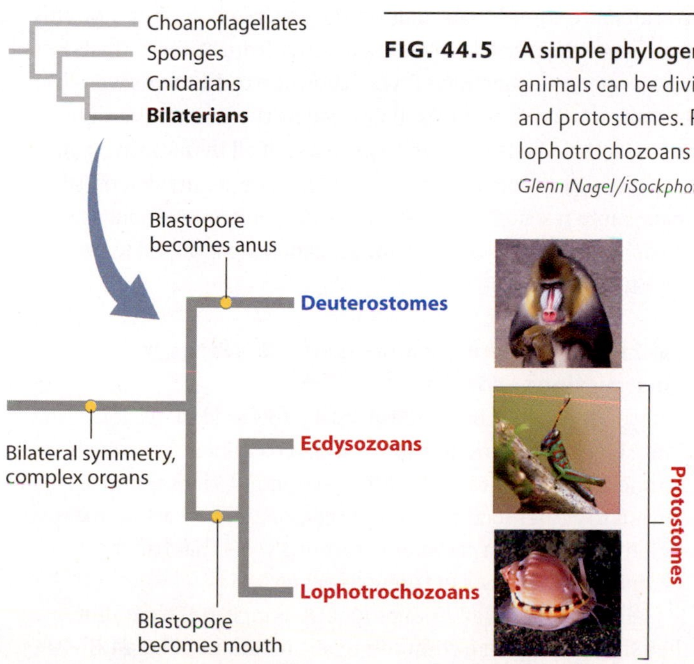

FIG. 44.5 A simple phylogenetic tree of bilaterian animals. Bilaterian animals can be divided into two major groups, the deuterostomes and protostomes. Protostomes can be further subdivided into lophotrochozoans and ecdysozoans. *Photo sources: (top to bottom) Glenn Nagel/iSockphoto; Lal/Getty Images; Franco Banfi/WaterF/age fotostock.*

digestion. Most dramatically, multicellularity permits sponges to extend above the seafloor, giving them access to food suspended in water currents.

The sponge body plan resembles a flower vase, with many small pores along its sides and a larger opening at the top (**Fig. 44.6**). The cells on the outer surface of the sponge are tough and act as the sponge's skin. The interior surface is lined by cells called **choanocytes,** which have flagella and function in nutrition and gas exchange. Choanocyte cells have a collar of

FIG. 44.6 Anatomy of a sponge. *Photo source: imagebroker/Alamy.*

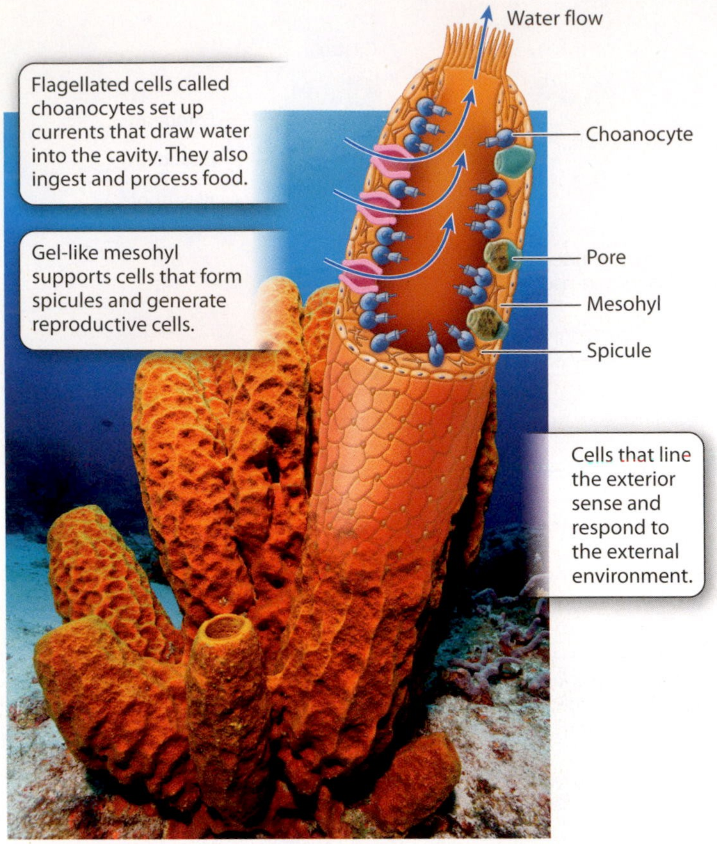

small cilia around their flagellum, much like the cells of animals' closest unicellular relatives, the choanoflagellates (Chapter 27). Between the interior and exterior cell layers lies a gelatinous mass called the **mesohyl** (Fig. 44.6). Mesohyl is mostly noncellular, but it contains some amoeba-like cells that function in skeleton formation and the dispersal of nutrients.

Sponges gain nutrition by intracellular digestion (Chapter 40). The choanocytes that surround the interior chambers and passageways of the sponge beat their flagella, creating a current that draws water from outside the body, through the pores in its walls and upward through the central cavity of the sponge, where it exits through the large opening at the top. The circulating water contains food particles and dissolved organic matter, which cells lining the cavity capture by endocytosis. The individual cells then metabolize the food in their interiors, much the way that protozoans feed.

From this observation, we might conclude that sponges are simply a group of uncoordinated cells, each working for itself, but that isn't the case. The choanocytes beat their flagella in a coordinated pattern, helping to draw water into the body interior through the pore system and outward again through the vase opening. Moreover, the shape of the sponge body itself directs water movement across feeding cell surfaces. These features effectively move water across the cells, facilitating food uptake. Sponge cells require oxygen for respiration and must get rid of the carbon dioxide respiration generates. Gas exchange occurs by diffusion, aided by the movement of water through the sponge cavity.

Sponges don't have highly developed reproductive organs. Instead, cells recruited from the choanocyte layer migrate into the mesohyl, where they undergo meiotic cell division and differentiate as sperm or eggs. Sperm released into the water fuse with eggs in the mesohyl of other sponges.

Many sponges build skeletons of simple structures called spicules. Some sponges precipitate spicules of glasslike silica (SiO_2). Others, however, make their spicules—and sometimes massive skeletons—of calcium carbonate ($CaCO_3$). Still other sponges have a skeleton made up of proteins (these sponges are often used in the bath). So we see that sponges are intermediate in the sophistication of their bodies: Their cells function much as single-celled protozoa do, but they coordinate their activities and so are more efficient in extracting food and oxygen from seawater.

Despite their anatomic simplicity, sponges are major contributors to seafloor communities. Approximately 9000 species of sponges have been described, most of them ocean dwelling, but also including a few that live in freshwater lakes. Many sponges obtain at least part of their nutrition from symbiotic microorganisms living within their bodies. In fact, sponge bodies are commonly full of bacterial cells. At least some of these bacteria are probably symbionts, but only a few experiments have demonstrated their function.

Until recently, most biologists regarded sponges as a monophyletic group, one that includes all the descendants of a common ancestor. Some molecular data, however, now suggest that sponges are paraphyletic—that is, that they contain some but not all the descendants of their common ancestor (Chapter 23). **Fig. 44.7** shows these two possibilities. If the tree on the left in the figure is correct, the two main limbs of the animal tree contain very different organisms—the simple sponges on one branch and all more complex animals on the other. As noted earlier, this figure indicates that sponges are distinct from other animals but leaves open the question of what the ancestral animals were like. In contrast, the hypothesis illustrated by the tree on the right suggests that the last common ancestor of all living animals, in essence, *was* a sponge. At present, biologists are divided on this issue; more research is needed to demonstrate how sponges relate to each other and to other animals, which are grouped together as the **Eumetazoa.**

Cnidarians are the architects of life's largest constructions: coral reefs.

Many of us have encountered at least a few cnidarians. Jellyfish flourish in marine environments from coastlines to the deep sea; sea anemones cling to the seafloor, extending their tentacles upward to gather food and deter predators; corals secrete massive skeletons of calcium carbonate, forming reefs that fringe continents and islands in tropical oceans.

Jellyfish and sea anemones look strikingly different, but they share a body plan common to all cnidarians (**Fig. 44.8**). At

FIG. 44.7 Two hypotheses for the phylogenetic relationships of sponges to other animals.

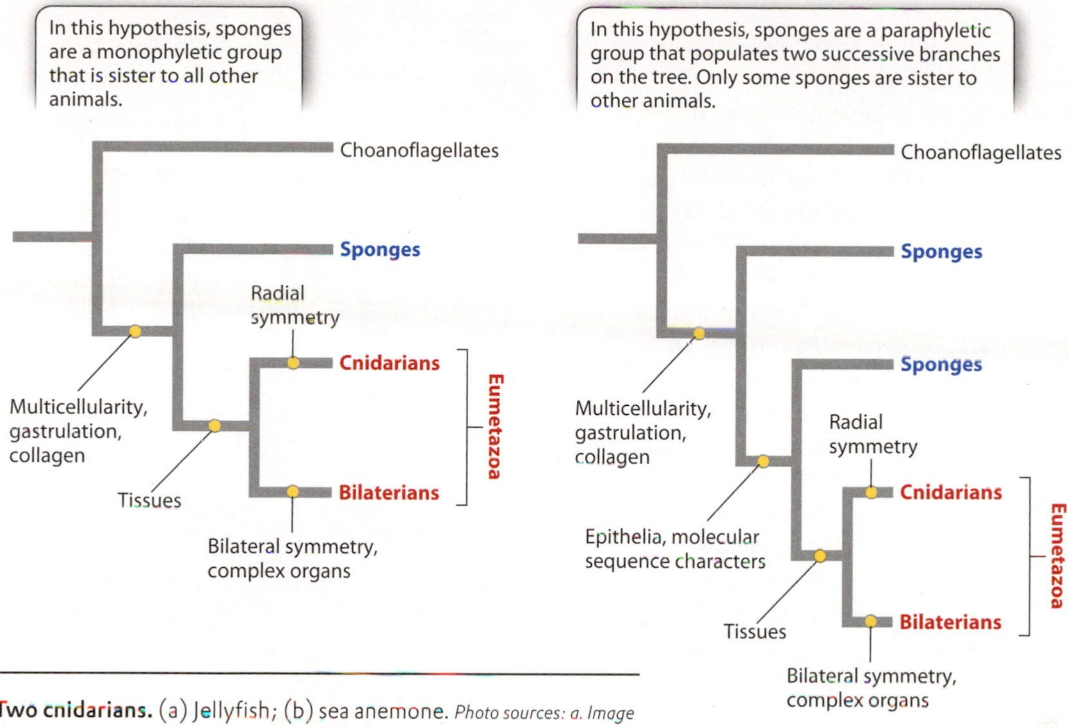

FIG. 44.8 Two cnidarians. (a) Jellyfish; (b) sea anemone. *Photo sources: a. Image Source/Getty Images; b. Andrew J. Martinez/Science Source.*

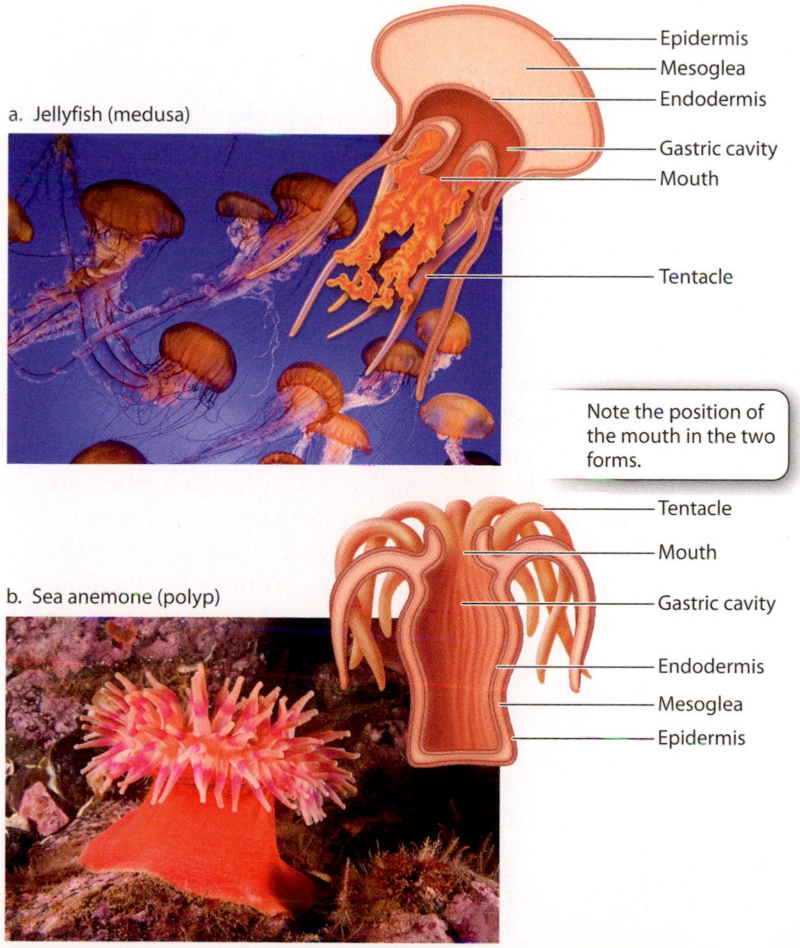

one end of the radially symmetrical body is a mouth surrounded by tentacles armed with stinging cells that subdue prey and defend against enemies. An anemone is like a jellyfish stuck upside down onto the seafloor with its tentacles and mouth facing upward. We call a free-floating jellyfish a medusa, and the stationary, or sessile, form of an anemone a polyp. Some cnidarians develop into one form or the other, but many have life cycles that alternate between the two.

The mouth of both medusa and polyp opens into a closed internal **gastric cavity,** the site of extracellular digestion and excretion. Rather than digesting food particles inside individual cells as sponges do, cnidarians receive both food and digestive enzymes in the gastric cavity. (The digestive enzymes are secreted by cells that line the walls of the cavity.) This arrangement permits cnidarians to digest large food items, such as a whole fish (paralyzed by the stinging cells on the tentacles), whose dissolved nutrients can then be absorbed through the walls of the cavity. Cnidarians, then, can consume many types of food that are unavailable to sponges.

The cnidarian body develops from a diploblastic embryo (one with two germ layers). It has an outer layer, the **epidermis,** which develops from the ectoderm, and an inner lining, the **endodermis,** that derives from the endoderm. These tissues enclose a gelatinous mass called the **mesoglea** (the "jelly" of jellyfish). This organization sounds a bit like the sponge body plan, but there are important differences, many of which are shared characters between cnidarians and the rest of the eumetazoa.

First, in cnidarians, the cells that form the epidermis and endodermis occur as closely packed layers of cells embedded in a protein-rich matrix, forming an epithelium (Chapter 10). In all animals that have an epithelium, it is formed from an unusually tight layer of specialized cells, bounded by a matrix composed of protein and connected to one another by junctions that regulate the passage of ions or other molecules. Epithelial layers line compartments within animals, often absorbing or secreting substances, and therefore making possible organs such as those of the digestive system (Chapter 40). For many years, most biologists accepted that sponges do not form epithelia. Now careful anatomical studies show that more complex sponges can form epithelia, although these are somewhat simpler in terms of cellular connections than those of other animals.

Second, cnidarians have a wider array of cell types than sponges do, permitting more sophisticated tissue function. Musclelike cells allow jellyfish to swim through the ocean. A simple network of nerve cells permits cnidarians to sense their environment and respond to it by directional movement. No such cells occur in sponges. Cnidarians don't have a brain, but at least a few have light-sensitive cells that function as simple eyes.

Moreover, whereas sponges filter water to gain food, cnidarians are predators, capturing prey with their tentacles and digesting it in the gastric cavity. Specialized cells on the tentacles contain a tiny harpoonlike organelle called a nematocyst, often tipped with a powerful neurotoxin that greatly aids prey capture and defense against other predators (**Fig. 44.9**). Other specialized cells lining the gastric cavity secrete digestive enzymes that break down ingested food into molecules that can be taken up by cells lining the cavity by endocytosis. There is no specialized passage for waste removal. Instead, waste is excreted back into the gastric cavity and leaves by way of the mouth. Oxygen uptake and carbon dioxide release occur by diffusion.

Many cnidarians reproduce asexually to form colonies. For example, corals form extensive rounded, fan-shaped, or hornlike colonies by budding, building potentially massive structures on the seafloor (**Fig. 44.10a**). In the Portuguese Man-of-War, different individuals in the same colony develop distinct morphologies, some specialized for flotation, others for prey capture, and still others for reproduction (**Fig. 44.10b**). Among animals without complex organs, such colonies represent the height of morphological complexity.

About 9000 species of cnidarians live in the oceans. Reef corals are remarkable not only for the size of their skeletal colonies, but also for their mode of nutrition. Most have lost the ability to capture prey and instead gain nutrition from symbiotic algae in their surface tissues.

→ **Quick Check 2** How do muscle cells enable cnidarians to function in ways that sponges cannot?

FIG. 44.9 Nematocysts. Nematocysts are harpoonlike organelles that capture prey.

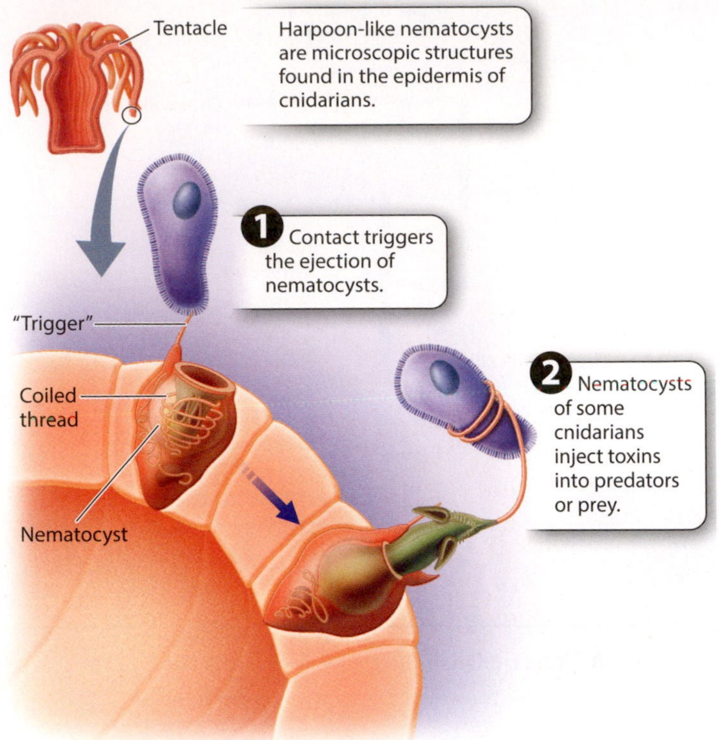

Ctenophores and placozoans represent the extremes of body organization among early-branching animals.

Our exploration of nonbilaterian animals ends with ctenophores, also called comb-jellies, and placozoans. Neither of these groups is diverse (just 100–150 species of comb-jellies have been named, and so far only one species of placozoan), but they may reveal a significant amount about the relationship between bilaterians and earlier-branching animals.

Ctenophores resemble cnidarians in body plan, and for many years most biologists thought that the two were close relatives. Like cnidarians, ctenophores have radial symmetry, with an outer epithelium and an inner endodermis that enclose a gelatinous interior (**Fig. 44.11**). Ctenophores also have a simple nerve net, as well as rudimentary gonads. They are predators that feed by ingestion, digesting prey within their gastric cavity by enzymes secreted from the cells lining the gut cavity. Gas exchange occurs by diffusion. However, there are important differences between cnidarians and ctenophores. Ctenophores propel themselves through the oceans by the coordinated beating of cilia that extend from epidermal cells. These cilia are usually arranged in comblike groups, hence the common name for this phylum. Unlike cnidarians, digestive wastes generated by ctenophores move through a gut cavity for elimination through an anal pore opposite the mouth.

FIG. 44.10 **Colony formation in cnidarians.** (a) In soft corals, the individual members of the colony are similar in form and function. (b) In the Portuguese Man-of War, the members of the colony differentiate to serve distinct functions, such as floating, feeding, and reproduction. *Sources: a. ifish/Getty Images; b. George G. Lower/Science Source.*

FIG. 44.11 **Ctenophores, or comb-jellies.** *Photo source: David Wrobel/SeaPics.com.*

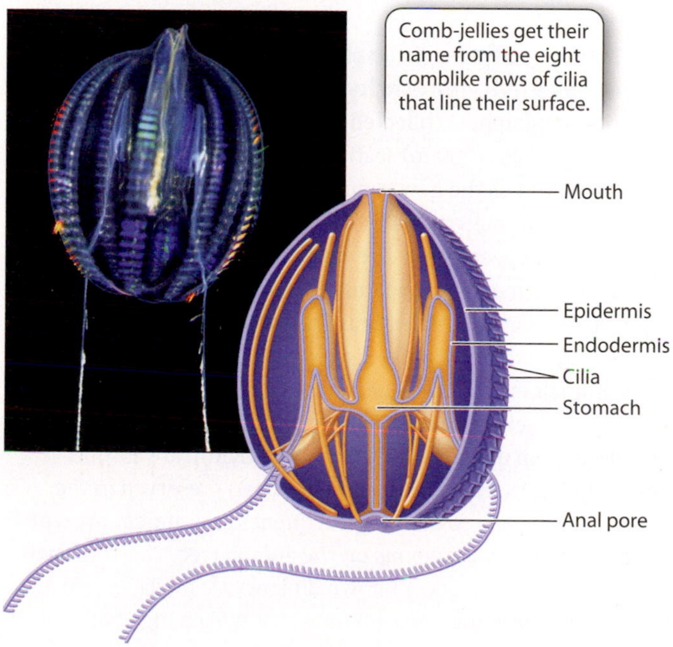

If ctenophores are the most complex animals discussed so far, **placozoans** are the simplest. These tiny (millimeter-scale) animals each contain only a few thousand cells arranged into upper and lower epithelia that sandwich an interior fluid crisscrossed by a network of multinucleate fiber cells (**Fig. 44.12**). They have no specialized tissues and few differentiated cell types. Cilia on cell surfaces allow movement, and gas exchange occurs by diffusion. Placozoans can absorb dissolved organic molecules, but commonly feed by surrounding food particles and secreting digestive enzymes to break them down. Despite their morphological simplicity, however, placozoans have a genome that contains many of the genes for transcription factors and signaling molecules that are present in cnidarians and bilaterian animals. What roles these genes play in placozoan biology remain unclear.

The order of early branches on the animal tree remains uncertain.

Molecular, anatomical, and embryological data agree that bilaterally symmetrical animals form a monophyletic group whose sister is the radially symmetrical Cnidaria. Increasingly, DNA sequence data place placozoans as sister to the cnidarians

FIG. 44.12 **The placozoan *Trichoplax adhaerens*.** *Photo sources: Reprinted by permission from Macmillan Publishers Ltd: NATURE 454, 955–960 (21 August 2008) M. Srivastava et al., The Trichoplax genome and the nature of placozoans. Copyright 2008. Image courtesy of Ana Signorovitch.*

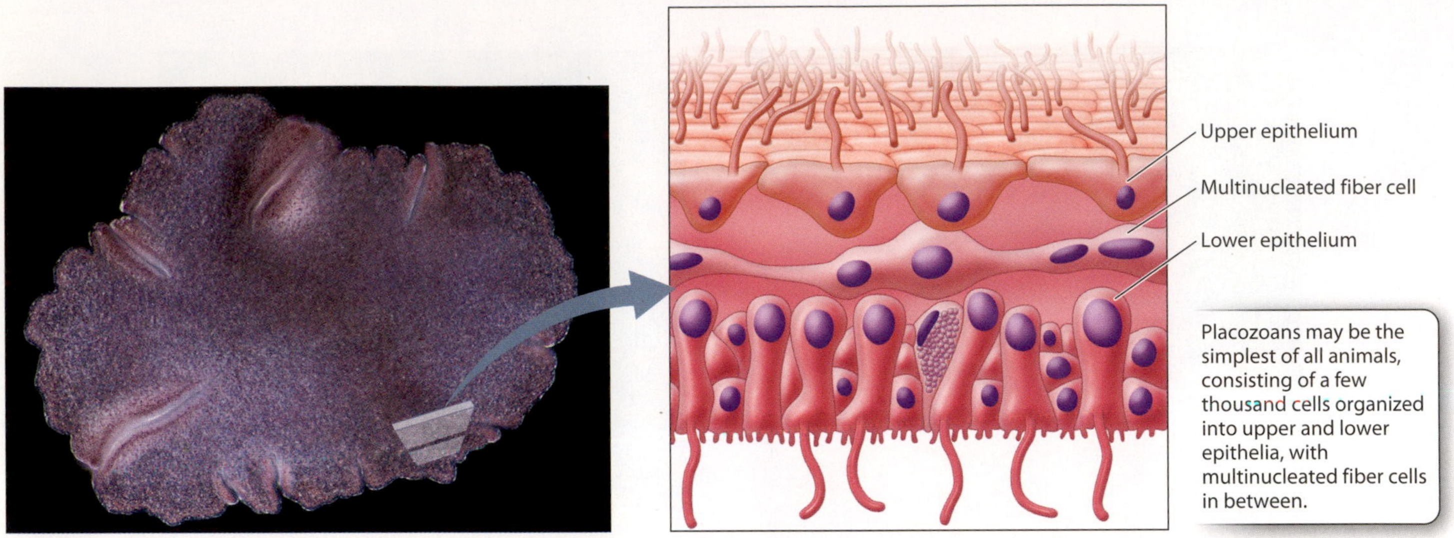

and bilaterian animals (**Fig. 44.13**). This placement suggests that feeding by the absorption of organic molecules evolved before gut cavities specialized for this function.

Where to place ctenophores in this tree is less clear. One recent hypothesis, based on DNA sequence comparisons, suggests rather surprisingly that ctenophores are the sister group to all other animals. If this hypothesis is correct, it means either that the complexity of ctenophores evolved independently of that found in eumetazoans or that the morphological simplicity of living sponges reflects loss of features retained in other animals. Other data, however, favor the more traditional placement of ctenophores as eumetazoans. DNA sequence data indicate that wherever ctenophores belong on the animal tree of life, their nervous system evolved independently of those found in cnidarians and bilaterian animals. At present, we simply don't know the correct branching order of the animal tree, although there is reason to believe that the earliest animals were much like present-day sponges in terms of body organization and function.

Fig. 44.13 shows one current view of the phylogenetic relationships among animals, with placozoans as the sister group to cnidarians and bilaterians, and uncertainty in the branching order of sponges and ctenophores. We use this tree to guide further discussion, but it should be regarded as a set of hypotheses to be tested against new data rather than an unambiguously "correct" tree. See also **Table 44.1** for a list of all currently recognized animal phyla.

FIG. 44.13 **A phylogenetic tree of animals, with placozoans and ctenophores.** The placements of sponge subgroups, placozoans, and ctenophores remain uncertain and should be treated as hypotheses to be tested.

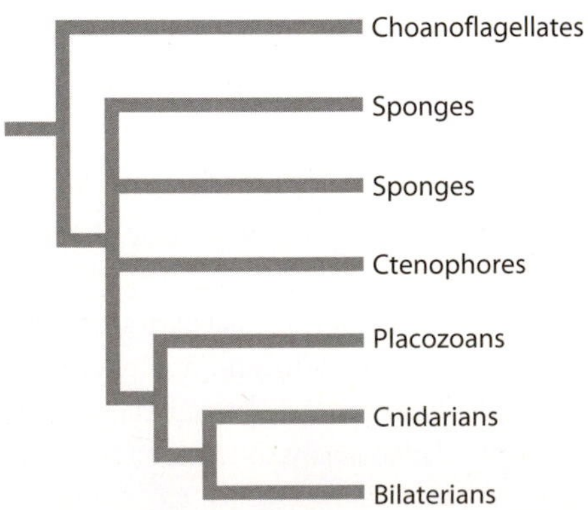

The discovery of new animals with a unique body plan complicates phylogenetic hypotheses still further.

In 1986, oceanographers collected some unusual organisms from the deep (400–1000 m) seafloor off the coast of Tasmania. Described in 2014 as the new genus *Dendrogramma,* these are diploblastic animals only a few millimeters long, with a mushroomlike appearance: a disc connected to a stalk (**Fig. 44.14**). The stalk bears a mouthlike opening at its end, connected by a tubular passageway to a highly branched series of canals in the disc that are thought to function in digestion and, perhaps, circulation.

The animals exhibit some characters that resemble cnidarians and others that resemble ctenophores, but features that might place them unequivocally in either phylum are absent. Unfortunately, DNA appears not to have been preserved in the animals collected and so molecular sequencing cannot at present be used to place *Dendrogramma* on the animal tree. These animals may be aberrant members of known diploblastic phyla, or—just maybe—they represent a new phylum that will change our view of early branches on the animal tree.

TABLE 44.1 Animal Phyla.

LIMB OF THE ANIMAL TREE	PHYLUM	COMMON NAME	APPROXIMATE NUMBER OF DESCRIBED SPECIES
Basal branches	Porifera	Sponges	9000
	Cnidaria	Corals, jellyfish, etc.	9000
	Ctenophora	Comb-jellies	100–150
	Placozoa	*Trichoplax adhaerens*	1
Basal bilaterians	Acoelomorpha	Acoels	350
Protostomia/Lophotrochozoa	Acanthocephala	Thorny-headed worms	1000
	Annelida	Segmented worms	15,000
	Brachiopoda	Lamp shells	350
	Bryozoa	Moss animals	5000
	Chaetognatha	Arrow worms	100
	Cycliophora	*Symbion*	at least 3
	Echiura	Spoon worms	150
	Entoprocta	Goblet worms	150
	Gastrotricha	Hairy backs	700
	Gnathostomulida	Jaw worms	100
	Micrognathozoa	*Limnognathia maerski*	1
	Mollusca	Mollusks	80,000
	Nemertea	Ribbon worms	1200
	Phoronida	Horseshoe worms	20
	Platyhelminthes	Flatworms	25,000
	Rhombozoa	—	75
	Rotifera	Rotifers	2000
	Sipuncula	Peanut worms	350
Protostomia/Ecdysozoa	Arthropoda	Arthropods	800,000–1,000,000
	Kinorhyncha	Mud dragons	150
	Loricifera	Brush heads	120
	Nematoda	Roundworms	20,000
	Nematomorpha	Horsehair worms	320
	Onychophora	Velvet worms	75
	Priapulida	Priapulid worms	17
	Tardigrada	Water bears	More than 600
Deuterostomia	Chordata	Chordates	50,000
	Echinodermata	Echinoderms	7000
	Hemichordata	Acorn worms, pterobranchs	90
	Xenoturbellida	—	2

FIG. 44.14 ***Dendrogramma enigmatica.*** Reported for the first time in 2014, this is a diploblastic animal that is not easily placed within any existing phylum. *Source: Jørgen Olesen.*

A mouthlike opening on the stalk connects to a highly branched system of internal canals for digestion and perhaps circulation.

44.3 BILATERIAN ANIMALS

The phyla discussed in the preceding section include about 20,000 species, a tiny fraction of all animal diversity. The final stop in this broad tour is the bilaterian branch, on which most animal species reside. Earlier, we pointed out that basic patterns of embryo formation divide bilaterians into two major groups, the Protostomia and the Deuterostomia (see Fig. 44.5). Molecular sequence data support the protostome–deuterostome groupings of bilaterian animals and show that the protostome animals can be further divided into two groups: the Lophotrochozoa (which include mollusks and annelid worms) and the Ecdysozoa (which include insects and other arthropods).

As noted above, bilaterian animals have bilateral symmetry and complex organs that develop from a triploblastic (three-germ layer) embryo. The anatomical complexity of bilaterian animals makes possible types of locomotion, feeding, gas exchange, behavior, and reproduction that are unknown in earlier branching groups. These capabilities underpin the remarkable diversity of bilaterian animals.

Lophotrochozoans make up nearly half of all animal phyla, including the diverse and ecologically important annelids and mollusks.

The Lophotrochozoa are a group of animals first recognized on the basis of unique molecular characters and now known to include several very different types of animal. On the one hand, there are the brachiopods, bryozoans, and related phyla that have a tentacle-lined organ for filter feeding called a lophophore. Then, there are the mollusks, annelid worms, and their close relatives, characterized by a type of larva called a trochophore. The odd name "lophotrochozoan" is a contraction of the terms "lophophore" and "trochophore," with "zoan" added to indicate that these organisms are animals (*zoon* is Greek for "animal"). The majority of lophotrochozoan species also have a distinctive form of spiral cleavage early in development.

The Lophotrochozoa contain 17 phyla, mostly small marine animals of limited diversity, but also the diverse and ecologically important **annelid worms** and **mollusks** (**Fig. 44.15**). We'll

FIG. 44.15 **A phylogenetic tree of lophotrochozoan animals, showing all 17 phyla in this group.** Phyla discussed in the text are shown in red. Uncertainties are shown by nodes with more than two branches. *Photo sources: (top) Valerie Giles/Science Source; (bottom) Phillip Colla/Oceanlight.*

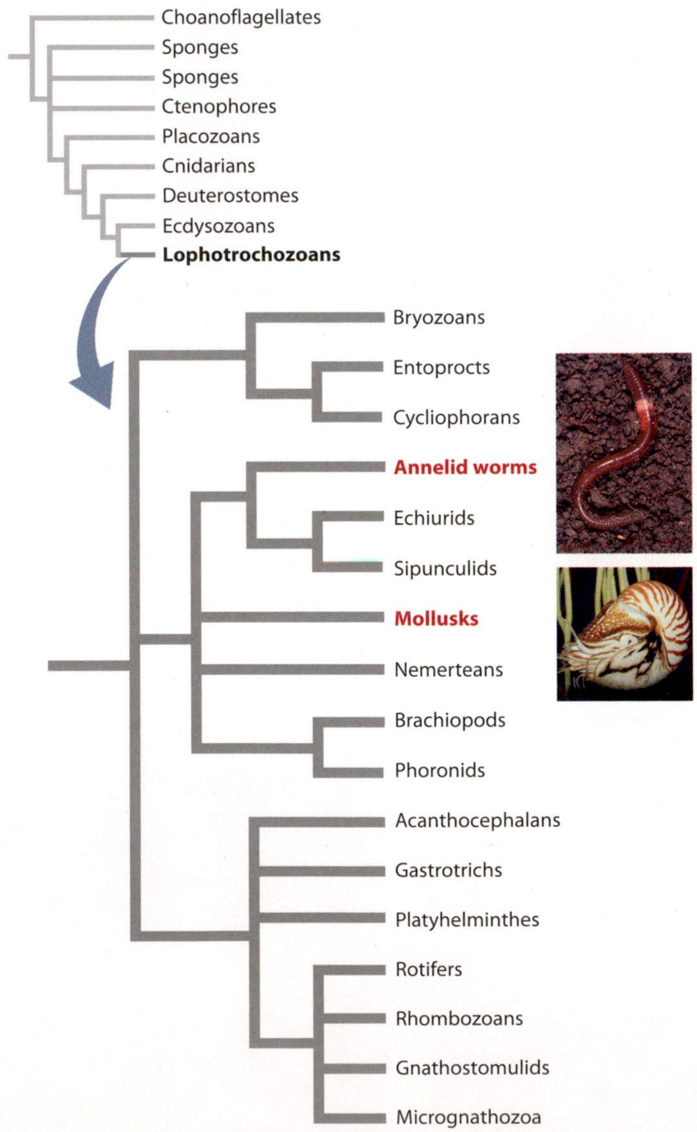

focus on these two phyla, which are closely related but which have nonetheless evolved along quite different anatomical trajectories in the half billion years since their origins.

The annelids most familiar to us are the earthworms, which, as their name suggests, are commonly found in soil (**Fig. 44.16a** and **44.16b**). Most of the 15,000 known annelid species, however, live in the oceans (**Fig. 44.16c**). All annelids have a cylindrical body with distinct segments, and they illustrate the advantages of a bilaterian body plan. At one end, the head has a well-developed mouth and, internally, a cerebral ganglion (a collection of nerve cells) that connects to an extensive nervous system. A digestive system extends through the body from the head to an anus, with a sequence of specialized organs for crushing, then digesting, and finally excreting ingested food, much like the digestive system in our own bodies. Working together, these features enable annelids to move through their environment, actively searching for food and digesting it efficiently after ingestion.

Aquatic annelids have gill-like organs for gas exchange, but terrestrial earthworms exchange gases through their skin. In both cases, a closed circulatory system moves dissolved gases through the body. Annelids have waste-filtering organs called nephridia, gonads (repeated in most segments), and a fluid-filled coelom, or body cavity. Fluids in the coelom form a hydrostatic skeleton that works in coordination with paired muscles in each segment to direct movement.

Many annelids are predators that capture and ingest prey, but some, like earthworms, ingest sediment, digesting the organic matter it contains and excreting mineral particles. A few marine annelids have evolved tentacles that enable them to filter food particles from water, while leeches, a specialized group of freshwater annelids, attach themselves to vertebrates to suck out a meal of blood (**Fig. 44.16d**). And, remarkably, the meter-long vestimentiferan worms that live around hydrothermal vents in the oceans have given up ingestion altogether (**Fig. 44.16e**). Without mouths, these enormous worms gain nutrition from chemosynthetic bacteria that live within a collar of specialized tissue. Discovered in the 1970s, vestimentiferan worms were originally placed in their own phylum, but molecular sequence comparisons indicate that they are annelids.

Mollusks are the second major phylum of the Lophotrochozoa, and include such familiar groups as snails, clams, mussels, squids, and octopuses. Mollusks are distinguished from other lophotrochozoans by a unique structure called the mantle that plays a major role in breathing and excretion, and which forms their shells (when shells are present). The mantle lines the visible body of a squid as well as what we commonly but mistakenly

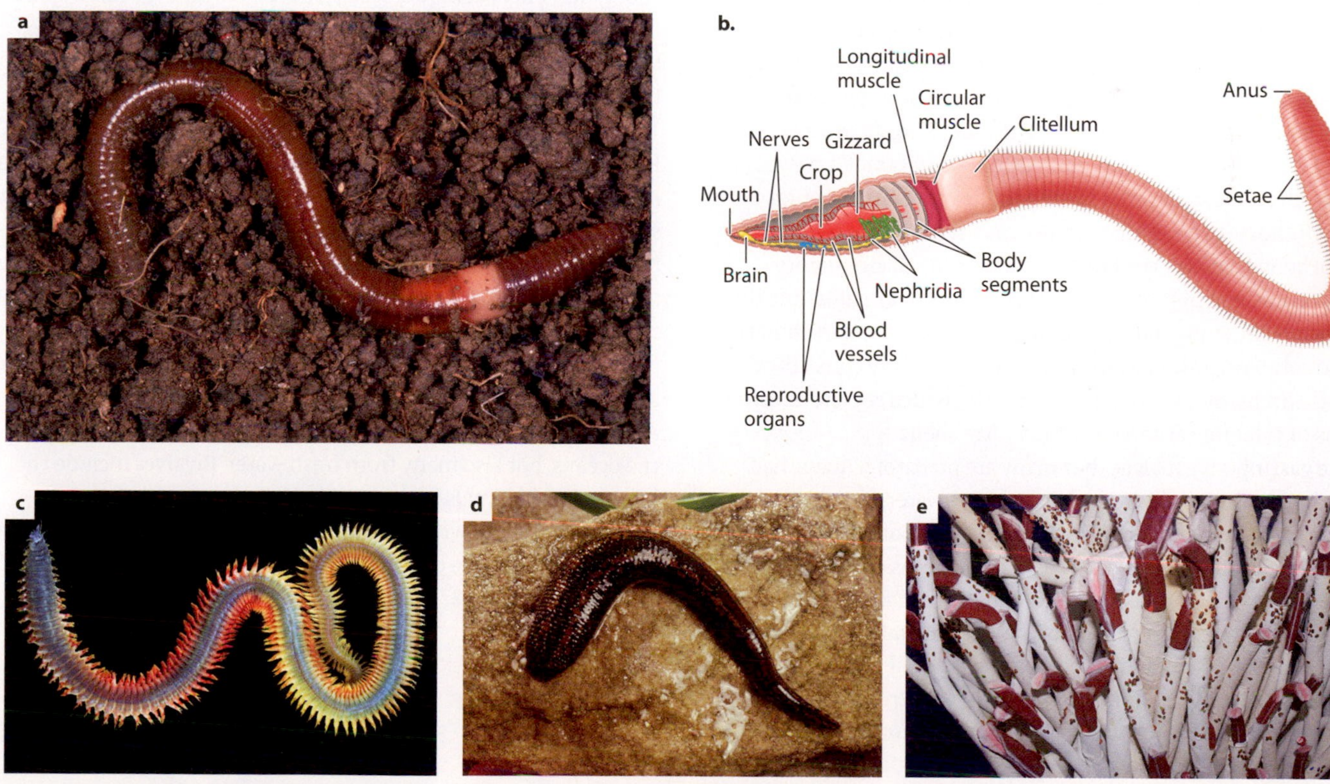

FIG. 44.16 **Annelids.** (a and b) Earthworms typify the segmented body plan found throughout the Annelida. (c) The most diverse annelids are polychaete worms, which live in the oceans. (d) Leeches are specialized for ingesting blood from mammalian hosts. (e) Vestimentiferan worms live along hydrothermal rift vents in the oceans, gaining nutrition from chemosynthetic bacteria in specialized organs (red in the image). *Photo sources: a. Valerie Giles/Science Source; c. Alexander Semenov/Science Source; d. Volk, F/age fotostock; e. Susan Dabritz/SeaPics.com.*

FIG. 44.17 A gastropod, showing a body plan typical of the phylum Mollusca. *Photo source: ARCO/W Rolfes/age fotostock.*

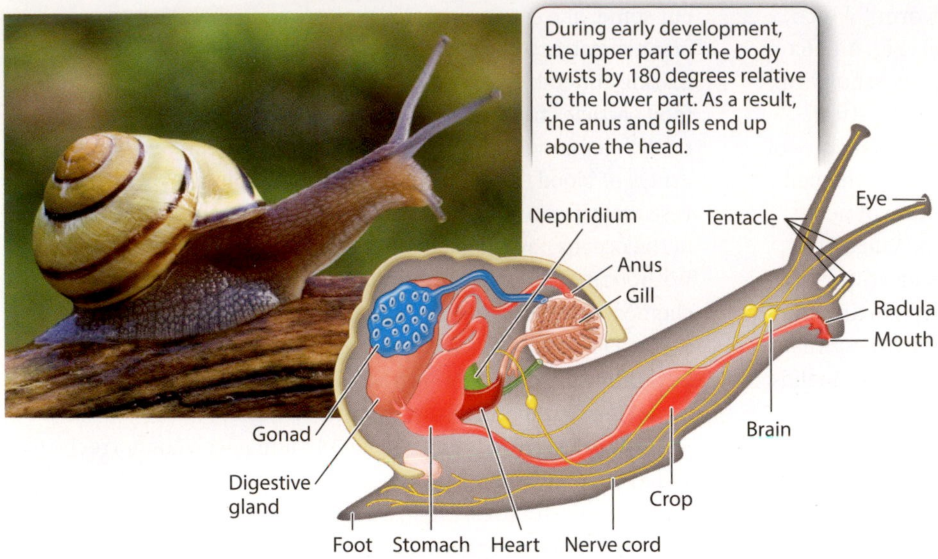

view as the head of an octopus. Mollusks also develop a distinctive larva called a trochophore that has a tuft of cilia at its top and additional cilia bands around its middle. Marine annelids develop from trochophores as well, and the presence of this unique form of larva in both phyla suggests a close evolutionary relationship between the two. This hypothesis is supported by molecular sequence comparisons. Nonetheless, mollusks and annelids share very few features of the adult body plan. More than 80,000 species of mollusks have been described, most of them gastropods.

Gastropods—snails and slugs—provide an illuminating example of the mollusk body plan (**Fig. 44.17**). Gastropods have a head with a well-developed mouth that contains a toothlike structure called a radula for feeding. The mouth connects to a gut cavity that extends to an anus. Featherlike gills facilitate gas exchange, and a muscular foot is used for locomotion. A neural ganglion in the head coordinates a nervous system that extends through a body that also contains a well-developed circulatory system, gonads, and nephridia. Gastropods are coelomates, but the body cavity is generally reduced to small pouches that surround the heart and other organs. The outer surface of the body consists of the mantle. In many gastropods the mantle tissues secrete external skeletons of calcium carbonate, which form shells.

Some gastropods eat algae, but many are predators. About half occur in the ocean, and the other half are freshwater and terrestrial species that primarily feed on plants. In land snails and slugs, the only terrestrial mollusks, the gills have been lost, and gas exchange occurs in an internal cavity that has been modified to function as a lung.

The second great class of mollusks is the **cephalopods**—about 700 species of squid, cuttlefish, octopus, and the chambered nautilus (**Fig. 44.18**). Cephalopods share a number of features with gastropods, including much of the internal anatomy, featherlike gills, and mantle. But they also have distinctive adaptations for their unique modes of life. Most obviously, cephalopods have muscular tentacles that capture prey and sense the environment. These leglike appendages are found on the head region rather than on the sides of the body as in other animals.

Cephalopods are superb swimmers, able to dart through the water by means of a jet propulsion system that forces water through mantle tissue that is fused to form a siphon. It is the combination of tentacles and rapid locomotion that makes cephalopods important predators in the oceans. Moreover, cephalopods have exceptional eyesight and exhibit the most complex behavior of any invertebrate animals: They are able to learn visual patterns and solve puzzles to gain food. The chambered nautilus secretes a coiled shell of calcium carbonate (Figs. 44.18a and 44.18b), but in other cephalopods, like squid and octopus, the shell is reduced or absent (Fig. 44.18c). In oceans of the Mesozoic Era (252–66 million years ago), however, shelled cephalopods called ammonites were among the most abundant and diverse predators.

A third major group of mollusks, the **bivalves,** includes the familiar clams, oysters, and mussels. Once again, these animals have anatomical features that point to their close relationship to snails and squids, but also have distinctive features that set them apart (**Fig. 44.19**). Notably, bivalves have (literally) lost their heads. As their name implies, bivalves have evolved a skeleton in which two hard shells called valves are connected by a flexible hinge.

Most bivalves obtain food by filtering particles from seawater. However, many bivalves live within marine sediments, burrowing into sand or mud with their muscular foot. How can bivalves filter seawater within the sediment? Once again, this group of mollusks has a modified mantle that allows a new function. Flaps of mantle tissue have fused to form a pair of siphons that extend upward from the bivalve's body to the surface of the seafloor above it. One siphon draws water containing food and oxygen into the body. The second siphon then returns water and waste materials to the environment. About 2000 species of bivalves have been described, mostly from the oceans, but also many from fresh water. Bivalves include the longest-lived animals in the world, the mahogany clams of the North Atlantic, for which the record is over 400 years of age. These bivalves would make excellent subjects for study of rates of genetic mutation (Chapter 22).

→ **Quick Check 3** From the shared traits of present-day snails, squids, and clams, what features do you think the common ancestor of all three had?

FIG. 44.18 **Cephalopods.** (a and b) *Nautilus,* (c) octopus, and squid (not shown) are ecologically important predators in the oceans. *Photo sources: a. Phillip Colla/Oceanlight; c. Michael Carey/SeaPics.com.*

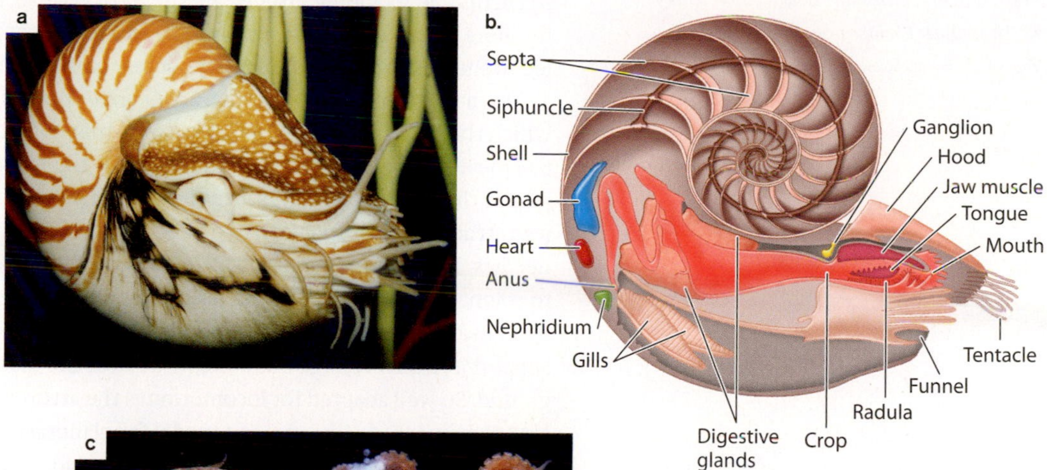

Ecdysozoans include nematodes, the most numerous animals, and arthropods, the most diverse.

The second major group of protostome animals is the Ecdysozoa (**Fig. 44.20**), which get their name from the process of ecdysis, or molting. Uniquely among animals, all ecdysozoans secrete a protein- or polysaccharide-rich cuticle that covers their bodies. This tough cuticle is like a suit of flexible, lightweight armor, protecting bodies from injury and physical challenges of the environment such as drying. This "armor" has a special property: The hard cuticle can be used to form appendages that function as tools, weapons, or even wings. Like a suit of armor, the cuticle is always a perfect fit for the wearer but does not stretch, and so it must be exchanged

FIG. 44.19 ***Mercenaria mercenaria,*** **a bivalve mollusk commonly harvested from coastal waters of eastern North America.** *Photo source: Glenn Price/Dreamstime.com.*

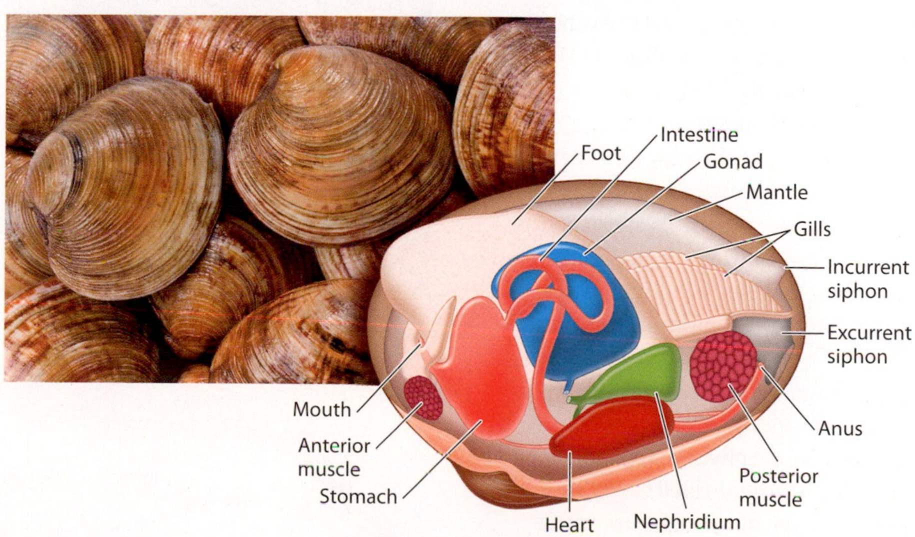

FIG. 44.20 A phylogenetic tree of ecdysozoan animals, showing all eight phyla in this group. Phyla discussed in the text are shown in red. Uncertainties are shown by nodes with more than two branches. *Photo sources: (top) London School of Hygiene & Tropical Medicine/Science Source; (bottom) Cindy Singleton/iStockphoto.*

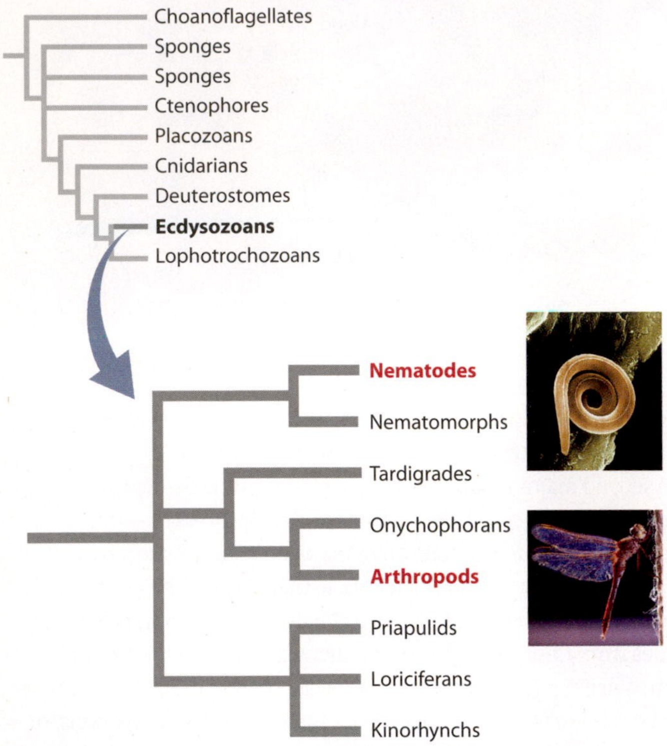

episodically during growth for a larger size to fit a larger, growing body. Ecdysozoans, then, are animals that molt their external cuticle during growth, producing a very soft, larger replacement underneath that soon hardens into a new protective covering.

The Ecdysozoa include eight phyla, but two dwarf the others in abundance and diversity. The most numerous of all animals are the small roundworms known as **nematodes** (**Fig. 44.21**). Most nematodes are tiny and so you've probably never noticed them, but they occur by the billions all around you and may well be inside your body as you read this. About 20,000 nematode species have been described, but estimates of their true diversity run as high as half a million. Most are microscopic, preying on microorganisms in soils and marine sediments. Their elongate cylindrical bodies have a mouth at one end, an intestine for digestion, and an anus in the rear. Other organs are simple: Nerve cords extend along the main axis from a ring in back of the mouth, rudimentary muscles line the body, and gonads expel sperm and eggs for reproduction. Nematodes are the main pseudocoelomate phylum, their digestive and reproductive organs surrounded by a fluid-filled cavity. As in other ecdysozoans, the body is covered by a tough cuticle that molts during growth. Parasitic nematodes are common and well known to doctors; hookworms in particular are major sources of illness in the tropics.

If nematodes are the most numerous animals, the **Arthropoda,** including insects, are the most diverse, accounting for more than half of all known animal species. What features have led to the remarkable evolutionary success of this branch?

Perhaps the most obvious feature of arthropods is the one for which they are named: jointed legs (*arthro* means "jointed," and *pod* means "foot" or "leg"). The jointed leg is the most versatile part of the arthropod body; it has been modified through evolution into structures that function as paddles, spears, stilts, pincers, needles, hammers, and more. The jointed leg allowed arthropods to colonize land by providing a multipoint, independent suspension—much like that of an all-terrain vehicle—able to support a relatively large body and move over the most uneven ground. So well adapted for locomotion is the arthropod leg that NASA engineers are using it as a model for planetary exploration vehicles. Researchers have conducted many studies to understand how this one type of limb could give rise to so many diverse structures with different functions (**Fig. 44.22**).

The arthropods' other defining characteristic is the material that forms their hard external skeleton: a strong, lightweight and nearly indestructible polysaccharide called **chitin.** Chitin, you may recall from Chapter 34, also forms the wall of fungal cells. Next to the cellulose that forms the wall of plant cells, chitin may be the most abundant biomolecule on Earth.

FIG. 44.21. A nematode. Nematodes are small but remarkably abundant animals. *Photo source: Gerd Guenther/Science Source.*

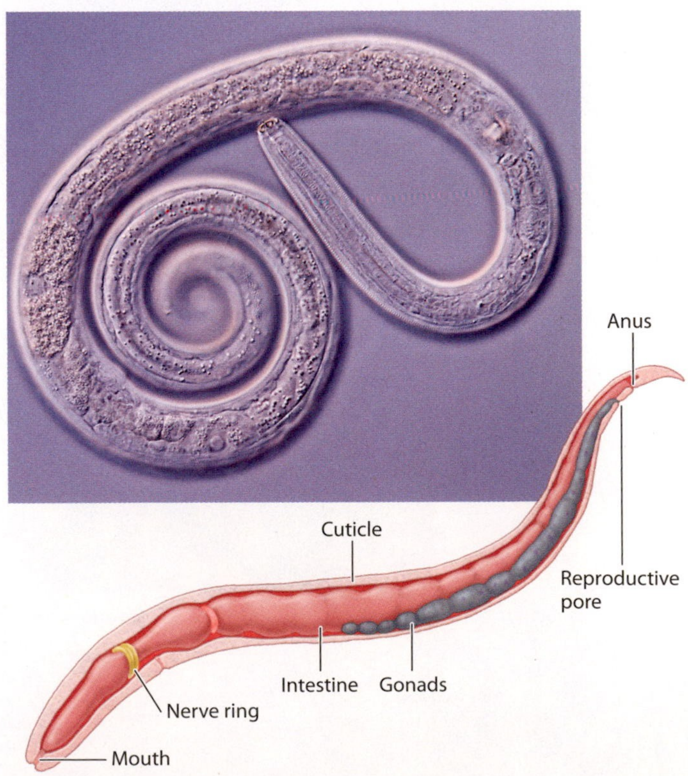

HOW DO WE KNOW?

FIG. 44.22

How did the diverse feeding appendages of arthropods arise?

BACKGROUND Arthropod species diversity is matched by diversity of form and function, challenging scientists to recognize and relate the equivalent parts of the very different front ends of, for example, spiders, crabs, and centipedes.

HYPOTHESIS Researchers hypothesized that the diverse front appendages of the different classes of arthropods develop from similar limb buds clearly visible in developing embryos, much as limb buds leading to bird wings, whale flippers, and human arms are present in vertebrate embryos.

EXPERIMENT The early stages of arthropod embryos have a definite head region and several body segments, each with a pair of appendages. *Hox* genes specify the identity of each body segment (Chapter 20). The expression patterns of different *Hox* genes can be visualized by means of labeled probes on whole embryos. In this way, the development of appendages on the same segment in different types of arthropod can be compared.

The dark bands show the expression of the *Hox* gene in the paired limb buds that can be seen in the embryo of the spider *Cupiennius salei.*

Photo source: Damen, W. G. M.; Tautz, D. (1998), "A Conserved Mode of Head Segmentation in Arthropods Revealed by the Expression Pattern of Hox Genes in a Spider," Proceedings of the National Academy of Sciences, USA 95 (18): 10665–10670. Copyright 1998 National Academy of Sciences, U.S.A.

RESULTS Study of the embryos revealed that, for example, the jawlike mandibles of insects and crustaceans form from appendages on the same segments of the developing body, whereas this segment on spiders and scorpions bears a pair of leglike feelers. Behind the mandibles are two pairs of feelers that are represented by legs in spiders and scorpions. On the other hand, the fangs of spiders and their counterparts in scorpions arise from a segment closer to the front end than the mandibles.

Feeding structures in the head regions of arthropods develop from paired appendages in the embryo.

Photo source: Piotr Naskreck.

Blue segments are in the head region, and the white ovals indicate the position of the mouth.

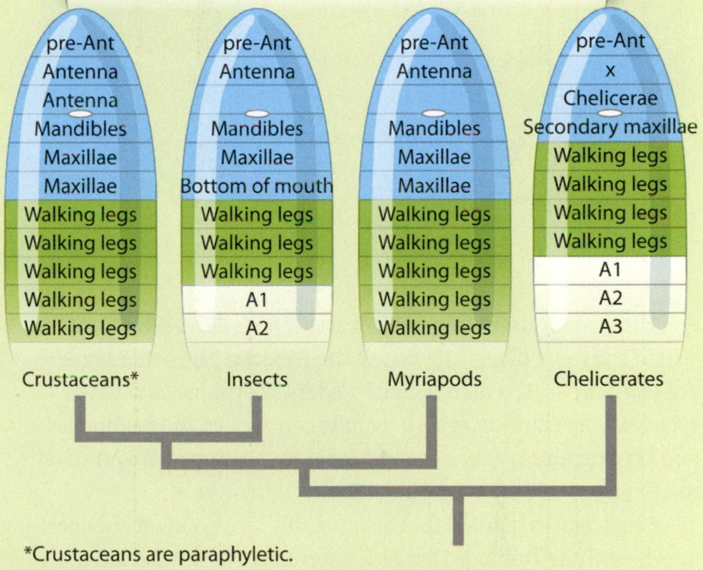

*Crustaceans are paraphyletic.

CONCLUSIONS The diverse feeding structures of arthropods arise from limb buds in the first segments of the developing embryo. The mandibles of crustaceans and insects arise from the same body segment and indicate the common ancestry of these animals, while spiders, scorpions, and their relatives are more distantly related. Evolution has modified simple legs into the remarkable diversity of forms, enabling structures as specialized as a spider's fangs and the grinding mandibles of crabs.

FOLLOW-UP WORK There has been recent controversy over the front appendages of sea spiders and whether or not they are equivalent to the great appendage on some Cambrian arthropod fossils.

SOURCE Damen, W. G. M., et al. 1998. "A Conserved Mode of Head Segmentation in Arthropod Revealed by the Expression Pattern of HOX Genes in a Spider." *Proceedings of the National Academy of Sciences USA* 95:10665–10670.

FIG. 44.23 **Arthropods.** This diverse phylum includes four main groups: chelicerates (spiders, scorpions, and ticks), myriapods (centipedes and millipedes), crustaceans (including lobsters, shrimp, and crabs), and insects.
Photo sources: (top to bottom) B. G. Thomson/Science Source; Tom McHugh/Science Source; Michael S. Nolan/SeaPics.com; Christian Hütter/imageBROKER/age fotostock.

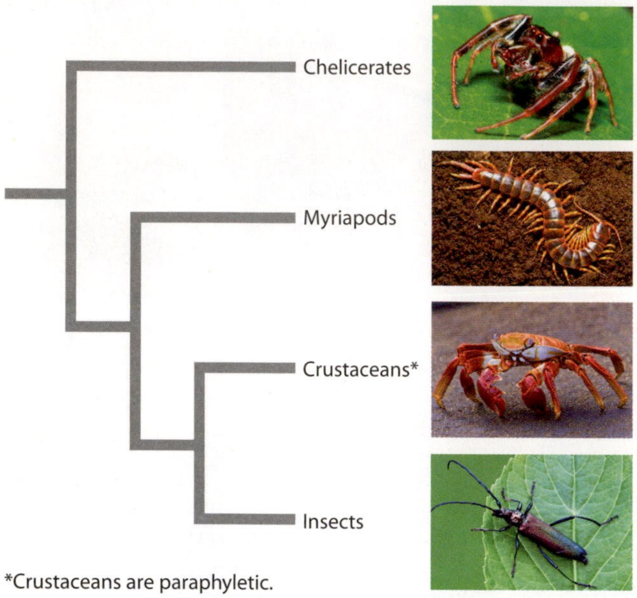

There are four main groups of arthropods (**Fig. 44.23**), of which the most diverse by far are the **insects.** The other three are **chelicerates,** which include spiders, scorpions, and their relatives; **myriapods,** which include centipedes and millipedes; and **crustaceans,** which include lobsters, shrimp, crabs, and their relatives.

Chelicerates, named for the pincerlike claws called chelicerae, are the only arthropods that lack antennae. The chelicerates are mostly carnivores, except for plant-eating mites. Scorpions and spiders are known for their venoms, used to subdue prey. In spiders, the venom is injected through chelicerae modified to form fanglike appendages, while in scorpions the venom is injected through the curved tail segment of their abdomen. Other familiar chelicerates include the harvestmen (often called daddy long-legs in the United States), ticks, and horseshoe crabs. Giant sea scorpions called eurypterids, up to 2 m long, were dominant predators in warm shallow seas 450–250 million years ago. The chelicerates were among the very first animals to invade land, about 400 million years ago, aided by their protective cuticle and supported by their arrays of jointed legs.

Myriapods are named for their many pairs of legs, which make them easy to recognize. Centipedes ("hundred legs") may have up to 300 pairs of legs and are fast-moving predators with front legs modified into venomous, fanglike organs. Millipedes ("thousand legs") may have several hundred pairs of legs (though never a thousand) and are slow-moving herbivores and detritivores (eaters of dead organisms). Millipedes often have glands lining their long bodies that produce a substance that contains cyanide, which they secrete to defend against predators.

Crustaceans also have distinctive appendages, notably their branched legs. They navigate their world with the aid of two pairs of highly sensory antennae, the larger of which has two branches. The unique crustacean larva, called a nauplius, swims with the aid of the smaller antenna and has a single eye that typically disappears in the adults. In the sea, these versatile arthropods fill many of the ecological roles that insects play on land, eating plants, other animals, or detritus. The tiny shrimplike plankton known as krill occur in vast numbers in polar seas, supporting ecosystems of fish, birds, and mammals such as whales. One Antarctic krill species is estimated to total over 500 million tons of biomass, more than the global total for humans. Lobsters and crabs are largely detritivores, scouring the ocean bottoms for dead and dying animals. Some crustaceans are even parasites, attaching themselves to the skin or inside the mouths of fish or sea mammals in order to feed. The strange creatures known as barnacles begin life resembling shrimp but soon settle on a surface for a life of filter-feeding. As the barnacles mature, their bodies become covered with thickened plates of armor and their legs become long and develop long brushes that sweep food into their mouths.

Insects were the first animals—and the only arthropods—to evolve wings, nearly 350 million years ago, and modern-day mayflies and dragonflies still bear structural similarities to these first fliers. The earliest-branching members of this group, the silverfish and rockhoppers, are referred to as primitively wingless because they evolved before the origin of wings. Some other insects, like fleas, have lost wings that were present in their common ancestor, presumably because their parasitic lifestyles are better served by jumping than flying. Like all arthropods, insects repeatedly shed their exoskeleton as they grow into adulthood. No insect can molt wings, however, and so only adults can fly.

Within the insects, the main evolutionary dividing line is between those that undergo **metamorphosis,** a major change in form from one developmental stage to another, and those such as grasshoppers and water bugs that do not. Insects that do not undergo metamorphosis look much like miniature adults when they hatch, and the main change in body form involves the appearance of wings. Some, like dragonflies, may also change the form of their legs or eyes as they leave the aquatic environment.

However, for those insects undergoing metamorphosis, the body changes from a wormlike larva specialized for feeding to a **pupa** (**Fig. 44.24**). During the pupa stage, the body tissues undergo a remarkable transformation from the relatively simple larva to a very different looking adult such as a fly, butterfly, wasp, or beetle, usually specialized for reproduction. Insects undergo a nearly complete dissolution of their muscles and other body parts, and so require the immobile, coffinlike pupa in which to transform their bodies from something as simple as a maggot, which lacks even an obvious head or tail, into an animal as sophisticated as a fly, able to walk on water, hover motionless in front of a flower, or

FIG. 44.24 Development in butterflies. Butterflies are a group of insects that undergoes metamorphosis. *Source: Ralph A. Clevenger/Corbis.*

fly at 60 mph. This evolutionary step has little parallel elsewhere in life and is one of the keys to insect diversification.

Unlike other terrestrial arthropods, all insects, whether or not they undergo metamorphosis, have highly specialized eggshells that can withstand desiccation while still allowing gas exchange. This shell allows insect eggs to survive in dry settings such as the surfaces of leaves, and so they can occupy habitats that are difficult for other terrestrial arthropods to access.

Insects have evolved one other critical adaptation for life on land. Aquatic arthropods obtain oxygen through gills, which don't work well in air because their large surface area dries out. Some terrestrial chelicerates have evolved simple lungs, called book lungs: modified appendages filled with hemolymph that have a large surface area for gas exchange, but are kept moist inside a protective pocket. Some crabs are able to make extended forays on land by keeping their gills moist underneath their carapace. In contrast, insects exchange gases through small pores in their exoskeletons called **spiracles** (Chapter 39). The spiracles connect to an internal system of tubes, the **tracheae,** that directs oxygen to and removes carbon dioxide from respiring tissues, much like air exchange ducts in a building.

Insects are extraordinarily diverse, with approximately 1 million described species making up 80% or more of all known animal species. Their success may be a function of the three adaptations that allow them to live in diverse habitats: dessication-resistant eggs, wings, and metamorphosis. The ability of insect eggs to resist drying out means that they can be placed on virtually any surface. Their wings allow insects to reach habitats and feed in ways other animals cannot. Metamorphosis allows them to have two bodies in their life cycles: a larval form specialized for feeding and an adult form specialized for mating and reproduction. These three adaptations enabled insects to feed directly on flowering plants. In turn, many insects pollinate plants, and so pollinating insects are in part responsible for flowering plant diversity (Chapter 33).

Deuterostomes include humans and other chordates, and also acorn worms and sea stars.

Having introduced diversity within protostome animals, we now turn our attention to the other great bilaterian group, the Deuterostomia. Deuterostome animals (**Fig. 44.25**) include three

FIG. 44.25 A phylogenetic tree of deuterostome animals, showing all phyla (and the major subdivisions of the chordates) in this group. Phyla discussed in the text are shown in red. *Photo sources: (top to bottom) Darlyne A. Murawski/Getty Images; Azure Computer & Photo Services/Animals Animals; © www.konig-photo.com; Reinhard Dirscherl/WaterFrame/Getty Images/Dwight Kuhn.*

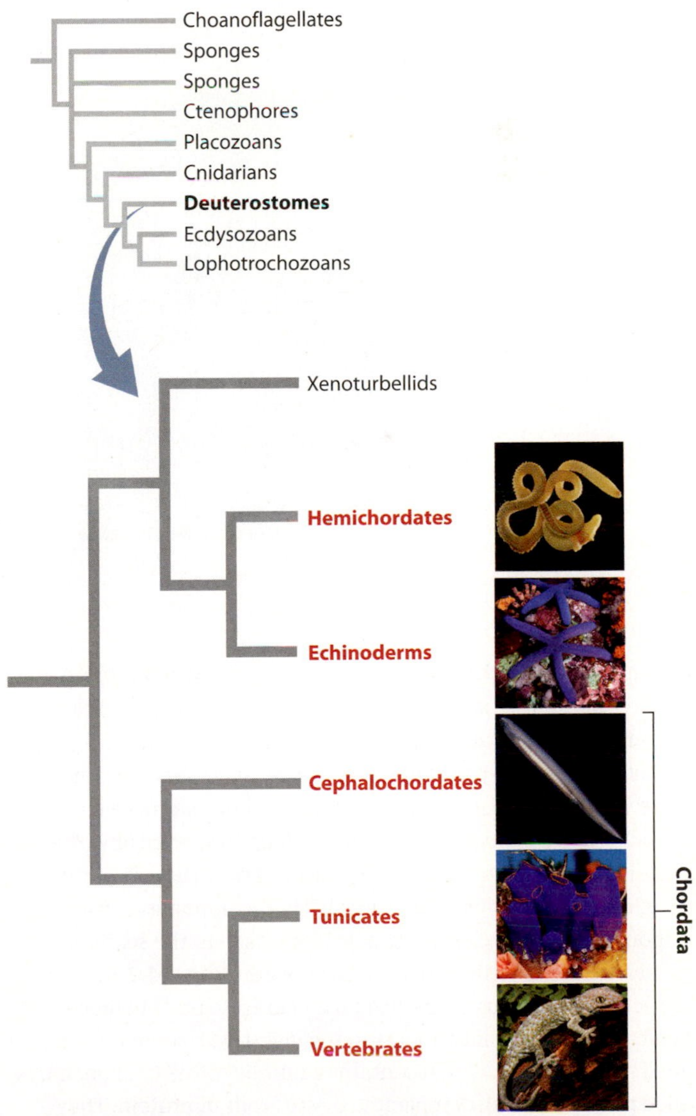

FIG. 44.26 **Hemichordates.** Hemichordates include (a) acorn worms, with pharyngeal slits, and (b) pterobranchs, with filter-feeding tentacles. *Photo sources: a. Darlyne A. Murawski/Getty Images; b. D. P. Wilson/FLPA/Science Source.*

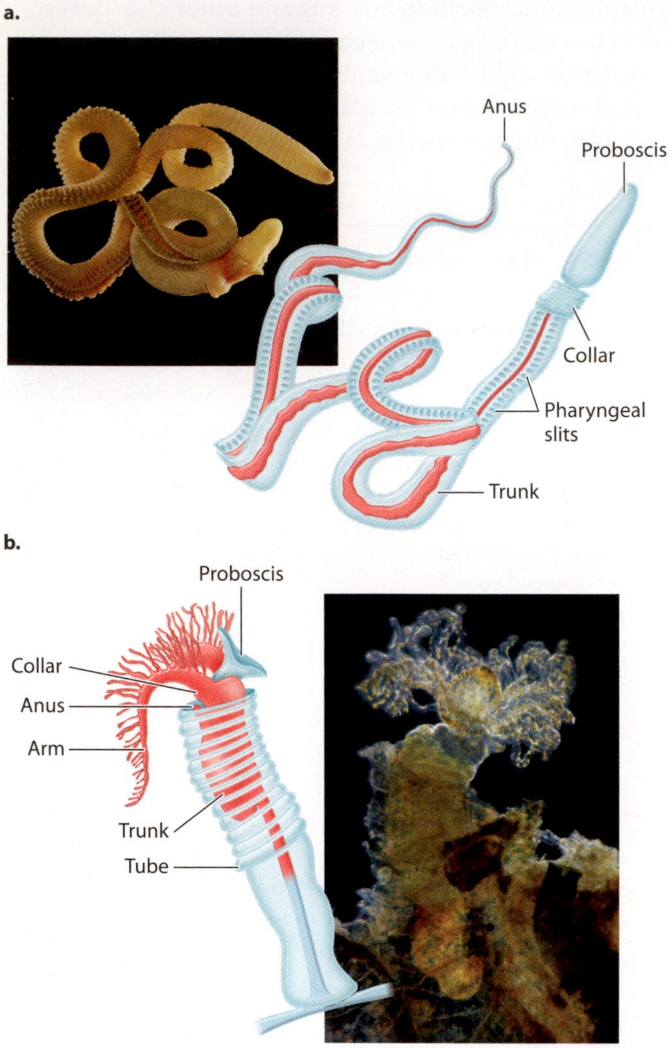

major phyla: **Hemichordata** (acorn worms), **Echinodermata** (sea urchins and sea stars), and **Chordata** (vertebrates and closely related invertebrate animals such as sea squirts).

Let's begin with the deuterostomes most distantly related to the vertebrates, the hemichordates and echinoderms. Hemichordates include acorn worms, about 75 species of wormlike animals that move through seafloor sediments in search of food particles. Pterobranchs, which are also hemichordates, consist of about a dozen species of animals that attach to the seafloor and use tentacles to filter food from seawater (**Fig. 44.26**). Hemichordates all have a mouth on an elongate protuberance called a proboscis that connects to the digestive tract by a tube called the **pharynx,** which contains a number of vertical openings called **pharyngeal slits** separated by stiff rods of protein. They also have a **dorsal nerve cord.** Additional characters that relate hemichordates to other deuterostomes are apparent in embryological and larval development. Nonetheless, adult hemichordates have body plans quite distinct from those of chordates and echinoderms.

Another major group of deuterostomes is the echinoderms, among the most distinctive of all animal phyla. Echinoderms include sea stars and sea urchins (including sand dollars), as well as brittle stars, sea cucumbers, and sea lilies—all of which share a unique fivefold symmetry on top of their basic bilaterian organization (**Fig. 44.27**). Echinoderms form distinctive skeletons made of interlocking plates of porous calcite, a form of calcium carbonate. This is why they generally feel hard and stony, and why they are well protected from most predators. The third unique feature of echinoderms is the **water vascular system,** a series of fluid-filled canals that permits bulk flow of oxygen and nutrients. **Tube feet,** small projections of the water vascular system that extend outward from the body surface, allow echinoderms to crawl, sense their environment, capture food, and breathe. Sea stars can use the tube feet on their powerful arms to pull open the valves of clams. About 7000 echinoderm species reside in present-day oceans, but the fossil record suggests that this phylum was more diverse in the past.

Zoologists have long known that echinoderm and hemichordate larvae share many features. Molecular sequence data now confirm that echinoderms are the closest relatives of hemichordates, with chordates a sister group to the group containing echinoderms and hemichordates (see Fig. 44.25).

Chordates include vertebrates, cephalochordates, and tunicates.

The other great branch of the deuterostome tree is the chordates, the phylum that includes vertebrate animals. Within the phylum Chordata, there are three subphyla: the **cephalochordates,** the **tunicates,** and the **vertebrates** (also called **craniates**). Like hemichordates, chordates all have a pharynx with pharyngeal slits (illustrated by the invertebrate chordate amphioxus, shown in **Fig. 44.28**). In fish, these pharyngeal slits form the gills, but in terrestrial animals like humans, these slits can be seen only in developing embryos. The **notochord,** a unique feature of chordates, is a stiff rod of collagen and other proteins that runs along the back, providing resistance for the muscles along each side in some chordates. In vertebrates, the notochord is apparent only during early embryogenesis and is replaced functionally by a **vertebral column** later in development. Also forming during early development is the **neural tube,** a cylinder of embryological tissue that develops into a dorsal nerve cord. Body musculature is organized into a series of segments called **myotomes,** most visible in fish as the flakes of muscle that become separated when the fish is cooked. Chordates have a tail (reduced but still internally present in humans and other apes) that extends posterior to the anus, and some chordates have muscularized appendages that include the fins of fish and legs of terrestrial vertebrates.

FIG. 44.27 Echinoderms. (a) Sea stars are familiar sights on the shallow seafloor; (b) sea urchins are motile echinoderms found on most seafloors. *Photo sources: a. Azure Computer & Photo Services/Animals Animals; b. V & W/Ricardo Fernandez/SeaPics.com.*

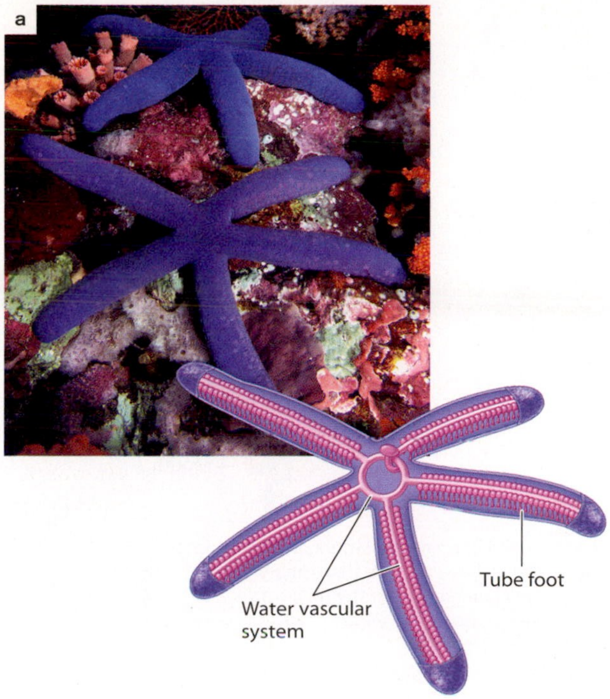

Amphioxus (Fig. 44.28) belongs to the cephalochordates. These animals share key features of body organization with vertebrates but lack a well-developed brain and eyes, have no lateral appendages, and do not have a mineralized skeleton. Nonetheless, their many similarities to vertebrates suggest they are closely related, and molecular sequence comparisons confirm this hypothesis.

FIG. 44.28 Amphioxus, a cephalochordate. Amphioxus has many features in common with vertebrates, but it lacks a well-differentiated head and has no cranium or vertebrae. *Photo source: © www.konig-photo.com.*

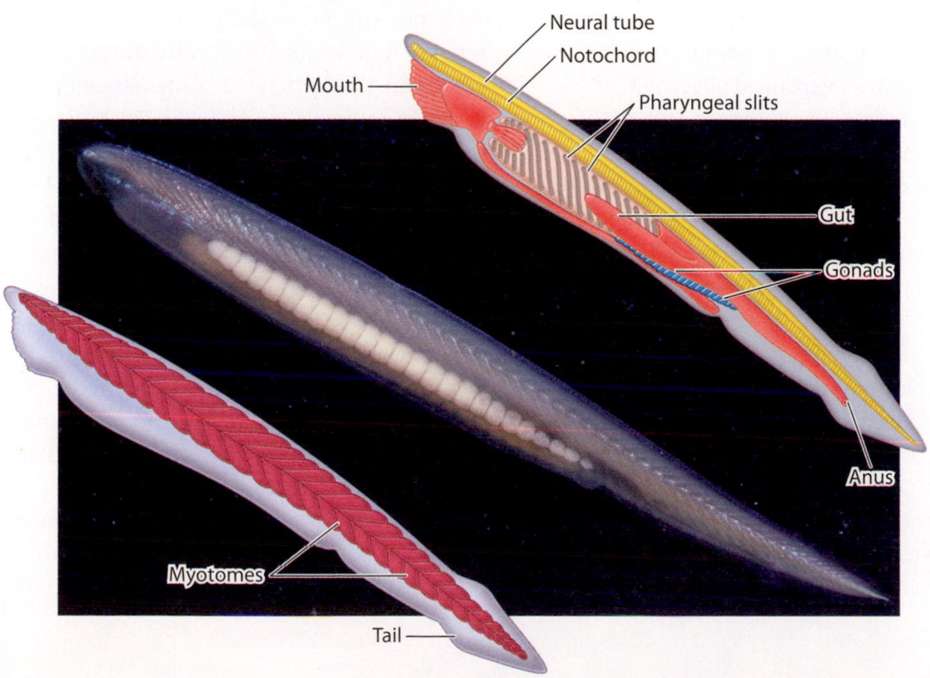

FIG. 44.29 **Tunicates.** (a and b) Adult form; (c) larval form. *Photo source: a. Reinhard Dirscherl/WaterFrame/Getty Images.*

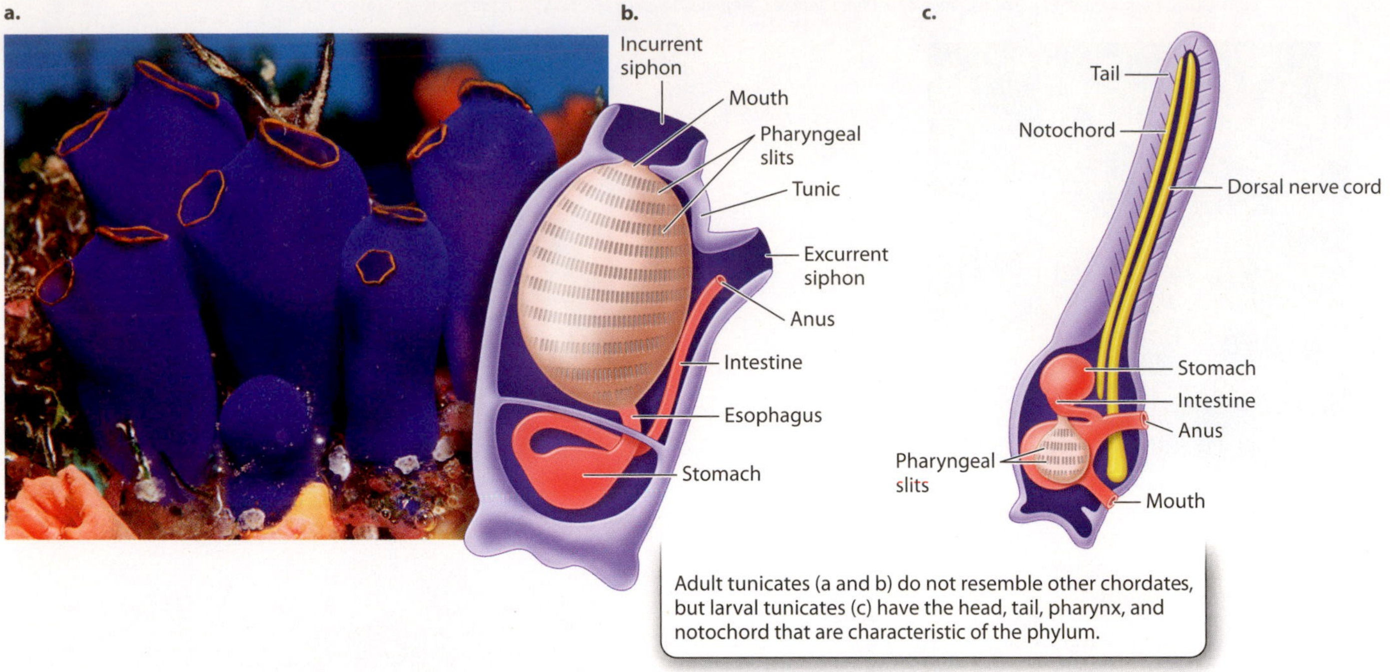

The tunicates also share key features of the chordate body plan during early development, but they have a unique adult form (**Fig. 44.29**). Tunicates include about 3000 species of filter-feeding marine animals, such as sea squirts anchored to the seafloor and salps that float in the sea. The adult tunicate body is a basketlike structure that is highly modified for filter-feeding, functionally not unlike the shapes of bivalves and barnacles, which are also adapted to this feeding mode.

A tunicate might be mistaken for a sponge, but its complex internal structure shows that it lies far from sponges on the animal tree. The tunicate's body wall, or tunic, has a siphonlike mouth at one end that draws water through an expanded pharynx that captures food particles and exchanges gases. Water and wastes are expelled through an anal siphon. In the adult tunicate, the only obvious similarities to other chordates are the pharynx and its pharyngeal slits. Larval tunicates, however, resemble other chordates like fish or tadpoles with a more typical chordate body plan, including a notochord, neural tube, and a long tail with muscles arranged in myotomes.

For many years, zoologists considered cephalochordates to be the closest relatives of vertebrate animals, but molecular data now suggest that, in fact, the tunicates are our closest invertebrate relatives. This means that the common ancestor of tunicates, amphioxus, and vertebrate animals was a small animal much like amphioxus in organization. After the divergence of the three major groups of chordates, however, the tunicates' adaptation for filter-feeding led to the unique anatomy they have today.

→ **Quick Check 4** You do not have a notochord, a tail, or pharyngeal slits. How do we know that you are a chordate?

44.4 VERTEBRATE DIVERSITY

In many ways, the last stop on our tour of animal diversity is the most familiar because it includes our own species, *Homo sapiens.* The animals known as vertebrates are named for their jointed skeleton, which runs along the main axis of the body, forming a series of hard segments collectively termed **vertebrae** (singular, **vertebra**). In addition to features shared with other chordates, vertebrate animals are distinguished by a cranium that protects a well-developed brain, a pair of eyes, a distinctive mouth for food capture and ingestion, and an internal skeleton commonly mineralized by calcium phosphate (**Fig. 44.30**). The vertebrates also have a coelom in which the organs are suspended and a closed circulatory system. Like the other chordates, vertebrates have pharyngeal slits in at least the early embryonic stages.

Many of the features that separate vertebrates from invertebrate chordates can be found in the head, including a bony cranium that protects the brain, joined in most vertebrates by the mandible (or jawbone) to form a skull. This protective helmet for the delicate neuronal tissues permits the development of a larger brain than would otherwise be possible. The mandible permits eating foods such as other living animals, themselves often protected by hard shells, and plants containing tough cellulose

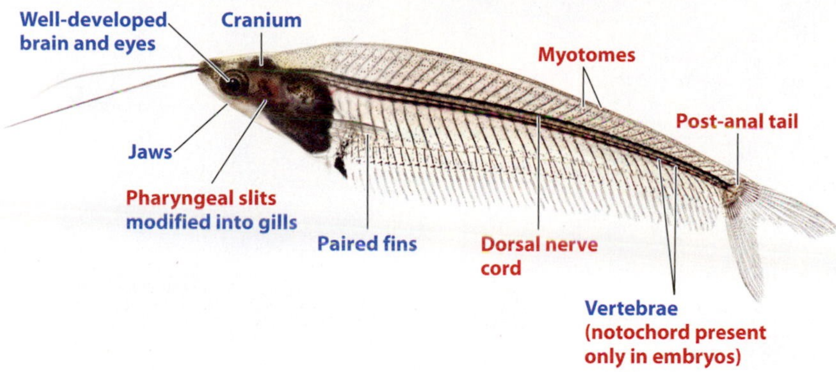

FIG. 44.30 **Vertebrate characters.** Anatomical features of vertebrates that are characteristic of chordates are shown in red, and those that are unique to vertebrates are shown in blue. *Source: Stockbroker xtra/age fotostock.*

and lignin. Because the bottom scavenger hagfish have a cranium but not vertebrae (or a mandible), some biologists use the name "Craniata" for this group, rather than the traditional "Vertebrata."

Fish are the earliest-branching and most diverse vertebrate animals.

We sometimes think of fish as one group, but the aquatic animals we commonly call "fish" include four distinct groups of aquatic vertebrates, not all of them monophyletic (**Fig. 44.31**).

The earliest-branching craniates are the **hagfish** and **lampreys** (**Fig. 44.32**). These animals have a cranium built of cartilage but lack jaws. Lampreys also have a vertebral column built of cartilage. Hagfish lack vertebrae, though current research suggests that this trait existed in a common ancestor of hagfish and lampreys but was lost in hagfish. These eel-like organisms feed on soft foods without the aid of jaws because they diverged before jaws evolved from pharyngeal slits. Hagfish feed on marine worms and on dead and dying sea animals, while lampreys live parasitically, sucking body fluids from fish prey. Both hagfish and

FIG. 44.31 **A phylogeny of the vertebrates.** This phylogeny is favored by recent molecular data, but, like all phylogenies, is a hypothesis. The placements of hagfish relative to other vertebrates and turtles relative to other sauropsids have provoked debate for many years.

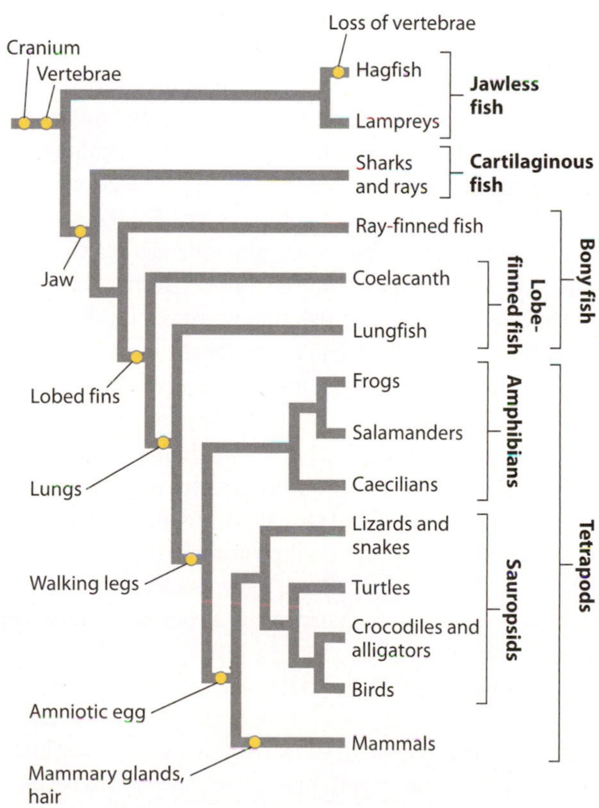

FIG. 44.32 **Hagfish and lampreys.** Commonly grouped together as jawless fish, (a) hagfish and (b) lampreys are the earliest- branching vertebrates alive today. *Sources: a. Tom McHugh/Science Source; b. F Hecker/age footstock.*

FIG. 44.33 Cartilaginous fish. The skeletons and jaws of sharks are made of cartilage. *Source: Michael Patrick O'Neill/Alamy.*

lampreys have a series of gill slits through which water enters to bring oxygen to the gills.

Biologists have long debated the phylogenetic relationships among hagfish, lampreys, and other vertebrate animals. The presence of a vertebral column in lampreys convinced many biologists that lampreys are the sister group to other vertebrates, and hagfish the sister group to all vertebrates, including lampreys, but several lines of molecular data now favor the view shown in Fig. 44.31 that hagfish and lampreys together form the sister group to all other vertebrates. About 65 species of lampreys and hagfish live in fresh water and coastal oceans. Fossils record the existence of many extinct species of jawless fishes. When we include fossils, the jawless fishes as a whole are paraphyletic because one of their groups gave rise to fish with jaws.

Chondrichthyes, or cartilaginous fish, form the next deepest branch on the vertebrate tree. This monophyletic group includes about 800 species of sharks, rays, and chimaeras, all of which have jaws and a skeleton made of cartilage (**Fig. 44.33**). These fish deposit calcium phosphate minerals only in their teeth and in small toothlike structures called denticles embedded in the skin. The best-known cartilaginous fish are the sharks, many of which are predators that put those mineralized teeth to good use. On the other hand, the group also includes Whale Sharks, filter-feeding animals that, at a length up to 40 feet (12 m), are gentle giants of the sea. Rays are closely related to sharks but have a flattened body adapted for life on the seafloor, where they feed on clams and crabs. The cartilaginous fish can retain levels of urea in their tissues that would poison other vertebrates, a physiological adaptation that helps them to maintain their salt balance without well-developed kidneys (Chapter 41). Nearly all Chondrichthyes occur in the oceans, but Bull Sharks (*Carcharhinus leucas*) commonly swim into rivers—they've been recorded well up the Mississippi—and there are several freshwater species of rays.

Moving up one more branch in the vertebrate tree, we encounter the **Osteichthyes,** or bony fish (see Fig. 44.31). Bony fish have a cranium, jaws, and bones mineralized by calcium phosphate. Numbering about 20,000–25,000 freshwater and seawater species, these are the fish that we most commonly encounter (**Fig. 44.34**). Bony fish are by far the most diverse group of vertebrates, possessing several unique features that facilitate their occupation of diverse niches. First, they have a system of movable elements in their jaws that allows them to specialize and diversify their feeding on many different types of food. Second, they possess a unique gas-filled sac called a swim bladder that permits exquisite control over their position in the water column through changes in buoyancy. Third, bony fish have kidneys that allow them to regulate water balance and so occupy waters over a wide range of salinity.

The swim bladder plays a special role in vertebrate evolution. Early vertebrates evolved a gut sac that enabled them to gulp air to obtain additional oxygen. In some fish, this sac evolved into the swim bladder, but in one group the sac was modified to become

FIG. 44.34 **Bony fish.** The char, well known to those who fish in fresh water, illustrates the principal anatomical features of bony fish: a cranium, jaws, and mineralized bones. *Source: JOEL SARTORE/National Geographic Creative.*

a lung. Charles Darwin was among the first to recognize that the swim bladder and the vertebrate lung are homologous organs.

The common ancestor of tetrapods had four limbs.

Most animal phyla occur in the oceans, and many are exclusively marine. A subset of animal phyla has successfully colonized fresh water, and a smaller subset has radiated onto the land. Eleven groups on the animal phylogenetic tree contain both aquatic *and* terrestrial species: nematodes, flatworms, annelids (earthworms and leeches), snails, tardigrades (microscopic animals with eight legs, sometimes called water bears), onychophorans (velvet worms), four groups of arthropods (millipedes and centipedes, scorpions and spiders, land crabs, and insects), and vertebrates. In each of these groups, the aquatic species occupy the earliest branches, and the terrestrial species occupy later branches. No two of these groups share a last common ancestor that lived on land; thus, they all made the transition independently. Earlier, we outlined the extraordinary diversity of terrestrial arthropods. Here, we focus on vertebrates as the other great animal colonists of the land.

Most bony fish have fins supported by a raylike array of thin bones. About half a dozen closely related species, however, have pectoral and pelvic fins that extend from a muscularized bony stalk. These fish are called **lobe-finned fish** and include the **coelacanth** and **lungfish** (**Fig. 44.35**). Although these animals resemble other fish, the coelacanth and lungfish are the nearest relatives of tetrapods, four-legged animals. Lungfish can survive periods when their watery habitat dries by burying themselves in moist mud and breathing air. Coelacanths first appeared over 400 million years ago and were thought to have been extinct for over 80 million years until a specimen was discovered in a fisherman's catch off South Africa in 1938. Hence, they are sometimes referred to as living fossils. There are two living species of coelacanths and three species of lungfish, one each in South America, Africa, and Australia. Because the fleshy-finned fish are the closest relatives of tetrapods, there is no monophyletic group that corresponds to our popular notion

FIG. 44.35 **Lobe-finned fish.** The fish most closely related to tetrapods are the lobe-finned fish: (a) the coelacanth and (b) the lungfish. These fish have paired fins with an anatomical structure similar to that of tetrapod legs. *Sources: a. Gerard Lacz/Animals Animals–Earth Scenes; b. D. R. Schrichte/SeaPics.com.*

FIG. 44.36 The amphibian life cycle.

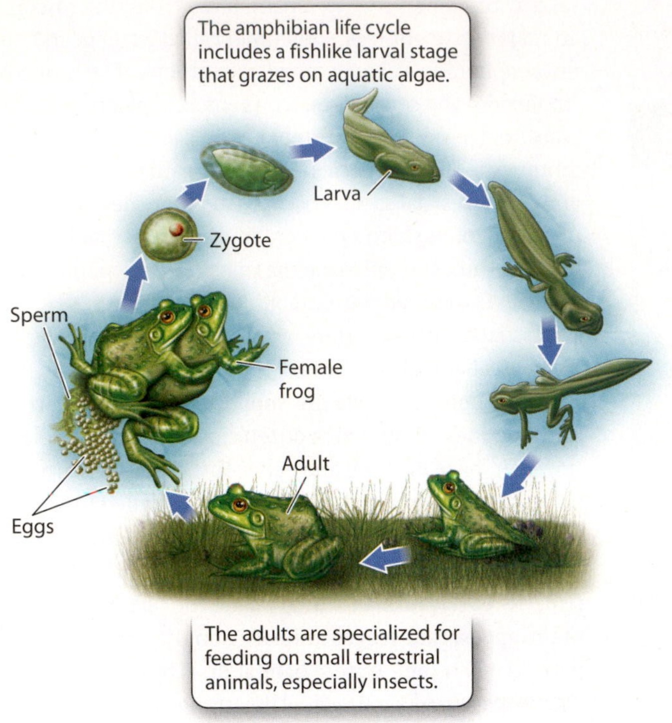

of "fish." To be monophyletic, such a group must include the tetrapods.

Fossils document the anatomical transition in vertebrate animals as they moved from water to land, showing changes in the form of the limbs, rib cage, and skull (see Fig. 23.20). Living tetrapods include amphibians, such as frogs and salamanders; lizards, turtles, crocodilians, and birds; and mammals. The last common ancestor of all these animals had four limbs, hence the name **Tetrapoda** ("four legs"). Some, however, like snakes and a few amphibians and lizards, have lost their legs in the course of evolution.

There are over 4500 species of **Amphibia** ("double life"), ranging in size from tiny frogs a few millimeters in length to the Chinese Giant Salamander, which is over 1 m long. Their name reflects their distinctive life cycle (**Fig. 44.36**). Most species have an aquatic larval form with gills that permit breathing under water and a terrestrial adult form that usually has lungs for breathing air. Amphibians must reproduce in the water or moist habitats, and so are not completely terrestrial. Whereas many amphibian larvae graze on algae, the adults are predators and often have a muscular tongue for capturing prey. Many are protected by toxins they secrete from glands in their skin, sometimes advertised by their brilliant color patterns. Most amphibian adults require moist skin for breathing and consequently have small lungs, with toads and red efts being notable exceptions. This is why frogs in particular have fallen victim to a fungus that affects the skin's ability to breathe (Chapter 49).

Amniotes evolved terrestrial eggs.

Like insects, some vertebrates evolved an egg adapted to tolerate dry conditions that accompany life on land. The amniotic egg has a desiccation-resistant shell and four membranes that permit gas exchange and management of waste products produced by the embryo (Chapter 42). These eggs must be fertilized internally before the eggshell is produced by the female because sperm cannot penetrate the shell. Amniotic eggs permit long development times, with the embryo's nutrition supported by a large yolk or a placenta. They keep wastes separated from the embryo so they do not poison it, giving the embryo time to build more complex bodies than otherwise would be possible.

The **amniotic egg** can exchange gases while retaining water, and so permits the group of vertebrates known as **amniotes** to live in dry terrestrial habitats that amphibian eggs cannot tolerate. Amniotes include lizards, snakes, turtles, and crocodilians (about 6000 species of scaly animals commonly referred to as reptiles), as well as birds and mammals (**Fig. 44.37**). Most mammals, many lizards, and some snakes have evolved live birth rather than laying eggs. Nonetheless, they retain the specialized membranes characteristic of all amniotes. Instead of being wrapped around the embryo inside a tough egg shell, these membranes surround and protect the embryo inside the womb.

As shown in Fig. 44.31, amniotes form a monophyletic group with two major branches. One are the mammals, and the other contains turtles, birds, and the amniotes traditionally grouped together as reptiles. Like fish, however, reptiles are not monophyletic —birds are descended from reptiles, although you have to look closely to see the similarities between the two groups. Birds lack teeth, most of their scales have been modified into feathers (except on the legs and toes), and they are generally adapted for flight. Adaptations for flight include hollow bones that are light but remain strong, and a method of breathing that extracts much oxygen from each breath, permitting high-performance flight (Chapter 39).

One consequence of this streamlined, lightweight body is a lack of the heavy jaws that many other tetrapods use to grind up plant leaves or other low-nutrition foods. Instead, birds have their grinding organ, the gizzard (also found in other reptiles), inside their bodies near their center of gravity, where it interferes less with flight. Only one kind of bird, the weak-flying Amazonian Hoatzin, can actually feed on leaves. As discussed in Chapter 23, fossils show that birds diverged from dinosaur ancestors, and molecular sequence comparisons indicate that crocodiles are the closest living relatives of birds.

→ **Quick Check 5** Why do bird feet have scales like those found on snakes and lizards?

All **mammals** are covered with hair and feed their young milk from the mammary glands for which the class Mammalia is named. Like many other groups of animals and plants, the

FIG. 44.37 **Amniotes.** (a) Turtles, (b) lizards, (c) mammals, and (d) birds. *Sources: a. Gary Meszaros/Science Source.; b. François Dorothé/Getty Images; c. Sarah Peters/Getty Images; d. NASA/Jim Grossmann.*

early-branching mammals show intermediate stages in the evolution of the body plans that dominate on Earth today. The first fossils that clearly show hair appeared about 210 million years ago, during the Triassic Period. Like feathers in birds, hair permits retention of body heat. The earliest-branching living mammals, the monotremes, lay eggs, like birds or lizards (**Fig. 44.38**), but their hatched young drink milk secreted from pores in the skin of the mother's belly. Among these are the Australian platypus and the echidna, or spiny anteater.

The first mammals that gave birth to live young appeared in the Jurassic Period, about 160 million years ago. These animals gave rise to the two major living groups, marsupial and placental mammals. **Marsupials** include kangaroos, koalas, and related groups native to Australia, as well as the opossums found in the Americas. Their young are born at an early stage of development, and the tiny babies must crawl to a pouch where mammary glands equipped with nipples provide them with milk. Only later do fully formed juveniles emerge from the pouch and begin life on the ground.

The **placental mammals** are named for a temporary organ called the **placenta** that develops in the uterus along with the embryo, providing nutrition that enables the offspring to be larger and more quickly independent when born (Chapter 42). Most living mammals fall into this group. Placental mammals include the carnivores, such as lions and weasels; the primates, including monkeys, apes, and humans; and the hooved mammals, which include cattle, pigs, deer, and—perhaps surprisingly—their marine relatives, the whales.

All together, the familiar mammals mentioned so far number about 1000 species. The most diverse mammals, however, belong

FIG. 44.38 A phylogenetic tree of mammals. There are three major groups of mammals: monotremes, marsupials, and placental mammals. *Photo sources: (top to bottom) Dave Watts/NHPA/Photoshot; C. Wermter/age footstock; David Macias/Science Source.*

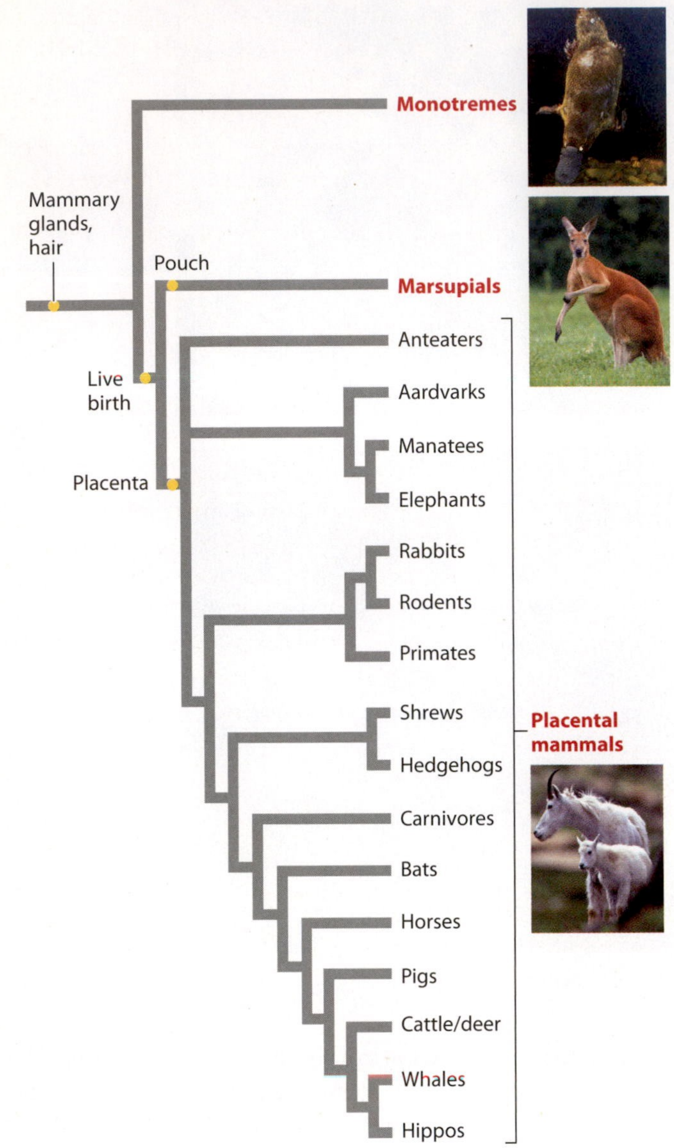

to two groups of mostly small animals, the bats and the rodents. Bats, which evolved flight and a form of sonar called echolocation to find food at night, include more than 1000 species. The largely plant-feeding rodents are as diverse as the large mammals and bats combined, numbering more than 2000 species. Beavers, gophers, rats, mice, hamsters, and squirrels are all rodents. Their evolutionary success is often attributed to innovations in tooth development. Their teeth are unusual in that they grow continually throughout life. This constant growth permits constant wear, and so allows these animals to gnaw through hard protective coverings to obtain nutrition from a remarkably wide variety of otherwise hard-to-access sources like the contents of seeds.

44.5 THE EVOLUTIONARY HISTORY OF ANIMALS

Because many animals form mineralized skeletons, sedimentary rocks deposited throughout their evolutionary history over the past 541 million years contain a rich fossil record of shells and bones. Animals also leave a sedimentary calling card in their tracks, trails, and burrows, again widespread in sedimentary rocks formed after 541 million years ago. Not all animal phyla are well represented in the fossil record—fossil annelids, for example, are rare and flatworms are essentially unknown as fossils because they lack mineralized body parts (Chapter 23). In contrast, the fossilized shells and bones of bivalve mollusks, brachiopods, echinoderms, and mammals preserve an excellent record of evolutionary history within these groups.

Phylogenies make predictions about the evolutionary history of animals, and fossils provide more than a confirmation of comparative biology. The remains of ancient animals record when innovations in morphology and function first appeared, preserving the remains of now extinct species that resemble the ancestors of modern groups. They often show intermediate combinations of traits not seen in modern animals and chronicle the relationship between Earth and life through time. And they underscore the evolutionary importance of extinctions, including a small number of events that removed a majority of existing species, paving the way for renewed diversification among survivors. If we step back and look at the big picture, what do fossils tell us about the evolutionary history of the animal kingdom?

Fossils and phylogeny show that animal forms were initially simple but rapidly evolved complexity.

Phylogeny suggests that animals are relative latecomers in evolutionary history, and the fossil record confirms this hypothesis. Life originated more than 3.5 billion years ago, but, as discussed in previous chapters, microorganisms were the only members of ecosystems for most of our planet's history. As noted in Chapter 28, macroscopic fossils of organisms thought to be animals first appear in rocks deposited only 579 million years ago. Called Ediacaran fossils after the Ediacara Hills of South Australia where they were discovered, these fossils have simple shapes that are not easily classified among living animal groups (**Fig. 44.39;** see also Fig. 28.13).

Phylogeny suggests that we should look for sponges, cnidarians, and other diploblastic animals among the oldest animal fossils, but sponges, at least, are rare among Ediacaran

FIG. 44.39 ***Dickinsonia,* an Ediacaran fossil.** The structural simplicity of *Dickinsonia* and similar fossils suggests that they record early branches of the animal tree. *Source: O. Louis Mazzatenta/National Geographic/Getty Images.*

fossils. This is possibly as much because of preservation (or the lack of it) as evolution, because sponges that do not make mineralized spicules have left few fossils. Instead, a majority of Ediacaran fossils show simple, fluid-filled tubes, without identifiable mouths or other organs. Many may have formed colonies, gaining complexity through colonial growth and differentiation, as some living cnidarians do. These early animals had epithelia and are thought to have obtained food by taking in dissolved organic matter or phagocytosing small particles while exchanging gases by diffusion, somewhat like living placozoans. Most Ediacaran fossils probably form an early branch or branches on the animal tree, falling among the limbs whose survivors include cnidarians, ctenophores, and placozoans. A few even resemble *Dendrogramma,* the newly discovered diploblastic animal introduced earlier.

Why did the first animal radiation occur so late in the evolutionary day? Scientists continue to debate this question, but part of the answer appears to lie in Earth's environmental history. Geochemical data suggest that only during the Ediacaran Period did the atmosphere and oceans come to contain oxygen in quantities sufficient to support the metabolism of large, active animals.

The animal body plans we see today emerged during the Cambrian Period.

Ediacaran fossils differ markedly from the shapes of living animals, but in the next interval of geologic history, the Cambrian Period (541–485 million years ago), we begin to see the remains of animals with familiar body plans (**Fig. 44.40**). Cambrian fossils commonly include skeletons made of silica, calcium carbonate, and calcium phosphate minerals, and these record the presence of arthropods, echinoderms, mollusks, brachiopods, and other bilaterian animals in the oceans. Spicules made by sponges are common in Cambrian rocks as well. The rocks also preserve complex tracks and burrows made by organisms with hydrostatic skeletons, muscular feet, and the jointed legs of arthropods. And in a few places, notably at Chengjiang in China and the Burgess Shale in Canada, unusual environmental conditions have preserved a treasure trove of animals that did not form mineralized skeletons (Chapter 23).

These exceptional windows on early animal evolution show that, during the first 40 million years of the Cambrian Period, the body plans characteristic of most bilaterian phyla took shape in a

FIG. 44.40 **Fossils of the Cambrian Period.** Cambrian rocks preserve early representatives of present-day phyla, including (a) early arthropods called trilobites, (b) annelid worms, (c) early deuterostomes, and (d) primitive vertebrates. *Sources: a. Sinclair Stammers/Science Source; b. Smithsonian Institution–National Museum of Natural History. Image: Jean-Bernard Caron; c. Bengtson, S. 2000, Teasing Fossils out of Shales with Cameras and Computers Palaeontologia Electronica, vol. 3, issue 1, art. 4:14 pp., 7.7 MBhttp://palaeo-electronica.org/2000_1/fossils/issue1_00.htm. Copyright: Palaeontological Association, 15 April 2000; d. Reprinted from Palaeoworld, 20, Jun-Yuan Chen, The origins and key innovations of vertebrates and arthropods. Fig. 5d. Copyright 2011 with permission from Elsevier.*

transition sometimes called the **Cambrian explosion.** Sponges and cnidarians radiated as well, producing through time the biomass- and diversity-enhancing habitats of reefs and imparting an ecological structure to life in the sea broadly similar to what we see today.

Scientists sometimes argue about whether the name "Cambrian explosion" is apt. The fossil record makes it clear that bilaterian body plans did not suddenly appear fully formed, so the event was not truly "explosive." Rather, fossils demonstrate a huge accumulation of new characters in a relatively short period of time, during which the key attributes of modern animal phyla emerged. For example, living arthropods have segmented bodies with a protective cuticle, jointed legs, other appendages specialized for feeding or sensing the environment, and compound eyes with many lenses. Cambrian fossils include the remains of organisms with some but not all of the major features present today in arthropods. In short, the first 40 million years of Cambrian animal evolution ushered in a world utterly distinct from anything known in the preceding 3 billion years.

FIG. 44.41 Genus-level diversity of marine animals through time, compiled from the fossil record. Arrows indicate mass extinctions. *Data from Sepkoski Online Database, http://strata.geology.wisc.edu/jack.*

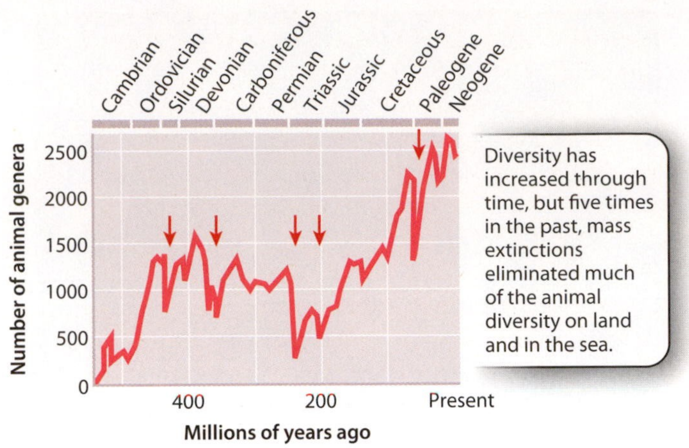

Tabulations of described fossils show that animal diversity has been shaped by both radiation and mass extinction over the past 500 million years.

The wealth of paleontological data we now have shows us a dynamic history of animal radiations and extinctions through time (Chapter 23). The graph shown in **Fig. 44.41** is a tabulation of fossil occurrences through time, compiled from paleontological literature by American paleontologist Jack Sepkoski since the 1970s. In this figure, we see that despite the burst of body plan evolution recorded by Cambrian fossils, there were still relatively few species at the end of the Cambrian Period.

The following Ordovician Period (485–443 million years ago) was a time of renewed animal diversification, especially the evolution of heavily skeletonized animals in the world's oceans. The number of genera recorded by fossils increased fivefold, suggesting that species diversity might have increased by an order of magnitude (most genera contain multiple species). The Ordovician radiation established a marine ecosystem that persisted for more than 200 million years. Interestingly, however, if you had walked along an Ordovician beach, the shells washing about your feet would have been far different from the ones you see today. The dominant shells were those of brachiopods, not clams. Broken corals in the surf were the skeletons of now-extinct cnidarians only distantly related to modern reef-forming corals. And arthropod shells, molted during growth, were those of now-extinct trilobites, not lobsters or crabs. Why was the Paleozoic world so distinct from our own?

Another look at Fig. 44.41 provides the answer. At the end of the Permian Period, 252 million years ago, environmental catastrophe eliminated most genera in the oceans. Paleozoic coral-like cnidarians become extinct, as did the trilobites. Brachiopods survived as a group, but most species disappeared. As noted in Chapter 23, the trigger for this devastation was massive volcanism that unleashed global warming, ocean acidification, and oxygen loss from subsurface oceans.

As ecosystems recovered from this mass extinction, they came to be dominated by new groups descended from survivors of the extinction: Bivalves and gastropods diversified; new groups of arthropods radiated, including the ancestors of the crabs and shrimps we see today; and surviving cnidarians evolved a new capacity to make skeletons of calcium carbonate, resulting in the corals that build modern reefs. In short, mass extinction reset the course of evolution, as it did four other times during the past 500 million years (Fig. 44.41).

Animals began to colonize the land 420 million years ago.

The colonization of land by animals began only after plants had established themselves there. Land plants evolved during the Ordovician Period, about 460 million years ago, and arthropods followed onto the land by the time of the Silurian Period, about 420 million years ago, especially chelicerates that included the ancestors of spiders, mites, and scorpions. Insects, descended from crustacean ancestors that independently gained access to the land surface, appeared at about the same time.

The radiation of the major groups of insects, however, began 360 million years ago with a marked diversification of dragonflies and the ancestors of cockroaches and grasshoppers. Some of the dragonfly-like insects that darted among the plants of Carboniferous Period coal swamps had bodies the size of a lobster and 75-cm wing spans! You might well wonder what prevents insects from attaining this size today. These immense insects flew through an atmosphere that contained more oxygen than is present in the atmosphere today, and this higher level of atmospheric oxygen is thought to have been a necessary condition for insect gigantism. After some time, levels of atmospheric oxygen declined to levels closer to those seen today, and the giant insects disappeared.

Flies, beetles, bees, wasps, butterflies, and moths radiated later, beginning in the early Mesozoic Era. Their rise in diversity parallels that of the flowering plants and reflects coevolution of flowering plants and the insects that both pollinate and feed on them.

Tetrapod fossils first appear in sedimentary rocks deposited near the end of the Devonian Period, about the same time as the early insect radiations 360 million years ago. As discussed in Chapter 23, the fossil record documents in some detail the shifts in skull, trunk, and limb morphology that allowed vertebrate animals to colonize land. These include the evolution of muscled, articulated legs from fins, together with a set of strong, articulated digits where the limbs meet the ground; lungs and a rib cage to support the muscles that control breathing; and an erect, elevated head with eyes oriented for forward vision. The appearance of amphibians and then reptiles follows the predictions of phylogenies based on comparative biology.

Mass extinction has influenced evolution on land as well as in the sea. As new groups of invertebrate animals spread through Triassic oceans, new types of tetrapod diversified on land, including small bipedal reptiles that gave rise to the most remarkable animals ever to walk on land: the dinosaurs. From their beginnings about 210 million years ago, dinosaurs radiated to produce many hundreds of species, dominating terrestrial ecosystems until the end of the Cretaceous Period, 66 million years ago, when another mass extinction, caused by a catastrophic asteroid impact, eliminated nearly all dinosaur species (Chapter 1). We say "nearly all" because the fossil record documents the evolutionary divergence of birds from a specific subgroup of dinosaurs, about 150 million years ago.

Mammals have been the dominant vertebrates in most terrestrial ecosystems since the extinction of dinosaurs, but the group originated much earlier, at least 210 million years ago. During the age of dinosaurs, most mammals were small nocturnal or tree-dwelling animals that stayed out of the way of large dinosaurs, although fossils from China show clearly that the largest mammals of this interval ate small dinosaurs and their eggs. Mammals are dominant components of terrestrial ecosystems today in part because they survived the end-Cretaceous mass extinction.

We summarize the biological diversity of all organisms, including animals, along with a timeline of Earth's history, in **Fig. 44.42.**

? CASE 8 BIODIVERSITY HOTSPOTS: RAIN FORESTS AND CORAL REEFS

How have reefs changed through time?

Today, coral reefs are among the most diverse ecosystems in the ocean, but they are under threat from human activities (Chapter 49). Is this unusual, or have reef systems bloomed and collapsed in the past? Reefs have existed through most of Earth's recorded history, but the kinds of creatures that compose them have changed through time because of both evolutionary innovation and mass extinction. Ecological opportunities are created by the appearance of a new habitat or place, or the origin of a novel innovation that grants new access to resources, or by the extinction of earlier occupants of the habitat or place. Diversification of one group of organisms often spurs the diversification of others that depend on them, and so the history of life is as much about new opportunities created by building on adaptive radiations as by their extinctions.

Bacteria built the first reefs, layered mounds of limestone that rose above the seafloor. With the Cambrian diversification of animals, calcified sponges rose to ecological prominence as participants in reef construction, but these communities were short lived, as they were wiped out later in the Cambrian Period, perhaps 515 million years ago. Subsequent renewed diversification of sponges, bryozoans, and cnidarians with massive calcium carbonate skeletons resulted in new and different reef communities beginning in the Ordovician Period, but these, too, were largely eliminated by renewed mass extinction 375 million years ago, during the Devonian Period. Reefs were widespread again by the Permian Period (300–252 million years ago), dominated this time by different groups of sponges and bryozoans, along with skeleton-forming green algae. The devastating extinction at the end of the Permian Period removed all reef-building animals, and only later did modern corals begin to build reefs. Indeed, reefs like those of the present day have developed only over 40 million years of Earth's history.

Reefs illustrate how evolution and extinction have repeatedly reshaped animal diversity through time. As discussed in Chapter 49, we may be nearing another great change in reef biology as global climate change threatens reef communities in many parts of the world. Today's coral species are at their tolerance limits of ocean temperature and pH, and both conditions may be changing more rapidly than species can adapt.

Humans, then, are influencing reefs and the biological diversity they sustain in ways matched only by massive volcanism and meteorite impact in the past. From the perspective of Earth history, the loss of coral reefs would simply be one more in a long series of extinction episodes. For humans, however, the commercial, aesthetic, and ecological consequences would be substantial and permanent. A similar, if simpler, history of evolution and extinction has shaped plant diversity in rain forests and other terrestrial ecosystems, which are also threatened today by the pace of habitat destruction and changing climate. ■

VISUAL SYNTHESIS Diversity Through Time

FIG. 44.42

Integrating concepts from Chapters 26, 27, 33, 34, and 44

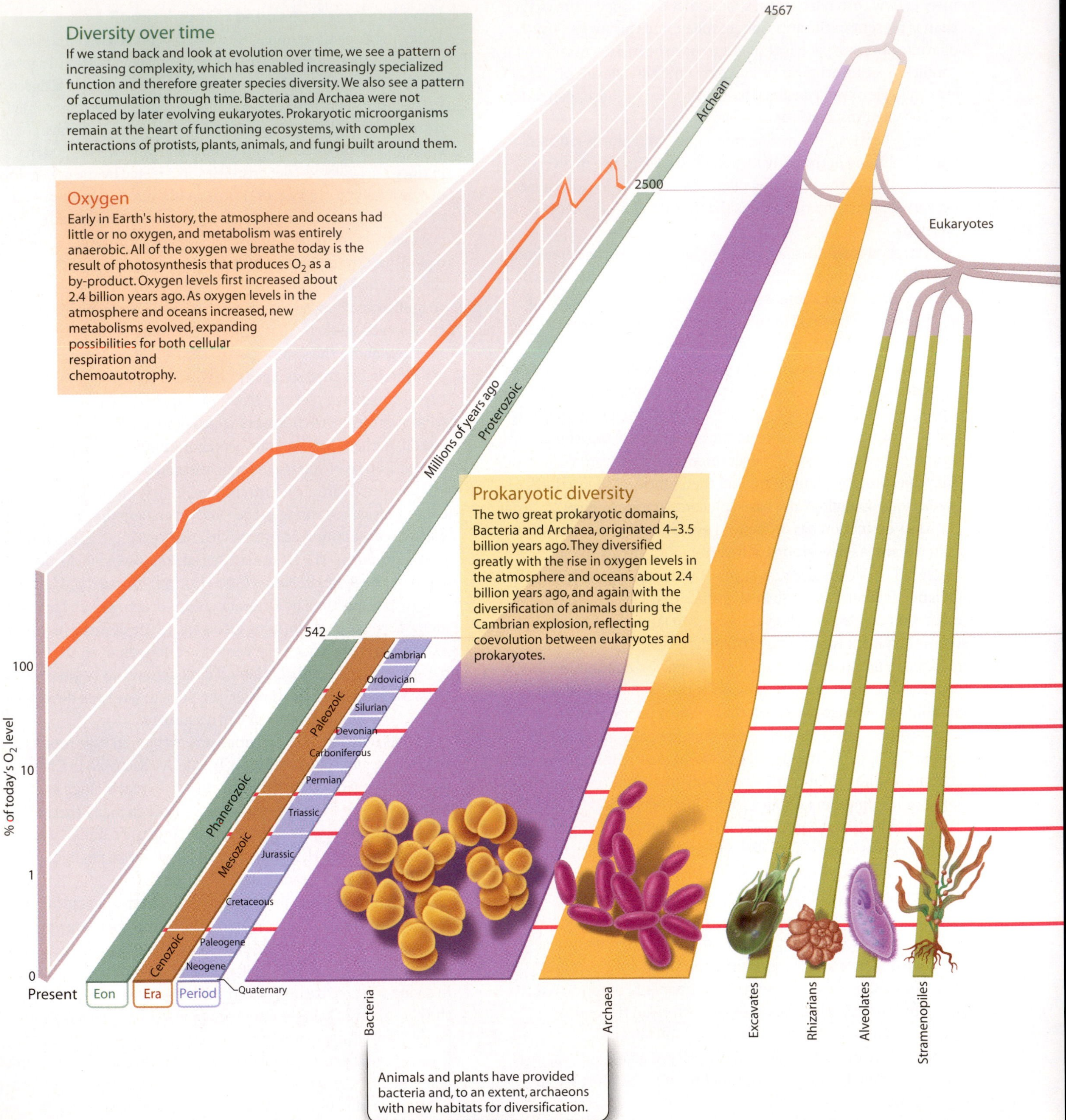

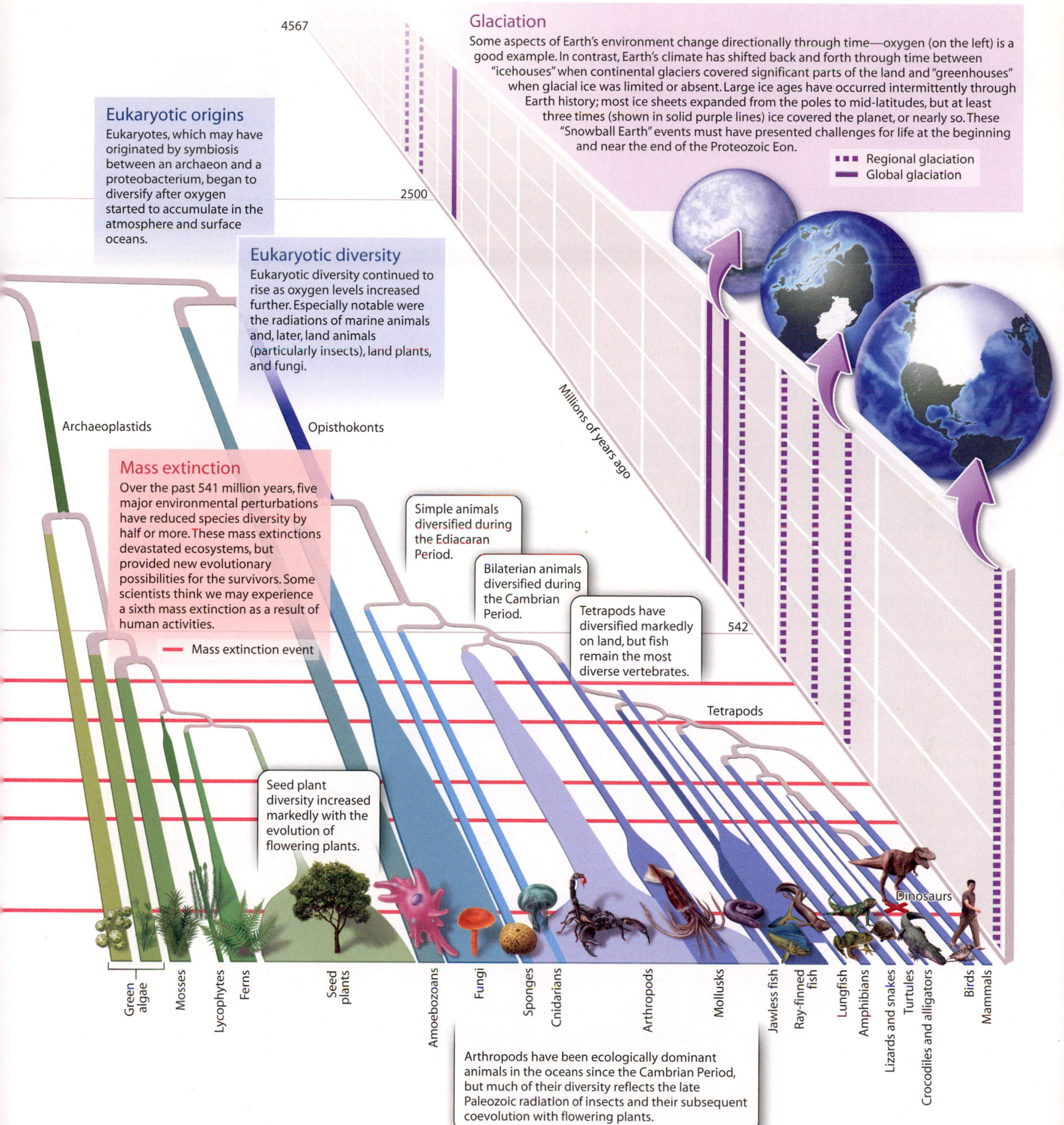

Glaciation
Some aspects of Earth's environment change directionally through time—oxygen (on the left) is a good example. In contrast, Earth's climate has shifted back and forth through time between "icehouses" when continental glaciers covered significant parts of the land and "greenhouses" when glacial ice was limited or absent. Large ice ages have occurred intermittently through Earth history; most ice sheets expanded from the poles to mid-latitudes, but at least three times (shown in solid purple lines) ice covered the planet, or nearly so. These "Snowball Earth" events must have presented challenges for life at the beginning and near the end of the Proterozoic Eon.
Regional glaciation
Global glaciation
4567
2500
542
Millions of years ago
Eukaryotic origins
Eukaryotes, which may have originated by symbiosis between an archaeon and a proteobacterium, began to diversify after oxygen started to accumulate in the atmosphere and surface oceans.
Eukaryotic diversity
Eukaryotic diversity continued to rise as oxygen levels increased further. Especially notable were the radiations of marine animals and, later, land animals (particularly insects), land plants, and fungi.
Archaeoplastids
Opisthokonts
Mass extinction
Over the past 541 million years, five major environmental perturbations have reduced species diversity by half or more. These mass extinctions devastated ecosystems, but provided new evolutionary possibilities for the survivors. Some scientists think we may experience a sixth mass extinction as a result of human activities.
Mass extinction event
Simple animals diversified during the Ediacaran Period.
Bilaterian animals diversified during the Cambrian Period.
Tetrapods have diversified markedly on land, but fish remain the most diverse vertebrates.
Tetrapods
Seed plant diversity increased markedly with the evolution of flowering plants.
Dinosaurs
Green algae
Mosses
Lycophytes
Ferns
Seed plants
Amoebozoans
Fungi
Sponges
Cnidarians
Arthropods
Mollusks
Jawless fish
Ray-finned fish
Lungfish
Amphibians
Lizards and snakes
Turtles
Crocodiles and alligators
Birds
Mammals
Arthropods have been ecologically dominant animals in the oceans since the Cambrian Period, but much of their diversity reflects the late Paleozoic radiation of insects and their subsequent coevolution with flowering plants.

Core Concepts Summary

44.1 Phylogeny charts the evolutionary relationships among more than a million animal species.

Early biologists grouped animals on the basis of shared features of adult bodies, such as symmetry and type of body cavity. page 946

Embryology provided additional information on animal phylogeny, and, in the age of molecular biology, molecular sequence comparisons have provided still more information. page 948

44.2 The earliest branches of the animal tree include sponges, cnidarians, ctenophores, and placozoans.

Sponges have a simple anatomical organization, feed by drawing water containing dissolved food particles into their interiors, and are widespread in the oceans. page 948

Cnidarians form true tissues, differentiate nerve and musclelike cells, and act as predators by means of specialized cells called nematocysts. page 950

Ctenophores, or comb-jellies, resemble cnidarians but move by the beating of cilia, have an anal pore for waste excretion, and have a rudimentary mesoderm germ layer. They remain a difficult limb to locate on the animal tree of life. page 952

Placozoans have a very simple organization, but genomic data support the hypothesis that they are the sister group to all other eumetazoans. page 953

The evolutionary relationships among sponges, placozoans, and ctenophores remain a subject of research. page 953

44.3 Bilaterians, including protostomes and deuterostomes, have bilateral symmetry and develop from three germ layers.

Protostomes are divided into lophotrochozoans and ecdysozoans. page 956

Lophotrochozoans consist of 17 phyla, including annelid worms and mollusks. page 956

Annelid worms include earthworms and leeches, but most species of annelids live in the oceans. page 956

Mollusks include gastropods (snails, slugs), cephalopods (squid, octopuses), and bivalve mollusks (clams, oysters). page 957

Ecdysozoans include several phyla that molt their cuticle, notably the nematodes and arthropods. page 959

Arthropods can be divided into four main groups—insects, chelicerates (spiders, scorpions), myriapods (centipedes, millipedes), and crustaceans (lobsters, shrimp). page 962

Deuterostomes include three main phyla—Chordata, Hemichordata, and Echinodermata. page 963

Well-known echinoderms are sea stars and sea urchins, which have outward fivefold symmetry. page 964

Chordates include vertebrates, cephalochordates, and tunicates. page 964

44.4 Most vertebrates have a bony cranium and vertebral column; vertebrates include fish, amphibians, reptiles, birds, and mammals.

The group commonly known as "fish" consists of four distinct groups of aquatic vertebrates: hagfish and lampreys, cartilaginous fish, bony fish, and lobe-finned fish. page 967

Lungfish are the closest relatives of tetrapods, which include amphibians, lizards, turtles, crocodilians, birds, and mammals. page 969

Amphibians have an aquatic larval form and a terrestrial adult form. page 970

Amniotes, such as lizards, snakes, crocodiles, birds, and mammals, have an amniotic egg, permitting movement into dry, terrestrial habitats. page 970

Mammals are covered with hair and feed their young milk from mammary glands. page 970

44.5 Animals first evolved more than 600 million years ago in the oceans, and by 500 million years ago the major structural and functional body plans of animal phyla were in place.

Ediacaran fossils from 579 million years ago provide evidence of early animals. page 972

The Cambrian explosion was an interval of rapid diversification beginning 541 million years ago, during which time most of the animal body plans we see today first evolved. page 973

Mass extinctions have repeatedly changed the trajectory of animal evolution during the past 500 million years. page 974

Chelicerates and insects were the first animals to colonize the land, sometime after 420 million years ago. page 974

Tetrapods first appear in the fossil record about 360 million years ago. page 975

Mammals originated at least 210 million years ago, but became dominant only after the extinction of the nonavian dinosaurs 65 million years ago. page 975

The fossil record of reef ecosystems reflects a more general pattern of repeated evolutionary innovation and mass extinction. page 975

Self-Assessment

1. Draw a simplified animal tree of life, indicating the relationships among sponges, cnidarians, protostomes, and deuterostomes.
2. What features distinguish sponges, cnidarians, protostomes, and deuterostomes? Place these features on the phylogeny drawn for Self-Assessment question 1.
3. Compare and contrast the anatomical organization of sponges and cnidarians. How do these differences shape their functional biology, especially feeding mode?
4. Describe feeding habits that do not require a head.
5. Name three features that account at least in part for the success of insects.
6. Draw a phylogeny of deuterostomes, indicating the most likely relationships among chordates, hemichordates, and echinoderms, with common examples of each group.
7. List features common to bony fish, amphibians, amniotes, and mammals.
8. Describe the evolutionary significance of the Cambrian explosion.

Log in to LaunchPad to check your answers to the Self-Assessment questions, and to access additional learning tools.

Dan L. Perlman/EcoLibrary.org.

CHAPTER 45

Animal Behavior

Core Concepts

45.1 For any behavior, we can ask what causes it, how it develops, what adaptive function it serves, and how it evolved.

45.2 Animal behavior is shaped in part by genes acting through the nervous and endocrine systems.

45.3 Learning is a change of behavior as a result of experience.

45.4 Orientation, navigation, and biological clocks all require information processing.

45.5 Communication involves an interaction between a sender and a receiver.

45.6 Social behavior is shaped by natural selection.

There is no Nobel Prize for biology. It is the Prize for Physiology or Medicine that is awarded for biological research. Biologists specializing in other areas, such as behavior and ecology, are not even in the running for science's ultimate prize—except, that is, in 1973, when the Nobel Committee bent its own rules and awarded the Prize for Physiology or Medicine to three scientists with no physiological or medical credentials. The three were Niko Tinbergen, Konrad Lorenz, and Karl von Frisch. The award was made in recognition of their roles in pioneering the study of animal behavior under natural conditions.

Why "under natural conditions"? In the early 1900s, Carl von Hess, a prominent visual physiologist, claimed that bees are color blind, a conclusion based on experiments in laboratory settings. However, by training bees to go to food sources associated with differently colored cards in a natural context, the Austrian behavioral biologist Karl von Frisch demonstrated that, in fact, bees have excellent color vision. He emphasized this point at a scientific meeting where he was showing bees that had been trained to associate blue cards with food. By chance, the audience was wearing blue name tags, which proved highly popular with the blue-fixated bees! Von Frisch showed that behavior cannot be studied in isolation, but can only be fully understood when studied in a natural context—in this case, foraging for food.

In this chapter, we explore animal behavior—how animals learn, communicate with one another, form social groups, and choose mates. In each case, animal behavior can be considered on multiple levels. On the individual level, behavior is determined in part by a complex interplay of the nervous and endocrine systems. Behavior is also influenced by the environment, as we saw with von Frisch's bees. Finally, behaviors, like all traits, are shaped by natural selection.

45.1 TINBERGEN'S QUESTIONS

We begin our exploration of animal behavior by asking a simple question: Why does an animal exhibit a particular behavior? The Dutch behavioral biologist Niko Tinbergen divided this overarching question into four separate questions, each one focusing on a different aspect of the behavior. Let's consider an example: Why does a bird sing? While apparently straightforward, this question can be broken down into more specific questions and then answered in a number of different ways:

1. *Causation.* What physiological mechanisms cause the behavior? This question can have multiple answers. A bird sings because its hormone levels have changed in response to changes in day length. Or, more immediately, a bird sings because air passing through its specialized singing organ, the syrinx, causes membranes to vibrate rhythmically.

2. *Development.* How did the behavior develop? Here, the focus is on the role of genes and the environment in shaping the development of the behavior. In birds, typically it's the male that sings, and he has learned the song from his father or other males.

3. *Adaptive function.* How does the behavior promote the individual's ability to survive and reproduce? In this case, the answer to the question might be that a male bird sings in order to attract a mate and then reproduce.

4. *Evolutionary history.* How did the behavior evolve over time? Complex bird songs may have evolved from vocalizations made by ancestors that became increasingly standardized over time, so much so that we can often identify a bird species simply by hearing it sing. A behavior may have originated to fulfill a function different from the one it currently serves. For example, the song of a particular species may have first evolved to claim a territory but now is used to attract mates.

Tinbergen's analysis allows us to see that different answers to the question "Why does a bird sing?" can all be correct. The first two, causation and development, are mechanistic explanations for behavior; these are referred to as *proximate* explanations of behavior because they deal with *how* the behavior came to be. The second two, adaptive function and evolutionary history, provide evolutionary explanations of how natural selection has shaped a behavior over time; these are referred to as *ultimate* explanations of behavior because they deal with *why* the behavior came to be. Tinbergen's four questions are complementary ways of looking at the same problem, and are a good starting point for the analysis of behavior.

The answers to Tinbergen's questions rely on an interplay between genes and the environment. The influence of genes is especially clear in **innate** behaviors, those that are instinctive and carried out regardless of earlier experience. Male silkworm moths of the genus *Bombyx* exhibit an innate behavior when they fly upwind toward the source of a female-produced pheromone, an airborne chemical signal that communicates with members of the same species, in this case to attract the opposite sex (**Fig. 45.1**). Pheromones released from a female's abdominal gland are sensed by small hairs on the male's antennae. In the presence of pheromones, the antennal sensory hairs fire action potentials (Chapter 35). When about 200 hairs are being activated per second, a male flies upwind distances of a kilometer or more, tracking the increasing pheromone concentration until he finds the female. The male does not need to learn this behavior; he performs it spontaneously. His genes encode molecular receptors to which the pheromone binds, and this binding triggers a cascade of events that result in the moth's heading up the pheromone's concentration gradient.

A particular behavior may depend on an individual's experience, in which case the behavior has been **learned.** As we will see, even neurologically simple organisms have a considerable capacity to learn. Fruit flies, for example, learn to avoid certain

FIG. 45.1 An example of innate behavior. A male moth flies toward the source of a female-produced pheromone. *Photo sources: (left to right) Leal Lab, Department of Molecular and Cellular Biology/UC Davis/Regents of the University of California; Courtesy of Prof. Dr. R. A. Steinbrecht. Cell Tissue Res (1995) 282:203-217; Fabio Pupin/FLPA/Science Source.*

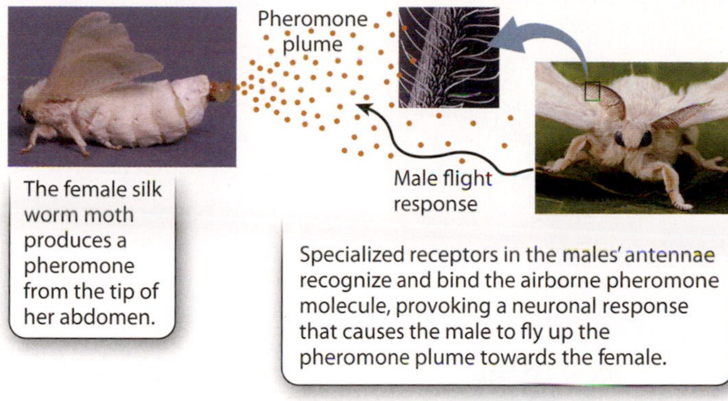

locations or substances if they associate them with an unpleasant experience.

We can consider, then, the extent to which a particular behavior is genetically encoded (part of the animal's nature) and the extent to which it is conditioned by the environment in which the animal develops (part of the animal's nurture). However, as we discussed in Chapter 18, nature and nurture are inextricably linked. For example, most human behaviors are the product of an interaction between nature and nurture. Consider language in the context of Tinbergen's second question concerning the development of behavior: We have an innate ability for acquiring language, but the specific language that we acquire depends entirely on our nurture. A baby in Italy learns Italian, and one in Finland learns Finnish.

45.2 DISSECTING BEHAVIOR

Behavior can be complex and therefore influenced by a large number of factors, both genetic and environmental. For that reason, complex behaviors can sometimes defy analysis. By choosing simple behaviors that can be more easily studied and understood, scientists can identify, isolate, and investigate the separate components of behavior.

The fixed action pattern is a stereotyped behavior.

Some of the first behaviors to be carefully analyzed were **displays,** which are patterns of behavior that are species specific and tend to follow the same sequence of actions whenever they are repeated and in a way that is similar from one individual to the next. Because these behaviors are so similar whenever they are produced, they are considered "stereotyped." Presumably, natural selection has favored display behaviors that are unmistakable in their function as signals. Consider, for example, courtship displays in birds that are preparing to mate (**Fig. 45.2**). Experiments have shown that, in some species, a bird raised in complete isolation from other members of its species will still perform a courtship

FIG. 45.2 Courtship display in the Raggiana Bird-of-paradise. The male (top) is performing a highly repeated set of behaviors to entice the female to mate with him. *Source: Bruce Beehler/age fotostock.*

FIG. 45.3 **A fixed action pattern.** A goose displays and completes the behavior to retrieve an egg (a), even if the behavior is interrupted (b) or the stimulus altered (c).

display with great precision, suggesting that the behavior is genetically encoded, that is, instinctual rather than learned.

Many displays are **fixed action patterns (FAPs),** meaning that they consist of a sequence of behaviors that, once triggered, is followed through to completion. A classic example, originally studied by Tinbergen, is the response of a goose to an egg that has fallen from its nest (**Fig. 45.3**). The stimulus that initiates the behavior is the **key stimulus** and in this case it is the sight of the misplaced egg. This sight provokes in the goose an egg-retrieval FAP, which consists of rolling the egg back to the nest with the underside of its beak (Fig. 45.3a). This response cannot be broken down into smaller subunits and it is always carried out to the end, even if it is interrupted. In fact, the goose will persist in its efforts, even if the researcher ties a string around the misplaced egg and removes it while the goose is in mid-action. The goose will still continue the task of rolling the now-absent egg (Fig. 45.3b).

It is possible to understand this behavior by varying attributes of the key stimulus (in this case, the misplaced egg). A remarkable finding is that many birds respond most strongly to the largest round object provided, even a soccer ball, as illustrated in Fig. 45.3c. A soccer ball in fact not only elicits the egg-retrieval FAP, but it does so even more strongly than does a normal egg. The soccer ball is considered a **supernormal stimulus** because it is larger than any egg the goose would naturally encounter and elicits an exaggerated response. Natural selection has likely favored geese that recognize and respond to large eggs that have rolled outside the nest, but since eggs never grow to the size of soccer balls, selection has not shaped an appropriate response to unrealistically large egg-shaped objects.

The nervous system processes stimuli and evokes behaviors.

How do animals recognize the stimulus that leads to a particular behavior? In the example of the goose retrieving an egg outside its nest, how does it recognize that the object in question is sufficiently egg-like to elicit the behavior? At its root, this is an extraordinarily difficult problem. The world is full of stimuli and there are many things that look like an egg. Thus, the challenge for the animal is to filter out the correct signal (the egg) from the surrounding noise (everything else). An animal has to process many different kinds of information in three dimensions.

We know that stimulus recognition is often carried out by **feature detectors,** specialized sensory receptors or groups of sensory receptors that respond to important signals in the environment. Frogs provide a good example. Male frogs call to attract females and warn off other males. A single pond may contain many species of frog, each with its specific advertisement vocalization, or call. A heavily populated frog pond can be a noisy place, but it is critical to an individual frog that it be able to distinguish the call of other members of its own species from those of other species. A further complication is that different species of frog can have similar calls, often virtually indistinguishable to the human ear.

One approach to the problem of understanding how frogs recognize their species-specific call has been to experiment with recordings of frog calls. By recording frog calls, altering aspects of the call in specific ways, playing them back, and then monitoring the responses of frogs, we can identify the impact of different components of the call—pitch, duration, and pulse frequency, for example. Studies like these have demonstrated that the frog's auditory nerves can act as a feature detector, each one "tuned" to a specific component of the call. The combination of different feature detectors stimulated by different components adds up to the recognition by the frog of a specific call, as shown in **Fig. 45.4.** Once the sound has been correctly identified, the appropriate behavior follows.

Hormones can trigger certain behaviors.

The effect of a neuron is typically short lived and local: A neuron may, for example, connect to another neuron or cause a muscle to contract (Chapter 35). One way in which a stimulus may provoke a more widespread and prolonged effect is through the production of hormones. Hormones can affect multiple cells in target organs simultaneously (Chapter 38).

Anolis lizards demonstrate the important role of hormones in sexual behavior, showing both how social stimuli can affect the release of hormones and how hormones can affect behavior (**Fig. 45.5**). If females of *Anolis carolinensis* are isolated from males during the breeding season, about 80% of individuals have active egg follicles in their ovaries (Fig. 45.5a). If one male is added to a group of females, that figure increases to 100% (Fig. 45.5b). The courting behavior of the male lizard stimulates the females to produce hormones that cause the full development of the ovaries, making the females reproductively active. If a group of males instead of a single male is added to an all-female population, on the other hand, only about 40% of the females undergo ovarian development (Fig. 45.5c). This unexpected result seems to occur because the males fight among themselves rather than court the females, and therefore the courtship stimulus is lacking. Castrated

FIG. 45.4 A feature detector. A frog in its natural environment is subjected to a barrage of sound stimuli. Each species has a feature detector tuned to its species-specific call, allowing it to distinguish the call of its own species (red arrow) from the calls of other frogs (blue arrows).

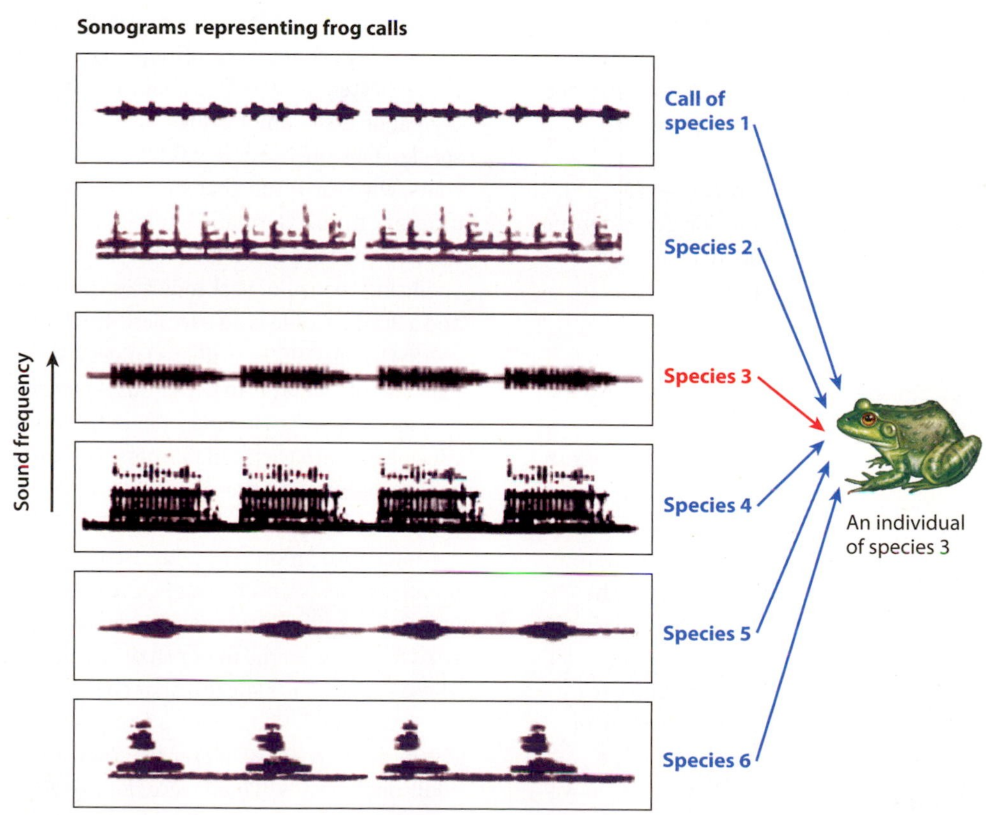

FIG. 45.5 Role of hormones in behavior, demonstrated by groups of *Anolis* lizards.

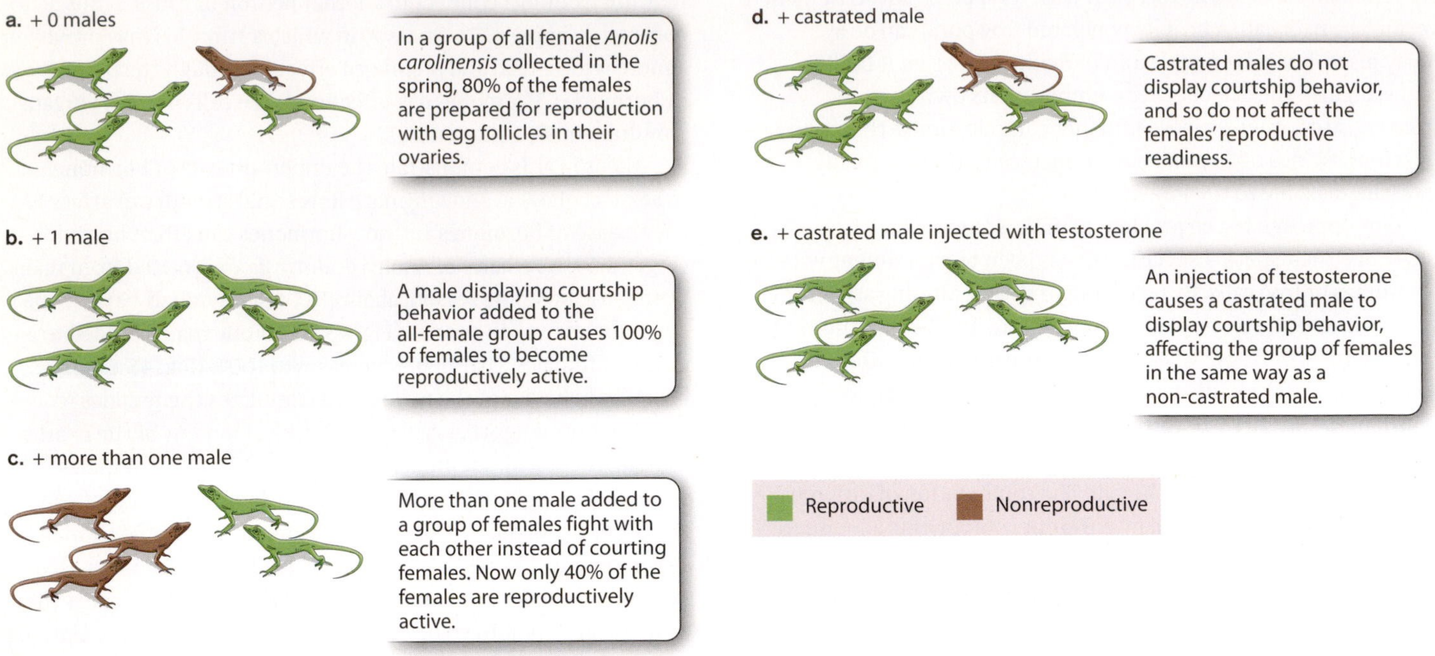

males (which do not produce testosterone) that are added to a group of females have no significant effect on rates of ovarian development, which remains about 80% (Fig. 45.5d). They fail to court because the behavior is caused in part by testosterone. However, a castrated male that is injected with testosterone does display courting behavior and has the same effect as the presence of a single male, inducing all the females to undergo ovarian development (Fig. 45.5e).

These simple laboratory experiments demonstrate the complex interplay between hormones and behavior. Testosterone mediates both male–male interactions and the male–female interactions that cause females to produce hormones that promote reproductive development.

Breeding experiments can help determine the degree to which a behavior is genetic.

The nervous and endocrine systems that in part shape behaviors are the product of genetic instructions. If you take an extreme view, then, all behaviors have a genetic component. After all, it is the instructions in the genome that build the nervous system, muscles, and glands whose actions lead to a behavior. On the other hand, all behavior can be considered environmental since, without the appropriate environment, an organism would not develop normally and would therefore lose the ability to perform the behavior. On a very basic level, an organism in an environment without appropriate nutrients would lack the necessary energy to perform a given behavior. How, then, do we study the genetic basis of behavior? Modern approaches harness the power of molecular genetics, but more traditional analyses are also highly informative.

Artificial selection (Chapter 21), in which humans breed animals and plants for particular traits, provides strong evidence of the role of genetics in influencing behavior. Dogs were domesticated from wolves about 10,000 years ago. Today, there is great diversity among dogs, although they are all the same species. Obviously, there has been extensive selection on physical traits, with the result that a Dachshund, for example, looks quite different from a German Shepherd or Great Dane. Selection has also been applied to behavior: A Pointer has extraordinary "pointing" behavior that indicates the location of a hunter's prey, and a Border Collie is an excellent herder (**Fig. 45.6**).

In the late 1950s, William Dilger studied the nest-building behavior of lovebirds in the genus *Agapornis*. Some species transport their nesting material in their beaks, whereas others tuck pieces into their tail feathers. Is there a genetic basis to the nesting behavior of these birds? To answer this question, Dilger set up crosses between species with different nest-building techniques to produce hybrids. He then observed how the hybrids built their nests. Interestingly, the hybrid offspring showed nest-building behavior intermediate between that of the parents: They tried to tuck material under their feathers, but were not successful. These experiments suggest that this behavior has a genetic basis.

It was not possible to do a true Mendelian analysis of the lovebirds' behavior (Chapter 16). Such an analysis would require additional crosses, which were not possible because the hybrids were sterile. Like most behaviors, the lovebirds' nest-building

FIG. 45.6 **Artificial selection.** Artificial selection can alter behavior, as shown by the remarkable differences in behavior we see among breeds of dogs, such as (a) a Pointer and (b) a Border Collie. *Sources: a. Adriano Bacchella/naturepl.com; b Fred Lord/Alamy.*

behavior is probably controlled by a large number of genes, each with a relatively small effect (Chapter 18). Therefore, it is difficult to identify which genes are responsible for the behavior. Even with molecular methods, the task of mapping and identifying the underlying genes when a large number of different genes govern a particular trait is daunting.

→ **Quick Check 1** Why is it typically difficult to identify individual genes that govern particular behaviors?

Molecular techniques provide new ways of testing the role of genes in behavior.

Molecular biology is changing the way we approach behavioral genetics. Some studies have identified complex behaviors that are strongly influenced by a single gene. One well-understood instance of a single gene's effect on behavior comes from the fruit fly *Drosophila.*

In *Drosophila,* there are two different alleles of the *foraging* (*for*) gene—*for*^{s} and *for*^{R}—that are common in populations. These two alleles have different effects on the behavior of *Drosophila* larvae. The gene encodes an enzyme expressed in the brain that affects neuronal activity and alters behavior. In the absence of food, both "sitter" (*for*^{s}) and "rover" (*for*^{R}) larvae move about in search of food. In the presence of food, however, sitters barely move, feeding on the patch on which they find themselves, whereas rovers move extensively both within a patch of food and between patches (**Fig 45.7a**).

Both forms are present in natural populations (**Fig. 45.7b**); typically, 70% are rovers, 30% are sitters. That both strategies are

FIG. 45.7 **The effect of genotype at the foraging locus on the feeding behavior of *Drosophila melanogaster* larvae.** (a) When food is present, rover larvae travel farther than sitter larvae. (b) Both types of larvae exist in natural populations. *Source: M. B. Sokolowski, 2001, "Drosophila: Genetics Meets Behaviour," Nature Reviews Genetics 2: 879–890, doi:10.1038/35098592.*

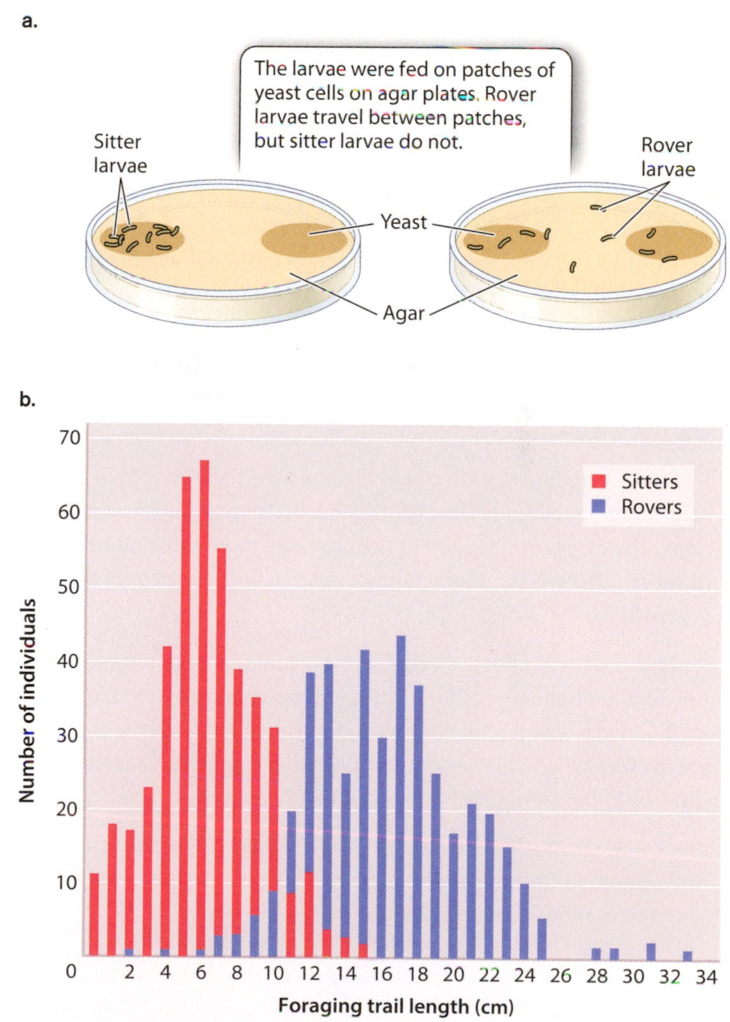

HOW DO WE KNOW?

FIG. 45.8

Can genes influence behavior?

BACKGROUND In fruit flies (*Drosophila melanogaster*), variation at the *foraging* (*for*) gene affects feeding behavior of larvae. Two alleles of the for gene—for^s and for^R—exist in natural populations; for^s larvae tend to stay on a patch of food ("sitters") and for^R larvae tend to move from patch to patch ("rovers"). Honeybees (*Apis mellifera*) also forage for food, but, in contrast to fruit flies, this behavior changes with age. Young honeybees stay at the hive ("nurses"), and older ones forage for nectar ("foragers"). The *for* gene is present in honeybees, raising the possibility that this gene is involved in the developmental change in foraging in honeybees.

HYPOTHESIS Foraging in honeybees is influenced by the expression of the *for* gene, and nurses have lower expression levels than foragers.

EXPERIMENT 1 Levels of *for* mRNA were measured in honeybee nurses and foragers.

RESULTS Levels of *for* mRNA are significantly higher in foragers than in nurses (Fig. 45.8a). However, foragers are older than nurses, so it was unclear whether the increased expression in foragers compared with nurses is associated with differences in age or differences in behavior. Therefore, the experiment was repeated with honeybees that forage at the age at which bees are normally nurses. These young foragers also have higher *for* mRNA levels than do nurses (Fig. 45.8b).

EXPERIMENT 2 The *for* gene encodes a cGMP-dependent kinase that phosphorylates other proteins (Chapter 9). Honeybee nurses were treated with cGMP to activate the kinase and their subsequent behavior was monitored. A related compound, cAMP, with a chemical structure similar to cGMP but which does not affect the kinase, was used as a control, to ensure that any effect observed was specific for the pathway and not the result of the treatment.

RESULTS Treatment with cGMP changed the behavior of nurses, causing them to forage (Fig. 45.8c). Furthermore, cGMP acted in a dose-dependent fashion: The higher the dose, the more foraging behavior was observed. No effect on foraging was seen in the control treatment (Fig. 45.8d).

CONCLUSION The same gene that is involved in two different behavioral phenotypes in fruit flies (sitters versus rovers) is also involved in a developmental behavioral change in honeybees (nurses versus foragers). This finding suggests that a gene can have related but different functions in different organisms.

FOLLOW-UP WORK Researchers have looked at the expression of many other genes in the honeybee genome to learn which genes are associated with the change of foraging behavior of honeybees.

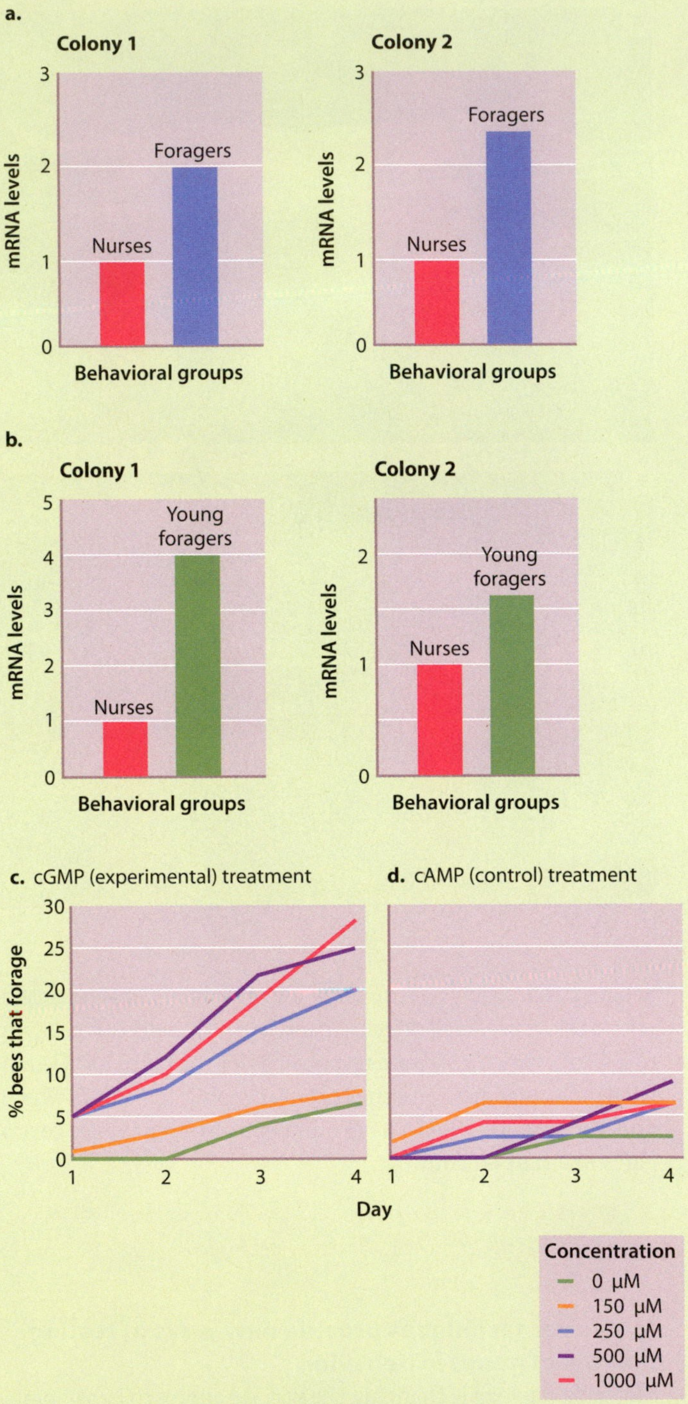

SOURCE Ben-Shahar, Y., et al. 2002. "Influence of Gene Action Across Different Time Scales on Behavior." *Science* 296:741–744.

present in populations suggests that both are adaptive because, if one strategy were markedly inferior, we would expect natural selection to eliminate it. Furthermore, studies have shown that rovers are selected for in crowded environments in which there is an advantage to seeking new food sources, and sitters are selected for in less crowded ones because they take maximum advantage of their current food source. In this case, variation at a single gene affects a complex behavior in fruit flies.

The *foraging* gene is present in other insects as well, including honeybees, suggesting that it has been evolutionary conserved. Honeybees with low levels of *for* expression in their brain tend to stay in the hive, whereas those with high levels of *for* expression are likely to be foragers (**Fig. 45.8**).

Studies of the North American vole, in the genus *Microtus*, have also revealed how genes influence behavior (**Fig. 45.9**). The prairie vole is monogamous: Couples pair bond for life, meaning that they mate only with each other. By contrast, the prairie vole's close relative, the montane vole, is promiscuous: A male mates and moves on, and a female over her lifetime produces litters by several different males. What differences underlie these different behaviors?

Hormones and their receptors provide an answer. In all mammals, antidiuretic hormone (ADH) controls the concentration of urine (Chapter 41). The receptor for ADH is not only expressed in the kidney, but also in the brain, where it affects behavior. Both the monogamous prairie vole and the promiscuous montane vole produce ADH, but the hormone receptors in the two species are different. The American neuroscientist Thomas Insel analyzed the genes that encode the ADH receptor in the two vole species. He found a difference in the region of the gene that determines under what conditions and in what cells it is expressed. Because of this difference, the distribution of ADH receptors in the brain is different in the prairie vole and in the montane vole.

FIG. 45.9 **Differing mating behavior in voles.** The different behaviors in (a) monogamous prairie voles and (b) promiscuous montane voles results from variation at a single gene. *Sources: a. Todd Ahern/Quinnipiac University; b. Rick & Nora Bowers/Alamy.*

Does the difference in gene expression explain the different sexual behaviors of the two species? The answer seems to be yes. Insel and his colleague Larry Young inserted the prairie vole ADH receptor gene, complete with its regulatory region, into a laboratory mouse (a promiscuous species, like the montane vole). Interestingly, Insel and Young observed a marked change in the mouse's behavior. Rather than mating with a female and then moving on to another mate, the transgenic male mouse pair bonded with the female. In short, the addition of a single gene affected a complex behavior such as pair bonding, making the otherwise promiscuous mouse more likely to pair bond.

45.3 LEARNING

Genes are not the only determinants of behavior. We all know that experience often leads to changes in behavior. We call this process **learning.** Humans are extraordinary learners, but even many relatively simple organisms respond to experience. Researchers have identified several types of learning. The categories are not strict, and there is some overlap among them, but they are a useful way to organize our thinking and give us deeper insights into the subject.

Non-associative learning occurs without linking two events.

Non-associative learning is learning that occurs in the absence of any particular outcome, such as a reward or punishment. One type of non-associative learning is **habituation,** which is the reduction or elimination of a behavioral response to a repeatedly presented stimulus. Chicks presented with silhouettes flying overhead provide an example. Initially, any overhead silhouette provokes a defensive crouching posture, but eventually, chicks habituate to overhead silhouettes that have proved not to be threatening. In the absence of any consequences, chicks no longer crouch in response to a harmless silhouette passing repeatedly overhead.

Sensitization is another form of non-associative learning. Sensitization is the enhancement of a response to a stimulus that is achieved by presenting a strong or novel stimulus *first.* This first stimulus makes the animal more alert and responsive to the next stimulus. The sea slug, *Aplysia,* a model organism among scientists interested in the neuronal basis of behavior, exhibits sensitization. *Aplysia* withdraws its gills in response to a touch on the siphon, the tube through which it draws water over its gills. Interestingly, if the animal is given a weak electric shock first, its response to a touch on its siphon is much more rapid, suggesting that the shock has made the slug more sensitive to later stimuli.

Associative learning occurs when two events are linked.

Associative learning (also called **conditioning**) occurs when an animal learns to link (or associate) two events. Perhaps the most famous example of conditioning is the response of Ivan Pavlov's dogs. Pavlov first presented the dogs with meat powder, and they salivated in response. He then presented the dogs with an additional cue, a ringing bell, whenever he presented the dogs with meat powder. After repeatedly experiencing the two stimuli together, the dogs salivated at the sound of the bell alone, in expectation of the meat reward. This form of conditioning, in which two stimuli are paired, is called **classical conditioning.** In this case, a stimulus that leads to a behavior (the meat powder) was paired with a neutral stimulus (the bell) that initially had nothing to do with salivation. Eventually, a novel association is made between the sound of the bell and food, and the sound of the bell alone elicits salivation.

A second form of associative learning links a behavior with a reward or punishment. If the behavior is rewarded (positively reinforced), it is more likely to occur the next time around. It is not just out of the goodness of her heart that the trainer at Sea World gives a sea lion or dolphin a fish reward at the end of each trick that the animal performs. However, if the behavior is punished (negatively reinforced), the response becomes less likely. This form of associative learning is called **operant conditioning.**

In classical conditioning, an association is made between a stimulus and a behavior, whereas in operant conditioning, an association is made between a behavior and a response. Operant conditioning involves a novel, undirected behavior that becomes more or less likely over time. Consider a rat in a cage with some levers in it, each a different color. The rat has no impulse to press any of the levers, but by chance, it presses the red one. A food reward tumbles into the cage. Eventually, the rat learns to associate pressing the red lever with the food reward; it has been operantly conditioned. It is thought that play in young animals provides them with the opportunity to explore their environment much in the same way as the rat does when first placed in the cage. Like the rat, they may find that some behaviors are more rewarding or more punishing than others.

→ **Quick Check 2** Both classical and operant conditioning involve learning to associate two events. What then is the difference between the two kinds of conditioning?

Learning is an adaptation.

The capacity to learn often has adaptive functions. Much learning, including human learning, is based on **imitation:** One individual copies another. Learning by imitation was famously observed in the days when people in Britain had milk deliveries left outside on the doorstep. As we saw in Chapter 24, a number of birds, including Great Tits and Blue Tits, learned to open foil-topped milk bottles, and this behavior spread rapidly through populations of birds as other birds copied the successful birds' approach. Individuals can also learn by imitating individuals of a different species. An octopus can learn to open a jar with a reward inside by watching a human or another octopus do it.

The capacity to learn a particular task seems in many instances to be innate. Female digger wasps are foragers. Tinbergen showed that a digger wasp learns the landmarks around her nest and then uses this information to find her way back to it from hunting. In his experiment, illustrated in **Fig. 45.10,** Tinbergen placed a ring of pine cones around a wasp's nest and, after she had flown out and returned a few times, moved the circle of pine cones to the side of the nest. The returning wasp now flew past her own nest and went directly to the center of the ring. This innate ability to learn landmarks is adaptive: It improves the survival or reproductive success of the individual.

This adaptive aspect of animal learning is also revealed by taste aversion experiments, in which an animal typically learns to avoid certain flavors associated with a negative outcome. Rats

HOW DO WE KNOW?

FIG. 45.10

To what extent are insects capable of learning?

BACKGROUND European digger wasps, *Philanthus triangulum,* live in the sand. These wasps are sometimes called "bee wolves" because they specialize on hunting for honeybees to feed their developing young. After mating, each female digs a long burrow with a few chambers at the end where she lays her eggs. She then forages for honeybees that she brings back to these chambers for her larvae to eat. The wasp faces a navigational challenge: Having captured her prey, how can she find her way back to and recognize her nest? Niko Tinbergen noticed that wasps lingered briefly near a new nest before heading off to hunt, and thought that they were learning local landmarks associated with the nest.

HYPOTHESIS The wasp learns visual cues around her nest to help her locate it upon her return.

METHOD Tinbergen recognized that a good test of the learning abilities of an insect should take place in its natural environment. He combined his skills as a naturalist and as an experimentalist to devise an elegant demonstration of the way in which female wasps learn landmarks for navigation. Tinbergen's approach was simple. He placed a ring of pine cones around the nest of a wasp, and then, once she had left to hunt, he shifted them to a new location away from the nest entrance. If visual cues are key, the wasp would return to the displaced pine cone ring. On the other hand, if cues are, for example, olfactory, she would return directly to the nest.

a.

b.

RESULTS Females carried out a brief landmark-learning flight on departure from the nest (Fig. 45.10a). When the landmarks were displaced, the females returned to the wrong location (Fig. 45.10b).

CONCLUSION Female digger wasps learn and then use local landmarks, such as the experimental ring of pine cones, as cues to the nest location.

SOURCE Tinbergen, N. 1958. *Curious Naturalists.* New York: Basic Books.

learn to avoid flavored water if consuming it is associated with nausea. However, they do *not* learn to avoid flavored water if it is associated with a different kind of negative reinforcement, such as a mild electric shock. Rats, then, can make some associations, but not others. The ones they can learn are the biologically meaningful ones, those that favor survival. In the course of evolution, the rat's ancestors encountered poisoned food that resulted in nausea and the aversion response evolved. Until

FIG. 45.11 **Imprinting on (a) Konrad Lorenz and (b) an ultralight aircraft.** By manipulating newly hatched birds' first sight of a moving object, it is possible to induce powerful attachments in the birds. *Sources: a. Nina Leen/Time Life Pictures/Getty Images; b. Jeffrey Phelps/Getty Images.*

humans started doing experiments on them, rats had never encountered a meal that resulted in an electric shock. It is not surprising, then, that the ability to pair these two phenomena never evolved. These experiments also show that the specific evolutionary history of the species in question matters when analyzing animal behavior.

In addition to adaptive predispositions for *what* can be learned and not learned, many species exhibit predispositions for *when* learning takes place. This is particularly evident in imprinting, a form of learning typically seen in young animals in which they acquire a certain behavior in response to key experiences during a critical period of development. Konrad Lorenz made imprinting famous by exploiting the observation that newly hatched goslings and ducklings rapidly learn to treat any sufficiently large moving object they happen to see shortly after hatching as their mother. Lorenz found that, if he was the first person the hatchlings saw, they would follow him as though he were their mother (**Fig. 45.11**).

This behavior is adaptive because the first being that a hatchling normally sees is a parent and the resulting close parent–offspring association ensures that the parent can provide care and protection. This type of imprinting is called **filial imprinting,** and it is most common in species whose offspring leave the nest and walk around while still young, like chicks and ducklings. Filial imprinting is rare in species of birds whose young stay in the nest until they are able to fly away.

Experiments have shown that filial imprinting typically occurs during a specific, sensitive period in the animal's life and that the results are usually irreversible. After Lorenz's baby ducks had imprinted on him, they would not change their minds about who their parent was even when presented with their real mother duck. The timing of the sensitive period varies from species to species. In some cliff-nesting sea birds, it has been shown that imprinting on auditory stimuli (like the call of the parents) begins while the chick is still in the egg.

45.4 ORIENTATION, NAVIGATION, AND BIOLOGICAL CLOCKS

Learning is a form of information processing in which experience shapes behavior. At a cellular level, it typically involves a change in the strength of connections between neurons. The central nervous system of even relatively simple animals is capable of extraordinary feats of information processing. The ways in which animal nervous and endocrine systems and the input of experience are integrated to generate adaptive behaviors are remarkable. Some of these adaptive behaviors include ways of moving, navigating, and keeping time, which we examine here.

Orientation involves a directed response to a stimulus.

Even the simplest bacteria and protozoa are capable of moving in response to stimuli. A *Paramecium* that finds itself in an unfavorable environment, such as water that is too warm or too salty, increases its speed and begins to make random turns. When it finds favorable conditions, such as cooler water, it slows and reduces its turning rate. These random, undirected movements are termed **kineses** (singular, **kinesis**).

In contrast, **taxes** (singular, **taxis**) are movements in a specific direction in response to a stimulus. An interesting example of a taxis is movement oriented to a magnetic field, a behavior called magnetotaxis. The term was first used by the American microbiologist Richard Blakemore in 1975. Blakemore had found that anaerobic bacteria in the genus *Aquaspirillum,* which swim by means of flagella, tend to swim toward magnetic north. These bacteria can be attracted to the side of a dish with a bar magnet. Little bits of magnetized iron oxide, arranged in a row inside the bacterial cell, allow the bacteria to sense the magnetic field. Blakemore hypothesized that the bacteria swim north in order to swim deeper. In the Northern Hemisphere, the north magnetic pole is inclined downward. At Woods Hole, Massachusetts, where these bacteria were found, the magnetic pole is at about a 70-degree incline. Since the bacteria are anaerobic, they must stay buried in sediment, and they remain buried by moving downward along the magnetic gradient.

To test his hypothesis, Blakemore looked for and found bacteria in New Zealand that exhibited similar behavior to the Woods Hole bacteria except that they swam toward the *south* magnetic pole, bringing them downward in the Southern Hemisphere. Finally, he took New Zealand cultures back to Woods Hole, where the bacteria responded inappropriately to the unfamiliar magnetic field of the Northern Hemisphere, swimming up into the oxygen and dying. These experiments demonstrated that bacteria are able to sense a magnetic field and move in a directed fashion relative to that field.

Navigation is demonstrated by the remarkable ability of homing in birds.

Many animals use environmental cues to migrate long distances. The navigational achievements of homing pigeons are legendary. They can home—that is, find their way back to the place where they are housed and fed—over extremely long distances, even of more than a thousand miles. It seems that they use a wealth of cues when homing. A compass may tell you which way is north, but you must also have **map information**—you must know where you are with respect to your goal—if the compass is to be useful for finding a particular location. So pigeons must have both compass and map senses. That map sense is presumably based on landmarks. Given the long distances the pigeons travel, these landmarks must vary in their type and probably include olfactory as well as visual cues.

The pigeon compass relies on different kinds of information. For example, pigeons can navigate during the day using the sun as a compass, and at night using the stars. And, like *Aquaspirillum,* they can detect Earth's magnetic field. The magnetic navigation system may be important to pigeons on cloudy days. Researchers have performed experiments with pigeons wearing magnetic helmets, which disrupt the pigeons' ability to detect Earth's magnetic field. On cloudy days, the pigeons wearing the helmets are unable to navigate home. On sunny days, however, the pigeons' sun compass overrides the erroneous magnetic information of the helmets.

The sun compass requires information about time as well. Every hour, the sun moves 15 degrees through the sky. To determine where north is, a pigeon needs information from the sun as well as some way to keep time. It turns out that pigeons have a clock—a biological one.

Biological clocks provide important time cues for many behaviors.

Like us, other animals live in space *and* time. So far, we have focused on their behavior in space. For many species, time is a life-or-death matter. When to migrate or mate is a critical decision in a seasonal environment. Researchers are beginning to unravel the neural and genetic underpinnings of the clocks in some species. In model organisms like *Drosophila,* for example, researchers have identified a number of genes that, when mutated, cause the clock to run slow or fast.

A biological clock is produced by a set of interacting proteins that cycles on its own to provide a regular rhythm. Different clocks work on different timescales. Daily cycles are governed by a **circadian clock.** Circadian clocks regulate many daily rhythms in animals, such as those of feeding, sleeping, hormone production, and core body temperature. Some species, especially seacoast species living in habitats where the tides are important, time activities by a **lunar** (moon-based) **clock.** There are also **annual** (yearly) **clocks.** For example, periodical cicadas are insects in the genus *Magicicada* that have a generation time of either 13 or 17 years: In a given location, there will be a cicada outbreak every 13 or 17 years.

Circadian clocks are observed in many organisms, including plants, fungi, and animals. Humans have circadian clocks, as jet lag never fails to remind us. The circadian clock is based on a set of "clock genes" whose protein products oscillate through a series of feedback loops in a roughly 24-hour cycle. Thus, when animals that, like us, are active during the day and inactive at night are placed in artificial conditions that are always lit, they continue to follow a basic day–night, active–inactive cycle. The clocks are not perfect, though: As the period spent in constant light is prolonged, the circadian clock drifts slowly until the animal is eventually no longer synchronized with the true day–night cycle.

Biological clocks remain synchronized with the day–night cycle because they are often reset by external inputs. For the circadian clock, light is the primary input, so the natural light–dark cycle keeps the clock from drifting. For clocks related to the seasons, day length (known as **photoperiod**) is the critical input because it is a good indicator of the time of the year (Chapter 30). For example, many mammals produce their offspring in the spring so that they can grow over the summer, when resources

HOW DO WE KNOW?

FIG. 45.12

Does a biological clock play a role in birds' ability to orient?

BACKGROUND One suggestion for how pigeons home is that they use a sun compass. If you are in the Northern Hemisphere and you know the time is 12 noon, then the sun is due south of you. Orienting yourself by this method is possible only if you know the time, so the question arises whether homing birds have the ability to tell the time. One way to answer this question is to "clockshift" the birds. Researchers raise birds in an artificial day–night cycle that is out of sync with the actual one. When released into a sunlit environment, these birds' sense of time is shifted by a set number of hours.

event	dawn	dusk
actual time	6 am	6 pm
clockshifted time	12 noon	12 midnight

HYPOTHESIS If a bird's ability to home is dependent on an internal clock, clockshifting should affect the bird's homing ability in a predictable way. Given that the sun travels 360 degrees in 24 hours, a 6-hour clockshift will result in a 90-degree error in homing direction because $360/(24/6) = 90$.

EXPERIMENT Birds were clockshifted by raising them in a chamber under an artificial light. Birds from a home loft in Ithaca, New York, were released on a sunny day at Marathon, New York, about 30 km east of Ithaca. Release on a sunny day made it possible for the birds to use the sun's position to navigate.

RESULTS As expected, the control birds (those that were not clockshifted) were usually good at picking the direction of their home loft, heading approximately westward toward Ithaca. The results for the clockshifted birds were very different: They miscalculated the appropriate direction. These birds headed approximately northward, as shown by the positions of the red triangles on the compass in the figure.

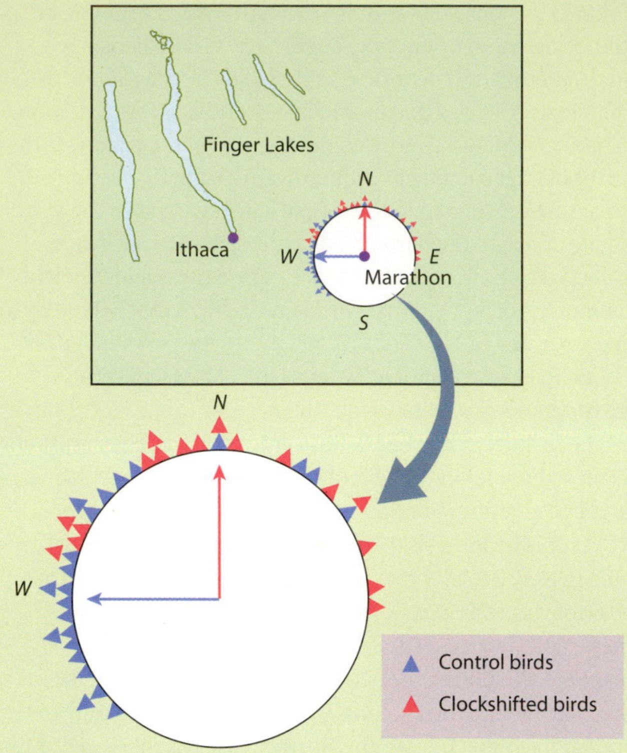

INTERPRETATION Assume the birds are released at 12 noon, when the sun is due south. The control birds know to fly in a direction 90 degrees clockwise from the direction of the sun, but the clockshifted birds "think" it is 6 p.m., so they expect the sun to be in the west. Their 90-degree clockwise correction, then, has them flying due north.

CONCLUSION The clear difference between control and clockshifted birds in the experiment shows that an internal time-based sun compass is an important component of the birds' homing abilities. However, the scatter of points (for both experiment and control) suggests that other factors are also important. This conclusion is reinforced by the observation that birds home well on cloudy days, when they cannot use a sun compass, suggesting that birds use multiple cues and navigational systems when they are homing.

SOURCE Keeton, W. T. 1969. "Orientation by Pigeons: Is the Sun Necessary?" *Science* 165:922–928.

are most abundant. These mammals have been selected over many generations to synchronize their reproduction at a time when the young have the greatest chance of survival. Photoperiod determines the timing of many kinds of behaviors, including migration, development, and reproduction.

The importance of the sun compass coupled with the biological clock in homing pigeons can be demonstrated by experimentally disturbing the birds' clock, as shown in **Fig. 45.12.**

→ **Quick Check 3** Why is photoperiod such a widely used time cue?

45.5 COMMUNICATION

Up to this point, we have focused on individual behaviors. Behaviors often depend on environmental cues, as we saw with the examples of navigation and keeping time. In addition, behaviors are often shaped by interactions with other individuals. Learning sometimes depends on communication between a teacher and student, and, as anyone knows who has ever watched birds tending their young, communication is central to parenting.

Communication is the transfer of information between a sender and receiver.

The sophistication of animal communication varies enormously, as can be seen by the variety of types of signal illustrated in **Fig. 45.13.** Even our closest relatives, the chimpanzees, cannot rival the human facility with language, but a number of species have evolved forms of communication that convey remarkably complex and specific information. For example, a vervet monkey that perceives a threat to its group will utter an alarm call that not only warns of a threat, but actually specifies the nature of the threat—whether it is a hawk, leopard, snake, or some other type of predator.

The simplest definition of **communication** is the transfer of information between two individuals, the **sender** and the **receiver.** The sender supplies a signal that elicits a response from the receiver. For example, the bright petals of a flower signal to an insect that nectar and pollen are available. This definition, however, has its problems. An owl hears the rustling of a mouse and responds accordingly. Most people would agree that the owl is not really communicating with the mouse (nor the mouse with the owl). For this reason, some biologists prefer to define communication as attempts by the sender to manipulate in some way the behavior of the receiver.

How has communication evolved? It is thought that communication has often evolved through co-opting and modifying behaviors used in another context. This process is called **ritualization,** and it involves (1) increasing the conspicuousness of the behavior; (2) reducing the amount of variation in the behavior so that it can be immediately recognized; and (3) increasing its separation from the original function. The scent markings with which mammals mark their territory provide an example. The original function was simply the elimination of waste, but this function has been modified through evolution: Strategically placed marks communicate with other individuals, indicating, for example, the extent of an individual's territory. The communication advantage is that an intruder detecting a territorial scent may be less likely to invade, thereby avoiding what could be an expensive fight.

FIG. 45.13 **Types of signal in different forms of communications.** Signals can be (a) visual, as in the male frigatebird's display, (b) auditory, such as a male frog's call, (c) electrical, as in African electric fish, which generate species-specific electric pulses, (d) chemical, such as the pheromones with which a queen termite controls her colony, and (e) mechanical, as when spiders "twang" webs to interact. *Sources: a. Dan L. Perlman/EcoLibrary.org; b. ANT Photo Library/Science Source; c. © Matthew E. Arnegard; d. Dan L. Perlman/EcoLibrary.org; e. Ron Rowan Photography/Shutterstock.*

Many forms of communication have evolved that prevent animals from coming to harm in what has sometimes been called a limited-war strategy. Two males, instead of battling it out, may engage in elaborate displays to size each other up, either literally by standing side by side or through displays of physical prowess such as roaring, in order to determine who is dominant. Fighting would be disadvantageous to both since even the winner might be seriously injured. In this case, communication may have evolved as a way for individuals to assess one another.

In some cases, communication in the natural world can be deceitful. A male may attempt to convince other males (or a female) that he is bigger (and stronger) than he really is. Possibly one reason that dogs circling in a fight raise the hair along their back is to inflate their apparent size. A potential prey may attempt to convince a predator that it is not, in fact, prey. This form of deceit is especially evident in some species of butterfly that mimic leaves or bark when their wings are at rest. Alternatively, predators may emit deceitful signals to entice their prey. Females of one species of firefly, for instance, mimic the flashes of the females of another species in order to lure males of the second species, which they then eat.

Some forms of communication are complex and learned during a sensitive period.

Bird song is one of the best-known and richest forms of animal communication. Because of the clear connection between what birds hear and how they respond, songbirds have served as a model system in the study of learning and communication. These songs are complex sequences of sounds, often repeated over and over. Like cricket and frog calls, bird songs are **advertisement displays,** behaviors by which individuals draw attention to their status. For example, the displays might indicate that they are sexually available or are holding territory. Bird songs are typically produced by males in the breeding season.

This sex difference in patterns of communication often results from intersexual selection (Chapter 21), and can extend to other traits. For example, male birds usually have bright plumage and elaborate displays, and females do not. Why do we see such differences between males and females? The key is the level of investment that an individual makes in an offspring. The eggs a female bird produces are metabolically expensive, whereas the sperm a male bird produces are metabolically cheap. Because of this and other costs associated with reproduction, a female typically only has one opportunity or a few opportunities to reproduce. It is therefore in her interest to ensure that she has chosen a high-quality male—one that sings well or has especially fine plumage. By contrast, it is in the male's interest to maximize his access to females—that is, to be as sexually attractive as possible. Sexual selection therefore tends to produce showy, attractive males and females that choose males on the basis of their performance or appearance.

In many species of birds that sing, some or all of the song is learned, often during a specific, sensitive period. Detailed studies of the White-crowned Sparrow, shown in **Fig. 45.14,** pioneered by neurobiologist and ethologist Peter Marler, and extended by Luis Baptista, have yielded the following general picture of the process.

Song in White-crowned Sparrows is learned by imprinting: The young male hears adult song during a sensitive period, 10–50 days after hatching. Then, shortly afterward, young male White-crowned Sparrows produce unstructured twittering sounds, known as a subsong, comparable to the babbling of human babies. If deprived of hearing adult song, for example by isolation during the sensitive period, the bird sings for the rest of his life a song not much different from unstructured twittering, even if he hears another male sing both before and after isolation.

Between about 250 and 300 days, the male sings an imperfect copy of the song (known as plastic song), and by about 300 to 350 days after hatching, song acquisition is complete (this song is known as structured song). At this point, even if the bird is deafened in the lab, he will still sing correctly. The song he produces is a precise copy of the one he learned, typically by hearing another male's song during the sensitive period, complete with most of its individual as well as species-specific characteristics.

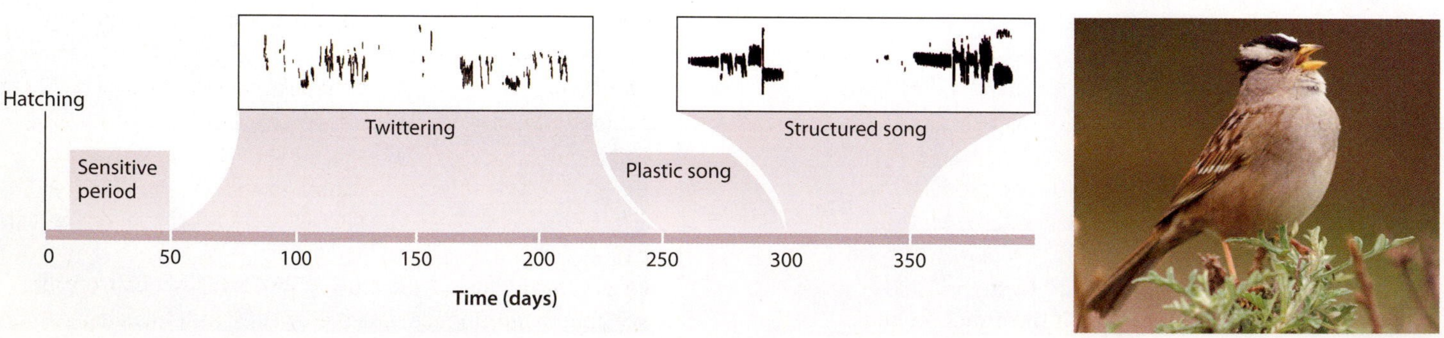

FIG. 45.14 Song acquisition in the White-crowned Sparrow. Sonograms show differences between the unstructured twittering of early life and the final, structured song. *Sources: After T. J. Carew, 2000, Behavioral Neurobiology, Sunderland, MA: Sinauer Associates, p. 244; Photo: Paul Sutherland/ Getty Images.*

FIG. 45.15 The waggle dance of honeybees. This form of communication provides information to other foragers about the distance to and direction of a food source. *Source: After M. B. Sokolowski, 2001, "Drosophila: Genetics Meets Behaviour," Nature Reviews Genetics 2:879–890, doi:10.1038/35098592.*

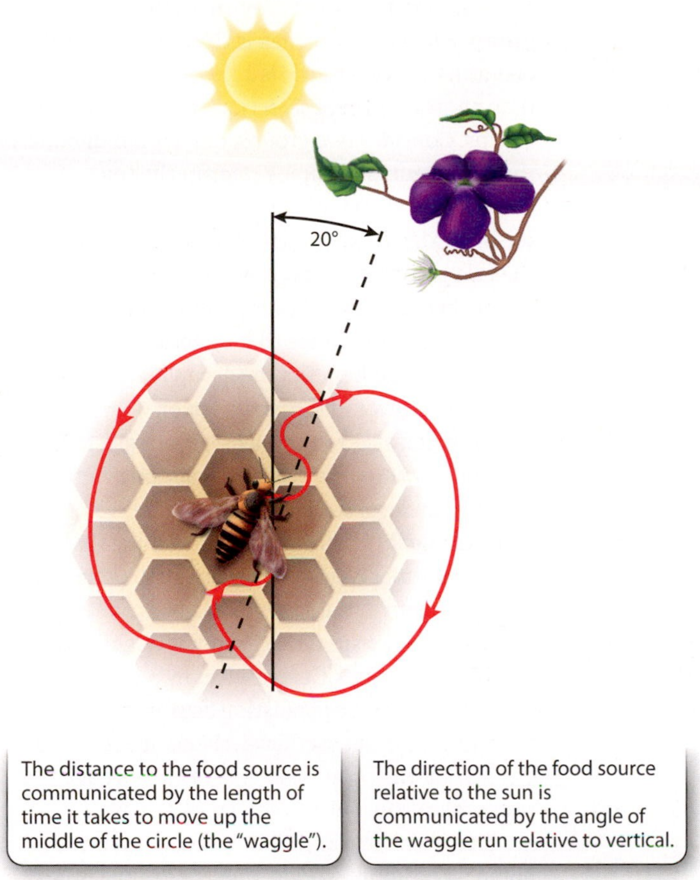

So White-crowned Sparrows have a programmed predisposition for *when* song learning takes place. Even more interesting, *what* can be learned is similarly constrained. If a tape of another species' song is played during the sensitive period, even the song of closely related Swamp Sparrows, the White-crowned male cannot learn that song; his adult song ends up not much different from his unstructured twittering. However, if he is provided with a live tutor, such as a live male Swamp Sparrow rather than a tape, his ability to learn the other bird's songs is greatly improved. Furthermore, if he hears tapes of both the song of his own species and that of closely related species played together, he can pick out the correct elements and sing a perfect White-crowned Sparrow song. Thus, birds of this species preferentially learn their species-specific song, but they cannot sing a normal song without learning it from another bird.

Song learning in White-crowned Sparrows has been well studied, and provides just one example of song learning in birds. Some species, such as mockingbirds, starlings, and canaries, learn a wider range of songs than do White-crowned Sparrows, and have a longer period of song learning, in some cases probably lifelong. By contrast, other species, such as many New World flycatchers, apparently have hard-wired (innate) songs. Other sound-based forms of communication, such as those seen in dolphins and whales, are only beginning to be decoded and understood.

Other forms of communication convey specific information.

Honeybees have an elaborate means of communication that is quite different from the sound-based communication of birds. Bees travel far and wide to forage for food for their hive and, upon returning home, use a variety of cues to inform outgoing foragers about the location and nature of the food resources they have found.

Returning workers pass samples of the collected food to other workers that gather around the forager when she returns to the hive. The returning foragers then begin a series of movements in specific patterns that encode information about the direction and distance of food sources. This is the dance language of honeybees, originally discovered by Karl von Frisch.

If the food source is more than about 50 m away, the incoming forager does a waggle dance. The steps of the dance are illustrated in **Fig. 45.15.** The forager moves quickly back and forth ("waggles") as it moves forward, circles back and repeats the waggle, and then circles back the other way, and so on. This dance usually goes on for several minutes and may take much longer. The distance to the food source is conveyed by the length of time it takes to waggle up the middle of the circle. The direction, with reference to the position of the sun, is conveyed by the angle of the line of the waggle run: If the food source is in a direct line with the sun, the dancer's waggle line points straight up; if food is at an angle of 45 degrees to the left of the sun, the dancer's line is 45 degrees to the left of vertical. Short of human language, the bees' dance language is among the natural world's most extraordinary and effective methods of communication.

45.6 SOCIAL BEHAVIOR

Returning to Tinbergen's questions, we can consider further the adaptive function and evolutionary history of animal behavior. The evolutionary advantages of most behaviors are readily apparent. The imprinting of Lorenz's goslings on the first moving object they encountered makes sense when you consider that that object is usually their parent (not Konrad Lorenz!) and that sticking close to their parent is an excellent survival strategy in a world filled with predators.

Social behaviors are also evolutionary advantageous. Consider the cooperative hunting seen in killer whales. Killer whales

FIG. 45.16 Mutually beneficial cooperation. Killer whales cooperate to create a wave to wash prey off an iceberg. *Sources: (top) John Durban/NOAA; (bottom) R. Pitman/NOAA SWFSC.*

participate in a behavior called wave washing, in which a pod of whales collectively creates waves to wash a prey species—typically a seal—off an iceberg (**Fig. 45.16**). This behavior is collaborative, and it is clear why each individual participates. By itself, a single killer whale is incapable of generating a big enough wave, but, through cooperation, an individual can contribute to prey capture and gain some portion of the prey.

Sometimes, however, how a behavior enhances survival and reproduction is less clear. Perhaps the most prominent examples of this kind of behavior are sacrifices made by individuals apparently for the good of others. Formally, we describe an act of self-sacrifice that helps another as **altruistic.** Such acts seem to fly in the face of natural selection, which selects for traits that increase fitness and selects against traits that decrease fitness. But altruistic acts, by definition, decrease fitness and therefore should be selected against. This contradiction is particularly stark in the case of species like honeybees: Honeybee workers, which are females, do not reproduce but instead assist the reproduction of the queen. Darwin confessed that such behaviors caused him sleepless nights. How, Darwin wondered, could altruism arise?

Group selection is a weak explanation of altruistic behavior.

Darwin himself suggested a solution to the altruism problem, a solution that today we call **group selection.** The idea is that natural selection operating on individuals is a less powerful force than another form of selection that operates on groups. Consider two groups, one with individuals that are altruistic, helping one another out, and another with individuals that act entirely selfishly. Darwin argued that, in a conflict between the two groups, the first group would beat the second because the first group works better as a unit. Here, then, we see that selection in favor of altruism that benefits the group has trumped selection in favor of selfish behavior.

Although an attractive idea, group selection is probably not important in the evolution of altruism. Altruism under group selection is typically not an **evolutionarily stable strategy,** meaning that this kind of behavior can be readily driven to extinction by an alternative strategy. It used to be thought that lemmings, a species of Scandinavian rodent, would collectively kill themselves by plunging into the ocean when the population became too large to sustain. This self-sacrificial, altruistic act would ensure that the population did not exhaust available resources and would prevent a population crash.

This is a nice story, but untrue. Lemmings do not in fact commit suicide; they take to the water to swim to new territories during migration. And we can see the evolutionary logic prohibiting the evolution of such altruistic behavior. An individual that "cheats" by not performing the altruistic act will survive and breed, and its altruistic companions will not. Now consider that this behavior—to kill oneself or not—is genetically encoded. A mutation that arises in a population that causes the individual to "cheat" and not kill itself will spread rapidly through the population by natural selection until, after many generations, it has completely replaced the altruistic allele that causes the individual to kill itself. The lemming behavior, then, is not an evolutionarily stable strategy because natural selection favors an alternative strategy.

This description is oversimplified because there is unlikely to be a simple genetic basis to traits such as these, but the point stands: Group selection is in general not an evolutionarily stable strategy because it can readily be overthrown by "selfish" strategies.

→ **Quick Check 4** It's sometimes said that the reason an animal does something is that it's "for the good of the species." Why is this argument incorrect?

Reciprocal altruism is one way that altruism can evolve.

If group selection does not provide a strong explanation for the evolution of altruism, how can we explain this behavior? **Reciprocal altruism,** whereby individuals exchange favors, is one way that altruism can evolve. This idea is perhaps best summarized as "You scratch my back, and I'll scratch yours." For this kind of behavior to work, individuals have to be able to recognize each other and remember previous interactions. If I did you a favor, I need to be able to recall who you are so that I can be sure that I get the return I expect.

A famous example of reciprocal altruism comes from an unlikely source—vampire bats. Adult female vampire bats live in groups of 8 to 12, and often remain in these groups for years: In one study, two tagged females shared the same roost for 12 years. The female groups are occasionally joined by unrelated females, so they are not all closely related. Because they have long-term associations, there is opportunity for reciprocal interactions.

Feeding for these bats is a "boom or bust" activity—either a bat has found and successfully fed on a blood source (boom), or a bat has failed to feed (bust). As a result, bats are often either overfed or underfed. Underfed bats are at risk of dying by starvation. Bats returning from a successful feeding expedition often regurgitate blood to unsuccessful ones, on the expectation that the favor will be returned at a future date.

The concept of kin selection is based on the idea that it is possible to contribute genetically to future generations by helping close relatives.

In the 1960s, we came to understand that the relatedness of participating individuals has played an important role in the evolution of altruism. This understanding lay in the insight that there are at least two ways to contribute genetically to the next generation: Either an individual can produce its own offspring, or it can produce no offspring and help close relatives reproduce instead. The latter strategy can become established in a population through a process called **kin selection,** a form of natural selection that favors the spread of alleles that promote behaviors that help close relatives, or kin.

What relationships qualify as kin? Consider the family depicted in **Fig. 45.17,** with one animal (indicated by gray shading in the figure) that has an allele A at a particular locus. If the animal reproduces, the probability of each offspring inheriting an A allele is 0.5. But what if the animal does not reproduce? According to Mendel's laws, there is a 50% chance that the animal's parents

FIG. 45.17 **Computing relatedness.** One's relatedness (r) to another individual is the probability that one shares, through inheritance, a given segment of DNA with that individual. Here, relatedness is calculated with reference to the individual shown on the gray background.

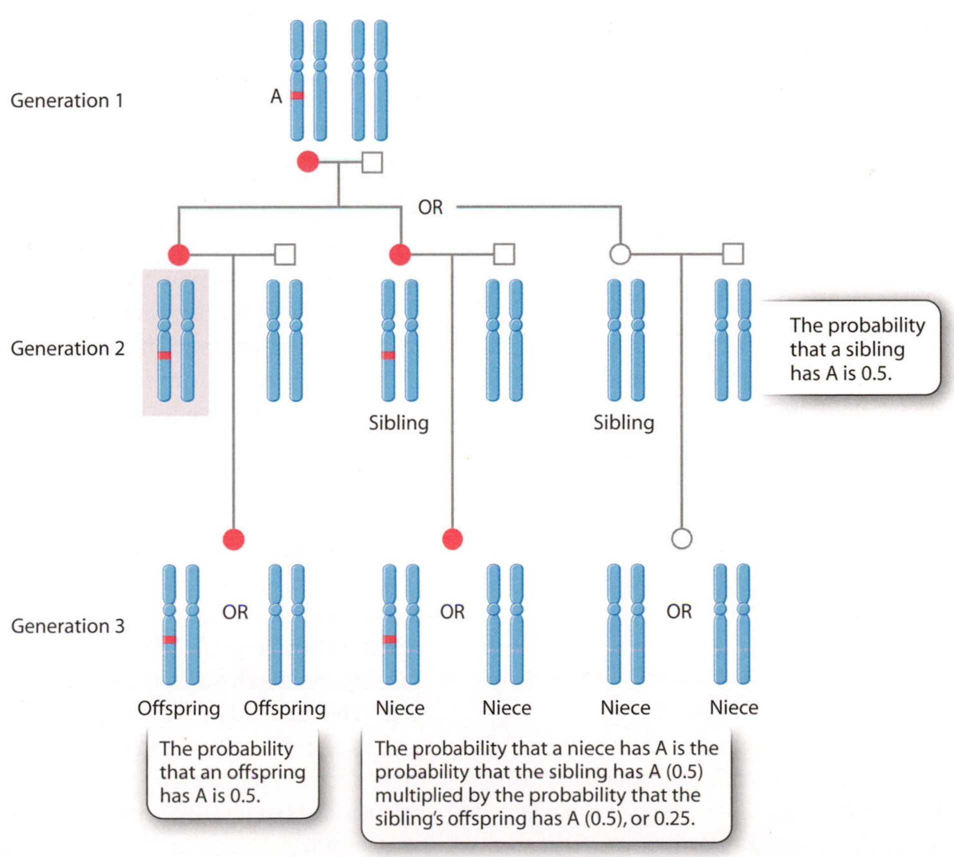

FIG. 45.18 Relatedness in haplodiploid insects, such as ants. Sisters are more related to each other (0.75) than they would be to their own offspring (0.5) if they could reproduce. *Photos source: Alex Wild/alexanderwild.com.*

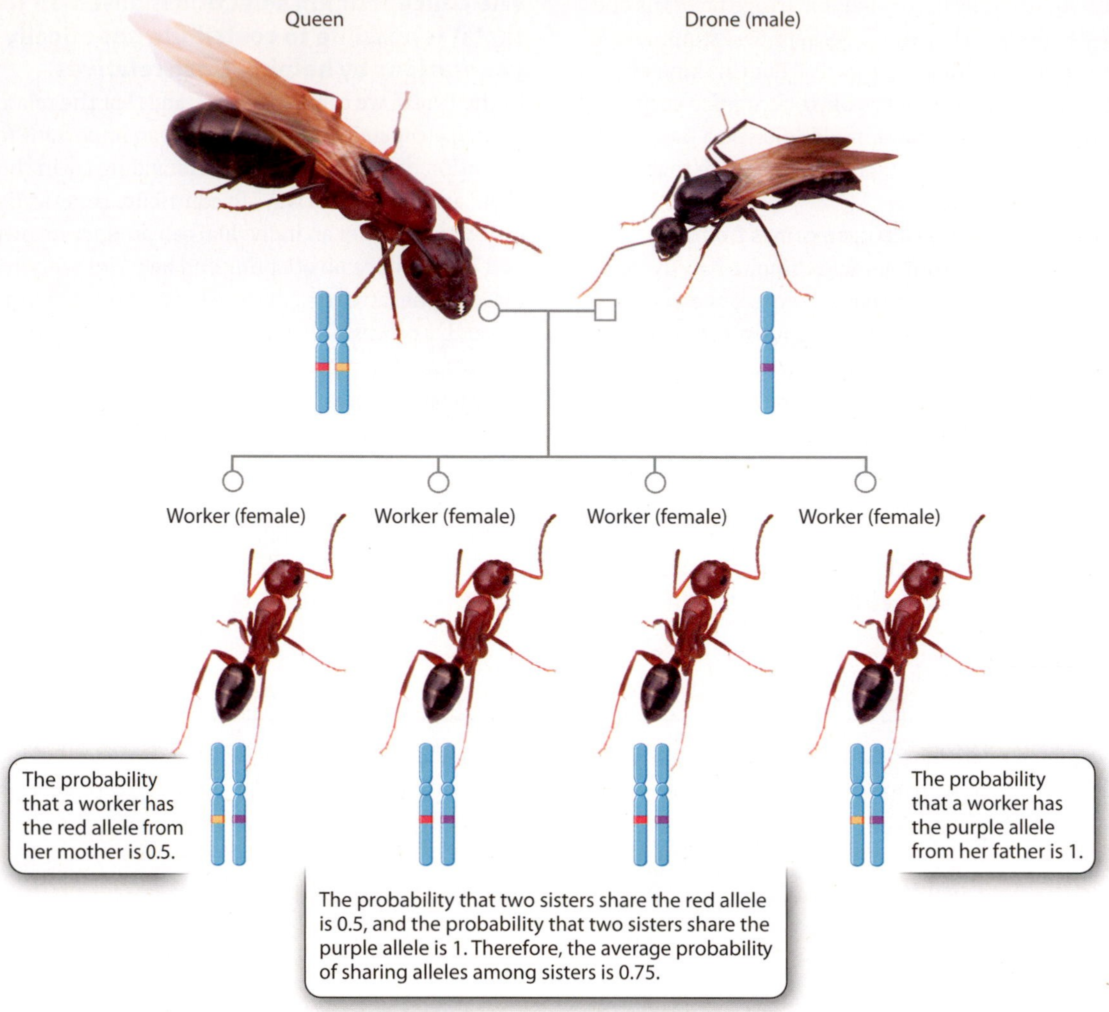

passed on the same *A* allele to the animal's sibling (Chapter 16). If the sibling does, in fact, inherit allele *A*, the probability that it passes allele *A* to each of its offspring is also 0.5. Given the two probabilities (that the sibling has an *A*, and that the allele passed on to an offspring), the probability that the animal's niece or nephew has *A* is 0.25. In other words, a diploid animal (such as humans) is twice as closely related to its own offspring than to its nephew or niece. Therefore, if the animal does not reproduce in order to help its sibling have two offspring, the animal will make the same net genetic contribution to the next generation as it would having one of its own offspring.

→ **Quick Check 5** How many cousins are genetically equivalent to a single offspring?

This logic was formalized in 1964 in a famous paper by British evolutionary biologist William D. Hamilton in terms of three key variables: *B*, the benefit of the behavior to the recipient (how many individuals does the behavior save?); *C*, the cost of the behavior to the donor (how many offspring are lost?); and *r*, the degree of relatedness between the recipient and donor. Hamilton asserted that as long as *rB* exceeds *C*, altruism can evolve.

How in practice does kin selection work? Let's start with an extreme case of relatedness: clonal organisms that are genetically identical to each other (meaning that the probability of a sibling's having allele *A* is 1). Many aphid species have a phase of their life cycle in which a "mother" aphid reproduces parthenogenetically to produce a small colony of clones of herself (Chapter 42). Any aphid in the colony is capable of reproducing in this way, but in a case of reproductive altruism, some individuals do not reproduce at all. They are "soldier" aphids that defend the colony against predators, typically other insects. As specialized defenders, they do not have the ability to reproduce. In this case, the Hamilton calculation is simple. By helping their clone-mates survive, the sterile soldier aphids are ensuring that their own genes survive to be passed on to the next generation.

The most famous cases of reproductive altruism occur in the insect group Hymenoptera, which includes ants, bees, and wasps. Many species of Hymenoptera are **eusocial,** meaning that they have overlapping generations in a nest, cooperative care of the young, and clear and consistent division of labor between reproducers (the queen of a honeybee colony) and nonreproducers (the workers). Often called social insects, these species are one of evolution's most extraordinary success stories.

What is it about Hymenoptera that predisposed them to eusociality? E. O. Wilson argues that natural selection has strongly favored group behavior in these insects. They typically receive group benefits from living in and defending a nest. For example, they can easily communicate with one another about the location of food resources such as nectar.

William Hamilton also noted that most Hymenoptera have an unusual mode of sex determination: Females (the queen and workers) are diploid, whereas males (drones) are haploid, produced from unfertilized eggs. As a result, the degree of relatedness, *r*, is higher for the sister–sister relationship (0.75) than for the mother–daughter one (0.5), as shown in **Fig. 45.18.** Put another way, a female is more closely related to her sister than she is to her daughter. Therefore, it makes more evolutionary sense, in terms of genetic representation in the next generation, for a worker to help her mother, the queen, to produce more sisters than for her to produce her own offspring.

This form of social organization can result in a huge colony of related individuals all working collectively, much as different cells work collectively within a body. Often, we see specialization into castes: Some ants are soldiers, defending the colony, others are foragers, and so on. If we extend the analogy of individuals acting like cells in a body, the multiple castes are equivalent to organs, all coordinating and working together. The term "superorganism" is sometimes used to describe a single social insect colony. ■

Core Concepts Summary

45.1 For any behavior, we can ask what causes it, how it develops, what adaptive function it serves, and how it evolved.

Tinbergen pointed out that questions about animal behavior can be asked at four different levels of analysis, focusing on cause, development, function, and evolution. page 982

The four questions have different answers, and all may be correct explanations of the behavior. page 982

Behavior can be innate (it is carried out in the absence of experience), or learned (it depends on experience). page 982

45.2 Animal behavior is shaped in part by genes acting through the nervous and endocrine systems.

Most behavior is produced by the interaction between genes (nature) and environment (nurture). page 983

A fixed action pattern is a stereotyped sequence of behaviors that, once initiated by a key stimulus, always goes to completion. page 983

Stimuli are recognized by feature detectors, which are specialized sensory receptors of the nervous system. page 984

Hormones can affect behavior through their wide-ranging and often long-term effects on multiple body systems. page 985

The extent to which genes influence a behavior can be determined by crossing closely related species having different behaviors and analyzing the behavior of the offspring. page 986

Most behaviors are influenced by many genes, but a few are strongly influenced by a single gene. page 987

Molecular studies are providing new ways to understand the role of genes in behavior in laboratory animals. page 987

45.3 Learning is a change of behavior as a result of experience.

Non-associative learning is a form of learning that occurs without linking two events. page 990

Habituation, a form of non-associative learning, is the reduction or elimination of a response to a repeated stimulus. page 990

Sensitization, another form of non-associative learning, is the enhancement of a response to a stimulus that is achieved by first presenting a novel stimulus. page 990

Associative learning occurs when an animal links two separate events. page 990

Classical conditioning is a form of associative learning that occurs when two stimuli are paired and a novel association is made between a formerly neutral stimulus and a behavior. page 990

Operant conditioning is a form of associative learning that occurs when a behavior is rewarded or punished, making the behavior more likely or less likely to occur, respectively. page 990

Imitation and imprinting are forms of learning. page 990

45.4 Orientation, navigation, and biological clocks all require information processing.

Kineses are undirected movements, whereas taxes are directed movements. page 992

Long-distance navigation may require the processing of many external cues, including the position of the sun, Earth's magnetic field, and information about time. page 993

Biological clocks control some physiological and behavioral aspects of animals. page 993

45.5 Communication involves an interaction between a sender and a receiver.

Communication is an interaction between two individuals, the sender and the receiver, and often involves attempts by the sender to manipulate the behavior of the receiver. page 995

Communication can evolve by the modification of noncommunicative aspects of an animal's behavior. page 995

Bird song is a rich form of communication that is often learned during a sensitive period. Acquisition of bird song proceeds through distinct stages. page 996

The dance language of honeybees communicates information about the location and direction of a food source. page 997

45.6 Social behavior is shaped by natural selection.

One of the most perplexing behaviors from the standpoint of natural selection is altruism, in which an organism reduces its own fitness by acts that benefit others. page 997

Group selection is the idea that altruistic behaviors can evolve if they benefit the group, not the individual, but this idea is not widely accepted. page 998

Reciprocal altruism requires that individuals interact regularly and remember each other so that they can "pay back" past favors. page 999

Kin selection is the concept that altruistic behaviors can evolve if they benefit close relatives. page 999

Self-Assessment

1. Choose a behavior described in this chapter and ask the four kinds of question that Tinbergen might have asked about it.
2. Explain the difference between innate and learned behavior and provide an example of each.
3. Explain, with an example, how crosses between closely related species can help us understand the genetic basis of behavior.
4. Distinguish between associative and non-associative learning and provide an example of each form of learning.
5. Differentiate between a kinesis and a taxis.
6. Explain why many biologists would not consider just any transfer of information between a sender and receiver to be a form of communication.
7. Describe how an altruistic behavior might evolve by kin selection.

Log in to LaunchPad to check your answers to the Self-Assessment questions, and to access additional learning tools.

CHAPTER 46

Population Ecology

Eladio Fernandez.

Core Concepts

46.1 A population consists of all the individuals of a given species that live and reproduce in a particular place.

46.2 Population size can increase or decrease over time.

46.3 The age structure of a population helps ecologists understand past changes and predict future changes in population size.

46.4 The dynamics of populations are influenced by the colonization and extinction of smaller, interconnected populations that make up a metapopulation.

In 2004, undergraduates on a field trip to the Dominican Republic, on the island of Hispaniola, swept up an unfamiliar black-and-white butterfly in their nets. That butterfly was a Lime Swallowtail (*Papilio demoleus*), the very first of its species, native to Asia, to be captured in the New World.

Soon, Lime Swallowtails spread to all corners of Hispaniola, a serious concern because their caterpillars strip the leaves off young lime and orange trees. Within 3 years, the Lime Swallowtail had also established itself in nearby Puerto Rico and Jamaica, and it has now reached Cuba. From there, it is just a short flight to Florida and the multibillion-dollar citrus industry of the United States. Can we enlist basic principles of biology to help understand this butterfly's Caribbean expansion and predict its future as it continues to spread?

Ecology is the study of the relationships of organisms to one another and to the environment. Ecological relationships sustain a flow of energy and materials from the sun, Earth, and atmosphere through organisms, with carbon and other elements eventually returning to the environment (Chapter 25). It is these interactions that determine the shifting ranges of Lime Swallowtails and other species on Earth.

In previous chapters, we discussed the anatomical, physiological, reproductive, and behavioral characteristics of organisms. These attributes are the products of evolution, and they determine the ways in which individual organisms interact with their physical and biological environment. Physical, or **abiotic,** factors that influence evolution include climate and nutrient availability. Biological, or **biotic,** factors include competitors, predators, parasites, and prey, as well as organisms that provide shelter or food.

In the following chapters, we develop the ideas of ecology in the context of interactions between organisms and species, and of the different strategies they can follow to ensure their survival and fitness. We trace the processes that govern interactions from the level of the evolved traits of individual organisms to the continent-spanning patterns of diversity and ecosystems. Ecology and evolution, then, are closely intertwined.

46.1 POPULATIONS AND THEIR PROPERTIES

When the first United States census was completed in 1790, the British scholar Thomas Malthus supposed that the remarkable recent growth of the American population, which had doubled in the previous 25 years, was a consequence of ample food supply and the active encouragement of marriages, neither of which characterized Europe at the time. In his 1798 volume *An Essay on the Principle of Population,* Malthus argued that a similar periodic doubling of population in the British Isles, on the other hand, would outstrip the more slowly increasing food supply in 50 years and lead to starvation. That is, increasing the number of births per year would eventually result in an increased number of deaths

FIG. 46.1 **(a) Size, (b) range, and (c) density of a population.** The places where populations occur can be mapped, and the numbers of individuals within those populations can be estimated. This information helps us understand what factors limit population distributions. *Sources: Data from A. Atkinson, V. Siegel, E. A. Pakhomov, M. J. Jessopp, V. Loeb, 2009, "A Reappraisal of the Total Biomass and Annual Production of Antarctic Krill," Deep-Sea Research I 56:727–740. Photos: Richard Hermann/SeaPics.com; (inset) British Antarctic Survey/Getty Images.*

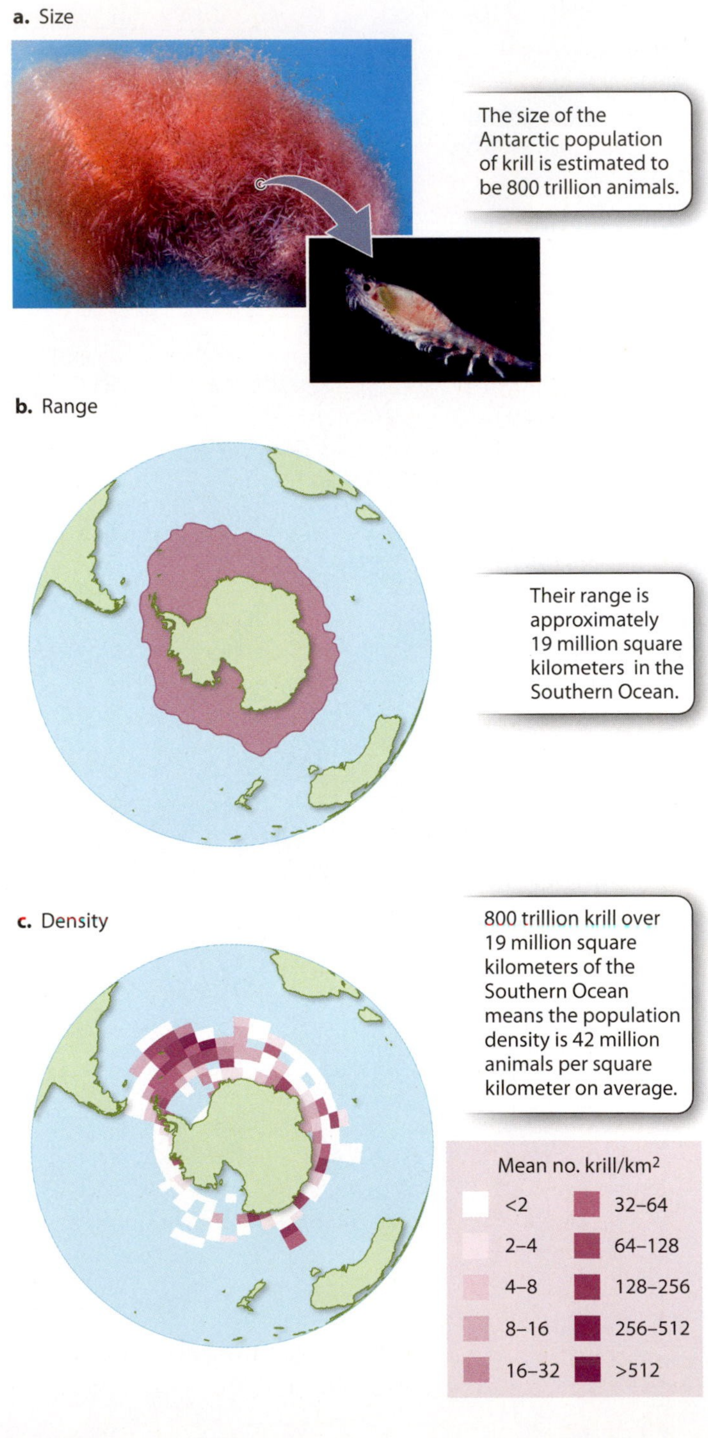

because the amount of food the land could produce was limited. This view of the human condition influenced the thinking of Charles Darwin, who as an inquisitive student read the most exciting works of the day. Darwin realized that the tension between population growth and resource limitation applies to all species, leading to a struggle for existence that results in adaptations that enhance survival (Chapter 21).

A population includes all the individuals of a species in a particular place.

In Chapter 21, we introduced the concept of population as a fundamental unit of evolution—it is populations, not individuals, that evolve. Here, we return to populations, this time as equally fundamental units of ecology. A **population** consists of all the individuals of a given species that live and reproduce in a particular place. Natural selection enables populations to adapt to the abiotic and biotic components of their environment. It is the traits evolved under natural selection that determine whether a population will grow or shrink under a given set of physical and biological conditions, perhaps disappearing entirely from a local environment.

How are populations characterized? What controls changes in population size? We first consider key features of populations. Then we look at what happens when there are more births than deaths, and what happens when resources such as food start to run out. We consider how individuals of different ages contribute to population growth, and how local populations are connected to one another. In the next chapter, we introduce interactions with other species and see how they shape the growth of populations. All these factors directly influence the flow of energy and materials through producers and consumers and, ultimately, the flow of genes through a population.

Three key features of a population are its size, range, and density.

Populations are defined by three key features: size, range, and density (**Fig. 46.1**). **Population size** is simply the number of individuals of all ages alive at a particular time in a particular place. For example, the population of krill in the cold waters around Antarctica is estimated at 800 trillion, making them by mass the most abundant animals on Earth (Fig. 46.1a).

A second key feature of a population is its geographic **range:** how widely a population is spread. Fig. 46.1b, for example, shows that the Antarctic krill population lives in an area of 19 million square kilometers around Antarctica. The range of any species reflects the range of climates a population can tolerate and determines how many other species the population encounters. The range of a plant population, for example, influences how many kinds of insect eat it, and the range of an insect population determines how many kinds of plant it encounters.

Taken together, population size and range contribute to a third key attribute of a population, its density. **Population density** is defined as a population's size divided by its range. In effect, it tells us how crowded or dispersed the individuals are that make up the population. For the Antarctic krill population, 800 trillion animals spread over 19 million square kilometers yields a density of 42 million animals per square kilometer (Fig. 46.1c). As we will see, population density can affect population size.

This way of characterizing population density, as average density across the range of a population, is often useful, but it can be misleading depending on how the organisms are distributed (**Fig. 46.2**). For example, the individuals within a population may be distributed randomly. That is, a new individual has an

FIG. 46.2 Distribution of individuals within populations. A population may be (a) randomly distributed, (b) clumped, or (c) uniform.

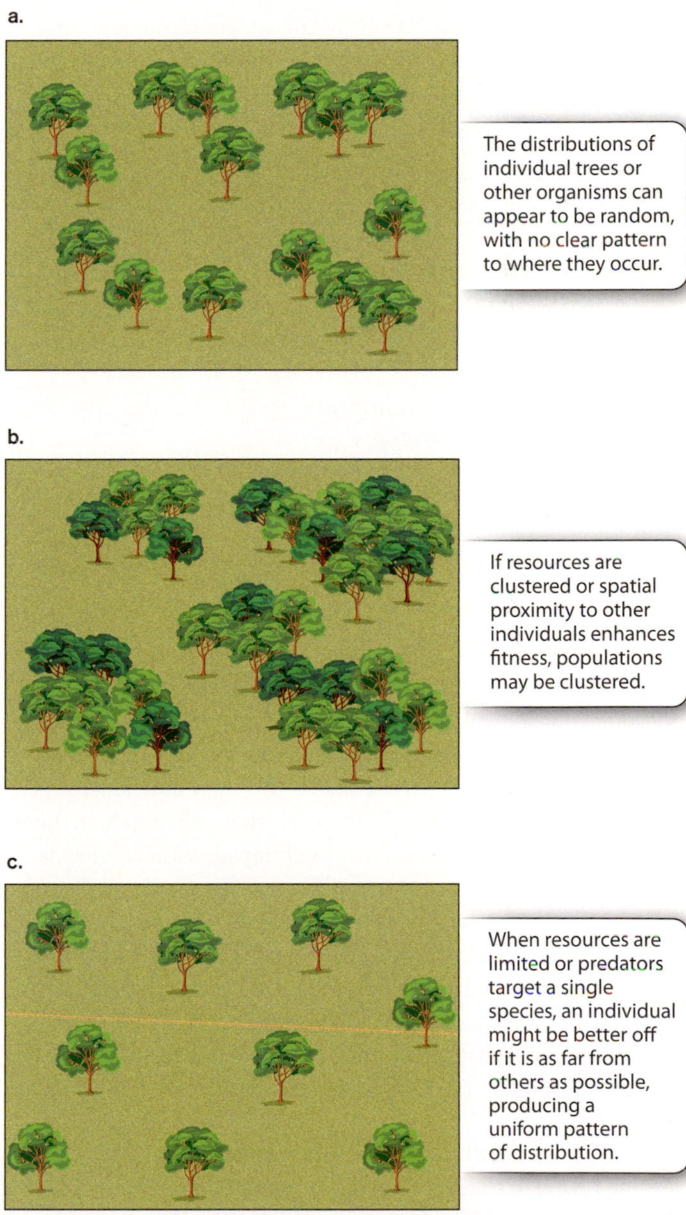

HOW DO WE KNOW?

FIG. 46.3

How many butterflies are there in a given population?

Photo source: AP Photo/Orlin Wagner.

BACKGROUND Individuals in many populations of butterflies (and other animals) most often live and die within the patch of habitat where they were born. Population sizes of such animals can be estimated by a technique called mark-and-recapture, in which organisms are captured and released on two successive days. We assume that the population size is about the same on the second day as the first day. We also assume that the captured and released animals mixed with the population and do not avoid recapture on the second day.

METHOD Butterflies, such as the Monarch, are captured in an area, marked with an identification sticker on the wing as shown here, and then released. The number of marked butterflies is recorded. The next day, butterflies are caught and the number of marked and unmarked butterflies is recorded. Butterflies with marks are the recaptures.

To find the population size (N), we take the total number of marked butterflies and unmarked butterflies caught on the second day (C) and divide by the number of recaptures (R). We multiply this number by the number of butterflies marked on the first day (M), as follows:

$$N = (C/R) \times M$$

RESULTS Let's say we capture 100 butterflies on the first day, mark them, and release them. On the second day, we capture 120 butterflies, 30 of which are marked and the rest unmarked. We conclude that the population size is $(120/30) \times 100$, or 400 butterflies total in the population sampled.

RELATED METHODS Mark-and-recapture techniques have proved extremely useful in estimating population sizes of organisms from polar bears to disease victims. There are refinements to the mark-and-recapture approach that take into account the movement of individuals between populations. These methods involve taking multiple samples and adding variables to the equation. Many animals have distinctive color markings, such as the blotches on humpback whale tailfins and frog skin-spotting, and so can be recorded without having to be marked.

SOURCE Lincoln, F. C. 1930. "Calculating Waterfowl Abundance on the Basis of Banding Returns." *Cir. U.S. Department of Agriculture* 118:1–4.

equal chance of occupying any position within the range, and the location of one individual has no influence on where the next will occur (Fig. 46.2a). Sometimes, however, resources are distributed patchily within the range, or the chances of a new individual's survival are enhanced by the presence of other individuals. In these circumstances, the population may be clumped, with individuals clustered together (Fig. 46.2b). In still other cases, one individual of a population might prevent another from settling nearby, for example if all individuals depend on the same set of resources. Or predators, having located one prey individual, may search for other members of the same species nearby. Under these conditions, individuals may be distributed more evenly than would be predicted by chance. Ecologists call such populations uniform (Fig. 46.2c).

Ecologists estimate population size by sampling.

Imagine trying to calculate the size of a population of Lime Swallowtails by counting every individual in the Dominican Republic. The task is hopeless. Like that of the Lime Swallowtail, the populations of most species are either too large or too widely dispersed for every individual to be counted. Instead, ecologists estimate population size by taking repeated samples of a population and from these samples estimating the total number of individuals. With an estimate of population size and range, ecologists can then determine population density.

For sessile, or sedentary, organisms like milkweeds in a field or barnacles on a rocky shoreline, ecologists commonly lay a rope across the area of interest and count every individual that touches it, or they throw a hoop and count the individuals inside it. Each such count gives one sample. They then take a number of samples and from them estimate the number and distribution of individuals in the population. For example, suppose 1-m^2 hoops are placed in three different spots in a field, encircling respectively 4, 5, and 3 individual milkweeds. We can multiply the average number of individuals (in this case, 4) per 1-m^2 sample by the total square meters in the area of interest (in this case, the size of the entire field) to give an estimate of population size.

It is relatively easy to count sedentary organisms along a straight line or within a grid area, but many species move around. How do ecologists estimate population size for turtles in a pond, or mice in a forest? In these cases, a method called **mark-and-recapture** is useful. Biologists capture individuals, mark them in a way that doesn't affect their function or behavior, release them back into the wild, and then capture a second set of individuals, some of which were previously marked. **Fig. 46.3** illustrates how the mark-and-recapture method can be used to estimate the number of butterflies in an area. If a large number of butterflies are caught, marked, released, and then recaptured, it is possible to make an estimate of the total population size by extrapolating from the proportion of the second sample that is marked.

These methods are simple, and they make a number of assumptions about the population. For example, we assume that the population has not changed in size between our first and second samples. Nonetheless, mark-and-recapture methods have proved effective in estimating the sizes of populations in nature. More sophisticated analyses are available for more complicated situations, as when more than two samples are caught, or when immigration or emigration occurs during the time between samples.

46.2 POPULATION GROWTH AND DECLINE

The size of a population may increase, decrease, or stay the same over time. Trends in population size are a key focus of ecological research and its application to conservation biology. A decrease in size through time, for example, might signal declining health in a local population—a matter of practical concern when the population in question is a commercially harvested fish or an endangered species. In the 1990s, catastrophic declines in cod populations led to a ban on commercial cod fishing on the Grand Banks of Newfoundland. The decline affected both cod and cod fishers, and it also led to changes in Grand Banks ecosystems: Crab and lobster populations have increased as there are fewer cod to prey on them.

Population size is affected by birth, death, immigration, and emigration.

Many factors affect population size. An obvious process that contributes to population increase is birth. Another is immigration: Individuals may arrive in a population from elsewhere. In contrast, mortality (death) and emigration, the departure of individuals from a population, both act to decrease population size (**Fig. 46.4**).

We can describe the changes in a population's size mathematically, defining population size as N and the change in population size over a given time interval as ΔN, where the Greek letter Δ (delta) means "change in." ΔN is the number

FIG. 46.4 Factors affecting the size of a population: birth, mortality, immigration, and emigration. *Photo source: Richard Hermann/SeaPics.com.*

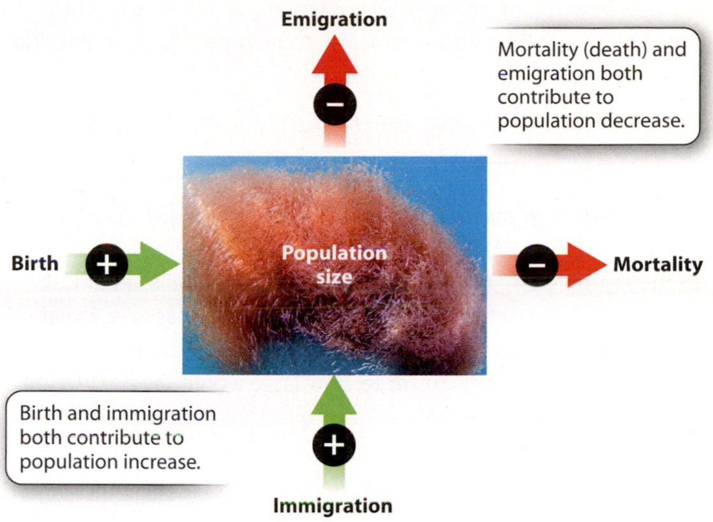

of individuals at a given time (time 2) minus the number of individuals at an earlier time (time 1), which is notated as $N_2 - N_1$. The processes leading to a change in population size through time include births (B), deaths (D), immigration (I), and emigration (E), so we can quantify ΔN as:

$$\Delta N = N_2 - N_1 = (B - D) + (I - E)$$

Commonly, ecologists want to know not just whether population size is increasing or decreasing, but also the rate at which population size is changing—that is, the change in population size (ΔN) in a given period of time (Δt), or $\Delta N/\Delta t$. Let's say a population starts with 80 individuals and after 2 years has grown to 120 individuals. The population, then, has gained 40 individuals in 2 years, for a rate of $40/2 = 20$ individuals per year.

The importance we place on such a number strongly depends on the actual population size. An increase of 20 individuals in a year is a big deal if the starting population had 80 individuals, but it is small if the starting population numbered 10,000. Usually, therefore, we are most interested in the proportional increase or decrease over time, especially the rate of population growth per individual, or the per capita growth rate (*capita* in Latin means "heads"). The per capita growth rate is commonly symbolized by r and is calculated as the change in population size per unit of time divided by the number of individuals at the start (time 1):

$$r = (\Delta N/\Delta t)/N_1$$

In our example, r equals 20 individuals per year divided by 80 individuals, for a per capita growth rate of 0.25 per year.

It is important to realize that, while rates are important in ecology, the data we usually have in hand actually consist of the numbers of individuals that we have counted, and we often do not know exactly how large a natural population really is and so must rely on estimates. In the previous section, we considered how we make such population size estimates, but for the rest of this discussion we focus on the per capita increase (or decrease) in a population over time and assume we know the population sizes for species of interest.

→ **Quick Check 1** If a population triples in size in a year, what is the per capita growth rate?

Exponential growth is characterized by a constant per capita growth rate.

In the example discussed in the preceding paragraphs, an initial population of 80 individuals increased by 20 per year over 2 years, for a per capita rate of increase of 0.25 per year. If this per capita rate of increase were to continue, how large would the population be after 10 years? We can answer this question using the following formula:

$$N_t = N_1 (1 + r)^t$$

Solving the equation for our example shows that in 10 years the population size would be 745 (1.25 raised to the 10th power, times 80). After 15 years, the population would swell to 2273.

This example illustrates **exponential growth,** the pattern of population increase that results when *r* is constant through time. A prominent feature of exponential growth is that the number of individuals added to the population in any time interval is proportional to the size of the population at the start of the interval. That is, while the per capita growth rate remains constant, the actual number of individuals added per time unit increases as the population grows. This is because the individuals added also reproduce, so through time more and more individuals contribute to the growing population. **Fig. 46.5** shows the characteristic shape of an exponential growth curve.

A second example brings home the remarkable consequences of exponential growth. Suppose we place one grain of wheat on the first square of an 8 × 8 chessboard that has 64 squares, two on the second square, four on the third, and keep doubling the number at each successive square. By the time we reach the 64th square, the total number of grains on the board is 18,446,744,073,709,551,615, more than 80 times what Earth could produce in a year if wheat were planted on every square inch of farmland, and about a million times the number of stars in the Milky Way galaxy.

→ **Quick Check 2** Would you rather have a million dollars today, or start with one cent and double the amount each day for a month? Why?

The per capita growth rate, *r*, is commonly called the **intrinsic growth rate** of a population, the maximum rate of growth when no environmental factors limit population increase. Intrinsic growth rates in mammals range from 82% per year in muskrats to 50% in snowshoe hares and 10% in moose.

FIG. 46.5 Exponential growth. When per capita growth rate, *r*, is constant, the curve shows ever-larger changes in population size over time.

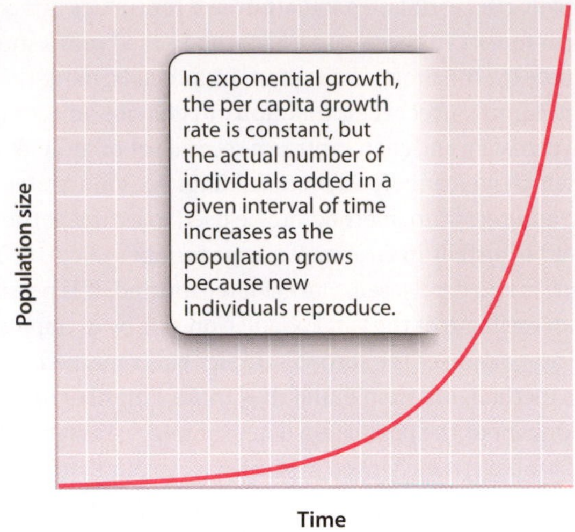

Malthus recognized that exponential growth cannot continue for long in nature because the number of individuals eventually outstrips the resources available to support the population. Darwin, in turn, recognized that when the number of young produced at birth exceeds the number of adults that can be supported by available resources, individuals within the population compete to obtain the resources needed for growth and reproduction. This **intraspecific** (within-species) **competition** results in natural selection (Chapter 21).

Individuals can also compete for resources with individuals from different species. Such **interspecific** (between-species) **competition** can result in an increase or decrease in population size. We focus for the most part on intraspecific competition in this chapter, reserving a detailed discussion of the effects of interspecific competition until Chapter 47.

Carrying capacity is the maximum number of individuals a habitat can support.

As a population grows, crowding commonly results in a decrease in the availability or quality of resources required for continuing growth. As a consequence, growth rate declines. On Hispaniola, for example, Lime Swallowtails have probably reached the limit of available citrus groves. Individual females have to search longer to find suitable plants on which no other females have already laid eggs. With that delay, and the pressure of eggs that are continuing to mature inside their bodies, females may end up laying too many eggs in one place or on poor-quality food, so there is not enough food for all larvae to survive. All of these factors lead to fewer births per female in the Lime Swallowtail population than occurred immediately after they first arrived in Hispaniola, when resources were not limiting.

For many organisms, crowding affects the death rate as well as the birth rate. For example, as the fish population of a pond increases, there is less food available for each individual, and oxygen may be used up by microbes that proliferate because of nutrients in the fish waste.

In other words, birth and death rates are themselves affected by population density, which we defined earlier as the number of individuals per unit of area or (for aquatic species) volume of habitat. Increasing density is not always detrimental to a population. For example, eggs of a salmon or sturgeon are more likely to be fertilized in rivers or streams swarming with their species than when there are few reproducing adults. Generally, however, at some point increasing population density causes the environment to deteriorate because of a scarcity of resources. When the density of a growing population reaches this point, the rate of growth slows and eventually levels off as the death rate increases to equal the birth rate.

We call the maximum number of individuals a habitat can support its **carrying capacity (K).** K represents the interplay between the requirements of individuals for growth and reproduction and the environmental resources such as food and space available to support these needs.

Many factors can keep a population below K. Predation and parasitism are two of them. For example, increasing numbers of Lime Swallowtails on Hispaniola have no doubt attracted predators and parasites that feed on them. Similarly, a high population density of fish in a pond increases the likelihood of infection by viruses, bacteria, and fungi because these agents are transmitted from fish to fish. Predation and parasitism can reduce the population size below K, but they do not change K itself.

Logistic growth produces an S-shaped curve and describes the growth of many natural populations.

When a population is small, resources are not limiting and the population can grow at or near its intrinsic growth rate, r. As the population increases, however, drawing closer to K, the worsening conditions due to crowding slow the rate of growth. For any given population at any given moment, such as deer today in New England forests or striped bass in Chesapeake Bay, we may wish to know what fraction of the carrying capacity is available for further population growth. The answer, expressed mathematically, is $(K - N)/K$.

The effects on population growth as a habitat fills up with individuals can be seen in **Fig. 46.6.** Growth is exponential at first and then slows as the population size approaches its maximum sustainable size, K. The resulting pattern of population growth, with its characteristic S-shaped curve, is termed **logistic growth.** Population size at any time (N_t) can be calculated using the equation:

$$N_t = N_1 + rN_1\,[(K - N_1)/\,K]$$

Logistic growth curves are S-shaped because they actually reflect two sets of processes, each related to population size. At first, growth reflects only the rate at which individuals can reproduce themselves (when N is very small relative to K), so the curve rises steeply. The second part of the curve, where it starts to level off, reflects the onset of additional factors that become important above a certain population size, such as decreasing availability of food and space.

FIG. 46.6 **Logistic growth.** The graph shows the characteristic S-shaped logistic growth curve, which is determined by r and K.

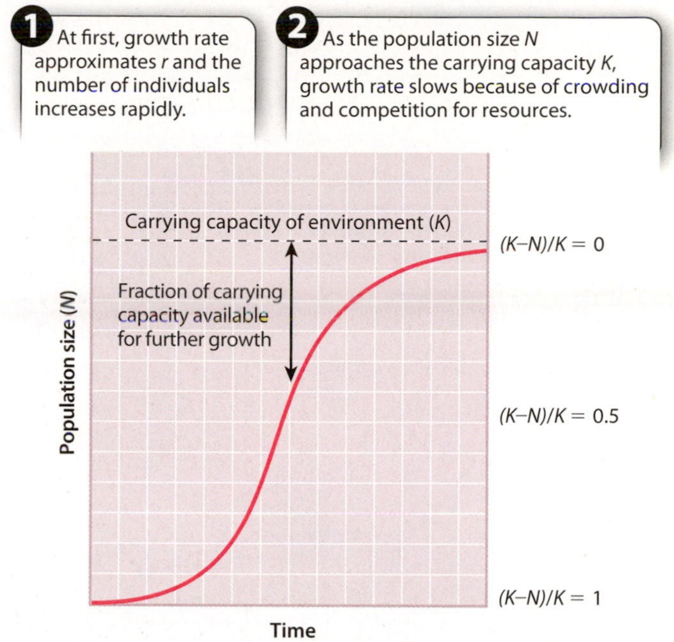

Many natural populations grow this way. Logistic curves also apply to many other phenomena in biology, as well as in such diverse fields as chemistry, sociology, political science, and economics. For example, a logistic curve describes the rate that oxygen binds to a population of hemoglobin molecules, as seen in Chapter 39. Similarly, the number of fertilized salmon eggs goes up with the amount of sperm (which reflects the number of males in the population) deposited in a stream by males but slows again as the percentage of unfertilized eggs available for the sperm to fertilize drops.

➔ **Quick Check 3** Why is the growth curve for most populations S-shaped?

Factors that influence population growth can be dependent on or independent of its density.

In the preceding discussion, the factors that limit population size depended on the density of the population. At low density, food and other resources do not limit growth, but as density increases, they exert more and more influence on population growth, spurring competition for available resources. Similarly, at high population density, individuals may be more vulnerable to predation or to infection. Factors such as resource availability and predation are called **density-dependent** factors.

Not all changes in population growth rates result from population density, however. **Density-independent** factors

FIG. 46.7 Density-independent effects on population size. Independent of population density, extreme weather (such as drought) lowers the quality of habitat for Song Sparrows, reducing the number of territories occupied. *Source: Data from M. K. Chase, N. Nur, and G. R. Geupel, 2005, "Effects of Weather and Population Density on Reproductive Success and Population Dynamics in a Song Sparrow (Melospiza melodia) Population: A Long-Term Study," The Auk 122(2):571–592.*

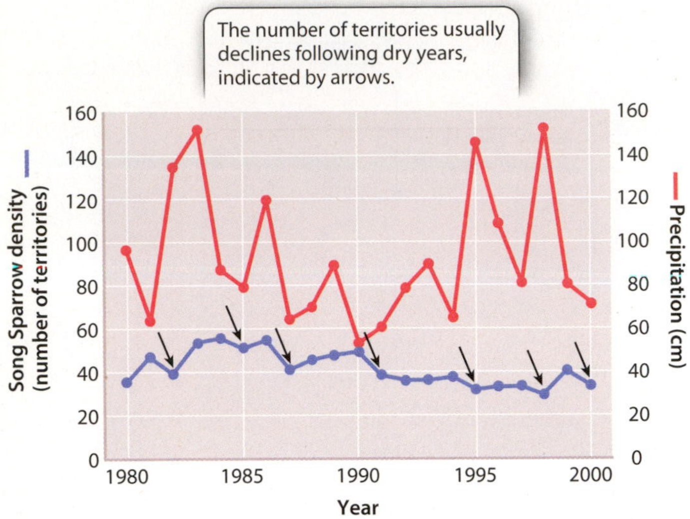

influence population size without regard for the population's density. These factors include events like severe drought or a prolonged cold period, either of which can cause widespread mortality independent of population density. A graph of density-independent effects on populations shows instantaneous, rather than gradual, drops in population size, and these changes can occur at different starting densities.

For example, population growth in Song Sparrows is a function of the number of nesting territories established by males each year, and of their young's success in those territories. **Fig. 46.7** shows that the number of territories generally increases for several years until an unusually dry spring occurs, which greatly decreases the amount of vegetation available to support the insects on which the sparrows feed. The number of territories then plummets, regardless of territory numbers the previous year.

46.3 AGE-STRUCTURED POPULATION GROWTH

Direct counting and mark-and-recapture methods permit ecologists to assess the size and density of populations in a particular place. However, not all organisms in a population contribute equally to population growth because some reproduce

FIG. 46.8 Age structure of two populations, with data from 2010. The data are depicted as histograms, bar graphs in which the length of each bar indicates the number of individuals in the group (here, age group). *Source: Data from United Nations Department of Economic and Social Affairs, Population Pyramids (France and India), http://esa.un.org/wpp/populationpyramids/population-pyramids.htm.*

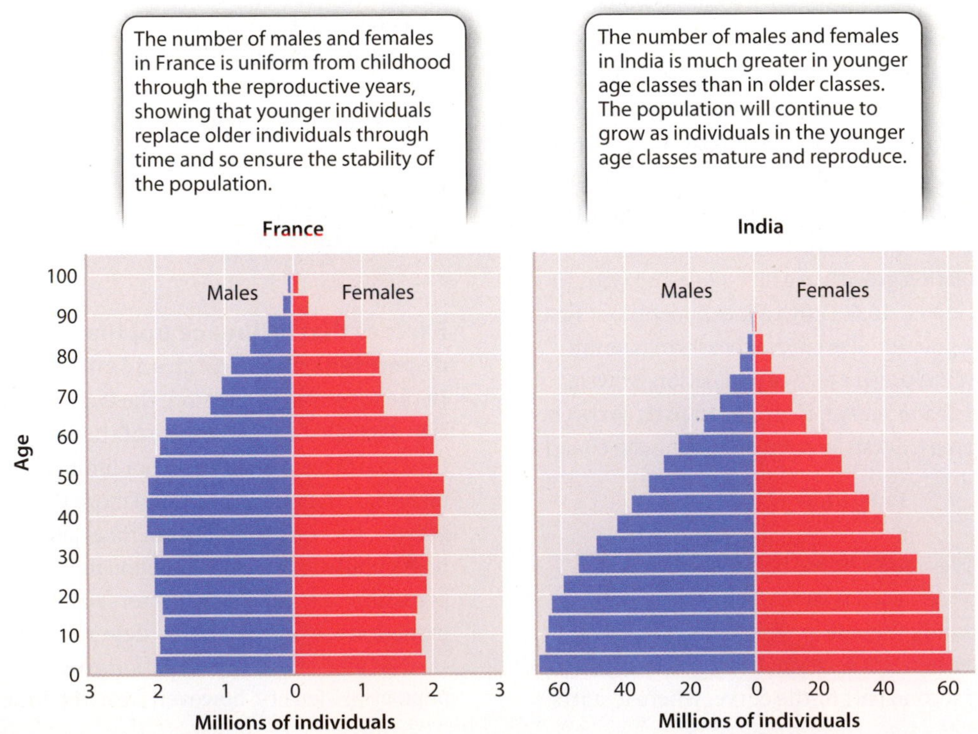

more than others. Therefore, it is also useful to take into account individuals that have differing reproductive capacities, variation that most often depends on differences in age.

Birth and death rates vary with age and environment.

To understand past and predict future changes in populations, ecologists use birth rates, expected longevity of individuals, and the proportion of individuals in a population that is able to reproduce. Insights into these parameters come from the **age structure** of a population, the number of individuals within each age group of the population studied. A population in which most individuals are past their age of greatest reproduction (because for some reason young are not surviving) or a population mostly composed of individuals that has not yet reached peak reproductive age (because older individuals are not surviving) will not increase as rapidly as one dominated by individuals capable of greatest reproductive output.

To reconstruct the age structure of a population, researchers commonly divide the population, or a sample of the population, into age classes, for example individuals born within a single year or within an interval of several years. Age can often be determined by measuring an aspect of the organism's anatomy (for example, counting growth rings in fish scales or in a core drilled into a tree trunk) or estimated by overall size (for example, length of a fish or tree height).

Estimating the number of individuals of differing ages enables ecologists to predict whether the size of a population will increase or decrease. A growing population usually shows a pyramid-shaped age distribution, with the youngest classes much more abundant than older classes, whereas a stable population shows a more even distribution of age classes. **Fig. 46.8** shows the difference in age structure between France, a country with a stable population, and India, a country with a growing population.

In humans, such differences reflect, in part, socioeconomic variation among countries, but differences in the age structure of other species can reveal threats from various sources. **Fig. 46.9** shows the effects of overfishing on the age structure of a species of long-lived Pacific rockfish, determined by comparing the ages of fish sampled in unfished waters with those in heavily fished waters. In the unfished population, the curve shows many individuals close to and past the age of 10 years, when reproductive ability starts to climb. In the fished populations, the curve shows a sharp decline of individuals older than 6 years because mostly older and larger fish are caught. Because these older fish produce most of the eggs and young fish, the annual reproductive effort of fished populations is severely affected by their removal.

Survivorship curves record changes in survival probability over an organism's life-span.

The age structure of a population reflects death rates as well as birth rates. **Demography** is the study of the size, structure, and distribution of populations over time and includes changes in response to birth rates, aging, migration, and death. The demography of the invasive Lime Swallowtail butterfly has been studied in Southeast Asia, where it is native, and this investigation may be helpful in predicting the population dynamics of this species in coming years as it spreads from Hispaniola across the Americas.

The factors that cause insect mortality commonly differ from one life stage to the next, and knowing their effects can help us understand the physical and biological influences on population

FIG. 46.9 Age structure in Pacific rockfish populations in fished and unfished waters. Harvesting larger and older individuals reduces the population's capacity for growth, jeopardizing commercial stocks. *Source: Data from S. A. Berkeley, M. A. Hixon, R. J. Larson, and M. S. Love, 2004, "Fisheries Sustainability via Protection of Age Structure and Spatial Distribution of Fish Populations," Fisheries 29(8): 23–32.*

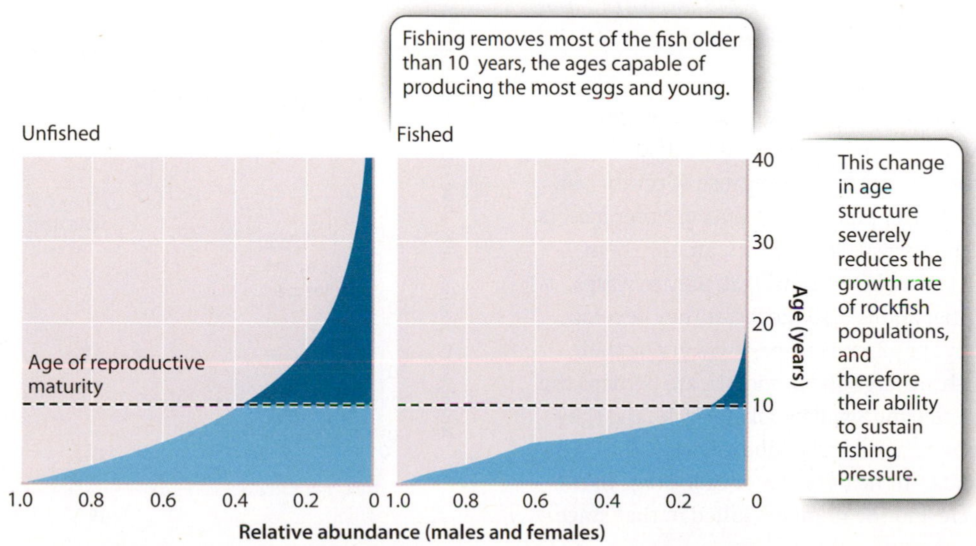

FIG. 46.10 Lime Swallowtail life cycle and natural enemies. The factors that cause insect mortality commonly differ from one life stage to the next.

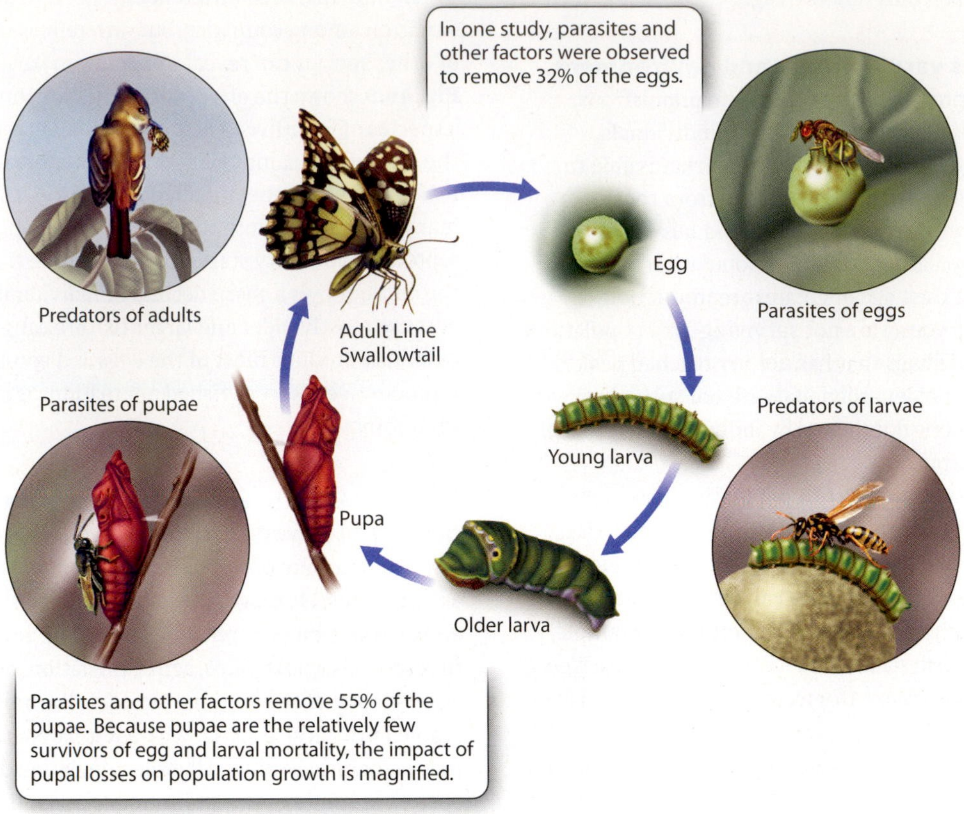

FIG. 46.11 Life table and survivorship curve. (a) A life table summarizes the number of individuals alive in each cohort of a population and can be represented graphically by (b) a survivorship curve, which plots the proportion of the original cohort alive at the end of each stage or time period. *Source: Data from M. Pathak, and P. Q. Rivsi, 2002, "Age Specific Life Tables of Papilio demoleus on Different Hosts," Annals of Plant Protection Sciences 10(2):375–376.*

a.

Stage	Approximate duration	Number alive	Survivorship
Egg	5 days	250	0.68
Larva	18 days	170	0.93
Pupa	15 days	158	0.45
Adult	5 days	71	

b.

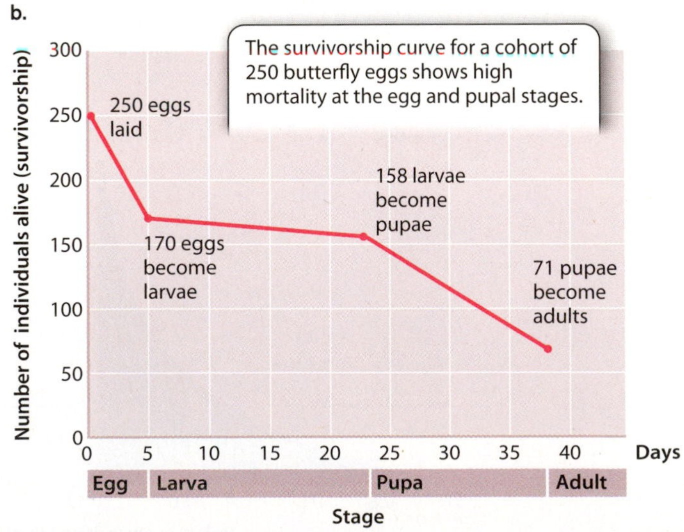

growth. **Fig. 46.10** shows the different dangers to each stage of the Lime Swallowtail's life. These butterflies each lay 30 or more eggs, but only a fraction of the eggs survive because of parasitism, drying out, falling from leaves, or predation by ants or other insects. The larvae have many defenses against predators, among them camouflage and chemical toxins, but they still fall prey to wasps, as well as to lizards and other animals, so the faster they develop into adults, the more likely they are to escape discovery by their enemies. The pupae, which are large and immobile, are perhaps the most vulnerable stage in the Lime Swallowtail life cycle, and they are often highly parasitized by a particular kind of wasp.

Fig. 46.11 presents data on the duration of each stage in the Lime Swallowtail life cycle and the numbers killed in that stage.

We can consider the individual eggs that begin our study to form a **cohort,** a group defined as the individuals born at a given time, and we can trace their survival as they age. The information about how many individuals of a cohort are alive at different points in time can be summarized in a **life table,** from which an ecologist can identify patterns of survival in a population. The proportion of individuals from the initial cohort that survive to each successive stage of the life cycle is called **survivorship.** When survivorship curves are plotted through time, they show an ever-declining fraction of the starting number of eggs (Fig. 46.11).

We can see in the life table and the survivorship curve for this example that survival is lowest among eggs (68% survive to the next stage) and pupae (only 45% survive to the next stage), which of course cannot move to escape harm. If we divide the number lost at each stage by the number of days at that stage, we see that the egg stage loses 16 per day, the pupal stage loses 6 per day, but the larval stage loses only 0.6 per day. There is therefore strong selective pressure to decrease the length of each stage, especially those stages most vulnerable to mortality. Nevertheless, if we remember that each adult female can lay 30 or more eggs, only a very few eggs need to survive to the adult egg-laying stage to ensure the growth of the population. In warm climates like the Caribbean, these butterflies may reproduce eight or nine times in a year. Most of their eggs will not survive to become adults. If they did, available resources would soon be depleted.

Patterns of survivorship vary among organisms.

Populations differ in the ages at which they experience the most mortality. In some species, many offspring are produced, but high mortality early in life means that few survive to reproductive age. In other species, mortality is highest in older individuals, while in still other species, mortality occurs with equal frequency at all ages. **Fig. 46.12** illustrates the three principal types of survivorship curve—Type I, Type II, and Type III— which correspond to the pattern of mortality: whether most mortality occurs late in life, is even throughout life, or is concentrated early in life, respectively.

Human populations in affluent countries experience most mortality late in life, and so typically show a Type I survivorship curve (Fig. 46.12a). Other large animals, such as elephants and whales, show similar survivorship curves. In contrast, birds and many small mammals commonly experience consistent levels of mortality throughout life, which can be graphed as a declining straight line. These animals, for which the probability of death does not change much through the life cycle, exhibit a Type II survivorship curve (Fig. 46.12b). Finally, organisms like Turtle Grass show high mortality at the earliest stages of their life cycles, resulting in a sharply declining survivorship curve that levels off (Fig. 46.12c). This pattern is characteristic of a Type III survivorship curve, which is typical of small herbaceous plants, small animals such as mice, insects like the Lime Swallowtail, and other organisms that grow fast and reproduce early.

FIG. 46.12 Three different types of survivorship curve. *Sources: a. Data from R. Pearl, 1978, The Biology of Population Growth, 3rd ed., North Stratford, NH: Ayer Company Publishers; b. data from E. S. Deevey, Jr., 1947, "Life Tables for Natural Populations of Animals," Quarterly Review of Biology 22:283–314; c. data from J. E. Kaldy and K. H. Dunton. "Ontogenetic Photosynthetic Changes, Dispersal and Survival of Thalassia testudinum (Turtlegrass) Seedlings in a Sub-tropical Lagoon," Journal of Experimental Marine Biology and Ecology 240(2):193–212.*

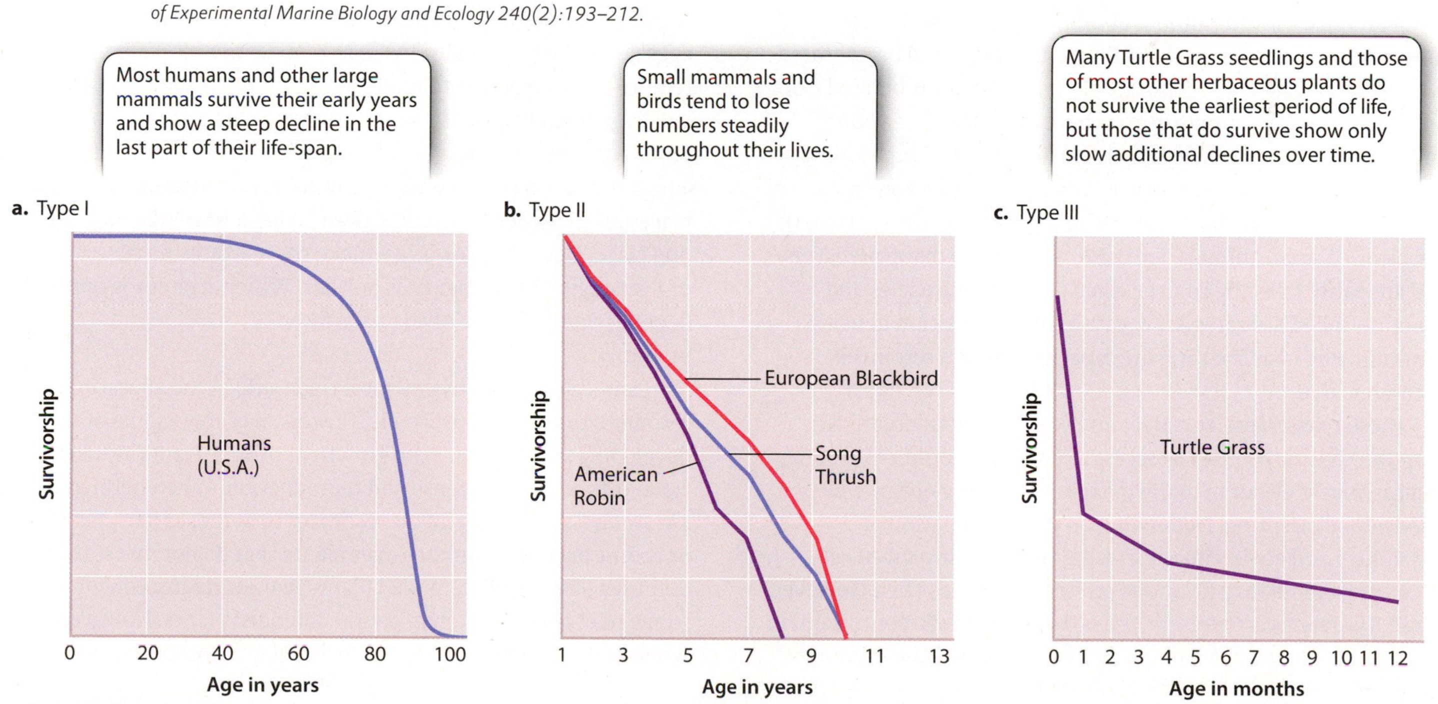

Reproductive patterns reflect the predictability of a species' environment.

In the preceding sections, we discussed population size and growth curves, and we outlined age structure and survivorship. These topics converge when we consider patterns of reproduction. Some species produce large numbers of offspring but provide few resources for their support. Such species are described as ***r*-strategists** (Chapter 42). In contrast, some species produce relatively few young but invest considerable resources in their support. These are termed **K-strategists.** How do these concepts relate to *r* and *K* in population growth?

In general, species that live in unpredictable environments produce many young. When the environment is full of predators watching for vulnerable prey, such as gulls that watch for young turtles on a beach, or when resources are patchily distributed and abundant for only a short time, the chances of a young organism's surviving to adulthood are low. Laying an egg, then, is like buying a lottery ticket—there is a limited chance of success. It is more advantageous to buy many tickets cheaply (that is, to have many offspring and make little parental investment in any of them) than to pay a lot for one. Thus, salmon and other fish produce many thousands of young, each with an extremely small chance of survival.

As we saw earlier, organisms like insects and plants that display Type III survivor curves also experience high mortality in the first stages of life. Because their reproductive success depends on the small chance that a few of their young will survive all the hazards they encounter, parents produce many offspring and invest relatively little in any of them. We refer to species like most fish as *r*-strategists because they reproduce at rates approaching *r*, the intrinsic rate of growth.

In contrast, species living where resources are predictable often produce a lower number of offspring and invest their reproductive resources in just a few, often larger and better cared for, offspring. These offspring are therefore better protected from predators, able to compete with others when they are young, and more likely to survive periods of bad weather. Large animals, such as eagles and bears, produce only two or three young, each of which is much more likely to survive to reproductive age than the offspring of small animals like mice, which have many predators and are easily killed by cold weather. Animals like eagles and bears are called *K*-strategists because their population densities commonly lie near the carrying capacity *K*, when resources become scarce.

These two strategies reflect an evolutionary trade-off. A reproductive individual has access to a finite amount of resources and can invest either in many inexpensive young or few well-provisioned offspring. Depending on the combination of physiology and environment, natural selection in a given set of circumstances may favor one strategy over the other. However, *r*- and *K*-strategists represent the opposite ends of a continuum in reproductive pattern. Many species lie between these extremes.

The life history of an organism shows trade-offs among physiological functions.

We have noted that population growth can be limited by resource availability. How an organism makes use of the resources that are available can also affect population growth. As we saw earlier in this chapter, each individual must devote some of its available food and energy intake to growth, some to the maintenance of cells and tissues, and some to reproduction. This allocation is very much like a household budget, in which incoming resources are divided among the day-to-day expenses of acquiring food and paying for housing and transportation, meeting present and future medical and educational expenses, and savings held in reserve for emergencies. When one category goes up in cost—it needs more money, or more resources—another must come down.

In nature, individual organisms strike a balance in how they allocate resources to the basic needs of growth, maintenance, and reproduction, a balance that has evolved to minimize waste while retaining the flexibility to use resources in different ways depending on environmental circumstances. These kinds of trade-off evolve through natural selection and are the rule in ecology, explaining in part why species tend to produce either many small offspring or a few larger ones.

The typical pattern of resource investment in each stage of a given species' lifetime is called its **life history.** The amount of energy available for reproduction varies over an individual's lifetime, and the degree of control an individual has over resource allocation also varies from species to species. Lizards, for example, show considerable flexibility in allocating resources to reproduction. Many lizards show a physiological trade-off from year to year between the number of eggs produced and the average size of each egg. There is a finite amount of resources to put into offspring, and lizards adjust the number and size of eggs depending on conditions in the environment. Larger eggs require more resources to build, leaving less for additional eggs, and larger eggs hatch into larger offspring that are better able to fend for themselves. But if, for example, the environment carries a high predation risk, then laying more eggs scattered in different places may be a better strategy than laying a few larger eggs in one place. Does this flexibility make these lizards *K*-strategists or *r*-strategists? The answer is neither. Which strategy is optimal depends on the nature of the risks faced by young.

Plants also show physiological trade-offs in the allocation of resources to seeds. For example, flowers may abort some of their developing seeds to favor the growth of others. Plants also make trade-offs between defense and reproduction. When herbivores are around, plants that invest resources in defenses like poisons or thorns have an advantage over plants that do not invest in defenses. These well-defended plants make a greater contribution to the next generation than poorly defended plants. However, when herbivores are absent, defensive plants are commonly at a competitive disadvantage. Individuals that do not invest in defenses can allocate more resources to reproduction. Such trade-offs in plant defense are especially important on nutrient-poor soils, where the cost of replacing valuable tissues lost to herbivores is higher than in nutrient-rich soils (Chapter 32).

The key point is that the various ways in which animals and plants allocate resources to growth and reproduction is rooted in their physiology, and their physiological traits are, themselves, evolved traits. Thus, ecology both reflects and determines how organisms function in nature.

→ **Quick Check 4** Why must a plant or animal make trade-offs among growth, reproduction, defenses, and other adaptive traits?

46.4 METAPOPULATION DYNAMICS

In nature, the populations of many, if not most, species are patchily distributed across the landscape, just as they are among islands in an archipelago (Chapter 22). Deer occur only where there are forest patches of suitable size, milkweeds occur only in open fields, and ducks need to be near water. Local populations are therefore usually separated from one another in space and are connected only occasionally by individuals that migrate between habitat patches. Moreover, some habitat areas that could, in principle, support populations of a given species are unoccupied. How does this kind of distribution influence the dynamics—that is, the change in abundances—of populations in nature?

A metapopulation is a group of populations linked by immigrants.

Groves of trees, separated from other groves by fields, are patches of habitat for the birds, insects, and fungi that rely on them for food and shelter. Similarly, humans are patches of habitat for infectious viruses and certain bacteria; coral reefs are patches for fish; and mountaintops are patches for mountain goats and other high-altitude species. For nuthatches or bark beetles, crossing between groves of trees is a risky business, marked by predatory birds and the possibility of not finding another suitable grove. Similarly, populations of reef corals are often separated by open waters full of predatory fish. In effect, all of these habitats are islands separated by uninhabitable territory. A **patch,** then, is a bit of habitat that is separated from other bits by inhospitable environments that are difficult or risky for individuals to cross.

The nuthatches that occupy a particular tree grove interact with one another more frequently than they do individuals of other tree groves, and so each group is considered a local population. Occasionally, a few migrants move from grove to grove, enabling gene flow between local populations. We can think of local populations of nuthatches in separate tree groves as parts of a larger, more inclusive group called a metapopulation. A **metapopulation** is a large population made up of smaller populations linked by occasional movements of individuals between them (**Fig. 46.13**).

In order to understand the dynamics of populations, then, we must consider not only the birth and death of individuals, but also the colonization and extinction of local populations. Local populations within a metapopulation have independent fates and commonly become extinct because of their relatively small sizes. In general, small populations become extinct more often than large populations because they are more susceptible to random factors that increase mortality, such as weather, natural disasters, or predation. For example, a nest of predators can eliminate a prey species locally; or, if population density falls too low in a small local population because of disease or lack of resources, survivors may not be able to find each other for mating. However, as long as some local populations remain—if they do not all become extinct at the same time—these populations can recolonize old habitats or find new habitat patches, thus ensuring survival of the metapopulation.

Experiments with patches of moss growing on boulders showed that the size and connectedness of the patches of moss

FIG. 46.13 **Metapopulations.** A metapopulation is a group of independent populations connected by occasional immigrants, illustrated here by populations of nuthatches in groves of trees. *Photo source: Charles O'Rear/Corbis.*

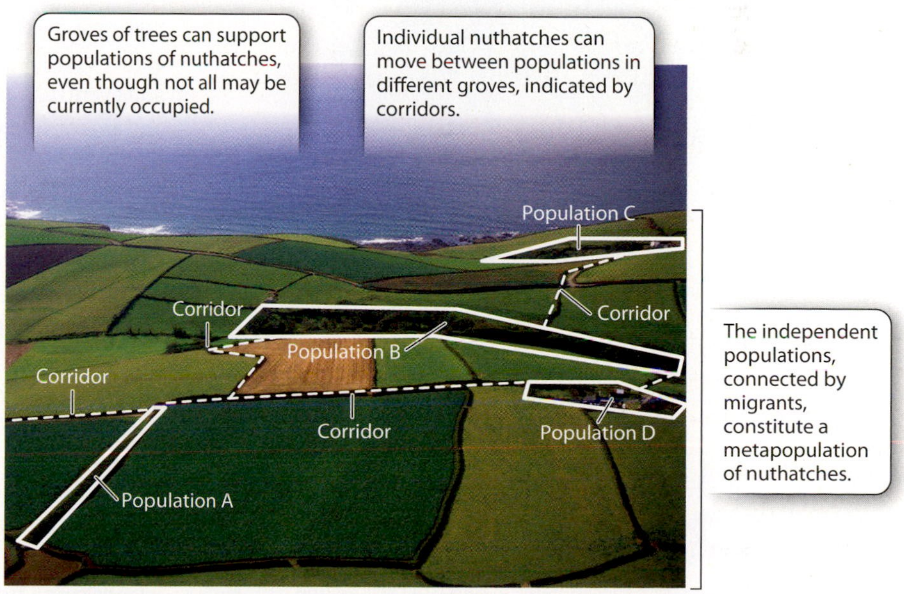

affected the probability that species of arthropods inhabiting the moss would become extinct (**Fig. 46.14**). Researchers cut large patches of moss into several small fragments and sampled the number of insect species in each patch over time. As expected, isolated patches showed decreases in species diversity, up to 40% in 1 year; some species became extinct. However, when nearly isolated moss patches were left with a narrow corridor of moss connecting them, the rate of extinction was lower. Even narrow corridors of suitable habitat can allow species to move between patches of habitat and prevent extinction, or serve to reintroduce species that become locally extinct.

In a world increasingly marked by habitat fragmentation caused by human activities, studies of metapopulations are playing an important role in conservation biology (**Fig. 46.15;** Chapter 49). Species survival depends not only on the health of local populations, but also on the ability of individuals to colonize new habitat patches. As the simple experiment with mosses shows, corridors that connect endangered populations can dramatically increase the probability that a species will survive. However, corridors are important only to species that use them to cross barriers between patches. For species like ferns that produce spores able to float between patches on breezes, corridors may not be as important to patch colonization. Dispersal ability differs among organisms, strongly influencing their capacity to colonize distant patches.

FIG. 46.14 Experimental patches of habitat. This study investigated the effects of patch size and connections in moss and the rates of colonization and extinction of the tiny arthropods that live there.

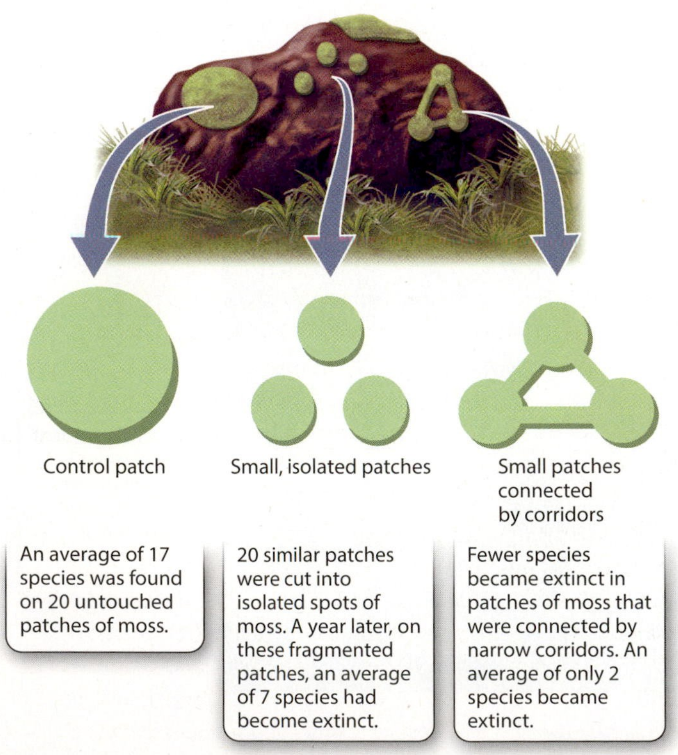

FIG. 46.15 The Biological Dynamics of Forest Fragments Project. Researchers have monitored nine forest patches near Manaus, Brazil, for more than 30 years and have found that fragmentation strongly affects forest biodiversity, providing clues to the design of effective refuges. *Source: Richard Bierregaard.*

Some patches or islands may be so far separated (impairing the ability to disperse) that their populations acquire new mutations, and eventually become reproductively isolated from one another and are considered separate species (Chapter 22).

Island biogeography explains species diversity on habitat islands.

Islands represent extremes in habitat patchiness for colonizing species—organisms that live on a mainland must successfully cross a barrier of water to establish a new population on an island. Ecologists commonly expand the use of the term "island" to include any habitat patch that is surrounded by a substantial expanse of inhospitable environment. While this use certainly includes parcels of land surrounded by water, it can also apply to a body of water surrounded by land, an expanse of forest surrounded by grasslands, or a mountaintop alpine field surrounded by forests at lower elevation. Where do colonists of these habitat islands come from, and what determines how many succeed in establishing a new population?

FIG. 46.16 **Island species diversity, a balance of colonization and extinction.**

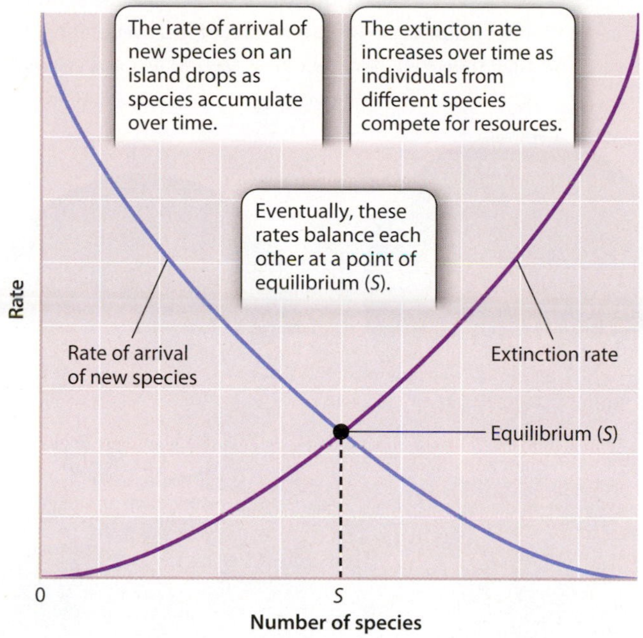

Species diversity reflects both the rate at which new species arrive on the island and the rate at which species already on the island become extinct (**Fig. 46.16**). As species arrive on an initially uninhabited island, the colonization rate is high because there are no predators yet and the small number of colonists means that there is little competition for resources. As more and more species arrive and successfully establish themselves on the island, the rate of colonization goes down because competition and predation increase, making successful colonization more difficult, and because of the diminishing number of species left in the pool of potential colonists. In contrast, extinction rate goes up as newly arrived species compete.

At some point, there is a balance between the arrival of new species and the loss of species by extinction. At this point, diversity is said to be at equilibrium (Fig. 46.16). This does not mean that no new species will colonize the island, but instead that each new colonist species will be accompanied by the extinction of one already in place. There is generally turnover in species composition through time.

The **theory of island biogeography,** articulated in 1967 by the American biologists Robert MacArthur and E. O. Wilson, states that the number of species that can occupy a habitat island depends on two factors (**Fig. 46.17**). The first is the size of the island. Because of their size, larger islands receive more colonists than smaller ones. Furthermore, larger islands can support more species than smaller ones can, so extinction rate is lower. For these reasons, the equilibrium number of species for larger islands is greater than that for smaller islands (Fig. 46.17a).

FIG. 46.17 **Effect of (a) island size and (b) distance from source populations on the equilibrium number of species on an island.**

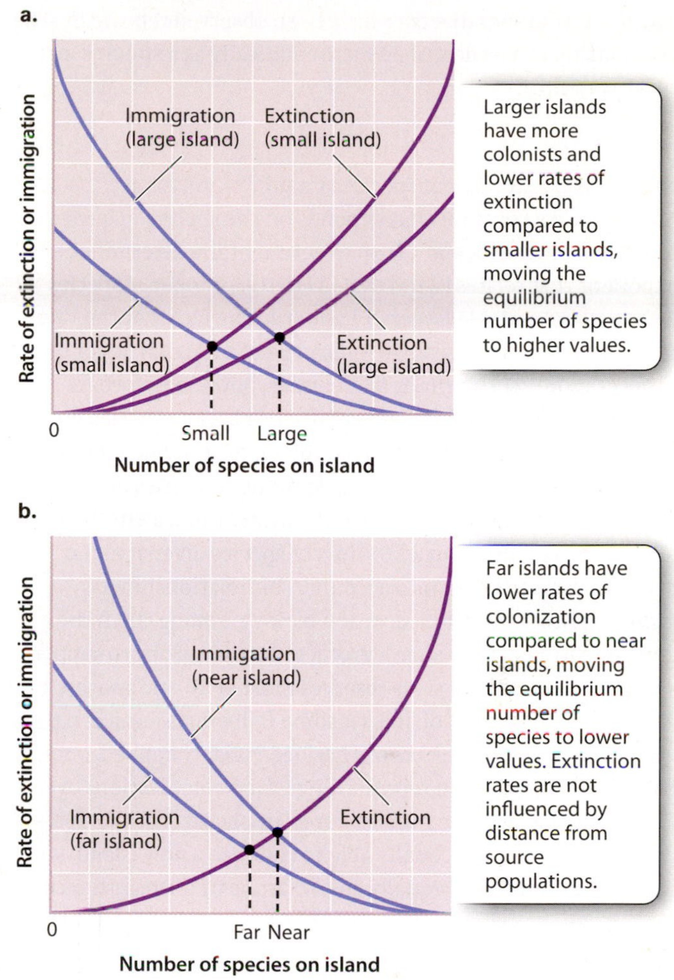

The second key factor is the distance of the island from a source of colonists. More distant islands have lower rates of colonization, moving equilibrium species diversity toward lower values (Fig. 46.17b). The theory of island biogeography therefore predicts that the equilibrium number of species on an island reflects two processes, competition and colonization. If one or the other of these processes is not important, then observations on the diversity of species on islands will not correspond to prediction based on the theory.

Wilson and his student Daniel Simberloff tested the theory of island biogeography in a novel way. Working among small mangrove islands off the Florida Keys, they used short-term insecticides to remove all the arthropods living on individual islands. Over the next few years, new arthropod populations established themselves on the mangrove islands. The researchers found that the species diversity of islands differing in size and distance from shore corresponded to those predicted by the theory. Larger islands had more successful colonists, whereas smaller and more distant islands had fewer.

Fig. 46.18 shows the relationship between island size and species number. Similar relationships between island size and equilibrium species diversity have been observed repeatedly. The relationship can be described mathematically as a **species–area relationship:**

$$S = cA^x$$

where S is the number of species at equilibrium, c is a mathematical constant that depends on the species in question, A is the habitat area, and x is an experimentally determined exponent that relates island size (A) to the number of species (S) on the island and is between 0.1 and 0.4. The fact that the value of x falls in this narrow range indicates not only that larger islands support more species than smaller ones, but also that there is a quantitative relationship between island size and species number.

A useful back-of-the-envelope calculation is that a tenfold increase in island size results in a doubling of species diversity. Perhaps more relevant for issues of conservation, a 90% reduction of habitat area results in a 50% loss of species diversity if no other factors are taken into account. This relationship has many applications in conservation biology, among them the determination of how large a protected area must be to support some species and how close reserves must be to one another to ensure the movement of at least some individuals between them.

Island biogeography theory has been widely applied, but it has some limitations in representing island colonizations over long evolutionary time frames. For example, the theory treats all potential colonists as equally able to colonize a new island, when in fact they are not all equally able to arrive or form an association

FIG. 46.18 **Island size and number of species.** The diversity of amphibian and reptile species on Caribbean islands fits the predictions of island biogeography theory. *Source: After R. H. MacArthur and E. O. Wilson, 1967, The Theory of Island Biogeography, Princeton, NJ: Princeton University Press.*

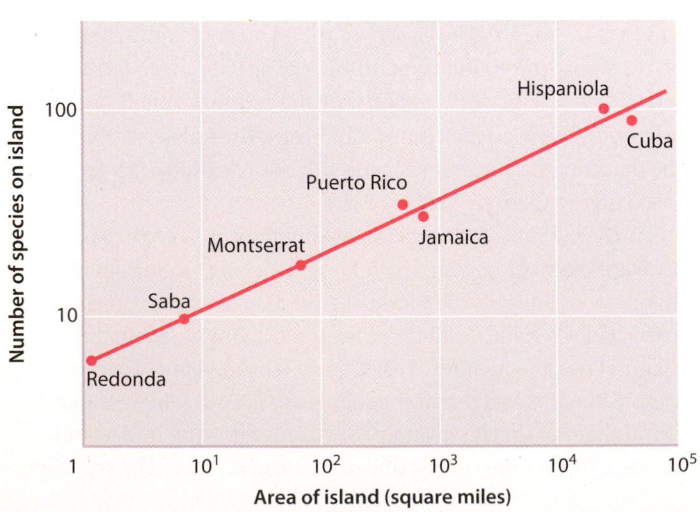

FIG. 46.19 ***Anolis*** **lizards in Hispaniola.** These lizards arrived as colonists from elsewhere in the Caribbean and subsequently diversified to produce a variety of species reflecting specialization in their insect-hunting in different parts of local vegetation. *Photo sources: (top to bottom) Eladio M. Fernandez; Eladio M. Fernandez; Eladio M. Fernandez; © 2013 Jake Scott; Stan Osolinski/Getty Images; Eladio M. Fernandez.*

with species already in place. These abilities are often determined by a species' particular evolutionary history. For example, flying animals like bats and birds more frequently arrive on oceanic islands than nonflying animals like rodents, lizards, and frogs. In contrast, the nonflying animals that colonize successfully more often evolve into multiple new species on islands. If these new, closely related species cannot themselves colonize another area, they will be found nowhere else in the world; these are called **endemic** species.

→ **Quick Check 5** Two islands lie off the coast of Florida. One is larger and the other is closer to the mainland. What determines which one will have more species?

? CASE 8 BIODIVERSITY HOTSPOTS: RAIN FORESTS AND CORAL REEFS

How do islands promote species diversification?

Anolis lizards provide a classic and dramatic example of evolutionary diversification on the island of Hispaniola, where many dozens of species of closely related species coexist. Phylogenies suggest that *Anolis* lizards colonized Hispaniola in five or six separate events, but the dozens of species present today indicate marked evolutionary diversification after colonization.

All *Anolis* lizards feed on insects and other invertebrates, but species differ in how and where they search for prey. Some with long legs run on the ground, while others with shorter legs and long tails climb on vegetation. Still others are large and robust and live in the high canopies of trees. Altogether, *Anolis* lizards have evolved six different feeding strategies with corresponding morphological traits (**Fig. 46.19**).

All six types can coexist in one place because they hunt insects in different ways and in different parts of the vegetation. Careful research is revealing the pattern of population expansion and fragmentation, followed by natural selection within different subpopulations, that has resulted in the diversity observed on Hispaniola today. Through evolution and complex ecological interactions, then, habitat diversity can depart from the simple predictions of island biogeography. Nonetheless, the theory remains a powerful way of thinking about species diversity and an important consideration in conservation biology. ■

Core Concepts Summary

46.1 A population consists of all the individuals of a given species that live and reproduce in a particular place.

A population is characterized by its size, range, and density. page 1004

Population size is the number of individuals at a particular time and place. page 1005

Population range is the total area in which a population lives. page 1005

Population density is the population size divided by its range. page 1005

Population size can be estimated by sampling and mark-and-recapture techniques. page 1006

46.2 Population size can increase or decrease over time.

Population size is influenced by births, deaths, immigration, and emigration. page 1007

The population growth rate is the change in population size, commonly expressed as the per capita rate of growth—the rate of growth (or decline) per individual in the initial population. page 1007

Initially, growth rates are often exponential, in which the number of individuals added to the population increases over time. page 1008

Population density often limits population growth because of decreasing resources. This limit defines the carrying capacity of the environment. page 1008

The growth curve of populations is commonly logistic, producing an S-shaped curve in which growth is exponential at first but tapers off as population size approaches the carrying capacity of the environment. page 1009

Population density can be limited by density-dependent factors, such as resources and predation, and by density-independent factors, such as drought and fire. page 1009

46.3 The age structure of a population helps ecologists understand past changes and predict future changes in population size.

The age structure of a population is the number of individuals of each age group and is often depicted as a histogram. page 1011

A growing population shows a pyramid-shaped age distribution, with the youngest classes more abundant than older classes, whereas a stable population shows an even distribution of age classes. page 1011

Survivorship is the proportion of individuals from an initial cohort that survives to each successive age or stage of the life cycle. page 1011

Type I, Type II, and Type III survivorship curves correspond to mortality that occurs predominantly late in life, is even throughout life, or is concentrated early in life, respectively. page 1013

r-strategists are organisms that live in unpredictable environments, producing many young but investing relatively little in each offspring. page 1014

K-strategists live where resources are predictable; they produce few offspring but invest a lot in each. page 1014

r- and *K*-strategists represent extremes of a continuum of reproductive strategies. page 1014

The life history of an organism is the evolved pattern of resource investment in each stage of a given species' lifetime. page 1014

Organisms often exhibit trade-offs between reproduction and other physiological functions. page 1014

46.4 The dynamics of populations are influenced by the colonization and extinction of smaller, interconnected populations that make up a metapopulation.

The size of local populations (patches) and the degree of connectedness between patches determines in part whether a given patch will persist or not. page 1015

An ecological island may be a true island surrounded by water or any habitat surrounded by uninhabitable areas. page 1016

The theory of island biogeography proposes that the number of species that can occupy an island depends on the size of the island and its distance from the mainland. page 1016

Species diversity on habitat islands reflects both the rate at which new species arrive and the rate at which colonist species become extinct. page 1019

Self-Assessment

1. Name the four factors that affect population size.
2. Draw an exponential and a logistic growth curve, and explain what accounts for their different shapes.
3. Name two density-dependent and two density-independent factors that can limit the size of a population.
4. Draw a graph showing the age structure of a population that is growing rapidly and a graph of one that is not.
5. Plot a survivorship curve for a species with high rates of predation early in life and one for a species with high mortality late in life. Name the types of survivorship these species display.
6. Explain how *r* and *K* strategies relate to the predictability of the environment and in what kinds of environment each strategist would be more successful.
7. Describe what is meant by a trade-off in physiological functions, and give an example.
8. Give three examples of a habitat island, and explain what makes it an island.
9. Name factors that determine the diversity of species on a habitat island, and explain the relevance of these factors in managing the habitat of an endangered species.

Log in to LaunchPad to check your answers to the Self-Assessment questions, and to access additional learning tools.

Charles J. Smith.

CHAPTER 47

Species Interactions, Communities, and Ecosystems

Core Concepts

47.1 The niche encompasses the physical habitat and ecological role of a species.

47.2 Competition, predation, parasitism, and herbivory are interactions in which at least one species is harmed.

47.3 Mutualisms are interactions in which both species benefit.

47.4 Communities are composed of local populations of multiple species that may interact with each other.

47.5 The cycling of materials and the flow of energy through ecosystems reflect the ways that species interact in communities.

The chocolate tree, *Theobroma cacao,* known simply as cacao, is native to tropical America. Cacao grows where the environment is consistently warm and humid. Physical environment alone, however, does not determine where cacao can live. Cacao reproduction, and ultimately distribution, is limited by the requirement that tiny flies called midges pollinate them (**Fig. 47.1**). The midges' eggs and young need deep shade and a thick layer of moist rotting vegetation, whether from the natural fall of leaves onto the forest floor or from banana plants left to decay in cacao plantations. The midges' needs mean that cacao cannot reproduce in areas with full sun or on clean ground that otherwise would support their growth—at least not without human help. The natural distribution of cacao plants, therefore, depends on interactions with both the physical environment and other species.

The reliance of chocolate trees on midges is not an unusual arrangement. Every species interacts with other species in ways that affect growth and reproduction. These other species may be the insects that help a plant pollinate its flowers or ones that eat its leaves. They may be a fungus that helps a tree gain nutrients or another that attacks it. They may be other species that compete for resources, predators and pathogens that reduce population size, prey species that provide a specialized food resource, or microbes that provide or take nutrients while living inside their hosts (Case 5: The Human Microbiome). Associations among species, whether positive, negative, or neutral, form a complex web of interactions in which each species influences the population size and distribution of other species, whether they are partners, enemies, or food.

In this chapter, we explore where and how species live, how they interact with one another in a community, and how different species and the physical environment form ecosystems seen around the world.

47.1 THE NICHE

Why do species live where they do? As discussed in Chapter 46, a population persists only in places where its members can tolerate the physical environment. Plants, for example, are sensitive to rainfall, freezing, and nutrient levels in the soil. The habitat occupied by a species usually contains other species, and the distribution of each species will also reflect interactions with the others, as the example of cacao illustrates. Competition between species can occur when different species use the same resources in the same place. Distinct species can coexist within shared habitat only if they use different resources, if they are active at different times, or if the local populations of one or both have so few individuals that resources are not limiting to their survival and reproduction. In other cases, like cacao, species persist only when other species that benefit their growth or reproduction are also present.

The niche is a species' place in nature.

This combination of a species' physical habitat and its ecological role in that habitat defines its **niche,** though the meaning of this term has changed over the years. Nearly a century ago, the American ecologist Joseph Grinnell defined the niche as the sum of the habitat requirements needed for a species' survival and reproduction. Later, ecologist Charles Elton redefined the niche as the role a species plays in a community, switching emphasis from the habitat to the species itself. By the middle of the twentieth century, ecologist G. Evelyn Hutchinson had combined these ideas, popularizing the concept of the niche as a multidimensional habitat that allows a species to practice its way of life.

FIG. 47.1 A mutually beneficial interaction between two species. Development of (a) seed-rich cacao fruits, whether for the plant's reproduction or to provide raw material for chocolate, depends on pollination of (b) the cacao flower by (c) midges. *Sources: a. Exotica im/age fotostock; b. A. Jagel/age fotostock; c. Brian V. Brown.*

FIG. 47.2 Fundamental and realized niches. The fundamental niche includes all the habitat potentially available to a population, whereas the realized niche is the subset of that habitat actually occupied by the population.

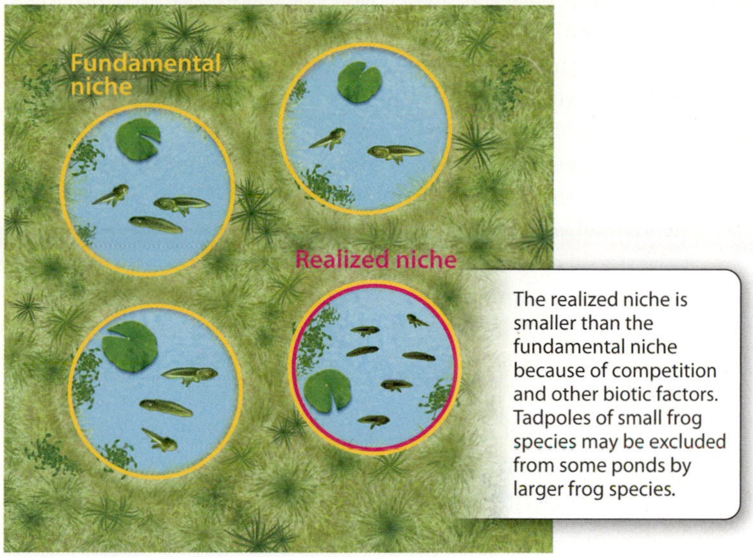

The niche is therefore determined by both physical (abiotic) factors such as climate and soil chemistry, and biological (biotic) factors based on interactions with other species. These factors lead to adaptations, the products of natural selection. Adaptations in turn enable different species to coexist by exploiting different combinations of resources. At the same time, species affect the resources and other species around them.

Species are commonly associated with a specific habitat, but "niche" and "habitat" are not interchangeable. Niches have a dual nature: They reflect both where organisms occur and what they do there. Furthermore, the niche refers to the ways that organisms respond to, and also affect, the resources and other species found in the habitat.

The realized niche of a species is more restricted than its fundamental niche.

The **fundamental niche** includes the full range of climate conditions and food resources that permits the individuals in a species to live. In nature, however, many species do not occupy all the habitats permitted by their anatomy and physiology because of interactions with other species that can reduce the range actually occupied through competition or predation. This actual range of habitats occupied by a species is called its **realized niche**.

For example, tree frogs can reproduce in ponds near forests throughout eastern North America, their fundamental niche (**Fig. 47.2**). However, tree frogs are excluded from ponds that also have larger frogs because the larger frogs eat most of the algal food available, limiting the tree frogs' realized niche to ponds without larger frogs (Fig. 47.2). Therefore, competition with other species can determine the size of the realized niche.

→ **Quick Check 1** Which is larger—the fundamental or the realized niche?

Niches are shaped by evolutionary history.

The niches of nearly all species are narrow in at least one and often several ways. *Anolis* lizards eat only live arthropods, as we saw in Chapter 46, and not plants, fungi, or dead organisms, though these resources are available in their habitats. Similarly, among the small birds that live in woodlands, there are warblers that hunt insects among the leaves in the trees, thrushes that seek insects in the soil underneath, and woodpeckers that find insects under bark. Each kind of bird is a member of a diverse family of related species that all seek insect prey.

Usually, some aspect of the niches of closely related species is similar (all *Anolis* lizards are alike in their dietary preferences for small live prey), and so are evolutionary conserved. This observation is called **phylogenetic niche conservatism.** It is commonly seen and speaks to the strong influence of evolutionary history in shaping at least some aspects of the niches of organisms.

47.2 ANTAGONISTIC INTERACTIONS BETWEEN SPECIES

Comparing the realized niche of tree frogs with their fundamental niche makes it clear that interactions with other species play key roles in the distribution and abundance of populations. Virtually every organism is involved in interactions with other species—think of plants and their pollinators or predators and their prey. These interactions often narrow the geographic distributions of species within the larger boundaries of their physical requirements, and so help shape their realized niches.

All interactions combine costs (often in resources) and benefits (usually increases in reproduction). Some interactions are direct, where species physically interact (such as plants and their pollinators). Others are indirect, where they influence one another through competition for a shared resource, such as food. In many interactions, the benefits are one sided: One organism actually consumes another or competes with it for food or some other resource. In this section, we focus on such one-sided interactions, in which at least one participant loses more than it gains.

Limited resources foster competition.

Whenever there are fewer resources than individuals seeking them, there can be competition. **Competition** is an interaction in which the use of a mutually needed resource by one individual or group of individuals lowers the availability of the resource for another individual or group. Different species of sea anemones compete for space in rocky shallows; swallows compete with one another for nesting holes in tree or cliffs; and many different species of dung beetles compete for animal droppings (some even cling to monkeys, waiting for feces to emerge). Males of many animal species commonly compete for female mates.

In Chapter 46, we saw that in many populations the tension between intrinsic growth rate and environmental carrying capacity results in **intraspecific competition,** competition between individuals of a *single* species. The increase in competition between individuals of a species that accompanies increased population density is a main reason why population growth slows as the species' environment approaches its carrying capacity.

Competition can also occur between individuals of *different* species, in which case it is **interspecific competition.** Whether competition occurs between individuals of one species or two, it is a lose–lose situation: Each side spends energy it would not spend in the absence of the other. Indeed, Darwin recognized that the "struggle for existence," as he called it, is a primary driver for natural selection, leading to the evolution of traits that aid survival and reproduction by reducing competition in one way or another.

Competitive exclusion can prevent two species from occupying the same niche at the same time.

In general, when two species have overlapping niches (if, for example, they share a physical habitat and overlap strongly in resource use), one will either become extinct in that place or change its niche. Coyotes and foxes eat the same small prey, but when they meet, a coyote will waste energy by chasing the fox, and the fox loses energy by running away from the coyote. In regions where foxes and coyotes occur together, the foxes avoid interactions with the larger coyotes by leaving the area, placing their dens in locations where coyotes do not occur.

This result of such an interaction, in which competition between two species prevents one from occupying a particular niche, is called **competitive exclusion.** It is another way in which a fundamental niche can be reduced to a smaller realized niche. It is important to recognize that even if two species overlap substantially in resource use, competitive exclusion may not take place unless the population sizes of both species are large enough to encounter each other or reduce the resources available.

Competitive exclusion can be more complicated than competition between two species that forces the two groups to use two different resources. Consider for example the variety of insect groups that can make use of different resources on just one species of plant. The insects feeding on a single patch of milkweed usually include some species of beetle that nibble roots and others that chew on leaf edges along with caterpillars. Some aphids suck phloem contents from the stems, and other beetle species feed on the plant's seeds (**Fig. 47.3**).

We may interpret this specialization on different parts of the milkweed as competitive exclusion, but it's possible that the

FIG. 47.3 **Insects feeding on different resources of the same host.** Different groups of insects feed on different parts of the milkweed plant. *Sources: a. Yogesh S. More/age fotostock; b. SimplyCreativePhotography/Getty Images; c. Eric Bean/Getty Images; d. strathroy/Getty Images; e. Photo by Sergio Rasmann.*

FIG. 47.4 Adaptations of *Anolis* lizards for feeding. (a) *Anolis* species that feed on twigs have short limbs and a prehensile tail that they use to support themselves when they sleep; (b) those that feed on trunks have long limbs. *Source: Eladio M. Fernandez.*

insects feed on different parts of the plant either because they competed for milkweeds in the distant evolutionary past and have long since diverged in niches, or because each has evolved its unique association with a particular plant part long before they were associated with milkweeds. In either case, there is no direct competition today among the current species.

? CASE 8 BIODIVERSITY HOTSPOTS: RAIN FORESTS AND CORAL REEFS

Can competition drive species diversification?

In Chapter 46, we discussed the *Anolis* lizards on the island of Hispaniola and other islands of the Caribbean. Within their fundamental niche of tropical forests, different *Anolis* species eat insects found in distinct parts of forest trees (see Fig. 46.19). That is, over time, they have evolved different adaptations for feeding (**Fig 47.4**). How did these different species and different feeding strategies evolve?

It is likely that competition for similar resources promoted species diversification in this group. That is, competitive exclusion, played out over time, resulted in initially similar species diverging in habitat use and feeding strategies. This process is known as **resource partitioning,** in which species whose niches overlap diverge and become different subspecies or species. The result is that the overlap between closely related groups is minimized.

Interactions with predators such as lizard cuckoos and curly-tailed lizards further influence, and limit, the distributions of different *Anolis* species. Moreover, there is reason to believe that the ability (or inability) to migrate contributes to the distributions of the individual species: When *Anolis* species are introduced by accident or deliberately to places beyond their dispersal capabilities, they commonly thrive. Therefore, competition, predation, and dispersal are all factors that influence the distribution of species.

→ **Quick Check 2** What is the difference between competitive exclusion and resource partitioning?

Species compete for resources other than food.

Food is the resource we most often think of as the limiting factor for population growth and reproduction, but many species just require a place to stand, so to speak, or suitable conditions for a nest. Barnacle and mussel larvae compete for a place to attach on submerged rocks. Their food, tiny plankton, is almost always abundant and therefore is not as likely to limit population growth as attachment space is.

Similarly, plants compete for space in fields and forests so they can collect sunlight and obtain water and nutrients from the soil. Sunlight cannot itself be used up, but room to collect sunlight on the ground below can be filled. Nesting sites for birds that lay their eggs in tree hollows or on cliff faces can also be strongly limited by the availability of space, as can the settling of corals on reef surfaces (**Fig. 47.5**).

Predation, parasitism, and herbivory are interactions in which one species benefits at the expense of another.

In contrast to competition, some interactions benefit one participant and harm the other. **Predation** is a type of interaction in which one organism consumes another, its prey (Case 7: Predator–Prey). In this interaction, the predator benefits at the expense of the prey. Early experiments by the Russian biologist Georgii Gause showed that a simple system with one predator and one prey population is inherently unstable. The predator overexploits the prey, driving it to extinction, and then becomes extinct itself.

In 1958, the American scientist Carl Huffaker demonstrated that if the prey had refuges where some individuals could escape from predators, they could persist while predator populations declined. Through time, the prey population would recover and

FIG. 47.5 Competition for space. (a) Herons, egrets, and cormorants compete for nest sites in wetland trees. (b) In the sea, corals compete for space on reef surfaces. *Sources: a. Courtesy of Jim Maddox, Galloway, NJ; b. Franco Banfi/Getty Images.*

a

b

HOW DO WE KNOW?

FIG. 47.6

Can predators and prey coexist stably in certain environments?

BACKGROUND In the 1950s, it was not clear whether a simple system of predators and prey could be stable and coexist indefinitely or whether both groups of species would become extinct as predators consumed all available prey. The results of experiments up to that time were equivocal, and ecologists knew that predators introduced to islands could hunt their prey to extinction.

HYPOTHESIS Ecologist Carl Huffaker hypothesized that predators and prey could stably coexist if temporary refuges were available for the prey.

EXPERIMENT Huffaker studied two kinds of mites: one that feeds on the surface of oranges, and another that preys on the orange-eating mite. Huffaker put two sets of oranges on a table, separating them with barriers of petroleum jelly, essentially making each orange its own habitat patch. He added toothpicks to one set of oranges so that the orange-eating mites could let out silk and float over the petroleum jelly barriers and, at least temporarily, escape predation.

RESULTS In the setup in which prey could not escape, a single cycle of increase and decline was observed: The population size of prey increased but was closely tracked by rising populations of their predators, and eventually both declined to extinction.

In experiments in which toothpicks were provided, the orange-feeding mites climbed the toothpicks and dispersed on silken threads they produced, moving away from predators and toward other oranges without mites. These populations of predators and prey went through three cycles of increase and decline, and cycles would probably have continued if oranges continued to be supplied.

CONCLUSION Predator–prey systems can be stable if there are sufficient areas available where prey can escape predators, at least temporarily.

SOURCE Information from Huffaker, C. B. 1958. "Experimental Studies on Predation: Dispersion Factors and Predator–Prey Oscillations." *Hilgardia: A Journal of Agricultural Science* 27: 795–834.

population density would rise to a point where predators would again expand and cause prey abundance to decline—and then decline again themselves (**Fig. 47.6**). Huffaker showed that predators and prey cycle repeatedly through periods of increasing and then decreasing density, as predators track their prey and some prey escape predation. In other words, a long-term, stable oscillation pattern can be achieved.

Predators can limit the population sizes of their prey, preventing prey populations from increasing to the level where competitive exclusion occurs. We have seen that when species overlap in resource use and their populations increase to the point where they compete for resources, they must be separated in space or time or risk local extinction through competitive exclusion when resources become limited.

If, however, populations do not rise to densities at which resources are limiting, competition is reduced and species can overlap in niches without excluding one another. For example, in ponds throughout the southern United States, tadpoles of the large Southern Toad, the Eastern Spadefoot, and the tiny Spring Peeper all use the same food source, grazing on algae as do other tadpoles. In the presence of the two larger species, Spring Peeper

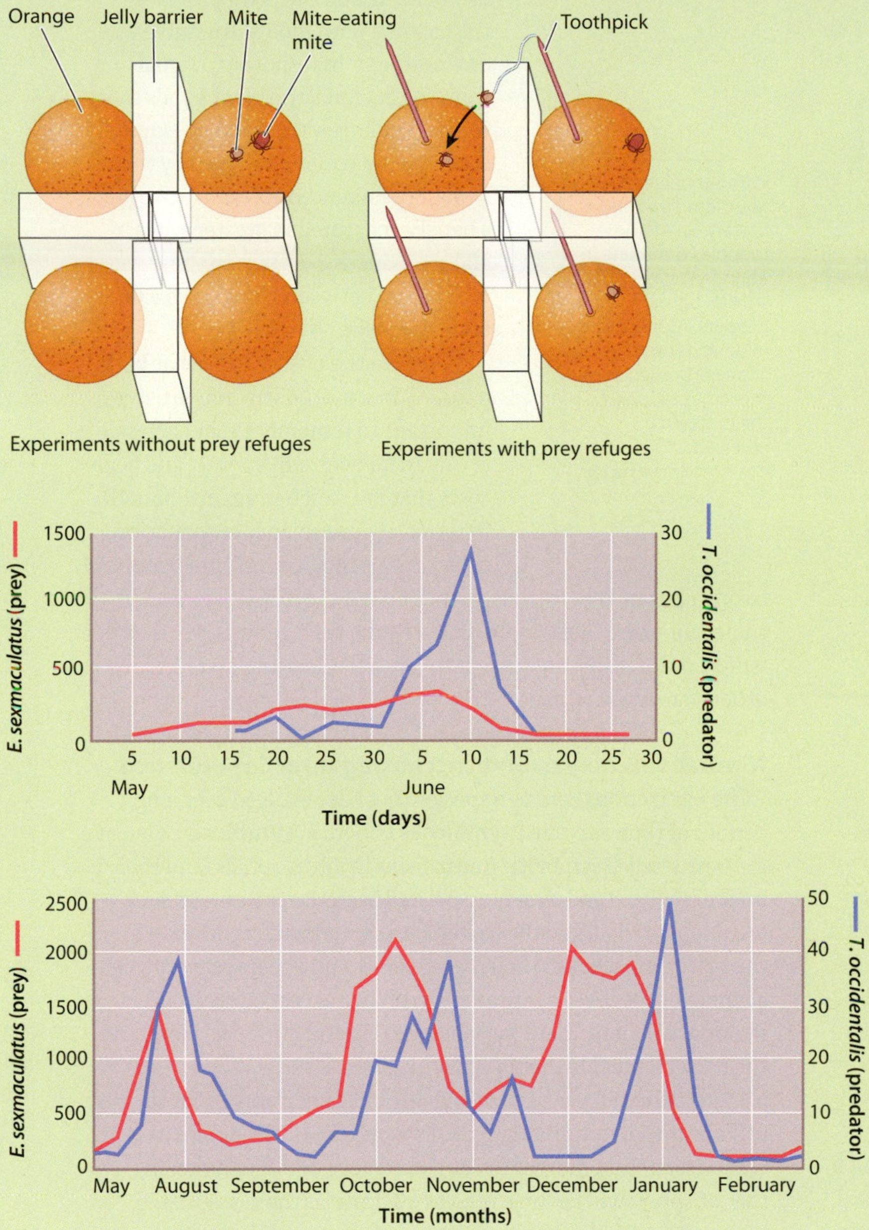

tadpoles compete poorly for food (**Fig. 47.7**). As a result, they have low survival rates, and those individuals that do make it to maturity are often undersized because as tadpoles they did not have access to adequate nutrition.

In the presence of a predatory Red-spotted Newt, however, the tables are turned. The newt prefers to eat the larger tadpole species, and so prevents them from becoming sufficiently abundant to outcompete the smaller Spring Peeper tadpoles. The Spring Peeper tadpoles, then, have the highest survival rates. With competition for food reduced, these small tadpoles grow to their maximal body size, and become the most abundant tadpole species in the ponds.

Parasites live in close association with another species, gaining nutrition by consuming their hosts' tissues. Unlike predators, parasites commonly do not kill their hosts, but they can reduce host fitness by tapping its resources. In this way, parasites can limit the population size of their host, keeping numbers well below the carrying capacity of the environment.

An extreme example is the fungus that infects the American chestnut, a tree once dominant in eastern North American forests. The fungus attacks the vascular system of young chestnut trees, which usually succumb when they reach 2 to 3 m in height. Today, there are only a few isolated patches of large chestnut trees in Vermont and a handful of other places that have escaped the fungus. As chestnut populations have declined, populations of oak, beech, and other trees that grow in the same forest have increased.

Herbivory, the consumption of plant parts, benefits herbivorous animals by providing some nutrients, and harms plants by directly affecting the products of photosynthesis. Plants are generally not as nutritious for animals as other animals are, and so herbivorous animals must eat more plant material to obtain the nutrients they need for survival and reproduction. As we saw in Chapter 32, plants are not passive victims of herbivores; most are well defended with chemical or physical deterrents to herbivory.

Facilitation can occur when two species prey on a third species.

Up to now, we have focused on interactions between two species. What happens when there are more than two species? Consider, for example, what occurs when two different organisms both attack another organism. In this case, the attack of one organism can sometimes aid the attack of the other. Such an indirect interaction between the two attacking species that benefits them both is called **facilitation.**

For example, both fungi and beetles attack pine tree phloem, and the attack of each is facilitated by the other. Bark beetles facilitate fungi that attack pine trees by providing entryways through their tunnels into the tree, while fungi facilitate bark beetles by lowering the amount of the pine's resinous defenses. It is clear that a single species can have both positive and negative effects on others that broadly overlap in a niche.

Like insects, viruses and other pathogens also usually attack only one particular type of host tissue such as the gastrointestinal, respiratory, nervous, or circulatory system. Severe infections can

FIG. 47.7 Effect of predators on the outcome of competition. Newt predators in ponds prefer to eat bigger tadpoles that compete well for food, enabling smaller tadpoles that are poorer competitors to survive to adulthood. *Data from P. J. Morin, 1981, "Predatory Salamanders Reverse Outcome of Competition Among Three Species of Anuran Tadpoles," Science 212:1284–1286.*

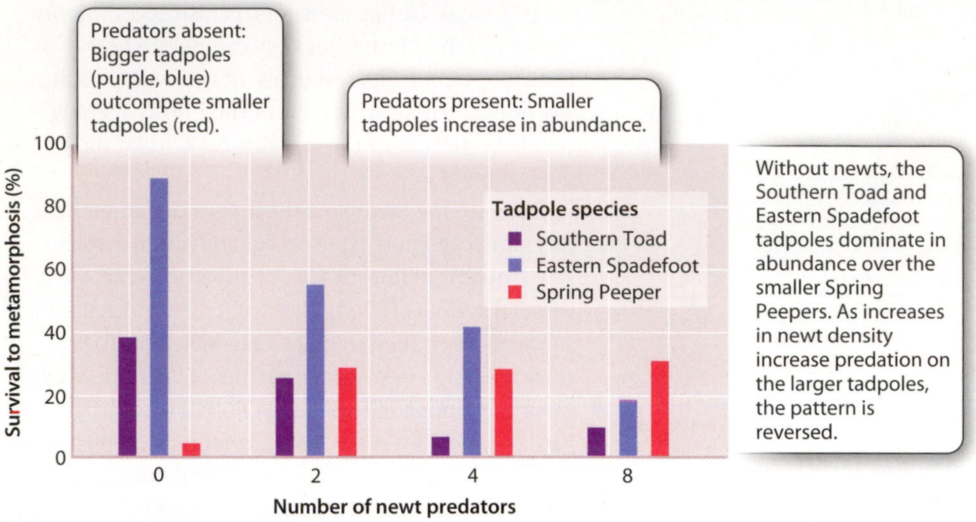

sometimes cause greater susceptibility of their hosts to infection by other pathogens.

47.3 MUTUALISTIC INTERACTIONS BETWEEN SPECIES

Interactions between species do not necessarily pair a gain for one participant with a loss for the other. As the example of cacao and midges shows, some interactions benefit both participants, though usually there are still some costs associated with each participant's benefits. Typical benefits include access to nutrients, shelter from enemies or weather, and direct help in reproduction, but all benefits are ultimately measured by natural selection in terms of reproductive output. Some typical costs include the proteins, fats, and carbohydrates that are invested in building structures such as flowers that attract pollinators, or specialized tissues that house bacteria or algae. Costs can also be energy-consuming activities such as the transport of pollen or seeds, or the loss of food resources used by a partner. When the benefits for each participant outweigh their costs, the interaction is a **mutualism.**

Mutualisms are interactions between species that benefit both participants.

The midges in our opening example obtain the benefit of food from the cacao blossoms at the cost of unwitting pollen transport between flowers; the chocolate tree obtains the benefit of pollination at the cost of producing the sugars and amino acids in the flowers' nectar.

The nitrogen-fixing bacteria that live in nodules on the roots of soybeans (Chapters 26 and 29) provide another illuminating example. The bacteria occur naturally in the soil and are attracted to the plant roots, where they take up residence in nodules produced by the plant. The bacteria provide their host plant with nitrogen in a biologically useful form, and the plant provides the bacteria with food and a stable environment. In response to signals from the bacteria, the plant builds the bacteria's home—the nodules—out of root tissue, a cost outweighed by the benefit of enhanced access to nitrogen, a benefit measured in greater growth and reproductive output. In the interaction between plant and bacteria, both sides win.

Despite the observation that both partners in a mutualism benefit, it is important to remember that each side is acting in its own self-interest and bears costs that are weighed against benefits in terms of growth and reproduction. Associations that, overall, are beneficial to one partner are often also beneficial to the other partner and, therefore, more often represented in the next generation. In other words, mutualisms are subject to natural selection, just like any other adaptation (Chapter 21).

Mutualisms may evolve increasing interdependence.

Close interactions between species that have evolved over long periods of time are called **symbioses** (singular, **symbiosis**). One of the best-studied symbiotic mutualisms involves aphids (and their insect relatives) and closely associated bacteria. The insects suck plant sap for food, and the bacteria live in tissues that develop in the insects adjacent to their digestive system. The insects provide a home for the bacteria at some physiological cost but gain an important benefit: Plant sap contains relatively few nutrients, and the bacteria provide their hosts with essential amino acids. The bacteria benefit from a stable and favorable environment at the cost of some loss of nutrients supplied to the aphids. The aphids pass the bacteria from mother to daughter in egg cells, guaranteeing that offspring will have bacteria as effective as those associated with earlier aphid generations. When the interaction between species drives reciprocal adaptations in both participants, there is a possibility for long-term coevolution because the descendants of each side will also be associated with each other, just as were past generations. Any mutations that arise can affect both sides of the partnership because neither aphids nor bacteria are found without the other.

DNA evidence suggests that aphids and their bacteria have been associated for 100 million years or more, and that the bacterial genomes have been gradually losing genes they would need for life outside the insects (**Fig. 47.8**). Some bacterial genes have in fact been acquired by the insects. We saw a similar process of

HOW DO WE KNOW?

FIG. 47.8

Have aphids and their symbiotic bacteria coevolved?

BACKGROUND Aphids are small insects that are common pests of garden plants. Specialized cells in the aphids harbor populations of the bacterium *Buchnera* that provide their host with amino acids essential for growth. Mother aphids pass the bacteria through their eggs to their daughters.

HYPOTHESIS Aphids and *Buchnera* bacteria have coevolved for millions of years.

EXPERIMENT Evolutionary ecologist Nancy Moran sequenced the DNA of bacteria and the aphids they come from to establish the phylogenetic relationships of aphid species and their symbionts. She reconstructed the phylogeny of both groups and compared their relationships. If the two phylogenies matched each other, the hypothesis of coevolution would be supported.

RESULTS The phylogenetic trees of aphids and their associated bacteria matched perfectly, just as if the bacteria were a gene of the aphids instead of a separate group. This matching of phylogenetic trees was seen both among aphid species in different genera and for aphid populations within a single species. A DNA-based molecular clock further showed that aphids, and the insects related to them, have all coevolved with these bacteria over nearly 200 million years.

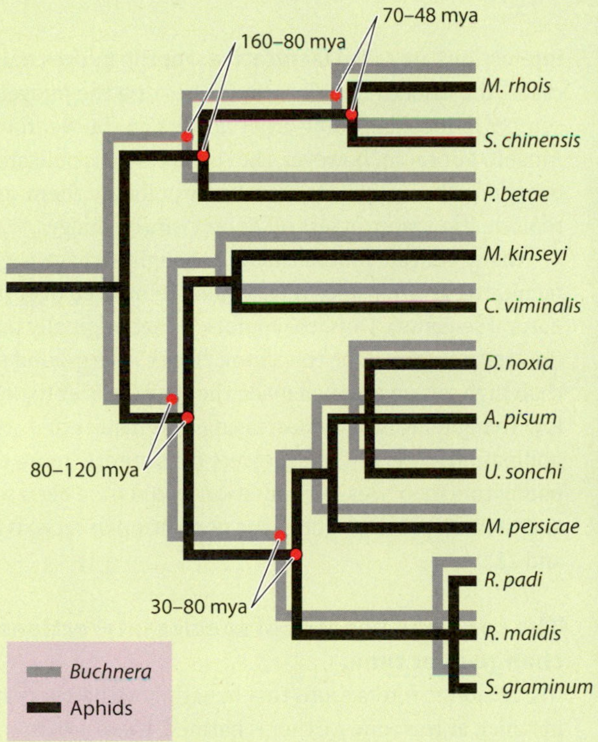

Photo source: Nigel Cattlin/Science Source.

CONCLUSION A single group of *Buchnera* has been passed down through the generations and the diversification of its aphid hosts for nearly 200 million years. Host–symbiont systems can be stable over long time intervals and show long-term coevolution.

FOLLOW-UP WORK Researchers are studying the changes in the genomes of bacteria living inside aphids to see how they lose genes necessary for living independently of their aphid hosts.

SOURCE Moran, N. A., et al. 1993. "A Molecular Clock in Endosymbiotic Bacteria Is Calibrated Using the Insect Hosts." *Proceedings of the Royal Society. London, Series B*, 253:167–171.

coevolution in Chapter 26 in the example of mitochondria and chloroplasts, which are probably nature's most dramatic examples of coevolved mutualism.

While some mutualisms involve close interactions between specific species, others are less particular. For example, sweet-tasting fruits evolved in flowering plants, attracting mammals and birds that disperse their seeds, but not in response to any particular bird or mammal species. Similarly, flowers evolved in response to insect pollinators such as bees and flies, whose own adaptations for visiting flowers evolved in response to flower availability.

Digestive symbioses recycle plant material.

In Case 5: The Human Microbiome, we saw that diverse microorganisms in our digestive tract provide important services to our own well-being. Humans aren't alone in hosting microbial symbionts; in fact, most animals have microbiomes. All herbivorous animals studied to date, from cows and sheep to ants, beetles, and termites, can feed on plant tissues only with the help of microbial symbionts. Much of Earth's biomass consists of wood and other plant tissues rich in cellulose and lignin. Few organisms have the ability to digest these molecules, but some microorganisms, mainly bacteria and fungi, can. When these organisms reside in an animal gut, for example the rumen or large intestine of a cow, they enable their host to benefit from plant tissues that the animal itself cannot break down (**Fig. 47.9**). Both host and symbiont therefore benefit from the association.

Not all of these plant-digesting symbionts live in the digestive tracts of their hosts. Consider leaf-cutting ants. Colonies of these ants in a Central American forest can consume about 14% of the leaves of nearby trees, hauling each in pieces down into their subterranean nests. In terms of kilograms of leaves consumed per hectare, a single colony of leaf-cutting ants is the functional equivalent of an herbivore the size of a sheep. The ants do not feed on the leaves directly, but rather use them to feed gardens of fungi that they tend carefully. The fungi, in turn, provide food for the ant colony. In contrast, bark beetles all over the world bring their fungi to the trees themselves, cultivating them deep inside the trunks to feed their young.

These insect–fungi symbioses can be thought of as a form of agriculture that, like human agriculture, has a large impact on the biosphere through their efficient breakdown of plant materials. Insect–fungus mutualisms are dramatic, but, like the human microbiome, they are more the rule than the exception: Nearly all eukaryotes depend on microbial symbionts with which they have coevolved (Chapters 26 and 27, Case 5).

Mutualisms may be obligate or facultative.

When one or both sides of a mutualism cannot survive without the other, the association is said to be **obligate.** The association of aphids and bacteria is obligatory for both sides: Bacteria cannot live without the shelter the aphids provide, and aphids cannot live without the nutrients provided by the bacteria.

Many associations, however, are not so tightly intertwined, and one or both participants can survive without the other. These interactions are called **facultative.** For the midges that pollinate cacao, the association is facultative because the midges have other sources of food, including other flowers, in the wet forests they inhabit. For cacao, however, the relationship is obligate because bees and other insects do not usually pollinate them, and so their reproduction more or less relies on visits by midges.

Many obligate mutualisms are thought to have begun as facultative relationships that became reinforced over time by natural selection. Thus, the mutation that originally made a cacao flower sprout from the tree trunk close to the ground rather than high out on a branch made the flower easier to pollinate by low-flying midges while decreasing the chances of discovery and pollination by bees. If midges were consistently more reliable pollinators than bees, the mutation would have been selected and then spread throughout the population of cacao (Chapters 21 and 22).

FIG. 47.9 Microbial mutualisms. Symbiotic mutualisms between cows and the bacteria that live in their rumen and large intestine help digest cellulose, the most abundant biomolecule on Earth.

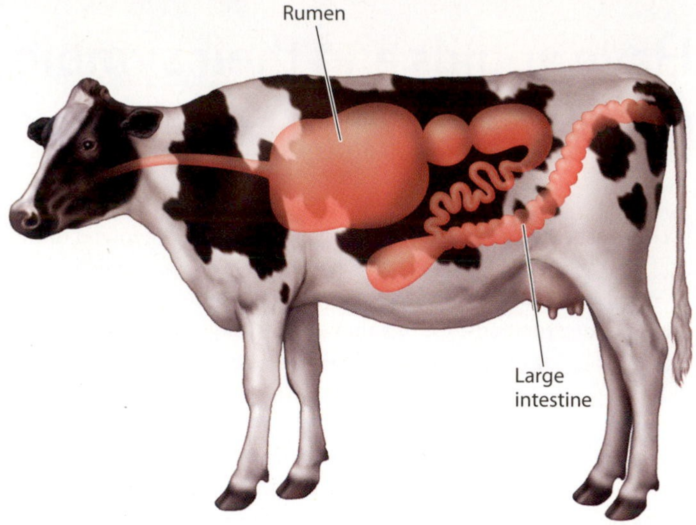

The costs and benefits of species interactions can change over time.

In addition to interactions that benefit both partners and those in which at least one partner is harmed, there are interactions in which one partner benefits with no apparent effect on the other. These interactions are called **commensalisms.** For example, Grey Whales in the Pacific Ocean are commonly covered with barnacles. The barnacles benefit from the association, obtaining both a substrate for growth and a free ride through waters rich in planktonic food, but the presence or absence of barnacles doesn't seem to affect the whales one way or the other.

Some ecologists also recognize another class of interaction, called amensalism, in which one partner is harmed with no

TABLE 47.1 Major Types of Species Interactions

INTERACTION	EFFECT ON EACH SPECIES	EXAMPLE
Competition	−/−	A grass and a wildflower: Each species loses the water, nutrients, and access to sunlight that the other takes.
Predation	+/−	Arctic Foxes and lemmings: Foxes benefit from eating lemmings; lemmings lose opportunities to reproduce.
Herbivory	+/−	Bison and grass: Bison benefit from eating grass; grass loses biomass that is eaten.
Parasitism	+/−	Tapeworms and humans: Tapeworms benefit from absorbing nutrients in human intestine; humans lose nutrients.
Mutualism	+/+	Flowers and bees: Flowers gain pollination; bees gain nectar and some pollen.
Commensalism	+/0	Egrets and cattle: Egrets benefit from insects stirred up by cattle; cattle are unaffected by egrets.

apparent effect on the other. A commonly cited example is the coconut palm and Brazil nut trees, tropical trees that produce heavy nuts that harm animals and plants they fall on. The spectrum of costs and benefits associated with interactions among organisms is summarized in **Table 47.1.**

Associations are not fixed—they can change over time. A mutualism can in some cases become antagonistic if one of the partners "cheats" by imposing a larger cost than benefit on the other. In fact, mutualisms that are loose associations among changing partners can become one sided rather quickly. For example, many plants have evolved tubular flowers that guide bees past their anthers or stigma on the way to the nectar at the base, taking just a little more time from bees who try to visit as many

FIG. 47.10 **Commensalism and mutualism.** Egrets feed on insects stirred up by water buffalo, which are usually not affected by the egrets—a commensalism. If, however, the egrets spot lions and the poor-sighted buffalo take warning, the relationship becomes a mutualism. *Source: Biosphoto/Gérard Lacz.*

flowers as possible. Some bees have short-circuited this plant mechanism by nipping the flower base from the outside and then drinking the nectar without pollinating the flower. Most plants have enough successful pollination to ensure seed production, but losses from cheating can still be costly.

Some interactions begin as a benefit to one species, with no benefit or cost to the other. For example, cattle egrets follow water buffalo to pick up the insects stirred up by the buffalo passing. The egrets gain food, but the buffalo are not affected either way—an example of commensalism (**Fig. 47.10**). However, cattle–egret commensalism can become a mutualism over time if the egrets give early warnings of nearby predators such as lions or eat insects like tsetse flies that carry buffalo diseases.

→ **Quick Check 3** What are some of the costs and benefits to an apple tree and to the honeybee that pollinates it?

47.4 ECOLOGICAL COMMUNITIES

We have been discussing how species interact and how these interactions can shape the distributions of populations, limiting distributions in one way or another. Now we look collectively at all the species that occur together in the same place. How, and to what extent, do they interact? And what are the properties of these interacting sets of species?

Species that live in the same place make up communities.

A **community** is the set of all populations found in a given place. This definition seems straightforward, but it raises some large questions about the nature of communities. Two different views were proposed nearly a century ago, and ecologists have debated their merits ever since. Frederic Clements, a prominent American ecologist of the early twentieth century, likened plant communities to a "superorganism" in which species interact strongly and predictably, like the organs within a body. In contrast, Henry Gleason, his contemporary, viewed plant communities as simply the products of species' acting individually in time and space. Not surprisingly, research inspired by their disagreement suggests that the plants in most communities lie somewhere between these extremes.

Populations in a community, whether a salt marsh, mountaintop, or desert, are tied together by the various interactions that secure their spot in a food web (Chapter 25), as well as by their physical location. The populations in a local community share the same rainfall, climate, and soil, and often encounter one another physically as they grow, reproduce, and die. Some provide food or shelter for others, and all end up being recycled by the many small decomposers in the soil. We can often characterize communities by the principal plants and animals they harbor, such as the cacti and lizards of deserts, the tiny, rock-loving mosses and marmots of a mountaintop, or the marsh grasses and mosquitoes of the coastal salt marsh. Indeed, communities often seem quite distinctive to us and easily recognized, and they are important as the local instances of biodiversity. At first glance, it would even appear that the plants and animals and fungi that live in one kind of place might compete for the very same local resources.

When we look at the details of where particular species occur, almost no two species have exactly the same geographic distribution. This is partly a result of competitive exclusion (unless the species are obligate associates of each other, like aphids and their bacteria). The northern bogs of New England may appear to have distinctive species, such as Black Spruce and Bog Club Moss, but while these and other bog-loving species overlap in their mutual preference for the water-saturated, acidic, sandy soils found in northern bogs, they do not occur in the same places over most of their geographic ranges. In contrast to populations or species, then, communities don't have clearly defined boundaries.

The activities of different populations in a community also vary in time. For example, flies are active during the day, pollinating flowers that open in the morning and falling prey to dragonflies, birds, and other predators. In contrast, moths are important pollinators at night, when spiders and bats that prey on them are also active. Similarly, in coastal marine communities, oysters, crabs, and many other animals follow the daily rise and fall of tides. And, of course, the activities of many species change seasonally in responses to sunlight, temperature, and rainfall. Protein-rich insects that birds prey on to feed their young are most abundant in the spring, and fruits and seeds rich in the oils that migratory birds need to supply fat reserves are mostly available in late summer and fall.

? CASE 8 BIODIVERSITY HOTSPOTS: RAIN FORESTS AND CORAL REEFS

How is biodiversity measured?

A key feature of any community is its **biodiversity,** or biological diversity. Biodiversity is often taken to mean the number of species, but we can also think of it more broadly and at many levels, including the variety of different genetic sequences, cell types, metabolisms, species, life histories, phylogenetic groups, communities, and ecosystems. We can also consider biodiversity for a particular place, such as the island of Hispaniola, or a tropical rain forest on the island of Hispaniola, or the entire planet.

Biodiversity is usually measured by counting the number of species present in a particular area. Because of limitations in time and expertise, scientists often focus on a subset of the taxonomic groups present. In terrestrial ecosystems, a good predictor of species diversity is the number of plant species, which reflects local moisture and temperature conditions. From this number, scientists can then estimate the numbers of other species, including the arthropods and other small animals that live on or near the plants, the birds and other vertebrates that prey on the insects and plants, the larger predators that prey on these, and the fungi and other decomposers that return the nutrients in dead bodies to the soil.

As a result, the number of species of flowering plants is among the best known, while the number of species of their ecological associates, the insects and fungi, is among the least known. On the island of Hispaniola, nearly every flowering plant species, approximately 6000 in all, is known. Because scientists estimate that there are about 2–3 times as many kinds of insects as flowering plants at any given place, the number of insect species will likely total about 12,000–18,000 once all are known. Each year studies of different insect groups are published that report a doubling or tripling of the known species. For example, recent studies of fireflies on the island reported a new total of 69 species, up from the 17 previously known. Other insects show the same pattern, as do other animals and fungi.

Tropical islands such as Hispaniola, Hawaii, and the Galápagos are missing elements of biodiversity that have not colonized them from the nearest mainland. Only a few of the South American frog and lizard groups are found on Hispaniola, and none but large iguanas made it to the Galápagos, which are farther away from the mainland. Nevertheless, the presence of many still unknown species of most animals and fungi makes ecological studies of tropical regions difficult. For this reason, scientists have turned to simpler communities such as those in extreme environments, where the ecological interactions can often be more completely understood.

One species can have a great effect on all other members of the community.

Predators and prey interact in settings as different as a tropical rain forest or Arctic tundra. In habitats with many different predator and prey species, identifying the ways that each species affects others can be challenging. For this reason, much research on how predators and prey influence community structure has been carried out in the Arctic, where communities tend to have relatively few species.

One well-studied community is found on Bylot Island in the Canadian territory of Nunavut. There, two species of lemming, the main herbivores in tundra communities, affect the populations of other herbivores and predators. Snow Geese are herbivores that live alongside the lemmings, and both geese and lemmings eat the grasses and other low plants that grow on the island. The main predators of lemmings and Snow Geese are Arctic Foxes and Snowy Owls. Lemmings, foxes, Snowy Owls, geese, and plants, then, are linked by a set of interactions, as shown in **Fig. 47.11.** Changes in one species can affect all of the others.

Lemming populations rise and fall every 4 years or so. When lemming populations decline, foxes turn to other prey, including the eggs of Snow Geese (while the now harder-to-find lemmings escape and their populations begin to increase again). Snow Geese, therefore, are indirectly influenced by lemming population densities. The population density of foxes also reflects that of lemmings—as lemming populations increase to the point at which they can again become a significant part of fox diets, the foxes have access to enough resources to have more pups, increasing the fox population. In time, the additional predators again cause the lemming population to decline, and fox reproduction drops as a result. The population density of lemmings also influences grass abundance through its indirect effect on whether or not foxes eat Snow Goose eggs; young geese are heavy grazers in years when goose reproduction is high.

FIG. 47.11 Predator–prey relationships in the Arctic grassland community on Bylot Island, Canada. Species have either a positive effect (+) or a negative effect (−) on other species.

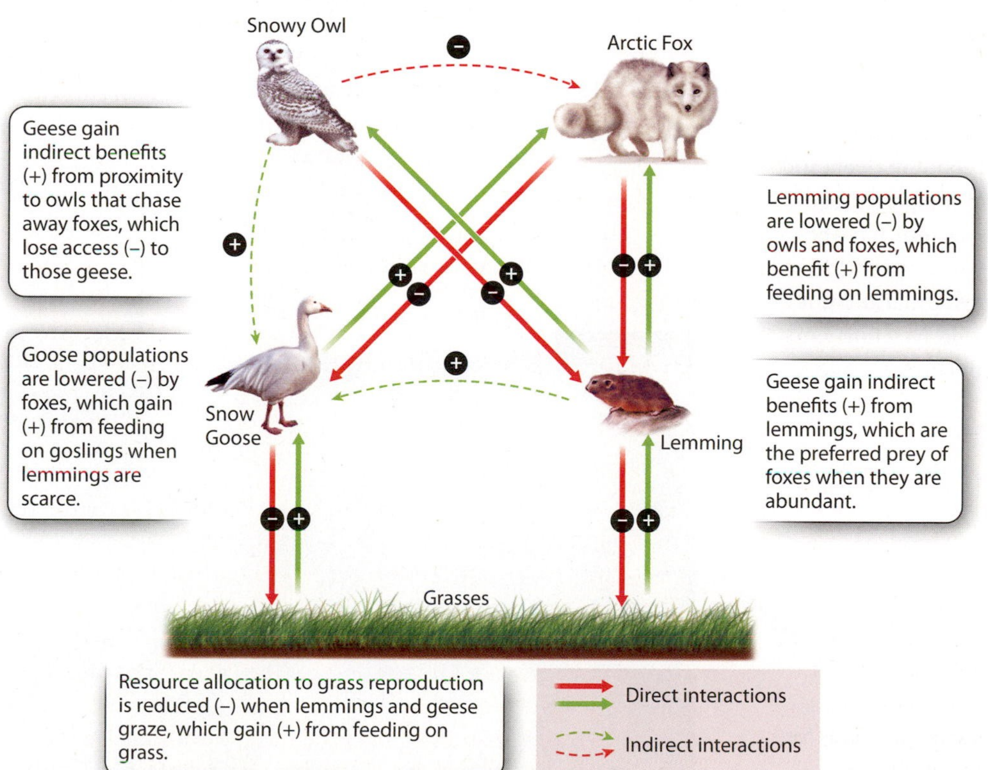

Keystone species have disproportionate effects on communities.

While all or nearly all species in a given place have some effect on one another, the integrity of a community may depend on a single species. This situation can occur when one species influences the transfer of a large proportion of biomass and energy from one trophic level to another or when the species modifies its physical environment, as beavers do when they make ponds. Such pivotal populations are called **keystone species** because they support a community in much the way an architectural keystone supports an arch. Though all species contribute to the structure of a community, a keystone species affects other members of the community in ways that are disproportionate to its abundance or biomass.

A classic experiment by American zoologist Robert Paine gave birth to the concept of keystone species. Along the northwestern coast of the United States, mussels adhere to rocks in the intertidal zone, where they coexist with about two dozen other species of invertebrate animals and seaweeds. Mussel populations are regulated by their principal predator, a sea star called *Pisaster ochraceus.* When Paine removed the sea star from experimental plots, mussel populations exploded, outcompeting other animals and seaweeds for space and so eliminating them from the rocky shoreline. Control plots in which the sea star remained showed no such change. From these results, Paine concluded that the activities of the predatory sea star had a singularly strong influence on the diversity and distribution of other populations in the rocky intertidal community. *Pisaster ochraceus* is a keystone species.

Another experiment, this time unintended, cements the point. Along the west coast of the United States and Canada, giant kelp grow more than 50 m tall, providing a home for a remarkable diversity of fish and invertebrates (**Fig. 47.12**). The kelp community depends strongly on a single predator: the sea otter. Sea otters are the predators of sea urchins, which graze on the rootlike holdfasts that anchor kelp to the seafloor. When the otters were removed by overhunting decades ago, sea urchin abundance soared and the kelp nearly disappeared, along with many other populations that live in kelp forests. Now, where otters are protected they once again keep urchin populations in check, enabling the kelp to return, and along with it, the diversity of fish and other animals for which kelp provides a habitat.

FIG. 47.12 Keystone species in Pacific kelp forests. The activities of keystone species strongly influence the communities in which they live. *Source: Carl Gwinn, blackcormorant.net.*

Keystone species called ecosystem engineers actively shape the physical environment, creating habitat for others. The classic example is beavers, which produce ponds by damming streams in forests, creating habitat that would not otherwise exist (**Fig. 47.13**). Hippos provide another good example, only recently recognized. In East Africa, hippos loll about in rivers and lakes during the day, but venture onto the land at night to feed on grasses and other low-lying plants. The hippos return to the water before the sun rises and—importantly—defecate in their watery home. In this way, hippos transfer a substantial amount of nutrients from land to river, and these feces-derived nutrients support a community that is more abundant and diverse than would be possible in the absence of hippos.

→ **Quick Check 4** Lemmings affect many other species in the Bylot Island community. Why are lemmings not considered a keystone species?

Disturbance can modify community composition.

We have seen how predation, herbivory, and parasitism can reduce the effects of competition. Severe physical impacts on a habitat, such as those caused by storms, earthquakes, or road building, can also dramatically lower the abundances of some species, as for example when an early frost kills mosquitoes and other insects. Such events are known as **disturbances,** and they have effects on populations of interacting species that are independent of their densities, as we saw in Chapter 46. Disturbances often affect multiple species in the same community and so can exert a strong influence over community composition. We usually think of disturbances as abiotic, but the line is not so clear with road-building by humans, for example, or dam building by beavers. Such changes are not uncommon, but are usually unpredictable.

Severe weather in forests demonstrates how disturbance can affect a community. When fire or a hurricane takes down tall trees in a forest, it also takes away the shade that had prevented some species from establishing new seedlings. Fires are frequent sources of disturbance in regions with a seasonally dry climate, such as parts of western North America and southwestern Australia. Like storms, fires remove existing plants and animals, opening up habitat for recolonization. Some species have adapted to frequent fires. For example, several pine species have evolved thick bark and high branches, traits that help protect them from fire. Others, which have evolved cones that open only in the wake of fire, are adapted to use the habitat space opened up by the fire and the nutrients that dead plant matter releases into the soil.

Noting the influence of storms on community composition in forests and coral reefs, American ecologist Joseph Connell proposed that the diversity of species in a given community reflects, at least in part, the frequency and intensity of disturbance. If disturbance is intense, few species can tolerate the physical conditions of the habitat and diversity will be low. On the other hand, if disturbance

FIG. 47.13 A beaver dam visible from space. This NASA satellite image shows a beaver dam in northern Alberta, Canada. *Source: DigitalGlobe/ScapeWare3d/Getty Images.*

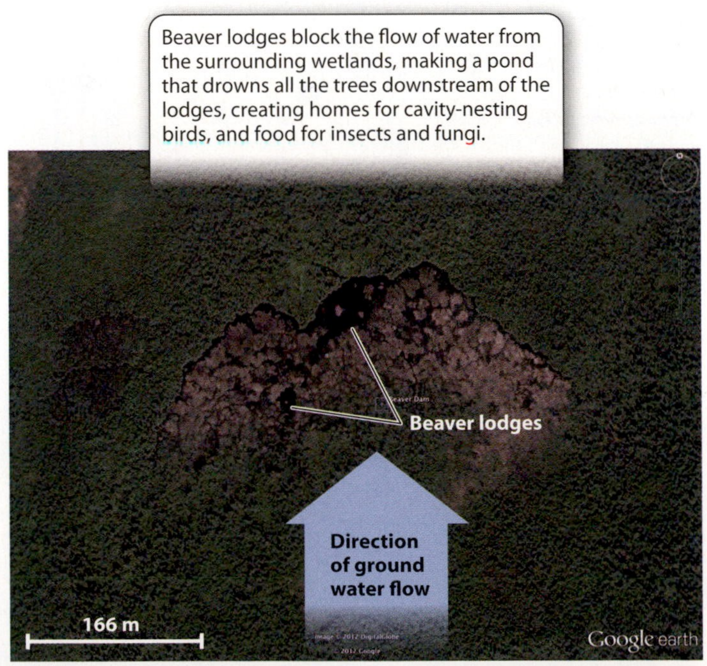

is rare or weak, competition may take over and only a few, stronger, competitors may remain. Species diversity tends to be highest when disturbance is frequent or intense enough to inhibit competition, but not so strong as to limit the number of species that can tolerate the environment.

Succession describes the community response to new habitats or disturbance.

Following disturbance or the appearance of a new habitat, a community goes through a series of changes. These changes involve a predictable sequence of colonization of species that then transform the community, in what can appear to be a linear process of maturation. For example, when a lava flow cools, the first organisms to colonize the rock are lichens and mosses that begin the slow process of soil formation. They are followed by grasses and other small plants, and by tiny arthropods and eventually larger animals and woody plants that eventually shade out the mosses and lichens that started the process of colonization. Similarly, following a large storm or fire, newly vacated habitat is commonly colonized by plant species adapted for rapid growth and strong sun. Eventually, these species are replaced by others able to grow in increasing shade. The pace of this transformation may be rapid, occurring over a few decades, or it may be much slower, depending on the environment.

This process of species replacing each other in time is called **succession** (**Fig. 47.14**). Clements viewed the pattern of species replacement in succession as strong evidence in support of his view that communities are highly integrated. Gleason, in contrast, explained succession purely in terms of the habitat preferences of individual populations.

The Pin Cherry, a resident of New England forests, illustrates how the life histories of individual species can contribute to succession. If you walk through a forest in New Hampshire, you will see oak trees and beeches, pine and hemlock. A mature forest may contain few or no Pin Cherry trees, but if you examine seeds lying dormant within the forest soil, you will see numerous Pin Cherry seeds. How did they get there?

Years earlier, disturbance from a storm or fire opened up the forest floor and Pin Cherries established themselves in the sunlit soils, growing from seeds left in the droppings of birds that had consumed the cherries elsewhere. The seedlings of Pin Cherries grow well in full sunlight, so the trees grew rapidly to reproductive maturity. A forest canopy was eventually reestablished, and Pin Cherry seeds do not germinate in the shade. But oak seedlings prosper in shade, so, through time, oaks and other species came to dominate this patch of forest. They will continue to dominate it until the next disturbance, when the removal of mature trees will once more expose the forest floor to bright sunlight, literally giving Pin Cherries another day in the sun.

The succession of species reflects the adaptations of each to the changes in sunlight, soil, water, and space that occur as the first organisms arrive, mature, and die. These organisms' bodies influence their habitat in both life and death, whether aquatic or terrestrial. Succession over thousands of years can turn bare rock into forest. Lichens are often the first organisms to colonize bare rock newly formed by cooling lava or exposed by landslides. The lichens may start off with only a few mites and other tiny invertebrates living on them. As the stone weathers, and microorganisms feed on the lichen and one another, they form particles of soil that accumulate in cracks and crevices. Eventually,

FIG. 47.14 **Succession.** In new habitats or following disturbance, species colonize and transform the habitat in a predictable sequence that reflects the adaptations of participating species.

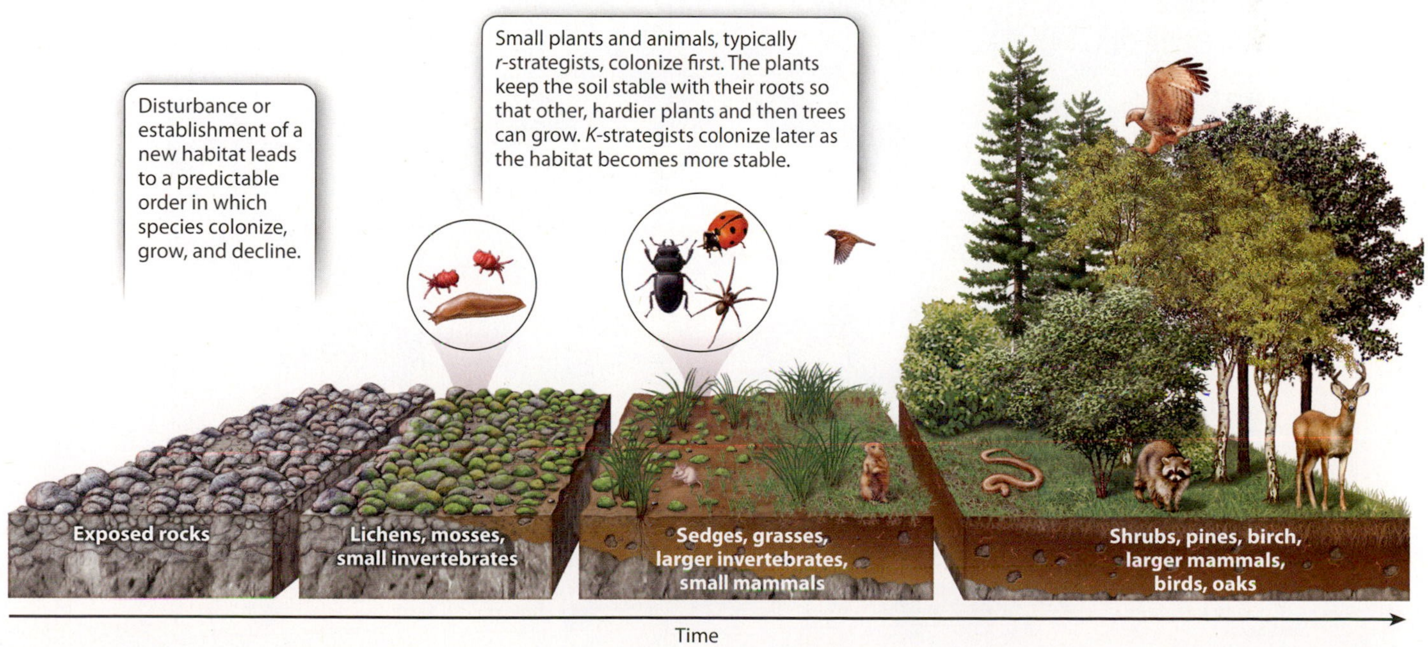

small plants, such as sedges and grasses, take root, and the growing plant community begins to gain height and diversity as other plants establish themselves.

Soon, larger insects, spiders, and other invertebrates join the community, as well as the first small mammals and birds. Many of these early colonizing species are *r*-strategists (Chapter 46), growing and reproducing quickly, and spreading their young far and wide. Birds carry the seeds of shrubs and other plants in their droppings, and the gradually deepening soil eventually permits the establishment of shrubs that in turn provide habitat for larger mammals and birds, and perhaps snakes and salamanders.

The first sun-loving trees, such as birches and pines, appear next, providing shade for the seedlings of later colonizing trees like oaks and maples. In many forests, these stable, long-lived *K*-strategists (Chapter 46) make up a final stage in the succession, forming a mature assembly commonly called a **climax community.** A climax community is one in which there is little further change in species composition.

47.5 ECOSYSTEMS

A community of organisms and the physical environment it occupies together form an **ecosystem** (**Fig. 47.15**). The physical and biological components of ecosystems are linked by the processes that cycle nutrients and transfer energy through the system. In Chapter 25, we introduced the concepts of the food web as a depiction of carbon cycling through ecosystems. Now we can revisit this concept and add to it the idea of energy flow through a community, understanding both as reflections of the functional properties of individual species and the interactions among species within a community.

→ **Quick Check 5** What is the difference between a community and an ecosystem?

Species interactions result in food webs that cycle carbon and other elements through ecosystems.

Of all the interactions observed among species, none are more prominent than predation and herbivory. Heterotrophic organisms, from bacteria to blue whales, obtain carbon and other elements needed for growth from other organisms. Thus, when a sea star preys on a mussel, we can view this interaction both as predation and as a link in the carbon cycle. Heterotrophic organisms commonly consume other heterotrophs, but, ultimately, ecosystem function depends on autotrophs—photosynthetic or chemosynthetic organisms that can form organic molecules by the reduction of carbon dioxide available in the physical environment (Chapter 6).

In ecology, photosynthetic and chemosynthetic organisms are known as **primary producers,** organisms that take up inorganic carbon, nitrogen, phosphorus, and other compounds

FIG. 47.15 **A savanna ecosystem in Africa.** The ecosystem includes the grasses, trees, animals, and microorganisms in the community, as well as the climate, soil chemistry, and other features that define its physical environment. *Source: Dan L. Perlman/EcoLibrary.org.*

FIG. 47.16 **A food web.** A food web diagrams the cycling of carbon and other elements from the environment through a succession of organisms and back to the environment.

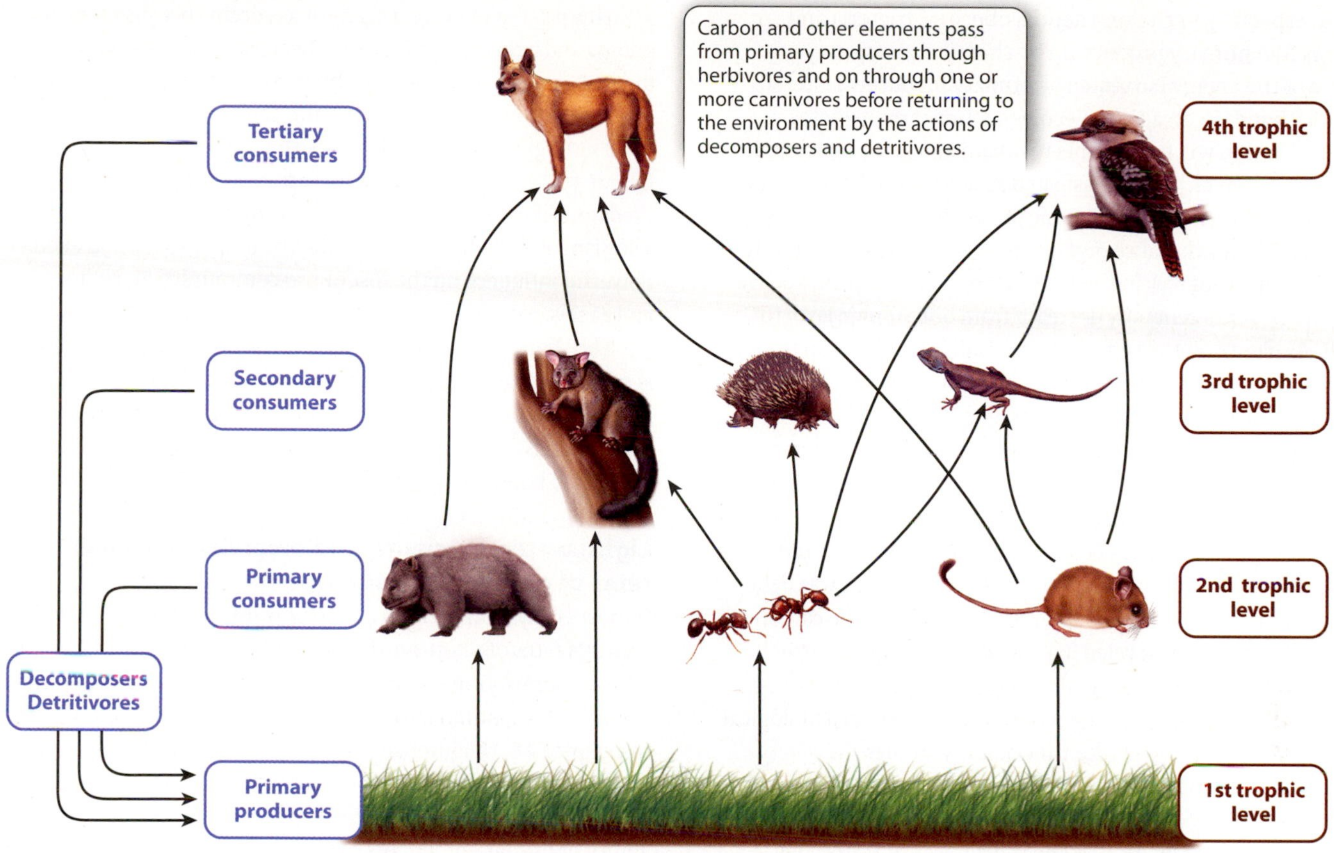

from the environment and convert them biochemically into proteins, nucleic acids, lipids, and more (Chapter 25). **Consumers,** heterotrophic organisms of all kinds, depend on primary production, directly consuming primary producers or consuming those that do. Predators hunt and eat prey, herbivores eat plants, and parasites infect their hosts. **Decomposers** such as certain bacteria and fungi are a special class of consumer, breaking down the molecules of primary producers and consumers and returning the carbon they contain to the atmosphere through respiration.

The various species present in an ecosystem can therefore be placed in an order that describes how one organism feeds on another, moving carbon (and other nutrients) through the system. This order is commonly depicted as linear, and it is called a **food chain.** In nature, however, every species is connected to many others. Many ecologists also stress that most communities contain species that are omnivorous, eating both plants and animals. For these reasons, the movement of carbon through an ecosystem is more realistically thought of as a **food web** (**Fig. 47.16**).

An organism's typical place in a food web is its **trophic level** (Fig. 47.16). The first trophic level consists of primary producers; consumers occupy the levels above the first trophic level. At the second trophic level are primary consumers; these are the herbivores, organisms that eat plants or other primary producers. Predators, organisms that eat other consumers, are secondary consumers, at the third trophic level and potentially higher levels as well. Top or apex predators, which have few, if any, natural predators, occupy the highest trophic level. Parasites exploit all levels of the system, obtaining carbon and other nutrients from primary producers and consumers alike. Finally, there are the all-important decomposers and detritivores. Decomposers feed on the dead cells or bodies of other organisms, and detritivores do much the same, consuming partially decomposed plants or animals in soil or sediment, as well as organic compounds in feces. These organisms return carbon dioxide and other inorganic compounds to the environment, completing the cycle of elements through the ecosystem.

Species interactions form trophic pyramids that transfer energy through ecosystems.

The interactions among species transfer energy as well as carbon and other elements through ecosystems. Unlike carbon, energy does not cycle through an ecosystem, and so new energy must continually be harvested from the environment to sustain the community. In essentially all ecosystems where sunlight is available, photosynthesis lies at the heart of ecosystem function. Plants, algae, and photosynthetic bacteria capture energy from the sun and use it to synthesize organic molecules. Where sunlight is absent, especially in the vast depths of ocean, bacteria and archaeons fuel primary production by chemical reactions.

Some of the energy harvested by primary producers is stored as chemical bonds in organic molecules such as carbohydrates (Chapter 8) and so will be available to organisms at the next trophic level. Some, however, will be dissipated as heat or used to do work (Chapter 6). Because building organic molecules is not 100% energy efficient, and because organisms at one trophic level rarely consume all the resources in the level below them, biomass and the energy it represents generally decrease from one trophic level to the next. As plants capture energy from sunlight and carbon from the atmosphere, they build bonds that store energy in molecules of carbon and other nutrients, converting radiant energy into chemical energy. Organisms that consume these organic molecules store some of the energy in other molecules such as fat and starch, but they also use much of the food energy to do work or generate heat. The energy expended as heat or work cannot be captured again by photosynthetic organisms (which require energy as light of only certain wavelengths, Chapter 8). Therefore, while carbon and other elements are recycled in ecosystems, energy transfer is unidirectional.

As we saw, the various producers and consumers in a biological community can be grouped into functionally similar classes, the trophic levels. Diagrams showing the biomass and therefore energy available at each level to feed the next are called **trophic pyramids** (**Fig. 47.17**). These diagrams of terrestrial biomass have a pyramid shape because the biomass of primary producers (usually plants) is generally more than that of primary consumers (herbivores), whose biomass is more than that of secondary consumers (the predators of herbivores), and so forth.

This pattern of decreasing biomass occurs because transfer of energy and biomass is inefficient. Because of wastes, work, and heat dissipation, only about 10% of the energy and biomass available at one trophic level is incorporated into the next level. For example, 1000 kg of grass biomass will support about 100 kg of lemmings (about 400 animals), and that, in turn, will support 10 kg of small predators like foxes. At the apex of the pyramid, we find a single kilogram of eagle tissue. In this way, primary production exerts a powerful influence on the rest of the community. To get more foxes and eagles, we would need more photosynthesis.

The shape of the biomass diagram for marine ecosystems is often inverted because the plants are all tiny plankton that reproduce much more rapidly than their consumers do, and so their biomass at any one time may be less than that of consumers present (Chapter 48).

FIG. 47.17 A trophic pyramid. A trophic pyramid graphically shows the transfer of energy through an ecosystem.

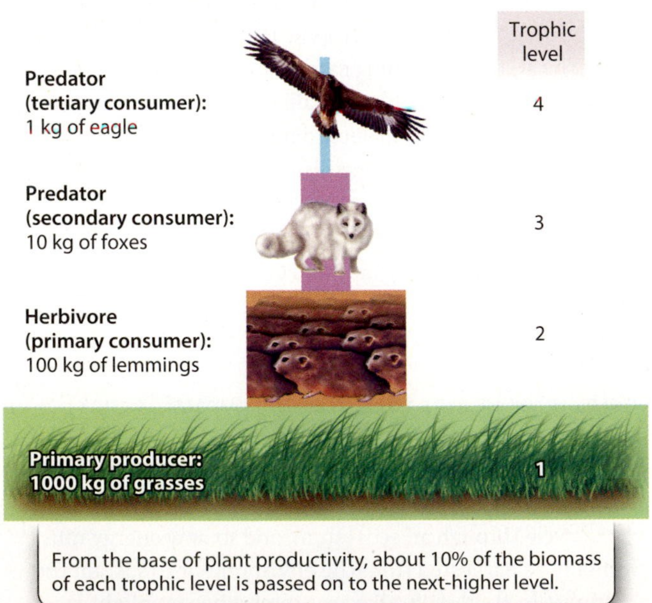

Light, water, nutrients, and diversity all influence rates of primary production.

If most ecosystems rest on a base of primary production, what controls rates of photosynthesis? Sunlight plays a role. Although all ecosystems, from polar deserts to equatorial forests, receive the same total number of hours of sunlight annually, as we discuss in Chapter 48, the intensity of solar radiation declines from the equator to the poles because of Earth's curvature. Seasonality also becomes more pronounced toward the poles, influencing the ability of primary producers to sustain growth throughout the year at higher latitudes.

In addition to sunlight, water and nutrients govern rates of primary production in most ecosystems. In Chapter 29, we saw that plants perform a physiological balancing act, opening their stomata to allow carbon dioxide into leaves but losing water vapor as a result. Because photosynthesis requires both carbon dioxide and water, photosynthetic rate—the amount of carbon dioxide fixed into organic matter in a given time interval—is closely tied to water availability. Lack of regular access to water limits the productivity of photosynthetic organisms in deserts and seasonally arid landscapes, explaining why irrigation increases crop yields in places like southern California, northern Africa, and the Middle East.

Even where water is not limiting, rates of photosynthesis are constrained by the availability of nutrients. This is why fertilizer generally increases crop yields (Chapter 29). Plants, algae, and photosynthetic bacteria all need nitrogen, phosphorus, sulfur, iron, and an array of trace elements to grow, and any one of these can limit primary production. In forests, nitrogen is commonly the soil nutrient that is least abundant relative to the needs of plants. In the open ocean, nitrogen can also be limiting, but usually it is a scarcity of phosphorus that limits photosynthesis. And in some parts of the ocean, nitrogen and phosphorus are both present in abundance, but primary production is still limited—in this case by iron, needed for many enzymes and electron transport molecules. The idea that

HOW DO WE KNOW?

FIG. 47.18

Does species diversity promote primary productivity?

BACKGROUND The different plant species found in communities have combinations of leaves, stems, and roots that tap different sources of water and nutrients in the soil and capture light at different levels of sun and shade. Do the differences among plant species result in higher levels of primary production than would be possible with fewer species?

HYPOTHESIS Plant diversity promotes primary production within a community.

EXPERIMENT Ecologist David Tilman conducted a long-term experiment in Minnesota, seeding each of 11 plots with 1, 2, 4, or 16 grassland species, and controlling each plot's amount of nutrients, water, and carbon dioxide, as well as manipulating the plots' exposure to herbivory and disturbance by fire. The primary production of each plot was measured annually for up to 23 years.

RESULTS The graph shows that the addition of nitrogen fertilizer [quantified as kg of N per hectare (ha)] increased primary production significantly, and additional water, higher CO_2, herbivory, and disturbance also resulted in at least a modest increase in productivity. The difference in primary production between high-diversity (16 species) and low-diversity (1, 2, or 4 species) plots was at least as great as the largest additions of fertilizer—and much greater than any of the other treatments.

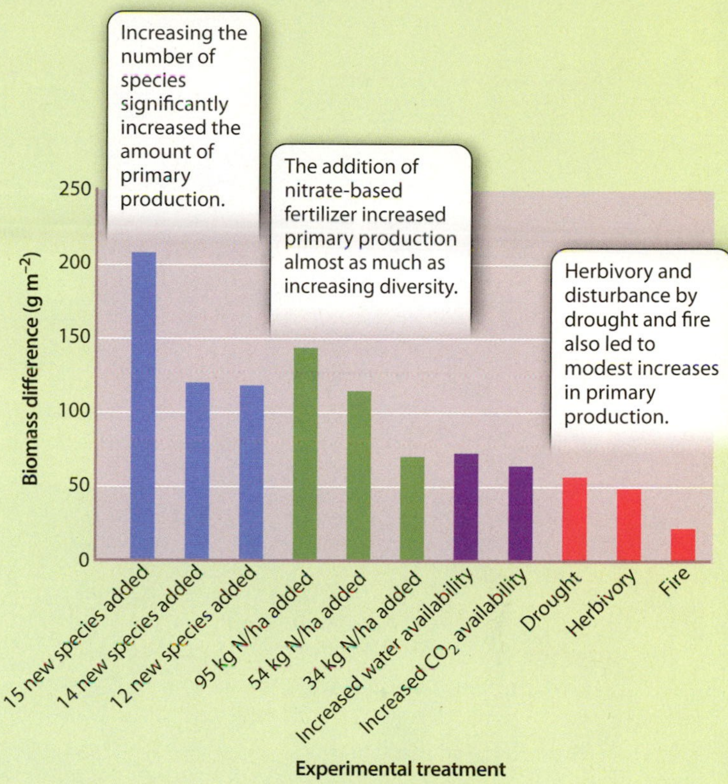

CONCLUSION The hypothesis was supported: In the grassland community under study, species diversity promoted primary production.

FOLLOW-UP WORK Researchers continue to study the ecosystem effects of species diversity, steadily accumulating data that show the importance of conserving the species diversity found in nature.

SOURCE Data from Tilman, D., P. B. Reich, and F. Isbell. 2012. "Biodiversity Impacts Ecosystem Productivity as Much as Resources, Disturbance, or Herbivory." *Proceedings of the National Academy of Sciences USA*, 109:10394–10397.

primary production is limited by the nutrient that is least available relative to its use by primary producers is called **Liebig's Law of the Minimum,** after Justus von Liebig, a nineteenth-century German scientist who first popularized the concept.

The biodiversity of a given region reflects its level of primary productivity. We have seen that the population sizes of predators depend on prey population size, which depends on primary production, and greater primary production also means a greater variety of plants of different sizes and kinds that can support a greater variety of herbivorous animals. While this suggests that high productivity can foster high diversity, recent experiments show that the reverse may also be true: Biodiversity may itself play an important role in ecosystem productivity (**Fig. 47.18**). This is a result of the many different ways that different plants and animals may use the resources presented by soil, water, and sunlight in a particular place. Collectively, a diverse collection of species will be more productive than a single species because no species is adapted for all functions in a given set of conditions.

The diversity of species on Earth or in a specific place is ultimately the result of the evolutionary, physiological, behavioral, and ecological processes we discussed in the preceding chapters. We follow changes over time in an aquatic and terrestrial habitat to summarize key concepts about population growth, species interactions, communities, and ecosystems in **Fig. 47.19.** We also introduce ways in which humans affect these processes, a topic that we revisit in Chapter 49. ■

VISUAL SYNTHESIS

FIG. 47.19

Succession: Ecology in Microcosm

Integrating concepts from Chapters 46 and 47

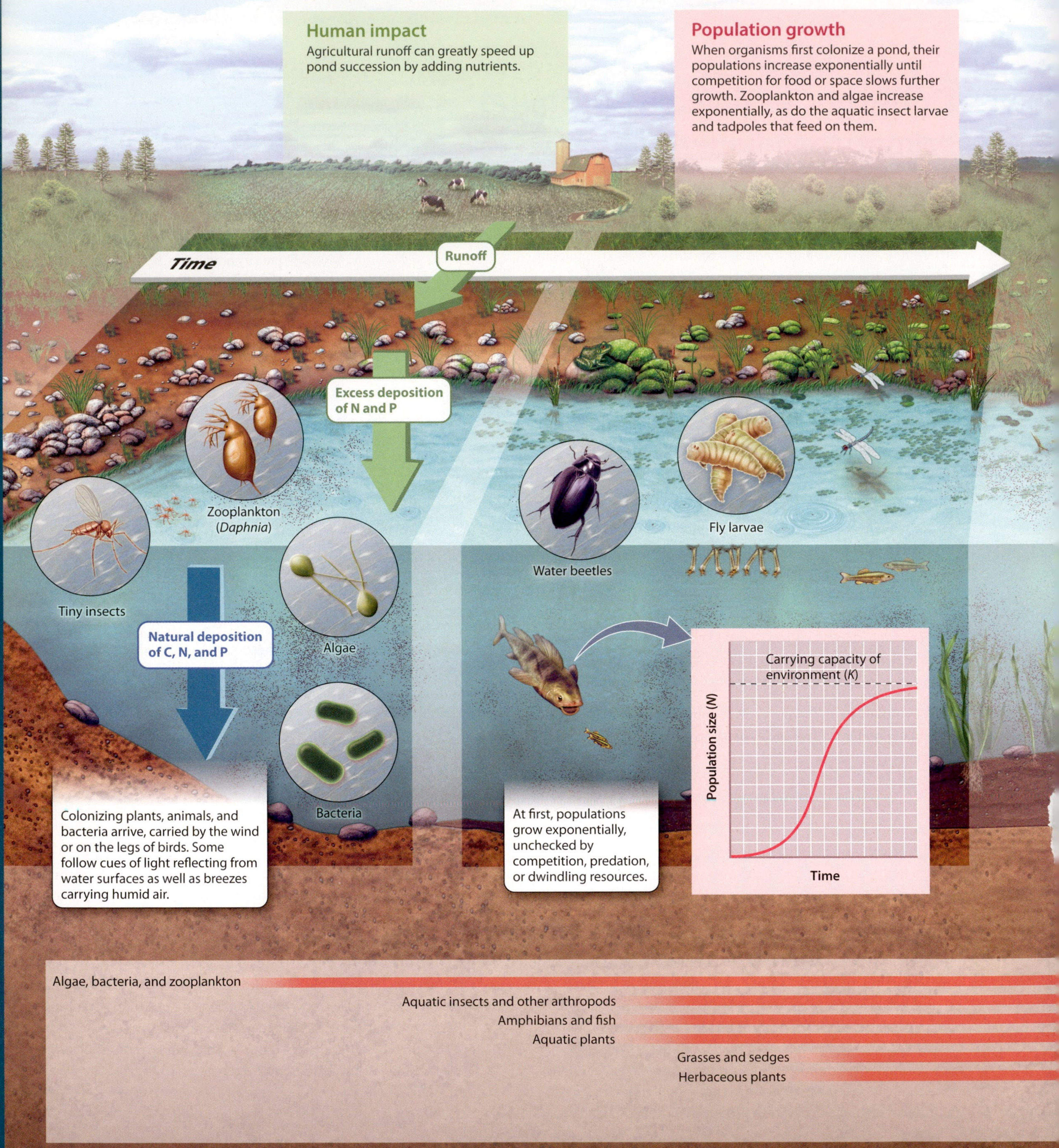

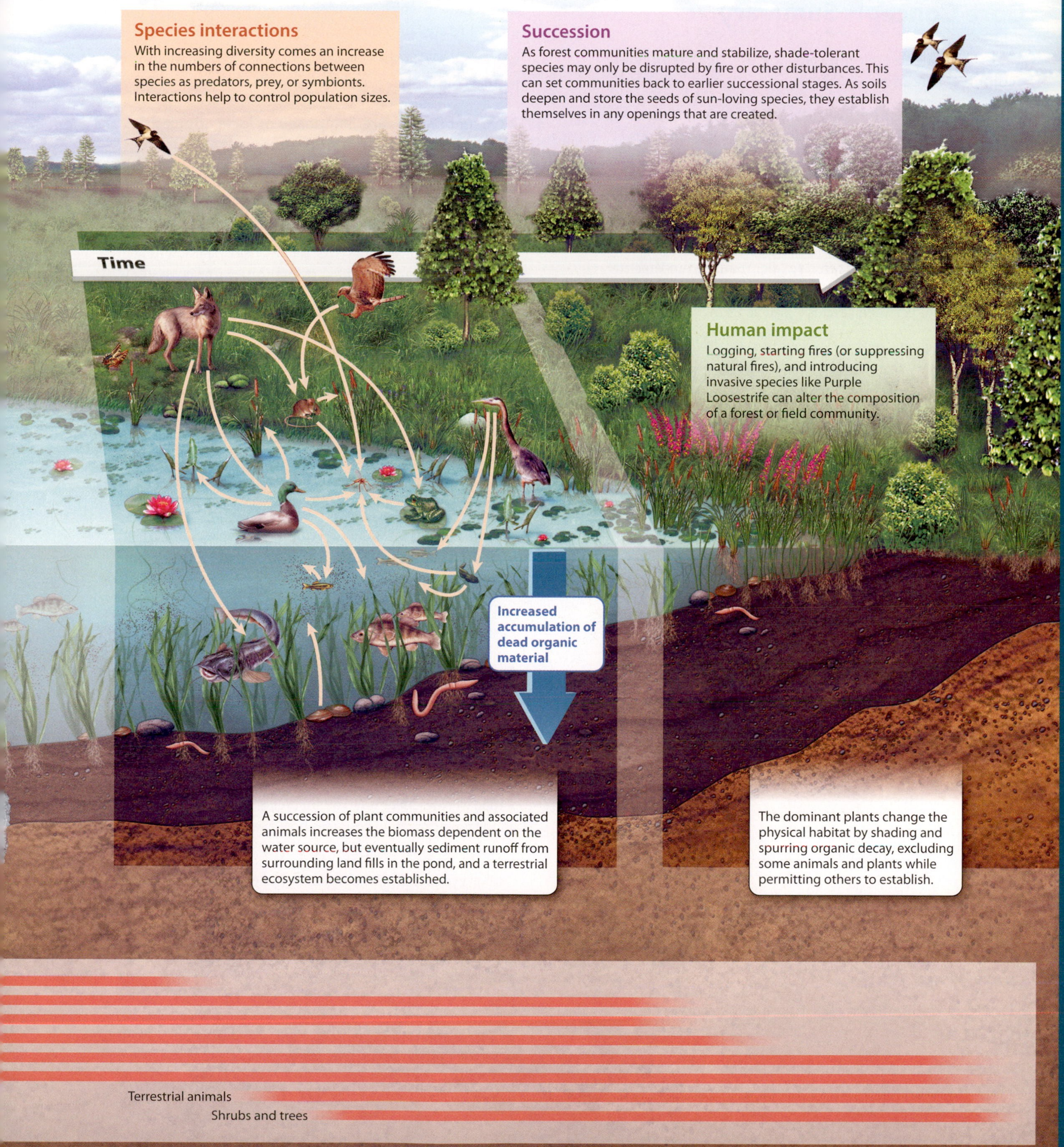
Species interactions
With increasing diversity comes an increase in the numbers of connections between species as predators, prey, or symbionts. Interactions help to control population sizes.
Succession
As forest communities mature and stabilize, shade-tolerant species may only be disrupted by fire or other disturbances. This can set communities back to earlier successional stages. As soils deepen and store the seeds of sun-loving species, they establish themselves in any openings that are created.
Time
Human impact
Logging, starting fires (or suppressing natural fires), and introducing invasive species like Purple Loosestrife can alter the composition of a forest or field community.
Increased accumulation of dead organic material
A succession of plant communities and associated animals increases the biomass dependent on the water source, but eventually sediment runoff from surrounding land fills in the pond, and a terrestrial ecosystem becomes established.
The dominant plants change the physical habitat by shading and spurring organic decay, excluding some animals and plants while permitting others to establish.
Terrestrial animals
Shrubs and trees

Core Concepts Summary

47.1 The niche encompasses the physical habitat and ecological role of a species.

The niche reflects the biological and nonbiological factors that determine where an organism lives and how it functions. page 1022

The fundamental niche is the full range of physical conditions and food resources that permit an organism to live. page 1023

The realized niche is the actual range of habitats where an organism lives, and it is often smaller than the fundamental niche because of competition, predation, and other interactions between species. page 1023

Many closely related species have broadly similar niches as a result of their shared evolutionary history. page 1023

47.2 Competition, predation, parasitism, and herbivory are interactions in which at least one species is harmed.

Competition is a form of interaction that occurs when two individuals of the same or different species use the same limited resource. page 1023

Intraspecific competition is competition between individuals of a single species; interspecific competition is competition between individuals of different species. page 1024

Competitive exclusion prevents two species from occupying the same niche at the same time. page 1024

Over time, competitive exclusion can lead to resource partitioning, in which two species evolve in such a way that they occupy different niches. page 1025

Individuals can compete for food, space, mates, or other resources. page 1025

Predation, parasitism, and herbivory are interactions in which one species benefits and one is harmed. page 1025

Two species that attack a third can help each other in a process called facilitation. page 1027

47.3 Mutualisms are interactions in which both species benefit.

In mutualisms, the benefits to both species outweigh the costs of participation. page 1028

Mutualisms can be obligate, meaning that they are required for survival or reproduction, or facultative, meaning that they are not required. page 1030

Interactions in which one species benefits and the other is unaffected are called commensalisms. page 1030

Species interactions can change over time. For example, partners can become parasites. page 1031

47.4 Communities are composed of local populations of multiple species that may interact with each other.

An ecological community consists of all the organisms living in one place at a given time. page 1031

Biodiversity, or biological diversity, includes the diversity of genetic sequences, cell types, metabolisms, species, life histories, phylogenetic groups, communities, and ecosystems. page 1032

Biodiversity is often measured by the number of species present in a given area. page 1032

Extreme environments can provide useful settings for the study of species interactions because of the relatively few number of species present. page 1032

Species that have a large influence on the composition of a community disproportionate to their numbers are known as keystone species. page 1033

Physical disturbances, such as a storm or drought, can reshape communities. page 1034

Succession is the predictable order of species colonization and replacement in a new or newly disturbed patch of habitat. page 1035

47.5 The cycling of materials and the flow of energy through ecosystems reflect ways that species interact in communities.

An ecosystem is the community and physical habitat in which the organisms live. page 1036

A food web describes how carbon cycles through ecosystems, and a trophic pyramid describes how energy flows through ecosystems. page 1036

Carbon cycles through ecosystems, but energy is used and dissipated. page 1037

Because of energy lost to heat, work, and the inefficiencies of consumers, each trophic level in an ecosystem generally has about 10% of the biomass found at the next-lower level. page 1037

In most terrestrial ecosystems, water and nutrients limit rates of primary production. page 1038

In the sea, primary production is commonly limited by nutrients. page 1038

Self-Assessment

1. Choose an organism, such as an oak tree, and define its niche.
2. Give an example of mutualism and predation, and in each case describe the benefits and costs to the participants.
3. Name three factors that help determine the species composition of a community.
4. Describe how a physical disturbance, such as a drought, can affect community composition.
5. Explain what is meant by "ecological succession" and give an example.
6. Describe how herbivores can affect the abundances of organisms at higher and lower trophic levels.

Log in to LaunchPad to check your answers to the Self-Assessment questions, and to access additional learning tools.

CHAPTER 48

Biomes and Global Ecology

Mike Hill/Getty Images.

Core Concepts

48.1 Solar radiation, wind and ocean currents, and topography determine the distribution of major climatic zones on Earth.

48.2 Biomes are broad, ecologically uniform areas whose characteristic species reflect regional climate.

48.3 Biologically driven cycles of carbon and other essential elements shape ecology and reflect evolution.

48.4 Global patterns of biological diversity reflect climate, history, and ecological interactions among species.

Watching a movie or video, you often do not need a map to figure out where the action is taking place. Cactus and scraggly shrubs suggest the deserts of Mexico or the American Southwest; dense stands of low-growing conifers bring to mind central Canada or Siberia. The deciduous forests of eastern North America and Europe are equally distinctive, as are prairies and tropical rain forests. Plants conspicuously show that biological communities vary from place to place, reflecting the adaptations of the species that live there to local conditions of temperature, rainfall, and nutrient availability. If we take the time to survey the animals in these diverse communities, we will find that they are distinctive as well.

Biomes are broad geographic areas with similar sets of communities. Climate exerts a major influence on the nature and distribution of biomes across the globe, but before we consider this interaction between Earth and life, we must first ask what climate is and why it varies from place to place.

48.1 THE PHYSICAL BASIS OF CLIMATE

Climate can be thought of as long-term average weather. In Boston, any given day in March may be warm or cold, dry or rainy, or even snowy. Records compiled over decades, however, show that, on average, March days in Boston vary from about 42°F (5°C) at the beginning of the month to 48°F (9°C) at its end. March brings an average of 4.3 inches (110 mm) of precipitation, at least some of it typically snow. Montreal is colder and snowier in March; Phoenix much warmer and drier. What controls this broadly predictable distribution of average weather across the planet? Climate in Boston, and around the world, is determined by solar radiation, global patterns of wind and ocean circulation, and Earth's varying topography, as we discuss below.

The principal control on Earth's surface temperature is the angle at which solar radiation strikes the surface.

From the perspective of our solar system, Earth stands out as the planet with life. Our nearest neighbors, Venus and Mars, differ in being respectively too hot for liquid water to persist and too cold. Earth is the Goldilocks planet: just the right distance from the sun so that, in conjunction with the greenhouse effect of the atmosphere, solar radiation maintains surface temperatures that can keep water in its liquid form. As we look across Earth, we see that climate varies markedly from one region to another. This variation also reflects the relationship between Earth and the sun, modified by the physical features of Earth and the circulation of air and seawater.

Fig. 48.1 shows how mean annual temperature is distributed across Earth's surface. At first glance, we see that Earth is hot near the equator and cold at the poles. This temperature distribution is a result of incoming solar radiation. Sunlight strikes equatorial regions directly, but, at higher latitudes, Earth's surface is tilted at an angle to incoming radiation. To see what a difference this makes, shine a flashlight directly on a sheet of paper placed perpendicular to the beam of light. Now, tilt the paper at 45 degrees to the light. You'll

FIG. 48.1 Solar radiation and climate. The angle at which solar radiation strikes our planet's curved surface is the primary determinant of Earth's surface temperatures.

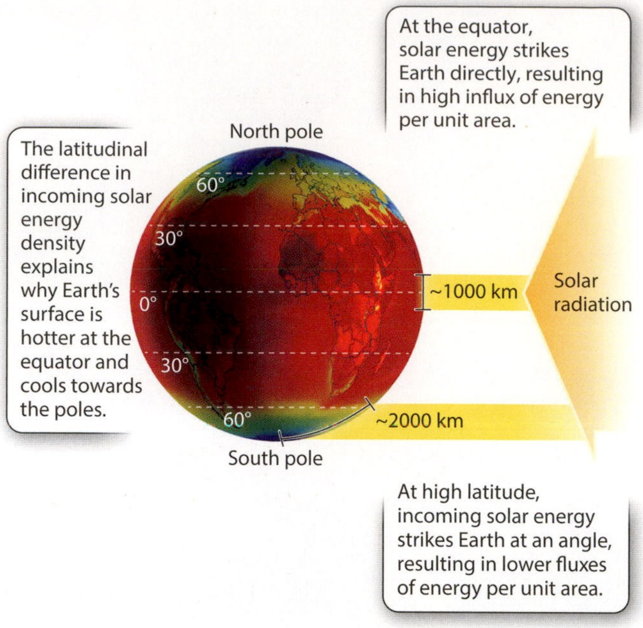

notice that the same amount of light (and so the same amount of energy) is now distributed over a larger area. On Earth, the amount of solar energy striking a given area in the tropics is about twice that found near the poles, and so tropical regions are warmer.

At higher latitudes, temperature is not only lower, it also exhibits greater variation through the year. Seasonality reflects a second feature of Earth as a planet: Earth's axis of rotation is not oriented perpendicular to incoming sunlight but rather tilts at an angle of 23.5 degrees (**Fig. 48.2**). Because of this tilt, solar radiation strikes North America and Europe more directly in July than it does in January—and so it is warmer in summer and colder in winter. At the extreme, above the Arctic Circle, incoming solar radiation can be absent for days to months during the winter. A similar pattern of seasonality characterizes the Southern Hemisphere, but with winter in July and summer in January. As we will see, patterns of temperature variation through the year go a long way toward explaining the geographic distribution of biomes across the globe.

→ **Quick Check 1** How would climates be different if Earth's axis of rotation were perpendicular to incoming solar radiation, rather than tilted at 23.5 degrees?

Topography, or the physical features of Earth, also contributes to global temperature patterns. Mountaintops can be glaciated, even at low latitudes, reflecting the fact that temperature declines with increasing elevation. In general, temperature drops about 4°F for every 1000 feet in elevation (6.5°C per kilometer). Mountains therefore exhibit a pattern in which climate and biomes change as elevation increases, mirroring the spatial pattern from lower to higher latitudes.

FIG. 48.2 Seasonality. Because Earth's axis of rotation tilts at an angle of 23.5 degrees, most parts of the globe experience seasonal variations in temperature.

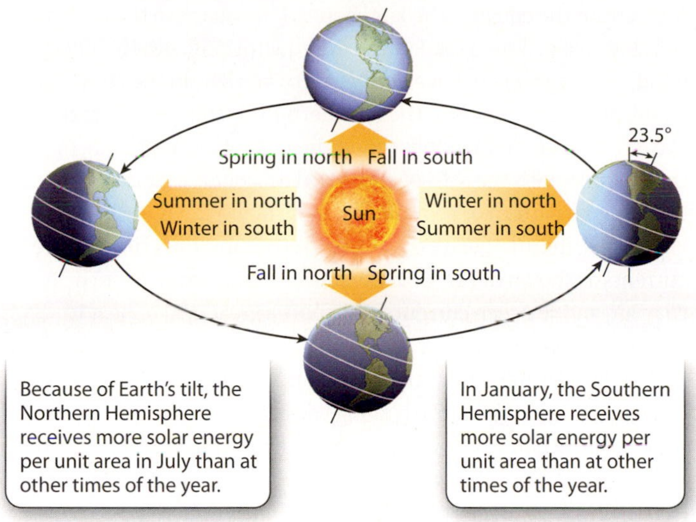

Heat is transported toward the poles by wind and ocean currents.

Temperatures at the equator are actually a bit cooler than we might predict based on how directly solar radiation strikes Earth at that latitude; and the poles are somewhat warmer than a diagram of solar radiation like Fig. 48.1 would predict. To reconcile predicted and measured temperatures, we need to consider another set of processes, those that transport heat from low to high latitudes. This heat transport is carried out by wind and ocean currents.

The gas molecules that make up air are constantly in motion, and when they are heated they move faster, and (unless confined) the volume of the air expands. For this reason, warm air is less dense than colder air and so it rises through the atmosphere. Not surprisingly, the rise of warm air is particularly strong at the equator (**Fig. 48.3**). The air cools as it rises, and, once it reaches an altitude of 10–15 km above Earth's surface, the air no longer continues to rise but spreads toward the poles, continuing to cool as it goes. Eventually, at about 23 to 30 degrees north and south latitude, this cooling air becomes dense enough to sink back to the surface. This air continues to move along Earth's surface, some of it moving toward a pole, while the rest moves back toward the equator. In either case, the air will warm again as it moves across Earth's surface, to rise once more at the equator or at about 60 degrees north and south latitude. As shown in Fig. 48.3, Earth's lower atmosphere is organized into cells of rising and falling air masses, and these transport heat from the equator toward the poles.

As warm air expands upward, cool air moves in to replace it. We experience this movement as wind, and as sailors have known for thousands of years, the prevailing direction of the wind differs from one latitude to another (Fig. 48.3). Prevailing winds reflect those cells of rising and falling air, but to understand them more fully, we have to consider one further aspect of our planet: its rotation about an axis.

FIG. 48.3 Wind currents. The rise of warm air through the atmosphere and the descent of colder air govern patterns of prevailing wind and transport heat from the equator toward the poles.

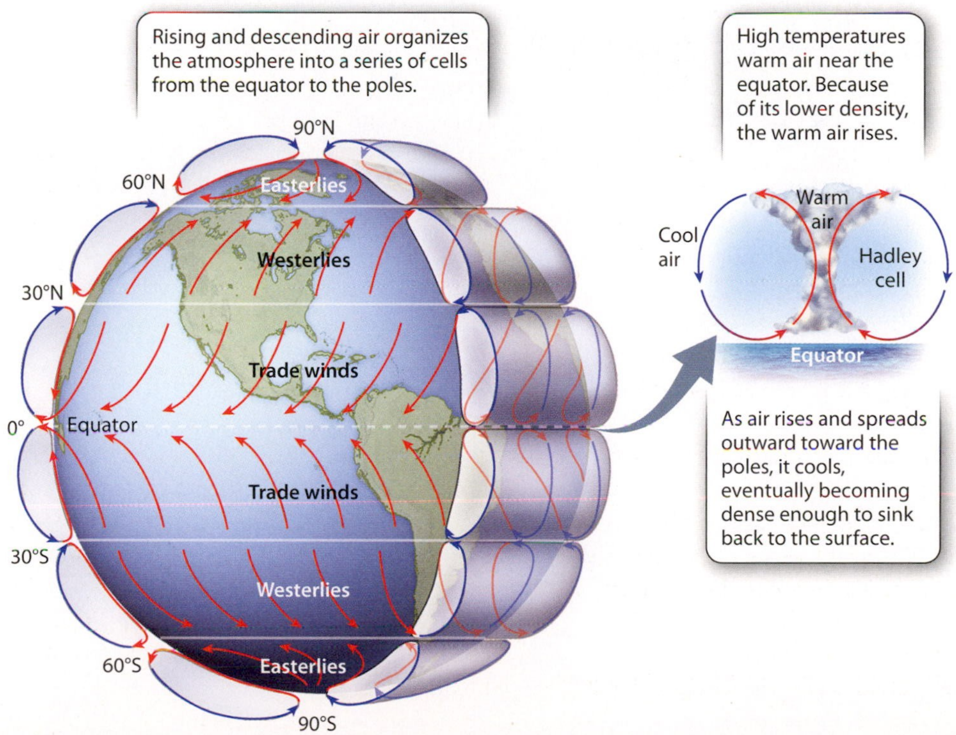

FIG. 48.4 The Coriolis effect. Because of Earth's counterclockwise rotation about its axis, winds in the Northern Hemisphere deflect to the right; those in the Southern Hemisphere deflect to the left.

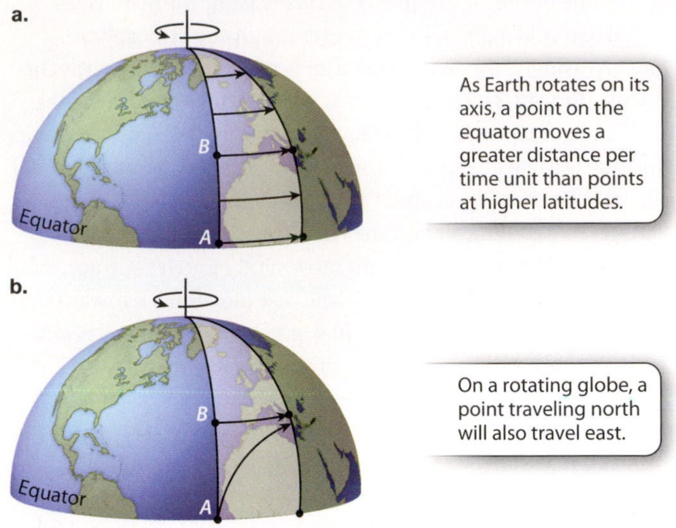

Earth rotates in a counterclockwise direction, moving from west to east. In the course of one daily rotation, a spot at the equator will move through a distance equivalent to our planet's circumference—nearly 25,000 miles, or 40,000 km. Points at higher or lower latitude travel a shorter distance in a single rotation (**Fig. 48.4a**). So, in a period of 24 hours, a point at the equator will have traveled farther than a point at higher or lower latitude, which is to say that the point at the equator moves more quickly than the point closer to the poles. As the wind moves air north and south from the equator, the land beneath it rotates to the east, but at a slower speed than the land at the equator (**Fig. 48.4b**). The wind, then, appears to have deflected to the right in the Northern Hemisphere, and winds in the Southern Hemisphere deflect to the left. This phenomenon is called the **Coriolis effect,** and, in conjunction with the cells of rising and descending air masses, it explains the pattern of prevailing winds shown in Fig. 48.3.

Winds, in turn, push on water in the oceans, directing surface currents as shown in **Fig. 48.5.** Water can carry much more heat than air, and so ocean currents transport a great deal of heat to higher latitudes. For example, the ocean current known as the Gulf Stream (and its northeasterly extension, the North Atlantic Drift) transports warm water north along the eastern margin of North America and then east toward Europe. The heat it carries helps to keep northwestern Europe much warmer than land at equivalent latitudes in North America. Subtropical plants ornament gardens in coastal Scotland but would be unthinkable in Newfoundland, Canada, which is at approximately the same latitude.

Just as colder air sinks below warmer and less dense air, cold waters sink beneath less dense water masses. At high latitudes, the sinking cold waters begin to move slowly along the seafloor toward the equator. The result is a complex three-dimensional circulation of water through Earth's oceans that plays an important role in the determination of regional climates.

FIG. 48.5 Ocean currents. Prevailing ocean currents reflect the winds that drive the circulation of water at the ocean's surface and transport a large amount of heat from the equator toward the poles.

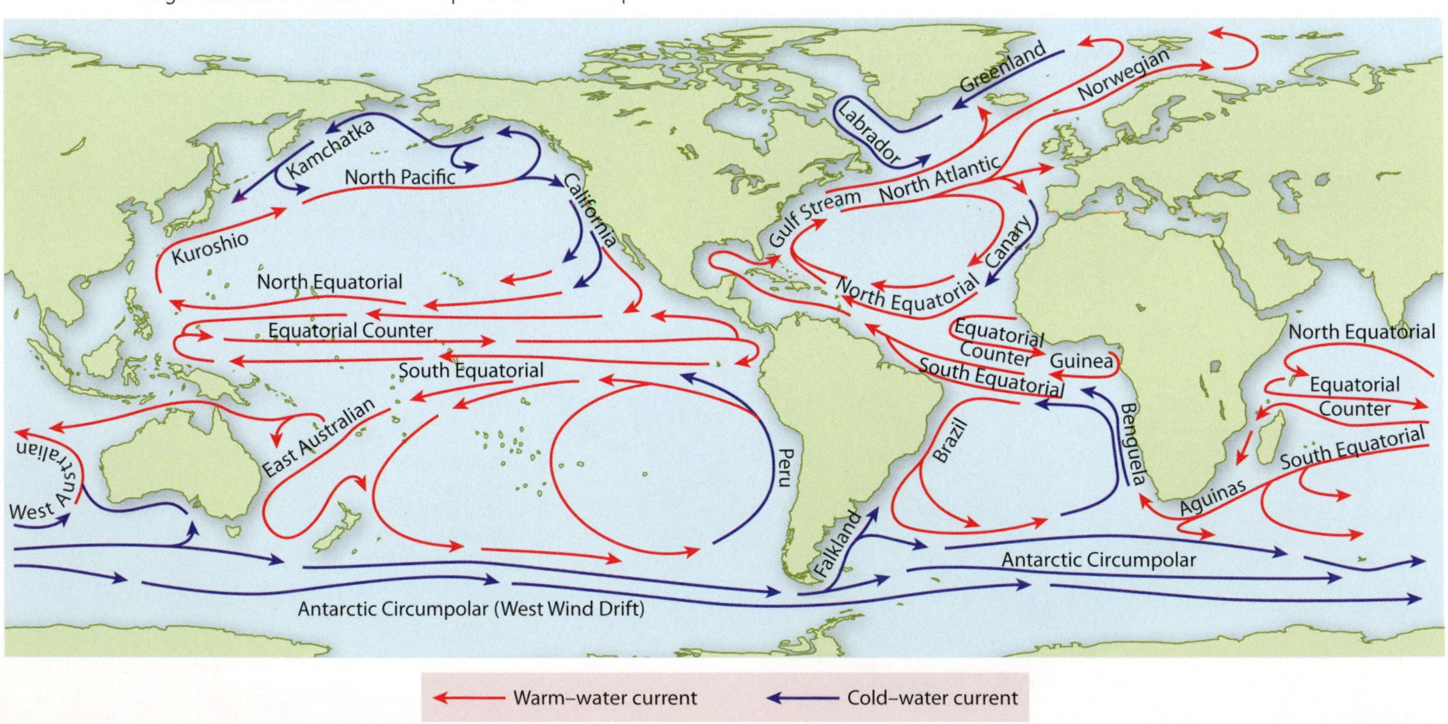

Wind and water, then, transfer heat from the equator toward the poles, modifying the direct effects of solar radiation. Together, these processes result in the temperature distribution observed across the planet, explaining why palm trees that cannot tolerate freezing are confined to low latitudes, while deciduous trees that lose their leaves every fall occur mostly in the seasonal climates found at mid-latitudes. At the same time, however, if you drive due west from New York to northern California, you will pass sequentially through deciduous forest, grass-dominated prairies, and montane stands of conifers. There is more to the distribution of biomes than temperature alone.

Global circulation patterns determine patterns of rainfall, but topography also matters.

To fully understand the features and distribution of biomes around Earth, we must introduce another physical principle: Warm air can carry more water vapor than can cold air. Armed with this insight, we can return to the global circulation of air. Warm air at the equator carries a great deal of water vapor, but, as it rises and the air cools, the amount of water vapor it can carry decreases. As a result, the vapor condenses, and it rains. Conversely, the air descending at about 23 to 30 degrees latitude north and south warms as it sinks and so takes up water vapor rather than releasing it; aridity results. As shown in **Fig. 48.6,** zones of high and low annual rainfall follow this broad pattern of rising and descending masses, with the highest rainfall near the equator and the lowest rainfall in belts between 23 and 30 degrees latitude north and south.

Regional precipitation (rain or snow) patterns are actually more complicated than this simple picture would suggest, as seen in the precipitation map for Washington state (**Fig. 48.7**). From the coast eastward into the Cascade Mountains, Washington and neighboring British Columbia are wet all year long, supporting magnificent conifer forests. However, east of the Cascades and the Coast Mountains that extend to the north, the land is dry. This observation suggests that the mountains themselves impose a regional pattern of rainfall. The effect is called a **rain shadow,** and it reflects the same general principles of rising and descending air masses. As wet air moves from the ocean up into the mountains, it cools, releasing its moisture as precipitation. Moving past the mountains, air masses descend, warming as they go and taking up water vapor. For this reason, lands in the rain shadow of the mountains are arid (Fig. 48.7).

In general, then, regional climate reflects the interactions among solar radiation, global patterns of circulation, and Earth's varying topography. Biomes, in turn, reflect these variations in climate.

FIG. 48.6 Global rainfall patterns. The rise and fall of air masses govern the global distribution of rainfall.

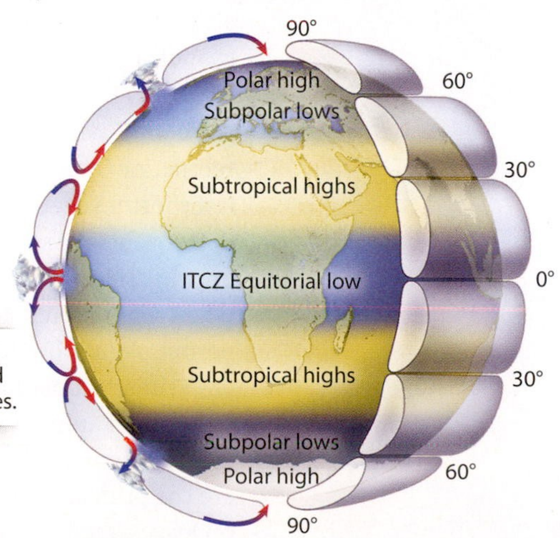

FIG. 48.7 Topography and rainfall. Mountains can impose a regional pattern on climate because of the rain shadow, which results in differing amounts of rain on the wind-facing and opposite sides of mountain ranges. *Data for top map from: PRISM Group and Oregon Climate Service, Oregon State University.*

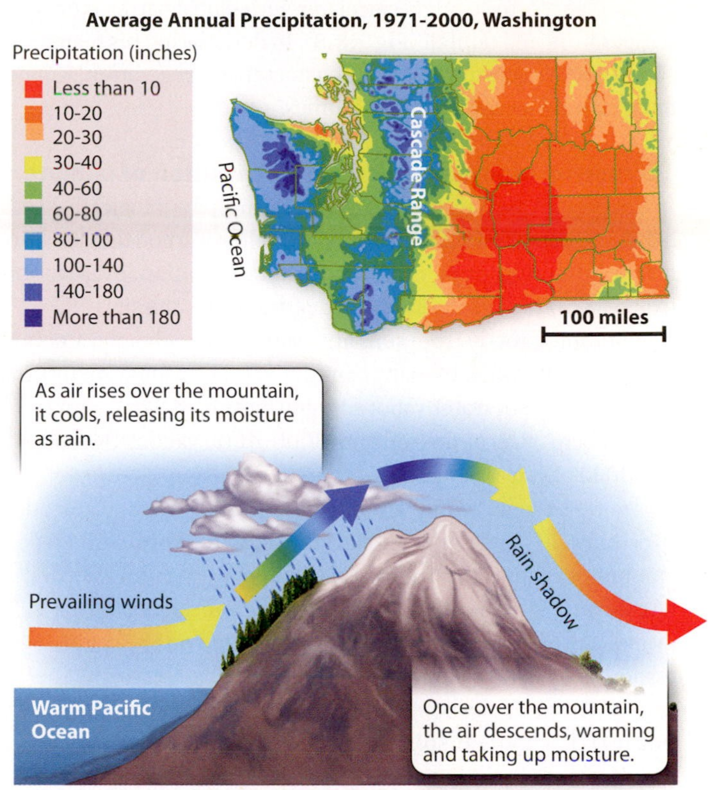

48.2 BIOMES

We're all familiar with the dense canopies of tropical rain forests and the rolling grasslands of prairies and steppes. As we noted at the start of the chapter, these and other terrestrial biomes are recognized by their characteristic communities—most visibly their vegetation. The vegetation reflects the evolutionary adaptation of plant form and physiology to climate. How close is the relationship between biome and climate? In 1884, Wladimir

Köppen published a climate map that, with relatively few changes, is still used today. You might wonder just how much information was available to Köppen about temperature and rainfall in, for example, the Amazon rain forest or central Asian steppes. The answer is, not much. In fact, Köppen's diagram actually charts global vegetation; the reason it works so well as a climate map is that plant form strongly correlates with global patterns of temperature and rainfall (**Fig. 48.8**). In designating his major climate zones according to their characteristic vegetation, Köppen mapped out Earth's terrestrial biomes.

Terrestrial biomes reflect the distribution of climate.

In Chapter 29, we saw that plants perform a physiological balancing act, opening their stomata to allow carbon dioxide into leaves, but losing water vapor through transpiration (evaporation from leaves) as a result. **Evapotranspiration** is the sum of evaporation directly from soils and water bodies plus the amount transpired by plants. In many communities, transpiration returns a large amount of water vapor to the atmosphere. For example, a single acre of corn can transpire 3000–4000 gallons (10,000–15,000 L) of water per day.

As we've seen, temperature varies with latitude and altitude, as do humidity and wind. **Potential evapotranspiration** is the amount of evapotranspiration that temperature, humidity, and wind would cause if water supply weren't limiting. Potential evapotranspiration, then, reflects the demand on the water resources of an ecosystem, just as precipitation represents the supply. The ratio of water demand to supply is known as the **potential evapotranspiration ratio,** and it effectively tells us what type of vegetation can be supported in a given region.

FIG. 48.8 Global distribution of biomes. Distributions of biomes around the world reflect regional climates. *Source: After D. M. Olson et al., 2001, "Terrestrial Ecoregions of the World: New Map of Life on Earth," Bioscience 51:933–938.*

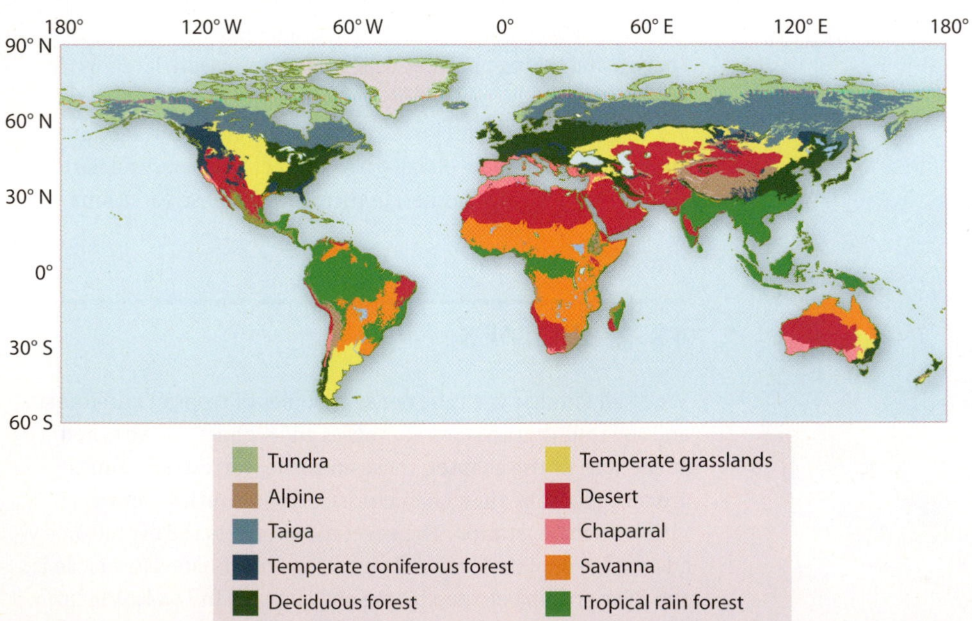

Deserts, with low annual precipitation, have a high potential evapotranspiration ratio (up to 20 or more): The limited moisture in soils evaporates rapidly, and plants must keep transpiration (and therefore photosynthetic rate) to a minimum. The plants' ability to store water in their tissues provides a hedge defense against drought, and so succulents—plants like cactus, which store water in their stems—are common. At the other extreme, rain forests have potential evapotranspiration ratios as low as 0.2: High precipitation permits nearly unlimited transpiration (and therefore high photosynthetic rates), enabling the development of wet forests with high biomass. Intermediate potential evapotranspiration ratios characterize the temperate zones that encompass much of the United States, Europe, and parts of eastern Asia, as well as mid-altitudes, below the highest mountain peaks but well above the coastline. Both latitude and altitude influence the potential evapotranspiration ratio through their effects on temperature and precipitation, and the combined effects of these factors are reflected in the distribution of biomes around the world (**Fig. 48.9**).

→ **Quick Check 2** Why can wet soils occur both near the equator, where rainfall is high, and at high latitudes, where rainfall is low?

In similar biomes on different continents, dominant plants generally look much the same, but commonly they are not closely related. Plants have independently evolved adaptations to particular climatic regimes by convergence. For example, the cacti that grow in deserts in western North America look very similar to plants in arid regions of Africa, but the plants in Africa belong to a different family entirely (**Fig. 48.10**). Convergent evolution is also seen in trees found in rain forests in different regions of the world; distantly related trees have evolved leaves with long tips that allow excess water to drip away, a reddish color in new leaves that protects them from intense sunlight, and roots with high buttresses that support tall trees atop thin soil.

In terrestrial biomes, vascular plants dominate primary production, although mosses, lichens, algae, and cyanobacteria also contribute. Plants also provide a physical structure for the biome, with high canopies in wet temperate forests and tropical forests, but shorter vegetation where water or nutrients are scarce, or where the growing season (the part of the year when temperature and precipitation permit plant growth) is short.

Primary consumers include a variety of animals, but in many environments, mammals and insects are the principal herbivores, grazing on leaves, fruits, seeds, and stems.

FIG. 48.9 Evapotranspiration and precipitation. The amounts of both potential evapotranspiration and precipitation change with latitude and altitude, and the ratio between the two is a primary determinant of vegetation type, primary production, and species diversity. *Data from: International Water Management Institute Global Monthly Dataset.*

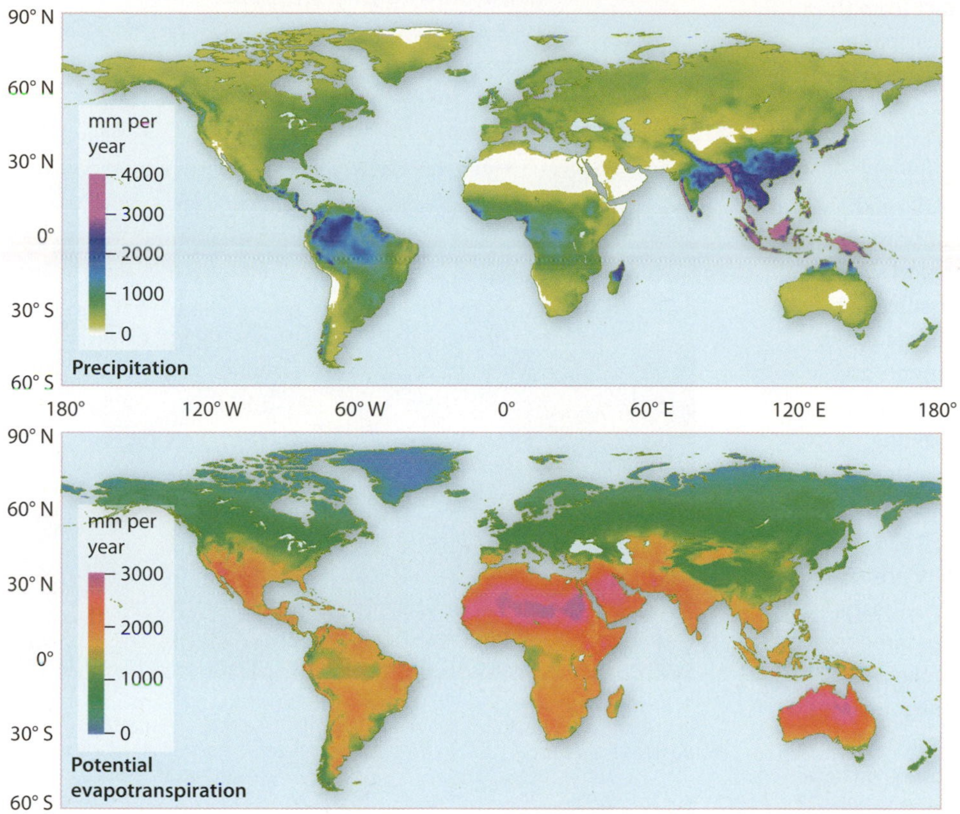

Insects and vertebrates also play major roles as secondary consumers. Predatory insects in various terrestrial biomes include hornets, the praying mantis, and larval ant lions that trap their prey in small sandy pits. Terrestrial biomes are often characterized by the carnivorous vertebrates that live there, which are among the most charismatic of all animals—for example, lions, tigers, and wolves. In all terrestrial biomes, despite the activities of consumers, detritus from both plants and animals ends up on the ground and may accumulate in soil.

Decomposers, especially fungi and bacteria, feed on this detritus, returning CO_2 to the atmosphere and nutrients to the soil. Still more protists and animals feed on these decomposers, populating a detritus-based wood web within the soil. Soil animals include minute nematodes (Chapter 44) as well as larger earthworms that eat both detritus and decomposers, ventilating the soil as they tunnel in search of food.

The principal biomes on land are summarized in **Fig. 48.11**, pages 1052–1057.

FIG. 48.10 Convergence of plant form. (a) Cacti in the deserts of North America and (b) euphorb plants in Africa look similar because they are both adapted to conserve and store water, but they are not closely related. *Sources: a. McDonald Wildlife Photography/Animals Animals; b. Michele Westmorland/Nature Picture Library.*

Aquatic biomes reflect climate, and also the availability of nutrients and oxygen and the depth to which sunlight penetrates through water.

Life, of course, is not limited to the land; diverse communities thrive in both freshwater and marine environments. Organisms that live in water face functional challenges distinct from those on land, and this helps to explain the nature of aquatic biomes. From the point of view of evolution, terrestrial organisms evolved from ancestors living in water, but to understand ecology, it is helpful to go from familiar terrestrial biomes to aquatic examples we generally experience only indirectly.

The most obvious distinction between terrestrial and aquatic biomes is that water is denser than air. For primary producers, buoyancy in water can provide mechanical support, reducing or eliminating the need for structural tissues such as wood. Also, primary producers on land are tethered to the ground both by gravity and the availability of nutrients and water, but this is not the case in aquatic biomes. In aquatic environments, nutrients are dissolved in encompassing waters, and by controlling their buoyancy, aquatic primary producers can live far from shore in surface waters bathed in sunlight.

Just as on land, the structure and distribution of primary producers governs patterns of consumer organisms. Aquatic plants and macroscopic algae are important primary producers in shallow aquatic environments where they can take advantage of nutrients transported from land while remaining anchored to the riverbed, lakebed, or seafloor. In larger lakes and the open ocean, however, phytoplankton—photosynthetic microorganisms that float in sunlit waters—are predominant primary producers, combining high rates of primary production with low biomass.

In shallow aquatic environments such as lake margins and coastal marine settings, much plant, algal, and animal debris accumulates in sediments, supporting luxuriant growth by bacteria. As on land, these decomposers provide food for many larger consumers, including the larvae and juveniles of commercially important fish and invertebrate species. Fungi, so important in terrestrial biomes as decomposers, are generally thought to play only a minor role in aquatic environments, although recent surveys of marine sediments show that where organic matter is abundant and oxygen is absent, microscopic fungi can be major components of the decomposer guild. Bacteria also play a critical role in cycling the remains of phytoplankton and other open-water organisms, both within the water and on deep lakebeds or seafloors.

Grazers such as snails and fish feed on marginal plants and algae, while single-celled protists and microscopic animals called zooplankton consume phytoplankton that floats in the water. As on land, secondary consumers such as other fish or squid feed on the grazers, adding complexity to food webs. Plankton, whether photosynthetic or heterotrophic, generally have only limited powers of locomotion and so drift

FIG. 48.11 **Some common land biomes.** *Sources: (Tundra) Helge Schulz/Picture Press/Getty Images; (Alpine) Rolf E. Kunz/age fotostock; (Taiga) Michael DeYoung/Design Pics/Getty Images; (Temperate coniferous forest) Charlie Ott/Science Source.; (Deciduous forest) NatureByDarrellYoung/Alamy; (Temperate grassland) Robert and Jean Pollock/Science Source; (Desert) Wayne Lynch/Getty Images; (Chaparral) Amit Basu Photography/Getty Images;, (Savanna) Hoffmann Photography/age fotostock; (Rain forest) Berndt Fischer/Getty Images.*

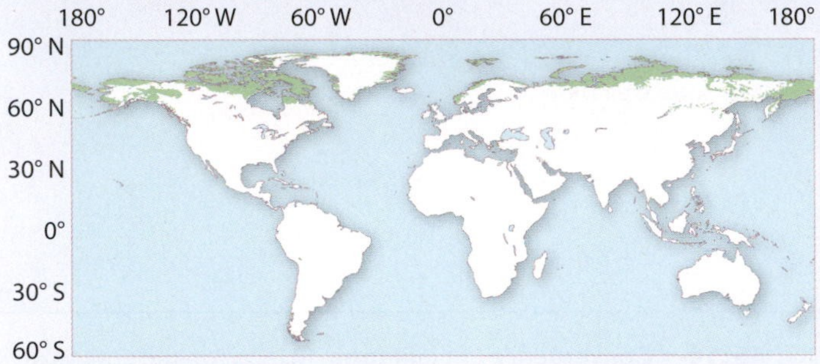

Tundra

Tundra is the coldest biome, and short days in winter limit the growing season. Tundra occurs close to the North Pole, above 65° N. The South Pole is largely surrounded by the ice and seas of Antarctica, and so there is very little area with plants. Precipitation is low, but because rates of evaporation are also low and because drainage is commonly poor, the ground is usually waterlogged and permanent ice occurs below a few centimeters of soil. Primary producers are mostly mosses, lichens, herbs, and low shrubs. Grasses and sedges occur in drier places, as do other flowering plants. Plant diversity is low, and most plants are small. Caribou are conspicuous grazers, but other primary consumers, including rabbits, birds, and insects, occur as well. Wolves and foxes are key predators. Fungi and bacteria decompose organic detritus, but low temperatures and water-logged soils limit rates of respiration.

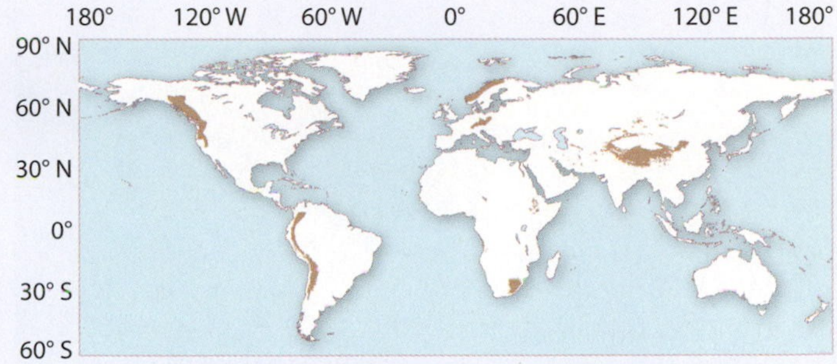

Alpine

The Alpine biome is similar to tundra but lacks permanent ice below the soil, and the temperatures vary more widely. Alpine areas occur throughout the world, often at about 10,000 feet (3000 m) at lower latitudes, but always just below the snow line. Because of their altitude, these are windy, cold places. The thin atmosphere provides only limited protection from UV radiation. Many alpine plants are therefore low and slow growing. Grasses are abundant, as are herbs that can provide spectacular wildflower displays during the short growing season. Alpine grazers include mountain goats, llamas, yaks, and marmots, as well as seasonally active insects. Predators include wolves and cats such as the Himalayan snow leopard.

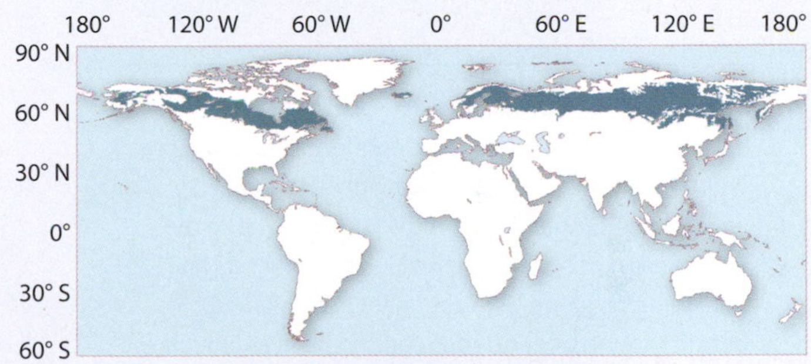

Taiga

These cool, moist forests occur from 50° N to 65° N. The short summer brings rain, and most of the plants are low-growing conifers like spruce, fir, larch, and pine, with an understory dominated by shrubs in the blueberry and rose families. Mosses commonly form an extensive and soggy groundcover. The soils are deep with accumulated organic matter because the low temperatures result in slow decomposition, but they are also acidic and poor in nutrients, helping to explain the limited stature of most plants. Birds, insects, and other invertebrate animals are more abundant and diverse than those of tundra biomes, but mammals are the most conspicuous consumers. Grazers include elk, moose, caribou, porcupines, hares, and a diversity of rodents. Bears, lynx, wolverines, weasels, mink, wolves, and foxes are well-known predators.

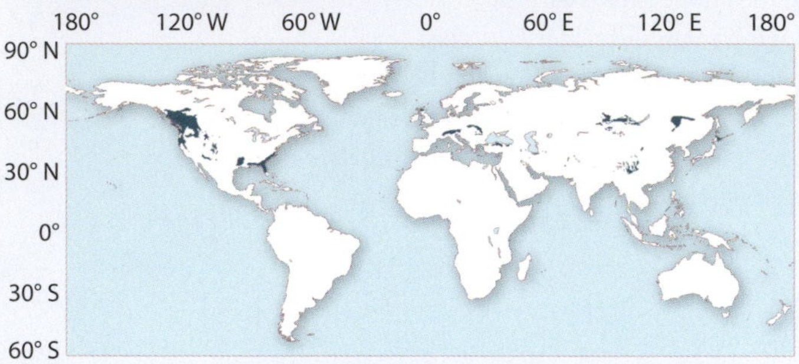

Temperate coniferous forest

Two broad areas of temperate coniferous forest occur below 50° N in North America, northern Japan, and parts of Europe and continental Asia. Along the Pacific coast of North America, warm summers and mild winters, along with abundant precipitation, permit growth of enormous conifers such as Douglas Fir, Red Cedar, Sitka Spruce, and redwoods. Much of the undergrowth consists of grasses, ferns, and members of the blueberry family. Understory shrubs are uncommon. In the interior of North America, much less precipitation and colder winter temperatures support drought-resistant conifers such as Ponderosa and Lodgepole Pines and Englemann Spruce. Diverse invertebrate species live in the soils, supported by active microbial decomposers. Insect and vertebrate diversity are all higher than in taiga forests.

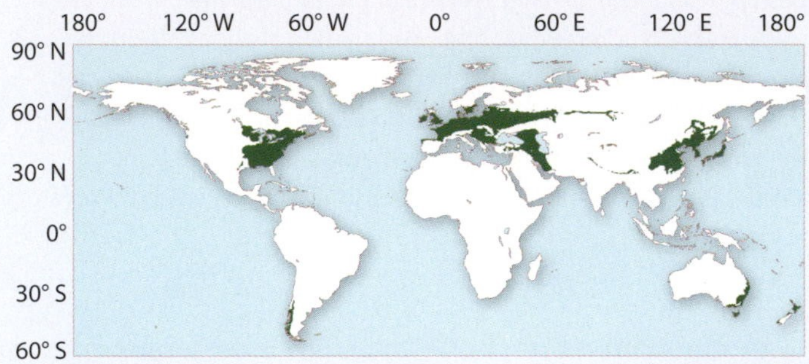

Deciduous forest

A moderate climate and dominance of hardwood deciduous trees occur across much of eastern North America, Europe, and Asia. Much of this biome has been subjected to human disturbance for agriculture and urban development. There are usually 15 to 25 tree species, including maples, oaks, poplars, and birches. Springtime sun passes through seasonally leafless trees to reach a diverse understory flora. Soils are rich in nutrients from annual leaf fall, and the moderate temperatures and precipitation promote decomposition, while the cool winters promote accumulation of organic materials. Soils contain a diversity of microbes and fungi, along with the protists and invertebrate animals that feed on them. Insects, birds, and mammals are relatively diverse, joined by snakes, lizards, and amphibians.

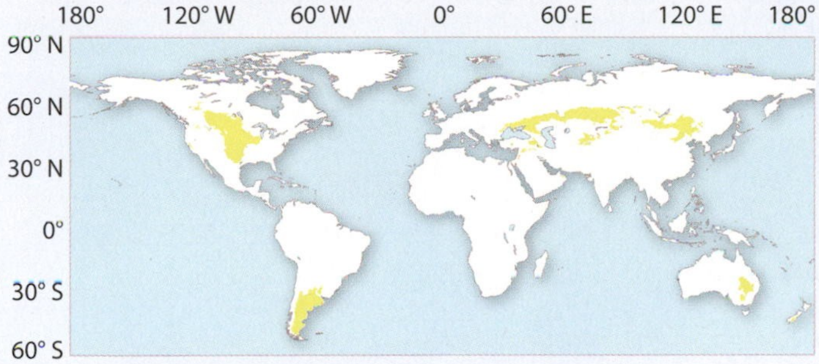

Temperate grassland

Before human settlement, this biome occupied most of the midwestern United States and south central Canada, dominated by blue-stem and buffalo grasses. Disturbance by fire helps to maintain grass populations in this biome. Lack of precipitation also prevents many species of tree from growing, and those trees that grow are usually in low areas with more available moisture. Because of active decomposers, soils accumulate nutrients, providing some of the most productive agricultural lands in the world. Before colonization by humans, temperate grasslands of North America supported abundant and diverse grazers, including bison, pronghorns, horses, and mammoths. Horses and mammoths became extinct 11,000 years ago, and herds of bison and pronghorn dwindled in the nineteenth century. Burrowing grazers such as prairie dogs remain relatively common, although population sizes have fallen; these species support predators such as ferrets, badgers, foxes, and birds of prey.

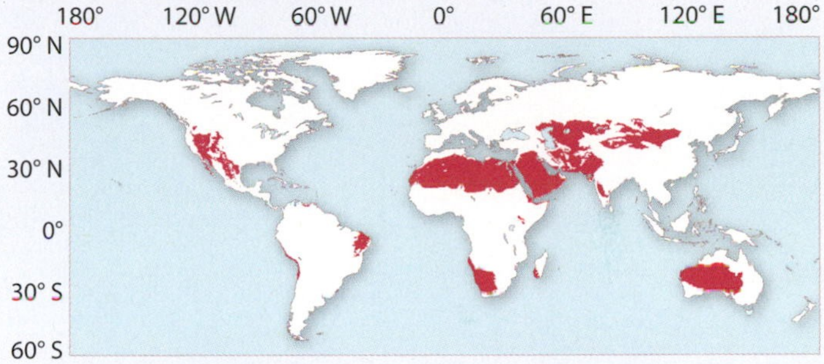

Desert

Desert occurs in continental interiors around the world north and south of the equator from 15° to 35° latitude. Wind patterns prevent this biome from receiving more than a few centimeters of precipitation annually. Deep-rooted plants, like cactus and euphorb plants, are adapted to store water. Primary production is low, and soils are poor in nutrients but may have high surface salt due to evaporation. Given the low rates of primary production, primary consumers tend to be small, and include diverse lizards as well as rodents. Predators include snakes, cat species, coyotes, and birds such as owls, hawks, and eagles.

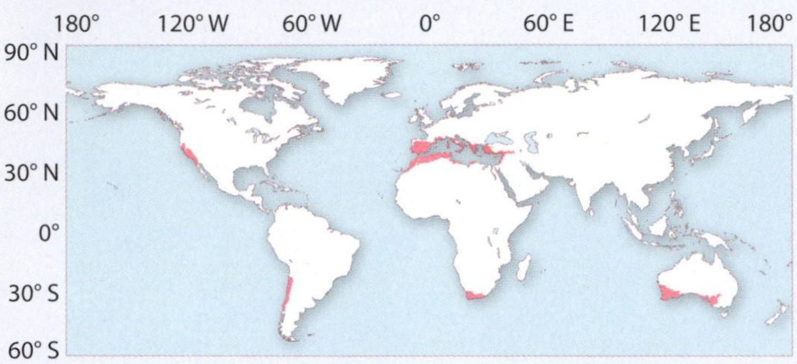

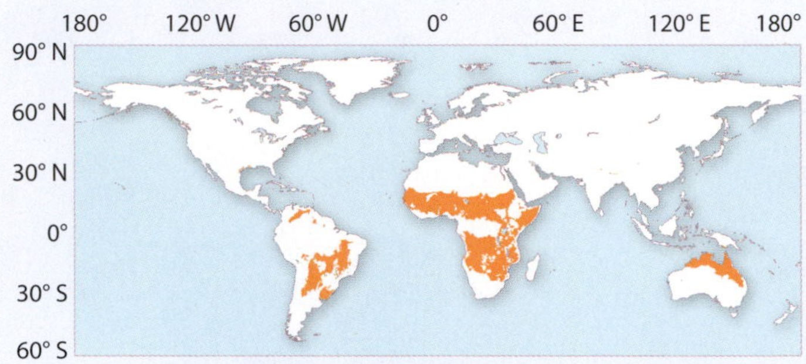

Chaparral

Like that of deserts, the distribution of chaparral reflects a narrow range of climate conditions and occurs on the western edge of continents from 32° to 40° latitude north and south of the equator. Precipitation ranges from 30 to 75 cm per year, usually falling in a period of 2 to 4 months. Typical plants are herbs that die and reseed themselves every year, evergreen shrubs, and small trees. Olives, eucalyptus, acacia, and oaks are typical woody species, always drought resistant and often adapted to withstand fire. Limited precipitation means soils are not rich in organic materials.

Savanna

Tall, perennial grasses dominate this biome, which occurs in warm, relatively dry regions of eastern Africa, South America, and Australia. Rain is seasonal and ranges from 75 to 150 cm per year. Scattered trees and shrubs usually drop their leaves in the dry season to conserve moisture. As in temperate grasslands, fire plays a key role in maintaining this biome. Large mammalian grazers are abundant and diverse; these include the migrating antelopes, zebras, and giraffes well known from Africa, but also kangaroos and other marsupials in Australia and large rodents in South America. Predator diversity can also be high, as exemplified by lions and other cat species, hyenas, and wild dogs. Dingos are important predators in Australian savannas, but were introduced only a few thousand years ago from southeastern Asia. Before that, the top carnivores included large, now-extinct lizards, as well as snakes and smaller lizards still present in the biome.

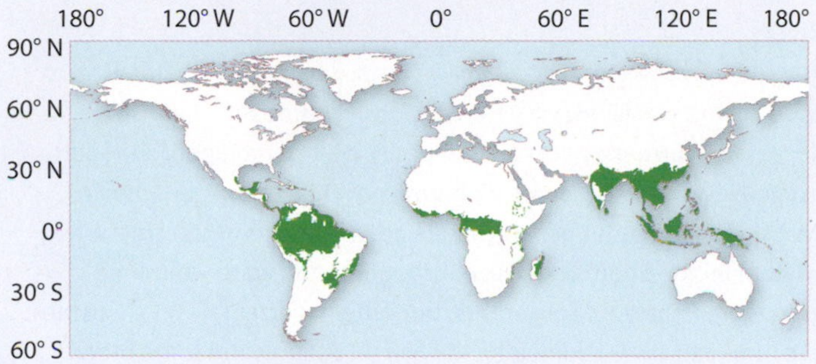

Tropical rain forest

The most diverse of all terrestrial biomes, tropical rain forests extend north and south of the equator from 10° N to 10° S. Annual precipitation is commonly more than 250 cm, and tree diversity alone often exceeds 300 species per hectare. Trees grow tall, and many have buttressed roots for support. Lianas and other epiphytic plants are common. Most leaves are evergreen and leathery and many have long pointed tips that facilitate drainage of excess moisture. Because of the high temperatures and heavy rains, decomposition by fungi and bacteria is rapid, preventing the accumulation of organic materials in clay-rich or sandy soils. There are few large grazers, but smaller mammals such as primates, bats, and rodents are highly diverse, as are birds, snakes, and lizards. Insects are especially abundant and diverse. Ants alone make up as much as 30% of animal biomass in rain forests, and they are the principal grazers on rain forest trees.

with the current. In contrast, secondary consumers collectively called nekton—fish, squid, whales, and others—can move actively through their watery surroundings. Because oxygen gas dissolves in water only to a limited extent, surface waters directly beneath the atmosphere will have an O_2 concentration about 20 times lower than that of the air. This compels larger animals to be moving constantly to keep their gill systems well oxygenated, or to live in places such as cold running streams and rocky marine coastlines where currents naturally produce well-oxygenated water.

The main freshwater biomes are lakes and rivers (**Fig. 48.12**). These vary tremendously in climate, and also in nutrient input and oxygen availability. In general, the turbulent flow of water ensures that rivers are well oxygenated. Some lakes, however, are physically stratified, meaning that the density of bottom waters keeps them from mixing upward with shallow waters. Organic remains that sink to deep waters provide food for respiring bacteria, and this draws down oxygen levels. Where nutrient influx is high, such lakes can become anoxic in their deeper parts; this is particularly common where human activities have increased nutrient loads entering the lake (Chapter 49).

Only about 2.5% of Earth's water is fresh, and most of this water occurs in glaciers, permafrost, and groundwater within soils. Overwhelmingly, the water on our planet is seawater. Covering 71% of Earth's surface and reaching a depth of nearly 11,000 m (almost 7 miles), the oceans constitute Earth's largest biomes. In fact, the oceans are so large and so diverse in their biology that ecologists commonly subdivide them into distinct realms based on depth and proximity to the shoreline (**Fig. 48.13**).

The dimensions of the oceans matter because they determine the distribution of nutrients, oxygen, and light. Runoff from the continents (supplemented by nitrogen fixation by marine cyanobacteria) provides most of the nutrients required by primary producers in the sea, but because sunlight rarely penetrates much deeper than about 200 m (the **photic zone**), most of the ocean's volume is off limits for photosynthetic organisms. Life in the deep sea is largely sustained by detritus that rains downward from surface waters, although chemoautotrophic bacteria and archaeons contribute as well. Particle- and filtering-feeding animals, as well as microbes, respire this rain of detritus, returning phosphorus and nitrogen to the water. These nutrients eventually return to the surface in upwelling waters, mostly along continental margins, where they support locally high rates of photosynthesis. Upwelling regions make up only about 1% of the ocean's surface area, but by some estimates, they support more than 20% of all commercial fish populations.

Many different groups of algae contribute to primary production on continental shelves, with diatoms, coccolithophorids, and dinoflagellates dominating in most regions. Because these algae turn over rapidly—most are eaten

FIG. 48.12 **Representative freshwater biomes.** *Sources: (Lakes) Ken Gillespie/Getty Images; (Rivers) Michael Runkel/Getty Images.*

Lakes

Lakes and ponds range in size from tiny temporary pools to the Great Lakes of North America, which contain more than 20% of all freshwater habitats in the world. They range geographically from the equator to above 80° N. Their chemistries vary with the types of terrain in which they are found, and nutrient levels can be low to extremely high. Aquatic plants and macroscopic green algae live along the shallow margins of lakes, and both algae and cyanobacteria are primary producers throughout the photic zone. Zooplankton include small arthropods and rotifers, whereas the nekton is dominated by fish. Turtles and birds also play an important role as grazers and predators in many lakes.

Rivers

Rivers are freshwater biomes characterized by moving water. Rivers and streams, like lakes, vary tremendously in size and chemistry. Because of currents, rivers are generally well oxygenated, although oxygen levels may be lower in slowly moving rivers on floodplains. As with lakes, plants and large algae grow along river margins, while phytoplankton photosynthesize throughout the water column. Insects are important consumers, but other invertebrates, fish, turtles, and birds can all be abundant and diverse. Salmon and some other fish spend much of their lives in the oceans but swim up rivers to reproduce.

FIG. 48.13. **Biomes in the sea.** Marine biomes reflect distance from the shore and the depth of ocean basins.

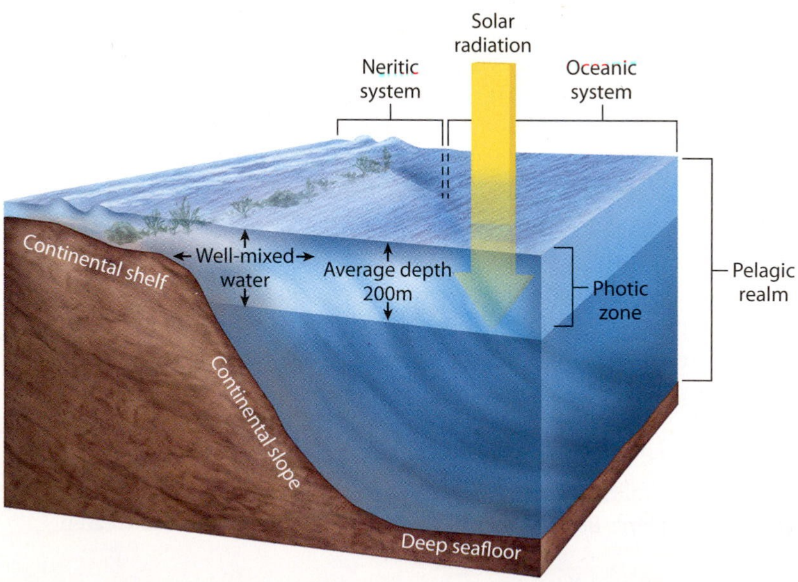

by grazers or lysed by viruses within a few days—their total biomass is commonly lower than that of grazing zooplankton, inverting the trophic pyramid characteristic of terrestrial ecosystems (Chapter 47). In the middle of large ocean basins, broad areas called gyres that are bounded by major currents, lie far from continental and upwelling sources of nutrients; their low primary productivity is dominated by cyanobacteria and tiny algae discovered only recently.

Viruses play a particularly large role in marine biomes, lysing up to 30% of phytoplankton daily in coastal ecosystems and releasing large amounts of organic molecules into the water column. This organic matter supports large populations of respiring bacteria, forming a microbial loop in the carbon cycle (Chapter 25). In turn, bacterial populations in the microbial loop support diverse populations of larger consumers.

Earlier in this chapter, we noted that winds drive ocean circulation, but winds affect only the uppermost waters of the sea. The surface layer that is mixed by the wind, broadly equivalent to the photic zone, is well oxygenated, but in deeper waters, oxygen can be depleted by respiration as microbes feed on sinking organic matter. In a few places, where surface primary production is high, the oxygen level in water masses just beneath the mixed layer declines to zero, forming an oxygen-minimum zone with distinctive microbial populations and only a limited presence of zooplankton and animals. On the deep seafloor and in the waters above it, oxygen is present, but food is sparse, consisting only of a slow rain of particles from shallower depths. Life persists in this biome as scattered populations of animals (especially fish, brittle stars, and polychaete worms), heterotrophic protists, and microbes. The exception occurs at hydrothermal vents, where high concentrations of sulfide, hydrogen, and methane support primary production by chemosynthetic bacteria and archaeons. These, in turn, support locally dense populations of animals.

Key marine biomes are summarized in **Fig. 48.14**, pages 1059–1060.

FIG. 48.14 Representative marine biomes. *Sources: (Intertidal) Dan L. Perlman/EcoLibrary; (Coral reefs) DIRSCHERL Reinhard/hemispicture.com/Getty Images; (Pelagic realm) Michael Aw/SeaPics.com; (Deep sea) Barrett Lia/SeaPics.com.*

Intertidal

The part of the neritic (shallow-water) system closest to shore is the intertidal zone, with its distinctive species associations. The intertidal zone lies along coastlines between mean high and low tides, meaning that organisms in this biome are exposed to the atmosphere on a daily basis. Intertidal organisms are thus challenged by periodic desiccation as well as by temperature extremes that accompany exposure to direct sunlight. In sandy intertidal zones, or tidal flats, many animals burrow into the wet sand to escape exposure at low tide. In rocky intertidal zones, organisms may close their shells tightly—for example, mussels and barnacles. Waves can cause mortality and so, especially along rocky coastlines, both algae and animals are securely attached to the substrate. Nutrient levels can be high, favoring strong algal growth. Consumers include a variety of sea stars, sea urchins, mollusks, barnacles, and corals. Intertidal biomes exhibit the greatest diversity where tidal exposure is relatively short. Tidal pools can harbor diverse animal and algal species.

Coral reefs

Coral reefs form another neritic biome, indeed, the most diverse biome in the oceans. The best-known reefs occur in shallow, tropical to subtropical environments where there is little water movement or accumulation of sand or mud, with the accumulating skeletons of corals building a structure above the seafloor. Nutrient levels are commonly low, but primary production is high and mostly tied to dinoflagellate algae that live within the tissues of corals. Rapid recycling of nutrients between the corals and their algal symbionts ensures high levels of primary production. Free-living algae occur in reefs, but are kept at low abundance by grazing fish. Many species of invertebrates live within the reefs, often in nooks and crannies among coral heads that provide shelter from predators. Fish diversity can be especially high. Coral reefs also occur in the deep sea, but build gradually, as coral growth is slow in the absence of symbiotic algae.

Pelagic realm

The pelagic realm—the part of the ocean that is neither close to shore nor close to the seafloor—forms the bulk of the oceanic system. Organisms live within the water column, either as plankton or nekton. In the upper 200 m or so of the pelagic realm, sunlight permits photosynthesis, with diverse algae and cyanobacteria as primary producers. In deeper waters, however, sunlight is absent and life is sustained by the rain of organic particles from surface waters. Minute arthropods dominate the zooplankton, while fish and cephalopods are key components of the nekton. This is true in both surface and deep waters, but animals are less abundant at depth. Fish and squids are conspicuous consumers, but recent research shows that, in the upper water column, heterotrophic protists may actually be more abundant and diverse. Bacteria and archaeons occur throughout the pelagic realm.

Deep Sea

Kilometers below the surface, deep seawaters are cold and dark, but they are not sterile. Sinking detritus supports sparse populations of animal consumers, as well as bacteria and archaeons that feed on organic particles. Despite their limited biomass, deep-sea biomes exhibit high species diversity, as high as those of shallow marine communities. However, the composition of communities in deep-sea biomes differs relatively little from place to place, and so the total diversity of these environments is lower than that found within the photic zone. Dense animal populations occur locally on the deep seafloor, where hydrothermal vents expel fluids containing high abundances of hydrogen, hydrogen sulfide, and methane. These compounds support locally high levels of primary production by chemosynthetic bacteria, many of them present as symbionts within the tissues of vent animals.

48.3 GLOBAL ECOLOGY: CYCLING BIOESSENTIAL ELEMENTS

In Chapter 25, we introduced the carbon cycle as a framework for understanding organisms, ecology, and diversity. The cycles of carbon and other biologically important elements are called **biogeochemical cycles** because they link organisms and their environment in the chemical cycling of the raw materials needed for life. As noted in Chapter 25, these cycles provide the fundamental ecological circuitry of life, sustaining life in all its diversity over long timescales. The following discussions bring us full circle, showing how the metabolic and functional diversity discussed in earlier chapters of Part 2 link to cycle carbon and nitrogen, structuring biomes on land and in the sea. We also introduce the phosphorus cycle, as phosphorus, like nitrogen, can limit rates of primary production.

The biological carbon cycle shapes ecological interactions and reflects evolution.

Fig. 48.15 shows how organisms cycle carbon through terrestrial and marine biomes, integrating concepts drawn from metabolism and ecology. All the ecological complexity of ecosystems and biomes begins with primary producers, the autotrophic organisms that convert carbon dioxide into organic molecules (Chapter 6).

On land, primary production is dominated by a single phylogenetic group of photosynthetic organisms, vascular plants. The tremendous variety of forms observed among plants in different biomes reflects evolutionary adaptation to differing environments. In the oceans, many distinct groups of algae and cyanobacteria convert carbon dioxide to organic molecules, and again the diversity of these primary producers reflects adaptations to the physical and chemical features of different water masses. In the dark waters of the deep sea, primary production by

FIG. 48.15 The biological carbon cycle. Carbon cycles, both (a) terrestrial and (b) marine, link metabolism and ecology into functioning ecosystems and biomes.

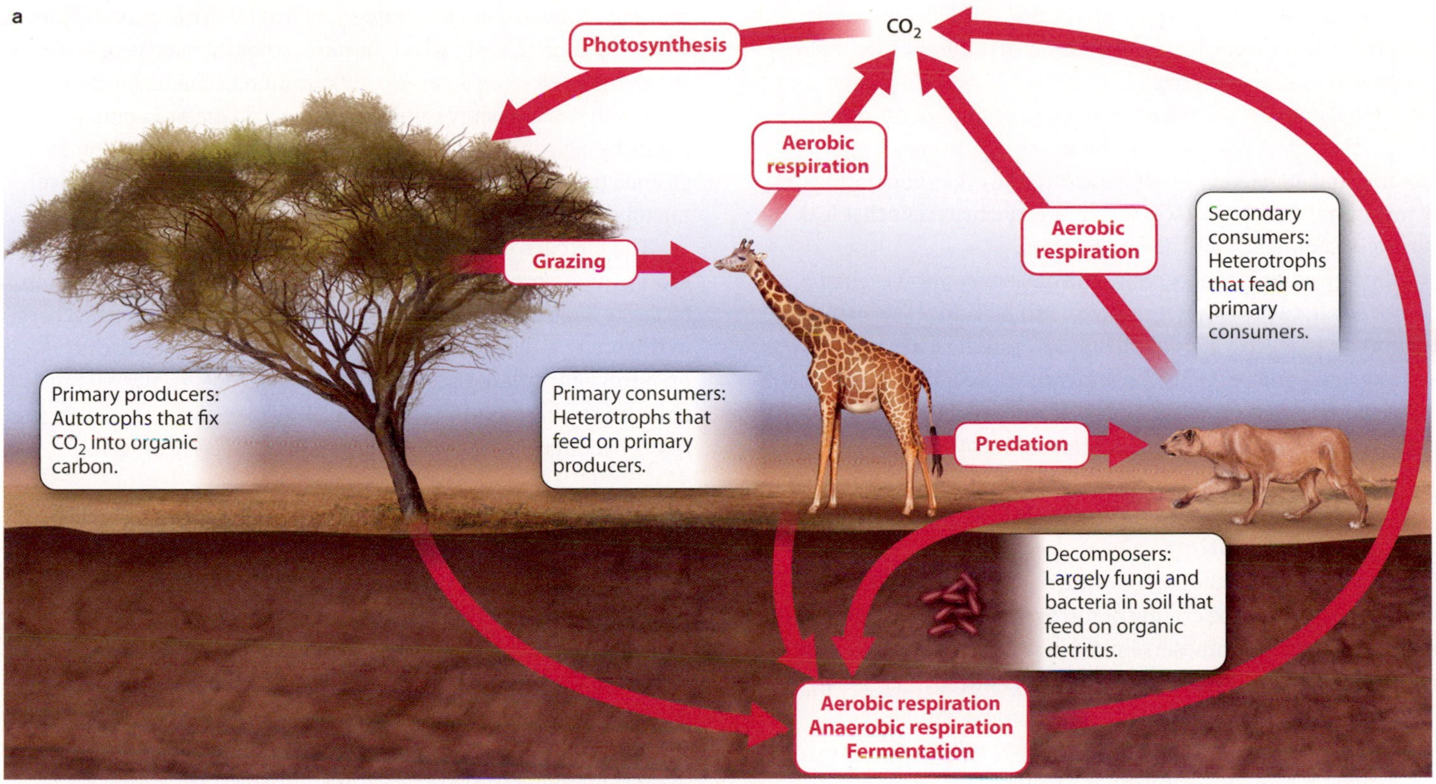

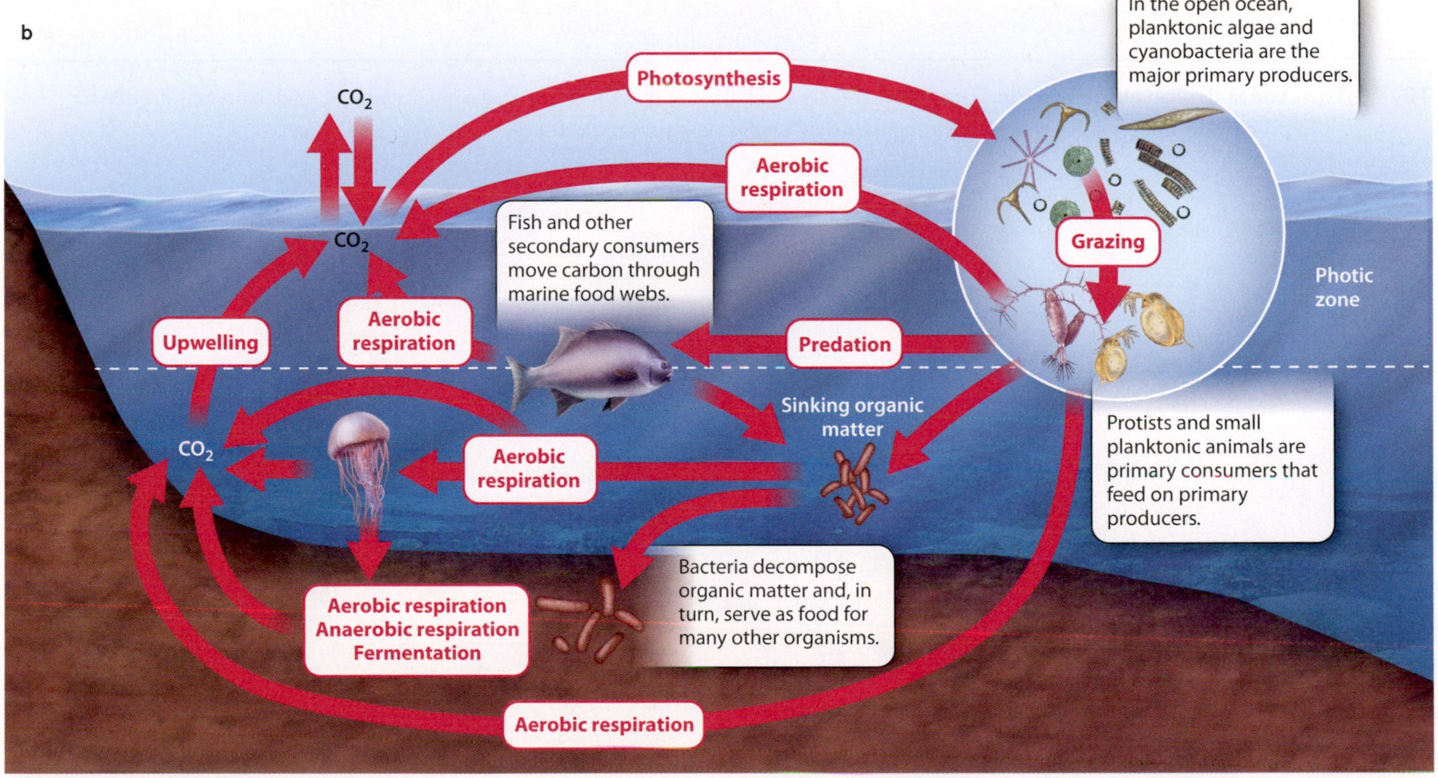

chemoautotrophic bacteria can contribute in important ways to the carbon cycle.

Consumers, from animals to bacteria, live on organic matter originally generated by primary producers, their diverse forms again reflecting evolutionary adaptation to different food sources and predators. As emphasized in Chapters 25 and 26, plants and animals may be the most conspicuous components of the biological carbon cycle, but it is the microorganisms—fungi, bacteria, and archaeons—that complete the cycle, generating CO_2 by respiration or fermentation from organic materials that leak into soils, sediments, and deep seawaters.

The key point is that ecology and biogeochemical cycling are not separate processes. Ecological interactions cycle carbon rapidly through ecosystems and biomes, and the leaks in this cycle, with some organic remains accumulating in sediments, link the biological carbon cycle to the longer-term geological cycling of carbon that governs climate (Chapters 25 and 49).

The nitrogen cycle also reflects the interplay between ecology and evolution.

Fig. 48.16 illustrates the same terrestrial ecosystem as Fig. 48.15, but now highlights the cycling of nitrogen. The primary producers, consumers, and decomposers linked in the carbon cycle also cycle nitrogen. Indeed, the biological carbon and nitrogen cycles are closely connected by both ecology and metabolism.

Once again, primary producers play a key role, as these are the organisms that can assimilate the biologically usable nitrogen found in soil or water. Primary consumers obtain nitrogen from the primary producers at the same time that they obtain carbon from these organisms, and through consumption both nitrogen and carbon travel through the food web. Decomposers return N_2 to the atmosphere by metabolic processes such as denitrification (a form of anaerobic respiration in which nitrate, rather than oxygen, serves as the terminal electron acceptor), nitrification (a chemoautotrophic process that uses energy gained from the oxidation of ammonia or nitrite by oxygen), or anammox (chemoautotrophy, with energy gleaned from the reaction of ammonia and nitrite)—all exclusively microbial processes (Chapter 26).

The nitrogen cycle differs from the carbon cycle in a key way: Most primary producers cannot assimilate nitrogen directly from by far its largest reservoir, N_2 in the atmosphere. As introduced in Chapter 26, the biological nitrogen cycle depends critically on nitrogen fixation, the reduction of N_2 to biologically useful ammonia by bacteria and archaeons. By extension, then, nitrogen-fixing bacteria are fundamental to the carbon cycle as well.

Phosphorus cycles through ecosystems, supporting primary production.

In Chapter 47, we introduced the concept of limiting resources, including nutrients. Photosynthesis is rarely limited by the availability of CO_2. Instead it is the availability of other elements needed to make proteins, nucleic acids, and membranes that determines the rate of primary production.

Phosphorus and nitrogen are the key limiting nutrients in most biomes, but their cycles are quite distinct. While nitrogen is found largely in the atmosphere, phosphorus resides

FIG. 48.16 The biological nitrogen cycle. The nitrogen cycle also links metabolism and ecology, governing the availability of nitrogen for primary producers and so exerting strong influence on the biological carbon cycle.

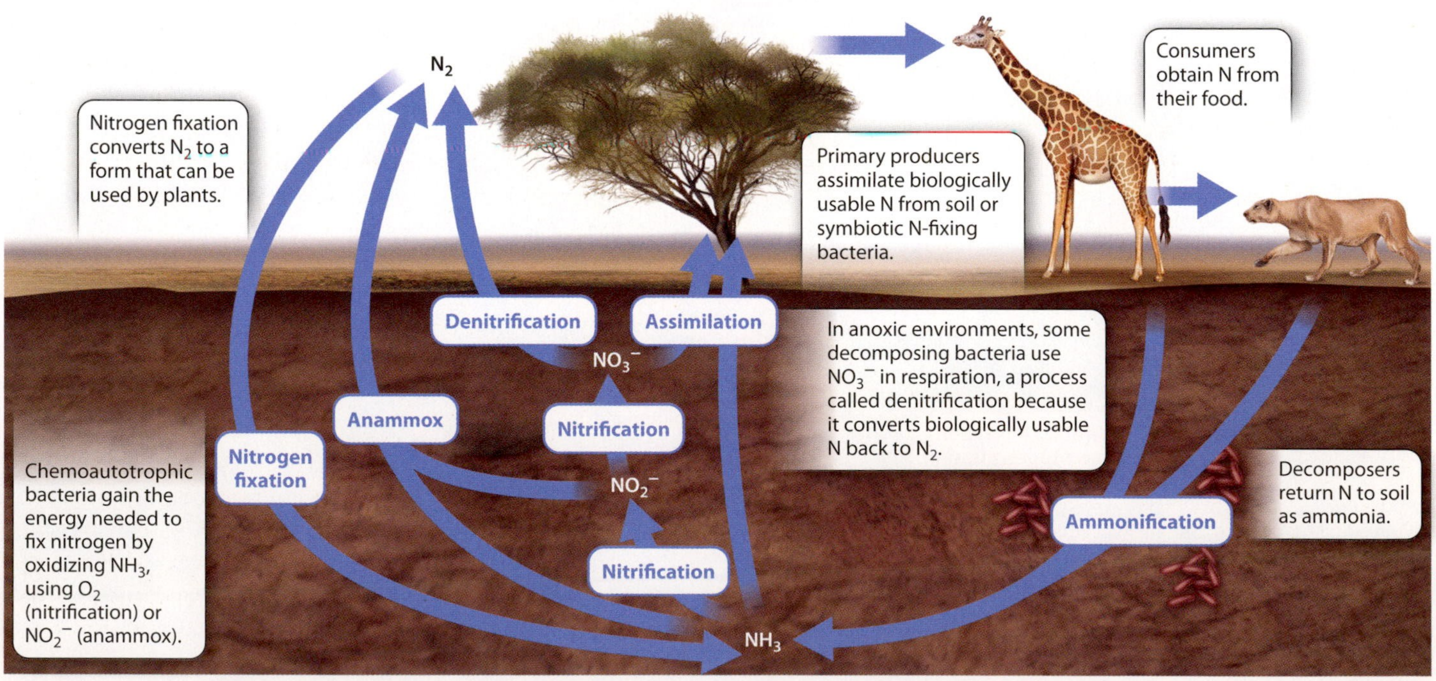

FIG. 48.17 The phosphorus cycle. Phosphorus, another key nutrient, is weathered from rocks and eventually returns to its geological reservoir through deposition in sediments. Before it does, organisms repeatedly cycle phosphorus through ecosystems and biomes.

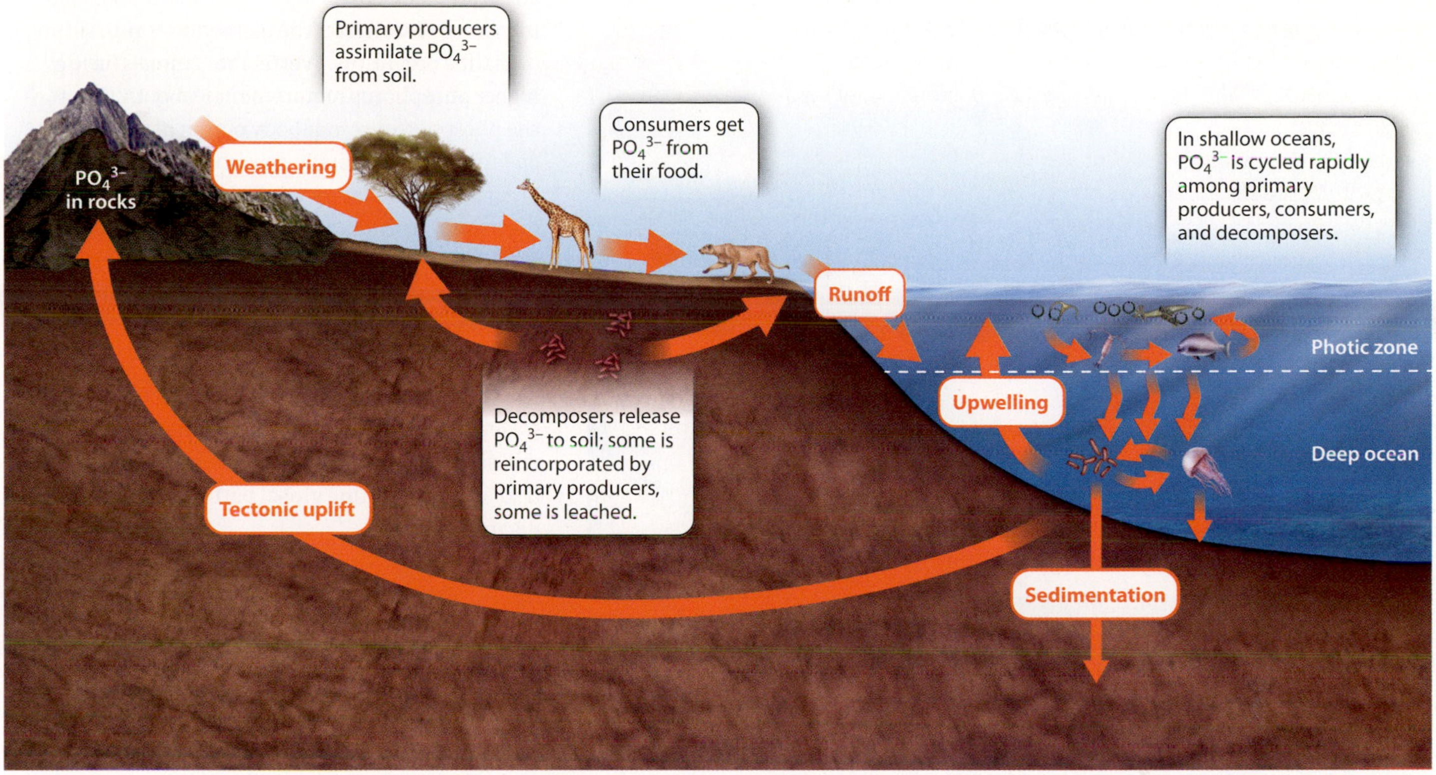

overwhelmingly in rocks. And though nitrogen and phosphorus are both assimilated by primary producers, phosphorus does not provide electron donors or electron acceptors for energy metabolism. It simply enters the food web as phosphate ion (PO_4^{3-}), released from rocks by chemical weathering, is taken up by primary producers, and then is transferred from one organism to another through cycles of consumption and decomposition, eventually returning to its geological reservoir by accumulation in sediments. The amount of phosphorus needed for growth is determined by the molecular composition of cells, mostly in RNA but also in DNA and membranes.

Fig. 48.17 shows how phosphorus cycles through terrestrial and marine biomes.

→ **Quick Check 3** In what way do the nitrogen and phosphorus cycles regulate Earth's carbon cycle?

Global patterns of primary production reflect variations in climate and nutrient availability.

Our review of biogeochemical cycles underscores the importance of primary production to ecosystems and biomes, but what controls rates of primary production? In Chapter 47, we saw that climate and nutrient limitation play important roles in determining the primary productivity of individual ecosystems. It stands to reason, then, that global patterns of primary production reflect planetary-scale variations in climate and nutrient availability.

Fig. 48.18 shows the global distribution of primary production on land. The similarity of this figure to maps of global evapotranspiration and precipitation (see Fig. 48.9) shows the strong control of climate on patterns of primary production. Warm and wet climates near the equator support highly productive forests, whereas dry habitats and cold environments with limited growing seasons have lower levels of primary productivity.

In contrast, examine **Fig. 48.19,** which shows how rates of photosynthesis vary in the marine realm. In the oceans, the relationship to global climatic zones is not so clear, which is unsurprising because, in aquatic ecosystems, water is not a limiting factor for photosynthesis. The principal pattern shown in Fig. 48.19 is one of high photosynthetic rates along the margins of continents and especially low rates in the central gyres of circulating oceans. As noted earlier in this chapter, rates of primary production are high near continents because runoff of nitrogen and phosphorus from the continents supplies abundant nutrients to the sea. Upwelling zones, where nutrient-rich deep waters return to the surface, are also (mostly coastal) regions of high primary production. In contrast, the interiors of large ocean basins contain little biologically available nitrogen or phosphorus, and so rates of primary production are low.

The relative amounts of carbon, nitrogen, and phosphorus in cells vary among organisms, but within relatively narrow limits.

FIG. 48.18 Primary production on land. High rates of primary production occur where climates are warm and wet. *Data from W. Cramer et al., 1999, "Comparing Global Models of Terrestrial Net Primary Productivity (NPP): Overview and Key Results," Global Change Biology 5 (Supplement 1): 1–15.*

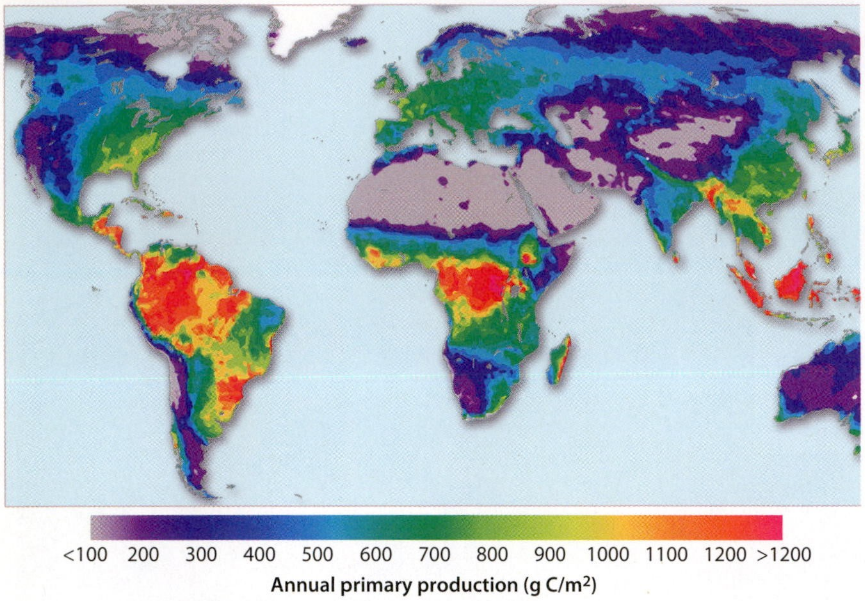

On average, cells contain 1 atom of phosphorus and 16 atoms of nitrogen for every 100 atoms of carbon, a correspondence called the **Redfield ratio** after the American biologist Alfred Redfield, who pioneered research on the elemental composition of marine organisms. (Vertebrate animals have a higher phosphorus requirement—about 80% of the phosphorus in your body occurs in your bones and teeth.) Interestingly, in some regions of the ocean, rates of primary production are relatively low even though nitrate and phosphate abundances are high. This observation suggests that another factor limits primary production. **Fig. 48.20** discusses experiments that point to iron as the limiting nutrient in these waters.

Global patterns of primary production help to determine the properties of biomes, and, as we will see in the next section, primary production also helps determine global patterns of biological diversity.

→ **Quick Check 4** Why do patterns of global primary production on land show strong latitudinal variation, whereas primary production in the oceans varies largely with distance from shore?

FIG. 48.19 Primary production in the oceans. In the sea, nutrient availability governs rates of primary production. *Data from K. Sherman, P. Celone, and S. Adams, 2004, NOAA Fisheries Service's Large Marine Ecosystems Program: Status Report. NOAA Tech Memo NMFS NE 183; 21 p.*

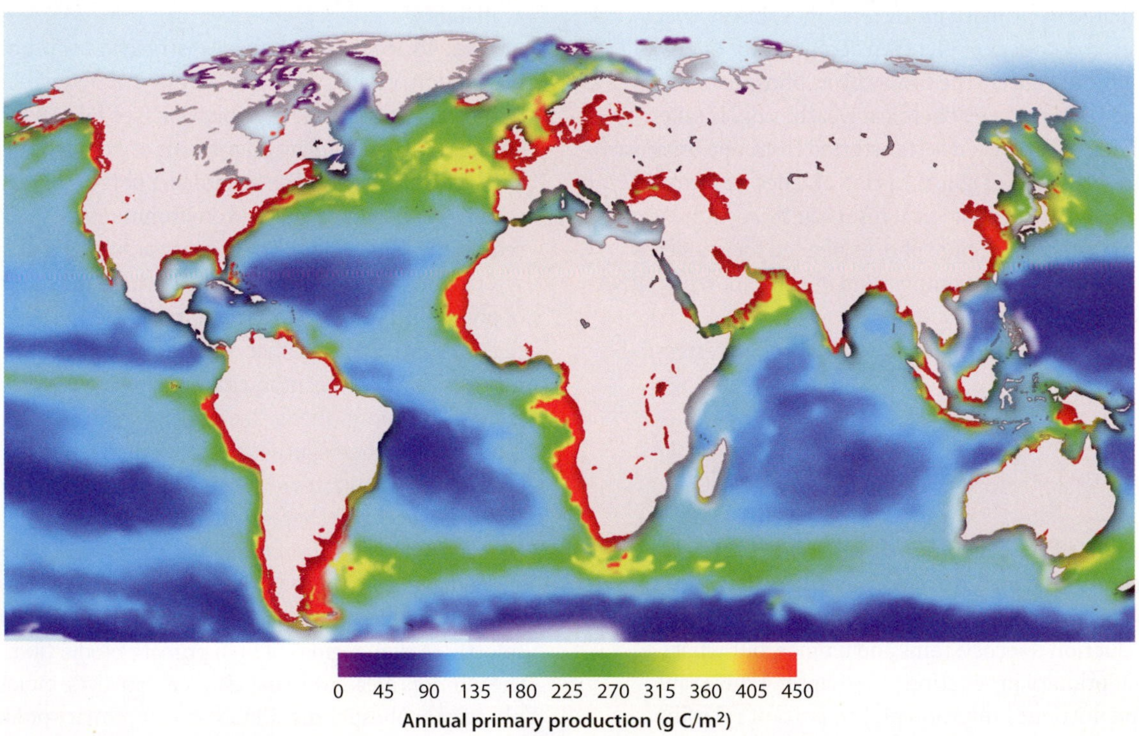

FIG. 48.20

Does iron limit primary production in some parts of the oceans?

BACKGROUND Nitrogen and phosphorus are nutrients that are thought to limit primary production in the oceans, and, consistent with this hypothesis, the concentrations of nitrate and phosphate in surface oceans are generally very low. Regionally, however, nitrogen and phosphorus concentrations are high, indicating that primary producers are not making use of all available nutrients.

HYPOTHESIS In some parts of the oceans, iron—required for the functioning of enzymes critical to photosynthesis and nitrogen uptake—is the nutrient in lowest supply relative to the needs of primary producers and so limits rates of photosynthesis.

EXPERIMENT Biologists John Martin and Steve Fitzwater sailed to subarctic waters of the northeastern Pacific Ocean, where high nitrogen and phosphorus concentrations suggested limitation by another nutrient. They collected samples of seawater, along with its phytoplankton, and added small amounts of iron to different samples: 1, 5, or 10 nanomoles per kilogram of sample (in the control sample, no iron was added). As photosynthesis proceeded over the course of a week, the researchers made daily measurements of chlorophyll concentration (an indicator of photosynthetic rate), as well as of the concentrations of nitrate and phosphate in the sample treatments.

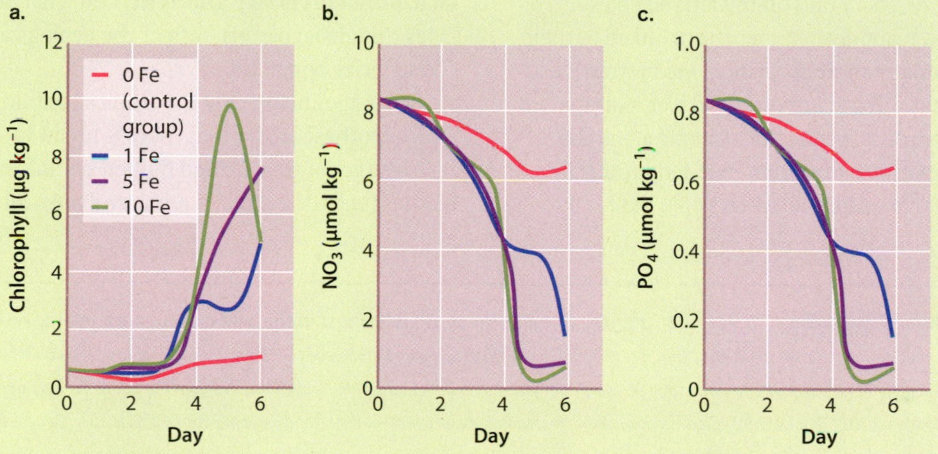

RESULTS As the figure shows, over the course of the experiments, samples to which iron was added showed increased chlorophyll concentration and decreased levels of nitrogen and phosphorus, with higher additions of iron generally showing more pronounced effects. Controls with no added iron showed little change over the course of the experiment.

CONCLUSION In samples from the northeastern Pacific Ocean, addition of iron increased rates of photosynthesis, with more uptake of nitrogen and phosphorus. This supports the hypothesis that, in these waters, iron and not nitrogen or phosphorus limits primary production.

FOLLOW-UP WORK Since this initial experiment, many researchers have done similar research in areas of the surface ocean with relatively high levels of nitrogen and phosphorus, with consistent results. Some researchers have suggested that seeding these ocean waters with iron could provide a sink for atmospheric carbon dioxide, helping to alleviate global change associated with anthropogenic CO_2 emissions. Results, however, have been mixed, with many experiments showing a spike of increased photosynthesis followed by an interval of enhanced respiration—with little carbon sinking into the deep sea.

SOURCE Martin, J. H., and S. E. Fitzwater. 1988. "Iron-Deficiency Limits Phytoplankton Growth in the Northeast Pacific Subarctic." *Nature* 331: 341–343.

48.4 GLOBAL BIODIVERITY

On a half-hour walk through a deciduous forest—for example, in the northeastern United States or southeastern Canada—we can expect to see maybe a dozen different kinds of tree: two or three kinds of oak perhaps, a couple of maple species, birch, beech, hornbeam, ironwood, one or two pine species, and possibly a sycamore, poplar, or spruce, depending on just where we are. Standing in front of a tree and turning around, we can almost always see several more trees of the same kind. In contrast, on a similar walk through a forest in the Amazon or in Borneo, we would pass 300 or 400 different tree species, with almost no near neighbors of the same kind. The great Victorian naturalist Alfred Russel Wallace took note of the low population density of trees in tropical forests and observed, "If the traveller . . . notices a particular species and wishes to find more like it, he may often turn his eyes in vain in every direction."

This intuitive feel for biodiversity is reflected in the global distribution of plant species richness shown in **Fig. 48.21.** Tropical rain forests harbor the highest number of plant species, and in general the amount of plant diversity tails off toward the poles. Deserts have particularly low plant diversity, a pattern linked to their low primary productivity. Perhaps more surprising, Mediterranean climates—temperate regions with dry summers and wet winters—can support large numbers of plant species, for example along the southern tip of Africa, southwestern Australia, and Mediterranean Europe. These regions show that additional factors only indirectly linked to climate influence regional plant diversity, including habitat diversity (in Mediterranean regions, a mixture of woodland, savannahs, and grassland) and frequent fires (disturbance, discussed in Chapter 47).

It isn't just trees. Mammals, birds, insects, and many other kinds of animal show similar patterns of species diversity. **Fig. 48.22,** for example, shows the global distribution of mammal species. To get a sense of animal diversity in tropical rain forests, consider that one survey found more than 700 species of leaf beetles in a single small set of trees in the Amazon, a number equal to the total diversity of leaf beetles in all of eastern North America. A single tree in this same forest harbored 43 species of ants, exceeding the total for all of Great Britain.

? CASE 8 BIODIVERSITY HOTSPOTS: RAINFORESTS AND CORAL REFFS

Why does biodiversity decrease from the equator toward the poles?

When we compare biomes across the continents, the pattern just described stands out clearly: Species diversity generally peaks near the equator and declines toward the poles. What are causes of this **latitudinal diversity gradient,** and what are its consequences? Does this global pattern reflect the ecological processes that characterize populations?

Ideas abound on why the tropics contain so many species. One hypothesis, for which there is broad support, looks to the relative ages of tropical and higher-latitude biomes. Tropical biomes tend to be older, having evolved over tens of millions of

FIG. 48.21 Latitudinal diversity gradient in vascular plants. Species diversity is highest in the wet tropics and declines toward the poles. *Data from Jens Mutke and Wilhelm Barthlott, 2008, "Biodiversität und ihre Veränderung im Rahmen des Globalen Umweltwandels: Biologische Aspekte" (Biodiversity and its change in the context of global environmental change: biological aspects) In Dirk Lanzerath, Jens Mutke, Wilhelm Barthlott, Stefan Baumgärtner, Christian Becker, and Tade M. Spranger, 2008, Biodiversität (Biodiversity.) (Ethik in den Biowissenschaften–Sachstandsberichte des DRZE, Bd.5). (Series: Ethics in the life sciences–DRZE expert reports, vol. 5). Freiburg i.B.: Alber: 63.*

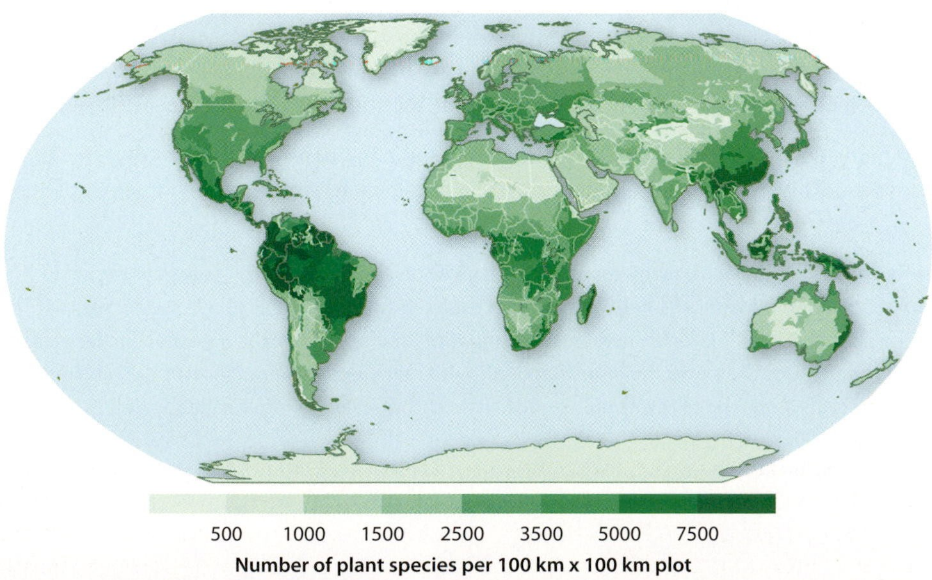

FIG. 48.22 Latitudinal diversity gradient in mammals. The local number of mammal species decreases from the equator to the poles. *Source: After D. M. Kaufman, 1995, "Diversity of New World Mammals: Universality of the Latitudinal Gradients of Species and Bauplans," Journal of Mammalogy 76:322–334.*

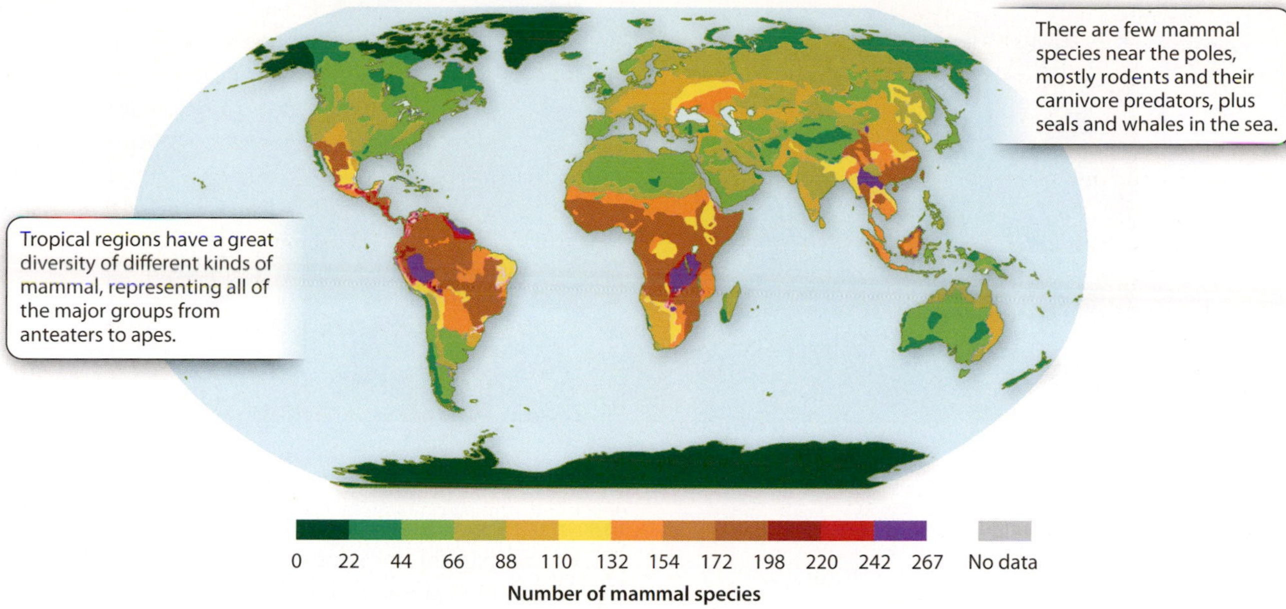

years, whereas biomes at higher latitudes have been shaped by climatic change, and ultimately an ice age, within the past few million years. Because tropical habitats have existed much longer than temperate ones, biologists hypothesize that there has been more time for species to evolve and diversify at low latitudes. Several studies of diversification rate, in frogs for example, show comparable rates of diversification in temperate and tropical biomes, supporting the hypothesis that the higher tropical diversity reflects the longer interval through which diversification has taken place.

Time, however, is just one of several factors that influence diversity. Another hypothesis is that temperate communities have fewer species because it is relatively difficult to adapt successfully to cold, dry winters and the range of weather typically experienced at higher latitudes. A tree growing in a temperate area like New England experiences a wide range of temperatures over a year (and sometimes even over a single day) as winds blow and snow falls, or heat waves move through the area. Because species living at high latitudes are adapted to a greater range of conditions on average than those nearer the equator, they tend to have larger geographic ranges than tropical species, which are adapted to the narrow range of environmental variation experienced in any particular place.

It has also been suggested that the many insects and fungi specialized to attack particular kinds of tree are generally more abundant in wet tropical forests, and so trees of a given species tend to be as far apart as they can be, thus reducing the chance of being found by their enemies. This means that there would be more room for different species in any given place. It has also been noted that there is more land surface area at low latitudes, so the species–area curve (Chapter 46) plays a role in the latitudinal diversity gradient.

Very likely, the latitudinal diversity gradient reflects all these influences and more not yet appreciated. What is clear, however, is that where there are more species of trees, there are also more species of animals and other consumers. Tropical temperatures and high humidity have permitted a wide range of plant species to evolve over the past 100 million years, creating many niches for the animals that depend on them.

The high diversity—and therefore the low population density—of many tropical forests prompts another question. What determines how far apart individual trees can be and still maintain a viable population? One important consideration goes back to mutualisms: Animal pollination ensures that pollen from one tree can reach another individual of the same species if it is within the foraging range of a pollinator carrying its pollen. Of course, if trees are farther apart than any single pollinator can visit within a foraging excursion, they will not be pollinated. Animal pollination, therefore, contributes to the diversity differences between tropical trees and higher-latitude species, which are more likely to be pollinated by wind.

Both on land and in the sea, biodiversity hotspots occur predominantly in tropical to subtropical environments. In part, this simply reflects the higher species diversity typical of lower latitudes, but it also highlights that tropical communities contain a relatively high proportion of **endemic species,** species found in one place and nowhere else. Continued speciation coupled with a limited ability to colonize new areas successfully underpins the distribution of hotspots. Because of the way that biodiversity

FIG. 48.23 **Fossils in 30–23-million-year-old Dominican amber.** These fossils document the evolutionary history of (a) orchid-pollinating stingless bees, now extinct on Hispaniola, and (b) *Anolis* lizards that, like their modern descendants, hunted insects in trees.
Source: Courtesy George Poinar

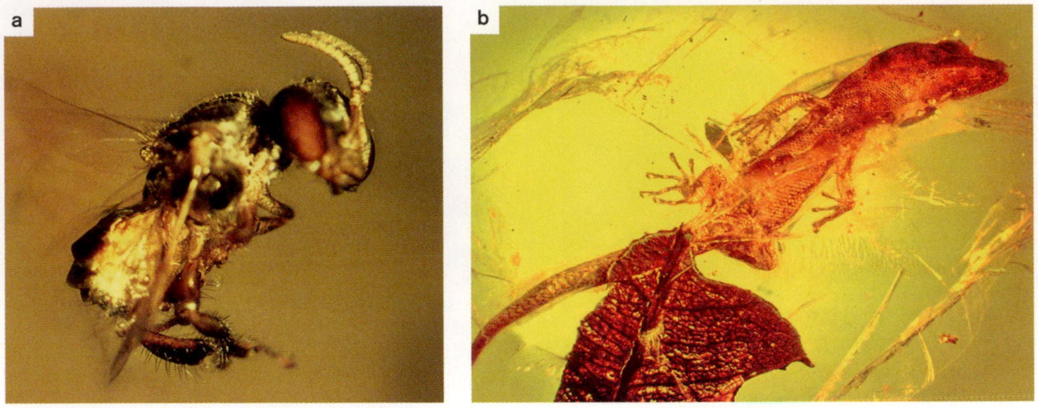

hotspots are distributed, the burden of conservation efforts falls disproportionately on developing countries.

→ **Quick Check 5** From what you know about how species diversity changes with latitude, predict how species diversity changes with altitude.

Evolutionary and ecological history underpins diversity.

Earth's biomes are the historical outcome of natural selection and environmental change through time. For this reason, the fossil record can shed light on how current patterns of diversity came to be. In general, fossils tell us only a little about the origins of tropical forests, but, on the island of Hispaniola, there is a remarkable exception: nuggets of amber that provide an extraordinary window on life in the island's rain forest 30 to 23 million years ago.

Amber is fossilized tree resin, and in Hispaniola it comes from the tree *Hymenea,* a member of the pea family. *Hymenea* trees grew in the Hispaniola rain forest 30 to 23 million years ago, much as they do today. Lizards, tree frogs, insects, plants, and fungi were all captured in their sticky sap, which hardened into amber, preserving a record of forest diversity and ecology (**Fig. 48.23**).

Studies of these fossils reveal ecological relationships similar to some observed on Hispaniola today. Not surprisingly, some of the most abundant animals in the amber are insects whose modern counterparts have ecological associations with *Hymenea.* Ants that patrolled the bark of *Hymenea* and other tropical trees in search of food, beetles that fed on trees that secrete resin as a defense mechanism, and stingless bees that gathered resins for their nests are all preserved intact in the amber as if they were just caught. Many of the insects seen in this amber, perhaps even the majority, are practically indistinguishable from species that continue to exist today on Hispaniola.

Dominican amber also reveals species that are no longer found on Hispaniola, for example, many of the more specialized kinds of ant found in the amber. A kind of beetle that feeds on palm trees in South America is found in amber but not on the island today, even though there are plenty of palms. And there are no longer stingless bees on Hispaniola, although the flowers they visit persist, and the bees occur on nearby Jamaica.

Vertebrate fossils are rare in Dominican amber, but they include *Anolis* lizards closely related to ones specialized for life in the canopies and high on the trunks of forest trees. In other words, resource partitioning began long ago in the Hispaniola rain forest. More vertebrate fossils occur in other settings, and these show that most of the birds and mammals found 25 million years ago are quite different from those in modern forests.

Such fossils allow us to add a deeper dimension of time to our understanding of how ecosystems have come to be. Whether forest, reef, or grassland, every ecosystem contains the present-day survivors of groups representing a broad range of origins in evolutionary time and geographic space. The amber fossils remind us that what G. Evelyn Hutchinson called "the ecological theater and the evolutionary play" have been intertwined throughout our planet's history, producing the diversity of species we see today. ■

Core Concepts Summary

48.1 Solar radiation, wind and ocean currents, and topography determine the distribution of major climactic zones on Earth.

Equatorial climates are warm and polar climates are cold because of the angle at which solar radiation strikes Earth's curved surface. page 1046

The tilt of Earth on its axis means that solar radiation strikes the Northern and Southern Hemispheres unevenly at different parts of Earth's orbit, causing seasonality. page 1046

Wind and ocean currents transport heat from the equator toward the poles. page 1047

Topographical features, like mountains, regionally modify global circulation patterns. page 1049

48.2 Biomes are broad, ecologically uniform areas whose characteristic species reflect regional climate.

The distribution of the plants that characterize terrestrial biomes reflects in large part abiotic factors such as climate and soil. page 1050

Regional climate determines the ratio of potential evapotranspiration to precipitation, a key determinant of vegetation type, primary production, and biodiversity. page 1050

Aquatic biomes also reflect abiotic factors, especially the spatial distribution of nutrients and the depth to which sunlight penetrates water. page 1052

48.3 Biologically driven cycles of carbon and other essential elements shape ecology and reflect evolution.

The evolved metabolisms of organisms cycle carbon through the environment and give structure to the food webs characteristics of biomes. page 1060

Nitrogen and phosphorus limit primary production in most biomes. page 1062

Nitrogen in the atmosphere is fixed by bacteria and archaeons, and then cycles through primary consumers, consumers, and decomposers, so is closely linked to the carbon cycle. page 1062

Phosphorus is found mostly in rocks, enters the food web as phosphate ion by chemical weathering, and cycles through ecosystems, supporting primary production. page 1062

Nutrient levels and climate determine global patterns of primary production. page 1063

48.4 Global patterns of biological diversity reflect climate, history, and ecological interactions among species.

Species diversity generally declines from the equator toward the poles, a pattern known as a latitudinal diversity gradient. page 1066

Earth's biomes are the historical outcome of environmental change through time and natural selection. page 1068

Self-Assessment

1. Explain how incoming solar radiation exerts a direct influence on the distribution of temperature across Earth's surface.
2. Why are equatorial latitudes generally wet, but land masses at about 23 to 30 degrees north and south latitude generally dry?
3. Why are the sides of tall mountains that face oncoming wind generally wetter than their opposite slopes?
4. Choose five terrestrial biomes, describe their climate and vegetation, and explain why they differ from one another.
5. How do marine biomes differ between the surface ocean and the deep seafloor?
6. Describe two ways that the biological nitrogen cycle interacts with the biological carbon cycle.
7. Compare and contrast biomes on land and in the sea in terms of their global patterns of primary production.
8. Describe the general pattern of diversity from the equator to the poles. Provide two hypotheses to explain this pattern of diversity.

Log in to LaunchPad to check your answers to the Self-Assessment questions, and to access additional learning tools.

Source: Michael Utech/Getty Images.

CHAPTER 49

The Anthropocene

Humans as a Planetary Force

Core Concepts

49.1 Some scientists call the time in which we are living the Anthropocene Period to reflect our significant impact on the planet.

49.2 Humans have a major impact on the carbon cycle, primarily through the burning of fossil fuels, which returns carbon dioxide to the atmosphere.

49.3 Humans have an important impact on the nitrogen and phosphorus cycles, primarily through the use of fertilizer in agriculture.

49.4 The impact of humans on the environment is changing the setting in which evolution acts.

49.5 Conservation biology aims to safeguard biodiversity for future generations.

49.6 In the 21st century, biologists, doctors, engineers, teachers, and informed citizens have vital roles to play in understanding our changing planet and making wise choices for our future.

By almost any measure, you were born into humanity's golden age. Beginning with the invention of the steam engine, humans have mastered the ability to harness Earth's store of chemical energy, dramatically extending the power of muscles. You probably don't find it remarkable to drive across a continent in less than a week or fly across an ocean overnight, but your great-great-grandparents would have been astonished. At the same time, advances in antibiotics and public health have drastically reduced infectious disease in many parts of the world. Smallpox, long one of life's deadliest realities, has been eliminated, and polio may eventually follow suit. And the Green Revolution of the last 50 years provides unprecedented amounts of food for a hungry planet (Case 6: Agriculture).

One consequence of this success is an increasing human population: We recently crossed the 7-billion mark. Another is that our actions affect the world around us directly or indirectly. In this chapter, we look at some of these impacts, and what we can do about them. Can we sustain or control current patterns of population growth and resource use? Can we ensure that future generations will be able to enjoy and benefit from the diversity of life we see today? As discussed in the following sections, there is reason both for concern and optimism. Together, scientific insight, technical ingenuity, and a wise citizenship provide our best hope of conserving our natural world while sustaining hard-won gains in food, energy, and health through the 21st century and beyond.

FIG. 49.1 Growth through time of the human population. *Data from "Historical Estimates of World Population," United States Census Bureau International Data Base, last modified June 2012, http://www.census.gov/population/international/data/worldpop/table_history.php.*

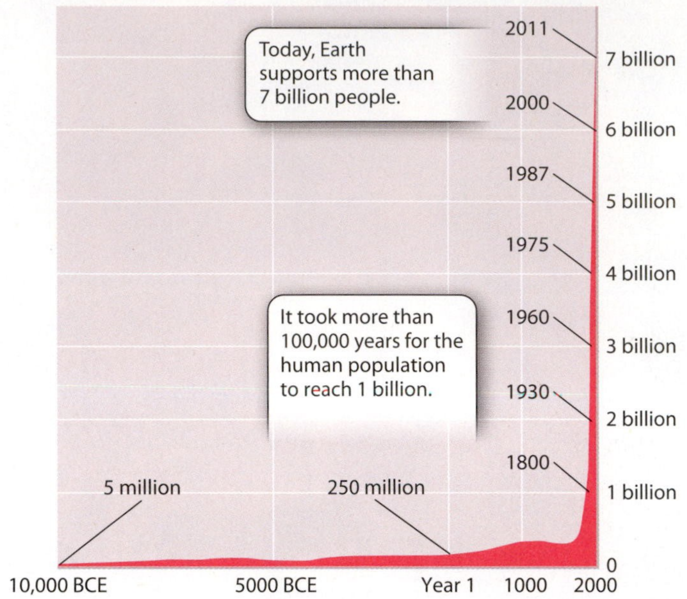

49.1 THE ANTHROPOCENE PERIOD

Increasingly, scientists refer to the modern era as the **Anthropocene Period,** emphasizing the dominant impact of humans on Earth. Our impact is partly a consequence of our sheer numbers, which in turn drive our energy and land use, but it also reflects the individual and societal choices we make.

Humans are a major force on the planet.

There are more humans than ever before, and our numbers are climbing rapidly. A world that had about 500 million inhabitants when Christopher Columbus set sail in 1492 now supports more than 7 billion people (**Fig. 49.1**). By the time you retire, our numbers will probably exceed 9 billion.

→ **Quick Check 1** Take a look at Fig. 49.1. How long did it take our numbers to double, from 1 billion to 2 billion? How about from 2 billion to 4 billion? What does this pattern say about the rate of population growth?

Of course, population size doesn't tell the whole story. Whether driving a car, living in a heated apartment, plugging in a computer, or eating fruit shipped from another hemisphere, you use a great deal more energy than your ancestors did. To give just one example, it has been suggested that the electrical outlets in our homes let us accomplish about the same amount of work performed by 30 servants in a Roman villa. **Fig. 49.2** shows annual per capita energy use in selected countries. Energy use by individuals varies greatly from one part of the world to another. Americans, Canadians, Europeans, and Australians live energy-intensive lives, whereas populations in rural Africa and parts of Asia consume relatively little energy per person.

Our direct use of energy, however, reflects only part of our impact on the planet. Nearly everything we use requires both land and energy. All of us consume food grown in fields, fertilized and harvested through energy, and shipped to a local store. We live in houses built of wood, brick, and stone, and use clothing, furniture, and appliances fashioned from plant materials grown in fields and forests or from minerals extracted from Earth by mining. It takes land to grow plants, and energy to sow seeds, make and distribute fertilizer, harvest useful materials, and manufacture, ship, and clean the finished products.

The concept of the **ecological footprint** is an attempt to quantify our individual claims on global resources by adding up all the energy, food, materials, and services we use and estimating how much land is required to provide those resources. **Fig. 49.3** plots the average ecological footprint per person against a ranking of the Human Development Index, a measure of standard of living maintained by the United Nations, for many countries. Not surprisingly, developed countries tend to have large ecological footprints—it takes about 7 hectares of land to support an average American. By contrast, in many parts of Asia and Africa, only a single hectare supports the average citizen.

In many countries, living standards have risen over the past 50 years, nearly always increasing the mean ecological footprint of their citizens. By some estimates, the total ecological footprint

FIG. 49.2 Energy use by individuals among countries. Energy use is distributed unevenly among countries. *Source: Data from United Nations Department of Economic and Social Affairs, 2009, World Economic and Social Survey 2009: Promoting Development, Saving the Planet, New York: United Nations, p. 44, http://www.un.org/en/development/desa/news/policy/wess-2009.shtml.*

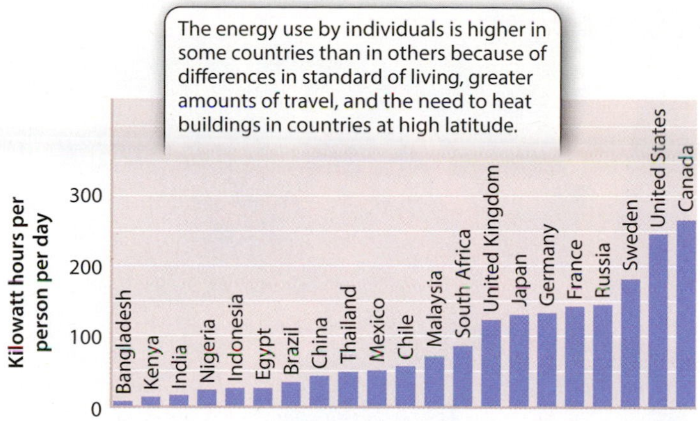

of humanity now exceeds the actual surface area of Earth. Taken together, then, trends in population size and the ecological footprint highlight both the remarkable technological successes of the past century and the challenges we face as citizens, policy makers, scientists, and engineers in the decades to come.

FIG. 49.3 Our ecological footprint. The graph shows the number of hectares needed to support an average person compared with a ranking of the Human Development Index, a measure of standard of living. *Source: European Environmental Agency (http://www.eea.europa.eu/data-and-maps/figures/correlation-of-ecological-footprint-2008) Data from Global Footprint Network, 2008; United Nations Development Programme, 2012.*

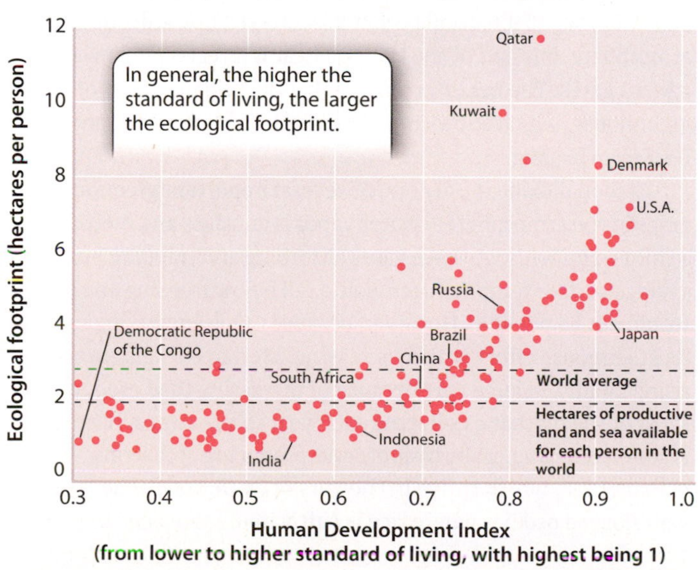

49.2 HUMAN INFLUENCE ON THE CARBON CYCLE

In Chapter 25, we discussed the carbon cycle, explaining how photosynthesis removes carbon dioxide from the atmosphere and respiration replaces it. The remarkable data plotted in the Keeling curve (see Fig. 25.1) show that atmospheric CO_2 levels have increased annually over the past 60 years, and analyses of air bubbles trapped in glacial ice show that this pattern of persistent increase began earlier, with the onset of the Industrial Revolution in the nineteenth century (see Fig. 25.3).

There is no longer any doubt that humans have become major contributors to the carbon cycle, particularly the part of the cycle in which ancient organic matter, or fossil fuel, is oxidized to carbon dioxide (CO_2), returning CO_2 to the atmosphere (**Fig. 49.4**). As discussed in Chapter 25, the amount of CO_2 we add to the atmosphere each year by burning fossil fuel is about 100 times that

FIG. 49.4 Human-generated sources of carbon dioxide. Every year, humans convert organic matter to carbon dioxide at rates far higher than those found in nature, predominantly by (a) burning fossil fuels and (b) biomass burning, as seen in this Amazon forest. *Sources: a. Tim McCaig/iStockphoto; b. Stockbyte/Getty Images.*

produced by all Earth's volcanoes together. And, as we also saw in Chapter 25, measurements over the past several decades of the relative abundances of ^{12}C, ^{13}C, and ^{14}C in atmospheric CO_2 make it clear that the CO_2 now being added to our air comes primarily from the combustion of fossil fuels (see Fig. 25.4). Clearing forests for agriculture generates still more carbon dioxide (Fig. 49.4). When land is cleared for agriculture, the existing vegetation is generally burned off, converting much of the organic carbon in biomass and soil organic matter to CO_2. No counteracting process removes CO_2 at comparable rates.

As atmospheric carbon dioxide levels have increased, so has mean surface temperature.

For the past century, scientists, sailors, and interested citizens have monitored temperature at weather stations around the world. More recently, satellites have enabled us to measure temperature in places as remote as the high Arctic and the middle of the ocean. The results are clear: In most parts of the world, mean annual temperature during the decade 1999–2008 was warmer than 1940–1980 averages (**Fig. 49.5**). In some places, the temperature change has been slight, but in others, especially at high latitudes, the increase has been as much as 2°C.

We can measure CO_2 levels in the atmosphere, and they are increasing. We can measure global temperature, and it is increasing. Is increasing CO_2 responsible for observed temperature changes? To address this question, we must understand that carbon dioxide is a **greenhouse gas**—that is, a gas that absorbs heat energy and then emits it in all directions. As shown in **Fig. 49.6,** solar radiation passes freely through the atmosphere, from top to bottom. Some incoming radiation is reflected from Earth's surface, and the rest is absorbed by the land and sea. In turn, some of the energy absorbed by Earth's surface is radiated back again as infrared radiation, or heat. Greenhouse gases in the atmosphere absorb the infrared radiation reflected up from Earth's surface and emit it in all directions—some is directed upward, out of the atmosphere, but half of the trapped heat is directed downward, toward Earth. The net effect is like that of the glass panes of a greenhouse, which allow sunlight into the structure but prevent heat from leaving.

Carbon dioxide is only one of several important greenhouse gases in the atmosphere—water vapor is another, and methane a third—and without these gases absorbing and trapping heat, average surface temperatures would fall below freezing and life would not be possible. However, because CO_2 is increasing rapidly, its greenhouse effect is also increasing. Methane levels are also rising rapidly, in large part because of increasing food production. Most of the methane delivered each year to the atmosphere is generated biologically by methane-producing archaeons (Chapter 26), which thrive in the guts of cattle and in the waterlogged paddies where rice is cultivated. More beef and

FIG. 49.6 The greenhouse effect. Carbon dioxide is a greenhouse gas: It traps heat from the sun, warming the planet. *Photo source: Fotosearch Stock Images.*

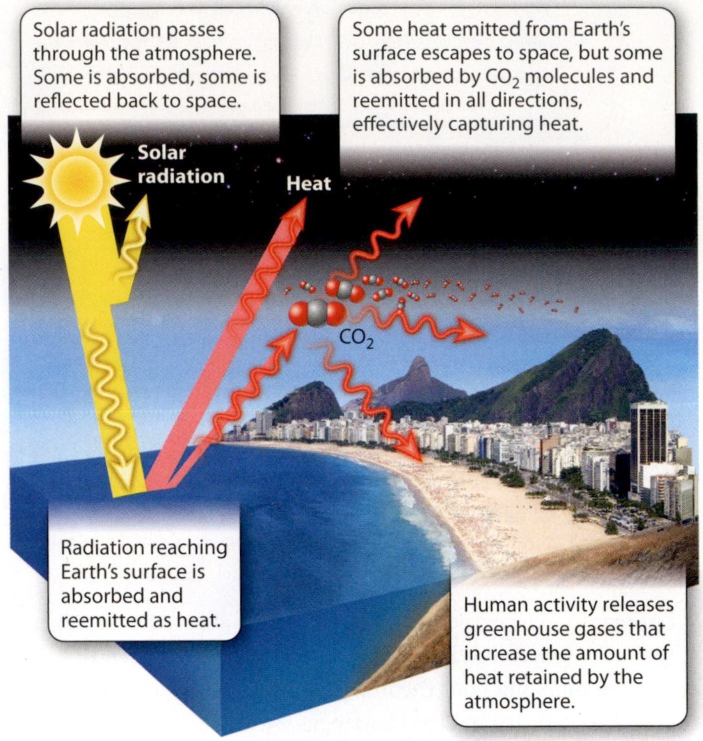

FIG. 49.5 Global warming. *Source: Data from R. A. Rohde, Global Warming Art, http://www.globalwarmingart.com/. Based on data from the Hadley Centre Coupled Model version 3 (HadCM3), http://badc.nerc.ac.uk/view/badc.nerc.ac.uk__ATOM__dpt_1162913571289262.*

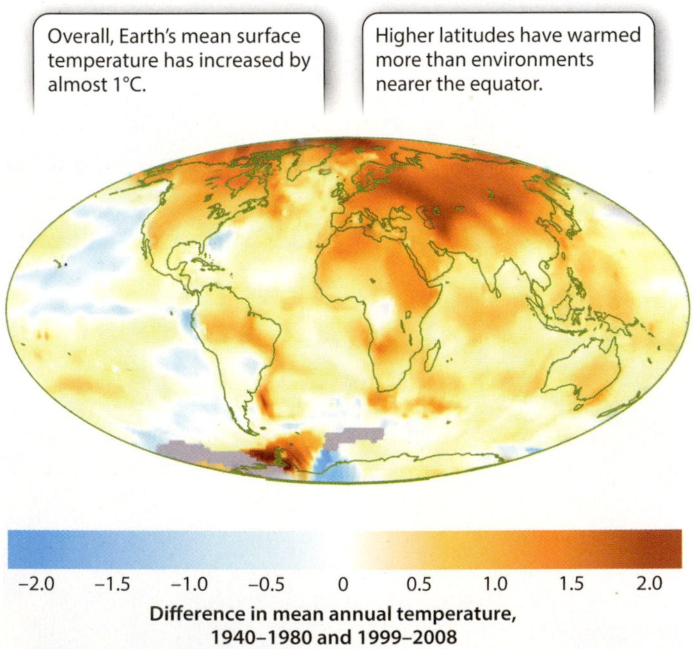

expanded rice paddies translate into higher rates of methane production. The thawing of permafrost at high latitudes releases additional methane that was trapped in frozen soils when the ice formed long ago.

Physics can help us determine how much atmospheric warming is due to increases in greenhouse gases. Each molecule of greenhouse gas absorbs and emits a specific amount of heat, and calculations show that the increases in atmospheric greenhouse gases measured over the past 50 years have increased the difference between incoming (solar) and outgoing radiation by about 2.5 watts per square meter. This difference is what adds heat to the oceans and atmosphere. Scientific consensus, reflected in reports from the Intergovernmental Panel on Climate Change, is that this greenhouse effect is the principal cause of observed twentieth-century temperature change. Scientific consensus also holds that global temperature will continue to rise as atmospheric CO_2 continues to increase in this century.

If this view is correct, human activities are changing the world. Can we, however, eliminate the possibility that the observed increases in greenhouse gases and temperature have natural causes? After all, as we saw in Chapter 25, the long-term geologic record indicates that climate and atmospheric composition have changed dramatically and repeatedly throughout our planet's history. Perhaps volcanic emissions have increased, driving the observed changes. However, measurements of the isotopic composition of atmospheric CO_2 effectively eliminate volcanic eruptions as a principal source of rising CO_2 (see Fig. 25.4).

Moreover, we can monitor the effects of volcanoes as they occur, and we can gauge the effects of past eruptions because volcanic ash accumulates along with the ice in continental glaciers. Historically, the major effect of large volcanic eruptions has been to *decrease* temperature because volcanic ash and aerosols reflect incoming solar radiation back into space. What's more, the impact of volcanic eruptions lasts only a few years. It doesn't drive the century-long temperature increase that we are currently observing.

What about the possibility that the amount of solar radiation entering the atmosphere has varied through time? The sun's output oscillates, and we know from direct measurements how solar radiation has varied during the past 50 years. The effect of this variation can be calculated, and it is small relative to greenhouse effects. It does not seem that variable solar radiation can account for the amount of temperature change that we have observed. In short, natural processes alone cannot explain the temperature increases observed in recent decades (**Fig. 49.7**).

As a result, there is overwhelming agreement among scientists that atmospheric CO_2 levels are rising, that temperature is increasing, and that the physics of greenhouse gases relates the two effects. Beyond these facts, however, it is difficult to say with certainty how human-induced global trends in climate will affect a given area.

Why don't we know for certain? Modeling future climatic conditions is difficult because of the many complex interactions that contribute to climate. Climate models are attempts to understand how climate works by fashioning equations that relate a simplified set of variables and interactions. Using mathematical models that accurately approximate current climate, scientists can change inputs into the model, adding CO_2 to the atmosphere, for example, or changing the extent of forest cover. The model then generates a set of results that can be used to predict future climate. All models must be checked against actual observations, and most are sensitive to assumptions made in constructing the model. That said, climate models do a pretty good job of explaining global-scale features of climate and climatic change.

Nevertheless, questions remain. How, in detail, might cloud cover change over the next century, and what effect would this change have on temperature and precipitation? What will be the effect of additional air pollutants, like the black carbon particles released into the air when coal or wood burns incompletely? Will black carbon increase warming by absorbing solar radiation, or decrease it by reflecting radiation back into space? What will be the quantitative effects of feedbacks such as the release of

FIG. 49.7 Two models for observed changes in mean global temperature over the past century. The model that incorporates both human and natural inputs (pink) closely matches the observed record (black), but the model based only on natural inputs (blue) does not. *Data from Meehl et al. (2004) Journal of Climate 17: 3721-3727, Fig. 2d.*

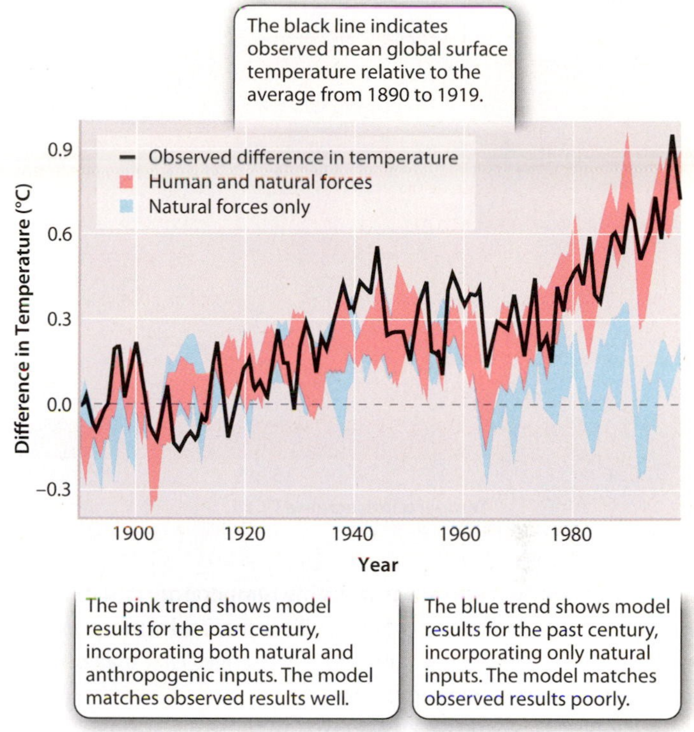

FIG. 49.8 A prediction for global warming by the end of the 21st century based on current rates of CO_2 increase. *Source: Adapted from R. A. Rohde, "Global Warming Predictions," from Global Warming Art, JPEG image, http://www.globalwarmingart.com/wiki/File:Global_Warming_Predictions_Map_jpg. Based on data from the Hadley Centre Coupled Model version 3 (HadCM3), http://badc.nerc.ac.uk/view/badc.nerc.ac.uk__ATOM__dpt_1162913571289262.*

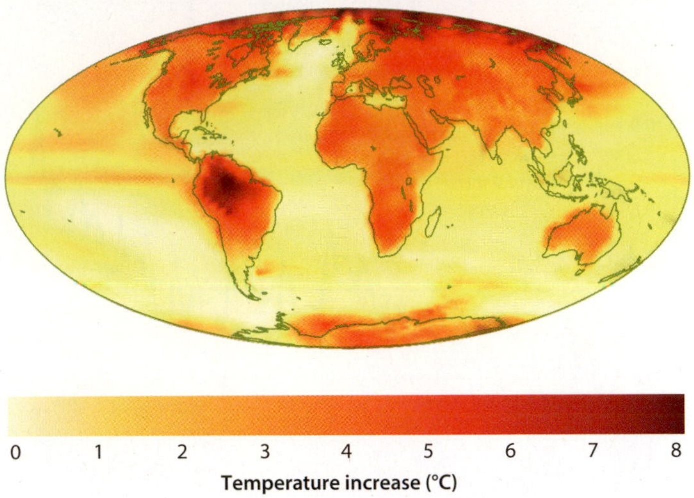

methane and carbon dioxide as warming temperatures melt permafrost at high latitudes? How will oceanic and atmospheric circulation patterns change, and will any changes that may occur enhance or dampen climatic change?

We still seek definitive answers for all these questions. Nonetheless, most climate models suggest that mean global temperature will increase 2°C–5°C during the 21st century (**Fig. 49.8**). That increase is not expected to occur uniformly throughout the globe. As has been true during the past 50 years, changing circulation patterns may cool some places and warm others. Likewise, rainfall is likely to increase in some areas, and decline in others.

As discussed in Chapter 48, climate can be considered as average weather over a long time interval. From one year to the next, however, weather is tremendously variable—there will be cold years and warm years, wet ones and dry ones. Because weather is so variable, evidence for climate change comes not from individual weather events but from records kept over decades. We can draw few conclusions from the single observation that May 2015 was Earth's warmest May on record, but the fact that May 2015 was the 361st month in a row that was warmer than its twentieth-century average tells us the world is warming.

Models predict that rainfall patterns should change as Earth warms. To test this, oceanographers analyzed 1.7 million measurements of seawater salinity taken over the past century. The oceans were chosen because they contain 97% of our planet's water and receive 80% of its rainfall. The salinity of surface seawater reflects both the addition of fresh water by rain, which decreases salinity, and evaporation, which increases it. For this reason, changing salinity can indicate whether the balance of rainfall and evaporation is shifting over broad regions of Earth. Consistent with many model predictions, wet areas of Earth are becoming wetter and dry regions drier.

→ **Quick Check 2** What is the difference between global warming and the greenhouse effect?

Changing environments affect species distribution and community composition.

What will be the consequences of climate change? Without question, some locations will benefit. For example, temperature increase in New England and Scandinavia will mean longer growing seasons. Other regions, however, will suffer. As precipitation patterns change, many places will become drier. Indeed, a number of climate models predict that some of the strongest declines in rainfall will occur in regions that currently produce much of the corn and wheat that feed the world. Already, farmers in southeastern Australia have experienced the worst droughts in a century, and with them unprecedented damage from brushfires. Over much of North America, 2012 was also the driest summer in many decades, and extreme drought returned in the western half of the continent in 2014. Such events are consistent with model predictions, but because of the great natural variability of weather, they do not by themselves confirm climate models. Scientists will be watching carefully to see whether such extreme events increase in frequency in coming years.

To understand how different species will respond to global climate change, we return to basic principles of ecology and evolution. Natural selection describes the changes in allele frequencies that result from the interactions between individuals of a population and their environment. We tend to think about natural selection in terms of individuals' adapting to a constant or slowly changing environment. In the 21st century, however, the environment will probably be a rapidly moving target, and so genotypes that conferred high fitness in the past may not be the most advantageous in the future.

Plants illustrate the complicated ways in which organisms respond to environmental change. Meticulous notes taken during the 1840s by Henry David Thoreau provide an inventory of plant species diversity in woodlands west of Boston, Massachusetts. Climate records show that mean annual temperature in this area has increased by about 2.5°C since Thoreau began his survey, and in response, many but not all species now flower a week earlier than they did in the nineteenth century (**Fig. 49.9**). A number of species have declined in abundance or even disappeared since Thoreau's time, and these tend to be the plants least able to change their flowering time (Chapter 30). Earlier onsets of leaves, flowers, and fruit have been documented in locations distributed widely throughout the temperate zones of North America, Europe, and Asia. The Massachusetts example highlights the simple but important observation that different plant species respond in distinct ways to changing climate, favoring some populations over others.

FIG. 49.9 Plant responses to warming temperatures. Observations over a century and a half show that (a) the time of first flowering has become earlier as (b) regional temperatures have become warmer in the woods of Concord, Massachusetts. *Source: Data from A. J. Miller-Rushing and R. B. Primack, 2008, "Global Warming and Flowering Times in Thoreau's Concord: A Community Perspective," Ecology 89(2):332–341, doi: 10.1890/07-0068.1.*

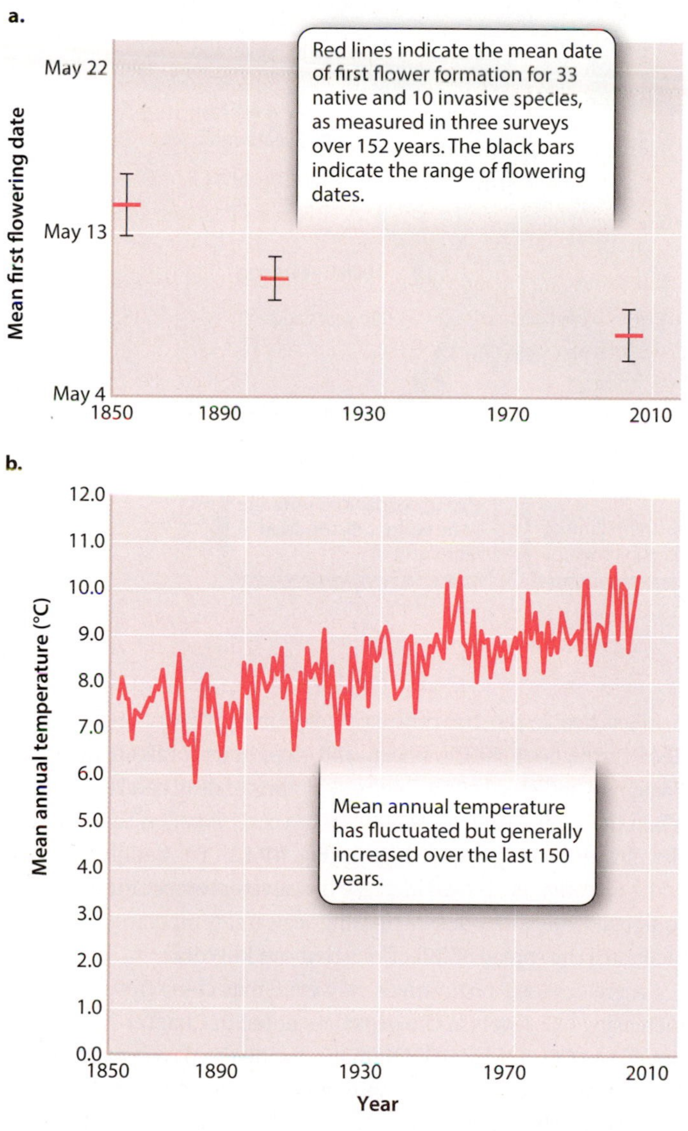

➔ **Quick Check 3** Why should the capacity to shift flowering time influence the distribution of a species during the 21st century?

The example of Thoreau's woods shows that as climate changes, some plants can modify flowering time—a physiological response to environmental cues that now occur earlier in the year. Experimental evidence indicates that many plants also show a direct growth response to elevated CO_2. Where nitrogen in the soil is abundant, increased carbon dioxide levels tend to stimulate growth, providing a negative feedback mechanism that might help to mitigate the effects of anthropogenic CO_2 emissions. In most terrestrial environments, however, soil nitrogen levels are low, limiting any growth response to elevated CO_2. Moreover, increased drought stress and more frequent outbreaks of insect pests may counteract any capacity for accelerated growth.

Plants grown in air with CO_2 levels like those predicted for the end of this century often change the way that they distribute the products of photosynthesis to different organs: The amount of nitrogen in plant tissues commonly declines, as does the proportion of resources devoted to reproduction. Again, this doesn't reflect genetic change, just a physiological response to altered environmental conditions. But if environmental change persists for a long time, the varying abilities of different plants to respond physiologically will result in changing allele frequencies, as some variants survive and reproduce better than others. That is, plant populations may evolve by natural selection in response to climate change.

Evolution is one response to global change, but, given the pace at which atmospheric composition and climate are changing, many populations may not have time to adapt. In this case, populations will either migrate or become extinct. Fossils deposited as Earth's climate warmed at the end of the last ice age show that many plant species dramatically changed their geographic distributions in response to changing climate. As discussed in Chapter 30, trees don't move, but they can migrate by dispersing seeds to new areas. For example, between 18,000 and 8000 years ago, Red Pine and Jack Pine migrated thousands of kilometers to the north from the Gulf of Mexico, where they lived during the ice age, to their current distributions in the northern United States and Canada (**Fig. 49.10a**). Hickory also migrated northward, expanding its range rather than simply shifting from one place to another as ice retreated and climate warmed (**Fig. 49.10b**).

Migration is a potentially important response to environmental changes projected for the 21st century, but it requires a continuous route to get from one place to another, and it is a challenge to find these direct paths on continents broken up by agricultural lands and cities. For this reason, some biologists advocate "assisted migration": the deliberate transplantation of plant populations from existing habitats to new ones more favorable to growth. To make sure that assisted migration does more good than harm, we need to do more research to understand how plants grow and compete in new ecological situations and under different environmental conditions.

? CASE 8 BIODIVERSITY HOTSPOTS: RAIN FORESTS AND CORAL REEFS

How has global environmental change affected coral reefs around the world?

Increasing atmospheric CO_2 is not just influencing life on land. It is also having an observable impact on organisms in the oceans, not least on coral reefs that provide hotspots of biological diversity in the marine realm. The Great Barrier Reef is visible from space, its wave-washed necklace of limestone stretching more than

FIG. 49.10 Plant migration in response to climate change. Records of fossil pollen over the 13,000 years since the last ice age document (a) the northward migration or (b) expansion of selected tree species. *Source: After M. B. Davis, 1976, "Pleistocene Biogeography of Temperate Deciduous Forests," Geoscience and Man 13:13–26.*

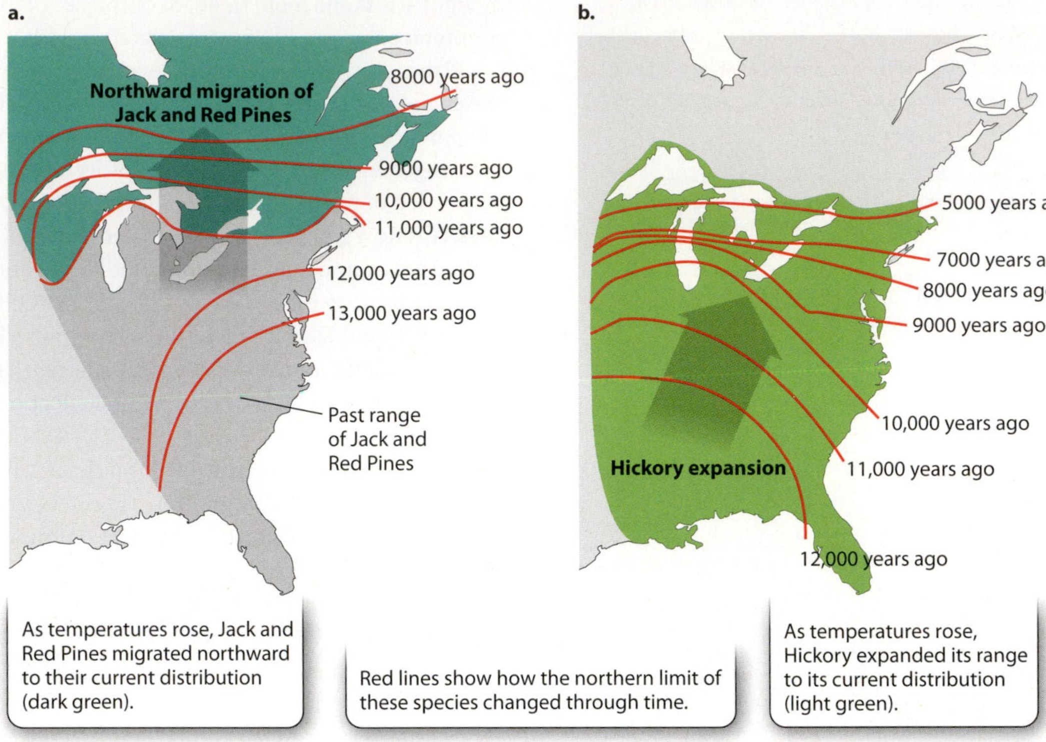

2000 km along the northeastern coast of Australia (**Fig 49.11**). A global diversity hotspot, the reef's 1500 fish species (a tenth of all known fish), more than 4000 species of clams and snails, and myriad other species prompt basic questions about ecology and diversity, many of which we have addressed in previous chapters. Lately, however, one question has come to dominate scientific discussion of the Great Barrier Reef: Will it exist a century from now?

Corals are the principal architects of the Great Barrier and most other reefs (Chapter 44), and so coral biology is the key to understanding both the reef's unusual diversity and its vulnerability in the face of climate change. Most reef corals gain nutrition from unicellular algae that live within their tissues. The algal symbionts in reef corals provide food for their host, in return receiving nutrients, a means of waste disposal, and a stable environment. It is the efficient cycling of carbon, nitrogen, and phosphorus between corals and their symbionts that underpins the biological richness of coral reefs. Reef corals accomplish another noteworthy feat: They make skeletons of calcium carbonate ($CaCO_3$) that through time build the reef's three-dimensional framework.

The very processes that facilitate reef growth—biomineralization and microbial symbiosis—are exposing their Achilles' heel in the 21st century. Today, on the Great Barrier Reef, throughout the Caribbean, and elsewhere, corals are dying. Bleaching, indicated by the white skeletons of dead corals, occurs when the symbiotic algae that feed the corals abandon their hosts, thereby sentencing them to death (**Fig. 49.12**). The details are still being investigated, but an increase in seawater temperature appears to signal the algal exodus. Increasing seawater temperature also facilitates the spread of infections that can kill coral.

A greater long-term threat, however, may come from increasing CO_2 levels in the ocean. As noted in Chapter 25, not all the CO_2 produced by humans during the past century has accumulated in the atmosphere: About one-third has been absorbed by the ocean. That's good news for slowing climate change because it means there is less CO_2 in the atmosphere to trap heat, but it is potentially bad news for marine habitats because CO_2 dissolves in seawater to form carbonic acid (H_2CO_3) (Chapters 6 and 39). Unlike climate change, the effects of increasing CO_2 on ocean pH are straightforward: Increasing the abundance of CO_2 in the oceans causes the pH of seawater to go down, a phenomenon known as **ocean acidification.**

In the past 50 years, as the concentration of CO_2 in the atmosphere has grown from about 315 ppm to about 400 ppm, the pH of surface oceans has correspondingly dropped by 0.1 unit

FIG. 49.11 **The Great Barrier Reef.** The world's largest coral reef, lying off the coast of Australia, is visible from space. *Source: NASA.*

FIG. 49.12 **Coral bleaching.** This coral shows a pattern of whitening that occurs when corals lose their algal symbionts. *Source: Scubazoo/Science Source.*

FIG. 49.13 **Trends in atmospheric CO_2 levels (red), seawater CO_2 levels (blue), and seawater pH (purple) in the central Pacific Ocean.** *Data from: Doney, S.C. et al., Ocean Acidification: The Other CO_2 Problem. Annual Review of Marine Science 1: 169–192.*

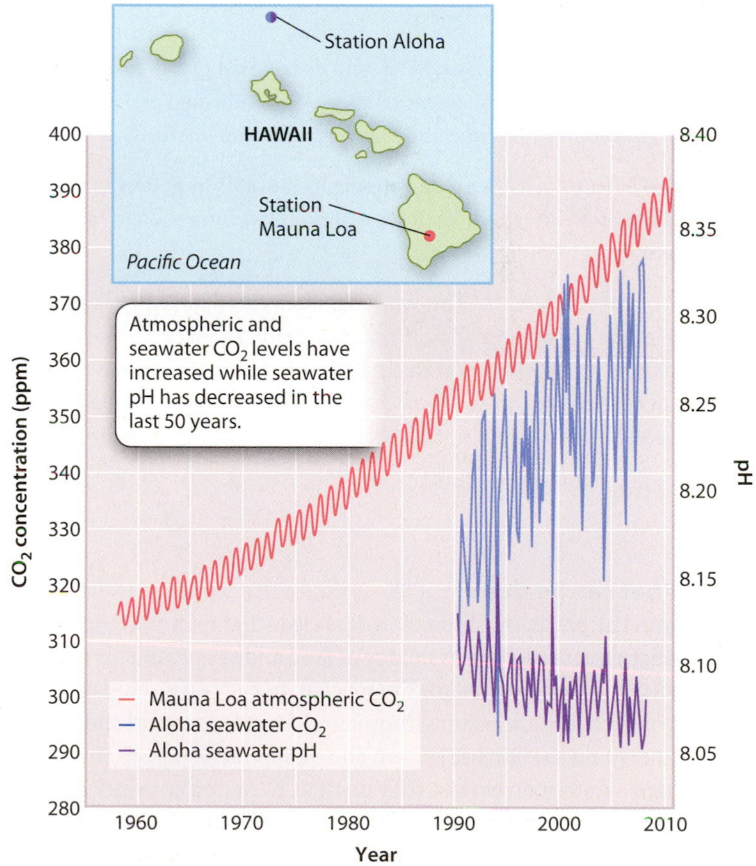

(**Fig. 49.13**). That may not seem like a lot, but remember that the pH scale is logarithmic, and so the 0.1 unit drop represents an increase in hydrogen ions of about 30%. In turn, ocean acidification causes carbonate ions in seawater to decrease, making it more difficult for some marine algae and corals to build their $CaCO_3$ skeletons (**Fig. 49.14**).

Essentially, all attempts to model our environmental future predict that, by the end of this century, atmospheric CO_2 will rise from its current level to more than 500 ppm, and changing pH in many parts of the ocean will limit the ability of corals to build skeletons. Some corals can survive this change. For example, experiments on two Mediterranean corals showed that at a pH level 0.8 units below that of present-day seawater, the corals grew naked—essentially as sea anemones—and recovered their skeletons upon return to normal seawater. But, if ocean acidification persists over timescales much longer than coral generations, the structural framework for tropical reefs may be irrevocably lost.

FIG. 49.14

What is the effect of increased atmospheric CO_2 and reduced ocean pH on skeleton formation in marine algae?

BACKGROUND It is well established that atmospheric CO_2 levels are increasing, which in turn decreases the pH of ocean water. In the late 1990s, experiments showed that a decrease in pH affected the ability of some marine organisms to build skeletons made of calcium carbonate ($CaCO_3$). The German biologist Ulf Riebesell and his colleagues carried out experiments to investigate whether ocean acidification affects algae called coccolithophorids, which account for the majority of carbonate precipitation in the open ocean.

HYPOTHESIS From the effects of decreased pH on $CaCO_3$ skeleton formation in other marine organisms, Riebesell hypothesized that increasing CO_2 would interfere with skeleton formation in coccolithophorids.

EXPERIMENT Riebesell and his colleagues studied two species of coccolithophorids, *Emiliania huxleyi* and *Gephyrocapsa oceanica*. They grew both species in the laboratory under conditions of increasing CO_2, from pre-industrial levels, to present-day levels, and then to levels predicted for the future. They measured pH and rates of calcification and used scanning electron microscopy to observe skeleton formation.

RESULTS With increasing CO_2 and decreasing pH, they observed decreasing rates of calcification as well as deformed and incomplete coccolithophorids under the scanning electron microscope.

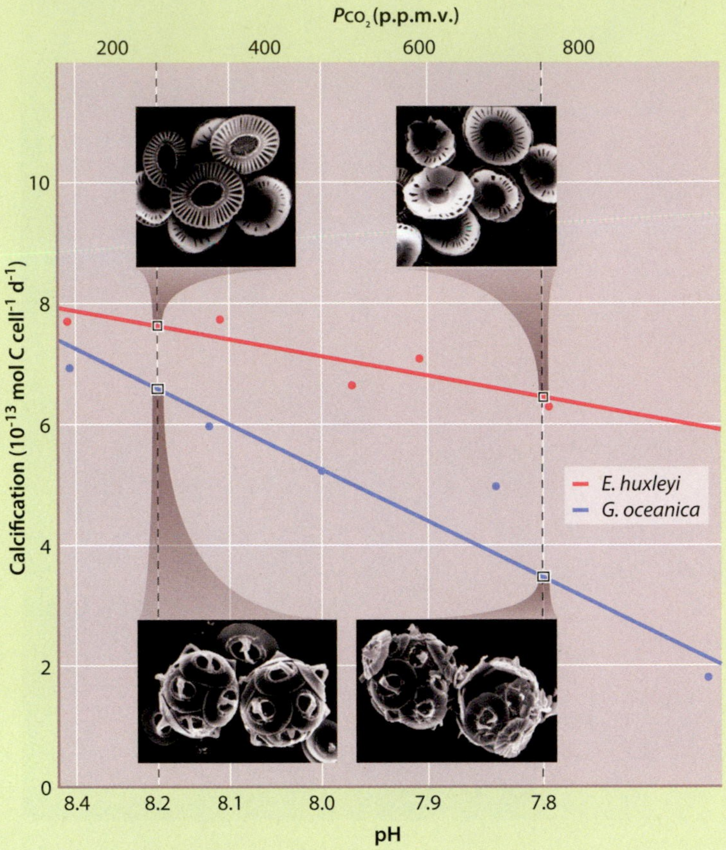

CONCLUSION The results support Riebesell's hypothesis: Ocean acidification interferes with normal skeleton formation in marine plankton.

FOLLOW-UP WORK This work has been extended to investigate the effects of ocean acidification on skeleton formation in corals. For example, the Mediterranean coral *Oculina patagonica* grows normally at pH 8.2. When grown at pH 7.4, however, the coral skeletons dissolve. Upon restoration of normal conditions, the corals resume skeleton formation. This experiment shows that corals might survive ocean acidification, although the ecosystem benefits provided by reefs would be lost.

SOURCES Riebesell, U., et al. 2000. "Reduced Calcification of Marine Plankton in Response to Increased Atmospheric CO_2." *Nature* 407: 64–367; Fine, M., and D. Tchernov. 2007. "Scleractinian Coral Species Survive and Recover from Decalcification." *Science* 315:1811–1811.

What can be done?

From the preceding discussion, it is clear that rising CO_2 has both direct effects on the physiology of land and sea organisms and indirect effects related to climate change and ocean chemistry. In order to ensure our environmental future, scientists must understand the climate system and biological responses to global change much better than they do now. Because the principal source of rising CO_2 is the burning of fossil fuels, the solutions we develop as citizens will clearly focus on energy—how we obtain it and how we use it. As the global population continues to increase and people aspire to higher standards of living, human impact on the carbon cycle will continue to grow, at least if we continue business as usual. Business as usual, however, may not be a viable option for the long term because population growth may simply outstrip resource availability, especially for petroleum. Ultimately, our success in both slowing rates of climate change and sustaining

FIG. 49.15 Alternative energy sources, including (a) solar and (b) wind energy. *Sources: a. euroluftbild.de/dpa/Corbis; b. Travel Pix/Robert Harding.*

energy supplies will depend on our ability to use energy more efficiently and generate new forms of power, most likely from the sun, wind, and nuclear sources.

Half a century ago, geologist M. King Hubbard predicted from patterns of oil use and the rate at which new oil fields are discovered that petroleum production in the United States would peak in the 1970s and decline thereafter. Global production, he suggested, would peak later, in the early decades of the 21st century. As King predicted, conventional petroleum production in the United States peaked in 1972 and has fallen since then to about half its peak rate of production. On the other hand, unconventional petroleum resources, made available in no small part by hydraulic fracturing, or fracking, is expected to restore total petroleum production in the USA to near-1970s levels by 2020. (Fracking is a technique in which large volumes of fluid and sand are injected into shale rock, fracturing the rock and thereby releasing trapped oil or gas.) After 2020, petroleum production will resume its decline, both in the United States and globally. Coal, natural gas, and unconventional reserves such as the oil sands found widely in western Canada can quench the world's thirst for fossil fuels for another century or more, and biofuels—fuels generated from plant biomass—may also help conserve petroleum, but the energy produced will be expensive, and climate change will continue.

In particular, natural gas (methane, CH_4) looms large in estimates of 21st century energy generation. Fracking has revolutionized natural gas extraction; in the United States alone, shale gas production has increased by 700% since 2007. Of all fossil fuels, natural gas generates the most energy for each mole of CO_2 produced by combusting it; thus, generating electricity by natural gas has advantages over burning coal. CO_2 emissions do continue, however, and gas leaks from expanding wells and pipelines add more methane, a strong greenhouse gas, to the atmosphere. Scientists and citizens debate whether the toxic fluids used in fracking will contaminate ground water in regions where shale gas is produced.

Like fossil fuels, alternative energy sources—wind, the sun, tidal energy, and nuclear power (**Fig. 49.15**)—all have important pros and cons, not the least of which, in the case of nuclear power, is the problem of long-term storage of radioactive waste. On the other hand, all these sources have the potential to benefit humanity at least three ways: by reducing the amount of CO_2 emitted per kilowatt of energy generated, by conserving finite natural resources, and by helping to contain the cost of energy production. The development of more fuel-efficient cars, airplanes, and buildings will also help greatly to maintain energy-intensive lifestyles while decreasing their environmental consequences. As individuals, we can choose energy-efficient transportation and appliances and develop habits that decrease energy waste. When one person turns off the light in an empty room, the energy saved is tiny, but when millions of people do it, the benefits add up.

Wise choices by citizens and consumers can slow the growth of atmospheric CO_2, but the CO_2 already in the atmosphere isn't going to go away quickly. As noted earlier, we have dramatically increased the rate at which CO_2 is added to the atmosphere, but we have not changed the rate at which it is removed. The geologic processes that control atmospheric CO_2 work on longer timescales. Therefore, even if we could cut CO_2 emissions by 90% tomorrow, CO_2 would not return to nineteenth-century levels for hundreds or even thousands of years.

Another possibility is to actively remove CO_2 from the air. Reforestation of previously cleared landscapes removes carbon from the atmosphere because plants build biomass from CO_2 during photosynthesis (Chapter 8). It is also possible to capture CO_2 as it rises upward through smokestacks, although it is currently expensive to do so. We do not yet know how to scrub CO_2 directly from the atmosphere, but if we could do so at reasonable cost, it would apply a significant brake to climate change.

Clearly, forestry and 21st-century technology have potentially important roles to fill in undoing the consequences of the past hundred years. For the most part, however, gaining control over our environmental future will entail wise personal and governmental choices about how we obtain and use energy.

49.3 HUMAN INFLUENCE ON THE NITROGEN AND PHOSPHORUS CYCLES

Human activities influence the environment not because we operate outside the carbon cycle, but because we have become an important part of it. Our interest in improved crop production also results in unprecedented human impact on the nitrogen and phosphorus cycles. Once again, environmental changes tied to human nitrogen and phosphorus use are altering the ecological landscape for species in lakes and the sea.

Nitrogen fertilizer transported to lakes and the sea causes eutrophication.

As discussed in Chapter 26, all organisms require nitrogen to synthesize proteins, nucleic acids, and other molecules. Although nitrogen is plentiful in the atmosphere as N_2 gas, this form is not biologically available to animals and plants. Nitrogen-fixing bacteria and archaeons convert N_2 to biologically available forms of nitrogen, called fixed nitrogen, such as ammonia (NH_3) and nitrate (NO_3^-). Plants then obtain nitrogen from ammonia and nitrate in the soil, and animals get the nitrogen from the food they eat. Crops, harvested year after year in the same fields, deplete available nitrogen, and so high and sustained yields require that farmers add biologically available nitrogen to the soil as fertilizer.

Through the use of fertilizer and other activities, humans add about 140 million tons of fixed nitrogen to the biosphere each year, an amount comparable to that contributed by microbial nitrogen fixation. The majority of this fixed nitrogen is created industrially by reacting nitrogen gas and hydrogen in the presence of a catalyst to produce ammonia. Most commercially produced ammonia is further processed to nitrate, which is spread across fields as fertilizer to increase crop production. Success in feeding the world's people owes much to fertilization with nitrates, but in fact, only about 10% of the nitrogen added to croplands ends up in food. Much of the nitrate fertilizer leaves fields as surface runoff and travels by rivers to lakes or the ocean. Denitrifying bacteria also use some of the added nitrate for respiration, returning N_2 to the atmosphere—and also generating N_2O, yet another potent greenhouse gas.

In lakes and oceans, the fertilizing effects of agricultural runoff are finally realized, but in unintended ways. The added nutrients lead to a great increase in the populations of algae and cyanobacteria in a process called **eutrophication.** These algal and cyanobacterial masses eventually sink to the bottom, where heterotrophic organisms, mostly bacteria, feed on them, fueling high rates of aerobic respiration. The bacteria's demand for oxygen can completely deplete O_2 in bottom waters, with potentially catastrophic consequences for animal life on the seafloor or lakebed.

In the Gulf of Mexico, nutrients leached from croplands and transported down the Mississippi River result in an ominously named dead zone of oxygen-depleted bottom water. This dead zone expands seasonally from the mouth of the river west toward Texas, covering an area about the size of New Jersey (**Fig. 49.16**). Within the dead zone, fish and seafloor invertebrates die in masses. Several hundred dead zones have been identified in coastal oceans around the world, and most are expanding year by year. Those in the Gulf of Mexico and the Baltic Sea are among the largest.

Nitrogen applied as fertilizer but spread by runoff and streamflow can affect the full range of ecological interactions discussed in Chapters 46 through 48. We benefit tremendously from increased crop production but pay a price in the eutrophication of lakes and oceans. We pay another price as well—as much as 1% to 2% of all fossil fuel consumption goes to make nitrogen fertilizer.

Phosphate fertilizer is also used in agriculture, but has finite sources.

The other major nutrient added to fields as fertilizer is phosphate (PO_4^{3-}), and here the problem is a different one. Like nitrate, much of the phosphate added to fields as fertilizer leaves as runoff, and phosphate runoff can be an important contributor to eutrophication. The Everglades of south Florida provide a good example. This ecosystem is built of plants that have adapted to low phosphate levels. However, phosphate levels in parts of the Everglades have increased up to a hundredfold. Chemical tracers show that these increased phosphate levels stem from phosphate fertilizer used on agricultural lands to the north. As a consequence, phosphate-loving cattails and other introduced plants are expanding at the expense of native plants, decreasing the diversity of this remarkable ecosystem.

There's no question that overuse of phosphates has detrimental effects on certain natural ecosystems, and yet we need

FIG. 49.16 **A dead zone, visible in this NASA satellite image as billows of green.** *Source: Image courtesy of Liam Gumley, Space Science and Engineering Center, University of Wisconsin–Madison and the MODIS Science Team.*

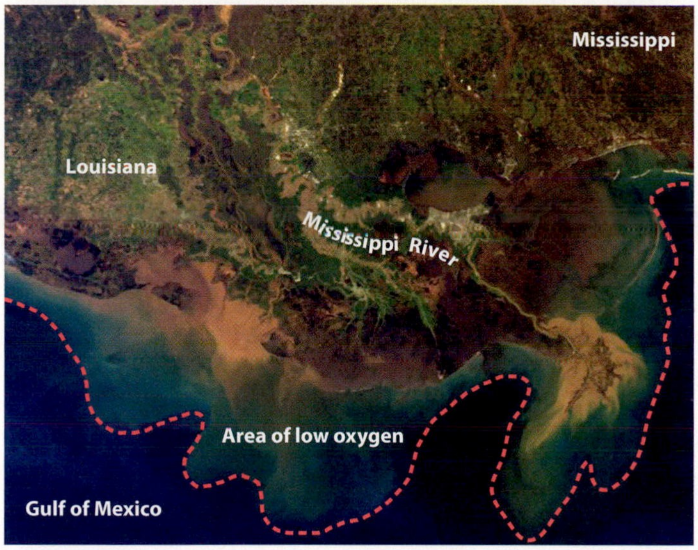

to be concerned that we will eventually run out of phosphates to use as fertilizers. Unlike for nitrate, there is no industrial process to produce phosphate. Thus, the only source of phosphate fertilizer is mining. Only a few places in the world have sedimentary rock with phosphates in high enough concentration to be worth mining, and these stores are being depleted.

Globally, phosphate production peaked about 1990, raising a serious question of how the world's need for phosphate fertilizer will be satisfied in the future. In many ways, the problem of phosphate availability parallels that of petroleum production. As abundant sources become depleted, we will have to mine poorer and poorer resources, raising production costs and ultimately calling into question our capacity to sustain agricultural production.

Human encroachment on the nitrogen cycle and long-term phosphate availability, then, converge on a single question: Can we manage future agricultural production in ways that feed the growing human population in a sustainable fashion?

What can be done?

The Green Revolution in agriculture began in the 1940s with the introduction of high-yield, disease-resistant varieties of wheat to Mexico (Case 6: Agriculture). In subsequent decades, new strains of corn, wheat, and rice, widespread application of industrial fertilizer, and the spread of farm machinery increased global crop production to levels that could scarcely have been imagined at the beginning of the twentieth century. As population growth has accelerated, however, the need to grow ever more food remains strong. And other factors, notably rising meat consumption in many countries and the increasing use of corn to generate biofuel, now place still greater demands on global agriculture. It is estimated that, over the next decade alone, global corn and wheat production will have to increase by 15% to meet demand.

How can we meet the needs of a growing population? Conventionally, agronomists see two options: We can devote more land to crops, or we can increase the yield of fields already in place. **Fig. 49.17** shows that globally both the amount of land devoted to growing corn and corn yield have increased steadily over the past 50 years. It is possible to increase yield further, as the yield per hectare in the United States, Canada, China, and western Europe far exceeds that of most other regions. Increasing yields, however, will require increasing the use of fertilizer and the use of fossil fuels to power

FIG. 49.17 **Corn yield and area in the United States and globally from 1961 to 2007.** Global corn yield and area have increased at about the same rate, whereas U.S. corn yield has risen much more quickly than corn area because of energy-intensive agriculture and use of fertilizer. *Source: Data from M. D. Edgerton, 2009, "Increasing Crop Productivity to Meet Global Needs for Feed, Food, and Fuel," Plant Physiology 149:9, http://www.plantphysiol.org/content/149/1/7.*

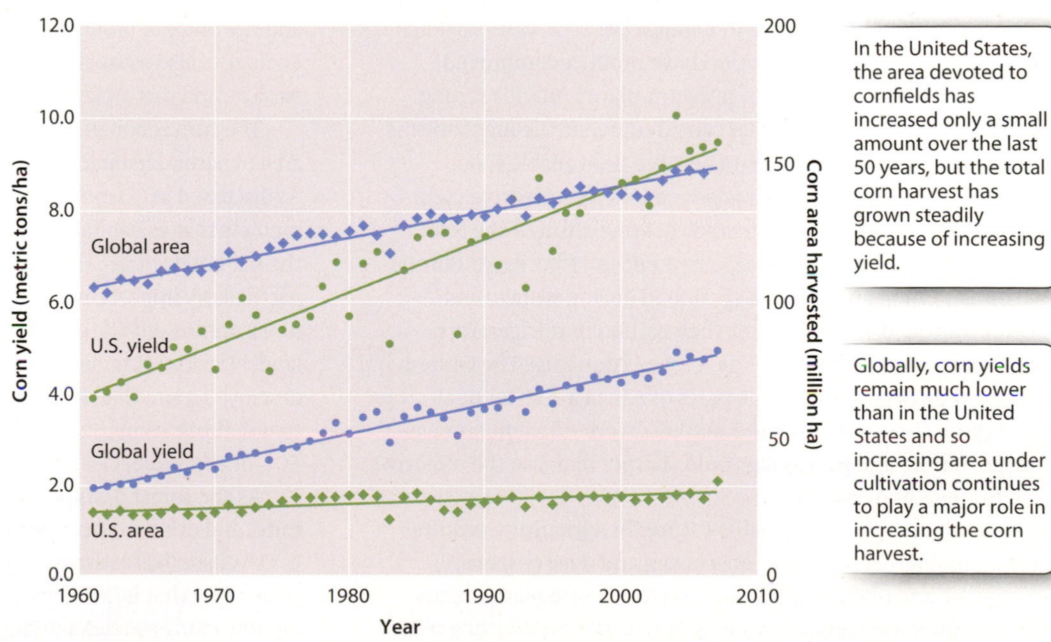

farm machinery, with all the attendant problems of climate change and eutrophication. Increasing crop area also comes at a cost, as forests and natural grasslands go under the plow, decreasing the biological storage of carbon and threatening biodiversity.

For more than a century, biologists have developed improved varieties of crop plants by breeding, the highly effective practice that Darwin called "variation under domestication" and used to frame his arguments about natural selection. Today, we have another tool available: genetic engineering. Biologists can now introduce genes for desirable traits into crop plants, improving their yield, disease resistance, or nutritional value (Chapter 32). The potential benefits of genetically modified crops are obvious, but some biologists see risks as well, many of which are detailed in Chapter 12.

Conventional breeding and genetic modification may improve crop yields, but, in a world of changing climate and evolving pests, these gains may have something of a "Red Queen" flavor to them (Chapter 42). In Lewis Carroll's *Through the Looking-Glass,* the Red Queen tells Alice, "Now, here, you see, it takes all the running you can do, to keep in the same place. If you want to get somewhere else, you must run at least twice as fast as that!" Much of evolution appears to work this way: As the physical or biological environment changes, continued adaptation is needed just to maintain fitness. Strains of wheat, modified to improve drought tolerance, may simply maintain high yields in a changing environment, not increase them. Increases in yields through genetic modification may be possible, but limited. Thus, massive fertilization by nitrate and phosphate fertilizers will continue to be a key component of agriculture.

Another type of biological intervention, the reduction of pests, may also help increase crop yields. We are not the only species with a nutritional interest in croplands. Fungi, bacteria, protists, nematodes, insects, birds, and mice all target cornfields, sharply reducing yield. In the United States alone, such unwanted guests cause about $1.5 billion worth of damage per year. Conventional breeding and genetic modification have produced improved resistance to pests, but fungus-resistant plants provide strong selective pressure for fungi that can circumvent the mechanisms of resistance, gaining access to otherwise unavailable food.

Food spoilage after harvest also reduces the effective yield of agricultural lands. It is estimated that one-third of the food grown in the United States is never eaten, and figures for Europe are similar. Fungi and bacteria spoil food during storage and transportation, on supermarket shelves, and in refrigerators. The problem also exists in less developed countries. The United Nations estimates that in some parts of Africa a quarter of all crops spoil before they ever reach the market. More efficient storage, transportation, and packaging could sharply increase the effective yield of fields and pastures.

As with energy, a sustainable future for agriculture requires creative biologists, ingenious engineers, and wise citizens. Geneticists and physiologists can help to develop plant breeds that improve crop yield while using less fertilizer. Working with microbiologists, engineers can discover new ways of storing and transporting food that increase the likelihood that it will be eaten before it spoils. As individuals, we make daily choices about the food we eat. We can choose to eat more plant products, for example. The basic ecological logic of the trophic pyramid (Chapter 47) tells us that a hectare of grain fed directly to humans feeds more people than the meat of cattle fed on the same crops. We can also choose to eat locally grown products where possible, saving energy while decreasing the likelihood of spoilage.

The entire carbon cycle, including its interactions with the N and P cycles and human impacts, is summarized in **Fig. 49.18,** pages 1086–1087.

49.4 HUMAN INFLUENCE ON EVOLUTION

It has sometimes been claimed that, as humans exert increasing influence on nature, evolution will grind to a halt. Nothing could be further from the truth. As examples in the preceding sections show, what is changing, and what will continue to change as the human footprint grows, is the selective landscape within which evolution operates. Some species, for example rats and cockroaches, thrive in the urban environments we have constructed for our own benefit. Others will expand or decline as we replace prairies and forests with crop- and pastureland. What are our conscious and inadvertent effects on evolution and biodiversity?

Human activities have reduced the quality and size of many habitats, decreasing the number of species they can support.

As noted in Chapter 1, it has been estimated that humans make use of nearly 25% of the entire photosynthetic output on land through the crops and livestock we harvest, the biomass we burn, and the primary production we eliminate when we clear-cut forests. Ecologists also estimate that crops cover 10% to 15% of Earth's land surface, and pastureland another 6% to 8% (**Fig. 49.19**).

The conversion of natural ecosystems to cropland decreases the area of natural habitats, as does the expansion of cities and towns. As discussed in Chapter 46, the theory of island biogeography predicts that as habitable area decreases, species diversity within the habitat declines. Habitat loss in tropical rain forests poses a particularly important threat to biodiversity (**Fig. 49.20**). These forests represent the greatest concentrations of species diversity on land, but clear-cutting to make room for grazing and croplands has destroyed more than half of their pre-industrial area. It is estimated that deforestation over the next century could eliminate 40% to 50% of all tree species in the Amazon forest.

As the forest disappears, so, too, will untold numbers of insect, mite, and other animal species. The renowned biologist E. O. Wilson has estimated that Earth is losing 0.25% of its species annually—that is, 5000 to 25,000 species per year, depending on how many species actually exist. And even where ecosystems

FIG. 49.19 **Global agriculture land.** Through time, land under cultivation has expanded, diminishing natural landscapes and the species diversity they support. *Source: After Millennium Ecosystem Assessment, 2005, Ecosystems and Human Well-being: Synthesis, Washington, DC: Island Press).*

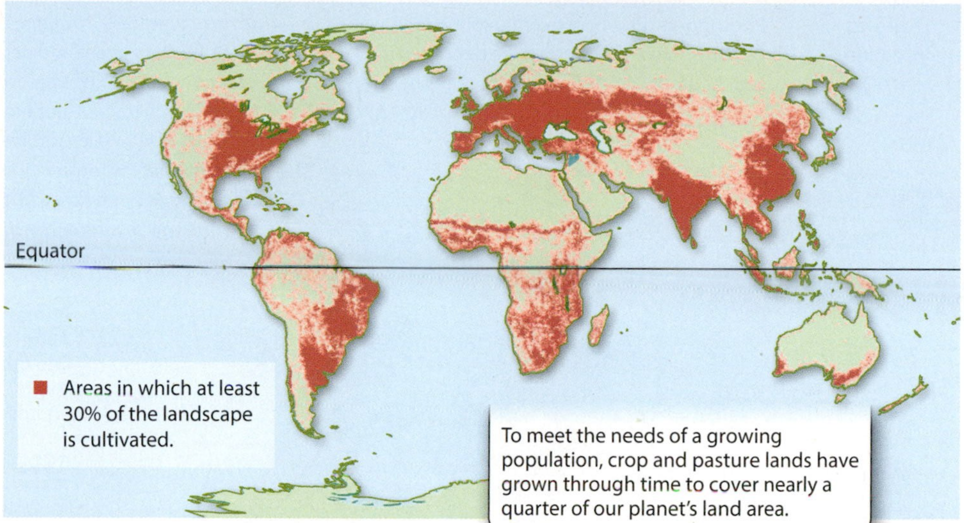

remain physically intact, pollution can erode populations. For example, recent studies in Europe show that levels of neonicotinoid pesticides in soil and water accurately predict patterns of population decline in bees and birds.

In the words of another distinguished biologist, Paul Ehrlich, "The fate of biological diversity for the next 10 million years will almost certainly be determined during the next 50–100 years by the activities of a single species."

Overexploitation threatens species and disrupts ecological relationships within communities.

For thousands of years, humans have harvested protein from the sea, tapping a resource that long seemed both free and endless. Citizens of Newfoundland, however, know better. Journals from John Cabot's voyage of 1498 tell of cod so abundant that a person "could walk across their backs," and for two centuries cod supported the Newfoundland economy. The amount of cod caught each year increased markedly in the 1960s—surveys estimated that as much as 60% of adult cod was being harvested. By the late 1980s, however, harvests had plummeted

FIG. 49.20 **Forests under threat.** (a) The map shows the extent of deforestation worldwide. (b) This desolate scene in Amazonia was once an expanse of tropical rain forest. Nearly 20% of the Amazon rain forest has been cleared for crops, pasture, or mining. *Photo source: AP Photo/Andre Penner.*

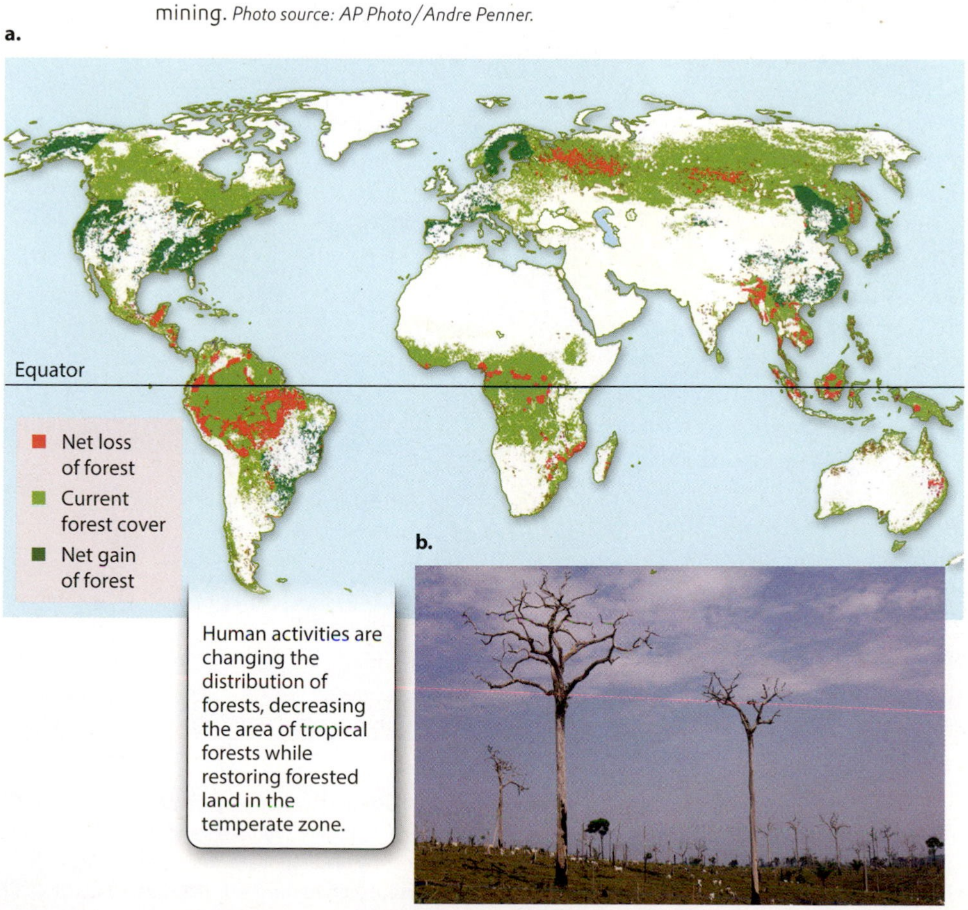

VISUAL SYNTHESIS

Flow of Matter and Energy Through Ecosystems

FIG. 49.18

Integrating concepts from Chapters 47–49

Terrestrial carbon cycle

Carbon (C) enters the cycle through photosynthesis and is transferred through food webs by consumers. Decomposers complete the return of carbon to the atmosphere as CO_2.

Terrestrial phosphorus cycle

Phosphate ions released from rocks by weathering are incorporated into biomass by plants. Phosphorus (P) is transferred through food webs by consumers, eventually to be deposited in sediments or transferred to the oceans by rivers.

Terrestrial nitrogen cycle

Nitrogen (N) enters the cycle through nitrogen fixation, as bacteria in soils reduce N_2 to the biologically available form of ammonia. Nitrogen is transferred through food webs by consumers, and decomposers return nitrogen compounds to soils, where ammonia can be oxidized to nitrate by chemoautotophic bacteria, and nitrate can be reduced to N_2 (which returns to the atmosphere) by bacteria that respire anaerobically. ***Go to the Visual Synthesis Map on LaunchPad to explore how the nitrogen cycle interacts with the cycles shown here.***

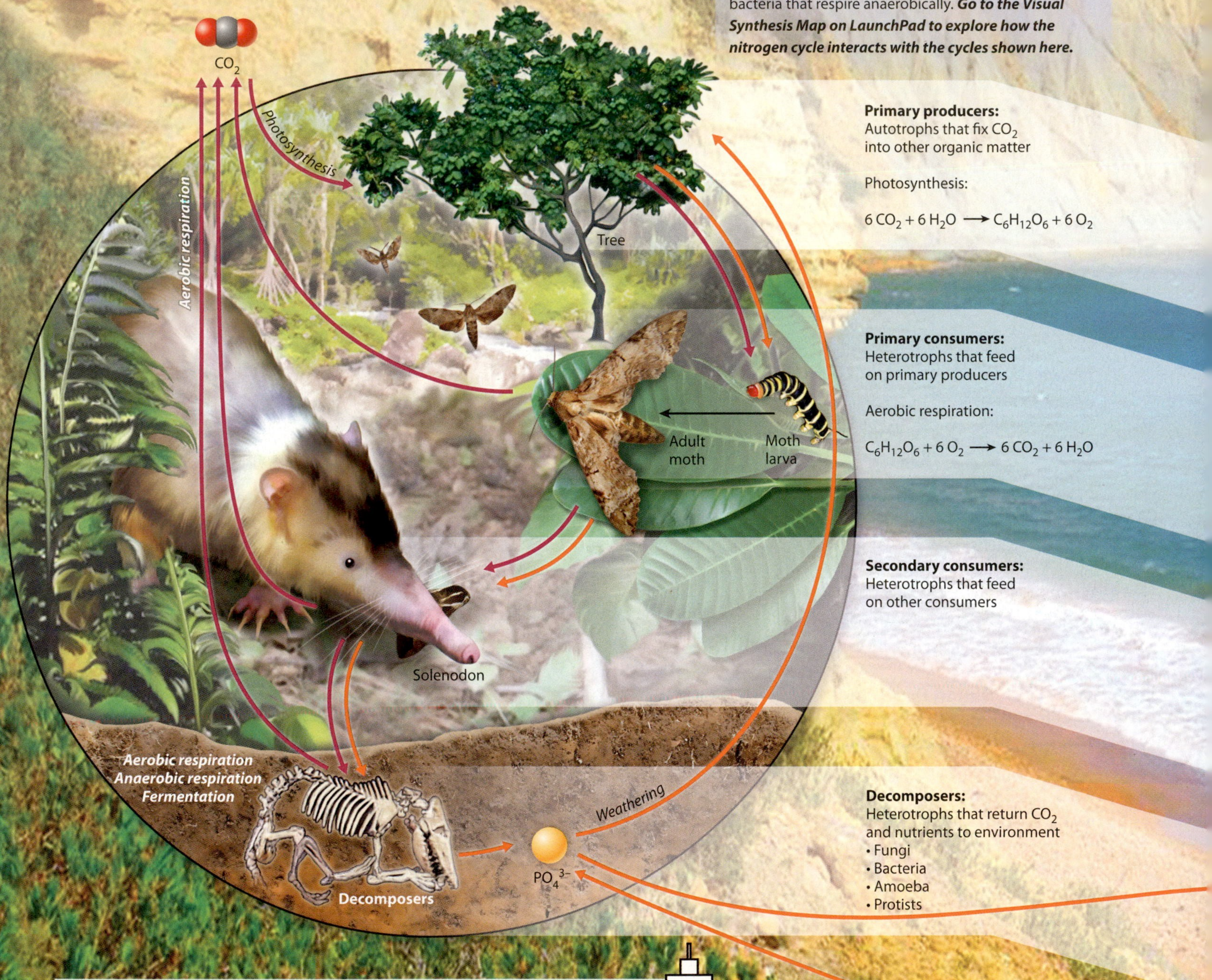

Trophic pyramids

As organisms cycle carbon, they also transfer energy, as illustrated by trophic pyramids. Some of the energy stored chemically in the sugars synthesized by primary producers and the biomolecules consumers eat is converted to heat and work by respiration. For this reason, only a portion of the energy taken in is available to consumers further up the food web. ***Go to the Visual Synthesis Map on LaunchPad to explore how trophic pyramids reflect the carbon cycle.***

Marine carbon cycle

As in the terrestrial carbon cycle, carbon enters food webs through photosynthesis and is transferred from species to species by consumers. In the oceans, microorganisms excrete carbon compounds and viruses lyse cells, providing food molecules for bacteria. The bacteria, in turn, provide food for protists and animals.

Marine phosphorus cycle

Phosphorus enters the oceans as erosional run-off from land and is taken up by primary producers. Phosphorus is transferred through food webs by consumers, but organic particles eventually sink into deeper waters, where phosphorus is released by microbial consumers and returned to the surface by upwelling water masses.

Marine nitrogen cycle

Nitrogen enters the oceans as run-off from the continents or through nitrogen-fixing microorganisms. Fixed nitrogen is taken up by primary producers and transferred through food webs by consumers. Microbial consumers feed on organic particles that sink into deeper waters, returning nitrate to the surface by means of upwelling. ***Go to the Visual Synthesis Map on LaunchPad to explore how the nitrogen cycle interacts with the cycles shown here.***

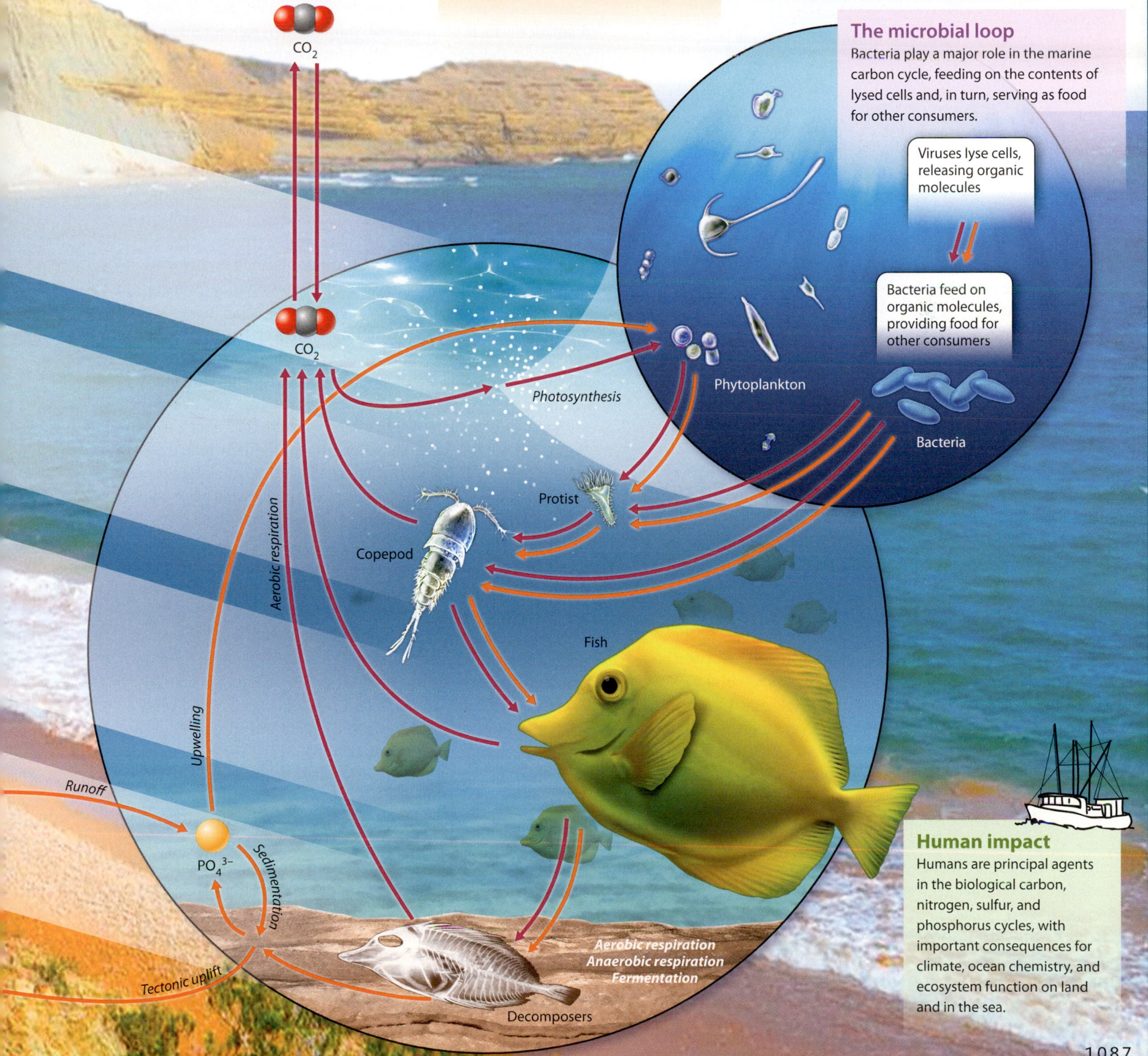

The microbial loop

Bacteria play a major role in the marine carbon cycle, feeding on the contents of lysed cells and, in turn, serving as food for other consumers.

Human impact

Humans are principal agents in the biological carbon, nitrogen, sulfur, and phosphorus cycles, with important consequences for climate, ocean chemistry, and ecosystem function on land and in the sea.

FIG. 49.21 Collapse of cod fishing in Newfoundland. Persistent harvesting of cod at rates higher than those at which stocks could be replenished led to the collapse of cod stocks. *Data from Millennium Ecosystem Assessment.*

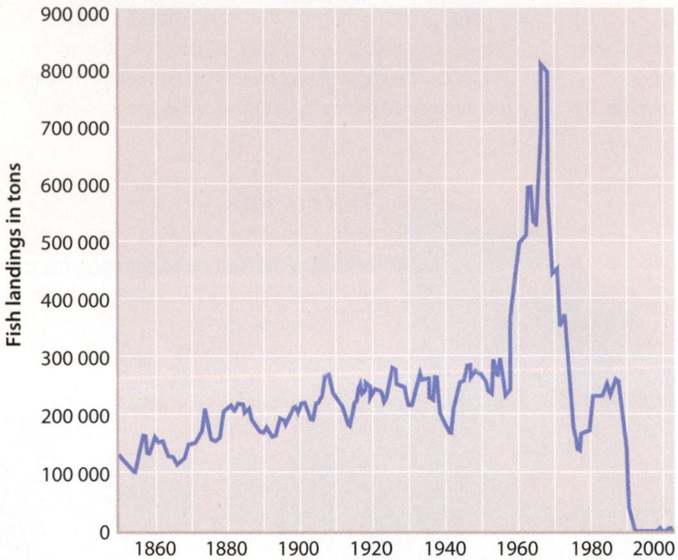

(**Fig. 49.21**). By 1992, cod stocks were estimated at 1% of 1960 levels (which were themselves well below those in Cabot's time) and so, with fish stocks on the verge of collapse, cod fishing was halted.

Unfortunately, the tragedy of cod in Newfoundland is not an isolated story. By 2003, nearly 30% of open ocean fisheries had declined by 90% or more, a percentage that will continue to rise unless sustainable fishing can be achieved. With collapsing fish stocks comes a loss in food production, economic decline, and, hidden beneath the surface of the sea, a shift in community structure, as species that compete with or depend on exploited fish species respond to changed conditions of natural selection.

Overexploitation occurs on land as well, perhaps most conspicuously affecting the majestic mammal species that have graced the African savanna for millennia. A global appetite for ivory has encouraged poaching, contributing to the reduction of elephant populations from an estimated 3–5 million in the 1930s to less than 10% of that today. Fueled by a (biologically unfounded) demand in parts of Asia for rhino horn as an aphrodisiac or folk medicine, poaching has similarly reduced populations of most rhinoceros species, limiting these once widespread animals to protected areas in both Africa and southeastern Asia. Note that habitat loss and overexploitation do not occur independently in these ecosystems; their effects are additive and, in some cases, synergistic, meaning that overexploitation exacerbates the harmful effects of habitat loss and vice versa.

Humans play an important role in the dispersal of species.

Biodiversity loss has many causes. In addition to habitat destruction and overexploitation, one of the most insidious is the introduction of non-native species. Non-native species that become established in new ecosystems are commonly termed **invasive species.** Removed from natural constraints on population growth, they can expand dramatically when introduced into new areas, sometimes with negative consequences for native species and ecosystems.

Species can sometimes be transported long distances by natural processes. For example, the Gulf Stream and North Atlantic Drift occasionally carry coconuts from the Caribbean to the coast of Ireland. Migratory birds can transport microorganisms across continents. In the Anthropocene Period, however, humans have dramatically increased dispersal. Ships fill their ballast tanks with seawater in Indonesia and empty it into Los Angeles Harbor, carrying organisms across the Pacific Ocean. Insects in fruit from South America wind up on a dock in London. Humans have become major agents of dispersal, adding new complexity to 21st-century ecology.

It has been estimated that 49,000 non-native species have been introduced into the United States (**Fig. 49.22**). Many of these provide food or pleasure (for example, ornamental plants), and only a few have escaped into natural ecosystems. Some of the invasive species that have become established in their new surroundings do little beyond increasing community diversity, but others put strong pressure on native species.

Kudzu, for example, a plant imported from Japan to retard soil erosion, now covers more than 3 million hectares in the southeastern United States, displacing native plants through competition for space and resources (Fig. 49.22a). The Zebra Mussel, a European bivalve that hitchhiked to the Great Lakes in the ballast water of cargo ships, has also multiplied dramatically in its new habitat, displacing native species and clogging intake pipes to power plants (Fig. 49.22b). Nearly 500 introduced plant species have become weeds, contributing disproportionately to crops lost each year to competition from invasive plants. And in Australia, more than 10% of all indigenous mammal species have become extinct in the past 200 years, largely because European colonists introduced cats and red foxes.

Invasive species have particularly devastating effects on islands because island species have commonly evolved with relatively few competitors or predators. Bones in Hawaiian lava caves record more than 40 bird species that existed 1000 years ago but are now extinct, eliminated by rats and pigs introduced by Polynesian colonists, as well as by landscape alteration. Introduced predators have had a similar impact on the island of Guam. Brown tree snakes introduced from Australasia since World War II have reduced native bird and reptile diversity by 75% (Fig. 49.22c).

Such introductions contribute to the changing diversity and ecology of 21st-century landscapes, but it may be the spread of modern diseases that should impress (and concern) us most. In the world of airplanes and oceangoing ships, bacteria and viruses spread much more rapidly than they did in our pre-industrial past. The H1N1 (swine) flu, a global concern in 2009, is a recent example. The first diagnosed case was identified on April 13

FIG. 49.22 **Invasive species.** (a) Kudzu (*Pueraria lobata*), a Japanese vine, was introduced to the United States in 1876 and now covers more than 3 million hectares. (b) The Zebra Mussel (*Dreissena polymorpha*), introduced into the Great Lakes from Europe in 1985, disrupts lake ecosystems, threatening native biological diversity. (c) The Brown Tree Snake (*Boiga irregularis*), introduced into Guam from Asia and Australia shortly after World War II, has decimated indigenous bird populations across the island. *Sources: a. James K. York/Dreamstime.com; b. Dr. David J. Jude, School of Natural Resources and Environment, University of Michigan; c. AP Photo/U.S. Department of Agriculture; c. John Mitchell/Science Source.*

in Mexico, and by mid-May this flu had been reported on five continents.

➔ **Quick Check 4** How do invasive species affect species diversity of communities and ecosystems?

Humans have altered the selective landscape for many pathogens.

From the Black Death of the fourteenth century to the flu pandemic of 1918, disease has played a major role in human history. In the Anthropocene Period, we have unprecedented opportunity to control or even eradicate some of the great diseases of the past and present. At the same time, however, the strong selective pressures placed on pathogens by antibiotics pose a key challenge for 21st-century medicine.

Malaria, discussed in Case 4, provides an illustrative example. In any given year, malaria infects an estimated 500 million people, causing a million or more deaths. Infection is particularly severe in parts of Africa, where the direct and indirect effects of malaria account for almost half of all childhood mortality. Insecticide-soaked bed nets have proven highly effective in affected areas, reducing transmission by as much as 90%. Natural selection, however, has begun to chip away at success in controlling the disease: Mosquito strains have begun to emerge that are resistant to DDT and another family of insecticides called pyrethroids.

Quinine, a chemical extracted from the bark of the cinchona tree, was the first effective treatment for malaria, and remained in widespread use until the 1940s. Since that time, a small arsenal of antibiotics has been developed, including chloroquine, mefloquine, primaquine, and doxycycline. Their usefulness, however, has declined through time as resistance to each of these drugs has evolved in populations of *Plasmodium*, the infectious agent of malaria. More recently, artemisinins, antibiotics based on a traditional Chinese plant remedy, have proved effective in treatment. Again, however, resistant strains of malaria have been identified.

→ **Quick Check 5** The interplay of pathogens and antibiotics again brings up the problem of the Red Queen. How does the Red Queen hypothesis apply to these kinds of interaction?

The arms race between antibiotics and pathogens is not limited to malaria. It characterizes nearly the whole range of diseases caused by viruses, bacteria, protists, and fungi that infect humans. For example, tuberculosis, a lung infection spread by inhalation of the bacterium *Mycobacterium tuberculosis,* causes 1.8 million deaths annually, mostly in sub-Saharan Africa, where many people are also infected with HIV (Chapter 43). Antibiotic treatment is effective, but many who are infected do not or cannot comply with the strict regimen of daily treatment for 6 months, hastening the evolution of drug-resistant *M. tuberculosis* strains. This occurs because an incomplete course of antibiotics does not eradicate all the bacteria, selecting for those bacteria that are least susceptible to the antibiotic. In 2012, nearly half a million new cases of antibiotic-resistant tuberculosis were reported. More troubling, a small percentage of these infections have been characterized as "extremely drug resistant," meaning that the pathogens resist both standard treatments and alternative antibiotics developed over the past 20 years (**Fig. 49.23**). In effect, extremely drug-resistant tuberculosis is untreatable.

Because bacteria can pass genes horizontally (Chapter 26), antibiotic resistance evolved in one species can jump to another. Thus, the bacterial incubators for drug resistance can arise anywhere and spread rapidly.

Clearly, despite the remarkable gains of the twentieth century, public health remains a key issue in the decades ahead. Improved sanitation—clean drinking water and safe food—can do much to reduce transmission of infectious diseases, and the search continues for vaccines that can further reduce or even eliminate some historic scourges. But the ongoing evolution of drug resistance requires the continuing efforts of scientists to stay one step ahead of pathogens.

Are amphibians ecology's "canary in the coal mine"?

In the ninetheenth and well into the twentieth century, British, American, and Canadian coal miners sometimes took canaries with them down long mine shafts. Canaries are more sensitive than humans to carbon monoxide and other toxic gases that can accumulate in mines. When the canaries passed out, the miners knew it was time to leave. Because many amphibian species have narrow environmental tolerances and exchange gases through their skin, they can be especially sensitive to environmental disruption. For this reason, some biologists see them as "canaries" for the global environment.

If amphibians are indeed our early-warning system, we should be concerned. Fully one-third of all amphibian species are threatened with extinction, and more than 40% have undergone significant population declines over the past decade (**Fig. 49.24**). What accounts for this pattern? Just about all the environmental issues discussed in this chapter come into play. Amphibian habitat is being destroyed in many parts of the world, especially old-growth forests that provide damp, food-rich environments in abundance. Pesticides and other toxins of human manufacture take their toll as well. For example, the common herbicide atrazine has been shown experimentally to impair development in frogs. Studies on placental cells grown in the laboratory suggest that atrazine in our food may also affect human reproduction. When atrazine was applied to human placental cells, it caused overexpression of genes associated with abnormal fetal

FIG. 49.23 Distribution of extremely drug-resistant tuberculosis. By 2012, confirmed cases had been reported in more than 50 countries. *Data from Global Tuberculosis Report 2013. WHO, 2013.*

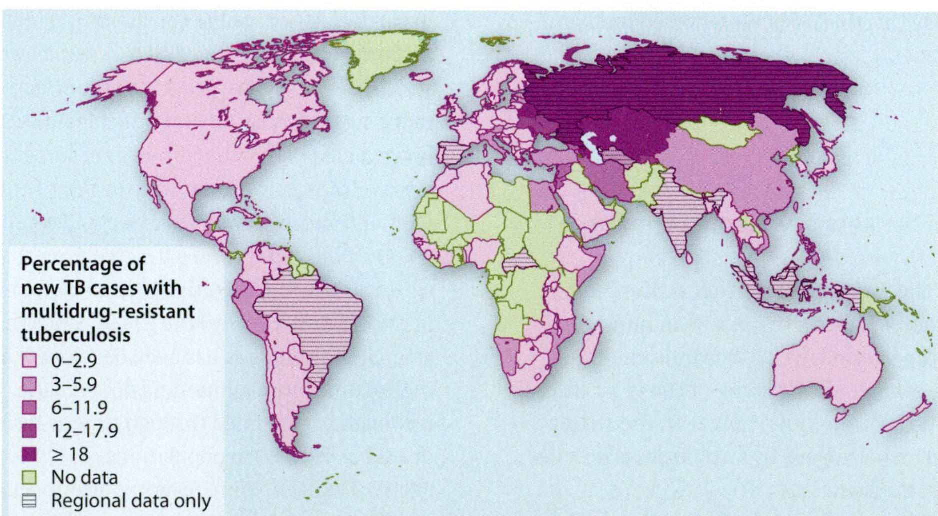

FIG. 49.24 **Amphibians under threat.** The Harlequin Toad (*Atelopus varius*), from Costa Rica and Panama, has been classified as critically endangered. *Source: Marie Read/Science Source.*

development. Because of concerns about public health, atrazine has been banned in several European countries.

Recently, infection has also been implicated in amphibian decline. The fungus *Batrachochytrium dendrobatidis*, or *Bd*, infects the skin of frogs and impairs many of the skin's functions. Recent studies of *Bd* show the classic features of epidemics—rapid geographic spread and high mortality—along the expanding front of fungal populations. How *Bd* spreads and whether it acts with other aspects of global change remain controversial, but there can be no doubt that disease has joined habitat destruction and pollution as a major cause of amphibian decline.

49.5 CONSERVATION BIOLOGY

Biological diversity, or **biodiversity,** refers to the summed variation of life on Earth (Chapter 47). This includes genetic and population variation within species; the number of species in local communities, biomes, or the planet as a whole; and the variations in species composition among ecosystems across the planet. We don't know how many species exist on our planet, although the number certainly runs into the millions and may exceed 10 million. Whether local or global, patterns of biodiversity reflect all of the processes discussed in earlier chapters: natural selection, functional biology, ecological interactions among species, and the distribution of physical environments. Increasingly, biodiversity also reflects human activities.

As discussed earlier in this chapter, habitat destruction, overexploitation, and invasive species all pose threats to biodiversity, and in the 21st century, global change threatens biodiversity as well. Earth history is punctuated by mass extinctions, but unlike the great extinctions of the past, current threats come not from meteorites or massive volcanism, but from us. Importantly, as human activities lie at the heart of threats to biodiversity, humans can make choices that will help us to preserve and protect the natural world. **Conservation biology** addresses the challenge of sustaining biodiversity in a changing world crowded with people.

? CASE 8 BIODIVERSITY HOTSPOTS: RAIN FORESTS AND CORAL REEFS

What are our conversation priorities?

Before examining how biologists and policy makers are working to conserve biodiversity, it is worth taking a moment to consider what we wish to preserve. At the broadest level, most biologists would hope to sustain our shared heritage of global species richness—a tall order on a planet increasingly marked by human activities. Not all ecosystems are equal in terms of biodiversity, and, following the lead of British conservationist Norman Myers, biologists have identified about two dozen **biodiversity hotspots** on land (Case 8). Hotspots are relatively small areas that have unusually high numbers of endemic species and that are under threat from human activities. (By definition, hotspots contain at least 1500 plant species found nowhere else and have lost 70% of their original habitat area.) Biodiversity hotspots represent only about 2% of the land surface, but they support more than half of the world's endemic plant species and nearly 43% of endemic bird, mammal, reptile, and amphibian species. Therefore, biodiversity hotspots are high priorities for conservation efforts.

Biodiversity reflects both species richness within local communities and the variations in species composition from one community to another, and from one biome to the next. Biodiversity hotspots reflect the variety of biomes found on land, but, for many reasons, conservation biologists wish to conserve species richness in local communities and regional ecosystems throughout the world. There is good evidence that biodiversity provides important **ecosystem services**—that is, benefits to humans—including cleaner air and water, greater primary productivity and improved resilience to environmental disruption, and untapped sources of food and molecular compounds for use in medicine and agriculture.

As discussed in Chapter 15, genetic variation exists within as well as among species, and this variation is likely to play an important role in conserving species in a time of global change. Genetic variants in plants provide the raw materials for breeding crops able to thrive in habitats that may become warmer or drier than at present. And within marine species vulnerable to global change, geneticists are beginning to find local populations with enhanced capacity to evolve tolerance to warmer temperature or decreased pH. Such variants may help to sustain the commercial shellfishing in the 21st century and provide robust coral populations able to sustain tropical reefs.

Conservation biologists have a diverse toolkit for confronting threats to biodiversity.

Fig. 49.25 makes it clear that the scope of the problem is large, requiring a variety of conservation strategies. Some programs focus on individual species, targeting specific threats that have reduced or could reduce populations below sustainable levels. Widely publicized efforts to protect elephants and rhinos provide illustrative examples. By protecting and restoring habitat where these animals thrive, by expanding efforts to curb poaching, and by working to shut down trade in elephant ivory and rhinoceros horn, conservationists hope to build and maintain sustainable populations. In some cases, for example that of the Arabian Oryx, long extinct in the wild but recently reintroduced to the deserts of the Arabian Peninsula, zoos and captive breeding programs have played a critical role in rebuilding populations. As the Arabian Oryx illustrates, species-specific programs are often successful. Similarly, the Southern White Rhino, known from a single local population a century ago, has prospered in protected reserves and now numbers more than 20,000 individuals. Whooping Cranes, majestic Northern American birds reduced to just 23 known individuals in 1940, now number more than 600.

FIG. 49.25 **Biodiversity at risk.** A substantial proportion of vertebrate species has been classified as at risk (likely to become endangered in the absence of intervention), endangered (likely to become extinct in the absence of intervention), or critically endangered (very high risk of extinction in the wild). In the 21st century, these proportions are increasing year by year. *Data from: Hogan, C. (2014). Endangered species. Retrieved from http://www.eoearth.org/view/article/51cbedac7896bb431f69323a.*

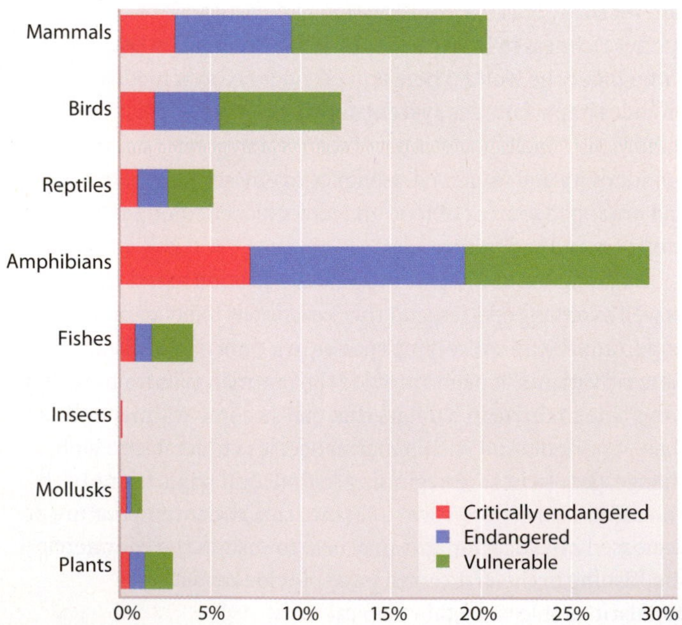

While conservation strategies focused on specific species are appropriate and necessary, they are simply not practical for the greater diversity of species at risk. Fortunately, one key component of targeted conservation efforts acts to conserve many species at once: protecting and restoring ecosystems.

In 1872, Yellowstone National Park was established as the world's first national park. The idea was to set aside a wilderness region of outstanding beauty and preserve it for all time. Since then, national parks and other reserves have been established throughout the world, and they have played a key role in conserving biodiversity. Reserves speak to the three major threats to biodiversity introduced earlier in this chapter. They protect habitat area, and, in many cases, have provided land to restore degraded habitats. They provide managed areas within which overexploitation can be eliminated or minimized. And they provide a buffer against invasive species. Notably, although we may value reserves because they protect charismatic animals such as lions, rhinos, or bison, they actually help to conserve entire ecosystems.

A great deal of research has shown that some designs for reserves work better than others. Not surprisingly, bigger is better. Species differ in the area needed to support sustainable populations, with top carnivores requiring particularly large territories to thrive. Thus, one large reserve functions better than four smaller reserves of the same total size. Also, because conservation is commonly more challenging along the edges of reserves, round or square areas function better than elongated or fragmented areas because they minimize edge effects. Importantly, successful plans for reserves include **corridors** that provide routes for migration from one reserve to another.

Global change provides new challenges for conservation biology in the 21st century.

Between 1985 and 2012, the area of Australia's Great Barrier Reef covered by coral declined by about 50%. Three processes drove the observed decline: physical destruction by tropical cyclones, predation by Crown-of-Thorns Sea Stars, and bleaching. Over the same interval, reefs in the Caribbean Sea similarly declined at a rate of about 1.5% per year, mostly because of storms, disease, bleaching, and the spread of algae due to overexploitation of fish that keep algal populations in check.

The declining health, areal extent, and diversity of coral reefs reflect some of the same factors that challenge biodiversity in general: overexploitation of associated species and habitat alteration due to fertilizer runoff. Outbreaks of Crown-of-Thorns Sea Stars may reflect both the overexploitation of predators on the sea stars and increased nutrient influx that appears to increase the survival of sea star larvae. Increased storm frequency and intensity, as well as bleaching, reflect another factor—climate change. Human activities such as agriculture and fishing will continue into the future, but warming temperatures and ocean acidification are becoming increasingly important threats to reefs in coming decades. This shouldn't deter us from addressing threats to reef ecosystems from local disturbances, but underscore the probability that reef conservation and management will require strategies to mitigate the effects of global change, as well.

The same is true on land. It is necessary and important to maintain reserves that protect natural habitats and the species they support, but, on the timescale of the next century, the climate of protected areas will change, and with it the capacity of the reserve to support its native biodiversity. Under these conditions, corridors will play a particularly important role in enabling populations to move from one reserve to another with local climate more amenable for growth and reproduction. Many ecologists think that natural corridors to promote migration will not be sufficient—that conservation biologists will need to implement a carefully considered program of assisted migration to match populations with appropriate environments.

In southern Africa, conservation efforts have benefited substantially by grouping reserves across national borders and connecting them with corridors. For example, a broad conservation area now links national parks and other reserves in South Africa, Mozambique, and Zimbabwe, providing an enlarged area for conservation corridors that connect established reserves and permit seasonal migration (**Fig. 49.26**). A key point is that, within this conservation area, animals and people coexist. Many corridors, in Africa and elsewhere, necessarily pass through populated areas, so local support is essential. In southern Africa, people within the conservation area have been persuaded that the preservation of biodiversity is in their best economic interests.

FIG. 49.26. Linking nature reserves in southern Africa. In a multinational effort, national parks and other reserves in South Africa, Zimbabwe, and Mozambique are being linked by a broad conservation area that provides corridors for migration. Within this area, animals and people coexist. *Data from http://www.krugerpark.co.za/great-limpopo-transfrontier-park-south-africa.html.*

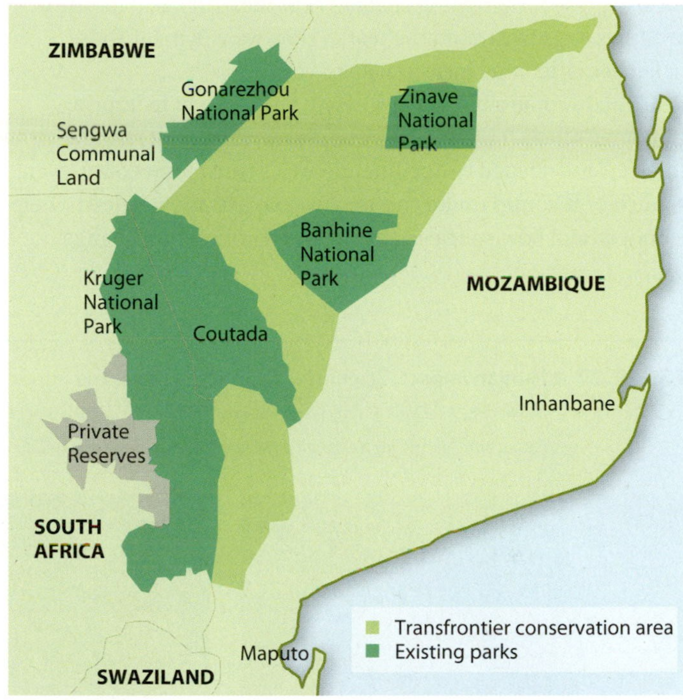

Sustainable development provides a strategy for conserving biodiversity while meeting the needs of the human population.

By now it should be clear that resource use and conservation biology are closely intertwined: The fate of many species will depend on the decisions we make about food and energy for a burgeoning human population. With this in mind, **sustainable development** has emerged as an important principle for trying to balance priorities for humans and the natural world. In principle, the concept is simple: The rate at which we use resources should be no higher than the rate at which they can be replenished.

Sustainable development is most easily understood in terms of species that grow and are harvested by humans. Thus, logging in the forests of western North America or the Amazon basin is sustainable only if tree harvest is matched by growth, keeping the forests dynamically intact. Similarly, fisheries are sustainable only if annual catch is no larger than the capacity of fish populations to grow and produce. By changing the way that catch is distributed among fishermen, the harvest of fish along the Pacific coast of the United States has become more predictable (as has the income of fishermen) and stocks of commercial species are increasing. Expanding marine reserves are also proving successful in rebuilding fish populations. At the same time, decisions by both fishermen and consumers are shifting the targets of commercial fishing away from vulnerable species and toward those that can be harvested sustainably.

It can be difficult to quantify sustainable yields for agriculture, as the cost of producing food includes the costs of fuel and fertilizer and, where episodic floods or droughts influence food supply, sustainable yield may vary from year to year. Despite these difficulties, the broader target of sustainable development—reducing our ecological footprint—has begun to influence agricultural practice and consumer choices. The task is formidable, but the goal is important: to ensure that meeting the needs of humans does minimal damage to Earth's biodiversity.

Without question, there will be fewer species when the 21st century ends than when it began. Lost species mean lost opportunities for the discovery of novel compounds for medical research. Diminished biodiversity may make communities less productive and less resilient to fires, hurricanes, or other environmental events. There is a pressing need to feed the world, and there are also many good reasons to conserve the biodiversity that has evolved over 4 billion years. Balancing these two imperatives will entail creative approaches to both science and policy, including thoughtful ways of using land that serves both people and conservation.

49.6 SCIENTISTS AND CITIZENS IN THE 21ST CENTURY

The decades ahead will be a time of great opportunity, but they will also present increasing challenges as our ecological footprint continues to grow around the world. How we will meet those challenges is, in large measure, up to you.

If you become a biologist, you will be able to help improve our planet's health in many different ways (**Fig. 49.27**). Field biologists are needed to document Earth's true biodiversity, even as much of it comes under threat. Physiologists are needed to help us understand how species respond to environmental change. Ecologists and population geneticists are needed to discover the principles that govern species interactions in communities so that we can understand how physiological responses to global change in one species will affect others. Only then will we be able to design effective ways to preserve the world's biodiversity. Molecular biologists will help us to combat disease in new ways, and geneticists will develop crops that yield more food. Perhaps most important, biologists will work with economists, lawmakers, and others to ensure that government policies reflect what we know about the natural world on which we all depend.

If you become a doctor, you will have many opportunities to improve the health of people in countries around the world. Through laboratory research, physicians will discover new ways to treat cancer, malaria, AIDS and other diseases. And, by treating patients in new ways, doctors will reduce the threat of

FIG. 49.27 **Human impact.** Scientists, doctors, teachers, and informed citizens can make a difference. *Sources: (Doctor) AP Photo/Sebastian John; (Science teacher) LWA/Dann Tardif/Getty Images; (Biochemist underwater) Alexis Rosenfeld/Science Source; (Testifying HHS Asst. Secretary) Matthew Cavanaugh/EPA/Newscom; (Earth Day tree planting) Chas Metivier/The Orange County Register/ZUMAPRESS.com.*

antibiotic-resistant pathogens. Working with government and private agencies, doctors will help to ensure clean water and adequate food for people in parts of Africa, Asia, and wherever else the basic needs of life pose an ongoing challenge.

If you become an engineer, you will have the opportunity to design energy-efficient machines or establish improved ways of slowing the spoilage of food. Novel solutions for managing resources and balancing resource use with conservation will benefit all people.

If you become a teacher, you will have a particularly important role in making sure that our children have the knowledge they need to meet the challenges of the Anthropocene Period. And, by choosing to be a responsible citizen, you can help to ensure that your family, workplace, and government make wise choices. In an age of environmental change, knowledge of Earth and life provides a key framework for personal choice. As citizens, our votes shape governments, and biological education can help to ensure wise leadership.

We cannot change the world we've been given, but we can and will influence the one our children inherit. It is easy to despair about the future, but as the novelist Philip Pullman put it, "Hope is not the name of a temperament, or of an emotion, but of a virtue." Hope inspires action, and our collective capacity for positive change is great. Knowledge of biology will not by itself ensure that humanity's golden age can be sustained in coming years, but it will surely be an important part of our solutions. ■

Core Concepts Summary

49.1 Some scientists call the time in which we are living the Anthropocene Period to reflect our significant impact on the planet.

The human population recently passed the 7-billion mark. page 000

The ecological footprint is the amount of land required to support an individual at an average standard of living. page 000

Our ecological footprint reflects all the activities that support our daily life, from growing and transporting the foods that we eat to supplying the materials and energy for our homes and the vehicles that provide our transportations. page 000

49.2 Humans have a major impact on the carbon cycle, primarily through the burning of fossil fuels, which returns carbon dioxide to the atmosphere.

Carbon dioxide is a greenhouse gas, trapping heat and thereby keeping the surface of Earth warm. page 000

Greenhouse gases play an important role in maintaining temperatures on Earth compatible with life, but increasing levels of greenhouse gases are resulting in higher mean surface temperatures. page 000

Changing climate has already affected the distribution of species and the composition of communities, and will continue to do so in the 21st century. page 000

Increasing levels of CO_2 are increasing the temperature and decreasing the pH of the oceans, affecting marine organisms, including corals. page 000

We can reduce CO_2 emissions, conserve finite natural resources, and contain the cost of energy by developing energy sources other than fossil fuels and improving energy efficiency. page 000

49.3 Humans have an important impact on the nitrogen and phosphorus cycles, primarily through the use of fertilizer in agriculture.

Humans add large amounts of industrially fixed nitrogen to the environment, which enters the biosphere as fertilizer to increase food production. page 000

Only a small percentage of fixed nitrogen is taken up by crops; the rest is returned to the atmosphere by soil microorganisms or ends up in lakes and oceans, where it results in eutrophication, a boost in primary production that depletes oxygen and leads to the death of fish and other aquatic organisms. page 000

A second major nutrient in fertilizer is phosphorus, which is mined rather than produced industrially. The finite supply of phosphorus raises questions about our ability to maintain agricultural production in the future. page 000

We have increased agricultural yields through the use of fertilizer, high-yield and disease-resistant plants, and farm machinery. page 000

To feed a growing population, we can devote more land to agriculture or increase productivity further, but each option brings with it costs. page 000

49.4 The impact of humans on the environment is changing the setting in which evolution acts.

Humans have altered the landscape in dramatic ways, leading to the expansion of some species, like corn and wheat, and the decrease and extinction of others. page 000

Habitat degradation and loss may cause extinction in some of Earth's most diverse ecosystems, such as tropical rain forests. page 000

Overexploitation of species by hunting, fishing, and logging, threatens exploited species and could change the ecology of ecosystems where these species live. page 000

Invasive species, spread by human activity, are major players in biodiversity loss. page 000

The use of antibiotics is one of the success stories of twentieth-century medicine, but has also led to the spread of antibiotic resistance, a major challenge for 21st-century biologists. page 000

Amphibian populations have declined dramatically because of a combination of habitat loss, pesticide use, and the spread of infection. page 000

49.5 Conservation biology aims to safeguard biodiversity for future generations

Reserves provide strong defense against habitat loss, overexploitation, and invasive species. page 000

Corridors that allow species to move between reserves strengthen conservation and will be particularly important as global change affects reserves. page 000

Sustainable development strives to match the rate at which we use resources to the rate at which they can be replenished. page 000

49.6 In the 21st century, biologists, doctors, engineers, teachers, and informed citizens have vital roles to play in understanding our changing planet and making wise choices for our future.

Human impact can be positive as well as negative. page 000

Self-Assessment

1. Describe what is meant by the term "ecological footprint."
2. Name three sources of atmospheric CO_2.
3. Explain the relationship between atmospheric CO_2 levels and mean temperature.
4. Provide evidence that indicates that human activities are responsible for increases in atmospheric CO_2 over the last century.
5. Describe the two ways organisms can respond to changes in CO_2 levels and temperatures.
6. Describe several possible solutions to the problem of increased atmospheric CO_2.
7. Describe the causes and consequences of eutrophication.
8. Describe several possible solutions to the problem of feeding a growing human population.
9. Give three examples of species that have benefited and three examples of species that have been harmed by human activity.
10. Explain why amphibians are sometimes called "canaries in a coal mine."
11. In what ways do habitat loss, overexploitation, and invasive species threaten biodiversity?
12. How do reserves and corridors help to conserve biodiversity?
13. What is the Red Queen hypothesis, and how does it pertain to 21st-century plants and animals?

Log in to LaunchPad to check your answers to the Self-Assessment questions, and to access additional learning tools.

QUICK CHECK ANSWERS

Chapter 21

1. The allele frequency of *a* was calculated as follows:

$$\text{Frequency}(a) = [2 \times (\text{number } aa) + 1 \times (\text{number } Aa)]/[2 \times (\text{total number of individuals})]$$

This equation can be rewritten as

$$\text{Frequency}(a) = [(\text{number } aa) + \tfrac{1}{2} \times (\text{number } Aa)]/(\text{total number of individuals})$$

Note that

$$\text{Number } aa/\text{total number of individuals} = \text{frequency}(aa)$$

and

$$\tfrac{1}{2} \times (\text{number } Aa)/\text{total number of individuals} = \tfrac{1}{2}\,\text{frequency}(Aa)$$

Therefore,

$$\text{Frequency}(a) = \text{frequency}(aa) + \tfrac{1}{2}\,\text{frequency}(Aa)$$

Stated in words, the frequency of allele *a* equals the frequency of *aa* homozygotes plus half the frequency of *Aa* heterozygotes.

By similar logic,

$$\text{Frequency}(A) = \text{frequency}(AA) + \tfrac{1}{2}\,\text{frequency}(Aa)$$

These equations are very useful for determining allele frequencies directly from genotype frequencies.

2. More. Take, for example, a protein variant that can be identified through protein gel electrophoresis. Now that we can sequence the DNA of multiple copies of that allele, we may find that there are several other amino acid differences that do not affect the mobility of the protein on a gel.

3. We can conclude that the population is evolving. What we cannot tell is what mechanism—selection, migration, mutation, genetic drift, or non-random mating—is causing it to evolve. To determine what mechanism(s) is (are) driving the process requires more-detailed population genetics analysis.

4. Sexual selection for bigger size, more elaborate weaponry, or brighter colors in males is an example of directional selection.

5. Adaptation is the fit between an organism and its environment. Of all the evolutionary mechanisms, only selection causes allele frequencies to change based on how they contribute to the success of an individual in terms of survival and reproduction. This means that allele frequencies in the next generation are ultimately governed by the environment. Because phenotype is in part determined by genotype, organisms become adapted to their environment under the influence of selection over time.

Chapter 22

1. Species change over time, making it difficult to craft a single definition that can be applied in all cases. Also, a species concept has to apply to such an astonishing variety of living and dead biological forms—everything from a living microbe to a long-extinct dinosaur—that it seems impossible to find an ideal one-rule-fits-all definition. Thus, though the BSC is useful in many cases, it is not applicable to asexual organisms and organisms known only from fossils.

2. Volcanic island chains offer many examples of adaptive radiation because they are dependent on the serendipitous process of colonization. This means that only some plants and animals are able to get there. In addition, because of the absence of competitors on these islands, there are often many available ecological opportunities for the colonizers. For example, there may be no insect-eating birds on the islands, providing an ecological niche to be filled.

3. If populations of fish became separated from other populations of fish in different ponds during dry periods, this scenario suggests that speciation was more likely to be allopatric rather than sympatric. The allopatrically evolved species became sympatric only when the lake flooded again, combining all the separate ponds into a single body of water.

Chapter 23

1. No. The two trees are equivalent. The node shared by humans and mice is simply rotated in one tree. In both trees, the closest relative of humans is mice because they share a node (where they split) not shared with lizards.

2. A group called "fish" is paraphyletic because it includes some, but not all, of the descendants of a common ancestor. The descendants of the common ancestor of fish also include amphibians, sauropsids, and mammals.

3. The traits are analogous, the result of convergent evolution. Both animals are adapted for swimming in water and converged independently on similar traits, including a streamlined body and fins. However, they are only distantly related, as one is a fish and the other is a mammal.

4. Character traits can be measured for fossils as well as living organisms, so you can examine the features of the skeleton and then construct a phylogenetic tree that takes into account the skeletal traits shared with known vertebrates.

Chapter 24

1. No. Modern humans and modern chimpanzees share a common ancestor. Changes have occurred along both lineages: from the common ancestor to humans, and from the common ancestor to chimpanzees.

2. The multiregional hypothesis suggests that the most recent common ancestor of all living humans lived about 2 million years ago, when *H. ergaster* first migrated from Africa to colonize Europe and Asia. The out-of-Africa hypothesis suggests that *H. sapiens* evolved much more recently in Africa and then moved out of Africa, replacing the remnants of populations originally established by *H. ergaster* when it left Africa. The out-of-Africa hypothesis suggests a much more recent common ancestor for all present-day humans, one that lived about 200,000 years ago.

3. Studies of mtDNA, which is maternally transmitted, indicated that there was no input of Neanderthal mtDNA into the modern human population. In contrast, studies of Neanderthal genomic DNA showed that 1% to 4% of non-Africans' genomes are composed of Neanderthal DNA, implying that there was interbreeding between the ancestors of non-Africans and Neanderthals, presumably when the *H. sapiens* population first left Africa. If there is evidence of genetic input but no evidence of a female-based genetic contribution, we can hypothesize that male Neanderthals were the key contributors.

4. It is inaccurate to label *FOXP2* the "language gene" for several reasons. First, the gene is expressed in many tissues, so its effects are not limited to speech and language. Second, many genes are required for language, not just *FOXP2*. Nevertheless, its association with vocal communication in multiple species is striking.

Chapter 25

1. The coincidence in timing between the Industrial Revolution and the observed increase in CO_2 levels—following a millennium of little change—suggests that human activities played a role in recent changes in atmospheric composition. Alternative hypotheses would focus on other processes that add CO_2 to the atmosphere. Perhaps, for example, increased volcanic activity added more CO_2 to the atmosphere, or perhaps warming induced by changes in solar radiation caused thawing of permafrost, facilitating increased respiration of soil organic matter at high latitudes. These hypotheses can be tested against the historical record of volcanic activity and measurements of solar radiation over the past century.

2. All else being equal, increasing the elevation of mountains should increase rates of chemical weathering and erosion. As chemical weathering of continental rocks consumes CO_2, atmospheric CO_2 levels should decline.

Chapter 26

1. In prokaryotic organisms, genes are transferred from one organism to another by horizontal gene transfer, facilitating the generation of new gene combinations.

2. Archaeal cell and genome organization are similar to those of Bacteria. It was only through the use of molecular sequence comparisons that Archaea were shown to occupy a distinct branch on the tree of life. Continuing research on membrane lipids, cell wall chemistry, and gene transcription and translation confirm that Archaea are distinct from bacteria.

3. Cyanoabcteria use much of the ATP they generate to drive the reduction of CO_2 into organic molecules used for growth and reproduction. In photoheterotrophs, the organic molecules needed for growth are taken in from the environment, leaving ATP available for other uses.

4. Primary production could be accomplished by anoxygenic photosynthesis, which does not generate oxygen. Oxidation of organic matter to carbon dioxide could be accomplished by anaerobic respiration (and fermentation), which does not use oxygen.

5. It wouldn't. Bacteria and archaeons cycle nitrogen between the atmosphere (nitrogen gas) and biologically useful forms. In the absence of these prokaryotes, only a limited amount of nitrogen would be fixed into biologically useful form by lightning, greatly limiting the extent of life.

6. Not necessarily. Today, hyperthermophiles live in hot springs and hydrothermal ridges on the ocean floor. These might have been the environments in which the last common ancestors of these organisms thrived.

Chapter 27

1. In general, eukaryotes employ a subset of the metabolic pathways used by bacteria. Most eukaryotes are capable of aerobic respiration; most can also gain at least some energy by fermentation; and some can also photosynthesize. In many ways, then, carbon cycling by eukaryotes is much like carbon cycling by aerobic prokaryotes. The novel contribution of eukaryotes is the ability to capture and ingest other cells, thus introducing predation into the carbon cycle.

2. No. Plants form one branch of the green algal tree, whose early branches are all occupied by unicellular forms. Similarly, the branch containing animals has choanoflagellates and other unicellular forms in its lower branches. Thus, animals and plants evolved multicellularity independently of each other. (Note that genetic evidence further supports this phylogenetic hypothesis: Plants and animals have different sets of genes that regulate multicellular development; see Chapter 28.)

3. Molecular sequences for genes in red and green algae show the close evolutionary relationship of these organelles to free-living cyanobacteria. This supports the hypothesis that eukaryotes initially gained the ability to photosynthesize by the symbiotic incorporation of cyanobacterial cells. Gene sequences in the chloroplasts of other photosynthetic eukaryotes (brown algae, diatoms, euglenids, and other algae) are very similar to those of red and green algal chloroplasts, indicating that photosynthesis spread throughout the eukaryotic tree by the symbiotic incorporation of green or red algal cells into other eukaryotes.

Chapter 28

1. Complex multicellular organisms have differentiated cells and tissues. Moreover, in simple multicellular organisms, all or nearly all cells are in direct contact with the environment. In complex multicellular organisms, most cells are completely surrounded by other cells.

2. Without mechanisms for bulk flow, the movement of oxygen, nutrients, and molecular signals through organisms is limited by diffusion. Rates of diffusion, in turn, limit the size and shape that an organism can achieve. Bulk flow allows key molecules to be transported over distances much greater than those possible by diffusion alone, making larger organisms possible.

3. All multicellular organisms have molecules that promote adhesion between cells and communication between cells, but plants and animals have largely distinct sets of molecules for these functions.

Plants and animals also exhibit distinct patterns of development, with stem cells called meristems commonly remaining active for life in plants, while tissues and organs differentiate early in animal development. Many of the differences between plants and animals reflect the fact that plant cells have cell walls, whereas animal cells do not.

Chapter 29

1. Plants transpire to obtain CO_2 from the atmosphere. Plants cannot completely seal themselves off against water loss because that would prevent them from obtaining the CO_2 needed for photosynthesis. As CO_2 diffuses into a leaf to reach the photosynthetic cells, water vapor evaporates from surfaces within the leaf and diffuses out of the leaf. Thus, transpiration is an unavoidable by-product of acquiring CO_2 by diffusion from the atmosphere.
2. CAM plants open their stomata at night when rates of evaporation are low and close them during the day to conserve water. At night, CO_2 is combined with 3-carbon PEP to form 4-carbon organic acids that are stored in the vacuole; during the day, these 4-carbon organic acids are converted back to PEP and CO_2, providing CAM plants with a source of CO_2 for photosynthesis, even though their stomata are closed. The production of 4-carbon organic acids by C_4 plants results in an increase in the concentration of CO_2 in the bundle-sheath cells. The high concentration of CO_2 in bundle-sheath cells makes it unlikely that rubisco will use O_2 as a substrate, preventing the losses in energy and reduced carbon associated with photorespiration.
3. The statement is correct in that the force that pulls water from the soil results from the evaporation of water from leaf cells, the energy for which comes from the heating of the leaf by the sun. It is incorrect in that there are structures within the plant that make water transport possible, notably the xylem conduits, and also the roots and the stem. The plant must expend energy to produce all these structures.
4. The carbohydrates in phloem move from source to sink. Sources can be actively photosynthesizing leaves, or any organ with a large amount of stored carbohydrates. Carrots are examples of roots with large amounts of stored carbohydrates. In contrast, sinks are organs that require an import of carbohydrates to meet their growth and respiratory needs. Examples of sinks include young leaves, developing fruits, and most roots. Typically, the shoot is the source and the roots are a sink. At the end of winter or following a severe drought, however, a plant may regrow its shoot by drawing upon carbohydrates that had been stored in the roots. In this case, the movement of carbohydrates is from roots to shoot.

Chapter 30

1. Both spores and gametes are unicellular and haploid. Gametes are short-lived and require hydration; they are capable of fusing only with another gamete. In contrast, spores can be long-lived and can survive exposure to air because of the presence of sporopollenin, a tough, resistant covering that allows spores to withstand environmental stresses such as ultraviolet radiation and desiccation. A spore, once dispersed, can grow into a new individual.
2. Both mosses and ferns exhibit an alternation of a haploid (gametophyte) generation and a diploid (sporophyte) generation, the release of swimming sperm, and the dispersal of spores. However, in mosses, the sporophyte is completely dependent on the gametophyte, whereas in ferns the sporophyte is only initially dependent on the gametophyte and eventually becomes free-living.
3. The male gametophyte develops within the spore wall, forming pollen. Following pollination, the male gametophyte produces a pollen tube that grows through the ovule to deliver male gametes to the egg. The female gametophyte grows to fill the entire sporangium and is never dispersed. Seed plant gametophytes are never free-living; all of the resources that fuel the growth of both male and female gametophytes come from the sporophyte generation.
4. Compared with spores, seeds can store more resources, slow down their metabolism, and exhibit dormancy, all of which aid their dispersal.
5. A flower consists of four whorls of organs: From outside in, they are the sepals, petals, stamens, and carpels. The sepals generally protect developing flowers; the petals commonly serve to attract pollinators; the stamens produce pollen; and the carpel protects the ovules inside that develop into seeds following fertilization.
6. A fruit forms from the ovary that encloses the seeds, and, in some plants, it also forms from other parts of the flower, including the petals and sepals.

Chapter 31

1. The shoot apical meristem is the site of cell division and thus is the source of all the cells that make up a stem. It also initiates leaves and produces new meristems, which allow stems to branch.
2. One of a plant's lateral meristems produces new vascular tissues; the other maintains an outer protective layer.
3. Both roots and stems grow from apical meristems, have regions of cell division, elongation, and differentiation, and respond to light and gravity. However, root apical meristems are covered by a root cap, the root has a single vascular bundle in the center, and branching occurs by the formation of new meristems from the pericycle; all these features are different from what is seen in stems.

Chapter 32

1. No. The component of the plant immune system that uses *R* genes can defend only against biotrophic pathogens, not necrotrophic ones, because this component of the immune system is based on receptor molecules within living cells. Necrotrophic pathogens kill cells before colonizing them.
2. In the hypersensitive response, plants block off an infected area. In SAR, plants send a signal to uninfected tissue, enabling it to mount a defense and prevent infection.
3. Plants that cannot synthesize jasmonic acid are unable to mount a defense throughout the plant body and as a result suffer more damage from herbivores.

4. The trade-off between growth and defense is a diversifying force that favors different species that are specialized for each habitat, rather than a single species that dominates across all soil types.

Chapter 33

1. The green algae are paraphyletic because the last common ancestor of all green algae also gave rise to land plants; the bryophytes are paraphyletic because the last common ancestor of mosses, liverworts, and hornworts also gave rise to vascular plants. Vascular plants, however, are monophyletic.
2. An alternative explanation is that stomata evolved once, in the common ancestor of mosses, hornworts, and vascular plants.
3. Xylem vessels evolved convergently in gnetophytes and angiosperms; insect pollination evolved convergently in cycads and angiosperms; fleshy tissues to facilitate seed dispersal evolved convergently in early seed plants (reflected today in ginkgo and cycad seeds), conifers such as juniper and yew, and flowering plants.
4. Monocots produce only a single cotyledon; monocot roots are produced continuously directly from the stem; many monocots produce creeping stems; and the strap-shaped leaves of many monocots elongate from a zone of cell division located at the base of the leaf, which surrounds the stem to form a sheath. Monocots do not form a vascular cambium; instead, radial growth occurs in a narrow zone just below the shoot apical meristem. In most monocot stems, the vascular bundles are distributed throughout the cross section, as opposed to being arranged in a ring.

Chapter 34

1. Hyphae enable fungi to seek out new food resources; they are able to penetrate large food bodies such as rotting logs or animal carcasses. Hyphae also transport nutrients from one part of the fungus to another.
2. A diploid cell has a single nucleus with two complete sets of chromosomes; a dikaryotic cell has two genetically distinct haploid nuclei.
3. Complex multicellular fruiting bodies evolved independently on two branches, the basidiomycetes and the ascomycetes.
4. You are eating the multicellular fruiting body built from dikaryotic hyphae.

Chapter 35

1. Nerve signals are transmitted electrically (in the form of an action potential) from one end of a neuron to the other. They are transmitted chemically (by neurotransmitters) across a synapse from one neuron to another.
2. Nerve impulses are slowed or completely stopped in response to loss of myelin surrounding nerve cells.
3. The presynaptic nerve cell has voltage-gated Ca^{2+} channels that open to cause neurotransmitter release from vesicles of the presynaptic cell. This results in neurotransmitter binding to postsynaptic ligand-gated ion channels, causing a change in the postsynaptic cell's membrane potential, after which the neurotransmitter is inactivated or taken up by the presynaptic cell.
4. Nerve gases that block acetylcholinesterase, the enzyme that inactivates acetylcholine, cause violent muscle contractions, leading to paralysis of the diaphragm (the main muscle involved in breathing) and death by asphyxiation.
5. The sensory neuron releases only one type of neurotransmitter, which is excitatory for the extensor muscle from which the stretch receptor was stimulated. An interneuron provides an inhibitory synapse to the flexor muscle.

Chapter 36

1. Prostaglandins depolarize nociceptors and increase their firing rate.
2. Excess mucus secretion blocks airborne odor molecules from reaching the chemosensory hairs of olfactory neurons, reducing their ability to bind odorants and therefore reducing the sense of smell. Taste is also affected because much of the sense of taste depends on smell.
3. Spinning causes the hair cells in the semicircular canals to accommodate to the new motion. When you stop spinning, the hair cells detect a change in angular motion of the head (relative to the motion when your head was spinning), which leads to a sense of imbalance when you stand or try to walk.
4. Rod cells are extremely sensitive to light, allowing an animal to see in dimly lit conditions. Cone cells provide color vision through opsins with absorption peaks at different light wavelengths, and they provide sharp vision.
5. The three areas of the brain that are used to remember and recognize a face are the occipital cortex for processing visual information, the hippocampus for forming long-term memory of a face, and the temporal lobe for facial recognition.

Chapter 37

1. Without newly synthesized ATP, myosin cross-bridges formed with actin cannot detach from their actin binding sites, so they remain in the bound state and make the muscle stiff.
2. Because acetylcholine is blocked and cannot depolarize the muscle cell membrane, calcium is not released from the sarcoplasmic reticulum, and the muscle is unable to contract.
3. Contracting at an intermediate length maximizes the overlap of actin and myosin filaments in the biceps fibers, allowing the greatest number of cross-bridges to form and produce force to lift a larger weight.
4. The force of contraction can be increased by temporal summation of action potentials transmitted by the motor neuron to the muscle (increased firing frequency) and by activating a greater number of

motor neurons to recruit more motor units (and thus, more muscle fibers).

5. Glass is composed of a single material (silica), which is rigid but can crack easily. Bone is a composite material, built of crystalline mineral reinforced by the protein collagen, enabling bone to absorb much more energy before breaking.

Chapter 38

1. Diabetes can result either from decreased insulin production by the pancreas (type 1 diabetes) or decreased effect of insulin on target cells (type 2 diabetes). Type 1 diabetes is an autoimmune disease in which the insulin-producing cells in the pancreas are attacked by the immune system. The result is inadequate insulin secretion. Daily injections of insulin can maintain stable normal circulating levels of blood glucose. In individuals with type 2 diabetes, insulin production is unaffected, but cells are not able to respond to normal circulating levels of insulin. Type 2 diabetes is commonly linked to obesity and is most effectively prevented by proper diet and exercise.
2. Hormones are chemical compounds that are secreted by a cell or gland that act on other cells in the body either locally or at distant sites (transported in the bloodstream). Hormones achieve specificity by binding to receptors that are present only on their target cells.
3. Peptide hormones bind cell-membrane receptors because they are hydrophilic and cannot diffuse across the cell membrane, activating signaling cascades within the target cell. Steroid hormones diffuse across the cell membrane and bind intracellular cytoplasmic or nuclear receptors, leading to changes in gene expression and protein synthesis within the target cell.
4. Paracrine signals are chemicals secreted by a cell that influence the activity of neighboring cells without entering the bloodstream. Pheromones are chemicals secreted into the environment and transmitted to other members of the same species, influencing their behavior. Hormones are released into the bloodstream and have effects on distant cells within the same individual.

Chapter 39

1. Diffusion is increased by a large surface area for exchange and a short diffusion distance.
2. In individuals with asthma, resistance to airflow increases, thereby decreasing the flow of air. Resistance increases because the diameter of the airways decreases as a result of smooth muscle contraction in the airway walls.
3. Air has a low density and viscosity and a high O_2 content relative to that of water.
4. Sickle-cell anemia is life threatening because transport of the deformed sickle-shaped red blood cells is limited in small blood vessels. Therefore, the supply of O_2 to tissues is reduced, cellular metabolism in the tissues supplied by the blood vessels is disrupted, and CO_2 builds up, possibly leading to organ failure.
5. A change in a vessel's radius affects blood flow most. For example, a twofold reduction in radius increases resistance to flow 16 times.
6. The atrioventricular valve closes when pressure in the ventricle exceeds the pressure in the atrium.

Chapter 40

1. Glycogen stores are used first, and then fat, and finally muscle protein.
2. No. Metabolic rate increases with increasing mass, but the relationship between the two is not linear. Instead, metabolic rate increases with mass raised to the ¾ power. This means that a dog that is twice as heavy as a cat has a metabolic rate that is less than twice as much as the cat.
3. "Essential," when used to describe dietary amino acids, minerals, or vitamins, does not mean that other nutrients are less important or useful. "Essential" in this context means that these nutrients must be obtained in the diet; the organism cannot synthesize them on its own. In other words, it is "essential" that these nutrients are consumed.
4. Rocks and sediments in gizzards break apart food, as teeth do in vertebrates.
5. By eating the products of the cecum, these animals can pass the partially digested food by the small intestine a second time, where nutrients are then absorbed.

Chapter 41

1. When a cell is placed in a hypotonic solution, water moves into the cell by osmosis and the cell swells or bursts. When a cell is placed in a hypertonic solution, water leaves the cell by osmosis and the cell shrinks.
2. Osmosis depends on the difference in total solute concentration on the two sides of a membrane. There is no net movement of water molecules across a selectively permeable membrane with a high concentration of a particular solute on one side of a membrane if this concentration is matched by a similar concentration of some other solute on the other side of the membrane. For example, the concentration of sodium ions may be high on one side of a selectively permeable membrane and low on the other, whereas the distribution of potassium ions may be just the reverse, so the total solute concentration is the same on the two sides of the membrane. The result is no net water movement.
3. Ammonia excreted by fish is rapidly diluted to nontoxic levels in the surrounding water. Mammals convert ammonia to urea, which is much less toxic than ammonia and therefore can be concentrated and stored before being eliminated.
4. Although the filtrate is dilute at the start and end of the loop of Henle, the loop creates a concentration gradient from the cortex to

the medulla, which is important for subsequent water reabsorption from the collecting ducts. The loops also leave urea as the main solute.

Chapter 42

1. Sexual reproduction generates genetically unique offspring by (1) chance mutations; (2) recombination between homologous chromosomes during meiosis I; (3) random segregation of homologous chromosomes during meiosis I; and (4) new combinations of chromosomes in the process of fertilization. Asexual reproduction, like sexual reproduction, is subject to chance mutations.
2. The vas deferens carries sperm from the epididymis to the ejaculatory duct. Cutting the vas deferens on both sides prevents sperm from entering the ejaculatory duct and leads to male sterility. This procedure, called a vasectomy, is a method of male contraception.
3. High levels of estrogen and/or progesterone lead to the growth of the uterine lining. The drop in these hormones while a woman is taking placebos allows the uterine lining to be shed each month. Placebos are usually taken instead of no pills in order to help keep track of the number of days before resuming hormone-containing pills.
4. In females, oogonia differentiate into primary oocytes before birth and therefore a female is born with all the primary oocytes she will have. In males, by contrast, spermatogonia do not differentiate into primary spermatocytes until puberty, but sperm can be produced throughout a male's lifetime. Only one of the four products of meiosis becomes a gamete in females, whereas all four products of meiosis become sperm in males.
5. In the process of development, cells differentiate into different cell types. The process of differentiation involves the turning on and turning off of specific genes. So, for example, genes that are expressed in a skin cell are different from those that are expressed in a liver or muscle cell. These changes in gene expression, once established, are for the most part stable through mitotic cell divisions.

Chapter 43

1. Damage to cilia in the respiratory tract increases the probability of lung infection by impairing the ability of the respiratory tract to sweep out microorganisms.
2. A second novel antigen injected on day 40 will produce a primary response, not a secondary response, because it is a novel antigen that has not been encountered before by the immune system.
3. Although a given B cell produces only one antibody, the entire population of B cells produces many different antibodies. Each B cell produces a different antibody as the result of a unique pattern of genomic rearrangement.
4. T cells are activated when their surface T cell receptors bind to an antigen in association with an MHC molecule. By contrast, B cells are activated when their surface antibodies bind to a free antigen.

Chapter 44

1. Cnidaria and Bilateria share a common ancestor that diverged from sponges and other early animals. They are, therefore, equally closely related to the sponges.
2. Muscle cells permit movement, enabling many cndarians to seek and capture prey.
3. The shared features among present-day snails, squids, and clams suggest that all mollusks descended from a common ancestor that had a head, radula, foot, mantle, and well-differentiated organ systems.
4. Young humans and other young animals often have features that they lose as they age. Our first "baby" teeth are an example. If you could look at a photograph of yourself when you were a 1-month-old, inch-long embryo, you would see a notochord, pharyngeal slits, and a tail—characters of the phylum Chordata.
5. Scales are found on all these animals because they were present in a common ancestor. Even feathers are modified scales.

Chapter 45

1. The connection between gene and behavior is mediated by multiple factors, including the neurons and hormones encoded by genes, so the mapping of behavior to a gene (or genes) is inevitably complex. Also, as we saw in Chapter 18, complex features of organisms, including behavior, are typically encoded by multiple genes. The effects of these genes are often influenced by environmental factors in the development of the organism, so the same genes will produce different outcomes when development occurs in different environments.
2. In classical conditioning, one stimulus is exchanged for another (in the case of Pavlov's dogs, the smell of food is exchanged for the sound of a bell). As a result, a novel stimulus evokes a behavior. In operant conditioning, a behavior is associated with a reward or punishment, so the behavior becomes more likely (with a reward) or less likely (with a punishment). Therefore, in the case of classical conditioning, an association is made between a stimulus and a behavior, and in operant conditioning, an association is made between a behavior and a response.
3. Photoperiod is a good marker for time of year, and therefore for season. Many species have season-specific behaviors—hibernation in bears, migration in birds—that must be timed appropriately.
4. Organisms do not act "for the good of the species" because natural selection operates on individuals: It is the individual that lives or dies, reproduces, or fails to reproduce. Traits that are disadvantageous to the individual are therefore selected against by natural selection, even if they are beneficial to the species as a whole.
5. With reference to Fig. 45.17, the probability of each offspring inheriting an *A* allele is 0.5. The probability of a cousin sharing an *A* allele is 0.25, so the probability of each of the animal's cousin's offspring having an *A* is $0.25 \times 0.5 = 0.125$. In this case, the animal would have to substitute four of its cousin's offspring for each of its own.

Chapter 46

1. The per capita growth rate, r, equals $(\Delta N/\Delta t)/N_1$. In this case, the change in time Δt is 1 year, so $r = \Delta N/N_1$. If the starting population is x, at triple the size it is $3x$. Therefore, ΔN is $3x - x$, or $2x$, and $r = 2x/x$, or 2. In general, when a population triples in size in a year, it grows by 200%, so the per capita growth rate is 2.
2. One cent per day, doubled each day (a rate of increase of 100%) for 30 days exceeds $10,000,000.
3. Logistic curves are S-shaped because exponential increase during early stages of population growth causes the first half of the curve to rise steeply, but then growth becomes limited by resource availability and so slows as population size approaches the carrying capacity of the environment.
4. Every organism invests all available energy in the ways that are most useful in the current environment. Increase in one area means decrease in another. To increase investments in all these areas, more resources in terms of food or energy would have to be acquired.
5. The closer island will initially receive more colonists than the farther one, but the larger island can eventually support more kinds of species.

Chapter 47

1. The fundamental niche is larger than the realized niche because the fundamental niche represents all the habitats where a species could possibly live, but competition among different species usually restricts a species to a smaller area, which is the realized niche.
2. In competitive exclusion, there is ongoing competition between two species for a particular resource, leading one to change its niche in the presence of the other. In resource partitioning, species evolve to use different resources, so there is no longer competition for that particular resource. Over time, competitive exclusion can lead to resource partitioning.
3. The cost to the apple tree is resources in the form of nectar, which takes energy to produce, and the benefit is pollination. The cost to the bee is the energy required to search for food and transfer pollen, and the benefit is food.
4. Although they can affect many other species in the Arctic, lemmings are not a keystone species because their effects are a function of their great abundance. Keystone species have effects on communities that are disproportionate to their numbers because of the particular roles they play, often as predators.
5. A community is the set of organisms that live in a given place. A community together with the physical environment in which the organisms live constitutes an ecosystem.

Chapter 48

1. Climates would not show strong seasonality.
2. Soil moisture reflects both rainfall and evaporation. Therefore, wet soils occur both near the equator where rainfall is high, and at high latitudes where rainfall is low but where evaporation is also low owing to cold climate, which limits the rate of evaporation of water from the soil.
3. In most biomes, nitrogen and phosphorus are limiting nutrients and so strongly constrain the rate of primary production.
4. Primary production on land is governed largely by temperature and rainfall, which vary by latitude. In contrast, primary production in the sea is strongly limited by nutrient availability; continental margins receive nutrients released by chemical weathering on the continents and transported to the ocean by rivers. They are also sites of upwelling, where nutrient-rich deep waters return to the surface.
5. A diversity gradient can be seen with altitude that is similar to the one seen with latitude: Typically, the higher the altitude, the fewer the number of species.

Chapter 49

1. In the 130 years from 1800 to 1930, our numbers doubled from 1 to 2 billion, and they doubled again in just the 45 years from 1930 to 1975. This pattern indicates that the *rate* of human population growth increased in this time.
2. Global warming is the measured increase in Earth's surface temperatures over the past 50 years. The greenhouse effect describes a process by which global warming can occur. The greenhouse effect is the result of the capacity of some molecules—especially carbon dioxide, methane, and water vapor—to absorb heat energy and then emit it in all directions. Without the greenhouse effect, Earth would not be habitable. However, in recent decades, increasing amounts of greenhouse gases in the atmosphere have resulted in global warming.
3. As ecosystems warm, plants able to open flowers earlier appear to have higher fitness, enabling them to maintain or increase their geographic ranges.
4. Invasive species may outcompete native species for available resources, diminishing the population size and, in time, the diversity of native species. Introduced animals may prey on native species, reducing their numbers. Or introduced species may simply increase local species diversity.
5. The Red Queen hypothesis suggests that organisms need to keep evolving (running) just to stay in the same ecological niche (place). As new antibiotics are developed, pathogens like malaria frequently evolve resistance, so researchers must continually strive to develop new medicines as natural selection erodes the effectiveness of the old.

GLOSSARY

10-nm fiber A relaxed 30-nm chromatin fiber, the state of the chromatin fiber in regions of the nucleus where transcription is currently taking place.

3′ end The end of a nucleic acid strand that carries a free 3′ hydroxyl.

30-nm fiber A chromosomal conformation created by the folding of the nucleosome fiber of DNA and histones.

3-phosphoglycerate (3-PGA) A 3-carbon molecule; two molecules of 3-PGA are the first stable products of the Calvin cycle.

5′ cap The modification of the 5′ end of the primary transcript by the addition of a special nucleotide attached in an unusual chemical linkage.

5′ end The end of a nucleic acid strand containing a free 5′ phosphate group.

ABC model A model of floral development that invokes three activities, A, B, and C, each of which represents the function of a protein or proteins that specify organ identity in each whorl.

abiotic factors Aspects of the physical environment, such as temperature, water availability, or wind, that affect an organism.

abomasum The fourth chamber in the stomach of ruminants, where protein digestion begins.

abscisic acid A plant hormone that triggers stomatal closure, stimulates root elongation, and maintains seed dormancy.

absolute temperature (T) Temperature measured on the Kelvin scale.

absorption The direct uptake of molecules by organisms, commonly to obtain food. In vertebrate digestion, it is the process by which breakdown products are taken up into the bloodstream.

accessory pigment A light-absorbing pigment other than chlorophyll in the photosynthetic membrane; carotenoids are important accessory pigments.

acclimatization An adaptive change in body function to a new environment.

acidic Describes a solution in which the concentration of protons is higher than that of hydroxide ions (the pH is lower than 7).

acoelomate A bilaterian without a body cavity.

acrosome An organelle that surrounds the head of the sperm containing enzymes that enable sperm to transverse the outer coating of the egg.

actin A protein subunit that makes up microfilaments; used by both striated and smooth muscles to contract and generate force.

action potential A brief electrical signal transmitted from the nerve cell body along one or more axon branches.

activation energy (E_A) The energy input necessary to reach the transition state.

activator A synthesized compound that increases the activity of an enzyme.

active site The portion of the enzyme that binds substrate and converts it to product.

active transport The “uphill” movement of substances against a concentration gradient requiring an input of energy.

adaptation In an evolutionary context, the fit between an organism and its environment that results from evolution by natural selection. In sensory reception, the process in which sensory receptors reduce their firing rate when a stimulus continues over a period of time.

adaptive (acquired) immunity The part of the immune system that is specific to given pathogens.

adaptive radiation A period of unusually rapid evolutionary diversification in which natural selection accelerates the rates of both speciation and adaptation in a single lineage.

addition rule The principle that the probability of either of two mutually exclusive outcomes occurring is given by the sum of their individual probabilities.

adenine (A) A purine base.

adenosine triphosphate (ATP) The molecule that provides energy in a form that all cells can readily use to perform the work of the cell. ATP is the universal energy currency for all cells.

adherens junction A beltlike junctional complex composed of cadherins that attaches a band of actin to the plasma membrane.

adrenal glands Paired glands located adjacent to the kidneys that secrete cortisol in times of stress.

adrenal medulla The inner part of the adrenal gland, which is stimulated by the sympathetic nervous system.

adrenaline (epinephrine) A hormone released by the adrenal gland that causes alertness and arousal.

advantageous mutations Genetic changes that improve their carriers’ chances of survival or reproduction.

advertisement display Behavior by which individuals draw attention to themselves or their status.

aerobic Utilizing oxygen.

aerobic metabolism Energy metabolism that uses oxygen gas to oxidize organic compounds, generating ATP; found in Bacteria, Archaea, and the mitochondria of eukaryotic organisms.

age structure The number of individuals within each age group of a population.

agonist muscles Muscle pairs that combine to produce similar motions.

aldosterone A hormone produced by the adrenal glands that stimulates the distal convoluted tubule and collecting ducts to take up more salt and water.

alga (plural, algae) A photosynthetic protist.

alkaloid Any one of a group of nitrogen-bearing compounds that damages the nervous system of animals, produced by some plants as a defensive mechanism.

allantois In the amniotic egg, a membrane that encloses a space where metabolic wastes collect.

allele frequency Among all the alleles of a gene in a population, the proportion that are of a specified allele.

alleles The different forms of a gene, corresponding to different DNA sequences in each different form.

allopatric Describes populations that are geographically separated from each other.

allosteric effect A change in the activity or affinity of a protein as the result of binding of a molecule to a site other than the active site.

allosteric enzyme An enzyme whose activity is affected by binding a molecule at a site other than the active site. Typically, allosteric enzymes change their shape on binding an activator or inhibitor.

alpha (α) carbon The central carbon atom of each amino acid.

alpha (α) helix One of the two principal types of secondary structure found in proteins.

alternation of generations The life cycle in which a haploid phase, the gametophyte, and a diploid phase, the sporophyte, follow one after the other.

alternative splicing A process in which primary transcripts from the same gene can be spliced in different ways to yield different mRNAs and therefore different protein products.

altruism Self-sacrificial behavior in which an individual's actions decrease its own fitness while increasing that of another individual.

Alveolata A eukaryotic superkingdom, defined by the presence of cortical alveoli, small vesicles that, in some species, store calcium ions.

alveoli (singular, alveolus) Clusters of tiny thin-walled sacs where gas exchange by diffusion takes place; found at the ends of very fine bronchioles.

amacrine cell A type of interneuron in the retina that communicates between neighboring bipolar cells and ganglion cells, enhancing motion detection and adjusting for changes in illumination of the visual scene.

amine hormone A hormone that is derived from a single aromatic amino acid, such as tyrosine.

amino acid An organic molecule containing a central carbon atom, a carboxyl group, an amino group, a hydrogen atom, and a side chain. Amino acids are the building blocks of proteins.

amino acid replacement A change in the identity of an amino acid at a particular site in a protein resulting from a mutation in the gene.

amino end The end of a polypeptide chain that has a free amino group.

amino group NH_2; a nitrogen atom bonded to two hydrogen atoms, covalently linked to the central carbon atom of an amino acid.

aminoacyl (A) site One of three binding sites for tRNA on the large subunit of a ribosome.

aminoacyl tRNA synthetase An enzyme that attaches a specific amino acid to a specific tRNA molecule.

amnion In the amniotic egg, a membrane surrounding a fluid-filled cavity that allows the embryo to develop in a watery environment.

amniotes The group of vertebrate animals that produces amniotic eggs; this group includes lizards, snakes, turtles, and crocodilians.

amniotic egg An egg that can exchange gases while retaining water, permitting amniotes to reproduce in dry terrestrial habitats that amphibian eggs cannot tolerate.

Amoebozoa A superkingdom of eukaryotes with amoeba-like cells that move and gather food by means of pseudopodia.

Amphibia A monophyletic group of vertebrates, including frogs and salamanders, with an aquatic larval form with gills and an adult terrestrial form that usually has lungs.

amphipathic Having both hydrophilic and hydrophobic regions.

amplified In PCR technology, an alternative term for "replicated."

amylase An enzyme that breaks down starch into smaller subunits.

anabolism The set of chemical reactions that build molecules from smaller units utilizing an input of energy, usually in the form of ATP. Anabolic reactions result in net energy storage within cells and the organism.

anaerobic metabolism Energy metabolism in the absence of oxygen.

analogous Describes similar characters that evolved independently in different organisms as a result of adaptation to similar environments.

anammox Anaerobic ammonia oxidation; energy metabolism in which ammonium ion is oxidized by nitrite, yielding nitrogen gas as a by-product.

anaphase The stage of mitosis in which the sister chromatids separate.

anaphase I The stage of meiosis I in which the two homologous chromosomes of each bivalent separate as they are pulled in opposite directions, but the sister chromatids remained joined at the centromere.

anaphase II The stage of meiosis II in which the centromere of each chromosome splits and the separated chromatids are pulled toward opposite poles of the spindle.

anchor A membrane protein that attaches to other proteins and helps to maintain cell structure and shape.

angiosperms The flowering plants; angiosperms are a monophyletic group of seed plants characterized by flowers, double fertilization, and fruits.

angiotensin II A hormone that causes the smooth muscles of arterioles throughout the body to constrict, which increases blood pressure as well as stimulates the release of the hormone aldosterone, increasing reabsorption of electrolytes and water by the kidneys.

annealing The coming together of complementary strands of single-stranded nucleic acids by base pairing.

annelid worms A phylum of worms (Annelida) that have a cylindrical body with distinct segments and a bilaterian body plan.

annual clock A biological clock that corresponds closely to a solar year.

anoxygenic Not producing oxygen; anoxygenic photosynthetic bacteria do not gain electrons from water and so do not generate oxygen gas.

antagonist muscles Muscle pairs that pull in opposite directions to produce opposing motions.

anterior pituitary gland The region of the pituitary gland that forms from epithelial cells that develop and push up from the roof of the mouth; it receives hormones from the hypothalamus that stimulate it to release hormones in turn.

anther In flowering plants, a structure at the top of a stamen consisting of several sporangia in which pollen is produced.

Anthropocene Period The modern era, so named to reflect the dominant impact of humans on Earth.

antibody A large protein produced by plasma cells that binds to molecules called antigens.

anticodon The sequence of three nucleotides in a tRNA molecule that base pairs with the corresponding codon in an mRNA molecule.

antidiuretic hormone (ADH) A posterior pituitary gland hormone that acts on the kidneys and controls the water permeability of the collecting ducts, thus regulating the concentration of urine that an animal excretes; also known as *vasopressin*.

antigen Any molecule that leads to the production of antibodies.

antigenic drift The gradual process in which a high rate of mutation leads to changes in the amino acid sequences of antigens, thus allowing a population of viruses to evolve over time and evade memory T and B cells.

antigenic shift Reassortment of RNA strands in the viral genome, leading to sudden changes in cell-surface proteins, thus making it difficult to predict from year to year which virus strains will be most prevalent and therefore what vaccine will be most effective.

antigenic variation The encoding of a protein at different times by any one of a number of different genes.

antigen-presenting cell A type of cell (including macrophages, dendritic cells, and B cells) that takes up an antigen and returns portions of it to the cell surface bound to MHC class II proteins.

antiparallel Oriented in opposite directions; the strands in a DNA duplex are antiparallel.

aorta A large artery through which oxygenated blood flows from the left ventricle to the head and rest of the body.

aortic body A sensory structure of the vertebrate aorta that monitors the levels of oxygen and carbon dioxide and the pressure of blood moving through the body.

aortic valve A valve beween the left ventricle and the aorta.

apical dominance The suppression of growth of axillary buds by the shoot apical meristem.

apomixis A process in some species of flowering plants in which seeds can develop even in the absence of fertilization.

appendicular Describes the part of the vertebrate skeleton that consists of the bones of the limbs, including the shoulder and pelvis.

appendix A narrow, tubelike structure that extends from the cecum. The appendix is a vestigial structure that has no clear function in nonherbivorous animals.

aquaporin A protein channel that allows water to flow through the plasma membrane more readily by facilitated diffusion.

aqueous Watery.

aqueous humor A clear watery liquid that fills the interior region in front of the lens of the vertebrate eye.

Archaea One of the three domains of life, consisting of single-celled organisms with a single circular chromosome and no true nucleus that divide by binary fission and differ from bacteria in many aspects of their cell and molecular biology.

Archaeplastida A eukaryotic superkingdom of photosynthetic organisms; includes the land plants.

Ardi A specimen of *Ardipethecus ramidus*, an important early hominin, dating from about 4.4 million years ago.

arteriole A small branch of an artery.

artery A large, high-pressure vessel that moves blood flow away from the heart to the tissues.

Arthropoda A monophyletic group of animals that includes insects and contains more than half of all known animal species; distinguished by their segmented bodies and jointed legs.

artificial selection A form of directional selection analogous to natural selection, but without the competitive element; successful genotypes are selected by the breeder, not by competition.

ascomycetes A monophyletic fungal subgroup of the Dikarya, making up 64% of all fungal species, in which nuclear fusion and meiosis take place in an elongated saclike cell called an ascus; also called *sac fungi*.

asexual reproduction The reproduction of cells or single-celled organisms by cell division; offspring are clones of the parent.

assimilation The process by which organisms take up nutrients from the environment.

associative learning (conditioning) Learning that two events are correlated.

astrocyte A type of star-shaped glial cell that contributes to the blood–brain barrier by surrounding blood vessels in the brain and thus limiting the size of compounds that can diffuse from the blood into the brain.

atom The basic unit of matter.

atomic mass The mass of the atom determined by the number of protons and neutrons.

ATP synthase An enzyme that couples the movement of protons through the enzyme with the synthesis of ATP.

atrioventricular (AV) node A specialized region of the heart containing pacemaker cells that transmit action potentials from the sinoatrial nodes to the ventricles of the heart.

atrioventricular (AV) valve A valve between the right atrium and the right ventricle and between the left ventricle and left atrium.

atrium (plural, atria) A heart chamber that receives blood from the lungs or the rest of the body.

auditory cortex The area of the brain that processes sound.

autocrine signaling Signaling between different parts of a cell; the signaling cell and the responding cell are one and the same.

autoimmune disease A disease in which tolerance is lost and the immune system becomes active against antigens of the host.

autonomic nervous system The involuntary component of the peripheral nervous system, which controls internal functions of the body such as heart rate, blood flow, digestion, excretion, and temperature.

autosome Any chromosome other than the sex chromosomes.

autotroph Any organism that is able to convert carbon dioxide into glucose, thus providing its own organic source of carbon.

auxin A plant hormone that causes shoots to elongate and guides vascular differentiation.

avirulent In plants, describes pathogens that damage only a small part of the plant because the host plant is able to contain the infection. In animals, describes nonpathogenic microorganisms.

axial Describes the part of the vertebrate skeleton that consists of the skull and jaws of the head, the vertebrae of the spinal column, and the ribs.

axillary bud A meristem that forms at the base of each leaf.

axon The fiberlike extension from the cell body of a neuron that transmits signals away from the nerve's cell body; the output end of a nerve cell.

axon hillock The junction of the nerve cell body and its axon.

B lymphocyte (B cell) A cell type that matures in the bone marrow of humans and produces antibodies.

Bacteria One of the three monophyletic domains of life, consisting of single-celled organisms with a single circular chromosome but no nucleus that divide by binary fission and differ from archaeons in many aspects of their cell and molecular biology.

bacteriochlorophyll A light-harvesting pigment closely related to the chlorophyll found in plants, algae, and cyanobacteria.

bacteriophage Virus that infects bacterial cells.

balancing selection Natural selection that acts to maintain two or more alleles of a given gene in a population.

ball-and-socket joint A joint that allows rotation in three axes, like the hip and shoulder.

basal lamina A specialized form of extracellular matrix that underlies and supports all epithelial tissues.

base A nitrogen-containing compound that makes up part of a nucleotide.

base excision pair A specialized repair system in which an incorrect DNA base and its sugar are both removed and the resulting gap is repaired.

base stacking Stabilizing hydrophobic interactions between bases in the same strand of DNA.

basic Describes a solution in which the concentration of protons is lower than that of hydroxide ions (the pH is higher than 7).

basidiomycetes A monophyletic fungal subgroup of the Dikarya, including smuts, rusts, and mushrooms, in which nuclear fusion and meiosis take place in a club-shaped cell called a basidium; also called club fungi.

basilar membrane The membrane that, with the cochlear duct, separates the upper and lower canal of the cochlea.

basophil A type of granulocyte that, along with eosinphils, defends against parasitic infections but also contributes to allergies.

behaviorally isolated Describes individuals that only mate with other individuals on the basis of specific courtship rituals, songs, and other behaviors.

beta (β) sheet One of the two principal types of secondary structure found in proteins.

beta-(β-)oxidation The process of shortening fatty acids by a series of reactions that sequentially remove two carbon units from their ends.

bilateral symmetry Symmetry on both sides of a midline; animals with bilateral symmetry have a distinct head and tail, marking front and back, with a single plane of symmetry running between them at the midline.

Bilateria The monophyletic group of animals with bilateral symmetry.

bilayer A two-layered structure of the cell membrane with hydrophilic "heads" pointing outward toward the aqueous environment and hydrophobic "tails" oriented inward, away from water.

bile A fluid produced by the liver that aids in fat digestion by breaking large clusters of fats into smaller lipid droplets.

binary fission The process by which cells of bacteria or archaeons divide.

binding affinity The tightness of the binding between the receptor and the signaling molecule.

biodiversity Biological diversity; the aggregate number of species, or, more broadly, also the diversity of genetic sequences, cell types, metabolism, life history, phylogenetic groups, communities, and ecosystems.

biodiversity hotspots Relatively small areas that have unusually high numbers of endemic species and that are under threat from human activities.

biogeochemical cycles The cycling of carbon and other biologically important elements through the biosphere.

biological species concept (BSC) As described by Ernst Mayr, the concept that "species are groups of actually or potentially interbreeding populations that are reproductively isolated from other such groups." The BSC is the most widely used and accepted definition of a species, but cannot be applied to Bacteria or Archaea.

biologist A scientist who studies life.

biology The science of life and how it works.

biome The distinctive and stable assemblage of species found over a broad region of Earth; terrestrial biomes are each recognized by their distinctive vegetation.

biomineralization The precipitation of minerals by organisms, as in the formation of skeletons.

biotrophic pathogen A plant pathogen that obtains resources from living cells.

biparental inheritance A type of inheritance in which the organelles in the offspring cells derive from those in both parents.

bipedal Moving by two feet and habitually walking upright.

bipolar cell A type of interneuron in the retina that adjusts its release of neurotransmitter in response to input from rod and cone cells.

biotic factors Any aspect of organisms that affects another organism.

bivalent The four-stranded structure consisting of two pairs of sister chromatids aligned along their length and held together by chiasmata.

bivalves A group of mollusks that includes clams, oysters, and mussels; they have an enclosing skeleton in which two hard shells are connected by a flexible hinge.

bladder A hollow organ in mammals and fishes for the storage and elimination of urine.

blastocyst A hollow sphere produced by cells in the morula that move in relation to one another, pushing against and expanding the membrane that encloses them. A blastocyst forms from the blastula, has an inner cell mass, and occurs only in mammals.

blastula A fluid-filled ball of undifferentiated cells formed after the fertilized egg has undergone several rounds of mitotic cell division following the morula stage.

blending inheritance The now-discredited model in which heredity factors transmitted by the parents become intermingled in the offspring instead of retaining their individual genetic identities.

blood The circulatory fluid in vertebrates.

bone marrow A fatty tissue between trabeculae and within the central cavity of a bone that contains many important cell populations.

bottleneck An extreme, usually temporary, reduction in population size that often results in marked genetic drift.

Bowman's capsule A membranous sac that encases the glomerulus.

brain The centralized concentration of neurons in an organ that processes complex sensory stimuli from the environment or from anywhere in the body.

brainstem Part of the vertebrate brain, formed from the midbrain, which activates the forebrain by relaying information from lower spinal levels.

bronchiole Any one of the fine branches of secondary bronchi.

bryophytes A paraphyletic group of nonvascular plants that includes the mosses, liverworts, and hornworts.

bud scale One of many like structures formed from leaf primordia that, together, protect shoot apical meristems from desiccation and damage due to cold.

budding A form of asexual reproduction in fungi, plants, and some animals in which a bud forms on the organism and eventually breaks off to form a new organism that is smaller than its parent.

bulbourethral glands Glands below the prostate gland that produce a clear fluid that lubricates the urethra for passage of the sperm.

bulk flow The movement of molecules through organisms due to pressure differences at rates beyond those possible by diffusion across a concentration gradient.

bundle sheath A cylinder of cells that surrounds each vein in C_4 plants in which carbon dioxide is concentrated in bundle sheath cells, suppressing photorespiration.

Burgess Shale A sedimentary rock formation in British Columbia, Canada, that preserves a remarkable sampling of marine life during the initial diversification of animals.

C_3 plant A plant that does not use 4-carbon organic acids to supply the Calvin cycle with carbon dioxide.

C_4 plant A plant in which carbon dioxide is incorporated into 4-carbon organic acids that are then used to supply the Calvin cycle.

cadherin A calcium-dependent adherence protein, important in the adhesion of cells to other cells.

calmodulin A protein that binds with Ca^{21} and activates the enzyme myosin kinase.

Calvin cycle The process in which carbon dioxide is reduced to synthesize carbohydrates, with ATP and NADPH as the energy sources.

Cambrian explosion A transition period in geologic time during which the body plans characteristic of most bilaterian phyla developed.

canine One of the teeth in carnivores specialized for piercing the body of prey.

capacitation A series of physiological changes that allow the sperm to fertilize the egg.

capillary A very small blood vessel, arranged in finely branched networks connected to arterioles or venules, where gases are exchanged by diffusion with surrounding tissues.

capsid The protein coat that surrounds the nucleic acid of a virus.

carbohydrate An organic molecule containing C, H, and O atoms that provides a source of energy for metabolism and that forms the starting point for the synthesis of all other organic molecules.

carbon cycle The intricately linked network of biological and physical processes that shuttles carbon among rocks, soil, oceans, air, and organisms.

carboxyl end The end of a polypeptide chain that has a free carboxyl group.

carboxyl group COOH; a carbon atom with a double bond to oxygen and a single bond to a hydroxyl group.

carboxylation The first step of the Calvin cycle, in which carbon dioxide absorbed from the air is added to a 5-carbon molecule.

cardiac cycle The contraction of the two atria of the heart followed by contraction of the two ventricles.

cardiac muscle Muscle cells that make up the walls of the atria and ventricles and contract to pump blood through the heart.

cardiac output (CO) The volume of blood pumped by the heart over a given interval of time, the key measure of heart function.

carnivores A monophyletic group of animals, including cats, dogs, seals, and their relatives, that consume other animals.

carotid body A chemosensory structure of the carotid artery that monitors the levels of oxygen and carbon dioxide and the pH of blood moving to the brain.

carpel An ovule-producing floral organ in the center whorl.

carrier A transporter that facilitates movement of molecules.

carrying capacity (*K*) The maximum number of individuals a habitat can support.

cartilage A type of connective tissue found, for example, in the walls of intervertebral discs and the joint surfaces between adjacent bones.

Casparian strip A thin band of hydrophobic material that encircles each cell of the endodermis of a root, controlling which materials enter the xylem.

catabolism The set of chemical reactions that break down molecules into smaller units and, in the process, produces ATP to meet the energy needs of the cell.

causation A relationship in which one event leads to another.

cavitation The abrupt replacement of the water in a conduit by water vapor, which blocks water flow in xylem.

cecum A chamber that branches off the large intestine; along with the colon, the site of hindgut fermentation.

cell The simplest self-replicating entity that can exist as an independent unit of life.

cell adhesion molecule A cell-surface protein that attaches cells to one another and to the extracellular matrix.

cell cycle The collective name for the steps that make up eukaryotic cell division.

cell division The process by which cells make more cells.

cell junction A complex of proteins in the plasma membrane where a cell makes contact with another cell or the extracellular matrix.

cell theory The theory that the cell is the fundamental unit of life in all organisms and that cells come only from preexisting cells.

cell wall A defining boundary in many organism, external to the cell membrane, that helps maintain the shape and internal composition of the cell.

cell-mediated immunity The ability of T cells, which do not secrete antibodies, to recognize and act against pathogens directly.

cellular blastoderm In *Drosophila* development, the structure formed by the nuclei in the single-cell embryo when they migrate to the periphery of the embryo and each nucleus becomes enclosed in its own cell membrane.

cellular respiration A series of chemical reactions that convert the energy stored in nutrients into a chemical form that can be readily used by cells.

cellulase The enzyme that breaks down cellulose.

central dogma The theory that information transfer in a cell usually goes from DNA to RNA to protein.

central nervous system (CNS) In vertebrates, the brain and spinal cord; in invertebrates, centralized information-processing ganglia.

centromere A constriction that physically holds sister chromatids together; the site of the attachment of the spindle fibers that move the chromosome in cell division.

centrosome A compact structure that is the microtubule organizing center for animal cells.

cephalization The concentration of nervous system components at one end of the body.

cephalochordates A subphylum of Chordata that shares key features of body organization with vertebrates but lacks a well-developed brain and eyes, has no lateral appendages, and does not have a mineralized skeleton.

cephalopods A monophyletic group of mollusks, including squid, cuttlefish, octopus, and chambered nautilus, with distinctive adaptations such as well-developed eyes and muscular tentacles that capture prey and sense the environment.

cerebellum Part of the vertebrate brain, formed from the hindbrain, which coordinates complex motor tasks by integrating motor and sensory information.

cerebral cortex Part of the vertebrate brain formed from a portion of the forebrain that is greatly expanded in mammals, particularly primates.

cerebrum The outer left and right hemispheres of the cerebral cortex.

cervix The end, or neck, of the uterus.

chain terminator A term for a dideoxynucleotide, which if incorporated into a growing daughter strand stops strand growth because there is no hydroxyl group to attack the incoming nucleotide.

channel A transporter with a passage that allows the movement of molecules through it.

chaperone A protein that helps shield a slow-folding protein until it can attain its proper three-dimensional structure.

character In the discipline of systematics, an anatomical, physiological, or molecular feature of an organism that varies among closely related species.

character state The observed condition of a character, such as presence or absence of lungs or arrangement of petals.

checkpoint One of multiple regulatory mechanism that coordinate the temporal sequence of events in the cell cycle.

chelicerates One of the four main groups of arthropods, including spiders and scorpions, chelicerates have pincer-like claws and are the only arthropods that lack antennae.

chemical bond Any form of attraction between atoms that holds them together.

chemical energy A form of potential energy held in the chemical bonds between pairs of atoms in a molecule.

chemical reaction The process by which molecules are transformed into different molecules.

chemoautotroph A microorganism that obtains energy from chemical compounds, not from sunlight.

chemoreceptor A receptor that responds to molecules that bind to specific protein receptors on the cell membrane of the sensory receptor.

chemiosmotic hypothesis The hypothesis that the gradient of protons across a membrane provides a source of potential energy that is converted into chemical energy stored in ATP.

chemotroph An organism that derives its energy directly from organic molecules such as glucose.

chiasma (plural, chiasmata) A crosslike structure within a bivalent constituting a physical manifestation of crossing over.

chitin A modified polysaccharide containing nitrogen that makes up the cell walls of fungi and the hard exoskeletons of arthropods.

chloride cell A type of specialized cell in the gills of marine bony fishes that counters the ingestion and diffusion of excess electrolytes into the animal by pumping chloride ions into the surrounding seawater; chloride cells in freshwater fishes have opposite polarity.

chlorophyll The major photosynthetic pigment contained in the thylakoid membrane; it plays a key role in the chloroplast's ability to capture energy from sunlight. Chlorophyll appears green because it is poor at absorbing green wavelengths.

chloroplast An organelle that converts energy of sunlight into chemical energy by synthesizing simple sugars.

chloroplast genome In photosynthetic eukaryotes, the genome of the chloroplast.

choanocyte A type of cell that lines the interior surface of a sponge; choanocytes have flagella and function in nutrition and gas exchange.

choanoflagellate One of a group of mostly unicellular protists characterized by a ring of microvilli around the cell's single flagellum.

cholecystokinin (CCK) A peptide hormone that causes the gallbladder to contract and thus release bile into the duodenum.

cholesterol An amphipathic lipid that is a major component of animal cell membranes.

Chondrichthyes Cartilaginous fish, a monophyletic group that includes about 800 species of sharks, rays, and chimaeras.

Chordata One of the three major phyla of deuterostomes, this group includes vertebrates and closely related invertebrate animals such as sea squirts.

chorion In the amniotic egg, a membrane that surrounds the entire embryo along with its yolk and allantoic sac.

chromatin A complex of DNA, RNA, and proteins that gives chromosomes their structure; chromatin fibers are either 30 nm in diameter or, in a relaxed state, 10 nm.

chromatin remodeling The process in which the nucleosomes are repositioned to expose different stretches of DNA to the nuclear environment.

chromosome In eukaryotes, the physical structure in which DNA in the nucleus is packaged; used more loosely to refer to the DNA in bacterial cells or archaeons.

chromosome condensation The progressive coiling of the chromatin fiber, an active, energy-consuming process requiring the participation of several types of proteins.

chytrid A single-celled aquatic fungus with chitin walls that attaches to decomposing organic matter.

cilium (plural, cilia) A hairlike organelle that propels the movement of cells or of substances within cells or out of the body; shorter than a flagellum.

circadian clock A biological clock, on a near-daily cycle, that can be set by external cues and regulates many daily rhythms. In animals, the circadian clock affects feeding, sleeping, hormone production, and core body temperature.

circular muscle Smooth muscle that encircles the body or an organ; in the digestive tract, a circular muscle layer contracts to reduce the size of the lumen. A circular muscle layer contracts alternately with longitudinal muscle to move contents through the digestive tract and to enable locomotion in animals with hydrostatic skeletons.

circulation The movement of a specialized body fluid that carries oxygen and carbon dioxide, and nutrients and waste products, through the body.

cis-regulatory element A short DNA sequence adjacent to a gene, usually at the 54′ end, that interacts with transcription factors.

cisternae (singular, cisterna) The series of flattened membrane sacs that make up the Golgi apparatus.

citric acid cycle The third stage of cellular respiration, in which acetyl-CoA is broken down and more carbon dioxide is released.

cladistics Phylogenetic reconstruction on the basis of shared evolutionary changes in characters, often called synapomorphies.

class In the Linnaean system of classification, a group of closely related orders.

classical conditioning Associative learning in which two stimuli are paired.

cleavage The successive mitotic divisions of the zygote after fertilization, in which the single large egg is divided into many smaller cells.

climax community A mature assembly, a final stage in succession, in which there is little further change in species composition.

clitoris The female homolog of the glans penis.

clonal selection A hypothesis proposing that the antigen instructs the antibody to fold in a particular way so that the two interact in a specific manner; now a central principle of immunology.

clone An individual that carries an exact copy of the nuclear genome of another individual; clones are genetically identical cells or individuals.

closed circulatory system A circulatory system made up of a set of internal vessels and a heart that functions as a pump to move blood to different regions of the body.

Cnidaria A phylum characterized by radial symmetry, two germ tissues in the embryo, a closed internal gastric cavity, and well-developed tissues but not organs; includes jellyfish, sea anemones, and corals.

cochlea A coiled chamber within the skull containing hair cells that convert pressure waves into an electrical impulse that is sent to the brain.

cochlear duct A fluid-filled cavity in the cochlea, next to the upper canal, that houses the organ of Corti.

codon A group of three adjacent nucleotides in RNA that specifies an amino acid in a protein or that terminates polypeptide synthesis.

coelacanth A genus of two species of lobe-finned fish found off the coast of Africa and thought to have been extinct for 80 million years but which is still living today; along with lungfish, the nearest relative of tetrapods.

coelom A body cavity surrounding the gut.

coenocytic Containing many nuclei within one giant cell; the nucleus divides multiple times, but the nuclei are not partitioned into individual cells.

coenzyme Q (CoQ) In respiration, a mobile electron acceptor that transports electrons from complexes I and II to complex III in the electron transport chain and moves protons from the mitochondrial matrix to the intermembrane space.

coevolution The process in which species evolve together, each responding to selective pressures from the other.

cofactor A substance that associates with an enzyme and plays a key role in its function.

cognition The ability of the brain to process and integrate complex sources of information, interpret and remember past events, solve problems, reason, and form ideas.

cohesion Attraction between molecules; one consequence of cohesion is high surface tension.

cohort A group of the individuals born at a given time.

collagen A strong protein fiber found in bone, artery walls, and other connective tissues.

collecting duct A type of duct in the vertebrate kidney where urine collects.

colon Part of the hindgut and the site of reabsorption of water and minerals; also known as the *large intestine.*

combinatorial control Regulation of gene transcription by means of multiple transcription factors acting together.

commensalism An interaction between species in which one partner benefits with no apparent effect on the other.

communication The transfer of information between two individuals, the sender and the receiver.

community The set of all populations found in a given place.

compact bone Dense, mineralized bone tissue that forms the walls of a bone's shaft.

companion cell In angiosperms, a cell that carries out cellular functions such as protein synthesis for sieve tubes.

comparative genomics The analysis of the similarities and differences in protein-coding genes and other types of sequence in the genomes of different species.

competition An interaction in which the use of a mutually needed resource by one individual or group of individuals lowers the availability of the resource for another individual or group.

competitive exclusion The result of an antagonistic interaction in which one species is prevented from occupying a particular habitat or niche.

complement system The collective name for certain proteins circulating in the blood that participate in innate immune function and thus complement other parts of the immune system.

complementary Describes the relationship of purine and pyrimidine bases, in which the base A pairs only with T and G pairs only with C.

complex carbohydrate A long, branched chain of monosaccharides.

complex trait A trait that is influenced by multiple genes as well as by the environment.

compound eye An eye structure found in insects and crustaceans that consists of a number of ommatidia, individual light-focusing elements.

concordance The percentage of cases in which both members of a pair of twins show the trait when it is known that at least one member shows it.

cone cell A type of photoreceptor cell on the retina that detects color.

conjugation The direct cell-to-cell transfer of DNA, usually in the form of a plasmid.

conservation biology The efforts by biologists and policymakers to address the challenge of sustaining biodiversity in a changing world crowded with people.

conserved Describes sequences that are similar in different organisms.

constant (C) Describes an unchanging region of the H and L chains.

constitutive Describes expression of a gene that occurs continuously.

constitutive defense A defense that is always active.

consumer An organism that obtains the carbon it needs for growth and reproduction from the foods it eats and gains energy by respiring food molecules; heterotrophic organisms of all kinds that directly consume primary producers or consume those that do.

contractile ring In animal cells, a ring of actin filaments that forms at the equator of the cell perpendicular to the axis of what was the spindle at the beginning of cytokinesis.

contractile vacuole A type of cellular compartment that takes up excess water and waste products from inside the cell and expels them into the external environment.

control group The group that is not exposed to the variable in an experiment.

cooperative binding The increase in binding affinity with additional binding of O_2.

copy-number variation (CNV) Differences among individuals in the number of copies of a region of the genome.

Coriolis effect The phenomenon in which, because of Earth's counterclockwise rotation about its axis, winds and oceanic currents in the Northern Hemisphere deflect to the right, and those in the Southern Hemisphere deflect to the left.

cork cambium Lateral meristem that renews and maintains an outer layer that protects the stem against herbivores, mechanical damage, desiccation, and fire.

cornea The transparent portion of the sclera in the front of the vertebrate eye.

corpus luteum A temporary endocrine structure that secretes progesterone.

correlation The co-occurrence of two events or processes; correlation does not imply causation.

corridors Areas that provide species with routes for migration from one reserve to another.

cortex In a stem, the region between the epidermis and the vascular bundles, composed of parenchyma cells. In the mammalian brain, the highly folded outer layer of gray matter, about 4 mm thick, made up of densely packed neuron cell bodies and their dendrites. In the mammalian renal system, the outer layer of the kidney.

co-speciation A process in which two groups of organisms speciate in response to each other and at the same time, producing matching phylogenies.

countercurrent exchange A mechanism in which two fluids flow in opposite directions, exchanging properties.

countercurrent multiplier A system that generates a concentration gradient as two fluids move in parallel but in opposite directions.

covalent bond A chemical bond formed by a shared pair of electrons holding two different atoms together.

CpG island A cluster of CpG sites on a DNA strand where cytosine (C) is adjacent to guanosine (G); the "p" represents the phosphate in the backbone.

cranial nerve In vertebrates, a nerve that links specialized sensory organs to the brain; most contain axons of both sensory and motor neurons.

craniates A subphylum of Chordata, distinguished by a bony cranium that protects the brain; also known as *vertebrates*.

crassulacean acid metabolism (CAM) A mechanism in plants that helps balance carbon dioxide gain and water loss by capturing carbon dioxide into 4-carbon organic acids at night, when transpiration rates are low, and then using it to supply the Calvin cycle during the day while stomata remain closed.

Crenarchaeota One of the three major divisions of Archaea; includes acid-loving microorganisms.

CRISPR (clustered regularly interspaced short palindromic repeats) A method of DNA editing in which any sequence in the genome can be replaced with any other sequence; the phrase describes the organization of the viral DNA segments in certain bacterial genomes from which the DNA editing method derives.

crisscross inheritance A pattern in which an X chromosome present in a male in one generation is transmitted to a female in the next generation, and in the generation after that can be transmitted back to a male.

Cro-Magnon The first known population of *Homo sapiens* in Europe, named for the site in France where specimens were first described.

crop The last part of the foregut, which serves as an initial storage and digestive chamber; also known as the *stomach*.

cross-bridge The binding of the head of a myosin molecule to actin at a specific site between the myosin and actin filaments.

cross-bridge cycle Repeated sequential interactions between myosin and actin filaments at cross-bridges that cause a muscle fiber to contract.

crosscurrent Running across another current at 90 degrees.

crossover The physical breakage, exchange of parts, and reunion between non-sister chromatids.

crustaceans One of the four main groups of arthropods, including lobsters, shrimp, and crabs; distinguished by two pairs of antennae and their branched legs or other appendages.

ctenophores Comb-jellies; species in this phylum have a radial body plan but a flow-through gut; they propel themselves by cilia arranged like a comb along the long axis of the body.

cultural transmission The transfer of information among individuals through learning or imitation.

cuticle In leaves, a protective layer of a waxy substance secreted by epidermal cells that limits water loss; also, an exoskeleton that covers the bodies of invertebrates such as nematodes and arthropods.

C-value paradox The disconnect between genome size and organismal complexity (the C-value is the amount of DNA in a reproductive cell).

cyanobacteria A monophyletic group of bacteria capable of oxygenic photosynthesis.

cyclic electron transport An alternative pathway for electrons during the Calvin cycle that increases the production of ATP.

cyclin A regulatory protein whose levels rise and fall with each round of the cell cycle.

cyclin-dependent kinase (CDK) A kinase that is always present within the cell but active only when bound to the appropriate cyclin.

cytochrome-*b6f* complex Part of the photosynthetic electron transport chain, through which electrons pass between photosystem II and photosystem I.

cytochrome c The enzyme to which electrons are transferred in complex III of the electron transport chain.

cytokine A chemical messenger released by phagocytes that recruits other immune cells to the site of injury or infection.

cytokinesis In eukaryotic cells, the division of the cytoplasm into two separate cells.

cytokinin A plant hormone that stimulates cell division.

cytoplasm The contents of the cell other than the nucleus.

cytosine (C) A pyrimidine base.

cytoskeleton In eukaryotes, an internal protein scaffold that helps cells to maintain their shape and serves as a network of tracks for the movement of substances within cells.

cytosol The region of the cell inside the plasma membrane but outside the organelles; the jelly-like internal environment that surrounds the organelles.

cytotoxic T cell One of a subpopulation of T cells, activated by cytokines released from helper T cells, that can kill other cells.

daughter strand In DNA replication, the strand synthesized from a parental template strand.

day-neutral plant A plant that flowers independently of any change in day length.

decomposer An organism that breaks down dead tissues, feeding on the dead cells or bodies of other organisms.

delayed hypersensitivity reaction Reactions initiated by helper T cells, which release cytokines that attract macrophages to the site of exposure, which is typically the skin.

deleterious mutations Genetic changes that are harmful to an organism.

deletion A missing region of a gene or chromosome.

demography The study of the size, structure, and distribution of populations over time, including changes in response to birth, aging, migration, and death.

denaturation The unfolding of proteins by chemical treatment or high temperature; the separation of paired, complementary strands of nucleid acid.

dendrite A fiberlike extension from the cell body of a neuron that receives signals from other nerve cells or from specialized sensory endings; the input end of a nerve cell.

dendritic cell A type of cell with long cellular projections that is typically part of the natural defenses found in the skin and mucous membranes.

denitrification The process in which some bacteria use nitrate as an electron acceptor in respiration.

density-dependent Describes processes affecting populations that are influenced by the number of individual organisms, such as the use of resources or susceptibility to predation or parasitism.

density-independent Describes factors such as severe drought that influence population size without regard for the density of the population.

deoxyribonucleic acid (DNA) A linear polymer of four subunits; the information archive in all organisms.

deoxyribose The sugar in DNA.

depolarization An increase in membrane potential

dermis The layer of skin beneath the epidermis, consisting of connective tissue, hair follicles, blood and lymphatic vessels, and glands. It supports the epidermis both physically and by supplying it with nutrients and provides a cushion surrounding the body.

desiccation Excessive water loss; drying out.

desiccation tolerance A suite of biochemical traits that allows cells to survive extreme dehydration without damage to membranes or macromolecules.

desmosome A buttonlike point of adhesion that holds the plasma membranes of adjacent cells together.

deuterostome A bilaterian in which the blastopore, the first opening to the internal cavity of the developing embryo, becomes the anus. The taxonomic name is Deuterostomia and includes humans and other chordates.

development The process in which a fertilized egg undergoes multiple rounds of cell division to become an embryo with specialized tissues and organs.

diabetes mellitus A disease that results when the control of blood-glucose levels by insulin fails.

diaphragm A domed sheet of muscle at the base of the lungs that separates the thoracic and abdominal cavities and contracts to drive inhalation.

diaphysis The middle region of a bone; blood vessels invading the diaphysis and epiphysis trigger the transformation of cartilage into bone.

diastole The relaxation of the ventricles, allowing the heart to fill with blood.

dideoxynucleotide A nucleotide lacking both the 2′ and 3′ hydroxyl groups on the sugar ring.

dietary mineral A chemical element other than carbon, hydrogen, oxygen, or nitrogen that is required in the diet and must be obtained in food; see also *minerals.*

differentiation The process in which cells become progressively more specialized as a result of gene regulation.

diffusion The random motion of individual molecules, with net movement occurring where there are areas of higher and lower concentration of the molecules.

digestive tract Collectively, the passages that connect the mouth, digestive organs, and anus; also known as the *gut.*

Dikarya A vast fungal group that includes about 98% of all described fungal species and in which dikaryotic cells are formed.

dikaryotic ($n + n$) Having two haploid nuclei, one from each parent, in each cell.

diploid Describes a cell with two complete sets of chromosomes.

directional selection A form of selection that selects against one of two extremes and leads over time to a change in a trait.

dispersal The process in which some individuals colonize a distant place far from the main source population.

display A pattern of behavior that is species specific and tends to be highly repeatable and similar from one individual to the next.

disruptive selection A form of selection that operates in favor of extremes and against intermediate forms, selecting against the mean.

distal convoluted tubule The third portion of the renal tubule, in which urea is the principal solute and into which other wastes from the bloodstream are secreted.

disturbance A severe physical impact on a habitat that has density-independent effects on populations of interacting species.

divergence The slow accumulation of differences between duplicate copies of a gene that occurs over time.

DNA fingerprinting The analysis of a small quantity of DNA to uniquely identify an individual; also known as *DNA typing.*

DNA ligase An enzyme that uses the energy in ATP to close a nick in a DNA strand, joining the 3′ hydroxyl of one end to the 5′ phosphate of the other end.

DNA polymerase An enzyme that is a critical component of a large protein complex that carries out DNA replication.

DNA replication The process of duplicating a DNA molecule, during which the parental strands separate and new partner strands are made.

DNA transposons Repeated DNA sequences that replicate and can move from one location to another in the genome by DNA replication and repair.

DNA typing The analysis of a small quantity of DNA to uniquely identify an individual; also known as *DNA fingerprinting.*

domain One of the three largest limbs of the tree of life: Eukarya, Bacteria, or Archaea.

dominant The trait that appears in the heterozygous offspring of a cross between homozygous genotypes.

donor In recombinant DNA technology, the source of the DNA fragment that is inserted into a cell of another organism.

dormancy Describes a time period in the life of an organism when growth, development, and metabolism slow down or stop; in plants, a state in which seeds are prevented from germinating.

dorsal nerve cord A nerve cord that develops in a location dorsal to the notochord; this embryonic feature is unique to chordates.

dosage The number of copies of each gene in a chromosome.

dosage compensation The differential regulation of X-chromosomal genes in females and in males.

double bond A covalent bond in which covalently joined atoms share two pairs of electrons.

double fertilization In angiosperms, the process in which two sperm from a single pollen tube fuse with (1) the egg and (2) the two haploid nuclei of the central cell.

double helix The structure formed by two strands of complementary nucleotides that coil around each other.

Down syndrome A condition resulting from the presence of an extra copy of chromosome 21; also known as *trisomy 21.*

downstream gene A gene that functions later than another in development.

duodenum The initial section of the small intestine, into which food enters from the stomach.

duplication A region of a chromosome that is present twice instead of once.

duplication and divergence The process of creating new genes by duplication followed by change in sequence over evolutionary time.

dynamic instability Cycles of shrinkage and growth in microtubules.

dynein A motor protein that carries cargo away from the plasma membrane toward the minus ends of microtubules.

eardrum In mammals, another name for the *tympanic membrane,* which transmits airborne sounds into the ear.

Echinodermata One of the three major phyla of deuterosomes, defined by five-part symmetry; this group includes sea urchins and sea stars.

echolocation Using sound waves to locate an object; bats find insect prey by emitting short bursts of high-frequency sound that bounce off surrounding objects and are reflected to the bat's ears.

ecological footprint The quantification of individual human claims on global resources by adding up all the energy, food, materials, and services used and estimating how much land is required to provide those resources.

ecological isolation Pre-zygotic isolation between individuals that specialize ecologically in different ways.

ecological niche A complete description of the role a species plays in its environment.

ecological species concept (ESC) The concept that there is a one-to-one correspondence between a species and its niche.

ecology The study of how organisms interact with one another and with their physical environment in nature.

ecosystem A community of organisms and the physical environment it occupies.

ecosystem services Benefits to humans provided by biodiversity, such as cleaner air and water, greater primary productivity, improved resilience to environmental disruption, and untapped sources of food and molecular compounds for use in medicine and agriculture.

ectoderm The outer germ layer, which differentiates into epithelial cells, pigment cells in the skin, nerve cells in the brain, and the cornea and lens of the eye.

ectomycorrhizae One of the two main types of mycorrhizae; ectomycorrhizae produce a thick sheath of fungal cells (hyphae) that surround, but do not penetrate, root cells.

ectotherm An animal that obtains most of its heat from the environment.

ejaculatory duct The duct through which sperm travel from the vas deferens to the urethra.

elastin A protein fiber found in artery walls that provides elasticity.

electrochemical gradient A gradient that combines the charge gradient and the chemical gradient of protons and other ions.

electromagnetic receptor A receptor that responds to electrical, magnetic, or light stimuli.

electron A negatively charged particle that moves around the atomic nucleus.

electron carrier A molecule that carries electrons (and energy) from one set of reactions to another.

electron donor A molecule that loses electrons.

electron transport chain The system that transfers electrons along a series of membrane-associated proteins to a final electron acceptor, using the energy released as electrons move down the chain to produce ATP.

electronegativity The ability of atoms to attract electrons.

electroreceptor A sensory receptor found in some fish that enables them to detect weak electrical signals emitted by other organisms.

element A pure substance, such as oxygen, copper, gold, or sodium, that cannot be further broken down by the methods of chemistry.

elongation The process in protein translation in which successive amino acids are added one by one to the growing polypeptide chain.

elongation factor A protein that breaks the high-energy bonds of the molecule GTP to provide energy for ribosome movement and elongation of a growing polypeptide chain.

embryo An early stage of multicellular development that results from successive mitotic divisions of the zygote.

endemic species Species found in one place in the world and nowhere else.

endergonic Describes reactions with a positive ΔG that are not spontaneous and so require an input of energy.

endocrine signaling Signaling by molecules that travel through the bloodstream.

endocrine system A system of cells and glands that secretes hormones and works with the nervous system to regulate an animal's internal physiological functions.

endocytosis The process in which a vesicle buds off from the plasma membrane, bringing material from outside the cell into that vesicle, which can then fuse with other organelles.

endoderm The germ layer that differentiates into cells of the lining of the digestive tract and lung, liver cells, pancreas cells, and gallbladder cells.

endodermis In plants, a layer of cells surrounding the xylem and phloem at the center of the root that controls the movement of nutrients into the xylem. Also, the inner lining of the cnidarian body.

endomembrane system A cellular system that includes the nuclear envelope, the endoplasmic reticulum, the Golgi apparatus, lysosomes, the plasma membrane, and the vesicles that move between them.

endomycorrhizae One of the two main types of mycorrhizae; endomycorrhizal hyphae penetrate into root cells, where they produce highly branched structures (arbuscules) that provide a large surface area for nutrient exchange.

endophyte A fungus that lives within leaves and that may help the host plant by producing chemicals that deter pathogens and herbivores.

endoplasmic reticulum (ER) The organelle involved in the synthesis of proteins and lipids.

endoskeleton The bony skeletal system of vertebrate animals, which lies internal to most of the animal's soft tissues.

endosperm A tissue formed by many mitotic divisions of a triploid cell, it supplies nutrition to the angiosperm embryo.

endosymbiosis A symbiosis in which one partner lives within the other.

endotherm An animal that produces most of its own heat as by-products of metabolic reactions.

energetic coupling The driving of a non-spontaneous reaction by a spontaneous reaction.

energy A property of objects that can be transferred from one object to another, and that cannot be created or destroyed

energy balance A form of homeostasis in which the amount of energy calories from food taken in equals the amount of calories used over time to meet metabolic needs.

energy intake Sources of energy.

energy use The ways in which energy is expended.

enhancer A specific DNA sequence necessary for transcription.

enthalpy (*H*) The total amount of energy in a system.

entropy (*S*) **envelope** A lipid structure that surrounds the capsids of some viruses.

environmental risk factor A characteristic in a person's surroundings that increases the likelihood of developing a particular disease.

environmental variation Variation among individuals that is due to differences in the environment.

enzyme A protein that functions as a catalyst to accelerate the rate of a chemical reaction; enzymes are critical in determining which chemical reactions take place in a cell.

eosinophil A type of granulocyte that, along with basophils, defends against parasitic infections but also contributes to allergies.

epidermis In mammals, the outer layer of skin, which serves as a water-resistant, protective barrier. In plants, the outermost layer of cells in leaves, young stems (lacking secondary growth), and roots. Also, the outer layer of the cnidarian body.

epididymis An organ that lies above the testes where sperm become motile and are stored prior to ejaculation.

epigenetic Describes effects on gene expression due to differences in DNA packaging, such as modifications in histones or chromatin structure.

epiglottis A flap of tissue at the bottom of the pharynx that prevents food from entering the trachea and lungs.

epiphysis The end region of a bone; blood vessels invading the epiphysis and diaphysis trigger the transformation of cartilage into bone.

epiphyte A plant that grows high in the canopy of other plants, or on branches or trunks of trees, without contact with the soil.

epistasis Interaction between genes that modifies the phenotypic expression of genotypes.

equational division Another name for meiosis II because cells in meiosis II have the same number of chromosomes at the beginning and at the end of the process.

esophagus Part of the foregut; the passage from the mouth to the stomach.

essential amino acid An amino acid that cannot be synthesized by cellular biochemical pathways and instead must be ingested.

estrogen A hormone secreted by the ovaries that stimulates the development of female secondary sexual characteristics.

estrus cycle A cycle in placental mammals other than humans and chimpanzees characterized by phases in which females are sexually receptive.

ethanol fermentation The fermentation pathway in plants and fungi during which pyruvate releases carbon dioxide to form acetaldehyde and electrons from NADH are transferred to acetaldehyde to produce ethanol and NAD^+.

Eukarya The eukaryotes; one of the three domains of life, in which cells have a true nucleus and divide by mitosis.

eukaryote An organism whose cells have a true nucleus.

Eumetazoa The monophyletic group of all animals other than sponges.

Euryarchaeota One of the three major divisions of Archaea; includes acid-loving, heat-loving, methane-producing, and salt-loving microorganisms.

eusocial Describes behavior most commonly observed in species of Hymenopteran insects, in which they have overlapping generations in a nest, cooperative care of the young, and clear and consistent division of labor between reproducers (the queen of a honeybee colony) and nonreproducers (the workers).

eutrophication The process in which added nutrients lead to a great increase in the populations of algae and cyanobacteria.

evaporation The amount of water evaporated from the Earth's surface, including ponds, rivers, and soil.

evo-devo The short name for evolutionary-developmental biology, a field of study that compares the genetic programs for growth and development in species on different branches of phylogenetic reconstructions.

evolution Changes in the genetic make-up of populations over time, sometimes resulting in adaptation to the environment and the origin of new species.

evolutionarily conserved Characteristics that persist relatively unchanged through diversification of a group of organisms and therefore remain similar in related species.

evolutionarily stable strategy A type of behavior that cannot readily be driven to extinction by an alternative strategy.

excitation–contraction coupling The process that produces muscle force and movement, by excitation of the muscle cell coupled to contraction of the muscle.

excitatory postsynaptic potential (EPSP) A positive change in the postsynaptic membrane potential.

excretion The elimination of waste generated by metabolism.

excretory tubule In renal systems, a type of tube that drains waste products and connects to the outside of the body.

exergonic Describes reactions with a negative ΔG that release energy and proceed spontaneously.

exhalation The expelling of oxygen-poor air by the elastic recoil of the lungs and chest wall.

exit (E) site One of three binding sites for tRNA on the large subunit of a ribosome.

exocytosis The process in which a vesicle fuses with the plasma membrane and empties its contents into the extracellular space or delivers proteins to the plasma membrane.

exon A sequence that is left intact in mRNA after RNA splicing.

exoskeleton A rigid skeletal system that lies external to the animal's soft tissues.

experimentation A disciplined and controlled way of learning about the world and testing hypotheses in an unbiased manner.

exponential growth The pattern of population increase that results when r (the per capita growth rate) is constant through time.

expressed Turned on or activated, as a gene or protein.

extension (PCR) A step in the polymerase chain reaction (PCR) for producing new DNA fragments in which the reaction mixture is heated to the optimal temperature for DNA polymerase, and each primer is elongated by means of deoxynucleoside triphophosphates.

extension (joint) The joint motion in which bone segments move apart.

external fertilization Fertilization that takes place outside the body of the female; in aquatic organisms, for example, eggs and sperm are released into the water.

extracellular digestion The process in most animals in which food is isolated and broken down outside a cell, in a body compartment.

extracellular matrix A meshwork of proteins and polysaccharides outside the cell; the main constituent of connective tissue.

extraembryonic membrane In the amniotic egg, one of several sheets of cells that extend out from the developing embryo and form the yolk sac, amnion, allantois, and chorion.

extravasation The process in which phagocytes that travel in the blood move from a blood vessel to the site of infection.

eyecup An eye structure found in flatworms that contains photoreceptors that point up and to the left or right.

F_1 generation The first filial, or offspring, generation.

F_2 generation The second filial generation; the offspring of the F_1 generation.

facilitated diffusion Diffusion through a membrane protein, bypassing the lipid bilayer.

facilitation A beneficial indirect interaction between organisms that are independently interacting directly with a third, as when two different organisms attack another organism, and the attack of each is aided as a result.

facultative In mutualisms, describes one in which one or both sides can survive without the other. In metabolism, describes a means of obtaining energy that is sometimes but not always used.

fallopian tube (oviduct) A tube from each ovary, through one of which a released oocyte passes.

family A group of closely related genera.

fast-twitch Describes muscle fibers that generate force quickly, producing rapid movements, but consume much more ATP than do slow-twitch fibers.

fatty acid A long chain of carbons attached to a carboxyl group; three fatty acid chains attached to glycerol form a triacylglycerol, a lipid used for energy storage.

feature detector A specialized sensory receptor or group of sensory receptors that respond to important signals in the environment.

fermentation A process of breaking down pyruvate through a wide variety of metabolic pathways that extract energy from fuel molecules such as glucose; the partial oxidation of complex carbon molecules to molecules that are less oxidized than carbon dioxide.

ferns and horsetails A monophyletic group of vascular plants that have leaves and disperse by spores.

fertilization The union of gametes to produce a diploid zygote.

fetus In humans, the embryo toward the end of the first trimester.

fiber In angiosperms, a narrow cell with thick walls that provides mechanical support in wood. In animals, a term for a muscle cell, which produces forces within an animal's body and exerts forces on the environment.

filament In animals, a thin thread of proteins that interacts with other filaments to cause muscles to shorten. In plants, the part of the stamen that supports the anther.

filial imprinting Imprinting in which newborn offspring rapidly learn to treat any animal they see shortly after birth as their mother.

filtration The separation of solids from fluids, as when circulatory pressure pushes fluid containing wastes through specialized filters into an extracellular space.

firing rate The number of action potentials fired over a given period of time.

first law of thermodynamics The law of conservation of energy: Energy can neither be created nor destroyed—it can only be transformed from one form into another.

first-division nondisjunction Failure of chromosome separation in meiosis I.

fitness A measure of the extent to which an individual's genotype is represented in the next generation

fixed Describes a population that exhibits only one allele at a particular gene.

fixed action pattern (FAP) A sequence of behaviors that, once triggered, is followed through to completion.

flagellum (plural, flagella) An organelle that propels the movement of cells or of substances within cells; longer than a cilium.

flexion The joint motion in which bone segments rotate closer together.

fluid mosaic model A model that proposes that the lipid bilayer is a fluid structure that allows molecules to move laterally within the membrane and is a mosaic of two types of molecules, lipids and proteins.

flux The rate at which a substance, for example carbon, flows from one reservoir to another.

folding domain A region of a protein that folds in a similar way across a protein family relatively independently of the rest of the protein.

follicle A type of cell that makes up the shell of cells surrounding an oocyte that supports the developing oocyte.

follicle-stimulating hormone (FSH) A hormone secreted by the anterior pituitary gland that stimulates the male and female gonads to secrete testosterone in males and estrogen and progesterone in females.

follicular phase The phase of the menstrual cycle during which FSH acts on granulosa cells, resulting in the maturation of several oocytes, of which, usually, only one becomes completely mature.

food chain The linear transfer of carbon from one organism to another.

food web A map of the interactions that connect consumer and producer organisms within the carbon cycle; the movement of carbon through an ecosystem.

force An interaction that changes the movement of an object, such as a push or pull by one object interacting with another object.

forebrain The region of the vertebrate brain that governs cognitive functions.

foregut The first part of an animal's digestive tract, including the mouth, esophagus, and stomach.

fossil The remains, impressions, or evidence of a once-living organism, preserved through time, most often in sedimentary rocks or tree resins.

founder event A type of bottleneck that occurs when only a few individuals establish a new population.

fovea The center of the visual field of most vertebrates, where cone cells are most concentrated; the region of greatest acuity.

fragmentation A form of asexual reproduction in which new individuals arise by the splitting of one organism into pieces, each of which develops into a new individual.

frameshift mutation A mutation in which an insertion or deletion of some number of nucleotides that is not a multiple of three causes a shift in the reading frame of the mRNA, changing all following codons.

fraternal (dizygotic) twins Twins that arise when two separate eggs, produced by double ovulation, are fertilized by two different sperm.

free energy The capacity to do work.

frequency of recombination The proportion of recombinant chromosomes among the total number of chromosomes observed.

frontal lobe The region of the brain located in the anterior region of the cerebral cortex, important in decision making and planning.

fruit In angiosperms, the structure that develops from the ovary, sometimes united with adjacent tissues, and serves to protect immature seeds and enhance dispersal once the seeds are mature.

fruiting body A multicellular structure in some fungi that facilitates the dispersal of sexually produced spores.

functional groups Groups of one or more atoms that have particular chemical properties of their own, regardless of what they are attached to.

fundamental niche The full range of climate conditions and food resources that permit the individuals in a species to live.

fungi An abundant and diverse group of heterotrophic eukaryotic organisms, principally responsible for the decomposition of plant and animal tissues.

G protein A protein that binds to the guanine nucleotides GTP and GDP.

G protein-coupled receptor A receptor that couples to G proteins, which bind to the guanine nucleotides GTP and GDP.

G_0 phase The gap phase in which cells pause in the cell cycle between M phase and S phase; may last for periods ranging from days to more than a year.

G_1 phase The gap phase in which the size and protein content of the cell increase and specific regulatory proteins are made and activated in preparation for S-phase DNA synthesis.

G_2 phase The gap phase in which the size and protein content of the cell increase in preparation for M-phase mitosis and cytokinesis.

gain-of-function mutation Any mutation in which a gene is expressed in the wrong place or at the wrong time.

gallbladder The organ in which bile produced by the liver is stored.

gamete A reproductive haploid cell; gametes fuse in pairs to form a diploid zygote. In many species, there are two types of gametes: eggs in females, sperm in males.

gametic isolation Incompatibility between the gametes of two different species.

gametogenesis The formation of gametes.

gametophyte In alternation of generation, describes the haploid multicellular generation that gives rise to gametes.

ganglion (plural, ganglia) A group of nerve cell bodies that processes sensory information received from a local, nearby region, resulting in a signal to motor neurons that control some physiological function of the animal.

ganglion cell A type of interneuron in the retina that synapses with bipolar cells and, if activated, transmits action potentials along the optic nerve to the visual cortex in the brain.

gap junction A type of connection between the plasma membranes of adjacent animal cells that permits materials to pass directly from the cytoplasm of one cell to the cytoplasm of another.

gas exchange The transport of oxygen and carbon dioxide between an organism and its environment.

gastric cavity In cnidarians, a closed internal site where extracellular digestion and excretion take place.

gastrin A peptide hormone produced in the stomach that stimulates cells lining the stomach to increase their production of HCl.

gastropods A group of mollusks consisting of snails and slugs.

gastrula A layered structure formed when the inner cell mass cells of the blastocyst migrate and reorganize.

gastrulation A highly coordinated set of cell movements in which the cells of the blastoderm migrate inward, creating germ layers of cells within the embryo.

gel electrophoresis A procedure to determine the size of a DNA fragment, in which DNA samples are inserted into slots or wells in a gel and a current passed through. Fragments move toward the positive pole according to size.

gene The unit of heredity; the stretch of DNA that affects one or more traits in an organism, usually through an encoded protein or noncoding RNA.

gene expression The production of a functional gene product.

gene family A group of genes with related functions, usually resulting from multiple rounds of duplication and divergence.

gene flow The movement of alleles from one population to another through interbreeding of some of their respective members.

gene pool All the alleles present in all individuals in a population or species.

gene regulation The various ways in which cells control gene expression.

general transcription factors A set of proteins that bind to the promoter of a gene whose combined action is necessary for transcription.

genetic code The correspondence between codons and amino acids, in which 20 amino acids are specified by 64 codons.

genetic drift A change in the frequency of an allele due to the random effects of finite population size.

genetic incompatibility Genetic dissimilarity between two organisms, such as different numbers of chromosomes, that is sufficient to act as a post-zygotic isolating factor.

genetic information Information carried in DNA, organized in the form of genes.

genetic map A diagram showing the relative positions of genes along a chromosome.

genetic risk factor Any mutation that increases the risk of a given disease in an individual.

genetic test A method of identifying the genotype of an individual.

genetic variation Differences in genotype among individuals in a population.

genetically modified organism (GMO) An organism that has been genetically engineered, such as modified viruses and bacteria, laboratory organisms, agricultural crops, and domestic animals; also known as a *transgenic organism.*

genome The genetic material transmitted from a parental cell or organism to its offspring.

genome annotation The process by which researchers identify the various types of sequence present in genomes.

genomic rearrangement The process of joining different gene segments as a B cell differentiates to produce a specific antibody.

genotype The genetic makeup of a cell or organism; the particular combination of alleles present in an individual.

genotype-by-environment interaction Unequaleffects of the environment on different genotypes, resulting in different phenotypes.

genus (plural, genera) A group of closely related species.

geographic isolation Spatial segregation of individuals.

geologic timescale The series of time divisions that mark Earth's long history.

germ cells The reproductive cells that produce sperm or eggs and the cells that give rise to them.

germ layers Three sheets of cells, the ectoderm, mesoderm, and endoderm, formed by migrating cells of the gastrula that differentiate further into specialized cells.

germ-line mutation A mutation that occurs in eggs and sperm or in the cells that give rise to these reproductive cells and therefore is passed on to the next generation.

gibberellic acid A plant hormone that stimulates the elongation of stems.

Gibbs free energy (*G*) The amount of energy available to do work.

gills Highly folded delicate structures in aquatic animals that facilitate gas exchange with the surrounding water.

gizzard In birds, alligators, crocodiles, and earthworms, a compartment with thick muscular walls in the digestive tract where food mixed with ingested rock or sediment is broken down into smaller pieces.

glans penis The head of the human penis.

glial cell A type of cell that surrounds neurons and provides them with nutrition and physical support.

glomeromycetes A monophyletic fungal group of apparently low diversity but tremendous ecological importance that occurs in association with plant roots.

glomerulus A tufted loop of porous capillaries in the vertebrate kidney that filters blood.

glycerol A 3-carbon molecule with OH groups attached to each carbon.

glycogen The form in which glucose is stored in animals.

glycolysis The breakdown of glucose to pyruvate; the first stage of cellular respiration.

glycosidic bond A covalent bond that attaches one monosaccharide to another.

Golgi apparatus The organelle that modifies proteins and lipids produced by the ER and acts as a sorting station as they move to their final destinations.

gonad In mammals, the part of the reproductive system where haploid gametes are produced. Male gonads are testes, where sperm are produced. Female gonads are ovaries, where eggs are produced.

gonadotropin-releasing hormone (GnRH) A hormone released by the hypothalamus that stimulates the anterior pituitary gland to secrete luteinizing hormone and follicle-stimulating hormone.

gram-positive bacteria Bacteria that retain, in their thick peptidoglycan walls, the diagnostic dye developed by Hans Christian Gram. (Bacteria with thin walls, which do not retain the dye, are said to be gram negative.)

grana (singular, granum) Interlinked structures that form the thylakoid membrane.

granulocyte A type of phagocytic cell that contains granules in its cytoplasm.

granuloma A structure formed by lymphocytes surrounding infected macrophages that helps to prevent the spread of an infection and aids in killing infected cells.

gravitropic Bending in response to gravity. A negative gravitropic response, as in stems, is growth upward against the force of gravity; a positive gravitropic response, as in roots, is with the force of gravity.

gray matter Densely packed neuron cell bodies and dendrites that make up the cortex, a highly folded outer layer of the mammalian brain about 4 mm thick.

greenhouse gas A gas in the atmosphere that allows incoming solar radiation to reach the Earth's surface, but absorbs radiation re-emitted as heat, trapping it in the atmosphere and causing the temperature to rise.

group selection Selection caused by the differential success of groups rather than individuals.

growth factor Any one of a group of small, soluble molecules, usually the signal in paracrine signaling, that affect cell growth, cell division, and changes in gene expression.

growth plate A region of cartilage near the end of a bone where growth in bone length occurs.

growth ring One of the many rings apparent in the cross section of the trunk of a tree, produced by decreases in the size of secondary xylem cells at the end of the growing season, that make it possible to determine the tree's age.

guanine (G) A purine base.

guard cell One of two cells surrounding the central pore of a stoma.

gustation The sense of taste.

gut Collectively, the passages that connect the mouth, digestive organs, and anus; also known as the *digestive tract.*

gymnosperms Seed plants whose ovules are not enclosed in a carpel; gymnosperms include pine trees and other conifers.

habituation The reduction or elimination of a behavioral response to a repeatedly presented stimulus.

hagfish One of the earliest-branching craniates, with a cranium built of cartilage but no jaws; hagfish feed on marine worms and dead and dying sea animals.

hair cell A specialized mechanoreceptor that senses movement and vibration.

half-life The time it takes for an amount of a substance to reach half its original value. Radioactive half-life is the time it takes for half of the atoms in a given sample of a substance to decay.

haploid Describes a cell with one complete set of chromosomes.

haplotype The particular combination of alleles present in any defined region of a chromosome.

Hardy–Weinberg equilibrium A state in which particular allele and genotype frequencies do not change over time, implying the absence of evolutionary forces. It also specifies a mathematical relationship between allele frequencies and genotype frequencies.

heart The pump of the circulatory system, which moves blood to different regions of the body.

heart rate (HR) The number of heartbeats per unit time.

heartwood The center of the stem in long-lived trees, which does not conduct water.

heavy (H) chains Two of the four polypeptide chains that make up the simplest antibody molecule.

helicase A protein that unwinds the parental double helix at the replication fork.

helper T cell One of a subpopulation of T cells that help other cells of the immune system by secreting cytokines, thus activating B cells to secrete antibodies.

hematocrit The fraction of red blood cells within the blood of vertebrates.

Hemichordata One of the three major phyla of deuterosomes, this group includes acorn worms and pterobranchs.

hemidesmosome A type of desmosome in which integrins are the prominent cell adhesion molecules.

hemoglobin An iron-containing molecule specialized for oxygen transport.

hemolymph The circulatory fluid in invertebrates.

hemophilia A trait characterized by excessive bleeding that results from a recessive mutation in a gene encoding a protein necessary for blood clotting.

herbivory The consumption of plant parts.

heritability In a population, the proportion of the total variation in a trait that is due to genetic differences among individuals.

heterokaryotic Describes a stage in the life cycle of some fungi, in which plasmogamy is not followed immediately by karyogamy and the cells have unfused haploid nuclei from both parents.

heterotroph An organism that obtains its carbon from organic molecules synthesized by other organisms.

heterozygote advantage A form of balancing selection in which the heterozygote's fitness is higher than that of either of the homozygotes, resulting in selection that ensures that both alleles remain in the population at intermediate frequencies.

heterozygous inherits different types of alleles from the parents, or genotypes in which the two alleles for a given gene are different.

hierarchical Describes gene regulation during development, in which the genes expressed at each stage in the process control the expression of genes that act later.

hindbrain Along with the midbrain, the region of the vertebrate brain that controls basic body functions and behaviors.

hindgut The last part of an animal's digestive tract, including the large intestine and rectum.

hinge joint A simple joint that allows one axis of rotation, like the elbow and knee.

hippocampus A posterior region of the limbic system involved in long-term memory formation.

histamine A chemical messenger released by mast cells and basophils; an important contributor to allergic reactions and inflammation.

histone code The pattern of modifications of the histone tails that affects the chromatin structure and gene transcription.

histone tail A string of amino acids that protrudes from a histone protein in the nucleosome.

homeodomain The DNA-binding domain in homeotic proteins, whose sequences are very similar from one homeotic protein to the next.

homeostasis The active regulation and maintenance, in animals, organs, or cells, of a stable internal physiological state in the face of a changing external environment.

homeotic gene A gene that specifies the identity of a body part or segment during embryonic development. One type of homeotic gene is known as a *Hox* gene.

hominins A member of one of the different species in the group leading to humans.

homologous Describes characters that are similar in different species because of descent from a common ancestor.

homologous chromosomes Pairs of chromosomes, matching in size and appearance, that carry the same set of genes; one of each pair was received from the mother, the other from the father.

homozygous Describes an individual who inherits an allele of the same type from each parent, or a genotype in which both alleles for a given gene are of the same type.

horizontal cell A type of interneuron in the retina that communicates between neighboring pairs of photoreceptors and bipolar cells, enhancing contrast through lateral inhibition to sharpen the image.

horizontal gene transfer The transfer of genetic material between organisms that are not parent and offspring.

hormone A chemical signal that influences physiology and development in both plants and animals; in animals hormones are released into the bloodstream and circulate throughout the body.

host cell A cell in which viral reproduction occurs.

host plant A plant species that can be infected by a given pathogen.

hotspot A site in the genome that is especially mutable.

housekeeping gene A gene that is transcribed continually because its product is needed at all times and in all cells.

human chorionic gonadotropin (hCG) A hormone released by the developing embryo that maintains the corpus luteum.

human genome The DNA in the chromosomes present in a human sperm or egg. The term is often used informally to mean all of the genetic material in a human cell or organism.

hybridization Interbreeding between two different varieties or species.

hydrogen bond A weak bond between a hydrogen atom in one molecule and an electronegative atom in another molecule.

hydrophilic "Water loving"; describes a class of molecules with which water can undergo hydrogen bonding.

hydrophobic "Water fearing"; describes a class of molecules poorly able to undergo hydrogen bonding with water.

hydrophobic effect The exclusion of nonpolar molecules by polar molecules, which drives biological processes such as the formation of cell membranes and the folding of proteins.

hydrostatic skeleton A skeletal system in which fluid contained within a body cavity is the supporting element.

hydroxyapatite The calcium phosphate mineral found in bone.

hypersensitive response A type of plant defense against infection in which uninfected cells surrounding the site of infection rapidly produce large numbers of reactive oxygen species, triggering cell wall reinforcement and causing the cells to die, thus creating a barrier of dead tissue.

hyperthermophile An organism that requires an environment with high temperature.

hyphae In fungi, highly branched filaments that provide a large surface area for absorbing nutrients.

hypothalamus The underlying brain region of the forebrain, which interacts with the autonomic and endocrine systems to regulate the general physiological state of the body.

hypothesis A tentative explanation for one or more observations that makes predictions that can be tested by experiments or additional observations.

identical (monozygotic) twins Twins that arise from a single fertilized egg, which after several rounds of cell division, separates into two distinct, but genetically identical, embryos.

IgA One of the five antibody classes, IgA is typically a dimer and the major antibody on mucosal surfaces.

IgD One of the five antibody classes, IgD is a monomer and is typically found on the surface of B cells; it helps initiate inflammation.

IgE One of the five antibody classes, IgE is a monomer that plays a central role in allergies, asthma, and other immediate hypersensitivity reactions.

IgG The most abundant of the five antibody classes, IgG is a monomer that circulates in the blood and is particularly effective against bacteria and viruses.

IgM One of the five antibody classes, IgM is a pentamer in mammals and a tetramer in fish and is particularly important in the early response to infection, activating the complement system and stimulating an immune response.

ileum A section of the small intestine that, with the jejunum, carries out most nutrient absorption.

imitation Observing and copying the behavior of another.

immediate hypersensitivity reaction A reaction characterized by a heightened or an inappropriate immune response to common antigens.

immunodeficiency Any disease in which part of the immune system does not function properly.

imprinting A form of learning typically seen in young animals in which a certain behavior is acquired in response to key experiences during a critical period of development.

in vitro fertilization (IVF) A process in which eggs and sperm are brought together in a petri dish, where fertilization and early cell divisions occur.

inbred line A true-breeding, homozygous strain.

inbreeding depression A reduction in fitness resulting from breeding among relatives causing homozygosity of deleterious recessive mutations.

incisor One of the teeth in the front of the mouth, used for biting.

incomplete dominance Describes inheritance in which the phenotype of the heterozygous genotype is intermediate between those of homozygous genotypes.

incomplete penetrance The phenomenon in which some individuals with a genotype corresponding to a trait do not show the phenotype, either because of environmental effects or because of interactions with other genes.

incus A small bone in the vertebrate middle ear that helps amplify the sound waves that strike the tympanic membrane.

induced pluripotent stem cell (iPS cell) A cell that has been reprogrammed to become pluripotent by activation of certain genes, most of them encoding transcription factors or chromatin proteins.

inducer A small molecule that elicits gene expression.

inflammation A physiological response of the body to injury that removes the inciting agent if present and begins the healing process.

inhalation The drawing of oxygen-rich blood into the lungs by the expansion of the thoracic cavity.

inhibitor A synthesized compound that decreases the activity of an enzyme.

inhibitory postsynaptic potential (IPSP) A negative change in the postsynaptic membrane potential.

initiation The stage of translation in which methionine is established as the first amino acid in a new polypeptide chain.

initiation factor A protein that binds to mRNA to initiate translation.

innate Describes behaviors that are instinctive and carried out regardless of earlier experience.

innate (natural) immunity The part of the immune system that provides protection in a nonspecific manner against all kinds of infection; it does not depend on exposure to a pathogen.

inner cell mass A mass of cells in one region of the inner wall of the blastocyst from which the body of the embryo develops.

inner ear The part of the vertebrate ear that includes the cochlea and semicircular canals.

insects The most diverse of the four main groups of arthropods.

instantaneous speciation Speciation that occurs in a single generation.

integral membrane protein A protein that is permanently associated with the cell membrane and cannot be separated from the membrane experimentally without destroying the membrane itself.

integrin A transmembrane protein, present on the surface of virtually every animal cell, that enables cells to adhere to the extracellular matrix.

intercostal muscles Muscles of vertebrates attached to adjacent pairs of ribs that assist the diaphragm by elevating the ribs on inhalation and depressing them during exhalation.

intermediate filament A polymer of proteins, which vary according to cell type, that combine to form strong, cable-like filaments that provide animal cells with mechanical strength.

intermembrane space The space between the inner and outer mitochondrial membranes.

internal fertilization Fertilization that takes place inside the body of the female.

interneuron A neuron that processes information received by sensory neurons and transmits it to motor neurons in different body regions.

internode The segment between two nodes on a shoot.

interphase The time between two successive M phases.

intersexual selection A form of sexual selection involving interaction between males and females, as when females choose from among males.

interspecific competition Competition between individuals of different species.

intervertebral disc A fluid-filled support structure found between the bony vertebrae of the backbone that enables flexibility and provides cushioning of loads.

intracellular digestion The process in single-celled protists in which food is broken down within cells.

intrasexual selection A form of sexual selection involving interactions between individuals of one sex, as when members of one sex compete with one another for access to the other sex.

intraspecific competition Competition within species.

intrinsic growth rate The per capita growth rate; the maximum rate of growth when no environmental factors limit population increase.

intron A sequence that is excised from the primary transcript and degraded during RNA splicing.

invasive species Non-native species; since they are removed from natural constraints on population growth, invasive species can expand dramatically when introduced into new areas, sometimes with devastating consequences for native species and ecosystems.

inversion The reversal of the normal order of a block of genes.

involuntary Describes the component of the nervous system that regulates internal bodily functions.

ion An electrically charged atom or molecule.

ion channels Cell-surface receptors that open and close, thereby altering the flow of ions across the plasma membrane.

iris A structure found at the front of the vertebrate eye, surrounding the pupil, that opens and closes to adjust the amount of light that enters the eye.

island population An isolated population.

isomers Molecules that have the same chemical formula but different structures.

isometric The generation of force without muscle movement.

isotopes Atoms of the same element that have different numbers of neutrons.

jejunum A section of the vertebrate small intestine that, with the ileum, carries out most nutrient absorption.

juxtaglomerular apparatus The structure formed by specialized cells of the efferent arteriole leaving the glomerulus of each nephron, which secretes the hormone rennin into the bloodstream.

karyogamy The fusion of two nuclei following plasmogamy.

karyotype A standard arrangement of chromosomes, showing the number and shapes of the chromosomes representative of a species.

key stimulus A stimulus that initiates a fixed action pattern.

keystone species Pivotal populations that affect other members of the community in ways that are disproportionate to their abundance or biomass.

kidneys In vertebrates, paired organs of the renal system that remove waste products and excess fluid; their action contributes to homeostasis.

kin selection A form of natural selection that favors the spread of alleles promoting behaviors that help close relatives.

kinesin A motor protein, similar in structure to myosin, that transports cargo toward the plus end of microtubules.

kinesis (plural, kineses) A random, undirected movement in response to a stimulus.

kinetic energy The energy of motion.

kinetochore The protein complexes on a chromatid where spindle fibers attach.

kingdom A group of closely related phyla.

Klinefelter syndrome A sex-chromosomal abnormality in which an individual has 47 chromosomes, including two X chromosomes and one Y chromosome.

knee-jerk extension reflex A reflex commonly tested by physicians to evaluate peripheral nervous and muscular system function.

***K*-strategist** A species that produces relatively few young but invests considerable resources into their support

labia majora Outer folds of skin in the vulva.

labia minora Inner folds of skin in the vulva that meet at the clitoris.

lactic acid fermentation The fermentation pathway in animals and bacteria during which electrons from NADH are transferred to pyruvate to produce lactic acid and NAD^1.

lagging strand A daughter strand that has its 5′ end pointed toward the replication fork, so as the parental double helix unwinds, a new DNA piece is initiated at intervals, and each new piece is elongated at its 3′ end until it reaches the piece in front of it.

lamellae (singular, lamella) The many thin, sheetlike structures spread along the length of each gill filament, giving gills an enormous surface area relative to their size.

lamprey One of the earliest-branching craniates, with a cranium and vertebral column built of cartilage but no jaws; many lampreys live parasitically, sucking body fluids from fish prey.

large intestine Part of the hindgut and the site of reabsorption of water and minerals; also known as the *colon.*

lariat A loop and tail of RNA formed after RNA splicing.

larynx The structure, above the trachea, that contains the vocal cords.

lateral inhibition Inhibition of a process in cells adjacent to the cell receiving a signal inducing that process, enhancing the strength of a signal locally but diminishing it peripherally.

lateral line system In fish and sharks, a sensory organ along both sides of the body that uses hair cells to detect movement of the surrounding water.

lateral meristem The source of new cells that allows plants to grow in diameter.

latex A white sticky liquid produced in some plants.

latitudinal diversity gradient The increase in species diversity from the poles to the equator.

leading strand A daughter strand that has its 3′ end pointed toward the replication fork, so as the parental double helix unwinds, this daughter strand can be synthesized as one long, continuous polymer.

leaf The principal site of photosynthesis in vascular plants.

leaf primordia A lateral outgrowth of the apical meristem that will eventually develop into a leaf.

learned Describes a behavior that depends on an individual's experience.

learning The process in which experience leads to changes in behavior.

lengthening contraction The contraction of a muscle against a load greater than the muscle's force output, leading to a lengthening of the muscle.

lens A flexible structure in the vertebrate eye through which light passes after entering through the pupil; it is controlled by ciliary muscles that contract or relax to the adjust the shape of the lens to focus light images.

lenticel A region of less tightly packed cells in the outer bark that allows oxygen to diffuse into the stem.

Leydig cell A type of cell in the testes that secretes testosterone.

Liebig's Law of the Minimum The principle that primary production is limited by the nutrient that is least available relative to its use by primary producers.

lichens Stable associations between a fungus and a photosynthetic microorganism, usually a green alga but sometimes a cyanobacterium.

life history The typical pattern of resource investment in each stage of a given species' lifetime.

life table A table that presents information about how many individuals of a cohort are alive at different points in time.

ligand Alternative term for a signaling molecule that binds with a receptor, usually a protein.

ligand-binding site The specific location on the receptor protein where a signaling molecule binds.

ligand-gated ion channel A receptor that alters the flow of ions across the plasma membrane when bound by its ligand.

light (L) chains Two of the four polypeptide chains that make up the simplest antibody molecule.

limbic system Inner components of the forebrain that control physiological drives, instincts, emotions, motivation, and the sense of reward.

linked Describes genes that are sufficiently close together in the same chromosome that they do not assort independently.

lipase A type of enzyme produced by the pancreas that breaks apart lipids, thus enabling their more effective digestion.

lipid An organic molecule that stores energy, acts as a signaling molecule, and is a component of cell membranes.

lipid raft Lipids assembled in a defined patch in the cell membrane.

liposome An enclosed bilayer structure spontaneously formed by phospholipids in environments with neutral pH, like water.

liver A vertebrate organ that aids in the digestion of proteins, carbohydrates, and fats in the duodenum by producing bile, which breaks down fat.

lobe-finned fish Species of fish with paired pectoral and pelvic fins that have a bone structure similar to that of tetrapod limbs.

logistic growth The pattern of population growth that results as growth potential slows down as the population size approaches *K*, its maximum sustainable size.

long-day plant A plant that flowers only when the light period exceeds a critical value.

longitudinal muscle Smooth muscle that runs lengthwise along a body or organ; in the digestive tract, a longitudinal muscle layer contracts to shorten small sections of the gut. A longitudinal muscle layer contracts alternately with circular muscle to move contents through the digestive tract and to enable locomotion in animals with hydrostatic skeletons.

loop of Henle The middle portion of the vertebrate renal tubule, which creates a concentration gradient that allows water passing through the collecting duct to be reabsorbed.

loss-of-function mutation A genetic change that inactivates the normal function of a gene.

Lucy An unusually complete early hominin fossil, *Australopithecus afarensis,* found in 1974 in Ethiopia.

lumen In eukaryotes, the continuous interior of the endoplasmic reticulum; in plants, a fluid-filled compartment enclosed by the thylakoid membrane; generally, the interior of any tubelike structure.

lunar clock A moon-based biological clock that times activities in some species, especially those living in habitats where tides are important.

lungfish Several species of lobe-finned fish that use a simple lung to survive periods when their watery habitat dries by burying themselves in moist mud and breathing air; along with coelacanths, the nearest relative of tetrapods.

lungs The internal organs for gas exchange in many terrestrial animals.

luteal phase The phase of the menstrual cycle beginning with ovulation.

luteinizing hormone (LH) A hormone secreted by the anterior pituitary gland that stimulates the male and female gonads to secrete testosterone in males and estrogen and progesterone in females.

lycophytes A monophyletic major group of spore-dispersing vascular plants that are the sister group to all other vascular plants.

lymph The fluid in the lymphatic system in which T and B cells ciculate.

lymphatic system A network of vessels distributed through the body with important functions in the immune system.

lysis The breakage, or bursting, of a cell.

lysogenic pathway The alternative to the lytic pathway; in the lysogenic pathway, a virus integrates its DNA into the host cell's DNA, which is then transmitted to offspring cells.

lysogeny The integration of viral DNA into a host cell's DNA, which is then transmitted to offspring cells.

lysosome A vesicle derived from the Golgi apparatus that contains enzymes that break down macromolecules such as proteins, nucleic acids, lipids, and complex carbohydrates.

lytic pathway The alternative to the lysogenic pathway; in the lytic pathway, a virus bursts, or lyses, the cell it infects, releasing new virus particles.

M phase The stage of the cell cycle consisting of mitosis and cytokinesis, in which the parent cell divides into two daughter cells.

macrophage A type of large phagocytic cell that is able to engulf other cells.

mainland population The central population of a species.

major groove The larger of two uneven grooves on the outside of a DNA duplex.

major histocompatibility complex (MHC) A group of proteins that appear on the surface of most mammalian cells; only the antigen associated with MHC proteins is recognized by T cell receptors.

malleus A small bone in the middle ear that helps amplify the waves that strike the tympanic membrane.

Malpighian tubule One of the tubes in the main body cavity of insects and other terrestrial arthropods through which fluid passes and which empties into the hindgut.

mammals A class of vertebrates distinguished by body hair and mammary glands from which they feed their young.

map information The knowledge of where an individual is in respect to the goal.

map unit A unit of distance in a genetic map equal to the distance between genes resulting in 1% recombination.

mark-and-recapture A method in which individuals are captured, marked in way that doesn't affect their function or behavior, and then released. The percentage of marked individuals in a later exercise of capture enables ecologists to estimate population size.

marsupials A group of mammals that includes kangaroos, koalas, and opossums; their young are born at an early stage of development and must crawl to a pouch where mammary glands equipped with nipples provide them with milk.

mass extinction A catastrophic drop in recorded diversity, which has occurred five or more times in the past 541 million years.

mast cell A cell that releases histamine.

maternal inheritance A type of inheritance in which the organelles in the offspring cells derive from those in the mother.

maternal-effect gene A gene that is expressed by the mother that affects the phenotype of the offspring, typically through the composition or organization of the oocyte.

mating types Genetically distinct forms of individuals of a fungus species that, by enabling fertilization only between different types, prevent self-fertilization and promote out-crossing.

mechanical incompatibility Structural configuration of the genitalia that prevents mating with another species.

mechanoreceptor A sensory receptor that responds to physical deformations of its membrane produced by touch, stretch, pressure, motion, or sound.

mediator complex A complex of proteins that interacts with the Pol II complex and allows transcription to begin.

medulla A part of the brainstem; also, the inner layer of the mammalian kidney.

meiosis I Reductional division, the first stage of meiotic cell division, in which the number of chromosomes is halved.

meiosis II Equational division, the second stage of meiotic cell division, in which the number of chromosomes is unchanged.

meiotic cell division A form of cell division that includes only one round of DNA replication but two rounds of nuclear division; meiotic cell division makes sexual reproduction possible.

membrane attack complex (MAC) A complex of complement proteins that makes holes in bacterial cells, leading to cell lysis.

membrane potential A difference in electrical charge across the plasma membrane.

memory cell A type of long-lived cell that contains membrane-bound antibodies having the same specificity as the parent cell.

menopause The cessation of menstrual cycles resulting from decreased production of estradiol and progesterone by the ovaries.

menstrual cycle A monthly cycle in females in which oocytes mature and are released from the ovary under the influence of hormones.

menstruation The monthly shedding of the uterine lining.

meristem A discrete population of actively dividing, totipotent cells; apical meristems are located at the tip of stems and roots and produce cells that allow plants to grow in length, while lateral meristems surround stems and roots and produce cells that allow growth in diameter.

meristem identity gene A gene that contributes to meristem stability and function.

mesentery A membrane in the abdominal cavity through which blood vessels, nerves, and lymph travel to supply the gut.

mesoderm The intermediate germ layer, which differentiates into cells that make up connective tissue, muscle cells, red blood cells, bone cells, kidney cells, and gonad cells.

mesoglea In cnidarians, a gelatinous mass enclosed in by the epidermis and endodermis.

mesohyl A gelatinous mass that lies between the interior and exterior cell layers of a sponge that contains some amoeba-like cells that function in skeleton formation and the dispersal of nutrients.

mesophyll A leaf tissue of loosely packed photosynthetic cells.

Messel Shale A sedimentary rock formation in Germany, preserving fossils that document fish, birds, mammals, and reptiles from the beginning of the age of mammals.

messenger RNA (mRNA) The RNA molecule that combines with a ribosome to direct protein synthesis; it carries the genetic "message" from the DNA to the ribosome.

metabolic rate An animal's overall rate of energy use.

metabolism The chemical reactions occurring within cells that convert molecules into other molecules and transfer energy in living organisms.

metamorphosis The process in some animals in which the body changes dramatically at key stages in development.

metanephridia A pair of excretory organs in each body segment of annelid worms that filters the body fluid.

metaphase The stage of mitosis in which the chromosomes are aligned in the middle of the dividing cell.

metaphase I The stage of meiosis I in which the meiotic spindle is completed and the bivalents move to lie on an imaginary plane cutting transverely across the spindle.

metaphase II The stage of meiosis II in which the chromosomes line up so that their centromeres lie on an imaginary plane cutting across the spindle.

metapopulation A large population made up of smaller populations linked by migration.

methanogens Euryarchaeotes that generate natural gas (methane, CH_4).

MHC class I (genes and proteins) MHC genes and proteins in vertebrates that are expressed on the surface of all nucleated cells.

MHC class II (genes and proteins) MHC genes and proteins in vertebrates that are expressed on the surface of macrophages, dendritic cells, and B cells.

MHC class III (genes and proteins) MHC genes and proteins in vertebrates that encode several proteins of the complement system and proteins involved in inflammation.

micelle A spherical structure in which lipids with bulky heads and a single hydrophobic tail are packed.

microfilament A helical polymer of actin monomers, present in various locations in the cytoplasm, that helps make up the cytoskeleton.

microfossil A microscopic fossil, including fossils of bacteria and protists.

microRNA (miRNA) Small, regulatory RNA molecules that can cleave or destabilize RNA or inhibit its translation.

microtubule A hollow, tubelike polymer of tubulin dimers that helps make up the cytoskeleton.

microvilli Highly folded surfaces of villi, formed by fingerlike projections on the surfaces of epithelial cells.

midbrain Along with the hindbrain, the region of the vertebrate brain that controls basic body functions and behaviors; it is a part of the brainstem.

middle ear The part of the mammalian ear containing three small bones, the malleus, incus, and stapes, which amplify the waves that strike the tympanic membrane.

midgut The middle part of an animal's digestive tract, including the small intestine.

migration The movement of individuals from one population to another.

mineral nutrients Elements that come from the soil or water that play essential roles in all metabolic and structural processes in plants and algae; see also *dietary minerals.*

minor groove The smaller of two unequal grooves on the outside of a DNA duplex.

mitochondria (singular, mitochondrion) Specialized organelles that harness energy for the cell from chemical compounds like sugars and convert it into ATP.

mitochondrial DNA (mtDNA) A small circle of DNA, about 17,000 base pairs long in humans, found in every mitochondrion.

mitochondrial genome In eukaryotic cells, the DNA in the mitochondria.

mitochondrial matrix The space enclosed by the inner membrane of the mitochondria.

mitosis In eukaryotic cells, the division of the nucleus, in which the chromosomes are separated into two nuclei.

mitotic spindle A structure in the cytosol made up predominantly of microtubules that pull the chromosomes into separate daughter cells.

Modern Synthesis The current theory of evolution, which combines Darwin's theory of natural selection and Mendelian genetics.

molar One of the teeth in the back of the mouth of mammals specialized for crushing and shredding tough foods such as meat and fibrous plant material.

molecular clock Estimates of the time when different taxa diverged, based on the amount of genetic divergence between them.

molecular evolution Evolution at the level of DNA, which in time results in the genetic divergence of populations.

molecular fossils Sterols, bacterial lipids, and some pigment molecules, which are relatively resistant to decomposition, that accumulate in sedimentary rocks and document organisms that rarely form conventional fossils.

molecular orbital A merged orbital traversed by a pair of shared electrons.

molecular self-assembly The process by which, when conditions and relative amounts are suitable, viral components spontaneously interact and assemble into mature virus particles.

molecule A substance made up of two or more atoms.

mollusks A monophyletic group distinguished by a mantle, which plays a major role in movement, skeleton-building, breathing, and excretion; includes clams, snails, and squid.

molting Periodic shedding, as of an exoskeleton.

monophyletic Describes groupings in which all members share a single common ancestor not shared with any other species or group of species.

monosaccharide A simple sugar.

morphospecies concept The idea that members of the same species usually look like each other more than like other species.

morula The solid ball of cells resulting from early cell divisions of the fertilized egg.

motor endplate The region on a muscle cell where acetylcholine binds with receptors.

motor neuron A neuron that, on receiving information from interneurons, effects a response in the body.

motor protein Any of various proteins that are involved in intracellular transport or cause muscle contraction by moving the actin microfilaments inside muscle cells.

motor unit A vertebrate motor neuron and the population of muscle fibers that it innervates.

mouth The first part of the foregut, which receives food.

mucosa An inner tissue layer with secretory and absorptive functions surrounding the lumen of the digestive tract.

multiple alleles Two or more different alleles of the same gene, occurring in a population of organisms.

multiplication rule The principle that the probability of two independent events occurring together is the product of their respective probabilities.

multipotent Describes cells that can form a limited number of types of specialized cell.

multiregional hypothesis The idea that modern humans derive from the *Homo ergaster* populations that spread around the world starting 2 million years ago.

mutagen An agent that increases the probability of mutation.

mutation Any heritable change in the genetic material, usually a change in the nucleotide sequence of a gene.

mutualism An interaction between two or more species that benefits all.

mycelium A network of branching hyphae.

mycorrhizae Symbioses between roots and fungi that enhance nutrient uptake.

myelin Lipid-rich layers or sheaths formed by glial cells that wrap around the axons of vertebrate neurons and provide electrical insulation.

myofibril A long rodlike structure in muscle fibers that contains parallel arrays of the actin and myosin filaments.

myoglobin An oxygen-binding protein in the cells of vertebrate muscles, related to hemoglobin, that facilitates oxygen delivery to mitochondria.

myosin A motor protein found in cells that carries cargo to the plus ends of microfilaments and is also used by both striated and smooth muscles to contract and generate force.

myotome In chordates, any one of a series of segments that organizes the body musculature.

myriapods One of the four main groups of arthropods, including centipedes and millipedes; distinguished by their many pairs of legs.

natural killer cell A cell type of the innate immune system that does not recognize foreign cells but instead recognizes and kills host cells that are infected by a virus or have become cancerous or otherwise abnormal.

natural selection The process in which, when there is inherited variation in a population of organisms, the variants best suited for growth and reproduction in a given environment contribute disproportionately to future generations. Of all the evolutionary mechanisms, natural selection is the only one that leads to adaptations.

Neanderthal *Homo neanderthalensis*, a species similar to humans, but with thicker bones and flatter heads that contained brains about the same size as humans'; present in the fossil record 600,000–30,000 years ago.

necrotrophic pathogen A plant pathogen that kills cells before drawing resources from them.

negative feedback Describes the effect in which the final product of a biochemical pathway inhibits the first step; the process in which a stimulus acts on a sensor that communicates with an effector, producing a response that opposes the initial stimulus. Negative feedback is used to maintain steady conditions, or homeostasis.

negative regulation The process in which a regulatory molecule must bind to the DNA at a site near the gene to prevent transcription.

negative selection Natural selection that reduces the frequency of a deleterious allele.

negatively selected Describes T cells that react too strongly to self antigens in association with MHC and are eliminated through cell death.

nematodes Roundworms, the most numerous of all animals; a phylum of the Ecdysozoa.

neoteny The process in which the timing of development is altered so that a sexually mature organism retains the physical characteristics of the juvenile form.

nephron The functional unit of the kidney, consisting of the glomerulus, capsule, renal tubules, and collecting ducts.

nerve A bundle of long fiberlike extensions from multiple nerve cells.

nerve cord A bundle of long fiberlike extensions from multiple nerve cells that serves as the central nervous system of invertebrates such as flatworms and earthworms.

nervous system A network of many interconnected nerve cells.

neural tube In chordates, a cylinder of embryological tissue that develops into a dorsal nerve cord.

neuron Nerve cell; the basic fundamental unit of nervous systems.

neurosecretory cell A neuron in the hypothalamus that secretes hormones into the bloodstream.

neurotransmitter A molecule that conveys a signal from the end of the axon to the postsynaptic target cell.

neutral mutations Genetic changes that have no effect or negligible effects on the organism, or whose effects are not associated with differences in survival or reproduction.

neutron An electrically neutral particle in the atomic nucleus.

neutrophil A type of phagocytic cell that is very abundant in the blood and is often one of the first cells to respond to infection.

niche The combination of traits and habitat in which a species exists.

nicotinamide adenine dinucleotide phosphate (NADPH) An important cofactor in many biosynthetic reactions; the reducing agent used in the Calvin cycle.

nitrification The process by which chemoautotrophic bacteria oxidize ammonia (NH_3) to nitrite (NO_2^-) and then nitrate (NO_3^-).

nitrogen fixation The process in which nitrogen gas (N_2) is converted into ammonia (NH_3), a form biologically useful to primary producers.

nitrogenous waste Waste in the form of ammonia, urea, and uric acid, which are toxic to organisms in varying degrees.

nociceptor A type of nerve cell with dendrites in the skin and connective tissues of the body that responds to excessive mechanical, thermal, or chemical stimuli by withdrawal from the stimulus and by the sensation of pain.

node In phylogenetic trees, the point where a branch splits, representing the common ancestor from which the descendant species diverged. In plants, the point on a shoot where one or more leaves are attached.

nodes of Ranvier Sites on an axon that lie between adjacent myelin-wrapped segments, where the axon membrane is exposed.

non-associative learning Learning that occurs in the absence of any particular outcome, such as a reward or punishment.

nondisjunction The failure of a pair of chromosomes to separate normally during anaphase of cell division.

nonpolar Describes compounds that do not have regions of positive and negative charge.

nonpolar covalent bond A covalent bond between atoms that have the same, or nearly the same, electronegativity.

non-random mating Mate selection biased by genotype or relatedness.

nonrecombinants Progeny in which the alleles are present in the same combination as that present in a parent.

nonself From the point of view of a given organism, describes the molecules and cells of another organism.

nonsense mutation A mutation that creates a stop codon, terminating translation.

non-sister chromatids Chromatids of different members of a pair of homologous chromosomes; although they carry the same complement of genes, they are not genetically identical.

nonsynonymous (missense) mutation A point mutation (nucleotide substitution) that causes an amino acid replacement.

nontemplate strand The untranscribed partner of the template strand of DNA used in transcription.

norm of reaction A graphical depiction of the change in phenotype across a range of environments.

normal distribution A distribution whose plot is a bell-shaped curve.

notochord In chordates, a stiff rod of collagen and other proteins that runs along the back and provides support for the axis of the body.

nuclear envelope The cell structure, composed of two membranes, inner and outer, that defines the boundary of the nucleus.

nuclear genome In eukaryotic cells, the DNA in the chromosomes.

nuclear localization signal The signal sequence for the nucleus that enables proteins to move through pores in the nuclear envelope.

nuclear pore One of many protein channels in the nuclear envelope that act as gateways that allow molecules to move into and out of the nucleus and are thus essential for the nucleus to communicate with the rest of the cell.

nuclear transfer A procedure in which a hollow glass needle is used to insert the nucleus of a cell into the cytoplasm of an egg whose own nucleus has been destroyed or removed.

nucleic acid A polymer of nucleotides that encodes and transmits genetic information.

nucleoid In prokaryotes, a cell structure with multiple loops formed from supercoils of DNA.

nucleolus A distinct, dense, non–membrane-bound spherical structure within the nucleus that contains the genes and transcripts for ribosomal RNA.

nucleoside A molecule consisting of a 5-carbon sugar and a base.

nucleosome A beadlike repeating unit of histone proteins wrapped with DNA making up the 10-nm chromatin fiber.

nucleotide A constituent of nucleic acids, consisting of a 5-carbon sugar, a nitrogen-containing base, and one or more phosphate groups.

nucleotide excision repair The repair of multiple mismatched or damaged bases across a region; a process similar to mismatch repair, but over a much longer piece of DNA, sometimes thousands of nucleotides.

nucleotide substitution A mutation in which a base pair is replaced by a different base pair; this is the most frequent type of mutation; also known as a *point mutation.*

nucleus (of an atom) The dense central part of an atom containing protons and neutrons.

nucleus (of a cell) The compartment of the cell that houses the DNA in chromosomes.

obligate Describes a mutualism in which one or both sides cannot survive without the other.

observation The act of viewing the world around us.

occipital lobe The region of the brain that processes visual information from the eyes.

ocean acidification An increase in the abundance of carbon dioxide in the oceans that causes the pH of seawater to go down.

Okazaki fragment In DNA replication, any of the many short DNA pieces in the lagging strand.

olfaction The sense of smell.

oligodendrocyte A type of glial cell that insulates cells in the brain and spinal cord by forming a myelin sheath.

omasum The third chamber in the stomach of ruminants, into which the mixture of food and bacteria passes and where water is reabsorbed.

ommatidia (singular, ommatidium) Individual light-focusing elements that make up the compound eyes of insects and crustaceans; the number of ommatidia determines the resolution of the image.

oncogene A cancer-causing gene.

oocyte The unfertilized egg cell produced by the mother; the developing female gamete.

oogenesis The formation of ova or eggs.

open circulatory system A circulatory system found in many smaller animals that contains few blood vessels and in which most of the circulating fluid is contained within the animal's body cavity.

open reading frame (ORF) A stretch of DNA or RNA consisting of codons for amino acids uninterrupted by a stop codon. In genome annotation, this sequence motif identifies the region as potentially protein coding.

operant conditioning Associative learning in which a novel behavior that was initially undirected has become paired with a particular stimulus through reinforcement.

operator The binding site for a repressor protein.

operon A group of functionally related genes located in tandem along the DNA and transcribed as a single unit from one promoter; the region of DNA consisting of the promoter, the operator, and the coding sequence for the structural genes.

Opisthokonta A superkingdom that encompasses animals, fungi, and related protists.

opsin A photosensitive protein that converts the energy of light photons into electrical signals in the receptor cell.

opsonization The binding of a molecule to a pathogen to facilitate uptake by a phagocyte.

optic nerve A cranial nerve that transmits action potentials from ganglion cells in the retina to the visual cortex of the brain.

orbital A region in space where an electron is present most of the time.

order A group of closely related families.

organ Two or more tissues that combine and function together.

organ of Corti A structure in the cochlear duct, supported by the basilar membrane, with specialized hair cells with stereocilia, that functions to convert mechanical vibrations to electrical impulses.

organelle Any one of several compartments in eukaryotes that divide the cell contents into smaller spaces specialized for different functions.

organic molecule A carbon-containing molecule.

organogenesis The transformation of the three germ layers into all the organ systems of the body.

origin of replication Any point on a DNA molecule at which DNA synthesis is initiated.

osmoconformer An animal that matches its internal osmotic pressure to that of its external environment.

osmoregulation The regulation of water and solute levels to control osmotic pressure.

osmoregulator An animal that maintains internal solute concentrations that differ from that of its environment.

osmosis The net movement of a solvent, such as water, across a selectively permeable membrane toward the side of higher solute concentration.

osmotic pressure The pressure needed to prevent water from moving from one solution into another by osmosis.

Osteichthyes Bony fish that have a cranium, jaws, and mineralized bones; there are about 20,000–25,000 species.

osteoblast A type of cell that forms bone tissue.

osteoclast A type of cell that secretes digestive enzymes and acid that dissolves the calcium mineral and collagen in bone.

outer ear The part of the human ear that includes the pinna, the ear canal, and tympanic membrane.

out-of-Africa hypothesis The idea that modern humans arose from *Homo ergaster* descendants in Africa before dispersing beyond Africa around 60,000 years ago.

oval window The thin membrane between the stapes of the middle ear and the cochlea of the inner ear.

ovary In plants, a hollow structure at the base of the carpel in which the ovules develop and which protects the ovules from being eaten or damaged by animals; in animals, the female gonad where eggs are produced.

oviparity Laying eggs.

ovoviviparity Giving birth to live young, with nutritional support of the embryo from the yolk.

ovule A reproductive structure in seed plants consisting of the female gametophyte developing within a sporangium and surrounded by outer protective layers. Ovules, when fertilized, develop into seeds.

ovule cone A reproductive shoot in gymnosperms that produces ovules.

ovum (egg) (plural, ova) The larger, female gamete.

oxidation reaction Describes a reaction in which a molecule loses electrons.

oxidation–reduction reaction A reaction involving the loss and gain of electrons between reactants. In biological systems these reactions are often used to store or release chemical energy.

oxidative phosphorylation A set of metabolic reactions that occurs by passing electrons along an electron transport chain to the final electron acceptor, oxygen, pumping protons across a membrane, and using the proton electrochemical gradient to drive synthesis of ATP.

oxygen dissociation curve The curve that results when blood pO_2 is plotted against the percentage of O_2 bound to hemoglobin.

oxygenic Producing oxygen.

oxytocin A posterior pituitary gland hormone that causes uterine contraction during labor and stimulates the release of milk during breastfeeding.

P_1 generation The parental generation in a series of crosses.

pacemaker Describes cardiac muscle cells that function as a regulator of heart rhythm.

palindromic Reading the same in both directions; describes sequence identity along the paired strands of a duplex DNA molecule; a symmetry typical of restriction sites.

pancreas A secretory gland that has both endocrine function, secreting hormones, including insulin, directly into the blood, and exocrine function, aiding the digestion of proteins, carbohydrates, and fats by secreting digestive enzymes into ducts that connect to the duodenum.

paracrine signaling Signaling by a molecule that travels a short distance to the nearest neighboring cell to bind its receptor and deliver its message.

paraphyletic Describes groupings that include some, but not all, the descendants of a common ancestor.

parasexual Describes asexual species that generate genetic diversity by the crossing over of DNA during mitosis.

parasites Organisms that live in close association with another species, consuming or gaining nutrition from their hosts' tissues, generally without killing them.

parasympathetic division The division of the autonomic nervous system that slows the heart and stimulates digestion and metabolic processes that store energy, enabling the body to "rest and digest."

parathyroid gland A gland adjacent to the thyroid gland that secretes parathyroid hormone (PTH), which, with calcitonin, regulates the actions of bone cells.

parenchyma In plants, describes thin-walled, undifferentiated cells.

parietal lobe The brain region, posterior to the frontal lobe, that controls body awareness and the ability to perform complex tasks.

parsimony Choosing the simplest hypothesis to account for a given set of observations.

parthenogenesis A form of asexual reproduction in which females produce eggs that are not fertilized by males but divide by mitosis and develop into new individuals.

partial pressure (p) The fractional concentration of a gas relative to other gases present multiplied by the atmospheric pressure exerted on the gases.

partially reproductively isolated Describes populations that have not yet diverged as separate species but whose genetic differences are extensive enough that the hybrid offspring they produce have reduced fertility or viability compared with offspring produced by crosses between individuals within each population.

patch A bit of habitat that is separated from other bits by inhospitable environments that are difficult or risky for individuals to cross.

paternal inheritance A type of inheritance in which the organelles in the offspring cells derive from those in the father.

pathogens Organisms and other agents, such as viruses, that cause disease.

peat bog Wetland in which dead organic matter accumulates.

pedigree A diagram of family history that summarizes the record of the ancestor-descendent relationships among individuals.

penis The male copulatory organ.

pepsin An enzyme produced in the stomach that breaks down proteins.

peptide bond A covalent bond that links the carbon atom in the carboxyl group of one amino acid to the nitrogen atom in the amino group of another amino acid.

peptide hormone A hormone that is a short chain of linked amino acids.

peptidoglycan A complex polymer of sugars and amino acids that makes up the cell wall.

peptidyl (P) site One of three binding sites for tRNA on the large subunit of a ribosome.

pericycle In roots, a single layer of cells just to the inside of the endodermis from which new root apical meristems are formed, allowing roots to branch.

periodic selection The episodic loss of diversity as a successful variant outcompetes others.

periodic table of the elements The arrangement of the chemical elements in tabular form, organized by their chemical properties.

peripatric speciation A specific kind of allopatric speciation in which a few individuals from a mainland population disperse to a new location remote from the original population and evolve separately.

peripheral membrane protein A protein that is temporarily associated with the lipid bilayer or with integral membrane proteins through weak noncovalent interactions.

peripheral nervous system (PNS) Collectively, the sensory and motor nerves, including the cranial and spinal nerves, and interneurons and ganglia.

peristalsis Waves of muscular contraction that move food toward the base of the stomach.

personalized medicine An approach in which the treatment is matched to the patient, not the disease; examination of an individual's genome sequence, by revealing his or her disease susceptibilities and drug sensitivities, allows treatments to be tailored to that individual.

peroxisomes Organelles in eukaryotic cells that contain many different enzymes and are involved in metabolic reactions.

petal A structure, often brightly colored and distinctively shaped, occurring in the next-to-outermost whorl of a flower; petals attract and orient animal pollinators.

phagocyte A type of immune cell that engulfs and destroys foreign cells or particles.

phagocytosis The engulfing of a cell or particle by another cell.

pharyngeal slit A vertical opening separated from other slits by stiff rods of protein in the pharynx of hemichordates.

pharynx The region of the throat that connects the nasal and mouth cavities; in hemichordates, a tube that connects the mouth and the digestive tract.

phenol Any one of a class of compounds, produced by some plants as a defensive mechanism.

phenotype The expression of a physical, behavioral, or biochemical trait; an individual's observable phenotypes include height, weight, eye color, and so forth.

pheromone A water- or airborne chemical compound released by animals into the environment that signals and influences the behavior of other members of their species.

phloem The vascular tissue that transports carbohydrates from leaves to the rest of the plant body.

phloem sap The sugar-rich solution that flows through both the lumen of the sieve tubes and the sieve plate pores.

phosphatase An enzyme that removes a phosphate group from another molecule.

phosphate group A chemical group consisting of a phosphorus atom bonded to four oxygen atoms.

phosphodiester bond A bond that forms when a phosphate group in one nucleotide is covalently joined to the sugar unit in another nucleotide. Phosophodiester bonds are relatively stable and form the backbone of a DNA strand.

phospholipid A type of lipid and a major component of the cell membrane.

photic zone The surface layer of the ocean through which enough sunlight penetrates to enable photosynthesis.

photoheterotroph An organism that uses the energy from sunlight to make ATP and relies on organic molecules obtained from the environment as the source of carbon for growth and other vital functions.

photoperiod Day length.

photoperiodism The effect of the photoperiod, or day length, on flowering.

photoreceptor A molecule whose chemical properties are altered when it absorbs light; also, photoreceptors are the sensory receptors in the eye.

photorespiration A process in which rubisco acts as an oxygenase, resulting in a release of CO_2 and a net loss of energy.

photosynthesis The biochemical process in which carbohydrates are built from carbon dioxide and the energy of sunlight.

photosynthetic electron transport chain A series of redox reactions in which light energy absorbed by chlorophyll is used to power the movement of electrons; in oxygenic photosynthesis, the electrons ultimately come from water and the terminal electron acceptor is $NADP^+$.

photosystem A protein-pigment complex that absorbs light energy to drive redox reactions and thereby sets the photosynthetic electron transport chain in motion.

photosystem I The photosystem that energizes electrons with a second input of light energy so they have enough energy to reduce $NADP^+$.

photosystem II The photosystem that supplies electrons to the beginning of the electron transport chain. When photosystem II loses an electron, it can pull electrons from water.

phototroph An organism that captures energy from sunlight.

phototropic Bending in response to light. A positive phototropic response, as in stems, is toward light; a negative phototropic response, as in roots, is away from light.

phragmoplast In dividing plant cells, a structure formed by overlapping microtubules that guide vesicles containing cell wall components to the middle of the cell.

phylogenetic niche conservatism The observed similarity in closely related species of some aspect of their niches, indicating its presence in their common ancestor.

phylogenetic species concept (PSC) The idea that members of a species all share a common ancestry and a common fate.

phylogenetic tree A diagrammed hypothesis about the evolutionary history, or phylogeny, of a species.

phylogeny The history of descent with modification and the accumulation of change over time.

phylum (plural, phyla) A group of closely related classes, defined by having one of a number of distinctive body plans.

phytochrome A photoreceptor for red and far-red light that switches back and forth between two stable forms, active and inactive, depending on its exposure to light.

pilus (plural, pili) A threadlike that connects bacteria, allowing plasmids to be transferred between them.

pineal gland A gland located in the thalamic region of the brain that responds to autonomic nervous system input by secreting melatonin, which controls wakefulness.

pinna (plural, pinnae) Small unit of the photosynthetic surface on fern leaves; also, the external structure of mammalian ears that enhance the reception of sound waves contacting the ear.

pit A circular or ovoid region in the walls of xylem cells where the lignified cell wall layer is not produced.

pith In a stem, the region inside the ring of vascular bundles.

pituitary gland A gland beneath the brain that produces a number of hormones, including growth hormone.

placenta In placental mammals, an organ formed by the fusion of the chorion and allantois that allows the embryo to obtain nutrients directly from the mother.

placental mammal A mammal that provides nutrition to the embryo through the placenta, a temporary organ that develops in the uterus; placental mammals include carnivores, primates, hooved mammals, and whales.

placozoans Possibly the simplest of all animals; each contains only a few thousand cells arranged into upper and lower epithelia that sandwich an interior fluid crisscrossed by a network of multinucleate fiber cells.

plasma (effector) cell A cell that secretes antibodies.

plasma membrane The membrane that defines the space of the cell, separating the living material within the cell from the nonliving environment around it.

plasmid In bacteria, a small circular molecule of DNA carrying a small number of genes that can replicate independently of the bacterial genomic DNA.

plasmodesmata (singular, plasmodesma) Connections between the plasma membranes of adjacent plant cells that permit molecules to pass directly from the cytoplasm of one cell to the cytoplasm of another.

plasmogamy The cytoplasmic union of two cells.

plate tectonics The dynamic movement of Earth's crust, the outer layer of Earth.

pleiotropy The phenomenon in which a single gene has multiple effects on seemingly unrelated traits.

pluripotent Describes embryonic stem cells (cells of the inner mass), which can give rise to any of the three germ layers and therefore to any cell of the body.

podocytes Cells with footlike processes that make up one of the three layers of the filtration barrier of the glomerulus.

point mutation A mutation in which a base pair is replaced by a different base pair; this is the most frequent type of mutation; also known as a *nucleotide substitution.*

Pol II The RNA polymerase complex responsible for transcription of protein-coding genes.

polar (molecule) A molecule that has regions of positive and negative charge.

polar body A small cell produced by the asymmetric first meiotic division of the primary oocyte.

polar covalent bond Bonds that do not share electrons equally.

polar transport The coordinated movement of auxin across many cells.

polarity An asymmetry such that one end of a structure differs from the other.

polarized Having opposite properties in opposite parts; describes a resting membrane potential in which there is a buildup of negatively charged ions on the inside surface of the cell's plasma membrane and positively charged ions on its outer surface.

pollen In seed plants, the multicellular male gametophyte surrounded by a sporopollenin-containing outer wall.

pollen cone A reproductive shoot in gymnosperms that produces pollen.

pollen tube A structure produced by the male gametophyte that grows through an opening in the sporopollenin-containing pollen wall and eventually conveys the sperm to the ovule.

pollination The process in which pollen is carried to an ovule.

poly(A) tail The nucleotides added to the 3′ end of the primary transcript by polyadenylation.

polyadenylation The addition of a long string of consecutive A-bearing ribonucelotideas to the 3′ end of the primary transcript.

polycistronic mRNA A single molecule of messenger RNA that is formed by the transcription of a group of functionally related genes located next to one another along bacterial DNA.

polymer A complex organic molecule made up of repeated simpler units connected by covalent bonds.

polymerase chain reaction (PCR) A selective and highly sensitive method for making copies of a piece of DNA, which allows a targeted region of a DNA molecule to be replicated into as many copies as desired.

polymorphism Any genetic difference among individuals sufficiently common that it is likely to be present in a group of 50 randomly chosen individuals.

polypeptide A polymer of amino acids connected by peptide bonds.

polyphyletic Describes groupings that do not include the last common ancestor of all members.

polyploidy The condition of having more than two complete sets of chromosomes in the genome.

polysaccharide A polymer of simple sugars. Polysaccharides provide long-term energy storage or structural support.

polyspermy In animals, fertilization by more than one sperm.

pons A part of the brainstem.

population All the individuals of a given species that live and reproduce in a particular place; one of several interbreeding groups of organisms of the same species living in the same geographical area.

population density The size of a population divided by its range.

population size The number of individuals of all ages alive at a particular time in a particular place.

positive feedback In the nervous system, the type of feedback in which a stimulus causes a response that leads to an enhancement of the original stimulus that leads to a larger response. In the endocrine system, the type of feedback in which a stimulus causes a response, and that response causes a further response in the same direction. In both cases, the process reinforces itself until interrupted.

positive regulation The process in which a regulatory molecule must bind to the DNA at a site near the gene in order for transcription to take place.

positive selection Natural selection that increases the frequency of a favorable allele.

positively selected Describes T cells that recognize self MHC molecules on epithelium cells and continue to mature.

posterior pituitary gland The region of the pituitary gland that develops from neural tissue at the base of the brain and into which neurosecretory cells of the hypothalamus extend that secrete releasing factors.

postreplication mismatch repair The correction of a mismatched base in a DNA strand by cleaving one of the strand backbones, degrading the sequence with the mismatch, and resynthesizing from the intact DNA strand.

posttranslational modification The modification, after translation, of proteins in ways that regulate their structure and function.

post-zygotic isolation Describes factors that cause the failure of the fertilized egg to develop into a fertile individual.

potential energy Stored energy that is released by a change in an object's structure or position.

potential evapotranspiration The amount of evapotranspiration that temperature, humidity, and wind would cause if water supply weren't limiting; the demand on the water resources of an ecosystem.

potential evapotranspiration ratio In an ecosystem, the ratio of water demand to supply.

power stroke The stage in the cross-bridge cycle in which the myosin head pivots and generates a force, causing the myosin and actin filaments to slide relative to each other.

predation An interaction between organisms in which one (the predator) consumes the other (the prey).

pregnancy The carrying of one or more embryos in the mammalian uterus.

premolar One of the teeth of mammals between the canines and molars that are specialized for shearing tough foods.

pre-zygotic isolation Describes factors that that prevent the fertilization of an egg.

primary active transport Active transport that uses the energy of ATP directly.

primary bronchi The two divided airways from the trachea, supported by cartilage rings, each airway leading to a lung.

primary consumers Herbivores, which consume primary (plant or algae) producers. Sometimes called grazers.

primary growth The increase in plant length made possible by apical meristems.

primary motor cortex The part of the frontal lobe of the brain that produces complex coordinated behaviors by controlling skeletal muscle movements.

primary oocyte A diploid cell formed by mitotic division of oogonia during fetal development.

primary producer An organism that takes up inorganic carbon, nitrogen, phosphorus, and other compounds from the environment and converts them into organic compounds that will provide food for other organisms in the local environment.

primary response The response to the first encounter with an antigen, during which there is a short lag before antibody is produced.

primary somatosensory cortex The part of the parietal lobe that integrates tactile information from specific body regions and relays it to the motor cortex.

primary spermatocyte A diploid cell formed by mitotic division of spermatogonia at the beginning of spermatogenesis.

primary structure The sequence of amino acids in a protein.

primary transcript The initial RNA transcript that comes off the template DNA strand.

primate A member of an order of mammals that share a number of general features that distinguish them from other mammals, including nails rather than claws, front-facing eyes, and an opposable thumb.

primer A short stretch of RNA at the beginning of each new DNA strand that serves as a starter for DNA synthesis; an oligonucleotide that serves as a starter in the polymerase chain reaction.

primordia (singular, primordium) An organ in its earliest stage of development; leaf primordia form near the tips of shoot apical meristems and develop into leaves.

principle of independent assortment The principle that segregation of one set of alleles of a gene pair is independent of the segregation of another set of alleles of a different gene pair.

principle of segregation The principle by which half the gametes receive one allele of a gene and half receive the other allele.

probability Among a very large number of observations, the expected proportion of observations that are of a specified type.

probe A labeled DNA fragment that can be tracked in a procedure such as a Southern blot.

procambial cell A cell that retains the capacity for cell division and gives rise to both xylem and phloem.

product Any one of the transformed molecules that result from a chemical reaction.

progesterone A hormone secreted by the vertebrate ovaries that maintains the thickened and vascularized uterine lining.

prokaryote An organism whose cell or cells does not have a nucleus. Often used to refer collectively to archaeons and bacteria.

prometaphase The stage of mitosis in which the nuclear envelope breaks down and the microtubules of the mitotic spindle attach to chromosomes.

prometaphase I The stage of meiosis I in which the nuclear envelope breaks down and the meiotic spindles attach to kinetochores on chromosomes.

prometaphase II The stage of meiosis II in which the meiotic spindles attach to kinetochores on chromosomes.

promoter A regulatory region where RNA polymerase and associated proteins bind to the DNA duplex.

proofreading The process in which DNA polymerases can immediately correct their own errors by excising and replacing a mismatched base.

prophase The stage of mitosis characterized by the appearance of visible chromosomes.

prophase I The beginning of meiosis I, marked by the visible manifestation of chromosome condensation.

prophase II The stage of meiosis II in which the chromosomes in the now-haploid nuclei recondense to their maximum extent.

prostate gland An exocrine gland that produces a thin, slightly alkaline fluid that helps maintain sperm motility and counteracts the acidity of the female reproductive tract.

protease inhibitor An antidigestive protein that binds to the active site of enzymes that break down proteins in a herbivore's digestive system.

protein family A group of proteins that are structurally and functionally related.

protein sorting The process by which proteins end up where they need to be in the cell to perform their function.

proteins The key structural and functional molecules that do the work of the cell, providing structural support and catalyzing chemical reactions. The term "protein" is often used as a synonym for "polypeptide."

proteobacteria The most diverse bacterial group, defined largely by similarities in rRNA gene sequences; it includes many of the organisms that populate the expanded carbon cycle and other biogeochemical cycles.

protist An organism having a nucleus but lacking other features specific to plants, animals, or fungi.

proton A positively charged particle in the atomic nucleus.

protonephridia Excretory organs in flatworms that isolate waste from the body cavity.

proto-oncogene The normal cellular gene counterpart to an oncogene, which is similar to a viral oncogene but can cause cancer only when mutated.

protostome A bilaterians in which the blastopore, the first opening to the internal cavity of the developing embryo, becomes the mouth.

protozoan (plural, protozoa) A heterotrophic protist.

proximal convoluted tubule The first portion of the renal tubule from which electrolytes and other nutrients are reabsorbed into the blood.

pseudogene A gene that is no longer functional.

pulmonary artery One of two arteries, left and right, that carries deoxygenated blood from the right ventricle to the lungs.

pulmonary capillary A small blood vessel that supplies the alveolar wall.

pulmonary circulation Circulation of the blood to the lungs.

pulmonary valve A valve between the right ventrical and the pulmonary trunk, which divides into the pulmonary arteries.

pulmonary vein A vein that returns oxygenated blood from the lungs to the left ventricle.

Punnett square A worksheet in the form of a checkerboard used to predict the consequences of a random union of gametes.

pupa The quiescent stage of metamorphosis in insects, during which the body tissues undergo a transformation from larva to an adult.

pupil An opening in the vertebrate eye through which light enters.

purine In nucleic acids, either of the bases adenine and gunanine, which have a double-ring structure.

pyloric sphincter A band of muscle at the base of the stomach that opens to allow small amounts of digested food to enter the small intestine.

pyrimidine In nucleic acids, any of the bases thymine, cytosine, and uracil, which have a single-ring structure.

quantitative trait A complex trait in which the phenotype is measured along a continuum with only small intervals between similar individuals.

quaternary structure The structure that results from the interactions of several polypeptide chains.

***R* gene** Any one of the group of genes that express the R proteins in plants.

R group A chemical group attached to the central carbon atom of an amino acid, whose structure and composition determine the identity of the amino acid; also known as a *side chain.*

R protein Any one of a group of receptors in plant cells, each expressed by a different gene, that function as part of the plant's immune system by each binding to a specific pathogen-derived protein.

radiometric dating Dating by using the decay of radioisotopes as a yardstick, including (for time intervals up to a few tens of thousands of years) the decay of radioactive ^{14}C to nitrogen and (for most of Earth history) the decay of radioactive uranium to lead.

rain shadow The area on the lee side of mountains, where air masses descend, warming and taking up water vapor; as a result, lands in the rain shadow are arid.

range The extent of the geographic area over which the populations of a species are distributed.

reabsorption In renal systems, an active or passive process in which substances that are important for an animal to retain are taken up by cells of the excretory tubule and returned to the bloodstream.

reactant Any of the starting molecules in a chemical reaction.

reaction center Specially configured chlorophyll molecules where light energy is converted into electron transport.

reactive oxygen species Highly reactive forms of oxygen.

reading frame Following a start codon, a consecutive sequence of codons for amino acids.

realized niche The actual range of habitats occupied by a species.

receiver The individual who, during communication, receives from the sender a signal that elicits a response.

receptor A molecule on cell membranes that detects critical features of the environment. Receptors detecting signals that easily cross the cell membrane are sometimes found in the cytoplasm.

receptor activation The "turning on" of a receptor, which often occurs when a signaling molecule binds to a receptor on a responding cell.

receptor kinase A receptor that is an enzyme that adds a phosphate group to another molecule.

receptor protein The molecule on the responding cell that binds to the signaling molecule.

recessive The trait that fails to appear in heterozygous genotypes from a cross between the corresponding homozygous genotypes.

reciprocal altruism The exchange of favors between individuals.

reciprocal cross A cross in which the female and male parents are interchanged.

reciprocal inhibition The activation of opposing sets of muscles so that one set is inhibited as the other is activated, allowing the movement of joints such as the knee.

reciprocal translocation Interchange of parts between nonhomologous chromosomes.

recombinant An offspring with a different combination of alleles from that of either parent, resulting from one or more crossovers in prophase I of meiosis.

recombinant DNA DNA molecules from two (or more) different sources combined into a single molecule.

recovery metabolism An animal's elevated consumption of oxygen following activity.

rectum The part of the hindgut where feces are stored until elimination.

Redfield ratio On average, cells contain atoms in the ratio 1 P:16 N:100 C, a correspondence first described by the American biologist Alfred Redfield.

reduction A reaction in which a molecule gains electrons.

reductional division An alternative name for meiosis I, since this division reduces the number of chromosomes by half.

refractory period The period following an action potential during which the inner membrane voltage falls below and then returns to the resting potential.

regeneration In the context of photosynthesis, the third step of the Calvin cycle, in which the 5-carbon molecule needed for carboxylation is produced.

regenerative medicine A discipline that aims to use the natural processes of cell growth and development to replace diseased or damaged tissues.

regression toward the mean With regard to complex traits, the principle that offspring exhibit an average phenotype that is intermediate between that of the parents and that of the population as a whole.

regulatory transcription factor A protein that recruits the components of the transcription complex to the gene.

reinforcement of reproductive isolation (reinforcement) The process by which diverging populations undergo natural selection in favor of enhanced pre-zygotic isolation to prevent the production of inferior hybrid offspring.

release factor A protein that causes a finished polypeptide chain to be freed from the ribosome.

releasing factor A peptide hormone that signals to the anterior pituitary gland through blood vessels, leading to a much larger release of associated hormones from that organ.

renal tubules Tubes in the vertebrate kidney that process the filtrate from the glomerulus by reabsorption and secretion.

renaturation The base pairing of complementary single-stranded nucleic acids to form a duplex; also known as *hybridization*, it is the opposite of denaturation.

renin An enzyme produced by the kidneys that converts angiotensinogen to angiotensin I and is involved in the regulation of blood volume and pressure.

repetitive DNA The collective term for repeated sequences of various types in eukaryotic genomes.

replication The process of copying DNA so genetic information can be passed from cell to cell or from an organism to its progeny.

replication bubble A region formed by the opening of a DNA duplex at an origin of replication, which has a replication fork at each end.

replication fork The site where the parental DNA strands separate as the DNA duplex unwinds.

repressor A protein that, when bound with a sequence in DNA, can inhibit transcription.

reservoir A supply or source of a substance. Reservoirs of carbon, for example, include organisms, the atmosphere, soil, the oceans, and sedimentary rocks.

residue In the context of protein synthesis, any of the amino acids that is incorporated into a protein.

resource partitioning A pattern in which species whose niches overlap may diverge to minimize the overlap.

responding cell The cell that receives information from the signaling molecule.

response A change in cellular behavior, such as activation of enzymes or genes, following a signal.

resting membrane potential The negative voltage across the membrane at rest.

restriction enzyme Any one of a class of enzymes that recognizes specific, short nucleotide sequences in double-stranded DNA and cleaves DNA at or near these sites.

restriction fragment length polymorphism (RFLP) A polymorphism in which the length of the restriction fragments is different in the two alleles.

restriction site A recognition sequence in DNA cutting, which is typically four or six base pairs long; most restriction enzymes cleave double-stranded DNA at or near these restriction sites.

reticulum The second chamber in the stomach of ruminants, which, along with the rumen, harbors large populations of anaerobic bacteria that break down cellulose.

retina A thin tissue in the posterior of the vertebrate eye that contains the photoreceptors and other nerve cells that sense and initially process light stimuli.

retinal A derivative of vitamin A that absorbs light and binds to rhodopsin, a transmembrane protein in the photosensitive cells of vertebrates.

retrotransposons Transposable elements in DNA sequences in which RNA is used as a template to synthesize complementary strands of DNA, a reversal of the usual flow of genetic information from DNA into RNA.

reverse transcriptase An RNA-dependent DNA polymerase that uses a single-stranded RNA as a template to synthesize a DNA strand that is complementary in sequence to the RNA.

rhizosphere The soil layer that surrounds actively growing roots.

ribonucleic acid (RNA) A molecule chemically related to DNA that is synthesized by proteins from a DNA template.

ribose The sugar in RNA.

ribosomal RNA (rRNA) Noncoding RNA found in all ribosomes that aid in translation.

ribosome A complex structure of RNA and protein, bound to the cytosolic face of the RER in the cytoplasm, on which proteins are synthesized.

ribulose bisphosphate carboxylase oxygenase (rubisco) The enzyme that catalyzes the carboxylation reaction in the Calvin cycle.

ribulose-1,5-bisphosphate (RuBP) The 5-carbon sugar to which carbon dioxide is added by the enzyme rubisco.

ring species Species that contain populations that are reproductively isolated from each other but can exchange genetic material through other, linking populations.

RISC (RNA-induced silencing complex) A protein complex that is targeted to specific mRNA molecules by base pairing with short regions on the target mRNA, inhibiting translation or degrading the RNA.

ritualization The process of co-opting and modifying behaviors used in another context by increasing the conspicuousness of the behavior, reducing the amount of variation in the behavior so that it can be immediately recognized, and increasing its separation from the original function.

RNA editing The process in which some RNA molecules become a substrate for enzymes that modify particular bases in the RNA, thereby changing its sequence and sometimes what it codes for.

RNA polymerase The enzyme that carries out polymerization of ribonucleoside triphosphates from a DNA template to produce an RNA transcript.

RNA polymerase complex An aggregate of proteins that synthesizes the RNA transcript complementary to the template strand of DNA.

RNA primase An RNA polymerase that synthesizes a short piece of RNA complementary to the DNA template and does not require a primer.

RNA processing Chemical modification that converts the primary transcript into finished mRNA, enabling the RNA molecule to be transported to the cytoplasm and recognized by the translational machinery.

RNA splicing The process of intron removal from the primary transcript.

RNA transcript The RNA sequence synthesized from a DNA template.

RNA world hypothesis The hypothesis that the earliest organisms relied on RNA for both catalysis and information storage.

rod cell A type of photoreceptor cell on the retina that detects light and shades ranging from white to shades of gray and black, but not color.

root apical meristem A group of totipotent cells near the tip of a root that is the source of new root cells.

root cap A structure that covers and protects the root apical meristem as it grows through the soil.

root hair A slender outgrowth produced by epidermal cells that greatly increases the surface area of the root.

root nodule A structure, formed by dividing root cells, in which nitrogen-fixing bacteria live.

roots A major organ system of vascular plants, generally belowground.

rough endoplasmic reticulum (RER) The part of the endoplasmic reticulum with attached ribosomes.

***r*-strategist** A species that produces large numbers of offspring but provides few resources for their support.

rumen The first chamber in the stomach of ruminants, which, along with the reticulum, harbors large populations of anaerobic bacteria that break down cellulose.

S phase The phase of interphase in which the entire DNA content of the nucleus is replicated.

saccharide The simplest carbohydrate molecule, also called a *sugar.*

saltatory propagation The movement of an action potential along a myelinated axon, "jumping" from node to node.

Sanger sequencing A procedure in which chemical termination of daughter strands help in determining the DNA sequence.

sapwood In long-lived trees, the layer adjacent to the vascular cambium that contains the functional xylem.

sarcomere The region from one Z disc to the next, the basic contractile unit of a muscle.

sarcoplasmic reticulum (SR) A modified form of the endoplasmic reticulum surrounding the myofibrils of muscle cells.

saturated Describes fatty acids that do not contain double bonds; the maximum number of hydrogen atoms is attached to each carbon atom, "saturating" the carbons with hydrogen atoms.

scaffold A supporting protein structure in a metaphase chromosome.

scientific method A deliberate, careful, and unbiased way of learning about the natural world.

sclera A tough, white outer layer surrounding the vertebrate eye.

scrotum A sac outside the abdominal cavity of the male that holds the testes.

second law of thermodynamics The principle that the transformation of energy is associated with an increase in the degree of disorder in the universe.

secondary consumers Predators or scavengers that feed on primary consumers.

second messenger An intermediate cytosolic signaling molecule that transmits signals from a receptor to a target within the cell. (First messengers transmit signals from outside the cell to a receptor.)

secondary active transport Active transport that uses the energy of an electrochemical gradient to drive the movement of molecules.

secondary growth The increase in plant diameter resulting from meristems that surround stems and roots.

secondary oocyte A large cell produced by the asymmetric first meiotic division of the primary oocyte.

secondary phloem New phloem cells produced by the vascular cambium, which are located to the outside of the vascular cambium.

secondary response The response to re-exposure to an antigen, which is quicker, stronger, and longer than the primary response.

secondary sexual characteristic A trait that characterizes and differentiates the two sexes but that does not relate directly to reproduction.

secondary spermatocyte A diploid cell formed during the first meiotic division of the primary spermatocyte.

secondary structure The structure formed by interactions between stretches of amino acids in a protein.

secondary xylem New xylem cells produced by vascular cambium, which are located to the inside of the vascular cambium.

second-division nondisjunction Disjunction in the second meiotic division.

secretin A hormone released by cells lining the duodenum in response to the acidic pH of the stomach contents entering the small intestine and that stimulates the pancreas to secrete bicarbonate ions.

secretion In renal systems, an active process that eliminates substances that were not previously filtered from the blood.

seed A fertilized ovule; seeds are multicellular structures that allow offspring to disperse away from the parent plant.

seed coat A protective outer structure surrounding the seed.

segmentation The formation of discrete parts or segments in the insect embryo.

segregate Separate; applies to chromosomes or members of a gene pair moving into different gametes.

selection The retention or elimination of mutations in a population of organisms.

selectively permeable Describes the properties of some membranes, including the plasma membrane, which lets some molecules in and out freely, lets others in and out only under certain conditions, and prevents other molecules from passing through at all.

self Describes an organism's own molecules and cells.

self-compatible Describes species in which pollen and ovules produced by flowers on the same plant can produce viable offspring.

self-incompatible Describes species in which pollination by the same or a closely related individual does not lead to fertilization.

self-propagating Continuing without input from an outside source; action potentials are self-propagating in that they move along axons by sequentially opening and closing adjacent ion channels.

semen A fluid that nourishes and sustains sperm as they travel in the male and then the female reproductive tracts.

semicircular canal One of three connected fluid-filled tubes in the mammalian inner ear that contains hair cells that sense angular motions of the head in three perpendicular planes.

semiconservative replication The mechanism of DNA replication in which each strand of a parental DNA duplex serves as a template for the synthesis of a new daughter strand.

seminal vesicles Two glands at the junction of the vas deferens and the prostate gland that secrete a protein- and sugar-rich fluid that makes up most of the semen and provides energy for sperm motility.

seminiferous tubules A series of tubes in the testes where sperm are produced.

sender The indivdual who, during communication, supplies a signal that elicits a response from the receiver.

sensitization The enhancement of a response to a stimulus that is achieved by first presenting a strong or novel stimulus.

sensory neuron A neuron that receives and transmits information about an animal's environment or its internal physiological state.

sensory organ A group of sensory receptors that converts particular physical and chemical stimuli into nerve impulses that are processed by a nervous system and sent to a brain.

sensory receptor cell A sensory neuron with specialized membranes in which receptor proteins are embedded.

sensory transduction The conversion of physical or chemical stimuli into nerve impulses.

sepal A structure, often green, that forms the outermost whorl of a flower with other sepals and encases and protects the flower during its development.

septum (plural, septa) In fungi, a wall that partially divides the cytoplasm into separate cells in hyphae.

sequence assembly The process in which short nucleotide sequences of a long DNA molecule are arranged in the correct order to generate the complete sequence.

sequence motif Any of a number of sequences or sequence arrangements that indicate the likely function of a segment of DNA.

serosa An outer layer of cells and connective tissue that covers and protects the gut.

Sertoli cell A type of cell in the seminiferous tubules that supports sperm production.

sex chromosome Any of the chromosomes associated with sex.

sex determination The factors that trigger development of male or female characteristics; in humans, sex determination occurs because of the Y chromosome.

sexual reproduction The process of producing offspring that receive genetic material from two parents; in eukaryotes, the process occurs through meiosis and fertilization.

sexual selection A form of selection that promotes traits that increase an individual's access to reproductive opportunities.

shell (of an atom) An energy level.

shoot The collective name for the leaves, stems, and reproductive organs, the major aboveground organ systems of vascular plants.

shoot apical meristem A group of totipotent cells near the tip of a stem or branch that gives rise to new shoot tissues in plants.

short-day plant A plant that flowers only when the day length is less than a critical value.

shotgun sequencing DNA sequencing method in which the sequenced fragments do not originate from a particular gene or region, but from sites scattered randomly across the molecule.

sickle-cell anemia A condition in which hemoglobin molecules tend to crystallize when exposed to lower-than-normal levels of oxygen, causing the red blood cells to collapse and block capillary blood vessels.

side chain A chemical group attached to the central carbon atom of an amino acid, whose structure and composition determine the identity of the amino acid; also known as an *R group.*

sieve plate A modified end wall with large pores that links sieve elements.

sieve tube A multicellular unit composed of sieve elements that are connected end to end, through which phloem transport takes place.

sigma factor A protein that associates with RNA polymerase that facilitates its binding to specific promoters.

signal sequence An amino acid sequence that directs a protein to its proper cellular compartment.

signal transduction The process in which an extracellular molecule acts as a signal to activate a receptor, which transmits information through the cytoplasm.

signal-anchor sequence In protein sorting, an amino acid sequence in a polypeptide chain that embeds the chain in the membrane.

signaling cell The source of the signaling molecule.

signaling molecule The carrier of information transmitted when the signaling molecule binds to a receptor; also referred to as a *ligand.*

signal-recognition particle (SRP) An RNA–protein complex that binds with part of a polypeptide chain and marks the molecule for incorporation into the endoplasmic reticulum (eukaryotes) or the plasma membrane (prokaryotes).

silencers DNA sequences that bind with regulatory transcription factors and repress transcription.

single-gene trait A trait determined by Mendelian alleles of a single gene with little influence from the environment.

single-lens eye An eye structure found in vertebrates and cephalopod mollusks that works like a camera to produce a sharply defined image of the animal's visual field.

single-nucleotide polymorphism (SNP) A site in the genome where the base pair that is present differs among individuals in a population.

single-strand binding protein A protein that binds single-stranded nucleic acids.

sink In plants, any portion of the plant that needs carbohydrates to fuel growth and respiration, such as a root, young leaf, or developing fruit.

sinoatrial (SA) node A specialized region of the heart containing pacemaker cells where the heartbeat is initiated.

sister chromatids The two copies of a chromosome produced by DNA replication.

sister groups Groups that are more closely related to each other than either of them is to any other group.

skeletal muscle Muscle that connects to the body skeleton to move an animal's limbs and torso.

sliding filament model The hypothesis that muscles produce force and change length by the sliding of actin filaments relative to myosin filaments.

slow-twitch Describes muscle fibers that contract slowly and consume less ATP than do fast-twitch fibers to produce force.

small interfering RNA (siRNA) A type of small double-stranded regulatory RNA that becomes part of a complex able to cleave and destroy single-stranded RNA with a complementary sequence.

small intestine Part of the midgut; the site of the last part of digestion and most nutrient absorption.

small nuclear RNA (snRNA) Noncoding RNA found in eukaryotes and involved in splicing, polyadenylation, and other processes in the nucleus.

small regulatory RNA A short RNA molecule that can block transcription, cleave or destabilize RNAA, or inhibit mRNA translation.

smooth endoplasmic reticulum (SER) The portion of the endoplasmic reticulum that lacks ribosomes.

smooth muscle The muscle in the walls of arteries, the respiratory system, and the digestive and excretory systems; smooth muscle appears uniform under the light microscope.

solubility The ability of a substance to dissolve.

solute A dissolved molecule such as the electrolytes, amino acids, and sugars often found in water, a solvent.

solvent A liquid capable of dissolving a substance.

somatic cell A nonreproductive cell, the most common type of cell in body.

somatic mutation A mutation that occurs in somatic cells.

somatic nervous system The voluntary component of the peripheral nervous system, which is made up of sensory neurons that respond to external stimuli and motor neurons that synapse with voluntary muscles.

source In plants, a region that supplies carbohydrates to other parts of the plant.

Southern blot A method for determining the size and number of copies of a DNA sequence of interest by means of a labeled probe hybridized to DNA fragments separated by size by means of electrophoresis.

spatial summation The converging of multiple receptors onto a neighboring neuron, increasing its firing rate proportionally to the number of signals received.

speciation The process whereby new species are produced.

species A group of individuals that can exchange genetic material through interbreeding to produce fertile offspring.

species–area relationship The relationship between island size and equilibrium species diversity.

spermatogenesis The formation of sperm.

spermatozoa (sperm) The smaller, male gametes.

spinal cord In vertebrates, a central tract of neurons that passes through the vertebrae to transmit information between the brain and the periphery of the body.

spinal nerve In vertebrates, a nerve running from the spinal cord to the periphery containing axons of both sensory and motor neurons.

spiracle An opening in the exoskeleton on either side of an insect's abdomen through which gases are exchanged.

spliceosome A complex of RNA and protein that catalyzes RNA splicing.

spongy bone Vertebrate bone tissue consisting of trabeculae, and thus lighter than compact bone, found in the ends of limb bones and within vertebrae.

spontaneous Occurring in the absence of any assignable cause; most mutations are spontaneous.

sporangium In plants, a multicellular structure in which haploid spores are formed by meiosis.

spore A reproductive cell capable of growing into a new individual without fusion with another cell.

sporophyte Describes the diploid multicellular generation in plants that produces spores.

sporopollenin A complex mixture of polymers that is remarkably resistant to environmental stresses such as ultraviolet radiation and desiccation.

stabilizing selection A form of selection that selects against extremes and so maintains the status quo.

stamen A pollen-producing floral organ.

stapes A small bone in the middle ear that helps amplify the waves that strike the tympanic membrane; the stapes connects to the oval window of the cochlea.

starch The form in which glucose is stored in plants.

Starling's Law The correspondence between change in stroke volume and change in the volume of blood filling the heart.

statocyst A type of gravity-sensing organ found in most invertebrates.

statolith In plants, a large starch-filled organelle in the root cap that senses gravity; in animals, a dense particle that moves freely within a statocyst, enabling it to sense gravity.

stem cell An undifferentiated cell that can undergo an unlimited number of mitotic divisions and differentiate into any of a large number of specialized cell types.

stereocilia Nonmotile cell-surface projections on hair cells whose movement causes a depolarization of the cell's membrane.

steroid A type of lipid.

steroid hormone A hormone that is derived from cholesterol.

stigma The surface at the top of the carpel, to which pollen adheres.

stomach The last part of the foregut, which serves as a storage and digestive chamber; also known as the *crop*.

stomata (singular, stoma) Pores in the epidermis of a leaf that regulate the diffusion of gases between the interior of the leaf and the atmosphere.

Stramenopila A eukaryotic superkingdom including unicellular organisms, giant kelps, algae, protozoa, free-living cells, and parasites; distinguished by a flagellum with two rows of stiff hairs and, usually, a second, smooth flagellum.

striated muscle Skeletal muscle and cardiac muscle, which appear striped under a light microscope.

strigolactone A hormone, produced in roots and transported upward in the xylem, that inhibits the outgrowth of axillary buds.

stroke volume (SV) The volume of blood pumped during each heart beat.

stroma The region surrounding the thylakoid, where carbohydrate synthesis takes place.

stromatolite A layered structure that records sediment accumulation by microbial communities.

structural gene A gene that codes for the sequence of amino acids in a polypeptide chain.

style A cylindrical stalk between the ovary and the stigma, through which the pollen tube grows.

suberin A waxy compound coating cork cells that protects against mechanical damage, the entry of pathogens, and water loss.

submucosa A tissue layer surrounding the mucosa that contains blood vessels, lymph vessels, and nerves.

subspecies Allopatric populations that have yet to evolve even partial reproductive isolation but which have acquired population-specific traits.

substrate (*S*) A molecule acted upon by an enzyme.

substrate-level phosphorylation A way of generating ATP in which a phosphate group is transferred to ADP from an organic molecule, which acts as a phosphate donor or substrate.

succession The replacement of species by other species over time.

sugar The simplest carbohydrate molecule; also called a *saccharide.*

sulci Deep crevices in the brain that separate the lobes of the cerebral hemispheres.

supercoil A coil of coils; a circular molecule of DNA can coil upon itself to form a supercoil.

superkingdom One of seven major groups of eukaryotic organisms, classified by molecular sequence comparisons.

supernormal stimulus An exaggerated stimulus that elicits a response more strongly than the normal stimulus.

surfactant A compound that reduces the surface tension of a fluid film.

survivorship The proportion of individuals from an initial cohort that survive to each successive stage of the life cycle.

suspension filter feeding The most common form of food capture by animals, in which water with food suspended in it passes through a sievelike structure.

sustainable development Use of natural resources at rates no higher than the rate at which they can be replenished.

symbiont An organism that lives in closely evolved association with another species.

symbiosis (plural, symbioses) Close interaction between species that live together, often interdependently.

sympathetic division The division of the autonomic nervous system that generally produces arousal and increased activity; active in the fight-or-flight response.

sympatric Describes populations that are in the same geographic location.

synapomorphy A shared derived character; a homology shared by some, but not all, members of a group.

synapse A junction through which the axon terminal communicates with a neighboring cell.

synapsis The gene-for-gene pairing of homologous chromosomes in prophase I of meiosis.

synaptic cleft The space between the axon of the presynaptic cell and the neighboring postsynaptic cell.

synaptic plasticity The ability to adjust synaptic connections between neurons.

synonymous (silent) mutation A mutation resulting in a codon that does not alter the corresponding amino acid in the polypeptide.

systemic acquired resistance (SAR) The ability of a plant to resist future infections, occurring in response to a wide range of pathogens.

systemic circulation Circulation of the blood to the body, excluding the lungs.

systole The phase of the vertebrate cardiac cycle when the ventricles contract to eject blood from the heart.

T cell receptor (TCR) A protein receptor on a T cell that recognizes and binds to an antigen.

T lymphocyte (T cell) A cell type that matures in the thymus and includes helper and cytotoxic cells.

tannin Any one of a group of phenols found widely in plant tissues that bind with proteins and reduce their digestibility.

taste bud One of the sensory organs for taste.

TATA box A DNA sequence present in many promoters in eukaryotes and archaeons that serves as a protein-binding site for a key general transcription factor.

taxis (plural, taxes) Movement in a specific direction in response to a stimulus.

taxon (plural, taxa) All the species in a taxonomic entity such as family or genus.

tectorial membrane A rigid membrane in the cochlear duct, against which the stereocilia of hair cells in the organ of Corti bend when stimulated by vibration, setting off an action potential.

telomerase An enzyme containing an RNA template from which complementary telomere repeats are synthesized.

telomere A repeating sequence at each end of a eukaryotic chromosome.

telophase The stage of mitosis in which the nuclei of the daughter cells are formed and the chromosomes uncoil to their original state.

telophase I The stage of meiosis I in which the chromosomes uncoil slightly, a nuclear envelope briefly reappears, and in many species the cytoplasm divides, producing two separate cells.

telophase II The stage of meiosis II in which the chromosomes uncoil and become diffuse, a nuclear envelope forms around each set of chromosomes, and the cytoplasm divides by cytokinesis.

template A strand of DNA or RNA whose squence of nucleotides is used to synthesize a compementary strand.

template strand In DNA replication, the parental strand whose sequence is used to synthesize a complementary daughter strand.

temporal isolation Pre-zygotic isolation between individuals that are reproductively active at different times.

temporal lobe The region of the brain involved in the processing of sound, language and reading, and object identification and naming.

temporomandibular joint A specialized jaw joint in mammals that allows the teeth of the lower and upper jaws to fit together precisely.

tendon A collagen structure that attaches muscles to the skeleton and transmits muscle forces over a wide range of joint motion.

termination In protein translation, the time at which the addition of amino acids stops and the completed polypeptide chain is released from the ribosome. In cell communication, the stopping of a signal.

terminator A DNA sequence at which transcription stops and the transcript is released.

terpene Any one of a group of compounds that do not contain nitrogen and are produced by some plants as a defensive mechanism.

tertiary structure The overall three-dimensional shape of a protein, formed by interactions between secondary structures.

test (of a protist) A "house" constructed of organic molecules that shelters a protist.

test group The experimental group that is exposed to the variable in an experiment.

testcross Any cross of an unknown genotype with a homozygous recessive genotype.

testis (plural, testes) The male gonad, where sperm are produced.

testosterone A steroid hormone, secreted by the testes, that plays key roles in male growth, development, and reproduction.

tetanus A muscle contraction of sustained force.

tetraploid A cell or organism with four complete sets of chromosomes; a double diploid.

Tetrapoda A monophyletic group of animals whose last common ancestor had four limbs; this group includes amphibians, lizards, turtles, crocodilians, birds, and mammals (some tetrapods, like snakes, have lost their legs in the course of evolution).

thalamus The inner brain region of the forebrain, which acts as a relay station for sensory information sent to the cerebrum.

thallus A flattened photosynthetic structure produced by some bryophytes.

Thaumarchaeota One of the three major divisions of Archaea; thaumarchaeota are chemotrophs, deriving energy from the oxidation of ammonia.

theory A general explanation of a natural phenomenon supported by a large body of experiments and observations.

theory of island biogeography A theory that states that the number of species that can occupy a habitat island depends on two factors: the size of the island and the distance of the island from a source of colonists.

thermoreceptor A sensory receptor in the skin and in specialized regions of the central nervous system that responds to heat and cold.

thick filament A parallel grouping of myosin molecules that makes up the myosin filament.

thin filament Two helically arranged actin filaments twisted together that make up the actin filament.

threshold potential The critical depolarization voltage of −50 mV required for an action potential.

thylakoid membrane A highly folded membrane in the center of the chloroplast that contains light-collecting pigments and that is the site of the photosynthetic electron transport chain.

thymine (T) A pyrimidine base.

thyroid gland A gland located in the front of the neck that leads to the release of two peptide hormones, thyroxine and triiodothyorine.

Ti plasmid A small circular DNA molecule in virulent strains of *R. radiobacter* containing genes that can be integrated into the host cell's genome, as well as the genes needed to make this transfer.

tidal ventilation A breathing technique in most land vertebrates in which air is drawn into the lungs during inhalation and moved out during exhalation.

tidal volume The amount of air inhaled and exhaled in a cycle; in humans, tidal volume is 0.5 liter when breathing at rest.

tight junction A junctional complex that establishes a seal between cells so that the only way a substance can travel from one side of a sheet of epithelial cells to the other is by moving through the cells by a cellular transport mechanism.

tissue A collection of cells that work together to perform a specific function.

tolerance The ability of T cells not to respond to self antigens even though the immune system functions normally otherwise.

toll-like receptors (TLRs) A family of transmembrane receptors on phagocytes that recognize and bind to molecules on the surface of microorganisms, providing an early signal that an infection is present.

topography The physical features of Earth.

topoisomerase Any one of a class of enzymes that regulates the supercoiling of DNA by cleaving one or both strands of the DNA double helix, and later repairing the break.

topoisomerase II An enzyme that breaks a DNA double helix, rotates the ends, and seals the break.

totipotent Describes cells that have the potential to give rise to a complete organism; a fertilized egg is a totipotent cell.

trabeculae Small plates and rods with spaces between them, found in spongy bone.

trace fossil A track or trail, such as a dinosaur track or the feeding trails of snails and trilobites, left by an animal as it moves about or burrows into sediments.

trachea The central airway leading to the lungs, supported by cartilage rings.

tracheae An internal system of tubes in insects that branch from openings along the abdominal surface into smaller airways, directing oxygen to and removing carbon dioxide from respiring tissues.

tracheid A unicellular xylem conduit.

trade-off An exchange in which something is gained at the expense of something lost.

trait A characteristic of an individual.

transcription The synthesis of RNA from a DNA template.

transcriptional activator protein A protein that binds to a sequence in DNA to enable transcription to begin.

transcriptional regulation The mechanisms that collectively regulate whether or not transcription occurs.

transduction Horizontal gene transfer by means of viruses.

transfer RNA (tRNA) Noncoding RNA that carries individual amino acids for use in translation.

transformation The conversion of cells from one state to another, as from nonvirulent to virulent, when DNA released to the environment by cell breakdown is taken up by recipient cells. In recombinant DNA technology, the introduction of recombinant DNA into a recipient cell.

transgenic organisms An alternative term for genetically modified organisms.

transition state The brief time in a chemical reaction in which chemical bonds in the reactants are broken and new bonds in the product are formed.

translation Synthesis of a polypeptide chain corresponding to the coding sequence present in a molecule of messanger RNA.

transmembrane proteins Proteins that span the entire lipid bilayer; most integral membrane proteins are transmembrane proteins

transmission genetics The discipline that deals with the manner in which genetic material is passed from generation to generation.

transpiration The loss of water vapor from leaves.

transporters Membrane proteins that move ions or other molecules across the cell membrane.

transposable element (TE) A DNA sequence that can replicate and move from one location to another in a DNA molecule; also known as *transposon.*

transposase The enzyme that cleaves a transposon from its original location in the genome and inserts it into a different position.

transposition The movement of a transposable element.

transposon A DNA sequence that can replicate and move from one location to another in a DNA molecule; also known as *transposable element (TE).*

triacylglycerol A lipid composed of a glycerol backbone and three fatty acids.

trimesters The three periods of pregnancy, each lasting about 3 months.

triose phosphate A 3-carbon carbohydrate molecule, produced by the Calvin cycle and exported from the chloroplast.

triploid A cell or organism with three complete sets of chromosomes.

trisomy 21 A condition resulting from the presence of three, rather than two, copies of chromosome 21; also known as *Down syndrome.*

trophic level An organism's typical place in a food web as a producer or consumer.

trophic pyramid A diagram that traces the flow of energy through communities, showing the amount of energy available at each level to feed the next. The pyramid shape results because biomass and the energy it represents generally decrease from one trophic level to the next.

tropic hormone A hormone that controls the release of other hormones.

tropism The bending or turning of an organism in response to an external signal such as light or gravity.

tropomyosin A protein that runs in the grooves formed by the actin helices and blocks the myosin-binding sites.

troponin A protein that moves tropomyosin away from myosin-binding sites, allowing cross-bridges between actin and myosin to form and the muscle to contract.

true breeding Describes a trait whose physical appearance in each successive generation is identical to that in the previous one.

trypsin A digestive enzyme produced by the pancreas that breaks down proteins.

tube feet Small projections of the water vascular system that extend outward from the body surface and facilitate locomotion, sensory perception, food capture, and gas exchange in echinoderms.

tubulin Dimers (composed of an α tubulin and a β tubulin) that assemble into microfilaments.

tumor suppressors A family of genes that encode proteins whose normal activities inhibit cell division.

tunicates A subphylum of Chordata that includes about 3000 species of filter-feeding marine animals, such as sea squirts and salps.

turgor pressure Pressure within a cell resulting from the movement of water into the cell by osmosis and the tendency of the cell wall to resist deformation.

Turner syndrome A sex-chromosomal abnormality in which an individual has 45 chromosomes, including only one X chromosome.

twitch A muscle contraction that results from a single action potential.

twofold cost of sex Population size can increase more rapidly in asexually reproducing organisms than in sexually reproducing organisms because only female produce offspring, and sexual females have only half the fitness of asexual parents.

tympanic membrane A thin sheet of tissue at the surface of the ear that vibrates in response to sound waves, amplifying airborne vibrations; in mammals, also known as the *eardrum.*

unbalanced translocation Translocation in which only part of a reciprocal translocation (and one of the nontranslocated chromosomes) is inherited from one of the parents.

unsaturated Describes fatty acids that contain carbon–carbon double bonds.

uracil (U) A pyrimidine base in RNA, where it replaces the thymine found in DNA.

urea A waste product of protein metabolism that many animals excrete.

ureter A large tube in the vertebrate kidney that brings urine from the kidneys to the bladder.

urethra A tube from the bladder that in males carries semen as well as urine from the body.

uterus A hollow organ within the reproductive tract of female mammals with thick, muscular walls that is adapted to support the developing embryo if fertilization occurs and to deliver the baby during birth.

vaccination The deliberate delivery in a vaccine of an antigen from a pathogen to induce a primary response but not the disease, thereby providing future protection from infection.

vacuole A cell structure that absorbs water and contributes to turgor pressure.

vagina A tubular channel connecting the uterus to the exterior of the body; also known as the birth canal.

valence electrons The electrons farthest from the nucleus, which are at the highest energy level.

van der Waals interactions The binding of temporarily polarized molecules because of the attraction of opposite charges.

variable (in experimentation) The feature of an experiment that is changed by the experimenter from one treatment to the next.

variable (V) region A region of the heavy (H) and light (L) chains of an antibody; the variable region distinguishes a given antibody from all others.

variable expressivity The phenomenon in which a particular phenotype is expressed with a different degree of severity in different individuals.

variable number tandem repeat (VNTR) A genetic difference in which the number of short repeated sequences of DNA differs from one chromosome to the next.

vas deferens A long, muscular tube from the scrotum, through the abdominal cavity, along the bladder, and connecting with the ejaculatory duct.

vasa recta The blood vessels in the kidneys.

vascular cambium Lateral meristem that is the source of new xylem and phloem.

vascular plant A plant that produces both xylem and phloem.

vasoconstriction The process in the supply of blood to the limbs is reduced by constriction of the arteries that supply the limbs.

vasodilation The process in which resistance in the arteries is decreased and blood flow increased following relaxation of the smooth arterial muscles.

vasopressin A posterior pituitary gland hormone that acts on the kidneys and controls the water permeability of the collecting ducts, thus regulating the concentration of urine that an animal excretes; also known as *antidiuretic hormone (ADH).*

vector In recombinant DNA, a carrier of the donor fragment, usually a plasmid.

vegetative reproduction Reproduction by growth and fragmentation. In plants, the production of upright shoots from horizontal stems, permitting new plants to be produced at a distance from the site where the parent plant originally germinated.

veins In plants, the system of vascular conduits within the leaf; in animals, the large, low-pressure vessels that return blood to the heart.

vena cava (plural, venae cavae) One of two large veins in the body that drain blood from the head and body into the heart.

ventilation The movement of an animal's respiratory medium—water or air—past a specialized respiratory surface.

ventricle A heart chamber that pumps blood to the lungs or the rest of the body.

venule A blood vessel into which capillaries drain as blood is returned to the heart.

vernalization A prolonged period of exposure to cold temperatures necessary to induce flowering.

vertebrae (singular, vertebra) The series of hard bony segments making up the jointed skeleton that runs along the main axis of the body in vertebrates.

vertebral column A skeletal structure in vertebrates that functionally replaces the embryonic notochord that supports the body.

vertebrates A subphylum of Chordata, distinguished by a bony cranium that protects the brain and (unless lost through evolution), a vertebral column; also known as *craniates.*

vesicle A small membrane-enclosed sac that transports substances within the cell.

vessel A multicellular xylem conduit.

vessel element An individual cell that is part of a xylem vessel for water transport in plants.

vestibular system A system in the mammalian inner ear made up of two statocyst chambers and three semicircular canals.

vestigial structure A structure that has lost its original function over time and is now much reduced in size.

vicariance The process in which a geographic barrier arises within a single population, separating it into two or more isolated populations.

villi Highly folded inner surfaces of the jejunum and ileum of the small intestine.

virulent Describes pathogens that are able to overcome a host's defenses and lead to disease.

virus A small infectious agent that contains a nucleic acid genome packaged inside a protein coat called a capsid.

visible light The portion of the electromagnetic spectrum apparent to our eyes.

visual cortex The part of the brain that processes visual images.

vitamin An organic molecule that is required in very small amounts in the diet.

vitreous humor A gel-like substance filling the large cavity behind the lens that makes up most of the volume of the vertebrate eye.

viviparity Giving birth to live young, with nutritional support of the embryo from the mother.

vocal cords Twin organs in the larynx that vibrate as air passes over them, enabling speech, song, and sound production.

voltage-gated channels Ion membrane channels that open and close in response to changes in membrane potential.

voluntary Describes the component of the nervous system that handles sensing and responding to external stimuli.

vulva The external genitalia of the female.

water vascular system A series of fluid-filled canals that permit bulk transport of oxygen and nutrients in echinoderms.

white blood cell (leukocyte) A type of cell in the immune system that arises by differentiation from stem cells in the bone marrow.

white matter Collectively, the axons of cortical neurons in the interior of the brain; it is the fatty myelin produced by glial cells surrounding the axons that makes this region of the brain white.

wild type The most common allele, genotype, or phenotype present in a population; nonmutant.

***X* chromosome** One of the sex chromosomes; a normal human female has two copies of the *X* chromosome; a normal male has one *X* and one *Y* chromosome.

xanthophyll Any one of several yellow-orange pigments that slow the formation of reactive oxygen species by reducing excess light energy; these pigments accept absorbed light energy directly from chlorophyll and convert this energy to heat.

***X*-inactivation** The process in mammals in which dosage compensation occurs through the inactivation of one *X* chromosome in each cell in females.

***X*-linked gene** A gene in the *X* chromosome.

xylem Vascular tissue consisting of lignified conduits that transports water and nutrients from the roots to the leaves.

***Y* chromosome** One of the sex chromosomes; a normal human male has one *X* and one *Y* chromosome.

yeast A single-celled fungus found in moist, nutrient-rich environments.

***Y*-linked gene** A gene that is present in the region of the *Y* chromosome that shares no homology with the *X* chromosome.

yolk A substance in the eggs of animals with external fertilization that provides all the nutrients that the developing embryo needs until it hatches.

Z disc A protein backbone found regularly spaced along the length of a myofibril.

Z scheme Another name for the photosynthetic electron transport chain, so called because the overall energy trajectory resembles a "Z."

zygomycetes Fungi groups that produce hyphae undivided by septa and do not form multicellular fruiting bodies; they make up less than 1% of known fungal diversity.

zygote The diploid cell formed by the fusion of two gametes.

Index

bold face indicates a definition
italics indicate a figure
t indicates a table

G